Laser in Forschung und Technik
Laser in Research and Engineering

Springer

Berlin
Heidelberg
New York
Barcelona
Budapest
Hongkong
London
Mailand
Paris
Santa Clara
Singapur
Tokio

Laser
in Forschung und Technik
in Research and Engineering

Vorträge des 12. Internationalen Kongresses
Proceedings of the 12th International Congress

Laser 95

Herausgegeben von/Edited by
W. Waidelich, H. Hügel, H. Opower, H. Tiziani
R. Wallenstein, W. Zinth

Mit 873 Abbildungen

Springer

Herausgeber:

Prof. em. Dr. W. Waidelich, Inst. Mediz. Optik, Universität München
Prof. Dr.-Ing. H. Hügel, Inst. Strahlwerkzeuge, Universität Stuttgart
Prof. Dr. H. Opower, DLR Inst. Techn. Physik, Stuttgart
Prof. Dr. H. Tiziani, Inst. Techn. Optik, Universität Stuttgart
Prof. Dr. R. Wallenstein, Fachb. Physik, Universität Kaiserslautern
Prof. Dr. W. Zinth, Inst. Medizinische Optik, Universität München

Die Deutsche Bibliothek - CIP-Einheitsaufnahme
(Laser in Forschung und Technik) Laser in Forschung und Technik, in research and
engineering: Vorträge des 12. Internationalen Kongresses Laser 95 / hrsg. von Waidelich ...
Berlin; Heidelberg; New York; Barcelona; Budapest; Hongkong; London; Mailand; Paris;
Santa Clara; Singapur; Tokio: Springer, 1996

NE: Waidelich, Wilhelm (Hrsg.); Internationaler Kongress Laser -12, 1996, München-;
Laser in research and engineering

Satz: Reproduktionsfertige Vorlagen der Autoren
SPIN: 10539718   62/3020 - 5 4 3 2 1 0 - Gedruckt auf säurefreiem Papier

ISBN 978-3-540-61316-9   ISBN 978-3-642-80263-8 (eBook)
DOI 10.1007/978-3-642-80263-8

Vorwort

LASER 95, die von der Messe München veranstaltete 12.Internationale Leitmesse für innovative und angewandte Optoelektronik, ermöglichte durch die Verbindung von Ausstellung und Kongreß einen generellen Überblick über den neuesten Stand und die Entwicklungstendenzen. Die Synthese der Ausstellung kommerziell erhältlicher Komponenten und Geräte (Messe) mit Vorträgen über Forschungsergebnisse und über Anwendungen (Kongreß) hat auf der LASER seit ihrer Gründung 1973 eine erfolgreiche Tradition. Als Folge der dynamischen Entwicklung der Optoelektronik und der Anwendungen des Lasers wurde der Kongreß 1995 in mehrere Einzelkongresse aufgeteilt.

Die auf die Schwerpunkte der Messe abgestimmten Kongreßthemen verfolgen ein zweifaches Ziel. Einerseits zeigen Erfahrungsberichte sowie Vorschläge für neue Anwendungen Grenzen und Möglichkeiten des Lasers und der Optoelektronik auf (Technologietransfer), andererseits regen neue Forschungsergebnisse zu weiterem Fortschritt in Forschung und Entwicklung an (Innovation).

Um die Ergebnisse von LASER 95 einem größeren Kreis von Interessenten zugänglich zu machen, sind im vorliegenden Band die Vorträge folgender Kongresse zusammengestellt:

LASER IN RESEARCH - W. Zinth
MODERNE FESTKÖRPERLASER - R. Wallenstein
OPTISCHE MESS- UND PRÜFTECHNIK - H.J. Tiziani
OPTISCHE KOMPONENTEN UND SYSTEME - H. Opower, H. Hügel
LASER IN DER MIKROSTRUKTURTECHNIK - H. Opower, H. Hügel

Das Buch enthält eine Fülle wertvoller Informationen aus den genannten Gebieten, die Grundlagen und Anregungen für Anwendungen sowie Impulse für Forschung und Entwicklung vermitteln.

Für das Zustandekommen dieses Kongressbandes sei allen Autoren, den Chairmen und Diskussionsrednern, der Messe München GmbH und dem Springer-Verlag verbindlichst gedankt.

Wilhelm Waidelich

Preface

LASER 95, the 12th International Congress and Trade Fair, was held at the Munich Trade Fair Center in June 1995.

Since 1973, the Messe Munich has been providing a survey of the state-of-the-art and trends in development through its LASER Opto-electronics Congress and Trade Fair. The successful synthesis of theory and practice made evident by the coordination of the congress and exhibition, has helped this high-tech information forum in Munich to gain an international reputation. This is the oldest and most important event of this kind.

To reflect the ongoing specialization in the development and application of Laser technology the Messe Munich organized 1995 a series of technical congresses according to major application sectors.

The congresses which are attuned to the Trade Fairs key points aim at a double goal. On the one hand critical reports from experience and suggestions for new applications show both limits and possibilitiies of Laser and the Opto-electronics; on the other hand results of new researches insite further progress in research and development.

To make the results accessible to all interested parties the lectures of the following congresses are put together in this volume:

LASER IN RESEARCH - W. Zinth
ADVANCED SOLID STATE LASERS - R. Wallenstein
OPTICAL MEASURING TECHNIQUES - H.J. Tiziani
OPTICAL COMPONENTS AND SYSTEMS - H. Opower, H. Hügel
LASER IN MICROSTRUCTURE TECHNOLOGY - H. Opower, H. Hügel

The book contains a wealth of information and ideas of the above mentioned branches and should thus stimulate application, research and development in the field of opto-electronics.

We would like to thank the authors, the Messe Munich GmbH and the Springer-Verlag for making this publication possible.

Wilhelm Waidelich

Inhaltsverzeichnis–Contents

Posters

MODERNE FESTKÖRPERLASER / ADVANCED SOLID STATE LASER
Session-Chairmen: R. Wallenstein

Session 1
Moderation: A. Borsutzky

Poster Session

Session 4
Moderation: R. Beigang

OPTISCHE MESS- UND PRÜFTECHNIK
OPTICAL MEASURING TECHNIQUES
Session chairman: H. J. Tiziani

Holographische- und Specklemeßtechnik
Application of Holography and Speckle Techniques
Moderation: W. Jüptner

3-D Information, Gewinnung und Verarbeitung
3-D Information, Aquisition and Analysis
Moderation: K. Biedermann

Qualitätssicherung und Kalibrierung der Messverfahren
Quality Assurance and Calibration of Measuring Techniques
Moderation: L. Mattson

OPTISCHE KOMPONENTEN UND SYSTEME
OPTICAL COMPONENTS AND SYSTEMS
Session Chairmen: H. Opower, H. Hügel

LASER IN DER MIKROSTRUKTURTECHNIK
LASER IN MICROSTRUCTURE TECHNOLOGY
Session Chairmen: H. Opower, H. Hügel, W. Pompe

Referenten - Contributors

Laser in der Forschung
Laser in Research
W. Zinth

Atoms in Solid Helium: A Sensitive Probe to Test Time Reversal Invariance

A. Weis

Max-Planck-Institut für Quantenoptik, D-85748 Garching

trw@ipp-garching.mpg.de

Abstract: We discuss the basic principle of experiments aimed at the detection of parity and time reversal invariance violating permanent electric dipole moments in atoms, and show that the use of paramagnetic atoms embedded in a matrix of solid ^{4}He is a promising alternative method to conduct a highly sensitive search for such moments. The recently developped preparation technique of these samples and their optical and magnetic properties are discussed and an outlook to the anticipated sensitivity to dipole moments is given.

1. Violation of discrete symmetries and permanent dipole moments

1.1 P-violation

The laws of nature are invariant under the combined symmetry operation CPT, where C is the charge conjugation operation (matter-antimatter exchange) , P the parity operation (spatial mirror inversion) and T the time reversal operation. This fact is known as the CPT theorem. Since the late 1950's it is known that these operations taken individually are not good symmetries, the first experimental evidence being the violation of parity (P) in disintegration processes governed by the weak interaction. Many examples of P-violation have since been observed in nuclear physics and in the desintegration and interaction of elementary particles. All of these processes are governed by so-called charged weak currents, mediated by the exchange of virtual charged W-bosons. In the 70's experiments revealed that parity was also violated in processes in which the interaction is mediated by the neutral Z-boson, the so-called weak neutral currents. In the 70's and 80's experiments showed that these currents make that parity is violated in atoms as well. All parity violating processes observed so far are well described by the standard model of electroweak interactions. In all of these processes T (time reversal) is a good symmetry.

4

1.2 T-violation

In the 60's experiments investigating the decay of the K^0 meson revealed an anomalous decay channel of the long lived component K_L, which is a manifestation of a violation of the combined symmetry CP. Together with the belief in the CPT-theorem, this was the first indication for a violation of T. Although an enormous amount of experiments have been carried out since with the aim of detecting T-violation in other systems, such as atoms, molecules, nuclei, the neutron, and others, no evidence for T-violation has been found in any of these systems. The fact that experiments so far have only revealed T-violation in one single system remains one of the biggest puzzles of modern physics.

1.3 Permanent electric dipole moments

One possible manifestation of T-violation is the coexistence in an elementary particle of electric and magnetic dipole moments. This coexistence also violates parity (P). Most modern experiments aimed at detecting T-violation in atoms or the neutron try to detect such a coexistence[1]. The only vector quantity which describes a particle at rest is its angular momentum $\vec{J}$ to which the 2 vector quantities $\vec{\mu}$ (magnetic moment) and $\vec{d}$ (electric moment) have to be parallel. The Hamiltonian of a particle with magnetic and electric moments in parallel static electric and magnetic fields then reads

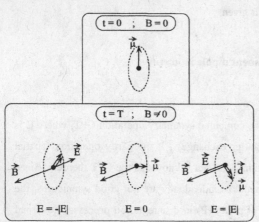

Fig. 1 Detection of EDM via E-field induced acceleration
    or deceleration of Larmor precession

$$H = -(\mu\,\vec{B} + d\,\vec{E}) \cdot \frac{\vec{J}}{J} \qquad (1)$$

In an experiment where the first term in Eq.1 is measured via the Larmor precession at $\bar{\omega}_L$, the effect of an electric dipole moment (EDM) can thus be detected as an acceleration or deceleration of the precession frequency (Fig.1):

$$\omega_{prec} = \omega_L \pm \frac{\vec{d} \cdot \vec{E}}{\hbar} \qquad (2)$$

The major criterium for a sensitive experiment is thus to have a medium which

    a) can hold a large electric field strength E, and

    b) allows a large precession angle $\Phi = \omega_{prec} T$ to be accumulated. Here T is either the finite interaction time of the particles with the static fields, or the spin relaxation time.

Atomic EDM experiments so far have been carried out in vapour cells or in atomic beams[1]. While the latter category allows the use of electric fields up to 100 kV/cm, the thermal atomic velocity usually limits the interaction time to a few ms. In vapour cells the use of buffer gas allows to achieve spin relaxation times on the order of 15 ms, electrical break-down however prevents the use of fields exceeding 4 kV/cm.

The motivation to use paramagnetic atoms trapped in cryogenic solid helium matrices is based on the following considerations:

    a) solid helium has electrical break-down fields exceeding 100 kV/cm, and

    b) spin relaxation times of the implanted atoms were expected, and have since been demonstrated to be very long, due to the long storage times and the isotropy and non-magnetic character of the environment of the local trapping site.

In this sense atoms in cryogenic matrices combine the advantages of both beam and vapour experiments by allowing for long precession times in very large electric fields [2,3]. Long relaxation times are a prerequisite to achieve narrow magnetic resonance linewidths and the common feature of all EDM experiments is the search for a shift of a narrow magnetic resonance line induced by (and proportional to) an applied static electric field.

In the next section we discuss the techniques by which this novel type of matrix isolated species can be prepared and review the optical and magnetic properties of atoms in ^{4}He matrices.

2. Atoms in solid helium matrices

Matrix isolation spectroscopy, i.e. the investigation of foreign atoms or molecules imbedded in cryogenic solid noble gas matrices is a well established technique [4] which has been limited in the past to the use of *heavy* noble gases such as Ne, Ar or Kr as matrix materials. The deposition techniques used to grow these matrices cannot be applied to He which is the only substance known not to solidify under cooling only. In order to solidify He it has to be pressurized to above 27 bar, while being cooled below 2K (Fig.2). In the past few years we have developped at the Max-Planck-Institute for Quantum Optics in Garching/Munich a technique [5,6] by which foreign atoms can be implanted into solid helium. The implanted atoms can be detected by the technique of laser induced fluorescence (LIF). Furthermore the absorption-reemission cycle of *polarized* resonance radiation can be used to transfer angular momentum to atoms which have a ground state with a non-vanishing angular momentum, a technique called optical pumping which allows the build-up (and detection) of atomic spin polarization [3].

2.1. Sample preparation

The experiments are performed in a copper pressure cell immersed in a double bath He cryostat cooled to approx.1.6 K by pumping helium off the bath. He gas is condensed in the cell and the pressure in the cell raised to above 27 bar, where solidification sets in. The pressure is controlled by a calibrated resistive pressure sensor. At low pressures He crystallizes in two phases, an isotropic bcc (body-cubic centered) phase and an anisotropic hcp (hexagonally-close packed) phase (Fig.2). The cell has optical access through five windows. The top window is a short focal length lens through which radiation from a frequency-doubled Nd-YAG laser is focused onto a metal target on the bottom of the cell, thereby melting the solid and laser ablating atoms into the liquified He. Upon resolidification atoms are trapped in the solid bulk and are detected via laser induced fluorescence. Typical atomic sample densities are 10^8 cm^{-3}.

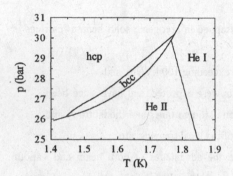

Fig. 2 Phase diagram of solid ^{4}He

2.2. Atomic bubbles

Electrons in liquid He have been known for a long time to form cavities from which helium is expelled; these defect structures are commonly called „bubbles" and have diameters of approximately 36 Å, which is ten times the average He-He separation in the bulk. Implanted atoms form similar structures in condensed He. The cavity in the case of Cs atoms has a diameter of approximately 12 Å. The repulsive interaction governing this bubble formation is the exchange potential between the valence electron of the impurity atom and the closed electronic shell of the He atoms (Pauli principle). The bubble size is determined by the balance of this repulsion with a weak van-der-Waals attraction and the surface tension of the bubble. Our studies of the optical properties of Ba atoms in superfluid and solid helium have shown that, as with trapped electrons, the structure of the trapping site in the *solid* phase has also a bubble structure, in contrast to the case of foreign atoms implanted in heavier noble gas matrices, where the interaction is too weak to deform the lattice and where multiple trapping sites were found.

2.3. Optical properties and optical pumping

The optical absorption-fluorescence cycle for atoms in He matrices is characterized by a strongly blue shifted and asymmetrically broadened excitation line and a less shifted and less broadened emission line [7]. For implanted Cs atoms the D_1 excitation line ($6^2S_{1/2} \rightarrow 6^2P_{1/2}$) peaks at 850 nm with a width of 10 nm, whereas emission occurs at 888 nm [8]. Due to the large absorption linewidth the hyperfine structure of the transition remains unresolved in the optical spectra. The Cs ground state has two hyperfine levels (F=3,4) and by pumping with $\sigma_\pm$ circularly polarized light most of the population is transferred to the „dark" |F=4, $M_F=\pm4$> sublevel of the ground state, which is decoupled from the light. A high degree of spin polarization, is thus characterized by a low fluorescence intensity.

We found that the degree of spin polarization strongly depends on the crystalline structure of the host matrix. When working in the bcc phase low light intensities are sufficient to transfer more than 90% of the population into the dark state [9]. On the other hand, when the atoms were implanted into the hcp phase even pumping with several 10 mW/mm^2 resulted in contrasts not exceeding 10%. It thus seems that in hcp the optical pumping process has to compete against a stronger relaxation mechanism than in bcc. This is supported by our measurement of the spin relaxation times in the two phases. Using the method of relaxation in the dark [10], we have determined the longitudinal spin relaxation times T_1. In the bcc matrix T_1 was found to be 1 - 2 seconds independent of the longitudinal magnetic holding field in the range 10mG - 2G [10], while in the hcp matrix T_1 increases monotonically with the holding field in the range 200mG to 7G with typical values of a few milliseconds. Although the matrix is basically polycrystalline, it seems that the crystalline symmetry still determines the symmetry of the trapping site. While the anisotropic hcp phase produces an electric field gradient at the local trapping site, which, by mixing of orbital angular momentum states and spin-orbit interaction depolarizes the ground state, the symmetric bcc phase preserves the spin polarization. For obvious reasons the understanding of these crystalline structure effects is of primary importance when considering the extension of these experiments to an EDM search.

2.4. Magnetic resonance and outlook to EDM experiment

Spin polarization in the ground state can be destroyed by irradiating the atoms with radiofrequency radiation. A resonant depolarization and hence change in fluorescence intensity appears when the frequency of the r.f field is equal to the Larmor precession frequency (lower curve of Fig.3). These double resonance technique is a very efficient way to optically detect magnetic resonance even with dilute samples. In the bcc phase we observed minimal linewidths of these double resonance signals on the order of 300Hz. These widths do not reflect the large spin lifetime, but rather the large magnetic field inhomogeneities due to the small size of the coils used.

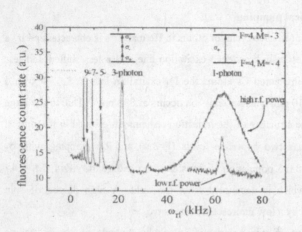

Fig. 3   R.f. multiphoton resonances in the ground state of Cs by absorption from the co- and counter-ratating r.f field. Processes involving up to 23 photons can be observed. These resonances appear only in the bcc phase.

Work is in progress to improve the field homogeneity and hence the ultimate linewidth of the magnetic resonance lines. In the hcp matrix we have observed additional magnetic resonances at frequencies corresponding to $|\Delta M_F| = 2$ and 3, which indicate a breaking of cylindrical symmetry (Fig.4). These resonances are yet another indication for the matrix effects discussed above.

The signature of an atomic EDM in magnetic resonance experiments is a shift of the resonance frequency proportional to an applied static electric field. Sensitive experiments require narrow resonance lines and large electric fields. We have shown previously[3] that for an optically detected magnetic resonance with a linewidth limited by optical power broadening a (shot-noise limited) sensitivity to atomic EDM's of the order of 10^{-25} e cm may be expected with the following experimental parameters: spin polarization = 50%, electric field strength = 100kV/cm, fluorescence collection efficiency = 10^{-3}, number of atoms = 10^7, pump time = 10 ms, and total integration time = 100 hours. This sensitivity is more than one order of magnitude higher than the lowest current experimental limit for the EDM of a paramagnetic atom[11]. These purely statistical arguments are clearly in favour of using paramagnetic atoms in solid He to search for EDM's.

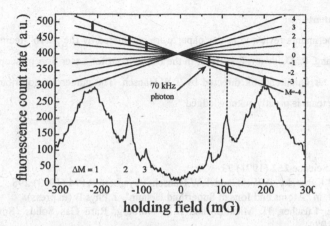

Fig. 4 Breaking of cylindrical symmetry in hcp phase evidenced by appearance of "forbidden" lines. These resonances are absent in the bcc phase.

3. Conclusion and Outlook

Although the He matrix is basically polycrystalline due to the stress introduced by the implantation process, it seems that the local atomic trapping sites conserve the symmetry of the phase corresponding to the experimental pressure and temperature values, i.e isotropic in the bcc phase and uniaxially anisotropic in the hcp phase. We thus find in hcp ⁴He a situation similar to the heavier rare gas solids (RGS), in which the anisotropic character of local trapping sites prevents the efficient optical pumping of implanted paramagnetic species. With our present knowledge however we can exclude the existence of multiple trapping sites, in contrast to the situation found in heavier RGS's. The matrix effects reported here are preliminary first results, and more effort is needed to get a more quantitative understanding. However, the fact that in the bcc phase no evidence for symmetry breaking effects could be detected so far is good news for an eventual EDM experiment.

Our main current efforts are oriented towards the improvement of the magnetic field homogeneity and the implementation of magnetic shields in order to measure the ultimate magnetic resonance linewidth. Furthemore we work on improving our detection efficiency and testing new ways for detecting the resonance. Finally, first steps will be undertaken to measure electrical breakdown voltages and leakage currents, and to test the properties of our sample in large electric fields. After completing these changes and studies we should be able to answer the question of whether paramagnetic atoms in solid Helium are really a promising novel sample to hunt for T-violation.

Acknowledgements

The experiments reviewed in this paper were conducted by the „cryo"-team (S.Kanorsky, M. Arndt, S. Lang, R. Dziewior, S.B.Ross) in the Department for Laser Spectroscopy at the Max-Planck-Institut für Quantenoptik directed by T.W.Hänsch. The numerous contributions from all members of the team is greatly acknowledged.

References

1. L.R.Hunter, Science **252** (1991) 73
2. M.Arndt, S.I.Kanorsky, A.Weis, and T.W.Hänsch, Phys.Lett. **A174** (1993) 298
3. A. Weis et al, in "Atoms and Ions in Superfluid Helium", Z.Phys.B (in press)
4. H.Coufal, E. Lüscher, H. Micklitz, and R.E. Norberg, Rare Gas Solids (Springer, Berlin, Heidelberg,1984)
5. S.I.Kanorsky,M.Arndt,R.Dziewior,A.Weis, and T.W.Hänsch, Phys.Rev.**B49** (1994) 3645
6. M. Arndt et al, in "Atoms and Ions in Superfluid Helium", Z.Phys.B (in press)
7. S.I.Kanorsky,M.Arndt,R.Dziewior,A.Weis, and T.W.Hänsch, Phys.Rev.**B50** (1994) 6296
8. S. Kanorsky et al, in "Atoms and Ions in Superfluid Helium", Z.Phys.B (in press)
9. S.Lang, S.I.Kanorsky, M.Arndt, S.B.Ross, T.W.Hänsch, and A.Weis, Europhys.Lett. **30** (1995) 233
10. M. Arndt, S.I. Kanorsky, A. Weis, and T.W. Hänsch, Phys.Rev.Lett. **74** (1995) 1359
11. E.D. Commins, S.B. Ross, D. DeMille, and B.C.Regan, Phys.Rev.**A50**, 2960 (1995)

Quantum Optics with Few Atoms and Photons

Axel Schenzle

Sektion Physik der Universität München,
Theresienstr. 37, 80333 München

Abstract

Macroscopic as well as microscopic systems are governed by the laws of quantum mechanics. The summation over the myriads of degrees of freedom characterizing a sizeable chunk of material, however, disguises the peculiar features of the quantum world and in most cases leaves us with phenomena that are in agreement with the predictions of classical physics. Experimenting with a few atoms and photons at a time, one expects that the characteristic properties of quantum mechanics must become clearly visible. The delicate measurements in this microscopic regime are confronted with the subtleties of the quantum measurement process.

1 Introduction

The only physical phenomena that display quantum features on a macroscopic scale are superconductivity and the superfluidity of He_4 and He_3, where a large fraction of the particles that constitute the system condense into a single quantum state. In atomic or solid state physics the coherence properties among different atoms are mostly limited to a small number of particles or to very small time intervals due to the randomizing nature of macroscopic systems. With the rapid development of experimental techniques like storage and cooling of single atomic particles, it has become possible to experiment with an individual atom over an extended period of time.

In what way is the physics different when looking at a single particle only? A thermally excited ensemble of atoms e.g. emitts fluorescence light with Poissonian statistics causing bunches of photon counting events. A single atom driven continuously by laser light is a weak but nevertheless observable light source. The photons detected from such a source are more regularly spaced than in the previous case - this is immediately obvious: a single atom can only emit a single photon at a time. In order to produce a second fluorescence quantum it must first be excited again, and this takes time. Therefore, the emission characteristics from a single atom shows a minimal separation of quanta, which is called anti-bunching. This is just one, but an especially intuitive example where the quantum laws become visible when single particles are involved. The anti-bunching is a unique quantum effect which is absent - as a matter of principle - in a classical description of a fluctuating electromagnetic field.

2 What is a non-classical field ?

Considerable confusion was created recently by the use of the term non-classical light. Are not all phenomena in physics non-classical, in the sense that we believe that quantum theory provides the proper description of microscopic as well as macroscopic physical processes. Loosely speaking, we call a state non-classical, when physical observables have properties that are classically impossible. We will explain this from different points of view in the following.

2.1 Phase-space densities

In 1932 Wigner constructed a phase-space description of quantum mechanics revealing a useful analogy between quantum mechanics and classical statistical mechanics. The classical positive phase density is replaced by the Wigner-function which serves exactly the same purpose as the classical probability density. For instance, quantum ensemble averages are calculated by integration over the Wigner-Function times the classical variables. The Wigner-Function $W(x, p, t)$ can be positive like a classical density, but is not restricted to positive values. In the first case, observables of the quantum system must have the same properties as a possible classical analog. In the latter case, however, no classical description can ever explain the properties of the quantum system consistently. Glauber [1] has discovered a very useful generalization of this concept originally developed in the field of quantum optics :

$$P(\alpha, \alpha^*, \beta, \beta^*, ..., t, s) \not\geq 0$$

where the continuous parameter s, ranging between -1 to 1, specifies different phase space densities. The close internal relationship between these concepts is obvious. For : $s = +1$: P = Glauber P-function, $s = 0$: P = Wigner function and $s = -1$: P = Husimi function, another widely used representation which is always positive. Positivity itself is an ambiguous property which cannot be used as an indicator for non-classical behaviour. We will see that it is the Gauber representation that plays the preferred role.

2.2 Inequalities of classical statistics

As a consequence of positivity of classical phase space densities, there exist a number of inequalities for the stationary correlation functions. These do not have to be obeyed by the corresponding quantum process. The following correlation functions for a classical variavble x(t) :

$$G_1 = \langle x(t)x(0) \rangle$$

and

$$G_2 = \langle x^2(t)x^2(0) \rangle$$

satisfy the inequalities:

$$G_1(t = 0) \geq G_1(t)$$
$$G_2(t) \geq G_1^2(t)$$

Quantum states that violate those inequalities have no classical correspondence and are therefore called non-classical states.

2.3 Squeezed states

The averages of many physical quantities vanish in the vacuum states but nevertheless experience random fluctuations about zero. Physical states that are more quiet than the vacuum states are called "squeezed" and require a non-positive Glauber P-function. An example is the quadrature operator of the electro-magnetic field:

$$\hat{E}(\phi) = \frac{1}{\sqrt{2}}(b^\dagger e^{i\phi} + b e^{-i\phi})$$

Squeezing is realized when :

$$\langle \Delta^2 \hat{E}(\phi) \rangle \leq \langle \Delta^2 \hat{E}(\phi) \rangle_{vacuum}$$

2.4 Photon counting statistics

The non-classical statistics of a light beam may be characterized by the fluctuations of the photo-electric current generated by a detector which is exposed to light for a given time interval T. A classical field with a perfectly stable intensity is associated with a Poissonian statistics of counting events. Any additional noise in the field intensity results in a broadening of the counting statistics. Observation of a more regular photon current with noise less than for the Poissonian case is a unique and unmistakeable indication of the quantum nature of the field.

3 Experiments with single atoms

In an ideal sense, experiments with single or few atoms can be carried out in ion or neutral atom traps, where individual particles can be stored over a substantial period of time by electromagnetic fields and can be irradiated by laser light [2]. In order to catch and store particles their kinetic energy must first be reduced by laser cooling. Examples are the Paul-trap for ions, magnetic quadrupol traps for neutral atoms and trapping potentials created by the gradient forces in standing light fields or optical molasses. The first experiments with single particles were made with atomic beams of such weak flux rates that at any given instant of time at most a single particle is present in the observation region [3]. Below we briefly summarize some experiments with few atoms that had a substantial impact on our understanding of matter field interaction, spontaneous processes and the measurement problem.

3.1 Quantum jumps

The apparent contradiction of the continuous time evolution as described by the Schrödinger equation and the discrete nature of spontaneous emission or radioactive decay was a puzzling aspect of early quantum mechnanics. However, it was a conceptual and not a practical problem as long as experiments always involved a large number of atoms. As Schrödinger argued himself [4]: "We never experiment with one electron or atom - in thought experiments we sometimes assume we do; this invariably entails ridiculous consequences" - like quantum jumps. That this is not as ridiculous as Schrödinger thought was suggested and demonstrated first by Dehmelt [5], [6] and analyzed theoretically by [7]. Here a single ion was monitored in a Paul-trap while it was irradiated by two light fields. The atom was approximated by a three level structure in V-configuration and the two excited states were reached by the two resonant external fields. One transition was optically allowed, while the other one was forbidden with a lifetime of the order of ten seconds. Fluorescence from the allowed transition of a single atom was observed as a faint but visible signal. As soon as the second laser succeeded to excite the electron into the long living state, fluorescence was quenched and only reappeared when the electron spontaneously returned to the ground state. In this way the individual jumps of the electron up and down the forbidden transition became observable even to the naked eye as an intermittend fluorescence signal. The three level configuration provided a unique amplification mechanism turning the single photon emitted in the forbidden transition into an avalange of visible photons. This is a unique single atom quantum effect which is quickly washed out when more that one atom is present.

3.2   Micro-maser

While in the previous example a single atom interacted with the multitude of vacuum modes, one can study the interaction with a single quantized mode by placing the atom inside a resonant high Q cavity. The fundamental mode frequency of a micro-wave cavity lies in the Giga-Herz regime, the corresponding resonant atomic transition can be realized by Rydberg states of alkali-atoms. In this case the higher harmonics of the field are so far from resonance that they can safely be neglected. An excited atom entering a cavity in its vacuum state may deposit its energy - depending on the time of interaction - and leave in the ground state. In case that the field energy is not already dissipated until the next atom enters, energy can be accumulated in the field and the device acts as a "single-atom maser" [8], [9]. By "single atom" we want to indicate that at a given instant in time at most one atom is participating in the interaction process and - most of the time the cavity is entirely empty. Such a system must have peculiar properties because raising the flux of atoms does not necessarily raise the photon number in the cavity. The atomic beam drives the field most effectively, when the atoms stay just long enough in the resonator to undergo half a Rabi-cycle and deposit all their energy. Thereby the photon number as well as the Rabi-frequency increases until the atoms undergoe a full cycle and leave the cavity in the excited state again - energy is no longer deposited and cavity damping gradually reduces the amplitude again. The photon statistics of the cavity field displays vaguely different characteristics, depending on the pumping conditions. By increasing the pump flux the variance of the photon number varies from strongly Super-Poissonian fluctuations i.e. $\langle \Delta^2 n \rangle \geq \langle n \rangle$ to situations with a sharp photon distribution $\langle \Delta^2 n \rangle \leq \langle n \rangle$. At higher pump rates, the distribution becomes doubly humped and the operation of the maser shows bistability and hysteresis [10]. All these features - classical or quantum - are observed while the cavity contains only 10 to 20 micro-wave photons.

3.3   Zeno-effect

Up to this very day the controversy on what a quantum measurement is still puzzles the minds of many scientists. Many feel uneasy with the traditional concept of measurement but take a pragmatic attitude: Quantum theory provides a set of rules by which we can predict - as it turns out with unprecedented accuracy - the outcome of all possible experiments. From an experimental point of view there is no need for a "new quantum theory" since the present one is in excellent agreement with measurements - it is the interpretation that is irritating. But who says that we must be comfortable with the findings of science? In the development of science one attempts to discover the rules by which nature operates, and nature does not care if we are content with what we find or not. In developing macroscopic physics it has been a useful guide line to follow pathes we feel comfortable with and which seem to satisfy our logical requirements, but this is not a necessary condition, especially not in the realm of microscopic processes where we have no intuition to start with.

While the idea of a classical measurement poses no basic or merely interesting questions, the outcome of a quantum measurement can only be predicted with a certain probability, even when the state prior to measurement is completely known in the sense of quantum mechanics. Even taking into account the detailed interaction mechanism of the system with the meter and the dynamics of the meter itself does not get around this problem. There is no way to calculate at which number of the dial the pointer will stop. And this is how it should be, since there is also no way to manipulate the experiment to ensure a specific meter reading. We therefore must conclude that the meter changes the state of the observed system in an irreversible way into the state consistent with the meter reading - knowledge of the previous state is lost for ever. It doesn't even make sense to talk about the state before it is measured. Repeating the measurement immediately after the first reading must result in an identical outcome. When repeating the measurement in rapid succession, the meter must keep reading the same value and therefore must bring the dynamics of the watched sytem to a complete halt [11] in analogy with Zeno's paradox where a watched arrow cannot move.

This effect can be demonstrated in a quantum optical experiment. Taking the previous three level atom, we can argue that the observation of the intermittend fluorescence field is a measurement of the population probability of the slow transition. A fluorescence signal indicates that the atom is in its ground state, while a dark period results when it occupies the metastable state. Watching the unitary time evolution along the slow transition by applying a rapid sequence of optical pulses must interfere with the Rabi-oscillation and bring the unitary time evolution to a halt [12]. This was observed in quantitative agreement with theory using an ensemble of Mg atoms stored in a Paul trap by Wineland et al.[13].

Zeno argued on a slowing down of the arrow in real space, not in configuration space. Just for curiosity one might want to see if this can also be realized experimentally. We might irradiate an atom in free flight by laser light. Resonance fluorescence from the internal degrees of freedom reveal the information about the position of the atom. The accuracy is given by the wavelength of light. A simple calculation, however, shows that this is not the case as long as the dynamical system is strictly linear. Placing the atom in a double well potential, where free motion is replaced by tunneling between "left" and "right", provides us with a non-linear process and quenching of motion is obtained. If the entire potential domain is irradiated, motion is only effected when the wavelength of the illuminating field is comparable or shorter than the well separation [14]. An inevitable side effect is momentum transfer by photon recoil. This phenomenon my easily mask the effect by heating the particle out of the well. But there is a fascinating way to get around this problem. If we illuminate only one of the wells and leave the other in the dark we can detect whether the particle is on the left or right and thereby make a possition measurement. If the particle is initially in the dark it will stay there indefinitely and the measurement prevents it from tunneling into the bright zone [15].

3.4 Quantum-non-destruction experiment

Until recently it appeared conceptually useful to differentiate between two kinds of measurements: Those which only determine the state of a system but otherwise leave the system intact, from measurements which destroy the system after the measurement has been taken. Examples of the first class are measurements of momentum or position of an electron or the population of atomic levels as in the previous examples. The obvious example of the second kind is the counting of optical photons. After a measurement with electrons or atoms we can still talk about the particles in the future. When counting photons all we can say is that there have been n photons before we detected them and sadly report that they are inevitably lost due to detector absorption. Is there atleast in principle a possibility to count photons with out destroying them? Then we can say confidently and not in past tense that after the measurement we still have n photons in a cavity. Traditional photon detectors derive their energy from the absorption of the photon. A non-destructive detector instead must be based on a dispersive effect.

Haroche et al. have proposed and partly realized such a device on the basis of the micro maser concept combined with the Ramsey interference technique [16]. The field to be measured is contained in a high-Q micro wave cavity. A weak beam of atoms passes through the cavity to count the number of photons. Prior to entering the cavity the atoms are excited from a Rydberg "ground state" $|g\rangle$ into a coherent 50:50 superposition with a higher Rydberg state $|e\rangle$. The superposition evolves freely and the phase difference accumulates while the atom passes through the cavity. In a second Ramsey zone at the end of the cavity the population is turned back to the ground state if the phase difference to the initial state is a multiple of 2π. If, however, the phase has been modified along the way due to the presence of the cavity field, the atoms can end up in any superposition state. A detector at the end will determine the ratio of atoms in excited and ground state. A single atom only contains the binary information "up" or "down" and can only carry a single bit of information. If we expect the cavity to contain any number of photons between 0 and n then we need at least to read out m atoms where $n = 2^m$, m is the number of binary digits needed to represent the possible photon numbers. In a real experiment the number can only be larger.

But how does one obtain the necessary information on the cavity field without changing the field state? One moment - that is not what we had been looking for; we only wanted to count photons without destroying them, the state of the field is inevitably modified in course of the measurement. The modification of the state is a basic consequence of the quantum measurement process. The modification in this case, however, must be a subtle one that leaves the photon number invariant. What is changed is the complementary property, the phase information of the state measured. In order to imprint the information of the photon number on to the atomic beam we consider a third Rydberg state above the two other ones. The transition frequency between the two excited states is chosen close to the resonance frequency of the occupied cavity mode, but detuned enough that no real transitions occur. But virtual transitions induced by the field shift the energy levels proportional to the intensity inside the cavity:

$$\Delta E = \hbar g^2(n+1)/\delta$$

where g is the dipole matrix element of the transition, δ the detuning between the field and the two excited states and n is the photon number. Shifting the middle atomic level changes also the transition frequency of the $|g\rangle$ to $|e\rangle$ transition, and thereby the phase accumulation of the superposition state at the end of the cavity. Assuming for a moment that in case of vacuum the total phase is a multiple of 2π then in the presence of a field the actual phase change is proportional to n, the number of photons. In this case the desired signal is derived from the dispersive part of the interaction and not from the absorptive one - this is the key to the non-destructive character of the measurement.

Dirac had speculated about complementarity in the case of the photon field. A particle's mechanical properties x and p can not be measured simultaneously due to the non-comutability of the representing operators. There should be a similar complementarity involving the photon number. Intuitively the variable complementary to the number is the phase: knowing the photon number wipes out all information on the phase of the field and vice versa. Unfortunately, there is no phase operator in the sense of a well defined mathematical object. We have an intuitive picture of the phase and of the states that possess a prefered phase, or of states that are phase invariant. The phase space representation discussed above directly visualizes the phase and amplitude properties of a field. $P(\alpha, \alpha^*, t)$ describes quantum properties of a single field mode. The complex variables α, α^* representing the complex field amplitudes $b, b^\dagger$, span a two dimensional space where $\alpha = r exp(i\phi)$ contains information on the modulus r and the phase ϕ of the field. Let us suppose that the initial state of the cavity has a certain amplitude r_0 and a certain phase ϕ_0, then it is represented by a density $P(\alpha, \alpha^*)$ localized around r_0, ϕ_0. The photon number measurement step by step redistributes this density and smears it over the entire range 0 and 2π leaving the "radius" i.e. the intensity unchanged: $r = r_0$. In an intuitive way this illustrates the Dirac conjecture about the complementarity of number and phase of the electro-magnetic field [17].

When simulating such an experiment by throwing dice for each outcome of the detector reading, according the calculated probabilities, one finds that it takes a large number of atoms to finaly determine the photon number inside the cavity. The reason is that in the simulation a fixed atom velocity and therefore a fixed interaction time had been used. This is not very realistic since an atomic beam, even with velocity selectors, has a finite spread. Repeating the simulation and keeping track of the the individual velocities cuts down the number of atoms needed. Making the calculation more realistic in this case has a very desirable effect: it shortens the time necessary for the measurement and therefore lifts the restrictions on the cavity storage time. The reason for this - unexpected - improvement is easy to see. Due to the perodicity of the Rabi-oscillations: $g\sqrt{n+1}\tau$, τ being the transit time of the atom, a fixed velocity makes the measurement sensitive to certain photon numbers and insensitive to others. A beam with a velocity spread, however, is testing the distribution much more efficiently, as long as we keep track of the velocity of every particle. But even with a finite velocity bandwidth, we do not reach the theoretical limit based on the binary information content of the beam.

In order to reach the absolute minimum of atoms needed, we have to optimize the data taking even further. According to the outcome of the first measurement and the correponding reduction of the wave function, we can calculate the expected entropy decrease in a second measurement as a function of the paricle velocity. It has been shown that choosing the optimum velocity each time a measurement is performed reduces the number of measurements to the suggested minimum. It might be argued that such a procedure is entirely academic since it seems not possible to select the atom velocity in an arbitary and fast way. For a setup with mechanical Fizeau selectors it is obviuosly impossible to select different groups. With Doppler selective excitation of the used Rydberg states it seems feasible to change the frequency of the exciting laser and thereby select the desired velocity from the beam with a wide velocity spread.

4 Atom-counting statistics

In his pioneering work Glauber [1] has formulated a quantum mechanically consistent theory of photon counting experiments and thereby generalized the existing semiclassical theory of counting events developed by Mandel [18]. On the basis of an elementary and idealized model of the photon detector, he showed how the operator ordering in quantum averages and quantum correlations is determined by the mesurement process. For a single mode of the field quantum correlations are defined as :

$$G_1(t) = \langle b^\dagger(t)b(0) \rangle$$
$$G_2(t) = \langle b^\dagger(0)b^\dagger(t)b(t)b(0) \rangle$$
$$..... = ...$$

The photon counting statistics i.e. the probability W of detecting n events during a time interval T was found to be :

$$W(n,T) = tr\hat{T}\rho\frac{1}{n!} \left(\eta \int_0^T b^\dagger(t)b(t)dt \right)^n exp(-\eta \int_0^T b^\dagger(t)b(t)dt)$$

where η, $0 \leq \eta \leq 1$ is the quantum efficiency of the detector and $\hat{T}$ an ordering convention that ensures time and normal ordering. The mathematical structure of the result suggests that all we need is the "no-count" probability $W(n = 0, T)$ since all other probabilities can be derived from there. By differentiating W(n=0,T) with respect to T associated statistical measures can be derived like the "waiting time" or "dark time" probability.

In recent experiments like the micro-maser and the quantum-non-destruction measurement, the entire information about the system is encoded into the level population statistics of the atomic beam that emerges from the interaction region. In analogy to the photon statistical tools summarized above, it would be desirable for a description of the micro-maser e.g. to develop similar tools for the atomic beam.

The probability of observing n atoms in a given time interval T represents merely the statistics of the beam source, since atoms are neither created nor destroyed in the non-relativistic interaction process. The probability is in good approximation Poissonian :

$$F(n,T) = \frac{1}{n!}(rT)^n exp(-rT)$$

This is the underlying classical statistics originating in the atomic oven. What we are interested in is the level selective statistics : the probability of observing n atoms in state $|a\rangle$ and m atoms in state $|b\rangle$ in a time T.

$$W(n, |a\rangle, m, |b\rangle, \eta, T)$$

where η is the quantum efficiency of the two state selective detectors. The redistribution of the electron over the two resonant states contains the information on the state of the cavity field. In analogy to the photon case we are primarily interested in the probability of observing *no* atom in state $|b\rangle$ irrespective of the number of atoms in state $|a\rangle$ that went by. More interesting from an experimental point of view is the waiting time probability : after the observation of an atom in state $|b\rangle$, how long does one have to wait until the next one appears in the same state? This probability is closely related to the above one like in the photon case - but not quite as simply [19], [20].

To illustrate the idea of the atom-counting statistics we apply the method to the micro-maser problem. An excited atom enters the cavity at time t_0 and starts to interact with the field through the Jaynes-Cummings Hamiltonian. During the short period the atom spends in the cavity, losses can safely be neglected. After the transit time τ the atom leaves the cavity and the total state of the combined system became entangled. A measurement of the internal state of the atom therefore projects the field into a different state, depending on the outcome of this measurement. Formally we write without the proper normalization:

$$\rho(t_0 + \tau) = \hat{O}_{|j\rangle}\rho(t)$$

where $j = a, b$. The operator $\hat{O}_{|j\rangle}$ has the form :

$$\hat{O}_{|j\rangle} = \langle j|exp(-i/\hbar H_{J-C}(\tau))\rho(t_0)|j\rangle$$

Between two successive atoms the cavity relaxes freely with the usual damping Liouvillian L.

$$\rho(t) = exp(L(t - t_0 - \tau))\rho(t_0 + \tau)$$

The probability that in the interval T no atom enters the resonator is $F(T) = exp(-rT)$ due to the Poissonian nature of the atomic beam. The probability for finding n atoms in state $|a>$ and no atoms in state $|b>$ under stationary conditions is given by:

$$W(n, |a>; 0, |b>; T) = tr\rho_{stst}G_n(T)$$

where the propagator $G_n(T)$ for the sub-ensemble is given by:

$$G_0(T) = e^{LT}F(T)$$

$$G_1(T) = r\int_0^T dt_1 e^{L(T-t_1)}F(T-t_1)\hat{O}_{|a>}e^{Lt_1}F(t_1)$$

$$G_n(T) = r^n \int_0^T dt1 \int_0^{t_1} dt_2 .... \int_0^{t_{n-1}} dt_n *$$

$$* \ e^{L(T-t_1)}F(T-t_1)\hat{O}_{|a>}.....\hat{O}_{|a>}e^{Lt_n}F(t_n)$$

ρ_{stst} is the steady state density operator for the field alone as it follows from the micro-maser theory. If we only use a $|b\rangle$ atom detector $|a\rangle$ atoms go by unnoticed and the probability of measuring no atom in $|b\rangle$ is given by summation over the unobserved events:

$$W(0, |b>, T) = \sum_{n=0}^{\infty} W(n, |a>; 0, |b>, T)$$

$$= tr\, exp(LT + r(\hat{O}_{|a>} - 1)T)\rho_{stst}$$

State selective atom detectors are notoriously insensitive, and the quantum efficiency η must be taken into account explicitly. Particles in state $|b\rangle$ are detected with probability η and they go unnoticed with probability $(1 - \eta)$. In a similar calculation we obtain the following result :

$$W(0, |b>, \eta, T) = tr\, exp(LT + r(\hat{O}_{|a>} + (1 - \eta)\hat{O}_{|b>} - 1)T)\rho_{stst}$$

This is now the atomic analogon of the zero-count probability $W(n = 0, T)$ for photons.

A closely related property is the probability density $P_2(|b>, \eta, T)$ for the time T one has to wait for the next atom in state $|b>$. The propagator for this probability measure is the same as in the previous case. We only have to guarantee that at the initial and the final instant of time an atom was detected in state $|b>$:

$$P_2\ (0, |b>, \eta, T) = \frac{tr\, \eta\hat{O}_{|b>}\, e^{LT + r(\hat{O}_{|a>} + (1 - \eta)\hat{O}_{|b>} - 1)T}\, \eta\hat{O}_{|b>}\rho_{stst}}{(tr\, \eta\hat{O}_{|b>}\rho_{stst})^2}$$

A useful tool for characterizing the fluctuations in a photon field are correlation functions. Something that comes to mind immediately here is the two atom correlation function $g_2(T)$ in correspondance to the two-photon correlation $G_2(t)$ in section 2. In contrast to the waiting time probability density which is an exclusive measure, the correlation function only relates two instants in time and does not notice what happens inbetween. This is easily achieved with our previous result, when we allow the detection efficiency to approach zero, provided we have used the proper normalization [19]. In the limit :

$$g_2(T) = \lim_{\eta \to 0} P_2(0, |b>, \eta; T)$$

we obtain the "two-atom in state $|b>$ correlation function"

$$g_2(T) = \frac{tr\hat{O}_{|b>}\, e^{LT + r(\hat{O}_{|a>} + \hat{O}_{|b>} - 1)T}\, \hat{O}_{|b>}\rho_{stst}}{(tr\hat{O}_{|b>}\rho_{stst})^2}$$

A mathematical detail should not go unnoticed: the two limits $\eta \to 0$ and $T \to \infty$ obviously do not commute since:

$$\lim_{T \to \infty} P_2(T) = 0$$
$$\lim_{T \to \infty} g_2(T) = 1$$

This is follows from the facts that a normalizable density must vanish at infinity, while a correlation function factorizes at large times due to the statistical independence of infinitely distant statistical events.

The atom counting measurements are the ony way to obtain information on the state of the field in a closed cavity like its photon number and its fluctuations. But even the time scale of phase fluctuations can be determined by detecting the correlations in the state distribution of the atomic beam [21].

References

[1] R.Glauber. *Quantum Optics and Electronics, Les Houches*, 1965.

[2] H.Dehmelt. *Angew.Chem.*, 102, 1990.

[3] R.Short L.Mandel. *Phys.Rev.Lett.*, 51, 1983.

[4] E.Schroedinger. *BPS*, 3, 1952.

[5] W.Nagourney J.Sandberg and H.Dehmelt. *Phys.Rev.Lett.*, 56, 1986.

[6] Th.Sauter W.Neuhauser R.Blatt and P.E.Toschek. *Phys.Rev.Lett.*, 57, 1986.

[7] A.Schenzle and R.G.Brewer. *Phys.Rev.A.*, 34, 1986.

[8] D.Meschede H.Walther and G.Mueller. *Phys.Rev.Lett*, 54:551, 1985.

[9] G.Rempe and H.Walther. *Phys.Rev.A*, 42, 1990.

[10] O.Benson G.Raitel and H.Walther. *Phys.Rev.Lett.*, 72, 1994.

[11] B.Misra and E.C.G.Sudarshan. *J.Math.Phys.*, 18, 1977.

[12] V.Frerichs and A.Schenzle. *Phys.Rev.A*, 44, 1991.

[13] W.M.Itano D.J.Heinzen J.J.Bollinger and D.J.Wineland. *Phys.Rev.A.*, 41, 1990.

[14] T.Altenmueller and A.Schenzle. *Phys.Rev.A.*, 49, 1994.

[15] I.Cirac A.Schenzle and P.Zoller. *Euro.Phys.Lett.*, 13, 1994.

[16] M.Brune S.Haroche V.Lefevre J.M.Raimond and N.Zagury. *Phys.Rev.Lett*, 65, 1990.

[17] R.Schack A.Breitenbach and A.Schenzle. *Phys.Rev.A*, 45, 1992.

[18] L.Mandel. *Proc.Phys.Soc.London*, 72:1037, 1958.

[19] C.Wagner A.Schenzle and H.Walther. *Opt.Commun.*, 107, 1994.

[20] H.-J.Briegel B.-G.Englert N.Strepi and H.Walther. *Phys.Rev.A.*, 49, 1994.

[21] Ch.Wagner R.J.Brecha A.Schenzle and H.Walther. *Phys.Rev.A*, 46, 1992.

Solid State Laser Intensity Noise Reduction

Charles C Harb, Timothy C Ralph[†], David E McClelland, Hans-A Bachor,
Ingo Freitag[‡], Andreas Tünnermann[‡] & Prof. Herbert Welling[‡]
Department of Physics, Faculty of Science,
The Australian National University, ACT 0200 Australia
[†]*Department of Physics, Private Bag 92019,*
University of Auckland, Auckland, New Zealand
[‡] *Laser Zentrum Hannover, Hollerithallee 8, D-30419 Hannover, Germany*

1   Introduction

Intensity stable laser sources are requires for both applied and fundamental research fields, such as telecommunications or gravitational wave antenna research. The aim of this collaboration between   the Australian National University and Laser Zentrum Hannover is to examine techniques for producing stable light sources that are operating at their fundamental limit as set by the quantum properties of the light.

Two types of techniques are possible for stabilising the intensity noise of a laser system, classical or non-classical intensity stabilisation. Classical intensity stabilisation covers the techniques that rely on linear behaviour of both the controlled and the controllers in the stabilisation process. Non-classical, on the other hand, depends on some non-linear behaviour of some section of the stabilisation process such as the χ^3 non-linear susceptibility of a medium. This latter case can produce lasers with outputs that are squeezed states, and hence have intensity noise that is lower than the quantum noise limit, QNL. Non-classical intensity control will be discussed by M. Taubmann in the talk entitled *'A reliable source of squeezed light, and an accurate theoretical model'* in these proceedings.

In this paper we will discuss active intensity control and injection locking intensity control which are both forms of classical intensity control.

2   Active Intensity Control

The basic design for an active intensity control system is shown in figure 1. This diagram illustrates the main components of the control system, namely, the laser to be controlled; an intensity detector in the form of a photodetector (PD); the control electronics that are used to ensure that the correct signals are used in the control loop; and some device that converts the electrical signals into intensity signals on the laser light. The effect of the intensity control system is monitored on a PD that has its signal suitably amplified and is connected to an RF Spectrum Analyser (SA) so that the RF spectral density intensity noise can be recorded.

This control system be analysed using conventional control theory [1]. The expected noise suppression is given by the simple expression:

$$S(f) = \left(\frac{1}{1 + G(f)} \right) \tag{1}$$

where $S(f)$ is the residual noise in the control loop and $G(f)$ is the complex open-loop gain.

In an ideal classical system $S(f)$ would continually decrease as $G(f)$ increased. However, the quantum noise of the light being detected by the PD in the control loop [2] limits the achievable noise reduction. This is because the quantum fluctuations of the light injects uncorrelated noise into the control loop, and hence can add intensity noise if $G(f)$ is too large. The minimum excess noise with respect to the quantum noise limit (QNL) is given by the relation:

$$\Delta Noise(dB) = 10 Log_{10}\left(1 + \frac{\imath_2}{\imath_1}\right) \tag{2}$$

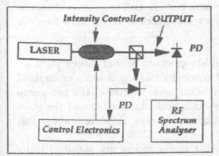

Figure 1. Experimental design for a basic intensity noise control system. In this diagram PD refers to a photodetector.

where $\imath_1$ is the photocurrent detected by the PD in the control loop and $\imath_2$ is the photocurrent detected by the detector monitoring the output of the control system. The lower limit for the intensity control is thus 3dB above this quantum noise limit when $(\imath_2/\imath_1) \approx 1$.

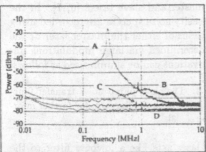

Figure 2. Experimental results for the suppression of intensity noise in an diode pumped Nd:YAG miniture ring laser system. A: relaxation oscillation; B: noise suppression; C: QNL; D: electronic noise level. For further information see ref. [2].

The advantages of using this type of control system is that large classical intensity oscillations, such as resonant relaxation oscillations [3], that occur in solid state laser systems, can be reduced by many orders of magnitude. This cannot be achieved easily by other methods. The disadvantage is that this technique will never completely reduce the noise to the QNL and in general there will be an increase in noise at certain frequencies of the intensity noise spectrum due to loop stability considerations.

3   Injection Locking Intensity Control

Injection locking is an established field of research dating back to the 1960s [4, 5, 6] but until recently it has been extremely difficult to investigate its properties for several reasons including: the lack of stable low noise light sources and of a suitable quantum theory that takes into account the full dynamics of the coupled cavity and laser systems. A conventioal injection locked laser system consist of a high power laser (the slave laser) which is locked to a low power, low noise laser (the master laser), see figure 3.

Injection locking is a phenomena that occurs when the injected master laser mode has a wavelength that lies simultaneously inside the gain bandwidth of the slave laser and is near a longitudinal mode of the slave cavity [3]. We are concerned primarily with the intensity noise spectrum of the injection locked system when the master and slave wavelengths are nearly identical.

We have developed a fully quantum mechanical model that can accurately describe the cascaded laser cavity system, and performed a set of experiments to test the validity of the model and to attempt to understand the properties of the injection locking process.

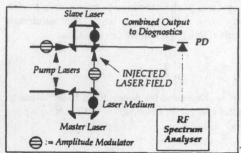

Figure 3. Schematic design of an injection locking system. In this diagram we have represented the master and slave lasers as ring lasers, as they are the case in the experiments performed. Shown in this diagram is the pump lasers and the injection of the master light into the slave cavity. Not shown is the diagnostics that are used to measure the intensity noise properties, such as the balanced detection system.

Our complete quantum optical theory [3, 5] allows us to predict the behaviour of the injection locked system by treating it as an example of cascaded optical systems. The theoretical model describes the noise spectra of the two pump lasers and it calculates how this pump noise is coupled into both the master and the slave lasers, and finally how this in turn generates the output noise spectrum. For details about this model see ref. [3].

All the results for the experiments testing the validity of our model are obtained using diode pumped nonplanar Nd:YAG ring lasers as both master and slave lasers [6, 7]. In the experiments two types of observations are made. The intensity noise spectra are recorded using a PD plus RF SA and the transfer functions for modulations of either the master laser or the pump of the slave laser through the entire system are determined via large modulations imposed using appropriate amplitude modulators. Our observations can be summarised as follows:

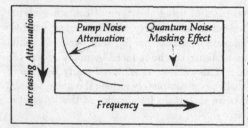

Figure 4. The slave pump noise transfer function. This is a schematic diagram, for further information see ref. [3].

The slave laser acts like a lowpass filter to the noise of its pump source. The transfer of modulation from the slaves' pump laser through to the injection locked slave output is characterised by no attenuation at DC (zero modulation frequency), followed by continuously increasing attenuation with the modulation frequency until the signals fall below the QNL of the slave laser. The rate of change of the attenuation, which determines the bandwidth of the lowpass filter, can be as high as 20 dB per decade for laser cavities with high atom-cavity coupling.

The transfer of modulation of the master lasers intensity through the injection locked system is more complicated, it can best be described by discussing three distinct frequency regions. At low frequencies, region (i) , the slave laser acts like a low impedance sink to the modulation of the master laser. That is, the dynamics of the slave responds too slowly, and hence any forced oscillation cannot be followed. Therefore, the master has small influence at DC but has greater influence as the frequency increases till it reaches the region (ii). In region (ii), the output signal is an amplified version of the spectrum of the master laser, and is thus called the amplification region. The modulation of the master lasers intensity is directly converted into a modulation of the slave laser, the amplification factor is determined directly by the ratio of the power of the slave laser to the power of the master laser. However, the width of this region is inversely proportional to the amplification factor and therefore large amplification factors may not be desirable due to the reduced influence of the master laser. It should be noted that in this region, and with the absence of classical modulations, the quantum noise of the master laser is amplified and dominates the noise spectrum of the injection locked system. In the final region, (iii), there is neither attenuation nor amplification of the signal from the master laser. Modulation from the master are reflected off the cavity of the slave laser without entering. The reflected master signal coherently beats with the strong new injection locked slave carrier mode and produces a signal at the output which is smaller in magnitude than the original by the power ratio. In region (iii) the noise floor is set by the quantum noise of the of the master laser beating with the slave mode. It is at the QNL of the slave if the master is QNL limited, but can be higher or lower if the master is super or sub-Poissonian.

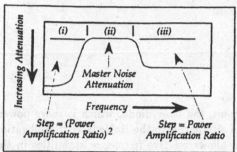

Figure 5. The master noise transfer function. This is a schematic diagram, for further information see ref. [3].

The advantage of our full quantum theoretical model, over earlier classical models, is that both modulation signals as well as the noise characteristic of the system at the quantum noise level can be predicted. The agreement with the experimental results is good qualitatively as the three distinct regions are observed experimentally.

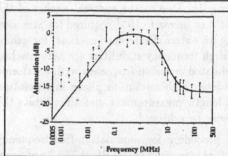

Figure 6. Experimentally measured master noise transfer function compared to theory. See ref. [3].

In regions (ii) and (iii), where the theory predicts the behaviour to with in 2dB of the measured results. However, in region (i) the theory predicts the genaral behaviour and the corner frequency between regions (i) and (ii), but the results disagree as to the magnitude of step shown in figure 5. Furthermore, the slave laser pump noise transfer function also disagrees qualitatively, but agrees extreemly well quantitatively with the theory. We believe that these discrepancies are due to excess master phase noise being amplified by the slave laser, and when the master field is off resonance with the slave cavity. As yet the effect of off resonance injection locking has not been modelled into the theory, but this will be examined in future work.

4   Acknowledgments

We would like to thank E. Huntington and G Newton for their assistance to this work.

References

[1] Dorf, *Modern Control Systems*, 5th ed. Cambridge, MA:, Addison Wesley, 1990.

[2] C. C. Harb, *et. al.*, IEEE Jour. of Quant. Elect., **30**, 2907 (1994); M. Taubman *et. al.*, JOSA B in print 1995.

[3] C. C. Harb, *et. al.*, To be submitted to Phys. Rev. A (1995).

[4] A. E. Siegman, *Lasers*, University Science: Mill Valley, Calif. 1986.

[5] T. C. Ralph, PhD Thesis, Australian National University (1993), unpublished; T. C. Ralph *et. al.*, Opt. Comm. in print (1995).

[6] I. Freitag *et. al.*, App. Phys. B, **58**, 537 (1994).

[7] T. J. Kane *et. al.*, Opt. Lett., **10**, 65 (1985); I. Freitag *et. al.*, Opt. Commun., **101**, 371 (1993).

Modulation-Free Frequency Stabilization of a Diode Laser: Zeeman-Effect

A. Abou-Zeid, A. Vogel
Physikalisch–Technische Bundesanstalt, Laboratory 5.23
Bundesallee 100, D–38116 Braunschweig

In interferometric precision length measurement, the laser diode, which is more powerful and more compact, is suitable for use as an alternative to the HeNe laser. Stabilization of the operating parameters of a laser diode allows relative frequency stabilities in the range of $1 \cdot 10^{-7}$ to be achieved for a short time [1]. Due to the spectral ageing of the diodes ($> 1 \cdot 10^{-5}$/Jahr [2]), the long–term stabilities below $1 \cdot 10^{-7}$ required for precision length measurement can be achieved only using an external reference such as, for example, an absorption line or a material measure. High frequency stabilities are here reached conventionally by linking-up of the slightly modulated emission frequency with an atomic absorption line. The modulation prevents, however, high-resolution phase interpolation techniques from being used in interferometric length measurement and thus limits the uncertainty of measurement which could otherwise be achieved.

Within the scope of the present work, a new procedure for modulation-free frequency stabilization of a laser diode to a hyperfine structure transition of the Rb at 780,2 nm using the Zeeman effect has been developed. The relative long-term stability and reproducibility of the vacuum wavelength of the stabilized diode laser will be described.

Previous stabilization of the laser frequency

Stabilization to a transition of the Rb–D_2 lines at 780,2 nm offers the advantage that a great number of commercial laser diodes are available. Rubidium vapour with the isotopic composition of $37,5\%$ ^{85}Rb and $62,5\%$ ^{87}Rb are used for stabilization. The D_2 transition ($5S_{1/2} \rightarrow 5P_{3/2}$) is composed of six hyperfine structure transitions per Rb isotope, three Hfs lines lying close together because of the modest splitting of the $5P_{3/2}$ state.

Fig. 1 shows the calculated distribution of the 12 Hfs lines of the isotope mixture over the centre frequency of the D_2 transition with the associated quantum numbers of the transitions [5]. When not only the natural line width of 6 MHz but also the Doppler broadening of each Hfs line at 293 K of 511 MHz (^{85}Rb) and 505 MHz (^{87}Rb) are taken into account, the dotted curve of the absorption coefficient in Fig. 1 is obtained. The measured curve was obtained with the temperature of the diode heat sink being stabilized, by variation of the injection current of a laser diode. The laser intensity through the Rb cell was of the order of 6 μW/cm^2; the increase in the optical output power of the laser diode with the current was compensated in front of and behind the cell by formation of the optical signal difference. At higher intensities, deviations from the calculated curve appeared, which is due to the beginning of saturated absorption (*optical pumping*). In frequency stabilization, intensities of 80 μW/cm^2 were required to achieve a good signal-to-noise ratio. At this intensity, the nonlinear effects in the area of the strongest absorption triplet used (F=2 $\rightarrow$ F'=1,2,3) were relatively small.

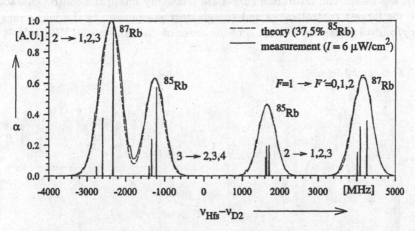

Figure 1: The continuous line shows the curve of the absorption coefficient $\alpha(\nu)$ measured over the four Rb–D$_2$ triplets at an intensity of 6 μW/cm^2. In addition, the position and the relative intensity of the 12 Hfs lines are represented with the natural line width.

For stabilization to a Doppler-broadened absorption triplet, the emission frequency of the laser diode should remain constant within ±100 MHz at least for a short time. For typical dependencies of the emission frequency on the heat sink temperature of -30 GHz/K and on the injection current of -3 GHz/mA, this requires a parameter constancy of ±3 mK and $\pm0{,}5$ μA.

Generation of the control signal

Stabilization without using a lock-in amplifier will be possible if the control signal is formed from the difference of two absorption signals whose frequency is shifted in opposite directions. When influenced by a longitudinal magnetic field, the absorpiton lines of the Rb–D2 transition for $\sigma^\pm$-polarized light are shifted with ±2 MHz/G [4]. The formation of the difference between the transmission signals of σ^+- and σ^+-polarized light yields the desired control characteristic with zero passages at the triplet maxima frequencies. Around these frequencies a linear control range is available which allows the frequency to be adjusted with the correct sign through the laser diode current. In the stabilization carried out, the opposite frequency shift was possible because two equidirectional beams pass through the cell in parallel with each other. The longitudinal magnetic field of 57 G was generated with the aid of a pair of coils.

Fig. 2 shows the curve of the differential signal above the laser diode current, the signal being generated with the aid of two photodiodes and a low-drift instrument amplifier control. Compared with the curve of the differential signal calculated on the assumption of linear absorption, at an intensity of 80 μW/cm^2, the deviations for the triplets with a smaller absorption coefficient were stronger due to nonlinear absorption effects. For the control the zero passage of the differential signal belonging to the triplet $(1 \rightarrow 0, 1, 2)$ at 56,55 mA was used. The saturation effects are relatively small; the control characteristic has here the largest control range and the greatest steepness. In the linear range, the frequency/control voltage coefficient is of the order of $\Delta\nu/\Delta U = -90$ KHz/mV.

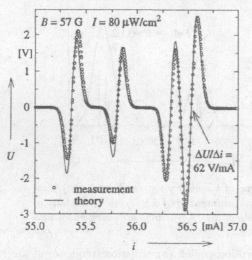

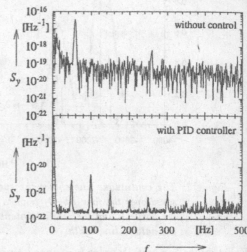

Figure 2: The circles show the output signal of the differential amplifier when the laser diode frequency is varied with the aid of the injection current above the Rb–D$_2$ lines.

Figure 3: Spectral FM power density $S_y(\nu)$ of the controlled (top) and of the uncontrolled laser diode (bottom) in the range 0 to 500 Hz.

With the aid of a PID controller, the output signal of the difference amplifier was used to adjust the laser diode current such that the emission frequency remains constant in the zero passage of the triplet employed. The control allowed the spectral FM noise density $S_y(\nu)$ to be reduced in the low–frequency range by the factor 10^3 (cf. Fig. 3).

Stability investigations

The initial assessment of the stability of the frequency control can be carried out with the aid of the output signal of the differential amplifier furnished by the coefficient $\Delta\nu/\Delta U$. The distributions of the deviations obtained in various long–time measurements over several hours showed relative stabilities σ/ν_o below $5 \cdot 10^{-11}$. Stabilities obtained only from the control signal must, however, be regarded critically as they do not allow for drift phenomena of the sensor of the control system, here the Rb cell with magnetic field and electronic receiving system. This is why a transportable lambdameter developed at the PTB was used to carry out independent long–time and reproducibility measurements of the vacuum wavelength. The absolute uncertainty of measurement of the lambdameter for wavelength measurements in the range 600 nm to 900 nm was of the order of $\pm 1 \cdot 10^{-7}$ and the resolution was $\pm 1 \cdot 10^{-8}$ [3].

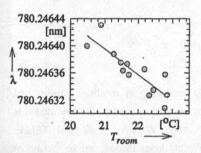

Figure 4: The measured values show the mean values of the vacuum wavelength determined in long–time measurements, above the mean values of the ambient temperature of the respective measurement.

Even with a stabilized laser diode, the total of 40 measurements of a duration of at least eight hours showed a dependence of the laser diode vacuum wavelength on the ambient temperature which could not be unambiguously explained by electronic or mechanical drift effects or temperature-dependent control point shifts as a result of edge asymmetries in the two absorption signals subtracted from one another. Fig. 4 shows the distribution of the mean vacuum wavelength from 12 measurements above the mean ambient temperature of the respective measurement. The relative stability σ/λ_o of each measurement was smaller than $6 \cdot 10^{-8}$. For the laser diode stabilized with the Zeeman effect to the transition $F = 2 \rightarrow F' = 1, 2, 3$ of the ^{87}Rb, reproducibility measurements yielded a vacuum wavelength of

$$\lambda_{Zeeman} = 780, 24636 \pm 0, 00003 \text{ nm}$$

References

[1] ABOU-ZEID, A. Emissionswellenlänge kommerzieller Laserdioden als Funktion ihrer Betriebsparameter. (Emission wavelength of commercial laser diodes as a function of their operating parameters) *PTB–Mitteilungen 94* (1984), 163.

[2] ABOU-ZEID, A. Spektrale Alterung kommerzieller GaAlAs–Diodenlaser. (Spectral ageing of commercial GaAlAs diode lasers) In: *Int. Kongreß für Laser–Optoelektronik und Mikrowellen, Munich* (1989), vol. 9.

[3] ABOU-ZEID, A., BADER, N. AND PRELLINGER, G. A transportable lambdameter for the calibration of laser wavelengths. *PTB–Mitteilungen 105* (1995), 89.

[4] TETU, M., VILLENEUVE, B., CYR, N., TREMBLAY, P., THERIAULT, S. AND BRETON, M. Multiwavelength sources using laser diodes frequency-locked to atomic resonances. *IEEE J. Lightwave Techn. 7* (1989), 1540.

[5] VOGEL, A. Entwicklung modulationsfreier Frequenzstabilisierungsverfahren von Laserdioden. (Development of procedures for modulation–free frequency stabilization of laser diodes). Dissertation submitted for diploma, Braunschweig Technical University, 1994.

Dynamic Optical Characteristics of High-Frequency Pulsed Laser Diodes

H. Wang, Th. Sodomann, H. Müller, D. Dopheide

Physikalisch-Technische Bundesanstalt (PTB), Department for Fluid Mechanics

Bundesallee 100, D-38116 Braunschweig

1. Introduction

Laser diodes are widely used, as they have been steadily improving in reliability, power, and wavelength coverage, while steadily decreasing in cost. One of the unique aspects of laser diodes is their modulation or pulsing capability [LENTH, 1984; TUCKER, 1985; WIEMAN et al, 1991; TELLE, 1993; IKEGAMI et al, 1995]. It has been shown that mono-mode laser diodes can be pulsed at frequencies up to 100 MHz and applied in the laser Doppler anemometry (LDA) advantageously, if they have mono-mode emissions [DOPHEIDE et al, 1989 & 1990 & 1993; WANG et al, 1994a]. However, pulsed laser diodes have sometimes multi-mode emissions [SELWAY et al, 1975; NAKAMURA et al, 1978; STRUNCK et al, 1992; DOPHEIDE et al, 1993; LOCHEY et al, 1994], and this in turn causes adjustment problems and measurement errors in LDA applications [WANG et al, 1994b]. Up to now, only the time-averaged measurements of emission spectra over many laser pulses [DOPHEIDE et al, 1993; WANG et al, 1994b; LOCHEY et al, 1994] or wavelength measurements of a single laser pulse with the pulse width of 1 µs and longer [CLARK et al, 1982] have been performed.

In this paper the dynamic emission characteristics of high-frequency pulsed (HF-pulsed) laser diodes have been studied 1) to measure the relationship between the optical output peak power of a HF-pulsed laser diode and the injection current pulses with a pulse width in the nanosecond range and with a pulse repetition frequency up to 100 MHz; 2) to investigate the spectral stability of each laser pulse using a dynamic wavelength measurement set-up; and 3) to examine the influence of the HF-pulsed laser diodes mode hopping and frequency jitter on the interferential fringe in LDA applications.

2. Operating parameters of HF-pulsed laser diodes

The optical characteristics of a laser diode can be described by the following parameters [WANG et al, 1994b]:

λ is the laser wavelength,

I_{th} is the minimum (threshold) bias current,

I_{bias} is the bias current when operating in the HF-pulsed mode,

I_{pulse} is the pulse current (or current step),

P_{cw} is the pure optical power of continuous laser light significant for I_{bias},

P_{pulse} is the peak power of HF-pulsed laser outputs,

η is the slope efficiency (i. e. $\Delta P / \Delta I$ — the output power change per unit injection current above

threshold, where $P = P_{cw} + P_{pulse}$ and $I = I_{cw} + I_{pulse}$),

f_{pulse} is the frequency of laser pulses (also of current pulses),

τ_{pulse} is the width of a laser pulse.

Duty cycle = $(1/f_{pulse} - \tau_{pulse})/\tau_{pulse}$ (1)

Unless otherwise specified, a sinusoidal pulse current at a frequency of 50 MHz (duty cycle = 1) has been used and the laser diode operating temperature was maintained at 22 °C during our experiments.

3. P - I characteristic lines of HF-pulsed laser

The laser diode was biased with a direct current and current pulses from a pulse driver (Fig. 1). The current monitor used has a frequency response up to 200 MHz, which allows each injection pulse current to be measured. The frequency of the oscillator can be adjusted up to 100 MHz. An avalanche photodiode (APD) connected to a transient recorder allows the light output of HF-pulsed laser diodes to be monitored. The optical output power was measured with a power meter.

Fig. 2 shows that the laser diode HL8325 can be pulsed at 50 MHz with various pulse currents. The slope efficiency of the laser diode is about 0.43 mW/mA in the CW and 0.50 mW/mA in the HF-pulsed operation mode. Similar results have been obtained with the f_{pulse} varying in the region between 1 MHz and 100 MHz.

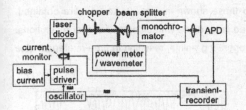

Fig. 1 Block diagram of the experimental set-up used for measuring P - I characteristic lines and dynamic wavelength tuning of HF-pulsed laser diodes. When measuring P - I lines, the monochromator was removed and a chopper frequency of 1 kHz was used to eliminate background influences.

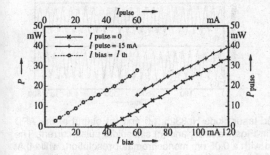

Fig. 2 Typical P - I characteristic lines of a HF-pulsed laser diode (HL8325 from HITACHI, I_{th} = 45 mA). The dashed line with circles has the axis I_{bias} against P_{pulse}, and the black lines have the axis P against I_{bias}. P_{pulse} is measured with an APD and calibrated using P_{cw}. The slope efficiency difference of the laser diode between CW and HF-pulsed modes needs to be further investigated.

4. Time-averaged emission spectra of many laser pulses

By using a wavemeter based on a Fabry-Perot interferometer and a CCD array (refer to Fig. 1 without the chopper, [WANG et al, 1994b]), the time-averaged emission spectra of a HF-pulsed laser diode

32

HL8325 over 5×10^6 laser pulses were measured and depicted in Figs. 3a to 3c. The multi-mode emissions here are due to the carrier effect [ITO et al, 1980 & 1981; KOBAYASHI et al, 1982]. The emission spectra of the laser diode HL8325 (40 mW) have broader bandwidths than those of HL7801 (5 mW) [WANG et al, 1994b]. This may be because the pulse current applied to HL8325 is stronger than that applied to HL7801, resulting in mode hopping and frequency jitter. Mono-mode emissions of the HF-pulsed laser diode can be obtained by selecting appropriate operating parameters (Fig. 3c).

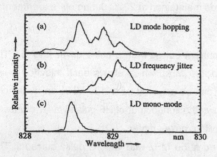

Fig. 3 Typical emission spectra of a HF-pulsed HL8325 laser diode at three different operating points (I_{th} = 45 mA):
  (a) I_{bias} = 0.5 I_{th}, I_{pulse} = 45 mA
  (b) I_{bias} = I_{th}, I_{pulse} = 75 mA
  (c) I_{bias} = 1.1 I_{th}, I_{pulse} = 60 mA
By using a digital optical wavemeter [WANG et al, 1994b], the emission spectra of the laser diode in the CW operation mode were measured and taken into account in selecting these operating points.

5. Dynamic spectral characteristics of a single laser pulse

<u>Dynamic wavelength change of HF-laser-pulses</u>   As shown in Fig. 1 (without the chopper), the output light of a laser diode was focused onto a monochromator and detected by an APD. By scanning the wavelength of the monochromator, spectra similar to those shown in Figs. 3a to 3c were obtained. Figs. 4a & 4b show the spectral stability of a HF-pulsed laser diode with different spectral resolutions. Mode hopping and frequency jitter occurred at some operating points, resulting in time-averaged wide band emission spectra as shown in Figs. 3a & 3b.

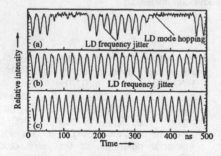
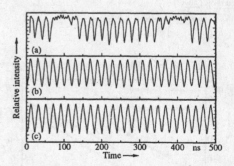

Fig. 4 Dynamic wavelength tuning of a HF-pulsed laser diode HL8325. (a) and (b) output of the APD with the operating points corresponding to those in Figs. 3a & 3b, and (c) sinusoidal pulse current. The data shown in the figure on the left were obtained with a 0.05 nm monochromator resolution, while that in the figure on the right were obtained with a 0.5 nm resolution, indicating that the influence of the frequency jitter becomes weaker if the slit width increases. Observations of the APD output with a very wide slit showed that the laser pulse intensity was constant over the whole measurement time. The central wavelength of the monochromator was set at 831.69 nm.

Laser chirping during a single HF-laser-pulse   As shown in Fig. 5, a modified Mach-Zehnder interferometer is used to measure the laser chirping. The laser light is split to two beams and one beam goes through a delay line. The two beams are crossed at an angle of 2φ instead of recombined parallelly like conventional interferometers. The interferential fringes are generated in the crossing region of two laser beams (called the measuring volume in LDA applications, refer to the chapter 6 of the paper). Here, light fibre optics were used because the optical delay L (optical path length difference between two laser beams) can be easily done by fibre couplers. A mono-mode light fibre (the same as a pinhole with a diameter of about 6 µm) was placed in the measuring volume, allowing the fringe motion past its end-face to be detected.

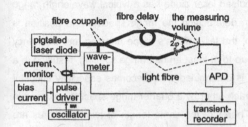

Fig. 5 Modified Mach-Zehnder interferometer (modified delayed self-homodyne method, IKEGAMI et al, 1995). In order to couple the light into the light fibre, arcsine $\langle\varphi\rangle$ (= 0.10 in our case) should be smaller than the numerical aperture of the light fibre (= 0.12). The influence of pigtailing on the spectrum can be reduced by using fibres with angled end-faces [WANG et al, 1994c].

Figs. 6a & 6b show that the optical frequency of the HF-pulsed laser diode changes even after the injection current becomes constant (laser chirping) because of the thermal effect [ITO et al, 1981; KOBAYASHI et al, 1982]. It takes relatively a long time for laser diodes to achieve final equilibrium and to show a smaller spectral change. Such changes increase if the I_{pulse} or the L increases. It has been shown that the frequency settling time may be as short as about 1 µs for a residual frequency error of 10^{-3} of the step height [MELMAN et al, 1981; WATTS et al, 1986], and the frequency response of laser diodes can be extended using modified injection current waveforms [ANDERSON et al, 1991 & 1993].

It is assumed that laser chirping superimposed onto the laser pulse would be observed by optimizing the I_{bias}, I_{pulse}, f_{pulse} and L. Due to the slow sampling rate of the transient recorder and limited bandwidth of the detector, however, such a laser chirping effect was not observed during the laser pulse at frequencies above 50 MHz (or laser chirping during the laser pulse with pulse widths smaller than 10 ns could not be time-resolved).

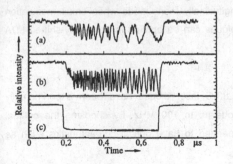

Fig. 6 Laser chirping measured in the measuring volume during one single laser pulse period.
(a) and (b) output of the APD with
I_{bias} = 1.1 I_{th} and L = 50 cm.
I_{pulse} = 45 mA in Fig. 6a and
I_{pulse} = 75 mA in Fig. 6b.
(c) square-wave current at 1 MHz.
Similar results were obtained if the I_{pulse} kept constant and the L changed.

6. Influence of the laser diode's frequency change on the fringe pattern

In order to investigate the influence of laser diodes' mode hopping and frequency jitter on the fringe pattern in LDA applications, the modified Mach-Zehnder interferometer in Fig. 5 was used, which actually shows a one-component HF-pulsed laser Doppler anemometer [WANG et al, 1994a] with the fringe spacing Δx defined as:

$$\Delta x = \lambda / (2 \bullet \sin(\varphi)) \tag{2}$$

Similar to the results shown in Figs. 4a & 4b, the intensity changes of the beat signals in Figs. 7a & 7b represent abrupt motions of interferential fringes in the measuring volume, indicating again the mode hopping and frequency jitter of HF-pulsed laser diodes at some operating points. Taking into account of Fig. 3c and Eq. (2), the emission spectrum of a HF-pulsed laser diode has a typical wavelength range of about 1 nm. In such a case, the relative error on determining the fringe spacing and position is negligible. However, with respect to Figs. 7a and 7b, the laser diode's mode hopping and frequency jitter could results in a shift of the interferential fringe up to at least an half fringe spacing, resulting in a relative error of nearly 50% when the fringe position is determined. This becomes especially critical if two equidistant fringe pattern locally shifted against each other by a quarter fringe spacing are used in directional LDA applications [WANG et al, 1994d]. Therefore the same length of two light fibres are required to impair such fringe motions.

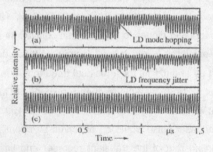

Fig. 7 Beat signals measured over 75 pulse periods in the measuring volume.
(a) and (b) output of the APD.
The operating parameters here correspond to those in Figs. 3a & 3b. $L = 1$ cm.
(c) sinusoidal pulse current.
As shown in Fig. 3a and here in Fig. 7a, the mode of the HF-pulsed laser diode kept the same over several pulse periods and then hoped to another mode over other several pulse periods, and so on. The reason for such mode hopping is not clear.

Besides, clear fringe patterns were observed if a laser pulse was superimposed onto the next following laser pulse (i.e. the pulse after one pulse period) and the modulation current is small ($I_{th} = 100$ mA, $f_{pulse} = 50$ MHz, $L = 6.5$ m, $I_{pulse} = 1$ mA). This means that laser pulses of HF-pulsed laser diodes are coherent with their neighborhoods and therefore longitudinally mono-mode when small modulation currents are biased above the threshold. Such techniques can be applied in frequency-shifted LDA applications [Müller et al, 1995].

7. Discussion and comments

The dynamic optical characteristics of HITACHI HL8325 laser diodes have been experimentally investigated in detail in the HF-pulsed operation mode up to 100 MHz. By choosing the optimal operating points, a HF-pulsed laser diode can be operated to have optical characteristics such as mono-mode emission, a stable wavelength from pulse to pulse, high optical output peak power and a fast pulse repetition frequency.

Inappropriate operating of HF-pulsed laser diodes results in mode hopping and frequency jitter. As a result, for LDA applications of the HF-pulsed laser diodes, optical adjustment becomes difficult, the fringe pattern becomes obscured, and two optical beams or fibres of exact the same length are required in LDA applications. Mono-mode operation of laser diodes is therefore recommended and can be achieved with the help of time-averaged measurements of emission spectra.

In this paper a mono-mode light fibre is used to measure the fringe motion Such a technique can also be extended to work as a beam scanner or to determine the dynamic measuring position in LDA applications [DURST et al, 1988; BERTRAND et al, 1993].

Acknowledgments   The work presented in this paper was supported by the Deutsche Forschungs-gemeinschaft (DFG) under contract No. DO 292/4-1. The authors are grateful to Dr. Strunck and Dr. Telle of PTB for their helpful suggestions, and to Mr. Federau from the firm Schäfter & Kirchhoff, Hamburg, for his support in pigtailing of laser diodes.

References   D. J. Anderson, J. D. Valera, J. D. C. Jones, Meas. Sci. Technol., 4, 982-987 (1993).
D. J. Anderson, J. D. C. Jones, P. G. Sinha, S. R. Kidd, J. Modern Optics., 38, 2459-2465 (1991).
C. Bertrand, P. Deevaux, J.P. Prenel, Experiments in Fluids, 16, 70 - 72 (1993).
G. L. Clark, L .O. Heflinger, C. Roychoudhuri, IEEE, J. Quantum Electronics, QE-18, 199-205 (1982).
D. Dopheide, H. J. Pfeifer, M. Faber, G. Taux, J. of Laser Applications, 1, 40 - 44 (1989).
D. Dopheide, V. Strunck, H. J. Pfeifer, Experiments in Fluids, 9, 309 - 316 (1990).
D. Dopheide, M. Rinker, V. Strunck, Optics and Lasers in Engineering. , 18, 135 - 145 (1993).
T. Ikegami, S. Sudo, Y Sakai, Artech House, Inc., (1995).
F. Durst, R. Müller, J. Jovanovic, Experiments in Fluids, 6, 105 - 110 (1988).
M. Ito, T. Kimura, IEEE, J. Quantum Electronics, QE-16, 910-911 (1980).
M. Ito, T. Kimura, IEEE, J. Quantum Electronics, QE-17, 787 - 795 (1981).
S. Kobayashi, Y. Yamamoto, M. Ito, T. Kimura, IEEE, J. Quantum Electronics, QE-18, 582 - 595 (1982).
W. Lenth, IEEE, J. Quantum Electronics, QE-20, 1045 - 1050 (1984).
R. A. Lockey, R. P. Tatam, The Proceedings of the 10th Optical Fibre Sensors Conference (Glasgow, Scotland), 408 - 410 (1994).
P. Melman, W. J. Carlsen, Appl. Opt., 20, 2694 - 2697 (1981).
H. Müller, V. Arndt, H. Wang, N. Pape, D. Dopheide, paper to be presented at the 4. Fachtagung Lasermethoden in der Strömungsmesstechnik: Aktueller Stand und neue Anwendungen, September 12th - 13th, 1995, Rostock University.
M. Nakamura, K. Aiki, N. Chinone, R. Ito, J. Umeda, J. Appl. Phys., 49, 4644 - 4648 (1978).
Selway, P. R., Goodwin, A. R., Electron. Lett., Vol. 12, pp. 25 - 26 (1976)
V. Strunck, D. Dopheide, M. Rinker, The proceedings of the 6th Intern. Symposium on Applications of Laser Techniques to Fluid Mechanics, (Lisbon, Portugal), I, 11.2.1 - 11.2.5 (1992).
R. S. Tucker, J. Lightwave Technol., LT-3, 1180- 1192 (1985).
H. R. Telle, Spectrochimica Acta Rev., 15, 301 - 327 (1993).
H. Wang, V. Strunck, H. Müller, D. Dopheide, Experiments in Fluids, 18, pp. 36 - 40 (1994a).
H. Wang, D. Dopheide, V. Strunck, The Proceedings of The 11th International Congress: LASER 93, Laser in Engineering (edited by W. Waidelich, Springer-Verlag Berlin Heidelberg), 313 - 319 (1994b).
H. Wang, H. Müller, V. Arndt, V. Strunck, D. Dopheide, The Proceedings of The 2nd International Conference on Fluid Dynamic Measurement and Its Applications: Modern Techniques and Measurements in Fluid Flows (Beijing, China), 126 - 129 (1994c)
H. Wang, H. Müller, V. Strunck, D. Dopheide, The Proceedings of The 7th International Symposium on Application of Laser Techniques to Fluid Mechanics (Lisbon, Portugal), I, 14.1.1 - 14.1.5 (1994d).
R. N. Watts; C. E. Wieman, Opt. Comm., 57, 45 - 48 (1986).
C. E. Wieman, L. Hollberg, Rev. Sci. Instrum., 62, 1 - 20 (1991).

Towards a Sealed-Off Excimer Laser

F. Voß, J. Bäumler, U. Rebhan, D. Basting
Lambda Physik GmbH
37079 Göttingen

1. Introduction

The limitation factors for sealed-off operation of excimer lasers are the loss of halogen (F2 or HCL), the built up of impurities in gas phase and the contamination of windows. The efforts of research and development in excimer laser technology in the last two years at Lambda Physik led to a new type of laser tube with longer gas lifetime, lower halogen consumption and reduced window contamination (Nova-Tube). This became possible by using selected materials inside the tube metal ceramic technology and improved discharge and preionisation technics.

2. Experiments with Laser tube

Many long run time experiments were done with different wavelengths and types of exicmer lasers equiped with Nova-Tube.

Fig.1 shows an ArF-Laser operated at 300 Hz (18 W average power at 193 nm) without cryogenic or chemical gas purification and without halogen injections. The dynamic gas lifetime exceeds 20 million pulses. This is an improvement of more than a factor of 10 compared to standard tube supported by cryogenic gaspurification.

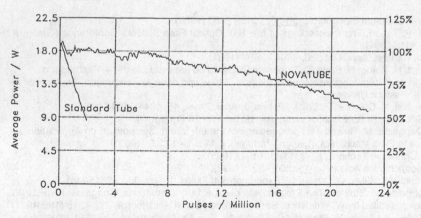

Fig. 1   An ArF laser with NOVATUBE™ was operated at 300 Hz without cryogenic gas
purification. The dynamic gas lifetime exceeds 20 million pulses compared to less than
2 million for a standard laser tube.

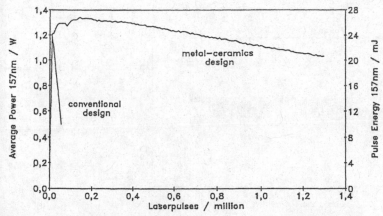

Fig. 2   F₂ test run with a COMPEX 205 at 50 Hz. The extrapolated dynamic gas lifetime exceeds
2.3 million pulses. The performance of a conventional F₂-laser is shown for comparison.

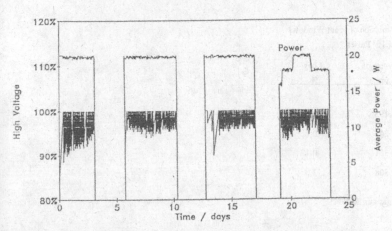

Fig. 3   Longterm test run of a 20W ArF laser over 140 million pulses with a single gas fill
without cryogenic gas purification.

A factor of more than 20 in improvement of gas lifetime was shown in case of F2-Laser (157 nm)
also without any gaspurification. This reduce number of gas fills and running costs dramatically and
makes this wavelength more interesting for applications (Fig. 2).

In Fig.3 a testrun of 20 W/100 Hz stabilized ArF- Laser is shown. This laser runs over 23 days with
breaks on weekends with halogen injections in a single gas fill. No kind of gas purification was
used.

A first result of KrF is shown in Fig.4. A 50 W-stabilized KrF-Laser at 600 Hz was running more
than 1 billion pulses with cryogenic gaspurification and halogen injections. Without any
gaspurification a KrF-Laser reached a gasliftime of more than 330 million pulses stabilized on 300
mJ at 200 Hz.

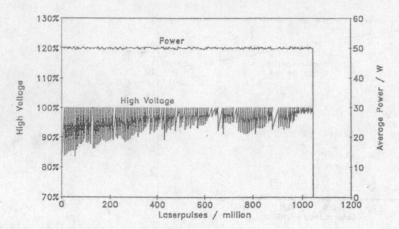

Fig. 4   Marathon test run over more than 1 billion pulses on a single gas fill with a 50 W KrF at 600 Hz.

Tab. 1 Loss of Total Transmission of Front Windows
after 10^7 Pulses

Gas Excimer	Wavelength nm	Loss %
ArF	193	0.9
KrF	248	0.2
XeF	351	0.12
XeCl	308	2.0

without cryogenic gas purification

For high power KrF-Lasers with beam size of about 30 * 15 mm2 and 200 W output power at 250 Hz (single pulse energy more than 1 J) more than 30 million pulses gas liftime was obtained. Also here no gaspurification was used.

The loss of of total transmission in frontwindows for the different excimers is shown in table 1. The loss due to contamination and ageing of optical bulk material is in the region of 1% for 10 million pulses. The above results obtained with several lasers at different wavelengths show a drastical improvement in gaslifetime and window contamination compared to older types of excimer-lasers.

3. Impurities in gas phase

For sealed-off operation it is helpful to know

1. What kind of impurities are in the gas inside the new Laser tube?
2. How much is the concentration of impurities inside Nova-Tube?

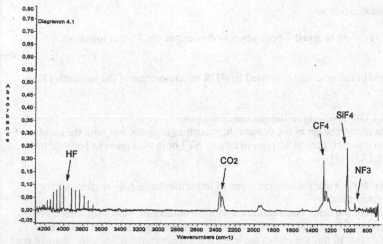

Fig.5

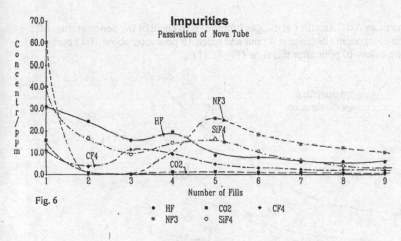

Impurities
Passivation of Nova Tube

Fig. 6

● HF   ■ CO2   ◆ CF4
✳ NF3   ○ SiF4

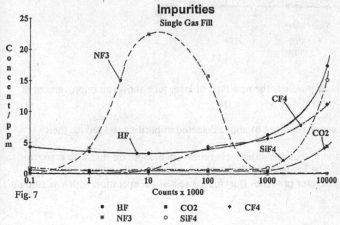

Impurities
Single Gas Fill

Fig. 7

● HF   ■ CO2   ◆ CF4
✳ NF3   ○ SiF4

40

First results of unpassivated tube are:

Concentration of H2O is lower or equal 2 ppm whereas concentration of O2 is lower or equal 0.4 ppm.(1)

The passivation process of the tube was controled by FTIR measurements of the impurities HF, NF3, SiF4, CF4, CO2.

Fig. 5 shows a typical FTIR-Spectrum of the gas inside the tube.
As seen from Fig.6 the concentration of the detectable impurities decrease fast with the number of fills (F2/He) from maximum 60 ppm to 10 ppm in case of NF3 or in the region to 5ppm or lower in case of HF, SiF4, CF4, CO2. (1)

In a fluorine passivated Nova-Tube the concentration of impurities in an ArF- mixture slightly increase with accumulated pulses.
Fig.7 shows the increase of impuritie concentrations with the number of pulses for an 300 Hz/ 30 W ArF-Laser in a single gasfill without gaspurification and halogen injection. The concentrations starts below 5 ppm and after 10 million pulses the highest concentration (HF) is lower than 18 ppm. The concentration of NF3 reach a maximum at about 10 thousand pulses and goes to zero at about 1 million pulses.(1)

A long stand by time of an ArF-Laserfill (static gas lifetime) also changes the concentration of impurities. The main component HF starts at 4 ppm and reach 15 ppm after about 100 hours. All other components are below 10 ppm after this time (Fig.8). (1)

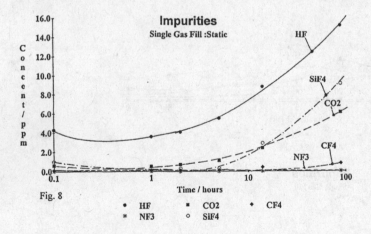

Fig. 8

Compared with other excimer-laser products the new type of laser tube shows an improvement in impurities of a factor between 5 and 50 (2).

Further investigations will find out the sources of above detected impurities and reduce their concentrations.
For long time sealed-off operation of excimer-lasers it is necessary to work on-line with cryogenic purification and halogen supply. The halogen supply is on-line possible without gascylinder by using halogen generator. This generator produces fluorine for sealed-off operation with low amount of impurities (Fig.9).

	fluorine generator
purity	> 99.98 %
contamination level	< 200 ppm
O_2	< 200 ppm*
N_2	< 40 ppm
HF	< 10 ppm
CF_4	< 10 ppm
COF_2	< 5 ppm
SiF_4	< 5 ppm

* resolution limit of available gas analyzer

Fig. 9 Halogen Gas Purity of the New Fluorine Generator

Conclusion

We have developed a new type of excimer-laser tube with excellent results in gaslifetime. Studies of impurities in gas give small values of concentrations. This show the way for further research and development in the direction to sealed-off eximer laser.

Acknowledgements

The autors would like to thank Reiner Hamm and Georg Schröder of Messer Griesheim Entwicklung Sondergase for their support in gas impurities measurements.

References

(1) Messer Griesheim, Entwicklung Sondergase (1995)
(2) G.M. Jursich et al. Applied Optics Vol.31, No.12 (1992)

Physical Conceptions of Developing High-Power Pulsed Chemical Layers on Photon-Branching Chain Reactions

V.I. Igoshin, V.A. Katulin, S.Y. Pichugin
Lebedev Physics Institute, Samara Branch.
Novo-Sadovaya str. 221, Samara 443011, Russia.

Abstract

Since 1983 our research group had pursued an approach to the laser on photon branching chain reaction. If a chain reaction leading to stimulated emission can be initiated by the same photons and emission of photons exceeds their expense to initiate the reaction the process on the whole becomes self-sustained. This type of reaction can be considered as a reaction with photon branching mechanism, as opposed to the well-known material and energetic branching mechanism. Basing on photon branching reactions it is possible in the future to develop high power pure chemical pulsed laser systems. This approach can avoid many difficulties encountered with traditional pulsed laser systems in which one must use the external pumping source.

Some our previous results on modeling of oscillators and amplifiers employing reactions with photon branching in chemically active media containing ultradispersed particles are given in this work. To increase the energy amplification in one stage we offer to fulfil the photon branching reaction in the multipass telescopic resonator and the geometro-optical calculations of the cavity parameters are given. The optical scheme of a three-stage system with the energy amplification as high as 10^6 is proposed. The technique of preparing the gas-dispersed media with necessary parameters is discussed. We also report on the first successful experiments on preparation of the hydrogen-fluorine working mixture containing ultradispersed passivated particles of aluminium.

1. Background and Significance

At present, it has been proved possible to create a purely chemical laser, operating only under continuous-wave condition. Obtaining high pulsed-laser energies from the existing chemical lasers involves expense of an energy comparable with the output laser energy to produce active centers. Thus, the existing laser systems to increase the energy of pulsed chemical lasers require a proportional increase in the initiation energy, which sets a limit on the energy rise in the pulsed systems. More strong problem arises in the case of nonchemical lasers. The main disadvantage of the existing projects of high-energy pulsed lasers is their high cost due to high cost of power supplies.

It would be possible to develop purely chemical high power pulsed laser systems based on so called photon branching chain reactions. In such type of reactions predicted theoretically in our works induced emission of photons in the course of the chain reaction exceeds their expense to initiate the reaction (to produce active centers). So in multiple-stage system Fig.1, the emission of preceding stage initiates the reaction in the next stages of a greater size and greater output energy. Experimental realization of this type of reaction is of great interest for development of high power pulsed laser systems.

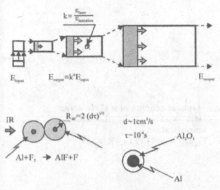

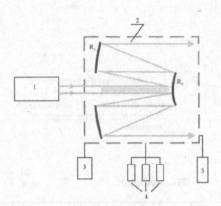

Fig.1   Top: A multiple-stage laser system on photon branching reaction; k is the ratio of specific volume laser energy ε_{laser} to specific volume initiation energy $\varepsilon_{initiation}$; E_{output} and E_{input} are output and input laser energies.
Bottom: Evaporation of dispersed phase by IR radiation; R_{dif} is the diffusion radius of evaporated atoms.

Fig.3.   The created setup with the multipass amplifier on the basis of unstable resonator.

2. Preliminary Studies

In previous works a well-argued and fundamentally new way to bring about photon branching chain reactions, based on nonresonant interaction between IR laser radiation developed and dispersed chemically active media, has been advanced [1-3]. It consists in introducing finely dispersed particles (e.g. passivated metal particles) into the active medium of HF or DF - CO_2 laser (Fig.1). Heating under the action of input laser radiation causes evaporation of the particles and initiation of the chain reaction. We performed numerical calculations of chemical kinetics equations, relaxation equation, heat balance equation, oscillator or radiation transfer equation, and also equations for the temperature and radius of the dispersed aluminium particles subject to evaporation in the IR radiation field. It was shown that specific output energy of the HF and DF - CO_2 laser with the mixture pressure of 1 atm which utilizes aluminium particles (coated with thin layer of Al_2O_3) with the concentration of $n \sim 10^9$ cm^{-3} and the radius of $r_0 \cong 0.1$ μm exceeds specific energy expenditures of the initiating radiation by 10 - 20 times Tabl., Fig.2)

Thus, it is theoretically possible to develop pulsed chemical IR laser amplifiers possessing a sufficiently high gain.

3. Research Design and Methods

The created setup is a complex of connected arrangements namely (Fig.3):
1 - the initiating laser oscillator,
2 - the laser amplifier with multipass unstable resonator,
3 - the source of condensed dispersed phase (CDP),
4 - the gases arrangement,
5 - the arrangement for exhausted gas neutralization and pump.

44

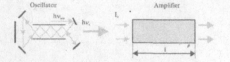

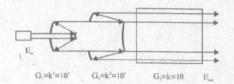

Fig.2. Optical configuration of oscillator and amplifier; I_0 is the intensity of initiating radiation, l is the length of amplifier.

$G_3=k'=10^2$ $G_2=k'=10^2$ $G_1=k=10$ E_{out}

Fig.4. Optical scheme of a three-stage laser system with the energy amplification
$E_{out}/E_{in}=G=G_1G_2G_3=K^6=10^6$

HF-oscillator:

P, atm	I_0, MWt/cm^2	n, cm^{-3}	r_0, mm	$t_{initiation}$, ms	E_L, J/liter	K
1	20	10^9	0,05	0,1	100	21

HF-amplifier:

P, atm	I_0, MWt/cm^2	E_{input}, J/cm^2	E_{output}, J/cm^2	K	l, m
1	14	4,2	36,7	8,7	4

Tabl.1. Some results of computer simulatios of the energy amplification in the oscillator and amplifier, p is total pressure of $H_2+F_2+O_2+He$ mixture.

Initiating laser oscillator gives the energy 1 - 10 J with pulse duration 1μs. Cavity for photon branching chain reaction has the volume of 3 liters and permits to obtain the output energy up to 1 kJ. The source of CDP enables to create the concentration of particles of the order of 10^9cm^{-3} with size of 0.1 μm. The gases arrangement enables to operate at pressure in cavity up to 1 atm and to change the mixture composition. The arrangement for exhausted gas pumping can pump the gas with the velocity of 1 - 3 m/s.

In this project the levitation method of obtaining ultradispersed particles will be used. According this technique the metal is evaporated from metal drop which is warmed and trapped in high frequency electromagnetic field. The evaporation takes place in laminar flow of Ar with small addition of O_2 for passivation of particles. These particles are heated in the cavity up to high temperature at time of 0.1μs under the action of IR input laser radiation. The evaporation of metal takes place. Then evaporated metal interacts with F_2 with the formation of free atoms F ($Al+F_2=AlF + F$). Thus, around each particle the formation of active centers diffusing in active media takes place. To achieve a homogeneous distribution of free atoms in the volume during time τ it is necessary to fulfil the condition R

$\leq 2 (d\tau)^{1/2}$, where R - the mean distance between particles, d - diffusion coefficient (fig.1, bottom).

Another technique of obtaining gas - dispersed medium is to introduce available powder in laser gas.

The reaction will be carried in multipass telescopic resonator (Fig.3) the pulse from oscillator initiates the reaction in central region. The radiation arising in this region initiates the reaction in the next ring region which has the volume in $\sim M^2$ times higher than the volume of the first region and so on (M is the magnification of resonator after 2 passes). In optical scheme with 5 passes and consequently 3 ring regions which is supposed to realize the total energy gain is $G=K^3= 10^3$, where $K\sim10$ is the volume coefficient of amplification equaled to the ratio of specific radiated energy to the specific absorbed energy. To do this optical process be possible it is necessary to fulfil the geometry-optical condition of self consistency $M\sim K^{1/2}$. The length of resonatore is chosen from the condition of effective absorption of the input pulse energy after two passes (Fig3). Then the radius of curvature of each mirror can be easily defined by the relations : $R_2=2l/(M-1)$, $R_1=MR_2$. The optical scheme of laser system with total energy amplification of the order of 10^6 times is presented on the Fig.4.

4. First results.

Our research group in collaboration with the group from Russian Scientific Center "Applied Chemistry" in St.Peterburg has conducted the first successful experiments on preparing the working hydrogen-fluorine gas mixture containing dispersed particles. For this purpose the stream of He containing passivated aluminum particles was injected in the gas mixture $H_2 - F_2 :O_2 : He = 0.1:0.15:0.015:0.035$ atm. The total gas pressure was equal to 1 atm and the length of reactor is equal to 1m. The dark reaction was registered by the formation of HF. The increase of dark reaction rate has not been revealed in comparison with that for pure gas mixture (4×10^{-2} torr of HF for a minute). The mixture was exposed by HF laser radiation with the specific laser energy of 4-8 J/cm^2 and pulse duration of ~1 μs. The measured ratio of output to input energy is equal to 0,8. This result can be explained if we suppose that the diameter of particles is equal to $1\mu m$ and their concentration is about 10^5 cm^{-3}. The initial particle diameter in powder which was used in experiment is equal to 0,3 μm and the increase of particle size in the working gas is due to their coagulation. To achieve photon branching regime it is necessary to diminish the residence time of particles in the working gas before laser action.

The research described in this publication was made possible in part by Grant No NRX000 from International Science Foundation.

References

1. Igoshin V.I., Pichugin S.Yu. A Substantiation of Feasibility of a Purely Chemical $H_2 - F_2$ Laser with Evaporation of Fine-Dispersed Particles upon Exposure to the IR Radiation. Kvantovaya elektronika, 1989,v.16,n.3,p.437-441, (in Russian).
2. Igoshin V.I., Katulin V.A., Pichugin S.Yu. Pulse Chemical HF and DF - CO_2 Lasers when Photon Branch Chain Reaction Induced by Thermal Decay of File-dispersed NaN_3 Particles Takes Place. Khimiya vysokikh energii, 1990, v.24, n.4, p.372-377, (in Russian).
3. Igoshin V.I., Pichugin S.Yu. Computer Simulation of Pure Chemical $H_2 - F_2$ amplifier. Trudy FIAN, 1993, v.217, p.136-145, (in Russian).

Parametrics Generation of Intense, Widely Tunable Ultrashort Pulses in an All-Solid-State Laser System

A. Seilmeier, T. Dahinten, M. Hofmann, J. Kaiser, H. Graener
Institute of Physics, University of Bayreuth
D-95440 Bayreuth, Germany

The generation of tunable pico- and femtosecond light pulses with parametric frequency conversion processes has attracted increased attention recently[1]. New crystal materials, as BBO, LBO, and KTP, with superior nonlinearity and high damage threshold are now available, and pump lasers of improved stability allow the reliable operation of optical parametric oscillators. Parametric devices pumped either by cw modelocked lasers or by single intense ultrashort pulses have been developed. In this contribution parametric systems are discussed which are pumped by intense pulse trains of a few microseconds duration and repetition rates up to 50Hz. This technique allows the operation of widely tunable optical parametric oscillators[2] (OPO) and subsequent difference frequency generators[3] producing tunable pulses of high peak power. The higher energy of the pump pulse trains has certain benefits for the operation of a synchronously pumped OPO, e.g., relaxed stability requirements, and the potential to work with large cavity losses (e.g. metal mirrors) allowing large tuning ranges with a single set of mirrors and high output-coupling factors for pulse shortening.

The advantage of this technique is the capability to generate widely tunable pulses of considerable output energy in a relatively simple solid-state system. As a pump source, e.g., a feedback controlled, additive pulse modelocked (FCM-APM) Nd:YLF laser[4] is used producing trains of several hundred pulses with constant amplitude and a duration of ~3ps. The energy per pulse amounts to approximately 1µJ, which is sufficient for an efficient pumping of an optical parametric oscillator. The performance can be even improved by modest amplification of the pulse trains, in

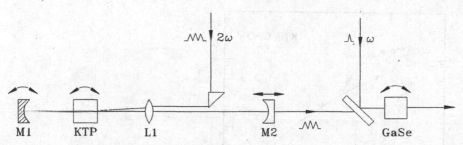

Fig. 1: Schematic of a single resonant OPO and a subsequent difference frequency generator. The OPO is pumped by a pulse train at the second harmonic of a Nd laser; an intense single pulse at the fundamental frequency serves as pump pulse in the difference frequency stage.

particular, when the second harmonic is used for pumping the parametric oscillator. From the pulse train a single pulse can be selected, amplified to energies of a few mJ, and used as a pump pulse for a subsequent difference frequency generation stage. In this way tunable mid-infrared pulses are generated.

Fig. 1 shows a schematic experimental setup of a parametric system consisting of an OPO pumped by the second harmonic frequency and a difference frequency stage. The simple OPO cavity consists of two curved mirrors M1 (r=-.2m, Au coating) and M2 (r=-2m, R=50%) and an uncoated f=.2m lens L1. The nonlinear crystal is a KTP crystal cut for type-II phase matching which is pumped in an off-axis geometry. Tuning of the idler resonant system is achieved by rotating the nonlinear crystal. For optimum performance the mirror M1, and the cavity length (position of M2) are simultaneously adjusted by computer controlled stepping motors. At the right hand side of Fig. 1 the difference frequency generator is shown. In a nonlinear crystal, e.g. GaSe, an intense single pulse at the fundamental frequency of the Nd laser is mixed with the output pulse train of the OPO at the idler frequency. In this way single pulses tunable from 2.5μm to 20μm are generated.

Fig. 2 shows the tuning range of the KTP-OPO pumped by the second harmonic of the Nd laser. The photon conversion efficiency is plotted as a function of the idler wavelength for two crystal cuts θ=78° and θ=62°. The energy per pump pulse is 500nJ. The first crystal cut is optimized for a subsequent difference frequency

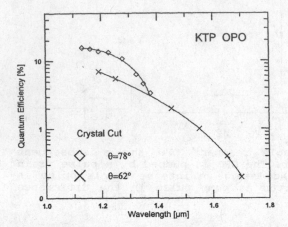

Fig.2: Photon conversion efficiency as a function of the wavelength for two KTP crystals cut at different phase matching angles. The idler resonant OPO is pumped by the second harmonic of the Nd laser.

mixing process in GaSe whereas the second cut was chosen for broad tunability. Conversion efficiencies exceeding 10% are observed. With pump pulses of 3ps duration tunable pulses of 1 to 2ps width are obtained.

The difference frequency step generates tunable pulses in the mid-infrared from 4µm to 20µm[5]. The output of the OPO at the idler frequency and an amplified single pulse of the Nd laser is mixed in a l=7mm GaSe crystal. Photon conversion efficiencies in the order of a percent are achieved despite of large reflection losses at the cleavage planes of the crystal oriented under ~40°. Fig. 3 shows a cross correlation curve between an idler pulse at λ=5µm and a pulse at the second harmonic frequency 2ω of the Nd laser of 3ps duration. The solid line represents a fit curve for Gaussian pulses. Deconvolution leads to a pulse duration of the 5µm pulse of 2.2ps. The gaussian shaped curve over 5 orders of magnitude is noteworthy. The bandwidth product was determined to be 0.5.

In summary it should be pointed out that different crystal materials have to be used for different frequency regions to obtain optimum output parameters. The technique discussed in this paper is reliable and has been used in many time resolved studies in semiconductor and molecular physics. Two examples for such applica-

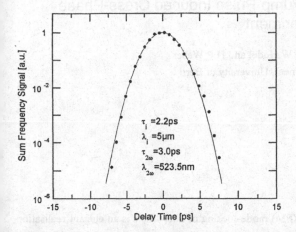

Fig. 3: Cross correlation between an IR pulse produced by difference frequency mixing in GaSe and the second harmonic of the Nd laser. The solid line is a fit curve for Gaussian shaped pulses.

tions are presented at this congress (paper by GRAENER et al. and paper by BAIER et al.). This all-solid state technique requires less complex laser systems. A flashlamp pumped modelocked Nd laser with an amplification stage is sufficient for the generation of intense, widly tunable pulses in the mid-infrared.

References

1 . for an overview see: special issue on optical parametric oscillation and amplification J. Opt. Soc. Am. **B 10** (1993) 2147
2 . R. Laenen, H. Graener, A. Laubereau, Opt. Commun. **77** (1990) 226
3 . R. Laenen, K. Wolfrum, A. Seilmeier, and A. Laubereau, J. Opt. Soc. Am. **B10** (1993) 2151
4 . P. Heinz, A. Reuther, and A. Laubereau, Opt. Commun. **97** (1993) 35
5 . T. Dahinten, U. Plödereder, A. Seilmeier, K.L. Vodopyanov, K.R. Allakhverdiev, and Z.A. Ibragimov, IEEE J. Quant. Electr. **QE-29** (1993) 2245
   I.M. Bayanov, R. Danielius, P. Heinz, and A. Seilmeier, Opt. Commun. **113** (1994) 99

Femtosecond Pulse Generation at 1.3 μ from a Pr³⁺-Doped Fiber Laser Modelocked by Pump Pulse Induced Cross-Phase Modulation: Theory and Experiment

M.Wegmüller, D.S.Peter, G.Onishchukov, W.Hodel and H.P.Weber
Institute of Applied Physics, Laser Department, University of Bern
Siedlerstr. 5,
CH-3012 Bern (Switzerland)
e-mail: wegmueller@iap.unibe.ch

Introduction: Cross-phase modulation (XPM) mode-locking of fiber-lasers is an elegant realisation of an all solid state, compact ultrashort pulse source - which also has great potential for all optical signal processing applications [PHILLIPS]. The technique was first demonstrated for Er-doped fibers [GREER, STOCK]. In these experiments, an additional, undoped piece of fiber was necessary in order to achieve a XPM sufficient for mode-locking. An improved version was demonstrated in our own experiment with a Pr-doped fiber, where XPM took place directly in the active fiber itself [PETER]. We obtained 570 fs pulses at 1.3 μm from this 'synchronously pumped' fiber laser.

In the following we describe this experiment, analyse in detail the pulse build up with the aid of numerical simulations and compare the theoretical and experimental results.

Experiment: The set-up of our experiment is shown in Fig.1. The pump signal was generated by a mode-locked Nd-YAG laser delivering pulses of 90 ps duration at a repetition rate of 82 MHz. The coupled average power was in the order of 1 W. The Pr-doped fiber (dopant level: 1000 ppm) was 8 m long and had a core diameter of 4.8 μm. At the wavelength of 1.3 μm the fiber had a relatively large normal group-velocity dispersion (GVD) of about 40ps/(km·nm) which was compensated for by an intracavity grating pair. The polarisation was varied with an intracavity λ/2 plate and had to be adjusted properly to obtain a stable train of short pulses (this point will be discussed in more detail later).

The autocorrelation and the spectrum of the resulting pulses are shown in Fig.2. Assuming a sech² pulse shape the pulse duration (fwhm) was 570 fs and the spectral width was 5.2 nm which is close to bandwidth limited. The pulses had the same repetition rate as the pump laser (82 MHz) and the average output power was typically 1mW.

Numerical results and comparison with the experiment: As gain (amplitude) modulation is practically absent in rare-earth doped fiber lasers due to the long upper state lifetimes involved, it was suggested that the phase-modulation caused by the pump pulses via XPM is the primary mode-

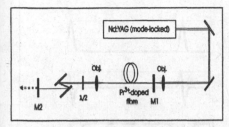

Fig.1: Experimental set-up of the 'synchronously pumped' fiber laser

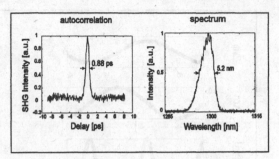

Fig.2: Measured autocorrelation and spectrum of the pulses

locking mechanism [GREER]. Hence, mode-locking of a fiber laser by 'synchronous pumping' can be regarded as the optical analogon to frequency modulation mode-locking. To confirm this and to analyse the pulse build up in detail we used numerical simulations. The model is described in detail in [WEGMÜLLER]. It contains the effects of dispersion (including third-order and gain dispersion), a frequency dependent gain and a constant loss of the cavity as well as the effect of XPM induced by the pump pulses. Note that by taking the gain to be constant along the fiber, pump pulse absorption and gain reduction by stimulated emission are neglected. The starting signal for the calculations consisted of spontaneous emission.

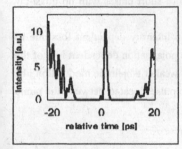

Fig.3: Calculated signal for the case of zero amplitude modulation

<u>*Role of XPM for the mode-locking:*</u> We first performed a series of calculations for different pump pulse parameters in order to see if pump pulse induced XPM is really sufficient for mode-locking the fiber laser. As expected, in each case a pulse is generated at the pump pulse location (t=0) with a duration that is only weakly dependent on the pump pulse parameters. A typical result is shown in Fig.3. The important point to note, however, is that in all cases a long 'pulse' (a quasi-cw signal) built up on either side of the short pulse as indicated in the figure.

The generation of this quasi-cw component can be explained by comparing a conventional (bulk) phase modulator with the 'optical phase modulator' of our model as illustrated in Fig.4. In the former case the induced phase modulation varies sinusoidally, i.e. a non-zero frequency-shift is induced on the signal at all times except when the derivative of the phase-shift is zero. The signal frequency components therefore experience a continuous shift unless they are located at (or in the immediate vicinity of) the maximum phase modulation amplitude. Since the gain in the laser cavity is highest for those components which are not continuously frequency shifted, pulses build up at the locations of maximum phase modulation amplitude (see left of Fig.4). For the case of XPM the situation is different. The phase modulation is strong only in the immediate vicinity of the pump pulses but is negligible in-between them. As a result a short pulse is generated at the location of the pump pulse and a long pulse is formed between the pump pulses (see right of Fig.4).

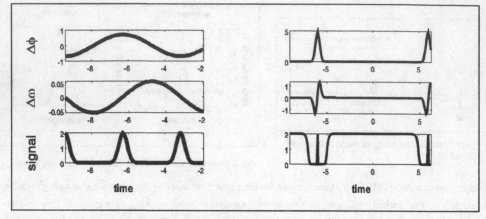

Fig.4: Phase modulation, frequency shift and resulting signal (all in a.u.) as a function of time for the case of a conventional phase modulator (left) and the 'optical phase modulator' by pump pulse induced XPM (right)

This argument explains why pump pulse induced XPM in itself is not sufficient to generate a 'clean' train of short pulses at the pump pulse repetition rate, and suggests that some additional amplitude modulation is necessary to suppress the quasi-cw signal. When such a mechanism was included in our numerical model, the simulations indeed showed the build up of short pulses with no quasi-cw signal present (inlet of Fig.5).

In the experiment this suppressing mechanism resulted from an intensity dependent loss due to nonlinear polarisation rotation in the fiber in combination with the polarisation dependent loss of the grating pair. In fact, without the λ/2 plate, i.e. with a substantially weaker amplitude modulation, the signal generated by the fiber laser consisted of 2 ns long, unstable pulses located between the pump

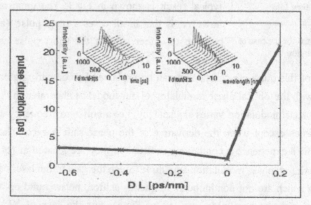

Fig.5: Calculated dependence of the pulse duration from the intracavity GVD (D·L, positive value = normal dispersion regime). The inlet shows the calculated pulse build up in time and frequency domain for the case of non-zero amplitude modulation.

pulses. Obviously, this result corresponds well with the numerically predicted presence of a quasi-cw signal for negligible amplitude modulation.

The influence of GVD: We further analysed the influence of the total intracavity GVD (D·L). Fig.5 shows the calculated pulse durations for several values of the intracavity GVD. The pulses were shortest for the case of zero GVD (i.e. third-order dispersion only). The difference between the measured (570 fs) and the calculated (1 ps) pulse duration is probably due to the influence of SPM in the experiment (SPM is not included in the numerical model). For increasing normal GVD (positive D·L), the pulse duration grew dramatically: for a value of D·L=0.2 the pulse duration increased by a factor of 20. On the other hand, anomalous intracavity GVD had only a small influence on the pulse duration: for D·L=-0.2 the pulse duration increased by a factor of 2 only.

In the experiment, the same dependence of the pulse duration on intracavity GVD was observed: for a value of D·L=0.3 the pulse duration increased by a factor of ~30; whereas for a similar amount of anomalous GVD the increase in pulse duration was not significant (factor of ~2). This demonstrates the importance of working in the anomalous GVD regime if short pulses are desired.

Conclusion: Mode-locking of a Pr-doped fiber laser by pump pulse induced XPM was analysed theoretically and experimentally. Both the experiment and the numerical simulations clearly showed that XPM is the main mode-locking mechanism but that it is in itself not sufficient to generate a 'clean' train of pulses. Only when introducing some amplitude modulation stable pulses are generated. In the experiment, 570 fs nearly bandwidth limited pulses were obtained for the case of close-to-zero intracavity GVD. Numerical analysis of the influence of overall intracavity GVD showed that normal GVD is very disadvantageous with respect to pulse duration, whereas the influence of anomalous GVD is relatively small. This result is in very good agreement with the experimentally observed behaviour.

References:
M. W. *Phillips*, A. I. Ferguson, D. B. Patterson, Optics Comm., **75**, 1990, pp. 33-38
E. J. *Greer*, K. Smith, Electron. Lett., **28**, 1992, pp. 1741-1743
M. L. *Stock*, L. M. Yang, M. J. Andrejco, M. E. Fermann, Optics Lett., **18**, 1993, pp. 1529-1531
D. S. *Peter*, G. Onishchukov, W. Hodel, H. P. Weber, Electron. Lett., **30**, 1994, pp.1595-1596
M. *Wegmüller*, W. Hodel, H.P. Weber, Optics Comm., **115**, 1995, pp.498-504

Widely Tunable Two-Color Femtosecond Ti:Sapphire Laser with Perfect Pulse Syncronization

A. Leitenstorfer, C. Fürst and A. Laubereau
Physik-Department E 11, Technische Universität München
D-85748 Garching, Germany

The development of the Kerr-lens mode-locked femtosecond Ti:sapphire laser [1] has had a great impact on ultrafast spectroscopy. Special versions have permitted the generation of pulses shorter than 10 fs [2,3]. Recently two pulse trains operating at different wavelengths were produced in a single laser system [4-7,9] and were reported to be synchronized to better than 50 fs. This two-color laser appears to be a promising tool for ultrafast experiments; unfortunately, the poor tunability of the reported systems turns out to be a serious hinderance for general applications in spectroscopy.

In this paper, we present the first widely tunable dual-wavelength Ti:sapphire laser [13]. The jitter between the two coupled pulse trains was found to be extremely small, less than 2 fs. In principle the laser consists of two asymmetric X-folded cavities (Fig. 1) both sharing the highly reflecting end mirror (EM), the folding mirrors (FM1, FM2) and the gain medium (G). The prism compressors (P1a, P1b, P2a, P2b) and output couplers (OC1, OC2) may be adjusted separately, providing fully independent wavelength tuning.

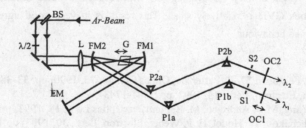

FIG. 1: Experimental setup of the dual-wavelength femtosecond Ti:sapphire laser. BS: beamsplitter; $\lambda/2$: half-wave plate for fine tuning of the absorbed pump power; L: focussing lens; EM: end mirror; FM1,FM2: folding mirrors; G: gain medium; P1a-P2b: dispersing prisms; S1,S2: adjustable slits for wavelength tuning; OC1,OC2: output couplers. OC1 and G are mounted on translation stages.

We start the discussion of the path of the two pulse trains at EM where the two laser beams are separated by 4 mm. Both beams are focused by FM1 crossing near the exit surface of the Brewster-cut laser rod. The folding mirrors have a focal length of 5 cm and are highly reflecting between 700 and 870 nm. The gain medium is a 6 mm, 0.15 %-doped Ti:sapphire rod. After FM2 the beams are again nearly parallel. The prism sequences (P1a, P1b, P2a, P2b) consist of fused-silica specimens that are separated by 60 cm in branch 1 and 85 cm in branch 2. Branch 1 is designed to have minimum third order dispersion [8] and permits pulse durations from 15 to 70 fs. The prism distance in branch 2 is chosen to be longer than necessary for dispersion compensation. In this way one is able to generate pulse durations between 40 and 120 fs by varying the amount of negative dispersion in the cavity [8].

The two pulse trains are synchronized when we adjust the round trip times to be equal by varying the cavity length of branch 1 by moving output mirror OC1 on a translation stage. The tolerance in cavity length for the cross coupled operation is found to be 1 μm. Interestingly, this syncronized mode exhibits a threshold behavior, hysteresis and bistability [14].

We find it essential for stable and tunable two-color operation that the Ar laser pumps both branches independently with separate pump beams (see Fig. 1). When the spatial overlap of the pumped regions becomes large the laser will turn to an unstable behavior owing to gain competition. On the other hand, the two beams must come close together for exact synchronization to ensure optimum pulse interaction, presumably by means of the third order nonlinear susceptibility.

The salient features of our laser system are as follows: In Fig. 2 the pulse durations routinely obtained in the two synchronized branches are depicted versus wavelength. The data are obtained with different output couplers OC1 and OC2 with minimum transmissions of 3 to 10%. For pulse durations of 70 fs and with a total pump power of 7 W we obtain output powers as high as 500 mW in each branch of the laser.

Interferometric autocorrelations for the laser operating in the coupled mode are presented in Fig. 3 for pulse durations as short as $\tau_1 = 16$ fs in branch 1 and $\tau_2 = 74$ fs in branch 2. We em-

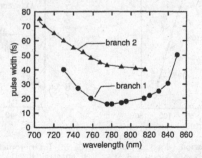

FIG. 2: Minimum pulse durations versus wavelengths in the cross-coupled mode of operation. The solid lines are to guide the eye.

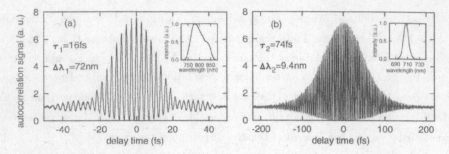

FIG. 3: Fringe resolved autocorrelations and spectra (inserts) of two coupled pulse trains. (a) shows the result for branch 1 and (b) for branch 2. Pulse durations are calculated assuming a sech^2 pulse shape for branch 1 and a Gaussian for branch 2.

phasize the very large wavelength distance (85 nm) between the two pulse trains corresponding to an energy difference of $1360 \, \text{cm}^{-1}$. The pulse spectra are depicted in the insets of Fig. 3. The bandwidth product of the 74 fs pulses is 0.42 which is close to the transform limit for Gaussian pulses. The shorter pulses in branch 1 exhibit a somewhat increased bandwidth product of 0.56. Figure 4 shows an intensity cross correlation of both pulse trains with pulse durations of $\tau_1 = 30 \pm 2 \, \text{fs}$ and $\tau_2 = 64 \pm 3 \, \text{fs}$ with corresponding laser spectra centered at $\lambda_1 = 830 \, \text{nm}$ and $\lambda_2 = 760 \, \text{nm}$, respectively. The data fit almost perfectly to a Gaussian shape over three orders of magnitude. It should be noted at this point that Gaussian pulse shapes are found in our laser system in autocorrelation and cross correlation measurements for pulse durations longer than 20 fs. Only if branch 1 is adjusted to shorter pulse durations, the autocorrelation exhibits slight wings and fits better to a sech^2. For the two Gaussian pulses with durations given above,

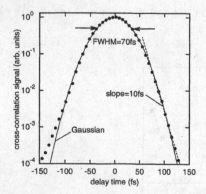

FIG. 4: Intensity cross-correlation (dots) of coupled pulses with $\tau_1 = 30 \, \text{fs}$ and $\tau_2 = 64 \, \text{fs}$ centered at $\lambda_1 = 830 \, \text{nm}$ and $\lambda_2 = 760 \, \text{nm}$. The solid line represents a Gaussian fit to the data.

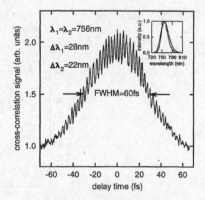

FIG. 5: Interferometric cross-correlation and related spectra (insert). Pulse durations are $\tau_1 = 42 \, \text{fs}$ and $\tau_2 = 45 \, \text{fs}$. The modulation depth of the fringes is consistent with a jitter of 1.7 fs between both pulse trains.

a cross-correlation width (FWHM) of 71±4 fs is expected for perfect synchronization. Within the experimental accuracy this value coincides with the 70±2 fs value we find in Fig. 4. Thus, we are not able to see any temporal jitter from these data.

To get quantitative information about the synchronization we perform an **interferometric cross-correlation** measurement. To this end we tune both pulse trains to the same wavelength (see insets in Fig. 5). Surprisingly, the data show a clearly resolved interferometric structure, indicating that the relative position of both pulse trains is stable within less than one cycle of light (Fig. 5)! Computer simulations for Gaussian pulses and a Gaussian distribution of varying time delays yield a timing jitter as small as 1.7 fs for this measurement. Interestingly, the interferometric pattern appears only if the two cavity lengths are matched within an interval of approximately 100 nm.

The sensitivity of the interferometric cross correlation leads us to the following explanation of the mechanism for the strong coupling of the two pulse trains: A self-sustained synchronization on a time scale below 10 fs necessitates some attractive interaction between the pulses. In the absence of such a mechanism even very small disturbances in cavity length (e.g. sound waves) would cause the temporal coupling to cease. It has been shown theoretically that a strongly phase-sensitive attractive interaction of solitary pulses can arise owing to cross-phase modulation in a Kerr medium [10]. Bound solutions of the nonlinear wave equations do exist even for pulses at different wavelengths and in the presence of group-velocity dispersion [10-12]. In our setup such an interaction takes place when the two pulses are copropagating through the laser crystal. Small mismatches in pulse positions, which may arise in every round trip through the laser cavities, can be compensated by this effect.

In conclusion we have developed a widely tunable dual-wavelength femtosecond source in the near infrared. For the first time to our knowledge a synchronization of two coupled pulse trains with an accuracy better than 2 fs was demonstrated. Because of its unique temporal resolution and high repetition rate our setup represents a powerful tool for ultrafast spectroscopy.

References

[1]   D. E. Spence, P. N. Kean, and W. Sibbett, Opt. Lett. **16**, 42 (1991)

[2]   A. Stingl, M. Lenzner, C. Spielmann, and F. Krausz, Opt. Lett. **20**, 602 (1995)

[3]   J. Zhou, G. Taft, C. Huang, M. M. Murnane, H. C. Kapteyn, and I. P. Christov, Opt. Lett. **19**, 1149 (1994)

[4]   M. R. X. de Barros and P. C. Becker, Opt. Lett. **18**, 631 (1993)

[5]   D. R. Dykaar and S. B. Darack, Opt. Lett. **18**, 634 (1993)

[6]   J. M. Evans, D. E. Spence, D. Burns, and W. Sibbett, Opt. Lett. **18**, 1074 (1993)

[7]   Z. Zhang and T. Yagi, Opt. Lett. **18**, 2126 (1993)

[8]   see C. Spielmann, P. F. Curley, T. Brabec, and F. Krausz,
        IEEE J. Quantum Electron. **QE-30**, 1100 (1994)
[9]   D. R. Dykaar, S. B. Darack, and W. H. Knox, Opt. Lett. **19**, 1058 (1994)
[10]   J. P. Gordon, Opt. Lett. **8**, 596 (1983)
[11]   V. V. Afanasjev, E. M. Dianov, and V. N. Serkin,
        IEEE J. Quantum Electron. **QE-25**, 2656 (1989)
[12]   M. Lisak, A. Höök, and D. Anderson, J. Opt. Soc. Am. **B7**, 810 (1990)
[13]   A. Leitenstorfer, C. Fürst, and A. Laubereau, Opt. Lett. **20**, 916 (1995)
[14]   C. Fürst, A. Leitenstorfer, and A. Laubereau, to be published

Single-Shot Absorption Spectroscopy with Chirped Femtosecond Laser Pulses

Harald Schulz, Ping Zhou, and Peter Kohns
Optikzentrum NRW
Abteilung Industrielle und wissenschaftliche Meßtechnik / Hochleistungsoptik
Universitätsstraße 142, D-44799 Bochum, Germany

The D_2 line of rubidium (Rb) was recorded using femtosecond (fs) laser pulses and a spectrally resolving autocorrelator. The interaction between photons and atoms lead to a significant change of the pulse correlation signal near the half of the D_2 wavelength.

Introduction

Narrow line spectroscopy and fs laser pulses seem to be incompatible. However, measuring the time dependend change of the pulse phase (chirp) allows investigations with high temporal and spectral resolution and opens new techniques such as microscopic LIDAR or atomic line spectroscopy as presented in this paper. Measuring the change of chirp when the fs pulse has passed a narrow bandwidth medium (e.g. an atomic gas) provides us enough information to identify the absorption species and estimate the absorption. We used a measurement technique similar to previously published FROG techniques [1,2]. The spectrally resolved autocorrelation (SRA) function [1] was used to reconstruct both the pulse phase and shape after crossing the medium.

Experimental setup

A self-modelocked Ti:Sapphire laser producing ultrashort pulses with a pulse duration of 100 fs and a pulse energy of 1 nJ was tuned to the D_2 absorption line of Rb ($\lambda \approx 780$ nm). The pulse bandwidth was about 10 nm. Using a dispersive delay line the chirp of the pulses could be adjusted accurately. The chirp was controlled using the measurement technique published in [2]. The pulses were sent through a Rb vapour cell (length $l_{cell} = 75$ mm) where the peak intensity was in the order of 10 kW/mm^2. The Rb concentration could be varied by the cell temperature from 10^{-7} Torr at 20° C to 10^{-4} Torr at 90° C due to the change of vapour pressure. The laser bandwidth exceeded the absorption linewidth of the Rb vapour which is

60

mainly due to Doppler broadening ($\gamma=2\pi\cdot300$ MHz) by four orders of magnitude. The peak intensity exceeded the Rb saturation intensity (50 μW/mm^2) by eight orders of magnitude. After passing the Rb cell each pulse was fed into a single-shot correlator [2,3], where it was split up into two replica and focussed by a cylindrical lens into a lithium borate crystal for second harmonic generation (SHG). The SH emitting linear area of the crystal was depicted onto the entrance slit of a grating spectrometer. The axis parallel to the entrance slit corresponded to the delay time τ between the pulse replica, whereas the axis of dispersion corresponded to the wavelength λ. The output of the spectrometer was monitored by a CCD camera. Thus the CCD output was the SRA signal described in [1].

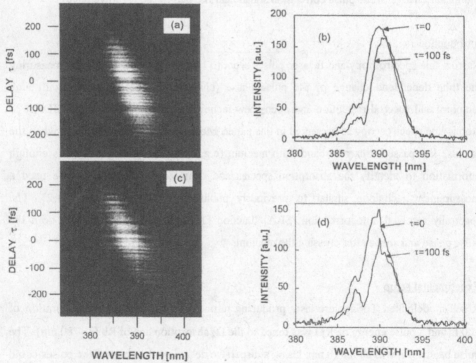

Figure 1: CCD output signals (see text). The side clouds at $\lambda = 387$ nm are due to a slight nonlinear chirp of the pulses

Experimental results

Fig. 1a) shows a typical CCD signal where the laser center wavelength was tuned near 780 nm and the Rb cell temperature was 20° C. Due to the low Rb pressure the SRA signal appeared as a cloud without a line structure at $\lambda=390$ nm, the half of the resonant wavelength of Rb (fig. 1b). The situation changed when the cell temperature was increased to 90° C (fig. 1c). The correlation signal increased at $\lambda=390$ nm as is indicated by the bright vertical line in fig. 1c, whereas the signal at other wavelengths did not change. Plots of the SRA shown in fig. 1c) at fixed delay times show a spike at 390nm (fig. 1d).

We note that neither the laser intensity nor the pulse spectrum measured directly behind the Rb cell depended on the temperature of the Rb cell where the accuracy of our measurements was in the order of 1%. We also note that the line structure in the SRA disappeared when the pulse spectrum was detuned from the Rb absorption line.

Theoretical explanation

As mentioned above our observations cannot be explained by the absorption of pulse energy at the transition line of Rb. On the other hand it is well known that the refractive index n of absorbing media near the resonance frequency ω_0 strongly depends on the laser frequency ω:

$$n(\omega) = 1 + \frac{N \cdot e^2}{4\varepsilon_0 \cdot m \cdot \omega_0} \cdot \frac{\omega_0 - \omega}{(\omega_0 - \omega)^2 + (\gamma/2)^2} \qquad (1)$$

Here N is the number density of illuminated particles, m the mass of the electron, e the elementary charge, and ε_0 the dielectric constant [4]. The formula is valid for two-level atoms and $n(\omega) \approx 1$. We note that the maximum absolute value of $n(\omega)-1$ is of the order 0.001 at the parameters used. Nevertheless the maximum of the phase shift $\Phi(\omega) = k \cdot l_{cell} \cdot (n(\omega) - 1)$ is of order 200π at a cell temperature of 90°C (k=wave vector $2\pi/\lambda$). Note that the dispersion of the Rb vapour changes the phase of the pulse yielding a strong phase modulation $d\Phi/d\omega$ only within the narrow spectral region of the atomic resonance and can be neglected outside. We assumed a gaussian shape of the laser pulses before passing the Rb cell:

$$E(t) = E_0 \cdot \exp[-(t/T)^2] \cdot \exp(-i\omega_L t) \cdot \exp(-i\alpha t^2). \qquad (2)$$

Here E(t) is the time-dependent electrical field of the pulse, E_0 its amplitude, T the pulse duration, ω_L the center frequency of the pulse, and α the chirp coefficient. The pulse spectrum can be obtained by a Fourier transformation:

$$E(\omega) \propto \int_{-\infty}^{\infty} E(t) \exp(i\omega t) dt. \qquad (3)$$

Thus the spectral dependence of the electrical field leaving the Rb cell can be described by

$$E'(\omega) = E(\omega) \cdot \exp[i\Phi(\omega)], \qquad (4)$$

Following PAYE et al. [1] the SRA signal of the pulse behind the cell can be calculated as

$$SRA(\tau, \omega) \propto \left| \int E'(t+\tau) \cdot E'(t) \exp(i\omega t) dt \right|^2, \qquad (5)$$

where E'(t) is the Fourier transformation of E'(ω).

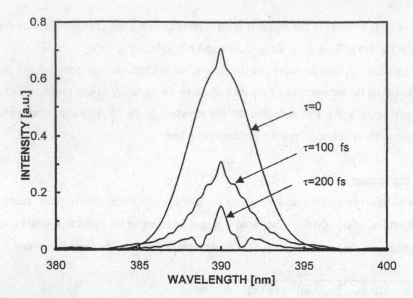

Figure 2: Results of our simulation

Because the strong phase modulation is located at the narrow atomic resonance a pronounced structure of the SRA signal at the resonance frequency ω_0 can be expected. We calculated the SRA signal for different delay times τ. The number density of the Rb atoms was chosen as $3 \cdot 10^{18}/m^3$ corresponding to the cell temperature of 90°C that we used in the experiment. Our calculated results at different delay times τ are plotted in fig. 2, which can be compared with the experimental results shown in fig. 1d. Each calculated spectrum contains a spike at half the resonant wavelength of Rb as observed in the experiment. The magnitude of observed and calculated spikes are within the same order of magnitude. The deviation of the simulated shape of the SRA to the experimental results can probably be attributed to the deviation of the real pulse shape from a gaussian shape that was used at our calculations. Additionally we neglected the real multilevel and two-isotope structure of the Rb atoms to simplify our theoretical estimation.

References:

[1] J.PAYE, M.RAMASWAMY, J.FUJIMOTO, and E.P.IPPEN, Opt.Lett.**18**,1946 (1993)
[2] P. ZHOU, H. SCHULZ, and P. KOHNS, "Correlator-measuring-systems for ultrashort laser pulses", Laser '95, congress F (1995)
[3] F. SALIN, P. GEORGES, G. ROGER, and A. BRUN, Appl. Opt. **26**, 4528 (1987)
[4] W.DEMTRÖDER, "Laserspektroskopie", Springer Verlag Berlin (1993)

High Repetition Rate Amplification of Femtosecond Pulses in the Ultraviolet Spectral Range

U. Stamm, J. Kleinschmidt

Lambda Physik GmbH, Hans-Böckler-Str. 12, 37079 Göttingen, Germany

O. Kittelmann, F. Noack, J. Ringling

Max-Born-Institut für Nichtlineare Optik und Kurzzeitspektroskopie,
Rudower Chaussee 6, 12489 Berlin, Germany

Abstract. UV optical pulses at 248 nm and 193 nm generated by frequency conversion of high-repetition rate Ti: sapphire laser pulses are amplified in commercial excimer amplifiers at up to 500 Hz repetition rate. At 248 nm maximum pulse energies exceeding 10 mJ at pulse durations of 350 fs have been obtained. At 193 nm average powers of 290 mW at 400 Hz with pulse durations of 300 fs have been generated.

Introduction

There is considerable interest in the development of excimer laser systems emitting high intensity, ultrashort optical pulses in the ultraviolet (1-3) for a variety of potential applications. All these systems are based on amplification of ultrashort seed pulses in excimer gain modules. Until now all fs amplification systems were restriced to rather low repetition rates ($\leq$30 Hz) which results for many applications in a decrease in signal-to-noise-ratio due to the low data aquisition rate.

Here we report on the amplification of fs pulses in commercial KrF and ArF amplifiers at repetition rates up to 500 Hz resulting in femtosecond pulse trains with single pulse energies above 10 mJ at 248 nm and 1 mJ at 193 nm.

Experimental

The generation of the femtosecond UV seed pulses from the Ti: sapphire regenerative amplifier system by frequency tripling and quadroupling has been reported earlier (4). The experimental arrangement is shown in Fig. 1.

The stable operation of the oscillator/amplifier system critically depends on the timing synchronization of its different components. The 41-MHz output of the driver electronics for the acousto-optic modulator of the fs-Ti:sapphire oscillator is divided down to the desired repetition rate (50 Hz-1 kHz) by the MEDOX-Pockels cell driver DR85-A. This TTL-signal is used for direct triggering the Q-switch of the Nd:YLF pump laser. Two programmable delays of the pockels cell driver define the moments to switch the stretched fs-pulse into the amplifier laser cavity and to switch out the amplified pulse. Additionally to this "normal" timing circuit the TTL-pulse also starts an adjustable electronic delay which triggers the amplifier gain module with the excimer controller used to stabilize the time between the start pulse and the discharge of the excimer gain module. The timing jitter between the seed pulses and the gain maximum is typically less than 1.5 ns r. m. s. This is considerably shorter than the temporal width of the gain profile allowing stable amplification of the seed pulses.

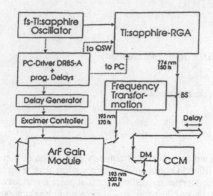

FIGURE 1 Experimental scheme, as example amplification at 193 nm.
BS: beam splitter; DM: dichroic mirror; CCM: cross-correlation
measurement; PC: pockels cell; QSW: Q-switch; RGA:
regenerative amplifier. Solid line: paths of the optical pulses,
dotted line: electronic signal connections for synchronization.

Femtosecond Pulse Amplification at 248 nm

The UV seed pulses at $\lambda = 248$ nm are generated by frequency tripling of the
745 nm radiation in two subsequent phase matched BBO crystals. The energy of
the seed pulses is typically 4 µJ. Amplification of these pulses was performed in a
commercial femtosecond KrF amplifier (based on the Lambda Physik excimer laser
LPX 140i).
In single pass geometry at repetition rates up to 500 Hz, the average power of the
amplified femtosecond pulses increases linearly with the repetition rate. A maximum average power of 1.0 W was achieved at 500 Hz corresponding to a single
pulse energy of 2 mJ.
Off-axis amplification (5) in a double pass geometry has been employed to reach
energy levels above 10 mJ with ASE background of less than 10^{-4}.

To determine the pulse duration of the amplified pulses the cross correlation
function was measured. To this aim the pulse energy of the difference frequency
signal at $\lambda = 372$ nm generated by mixing the amplified pulses at 248 nm and the
fundamental at $\lambda = 745$ nm in a 0.3 mm thick BBO crystal has been determined.
The duration of the amplified UV pulses was measured to 350 fs.

Amplification at 193 nm

The seed pulses were first tuned to the peak gain wavelength of ArF (193.4 nm) by
tuning the Ti:sapphire system to a wavelength of about 774 nm and adjusting the
BBO conversion stages (crystals) for optimum second, third and fourth harmonic
generation. The energy of the seed pulses was measured to be about 0.8 µJ and the
pulse duration was determined to be 170 fs by a cross-correlation technique (4).
Pulse energies up to 110 µJ at 100 Hz repetition rate could be obtained after a single
pass which corresponds to a small signal gain of about 140.

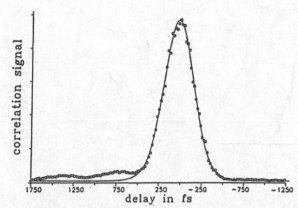

FIGURE 2 Cross-correlation signal as function of the relative delay between the fundamental and the amplified 193 nm pulse. Solid line: Calculated cross-correlation function of a 150 fs fundamental pulse and a 300 fs near VUV-pulse taking into account group velocity mismatch and assuming $sech^2$-shaped pulses.

The output after the first pass was reflected back through the discharge tube in a second pass resulting in ultrashort optical pulses with a short pulse energy content of ≈ 1 mJ at 100 Hz repetition rate. These pulses were accompanied by a copropagating amplified spontaneous emission (ASE) background of about 250 µJ emitted in about 15 ns. Already at this energy level saturation of amplification was observed.

At the highest repetition rate of 400 Hz an average power of more than 290 mW could be achieved. The pulse-to-pulse fluctuations have been measured to be less than 7 % r. m. s. The observed saturation behaviour in average power at higher repetition rates (>300 Hz) is typical for ArF amplifiers which gain strongly decreases when temperature gradients in the discharge occur (6).

The pulse duration of the amplified pulses both after the first and second pass through the gain module was determined by a cross-correlation measurement as described in Ref. 4. The cross-correlation traces show a temporal width of typically 350 fs and 420 fs (Fig. 2) after the first and second pass, respectively. Taking into account dispersion in the cross-correlator and assuming a $sech^2$-pulse shape pulsewidths of about 220 fs and 300 fs can be deduced from these measurements.

Inspection of the spectrum of the amplified pulses (Fig. 3a) reveals that its width is slightly broader than the spectrum of the ASE (Fig. 3b). Besides the pronounced dips in the spectral profile caused by absorption due to molecular oxygen the spectrum exhibits additional modulations. We believe that the amplified 193 nm pulses undergo strong self-phase modulation in atmospheric air leading to the observed spectral broadening (6). These pulses should be recompressible in time by compensation of the linear portion of the chirp in a proper compressor arrangement.

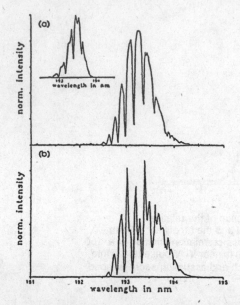

FIGURE 3: Spectra of the 193 nm pulses (20 shots accumulated).
(a) ASE of the ArF amplifier module, the inset shows a
spectrum of the seed pulses, (b) spectrum of the amplified
pulses after the second pass. The spectral resolution of the
optical multichannel analyzer system was 0.014 nm per
channel.

REFERENCES

1. Glownia, J.H., Kaschke, M., and Sorokin, P.P., *Opt. Letter* **17**, 337 (1992)
2. Momma, C., Eichmann, H., Jacobs, H., Tünnermann, A, Welling, H., and Wellegehausen, B., *Opt. Lett.* **18**, 516 (1993).
3. Szatmari, S., and Schäfer, F.P., *J. Opt. Soc. Am.* B **6**, 1877 (1989).
4. Ringling, J., Kittelmann, O., Noack, F., Korn, G., and Squier, J., *Opt. Lett.* **18**, 2035 (1993).
5. Szatmari, S., Almasi, G., and Simon, P., *Appl. Phys.* B **53**, 82 (1991).
6. Ringling, J., Kittelmann, O., Noack, F., Stamm, U., Kleinschmidt, J., and Voß, F., to be published in *Opt. Lett.*

A Multi-Pass Plane-Mirror Femtosecond Dye Laser Amplifier

M. Wittmann and A. Penzkofer

Naturwissenschaftliche Fakultät II - Physik, Universität Regensburg,

D-93040 Regensburg, Germany

Femtosecond dye laser amplifiers are reviewed by KNOX and DIELS and RUDOLPH. Multistage single-pass amplifiers are used with high power pump lasers. Bow-tie and confocal resonator multi-pass amplifiers amplify the femtosecond pulses in several passes within the pump pulse duration. Two-stage multi-pass amplifiers allow to adjust the pump pulse requirements to the small-signal preamplification and the final high energy extraction (CHAMBARET et al.).

Here a double-stage three-pass plane mirror amplifier system is described. The experimental system is shown in Fig. 1. The pulses to be amplified are generated in a linear cw passive mode-locked dispersion compensated femtosecond rhodamine 6G - DODCI dye laser (BÄUMLER and PENZKOFER). The laser amplifier system is pumped with a Q-switched Nd:YAG laser of 50 mJ second harmonic output, 5 ns pulse duration, and 20 Hz repetition rate. A synchronisation unit, SU, locks the Q-switched laser to the femtosecond laser with a temporal jitter of $\pm$ 0.5 ns (TURNER et al.). Partially coated plane mirrors, PM, form the multi-pass amplifier cavities. They allow loss-free femtosecond laser beam entrance and exit, while complete beam reflection occurs for the multiple

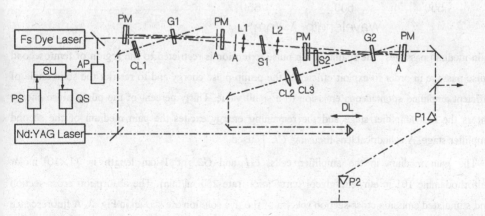

Fig. 1: Amplifier system. AP, avalanche photodiode; PS, laser power supply; QS, Q-switched system; A, apertures; DL, optical delay line.

68

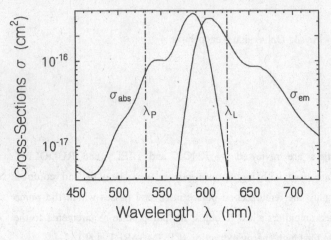

Fig. 2: Absorption and emission spectra of sulforhodamine 101 in ethylene glycol.

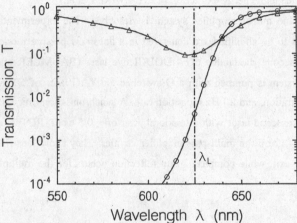

Fig. 3: Small-signal transmission through saturable filters (circles, RG645; triangles, malachite green).

gain medium passages. The spatial pump pulse extension is restricted to the region of femtosecond pulse passage in order to exploit efficiently the pump pulse energy and to restrict the solid angle of efficient amplified spontaneous emission to a small value. Thirty percent of the pump pulse energy enters the first amplifier stage and the remaining energy excites the gain medium of the second amplifier stage. A cylindrical lens focusing, CL, is used.

The gain medium in the amplifier cells, $G1$ and $G2$, of 1 cm length is 1.1×10^{-4} molar sulforhodamine 101 in ethylene glycol (circulation rate 250 ml/min). The absorption cross-section and stimulated emission cross-section spectra of the dye solution are shown in Fig. 2. A fluorescence quantum yield of $\phi_F = 0.93$ and a radiative lifetime of $\tau_{rad} = 5.6$ ns have been determined by spectro-fluorometry (PENZKOFER and LEUPACHER). At the pump laser wavelength of $\lambda_P = 532$ nm the groundstate absorption cross-section is $\sigma_P = 7.2 \times 10^{-17}$ cm^2 (Fig. 2) and the S$_1$-excited-state absorption is $\sigma_{ex,P} = 5.7 \times 10^{-17} \, \text{cm}^2$ (determined by non-linear transmission measurements, WITT-MANN and PENZKOFER). At the laser wavelength of $\lambda_L = 625 \, \text{nm}$ the stimulated emission cross-section is $\sigma_{em,L} = 1.85 \times 10^{-16} \, \text{cm}^2$ (Fig. 2) and the excited-state absorption cross-section is $\sigma_{ex,L} = 6 \times 10^{-17} \, \text{cm}^2$ (determined by small-signal amplification measurements, WITTMANN and PENZKOFER).

The wavelength of amplified spontaneous emission (ASE) approximately coincides with the femtosecond laser wavelength. The ASE generated in the first amplifier stage along the femtosecond pulse propagation direction is reduced by a 2 mm photodarkened Schott colour filter glass RG645, $S1$, which is placed between a telescopic lens arrangement, $L1, L2$. A detailed spectroscopic analysis of the filter is given by WITTMANN and PENZKOFER. The wavelength dependent small-signal transmission of the filter is shown by the circles in Fig. 3, and the intensity dependent transmission of the filter is shown by the circles in Fig. 4 for 240 fs pulses at 625 nm (absorption recovery time of dominant component is 25 ps). In the second amplifier stage a 1 mm cell, $S2$, of 6.4×10^{-5} molar malachite green in ethanol (absorption recovery time is 1.5 ps, MARCANO et al.) is incorporated for amplified spontaneous emission reduction. The wavelength dependent small-signal transmission of the solution is shown by the triangles in Fig. 3, and the intensity dependent transmission is shown by the triangles in Fig. 4. Experimental background energy measurements (WIEDMANN and PENZKOFER) gave an overall ASE content of about 5%.

In the first triple-passage amplifier the input pulses of 110 fs duration and 40 pJ single pulse energy are amplified a factor of 4×10^5. In the second amplifier gain saturation occurs. The output pulse energy of the second amplifier is 270 μJ. The pulse duration increases to 240 fs by group

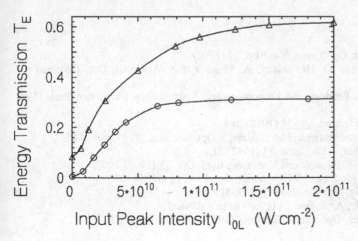

Fig. 4: Energy transmission through saturable filters (circles, RG645; triangles, malachite green).

70

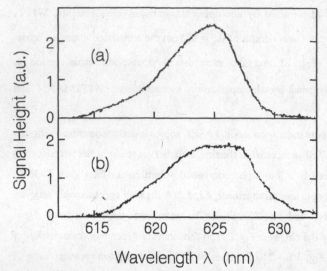

Fig. 5: Femtosecond laser spectra
(a) input spectrum,
(b) spectrum behind second amplifier.

velocity dispersion. The femtosecond pulse input spectrum and the amplified output spectrum are displayed in Fig. 5a and 5b, respectively. The spectral bandwidth - pulse duration products ($\Delta v_L \Delta t_L \approx 0.6$ before amplifier, $\Delta v_L \Delta t_L \approx 1.6$ behind amplifier) indicate self-phase modulation and group velocity dispersion (PENZKOFER et al.).

In a prism-pair compression stage (Schott glass SF59 prisms $P1$ and $P2$ at a distance of 35 cm) the amplified pulses are shortened to $\Delta t_L \approx 75\,\mathrm{fs}$ without a remarkable change of the spectral shape ($\Delta v_L \Delta t_L \approx 0.5$). The energy of the compressed pulses is 180 µJ.

In conclusion, a versatile double-stage three-pass plane mirror dye laser amplifier system has been built up with small output pulse energy fluctuation (< 10%) and weak ASE contribution (<5%). Numerical simulations of the amplification dynamics are given elsewhere (WITTMANN and PENZKOFER).

References
W. Bäumler and A. Penzkofer, Opt. Quant. Electron. 24 (1992) 313.
J.P. Chambaret, A. Dos Santos, G. Hamoniaux, A. Migus and A. Antonetti, Opt. Commun. 69 (1989) 401.
J.-C. Diels and W. Rudolph, Femtosecond Phenomena and Spectroscopy (Academic Press, New York, 1995).
W.H. Knox, IEEE J. Quant. Electron. QE-24 (1988) 388.
A. Marcano, L. Marquez, L. Aranguren and M. Salazar, J. Opt. Soc. Am. B7 (1990) 2145.
A. Penzkofer and W. Leupacher, J. Luminesc. 37 (1987) 61.
A. Penzkofer, M. Wittmann, W. Bäumler and V. Petrov, Appl. Opt. 31 (1992) 7067.
T. Turner, M. Chatelet, D.S. Moore and S.C. Schmidt, Opt. Lett. 11 (1986) 357.
J. Wiedmann and A. Penzkofer, Opt. Commun. 25 (1978) 226.
M. Wittmann and A Penzkofer, Appl. Opt. 34 (1995) to be published.
M. Wittmann and A. Penzkofer, Opt. Quant. Electron. (1995) to be published.

The Ultrafast Bleaching of CuS Microcrystals

V.P.Mikhailov, K.V.Yumashev, P.V.Prokoshin, A.M.Malyarevich
International Laser Center, 7 Kurchatov St., Minsk 220064, Belarus

Nonlinear optical processes in different systems with ultrasmall semiconductor particles are of significant interest in the connection with observed and expected peculiar optical and electrooptical features. These features are caused first of all that particles size becomes comparable with Bohr exciton radius and, second, properties of ultrasmall particles vary due to environment. Nonlinear optical phenomena of semiconductors like $A^{II}B^{YI}$ in different media are widely investigated now. These semiconductors have direct transition under excitation that results in the formation of excitons with considerably high binding energy. The nonlinear response features are attributed to two main effects: (1) band filling (energy level filling in the case of quantum-sized particles); (2) generation of electron-hole plasma under the more high pumping. Other mechanisms of nonlinear optical effects considered for such semiconductor particles are – two-photon absorption and induced absorption due to some trapping of carriers (defects, surface states, etc). In the most of cases the latter plays secondary role just affecting the relaxation pathways.

In present paper we report on the ultrafast bleaching of CuS microcrystals in PVA thin films and possible reasons of this phenomenon. Experiments have been performed on two types of CuS samples. One of them is the CuS quantum dots in transparent matrix (usual sample). The second – the same CuS microsryastals, but after certain treatment (modified sample). Due to this treatment the sample has new absorption band in the region 0.6 – 1.8 μm with maximum at 1.2 μm near the fundamental absorption edge of semiconductor. Figure 1 demonstrates spectra of linear absorption for both samples.

Thin films with ultrasmall CuS particles were prepared from acqueous colloids stabilized by polyvinyl-alcohol (PVA) as the result of deposition of certain volume of the colloid on an optical glass (transparent for λ>350 nm) and drying with formation of an uniform film with thickness 50-100 μm. According to the transmission electron microscopy data the mean diameter of CuS particles is about 5 nm.

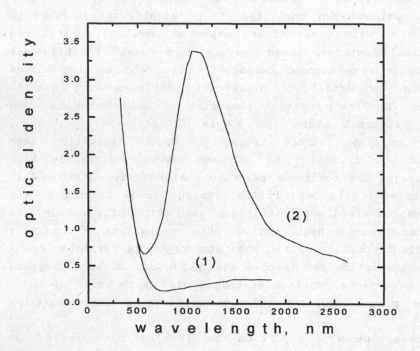

Figure 1. Linear absorption spectra of CuS microcrystals: (1) – usual, (2) – modified.

In order to obtain samples with additional absorption in the near-IR region, acqueous colloids were aged during 10 days. The aged colloids also were deposited on glass and dryed with formation the film containing CuS particles. The variation of the optical absorption in the result of such treatment is caused by the partial surface oxidation.

The transient absorption change spectra of CuS microcrystals as a function of delay time and pump intensity and the transmittance as a function of incident intensity have been studied using pump-probe laser spectroscopy method. Passively mode-locked $YAlO_3$:Nd laser generated the pulses of 15 ps duration and 1 Hz repetition frequency. First or second harmonic of laser radiation (λ=1.08 or 0.54 μm) was used as pump beam. Part of laser radiation was transformed into white continuum by means of D_2O cell and was used as probe beam. Its detection was performed in the range 300 - 900 nm by two photodiode arrays. All optical measurements were performed at room temperature.

For nontreated CuS samples we observed induced absorption in the range 0.5 - 1 μm. Relaxation of this effect has nonexponential type with character time much longer than 1 ns.

The nonlinear optical response of treated CuS microcrystals is rather unusual and consists in the combination of the bleaching effect in the low-energy range and induced absorption in the high-energy one (Figure 2). The former corresponds approximately to the

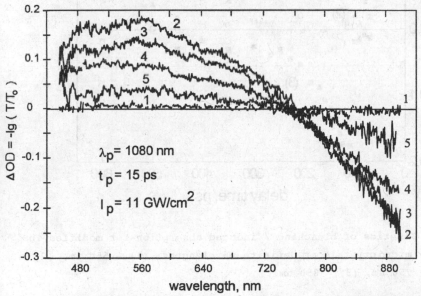

Figure 2. The differential absorption spectra of modified CuS micro-crystals for various delay times at room temperature at a pump wavelength λ_p=1.08μm. Delay times: (1) -30ps - base line, (2) 15ps, (3) 30ps, (4) 60ps, (5) 160ps.

position of the additional absorption band in the linear spectrum. The relaxation time observed is close for both effects and is measured to be ~ 100 ps (Figure 3).

We can connect these features with the existence of additional absorption band in band gap and peculiarities of electron excitation from valence band to this additional level.

The CuS and other semiconductor microcrystals, which have the above absorption band within the band gap due to partial oxidation, are very perspective as the saturable absorbers for pico- and femtosecond lasers.

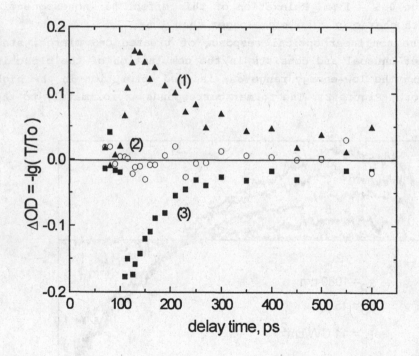

Figure 3. Kinetics of bleaching / induced absorption for modified CuS microcrystals at definite wavelengths: (1) - 550 nm, (2) - 750 nm, (3) - 850 nm.

The conditions of Selective Laser Sintering of Metal-Polymer Compositions

A.V. Levin, N.L. Kuprianov, A.L. Petrov, I.V. Shishkovsky

P.N.Lebedev Physical Institute, Samara branch

Novo-Sadovaja str. 221, 443011 Samara, Russia.

1 Introduction

Multiple layer SLS is a one from perspective method of such speed CAD modeling. Under selectivity we consider a possibility to scan of a laser beam over the free powder material surface by the some contour with account for the laser influence dose in each point. Specificity of such problem and her difference from the known technology of the laser cladding and soldering is joining with skill to obtain as a large as small layers of a sintering structures with a space resolution by the layer ~0.1 - 0.3 mm.

A powder materials of different nature use in SLS technology: ceramic systems, intermetallic compounds and polymers [1-2]. But there is a undeservedly limited place in literature given the questions of a uses of metal powders by the collected of compositions for the laser rapid phototyping. So in this work, the conditions of selective sintering had been studied in a volume of the powders for next powder materials: polyamid (PA), poly-carbonate (PC), polymetacrilate (PMA) and cladding powders on the basis of nickel (CP) and brass (BP).

2 Materials and equipment

In investigation as above mention powders, as these compositions took part. The SLS process was conducted by two scheme: as the scan laser beam from Nd-YAG and immovable powder, as immovable laser beam from cw CO_2 laser and moving powder. In both case a moving was created through personal computer. Laser on Nd-YAG worked in regimes with turned on (~3 kHz) and turned out (i.e. given from PC) internal resonator modulation. Diameter of laser beam was d = 50 and 100 mkm and laser power changed in range 1 - 16 W. The management of laser beam with computer had been used in range of field 50 x 50 and 100 x 100 mm (i.e. 1024 x 1024 pixels) and had allowed to conduct a scan by any contour. The duration of laser scanning determines the times of laser beam transition from point to point, a calm deflectors in each point and are changed us in range v = 4 - 1220 mm/s. The maximum laser radiation power of cw CO_2-laser was ~ 55 W and the divergence (total angle) was ~ 3 mrad. The unpolarized laser radiation as focused by ZnSe lens on the horizontal surface of a moving powder. The output laser power was modulated at a frequency of range 6-12 Hz by an attenuator, that reduced the average radiation power. We varied the laser radiation intensity from 10^2 to 10^3 W/cm^2. The duration of a single radiation pulse was varied from 1 ms to 1.4 ms. Optical microscopy was conducted on NIOPHOT - 30.

The powders are treated in free pour volume, significantly larger then a depth of sintered layer. We had been researched the conditions of SLS for one laser passage depend of laser power, scan time, degree of covering and a quantity of polymer in a mixture. It allows to determent the interval of sintering depths for one the laser passage under minimal deformation on the step.

3 Discussion

Figure 1 (plot 1,2) shows the results of laser sintering of pure CP and PA powders. Under small laser power the depth of sintering is such little, that sintered material spreads from a touch. With an increasing of power the size of sintering stripe is growing, while for CP it equal only $z = 0.7$ mm and stripe is sufficient brittle. Increasing of power for PA conducts at a full melting of powder. But a strong stresses changed initial form of sintering stripe. Compare of melting temperatures for these powders [3-5] allows to understand this situation. Indeed, the melt temperature of CP is sufficient highly and effective sintering are observed only for small particles from taken fraction under our laser powers. Treatment of PA on a lower powers doesn't gives any result because reflectivity of polymer is high. The growth of laser power leads in melt of polymer powder. Laser treatment of PC and PMA characterizes a full destruction of these powders. Those experiments showed that we can get an appropriate sintering conditions of PA and PC after a preliminary heating of the powder bulk only.

The mixing of polymer with metal powders improves a quality of sintering layers. From one hand, this allows to increase an absorption capability of the mixture composition on the low laser powers and to accelerate the sintering process. From second hand, there is observe an effectively coagulation of our compositions on a high laser powers. In result, energy of laser in wide range is sufficient for the sintering and remelting of a polymer component of a mixed composition. In this case, the polymer component plays a role of bunch. With account of high thermal diffusivity of metal powders in compare with polymer, the heat quickly turns off to the depth of composition. So the sintering layer practically have not deformation and polymer component doesn't decomposes on a monomers. The results of mixture treatment shown on fig. 1 (plot 3 - 6). An addition of metal powder in mixture softens, but polymer destruction process doesn't stops on the treated surface. It was expressed in form of soot particles for mixture PC+CP = 1:6 and flying monomer for PMA+CP = 1:6 on a regimes ~ 10 - 16 W. On the such laser powers we are continued to see a big shrink stresses on the edges of sintered stripe. The cohesion of compositions is good, expected mixture PMA+CP, where the size of PMA particles is more than CP fraction and sintering are observed in the places of arrangement of CP.

The change of proportions of mixed powders affects by size of sintering layer insignificantly for Nd-YAG laser beam (fig.1, plot 5,6). On fig. 2 dependence of the sintered depth are shown for mixture PC+BP = 1:8 in the volume from a laser power under a differences scan velocities. Obviously, that all curves go on saturation by the sintered depth, so there is optimally to treat of composition in interval of a moderate laser power, when deformation processes and destruction of polymer component are insignificantly.

Analysis of laser regimes with use Nd-YAG for mixtures are showing perspectively of the use for SLS the composition on the bases PC and PA with metal powders - BP, CP. Simultaneously, there is an important the possibility have a treatment as a fine layers of mixture for the precision sintering of some sections of a volume models, as a thick layers - for the acceleration of sintering process.

Microstructure of monolayer is homogeneous gray phase, in which a chaotic distributed pores present with the size of some hundreds of mkm (fig. 3). The roughness of surface for mixture PA+CP=1:8 equal 0.02 - 0.04 mm under $P = 4.5$ W and increases on order under $P = 11$ W, formed a continuous stripes along laser beam passage. Seen, that the unmelted great particles of cladding powder could saw under a big magnify, which reflect of light during survey. They have size 120 - 160 mkm and don't change during a treatment.

The experiments with use cw CO_2 - laser were carried out for BP to the polymer proportions from 12:1 to 20:1. Some results of a mono layer forming of a BP+PC composition are shown in fig.5 for the scanning velocity 400 mm/min. It was obviously from fig.5 that the properties of the mono layers of the 20:1 proportion considerably differed from that for the proportions of 12:1 and 16:1. For example, it is shown in fig.5 that

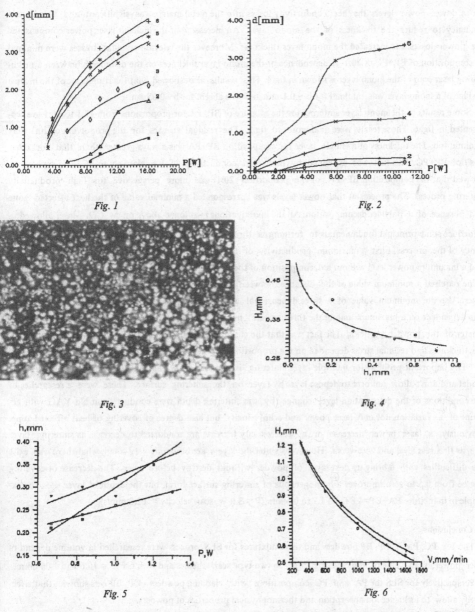

77

Fig. 1

Fig. 2

Fig. 3

Fig. 4

Fig. 5

Fig. 6

Fig. 1 Influence of laser power - P on sintering depth - Z for differences powder compositions: 1. - CP; 2. - PA; 3. - PMA+CP = 1:6; 4. - PC+CP = 1:6; 5. - PA+CP = 1:8; 6. - PA+CP = 1:12. (Regime with turn on external modulation, v = 46 mm/s., d = 50 mkm., covering ~ 0.5).

Fig. 2 Influence of laser power - P on sintering depth - Z for differences scan velocities of laser beam: 1. - 1200; 2. - 610; 3. - 244; 4. - 122; 5. - 61 mm/s. (regime with internal modulation, mixture PC+BP = 1:8, d = 50 mkm, covering ~ 0.5).

Fig. 3 Microstructire of sintering surface depth (conditions see fig.1, mixture PA+CP = 1:8, P=13,5 W, top view).

Fig. 4 Dependence of monolayer thickness H on the distance h between a neighbouring laser scan (cond. see fig.5).

Fig. 5 Dependence of sintering depth via laser power for cw CO_2 (v=400mm/min, PC+BP=20:1).

Fig. 6 Dependence of the monolayer depth H from scan velocity v (cw CO_2 laser power = 4 W, PA+BP=20:1).

for a lower power levels the heat conductivity losses into the metal matrix was significant and there was a tendency to reduce the thickness of the mono layer. An increasing of the incident laser power compensated the power losses and increased the mono layer thickness. Moreover the relevant residual stresses were minimal to composition of BP+PC of 20:1. A dependence of the mono layer thickness on the distance h between a neighboring laser scan of the mono layer is shown in fig.4. This results also showed, that the roughness of the inverse surface of a mono layer was minimal for h ~ 0,1 mm and was absent for h~ 0,05 mm.

Some results of the mono layer sintering for the mixture of BP+PA for proportions from 12:1 to 20:1 are represented in fig. 6. Those results were analogous to fig.5. The residual stresses for the proportion of 20:1 were minimal too. The thickness of a single layer for composition BP+PA was always greater than that for composition BP+PC. This fact was explained by high heat conductivity of PA relatively to PC. The experiment showed that the composition BP+PA of the proportion 20:1 was more perspective for high productivity sintering process. The process at that power levels was corresponded a minimal value of the heat affected zone and absence of a polymer decomposition. All the investigations mentioned above on cw CO_2 -laser allowed us to deduce some principal fundamentals to forming a three dimensional samples. There were two opposite tendency of the process. First, a maximum productivity of the process proposed a maximum scanning velocity and a maximum power level without a decomposition of the polymer. From the other side, a minimum roughness value required a minimum value of the distance between the neighboring laser scan within a single sintered layer. Also the minimum value of a three dimensional samples roughness and hence a maximum precision was dependent on a maximum among the three values: heat affected zone, laser focal spot and maximum diameter of the powder particles. The fact was that the characteristics of a mixture mono layer as, for example, thickness and residual stress depended on the proportions of the polymer and BP in the composition.

The important problem for multiple layer sintering is stresses. They change the form of a future volume model and don't allow uniform to depose layer by layer on the sintering surface. There were a researches of the conditions of the deformation level change by the sintering depth were conducted on Nd-YAG with account of a variation not only laser power and scan velocity, but also degree of covering of heat affected zone. Obviously, a laser power decrease or a scan velocity increase are conducted to decrease as sintering layer size, as tie stress level and vice-versa. However, sintering layers are becoming very small, what there was exist the difficulties with a bring up next layer of powder without destroy before layer. The decrease of covering degree from 0.5 to zero improves a homogeneous of sintering surface form, but the depth of layer dies, for example in in mixture PA+CP=1:8 from 1.7 to 0.5 mm (P = 3.9 W, scan velocity = 62.3 mm/s).

4 Conclusions

1. The PA, PC, PMA, CP, BP powders and these mixtures for SLS process were researched in volume depend of laser power, scan velocity and covering degree in two type scan schemes and with use λ = 10.6 and 1.06 mkm.
2. Perspectively for SLS the PA and PC compositions with cladding powders - CP, BP was shown , that effectively allow to influence on absorption and thermophysical properties of powders.
3. Experimental the intervals of maximal and minimal possible a sintering depths were determined for one laser passage under minimum deformation.

5 References

[1] Wu D.S.: Optical scanner design impacts rapid laser phototyping. Laser Focus World. 11 (1990) 99 - 106.
[2] Nott K.: The selective laser sintering process. Photonics spectra. 9 (1991) 102 - 104.
[3] Handbook by plastic masses. Editor: V. M. Kataev. 1. Moscow Chemistry (1975).
[4] Cladding materials of ECS countries. Handbook - Catalog. Kiev- Moscow (1979).
[5] Jakovlev A.D., Zdor V.F., KaplanU.I. Powder polymer materials and covers on these (1971).

Ultrafast Nonlinear Scattering and Deflection of Femtosecond Laser Pulses

Harald Schulz, Peter Kohns, Andreas Scheidt, and Ping Zhou
Optikzentrum NRW
Abteilung Industrielle und wissenschaftliche Präzisionsmeßtechnik / Hochleistungsoptik
Universitätsstraße 142, D-44799 Bochum, Germany

1. Introduction

Ultrafast processes in physics, chemistry, biology and material science are already investigated by time-resolved spectroscopy since picosecond (ps) and femtosecond (fs) laser sources became available [1]. Different interaction processes during the excitation and the relaxation of the illuminated matter were studied. The contribution due to the interaction of the photons with electrons, phonons, plasmons or other types of matter excitation can be separated with respect to the specific temporal and spectral characteristics [2].

In our experiments we used the time-resolved scattering and deflection of a light beam to investigate different excitation and decay processes. Scattering due to coherent electronic excitation and deflection due to coherent and incoherent phonon excitation on a time scale of 100 fs to 50 ps were studied. This is to our knowledge the first time where the transition from the electronic light scattering to the photothermal beam deflection is investigated.

2. Experimental setup

The amplified pulse of a Ti:sapphire fs laser was split into an intense pump and a weak probe pulse. As a sample a plate of glass and a NaCl single crystal were used, both of 1 mm thickness. The probe beam direction was chosen normal to the plane sample surface. Two different angles between pump and probe beam namely 5° (setup A) and 40° (setup B) were used. The interaction volume given by the spatial overlap of pump and probe beam was in the bulk of the sample. Measurements with setup A were carried out with equal waists of pump and probe beam at their crossing point inside the sample, whereas at setup B the probe beam waist was larger than the pump beam waist. The time delay τ between pump and probe pulse could be adjusted from zero to 50 ps in steps of 10 fs.

In the case of setup A the intensity of the scattered (deflected) probe beam and the deflection angle of the probe beam were detected as a function of τ using a linear CCD array and a pinhole in front of a photodiode. It should be noticed, that the scattering angle did not vary with τ. Therefore the detection of the scattered probe beam intensity with a fixed photodiode was possible.

In contrast to setup A at setup B only a fraction of the probe beam was passing through the excitation volume. Therefore the probe beam profile is changed at the position of the excitation volume and a local light intensity maximum occurs. Its position was studied as a function of τ.

3. Experimental results

The experimental results are collected in figures 1 and 2. Figure 1 shows the intensity of the scattered wave as a function of the delay τ using setup A. We observed a peak without any structure. The peak width is in the order of the pulse duration. Using setup B we observed the dependence of the probe pulse deflection on τ shown in figure 2. We obtained a peak similar to figure 1 accompanied by strong oscillations when the NaCl crystal was used. Using the glass plate a curve quite similar to figure 1 was obtained.

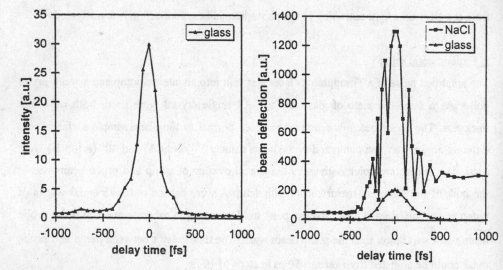

Fig.1: Scattered intensity as a function of τ (setup A)   **Fig.2:** Deflection of the probe beam as a function of τ (setup B)

4. Discussion

The observations made with type A experiments can be explained by elastic processes due to the interaction of the probe pulse with electronic excitations driven by the pump pulse. The scattered probe pulse occurs at a fixed angle which is independent of the delay time and intensity of the pump pulse. The intensity of the scattered probe pulse increases quadratic with the intensity of the pump pulse and varies with τ. The scattering angle is determined by the wave vectors of pump and probe pulse. These observations can be explained consistent by a degenerated four-wave mixing (DFWM) process similar to the observations described in [3]. The responsible process is a third order non-linear optical effect.

The same electronic process, the non-linear Kerr effect, gives rise to a change of the refractive index which effects the probe beam. If pump and probe pulse are not collinear the temporal and spatial change of the refractive index leads to the deformation of the probe beam profile and a deflection of the probe beam.

The deflection angle depends on both the pump beam intensity and τ. These observations are quite similar to the results reported in [4]. Nevertheless the experimental results presented in [4] were explained by photothermal phenomena whereas we believe that only an electronic process can be responsible for the observed beam deflection on the fs time scale. This is further confirmed by the fact that both the measured time-dependend probe beam deflection and the probe beam scattering intensity depend on the pulse duration.

In order to separate probe beam scattering due to DFWM from the probe beam deflection the angle between pump and probe beam was increased (setup B). As result DFWM is strongly supressed because the phase mismatch increases with the angle between pump and probe beam. On the other hand both beam deflection and beam profile deformation increase, if only a fraction of the probe beam spot is spatially overlapping with the volume of the matter that is excited by the pump beam.

The results of the type B experiment carried out with the glass sample showed a temporal behaviour of the beam deflection according to the beam scattering observed at the type A experiment. A third order correlation function of pump and probe pulse is observed that can be explained by an elastic interaction of the light with an instantaneous electronic excitation (non-linear Kerr effect).

In the case of the crystalline sample three types of light-matter interaction might be responsible for the observed temporal behaviour of the probe beam deflection:

Beside the already mentioned elastic interaction of the light with electronic excitation a part of the excitation energy is transferred to phonons having a characteristic oscillation period of about 50 fs. The observed extrema of the probe beam deflection during the temporal overlap with the pump pulse can be explained by a coherent interaction of phonons with the probe beam where the phonons yield a change of the refractive index.

As a further consequence the excitation of phonons transfers energy from the pump pulse to the matter that is stored in the sample resulting in photothermal probe beam deflection after the pump pulse has passed the sample. The photothermal beam deflection starts approximately 1 ps after the pump pulse and varies slowly up to 50 ps. The complete relaxation is expected to take place at a time scale of μs to ms depending on the thermal gradients produced by the pump pulse. We investigated this slow relaxation process in other experiments not shown here.

We believe that we observed an inelastic light matter interaction with the single crystal sample, where a coherent phonon excitation is followed by a phase decay of the coherent excitation on the picosecond time scale, that leads to the common photothermal beam deflection.

5. Conclusions

In the case of the single crystal sample, electronic and phonon induced refractive index changes seemed to be responsible for the probe beam deflection on the fs and ps time scale. The experiments with the glass sample yield an electronicly induced change of the refractive index that causes a beam deflection. In addition a scattered wave with constant scattering angle is observed. These processes can be explained by the nonlinear Kerr effect that describes an elastic electronic interaction of light and matter.

References:
[1] J.HERRMANN, B.WILHELMI, "Laser für ultrakurze Lichtimpulse", Physik-Verlag, Weinheim (1984)
[2] "Ultrafast Phenomena", Vol. 7, 1994 OSA Technical Digest Series (Optical Society of America, Washington, DC, 1994)
[3] H.SCHULZ, H.SCHÜLER, T.ENGERS, D.von der LINDE, IEEE J. Quantum Electron. 25, 2580-2585 (1989)
[4] P.HEIST and J.KLEINSCHMIDT, Opt.Lett.19, 1961-1963 (1994)

Investigation of a Pulsed EBCD CO-Laser Supplied from an LC Transmission Line

V.Kazakevich, K.Morozov, A.Petrov, G.Popkov
P.N.Lebedev Physical Institute, Samara branch, Russia

Summary

The output parameters of a free-running pulsed electron-beam-controlled discharge CO-laser (EBCD CO-laser) has been studied experimentally for the first time when collisions of the second kind (superelastic collisions) were of no importance in the active medium of the laser.

1 Introduction

It is known, that one of the important problems in the development of a repetitively pulsed EBCD CO-laser, just as for any other laser, is the choice of the active-medium excitation at which a high laser efficiency is reached. This is the vital problem for the EBCD CO-laser because only practical achievement of high output characteristics makes this technically complicated type of laser (at an average radiation power 15-20kW) attractive for use in industry.

As we suppose one of the possible reason which may decrease output parameters of a pulsed EBCD CO-laser are collisions of the second kind, i.e. the deactivation of excited vibrational levels of a CO molecule by electrons of discharge. We determined the excess of electronic temperature and electronic energy over the mean those, which are determined by the value E/N. The greatest excess of electronic energy over the mean was equal to $E_e =0.04eV$ at the end of a pump pulse for our experimental conditions at the pumping from a capacitor bank. In order to avoid the negative effect of the superelastic collisions on the output laser parameters we suggested to pump EBCD CO-laser by the discharge of pulse-forming line at constant value E/N[1].

2 Experimental setup and results

The experiments were carried out on a pulse-periodic LC EBCD CO-laser developed and constructed at the Samara branch of the Lebedev Physics Institute. It is experimentally shown that the discharge of an LC transmission line across a matched load at constant E/N allows increases output spectrum of the LC EBCD CO-laser in shortwave region more than 1.5 times and considerably increases the number of vibrational-rotational transitions in spectrum of oscillation in comparison with pumping from a capacitor bank. It is noted the differences of temporal characteristics of the radiation emitted for LC EBCD CO-laser and EBCD CO laser with a capacitor bank. The output power from the LC EBCD CO-laser reached its maximum and then remained constant throughout the pump pulse. A small (~5%) fall in the radiation power was observed for only the longest (~90 μs) pump pulses. This was qualitatively different from the EBCD CO-lasers with a capacitor bank when a strong (~100%) dip in the radiation power was observed at the end of a pump pulse for a wide range of pump-pulse durations. Under the same conditions the efficiency of the LC EBCD CO-laser (η =22% at T=140K for a CO:N2 =1:19) was higher than the efficiency of a laser with a capacitor bank (η = 15%).

3 Conclusions

In order to raise output parameters of a pulse high-pressure electron-beam controlled discharge carbon monoxide laser it is necessary to excite its active medium in conditions when superelastic collisions of the CO excited molecules and discharge electrons are absent. Technically it can be carried out by using the pulse-forming line for the main discharge.

4 References

[1] Igoshin V.I., Kazakevich V.S., Morozov K.V., Petrov A.L., Popkov G.N., Chernovaya V.B.:Calculation of the electron energy in the discharge plasma of a pulsed electron-beam controlled CO laser on the basis of an analysis of the temporal characteristics of its radiation. Quantum Electronics 24(5)395-398(1994)

Pulsed Operation of Laser Diodes in a Picosecond Time Regime

Gerald Kell, Rainer Erdmann, Uwe Ortmann, Michael Wahl, Edgar Klose
BiosQuant GmbH Berlin, Rudower Chaussee 6, 12484 Berlin;
Humboldt-Universität zu Berlin, Unter den Linden 6, 10099 Berlin

1. General problem and common solution

For many applications in the field of research and development it is important to use
a laser light source with characteristics similar to a dirac pulse. This contribution
shows the possibility for using common or special types of laser diodes to produce
very short laser pulses below 100ps. Typical applications are
- time resolved spectroscopy
- distance and speed measurements
- testing of opto-electronic devices and others.

The important questions are: How to get a maximum of electrical energy into the
active region of the laser diode as fast as possible? How long is the duration of the
shortest optical pulse, determined by the internal behaviour of the laser diode? And
how to design the electric circuit?

The first hand solution consists of a pulsed RF power amplifier and a biasing dc-
source. The signal fed to the laser diode is a superposition of both pulse and bias.
Problems are more difficult when electrical transmission lines are used. In the
common setup, the optical pulse shows a width (FWHM) of about 500ps. The bias
current is the reason for a permanent luminescence. Furthermore, a problem is the
optimal power control in connection with the parameter adjustment.

2. Analysis of the problem

The first step is constructing a usable model of the dynamic behaviour of the laser
diode.

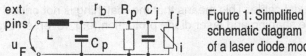

Figure 1: Simplified
schematic diagram
of a laser diode model

Fig. 1 shows a simplified example. It contains the parasitic reactances of the chip
characteristics, the bond wires and housing structure. However, it does not show the
chip internal inductance that can be important to calculate the maximum internal
resonance frequency.

Simulation calculations for the practical case have shown the dominance of the bonding inductance. Therefore, it is sufficient to use the simplified model. The non-linear resistance (r_j) represents the electrical junction properties. It can be approximated by the standard form of the pn-diode equation. The current (i) proportionally represents the laser intensity, approximately given by the static curve of the laser diode. This way, our simulation results are in good agreement with the measured curves of the laser pulse.

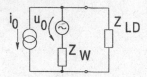

Figure 2: Line output equivalent circuit connected to the laser diode

Next step is to get an optimised matching of the laser diode to the 50-Ohms transmission line. Fig. 2 shows the output SPICE-model of the line connected to the model of the laser diode. Of course, an all-case precise matching is not possible due to the nonlinear behaviour of the laser diode. But in conjunction with a correct matching from the switch stage to the line input, the wave reflection will be swallowed there and does not cause any problems. The remaining question is the additional power loss caused by improper impedance matching.

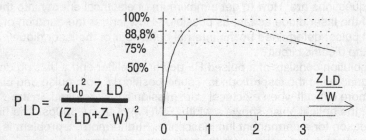

$$P_{LD} = \frac{4u_0^2 \; Z_{LD}}{(Z_{LD}+Z_W)^2}$$

Figure 3: Relative power fed to the LD versus ratio of LD impedance and characteristic line impedance

Generally, in fig. 3 the fundamental equation for the output power is given together with the basic curve. It shows, for example, that the power loss will be less than 25%, when the impedance of the laser diode is in the region of 17 Ohms to 150 Ohms, using a 50 Ohms coaxial cable. This means that the matching is not critical and can be optimised for pulse shaping and laser diode protection (e.g. reverse voltage).

3. Synthesis of a solution

First problem finding a solution is the creation of a matching circuit between line output and laser diode.
Basic solutions for the adaption between these elements are
 - a transformer circuit, e.g. by inductors or transmission lines,

- a lossy resistor network, or
- a nonlinear microwave resonance circuit including lines, resistors and diodes.

The last example results in a most difficult passive solution but shows the best performance. Similar to the structure of a differentiator, e.g. slew rates of more than 10V/200ps can be achieved from a primary pulse height of 25V at 800ps pulse width.

Another problem is the realisation of a very fast switch stage to generate this electrical pulse. The semiconductor technology offers two unusual ways for getting a shortest time and fast slopes:

Firstly, special gallium-arsenide FETs designed for RF-transmitters, yield switch-on and -off transients of less than 200ps. They allow highest repetition rates up to the GHz-range, however, the integral power generated this way may damage common laser diodes. Additionally, the voltage range above 12V requires special FET-types with high costs. Despite of these problems repetition rates up to 80MHz could be realised.

Secondly, by using silicon devices only the avalanche effect offers switch times down to 400ps at voltages up to 50V. The repetition rate is limited here due to the recombination time of typically more than 100ns. This is why the frequency in practice is less than 5MHz.

A general structure of the whole system and best results with different types of laser diodes are shown in the figures 4 and 5.

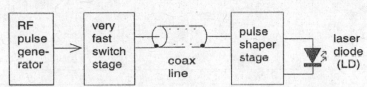

Figure 4: General structure of a laser pulse generator

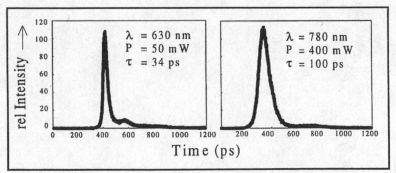

Figure 5: Measured laser pulses with different types of laser diodes and generators

4. Conclusions and further work

By modelling the behaviour of the laser diode, the development of a fast laser pulse generator will be well supported. The two mean problems are, on one hand, the generation of a powerful short electrical pulse and the transmission of this pulse into a coax line, on the other hand, the matching of the laser diode to the line output together with effective pulse shaping. Two versions of different switch stages can yield either a high repetition rate at a medium power or a medium repetition rate at higher voltages and currents.

In the future, laser pulses of durations below 100ps will be of increasing importance. For many applications it will be very useful to obtain repetition rates of more than 100MHz from diode lasers. A miniaturisation of the pulsed laser system with battery operation is an interesting aspect for further developments.

For optical peak powers above 10W further developments are necessary. Higher pulse energy offers the possibility of blue laser pulses e.g. with a frequency doubling crystal.

Technological Problems of Pulse YAG Lasers Integration in Manufacturing Sequences

A.L.Petrov, A.A.Gusev, S.V.Kayukov, P.V.Musorin

Lebedev Physics Institute, Samara Branch.
Novo-Sadovaya str. 221, Samara 443011, Russia

1.Introduction.

This paper is devoted to the specific problems of integration of new technological processes such as laser pulse welding in operating manufacturing lines

2.Laser welding of bearings protective washers.

Laser spot welding is used instead of traditional mechanical fastening to diminish the value of distorsion of rolling surfaces [1]. A scheme of the method is presented in fig.1. Appearance of the mechanical unit of laser welding installation is shown in fig.2. The following advantages were ensured by the laser welding:

- decreasing of distorsion of bearings external rings by the factor of 1.5;
- simplification of bearings construction;
- decreasing of labour input in bearings assembling process.

Fastening of protective washers is the finish operation in the technological sequence of closed bearings production. So the main requirements are as follows:

- productivity of laser welding must correspond to productivity of assembling process;
- appearance of spot welds must be rather good.

One welding mashine based on the laser of "Quant-15" tipe has an average productivity about 5 million bearings per a year, so a work-bay with 4 laser installations can fully provide necessities of rather great assembling workshop. Optimized regimes of laser welding ensure rather smooth surface of spot welds. Besides laser welding ensures enhanesment of fastening strength by the factor of 2.0 and encreasing of fastening reliability for a lubricant flowing out.

It has to be underlined that enhancement of bearing precise due to laser welding is followed by simplification of a bearing construction and by reduction of requirements to precision of bearing details with matched surfaces by the factor of 1.5.

Laser welding is preferable method of protective washers fastening for bearings of superlight and supermidget series.

3. Laser welding of aluminium casing of cables.

Communication cables have metal-plastic casing with thin aluminium layer on the internal surface. Bobbins with aluminium ribbon are installed at the small angle to the axes of cable drawing, fig.3. Necessity of joining of ribbon ends arrises when the ribbons exchange is carried out. The following requirements are made for joining of ribbon:

- strength coefficient of joining must be not less than 1.0 relatively to base material;
- joining must ensures good electrical contact;
- the characteristics 1 & 2 must be stable for a long period;
- joining place must be not thicker than threefold base value.

The best results according to stated requirements can be ensured with the help of welding but traditional methods are not applicable if a ribbon thick is less than 0.2 mm. The paper authors have developed technology and equipment of laser welding which is used in serial production of cables.

Lap welding is carried out by through melting of upper sheet by focused laser radiation, fig.4. Double weld seam is used to increase strength and reliability of a joining. It is formed at single run by special optical system. Welding is conducted without any flux and protective gases.

Double weld seam ensures reliable electrical contact and strength coefficient exceeding 1.0. Welds do not age and have

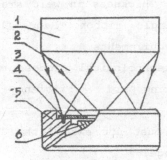

Fig.1. Scheme of laser welding of bearing protective washer: 1-optical system; 2-laser radiation; 3-protective washer; 4-spot weld; 5-external ring; 6-internal ring.

Fig.2. Appearance of the mechanical unit of laser welding of protective washer.

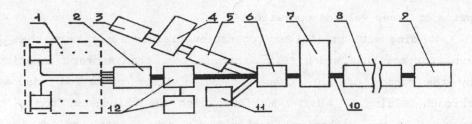

Fig.3. Scheme of production line of communication cables: 1-wire giving system; 2-Z turning; 3-giving unit of aluminium ribbon; 4-laser welding installation; 5-aluminium ribbon storage; 6-forming unit; 7-extruder; 8-water cooling line; 9-collector unit; 10-cable 11-giving unit of paper ribbon; 12-hydrophobic filler unit.

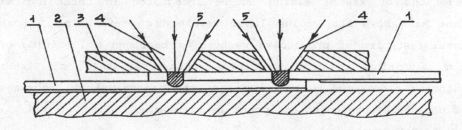

Fig.4. Lap welding of aluminium ribbon: 1-upper and down ends of aluminium ribbon; 2-down clamp plate; 3-upper clamp plate; 4-focused laser beam; 5-section of weld seams.

durability not less than base material. Thickness in weld area is 2.0-2.2 relative to thickness of single ribbon. Hence laser welding satisfies the above mentioned requirements.

Integration of laser welding in technological process makes additional requirements for ensuring of efficiency of technological line as a whole:

- time interval necessary for welding must not exceed the limit determined by capacity of a ribbon storage;
- laser welding regimes and mechanical control unit must be adjusted to ribbon with different thickness and width;
- working conditions of laser equipment must correspond to conditions of a cable workshop;
- working time of the laser equipment must ensures acceptable cost price of laser welding operation.

Welding with preliminery operations requires not more than 3 min. for a ribbon width 65 mm that is less than storage capacity by the factor of 2.0. Usage of pulse laser radiation ensures through melting of aluminium ribbon with the thickness from 0.05 mm to 0.22 mm. Mechanical control unit can be easy adjusted to ribbon width from 30 mm to 220 mm. Construction of the laser equipment ensures working time about 10 years that corresponds to increasing of the cost price of 1 km of cable about $ 0.25.

4. Conclusions.

Thus, the analisys of technological requirements for the operations of laser welding to be integrated in technological lines has shown that pulse laser welding is preferable in both cases: fastening of protective washers of bearings and joining of thin aluminium casing in cables. Characteristics of laser equipment correspond to parameters of the technological processes and use of the presented equipment is advisable.

5. References.

[1] A.L.Petrov, A.A.Gusev, S.V.Kayukov, I.G.Nesterov, V.A.Katulin. New Advances in Industry Applications of YAG Pulse Lasers. Proc.Int.Conf. LAMP'92, Nagaoka, Japan, p.993-997.

Focused Field Characteristics of Partially Coherent Light

Baida Lü   Bin Zhang   Bangwei Cai
Department of Opt0-Electronic Science & Technology,
Sichuan University, Chengdu 610064, China

In recent years much interest has arisen in partially coherent light, the most important class of which is Gaussian Schell-model (GSM) beams[1]. In this paper we deal with the focused field characteristics of partially coherent light. On the basis of the propagation law of partially coherent light and in consideration of the focus geometry, a theoretical model is presented to simulate GSM beams focused by a focus line array, where the complex degree of spectral coherence is used to describe the spatial coherence of GSM beams in the space-frequency domain. Then, numerical calculations are performed to illustrate the changes in spatial coherence properties along both the line focus length and width, and to illustrate the dependence of the irradiation uniformity on the coherence of incident light.

1. Simulation model

It is known that there is two types of array line focus system, i.e., the cylindrical lens array (CLA)[2] and the segmented wedge array (SWA)[3] which were proposed to improve the irradiation uniformity on the target in the X-ray laser experiment. Let us now consider the CLA illuminated by a GSM beam shown in Fig.1, which consists of a principal focusing lens with focal length f and an array of cylindrical lenslets with equal focal length f_{cy} and equal width d. The side width D of the array and the principal lens is equal, which reads

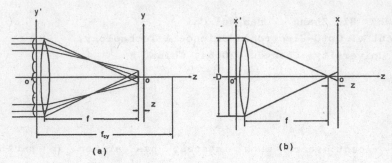

Fig.1   A schematic diagram of the CLA (a) a top view; (b) a side view.

$$D=Nd \tag{1}$$

with N being the number of lenslets. For definiteness, in the following calculations N is a positive odd number $N=1,3,5,\ldots$ and the distance between the lens and array can be neglected. Suppose that the waist of the incident GSM beam is located at the position $z=0$ of the principal lens, whose cross-spectral density function $W_0(x_1',y_1',x_2',y_2',0)$ is written as

$$W_0(x_1',y_1',x_2',y_2',0)=\exp(-\frac{x_1'^2+y_1'^2+x_2'^2+y_2'^2}{w_0^2})\exp[-\frac{(x_1'-x_2')^2+(y_1'-y_2')^2}{2\sigma_0^2}] \tag{2}$$

where w_0, σ_0 denote the beam waist radius and the correlation width, respectively. The corresponding intensity $I_0(x',y',0)$ and the complex degree of spectral coherence $\mu_0(x_1',y_1',x_2',y_2',0)$ are

$$I_0(x',y',0)=W_0(x_1'=x_2'=x',y_1'=y_2'=y',0)=\exp[-\frac{2(x'^2+y'^2)}{w_0^2}] \tag{3}$$

$$\mu_0(x_1',y_1',x_2',y_2',0)=\exp[-\frac{(x_1'-x_2')^2+(y_1'-y_2')^2}{2\sigma_0^2}] \tag{4}$$

If the two points for measuring spatial coherence are assumed to be positioned symmetrically about the z axis, Eq.(4) can be written as

$$\mu_0(x',y',-x',-y',0)=\mu_0(y',-y',0)\mu_0(x',-x',0) \tag{5}$$

and

$$\mu_0(y',-y',0)=\exp[-\frac{2y'^2}{w_0^2}(\beta^2-1)] \tag{6}$$

$$\mu_0(x',-x',0)=\exp[-\frac{2x'^2}{w_0^2}(\beta^2-1)] \tag{7}$$

where

$$\beta=(1+\frac{w_0^2}{\sigma_0^2})^{-1/2} \tag{8}$$

is called the coherent parameter.

According to the well-known propagation law of partially coherent light and under the quasi-monochromatic approximation,[4] after some lengthy algebras, we obtain the cross-spectral density function $W(x_1,y_1,x_2,y_2,z)$ of the GSM beam passing through the CLA

$$W(x_1,y_1,x_2,y_2,z)=\frac{k}{2\pi(f+z)}\sum_{m=-(N-1)/2}^{(N-1)/2}\sum_{n=-(N-1)/2}^{(N-1)/2}\int_{-d/2}^{d/2}\int_{-d/2}^{d/2}$$

$$\exp[-\frac{ik(y_1'^2-y_2'^2)}{2f_{cy}}]\exp\{-\frac{ik[(y_1'+md)^2-(y_2'+nd)^2]}{2f}\}$$

$$\exp[-\frac{(y_1'+md)^2+(y_2'+nd)^2}{w_0^2}]\exp[-\frac{(y_1'+md-y_2'-nd)^2}{2\sigma_0^2}]$$

$$\exp\{\frac{ik}{2(f+z)}[(y_1'+md)^2-(y_2'+nd)^2-2[y_1(y_1'+md)-y_2(y_2'+nd)]$$

$$+(y_1^2-y_2^2)]\}dy_1'dy_2'\frac{k}{2\pi(f+z)}\int_{-D/2}^{D/2}\int_{-D/2}^{D/2}\exp[-\frac{ik(x_1'^2-x_2'^2)}{2f}]$$

$$\exp(-\frac{x_1'^2+x_2'^2}{w_0^2})\exp[-\frac{(x_1'-x_2')^2}{2\sigma_0^2}]\exp\{\frac{ik}{2(f+z)}[(x_1'^2-x_2'^2)$$

$$-2(x_1x_1'-x_2x_2')+(x_1^2-x_2^2)]\}dx_1'dx_2'$$

$$=W(y_1,y_2,z)W(x_1,x_2,z) \tag{9}$$

where k is the wave number $k=\frac{2\pi}{\lambda}$ (λ—— wavelength) (10)

and $z=\mathit{l}-f$ (l is the propagation distance) (11)

Then the intensity $I(x,y,z)$ and the complex degree of spectral coherence $\mu(x,y,-x,-y,z)$ of the focused GSM beam are expressed as

$$I(x,y,z)=\frac{k}{2\pi(f+z)}\sum_{m=-(N-1)/2}^{(N-1)/2}\sum_{n=-(N-1)/2}^{(N-1)/2}\int_{-d/2}^{d/2}\int_{-d/2}^{d/2}\exp[-\frac{ik(y_1'^2-y_2'^2)}{2f_{cy}}]$$

$$\exp\{-\frac{ik[(y_1'+md)^2-(y_2'+nd)^2]}{2f}\}\exp[-\frac{(y_1'+md)^2+(y_2'+nd)^2}{w_0^2}]$$

$$\exp[-\frac{(y_1'+md-y_2'-nd)^2}{2\sigma_0^2}]\exp\{\frac{ik}{2(f+z)}[(y_1'+md)^2-(y_2'+nd)^2$$

$$-2y[(y_1'+md)-(y_2'+nd)]]\}dy_1'dy_2'\frac{k}{2\pi(f+z)}\int_{-D/2}^{D/2}\int_{-D/2}^{D/2}$$

$$\exp[-\frac{ik(x_1'^2-x_2'^2)}{2f}]\exp(-\frac{x_1'^2+x_2'^2}{w_0^2})\exp[-\frac{(x_1'-x_2')^2}{2\sigma_0^2}]$$

$$\exp\{\frac{ik}{2(f+z)}[(x_1'^2-x_2'^2)-2x(x_1'-x_2')]\}dx_1'dx_2'$$

$$=I(y,z)I(x,z) \tag{12}$$

$$\mu(x,y,-x,-y,z)=\mu(y,-y,z)\mu(x,-x,z) \tag{13}$$

$$\mu(y,-y,z)=\frac{W(y,-y,z)}{\sqrt{I(y,z)I(-y,z)}} \tag{14}$$

$$\mu(x,-x,z)=\frac{W(x,-x,z)}{\sqrt{I(x,z)I(-x,z)}} \tag{15}$$

Thus, Eq.(2), (3), (5) and (9), (12), (13) characterize the spatial coherence properties and intensity distribution of the incident and focused GSM beams, respectively.

2. Numerical calculation results

Numerical calculations were performed, using the Gauss-Legendre numerical integral method and the stationary-phase method. The calculation parameters are f=500mm, f_c=1300mm, $2w_0$=D=180mm, N=7 and λ=1.06μm. Typical results are compiled in Figs.2, 3, 4 and 5. Fig.2 gives the change in the complex degree of the incident GSM beam $\mu_0(x',-x',0)$ (or $\mu_0(y',-y',0)$) versus distance $|2x'|$ (or $|2y'|$) at the incident plane. As can be seen, μ_0 decreases monotonously with increasing $|2x'|$ (or $|2y'|$). The magnitude of μ_0 is dependent on β, the greater the coherent parameter β, the less μ_0 decreases with $|2x'|$ (or $|2y'|$), which is consistent with Ref.[1]. In Fig.3 the complex degree along the

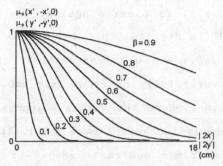

Fig.2 The complex degree of spectral coherence of the incident GSM beam as a function of the distance between two points (x' ,y' ,0) and (-x' ,-y',0) at the incident plane. Parameters for calculation are seen in the text.

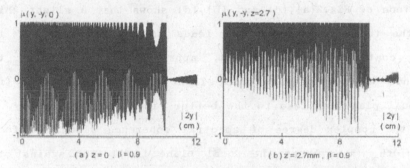

Fig.3 (a) (b) The complex degree of spectral coherence of the focused GSM beam for β =0.9 along the line focus length versus the distance between two points (x, y, z) and (x, -y, z) at the plane (a) z=0; (b) z=2.7 mm.

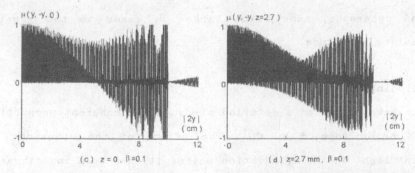

Fig. 3 (c) (d) The complex degree of spectral coherence of the focused GSM beam for β =0.1 along the line focus length versus the distance between two points (x, y, z) and (x, -y, z) at the plane (c) z=0; (d) z=2.7mm.

line focus length $\mu(y,-y,z)$ is plotted against the distance $|2y|$ at the focal plane $z=0$ (in Fig.3(a),(c)) and at the plane $z=2.7$mm (in Fig.3(b),(d)) for $\beta=0.1$ and 0.9, respectively, From which it follows that for the nearly fully coherent case $\beta=0.9$ more points are located in the region, where the values μ approaches 1 and -1 as compared with the nearly incoherent case $\beta=0.1$. It implies that in the former case the fringe contrast induced by the multibeam interference between lenslets becomes larger, which results in the large irradiation nonuniformity shown more clearly in Fig.4, where the intensity distribution $I(y,0)$ along the line focus length at the focal plane is given for $\beta=0.9$ and 0.1. In addition, a comparison of Fig.3(a),(c) and (b),(d) shows that a slight shift from the focal plane ($z=2.7$mm) leads to the decrement of the fringe contrast. As a results, appropriately decreasing the coherence of incident light and slightly moving the target from the focal plane can lead to the better line focus uniformity. In Fig.5 the complex degree of spectral coherence along the focal line width $\mu(x,-x,0)$ at the focal plane is plotted against the distance $|2x|$ for $\beta=0.1,0.3,0.5,0.7$ and 0.9. It is seen that starting from the value 1 corresponding to the point $|2x|=0$ at the z axis, $\mu(x,-x,0)$ oscillates with increasing $|2x|$, and the better spatial coherence, namely the larger β, leads to the greater oscillating amplitude.

3.Concluding remarks

We have presented a detailed study on the coherent properties of GSM beams focused by CLA. The effect of the coherence of incident light on the irradiation uniformity has been investigated from the view point of the complex degree of spectral coherence. It has been shown that an appropriate decrement of the coherence

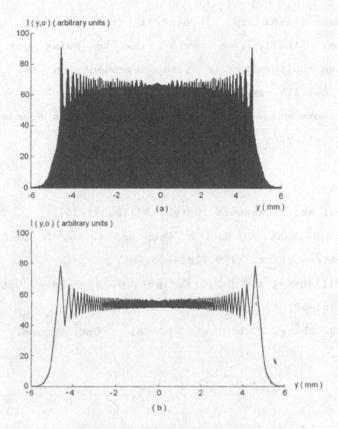

Fig.4 The intensity distribution $I(y,z)$ along the line focus length at the focal plane (a) β=0.9; (b) β=0.1

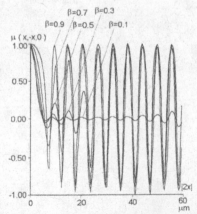

Fig.5 The complex degree of spectral coherence of the focused GSM beam along the line focus width for β=0.1,0.3,0.5,0.7 and 0.9 versus the distance between two points (x,y,z) and (x,-y,z) at the focal plane.

of incident light is of benefit for achieving the better uniformity. Finally, we would like to point out that the irradiation uniformity is also dependent on the number of lenslets, details can be found in Ref.[4].

This work was supported by National Hi-Tech Foundation under grant 416-2-11-2.

References

1.T. Shirai and T. Asakura, Optik, 94(1993)1-15.

2.W. Chen, S. Wang, C. Mao, B. Chen and A. Xu, In Conf. on Laser and Electro-Optics, (1990)282-285.

3.D. M. Villeneuve and G. D. Enright, H. A. Baldis, Opt. Commun., 81(1991)54-58.

4.B. Lü, B. Zhang, Z. Liu and B. Cai, Opt. Commun., 114(1995) 181-188.

Frequency-Stabilized CO (Δv=2) Laser System for Sub-Doppler Molecular Spectroscopy

M. Mürtz, M.H. Wappelhorst, R. Martini, P. Palm, B. Frech, W. Urban
Institut für Angewandte Physik der Universität Bonn
Wegelerstr. 8
D-53115 Bonn

Abstract

We report a frequency-stabilized CO-overtone laser system for precision spectros-copy applications. The laser frequency is stabilized in a frequency-offset locking scheme (FOL) on combination frequencies of two CO_2 laser standards. This FOL technique allows us to sweep the laser frequency in a well-defined and linear manner with very high frequency accuracy. The uncertainty is smaller than 40 kHz (corres-ponding to $\Delta\nu/\nu = 4\cdot 10^{-10}$ at $\nu = 100$ THz). This stabilized laser system enables us to record line shapes of molecular saturation signals in a computer-controlled experiment.

1. Introduction

In the mid-infrared region the most important light sources for high-resolution spectroscopy are tunable diode lasers (TDLs) and line-tunable molecular gas lasers. Gas lasers exhibit a much better spectral purity than standard TDLs but suffer from the lack of broadband tunability. Well-known gas lasers in the mid-IR are CO_2 lasers in the 9 to 12 μm region and CO lasers in the 5 to 8 μm region. They provide several hundred laser lines with frequency distances of typically 50 to 100 GHz. In recent years the development of the d.c. discharge CO overtone (Δv=2) laser in our laboratory has extended the wavelength domain covered by molecular gas lasers [1]. We achieved single-line laser operation in a liquid-nitrogen cooled plasma on CO overtone transitions between 2.6 to 4.1 μm (2450 to 3850 cm^{-1}) [2]. Over 330 laser lines are available with output powers up to 550 milliwatt.

Due to their high spectral purity molecular gas lasers are powerful instru-ments particularly for molecular spectroscopy with very high accuracy. During the past five years, a great number of very accurate sub-Doppler heterodyne frequency measurements have been performed utilizing CO or CO_2 lasers [3-8]. The combination of passive thermal and mechanical stability with sophisticated electronic servo loops for frequency control enabled measurements of molecular line positions with an accuracy of $\Delta\nu/\nu = 10^{-10}$ or better. Recently, the first sub-Doppler frequency measurements utilizing a CO overtone laser have been performed by our colleagues at the National Institute of Standards and Technology (NIST), Boulder, where a long-standing NIST-IAP collaboration was continued [9,10].

This paper describes a CO overtone laser system optimized for precision spectroscopy and its integration into the Bonn heterodyne spectrometer. The current performance of the frequency stabilization is reported and we demonstrate the use of the laser for saturated absorption spectroscopy around λ=3μm.

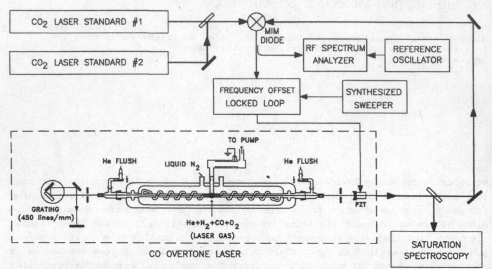

Fig.1. *Scheme of the experimental setup*

2. Experimental Setup and Results

Fig. 1 shows a diagram of the experimental setup. The laser gas is excited by a DC discharge which is split into two branches. The plasma consists of a mixture of He, N_2, CO and air which is pumped through the laser tube by a rotary pump. Detailed considerations on the laser operation can be found in [11,12]. A small amount of helium is added through inlets at the anodes in order to prevent soiling of anodes and Brewster windows from carbon. The gain tube is cooled by liquid nitrogen over a length of 90 cm. We have renounced of introducing ripples inside the laser tube and sonic injection of the laser gas into the plasma. These features would magnify the laser gain through the improved mixing and cooling of the plasma; on the other hand the laser linewidth might broadened due to the enhanced turbulence in the optical path of the laser beam.

The laser cavity is formed by a highly efficient gold-coated reflection grating (450 lines/mm) in Littrow mount and a 2% dielectric-coated output coupling mirror with 10 m radius of curvature. This mirror is mounted on a piezo-ceramic transducer to allow frequency control. In an effort to minimize the laser line-width we have acoustically capsuled the laser cavity. Moreover, we have inserted convection barriers to isolate the cooled segment from the regions between the Brewster windows of the gain tube and the resonator optics.

To stabilize and measure the frequency of the laser, we employ a frequency-offset-locking scheme based on a heterodyne technique. The reference frequency standard is provided by two CO_2 lasers which are saturation stabilized utilizing the Freed-Javan fluorescence technique [13]. Each laser provides a secondary frequency standard with an uncertainty smaller than 5 kHz [5]. Frequency mixing is achieved by heterodyning the three lasers on a MIM point-contact diode. An acousto-optical modulator was inserted into the CO laser beam to avoid trouble due to uncontrolled optical feedback from the MIM diode. The nonlinearity of the MIM diode permits frequency mixing up to the 100 THz region. Thereby we achieve a beat signal in fourth-order mixing at a frequency

$$\nu_{BEAT} = |\nu_{CO} - m\nu'_{CO2} - n\nu''_{CO2}|, \qquad\qquad m=1, \ n=2.$$

By an appropriate choice of the CO_2 laser transitions the beat frequency ν_{BEAT} falls into the frequency range below 5 GHz which can be easily manipulated by means of radio-frequency electronics.

In figure 2, a typical spectrum analyzer record of the beatnote is given. The signal-to-noise ratio was generally observed to be between 20 and 30 dB (not discernible here); the spectral width was about 100 kHz. The beatnote is employed to control the length of the CO laser resonator and consequently the laser frequency via an electronic servo loop. After down-conversion (by mixing with a variable microwave frequency ν_{SYN}) to an intermediate frequency of ν_{IF} = 160 MHz the signal is fed into a discriminator. This provides a frequency error signal which is applied to the PZT of the output coupler. When the servo loop is closed, the CO laser frequency ν_{CO} is locked to the CO_2 laser combination frequency with a variable offset, determined by the microwave synthesizer frequency ν_{SYN}. With this stabilization procedure we can transfer almost the full accuracy of the CO_2 fluorescence standard at 10 μm to the CO overtone laser at 3 μm. Moreover, the FOL technique allows us to sweep the laser frequency in a well-defined and linear manner with very high frequency accuracy.

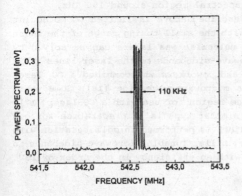

Fig.2. Beatnote CO laser ↔ CO_2 lasers (sweep time=200ms, res. bandw.=10kHz)

Fig.3. Lamb dip in 2f-detection of the OCS 10^01-00^00 P(27) transition

To demonstrate the application to precision molecular spectroscopy, we have used a pump-and-probe setup which allows us to observe saturated absorption signals. The measurement was made on the P(27) transition of the 10^01 - 00^00 band of carbonyl sulfide (OCS) which coincides with a CO (Δv=2) laser transition. We employed a derivative technique based on phase-sensitive 2f-lock-in detection. The details of the spectroscopy arrangement will be published elsewhere [14]. Fig. 3 shows the 2f-shaped OCS Lamb dip, obtained with a saturation power of 10 mW. The OCS pressure was near 1 Pa and the absorption length was 1m. The frequency intervall (4 MHz) was divided into 200 steps; the offset-locked CO laser frequency was swept over the interval and at each frequency step the phase-sensitive amplified absorption signal was recorded by the computer. Thus, we achieved digitized line shapes of the saturated absorption signals which have been fitted to a modulation-distorted 2f-Lorentzian profile with a non-linear least squares fitting procedure. In this way the center frequency of the Lamb dip was determined with a reproducibility of better than ±10kHz.

Systematic errors of the frequency calibration might arise from electronic offsets introduced by the discriminator or the loop filter. In order to correct for this, the center position of the beatnote has to be determined before and after the Lamb dip record. For this purpose, the beatnote is read out from the spectrum analyzer together with two marker signals from a RF synthesizer by the computer and analyzed by means of a center-of-mass method. A detailed description of this procedure can be found in [15]. The analysis of several measurements of this type proves that the stability of the CO overtone laser relative to the CO_2 laser frequencies is ±5 kHz over 30 minutes; the absolute accuracy is estimated to be ±40 kHz at 100 THz.

3. Conclusion and Outlook

In this paper we presented an offset-locked CO-overtone laser system which is suitable for very accurate sub-Doppler frequency measurements and line shape studies of calibration gas absorption lines in the 2.6 - 4.1 μm spectral region. Through the FOL stabilization scheme almost the full accuracy of the CO_2 laser standards at 30 THz is transferred to the spectral region around 100 THz.

Obviously, the type of measurement described above is limited to molecular transitions which have accidental overlap with the small tuning range of the laser lines. However, the tunability problem of molecular gas lasers can be solved by the generation of continously tunable microwave-sidebands to the laser lines. This was realized for the first time by Magerl and coworkers who combined a CO_2 laser with an electro-optic crystal mounted in a microwave waveguide [16]. Some years later, this technique was adapted to the 5μm region for use with a CO laser [17]. A tunable microwave sideband modulator of similar type is in preparation for our CO overtone laser system near 3μm. This project is performed in collaboration with the Laboratoire de Spectroscopie Hertzienne, Lille. Tunable microwave sidebands in the range of 8 to 18 GHz will be generated with an efficiency in the order of 10^{-3}.

4. Acknowledgements

We would like to thank Dr. J.S. Wells, on visit from NIST, Boulder, for helpful advice during the experiments. This work is supported by the Deutsche Forschungsgemeinschaft.

References
[1] M. Gromoll-Bohle, W. Bohle, W. Urban: Opt. Commun. **69**, 409 (1989)
[2] E. Bachem, A. Dax, T. Fink, A. Weidenfeller, M. Schneider, W. Urban: Appl. Phys. B **57**, 185 (1993)
[3] M. Schneider, K.M. Evenson, M.D. Vanek, D.A. Jennings, J.S. Wells, A. Stahn, W. Urban: J. Mol. Spectrosc. **135**, 197 (1989)
[4] Ch. Chardonnet, A. van Leberghe, Ch.J. Borde: Opt. Commun. **58**, 333 (1986)
[5] T. George, S. Saupe, M.H. Wappelhorst, W. Urban: Appl. Phys. B **59**, 159 (1994)
[6] T. George, M.H. Wappelhorst, S. Saupe, M. Mürtz, W. Urban, A.G. Maki: J. Mol. Spectrosc. **167**,419 (1994)
[7] A.G. Maki, C.-C. Chou, K.M. Evenson, L.R. Zink, J.-T. Shy: J. Mol. Spectrosc. **167**, 211 (1994)
[8] B. Meyer, S. Saupe. M.H. Wappelhorst, T. George, F. Kühnemann, M. Schneider, M. Havenith, W. Urban: Appl. Phys. B, in press

[9]   A. Dax, J.S. Wells, L. Hollberg, A.G. Maki, W. Urban: J. Mol. Spectrosc. **168**, 416 (1994)

[10] J.S. Wells, A. Dax, L. Hollberg, A.G. Maki: J. Mol. Spectrosc. **170**, 75 (1994)

[11] W. Urban: Infrared Lasers for Spectroscopy, in: Frontiers of Laser Spectroscopy (Eds.: A.C.P Alves, J.M. Brown, J.M. Hollas), Kluwer Academic Publishers (1988)

[12] W. Urban: The CO-Overtone Laser, in: Applied Laser Spectroscopy (Eds.: W. Demtröder, M. Inguscio), Plenum Press (1990)

[13] C. Freed, A. Javan: Appl. Phys. Lett. **17**, 53 (1970)

[14] M. Mürtz, M.H. Wappelhorst, B. Frech, P. Palm, W. Urban, A.G. Maki: Sub-Doppler heterodyne frequency measurements near 100 THz using a CO overtone sideband spectrometer, in preparation

[15] M.H. Wappelhorst, M. Mürtz, P. Palm, W. Urban: Sub-Doppler Line Shape Studies of CO and OCS with a Frequency Offset-Locked CO Laser, submitted to Appl. Phys. B

[16] G. Magerl, J.M. Frye, W.A. Kreiner, T. Oka: Appl.Phys. Lett **42**, 656 (1983)

[17] S.-C. Hsu, R.H. Schwendemann, G. Magerl: IEEE J. Quantum Electron. QE-**24**, 2294 (1988)

Quantum Mechanical Treatment of Micromaser Bistability

Hans-Jürgen Briegel*, Ullrich Martini[†] and Axel Schenzle[†]
* Texas A&M University [†] Ludwig-Maximilians-Universität München

A beam of twolevel-atoms is sent through a micromaser cavity. The atoms enter the cavity either in the excited state or in the ground state. The probability P for an atom to enter the cavity in the excited state depends on the laser intensity, the laser detuning and, because of the Doppler effect, the velocity of the atom. It has been shown experimentally that in this setup bistability is found when P is varied slowly [1]. The aim of this talk is to connect the bistability to the trapped states [2]. The Master equation which describes the interaction of the atom with the cavity, contains three parts: the interaction between atom and field in resonance and rotating wave approximation:

$$H = -\frac{\hbar g}{2}(a^\dagger \sigma_- + a\sigma_+),$$

cavity losses:

$$\mathcal{L}_f = -\frac{A}{2}(a^\dagger a\rho + \rho a^\dagger a - 2a\rho a^\dagger)$$

and spontaneous emission:

$$\mathcal{L}_a - \frac{B}{2}(\sigma_+\sigma_-\rho + \rho\sigma_+\sigma_- - 2\sigma_-\rho\sigma_+).$$

We transform the Master equation into its equivalent eigenvalue problem:

$$\mathcal{L}\rho_\Gamma = \Gamma\rho_\Gamma$$

This spectral decomposition may be used to obtain the dynamical properties of the system:

$$
\begin{aligned}
\rho(t) &= \sum_\Gamma c_\Gamma(t)\rho_\Gamma \\
\dot{\rho}(t) &= \sum_\Gamma c_\Gamma(t)\mathcal{L}\rho_\Gamma = \sum_\Gamma c_\Gamma(t)\Gamma\rho_\Gamma \\
c_\Gamma(t) &= c_\Gamma(0)e^{\Gamma t}
\end{aligned}
$$

thus we can propagate a given initial state of the system in time. The same method may be used to obtain correlation functions.

In order to find the eigenvalues and eigenstates of $\mathcal{L}$ we use the damping bases formalism [3]. The procedure is performed in four steps:

1. Diagonalize the damping operators L_a and L_f.

2. Express H in terms of eigenstates of L_a and L_f.

3. Transform the master equation into a continued fractions expression which may be solved easily for the eigenvalues.

4. Find the eigenstates from the eigenvalues.

The map summarizes the evolution between the moment t_o when one atom enters the cavity and the moment $t_o + T$ when the next atom enters the cavity. It consists of two parts: Interaction and damping while the atom crosses the cavity and then pure damping until the next atom enters the cavity. Therefore the map is defined by

$$\rho(T) = Pe^{\mathcal{L}_I(T-\tau)}\mathrm{tr}\{e^{\mathcal{L}\tau}\rho_\sigma(0)\rho(0)\} + (1-P)e^{\mathcal{L}_I T}\rho(0),$$

Here the time one atom spends in the cavity is denoted by τ. Imperfect velocity selection leads to an average over interaction times. With the help of the spectral decomposition of $\mathcal{L}$ this expression is easily evaluated.

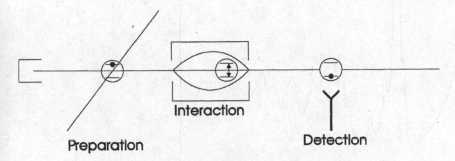

Preparation   Interaction   Detection

Figure 1: The experimental setup.

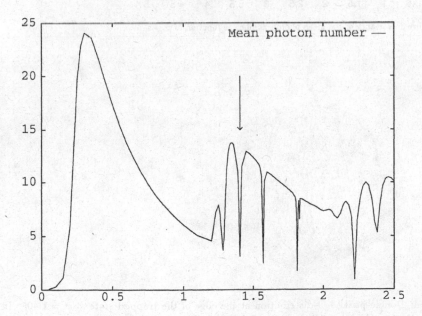

Figure 2: Mean photon number as function of τ

We are interested in the following information contained in the map:

- Dynamics of the system: Repeated applications of the map and its eigenvalues.

- Stationary state: The state which is reproduced by the map.

Fig. 2 shows the dependence of the mean photon number on the time the atom needs to cross the cavity, fig. 3 the variance of the mean photon number. From these plots it is clearly visible that the variance is large for values of τ corresponding to a trapped state. From now on we will concentrate on the trapped state marked with an arrow in fig. 2 and fig. 3. Fig. 4 shows

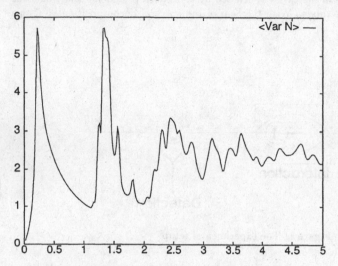

Figure 3: Variance of the mean photon number as function of τ.

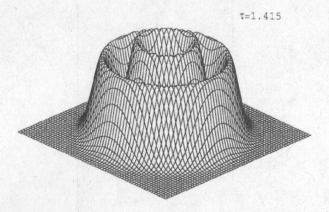

Figure 4: Double-peaked Q-distribution at the edge of the trapped state at $\tau = 1.405$. In this plot $\tau = 1.415$, which is in the middle of the ascending edge of the trapped state.

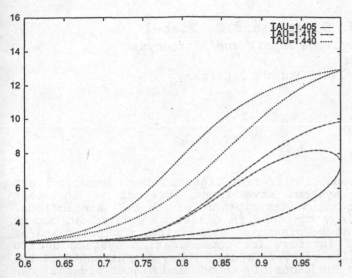

Figure 5: Hysteresis loops obtained by slowly varying the excitation probability between $P = 1$ and $P = 0.6$. All curves start at $P = 1$. $\tau = 1.405$ is in the center of the trapped state and shows no hysteresis. $\tau = 1.415$ is in the middle of the ascending edge of the trapped state and shows pronounced hysteresis. $\tau = 1.440$ is at the top of the edge and shows weak bistability.

the Q-function of the photon field in the middle of the ascending edge of this trapped state. An important clue to bistability are eigenvalues of the map with real part very close to one. The map has such eigenvalues, but one cannot tell if these eigenvalues correspond to relevant eigenstates. Our criterion for bistability will be the hysteresis loop. In order to obtain the hysteresis loop we start with the stationary state for $P = 1$. Then we calculate a new map with a slightly modified value for P and apply it to the stationary state to obtain a new state which is therefore not the stationary state for the new P. We repeat this procedure until $P = 0.6$, then increase P again until $P = 1$ and finally decrease P to $P = 0.6$. Fig. 5 shows the mean photon number versus P. The area between the upper and the lower branch of the hysteresis loop, which is a measure for the amount of bistability found, is large for values of P in the middle of the ascending edge of the trapped state, smaller at the top of the edge of the trapped state and very small in the center of the trapped state.

Thus, bistability is observed and connected to the trapped states. Different properties of bistable states have been used to ensure this result.

References

[1] O. Benson, G. Raithel, and H. Walther. Quantum jumps of the micromaser field: Dynamic behavior close to phase transition points. *Physical review Letters*, 72:3506, 1994.

[2] P. Filipowicz, J. Javainen, and P. Meystre. Theory of a microscopic maser. *Physical Review A*, 34(4):3087, 1986.

[3] H. J. Briegel and B. G. Englert. Quantum optical master equations: The use of damping bases. *Physical Review A*, 47:3311, 1993.

Stochastics in Some Quantum Optical Systems

C. Gerlach, F.-J. Schütte and R. Tiebel
Institut für Theoretische Physik und Astrophysik
Universität Potsdam
Postfach 601553, D-14415 Potsdam, Germany

1. **Introduction**

Nonlinear interactions of different light modes or between light fields and atomic systems have found interest in the last decades. In the quantum description of these interactions, states without analogy to those in classical optics can occur. Optical fields with sub-Poissonian statistics, squeezing and/or antibunching are of interest for communication systems due to their noise properties of lower order. Today measurements of the optical correlation functions of second and higher order(s) can be carried out with high accuracy to prove nonclassical behaviour.

In this context, the evolution of such systems in time is one of the important problems. However, only several approximative methods are available to determine the dynamics of nonlinear coupled systems.

We have investigated the dynamical behaviour of expectation values of certain optical quantum systems. Our interest refers to one or two optical cavity modes coupled nonlinearly in a medium characterized by susceptibility coefficients $\chi^{(2)}$ and $\chi^{(3)}$, respectively. A system of ring cavity modes may be coupled to an external optical monochromatic field with fixed amplitude and phase (driving field). Moreover, the cavity is coupled to a damping bath describing internal losses and outcoupled fields in the usual approximations.

We derive equations for the time evolution of the density operator. Then the corresponding (generalized) FOKKER-PLANCK equation for one of the quasi-probability densities can be formulated. Moments and correlation functions of ordered field operators follow by integration procedures.

If the FOKKER-PLANCK equation is one of the second order, a set of corresponding stochastic differential equations exists. By their integration, we obtain the interesting time dependent moments after averaging over a sufficient big number of trajectories. To overcome the problems connected with non-positive diffusion terms, we choose one of the generalized quasi distributions, namely the positive P-Representation. This function exists as positive function for any density operator. It means, however, a doubling of the phase space dimension. The results can be compared with those of the YUEN's method to introduce complex WIENER processes with negative variances. In the last treatment, a doubling of the number of independent field variables is needed also.

2. The model Hamiltonian for KERR interaction

As an example, we regard a self interaction of an one mode ring cavity field including coherent external driving and damping modelled by the Hamiltonian

$$\hat{H} = \hbar\omega\,\hat{a}^\dagger\hat{a} + \frac{1}{2}\,\chi\,\hat{a}^{\dagger 2}\hat{a}^2 + i\hbar f e^{-i\omega' t}\hat{a}^\dagger - i\hbar f^* e^{i\omega' t}\hat{a}$$

$$+ \hat{a}^\dagger\hat{\Gamma} + \hat{a}\hat{\Gamma}^\dagger, \tag{1}$$

where $\hat{a}$ and $\hat{a}^\dagger$ are the annihilation- and creation operators of the cavity mode, respectively. χ is proportional to the susceptibility of third order $\chi^{(3)}$ responsible for the KERR-interaction, and f means the amplitude of the driving field with frequency ω'. The last two terms describe the bilinear coupling with the damping bath assumed as a large system of harmonic oscillators as in the usual description of damping theory. The reservoir operators are $\hat{\Gamma}, \hat{\Gamma}^\dagger$.
From (1) it follows von NEUMANN's equation for the density operator $\hat{\rho}$

$$\frac{\partial\hat{\rho}}{\partial t} = \frac{\partial}{\partial\hat{a}}\left\{\left(\frac{\gamma}{2} + i\omega\right)\hat{a} - f\,e^{-i\omega' t}\right\}\hat{\rho} + h.a.$$

$$+ \chi\left\{\frac{i}{2}\frac{\partial^2\hat{\rho}}{\partial\hat{a}^{\dagger 2}}\hat{a}^{\dagger 2} + h.a.\right\} + \chi\left\{i\hat{a}^2\frac{\partial\hat{\rho}}{\partial\hat{a}}\hat{a}^\dagger + h.a.\right\} \tag{2}$$

$$+ \chi\left\{2i\frac{\partial\hat{\rho}}{\partial\hat{a}^\dagger}\hat{a}^\dagger + h.a.\right\} + \gamma\bar{n}\frac{\partial^2\hat{\rho}}{\partial\hat{a}\,\partial\hat{a}^\dagger}.$$

If a diagonal representation for the density operator $\hat{\rho}$ exists

$$\hat{\rho} = \int\Phi(\alpha,\alpha^*)\,|\alpha\rangle\langle\alpha|\,d^2\alpha, \tag{3}$$

then the quasi distribution $\Phi(\alpha,\alpha^*)$ obeys a differential equation of second order (FOKKER-PLANCK equation), which follows from (2):

$$\frac{\partial\Phi}{\partial t} = \frac{\partial}{\partial\alpha}\left\{\left(\frac{\gamma}{2} + i\,\delta\omega\right)\alpha - f + i\chi\alpha(\alpha\alpha^* - 2)\right\}\Phi + h.a.$$

$$- \frac{i}{2}\chi\alpha^2\frac{\partial^2\Phi}{\partial\alpha^2} + h.a. + \gamma\bar{n}\frac{\partial^2\Phi}{\partial\alpha^*\partial\alpha}, \tag{4}$$

where a transformation of α in a rotating frame with the optical driving frequency ω' is carried out, so that terms oscillating rapidly are eliminated. In (4), $\delta\omega$ means the difference between the cavity eigen frequency ω and the driving frequency ω'.

If we set in (4) formally $\alpha^* \to \beta$, we obtain the FOKKER-PLANCK equation for the positive P-Representation

$$\hat{\rho} = \int P(\alpha, \beta) \, \frac{|\alpha\rangle\langle\beta^*|}{\langle\beta^*|\alpha\rangle} \, d^2\alpha \, d^2\beta, \tag{5}$$

where we choose such a representation because a positive function $P(\alpha, \beta)$ in (5) always exists. The diffusion matrix in the FOKKER-PLANCK equation for this P-function is positive semidefinite.

Therefore, ITO stochastic differential equations (SDE) can be formulated equivalent to the above FOKKER-PLANCK equation:

$$\frac{d\alpha}{dt} = -\left(\frac{\gamma}{2} + i\delta\right)\alpha - i\chi\alpha^2\beta + f + \sqrt{i\frac{\chi}{2}}\,\alpha\,\xi_1(t)$$

$$\frac{d\beta}{dt} = -\left(\frac{\gamma}{2} - i\delta\right)\beta + i\chi\alpha\beta^2 + f^* + \sqrt{-i\frac{\chi}{2}}\,\beta\,\xi_2(t) \tag{6}$$

where $\xi_j(t)$ (j = 1,2) are independent delta-correlated stochastic forces.

We have integrated this system for fixed values of the driving field f with the ROSENBROCK method [2] and have made an average over a large number of trajectories. In each realization, the initial conditions describe the same coherent state.

So we are able to obtain statistically stationary solutions of the stochastic equations (6), if we choose appropriate system parameters. Furthermore, we compare the results with solutions resulting from the truncated set of ordinary differential equations (ODE) for the correlation functions up to the second order

$$\xi(t) = \langle\hat{a}\rangle, \; B(t) = \langle\hat{a}^\dagger\hat{a}\rangle - \langle\hat{a}^\dagger\rangle\langle\hat{a}\rangle, \; C(t) = \langle\hat{a}^2\rangle - \langle\hat{a}\rangle^2:$$

$$\frac{d\xi(t)}{dt} = -\left(\frac{\gamma}{2} + i\delta\right)\xi + f - i\chi(2B\xi + C\xi^* + |\xi|^2\xi)$$

$$\frac{dB(t)}{dt} = -\gamma B + \frac{\gamma}{2}\overline{n} + i\chi(C\xi^{*2} - C^*\xi^2) \tag{7}$$

$$\frac{dC(t)}{dt} = -(\gamma + 2i\delta)C - i\chi(C + \xi^2 + 6BC + 2B\xi^2 + 4C|\xi|^2).$$

3. Results

With the choosen paramaters (see fig. 1), we obtain a bistable behaviour for the photon number in dependence of the driving field. A similar behaviour was found by Escher et al.[3].

It is remarkable, that these special solutions are very similar to solutions of the system of deterministic differential equations for low-order moments (7), obtained in the GAUSS-approximation. Then, all moments of order higher than two can be reduced to those of second order [5,6]. It means, that the Gaussian property of the quasi probability distribution at the beginning will be true at all times.

So we can conclude, that under these conditions for system parameters and initial states the GAUSSIAN approximation is a appropriate description of the time behaviour of the moments in this optical nonlinear interaction.

Moreover, this model allows an analytical solution for the stationary moments of all orders using the complex P-Representation [3, 4]. From this formula we have calculated and the stationary photon number. It is represented also in Fig. 1.

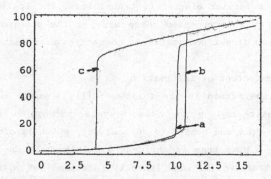

Fig.1. Plot of the stationary photon number n_{st} versus the amplitude f of the driving field (parameters $\gamma = 1$, $\delta = -3.41$, $\chi = 0.025$):
  (a) n_{st} calculated from of the stationary solution of the FOKKER-PLANCK equation for the complex P-Representation [3,4].
  (b) n_{st} calculated from the truncated system of ordinary differential equations (ODE) (7) for correlation functions.
  (c) n_{st} determined from the ITO stochastic differential equations (6) by averaging over 1000 trajectories.

References

[1] C.W. Gardiner, *Quantum Noise,* Springer-Verlag Berlin Heidelberg 1991.
[2] A. M. Smith and C. W. Gardiner, Phys. Rev. A39, 3511-3524 (1989). The ROSENBROCK method is similar to the iteration schemes descibed in that paper.
[3] F.Escher, F. Theilmann, U. Martini, A. Schenzle, Appl. Phys. **B** (1995);
Verhandlungen der Deutschen Phys. Gesellschaft 4/95, p.374;
Vortrag Q 8B.4 (Frühjahrstagung Innsbruck 1995).
[4] P. D. Drummond, D.F. Walls , J. Phys. A **13**, 725 (1980).
[5] J. Perina, J. Krzepelka, B. Horak, Z. Hradil and J. Bajer, *Czech. J. Phys.* **B37**, 1161-1173 (1987).
[6] R. Tiebel, F.-J. Schütte and M. Mareyen, *Chaos, Solitons & Fractals* **3**, No. 2, 177-184 (1993).

Non-Linear Polarization Spectroscopy in the Frequency Domain

D. Leupold, B. Voigt, J, Hirsch, J. Ehlert and F. Nowak

Max-Born-Institut für Nichtlineare Optik und Kurzzeitspektroskopie

D-12474 Berlin

Non-linear polarization spectroscopy in the frequency domain (NLPF) is a special variant of optical mixing, the third order coupling of optical fields which are in resonance with molecular electronic transitions. The application of NLPF to organic molecules in the condensed phase aims at the elucidation of ultrafast relaxation processes of excited states in connection with mechanisms of spectral band broadening.

The method and its content of information

The principle of the method is the following [1]: A weak, linearly polarized, highly monochromatic (ω_2) probe beam travels through a parallel oriented polarizer to the sample and behind to an analyzer, which is crossed with respect to the polarizer. A pump beam is incident simultaneously on the sample. It is also highly monochromatic (ω_1) and linearly polarized but with an angle of 45° with respect to the probe beam polarization. The signal behind the analyzer is the coherently generated light at ω_2. It is measured as a function of the frequency difference $\Delta = \omega_2 - \omega_1$ between the fixed probe frequency ω_2 and the pump frequency ω_1, which is scanned over the spectral range of interest.

In the following application both frequencies ω_1 and ω_2, resp., are in resonance with the first singlet transition S_0-S_1 of the sample, i.e. ω_1 is tuned over the red-most absorption band.

Basic parts of the experimental set-up, which is described in detail elsewhere [2,3], are excimer laser (LPX 105 Lambda Physik)- pumped dye lasers (LDL 205 Laser Analytical System for the pump and LPD 3000 Lambda Physik for the probe); their spectral width in less than 0.05 cm^{-1} and the pulse duration (fwhm) is 15 ns. Selected Glan-Thompson-polarizers (Halle Nachf.) with a block factor $\leq 10^{-8}$ are used. These parameters mainly determine the related time domain range of relaxation processes, which can be extracted from the NLPF signals by fitting with appropriate theoretical lineshape formulas.

The non-linear response of a sample in resonance with the two fields can be obtained by solving the equations of motion for the density matrix of an appropriate level system. The experimental NLPF lineshape is proportional to the square modulus of the lineshape of third order polarization, which is that of third order susceptibility. Theoretical expressions for two, three and four level systems with several relaxation channels have been published [1,3,4].

NLPF is especially well suited to characterize the different types of absorption band broadening [5]: In case of homogeneous broadening, i.e. if the spectral band profile of an electronic transition for each absorbing species is identical to the corresponding profile of the species in its entirety (with ω_0 as the center frequency), than the NLPF signal for a two level system has two maxima. The one is located at $\omega_1=\omega_2$ with a fwhm of $2\sqrt{2}\gamma$, and the other at $\omega_1=\omega_0$ with fwhm of approx. 1.3 Γ. (γ is the inverse of the population relaxation time T_1, Γ that of the phase relaxation T_2). - In case of extremely inhomogeneously broadened transitions, i.e. if the variation of the continuous distribution function of the center frequencies of all individual species is negligible within the homogeneous width of each of them, than the NLPF signal shows only one maximum at $\omega_1=\omega_2$ with fwhm of $2\sqrt{2}\gamma$. In this latter case the NLPF lineshape gives no indication of the overall absorption band maximum. - The usual case of a moderately inhomogeneously broadened band is that in-between these limits, here the width of the distribution function is an additional characteristic of the NLPF signal shape. - In case of heterogeneous broadening, where the overall absorption band profile is the envelope of a discrete number of distinct transitions (each of which is broadened according to one of the before mentioned cases), the NLPF lineshape is the square modulus of weighted sum of the individual contributions. Applications have shown, that NLPF is especially useful to characterize heterogeneity [3,6], if in the absorption spectrum the fine structure is hidden even at low temperatures. One further example will be describing in the following. All measurements were made at room temperature.

NLPF of cryptocyanine in alcoholic solutions

Earlier investigation of cryptocyanine (CC; 1,1´-diethyl-4,4´-carbocyanine iodide, cp.fig.1[1]) dissolved in methanol were interpreted as hints to extremely inhomogeneous broadening of the S_0-S_1 absorption band [7]. NLPF measurements clearly contradict this assumption. One example of a corresponding NLPF lineshape is shown in fig. 2 together with a theoretical NLPF curve calculated according to the literatur data of the inhomogeneous model. The discrepancy is obvious. Instead, this experimental lineshape and those obtained at other probe frequencies within the S_0-S_1 absorption band indicate heterogeneous broadening, which is still better resolved in case of the solvent iso-butanol. The latter is shown in fig. 3 together with a fit according to theoretical lineshape in case

---

[1]Notice that in the figures is representation of wavelength instead of frequency.

116

of heterogeneous broadening with four homogeneous subbands (The fit has been confirmed by χ^2 and F test investigations). For all subbands it holds $50 \leq T_2 \leq 25$fs. These subbands are hidden under the structureless absorption shape in the 680-730 nm range (cp.fig.1); there is also at 77 K not more fine structure in the absorption band [8,9].

These subbands are interpreted as belonging to vibrational transitions coupled to the electronic excitation (cp.fig.4). The contribution of a subband from S_{01} (cp.fig.4) to the overall band shape has the important consequence that the absorption cross section in the range of the S_0-S_1 transition is not simply proportional to the spectral shape of the corresponding optical density [9].

Acknowledgement

This work was supported be the Deutsche Forschungsgemeinschaft (grant Le729/2-1, 2-2).

References
1. J.J. Song, J.H. Lee and M.D. Levenson, Phys. Rev. A 17 (1978) 1439.
2. D. Leupold, B. Voigt, M. Pfeiffer, M. Bandilla and H. Scheer, Photochem. Photobiol. 57 (1993) 24.
3. D. Leupold, B. Voigt, F. Nowak, J. Ehlert, J. Hirsch, E. Neef, M. Bandilla and H. Scheer, Lietuvos Fizikos Zurnals 34 (1994) 339.
4. S. Seikan and S. Sei, J. Chem. Phys. 79 (1983) 4146.
5. S. Mory and E. Neef, Exp. Techn. Phys. 39 (1991) 385.
6. H. Lokstein, D. Leupold, B. Voigt, F. Nowack, J. Ehlert, P. Hoffmann and G. Garab, Biophys. J., in print.
7. G. Mourou, IEEE J. Quantum El. QE-11 (1975) 1.
8. D. Leupold, B. Voigt and R. König, J. Lumin. 5 (1972) 308.
9. H. Stiel et.al in preparation; cp. also D. Leupold, B. Voigt, J. Ehlert, J. Hirsch, H. Stiel and W. Sandner, contribution TUE-45 to ECAMP 5, European Conference on Atomic and Molecular Physics, Edinburgh, April 1995.

Figures

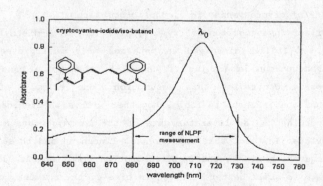

Fig.1 S_0-S_1 absorption band profile of CC in iso-butanol.

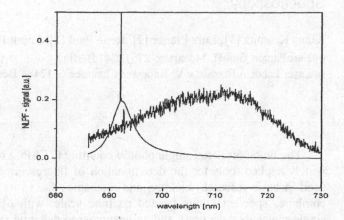

Fig. 2 Comparison of
the measured NLPF signal
of CC in methanol at probe
wavelength λ_2=692,578 nm
with a calculated NLPF
lineshape for extreme
inhomogeneous broadening
(Mourou model [7]).

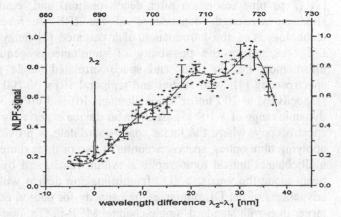

Fig. 3 Measured NLPF
signal of CC in isobutanol
at probe wavelength
λ_2=692,578 nm (crosses)
and curve of theoretical
fit from model of
heterogeneous broadening
with 4 subbands.

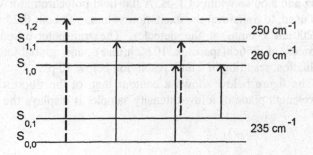

Fig. 4 Scheme of
transitions hidden in
the S_0-S_1 absorption band
of CC in iso- butanol,
and resolved by NLPF.

Time- and Space-Correlated Single Photon Counting Spectroscopy

Klaus Kemnitz [1], Lutz Pfeifer [2], René Paul [2], Frank Fink [2]

[1] EuroPhoton GmbH, Mozartstr. 27, 12247 Berlin

[2] Laser Labor Adlershof e.V, Rudower Chaussee 6, 12484 Berlin

The time-correlated single photon counting (TCSPC) method [1] is one of the most widely applied tools for the determination of fluorescence lifetimes and is intensively used in basic research, biology, and medicine, due to its unique mix of properties. It combines speed, i.e., the ps and ns time scale, with ultrasensitivity (epitomised by single-molecule detection), and an unsurpassed dynamic range, enabling the researcher to study very intensely and very weakly emitting species simultaneously. A TCSPC system typically consists of a high-repetition rate laser, electronic modules, and a fast and sensitive detector, such as a photomultiplier tube (PMT), or more recently, a microchannel-plate-(MCP)-PMT. The MCP-PMT is distinguished by its very small transit time spread, resulting in instrument response functions (IRF) well below 20 ps [2] (2 ps time resolution after deconvolution) and, equally important, by its small emission-wavelength dependence of the IRF. A recent advance in MCP-PMT technology, i.e., the introduction of a patented [3] delay-line (DL) anode by ELDY Ltd., resulted in the possibility of simultaneous acquisition of space and time information, i.e., in time- and space-correlated single photon counting (TSCSPC) spectroscopy [4], with spatial and temporal IRFs of 100 μm and 75 ps FWHM [5], respectively, at 200 linear space channels [6] and up to 350.000 cps through-put at a dynamic range of $> 10^5$ [5]. Such linear device is perfectly suited for emission-resolved spectroscopy, where the linear space-coordinate is provided by a polychromator. By applying fibre optics, space-resolution in two or three dimensions becomes feasible and multichannel optical tomography is within reach, and by scanning the exciting laser beam along the y-axis, a 2D lifetime imaging device will result. Due to the inherent advantages of the DL principle, the capacity for photon acquisition is about 200 times larger than with standard, single-channel MCP-PMTs, due to the dilution of the pile-up effect across many space channels.

We used a Ti-Saph laser system by Lexel Ltd. with SHG and THG, running at 76 MHz and a pulse width of 1 ps. A flat-field polychromator with cylindrical input optics was used, to transform the excitation volume of about 0.25 mm diameter into an image of 200μm x 6mm at the detector. The transputer-based 2D-MCA produced data matrices of 256ch(space) x 1024ch(time), and global analysis was used for data evaluation, resulting in a time resolution below 10 ps.

The figure below shows a contour map of the fluorescence intensity of the time-wavelength plane of a low-intensity sample. It displays the IRF at 400 nm, the Raman band of water, a solvent impurity and, at the red side of the spectral range, the fluorescence decay of rose bengal with a lifetime of about 100 ps.

The scope of applications of the present TSCSPC system ranges from basic investigations to applications in biology and medicine such as DNA-sequencing, microscope studies of living cells, and optical tomography. We present first data of potential applications in dental diagnosis and cell biology.

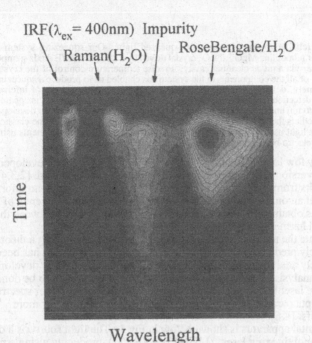

[1] D.V. O'Connor and D. Phillips, "Time-Correlated Single Photon Counting", 1984, Academic Press.
[2] K. Kemnitz, "Ultrafast TCSPC System", in preparation.
[3] US-patent No.5,148,031.
[4] M.R. Ainbund, O.E. Buevich, V.F. Kamalov, G.A. Menshikov, B.N. Toleutaev, "Simultaneous spectral and temporal resolution in a single photon counting technique", Rev.Sci.Instr. 63(1992)3274.
[5] K. Kemnitz, L.Pfeifer, R. Paul, F. Fink, "Time- and Space-Correlated Single Photon Counting Spectroscopy", in preparation.
[6] B.N. Toleutaev and H. Hamaguchi, "Symposium on Molecular Structure and Spectroscopy", Sep. 1992, Kyoto, Japan.

A Reliable Source of Squeezed Light, and an Accurate Theoretical Model

Matthew S. Taubman, Tim C. Ralph, Andrew G. White,
Ping Koy Lam, H. M. Wiseman, David McClelland, Hans-A. Bachor
The Department of Physics, The Australian National University,
ACT 0200, Australia

We demonstrate a reliable source of amplitude squeezed light. Our squeezing system is comprised of a linear monolithic MgO:LiNbO$_3$ crystal driven by a LIGHTWAVE diode-pumped monolithic ring laser via a mode cleaning cavity. Precise temperature control of the crystal and precise locking of all the components in the system has enabled us to produce a squeezing system which routinely delivers about 25 mW of light at 532 nm with 2.2 dB of inferred squeezing around a detection frequency of 12 MHz. The reliability of this system is apparent when it is observed to run unaltered for up to 5 hours. A theoretical model has been developed which accurately predicts the squeezing of the system by including the effects of the classical noise present on the light used to drive the crystal. We present results of experiments using this light, including electro-optic transfer of squeezing between two beams.

Recently, very low loss monolithic frequency doublers have been developed. [1] This has allowed high conversion efficiencies and good squeezing to be obtained. [2,3,4] In this letter we present results from such a second harmonic generator, producing an inferred squeezing of 2.2 dB at an output power of 25 mW, with a conversion efficiency of 50%. This level of squeezing was obtained by using a modecleaning cavity inserted between the laser source and the second harmonic generator.

We demonstrate the use of the Cascaded formalism [5] in developing a theoretical model which accurately predicts the level of squeezing from this system, as has been done in refs. [3,4]. An entirely analytic expression for the laser output spectrum is developed, allowing an analytic analysis of each successive component of the system to be done. The output spectrum of the laser becomes an input parameter for the modecleaner spectrum, which in turn becomes an input parameter for the second harmonic generator. For more details on this, see refs. [4,6].

The experimental apparatus is shown below in Fig 1. The light source is a diode-pumped Nd:YAG monolithic ring laser, (LIGHTWAVE 122), producing a single mode at 1064 nm. The half wave plate and Faraday isolator allow the laser beam to be attenuated and prevent any back reflection into the laser. The light is then incident on a mode-cleaning cavity, with 98% reflecting input and output mirrors. The cavity is 2.45 m long, has a linewidth of about 800 kHz, and a transmission of about 50%. The reflected light from the input mirror of the modecleaner is used to lock it. The output of the modecleaner passes through an electro-optic modulator, which adds the phase modulation necessary for locking the crystal, a glass plate and a dichroic beamsplitter. It is then incident on the monolithic crystal. The glass plate diverts a small amount of the infra-red light reflected from the crystal for locking purposes. The second harmonic produced in the crystal returns along the same

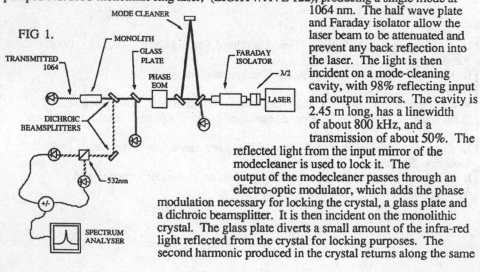

FIG 1.

MODE CLEANER
MONOLITH
GLASS PLATE
TRANSMITTED 1064
PHASE EOM
FARADAY ISOLATOR
λ/2
LASER
DICHROIC BEAMSPLITTERS
532nm
+/-
SPECTRUM ANALYSER

path as the incident fundamental, and is separated from it using two dichroic beamsplitters. It is then incident on a balanced detector, the total quantum efficiency of which is 65% ± 5%. The transmitted light from the crystal is also detected for diagnostic purposes. The second harmonic generator is a standing wave monolithic cavity, made from MgO doped $LiNbO_3$. The dimensions along the x, y and z crystal axes respectively are 5.5, 12.5 and 5.5 mm. The radius of curvature of the end faces is 14.25 mm. The waist size of the fundamental mode is 32.8 μm. The front and back coatings have reflectances at the fundamental wavelength of 99.60% ± 0.3% and 99.90% ± 0.03% respectively, while the reflectances at the second harmonic are ~4% and 99.9%

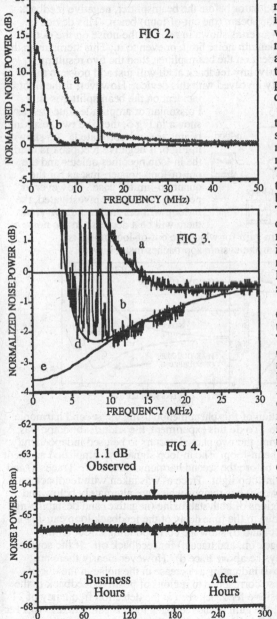

FIG 2.

FIG 3.

FIG 4.

1.1 dB
Observed

Business
Hours

After
Hours

respectively. The crystal is mounted inside a copper oven and maintained to within a precision of 0.001 K of its phase matching temperature, which is approximately 393 K. (Absolute temperature measurement to this precision is difficult and has not been done.)

The impact of the modecleaner on the noise characteristics of the light is shown in Fig 2. Trace a) is the laser noise spectrum, and trace b) is the output noise spectrum of the modecleaner. These traces were found by subtracting the "plus" and "minus" traces from a balanced detector (not shown in Fig 1). The optical power incident on this detector for both traces was 18 mW. The noise filtering action of the modecleaner is very clear. The laser noise spectrum is not quantum limited until beyond 50 MHz, while the modecleaner noise spectrum is quantum limited by 8 MHz.

A comparison between experimental squeezing results and theoretical predictions is shown in Fig 3. Traces a) and b) show squeezing obtained both without and with the modecleaner respectively. They have been corrected for the electronic noise level and efficiency of the balanced detector. Trace b) shows a maximum inferred squeezing of 2.2 dB at 11 MHz. Traces c) and d) are the theoretical predictions for the squeezing. The agreement is very good. The only changes made in the model in producing these last two traces were the inclusion of the modecleaner and an adjustment of the power incident on the second harmonic generator. This demonstrates the validity of the cascaded formalism and allows us to make predictions for other cascaded systems. Trace e) is the theoretical squeezing of the system if there were no laser noise. Comparison with this trace clearly demonstrates the action of the modecleaner. Fig 4 shows the results of a reliability test. The top and bottom traces are the "minus" and "plus" signals from the balanced detector in the second harmonic beam. They show a typical observed squeezing of 1.1 dB at a detection frequency of 11.16 MHz, which translates to about 2.2 dB inferred, and were taken over a period of 5 hours, clearly demonstrating the reliability of the squeezed source. The difference between business and after hours operation is clearly visible! (This is due to the difference in ambient noise levels).

We have used this source of squeezed light to demonstrate electro-optic transfer of squeezing between two squeezed beams as outlined in ref. [7]. When light with excess classical noise is incident on a beamsplitter, the fluctuations on the two resulting beams are correlated. Consequently, if one beam (the in-loop beam) is incident on a detector and the output amplified and fed to an intensity modulator before the beamsplitter, negative feedback will reduce the noise fluctuations on the other beam (the out-of-loop) beam. This device is commonly known as a "noise eater". However as shown in ref. [7], the noise on the out-of-loop beam cannot be reduced below the quantum noise limit, or even to it. This stems from the fact that if quantum-limited noise was incident on the beamsplitter, then the two resulting beams would be uncorrelated. Consequently, any feedback at all will just add noise to the out-of-loop beam. This is the noise penalty involved with this device. However, if the light

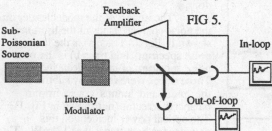

FIG 5.

incident on the beamsplitter is sub-Poissonian or amplitude squeezed as shown in Fig 5, then the two resulting beams are anti-correlated. As shown in Fig 6, if negative feedback is used, the in-loop becomes quieter and the out-of-loop noisier, just as for the quantum-limited case. However, if positive feedback is investigated, the in-loop will become more noisy, but there will be a decrease in the noise of

the out-of-loop beam. There is an optimum gain for which the out-of-loop noise is a minimum, after which it rapidly increases as the system approaches instability.

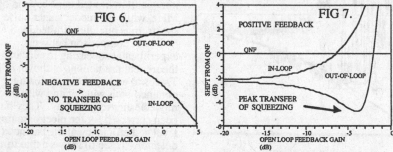

Fig 8 shows experimental verification of this phenomenon using the second harmonic generator as an amplitude squeezed source. To do this experiment, the adder/subtractor was removed from the balanced detector allowing the two photodetectors to be used independently, one for the out-of-loop and the other for the in-loop. The in-loop signal was amplified and sent to an amplitude modulator placed directly before the second harmonic generator. Traces a) and b) show the effect of the feedback on the in-loop light. Trace a) was taken with feedback on. Trace b) was taken with feedback off. Trace a) has regions which are alternately below and above trace b), indicating that there are regions of both stabilizing (negative) and destabilizing (positive) feedback respectively. This is due to the time delay of the feedback loop causing a frequency dependent phase shift. Traces c) and d) show the corresponding effect on the out-of-loop light, trace c) being for feedback on, and trace d) for feedback off. If the source was Poissonian, then trace c) would always be above trace d). However clearly there are regions in which trace c) falls below trace d) indicating a decrease in the noise in the out-of-loop, particularly the circled section. These correspond to regions of positive feedback in the in-loop traces. Judging by the level of positive feedback seen at the detection frequency of 21 MHz on the in-loop trace, (0.6 ± 0.1) dB, the gain of the feedback loop is estimated as (0.11 ± 0.02) dB. For the input squeezing level of 2.1 dB, this should give an enhancement in the

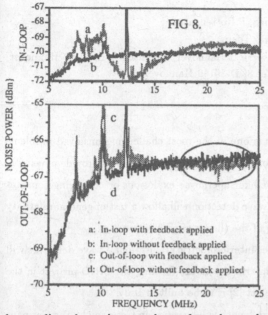

FIG 8.

a: In-loop with feedback applied
b: In-loop without feedback applied
c: Out-of-loop with feedback applied
d: Out-of-loop without feedback applied

electro-optic transfer of squeezing, see refs. [7,8].)

This technique, while only producing small changes for the levels of squeezing obtained from this experiment, can be applied to strongly squeezed sources to give a dramatic effect as shown in Fig 7. This could be used to reduce the coherent amplitude of a squeezed state while not making the sacrifice of squeezing which normally occurs with attenuation. This could be directly applied to laser diodes which, while capable of producing good squeezing, generally must be very bright to lift them far above threshold. We would also point out, that in theory, any quadrature squeezing may be treated in the same way, provided that one arranges to modulate the input field and detect the output fields in that same quadrature.

Note that this process cannot produce more squeezing than was present on the original beam before the beamsplitter, hence it cannot be used to enhance the squeezing of a given source, unless there is an auxiliary anti-correlated beam. This does happen in second harmonic generation experiments where the reflected or transmitted fundamental can be anti-correlated to the produced second harmonic. (In the second harmonic generation experiment we performed, it was found that these correlations were negligible, but this is not generally true). Thus in theory, these auxiliary beams can be used to improve the squeezing on the second harmonic (or visc versa) via this process as discussed in ref. [8].

References.

[1] Laser Zentrum Hannover.
[2] R. Paschotta, M. Collet, P. Kürz, K. Fiedler, H.-A. Bachor and J Mlynek, Phys. Rev. Lett., 72, 3807 (1994).
[3] T. C. Ralph, M. S. Taubman, A. G. White, D. E. McClelland and H.-A. Bachor, "Squeezed light from second harmonic generation; experiment versus theory", In print, Opt Lett.
[4] M. S. Taubman, T. C. Ralph, A. G. White, D. E. McClelland and H.-A. Bachor, "Reliable squeezing from second harmonic generation", in preparation.
[5] H. J. Charmichael, Phys. Rev. Lett., 70, 2273 (1993); C. W. Gardiner, Phys. Rev. Lett., 72, 2269 (1993).
[6] C. C. Harb, T. C. Ralph, I. Freitag, D. E. McClelland and H.-A. Bachor, "Properties of an Injection Locked Laser", in preparation.
[7] M. S. Taubman, H. Wiseman, D. E. McClelland and H.-A. Bachor, "Intensity feedback effects on quantum-limited noise" In print, JOSA B.
[8] H. M. Wiseman, M. S. Taubman and H.-A. Bachor, Phys. Rev. A, 51, 3227 (1995).

Stabilization of an Nd³⁺:YAG-Ring-Laser as a Light Source for Gravitational-Wave Observatories

B.Willke, K.Danzmann, I.Freitag[†], O.Jennrich, P.Rottengatter[†],
A.Tünnermann[†], H.Welling[†] and the GEO600 Team
Institut für Atom- und Molekülphysik, Universität Hannover, Appelstr. 2, D-30167 Hannover
[†]Laser Zentrum Hannover e.V., Hollerithallee 8, D-30149 Hannover

The direct detection of gravitational waves is one of the most challenging unsolved problems of experimental astrophysics. Observations of gravitational-wave-signals will provide new information about violent events in our Universe like supernovae explosions or coalescing compact binary systems . Additionally gravitational wave detection will allow a test of general relativity and will help to improve the understanding of the theory of gravitation.

Over the next six years three large scale laser-interferometric gravitational- wave detectors will be built worldwide. Two of these with 4 km arm-length are forming the LIGO-project in the USA [1]. A 3 km detector , the VIRGO-project [2] is to be built in Italy.

A medium-size interferometer with 600 m arm-length will be set up in Germany by the German-British GEO600 collaboration.

The GEO600 gravitational wave detector uses a novel variant of a Michelson interferometer [3] to sense the length difference of two orthogonal arms. A suitably orientated gravitational-wave causes a phase shift between the light which has passed the two arms and hence changes the output intensity of the interferometer.

The design target of GEO600 is to measure gravitational waves which cause a phase shift of less than 10^{-13} rad. Therefore the contribution of all different noise sources have to be kept below this value.

In the following part the noise introduced by the laser system is discussed and a setup to reach the desired specifications is presented.

First we will discuss the noise due to intensity and frequency fluctuations of the laser. If the arm lengths are slightly different, as it is planned for GEO600 to acquire some important control signals, a Michelson interferometer is sensitive to frequency and to intensity fluctuations. To reach the target sensitivity we have to keep the frequency noise spectral density below $10^{-4} \frac{\text{Hz}}{\sqrt{\text{Hz}}}$. The relative intensity stability of the light injected into the Michelson should be better then 10^{7}.

Further we have to consider the pointing noise or spatial jitter of the laser beam. If the beam splitter is misaligned by a small amount, the geometric fluctuations of the beam-direction will cause a change of the optical path length in the two arms. We have to keep the spectral density of the jitter at the beam-splitter below $10^{-14} \frac{m}{\sqrt{Hz}}$ to reach the desired sensitivity.

Even if all technical noise is kept below the limit, we have to consider the fundamental fluctuation in the light power due to the quantum nature of light. These fluctuations are known as shot noise and are proportional to the square root of the number of photons N. In comparison the output signal of a Michelson interferometer due to a gravitational wave is proportional to N and for that reason the signal-to-noise ratio will increase with the square root of the power inside the interferometer. GEO600 is planned to operate with a laser system of an output power of 10 Watt.

We expect that a injection-locked high-power laserdiode-pumped Nd^{3+}:YAG laser-system which is spatially filtered by two passive cavities may reach these specifications. A monolithic non-planar Nd^{3+}:YAG-ring-laser from the Laser Zentrum Hannover [4] will work as a stabilized master-oscillator for a injection-locked slave laser. By stabilizing the master laser on a high finesse cavity with a Pound-Drever scheme, we intend to reach the desired frequency-stabilization of 0.1 $\frac{mHz}{\sqrt{Hz}}$. As measured by Farinas et al. [5], no additional frequency noise is introduced by the injection locked slave at the low frequency region, where gravitational waves are expected to be detectable.

To reach the goal for the intensity stability we have to consider the master-slave behavior. We are able to reduce the power-fluctuations of the master laser to the $10^{-7} \frac{1}{\sqrt{Hz}}$ range [6]. For such a quiet oscillator the intensity noise in the low frequency region of the whole injection-locked system is dominated by the noise of the pumping power of the slave laser [7].

To suppress the pointing noise of the laser by 120 dB we will use two passive cavities. These so called mode cleaners with a finesse of 1600 and a length of 4m transmit the fundamental mode and reflect high order spatial contributions. Unfortunately these mode cleaners transfer a beam jitter into a intensity noise. As a consequence we are going to measure the intensity noise behind the second mode-cleaner-cavity and use this signal as the error-signal of a servo-loop which feeds back on the pumping source of the slave laser.

We think that this stabilization scheme will provide us with a frequency and intensity stabilized spatial filterred light source which allow us to reach the target sensitivity of GEO600.

References

[1] A. Abramovici et al. "LIGO: the laser interferometer gravitational wave observatory", Science **256**, 325 (1992)

[2] A. Giazotto et al. "The VIRGO Project", Proposal to CNRS and INFN (1989)

[3] K. Danzmann et al. "GEO600: Proposal for a 600 m Laser-Interferometric Gravitational Wave Antenna", Max-Planck-Institut für Quantenoptik MPQ190 (1994)

[4] I. Freitag et al. "Electrooptically Fast Tunable Miniature Diode-Pumped Nd:YAG Ring Oscillator", Opt. Commun.,**101**, 371 (1993)

[5] A.D. Farinas et al. "Frequency and Intensity noise in an injection-locked, solid-state laser", J.Opt.Soc.Am.B, **12**, 328 (1995)

[6] C.C. Harb et al. "Suppression of the Intensity Noise in a Diode-Pumped Neodymium:YAG Nonplanar Ring Laser", IEEE Jour. of Quant.Elect., **30** (1994)

[7] C.C Harb et al. "Solid State Laser Intensity Noise Reduction", Talk on this conference (1995)

Stabilization of Chaotic Outputs in an Acousto-Optic Q-Switched Nd:YAG Laser

Jeong-Moog Kim, Kwang-Suk Kim, Seung-Kyu Park, Seunghoon Lee, Cheol-Jung Kim, O.J. Kwon*, and Chil-Min Kim**
Korea Atomic Energy Research Institute, POB 105, Yusung-Gu,
Taejeon 305-600, Korea
*Dept. of Scien. Educ. Kongu Natl. Teacher's College, Chungnam314-060, Korea
** Dept. of Physics, Pai Chai University, Seogu, Taejeon 302-735, Korea.

Abstract

We have stabilized the chaotic output of the Acousto-Optic Q-switched Nd:YAG laser, by pertubing the modulation frequency at 5.5 kHz of modulation frequency and 11 A of discharge current. The pertubation is obtained by using return-map-based algorithm and the controlled laser has produced a period 2T output just like the stable Q-switched laser pulses.

I. Introduction

In recent, Roy et al reported a stabilization of the diode laser pumped Nd:YAG laser that contains a KTP doubling crystal within a cavity[1]. The laser was controlled by the method of occasional proportional feedback, related to the control scheme of Ott, Grebogi, Yorke (OGY)[2]. The method is based on the idea that an originally chaotic trajectory can be converted to a desired periodic orbit embedded within a chaotic attractor by applying small, precisely chosen temporal perturbations to an accessible parameter of the system. This controlling method has two major advantages; that it does not require a priori knowledge of the system equations and that one is flexibile to choose and control any of the unstable periodic orbits in the attractors. Although the method is very systematic, however it requires a permanent computer analysis to obtain the unstable periodic orbits and to determine stable manifold and unstable manifold near the unstable fixed points on Poincare surface in experiment. And the method can stabilize only those periodic orbits whose maximal Lyapunov exponent is small compared to the reciprocal of time interval between parameter changes, and can not control chaos to the unstable fixed points nor the unstable periodic orbits if they are not embedded within chaotic attractor.

After the Roy's report, some papers has been reported of controlling chaos in NMR laser[3] and Q-switched CO_2 laser[4] in experiment. To stabilize the lasers, OGY method in case of NMR laser and a delayed continuous feedback method in case of Q-switched CO_2 laser are used respectively. Lately many control methods have been developed theoretically. One of them which is easy and systematic is map-based algorithm[5],[6]. By using this method we have in experiment controlled the chaos appearing in Q-switched Nd:YAG laser.

II. Experimental Setup

The method was developed by Lai et al[5], and Peng et al[6]. The method uses the x_n vs x_{n+1} one-dimensional return map, and the desired unstable fixed point x_f for controlling is where the return map meet the diagnal. The chaos can be stabilized to that unstable fixed point on the Poincare surface by perturbing an accessible parameter systematically. Let the x_n vs x_{n+1} 1-dimensional return map be $f(x_n, \lambda_0)$ at the parameter λ_0. Then the perturbation at each pulse is that.

$$\Delta \lambda = - \frac{\partial f / \partial x}{\partial f / \partial \lambda} (x_n - x_f)$$

The Q-switched Nd:YAG laser used in this experiment has a 15 cm of Nd:YAG rod and an acouto-optic modulated Q-switching cell. The laser mirrors are coated at 1.06 μm and the cavity lenth is 75 cm. To detect the chaotic output a photodiode which is sensitive to IR light is mounted behind the back mirror with a birefringent and interference filters. To obtain the return maps and to control the laser output, the output power of the laser is integrated by a gate integrating circuit which is interfaced with a 486 DX-2 computer. A 12-bit A/D and a D/A converters are used to interface the laser output with computer and the Q-switching driver with computer respectively. The integrating time and the converting time of the A/D converter are about 20 μsec and 12 μsec respectively. These times restrict the modulating frequency in stabilizing the chaotic output. The output signals are measured by a LeCroy digital oscilloscope. To control the laser, we have perturbed the modulating frequency of the Q-switching driver by using a V/F converter. The signals from D/A converter drive the V/F converter which triggers Q-switching driver. The experimental diagram is given in Fig. 1. The experiment is conducted at 5.5 kHz of the modulating frequency and 11 A of discharge current.

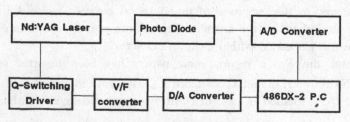

Fig. 1) Experimental Diagram.

III. Experimental Results

To control the chaotic output of the laser, we have obtained bifurcation diagram according to the modulating frequency at from 1kHz to 12kHz as shown in figure 2. The figure shows a typical bifurcation diagram with noise, which presents a transition from period 2T to chaos at 5 kHz, and from chaos to another 2T period at 6 kHz. We have controlled the laser in the caotic region of the diagram.

To control the laser we have obtained a return map at 5.5 KHz of modulation frequency as shown in Fig.3. The figure seems to be a closed cycle which consists of two lines and which has two unstable

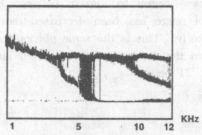

KHz

Fig. 2) Bifurcation diagram.

fixed points. To control chaos we have taken lower unstable fixed point as a desired control point. Although the lines of the return map is a little thick duc to the noise, we have obtained the factor, $\frac{\partial \lambda / \partial x}{\partial f / \partial x}$ (-4.75) on an average. As two lines appear on the map, we have perturbed the modulating frequency at every 2T period.

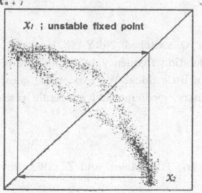

X_{n+1}

X_l ; unstable fixed point

X_2

X_n

Fig. 3) Return map.

On this experimental condition, chaotic laser outputs have been obtained as shown in Fig. 4-a). The temporal behavior shows a quasi-periodic signal. Considering the thickness of the return map, we have determined the control range $x_n - x_f$ to be four times the thickness of the return map. When the modulating frequency is perturbed, the laser has been controlled to 2T period as shown in Fig. 4-b). This time the perturbation signal is obtained as hown in Fig. 4-c). The 2T period is the minimum period; we can not control chaos to 1T period using this method, because x_n appears on the two lines alternately.

When the laser is controlled, the output also shows small fluctuations due to noise. As disscussed by Ott et al, the noise is injurious in controlling chaos. When the control range is smaller than the noise range, the controlled signal easily escapes from the unstable fixed point. To observe the noise effect, we have varied the control range. When the control range has been decreased slightly, chaotic bursts have begun to appear between the controlled signals irregularly. As the control range has been decreased more, the chaotic bursts have appeared more frequently. This is the same phenomenon that was observed in OGY method. But when the control range has been large enough, the laser has been well controlled to 2T period.

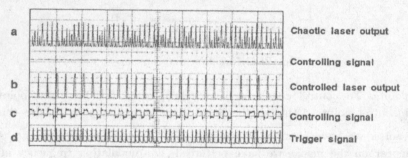

Fig. 4) Laser output and controlling signal

IV. Conclusions

In conclusion, the Q-switched Nd:YAG laser is well stablized to 2T period at 5.5 kHz of modulation frequency by using return map based algorithm, though the return map is a little thick. We expect the controlled Q-swiched laser will be used usefully industry, medicine and academic research as a stable light source.

References
[1] R. Roy, T.W. Murphy, Jr., T.D. Maier, and Z. Gills, Phys. Rev. Lett, 68, 1259(1992)
[2] E. Ott, C. Grebogi, and A. Yorke, Phys. Rev. Lett, 64, 1196(1990)
[3] C. Reyl, L. Flepp, R. Badii, and E. Burn, Phys. Rev, E47, 267(1993)
[4] S. Bielawski, D. Derozier, and P. Glorieux, Phys. Rev. E49, R971(1994)
[5] Y.-C. Lai, Computer Phys, 8, 62(1994)
[6] V. Petrov, B. Peng, and K. Showalter, J. Chem. Phys, 96, 7506(1992)

Spontaneous Raman Scattering with Picosecond Pulses

H. Graener and M. Hofmann
Physikalisches Institut, Universität Bayreuth
95440 Bayreuth, Germany

The first informations about vibrational relaxation processes in liquids were obtained by observing the time evolution of the spontaneous anti-Stokes Raman scattering from vibrational states with a non equilibrium population[1]. Due to the small raman scattering cross sections and the envolved relaxation time constants these experiment are rather inefficient and require a picosecond time resolution. With the increasing performance of ultrashort laser systems and detection setups it is now possible to use the whole potential of this technique especially to monitor the time evolution of complete (anti-Stokes) Raman spectra with sufficient resolution[2]. This allows to observe all Raman active vibrational states which carry a transient excess population. Exciting one single vibrational mode via resonant IR absorption rather complex relaxation pathways can be identified and analysed.

Experimentally we work with a APM-FCM Nd:YLF laser system[3] running with a repetition rate of 50 Hz. The oscillator produces a pulse train of approximatly 500 pulses (wavelength 1.047 μm, pulse duration 4 ps, single pulse energy 30 nJ), which is subsequently amplified and frequency doubled. This pulse train pumps synchronously an idler resonant OPO with a 9 mm KTP (yz-cut, θ = 62°) as nonlinear crystal[3]. The OPO (single pulse energy 20 nJ, pulse duration 1 ps) can be angle tuned between 1.2 and 2 μm. From the fundamental laser pulse train a single pulse is selected and amplified in two double pass Nd:YLF amplifier stages giving a single pulse energy of 1.5 mJ. The major part of this pulse serves as pump in a two crystal parametric down conversion stage (47°-cut $LiNbO_3$ crystals), which uses the OPO output as signal input, to produce the desired strong infrared excitation pulse (duration 4 ps, energy 40μJ). Tis pulse is tunable between 2.5 and 4 μm. The minor part of the laser fundamental is frequency doubled and time delayed; it serves as probe pulse (probing energy 20 μJ). From cross-correlation measurements a time resolution of 1.5 ps is obtained for the experimental setup. The IR and the green pulse are noncollinearly focussed into the sample. The scattered Anti-Stokes light is collimated with a f/1.4 objective, passes a notch filter to suppress the Rayleigh light and is focussed with an achromatic lens into the entrance slit of a f = 64 cm monochromator. The scattered raman light is detected by a 1100×330 pixel slow scan, back illuminated CCD array. To monitor a complete Raman spectrum a 150 l/mm grating is required giving a spectral resolution of 1 nm with a slit width of 100 μm (corresponding frequency resolution 50 cm^{-1} at 450 nm or 26 cm^{-1} at 620 nm). With a typical registration rate of 1 count/s in one channel an integration time of 20-100 s on the CCD array is sufficient to register a complete Raman spectrum; this corresponds to a summation of 1000-5000 laser shots.

First results are reported on liquid dichloromethane (CH_2Cl_2) at room temperature. Dichloromethane has nine fundamental vibrations, which are all more or less Raman and infrared active. Of special interest are three vibrations: the symmetric CH –

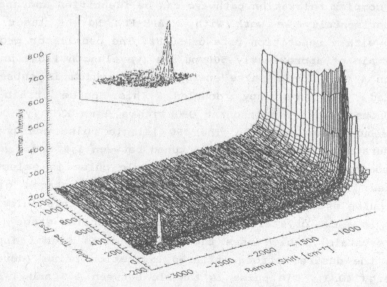

Fig 1.: 3D plot of the observed transient anti-Stokes Raman spectra for CH_2Cl_2; excitation frequency: 3050 cm^{-1}; the inset shows the short time spectra (-10ps < t_D < 70 ps) around the CH stretching vibration (2500 cm^{-1} < ν < 3400 cm^{-1})

stretch (v_1) at 2980 cm^{-1}, the sym. CCl stretch (v_3) at 705 cm^{-1} and the antisymmetric CH stretch (v_6) at 3050 cm^{-1}.

Exciting the strong IR activ v_6 vibration at 3050 cm^{-1} the anti-Stokes intensity as a function of delay time is monotored. Fig. 1 shows a 3D plot of this experiment for the frequency range of 650 - 3200 cm^{-1}. Around 3000 cm^{-1} a peak at early delay time can be observed, which is shown slightly enlarged and rotated as inset. This peak must be attributed to v_1 and v_6 population. Another Raman line which strongly varies in time is the sym. CCl stretching vibration at 705 cm^{-1}. At negative delay time (probe before pump pulse) a constant scattering intensity due to the 3.5 % thermal population is observered. Around delay time zero a first fast increase followed by a clearly slower one can be seen. Then the additional scattering decays on a several hundred picosecond time scale remaining on a slightly increased level still after 1.2 ns. Hardly visible in the 3D spectrum are additional anti - Stokes components between 1100 and 1450 cm^{-1}, which build up on a ten ps scale and decay within 200 ps. The anti-Stokes intensity from the 285 cm^{-1} CCl scissor vibration (not shown in Fig.1) remains nearly constant over the whole delay range.

To obtain the desired dynamical information one can integrate each spectrum over the interesting spectral regions and substract the corresponding integral without excitation pulse and normalize with respect to the input energies: the result is proportional to the population change of the corresponding vibrational state. Fig. 2a shows the short time behaviour of the three vibrations discussed above for an excitation frequency of 3050 cm^{-1}. Clearly visible is the fast rise of the anti-Stokes intensity for the pumped state (open circles), whereas the 2980 cm^{-1} scattering (open triangles) obviously increases slower and is thus populated via a relaxation process. For $t_D \gg 0$ the decay of both CH stretching vibrations can be described by a single exponential time constant $T_{eff} = 9 \pm 2$ ps. This result is consistant with previous measurements[4]. The scattering of the v_3 vibration (full points) shows a first fast rise with a time constant in order of T_{eff}, but the increase goes on even when the CH decay has finished. Fig. 2b shows the full time evolution of the 705 cm^{-1} sym. CCl stretch vibration. The decay can be described be a time sonstant of 350 $\pm$ 100 ps. The solid lines in Fig. 2 are the result of a rate equation model calculation. This model must include at minimum

134

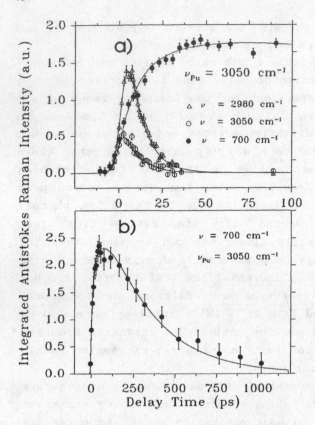

Fig. 2: Integrated
Anti-Stokes Intensity
as function of delay
time;
a) short time
behaviour of the
Raman lines centered
around 3050 cm^{-1} (open
circles), 2980 cm^{-1}
(open triangles) and
700 cm^{-1} (full
points);
b) long time
behaviour of the
700 cm^{-1} component;
experimental points,
calculated curves

five vibrational levels and six time constants. The equilibration of the vibrational energy in the CH stretch region occurs on a 2 ps time scale. The relaxation of the ν_6 state is directly to ν_3 with 8 ps, whereas the ν_1 relaxation populates an intermediate state, which decays with a 200 ps time constant to the ν_3. So two population pathways of ν_3 exsist, which explain the rather complicated scattering increase shown in Fig. 2a.

In summary it was shown that the time resolved anti-Stokes Raman spectroscopy after IR excitation is an extremely powerful technique giving detailed information about vibrational relaxation processes.

1. A. Laubereau, D. von der Linde, A. Kaiser, Phys. Rev. Lett. **28**, 1162 (1972)
2. A. Tokmakoff, B. Sauter, A.S. Kwok, M.D. Fayer, Chem. Phys. Lett. **221**, 412 (1994)
3. K. Wolfrum, P. Heinz, Opt. Comm. **84**, 290 (1991)
4. H. Graener, A. Laubereau, Appl. Phys. B **29**, 213 (1982)

Two Color Spectroscopy of Large Molecules with 20 fs Optical Pulses

S.H. Ashworth, T.Hasche, M. Woerner, E. Riedle, T. Elsaesser.
Max-Born-Institut für Nichtlineare Optik und Kurzzeitspektroskopie,
Rudower Chaussee 6, D-12489 Berlin, Germany

The photophysics of large molecules has been described empirically by Kasha's Rule that states the following: for any manifold of electronic states the first excited state of a given multiplicity will be the fluorescing state. For instance, numerous laser dyes show strong fluorescence emission from the S_1 state with a spectrum that remains unchanged for any excitation wavelength in the range of the absorption spectrum. Thus, higher lying electronic states relax predominantly by radiationless processes to the bottom of the S_1 state. This relaxation process is depicted schematically in the inset of Fig. 1 and occurs on a time scale much shorter than the S_1 lifetime which has values between hundreds of picoseconds and several nanoseconds.

The question we wish to address here is the timescale of the relaxation from S_n states to the S_1 state. To this end we chose the laser dye IR125 dissolved in ethylene glycol. Two Color pump-probe experiments with a very high time resolution of 20 fs were performed. The molecules were excited with a 30 fs pulse in the blue spectral range and the resulting transmission changes were detected with probe pulses that overlap both the S_1-S_0 absorption and fluorescence spectral regions. The absorption and emission spectra of IR125 in ethylene glycol are shown in Fig. 1 along with the spectra of both pump and probe pulses.

We used a home built Ti:sapphire laser with a cavity design similar to Ref. 1 which produces pulses of 20 fs duration tuned to a center frequency of 850 nm. The beam was divided and the larger portion of the output (90%) was doubled in a 400 μm BBO crystal to generate excitation pulses and chopped with a mechanical chopper before being recombined with the remainder of the fundamental (probe) beam and focused into the sample. Both pump and probe pulses were characterized at the sample position. The probe pulses were measured by non-collinear autocorellation in a 100 μm KDP crystal and corresponded to a 20 fs Gaussian pulse profile. The pump beam was characterized by recording a collinear cross-correlation with a 100 μm BBO crystal. Deconvolution of the cross-correlation allows an estimate of the blue pulse length of around 30 fs to be made. After interaction with the sample the pump radiation was filtered out and the change of transmission of the sample at the chopping frequency recorded as a function of delay time between the pump and probe beams.

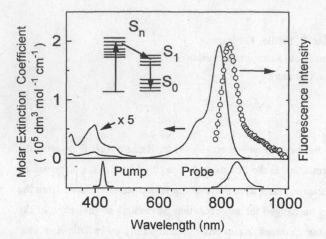

Figure 1. Absorption and emission spectra of Ithe laser dye IR125 dissolved in ethylene glycol. The spectra of both pump and probe pulses are given in the lower panel for comparison. The inset shows a simplified relaxation scheme.

This has been done both integrally and spectrally resolved [2]. Selected spectrally resolved scans are shown in Fig. 2. Because the probe pulses are 20 fs long the bandwidth is correspondingly large, hence it was possible to obtain spectrally resolved signals over a range of nearly 100 nm.

There are three components to the signal : A strongly damped oscillation at short times (< 80 fs) is followed by a level portion (< 200 fs) after which an exponential rise in transmission leads to a maximum transmission increase at around 3 ps. The signal then decays on a time scale of several hundreds of picoseconds by depopulation of the S_1 state. The first signal occurs on a time scale comparable to the vibronic dephasing time of the molecules, i.e. the laser pulses and the molecules

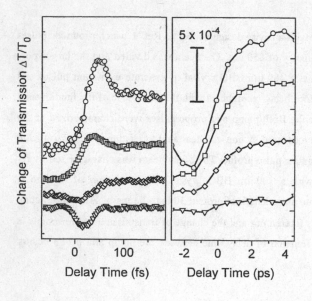

Figure 2. Selected scans from the spectrally resolved data. From above 800 nm (-○-), 820 nm (-□-), 840 nm (-◇-) and 890 nm (-▽-). Two different timescales are shown. For clarity the signals are offset vertically and both sets of data are plotted on the same vertical scale.

couple coherently. Our data show for the first time the coherent coupling of S_0-S_n and S_0-S_1 transitions which is mediated through the electronic ground state. The excitation pulse creates a coherent ground state motion via an impulsive stimulated Raman process [3,4]. The vibrational wavepacket propagating in the ground state potential leads to a transient absorption at long wavelengths (transient negative going signals in Fig. 2) and induced bleaching at short wavelengths (transient positive going signals in Fig. 2). The wavelength dependence of the signals contains information on the relative shift of the potential surfaces. The coherent motion is damped by the very rapid spreading of the vibrational wavepacket and disappears after about 80 fs.

The incoherent, step like signal which follows is due to bleaching of the ground state. This can be confirmed by comparing the signals after the coherence has decayed (at 80 fs) with the steady state absorption curve. Within our experimental accuracy this bleaching component quantitatively matches the steady state absorption.

The third component extending well into the picosecond regime is due to stimulated emission from the S_1 state. It reflects the delayed accumulation of molecules at the bottom of the S_1 state. Having subtracted the bleaching part of the signal the increase of transmission follows the stimulated emission spectrum at all delay times. That is to say that the stimulated emission develops in one spectral region and does not appreciably shift with time. The rise of the signal can be fitted to a single exponential and the resulting time constant is on the order of 1 ps. Several intramolecular relaxation processes are involved in this delayed accumulation. First, the states directly excited by the pump pulse are depopulated by coupling to a large manifold of other vibronic states. Second, redistribution within this manifold plays an important role for the population of the S_1 state. A separation of the different processes is not possible from the data of Fig. 2, giving the overall dynamics of relaxation.

To investigate the initial step of S_n relaxation, we turned our attention to the state initially prepared with the blue (pump) pulses and performed a one color pump-probe experiment that should allow us to investigate the relaxation of the S_n state directly. A non-collinear geometry was required in order to carry out this measurement. Using the same sample the one color pump-probe signal is plotted in Figure 3 (upper trace). Here only two contributions to the signal are apparent. A bleaching component which is constant to at least 10 ps and a more or less pulse limited peak at short delay times. This peak is a convolution of coherent signal and simulated emission and we are currently investigating the ratio between the two contributions.

From the one color pump-probe result we see that the molecules are lost on a timescale of tens of femtoseconds from the initially prepared state, i.e. much faster than they accumulate at the bottom of the S_1 state. This is clear from the comparison of the one color and the two color pump-probe results given in Fig. 3. The lower trace gives the spectrally integrated transmission change measured with

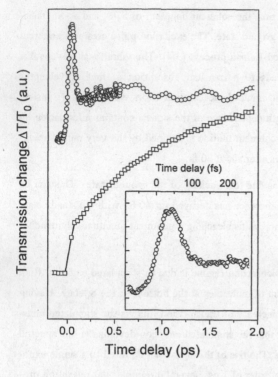

Figure 3. Spectrally integrated pump-probe signals. The curves are shifted relative to one another for clarity and are not plotted on the same vertical scale.
Upper: One color pump-probe signal (-○-)
Lower: Two color pump-probe signal (-□-)
Inset: Detail of one color pump-probe signal at short delay times.

pulses at 850 nm. The very rapid loss of S_n population may be through internal conversion to other singlet states or redistribution of energy within the S_n state. It is important to note that the depopulation of the optically coupled S_n levels is not the rate limiting step for S_n-S_1 relaxation. In contrast, our data suggest that the subsequent redistribution processes are responsible for the picosecond accumulation in S_1. Our aim now is to investigate this timescale "hole" using specially selected molecular systems and comparison with theoretical calculations to elucidate timescales of intermediate relaxation processes.

References

1. P.F. Curley, Ch. Spielmann, T. Brabec, F.Krausy, E.Wintner and A.J. Schmidt, *Opt. Letters*, **18**, 977 (1993)
2. G. Stock and W. Domcke, *Phys. Rev. A.*, **45**, 3032 (1992)
3. K.A. Nelson and I.W. Ippen, *Adv. Chem. Phys.*, **57**, 1 (1989)
4. M.S. Pschenichikov, K Duppen and D.W. Wiersma, *Phys. Rev. Letters*, **74**, 674 (1995)

Intersubband Relaxation and Carrier Transfer Within the Conduction Band of Quantum Well Structures

J. Baier, I. Bayanov, J. Kaiser, U. Plödereder and A. Seilmeier
Institute of Physics, University of Bayreuth
D - 95440 Bayreuth, Germany

Carrier scattering between subbands in semiconductor heterostructures is of interest for the development of sensitive quantum well infrared photodetectors /LEVINE/ and of intersubband lasers /FAIST/.

In this contribution intersubband scattering is investigated by an infrared bleaching technique /SEILMEIER (1987)/. An intense infrared pulse resonantly excites conduction electrons from the lowest subband to a higher lying subband. The recovery of the intersubband absorption is monitored by a weaker delayed pulse of the same frequency.

The infrared pulses with a duration of about 1 ps are generated by difference frequency mixing. A modelocked Nd:glass laser pumps a travelling wave dye laser. Subsequent generation of the difference frequency between the fundamental of the glass laser and the travelling wave dye laser in a GaSe crystal provides pulses tunable between 6 μm and 18 μm /DAHINTEN/.

The sample investigated is an n-modulation doped GaAs/$Al_{0.31}Ga_{0.69}As$ quantum well structure grown by MBE consisting of 50 periods. GaAs wells of 7.3 nm width are embedded between 36 nm wide AlGaAs barriers in which the central 8 nm are doped by Silicon ($c = 5.7 \times 10^{11}$ cm^{-2}). The sample prepared in a special waveguide geometry /SEILMEIER (1987)/ is mounted on the cold head of a closed-cycle He refrigerator which allows measurements between 10 K and 300 K.

Fig. 1 shows the subband structure of the sample which is the result of selfconsistent calculations. There exist three different kinds of subbands: i) subbands localized in the GaAs wells (solid lines, W1 and W2), ii) subbands mainly localized in the AlGaAs barriers with wavefunctions which exhibit a small amplitude also

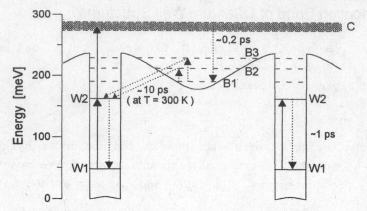

Fig. 1: Band structure of the investigated modulation doped sample. The excitation processes are indicated by solid arrows, the most significant relaxation channels by dotted arrows.

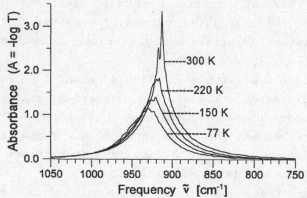

Fig. 2: Temperature dependence of the intersubband absorption (transition W1→W2) taken on the waveguide sample.

in the wells (dashed lines, B1 to B3) and iii) quasibound subbands in the continuum (e.g. hatched area C). In Fig. 2 the intersubband absorption at $\tilde{\nu} \sim 920$ cm^{-1} (transition W1→W2) is depicted for several temperatures. In the time resolved experiments the frequency of the pulses is always tuned to the maximum of the intersubband absorption.

The intersubband relaxation is systematically investigated as a function of the sample temperature. Fig. 3. shows results taken with a pump pulse intensity of ~70 kW/cm^2. In general we observe biexponential relaxation. At T≤77 K the fast component with the time constant τ_1 dominates the relaxation behavior. The amplitude of the second slower component with τ_2 rises with temperature and is prevailing at room temperature. There is indication that this

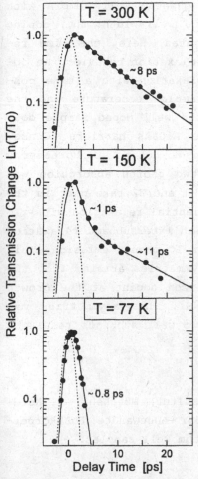

Fig. 3): Normalized transmission change as a function of delay time for different temperatures at a pump intensity of ~70 kW/cm². The dotted line represents the correlation function of the pump and probe pulses.

effect is due to the increasing intersubband absorption with rising temperature (see Fig. 2). The same correlation between the absorbed pump energy and the magnitude of the slowly relaxing component is found in measurements as a function of the pump intensity.

The findings can be explained by the following model: The infrared pump pulse excites electrons both to the first excited subband of the well (W2) and to a resonant continuum band (C) via two photon excitation (see Fig. 1, solid arrows). Efficient trapping processes occupate the states B1 to B3 which are mainly located in the barrier.

The fast relaxation component with a time constant of $\tau_1 \approx 1$ ps represents the direct intersubband relaxation W2→W1. This value is in good agreement with a calculated intersubband scattering time due to electron - LO phonon interaction /FERREIRA/.

The second slower relaxation component arises from the transfer of the trapped carriers in the barriers back into the quantum wells /SEILMEIER (1994)/. Inspection of the corresponding transfer rates shows that the most efficient path is phonon assisted excitation to the states B2 and B3 and a subsequent scattering B2→W2 and B3→W2, respectively, with a time constant of ~10 ps at room temperature. The temperature dependence of the activation process qualitatively explains the temperature behavior of the slower relaxation component; it increases from τ_2=8 ps at T=300 K to τ_2=11 ps at T=150 K.

142

For comparison, a well doped multiple quantum well sample with similar parameters (Al-concentration: 0.3, well width: 8nm, doping concentration: $7 \times 10^{11} cm^{-2}$) was investigated. Here, the time resolved data do not show such a complex relaxation behavior. We observe a monoexponential intersubband relaxation with a time constant of $\tau=3$ ps independent from the sample temperature and the pump intensity. The band structure of the well doped sample does not exhibit a potential minimum in the AlGaAs barriers. Consequently subbands localized in the barrier do not exist. Moreover a resonant continuum band for efficient two photon absorption was not found. Therefore only the two bands W1 and W2 take part in the relaxation process leading to a monoexponential relaxation with τ_1.

The time constant of 3 ps for the direct intersubband relaxation W2→W1 is longer than that calculated for LO phonon emission. This fact is believed to be due to doping asymmetries arising from the segregation and accumulation of the Silicon dopant at the growth surface during the MBE process /LARKINS/. This asymmetries may lead to a reduced overlap between the two well subbands resulting in larger intersubband relaxation times.

Acknowledgement

We thank G. Weimann (Walter-Schottky-Institut, München), P. Koidl and H. Schneider (Fraunhofer-Institut für Angewandte Festkörperphysik, Freiburg) for providing the quantum well samples.

References

Dahinten T., Plödereder U., Seilmeier A., Vodopyanov K., Allakhverdiev K. and Ibragimov Z.; IEEE J. Quant. Electron. **29**, 2245 (1993)

Faist J., Capasso F., Sivco D.L., Sirtori C., Hutchinson A.L. and Cho A.Y.; Science **264**, 553 (1994)

Ferreira R. and Bastard G.; Phys. Rev. B **40**, 1074 (1989)

Larkins E.C., Schneider H., Ehret S., Fleißner J., Dischler B., Koidl P. and Ralston J.D.; IEEE J. Trans. Electr. Dev. **41**, 511 (1994)

Levine B.F.; J. Appl. Phys. **74**, R1 (1993)

Seilmeier A., Hübner H.-J., Abstreiter G., Weimann G. and Schlapp W.; Phys. Rev. Lett. **59**, 1345 (1987)

Seilmeier A., Plödereder U., Baier J. and Weimann G.; *Quantum Well Intersubband Transition Physics and Devices* ed H.C. Liu *et al* (Academic Publishers) p 421 (1994)

Bloch Oscillations of Excitonic and Continuum States in Superlattices

P. Leisching[1], T. Dekorsy, H.J. Bakker, H.G. Roskos, K. Köhler[2], and H. Kurz
RWTH Aachen, Institute of Semiconductor Electronics II, Sommerfeldstr. 24, 52074 Aachen, Germany
[1]present address: Ecole Polytechnique, Laboratoire d'Optique Quantique du C.N.R.S., 91128 Palaiseau Cedex, France
[2]Fraunhofer-Institute of Applied Solid State Physics, Tullastr. 12, 79108 Freiburg, Germany.

The investigation of the coherent dynamics of optically excited wavepackets and the associated macroscopic polarizations in high quality GaAs/AlGaAs semiconductor superlattices has gained considerable interest during the last years [1]. It is both of fundamental and technological interest: this system is especially well suited for the investigation of the dephasing mechanisms of the phase coherence and the nonlinear Coulomb interactions observed in different polarizations. The coherent dynamics of the microscopic wavepacket dipoles might be a future practical source of tunable coherent radiation in the THz regime [2].

The *interband* polarization associated with the valence band to conductance band transitions can be monitored by four-wave-mixing (FWM). Due to the coherent superposition of different electronic states the wavepacket motion is also associated with an oscillating microscopic electric dipole in growth direction leading to a macroscopic *intersubband* or *intraband* polarization between the different continuum subbands. To measure the far-field THz radiation of this macroscopic oscillating electric field in the sample, an optically gated dipole antenna can be applied [2]. The internal electric field in the sample can be detected by using the transmittive electro-optic sampling (TEOS) technique [3].

Theory predicts a strong Coulomb coupling of both polarizations [4,5]. So far, most FWM and THz/TEOS experiments were performed separately [1-3], prohibiting a direct comparison of the complementary informations to be gained by the different techniques: it is difficult to use identical *optical* excitation conditions although that is required for the comparison of the results with theory. By using a combined FWM/TEOS set-up, differences in the dephasing and the beating frequencies of both polarizations were observed and explained in simple theoretical models [6].

In this paper we report on simultaneously performed FWM/TEOS experiments on an electrically biased high quality GaAs/$Al_{0.7}Ga_{0.3}As$ semiconductor superlattice [1] (electronic/heavy hole (hh) miniband width Δ: 19/1.3 meV). We investigate the *coupled* dynamics of both polarizations and the influence of dipole selection rules which govern the coherent dynamics. For the time-resolved experiments we use an Ar$^+$:ion-laser pumped Kerr-Lens mode-locked Ti:sapphire laser with a repetition rate of 76 MHz. It delivers 130 fs laser pulses with a spectral width ΔE_{laser} of obout 18 meV. The combined FWM/TEOS experiments are performed at 10 K using a modified two-beam pump-probe set-up, see Ref [3,6]. We take advantage of the unique electric field dependent properties of superlattices: by tuning the electric field, we are able to change the energy separation ΔE between two adjacent continuum WS states linearly with the applied electric field: $\Delta E = eFd$ (d denotes the superlattice period, F the electric field). At an intermediate electric field (1-5 kV/cm) we are able, to excite a coherent superposition of several of the continuum WS states and their excitonic transitions using a broadband ultrashort laser pulse with $\Delta E_{laser} > \Delta E$. The quantum interference of the different WS states leads to oscillating contributions in the signal. The fundamental frequency $\Delta \omega = \Delta E/\hbar$ corresponds to the Bloch oscillation period $\tau_B = h/\Delta E$ of the superlattice.

The experimental results at an excitation density of 2×10^9 cm^{-2} carriers per superlattice period and an applied voltage of -0.5 V (3.5 kV/cm) are depicted in Fig. 1, the laser pulse is centered at the excitonic hh$_0$ transition. The FWM data in the upper part of Fig. 1 reveal first an initial peak, followed by a pronounced beating structure in the decay of the signal. The peak is due to the fast interband dephasing (100 fs) of the excited continuum states. Without the strong modulation due to the excitonic quantum interference the purely excitonic residual signal decays monoexponential with the excitonic interband phase relaxation time T_2^{inter}. Assuming a purely homogeneous broadening leads to an estimation for T_2^{inter} of about 1 ps. We observe a beating period on the order of about 250 fs, corresponding to a WS splitting of about 16 meV. This result is in striking contrast to the Bloch period $\tau_B = 1$ ps expected at this field and can be explained with the existence of overtones, i.e. higher harmonics, of BO [1].

In the lower part of Fig. 1, the TEOS trace under *identical* experimental conditions is depicted. The signal is composed of a fast initial peak which is a coherent artifact. It is followed by a steplike function (instantaneous polarization) that is modulated with two frequencies: at the beginning a short living oscillation with the FWM frequency followed by a long living oscillation with about four times the FWM period. The longer period of 1 ps is in good agreement with the continuum WS splitting of 4 meV, the damping of the amplitude is assigned with the interband phase relaxation

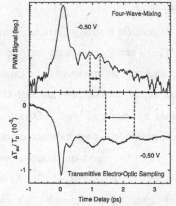

Figure 1: Upper part: semilogarithmic plot of the FWM data at - 0.55 V. Lower part: Linear plot of the TEOS data under identical excitation conditions.

time T_2^{intra}. A monoexponential fit to the data delivers a T_2^{intra} of 2.4 ps.

In order to give a *qualitative* theoretical description of the observed experimental differences in the interband and intraband polarizations we use the density matrix formalism approach for a multilevel quantum system, see Ref. [7]. For simple analytic expressions, we restrict ourselves to the case of equal spacing $\Delta\omega$ of the WSL and the same T_2^{inter} (T_2^{intra}) for each excitonic (continuum) WS level. The intensity of the diffracted FWM light in the direction of $2k_2-k_1$, assuming δ-shaped laser pulses #1 and #2 ($E_{laser} = E_1\delta(t+\tau) + E_2\delta(\tau)$) then accounts to:

$$|P^{(3)}_{2k_2-k_1}(\tau)|^2 \propto E_1^2 E_2^4 \times \theta(\tau)\exp(-2\frac{\tau}{T_2^{inter}}) \times \left\{\sum_k \mu_{0k}^4 + \sum_{k>j} \mu_{0j}^2\mu_{0k}^2\cos[\Delta\omega(k-j)\tau]\right\}^2 \quad (1)$$

with μ_{0k} given by the interband matrix elements of the envelope wave functions Φ_k^{WS}, denoted as $|k>: \mu_{0k} = <0|k>$. The oscillating terms of the signal strongly depend on the relative strength of the interband matrix elements. At low voltage (small WS splitting) compared to the spectral pulse width, one can observe overtones of the BO in the FWM signal [1] up to the fourth harmonic of the fundamental BO period, i.e., the signal is dominated by the $\cos(2\Delta\omega\tau)$ and $\cos(4\Delta\omega\tau)$ terms. The time evolution of the TEOS signal is given by the macroscopic longitudinal polarization $P_z^{(2)}$ and consists of two components: an step-like polarization generated by population matrix elements decaying with the relaxation time T_1 and a second oscillating part decaying with T_2^{intra}:

$$P_z^{(2)}(\tau) \propto E_1^2 \times \theta(\tau) \times \left\{\exp(-\frac{\tau}{T_1})\sum_k \mu_{kk}\mu_{0k}^2 + \exp(-\frac{\tau}{T_2^{intra}})\sum_{k>j} \mu_{jk}\mu_{0k}\mu_{j0}\cos[\Delta\omega(k-j)\tau]\right\} \quad (2)$$

In addition to the dependence on the interband matrix elements μ_{0k}, the occurrence of higher beating terms in the signal is governed by the *intraband* transition *dipole* matrix elements $<j|ez|k>$ between different electronic bands: $<j|ez|k> = \Delta/(4F)(\delta_{j,k+1} + \delta_{j,k-1})$ for the off-diagonal matrix elements and $<j|ez|j> = ejd$ for the diagonal matrix element (defining $z=0$ at the center of the hh n'$=0$ well). Therefore the dominant contribution to the signal is given by the $\cos(\Delta\omega\tau)$ terms. In order to give a more quantitative description, we evaluate Eq. 1 and 2 for a WS splitting of $\Delta\omega$ = 4 meV [7]. The calculated interband polarization $|P^{(3)}|^2$ in the upper part of Fig. 2 shows a pronounced modulation with the fourth harmonic of the BO (250 fs), in good agreement with the experiment. In the lower part of Fig. 2, the microscopic intraband dipole moment per excited electron-hole pair is depicted. The intraband polarization $P_z^{(2)}$ only reveals a beating with the first harmonic of BO (1 ps). Due to the different phase relaxation times, the decay of the intraband polarization is much slower.

Neglecting the Coulomb interaction between both polarizations we should observe only the *first harmonic* of the BO in the TEOS (see Fig. 1) and THz-emission experiments of Ref. [2]. These calculations are in contrast to the experimental results: for short delays there is still evidence for an occurance of the overtones in the TEOS data in Fig. 1, too.

Theoretical calculations including Coulomb coupling between both polarizations [4,5] predict a coupling between both polarizations which leads to oscillating components with the interband frequency in the intraband polarization. In contrast to the conventional difference of the excitonic

binding energy [6], in our system the observability of the coupling is strongly enhanced due to the large difference in the beating frequencies by a factor of four. The observed experimental behaviour is in good agreement with the theoretical predictions, assuming a difference of a factor of two in the phase relaxation times.

In conclusion, we presented a detailed experimental and theoretical investigation of the dephasing and Coulomb coupling of interband and intraband polarizations in a GaAs/AlGaAs superlattice. We used a combined FWM/TEOS setup that enables the simultaneous measurement of both complementary polarizations under *identical* optical excitation conditions. The experimental results of FWM and TEOS differ considerably: the

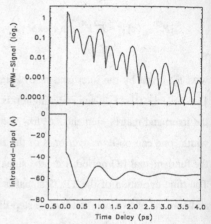

Figure 2: Upper part: semilogarithmic plot of the calculated FWM trace (Eq. 1). Lower part: linear plot of the intraband dipole moment (Eq. 2).

fundamental beating frequencies by a factor of four and the phase relaxation times by a factor of two. Additionally, there are also contributions in the TEOS signal with the FWM frequency.

This differences demonstrate, that the FWM signal is generated by the interband dynamics of excitonic states while the TEOS signal is dominated by the intraband dynamics of continuum states after the decay of the interband polarization.

The experimental data are in good agreement with numerical calculations using a density matrix formalism modeling the coherent dynamics: the existence of overtones, i.e. higher harmonics, in the interband polarization is strongly supressed in the intraband polarization due to dipole selection rules. For this reason one has to invoke the Coulomb coupling of both polarizations that leads to the occurrence of overtones also in the TEOS signal. The expected Coulomb coupling is in good agreement with the observed difference of a factor of two in the phase relaxation times.

This work has been supported by the Deutsche Forschungsgemeinschaft and the Stiftung Volkswagenwerk. One of us (P.L.) gratefully acknowledges financial support by the A. von Humboldt Foundation and the C.N.R.S.

[1] P. Leisching, P. Haring Bolivar, W. Beck, R. Schwedler, F. Brüggemann, H. Kurz, and K. Köhler, Phys. Rev. B **50**, 14 389 (1995).
[2] C. Waschke, H.G. Roskos, R. Schwedler, K. Leo, H. Kurz, and K. Köhler, Phys. Rev. Lett. **70**, 3319 (1993).
[3] T. Dekorsy, P. Leisching, H. Kurz, and K. Köhler, Phys. Rev. B **50**, 8106 (1994).
[4] T. Kuhn, E. Binder, and G. Mahler, Proceedings of the 22th International Conference on the Physics of Semiconductors, Vancouver 1994.
[5] T. Meier, G. von Plessen, P. Thomas, and S.W. Koch, Phys. Rev. Lett. **71**, 902 (1994).
[6] P. Leisching, T. Dekorsy, H.J. Bakker, H. Kurz, and K. Köhler, Phys. Rev. B **51**, 15.06.95.
[7] P. Leisching, T. Dekorsy, H.J. Bakker, K. Köhler, and H. Kurz, submitted to JOSA B (1995).

High Resolution CARS Spectroscopy of Polyatomic Molecules

H.W. Schrötter, S. Anders, T. Bican, D. Illig, J. Jonuscheit[1], T. Mangold,
K. Sarka[2], M. Stamova[3]

Sektion Physik der LMU München, Schellingstr.4, D-80799 München, F.R. Germany

The investigation and analysis of rotation-vibrational bands through high resolution infrared and Raman spectroscopy is an important source of information on the structure of polyatomic molecules. By cw-CARS (coherent anti-Stokes Raman scattering) an instrumental spectral resolution of less than 30 MHz can be routinely achieved. It is limited by the convoluted linewidths of the lasers used for excitation, namely a fixed frequency laser and a tunable dye laser. The CARS signal is generated when the frequency difference of the two lasers coincides with a Raman active rovibrational transition in the molecule.

In Munich FRUNDER et al. [1] have developed a cw-CARS spectrometer based on an argon ion ring laser. Recently the spectrometer was reconstructed around a new argon ion laser and equipped with a better stabilization for the tunable dye laser. The gaseous samples are contained in special sample cells with Brewster angle windows placed in the focal region of the argon ring laser which provides an intracavity power of between 100 and 200 W for excitation. A new data acquisition system allows the simultaneous recording of CARS and calibration spectra. An instrumental resolution of less than 0.001 cm^{-1} and a wavenumber accuracy of ± 0.006 cm^{-1} was achieved. A more detailed description of the spectrometer is published elsewhere [2] together with the high resolution CARS spectrum of ethane.

---

[1]Present address: Lehrstuhl für Technische Thermodynamik der Universität Erlangen, Am Weichselgarten 8, D-91058 Erlangen, F.R. Germany

[2]Permanent address: Department of Physical Chemistry, Faculty of Pharmacy, Comenius University, Odbojárov 10, SK-83232 Bratislava, Slovak Republic

[3]Permanent address: Institute of Solid State Physics, Bulgarian Academy of Sciences, Blvd. Tsarigradsko Chaussee 72, BG-1784 Sofia, Bulgaria

As further examples for the results obtained with this spectrometer we present here CARS spectra of isotopomers of ammonia and benzene. Some preliminary results were already published [3, 4]. Self pressure broadening in ammonia was investigated through the recording of the spectra of the Q-branches of the ν_1 bands of $^{14}NH_3$, $^{15}NH_3$, $^{14}ND_3$ and $^{15}ND_3$ at pressures from 1 kPa to 80 kPa. In a previous investigation on $^{14}NH_3$ by ANGSTL et al. [5] it was found that the spectrum can only be simulated by introducing individual linewidths $\Delta\nu(J,k)$ depending on the quantum numbers J and k of the symmetric top. Line positions and self broadening coefficients in the Q-branch of the corresponding infrared ν_1 band have been experimentally determined by MARKOV et al. [6]. Starting from these data in combination with microwave data [7] we have simulated the CARS spectra of $^{14}NH_3$ in the pressure range 1 to 5 kPa and found good agreement within our experimental accuracy. The linewidths of k = 0 transitions for which no experimental IR values exist were scaled down from the calculated values [6]. The positions of the k = 0 lines had to be shifted from the positions calculated from molecular constants derived from infrared spectra [5, 8] to obtain the best fit. This indicates that a slight revision of the molecular constants of the ν_1 state will result from inclusion of the CARS data in a future fit.

The CARS spectrum of the ν_1 band of $^{15}NH_3$ was assigned by combining the wavenumbers from infrared [9] and microwave [10] spectra. By comparison of unassigned lines in the IR spectrum with the CARS data 24 of these lines could be identified. The spectra were at first simulated by applying the same linewidth values $\Delta\nu(J,k)$ that were used for $^{14}NH_3$, but it was found that the pressure broadening coefficients had to be reduced by factors between 0.6 and 0.9 for isolated lines to obtain the best fit. Figure 1 shows part of the experimental and calculated CARS spectrum of $^{15}NH_3$, the lower trace representing the difference spectrum which indicates some remaining minor discrepancies. In the CARS spectra of the ν_1 bands of $^{14}ND_3$ and $^{15}ND_3$ the lines are much more closely spaced and could not be so well resolved. An assignment will only be possible after IR data of these bands become available.

In another series of measurements the CARS spectra of the CH- and CD-streching vibrations of several isotopomers of benzene (C_6H_6, C_6D_6, $1,3,5-C_6H_3D_3$,

150

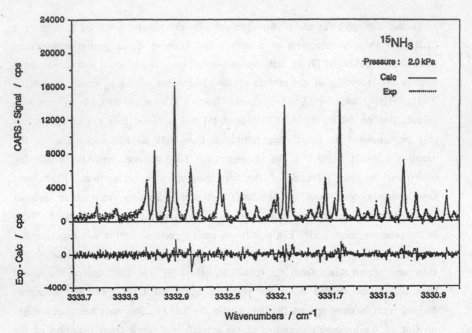

Fig. 1: Part of the CARS spectrum of $^{15}NH_3$ at a pressure of 2 kPa. Points: experiment, full line: simulation, lower curve: difference.

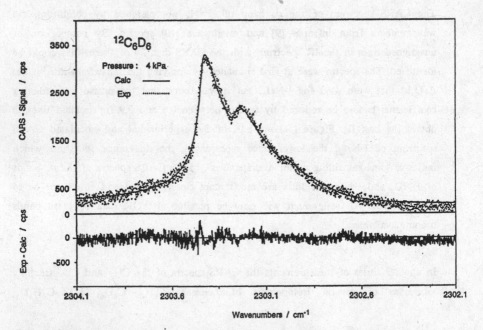

Fig. 2: CARS spectrum of the ν_1 band of C_6D_6 at a pressure of 4 kPa. Points: experiment, full line: simulation, lower curve: difference.

1,2-$C_6H_4D_2$, 1,3-$C_6H_4D_2$, 1,4-$C_6H_4D_2$, and C_6HD_5) were recorded. For C_6H_6 and C_6D_6 the spectra could be simulated as shown for the latter case in Fig. 2 and anharmonicity constants were determined from the resolved hot bands (C_6H_6: $x_{1,20} = -0.23$ cm^{-1}, $x_{1,18} = -2.62$ cm^{-1}; C_6D_6: $x_{1,20} = -0.20$ cm^{-1}). In the case of 1,3,5-$C_6H_3D_3$ the earlier results [4] were confirmed.

Acknowledgements

We thank F. Bauer and K. Sailer for preliminary measurements, R. Angstl and U. Lehner for the development of computer programs, D. Papoušek and Š. Urban for the communication of unpublished IR data of $^{15}NH_3$, and A. Ruoff and his collaborators at the University of Ulm for providing most of the deuterated benzene samples. K.S. gratefully acknowledges financial support from the Forschungszentrum Karlsruhe in the framework of Scientific-Technical Cooperation between Germany and the Slovak Republic and M.S. thanks for support of her stay in Germany by the Deutsche Forschungsgemeinschaft.

References

[1] H. Frunder, L. Matziol, H. Finsterhölzl, A. Beckmann, and H.W. Schrötter, J. Raman Spectrosc. 17, 143 (1986)

[2] T.S. Bican, J. Jonuscheit, U. Lehner, and H.W. Schrötter, J. Raman Spectrosc., in press.

[3] J. Jonuscheit, U. Lehner, D. Illig, S. Anders, K. Sarka, and H.W. Schrötter, J. Mol. Structure 349, 389 (1995).

[4] S. Zeindl, K. Sailer, F. Bauer, H.W. Schrötter, and A. Ruoff, J. Mol. Structure 350, 185 (1995).

[5] R. Angstl, H. Finsterhölzl, H. Frunder, D. Illig, D. Papoušek, P. Pracna, K. Narahari Rao, H.W. Schrötter, and Š. Urban, J. Mol. Spectrosc. 114, 454 (1985)

[6] V.N. Markov, A.S. Pine, G. Buffa, and O. Tarrini, J. Quant. Spectrosc. Radiat. Transfer 50, 167 (1993).

[7] R.L. Poynter and R.K. Kakar, Astrophys. J, Supp. 29, 87 (1974).

[8] G. Guelachvili, A.H. Abdullah, N. Tu, K. Narahari Rao, Š. Urban, and D. Papoušek, J. Mol. Spectrosc. 133, 345 (1989).

[9] D. Papoušek and Š. Urban, private communication (1984).

[10] H. Sasada, J. Mol. Spectrosc. 83, 15 (1980).

Laser-Photolysis of Organic Esters at 193 nm and 248 nm: Products and Quantum Yields

M.Wäsche, U.Brückner and E.Linke

Humboldt-Universität zu Berlin, Institute of Chemistry

D-12484 Berlin

Acetates, the esters of acetic acid, are excellent solvents of great technical importance. Due to their high vapour pressure, they are also atmospheric pollutants, which makes it necessary to develop ways for their destruction in the gaseous state. Efficient photochemical purification processes of gas mixtures represent an environmentally compatible alternative to thermal and plasmachemical degradation processes. Due to the precise control of the radiation parameters lasers can be readily adapted to specific problems. UV-photolysis is used to examine some general and specific aspects of excimer laser induced chemical reactions in the gas phase. The photolysis of different acetates with laser radiation at 193nm and 248nm is studied using FTIR-spectroscopy and photoionisation mass spectroscopy. In first experiments the formation of stable end products and the decomposition of parent molecules were analysed. From literature only laser photolysis experiments of acetates with IR multiphoton dissociation are known so far (HINTSA). The aim of this work is to give results of the UV-photodecomposition of methyl-, ethyl- and butyl acetate in dependence on laser wavelength, laser fluence, reactant pressure and buffer gas pressure.

Experimental

The gaseous samples were photolyzed in a closed stainless steel reaction cell at room temperature by using defocused pulsed radiation of a multigas excimer laser (EXC 200, ATL-Lasertechnik & Acc. GmbH, Berlin) operating at 248 nm and 193 nm, respectively. Pulse repetition rate and fluence are typically 1 - 10 Hz and 10 - 120 mJ/cm^2, respectively. The beam of the laser is directed into the reaction cell through a quartz window. The energy of the laser pulse before and after crossing the reaction cell is measured with pyroelectric detectors (PEM 50, Radiant Dyes Laser & Acc. GmbH). The pressure range in the reaction cell was varied between 50 Pa - 1500 Pa. The cell is equipped with NaCl windows for product and reactant analysis by means of FTIR spectroscopy. For a quantitative analysis the reactant and product concentrations were recorded for each gaseous compound. The FTIR spectrometer (M1200, MIDAC) typically operated with a resolution of 2 cm^{-1} and a DTGS detector.

Studies of the fragmentation mechanisms of the acetate molecules were carried out by electron impact and also by photoionisation mass spectroscopy in a commercial quadrupole mass spectrometer (Balzers QMG 421). For this purpose we used the focused excimer laser radiation for the ionisation of the molecules in the cross beam source of the spectrometer .

All acetates were obtained from commercial suppliers with purities better than 99,0% and were degassed by extensive freeze-pump-thaw cycles.

Results and discussion

The decomposition of the methyl-, ethyl- and butyl acetate in dependence on the number of laser shots at 193 nm is shown in Figure 1. The laser fluence was 75 mJ/cm^2. The reactant depletion was measured by the decrease of the ν_{10} (C - O stretch) mode of vibration signal from FTIR. There are no marked differences between the different acetates. The reactant depletion at 248 nm is not so strong, which can be explained by the lower absorption cross-sections and by the lower amount of excess energy. The measured integrated UV absorption cross sections of laser wavelengths 193 nm and 248 nm used in the overall quantum yield determination of the above molecules are summarized in Table 1.

Table 1: Integrated absorption cross-section values (cm^2 / molecule)

Molecule	193 nm	248 nm
methyl acetate	$2,1 \times 10^{-19}$	$0,7 \times 10^{-19}$
ethyl acetate	$2,2 \times 10^{-19}$	$0,6 \times 10^{-19}$
butyl acetate	$2,6 \times 10^{-19}$	$0,6 \times 10^{-19}$

Typically about 1000 laser shots were used to decompose about 32% of the acetate. The overall quantum yield (dissociated acetate molecules per absorbed photon) was found to be 0,5 in the initial range of the decomposition at 193 nm. Parallel secondary dissociation reactions and formation of products take place.

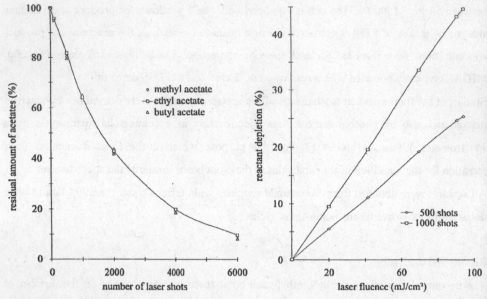

Figure 1. Decomposition of acetates at
193 nm, laser fluence 75 mJ/cm^2

Figure 2. Decomposition of ethyl acetate
at 193 nm in dependence on
laser fluence

To check for multiphoton absorption processes contributing to the quantum yields, the laser fluence was varied as shown in Figure 2. The decomposition efficiency varied linearly with the laser fluence, indicating in this region a single photon nature of the acetate decomposition at 193 nm.

From the FTIR and photoionisation mass spectroscopic measurements follows that each photon absorbed is able to break the C-H , C-C as well as the C-O bonds of the acetate molecules. Energetically this is not surprising, since a 193 nm photon carries more energy (148 kcal/mol, 6,42eV) than the heats of formation needed for the possible reactions. The present experiments are not able to decide which of the bond types breaks preferentially.

Information on the stable end products from the 193 nm and 248 nm photolysis of methyl-, ethyl- and butyl acetate were obtained by FTIR spectroscopy. We observed the following stable end products: CH_4, CO_2, CO, H_2O, for methyl acetate and CH_4, CO_2, CO, H_2O, C_2H_2, C_2H_4 for ethyl and butyl acetate. They give hints about possible photolysis mechanisms.

Addition of nitrogen as a buffer gas and oxygen as a reaction gas , respectively, has an influence on the photodecomposition. The gas pressure in the photolysis cell was changed from 1,3 kPa - 13,3 kPa. Nitrogen leads to a decrease, i.e. the acetate molecules are stablized by collisions with nitrogen. Oxygen, on the other hand, leads to an increase of the decomposition. By the addition of O_2 an extra photooxidative decomposition takes place. This effect is strongest for butyl acetate, i.e. the longer the chain the more effective is the attack by oxygen. The formation of oxy- and peroxyradicals in the presence of oxygen is possible. The stable end products are qualitatively the same as those produced without reaction gas. There are quantitative differences. CO_2 is the dominant product.

The photolysis fragments at 193 nm measured by laser ionisation in a quadrupole mass spectrometer suggest that the cleavage of the CH_3CO - OX bond (X stands for the alkyl radicals CH_3, C_2H_5, and C_4H_9) is the preferred dissociation pathway , as indicated by the intense signal at m/e=43 (CH_3CO^+). Studies of intermediates will permit a better understanding of the reaction mechanisms.

Work is in progress to complete the identification and the quantitative analysis of the end products and to investigate the primary reactions by time-resolved techniques.

References

HINTSA, E.J.; WODTKE, A.E.; LEE, Y.T.   J.Phys.Chem. **1988**, 92, 5379

Characterization of Internal Particle Structures by Dynamic Light Scattering

J. GIMSA, P. EPPMANN, B. PRÜGER, and E. DONATH
Inst. of Biology, Humboldt-University, Invalidenstr. 42, D-10115 Berlin, Germany
Correspondence should be sent to Jan Gimsa.

Summary

Available dynamic light scattering (DLS) methods determine only the size and surface charge of the particles since information is obtained from an analysis of the motion arising from thermal noise or electrophoretic force. We propose a new method combining DLS with dielectric single particle spectroscopy [GIMSA et al. 1995]. This method exploits the fact that also ac-fields of the radio frequency range may induce particle motion. In such fields the particle's polarization is a function of field frequency. Depending on the external field properties translation (dielectrophoresis) or rotation (electrorotation) may be induced. In electrorotation the dielectric properties of particles subject to a rotating electric field may be obtained from the frequency spectrum of individual particle rotation. Given a known geometry, dielectric constants and conductances of the particle's substructures can be obtained from the electrorotation spectrum.

Brief history

At the beginning of this century physicists studied effects of rotating electric fields intensively [LAMPA 1906]. Calculations were even extended to molecular dipole relaxations by BORN [1920] and experimentally proven by LERTES [1921]. 1982 rotating fields were introduced in biology as a dielectric cell spectroscopy method [ARNOLD and ZIMMERMANN 1982]. It was shown that electrorotation is suitable for determining specific properties of shells and internal media e.g. membrane capacitances and conductances, as well as internal conductivities and permittivities [FUHR et al. 1986]. The method was applied to investigate physiologically relevant phenomena. For a review on biological applications see [GIMSA et al. 1991, HUANG et al. 1992]. Only few investigations were carried out to characterize colloidal particles [ARNOLD et al. 1987].

Dielectric particle spectroscopy

For any particle suspended in a medium of different dielectric properties an external electric field induces a charge displacement described by the polarization vector, p. The dielectrophoretic force experienced by particles in a non-uniform electric field depends on field gradient and frequency. In case of electrorotation p rotates at the frequency of the external electric field. For sufficiently high field frequencies p follows the field with a delay. In this case the interaction of the induced polarization with the external field generates a torque depending on field frequency and relaxation time of the polarization process. It can be shown that the dielectrophoretic force and the electrorotational torque are proportional to the real and imaginary part of p, respectively. Since they are related by Kramers-Kronig-relation they principally carry the same information. In practice, the advantage of electrorotation is that the relaxation times of polarization processes are directly expressed as rotation maxima in the electrorotational spectrum. Analytical models have been developed for multi-shelled objects of spherical and cylindrical [FUHR et al. 1985, SAUER and SCHLÖGL 1985, PASTUSHENKO et al. 1985] as well as of ellipsoidal geometry [PAUL and OTWINOWSKI 1991]. Recently, micro-technological chambers were developed for particle characterization [FUHR et al. 1994].

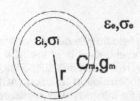

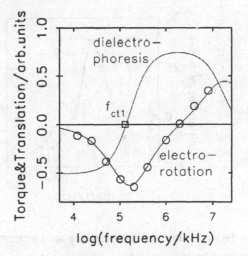

Fig. 1
Spherical single-shell model
of a membrane covered particle.

Fig.2
Electrorotational spectrum (O) measured microscopically on human erythrocytes. An external conductivity of σ_e=12 mS/m was used. For the fit a radius r=3.1 μm and dielectric constants of ε_e=80, and ε_i=50 were assumed. A membrane capacitance of C_m=6.2x10^{-3} S/m^2, a specific membrane conductance of g_m=10 S/m^2, and an internal conductivity of σ_i=0.17 S/m were obtained. The theoretical dielectrophoretic spectrum plotted for the obtained particle parameters is also presented. f_{ct1}, the frequency were dielectrophoretic movement ceases (□) was measured on 10 cells. Standard deviations in x-direction are too small to be shown.

Fig. 2 shows the dielectrophoretic and electrorotation spectrum for a single-shell model, a sphere covered by a low conductive membrane. Measured points are from human erythrocytes, suspended in an isotonic sucrose solution. Measurements were carried out as described in [GIMSA et al. 1991]. A field strength of 7 kV/m was used to induce a particle rotation speed in the order of one revolution per second. Negative rotation values reflect a sense of rotation opposite to that of the field.

Electrorotation measured by DLS

Microscopical measurements of electrorotation on single particles limit the particle size to the microscopic range. Long measuring times are required for statistical significance. We applied DLS to overcome these drawbacks [BERNE and PECORA 1976, BLOOMFIELD 1985]. For first measurements human red blood cells were used since they show significantly higher rotation speeds at the highest field strength available to date than non-biological particles. For electrorotation measurements the cell suspension was filled into an optical 4-electrode-chamber (for details see [PRÜGER et al., in this issue]). Since electrorotation adds an additional particle motion to Brownian motion a steeper decay of the ACF must be expected. We used this decay as a measure of the electrorotation speed.

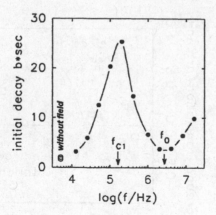

Fig.3
Rotating field frequency dependence of the initial decay, b, of the ACF of a human red blood cell suspension. The autocorrelation function (ACF) of the light intensity was measured at an angle of 10°. b was obtained from a straight line fitted to the first 30 ms of the logarithm of the ACF. A volume concentration of 0.02% ensured an independent movement of the particles. The field strength was about 14 kV/m and the external conductivity σ_e was 12 mS/m.

The decay values in Fig.3 resemble an electrorotational spectrum. The steepest decay corresponds to the anti-field rotation peak (compare to Fig.2). The frequency where the particle rotation ceases can be found around 2.5 MHz. In the range of the co-field rotation peak again steeper decays are found. The "decay value spectrum" agrees with results obtained from microscopic observation proving that the measured effect is clearly related to particle rotation. Therefore DLS measurements allow recalculation of dielectric particle properties. For a better understanding of the obtained correlation function first considerations for a simple particle model, carrying only one scattering point, were made [Gimsa et al. 1995]. We believe that the new method offers a potentially wide scale of applications on colloidal particles, vesicles, microemulsions, subcellular particles, cells, etc..

Acknowledgment
This study was supported by grant Gi 232/1-1 from Deutsche Forschungsgemeinschaft. Mrs. Ch. Mrosek and Ms. S. Klawitter provided excellent technical assistance.

References
Arnold, W. M., H.P. Schwan, U. Zimmermann, 1987. Surface conductance and other properties of latex particles measured by electrorotation. J. Phys. Chem. 91:5093.

Arnold, W. M., U. Zimmermann, 1982. Rotating field induced rotation and measurement of the membrane capacitance of single mesophyll cells of Avena sativa. Z. Naturforsch. 37c:908-916.

Berne, B. J., R. Pecora. Dynamic Light Scattering, Wiley, New York, 1976.

Bloomfield, V.A., Biological applications. in: R. Pecora (Ed) Dynamic Light Scattering, Plenum Press, New York, 1985, p. 363-416.

Born, M., 1920. Über die Beweglichkeit der elektrolytischen Ionen. Z. Phys. 1:221-241.

Fuhr, G., J. Gimsa, R. Glaser, 1985. Interpretation of electrorotation of protoplasts. studia biophysica 108:149-164.

Fuhr, G., R. Glaser, R. Hagedorn, 1986. Rotation of dielectrics in a rotating electric high-frequency-field. Model experiments and theoretical explanation of the rotation effect of living cells. Biophys. J. 49:395-402.

Fuhr, G., T. Müller, Th. Schnelle, R. Hagedorn, A. Voigt, S. Fiedler, W.M. Arnold, U. Zimmermann, B. Wagner, A. Heuberger, 1994. Radio-frequency microtools for particle and live cell manipulation. Naturwissenschaften 81:528-535.

Gimsa, J., R. Glaser, G. Fuhr, 1991. Theory and application of the rotation of biological cells in rotating electric fields (electrorotation). in: W. Schütt, H. Klinkmann, I. Lamprecht, T. Wilson (Eds) Physical Characterization of Biological Cells, Verlag Gesundheit GmbH, Berlin, p. 295-323.

Gimsa, J., B. Prüger, P. Eppmann, E. Donath, 1995. Electrorotation of particles measured by dynamic light scattering - a new dielectric spectroscopy technique. Coll. Surfaces. in press.

Huang, Y., R. Hölzel, R. Pethig, X.-B. Wang, 1992. Differences in the AC electrodynamics of viable and non-viable yeast cells determined through combined dielectrophoresis and electrorotation studies. Phys. Med. Biol. 37:1499-1517.

Lampa, A., 1906. Über Rotationen im electrostatischen Drehfelde. Wiener Ber. 115/2a: 1659-1690.

Lertes, P., 1921. Der Dipolrotationseffekt bei dielektrischen Flüssigkeiten. Zeitschr. Physik 6. 56-58.

Pastushenko, V.Ph., P.I. Kuzmin, Yu.A. Chizmadshev, 1985. Dielectrophoresis and electrorotation: a unified theory of spherically symmetrical cells. studia biophysica 110:51-57.

Paul, R., M. Otwinowski, 1991. The theory of the frequency response of ellipsoidal biological cells in rotating electrical fields. J. Theor. Biol. 148:495-519.

Sauer F.A., R.N. Schlögl, 1985. in: A. Chiabrera, C. Nicolini, and H.P. Schwan (Eds), Interactions between Electromagnetic Fields and Cells, Plenum Publishing Corp., New York, p. 203.

Dynamic Light Scattering Applied to Electrorotation: Measurement of Internal Particle Properties and Layer Capacitance

B. PRÜGER, P. EPPMANN, J. GIMSA, and E. DONATH

Inst. of Biology, Humboldt-University, Invalidenstr. 42, D-10115 Berlin, Germany

Correspondence should be sent to Bernhard Prüger.

Summary

Dynamic light scattering (DLS) is widely used to characterize the properties of colloidal particles. We combine DLS with electrorotation spectroscopy [see GIMSA et al. in this issue]. In electrorotation individual particle rotation is induced by a rotating radio-frequency field. Electrorotation is sensitive to the particle's internal dielectric structures. Our measurements on human red blood cells give the same results as electrorotational spectra measured microscopically. From characteristic points of the spectrum, values for the cell's membrane capacitance and internal conductivity can be obtained. We found the need of high field strength and low particle concentration to get a clear separation of the rotation from other field-induced particle motions.

Experimental setup

To perform electrorotation measurement by means of DLS, it was necessary to design a new measuring chamber, meeting the conditions of optical tranparency for the laser beam and allowing the application of driving signals for the rotating field (Fig. 1).

The chamber electrodes were driven by four, 90° phase-shifted, symmetrical square wave signals of maximum 20 V_{pp} amplitude and variable frequency. The distance between two opposed electrodes was about 1.4 mm. The special electrode arrangement induces a rotating field in a plane perpendicular to the optical axis. In the horizontal plane, the cuvet allowed detection of scattered light over a wide anglar range. The scattered light intensity at a fixed detection angle (θ, set by a pinhole) was determined by a photodiode detector. For a time dependent correlation analysis [BERNE and PECORA 1976, OSTROWSKI 1993] in the digital correlator card the intensity signal was converted into TTL-pulses by a voltage-controlled oscillator (VCO).

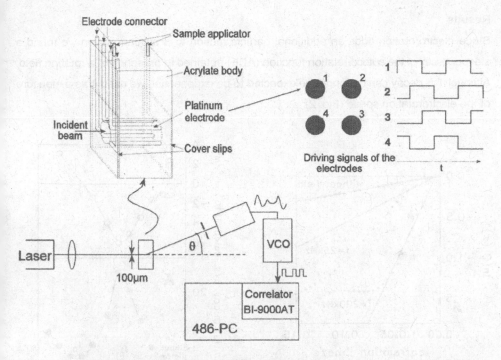

Fig. 1 Experimental setup with a view on the electrorotation chamber. The four 90° phase-shifted driving signals of the chamber electrodes create a rotating electrical field.

Measurements

Human red blood cells were suspended in a 300 mOsmol sucrose solution to a volume concentration of 0.02%. This dilution ensured independent movement of the red blood cells. For laser optical measurements, the red blood cell suspension was introduced into the chamber by a syringe. As known from microscopic observation, the oblate red blood cells orient in rotating electric fields. They rotate about the small semi-axis which aligns with the axis of the rotating field and, hence, with the incident beam. Therefore, their rotation axes are parallel to the beam. The effective detection angle was set to $\theta = 10°$. At this angle, the detected rotation effect was at maximum. The effective scatter volume of 0.04 mm^3 corresponds to a mean particle number of about 40. A runtime of 60 s was chosen, and the sample time, $\Delta\tau$, which denotes the smallest decay-step in the ACF, was set to 5 ms. Due to its channel number of 236, the correlator uses a maximum correlation depth of 1.18 s.

Results

Since electrorotation adds an additional particle motion to Brownian motion we found a steeper decay of the autocorrelation function (ACF) obtained in presence of a rotating field. Although this decay can no longer be expected to be exponential, we used it as a measure of the electrorotation speed (Fig. 2).

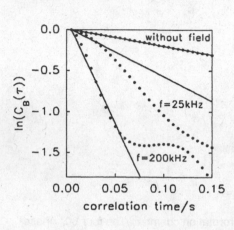
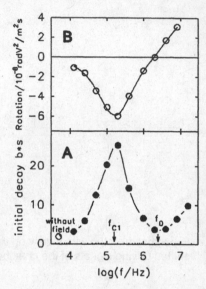

Fig.2 Experimental results on a suspension of human red blood cells (suspension conductivity $\sigma_e = 12$ mS/m) Left: Logarithm of baseline corrected correlation functions (dots) measured at two different field frequencies and without field. The field strength was about 14 kV/m and the detection angle was set to $\theta=10°$. The straight lines denote the initial decay, b, used as a measure for the particle rotation. Right: Rotating field frequency dependence of the initial decay (A) and the corresponding electrorotational spectrum measured microscopically (B), see [GIMSA et al. 1991].

The frequency dependent decay values in Fig.2 reflect the characteristic frequencies of the electrorotational spectrum. The steepest decay at $f_{c1} = 170$kHz corresponds to the anti-field rotation peak. The frequency where the particle rotation ceases can be found at $f_0 = 2.3$ MHz. Since the current setup cannot detect the rotation sence, in the range of the co-field rotation peak again steeper decays are found. Assuming a neglectible membrane coductivity, from the single-shell model [GIMSA et al. 1991], the following particle parameters can be obtained from the DLS-measurements: Membrane capacitance $C = 5.9 \times 10^{-3}$ F/m^2 and internal conductivity $\sigma_i = 0.18$ S/m.

For a better understanding of the ACF, theoretical considerations of a simple particle model, carrying only one scattering point were made [Gimsa et al. 1995]. These describe the rotational part of the ACF as a zero order Bessel term (similar to an expression found in [Racey et al. 1981]) and could qualitatively explain the rotation induced decay of the ACF. Theoretically an optimum detection angle could be derived for our particle size of 3.1µm. It agreed with the angle at which the influence of particle rotation on the measured ACF was at maximum. The model also allows to estimate a lower particle size limit of about 300 nm for the method. We believe, that more realistic models including rotational diffusion will allow a quantitative expression of the ACF and a direct determination of the rotation speed.

When particle concentrations higher then 0.02% were used the electrorotation peaks were not as clearly reflected in the decay-value spectrum. This was caused by particle-particle interaction induced additinal motions [SAUER 1985]. On the contrary, at lower concentrations, number fluctuations cause large variations to the measured correlation decay. These considerations suggest for every object an optimum particle concentration. To clearly observe particle rotation a high field strength (>10 kV/m) should be used.

Improvement of the method is needed to obtain the absolute speed and direction of the rotation.

Acknowledgment

This study was supported by grant Gi 232/1-1 from Deutsche Forschungsgemeinschaft.

References

Berne, B. J., R. Pecora. Dynamic Light Scattering, Wiley, New York, 1976.

Gimsa, J., P. Marszalek, U. Loewe, and T.Y. Tsong, 1991. Dielectrophoresis and electrorotation of neurospora and myrine myeloma cells. Biophys. J., 60:749.

Gimsa, J., B. Prüger, P. Eppmann, E. Donath, 1995. Electrorotation of particles measured by dynamic light scattering - a new dielectric spectroscopy technique. Coll. Surfaces. in press.

Ostrowski, N., 1993. Liposome size measurements by photon correlation spectroscopy. Chemistry and Physics of Lipids 64:45.

Racey, T. J., R. Hallett, B. Nickel, 1981. A quasi elastic light scattering and cinematographic investigation of motile Chlamydomonas Reinhardtii. Biophys. J., 35:557.

Sauer, F.A., 1985. in: A. Chiabrera, C. Nicolini, and H.P. Schwan (Eds), Interactions between Electromagnetic Fields and Cells. Plenum Publishing Corp., New York

One- and two Photon Excited Fluorescence of Motile Cells in the Optical Trap

K. König[1,2], B.J. Tromberg[2], M.W. Berns[2]

[1]Institute for Molecular Biotechnology, D-07708 Jena, [2]Beckman Laser Institute, Irvine 92715, California

ABSTRACT

Motile human spermatozoa were hold in a single-beam gradient force optical trap ("optical tweezers"). Xeno- and autofluorescence of optically-trapped cells were excited with 488 nm laser microbeams or 365 nm mercury lamp radiation (one-photon excitation). In addition, two-photon NIR excited fluorescence was induced by the cw trapping-beam (100 mW). Non-resonant two-photon absorption is responsible for pathophysiological effects.

INTRODUCTION

Healthy human sperm cells achieve velocities of more than 50 μm/s. In order to perform intracellular fluorescence measurements on these highly motile cells, optical traps can be employed to hold vital cells in a fixed position [1-3]. The trapping forces generated by the intracellular refraction of cw near infrared (NIR) microbeams have to be higher than the ATP-driven motility forces which are in the pN range. This condition requires laser powers of more than 100 mW. A highly focused cw 100 mW trapping beam at 760 nm (NA = 1.3) leads to intensities of about 40 MW/cm^2, and thus, to photon flux densities of 10^{27} photons cm^{-2}s^{-1}. With typical two-photon absorption cross sections of 10^{-48} to 10^{-50} cm^4 s photon^{-1} molecule^{-1} [4] these photon flux densities in optical traps are sufficient to induce nonlinear absorption effects including two-photon excited fluorescence.

MATERIALS AND METHODS

The NIR radiation (trapping beam) of an Ar$^+$-ion laser-pumped tunable cw Ti:Sapphire ring laser (Coherent, model 899-01) was introduced into a modified inverted confocal laser scanning microscope (CLSM, Axiovert 135M, Zeiss). The parallel beam was expanded to fill the back aperture of a 100x Zeiss Neofluar brightfield objective (NA=1.3). Laser scan images were obtained with 488 nm microbeams (2.2 μW) and a scanning time of 1s per image. 365 nm (mercury lamp) excited autofluorescence was detected with a slow-scan, thermoelectrically cooled CCD camera (Princeton Instr., model TE576/SET135), Fig. 1.

Semen was washed once to remove seminal fluid by 200g centrifugation. The pellet was layered with fresh human tubal fluid and allowed to swim up for 20 min. The supernatant containing sperm was then diluted in HEPES buffer containing 1% HSA. Spermatozoa were injected into microchambers. Experiments were performed within three hours following ejaculation at a room temperature of 29°C.

For viability analyzing, cells were incubated with the LIVE/DEAD FertiLight™ Sperm Viability Kit (Molecular Probes) 5 min before experiments. The kit contains the live-cell stain SYBR™14 (100 nM) with a 515 nm emission maximum and the dead-cell stain propidium iodide (12 µM) with a 636 nm peak.

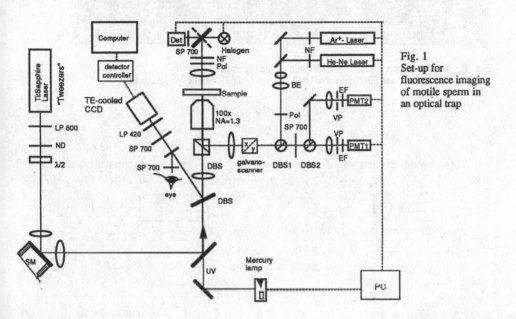

Fig. 1
Set-up for fluorescence imaging of motile sperm in an optical trap

RESULTS

One-photon excited fluorescence

365nm-excited autofluorescence images of sperm cells optically hold in 800 nm single-beam gradient force optical traps are seen in Fig. 2. The NADH-attributed autofluorescence of an intact motile sperm (5 s exposure time) is located in the mid-piece of the cell, the primary site of mitochondria. When exposed to the fluorescence excitation radiation (1 mW) for some minutes, the intensity increased and the sperm head became the major fluorescence site. Interestingly, 760 nm traps alone (no simultaneous UVA irradiation) induced similar fluorescence modifications.

Cell labeling with the Live/Dead kit allowed laser scanning fluorecence microscopy of optically-trapped spermatozoa. 488 nm excited fluorescence was recorded with photomultipliers in two spectral ranges 510-525 nm and 630-700 nm, respectively. Red fluorescence, indicating cell death, began to appear in the lower head section and spread with time throughout the entire head. Onset of red fluorescence occurred after a mean trapping time of 1 min in 760 nm traps (no UVA), after 5 min in 800 nm traps with simultaneous UVA exposure, and was not obtained for 800 nm traps alone (no UVA) up to 10 min.

Fig. 2   Autofluorescence images of an 800 nm trapped sperm cell
prior to (left) and after additional UVA exposure (right), middle: phase contrast

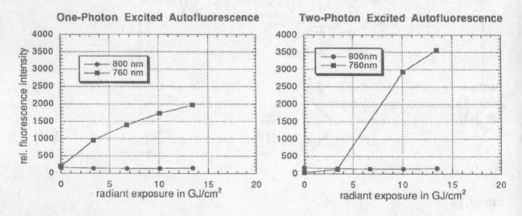

Fig. 3   Autofluorescence intensity of 760 nm-trapped sperm in dependence on NIR radiant exposure

Fig. 4   Images of 760 nm trapped sperm during white light illumination (left), UVA exposure (middle),
and with no external light source (NIR beam only, right)

Two-photon excited fluorescence

Optically-trapped cells exhibited a small visible fluorescence spot with sub-micrometer dimensions in the lower head part without additional excitation sources. The fluorescence was induced by the NIR trapping beam itself and can be explained by non-resonant two-photon excitation. Indeed, control measurements with different fluorophore solutions, such as Rhodamine 123, Acridine Orange, or Laurdan, confirm a squared dependence of fluorescence intensity on NIR laser power.

The spot position indicates the intracellular microbeam localization. We found changes in intracellular beam localization due to the superposition of trap-induced forces and variable motility forces.

Cell damaging effects of 760 nm traps were also detected by two-photon excited fluorescence. In 760 nm trapped cells labeled with the Life/Dead kit we observed a change in color of the fluorescence spot from green to red within 2 min of trapping.

Fig. 3 shows one-photon excited (UVA) and two-photon excited (NIR) autofluorescence intensity of 760 nm/800 nm trapped sperm cells in dependence on NIR radiant exposure (P = 105 mW). The two-photon intensity represents the maximum pixel signal, the one-photon intensity the maximum pixel signal out of the two-photon excited fluorescence region. Exposure time was 5 s. No fluorescence modifications occurred for 800 nm traps. In 760 nm traps, the one-photon fluorescence signal increased rapidly and achieved 5fold values at 5 GJ/cm^2 whereas the two-photon signal increased only 1.5fold. However, with increasing trapping time the two-photon excited autofluorescence became more intense.

Fig. 4 shows a transmission image (halogen lamp), the UVA-excited fluorescence image, and the NIR-induced fluorescence image of 760 nm trapped sperm incubated with SYBR[TM]14. The highly fluorescent spot in all images indicates the trapping beam induced two-photon excited fluorescence.

CONCLUSION

Optical trapping allows one- and two-photon excited fluorescence studies of single motile cells. NIR optical traps are thought to be harmless tools in cell micromanipulation and used in laser-assisted in vitro fertilization. However, short-wavelength traps may induce cell damage via nonlinear absorption. Excitation of biomolecules absorbing in the UVA/blue is possible with highly focused NIR continuous wave lasers.

REFERENCES
[1] Ashkin, A. et al. *Opt. Lett.* 11 (1986) 288-290.
[2] Tadir, Y. et al. *Fertil. Steril.* 52 (1989) 870-873.
[3] Greulich, K.O., Weber, G. *J. Microscopy* 167(1992)127-151.
[4] Jiang, S. *Prog. React. Kinet.* 15 (1989) 77-92.

Metal-to-Ligand Charge Transfer Spectra forTransition Metal Complexes of Chelated Aromatic Ligands: Theory and Experiment

J.-Y.Zhou,T.Luo,W.-L.She,X.-G.Huang,W.-J.Peng
Ultrafast Laser Spectroscopy Laboratory,Zhongshan University,
Guangzhou 510275,China.

J.-Z..Wu,L.-N.Ji
Department of Chemistry,Zhongshan University,Guangzhou 510275,China

1. Theory

Complexes of transition metal ions with aromatic.ligands,notably,Ruthenium(2+) complexes with tris-2,2'-pyridine(bpy) or tris-1,10-phenanthroline(phen), find important applications to photosensitization, photocatalization[1] and recently, to the determination of DNA secondary structures[2]. Metal-to-transfer(MLCT), in which an electron is optically excited from a metal orbital, d, to a ligand orbital, π^*, is a frequently encountered optical process involved in inorganic phtochemistry.However,theory has not been available to our knowledge to successfully explain the absorption spectra without using adjustable parameters to fit the experimental results.In this part, we discuss the nature of the electronic states of the metal complexes from a quantum mechanical approach.

An approximation of a single electorn in a non-zero periodic potential is made.The cyclic boundary conditions are used in the schrodinger equation to determine the eigenvalues of the electron.It is found that the spectral structures derived from the theory by using the metal complex structure data as input are in agreement with the main features of the absorption spectra of Ru $^{2+}$ complexes. We attribute that the agreement is due to consideration of electronic tunneling through metal ions and through the ligands.

To a first approximation, we assume that an electron be in a periodic square wave potential, i.e., Kronig-Phenney potential. The expression of the potential field is

$$V(x) = \begin{cases} 0, & \text{for } 0 \leq x \leq a \\ V_0 & \text{for } a \leq x \leq l \end{cases} \tag{1}$$

T where l=a+b, with a and b the length of the well and of the barrier, respectively. V_0 is the barrier height.

The electronic wavefunction, Φ is determined by

$$-\frac{\hbar^2}{2\mu} \frac{d^2\Phi}{dx^2} + V(x)\Phi(x) = \varepsilon\Phi(x) \tag{2}$$

with h the Planck contant, μ the mass of the electron, ε the electron energy,which can be greater or smaller than the barrier height.

The metal complexes studied in this work are Ruthenium(2+) complexes of tris-2,2-pyridine(bpy) and tris-1,10-phenanthroline(phen) i.e., Ru(bpy) $_3^{2+}$ and Ru(phen) $_3^{2+}$.Structure information of the metal complexes are used as input and the parameters a, b and V_0 are given for the following reasons. The length of the well is assumed to be slightly greater than twice of the covelent radius of Ru $^{2+}$, i.e.,a=(2*129)pm, as the nitrogen-Ru $^{2+}$ distance of the complex is slightly greater than the sum of the covelent radiuses of a nitrogen and of a Ru^{2+} ion. The depth of the well, V_0,is 2.1eV, obtained from the Ru^{2+} reduction potential[3] as well as potential well parameters.The barrier length is assumed to be equal to the shortest istance between the two chelating nitrogens in bpy or in phen, as the absorption wavelength does not depend so much on the derivitives of bpy or phen, but depends very much on the Ru^{2+} coordinating micro-structures of the ligands.Thus we have barrier lengths b=(3* 140+2*70)pm for bpy and phen as ligands.

The eigenstates determined with equation (1) and (2) have several unique characterizations. For an electron with the energy smaller than the barrier height, i.e., for the electron mainly with d electron character,there is only one state inside the well for given paramters discussed above.If the barrier length is varied, however, the eigenvalues change,indicating that ground state energy also depends on the length of the barrier.Thus the electronic tunneling effect may result in the splitting for ground eigenstate.For π^* electrons,i.e.,for the electrons mainly with the ligand character,there exists a number of eigenstates.The spectral structures of MLCT for $Ru(bpy)_3^{2+}$ and $Ru(phen)_3^{2+}$ can be determined by considering d and

π^* electron eigenstates. It is predicted that the firstmajor MLCT absorption peak should apear at the photon energy of 2.578eV, while the experimentally determined value is 2.7eV.Further, the theory predicts that the MLCT absorption should show two distinct absorption peaks in the visible with their separation being at 2270cm-1 The experimentally determined value is 1900cm-1.Thus the two prominent peaks appearing in the visible spectral region are arising from electronic transitions and they are not a vibronic structure accompanying a major electronic transition.The photon energy required to excite the MLCT to the next higher eigenstates are in ultraviolet spectral region, which is also in agreement with experimental data.

2.Experiment

We report the spectroscopic study for a series of mixed-ligand $[Ru(bpy)2L]^{2+}$ complexes. Here bpy=2,2'-bipyridine, L is imidazo[f]1,10-phenanthroline(IP) or 2-phenylimidazo[f] 1,10-phenanthroline(PIP) or PIP derivatives. The solvent used to dissolve the complexes is istilled water containing 5mM Tris,50mM NaCl,pH 7.0, which has been commonly regarded as a buffer in studying the binding of Ru\(II\) trischelates with DNA [2],Calf thymus DNA is used. All the spectroscopic measurement is conducted at room temperature.

The photophysical properties of the complexes are studied by ns and ps spectroscopy.The ns device involves an excimer laser pumped bye laser,a grating monochormator,a boxcar nd a computer-controlled data aquirezation system.The dye laser emission is at 450nm,which is at the absorption maximum of the MLCT transition for the Ru(II) complexes.

The study of the emission spectra with ns system reveals the wavelengths at the maximum intensity of the emission band,the intensities and the emissive lifetimes,which are listed in table 1 and are compared with those of $Ru(bpy)_3^{2+}$. It becomes obvious from table 1 that the effect of electronic donor group attached to the PIP ligand, Cl or -CH, is to increase the luminescence of the Ru(II) chelates while with an electronic acceptor group $-NO_2$ in the ligand, he luminecence is not detected when measured with ns laser system. We attribute the electron donor or acceptor dependent luminescence to the variation of electronic distribution for the antibounding π^* orbital in the ligand.

Table 1   Luminescence Properties of $[Ru(bpy)_2L]^{2+}$

L		bpy	IP	PIP	CLP	HOP
λ_o(nm)		617	625	615	616	606
$I_o \tau$	A	1.00	3.55	2.98	3.24	3.31
	B	1.11	17.33	21.28	21.46	14.12

λ_o is the emission wavelength. A represents the measured product of intensity I_o with the emission lifetime τ without DNA in solution and B represents that with DNA in solution at a concentration DNA/Ru=47. All the A and B values are calibrated in relative to $[Ru(bpy)_3^{2+}$M/L.

For all Ru(bpy)2L]$^{2+}$ complexes with an electronic donor group attached to L ligand the addition of calf-thymas DNA results in an increased emission intensities and excited state lifetimes,as shown in table 1.The product of intensity, I_0, with lifetime, τ_0, can be increased by up to an order of magnitude.Similar enhancement effect was observed in other mixed-ligand Ru(II) chelates and theincrease in the luminescence was attributed to the intercalating one ligand with DNA[4]. Bpy ligand has been demonstrated having no or little intercalation function, and we also observed the $I_0\tau$ value of Ru(bpy)$_3^{2+}$ only increased quite little in the presence of DNA compared to that in the absence of DNA, so the intercalating ligand is IP, PIP, CIP, HOP or NOP. The presence of an electronic donor group attached to the PIP ligand does not affect the intercalating binding mode of the Ru(II) complex to DNA.

References:

[1] K.Kalyanasundaram, Coord.Chem.Rev., 46(1982)159.
[2] C.V.Kumar,J.K.Barton,N.J.Turro,J.Am.Chem.Soc.,107(1985)5518.
[3] R.JWatts,J.Chem.Educ.,60(1983)834.
[4] J.Y.Zhou et.al.,Appl.spectr.,47(1993)2175.

Femtosecond Infrared Spectroscopy of Photosynthetic Reaction Centers

P. Hamm, M. Zurek, W. Zinth

Institut für Medizinische Optik

Ludwig Maximilians Universität München, Barbarastr. 16, 80797 München, Germany

Introduction:

The reaction center (RC) from purple bacteria is the key unit to perform the primary charge separation and energy storage in photosynthesis. Two basic experimental techniques have been applied to investigate the function of this unit: (i) The structure of bacterial RC's, a chromophore-protein complex, was determined by x-ray scattering with atomic resolution (ii) Time resolved spectroscopy in the visible or the near IR was applied to characterize the kinetics of the photoreaction as well as the electronic states of the chromophores. After light absorption, an electron is transferred from a dimer of bacteriochlorophyll molecules (special pair P) to a bacteriochlorophyll monomer B in ≈3-4 ps, to a pheophytin H in ≈1 ps and finally to a quinone Q in ≈200 ps. However, only one of the almost symmetric potential electron transfer (ET) branches seems to be essential for this reaction. This so-called 'unidirectionality of ET' is one of the remaining open questions of bacterial photosynthesis.

Therefore, in order to obtain supplement information on structural modifications of the chromophores as well as on protein residues surrounding the chromophores, on the reaction mechanism itself, and on the electronic structure of P* and P$^+$ time resolved vibrational spectroscopy in the mid-IR spectral range (1000-1800 cm^{-1}) was started only recently [1-4].

1. Experimental setup:

A laser system which enables femtosecond IR experiments in an extended frequency region (2000-1000 cm^{-1}) is described in detail in [5,6]. It is based on a standard Ti:sapphire regenerative amplifier system running at a repetition rate of 1 kHz. The RC's were excited by 150 fs pulses at 870 nm. The absorbance change of the sample was recorded by tunable probing pulses generated by difference frequency mixing between the output of the Ti:sapphire and that of a synchronized traveling wave dye laser in a $AgGaS_2$-crystal. The IR-pulses had a pulse duration of ≈300 fs and a spectral bandwidth of ≈60 cm^{-1}. For spectrally resolved detection, the probing pulses were dispersed in a grating spectrometer after passing the sample. A 10 element IR-detector array was used to cover the complete bandwidth of the probing pulses simultaneously.

2. Results and Discussion

A complete data set of the transient absorbance changes of the RC's from the purple bacterium *Rb. sphaeroides* covering the spectral range from 1000-1800 cm^{-1} and a temporal range of 0-1 ns was measured with a spectral resolution of 4-7 cm^{-1} and a temporal resolution of ≈300 fs. In the following, this data set is analyzed under two different aspects: (i) the temporal evolution of the absorbance changes at certain spectral positions and (ii) transient difference spectra at certain delay time positions.

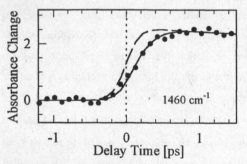

Fig. 1: Transient absorption signal measured at 1460 cm^{-1} in the maximum of an intense IR-absorption band. The absorption rise (solid line and dots) is delayed with respect to instrumental response function (dashed line). The delay can be modeled by a exponential function with a time constant of 200 fs.

2.1 Temporal evolution

Almost throughout the entire spectral range, an initial absorption increase occurs. However, a detailed analysis shows that in general, the signal rises slower than the temporal response function of the experimental setup [3]. This delay is particularly pronounced in certain spectral regions around 1460 cm^{-1} (see Fig. 1) and 1240 cm^{-1}. The delay can be fit well by assuming that the instantaneous absorption increase is negligible and rises with a subsequent time constant of 200 ± 100 fs; such a kinetic component was not detected before in VIS/NIR experiments.

The subsequent temporal evolution may be well described by two well known ET kinetic components: The dominant feature shows up the same time constant (3.5 ps) as the decay of the initial state P* while the $P^+H \rightarrow P^+Q^-$-ET step (220 ps) is observed with smaller amplitudes at certain spectral positions. The fast ET step from $P^+B^- \rightarrow P^+H^-$ (in 1 ps) which was observed in VIS/NIR measurements is not resolvable in the IR experiments. It can be concluded that the IR absorbance changes are mainly caused by modifications of the special pair P (or of the protein surrounding P) and not by the other chromophores. Within the dynamic range of the experiments, there is no evidence for further kinetic components which might be assigned to a slow protein relaxation due to a response of the protein on the changed charge distribution.

2.2 Transient difference spectra

In Fig. 2, transient difference spectra at distinct delay times at t_D=1 ps, 10 ps and 1000 ps are shown related to the intermediate states P*, P$^+$H$^-$ and P$^+$Q$^-$, respectively. The discussion of these spectra can be roughly divided into two spectral regions:

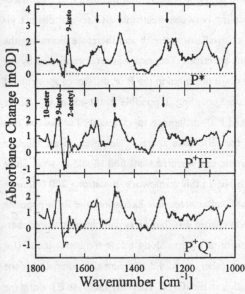

Fig. 2: Time resolved difference spectra at a delay time of 1 ps, 10 ps and 1000 ps where the RC's are predominantly in the states P*, P$^+$H$^-$, and P$^+$Q$^-$.

(i) C=O spectral range: In the spectral regions >1600 cm^{-1} absorption bands of the C=O groups of the chromophores and of the protein (Amide I band) are expected. Here, a detailed assignment of distinct absorption bands can be given [1,2]: In P* (1 ps), a red shift of the 9-keto carbonyl group is observed and the 10a-ester carbonyl group is not affected. However, already in the state P$^+$H$^-$, the 9-keto carbonyl group (1706 cm^{-1}/1684 cm^{-1}) and the 10a-ester carbonyl group (1752 cm^{-1}/1736 cm^{-1}) of the special pair are blue shifted just as observed in steady state FTIR spectra [1]. Also some weak bands corresponding to the 2-acetyl group can be identified around 1650 cm^{-1}. Only weak changes occur between P$^+$H$^-$ and P$^+$Q$^-$ in the carbonyl range. Due to the complex spectrum in the amide I region, no protein contribution can be identified unambiguously. Within the experimental error, the 1 ns difference spectrum and the steady state FTIR difference spectrum (according to a mean delay time of several milliseconds) are identical indicating that no further reactions occur.

(ii) The low frequency range: In the spectral regions <1600 cm^{-1}, the C=C, C-C, C-N stretching and C-H bending modes of the chlorophyll tetrapyrolle ring system are expected. Unfortunately, due to the lack of exact model calculations, a detailed assignment of difference bands to distinct modes is not yet possible. However, a striking feature is found especially in this part of the spectrum: a pronounced positive absorption background in connection with three strong broad positive bands which are observed in P* and P$^+$ at slightly different peak positions. In addition, as seen in section 2.1, these positive bands do not appear instantaneously after electronic excitation but within a delay time of ≈ 200 fs. Qualitatively, these observations are explained as follows: As a consequence of exitonic coupling between two chromophores to a dimer (the special pair P), low lying electronic transitions appear when the RC's populate P* or P$^+$ [3,4]. Due to the low excitation frequency of these electronic transitions (2000-3000 cm^{-1}), additional coupling between electronic and several distinct vibronic transitions is expected leading to a transfer of oscillator strength and therefore increasing the absorption cross section in the vibrational part of the spectrum. This explanation is confirmed by the observation, that a comparable absorption increase is not observed in the P$^+$/P-difference spectrum of the heterodimer mutants, where no effective excitonic coupling is possible in the dimer [3]. These arguments qualitatively explain the positive P*/P- and P$^+$/P-difference spectra as well as the fact that absorbance changes of the special pair dominate the difference spectra. However, in an absolute symmetrical special pair P, coupling between electronic and vibronic transition should be weak since no intramolecular charge separation within P is possible. In this framework, the strong 200 fs kinetic component can be understood as follows: Immediately after electronic excitation, the RC's populate an almost symmetric initial excited state with uniform charge distribution (small charge transfer character). Later on in parallel with the 200 fs component, a motion along a low frequency mode (for example a motion of both dimer halves against each other) occurs destroying the symmetry of the dimer and leading to an increased net charge separation in the dimer. This may facilitate ET along the active branch and therefore may be one reason for the 'unidirectionality of ET'.

References:
[1] P. Hamm, M. Zurek, W. Mäntele, M. Meyer, H. Scheer and W. Zinth, Proc. Natl. Acad. Sci. USA 92 (1995) 1826
[2] S. Maiti, G. C. Walker, B. R. Cowen, R. Pippenger, C. C. Moser, P. L. Dutton and R. M. Hochstrasser, Proc. Natl. Acad. Sci. USA 91 (1994) 10360
[3] P. Hamm and W. Zinth, submitted to Chem. Phys
[4] G. C. Walker, S. Maiti, B. R. Cowen, C. C. Moser, P. L. Dutton and R. M. Hochstrasser, J. Phys. Chem. 98 (1994) 5778
[5] P. Hamm, C. Lauterwasser and W. Zinth, Opt. Lett. 18 (1993) 1943
[6] P. Hamm, S. Wiemann, M. Zurek and W. Zinth, Opt. Lett. 19 (1994) 1642

Femtosecond Studies of Intramolecular Proton Transfer in the Condensed Phase

C. Chudoba, S. Lutgen, T. Jentzsch, M. Woerner, M. Pfeiffer, E. Riedle, and T. Elsaesser
Max-Born-Institut für Nichlineare Optik und Kurzzeitspektroskopie
D-12489 Berlin, Germany

Intramolecular proton transfer in the excited electronic state is an elementary photoreaction that often occurs on a subpicosecond time scale [1]. The compound 2-(2'-hydroxy-5'-methylphenyl)-benzotriazole (commercial name: TINUVIN P or TIN), a commercial ultraviolet stabilizer of polymers, performs an ultrafast reaction cycle after excitation to the S_1 state in the enol configuration as shown schematically in the insert of Fig. 1 (a) [2]. The enol-keto-type proton transfer in the excited state (time constant 100 fs) is followed by radiationless deactivation of the reaction product within 150 fs and proton back-transfer to the enol electronic ground state (600 fs). The total kinetics of the closed reaction cycle has been determined by femtosecond experiments at various wavelengths [3], whereas the rate constants pertaining to the keto-type states were previously only determined at a single wavelength. We now report for the first time spectrally and temporally resolved measurements over a wide wavelength range, providing the transient absorption and emission spectra of the keto-type reaction product. By the ultrafast proton transfer reaction and the following radiationless deactivation, a large amount of energy is transferred to the vibrational system of the reaction product. The broad spectral overlap of the keto-type absorption and emission bands between 600 and 700 nm gives direct evidence of non-equilibrium vibrational excitation of the keto-type tautomer. The proton transfer dynamics are determined by the shape of the potential surface on which the reaction proceeds. Varying the excitation from below the enol-S_0-S_1 transition to excess energies of about 5000 cm^{-1} does not influence the kinetics of the proton transfer cycle. This demonstrates a barrierless pathway of the reaction in the excited state.

Pump and probe pulses of 100 fs duration were derived from femtosecond white light continua generated with amplified pulses from a self-mode-locked Ti:sapphire laser. Frequency doubling of the continuum provides excitation pulses tunable between 350 and 400 nm, whereas probe pulses between 400 and 800 nm were generated by spectral selection and recompression. In Fig. 1, we present transient spectra measured with excitation at $\lambda_{ex} = 370$ nm, nearly resonant to the purely electronic S_0-S_1 transition of the enol tautomers at $v_{00} = 27\,260$ cm^{-1} [4]. On the left side of Fig. 1

(a), the steady state absorption spectrum of enol-TIN is plotted on a logarithmic scale. On the right side, the transient spectra of the keto-type reaction product are shown that have been derived from time-resolved data at the respective wavelength. The solid circles represent the small signal gain $G = \ln(T/T_0)$ of the probe pulses measured at a delay of 200 fs (T_0,T : transmission of the sample before and after excitation). After excitation of enol-TIN to the S_1 state, excited keto-type tautomers

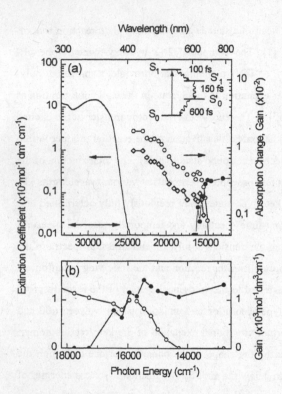

are created by proton transfer with a time constant of 100 fs. The population inversion between the excited state and the unpopulated ground state of the reaction product gives rise to the stimulated emission observed at early times. The emission band which agrees well with the steady-state emission spectrum, decays by internal conversion to the keto ground state with a time constant of 150 fs.

The open symbols in Fig. 1 (a) represent the induced absorption $\Delta A = -\ln(T/T_0)$ at delay times of 750 fs (circles) and 1250 fs (diamonds). The very broad absorption band between 400 and 700 nm originates from the keto-type ground-state S_0'. The shape of the absorption band does not vary for delay times greater than 750 fs.

The transient emission and absorption bands of the keto-type molecules show a strong spectral overlap. In Fig. 1 (b), the molecular extinction and gain coefficients of the induced absorption and stimulated emission in the overlap region are shown, respectively. We observe almost identical gain and extinction coefficients, which demonstrates that (1) the absorption and the emission originates from the same electronic transition ($S_1' - S_0'$) of keto-type TIN

Fig. 1 (a) Steady state absorption band of enol-TIN dissolved in cyclohexane (solid line), and spectra of transient absorption $\Delta A = -\ln(T/T_0)$ (open symbols) and gain $G = \ln(T/T_0)$ of keto-type TIN (T_0, T: transmission of the sample before and after excitation at $\lambda_{ex} = 370$ nm). The emission spectrum was recorded at a delay time of 200 fs, the transient absorption was measured with delays of 750 fs (circles) and 1.25 ps (diamonds). The arrow on the abscissa marks the range of excitation energies.
Insert: Schematic of the ultrafast reaction cycle of TIN.
(b) Molar coefficients of absorption and stimulated emission of keto-TIN on an extended energy scale (linear coordinates). Note the well pronounced overlap of absorption and emission spectra.

and (2) that absorption from the excited keto-type state to higher lying singlet levels is negligible in the range of the emission band. The energetic width of the spectral overlap amounts to approximate-

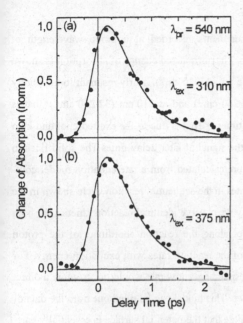

ly 2000 cm^{-1}. This fact is direct evidence for strong vibrational excitation of the keto-type tautomers, resulting in substantial populations of high lying vibrational states. This excess population leads to an enhancement of absorption at long wavelengths below the purely electronic transition and of emission in the short wavelength region. The excess energy of the vibrational system is supplied by the exciting photon and is transferred to the vibrations by the proton transfer reaction in the excited electronic state and the subsequent internal conversion to the ground state of the reaction product. Due to the unknown energy separation of the enol and the keto-type S'_0 states and some possible ultrafast energy transfer to the solvent, the energy content of the keto-type molecules cannot be determined exactly. However, an estimate of the maximum vibrational excess energies $E_{S1,max}$ in the S'_1

Fig. 2 Transient absorption at 540 nm showing identical kinetics with excitation (a) well above and (b) slightly below the 00-transition of enol-TIN at 367 nm (points). The solid lines are the result of numerical simulations.

state and $E_{S0,max}$ in the S'_0 state assuming isoenergetic ground states of the enol and the keto-type tautomers gives values of $E_{S1,max} = E_{ex}-E_{00}' = 11\,000$ cm^{-1} and $E_{S0,max} = E_{ex} = 27\,030$ cm^{-1}. $E_{ex} = 27\,030$ cm^{-1} represents the photon energy of the excitation pulses and $E_{00}' = 16\,000$ cm^{-1} is the approximate S'_1-S'_0 separation in the keto- type tautomer as determined from the transient spectra. From the maximum excess energy and a normal mode analysis of TIN [4], we calculate the maximum temperature of an equilibrated vibrational system in which the vibrational populations obey Bose statistics. This gives values of $T_{S1,max} = 600$ K and $T_{S0,max} = 1000$ K for the S'_1 and the S'_0 state, respectively. The corresponding thermal energies $kT_{S0,max} = 670$ cm^{-1} and $kT_{S1,max} = 400$ cm^{-1} are a measure for the energetic width of the corresponding vibrational distributions. It should be noted that the spectral overlap of more than 2000 cm^{-1} observed in our experiment is at least twice as broad as the sum of the maximum thermal energy widths. This finding demonstrates that the vibrational modes coupling to the S'_1 -S'_0 transition of keto-type TIN are populated much stronger than in thermal equilibrium, i.e. our data give evidence for nonequilibrium vibrational distributions of the keto-type molecules.

In Fig. 2, we present transient absorption measurements recorded at a probe wavelength of λ_{pr}=540 nm for two different excitation wavelengths. The normalized change of absorption is shown as a function of delay time for excitation at λ_{ex} = 375 nm (27030 cm^{-1}), nearly resonant to the purely electronic transition of the enol tautomer (ν_{00} = 27 260 cm^{-1}) and at 310 nm (32 260 cm^{-1}), that is 5000 cm^{-1} above ν_{00}. The absorption at early delay times (<1ps) is due to the excited enol and keto states, whereas the keto ground state gives rise to the signal at later delay times. The solid lines in Fig. 2 represent numerical fits of the data which were calculated from a rate equation model comprising the four electronic states and their transfer rates in the sequential reaction cycle shown in the insert in Fig. 1 (a). Within the experimental accuracy we observe identical reaction kinetics for both excitation wavelengths. A possible potential barrier along the reaction coordinate of the proton transfer would result in a well-pronounced change of the reaction rates with excitation energy. For small excess energies, proton tunneling with a correspondingly low rate would occur while a direct transfer process would be possible for higher energies. This model can be ruled out from the data of Fig. 2, showing identical rates. In contrast, we conclude that the potential surface is essentially barrierless along the reaction coordinate with a proton transfer rate determined by the vibrational degrees of freedom involved in the reaction.

[1] For a review see: Photoinduced proton transfer, Michael Kasha Festschrift, J. Phys. Chem. **95** (1991) 10215

[2] F. Laermer, T. Elsaesser, and W. Kaiser, Chem. Phys. Lett. **148** (1988) 119

[3] M. Wiechmann, H. Port, F. Laermer, W. Frey, and T. Elsaesser, J. Phys. Chem. **95** (1991) 1918

[4] K. Lenz, M. Pfeiffer, A. Lau, and T. Elsaesser, Chem Phys. Lett. **229** (1994) 340

Ultrafast Electron Transfer in Modified Photosynthetic Reaction Centers from Rhodobacter Sphaeroides Measured by Fluorescence Upconversion

T. Häberle[a], H. Lossau[a], G. Hartwich[a], H. Scheer[b], and M.E. Michel-Beyerle[a]

[a] Institut für Physikalische Chemie, TU München, Lichtenbergstr. 4, 85747 Garching. FRG
[b] Botanisches Institut der LMU München, Menzinger Str. 67, 80638 München

Abstract

In photosynthetic reaction centers determination of ultrafast electron transfer rates as a function of the thermodynamical driving force ΔG allows conclusions on the mechanism of primary charge separation. In this paper we present femtosecond time-resolved fluorescence measurements on the primary donor species P* in reaction centers from *Rhodobacter Sphaeroides* R26 after modifying the energetics of the neighboring pigment, a bacteriochlorophyll molecule. The modification was achieved by thermally replacing the pigment by a specially designed derivative, the Ni-bacteriochlorophyll. Room temperature rates for the primary charge separation in the modified reaction centers are the same as in the native ones. This supports a sequential two step mechanism also in native reaction centers.

Introduction

The first reactions involved in the conversion of light into chemical energy in photosynthesis occur in a protein (*reaction center*) spanning the membrane. Upon excitation of a bacteriochlorophyll dimer (P*) an electron is translocated across the membrane, thus providing the electrochemical potential gradient which drives the subsequent formation of energy-rich chemical products as sugar and oxygen.

One of the most exciting problems still under some debate is the role of a bacteriochlorophyll monomer (B_A) which in the X-ray structure [1-6] has been shown to be interlocated between the primary donor P*, and a bacteriopheophytin (H_A) at the active pigment branch (Fig. 1). The investigation of primary kinetics is complicated by a variety of factors, e.g. congested difference spectra, dispersive kinetics due to heterogeneous couplings and/or energetics, and coherence phenomena.

In this paper we study the mechanism of primary charge separation upon changing the energetics of the potential primary acceptor B_A. In the approach taken here we lower the free energy of $P^+B_A^-$ by 1000 cm^{-1}. This is achieved by thermally replacing the native bacteriochlorophyll bear-

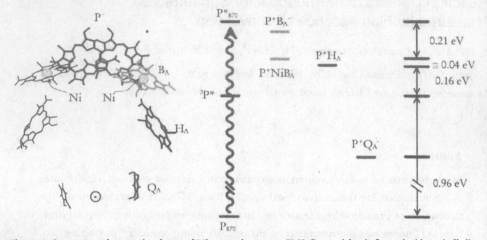

Figure 1: Structure and energetic scheme of Ni_{AB}-reaction center [7,8]. P=special pair (bacteriochlorophyll dimer), B=accessory bacteriochlorophyll, H=pheophytin, Q=quinone.

ing Mg^{2+} as central ion with the Ni^{2+} derivative (fig. 1). The functionality of the modified reaction centers (with both bacteriochlorophyll monomers exchanged, Ni_{AB}-RCs) has been demonstrated by steady state and time resolved fluorescence measurements [9,10], ns- and ps-time resolved difference absorption measurements [7,8], as well as highly sensitive determination of transmembrane charge separation to the quinone, forming $P^+Q_A^-$ [11]. Special access to the energetics is provided by the magnetic field dependence of the recombination dynamics [8].

In this paper discussion and predictions of electron transfer rates used are based on conventional electron transfer theory. There, the dependence of the electron transfer rate k on the free energy difference ΔG between the donor and the acceptor is given by [12]

$$k = \frac{2\pi}{\hbar} \frac{V^2}{\sqrt{4\pi\lambda_s k_B T}} \cdot \sum_{n=0}^{\infty} \exp(-S_V) \cdot \frac{S_V^n}{n!} \cdot \exp\left(-\frac{(\Delta G + \lambda_s + n\hbar\omega_V)^2}{4\lambda_s k_B T}\right) \qquad (1)$$

with V denoting the electronic coupling, λ_S the low frequency reorganization energy and T the temperature. The high frequency mode is represented by its energy $\hbar\omega_V$, the coupling parameter $S_V = \lambda_V/\hbar\omega_V$, and the reorganization energy λ_V. The excellent applicability of this formalism to ultrafast charge separation in a series of cyclophane-bridged porphyrin-quinone donor-acceptor compounds has been demonstrated recently over the so far widest range of energies [13]. In the context of this paper the most prominent feature of these studies is the weak dependence of the rate k on ΔG in the maximum of rate as well as the inverted region, where $-\Delta G \geq \lambda_s$. There the rate has been shown to be invariant over a width of ΔG over 4000 cm^{-1}.

Assuming this phenomenology to hold also for the primary charge separation in the photosynthetic reaction center, we expect the relative behavior of the primary rates in native and modified reaction centers (where the energy of $P^+B_A^-$ has been lowered) to depend on the specific mechanism in operation.

(i) If primary charge separation leads directly to the state $P^+H_A^-$ with $P^+B_A^-$ acting as *superexchange mediator*, lowering of the energetics of $P^+B_A^-$ is expected to diminish the energy denominator in the expression for the superexchange matrix element [14-18],

$$V_{super} = \frac{V_{DS}V_{SA}}{\Delta G + \lambda_s} \tag{2}$$

thus *increasing* the rate. Here, V_{DS}, V_{SA} are the electronic couplings between donor-superexchange state, and superexchange state-acceptor, respectively.

(ii) If, however, primary charge separation in native reaction centers populates in a kinetic sense the state $P^+B_A^-$, one would expect *constancy* or a slight *decrease* of the rate at room temperature when lowering the energetic level of $P^+B_A^-$.

It is the goal of this paper to discriminate between the two mechanisms in the native reaction center. This is achieved via conclusions from the primary electron transfer dynamics in the Ni_{AB}-reaction center by femtosecond time-resolved fluorescence measurements on the primary donor P^*.

Experimental

The photosynthetic reaction centers of *Rb. Sphaeroides* R26 were isolated by standard methods [19]. The transmetallated bacteriochlorophylls were obtained by the method described in HARTWICH et al. [20]; the exchange was done by the method of STRUCK et al. [21]. Exchange ratios were better than 95%. Quinone was removed by the method of GUNNER et al. [22]. The samples were solved in TL buffer. The measurements were performed in a 1 mm quartz kuvette cooled to 275 K by blowing with cold gaseous nitrogen.

The laser source was a commercial Ti:Al$_2$O$_3$-laser (Coherent Mira Basic), pumped by a fraction (8 W) of the output of an Ar$^+$-laser (Coherent Innova 400). The laser was tuned to the maximum of the reaction center dimer band at $\lambda = 866$ nm and delivered 450 mW of cw power. The autocorrelation width of the pulses was 200 fs (full width of half maximum, FWHM). Repetition rate was 76 MHz.

The upconversion setup (fig 2) is optimized for good time resolution following the guidelines in KAHLOW et al. [23] by minimizing the amount of glass in the optical path and using only quartz optics with its comparative low group velocity dispersion. The laser beam is splitted in two halves. One is focused into the sample, the other one (gate beam) passes a stepping motor driven

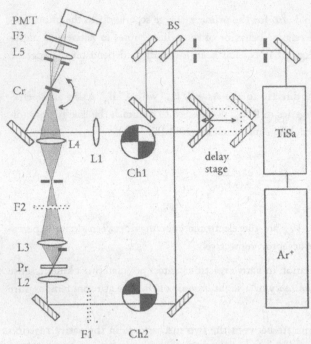

delay stage (resolution 16.7 fs, range 1.3 ns) and is joined with the fluorescence from the sample in a β-barium-borate (BBO) crystal (thickness 1 mm). By employing noncollinear noncritical type I phase matching, the fluorescence is detected parallel polarized with respect to the excitation. The sample cuvette is mounted in a holder rotating with up to 3000 rpm, in order to avoid thermal effects, re-excitation of long lived intermediates, and to distribute the stress of excitation to a larger sample volume. Sum frequency and background photons are detected by a photomultiplier (Hamamatsu R4220 P) and counted synchronously to the phase of light choppers in the excitation and gate path. Data acquisition and experi-

Figure 2: Optical setup of the upconversion apparatus. Ch=chopper, F=filter, PMT=photomultiplier tube, Cr=nonlinear crystal, L=lens, Ar^+ = argon ion laser, TiSa=Ti:Al$_2$O$_3$-laser, BS=beam splitter

ment control is automated using a personal computer. In order to reduce the influence of long term drifts of the laser and the degrading of the samples upon the decay curves, we repeat the measurement cycles up to 16 times and alter the direction of acquisition after each run. The fluorescence signal did not decrease significantly during a measurement cycle although it took up to several hours. The time window was at least 16 ps.

The instrument response function (IRF) is taken before each fluorescence measurement using scattered light from the sample. The alignment is not changed except for adjusting the phase matching angle of the crystal and removing a long pass filter after the sample. Time constants are extracted by fitting a convolution of the IRF and exponential decay functions [24] to the data using the Levenberg-Marquardt method [25,26]. Quality of fit is judged by observing the residuals and values of the reduced χ_v^2. We obtained values between 1.0 and 1.5 for the latter. By this deconvolution lifetimes down to a quarter of the IRF FWHM can be resolved.

Result and Discussion

The fluorescence decay of P* in Ni$_{AB}$-RCs excited at 866 nm and detected at 930 nm is shown in fig 3. Measurements were performed on quinone-depleted reaction centers at 275 K, where the triplet (^{3}P*) lifetime is shorter than 1 μs. The fitted time constants are 2.4 ps (79% amplitude)

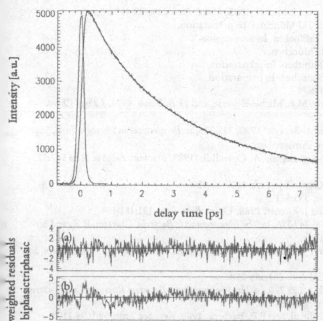

Figure 3: Decay of the dimer fluorescence in Ni_{AB}-RCs from *Rb. Sphaeroides* R26. Excitation wavelength 866 nm, detection wavelength 930 nm, $T = 295$ K. Fit parameters as shown in text. The IRF is shown also. (a) residuals for triphasic fit, (b) for biphasic fit.

and 8 ps (21%) when using a bi-exponential fit. The residuals show some structure below 1 ps, which can be removed by including a third, short time component in the fit; this leads to a shift of the other time constants to larger values yielding 0.6 ps (9%), 2.8 ps (78%), and 10 ps (13%). No longer components were observed within the experimental resolution in time windows up to 100 ps.

These time constants for the spontaneous emission in the Ni_{AB}-RCs at room temperature are the same as for quinone-depleted as well as quinone-reduced native RCs from *Rb. Sphaeroides* R26 [27-29]. This includes also ratios of amplitudes and even the qualitative appearance of the residuals when fitted with two time components [29]. This invariance of the charge separation rate upon lowering ΔG from -500 cm^{-1} to -1500 cm^{-1} follows from nonadiabatic electron transfer theory in the inverted region. The measured rate can be described with the identical set of parameters used to model the rate dependence on ΔG in a series of native and mutagenetically altered reaction centers [30].

Our conclusion on the sequential mechanism to be operative in native RCs of *Rb. Sphaeroides* is independently supporting the findings of ZINTH's group [31].

Acknowledgement

The work has been supported by the Deutsche Forschungsgemeinschaft (SFB143).

References

1. Deisenhofer, J., O. Epp, K. Miki, R. Huber, and H. Michel. 1984. *J. Mol. Biol.* 180:385-398.
2. Allen, J.P., G. Feher, T.O. Yeates, D.C. Rees, J. Deisenhofer, H. Michel, and R. Huber. 1986. *Proc. Nat. Acad. Sci. USA* 83:8589
3. Chang, C.H., J. Norris, M. Schiffer, U. Smith, J. Tang, and D. Tiede. 1986. *FEBS Lett.* 205:82-86.
4. Ermler, U., H. Michel, and M. Schiffer. 1994. *J. Bioenerg. Biomembrane* 26:5-15.
5. Ermler, U., G. Fritzsch, S.K. Buchanan, and H. Michel. 1994. *Structure* 2:925-936.

184

6. Deisenhofer, J., O. Epp, I. Sinning, and H. Michel. 1995. *J. Mol. Biol.* 246:429-457.
7. Langenbacher, T. 1995. Dissertation. TU München. In preparation.
8. Rousseau, G. 1995. Dissertation. TU München. In preparation.
9. Hartwich, G. 1994. Dissertation. TU München.
10. Friese, M. 1995. Dissertation. TU München. In preparation.
11. Müller, P. 1995. Dissertation. TU München. In preparation.
12. Jortner, J. 1976. *J. Chem. Phys.* 64:4860
13. Heitele, H., F. Pöllinger, T. Häberle, M.E. Michel-Beyerle, and H.A. Staab. 1994. *J. Phys. Chem.* 98:7402-7410.
14. Bixon, M., J. Jortner, and M.E. Michel-Beyerle. 1990. 325-336, in *Perspectives in Photosynthesis*, Jortner, J. and Pullman, B., eds. Kluwer, Amsterdam.
15. Bixon, M., J. Jortner, M.E. Michel-Beyerle, and A. Ogrodnik. 1989. *Biochim. Biophys. Acta* 977:273-286.
16. Hu, Y.M. and S. Mukamel. 1989. *Chem. Phys. Lett.* 160:410-416.
17. Marcus, R.A. 1987. *Chem. Phys. Lett.* 133:471
18. Michel-Beyerle, M.E., M. Bixon, and J. Jortner. 1988. *Chem. Phys. Lett.* 151:188-194.
19. Feher, G. and M.Y. Okamura. 1978. 349-386, in *The Photosynthetic Bacteria*, Clayton, R.K. and Sistrom, W.R. eds. Plenum Press. New York.
20. Hartwich, G., E. Cmiel, I. Katheder, W. Schäfer, A. Scherz, and H. Scheer. 1995. *J. Amer. Chem. Soc.* in press.
21. Struck, A. and H. Scheer. 1990. *FEBS Lett.* 261:385-388.
22. Gunner, M.R. and P.L. Dutton. 1988. 259-270, in *The Photosynthetic Bacterial Reaction Center - Structure and Dynamics*, Breton, J. and Verméglio, A. eds.
23. Kahlow, M.A., W. Jarzeba, T.P. DuBruil, and P.F. Barbara. 1988. *Rev. Sci. Instr.* 59:1098-1109.
24. Grinvald, A. and I.Z. Steinberg. 1976. *Anal. Biochem.* 75:260-280.
25. Bevington, P.R. and D.K. Robinson. 1992. Data Reduction and Error Analysis for the Physical Sciences. McGraw-Hill.
26. Press, W.H., S.A. Teukolsky, W.T. Vetterling, and B.P. Flannery. 1992. Numerical Recipes in C. The Art of Scientific Computing. Cambridge University Press, Cambridge.
27. Du, M., S.J. Rosenthal, X. Xie, T.J. DiMagno, M. Schmidt, D.K. Hanson, M. Schiffer, J.R. Norris, and G.R. Fleming. 1992. *Proc. Nat. Acad. Sci. USA* 89:8517-8521.
28. Hamm, P., K.A. Gray, D. Oesterhelt, R. Feick, H. Scheer, and W. Zinth. 1993. *Biochim. Biophys. Acta* 1142:99-105.
29. Stanley, R.J. and S.G. Boxer. 1995. *J. Phys. Chem.* 99:859-863.
30. Bixon, M., J. Jortner, and M.E. Michel-Beyerle. 1995. *J. Phys. Chem.* submitted.
31. Schmidt, S., T. Arlt, P. Hamm, H. Huber, T. Nägele, J. Wachtveitl, M. Meyer, H. Scheer, and W. Zinth. 1994. *Chem. Phys. Lett.* 223:116-120.

Sub-Picosecond Grating Measurements of the Resonant Energy Transfer in Cresyl Violet Solutions

R. Bierl, M. Seischab and S. Schneider

Institut für physikalische und theoretische Chemie, Universität Erlangen-Nürnberg
Egerlandstrasse 3, D-91058 Erlangen

1.Introduction

Electronic excitation transfer in disordered systems, such as solutions or amorphous solids, is of great importance in a large majority of fields, e.g. in the sensitization and quenching of photophysical an photochemical primary processes, or the collection of light energy in technical or natural antenna systems (photosynthesis). Therefore the dynamics and spatial mobility of excited-state energy transport in disordered systems has received considerable attention both with respect to experimental determination and theoretical description.

An alternative to pump-probe experiments for studying excited state kinetics is the grating experiment. In this a spatial modulation in material properties is induced due to absorption of light, whose intensity is modulated according to the interference pattern of two pump beams (Fig.1). The fringe spacing Λ of the thus produced optical grating varies with the crossing angle Θ: $\Lambda = \lambda/2n \sin\Theta/2$. A third (probe) beam is diffracted by the grating (Fig.1). The diffraction efficiency is proportional to the square of the modulation depth, i.e. the difference in concentration of excited molecules in the peaks and nulls of the interference pattern $(S \sim (N^*_{max} - N^*_{min})^2)$. It is obvious that the diffraction efficiency must decay like $S_0 * e^{-2t/\tau_0}$, if τ_0 denotes the lifetime of excited molecules in absence of energy transfer processes. In the presence of energy transfer which leads to an additional decrease in modulation depth, the decay of the scattered light intensity is given by:

$$S(t) \sim S_0 \cdot \exp\left[-2t\left(\frac{2\pi D}{\Lambda^2} + \frac{1}{\tau_0}\right)\right] \tag{1}$$

with D representing the energy diffusion constant.

The energy diffusion constant D of the resonant energy transfer is usually discussed in terms of FÖRSTER theory [1]. It yields:

$$D = A \cdot C^{4/3} R_0^2 \tau_0^{-1} = \text{const} \tag{2}$$

In its derivation, two important assumptions are made:

i. The energy absorbers are located on a regular lattice
ii. The donor's vibrational states are thermalized before energy transfer takes place.

Both assumptions can be questioned when studying solutions with high solute concentrations. It has been shown by several authors that due to the random distribution of solute molecules the energy diffusion constant is no longer constant in time (Förster model) but increases by about a factor of 5 for short times and sufficiently high concentrations. Furthermore, the rate of resonant energy transfer could be much higher for un-relaxed vibrational states with higher excess energy (better overlap betweeen absorption and (resonance) emission spectra) so one could get a solvent dependence in D.

FAYER [2] et al. allow the formation of dimers which act energy traps and terminate the energy migration process. This introduces a further decay channel and leads to a dependence of D like C^3.

2. Experimental Procedures

Pulses with about 100 fs fwhm and an energy of about 50 pJ at 625 nm are generated by a two-prism rhodamine 6G/DODCI CPM dye laser oszillator and amplified in a four stage excimer laser pumped Rh101 amplifier with a gain of 10^6. In order to avoid the effects of long-term drifts, averaging was done over ten laser shots per data point, but the scanning of the variable delay line was repeated ten times.

The two pump beams and the probe beam intersect the sample as shown in fig. 1. The angles between the probe beam and the pump beams are chosen with $\Theta_1 \neq \Theta_2$. This was done to avoid superimposition of diffracted beams of different origin.

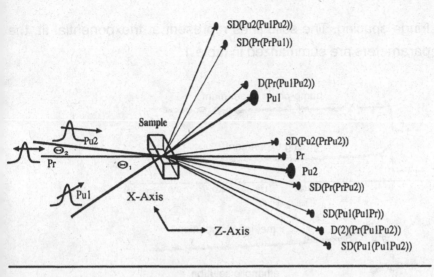

Pu1, Pu2	pump-beams
Pr	probe-beam
D	diffraction signal
SD	selfdiffraction signal

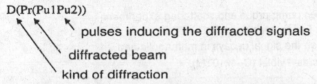

D(Pr(Pu1Pu2))

pulses inducing the diffracted signals

diffracted beam

kind of diffraction

Figure 1: Experimental arrangement for the three-pulse grating experiment.
The pump beams intersect the sample with the angles Θ_1 and Θ_2.
The unsymmetrical arrangement was chosen to avoid superpositon
of diffracted beams of different origin

The beam scattered into the first orders (see notation in fig.1) and the
probe beam were monitored by diodes whose signals are fed into a gated
integrator. The output of the latter is transferred to a computer.

3. Results and discussion

In Fig.3, the decay of the signal deflected by the induced transient grating
is displayed for two different situations. Firstly, the fringe spacing is varied
at constant dye concentration, secondly the dye concentration is varied at

constant fringe spacing. The solid lines represent at triexponential fit, the deduced parameters are summarized in table 1.

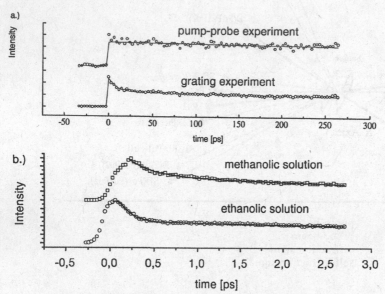

Fig.2 a.) Comparison between pump-probe and scattering experiment for longer delay times. Fringe spacing $\Lambda=10.0\mu m$.
b.) Comparison between the signal decays in methanolic and ethanolic solutions of cresyl violet ($C=4\times10^{-3}M$).

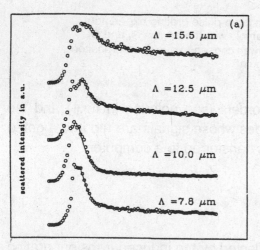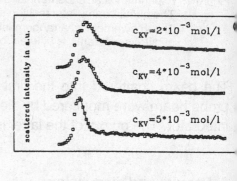

Fig. 3   Dependence of the intensity of the diffracted beam on delay time
a.)Influence of the fringe spacing ($C=4\times10^{-3}M$)
b.)Influence of the dye concentration ($\Lambda=10.0\mu m$)

Table 1
Parameters derived by the triexponential least-squares fit of scattered light intensity
displayed in fig.3. The decay times of component 2 and 3 were kept constant at 7ps
and 1.5ns. These values originate from the fit of the scattering curve shown in fig. 2a

C [mol/l]	Λ [µm]	τ_1 [ps]	A_1 rel.	A_2 rel.	A_3 rel.
4x10^{-3}	15.5	0.26	0.53	0.28	0.19
4x10^{-3}	12.5	0.14	0.61	0.29	0.10
4x10^{-3}	10.0	0.12	0.91	0.07	0.02
4x10^{-3}	7.8	0.11	0.83	0.14	0.03
2x10^{-3}	10.0	0.23	0.61	0.23	0.16
4x10^{-3}	10.0	0.12	0.91	0.07	0.02
5x10^{-3}	10.0	0.08	0.91	0.09	0.00

The inverse decay times ($1/\tau_1$) of the fast components are plotted versus
$1/\Lambda^2$ (fig. 4a) and versus c^n (fig. 4b) in order to test the relationships given
by equations 1 and 2.

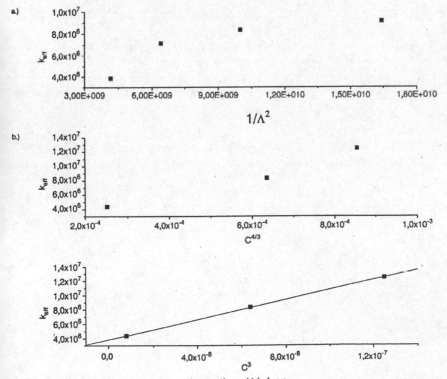

Fig.4 Dependence of the inverse decay time ($1/\tau_1$) on

a.) $1/\Lambda^2$ and   b.) C^n with n=4/3 and n=3

The lack of a linear relationship between $1/\tau_1$ and $1/\Lambda^2$ indicates that the variation of $1/\tau_1$ with fringe spacing is not governed by resonant energy transfer only.

In accordance with this conclusion is the fact that the decay rate does not vary with concentration like $c^{4/3}$(Förster limit). The data suggest a dependence like C^3. This seems to show good agreement with the energy trapping through dimers (Fayer). Since there is is a large difference in the signal decay (fig.2b) for ethanolic and methanolic solutions (under otherwise identical conditions) it is likely that also solute-solvent interactions are reflected in the different decay patterns. However, localized interactions should not give rise to a Λ dependence of the decay rate.

Degenerate Four-Wave Mixing on Dye Molecules and Photosynthetic Reaction Centers

T. Arlt, R. Heinecke, H. Penzkofer and W. Zinth

Institut für Medizinische Optik, Universität München

D-80797 München

Introduction

Investigations of the ultrafast processes in photosynthetic reaction centers are one of the most interesting subjects in biophysics today. The photosynthetic reaction center (RC) is a membrane protein complex containing different chromophores. Optical excitation starts an electron tranfer across this membrane protein, where the electron hops from one chromophore to the other. In the first step the electron is transferred from a pair of excitonically coupled chlorophylls called special pair P to a single chlorophyll molecule B within 3 picoseconds (ps) [1].

Optical pump-probe spectroscopy of the electron transfer reaction of photosynthetic reaction centers has been used during the last decade. However, it has not delivered complete information. As a consequence additional techniques are required. Degenerate four-wave mixing is an interesting coherent technique to characterize the nonlinear response of a medium. In the present paper it is used to investigate the Q_Y transition moments of chromophores in RCs of *Rhodobacter sphaeroides* in further detail ($Q_Y(P)$ at 865 nm and $Q_Y(B)$ at 802 nm). The data are compared with results on dye molecules.

The experimental setup is shown in Fig. 1. A lasersystem based on a CPM dye laser delivers femtosecond (fs) pulses in the range from 700 nm to 1000 nm depending on the dye used for the amplification (for details of the system see [2]). The instrumental response function of the four-wave mixing experiments has a width of about 90 fs (FWHM). Two pulses with wavevectors $\underline{k}_1$ and $\underline{k}_2$ are used for the experiment. The first laser pulse $\underline{k}_1$ creates a nonlinear polarization in the sample, which decays very rapidly in the case of an inhomogeneous broadened system. The second pulse $\underline{k}_2$, which is delayed by the time t, removes the phases of the different transitions and one finds the so called photon echo as a scattered signal in the direction $2\underline{k}_2 - \underline{k}_1$:

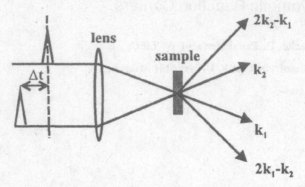

Fig. 1: Experimental setup used in the four-wave mixing experiments. The scattered signal in direction $2\underline{k}_2 - \underline{k}_1$ is observed as a function of the delay Δt.

$$I_{2\underline{k}_2-\underline{k}_1}(t) = I(0) \cdot \exp(-t/\tau) \sim \exp(-t/(T_2/4))$$

τ is the measured decay time of the scattered signal and T_2 the optical dephasing time. In the case of a simple two-level system T_2 is given by:

$$1/T_2 = 1/T_{2PD} + 1/(2 \cdot T_1)$$

Here T_{2PD} represents the pure dephasing time and T_1 the energy- or population relaxation time.

Results and Discussion

$Q_Y(B)$ band:

Fig. 2a shows the scattered signal of reaction centers (R 26) in direction $2\underline{k}_2 - \underline{k}_1$ at 804 nm. A model function using a sech²-function and a monoexponential decay with $\tau = 85$ fs describes the data quite well (solid line). In Fig. 2b measurements on the dye molecules IR 140 (Lambda Physics) and Kodak 9860 are plotted for comparison. For IR 140 a decay time of 75 fs could be detected, whereas for the Kodak dye no time constant could be observed within our time resolution. Since the RC can be treated as an inhomogeneous broadened system, the decaytime $\tau = 85$ fs refers to a dephasing time T_2 of about 340 fs. This value has to be compared with a time constant of some 100 fs found by pump-probe experiments for the energy transfer from the excited molecule B^* to P^* [3]. Without considering a pure dephasing time a T_1-time constant of 150 fs would lead to $T_2 = 300$ fs; i.e. it determines the decay of the nonlinear polarization in the $Q_Y(B)$ band at 804 nm.

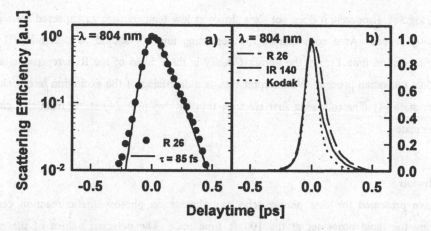

Fig.2: a) Scattered signal for reaction centers (R26) at 804 nm (points) and a model function using a sech²-function and a monoexponetial decay with 85 fs (solid line). b) For comparison the signal of the dyes IR 140 and Kodak 9860 is shown. Note the different scales used in a) and b).

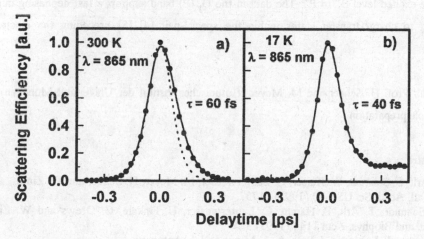

Fig.3: Scattered signal of reaction centers (R 26) in the Q_y (P) band at 865 nm for 300 K and 17 K. Points refer to the measured data, the dotted line represents the instrumental response function and the solid line to fit functions using exponential decay functions.

$Q_Y(P)$ band:

In Fig. 3 the scattered signal of reaction centers at 865 nm is plotted for 300 K and 17 K. The signals show a fast initial decay. At low temperatures a slow decaying background remains, which is probably due to an artefact of aggregating chlorophylls. The observed time constant τ of about 60 fs at room temperature (Fig.3a) does not change drastically at low temperatures

(40 fs, Fig.3b). Especially it does not slow down at low temperatures as expected for a pure dephasing process. As a consequence the dephasing time T_2 should be determined by an energy relaxation time T_1; i.e. the ultrafast decay in the P band of the RCs represents some electronic relaxation process. We interpret this as a dephasing of the excitation into a charge transfer state [4]. The following first electron transfer step probably starts from this charge transfer state.

Conclusion

We have presented the first photon-echo experiments on photosynthetic reaction centers resolving the rapid processes at the 100 fs time scale. The detected values of the phase relaxation times T_2 of the RCs are in the 200 - 400 fs range. Data at low temperatures show that population-relaxation processes (T_1-processes) determine the dephasing times. In the $Q_y(B)$ band the observed T_2 component of about 340 fs is probably related to energy transfer from the excited level B^* to P^*. The data in the Q_y (P) band support a fast dephasing due to coupling to charge-transfer states within the special-pair [4], [5] preceding the electronic transfer.

We thank Prof. H. Scheer and M. Meyer (Botanisches Institut der Universität München) for the sample preparation.

Literature

[1] T. Arlt, S. Schmidt, W. Kaiser, C. Lauterwasser, M. Meyer, H. Scheer and W. Zinth, Proc. Natl. Acad. Sci USA 90 (1993) 11757.
[2] S. Schmidt, T. Arlt, P. Hamm, C. Lauterwasser, U. Finkele, G. Drews and W. Zinth, Biochim. and Biophys. Acta 1144 (1993) 385.
[3] J. Breton, J.-L. Martin, J. Petrich, A. Migus and A. Antonetti, FEBS Letters 209 (1986) 37.
[4] P. Hamm, M. Zurek, W. Mäntele, M. Meyer, H. Scheer and W. Zinth Proc. Natl. Acad. Sci USA 99 (1995) 1826.
[5] S. R. Meech, A. J. Hoff and D. A. Wiersma, Proc. Natl. Acad. Sci USA 83 (1986) 9464.

Femtosecond Spectroscopy on Modified Bacterial Reaction Centers

T. Nägele, H. Huber, J. Wachtveitl, S. Schmidt, T. Arlt, W. Zinth,

Institut für Medizinische Optik

Ludwig-Maximilians Universität München, Germany

The solution of the molecular structure of photosynthetic reaction centers provided detailed information on the organisation of this pigment-protein-complex, in which a series of light induced electron transfer steps initiates bacterial photosynthesis. Optical exitation leads to the transfer of an electron from a pair of strongly coupled bacteriochlorophyll molecules to a quinone on the opposite side of the membrane. Only one of the two symmetrically arranged chains of chromophores is used for electron transfer. The primary electron donor (P) remains in the exited state for about 3 ps and the electron arives at the bacteriopheophytin (H_A) in nearly the same time, the quinone (Q_A) is reduced within 200 ps. The participation of the first chromophore in the chain, the monomeric bacteriochlorophyll B_A was under debate for a long time [1]. Former time resolved studies have shown that a population of the anion state B_A^- should be very small, short-livedand thus difficult to detect [2].

Here we present results on modified RCs of *Rhodobacter sphaeroides* (R26.1), where the bacteriopheophytin molecules BPhe-a are replaced by a plant pheophytins Phe-a (R26.Phe-a). This modification causes remarkable changes of the reaction process and yields important information on the energetics of the intermediate states within the RC.

The *in vitro* difference of the redox potential of BPhe-a and Phe-a of 130 meV suggests a comparable rise of the free energy of $P^+H_A^-$ in the modified RCs. The energy level of this radical pair state should lie in the range of the free energy of P^*. Provided that the electron transfer runs stepwise via B_A, a long-lasting population of the radical pair state $P^+B_A^-$ in thermal equilibrium with $P^+H_A^-$ should be detectable.

The extension of the spectral range up to probing wavelengths of 1100 nm (Fig. 1) allows us to measure in a region where only the BChl -a anion has a pronounced absorption band, i.e. no H_A^- band or any ground state absorption of the chromophores [3].

At wavelengths shorter than 1020 nm there is a strong stimulated emission decaying that practically vanishes at $\lambda = 1020$ nm. In the spectral region shown the absorption in R26.1 increases with a strong 0.9 ps component. The constant positive absorption difference at longer delay times is due to

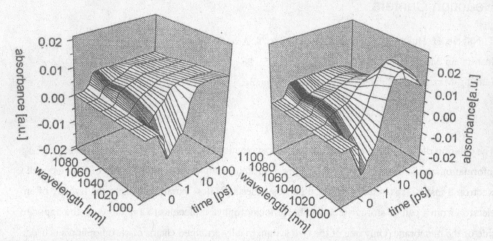

<u>Fig.1:</u> Time resolved absorption differences $\Delta A(t_D)$ of wild type (left) and R26.Pheo (right) RCs in the spectral region from 1000 nm to 1100 nm. In contrast to the native protein a long-living population of $P^+B_A^-$ can be seen with its maximum at a wavelength of 1020 nm at a delaytime of 30 ps in the modified RC.

P^+. In the modified RCs the absorption change is drastically different: it rises with 0.9 ps and 3.5 ps and subsequently decreases with a time constant of 380 ps. In this spectral region the difference can only be explained by a long-lived population of approximately 30 % of the BChl-a anion state $P^+B_A^-$. In both types of RCs only P^+ causes the absorption change at $\lambda = 1100$ nm.

Data obtained in other spectral regions allows the following conclusion [4]:

(i) a reduced formation of the radical pair state $P^+Q_A^-$ (70%) compared to 99 % in the native type.

(ii) a long-lived population of P^* of about 3 %.

Combinig these results, the microscopic rates and the free energy levels of the participating intermediates can be estimated within a stepwise electron transfer model (Fig. 2).

In the analysis for R26.Phe-a the recombination from $P^+B_A^-$ $(1/\gamma_{20})$ and the free energies of the first ET-intermediates - $\Delta G(P^+B_A^-)$ and $\Delta G (P^+H_A^-)$ - were varied as free parameters to reproduce the measured time constants and absorbance changes of the intermediates. We consider recombination from P^* and $P^+B_A^-$ to the ground state P in order to explain the reduced efficiency in quantum yield. The free energies G_i of the different states are used to correlate forward and backward rates by the principle of detailed balance :

$$\gamma_{ji} = \gamma_{ij} \cdot \exp(-(G_i - G_j) \backslash (k_B \cdot T))$$

Consistency is only obtained if the free energy level of the intermediate $P^+B_A^-$ is approximatly

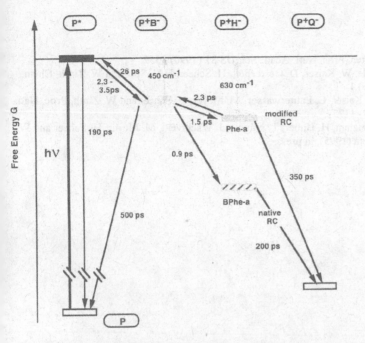

__Fig 2:__ Reaction scheme used to explain the primary steps of photosynthesis in bacterial reaction centers of _Rb. sphaeroides_. It is based on a stepwise ET along the pigment chain of the reaction center starting at the special pair P via B_A, H_A to Q_A.

450 cm^{-1} below P* and 180 cm^{-1} above $P^+H_A^-$. The rise of ΔG ($P^+H_A^-$) of 1400 cm^{-1} in R26.Pheo compared to the native type - this corresponds to a level of 630 cm^{-1} below P* - is in the range one would expect from the difference in the _in vitro_ redox potential. The value ΔG ($P^+B_A^-$) of 450 cm^{-1} can also be used as a first approximation for R26.1 since the modification is not expected to lead to pronounced changes in the energetics of this intermediate.

It can be concluded that the replacement of bacteriopheophytin a by pheophytin a leads to significant changes in the reaction kinetics. In the spectral region around 1020 nm a long-lived population of $P^+B_A^-$ is clearly seen, a proof that the monomeric bacteriochlorophyll B_A is a real electron carrier. With the calculated free energies of the intermediates the stepwise reaction scheme in wild-type RCs can be understood as the result of an optimizationprozess of the primary ET-steps. The fast decay of the state $P^+B_A^-$ works accelerating for the forward ET. Back reactions are slowed down by the large energy gap between $P^+B_A^-$ and $P^+H_A^-$. Both effects are very important to get a quantum yield of nearly one.

References

[1] C. Kirmaier and D. Holten, Proc. Natl. Acad. Sci. US 87 (1990) 3552.
[2] W. Holzapfel, U. Finkele, W. Kaiser, D. Oesterhelt, H. Scheer, H.U. Stilz and W. Zinth, Chem. Phys. Letters 160 (1989) 1.
[3] T. Arlt, S. Schmidt, W. Kaiser, C. Lauterwasser, M. Meyer, H. Scheer and W. Zinth, Proc. Natl. Acad. Sci. US 90 (1993) 11757.
[4] S. Schmidt, T. Arlt, P. Hamm, H. Huber, T. Nägele, J. Wachtveitl, M. Meyer, H. Scheer and W. Zinth, Spectrochem. Acta (1995), in press.

Primary Reactions in Bacterial Photosynthesis: Dynamics of Reaction Centers with Modified Electron Acceptors

H. Huber, T. Nägele, S. Biesdorf, J. Wachtveitl, W. Zinth

Institut für Medizinische Optik, Universität München, Germany

The photoinduced electron transfer in the bacterial reaction center from *Rb. sphaeroides* R26.1 has been studied extensively in the last years. It is now well established, that the electron transfer steps along the active branch predominantly follow a sequential mechanism, in which anions of the monomeric BChl $\underline{a}$ (B_A^-), BPhe $\underline{a}$ (H_A^-) and primary ubiquinone (Q_A^-) are formed with time constants of 3 ps, 0.9 ps and 200 ps, respectively. Strong evidence for the reaction via B_A came from experiments on pigment modified samples, where the state B_A^- is in equilibrium to an energetically increased state $P^+H_A^-$ and is populated to a much higher extent than in native RCs [1,2]. We report experiments on RCs where 3-acetyl-Phe $\underline{a}$ and 3-vinyl-BPhe $\underline{a}$ are introduced at the BPhe $\underline{a}$-sites $H_{A/B}$ [3]. These two pigments are especially interesting, since they both represent a structural link between Phe $\underline{a}$ used as electron acceptor in plant photosystems and the bacterial BPhe $\underline{a}$. The bacteriochlorin type pigment 3-vinyl-BPhe $\underline{a}$ differs from BPhe $\underline{a}$ only by its substituent at the C3-position (vinyl instead of acetyl), which results in similar absorption characteristics of the two pigments, with only small shifts of the main bands in the Q_y- and the Q_x-region of 3-vinyl-BPhe $\underline{a}$ to higher frequencies. The 3-acetyl-Phe $\underline{a}$ differs from BPhe $\underline{a}$ in the hydrogenation state of its ring II macrocycle and as a chlorin type pigment shows spectral features more similar to Phe $\underline{a}$. The absorption spectra of the corresponding RCs are shown in Fig. 1.

Time resolved experiments

The transient absorbance changes were recorded in pump-and-probe experiments with a temporal resolution of 150 fs. Light pulses for the measurements were generated by a Ti:Sapphire laser system operating at a wavelength of 870 nm, that has been described elsewhere [3].

Measurements of RCs containing 3-acetyl-Phe $\underline{a}$ taken at probing wavelengths of 907 nm (P*-decay), 850 nm (ground state bleaching) and 690 nm (maximum of the $H_{A/B}$-band) show no significant difference from WT-data in comparable wavelength regions [3]. It has been shown earlier for WT

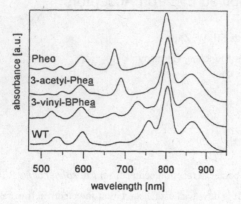

Fig. 1: Absorbance spectra of the different modified RCs with the incorporated chromophore at the sites $H_{A/B}$

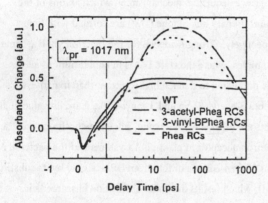

Fig. 2: Transient absorption changes of the modified RCs in the $P^+B_A^-$-band. The figure shows best fits to the measured data. RCs that contain a vinyl side-group at C3-position have a long lasting absorption due to thermal repopulation of the state $P^+B_A^-$.

RCs [4], that time resolved measurements in the BChl a-anion band (max. around 1020 nm) exhibit a pronounced subpicosecond kinetic component, which can be assigned to a signal originating from the reduced B_A^--intermediate. The 3-acetyl-Phe a RCs show qualitatively the same behaviour in this spectral range. However the absorption increase is somewhat slower than in wildtype RCs (Fig. 2, dashed-dotted line). The 1017 nm absorbance kinetics at long delay times are similar to wild type: Only a very weak relative absorbance decrease occurs on the 200 ps timescale.

We can conclude for RCs containing 3-acetyl-Phe a: The change of the double bond character of ring II - with the 3-acetyl-group being conserved as in BPhe a - mainly affects the absorption properties of the RCs. The dynamic characteristics, however, is very similar to the one found in WT. The kinetic data for the 3-vinyl-BPhe a RCs are completely different: Compared to 3-acetyl-Phe a RCs the changes in the absorbtion spectrum are less pronounced (Fig. 1), but the kinetic changes are drastic [3]. The data obtained at $\lambda_{pr} = 907$ nm show a slower initial decay of the gain signal, which can be fit with a time constant of 5.8 ps. The time resolved measurement in the P-band (at

λ_{pr} = 850 nm, [3]) demonstrates a slow decay at long delay times which is due to ground state recovery: In contrast to WT, the P$^+$-occupation is not stably maintained, since charge recombination processes lead to a 30 % repopulation of P within the timescale of the experiment. The time constant used to model this decay is 310 ps. This represents an eigenvalue of the rate equation system determined by the microscopic rates of the P$^+$-decay and the electron transfer from H$_A$ to Q$_A$. A long lived absorption due to the BChl a-anion-radical state could be detected at λ_{pr} = 1017 nm (Fig. 2, dotted line). After a small initial absorption decrease, caused by the stimulated emission of P*, the signal rises to a pronounced maximum at intermediate delay times. The curve subsequently decreases and a positive absorption difference due to the broad P$^+$-band remains at late times. The model curve shown in Fig. 2d (dotted line) is composed of a 1.5 ps and a 5.8 ps component of approximately equal amplitude for the absorption increase and a 310 ps component for the decay of the signal. In analogy to the Phe a containing RCs [1,2], the additional long lived absorption in 3-vinyl-BPhe a RCs on the 10 - 150 ps timescale must be due to the population of the state P$^+$B$_A$$^-$, since at a probing wavelength of 1017 nm absorption is only expected from P$^+$ or B$_A$$^-$.

Discussion

The interpretation of the observations follows non-adiabatic electron transfer theory. The electron transfer model introduces back rates, that are connected to the forward rates via the principle of detailed balance. It was described in detail elsewhere [2] . Estimates of the energetics of the intermediate states and the microscopic rates can be obtained by using the observed time constants, the efficiency of P$^+$Q$_A$$^-$-formation, the concentration of P* and P$^+$B$_A$$^-$ (in equilibrium with P$^+$H$_A$$^-$) within this reaction model. The model shows that the change in free energy of P$^+$H$_A$$^-$ - related with the exchange of a bacteriopheophytin - has drastic consequences on amplitudes and time constants. Qualitatively 3-vinyl-BPhe a and Phe a RCs show slower decay times of P* than WT and 3-acetyl-Phe a RCs. In addition for 3-vinyl-BPhe a and Phe a RCs time constants of 240 ps to 350 ps are observed (Fig 2). These long time constants are related with the decay of P$^+$B$_A$$^-$ and P$^+$H$_A$$^-$. The similarity of 3-vinyl-BPhe a and Phe a RCs suggests that the free energy of P$^+$H$_A$$^-$ is similar in both types of RCs. A value of 630 cm^{-1} (75 meV) below P* can be deduced by modelling the primary reaction. This value is $\approx$ 1470 cm^{-1} (175 meV) higher than in WT RCs and leads to a thermal repopulation of the state P$^+$B$_A$$^-$ from P$^+$H$_A$$^-$. In 3-acetyl-Phe a RCs no long-lived absoption of P$^+$B$_A$$^-$ could be observed (Fig 2). Here the free energy of P$^+$H$_A$$^-$ should be lower than in the case of 3-vinyl-BPhe a, so that no considerable thermal population of P$^+$B$_A$$^-$ is possible.

It is interesting to compare this changes in free energy with the changes in redox potential of the chromophores in solution [3]. The value of 3-vinyl-BPhe $\underline{a}$ is 180 mV (1500 cm^{-1}) lower than the one for BPhe $\underline{a}$. It is even slightly below that of Phe $\underline{a}$. Here it is assumed that this change in energy is conserved upon incorporation of the new pigment in the RC. These changes in redox potential agree well with the free energy values deduced from the time resolved experiments. The redox potential for 3-acetyl-Phe $\underline{a}$ is 70 meV (580 cm^{-1}) below the one of BPhe $\underline{a}$. Assuming that after incorporation of this pigment in the RC the free energy of P$^+$H$_A^-$ is shifted up by a similar value compared to WT, the reaction modell delivers a very small long lived population density P$^+$B$_A^-$ of approx. 1 % for RCs containing 3-acetyl-Phe $\underline{a}$. This value is at the detection limit of the measurements shown in Fig. 2 (dashed-dotted line) and thus no significant difference in long time kinetics to WT-RCs is observed. We can therefore conclude that the change in energetics that is caused by the incorporation of the different types of chromophores can be well described by the redox potential change of the pigments.

The direct comparison of all four types of RCs demonstrates the role of different functional groups of BPhe $\underline{a}$ in the ET process. The modifications of ring II strongly influence the absorption properties of the chromophores H$_{A/B}$ and hence should reflect the modified electron density in the molecule. The side-group at C3-position (acetyl or vinyl) determines the energetics of the intermediate P$^+$H$_A^-$ via the redox potential value. This alteration leads to pronounced changes in the reaction kinetics of the isolated RCs.

[1] S. Schmidt, T. Arlt, P. Hamm, H. Huber, T. Nägele, J.Wachtveitl, M. Meyer, H. Scheer and W. Zinth, Chem. Phys. Letters 223 (1994) 116

[2] S. Schmidt, T. Arlt, P. Hamm, H. Huber, T. Nägele, J.Wachtveitl, M. Meyer, H. Scheer and W. Zinth, Spectochim. Acta (1995), in press

[3] H. Huber, M. Meyer, T. Nägele, I. Hartl, H. Scheer, W. Zinth and J. Wachtveitl, Chem. Phys. (1995), in press

[4] T. Arlt, S. Schmidt, W. Kaiser, C. Lauterwasser, M. Meyer, H. Scheer and W. Zinth, Proc. Natl. Sci. USA 90 (1993) 11757

Intensity Effects on Ionization Pathways in K_2: Control of the Wave Packet Dynamics

E. Schreiber[1], S. Rutz[1], R. de Vivie-Riedle[2]

[1]Institut für Experimentalphysik, FU-Berlin,
 Arnimallee 14, D-14195 Berlin
[2]Institut für Physikalische und Theoretische Chemie, FU-Berlin,
 Takustraße 3, D-14195 Berlin

Abstract

Experimental and theoretical pump-probe studies were performed for the K_2 molecule. Special molecular spectroscopic properties combined with the induced dynamics by the 'femtosecond state preparation' allow the transition from pump-probe to pump-control spectroscopy. Hereby, the intensity of the laser field serves as a control parameter in the multi photon processes. We can distinguish between the effect of two main processes, i.e. multi photon ionization (MPI) and impulsive stimulated raman scattering (ISRS).

Introduction

Recently femtosecond techniques have been advanced in wide spectral regions (from ultraviolett to near infrared) /1/ to directly probe molecular motions in real time. Various examples of bound state dynamics /2-8/ have been investigated by experimental and theoretical groups over the last few years.

On the experimental side, the dynamics of these molecules have been studied successfully by means of pump-probe spectroscopy on a femtosecond time-scale /2-4/. Here, besides transient fluorescence spectroscopy the technique of transient multi photon ionization (MPI) was applied. The principle of this method is sketched in fig. 1a. A first ultrashort laser pulse prepares a coherent superposition of vibrational eigenstates in an excited electronic state. Hence, a vibrational wave packet is produced propagating on the potential energy surface of this excited state. The wave packet's motion is probed by a second time-delayed ultrashort laser pulse. This pulse is used to ionize the molecule. The oscillatory changes of the detected ion signal with respect to the delay time reflect the wave packet's propagation and reveal besides the vibrational period a detailed information about the transition pathways between the involved electronic states.

On the theoretical side, time-dependent quantum calculations allow to simulate these pump-probe ionization experiments. The calculations yield the ionization pathways as well as the

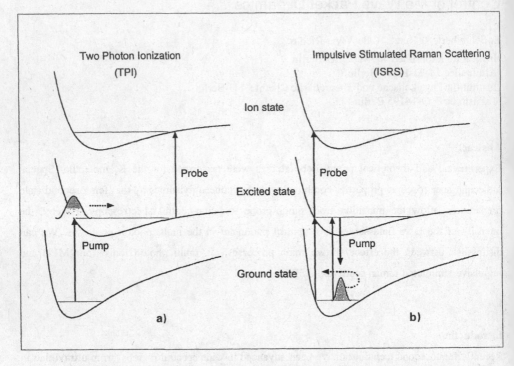

Fig. 1 *Schematic diagram for 2-photon ionization and impulsive stimulated resonant raman scattering. Ground, excited and ionic state of molecules as well as transitions between them after interaction with pump and probe laser pulses are shown (more details see text).*

transition mechanisms of the multi photon process. The interplay of theory and experiment then enabled an excellent understanding in the field of analysis of the MPI dynamics of small molecules /5/.

Beyond this, it is now of high interest to develop new far reaching concepts that use these special dynamics of the coherent superposition of states prepared by femtosecond laser pulses. The main goal is to influence a molecular system, either in the way to focus on vibrational modes in selected potential energy surfaces or to guide the system into distinctive reaction channels.

The control of reaction channels by femtosecond pulses was first suggested theoretically /9,10/ and was later on also observed in experiments /11,12/. In both cases the delay time between pump and probe (control) laser pulse was the control parameter. With some small modifications of the powerful pump-probe technique it is possible to drive molecular wave packets to a desired location on its potential energy surface from which selective dynamic processes (e.g. chemical reactions /11,12/) can be initiated.

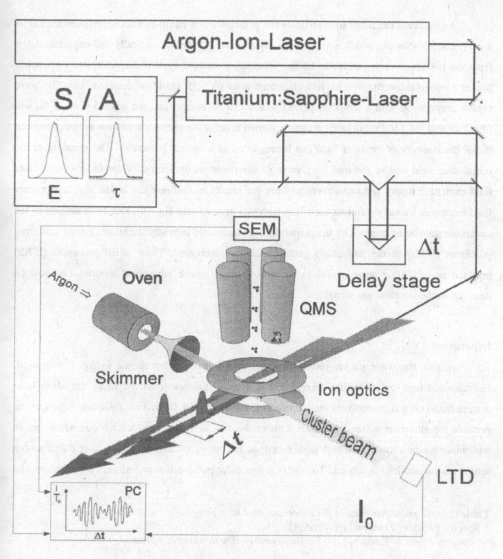

Fig. 2 *An Argon ion laser is used to pump a femtosecond laser (fs titanium:sapphire laser for moderate pulse power, ps titanium:sapphire laser with fiber compressor for high pulse power). 1% of the laser output is used to control the laser parameters by means of a spectrometer (S) and an autocorrelator (A). The laser is splitted with a 50% beam splitter, one part passes the delay unit. Both laser beams are recombined in the interaction region of the cluster chamber. Potassium vapor is produced in the oven and coexpanded with argon. The so produced cluster beam is collimated using a molecular beam skimmer. Potassium cluster ions produced by interaction of the laser beams with the cluster beam are focussed by the ion optics into the quadrupole mass filter (QMS) and continuously detected by means of a secondary electron multiplier (SEM). The Langmuir-Taylor detector (LTD) is used to control the cluster beam intensity. The intensity of ions is recorded as a function of the delay time ΔT.*

An excellent candidate to investigate the principles of a pump-control experiment should be a very simple molecule, which is relatively easy to handle, both, theoretically and experimentally. Here the potassium dimer seems to be the adequate molecular system. Experimentally the wave packet propagation in the first excited electronic state $(A^1\Sigma_u^+)$ could be detected /13/. The wave packet propagation with a temporal period which can be assigned to the ground state of K_2 was found as well /14,15/. In this paper it can be shown that for the excitation process in the potassium dimer the intensity of the laser field can be regarded as a control parameter. The intensity of the initial laser field will be the tool for control of the observed dynamics. While for moderate laser field excitation theory and experiment provide the results as outlined just before, for higher laser field excitation a totally different transition pathway appears (see fig. 1b). Due to saturation of the electronic state being pumped by the preparing high laser field intensity stimulated raman scattering processes (ISRS) to the molecule's ground state are increased. These ISRS processes (ISRS) produce an oscillating wave packet in the ground state whose vibrational motions can now be detected in the transient ion signal.

Experiment

To investigate the wave packet phenomena the experimental setup shown in fig. 2 was used. Femtosecond light pulses with a wavelength $\lambda_{central}$=840 nm are produced in an ultrashort laser system based on a regeneratively modelocked titanium:sapphire laser. Two different techniques to produce the ultrashort pulses were used. This enabled us to decide between a mode where pulses with moderate or a mode with high peak power of the pulses are available. The laser data for both setups are summarized in table 1. To realize a one color pump-probe experiment a Michelson-like

Table1: *Laser pulse parameters for moderate and high power.*
* Spectra Physics "Tsunami" (fs-Version);
** Spectra Physics "Tsunami" (ps-Version combined with pulse compressor)

	moderate power*	high power**
wavelength (nm)	840	840
pulse width (fs)	60 - 70	50 - 60
spectral width (cm^{-1})	190	260
peak power (10^9Wcm^{-2})	0.5	9

arrangement was used. The molecules were produced in a supersonic molecular beam. This provided vibrational and rotational temperatures $T_{vib} < 50$ K and $T_{rot} < 10$ K, respectively. After interaction with pump and probe laser pulses the generated ions were isotope selectively detected by means of a quadrupole mass spectrometer. The amount of so counted ions was recorded as a function of delay time between pump and probe pulses. A more detailed description of the experimental setup is given in /16/.

Theory

The time dependent Schrödinger equation is written and solved in the matrix representation:

$$
i\hbar \frac{\partial}{\partial t}
\begin{pmatrix} \Psi_1 \\ \Psi_2 \\ \vdots \\ \Psi_n \\ \Psi_{I(E')} \end{pmatrix}
=
\begin{pmatrix}
H_{11} & H_{12} & 0 & 0 & \cdots \\
H_{21} & H_{22} & H_{23} & 0 & \cdots \\
 & & \ddots & & \\
\cdots & 0 & H_{n-1n} & H_{nn} & H_{nI(E')} \\
\cdots & 0 & 0 & H_{I(E')n} & H_{I(E')I(E')}
\end{pmatrix}
\cdot
\begin{pmatrix} \Psi_1 \\ \Psi_2 \\ \vdots \\ \Psi_n \\ \Psi_{I(E')} \end{pmatrix}
\tag{1}
$$

Five relevant electronic states which are able to participate in the multi photon ionization process. he corresponding potential curves are the $X^1\Sigma_g^+$, $A^1\Sigma_u^+$, $4^1\Sigma_g^+$, $2^1\Pi_g$ states of the neutral and the K_2^+ ion ground state displayed in fig. 3. Crucial for the dynamical behaviour of the K_2 molecule is the close energetic position of the excited states $4^1\Sigma_g^+$ and $2^1\Pi_g$.

The matrix elements of the Hamiltonian (see eq. 1) describing the neutral molecular states are given by $H_{ii} = T_{ii} + V_{ii}$, where $T_{ii} = P^2/2m$ are the kinetic energy operators and V_{ii} are the potential energy surfaces. The off-diagonal elements describe the interaction with the laser field in the semiclassical dipole approximation and are given by $H_{ij} = -\mu_{ij} \cdot E(t)$ with μ_{ij} as the dipole transition moments. The electromagnetic field $E(t)$ is given by $E(t) = E_0 \cdot \cos\omega t \cdot s(t)$, where E_0 is the maximum amplitude of the electromagnetic field, ω the laser frequency and $s(t)$ a shape function. Gaussian functions are chosen as shape functions for both pump and probe pulse. The eigenfunctions Ψ_i symbolize the time dependent wave packets on the involved surfaces. The initial state of the system is assumed to be the lowest vibrational state of the electronic ground state. The basic foramlism used to propagate the wave packet in space is the FFT method /17,18/. The subscript I symbolizes the ion state. The electronic continuum due to the ejection of the electron with kinetic energy E_k is simulated by discretizing the corresponding energy range /19/.

The potential energy surfaces and the dipole transition moments μ_{ij} between the neutral states are obtained from ab initio data /20/. The transition moment into the ion state is assumed to be independent of E_k with an estimated constant value of $\frac{1}{3}\mu_{A\text{-}4\Sigma}$.

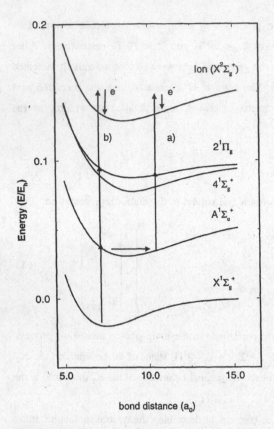

Fig. 3 *Ab initio potential energy surfaces of K_2 and transition pathways for*
a) resonant three photon ionization process
b) impulsive stimulated Raman scattering
(E_h =energy in hartree)

The parameters of the laser pulse were adjusted to the experimental values: pulse width 60 fs, wavelength 840 nm, intensity 0.8 GW/cm^2 and 9 GW/cm^2, respectively. The resulting fs pulses with the Gaussian shape have a spectral width of 245 cm^{-1}.

Results and Discussion

In the present five state model two main transition pathways can be distinguished (see fig. 3). Ionization pathway (a) is located at the outer turning point of the $A^1\Sigma_u^+$ state and involves the electronic states $X^1\Sigma_g^+$, $A^1\Sigma_u^+$, $2^1\Pi_g$, and the ion ground state continuum. Ionization pathway (b) which is located at the inner turning point of the $X^1\Sigma_g^+$ state involves the electronic states $X^1\Sigma_g^+$, $A^1\Sigma_u^+$, $4^1\Sigma_u^+$ and the ion ground state continuum /16/.

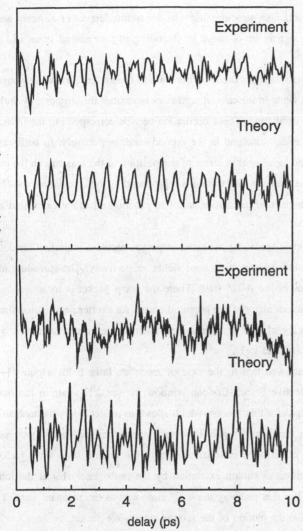

Fig. 4 *Comparison between measured pump-probe and calculated ionization signal as a function of delay time between the laser pulses. Panal a) moderate pulse intensities, panal b) high pulse intensities.*

Femtosecond pump-probe studies, experimental and theoretical, on the K_2 molecule were performed for moderate and high laser intensities. The comparison of experimental (upper curve) and theoretical total ion signal for delay times up to 10 ps is presented in fig. 4. The temporal evolution of the ion signal for moderate and high laser pulse intensities are shown in fig. 4a) and 4b), respectively. The obvious difference is that the transient ion signal for moderate laser intensity shows a distinct oscillation whereas for high laser field excitation the oscillation structure is rather purely observable. The disturbed oscillatory structure might be due to interference of different wave packet signals. The theoretical ion signals are obtained by including all four neutral states and the ion continuum in eq. 1.

In the case of moderate laser intensities the experimental and theoretical curves show strong oscillations with a period of 500 fs while in the case of high laser intensities the temporal evolution of the ion signal exhibits a 360 fs oscillation. These oscillation periods correspond to the induced motion of the wave packet in the $A^1\Sigma_u^+$ state and in the ground state, respectively. In both cases measurement and theory are in good agreement in terms of the position of the maxima. In the case of moderate laser field the envelope intensity modulation is reproduced by the calculation, too /16/. The beat structure in the case of high laser intensities is not so pronounced in the calculated ion signal.

In the investigated time domain spreading of the wave packet becomes important for delay times of about 7 ps for moderate and 9 ps for high laser fields, respectively. The spreading at a delay of 7 ps reflects the dynamics of the $A^1\Sigma_u^+$ state. There the wave packet is located around v=11 /16/. The influence of the anharmonicity of the potential occurs for shorter propagation times compared to the spreading at a delay of 9 ps for the wave packet in the ground state which is located at lower vibrational levels (around v=1).

Detailed calculations /21/ showed that in the case of moderate laser fields a pure (1+2) photon process takes place. A selective Franck-Condon window in the $2^1\Pi_g$ state at the outer turning point of the $A^1\Sigma_u^+$ state exists for this process which allows to reflect solely the motion of the $A^1\Sigma_u^+$ state in the ion state. In the case of higher laser intensities stimulated raman processes during the pump pulse induce an effective vibrational motion of the ground state wave packet, which opens the possibiltiy of a direct 3 photon excitation by the probe laser /16/ at the inner turning point of the ground state. For this pathway the $4^1\Sigma_g^+$ state acts as the resonant state. The resulting ion signal now mirrors also the motion of the ground state wave packet.

The existence of only one ionization pathway (a) in the case of moderate laser fields is reflected in the pure oscillations of the experimental and theoretical total ion signal. No interference with other pathways occurs. High laser intensities change the dynamical situation. Now, in principle two ionization pathways (a) and (b) are accessible. The oscillation period of 360 fs in the experimental signal and in the theoretical simulation, as well as, the results from the Fourier transform of the experimental ion signal /14/ indicate that in the case of high laser intensities the ionization pathway b) now dominates the pathway a). From the obtained theoretical results we can conclude that the contributions from the ionization pathway (b) cause the oscillation period, the contributions from process (a) the broadening of the oscillations in the total ion signal.

Conclusion

The intensity of the fs laser pulses can be used to control the ionization pathways in K_2. The laser field intensity influences the probability of multi photon processes, the efficiency of the ISRS processes as well as the Rabi oscillations.

During the transition from moderate to high laser fields the contributions from the energetically close lying $4^1\Sigma_g^+$ and $2^1\Pi_g$ states influence decisively the resulting ionization pathways and consequently the total ion signal. In the case of moderate laser fields the pathway (a) via the $2^1\Pi_g$ state is accessible, only. High laser fields, however, inforce the ISRR process and thus the contribution of higher vibrational levels in the ground state wave packet is increased. Its oscillation amplitudes are now large enough that its inner turning point is shifted versus shorter bondlength and reaches a nuclear conformation which now fulfills the resonance conditions with the laser frequency at the inner turning point via the $A^1\Sigma_u^+$ and $4^1\Sigma_g^+$ state and opens the pathway (b). The $4^1\Sigma_g^+$ state is now a resonant state in the 3 photon excitation scheme.

The possibility in the K_2 molecule to choose between different pathways, i.e. between different electronic states in the excitation ladder, opens the way from pump-probe to intensity controlled spectroscopy. The analysis shows that under certain spectroscopic conditions, as found in K_2, the intensity of the laser field serves as a control parameter. The induced femtosecond dynamics can probe selected molecular modes. Under this aspect the K_2 molecule can be regarded as a model system. We think that this example nicely demonstrates that femtosecond chemistry is providing exciting new discoveries in the field of control of chemical and physical processes.

Acknowledgment

We thank Prof. L. Wöste, B. Reischl and Prof. J. Manz for stimulated discussions. Generous financial support by the Deutsche Forschungsgemeinschaft DFG through project SFB 337 and a Habilitationsstipendium (for R. dV.-R.) is gratefully acknowledged.

References

[1]   W.S. Warren and M. Haner, in Atomic and Molecular Processes with short intense Laser Pulses, A.D. Bandrauk (Ed.), (Plenum Press, New York, 1988):
       W. Kaiser (Ed.), Ultrashort Laser Pulses and Applications: Topics in Applied Physics, Vol. 60 (Springer, Berlin, 1988).

[2]   L. R. Khundkar and A.H. Zewail, Ann. Rev. Phys.Chem. 41 (1990) 15;
       M. Gruebele, G. Roberts, M. Dantus, R.M. Bowmann and A.H. Zewail, Chem. Phys. Letters 166 (1990) 459;

M.H.M: Janssen, R.M. Bowman and A.H. Zewail, Chem. Phys. Letters 172 (1990) 99.

[3] T. Baumert, M. Grosser, R. Thalweiser and G. Gerber, Phys. Rev. Letters 67 (1991) 3753.

[4] J. Manz and L. Wöste (Eds.), Femtosecond Chemistry (Verlag Chemie, Weinheim, 1995).

[5] T. Baumert, V. Engel, C. Röttgermann, W.T. Strunz and G. Gerber, Chem. Phys. Letters 191 (1992) 639;

T. Baumert, V. Engel, Ch. Meier and G. Gerber, Chem. Phys. Letters 200 (1992) 488;

Ch. Meier and V. Engel, Chem. Phys. Letters 212 (1993) 691.

[6] V. Engel and H. Metiu, J. Chem. Phys. 92 (1990) 2317;

H. Metiu and V. Engel, J. Chem. Phys. 93 (1990) 5693.

[7] B. Hartke, R. Kosloff and S. Ruhman, Chem. Phys. Letters 158 (1989) 238.

[8] J. Manz, B. Reischl, T. Schröder, F. Seyl and B. Warmuth, Chem. Phys. Letters 198 (1992) 483.

[9] D. Tannor, R. Kosloff and S. Rice, J. Chem. Phys. 85 (1986) 5805.

[10] S. A. Rice, in Mode Selective Chemistry, J. Jortner et al. (Eds.), (Kluwer Academic Publishers, Netherlands,1991).

[11] E.D. Potter, J.L. Herek, S. Pedersen, Q. Liu, and A.H. Zewail, Nature 355 (1992) 66.

[12] T. Baumert and G.Gerber, Israel Journal of Chemistry, 34 (1990) 103.

[13] S.Rutz and E. Schreiber, Ultrafast Phenomena, IX, Springer Series in Chemical Physics, Vol. 60 (1994) 312;

S.Rutz, E.Schreiber, and L. Wöste, Surf. Rev. Lett., in press.

[14] K. Kobe, S. Rohland, S. Rutz, E. Schreiber and L. Wöste, submitted to J. Chem. Phys.

[15] E. Schreiber, in: Laser 94, Eds.: T. Goldmann and V.J. Cocroum, Society of Optical and Quantum Electronics.

[16] R. de Vivie-Riedle, K. Kobe, J. Manz, W. Meyer, B. Reischl, S. Rutz, E. Schreiber and L. Wöste, in preparation.

[17] R. Kosloff and D. Kosloff, J. Chem. Phys. 79 (1983) 1823;

D. Kosloff and R. Kosloff, J. Comput. Phys. 52 (1983) 35;

R. Kosloff, J. Phys. Chem. 92 (1988) 2087.

[18] C. Leforestier, R. Bisseling, C. Cerjan, M.D. Feit, R. Friesner, A. Guldberg, A. Hammerich, G. Jolicard, W. Karrlein, H.-D. Meyer, N. Lipkin, O. Roncero and R. Kosloff, J. Comput. Phys. 94 (1991) 59.

[19] R.S. Burkey and C.D. Cantrell, J. Opt. Soc. Am. B1 (1984) 169; B2 (1995) 485.

[20] Li Yan and W. Meyer, in preparation.

[21] R. de Vivie-Riedle and B. Reischl, Berichte der Bunsengesellschaft, in press.

Moderne Festkörperlaser
Advanced Solid State Lasers
R.Wallenstein

Q-Switching and Mode Locking of Solid-State Lasers by Semiconductor Devices

F. X. Kärtner and U. Keller
Institute of Quantum Electronics, Swiss Federal Institute of Technology
ETH Hönggerberg - HPT, CH - 8093 Zurich, Switzerland

Abstract:

We have demonstrated that an appropriately designed semiconductor saturable absorber can reliably start and sustain stable mode locking of solid-state lasers such as Nd:YAG, Nd:YLF, Nd:glass, Cr:LiSAF and Ti:sapphire lasers. Especially for solid-state lasers with rather long upper state lifetimes, previous attempts to produce self-starting passive mode locking with semiconductor saturable absorbers was always accompanied by self-Q-switching. Here, we derive criteria that characterize the dynamic behaviour of solid state lasers in the important regimes of Q-switching, Q-switched mode locking and continuous wave mode locking in the picosecond and femtosecond range for the pulsewidth. We demonstrate that semiconductor absorbers can be designed to predetermine the dynamic behaviour of a laser for a given solid state laser material and present an experimental verification. This allows for the development and design of robust, compact, pico- and femtosecond solid-state laser sources for scientific and industrial applications.

I. Introduction:

There is an intensive research towards an all-solid-state ultrafast laser technology. The availability of high-power diode lasers and laser arrays, which can be utilized to pump solid-state laser materials very efficiently, allows for more compact, reliable lasers with less utility requirements. New laser materials with extraordinary high intrinsic quantum efficiencies of more than 90% and an improved scalable cavity design make output powers in the kW-range from laser heads a size smaller than a shoe box possible in the near future [1]. Such powerful and compact laser sources will have applications in fundamental research like x-ray, plasma and higher harmonic generation as well as in medicine and industry.

To increase the laser peak power or to obtain ultrashort optical pulses one usually controls the laser dynamics by various Q-switching and continuous wave (cw) mode locking techniques which have been developed over the last three decades [2], [3]. Our research carries the idea of an all-solid-state ultrafast laser technology further by controlling the laser dynamics with compact scalable passive semiconductor devices. The semiconductor device essentially acts as a saturable absorber. It is well known that a saturable absorber can be used to passively Q-switch or modelock a laser. But there is a third regime where the laser is Q-switched-modelocked, i.e. modelocked pulses are underneath the Q-switched envelope. For most applications, this is an undesired behaviour.

It has been believed for a very long time that solid-state lasers like Nd:YAG or Nd:YLF with their long upper state lifetime can hardly be passively cw-modelocked using saturable absorbers without Q-switching or at least Q-switched mode locking [4]. However, in this paper we show that depending on the parameters of the saturable absorber, like recovery time of the absorption and saturation energy, pure passive Q-switching or pure continuous wave (cw) mode locking and, if desired, also Q-switched mode locking can be obtained. As a result full control over the saturable absorber device parameters is required to obtain well matched optimum values for a given solid-state laser material. Bandgap engineering and modern semiconductor growth technology in principle allow for the production of saturable absorbers with accurate control of the device parameters such as bandgap, saturation energy, and carrier lifetime. Semiconductor absorbers have an intrinsic bitemporal pulse response: intraband carrier-carrier scattering and thermalization processes which are in the order of 10 to several 100 fs as well as interband recombination processes which can be in the order of picoseconds to nanoseconds. This makes them ideal for the mode locking and Q-switching of solid state lasers in the picosecond and femtosecond regime.

In the first part of this paper we will investigate a basic laser model which describes the laser dynamics of solid-state lasers. Based on these model, we study the most important features of the laser dynamics analytically and derive criteria which characterize the laser dynamics with respect to Q-switching, self-starting of mode locking and Q-switched modelocking by a saturable absorber. We show that with the help of a properly designed saturable absorber even a solid state laser like Nd:YLF with its long upper state lifetime of 450 ms can be safely cw-modelocked without Q-switching [5, 6]. Saturable absorbers with relaxation times in the picosecond range can then generate stable picosecond pulses if the gain bandwidth is broad enough. Furthermore with a picosecond absorber one can even generate femtosecond pulses if soliton like pulse shaping is additionally employed [7]. This is a new regime of mode locking which we call soliton mode locking stabilized by a slow saturable absorber. It is extremely useful for diode pumped lasers where Kerr-Lens-Mode locking (KLM) is weak due to the large pump area in the laser crystal [8]. This mode locking technique also does not impose any requirements on the cavity design or operation of the cavity close to its stability limit [9].

II. Q-switching versus Q-switched mode locking and cw-mode locking in solid state lasers.

In this section we will briefly review the results derived by Haus on the parameter ranges for cw-Q-switching and cw-mode locking [4] to clarify the notation used and to extend these results to the case of Q-switched mode locking. We start from the most simple laser model of a single mode laser with homogeneously broadened laser medium and saturable absorber. Under the assumption that the transverse relaxation times of the equivalent two level models for the laser gain medium and for the saturable absorber are much faster than any other dynamics in our system we can use rate equations to describe the laser dynamics [10]. Further, we assume that the changes in the laser intensity, gain and saturable absorption on a time scale of the order of the roundtrip time T_R in the cavity are small, (i.e. less than 20%). Then we can write the rate equations of the laser as

$$T_R \frac{dP}{dt} = 2(g - l - q)P \tag{1}$$

$$T_R \frac{dg}{dt} = -\frac{g - g_0}{T_L} - \frac{g T_R P}{E_L} \tag{2}$$

$$T_R \frac{dq}{dt} = -\frac{q - q_0}{T_A} - \frac{q T_R P}{E_A} \tag{3}$$

where P denotes the laser power, g the gain per roundtrip, l the linear losses per roundtrip, q the saturable losses per roundtrip, g_0 the small signal gain per roundtrip, and q_0 the unsaturated but saturable losses per roundtrip. $T_L = \tau_L / T_R$ and $T_A = \tau_A / T_R$ are the upper state lifetime of the gain medium and the absorber recovery time normalized to the roundtrip time of the cavity. $E_L = \frac{h v A_{eff,L}}{2^* \sigma_L}$ and $E_A = \frac{h v A_{eff,A}}{2^* \sigma_A}$ are the saturation energies of the gain and the absorber respectively. The factor of 2^* in the definition of the saturation energies is due to the average over the standing wave effects in a linear cavity. If we use a ring cavity or if the laser is in pulsed operation and the media are much shorter than the equivalent pulselength we have to substitute 2^* by 1. For simplicity we also do not take into account effects due to spatial hole burning, which nevertheless can become important from a quantitative point of view, but will not change the overall qualitative picture [11, 12]. The factor of 2 in Eq.(1) is due to counting gain and loss with respect to amplitude.

For solid state lasers with relaxation times of the gain in the order of $\tau_L \approx 100 \mu s$ or more and cavity roundtrip times of $T_R \approx 10 ns$, we obtain $T_L \approx 10^4$. On the other side, we assume absorbers with recovery times much shorter than the roundtrip time of the cavity, i.e. $\tau_A \approx 1 - 100 ps$ so that typically $T_A \approx 10^{-4} - 10^{-2}$. This is achievable in semiconductors and can be engineered at will by low temperature growth of the semiconductor material [13, 14]. Thus as long as the laser is running cw and single mode, the absorber will follow the instantaneous laser power and we can eliminate the saturable absorption q adiabatically [15] by using (3) and we substitute in (1)

$$q = \frac{q_0}{1 + P / P_A} \quad \text{with} \quad P_A = \frac{E_A}{\tau_A} \tag{4}$$

where P_A is the saturation power of the absorber. At a certain amount of saturable absorption, the relaxation oscillations become unstable and Q-switching occurs.

II. A. Parameter ranges for cw- Q-switching

A linearized stability analysis of Eqs.(1) and (2) with (4) gives the stability criterion against Q-switching of the cw-running laser [4]:

$$-2P \frac{dq}{dP}\bigg|_{cw} < \frac{r}{T_L}\bigg|_{cw} \quad \text{with} \quad r = 1 + \frac{P}{P_L} \quad \text{and} \quad P_L = \frac{E_L}{\tau_L} \tag{5}$$

where r is the pump parameter which describes how many times above threshold the laser operates, and P_L is the saturation power of the laser gain. Inequality (5) has a simple physical explanation. The right side of (5) is the relaxation time of the gain into equilibrium at a given constant laser power. The left side is the decay time of a power fluctuation of the laser at fixed gain. If the gain can not react fast enough to fluctuations of the laser power, relaxation oscillations grow and result in passive Q-switching of the laser.

As can be seen from (4) and (5) we obtain

$$-2T_L P \frac{dq}{dP}\bigg|_{cw} = 2T_L q_0 \frac{\dfrac{P}{\chi P_L}}{\left(1 + \dfrac{P}{\chi P_L}\right)^2} < r\bigg|_{cw} \quad \text{with} \quad \chi = \frac{P_A}{P_L}. \tag{6}$$

where χ is an effective "stiffness" of the absorber against cw-saturation. As can be inferred from Eq. (6), the laser can never Q-switch if

$$-2T_L P \frac{dq}{dP}\bigg|_{cw} = \frac{2q_0 T_L}{\chi} < 1 \tag{7}$$

For solid-state lasers with long upper state lifetimes, very small amounts of saturable absorption, even a fraction of a percent, may lead to a large enough cw-Q-switching driving force (7) to Q-switch the laser. If we want to modelock the laser with the saturable absorber, we will try to leave the regime of cw-Q-switching. In principle one can always leave the regime of Q-switching if the laser can be pumped sufficiently to strongly saturate the absorber. Unfortunately, the cw-saturation power of an absorber is often higher than the damage threshold, so that the absorber can not be bleached with cw-radiation.

There is a trade-off: On one hand the cw-Q-switching driving force has to be small enough to prevent self-Q-switching and on the other hand the saturable absorption has to be large enough so that mode locking is self-starting.

II. B. Self-starting of mode locking due to a saturable absorber

We consider only one mechanism for self-starting of mode locking which is in our case the dominant mechanism. In an ideal homogeneously broadened laser without any spatial hole burning and without saturable absorber only one longitudinal cavity mode at the center of the gain is running and saturates the gain, so that all neighbouring modes are below threshold. Due to the saturable absorber two mode coupling [16] between the cavity mode at line center and the cavity mode m times the mode spacing $2\pi / T_R$ away from the line center occurs. This two-mode coupling provides additional gain for adjacent modes to reach threshold. The growth rate of these modes can be derived from Eqs.(18) to (20) of Ref. [4] and becomes in our notation:

$$\frac{1}{T_{mod}} = \left\{ \frac{2q_0}{\left(1 + \dfrac{P}{P_A}\right)^2 + (2\pi m T_A)^2} \frac{P}{P_A} - \frac{2g_0}{\left(1 + \dfrac{P}{P_L}\right)^2 + (2\pi m T_L)^2} \frac{P}{P_L} \right\}_{cw} \tag{8}$$

$T_{mod} = \tau_{mod} / T_R$, if positive, is the build-up time of neighbouring modes normalized to the roundtrip time. For a short normalized recovery time of the absorber, i.e. $2\pi m\, T_A \ll 1$ and long upperstate lifetime, i.e. $T_L \gg 1 + \dfrac{P}{P_L}$, we can simplify (8) for an enormous number of neighbouring modes and obtain for small cw-saturation of the absorber the approximate condition

$$\frac{1}{T_{mod}} = \frac{2q_0}{\left(1+\dfrac{P}{P_A}\right)^2}\frac{P}{P_A} - \frac{2g_0}{(2\pi m T_L)^2}\frac{P}{P_L} \approx \left(\frac{2q_0\, T_L}{\chi} - \frac{2g_0}{(2\pi m)^2 T_L}\right)\frac{P}{P_L T_L} \tag{9}$$

Thus, despite the fact that the absorber is too weak to lead to cw-Q-switching, see (7), it can easily be strong enough to excite neighbouring modes. These neighbouring modes grow until a new steady state is reached. With sufficiently large saturable absorption we can neglect effects like dispersion (i.e. unequal cavity mode spacing) and spatial hole burning which lead to frequency pushing and pulling of the individual cavity modes. The threshold for mode locking will be reached when the growth rate (8) is positive. Especially it is usefull to call the quantity $2q_0\, T_L / \chi$ the modelocking driving force, since it has to be large enough to overcome the gain saturation. Naively one would think that if (8) is positive and the absorber is hard enough to saturate so that (5) is satisfied, i.e. the modelocking driving force is less than one, the laser is always stable against Q-switching and would run purely cw-modelocked. However, what is often observed experimentally is a train of modelocked pulses under a Q-switched envelope (Fig. 1), referred to as Q-switched mode locking.

II. C. Parameter ranges for Q-switched mode locking

To understand the regime of Q-switched mode locking we reconsider the rate equations (1) to (3). Fig. 1 indicates that we can approximate the laser power as

$$P(T,t) = E_P(T)\sum_n f(t - nT_R) \text{ with } \int_{-T_R/2}^{T_R/2} f(t - nT_R)dt = 1 \tag{10}$$

where $E_P(T = nT_R)$ is the pulse energy of the n-th pulse which only changes appreciable over many cavity round-trips and $f(t)$ is the shape of the modelocked pulses which is not yet of interest. For simplicity we assume that the modelocked pulses are much shorter than the recovery time of the absorber, so that the absorber now saturates with the pulse energy and not any longer with the average intensity. In this case, we obtain for the loss in pulse energy per roundtrip

$$q_p(T) = \int_{-T_R/2}^{T_R/2} f(t)q(T,t)dt = q_0 \frac{1 - \exp\left[-\dfrac{E_P(T)}{E_A}\right]}{\dfrac{E_P(T)}{E_A}}. \tag{12}$$

With Eq. (12) for the loss in pulse energy per roundtrip we can set up rate equations for the slowly varying pulse energy and the laser gain on a coarse grained time scale in the order of the cavity roundtrip time. This rate equations are similar to Eqs.(1), (2) and (4) where P is replaced by pulse

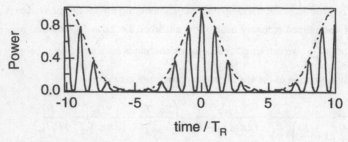

Fig. 1: *Time dependence of the laser output power in the Q-switched modelocked regime.*

energy E_P and the saturable losses by the losses for the pulse energy (12). Then the stability criterion (5) applies also to Q-switched mode locking if we replace the formula for cw-saturation of the absorber (4) by the formula for pulsed saturation (12)

$$-2E_P \frac{dq_P}{dE_P}\bigg|_{cw-mod} < \frac{r}{T_L}\bigg|_{cw-mod}$$

with

$$-2T_L E_P \frac{dq_P}{dE_P}\bigg|_{cw-mod} = 2T_L q_0 \frac{1 - \exp\left[-\dfrac{P}{\chi_P P_L}\right]\left(\dfrac{P}{\chi_P P_L} + 1\right)}{\left(\dfrac{P}{\chi_P P_L}\right)} \qquad (14)$$

where $\chi_P = \chi T_A$ describes an effective stiffness of the absorber compared with the gain when the laser is cw-modelocked at the same average power as the cw laser. Thus, similar to the case of cw-Q-switching it is useful to introduce the driving force for Q-switched mode locking:

$$-2T_L E_P \frac{dq_P}{dE_P}\bigg|_{cw} = \frac{2q_0 T_L}{\chi_P}. \qquad (15)$$

Comparing Eqs.(7) with (14) we obtain that in the case of cw-mode locking the absorber is more strongly saturated by a factor of $1/T_A$, which can easily be as large as 1000. Therefore, the tendency for Q-switched mode locking is significantly higher than for cw Q-switching. However, now it is much easier to strongly saturate the absorber with an average power well below the damage threshold of the absorber. Therefore, in the limit of strong pulsed saturation of the saturable absorber, we can design a saturable absorber which both produces a high mode locking driving force (9) to start the mode locking and is stable against Q-switching since the "pulsed stiffness" of the absorber is much lower.

The theory presented so far developed guidelines for the design of an absorber which prevents Q-switching instabilities but still self-starts mode locking. A saturable absorber alone can shorten a pulse until it is roughly on the order of the absorber relaxation time if, of course, the gain bandwidth is sufficiently large. The pulse formation process is essentially given by the fast saturable absorber mode locking model analysed by Haus [17]. This is the case for picosecond lasers where dispersion and self-phase modulation are not the dominant pulse shaping mechanisms. Recently, we have shown both theoretically and experimentally that soliton like pulse formation in combination with even a slow saturable absorber that has a recovery time much longer than the pulsewidth can stabilize the pulse [7]. This is in contrast to the traditional picture that the gain window has to close immediately before and after the passage of the pulse. This is possible in the soliton regime, because for the soliton the nonlinear effects due to SPM and the linear effects due to the negative GVD are in balance. In contrast, the noise or instabilities that possibly could grow within the longer time window are not intense enough to experience the Kerr nonlinearity and are therefore spread in time. When they spread in time they experience the higher absorption due to the slowly recovering absorber after passage of the much shorter soliton-like pulse. Thus the instabilities see less gain per roundtrip than the soliton and will decay with time.

III. The Antiresonant Fabry-Perot saturable absorber (A-FPSA) for controlling the Q-switching and mode locking in solid state lasers.

In the last section the parameter ranges for Q-switching, cw-mode locking, and Q-switched mode locking have been discussed. Given a laser with the key parameters intracavity pulse energy, loss, small-signal gain, upperstate lifetime of the laser material, and its saturation energy, it is important to be able to adjust the parameters of the absorber according to the conditions derived in section II in order to obtain stable, self-starting cw-mode locking or, if desired, operation in any other regime. In this section, we show how this can be achieved with the semiconductor antiresonant Fabry-Perot saturable absorber (A-FPSA) [18]. Antiresonance means that the intensity inside the Fabry-Perot is smaller than the incident intensity, which decreases the device loss and increases the saturation intensity. The A-FPSA is a nonlinear mirror, which simply replaces one of the laser cavity mirrors to passively mode-lock a cw pumped laser. The nonlinear reflectivity change in the A-FPSA is due to band-filling where the absorption is bleached with the photo-excited carriers because of the Pauli-exclusion principle. The impulse response shows a bitemporal behaviour, i.e. a slow time constant due to carrier recombination for efficient starting of the mode locking and the generation of ps-pulses as described in section II, as well as a fast time constant due to carrier thermalization for further pulse shortening and sustaining of sub-100-fs pulses utilizing soliton mode locking.

The samples we have built use a bottom mirror which consists of 16 to 25 pairs of GaAs- and AlAs-quarter wave layers forming a Bragg mirror with a center wavelength of either 830 nm (for Cr:LiSAF) or 1050 nm (for Nd-doped crystals or glasses) and approximately 100 nm bandwidth (see Fig. 2). On top of the mirror, the absorber is grown consisting of bulk (Al)GaAs for operation below 870 nm or a strained InGaAs/GaAs superlattice for operation at 1050 nm, where the Indium content in the wells determines the absorption edge which can be varied from 900 nm to potentially 2 μm. Finally, a top mirror consisting of 3 or 4 pairs of SiO_2- and TiO_2-quarter wave layers is evaporated on top of the absorber. This may increase the damage threshold for the device and adjustes the effective saturation intensity of the absorber to the laser. In addition, the absorber is grown at low temperatures (i.e. between 200°C and 400°C instead of > 600°C [13, 14]). The

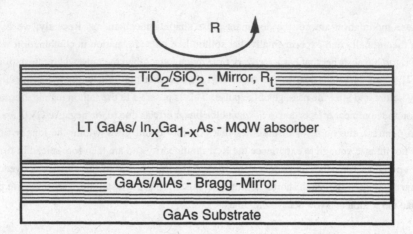

Fig. 2: *Structure of an Antiresonant Fabry-Perot Saturable Absorber (A-FPSA) designed for an operation wavelength of ≈ 1 μm. For operation at 800 nm, an AlGaAs bulk absorber was used.*

advantages of low temperature MBE-growth are twofold: first, incorporation of excess arsenic in form of interstitials and clusters leads to interband states which drastically reduce the lifetime of photogenerated carriers, and thus the absorber recovery time - an essential parameter for the mode locking performance [23], see also Eqs.(4) and (21). Second, the degradation in surface morphology of strained InGaAs/GaAs MQW's using normal growth conditions, which would result in high scattering losses, are reduced since the defects have dimensions much smaller than the optical wavelength and no cross-hatched roughness is present in our LT-grown MQW's.

IV. Experimental results

Samples from the same epitaxial runs with top mirror reflectivities of $96\% \pm 0.5\%$ have been used as A-FPSA's in various mode locking experiments with Nd:YAG, Nd:YLF and Nd:glass lasers [6]. To demonstrate the validity of the Q-switching and self-starting criteria developed above, we built a Nd:YLF laser with the gain medium in the middle of the cavity to minimize effects due to spectral hole burning [11]. The numerical evaluation of the stability condition for Q-switched modelocking Eq. (13) is shown in Fig. 3 for the different absorber parameters.

The value for the normalized upper state lifetime of the Nd:YLF laser was $T_L = 117 \cdot 10^3$. We used different samples which had about the same amount of saturable absorption $2q_o \approx 0.0045$ resulting in a critical product of $2q_oT_L = 530$, as shown in Fig. 3. For parameter values inside the corresponding loop in Fig. 3 the laser operates in the regime of Q-switched modelocking. The absorbers were grown at different temperatures and therefore had different carrier lifetimes $\tau_A = 3.8ps, 15ps$ and $22ps$. The sample with the lowest carrier life time exhibited the highest saturation intensity and therefore the lowest mode locking driving force and hence did not reach the self-starting condition for modelocking (9). The experimentally determined transition points from the regime of Q-switched mode locking to pure cw-mode locking for the other two absorbers are shown

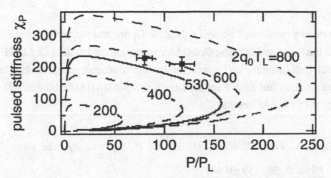

Fig.3: *Comparison between theory and experiment of the stability boundaries for Q-switched mode locking according to Eq.(13) for saturable absorbers with different growth temperature. The observed tansitions should be compared with the full line shown.*

in Fig. 3. At larger pump powers, where the absorber is more strongly saturated and with a high pulsed stiffness due to the high reflectivity of the top reflector the regime of Q-switched mode locking can be overcome and we obtain pure cw-modelocked pulses. A more detailed evaluation of this experiment can be found in [19].

There are several ways to produce a high stiffness of the absorber one is using a high top reflector, the other is to reduce the focussing on the saturable absorber. In general it will depend on the cavity design and the damage threshold of the absorber which technique will be used. Similar absorbers with different thickness of the absorber layer and top reflectivity have been used to modelock Cr:LiSAF [20], Nd:glass [8] and Ti:Sapphire lasers [9] in the femtocesond regime using the fast time constant of the semiconductor absorber and soliton formation in the laser.

V. Conclusion

Starting from a simple two level laser and absorber model we characterized the dynamics of solid state lasers modelocked and Q-switched by the absorber. Low-temperature grown semiconductors are an excellent material for saturable absorbers to modelock lasers since the carrier lifetime can be engineered to values well below typical roundtrip times in solid state laser cavities. We demonstrated experimentally that devices incorporating this material can control the laser dynamics of a large variety of solid-state lasers in the picosecond regime. In addition, by exploiting the bitemporal behaviour of the semiconductor devices and soliton formation due to negative GVD and SPM, we can use the same semiconductor absorbers to modelock the lasers in the femtosecond regime.

The saturable absorber techniques demonstrated do not interrelate the spatial profile of the laser mode with the laser dynamics as is the case for Kerr-Lens-Mode locking. Therefore, we believe that the saturable absorber control of the laser dynamics as shown here is easily scalable to the new emerging high power laser designs and will lead to less restricted design tolerances in general.

Acknowledgement:
This work was supported by the Swiss Priority Program in Optics and the Swiss National Fund No. 21-39362.93. Special thanks go to Lightning Optical for supplying the low loss Cr:LiSAF crystal, to Heng Chiu from AT&T Bell Laboratories for some of the semiconductor saturable absorbers and M. Moser at the Paul Scherrer Institut Zürich for the broadband AlGaAs/AlAs bottom mirrors which were used for some of the Cr:LiSAF results.

References:
1. A. Giesen, H. Hügel, A. Voss, K. Wittig, U. Brauch, H. Opower, *Applied Phys. B* **58**, 363 (1994)
2. E. P. Ippen, *Appl. Phys. B* **58**, 159 (1994)
3. A. Penzkofer, *Appl. Phys. B* **46**, 43 (1988)
4. H. A. Haus, *IEEE J. Quantum Electronics* **12**, 169 (1976)
5. U. Keller, D. A. B. Miller, G. D. Boyd, T. H. Chiu, J. F. Ferguson, M. T. Asom, *Optics Lett.* **17**, 505 (1992)
6. U. Keller, *Applied Phys. B* **58**, 347 (1994)
7. F. X. Kärtner, U. Keller, *Optics Lett.* **20**, 16 (1995)
8. D. Kopf, F. X. Kärtner, K. J. Weingarten, U. Keller, *Optics Lett.* **20**, 1169 (1995)
9. I. D. Jung, L. R. Brovelli, M. Kamp, U. Keller, to appear in Optics. Lett. (1995)
10. A. E. Siegman, *Lasers* (University Science Books, Mill Valley, California, 1986)
11. B. Braun, K. J. Weingarten, F. X. Kärtner, U. Keller, to appear in *Applied Physics B* (1995)
12. F. X. Kärtner, B. Braun, U. Keller, to appear in *Applied Physics B* (1995)
13. G. L. Witt, R. Calawa, U. Mishra, E. Weber, Eds., *Low Temperature (LT) GaAs and Related Materials* , vol. 241 Pittsburgh, 1992)
14. T. H. Chiu, U. Keller, M. D. Williams, M. T. Asom, J. F. Ferguson, *J. Electronic Materials* , submitted (1994)
15. H. Haken, *Synergetics: An Introduction* (Springer Verlag, Berlin, New York, 1983)
16. R. W. Boyd, *Nonlinear Optics* (Academic Press, New York, 1992)
17. H. A. Haus, J. G. Fujimoto, E. P. Ippen, *JOSA B* **8**, 2068 (1991)
18. U. Keller, D. A. B. Miller, G. D. Boyd, T. H. Chiu, J. F. Ferguson, M. T. Asom, in *Advanced Solid-State Lasers* , L. Chase, A. Pinto, Eds. (Optical Society of America, Washington, D.C., 1992), **13**, 98
19. F. X. Kärtner, L. R. Brovelli, D. Kopf, M. Kamp, I. Calasso, U. Keller, *Optical Engineering*, July (1995)
20. D. Kopf, K. J. Weingarten, L. Brovelli, M. Kamp, U. Keller, *Optics Lett.* **19**, 2143 (1994)

Passive Mode-Locking and Q-Switching of a 2.7 μm Fluoride Fiber Laser

Christian Frerichs, Joachim Urner
Institut für Hochfrequenztechnik
Technische Universität Braunschweig
P.O. Box 3329
D-38023 Braunschweig, Germany

Passive Q-switched and mode-locked operation of a 2.7 μm fluoride fiber laser is reported. Q-switched pulses containing mode-locked pulse trains were attained by the means of InAs epilayers used as saturable absorbers. The measured mode-locked pulse duration was 2 ns, which was the resolution limit of the detection setup. Average output powers up to 16 mW could be measured.

3 μm lasers have many potential applications especially in the medical field due to the strong absorption of water in this range. Fiber lasers at this wavelength combine long interaction lengths and high power densities in the active core with negligible thermal problems, thus leading to potentially low thresholds and high efficiencies. The standard fiber laser in this range is the the Er^{3+}-doped fluoride fiber laser at 2.7 μm based on the IR glass ZBLAN utilizing the transition $^4I_{11/2} ->$ $^4I_{13/2}$. This laser has been realized with threshold powers below 1 mW, with slope efficiencies exceeding 20% and has also been laser diode pumped [1].

Many applications need high peak powers in short pulses, e.g. to reach power levels required for tissue ablation. This can be attained by Q-switching or mode-locking the laser. In the 3 μm region, both techniques have been realized with crystal lasers both actively and passively [2, 3, 4]. Q-switching of a 2.7 μm fiber laser utilizing an acousto-optic modulator was reported recently [5]. Mode-locking of a 2.7 μm fiber laser was realized by the so called flying-mirror mode-locking technique [6]. In this paper the pulsed operation of a 2.7 μm fiber laser by saturable absorber mode-locking and Q-switching is presented for the first time. InAs epilayers on GaAs substrates served as saturable absorbers. This passive method previously yielded mode-locked and Q-switched pulses at 3 μm erbium crystal lasers [2, 3].

Fig. 1 shows the experimental setup. A 1000 ppm Er^{3+}-doped ZBLAN fiber with a core diameter of 30 μm and a numerical aperture of 0.15 made by Le Verre Fluoré was used, its length ranging

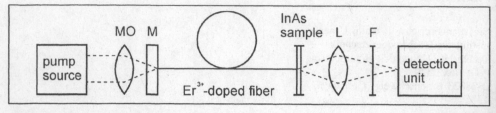

Fig. 1: Experimental setup

from 120 to 140 cm. The input mirror M with R(2.7 μm) > 99% was butted to the fiber endface. The laser was pumped by a 500 mW SDL diode laser or a Ti:Sapphire laser, both emitting at 792 nm and coupled into the fiber by a microscope objective MO. The InAs epilayers, MBE grown on 0.25 mm thick GaAs substrates [4], were directly butted against the rear fiber end, thus serving simultaneously as output mirror and as saturable absorber. Samples of 0.09, 0.1, 0.15, 0.2 and 0.4 μm thickness were used. The output radiation was focussed onto the detection unit by an IRGN6 lens L through a long pass filter F. The pulse shapes were monitored with an InSb photodiode, the signal of which was preamplified and displayed on a 400 MHz storage oscilloscope. With preamplifier the system response time was 2.5 ns, without less than 2 ns. Average power measurements were performed with a pyroelectric power meter.

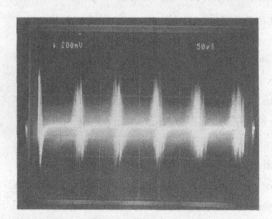

Fig. 2: Photograph of Q-switched pulses (0.1 μm layer, P_{launch} = 81 mW)

Starting the fiber laser with any InAs sample yielded Q-switched pulses containing mode-locked pulse trains. Fig. 2 shows a photograph of several superimposed Q-switched pulses measured with a 0.1 μm InAs layer. The negative ringing is due to the response of the preamplifier. The average distance between the pulses amounted to 100 μs. There was large time and also amplitude jitter in the pulse trains. The distance between the pulses (PRP) was pump power dependent: with increasing power the PRP decreased. At a launched power of 230 mW the average PRP with the 0.1 μm sample was only 12 μs. The average PRP was also dependent on the thickness of the InAs layer: at a launched power of 81 mW the largest PRP of about 900 μs was measured with the 0.4 μm sample, the shortest with the 0.09 μm sample amounted to 60 μs.

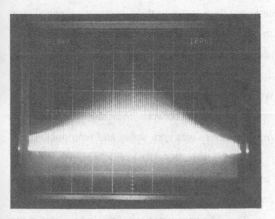

Fig. 3: Photograph of a Q-switched pulse
(0.4 μm layer, P_{launch} = 230 mW)

Fig. 3 shows a photograph of one Q-switched pulse recorded with a 0.4 μm sample at a launched pump power of 230 mW. In this case the output power was large enough to omit the preamplifier. The mode-locked pulses in the Q-switched envelope are clearly to be seen. The FWHM duration of the pulse was about 0.5 μs. The Q-switched pulse width was also a function of pump power and of layer thickness. Increasing the pump power yielded a decrease of the pulse duration. With the 0.4 μm sample the pulse width at a launched power of 81 mW amounted to 1.2 μs, at 230 mW 0.5 μs. At the maximum available power of about 250 mW a pulse duration of down to 0.4 μs could be measured. At a launched power of 81 mW the pulse widths varied between more than 10 μs with a 0.09 μm layer and 1.2 μs with a 0.4 μm layer. Q-switched operation with the 0.09 μm sample was very irregular, large time and amplitude jitter occurred.

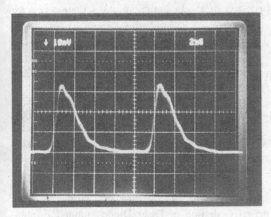

Fig. 4: Photograph of mode-locked pulses
(0.4 μm layer, P_{launch} = 230 mW)

Fig. 4 shows a photograph of two mode-locked pulses recorded with a 0.4 μm layer at a pump power of 230 mW. The FWHM pulse width was about 2 ns. At lower pump powers the preamplifier had to be used. In this case the mode-locked pulses always had FWHM durations of about 3.5 ns and fall times of about 2.5 ns. Hence it can be assumed that the pulse lengths measured are limited by the resolution limit of the detection setup and that the real pulse widths are much shorter at least at optimized conditions. The PRF was always dependent on the fiber length and increased from 70 to 90 MHz when using fiber lengths from about 140 to 110 cm. Due to a time base inaccuracy the time axis on the nanosecond scale shown in Fig. 4 has to be multiplied by a factor of 1.12.

The lasers emitted average powers comparable to cw fiber lasers in this wavelength range. Laser thresholds were measured in the range of 20 to 35 mW launched power. The thresholds increased with increasing layer thickness. The slope efficiencies reached 3.2 % with laser diode pumping and up to 9.4 % with Ti:Sapphire laser pumping. In the latter case the maximum average output power was 16.0 mW at a launched power of 232 mW, while with laser diode pumping only 1.58 mW at 81 mW could be achieved. There was no systematical relation between layer thickness and slope efficiency. The pulsed operation of the 2.7 μm fiber lasers was very stable and reproducible.

Generally, the results reported here meet the expected behaviour of the passively modulated fiber laser. Q-switching occurs due to the forced energy storage by the saturable absorbers leading to giant pulses when switching to high transmission. The short recovery time of the InAs layers in the order of 100 ps [4] allows for mode-locking. Therefore, the measured pulse widths could be in the order of the relaxation time of the InAs epilayers. Larger PRPs, shorter, more powerful pulses and higher threshold powers with thicker samples indicate that the switching threshold with these samples is higher than with thinner ones. There might be a great potential for optimization by adjusting layer thickness and fiber geometry. This passive method of mode-locking and Q-switching of 2.7 μm fiber lasers might hence be a simple, inexpensive and reliable technique for achieving short pulses in the 3 μm range.

The authors want to express their gratitude to Dr. Martin Frenz and Dr. Ludwig Wetenkamp for making the InAs samples available.

[1]   Ch. Frerichs, "All Optical Modulation of a 2.7 μm Erbium-Doped Fluorozirconate Fiber Laser", *OSA Proc. on Advanced Solid State Lasers* **15**, p. 399-402, (1993).

[2]   F. Könz, M. Frenz, V. Romano, M. Forrer, H. P. Weber, A. V. Kharkovskiy, S. I. Khomenko, "Active and Passive Q-Switching of a 2.79 μm Er:Cr:YSGG Laser", *Opt. Commun.*, **103**, pp. 398-404, (1993).

[3]   K.L. Vodopyanov, A.V. Lukashev, C.C. Phillips, I.T. Ferguson, "Passive Mode Locking and Q Switching of an Erbium 3 μm Laser Using Thin InAs Epilayers Grown by Molecular Beam Epitaxy", *Appl. Phys. Lett.*, **59**, pp. 1658-1660, (1991).

[4]   B. Pelz, M.K. Schott, M.H. Niemz, "Electro-optic Mode Locking of an Erbium:YAG Laser with a RF Resonance Transformer", *Appl. Opt.*, **33**, pp. 364-367, (1994).

[5]   Ch. Frerichs, T. Tauermann, "Q-switched Operation of Laser Diode Pumped Erbium-Doped Fluorozirconate Fibre Laser Operating at 2.7 μm", *Electron. Lett.*, **30**, pp. 706-707, (1994).

[6]   Ch. Frerichs, "Mode-Locking of a 2.7 μm Erbium-Doped Fluoride Fiber Laser", in: *Trends in Optical Fibre Metrology and Standards*, ed. O.D.D. Soares, NATO ASI Series, Vol. 285, pp. 840-841, (1995).

First Results of an Electrooptically Switched all-in-one Ti:Sapphire Laser

F.H. Loesel, C. Horvath, S. Ohlhauser, M.H. Niemz
University of Heidelberg, Dept. of Applied Physics
Albert-Ueberle-Str. 3-5, 69120 Heidelberg, Germany

Abstract: The potential of all-in-one techniques in modern femtosecond solid state lasers is being discussed. For first experiments we designed an electrooptically switched Ti:Sapphire laser. Q-Switching and dumping was performed at half-wave voltages. The generation of ultrashort pulses with energies at the μJ-level in one single cavity seems to be within the realms of possibility.

Modern solid state lasers, capable of generating ultrashort pulses in the sub-100 fs-range, are now commercially available. The dominating process for shaping the short pulses is Kerr lens modelocking (KLM) [1]. Pulse energies of these basic oscillator lasers are in the order of several nJ, typically. Higher energy levels can be reached by using an acoustooptic dumper to extract an increased fraction of the intracavity energy. RAMASWAMY et al. obtained 100 nJ pulses at a repetition rate of up to 950 kHz with this design [2]. Regeneratively amplified fs laser systems or multipass amplifiers yield output energies from several hundred μJ up to the mJ-level and are now also available from some manufacturers. Regarding the pulse energies, one can find a "gap" at the region of several μJ between the cavity dumped laser and the externally amplified systems, which is especially promising for the cold ablation of biological materials, using fs pulses and a plasmamediated process [3,4]. As the regenerative or multipass amplifiers are too complex and expensive for a future medical laser system, it is of high interest to investigate the potential of the all-in-one technique in the femtosecond range. This generation and amplification in just one single cavity already permits the construction of compact, reliable and powerful lasers in the picosecond region (e.g.[5]).

For the means of amplification and dumping of the pulses we inserted a Pockels cell (PC) and a polarizer (POL) in a standard KLM cavity (Fig.1). The laser contained a 10 mm long Ti:Sapphire crystal as the gain medium in a z-fold ($R_{M2,M3} = 100$ mm) and a pair of SF10 prisms $(P_{1,2})$ for compensation of the group velocity dispersion. The crystal was pumped through the lens L ($f_L = 100$ mm) with 6.5 W of visible multi-line emission from an argon-ion laser. The output coupler (OC) had a low transmission of 3%. A Brewster-cut acoustoop-

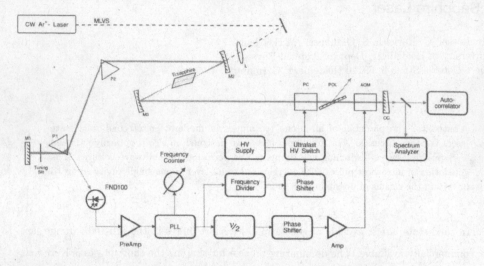

Figure 1: Schematic of the all-in-one laser system

tic modulator (AOM) served as the starting mechanism for the Kerr lens modelocking. The modulator was driven at an efficiency of about 1% by the phase-locked loop electronics of a regenerative feedback unit. Without the Pockels cell and the polarizer the laser produced pulses of 60 fs duration (F2 prisms), using gain guiding as a "soft" aperture [6]. For future use in a Cr:LiSAF laser the additional all-in-one optics were designed to show a minimum of passive losses. The Pockels cell consisted of a Brewster-cut KD*P crystal with a length of 18 mm and two 1.5 mm thick fused silica windows. The transmission was about 99%, and due to a z-cut geometry the passive birefringence of the cell was zero. The 2 mm thick polarizer was soft coated and had a very high transmission of 99.5% for p-polarized light and a reflectivity of about 85% for s-polarized light. Using ultrafast transistor switches the half-wave voltage of

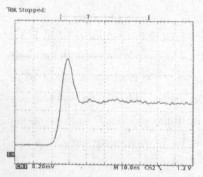

Figure 2: Optical response of the Pockels cell

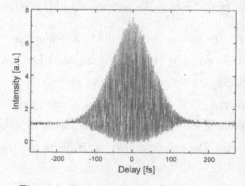

Figure 3: Autocorrelation of a 90 fs pulse

about 5.3 kV (at 800 nm) could be applied to the electrooptical cell. The optical rise time of the Pockels cell to the half-wave level was measured between crossed polarizers. Fig.2 shows the optical transmission of the setup when applying higher voltages. After 10 ns a peak marks the half wave voltage. Further increasing the voltage then reduces the portion of transmitted light as the polarization is rotated by more than 90°.

Without a high voltage over the Pockels cell we first investigated whether Kerr lens modelocking could still be observed, despite the high amount of optical material in the cavity. After varying the separation length of the prism pair we obtained stable KLM operation with pulse durations of about 90 fs (Fig.3). We measured a spectral width of 11 nm (FWHM), resulting in a pulse-bandwidth product of 0.45.

We then used the following switching scheme to investigate the all-in-one performance of the laser system: Applying the half-wave high voltage over the Pockels cell rotated the polarization of the light inside the cavity by 90°, and the initial p-polarization was reflected at the polarizer (the s-polarization experienced high losses at the Brewster-cut elements). Therefore, the cavity Q was low, lasing could not start, and the rate of inversion in the gain medium increased. Fast switching of the voltage to zero then suddenly increased the cavity Q, and a Q-switch pulse with 600 ns FWHM was generated. Without the regenerative feedback as a starting mechanism the Q-switch pulse was free of any modulation. When we started the PLL-driven acoustooptic modulator (at low efficiency) the Q-switch pulse showed some pulse-like modulation evolving under the envelop. Optimization of the phase shift in the PLL electronics and of the cavity configuration finally generated a fully modulated structure of short pulses within the Q-switch pulse (Fig.4). At the saturation limit the Pockels cell again was switched to its half-wave voltage within 10 ns and the short pulse was dumped out of the cavity at the polarizer (Fig.5).

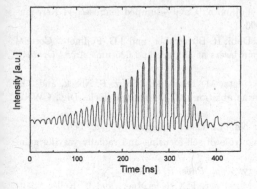

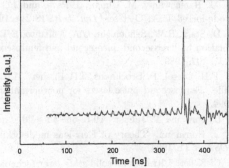

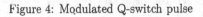

Figure 4: Modulated Q-switch pulse   Figure 5: Dumped pulse at max. gain

Switching of the Pockels cell was synchronized with the round trip frequency of the cavity to phase-lock the dumping process. The repetition rate was limited to 2 kHz by the HV power supply. We measured dumped pulse energies of about 300 nJ and noticed that damage at the soft coated polarizer occurred. Considering the extraction efficiency and the possibly reduced performance of the polarizer, we calculated the intracavity energy to be >650 nJ. Measurements after shortening the high-Q switch time by removing a preliminary series resistor for short circuit protection in the switches suggested a further increase in the energy by a factor of about 1.5. In our opinion, enhanced polarizer coatings, the integration of the polarizer as a window into the Pockels cell and higher pump powers could yield energies of the dumped pulses at μJ-level. Changing the position of mirror M_3 and of the prism P_2 resulted in pulse durations limited by the resolution of a fast photodiode (1ns) used for monitoring. Here, the quality of the modulation was very sensitive to variations of these elements. Nevertheless, we also observed fully modulated Q-switch pulses were the structure was not diode limited and the modulation seemed to be insensitive to changes. Autocorrelation measurements of the pulse duration could not be obtained during these first experiments, and it is still the question whether Kerr lens modelocking can generate ultrashort pulses during the Q-switch pulse. Alternatively, the use of solid state saturable absorbers [7] with their fast build-up time of the ultrashort pulse [8] seem to be a promising technique for the design of an all-solid-state all-in-one laser generating ultrashort pulses with energies at the μJ-level.

Frieder H. Loesel is grateful to the German National Scholarship Foundation for their support in many fields.

References

1. D.E. Spence, P.N. Kean, and W. Sibbett. 50-fsec pulse generation from a self-mode-locked Ti:sapphire laser. *Opt. Lett.*, 16(1), 1991.

2. M. Ramaswamy, M. Ulman, J. Paye, and J.G. Fujimoto. Cavity-dumped femtosecond Kerr-lens mode-locked Ti:Al$_2$O$_3$ laser. *Opt. Lett.*, 18(21), 1993.

3. D. Stern, R.W. Schoenlein, C.A. Puliafito, E.T. Dobi, R. Birngruber, and J.G. Fujimoto. Corneal ablation by nanosecond, picosecond and femtosecond lasers at 532 nm and 625 nm. *Arch. Ophthalmol.*, 107, 1989.

4. F.H. Loesel, P. Brockhaus, J.P. Fischer, M.H. Goetz, M. Tewes, M. Niemz, F. Noack, and J.F. Bille. Femtosecond pulse lasers for nonthermal tissue ablation. CLEO Europe Tech. Dig., CWN2, 1994.

5. L. Turi, and T. Juhasz. All-in-one all-solid-state Nd:YLF laser. Accepted by: *Opt. Lett.*

6. J. Herrmann. Theory of Kerr-lens modelocking: role of self-focusing and radially varying gain. *JOSA B*, 11, 1994.

7. U. Keller. Ultrafast all-solid-state laser technology. *Appl. Phys. B*, 58, 1994.

8. L.R. Brovelli, I.D. Jung, D. Kopf, M. Kamp, M. Moser, F.X. Kaerntner, and U. Keller. Self-starting soliton modelocked Ti-sapphire laser using a thin semiconductor saturable absorber. *Elect. Lett.*, 31(4), 1995.

Full Finite Element Analysis of Diode Pumped Lasers

K. Altmann
Micro Systems Design Dr. Altmann GmbH
Brunhildenstr. 9, 80639 München

Abstract

By the use of the analogy between the tranversal paraxial wave equation and the differential equation of a strechted membrane embedded on an elastic foundation with locally varying stiffness a full *Finite Element Analysis* (FEA) of thermo-mechanical effects on the formation of optical modes in an end-pumped laser rod was carried through. By an iteration procedure the extracted optical power and the beam profile have been taken into account in a dynamic heat balance.

Summary

The Parametric Design Language provided with the commercial FEA code ANSYS[1] was used to develop a simulation code in order to analyze the interaction between thermo-mechanical effects, material properties and optical mode formation in diode pumped solid state lasers.

Thermo-mechanical Simulation

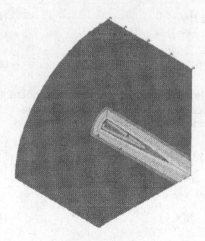

Fig. 1. Distribution of the pumping beam

The simulation of the thermo-mechanical behavior of the crystal is a straightforward FEA procedure. In ANSYS standard moduli are prepared to do the geometric model, the mesh and the solution. As inputs the material properties of the laser rod, the resonator configuration, the distribution of the pumping beam, and as a boundary condition the lay-out of the cooling arrangement are used. Fig. 1 shows a quarter section of a laser rod with a pumping beam distribution typical for frontal pumping. In this example, the rod is assumed to be cooled along the lateral surface, however any other boundary conditions for the temperature e.g. frontal cooling can be used. Figs. 2 and 3 show the temperature distribution and the thermal deformation obtained with these inputs by FEA.

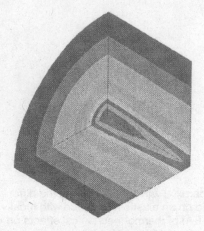 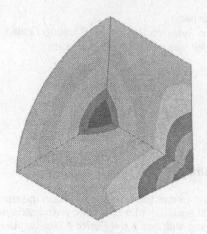

Fig. 2. Temperature distribution   Fig.3. Thermal deformation

Optical Simulation

The obtained temperature distribution and the thermal deformation are used as inputs for the computation of the transversal mode structure. The latter can be done by the use of an electromagnetic BPM code as shown in Ref. 2. However, in the present program this computation is also carried through by the use of FEA tools. For this purpose a mechanical model was developed whose vibrations are described by a differential equation which is formally identical to the transversal part of the scalar paraxial wave equation. The latter can be written in the form

$$\nabla_t^2 \psi + \left[\Theta - \Phi(x,y)\right]\psi = 0 \tag{1}$$

with

$$\Theta = k^2 n_0^2 - \beta^2. \tag{2}$$

Here ψ is the transversal distribution of the complex amplitude of any field component. k is the vacuum wave number, and β the scalar propagation constant. x and y are the transversal coordinates. In the case of a wave guide with the local index of refraction given by

$$n(x,y) = k^2 n_0^2 - \tilde{n}(x,y) \text{ with } \tilde{n} \ll n_0 \tag{3}$$

the function $\Phi(x,y)$ has the form

$$\Phi(x,y) = 2n_0 k^2 \tilde{n}(x,y). \tag{4}$$

In the case a monolithic laser, $\Phi(x,y)$ can be approximated in an analogous way by

$$\Phi(x,y) = 2\,n_0 k^2 \left[\frac{n_0}{L} \left(F_1(x,y) + F_2(x,y) \right) + \frac{1}{L}\int_0^L \tilde{n}(x,y,z)\,dz \right] \tag{5}$$

as will be shown in a separate paper in more detail[3]. In Eq.(5) $\tilde{n}(x,y,z)$ is the thermally induced modification of the index of refraction, $F_1(x,y)$ and $F_2(x,y)$ represent the deformation of the front faces 1 and 2 of the crystal respectively. L is the length of the crystal. It is also possible to introduce external optical elements like external mirrors into Eq.(5).

In order to solve Eq.(1), which has the form of the 2-D Schrödinger Equation, by use of the tools provided in ANSYS an analogous mechanical model was constructed by embedding a strechted membrane on an elastic foundation with locally varying stiffness. The vibrations of this model are described by the differential equation

$$S\nabla_t^2 \psi + \left[\rho\tilde{\omega}^2 - \varepsilon(x,y) \right] \psi = 0, \tag{6}$$

Here S is the tension of the membrane, ρ the weight per unit area. $\tilde{\omega}$ represents the frequencies of the vibrational modes, and $\varepsilon(x,y)$ the locally varying foundation stiffness. Since Eq.(6) is formally identical to Eq. (1) the vibrational modes of the membrane correspond to the optical modes of the resonator, if $\varepsilon(x,y)/S$ is replaced by $\Phi(x,y)$ according to Eq. (5). Thus an analysis of the effect of thermal lensing and deformation of the front faces F_i on the modal structure of the laser can be carried through by insertion of the results of the thermo-mechanical simulation into Eq.(5). Fig. 4 shows the TEM_{00} mode along with the pumping beam distribution. Fig. 5 shows a comparison of the FEA result versus an analytical solution obtained by a formula given in Ref. 2 for the TEM_{00} mode. In the present example the numerical mode shape is slightly broader than the analytical one. If the length of the resonator goes to zero, full agreement is obtained.

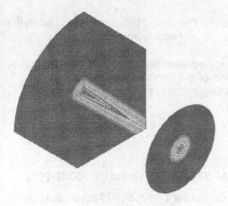

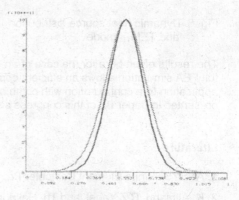

Fig. 4. Pumping beam distribution with TEM_{00} mode as obtained on a screen at the front end of the crystal

Fig. 5. Comparison: FEA result versus analytical solution for the TEM_{00} mode

The computational procedure described so far is incomplete since the power extracted by the laser beam is not taken into account. In order to include the latter in the simulation the power extracted locally per unit volume has to be subtracted from the local heat rate generated by the pumping beam. This is realized by an iteration procedure which uses the transversal beam profile obtained in a previous run to compute the locally extracted optical power for a new run which again starts with the thermo-mechanical FEA. Usually this procedure converges very well and after a few runs yields a dynamic heat source distribution, a dynamic temperature distribution, and a dynamic mode structure. Fig. 6. shows the dynamic heat source distribution along with the dynamic TEM_{11} mode. Please, notice the dip in the heat source distribution along the axis of the rod which is in contrast to Fig. 1. Fig. 7 shows the dynamic TEM_{00} mode and the analytical mode shape as already shown in Fig. 5. Since due to the optical power extraction the temperature distribution is flattened, the thermal lensing effect and also the deformation of the front faces are reduced. This results in a broadening of the mode shape as can be noticed by comparison of Fig. 7 with Fig. 5. In the present case, this effect is relatively small, however for longer crystals, it becomes more important.

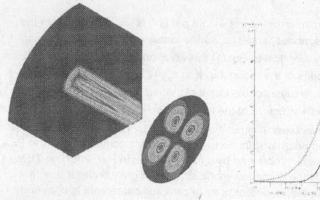

Fig. 6. Dynamic heat source distribution and TEM_{11} mode

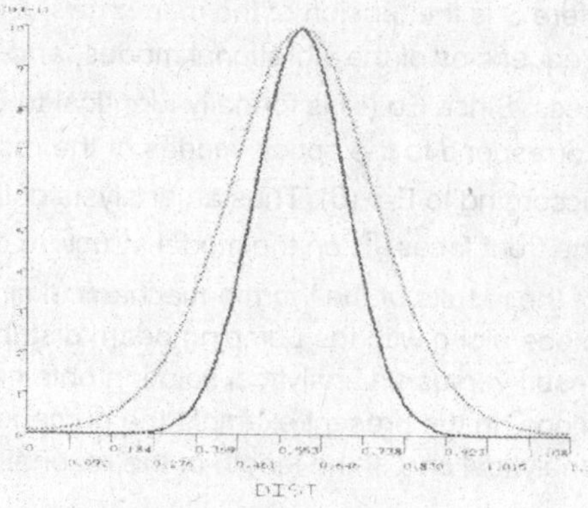

Fig. 7. Dynamic TEM_{00} mode and analytical approximation

The results given here for the case of an end-pumped laser rod show that the presented full FEA simulation allows an efficient computation of the dynamic mode structure. An application to a configuration with combined longitudinal and transversal pumping is presented in paper P2 of this congress session.

Literature

1. ANSYS® Finite Element Computer Program; ANSYS, Inc., Houston,PA 15342-9912

2. K. Altmann, R.Z.Yahel, and Th. Halldorsson, Computer Simulation of a Miniaturized End Pumped Laser Rod, Technical Digest of the CLEO' 94, Amsterdam

3. K. Altmann, paper in preparation

Modenausbildung in Festkörperlasern beim Pumpen mit Laserdioden

R. Axtner, K. Altmann, P. Zeller

Daimler Benz AG, Forschung und Technik, D-81663 München

Das sich ausbildende Strahlprofil eines Lasersystems ist neben dem Verstärkungsprofil im Lasermaterial von vielen weiteren Parametern abhängig.

Besonders bei diodengepumten Festkörperlasern kann sich die Komplexität durch die Vielzahl der möglichen Pumpkonfigurationen weiter steigern. Schließlich wird durch die Wahl einer Pumpkonfiguration auch die Fixierung und damit die Kühlung des Lasermaterials weitgehend festgelegt. Hinzu kommt, daß beim longitudinalen Pumpen häufig ein Endspiegel direkt auf den Laserstab aufgebracht wird, dessen Krümmungsradius dann nicht nur durch die geometrische Definition der mechanischen Bearbeitung festgelegt ist, sondern auch durch die thermischen Verhältnisse, die an der betreffenden Endfläche bei bestimmten Pumpleistungen herrschen.

Aus den beschriebenen Gründen ist es hilfreich, ein mathematisches Modell zu erstellen, das auch komplexe Verhältnisse in einem Resonator berücksichtigen kann. Ein solches Analyse-Werkzeug wurde unter Anwendung der Finiten Element Methode (FEM) geschaffen. Das Modell läßt sich an verschiedene Pumpkonfigurationen anpassen und berücksichtigt dabei folgende Parameter:

- Geometrie und Materialeigenschaften des Lasermaterials
- Pumpstrahlungsverteilung im Lasermaterial
- Wärmeabfuhr an der Oberfläche des Lasermatrials (gleichmäßig oder in beliebigen Segmenten)
- Energieabgabe durch den erzeugten Laserstrahl

Die aufeinanderfolgenden Berechnungsschritte sind aus dem Flußdiagramm (Abb.1) zu ersehen. Eine genauere Beschreibung der Berechnungsmethode kann aus [Lit.1] entnommen werden.

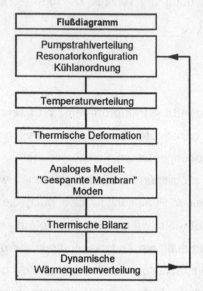

Abb. 1: Darstellung der Berechnungsschritte als Flußdiagramm

Wenn ein Laserstab von einem Ende (longitudinal) im Zentrum gepumpt wird, so ergibt sich am ehesten eine gute TEM_{00} - Effizienz. Abb. 2 zeigt die Berechnung der dynamischen Wärmequellenverteilung bei gleichmäßiger Kühlung des Laserstabes am Mantel und der daraus resultierenden Grundmode. Der Vergleich mit gemessenen Werten zeigt gute Übereinstimmung (Abb. 3).

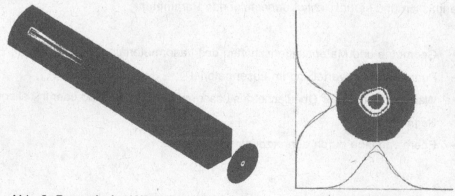

Abb. 2: Dynamische Wärmequellenverteilung   Abb. 3: Grundmode

Auch Anordnungen mit höherer Komplexität können berechnet werden. So ist in Abb. 4 die dynamische Wärmequellenverteilung bei gleichzeitigem Pumpen auf der Stabachse und transversal entlang der Mantelfläche in einem Segment dargestellt. Daß bei diesem Aufbau nur bestimmte Segmente der Mantelfläche gekühlt werden können, wurde berücksichtigt. Hier zeigt sich, daß die Tendenz zur Ausbildung von höheren Moden steigt, was u.a. zu der in Abb. 4 dargestellten TEM_{40} - Mode führt.

Durch ein Überlappungsintegral kann angegeben werden, mit welcher Wahrscheinlichkeit die verschiedenen Moden anschwingen (Abb. 5).

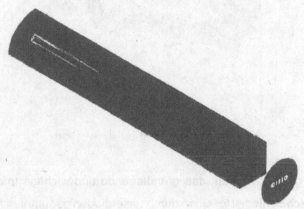

Abb. 4: Dynamische Wärmequellenverteilung bei longitudinalem und zusätzlichem transversalen Pumpen

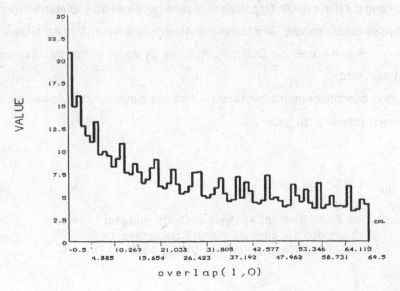

Abb. 5: Überlappungsintegral

240

Abb. 6 zeigt ein im Experiment aufgenommenes Strahlprofil dieser Pumpkonfiguration. Deutlich ist die Überlagerung der Intensitäten anschwingender Moden bis hinauf zur TEM$_{80}$ - Mode zu erkennen:

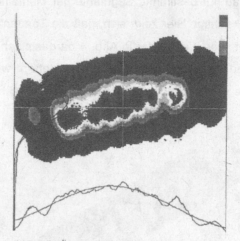

Abb. 6: Überlagerung höherer Moden

Es konnte gezeigt werden, daß grundlegende und wichtige Informationen aus dem mathematischen Modell für einen durch Laserdioden gepumpten Festkörperlaser abgeleitet werden können. Aufgrund der Tatsache, daß weitere Parameter (z.B. Krafteinwirkung auf den Kristall, Depolarisationseffekte) eine Rolle spielen können, die möglicherweise untereinander sehr komplexe Abhängigkeitsverhältnisse bilden, muß man davon ausgehen, daß zur Optimierung eines Systems immer das Experiment notwendig sein wird.

Die oben beschriebene mathematische Analyse hilft auch den Ansatzpunkt für Optimierungen schneller zu finden.

Literatur:

1. K. Altmann, Full Finite Element Analysis of Diode Pumped Lasers
      Vorträge des 12. internationalen Kongresses, LASER 95
      W. Waidelich (Hrsg.), Springer Verlag 1995, im Druck

Solid-State and Tunable Color Center Lasers with Nonlinear Crystalline Raman Shifter

Tasoltan T.Basiev, Peter G.Zverev.
General Physics Institute of Russian Academy of Sciences,
Moscow,Russia

The recent results on development of Solid State Raman Lasers for eye safe spectral region are presented. External and intra cavity schemes of Barium Nitrate Raman Shifters and Lasers pumped with 1.06 μm and 1.32 μm nanosecond neodimium laser are investigated.

With the help of the first Stokes scheme of Raman conversion we have obtained a number of new laser lines at 1.197 μm, 1.216 μm, 1.529 μm, 1.556 μm with conversion efficiency more then 50%. For the Second Stokes scheme the conversion efficiency for 1.369 μm and 1.394 μm lines was about 30%. The Third Stokes scheme provide us with the 1.598 μm and 1.632 μm wavelength of Raman oscillations with the efficiency near 20%.

Frequency Doubling of a Diode Pumped Nd:YAG Laser

Yutian Lu*, Keming Du and P. Loosen

FhG-ILT, Aachen, Germany; *Institute for Optics and Fine Mechanics CSA, Shanghai, China

A cw Nd:YAG laser pumped transversely with diode lasers was constructed. The laser diode emission was focused on to the Nd:YAG rod with elliptical cylinder mirrors. With this laser, intracavity frequency doubling was studied. Using a plane-plane resonator with the KTP crystal located immediately in front of the plane resonator mirror a maximum power is achieved 3.5W at 532nm. When a resonator with a telescope consisting of a lens and a curved end mirror is used and the KTP crystal is located at the common focus, the maximum power at 532nm is 5W. The maximum effective conversion efficiency is 53%. In this paper experimental results obtained with different configurations of frequency doubling are presented.

1   Introduction

Diode pumped high power Nd:YAG lasers are increasingly interesting for material processing and applications in basic research because of their high efficiency, low noise and long service interval [1]. Frequency doubling of this kind of laser offers attractive visible laser devices useful for a variety of applications in medicine, active imaging [2] and for pumping Ti:Sapphire in order to build up compact tunable all-solid-state-laser sources.

To study intracavity frequency doubling a cw Nd:YAG laser pumped transversely with laser diode arrays was constructed. Experimental investigations were carried out. The results are presented below.

2   Pumping arrangement

As shown in Fig. 1, six diode laser bars with elliptical cylinder mirrors were arranged symmetrically around the Nd:YAG rod. Each diode laser bar is located on the primary focus line of the mirror so that its emmision is focused on the secondary focus line of the elliptical cylinder mirror [3]. The lateral gain profile can be adjusted by changing the distance from the secondary focus line to the rod axis. Two such modules are used in our laser. In order to minimize the azimuthal gain modulations and the heat-induced

homogeneous inside the rod, the two pumping modules are rotated 30 degrees to each other.

Experimental results

The rod used is 4mm in diameter and 42mm in length. The resonator consists of two plane mirrors. The transmission of the output coupler is 9%. To check the pump efficiency a short plane-plane resonator with a length of 12cm was used. In this case the multimode output power is 60W for a total pump power of 190W. The optical slope efficiency is 37.5%. The resonator was then lengthened to 150cm. In TEM_{00} mode operation an output power of 22W is achieved for a pumping power of 168W (Fig. 2). The corresponding optical slope efficiency is 20%. The far-field intensity profiles in the multimode and TEM_{00} mode operations are shown in Fig. 3.

Intracavity frequency doubling was studied with this laser under cw-operation. The resonator was folded by a plane harmonic seperator (HR1064nm&HT532nm+AR532nm). Due to its large nonlinear coefficients, wide angular bandwidth, small walk-off, high optical damage threshold and lower temperature dependence, KTP crystal was chosen for our intracavity frequency doubling. The KTP crystal used is 10mm in length, is cut for type II phase matching under room temperature. Laser heating of the crystal was found still to seriously affect the stability of the green output power.

Two frequency doubling schemes were investigated. The first one is with internal telescope (Fig 4a). In this case the KTP-doubler was located immediately in front of the plane mirror. To increase the conversion efficiency a telescope resonator consisting of a lens ($f = 125mm$) and a curved resonator mirror ($R = 250mm$) was introduced. The KTP crystal was located at the focus of the telescope (Fig. 4b). The green output power in both cases was measured. Fig. 5 shows the output power in dependence of the pumping power. Without telescope a maximum output power of 3.5W at 532nm was obtained for a pumping power of 140W. The corresponding output power in TEM_{00}-mode at the fundamental wave length is 13.2W.

With the internal telescope a maximum green output power of 5.3W was obtained with a pumping power of 126W. With this pumping power the TEM_{00} mode power at 1064nm is 11.3W. The corresponding effective conversion efficiency is then 53%.

Intracavity doubling crystal works as a nonlinear output coupler. Its inhomogeneous birefringence will drastically degrade the fundamental mode performance of the laser and strongly reduce the second harmonic generated. Optical quality is extremely significant for the crystal used. We compared KTP crystals of the same specifications but from different company, a factor of 2 was found among their green power outputs. Although KTP crystal possesses lower temperature dependence, thermal gradient inside the crystal still seriously influences laser performance. A remarkable difference was observed for a crystal under

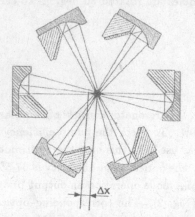

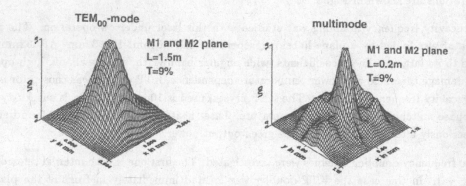

Fig. 1 The scheme of the diode pumping modul.

Fig. 2 The multimode and fundamental output power.

TEM₀₀-mode

multimode

Fig. 3 Far-field intensity profiles in multimode and fundamental mode operations.

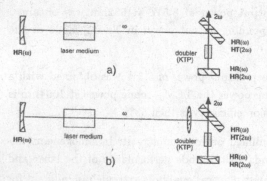

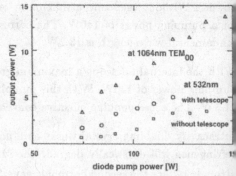

Fig. 4 Schemes of intracavity frequency doubling
a) without telescope; b) with telescope

Fig. 5 The output power in dependence of the
diode pumping power.

ifferent cooling conditions. When the thermal transfer was not very good, adjustment ecame very difficult and it was even impossible to get fundamental mode operation and ptimal frequency doubling at a high pumping level.

Conclusions

ntracavity frequency doubling was investigated with a Nd:YAG-laser pumped transver- ally with diode lasers. In the case without telescope 3.5W output power @532nm was btained. Using the telescospe resonator a maximum green output power of 5.3W was chieved with an effective conversion efficiency of 53%.

References

[1] FAN T. Y. AND BYER R. L.: Diode laser-pumped solid state lasers *IEEE J. of Quantum Electronics* **24** 6(1988) 895-912

[2] MARSHALL L. R. HAYS A. D. HAYS, KAZ A. AND BURNHAM R. L.: Intracavity doubled mode-locked and cw diode-pumped lasers *IEEE J. of Quantum Electronics* **28** 4(1992) 1158-63

[3] DU K.-M., LU Y.-T. AND LOOSEN P.: Operation performances of a diode pumped Nd:YAG-laser and efficient frequency doubling *Technical Digest CLEO'95, Baltimore* **CTuI64** (1995) 125-6

Frequency Tuning Characteristics of Single Longitudinal Mode in Nd:YVO₄ Laser Pumped by Laser-Diode

Lin Yueming   He Huijuan

Shanghai Institute of Optics & Fine Mechanics

Academia Sinica, Shanghai, 201800, China

Frequency tuning single longitudinal mode laser is very useful in many application such as spectroscopy, Quantum optics, Lidar and coherent communication. The laser frequency of solid state lasers can be tuned in different methods. There are heating, piezo-mechanic tuning, intercavity etalon tuning and electrooptically tuning. In this paper, the frequency of laser-diode pumped single longitudinal mode Nd:YVO₄ laser has been tuned by heating crystal and PZT controlling the length of cavity.

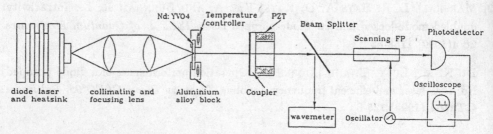

Fig.1 The schematic diagram of experimental Set-up.

Laser diode pumped single frequency Nd:YVO₄ laser has been reported in Ref[1]. The experimental setup is shown as fig.1. Laser-diode is a 1W CW SDL-2462-P1 with emitting area of $1 \times 200 \mu M$. The optics system consists of collimating lens and focusing lens. The sizes of Nd:YVO₄ is 3mm×3mm×1mm. The Nd:YVO₄ piece is high antireflection at 808nm and high reflection at 1064nm coated on one surface, forming the one mirror of the laser, and is antireflection at 1064nm coated on the other surface. A flat mirror with reflectivity 63% at 1064nm is used for coupler of laser. The temperature is varied by a small heater. The PZT used to control cavity length of laser is bonded on the coupler mirror.

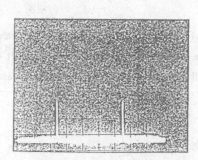

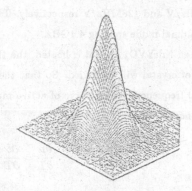

Fig.2 The oscilloscope trace of laser longitudinal modes.

Fig.3 CCD camera trace of the laser output mode pattern.

The output beem is splitted into two parts. One is used to monitor longitudinal mode spectrum by scanning confocal interferometer, the other is used to measure the laser wavelength by model WA-20 wavemeter. The optical spectra are monitored at 1064nm used a scanning confocal interferometer with 288 finesse and a 1.3GHz free spectral range. Fig 2 shows the longitudinal modes of the $Nd:YVO_4$ laser with 450mW pumping power. Fig 3 shows the CCD camera trace of the laser output mode pattern.

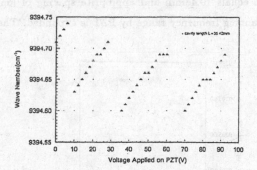

Fig.4 The relationship between oscillation frequency and the voltage applied on PZT at ambient temperature. Cavity length L=35.42mm. Tuning coefficient is up to 150MHz/V and tuning range up to 2.9GHz without mode hopping.

The oscillative frequency of laser as function of the voltage applied on PZT is shown in figure 4. The laser cavity length is 35.42mm and pumping power is 367mW. When the voltage is increased, the several stable regions of frequency are there. The mode hopping is discovered between two stable regions of frequency. There are slight deferences of tuning coefficient of every region due to nonlinear effect of PZT. The maximum and average tuning coefficient are

150MHz/V and 128MHz/V respectively. The maximum tuning region 2.9GHz is less than a longitudinal mode spacing 4.1GHz.

When Nd:YVO$_4$ crystal is heated, the thickness of crystal and it's mount, the refractive index of crystal will be varied. So that the variation of laser cavity length and the shift of central frequency of gain curve of active material will be happened. The expression of laser frequency as follow

$$\frac{\partial \nu}{\partial T} = \frac{\partial \nu}{\partial T}\Big|_{gain} + \frac{\partial \nu}{\partial T}\Big|_{cavity} \tag{1}$$

$$\frac{\partial \nu}{\partial T}\Big|_{cavity} = -\frac{\nu}{n_{YVO_4}d_{YVO_4} + L_{cavity}}\left(n_{YVO_4}\frac{\partial d_{YVO_4}}{\partial T} + d_{YVO_4}\frac{\partial n_{YVO_4}}{\partial T} + \frac{\partial L_{AL}}{\partial T}\right) \tag{2}$$

Where $\frac{\partial \nu}{\partial T}\big|_{gain}$ is temperature coefficient of frequency due to shift of central frequency of gain curve, $\frac{\partial \nu}{\partial T}\big|_{cavity}$ is temperature coefficient of frequency due to variation of laser cavity length. Lcavity is laser cavity length except the thickness d_{YVO_4} of Nd:YVO$_4$, $\frac{\partial L_{mount}}{\partial T}$ is thermal expansion coefficient of crystal mount, n_{YVO_4} is refractive index of Nd:YVO$_4$. The PZT used to scan the frequency of laser at different temperature value. We assume the central frequency of scanning range as the central frequency of gain curve. The practice cavity length of laser L equals 10.48mm and approprite spacing of longitudinal modes is 12.9GHz, so that the scanning frequency range by PZT is expanded. The measuring results

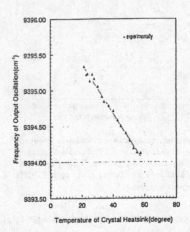

Fig.5 The oscillation frequency versus temperature of crystal heatsink for Nd:YVO$_4$ laser. Temperature co-efficient of output frequency is -1.04GHz/degree. At this time, cavity length L=10.48mm, the axial mode spacing is 12.9GHz. The maximum tuning range is up to 47.4GHz without mode hopping.

of oscillation frequency of Nd:YVO$_4$ laser versus temperature of crystal is shown in figure 5. The temperature coefficient of laser output frequency equals -1.04GHz/degree. The maximum tuning range is up to 47.4GHz without mode hopping.

The output power of laser with relation to scanning frequency in a axial mode spacing at temperature of crystal 26.6 degree is shown in figure 6. The variation of power is less than 6.6%.

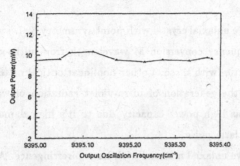

Fig.6 Output power in a axial mode spacing. The temperature of crystal is 26.6 degree and pumping power 250mW.

Summary

Frequency tuning of Nd:YVO$_4$ laser is realized using a piezo-electric transducer (PZT) mounted on the output coupler for cavity length variation and laser crystal heating. The single frequency laser-diode pumped Nd-doped 2% YVO$_4$ at 1064nm has been tuned over a range of 47GHz without mode hopping. Tuning coefficient is up to 150MHz/V for PZT and temperature coefficients is -1.04GHz/°C for the central frequency of gain curve.

Reference

1. Lin Yueming, He Huijuan, "Single frequency Nd:YVO$_4$ Laser", ACTA OPTICA SINICA, 14, 12(1994).

Tuning PS Ultra-Violet Light Generation by BBO AT 275-401NM

He Huijuan,   Lu Yutian

Shanghai Institute of Optics & Fine Mechanics

Academia Sinica, Shanghai 201800, China

β-$B_a B_2 O_4$ is a negative uniaxial crystal with point symmetry 3. It's a very useful nonlinear optical material for frequency conversion at wavelength from ultra-violet to mid-infrared. Comparism of its properties with those of other nonlinear optical materials shows that BBO is paticularly useful in the generation of ultra-violet radiation of wavelength as short as 200nm. BBO crystal has high power capacity due to it's high damage threshold, so that which is suitable for PS laser pulse.

β-$B_a B_2 O_4$ is a negative uniaxial material with large birefringence. A sellmeier equation of the form

$$n^2 = A + B/(1 - c/\lambda^2) \tag{1}$$

where λ is free space wavelength in μM,

for n_o	A=1.2365	B=1.4878	C=0.01451
for n_e	A=1.3061	B=2.6622	C=0.00635

Using the PS Nd:YAG laser and dye laser pumped by this PS Nd:YAG laser as the pumping sources, we have calculated the type Ⅰ phase-matching angles to obtain the second harmonic wavelength at 275–363nm as in figure 1. The phase-matching angles are tuning at 32.1°–45.4°. Then we calculated the phase-matching angles for sum frequency generation between the fundamental wave of PS Nd:YAG laser at 1064nm and dye laser pumped by Nd:YAG laser. The phase-matching angles are tuning at 28.3°–30.7° and the wavelength range generated by sum frequency is at 363–401nm as in figure 2.

According to the phase-matching curves, the tuning phase-matching angles for doubling frequency and sum frequency are 32.1°–45.4° and 28.3°–30.7° respectively. If the BBO crystal is cut at 36° and rotated from 28° to 45°. The doubling frequency and sum frequency are generated in same one piece of BBO crystal. The tuning angles amount to $\pm$8.5°.

The experimental set-up is shown in Fig.3. The active and passive mode-locking Nd:YAG laser is used to generate the 1064nm ps pulse. After doubling frequency crystal the 532nm

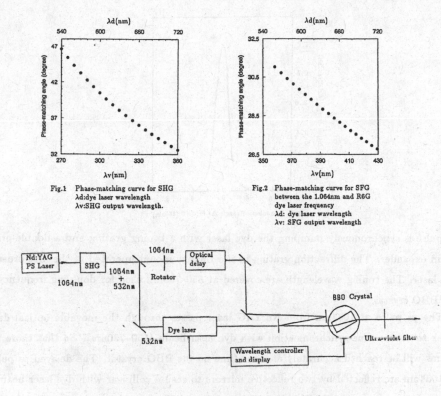

Fig.1 Phase-matching curve for SHG
λd:dye laser wavelength
λv:SHG output wavelength.

Fig.2 Phase-matching curve for SFG
between the 1.064nm and R6G
dye laser frequency
λd: dye laser wavelength
λv: SFG output wavelength

Fig.3 Schematic diagram of experimental set-up

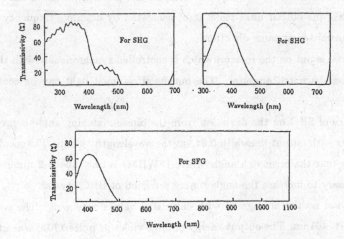

Fig.4 Transmissivity of ultra-violet filter.

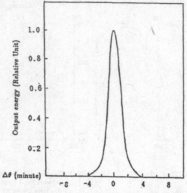

Fig.5 The mismatch angle $\Delta\theta$ for SHG (560nm).

ps pulse is synchronously pumping the dye laser with a tuning grating and a double-prism beam expander. The diffraction grating in the Littrow mount operated at the first order in dye laser. The tuning wavelength are covered at 550–730nm used for doubling frequency in the BBO crystal.

The ps pulse at 1064nm from Nd:YAG laser passes through the movable optical delay block to ensure time-synchronization with dye laser beam (550–730nm). So that those two beams will be reached accurately at same time on the BBO crystal. The delayed ps pulses at 1064nm are reflected by two reflective mirrors to ensure collinear with dye laser beam in the sum frequency crystal. There are three pieces of filters used to cut off the visible and infrared light with transmissivity <0.001. The transmissivity of ultra-violet filters are shown in figure 4. So that the output ultra-violet light generated by doubling frequency and sum frequency are transmitted at same direction.

The BBO crystal is put on the rotator, which is controlled synchronously with the tuning grating of dye laser by microcomputer. The output ultra-violet light can be continuously tuned.

The dependence of SHG on the deviation from the phase-matching angle is investigated using the dye laser with output linewidth 0.04Åat the wavelength 560nm. The results shown in Fig.5 indicates that the mismatch angle width (FWHM) is only about 2 nimutes of arc. So that it's necessary to increase the angle tuning precision of BBO crystal.

The experimental results are agreed with calculated values very well. The wavelength tuning range is 276–401nm. The output energy of ultra-violet ps pulse $170\mu j$ when the input energy of ps dye laser pulse is $805\mu j$. Therefore the maximum frequency conversion efficiency is up to 21%. The pulse duration is 40ps and the maximum output peak power is 1.5Mw. The tuning velocity for sum frequency and doubling frequency are $40–60°\text{Å/s}$ and $12–24°\text{Å/s}$ respectively.

Diodengepumpter regenerativer Verstärker

D.Hoffmann[1], H.G.Treusch[2] und E.W.Kreutz[1]

[1] Lehrstuhl für Lasertechnik der Rheinisch-Westfälischen Technischen Hochschule,
   Steinbachstraße 15, D-52074 Aachen, FRG

[2] Fraunhofer-Institut für Lasertechnik, Steinbachstraße 15, D-52074 Aachen, FRG

1. Einleitung

Für Untersuchungen zur Mikrobearbeitung mit Laserpulsen mit Pulsdauern kleiner 100 ps wird Laserstrahlung mit hoher Strahlqualität (TEM00-Mode, $M^2 < 1.1$), geringer zeitlicher Fluktuation der Laserleistung (< 2 %), sowie hoher Pulsleistung ($> 10^6$ W) benötigt.

Eine geringe zeitliche Fluktuation der Laserleistung wird mit einem diodengepumpten Festkörperlaser durch Temperaturstabilisierung und Versorgung der Pumplaserdioden mit einem zeitlich stabilen Strom erreicht.

Voraussetzung für die geforderte Strahlqualität ist eine homogene, an den TEM_{00}-Mode angepaßte, Pumplichtverteilung /1/.

Zur Erzeugung hoher Pulsleistungen bei Pulsdauern von 100 ps wird eine Anordnung bestehend aus einem modengekoppelten Oszillator und einem regenerativen Verstärker /2/ gewählt. Der Oszillator liefert ps - Pulse mit Pulsenergien im nJ - Bereich. Ein einzelner Puls zirkuliert im regenerativen Verstärker, bis die maximale Pulsenergie erreicht und mittels „cavity dumping" ausgekoppelt wird. Auf diese Weise wird eine Pulsenergie im mJ-Bereich und die geforderte Pulsleistung von 10^6 W erreicht. Der Vorteil des regenerativen-Verstärker-Prinzips besteht darin, daß eine niedrige Pumpleistung, in einer mit wenigen Diodenlasern erreichbaren Größenordnung, benötigt wird.

Der Aufbau und die Eigenschaften des regenerativen Verstärkers werden dargestellt. Bei der Konstruktion wurde die industrielle Einsetzbarkeit durch einen kompakten und robusten Aufbau berücksichtigt. Der Verstärker kann mit konventionellen Fertigungsverfahren hergestellt werden.

2. Anordnung

Aufgrund der für longitudinales Pumpen zu hohen Pumpleistung von 640 W wird eine transversale Pumpgeometrie gewählt /3//4/. Je 4 Diodenlaser werden zu einem Modul zusammengefaßt. Acht Module werden rotationssymmetrisch um den Laserstab angeordnet. Der, von einer „flow-tube" umgebene, zylindrische Laserstab wird mit Wasser gekühlt (Abbildung 1).

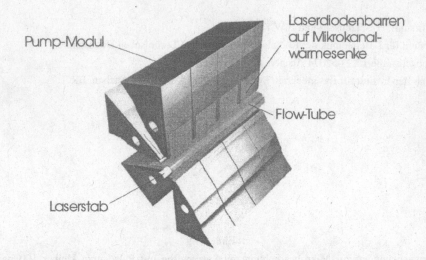

Abbildung 1: Schematische Darstellung der Anordung von Laserstab, Flow-Tube und 4
Pumpmodulen

Die Diodenlaser bestehen aus auf Kupfer-Mikrokanalwärmesenken montierten Laserdiodenbarren.
Die Stirnseiten der Mikrokanalwärmesenken sind zur Reflexion des Pumplichtes vergoldet. Auf
diese Weise ergibt sich eine 8-eckige Kavität für das Pumplicht, wodurch eine effiziente
Ausnutzung der verfügbaren Pumpleistung gewährleistet wird. Durch diese Anordnung und durch
die kompakte Bauweise kann auf Optiken zur Formung der Pumplichtverteilung verzichtet werden.

3. Konstruktion

Die Anforderungen an die Konstruktion des Pumpmoduls (Abbildung 2) beinhalten die Versorgung
der Laserdioden mit Strom und Kühlwasser, sowie die präzise Halterung relativ zum Laserstab.
Eine einfache Montage unter Ausschluß von Beschädigungen der Laserdioden ist sicherzustellen.
Zum Schutz vor zu hohen Spannungen in Sperrichtung, z.B. durch elektrostatische Entladungen,
wird jede Laserdiode durch eine parallel geschaltete Halbleiterdiode geschützt. Alle elektrischen
Kontaktflächen sind zur Reduzierung der Übergangswiderstände vergoldet.
Die Laserdioden sind zur Erzielung eines hohen Wirkungsgrades elektrisch in Reihe geschaltet. Die
Kühlkreisläufe sind parallel geschaltet, damit alle Laserdioden die gleiche Temperatur aufweisen.
Um eine Verschmutzung der Emitterfacetten zu verhindern, muß das Pumpmodul hermetisch
gekapselt werden.

3. Ergebnisse

In Abbildung 3 wird die gemessene Fluoreszenzlichtverteilung in einem Nd:YAG-Laserstab mit
einem Durchmesser von 3mm und einer Länge von 64 mm dargestellt. Da die Pumpintensitäten
unter der Sättigungsintensität des Nd:YAG-Materials liegen, besteht ein linearer Zusammenhang
zwischen der Intensität des Fluoreszenslichtes und der Verstärkung beim Einsatz als Laser, so daß
eine homogene Verstärkung des Laserstrahls erwartet wird. Der in Abbildung 2 dargestellte

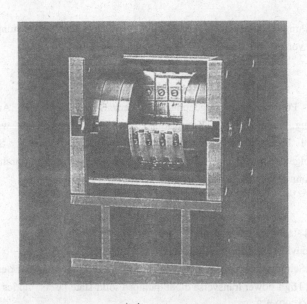

Abbildung 2: Realisierter Prototyp ohne hermetische Kapselung; durch Entfernen von 2 Pumpmodulen wird die Anordnung der Diodenlaser sichtbar.

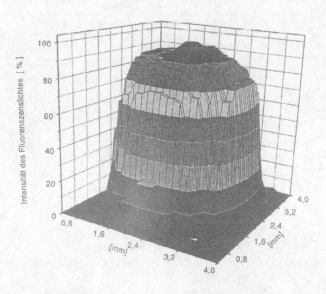

Abbildung 3: Gemessene Fluoreszenzlichtverteilung

Verstärker mit 8 Pumpmodulen mit je 4 Laserdioden erreicht eine Pumpleistung von 640 W bei einer Wellenlänge von 807 nm und einer Halbwertsbreite von 3 nm.

4. Zusammenfassung

Es wurde ein Prototyp des regenerativen Verstärkers, der die Anforderungen an Pumplichtverteilung, Pumpleistung, Kompaktheit und robustem Design weitgehend erfüllt, realisiert. Weitere Eigenschaften der gewählten Konstruktion sind die Skalierbarkeit der Pumpleistung infolge des modularen Aufbaus, sowie die einfache Wartung durch Austauschbarkeit der Pumpmodule ohne Demontage des Verstärkers.

5. Literatur

/1/ Köchner, Solid state laser engineneering, Springer-Verlag, Berlin (1992).

/2/ P. Bado, M. Bouvier, J.S. Coe, Optics letters, Vol.12, No.5 (1987) 319.

/3/ P.Peuser, N.P. Schmitt, Diodengepumpte Festkörperlaser, Springer-Verlag, Berlin (1995)

/4/ R.L. Burnham, High-power transverse diode-pumped solid-state lasers, Optics and Photonics News, August (1990) 4-8.

Efficient Flash-Lamp-Pumped Ti:Sapphire for Oscillator and Amplifier Operation

A. Hoffstädt

Festkörper-Laser-Institut Berlin GmbH, Optisches Institut, Technische Universität Berlin, Straße des 17. Juni 135,
D-10623 Berlin

Flash-lamp-pumped Ti:sapphire lasers attract much interest because of their high emitted pulse energies and wide tuning range, which can be expanded into the UV by frequency conversion techniques. There are so far only few applications of lamp-pumped Ti:sapphire lasers. Although efficient lamp-pumped Ti:sapphire operation has been demonstrated by ERICKSON [1] and BOQUILLON and MUSSET [2] as well as high emitted and stored energies by BROWN and FISHER [3], a system combining high average output power at high efficiency and good reliability, which could allow for new applications as surface treatment and fs-pulse amplification, was still outstanding. Design and first results of such a Ti:sapphire laser have been given in [4], here laser and amplifier performance data are presented.

Four flash-lamps series-connected (ILC L-6341, l_{arc} x d_i = 6" x 5 mm) are driven by a LC-discharge circuit including a Thyratron switch (CX1574C) and a 0.52-μF capacitor, which is loaded by a 10-kW power supply. A high current dc-simmer pre-ionizes the lamp plasma. At 20 kV loading voltage current pulse duration is 5.2 μs (10% width). The Czochralski-grown, 90°-cut, flat-flat polished and BBAR-coated laser rod from Union Carbide Corp.

(l x d = 6" x 10 mm) with a 0.10 wt% Ti_2O_3 doping has a FOM = 420 (= α @ 490 nm/ α @ 820 nm). Utilizing a high transfer efficiency reflector design [5], a high Xenon cold-fill pressure of 400 Torr and a simmer current of 5.0 A the laser oscillator with a short flat-flat resonator (output coupler

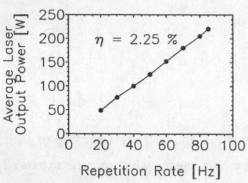

R = 66%) emitted at 764 nm with 6 nm FWHM 220 W average power (Fig. 1). Loading energy per pulse was 115 J, resulting in a total laser efficiency of 2.25% without the use of a fluorescent converter.

Fig.1 Average laser output power.

Employing the cerium-doped 1-mm quartz lamp wall and the Duran flowtube of the laser rod as spectral filter (cut-off wavelength λ_C = 370 nm) cannot prevent pump-light-induced broadband parasitic absorption in the Ti:sapphire crystal. Measurement of the transmission of another Czochralski-grown laser rod from Union Carbide Corp. (6" x 1/4", 0.10 wt% Ti_2O_3, FOM = 200) before and after lamp pumping with the same filter combination shows no additional absorption at

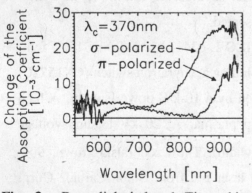

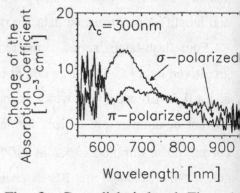

Fig. 2 Pump-light-induced Ti:sapphire absorption. Cerium doped quartz & Duran.

Fig. 3 Pump-light-induced Ti:sapph absorption. Duran filter only.

800 nm but a broad band at 950 nm (Fig. 2) [5]. The same measurements but with the 10-mm rod used in this work indicate a lesser absorption increase for higher FOM. Additional pumping after the experiments for Fig. 2 with a Duran filter only, results in further increased absorption (Fig. 3) [5].

Thermal lensing without laser action has been determined by the deflection of a He-Ne-laser beam displaced parallel to the optical axis. Fig. 4 and 5 show the results for two orientations of the Ti:sapphire rods c-axis to the four 200-Torr-lamps placed in a cylindrical Spectralon-reflector. The oscillatory behaviour of the refractive power per loading power is reproducable. The slight increase with loading power is, to a great extend, due to the temperature dependence of the Ti:sapphire refractive index and thermal conductivity.

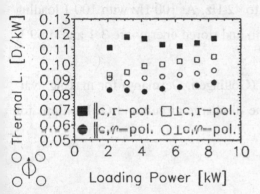

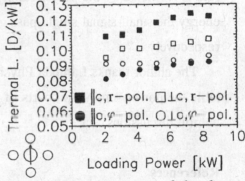

Fig. 4 Refractive power of the thermal lens per loading power.

Fig. 5 Refractive power of the therm lens per loading power.

Determining the loading power needed to obtain the same probe-beam deflection (thermal lens $\parallel$ c, r-pol., c-axis orientation as for Fig. 4) for different loading voltages gives the relative ratio of heating power to pumping power (Fig. 6). The decrease of laser rod heating with higher loadings is due to the blue-shift of the lamp emission spectrum which causes a lower spectral filter transmission and a smaller spectral overlap with the Ti:sapphire absorption.

The beam parameter product (waist radius · half far field angle) has been determined for the same pump chamber set-up as used for Fig. 4 to 29 mm

260

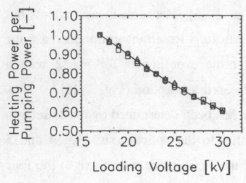

Fig. 6 Relative heating power per pumping power.

mrad ($\parallel$c) and 23 mm mrad ($\perp$c) at 8.3 kW loading power. The relation of the beam parameter products $\parallel$c and $\perp$c corresponds well with the multimode far field divergence calculated from the refractive powers given in Fig. 4.

Utilizing again the same pump chamber set-up, the small-signal single-pass gain and stored energy have been measured to 6.1 and 1.1 J, respectively, at maximum loading energy of 235 J. Maximum possible repetition rate is restricted by the power supply to 42 Hz. At 100 Hz with 100 J loading energy the small-signal single-pass gain and stored energy are 3.1 and 0.69 J, respectively.

The author thanks Lambda Physik (Göttingen, Germany) for making available the power supply and parts of the discharge circuit and G. Phillipps for helpful discussions.

References

1. E.G. Erickson, "Flashlamp-Pumped Titanium:Sapphire Laser", in Proc. of the OSA Top. Meet. on Tunable Solid State Lasers, North Falmouth, Cape Cod, MA, USA, 1.-3. May 1989, Vol. 5 of OSA Proceeding Series, 26-32 (1989)

2. J.P. Boquillon and O. Musset, "Flashlamp-pumped Ti:Sapphire laser: Influence of the rod figure of merit and Ti^{3+} concentration", Appl. Phys. B **59**, 357-360 (1994)

3. A.J.W. Brown and C.H. Fisher, "A 6.5 J Flashlamp-Pumped $Ti:Al_2O_3$ Laser", IEEE J. Quantum Electron. **29**(9), 2513-2518 (1993)

4. A. Hoffstädt, "High-average-power flash-lamp-pumped Ti:sapphire laser", Opt. Lett. **19**(19), 1523-1525 (1994)

5. A. Hoffstädt, "Blitzlampengepumpte Ti:Saphir-Laser", Laser Optoelectron. **3**, 65-74 (1992)

Passive Q-Switching and Mode-Locking of Solid State Lasers with V³⁺:YAG Crystal

V.P.Mikhailov, N.I.Zhavoronkov, N.V.Kuleshov, P.V.Prokoshin, K.V.Yumashev, V.A.Sandulenko
International Laser Center, Kurchatov str.7, Minsk 220064, Belarus

The saturation of the optical absorption in V-doped YAG is investigated. The absorption cross section of tetrahedrally coordinated V^{3+} ion at 1.08 μm is estimated to be 8.2 x 10^{-18} cm², and absorption recovery time of about 5 ns is measured. Using V^{3+}:YAG as a saturable absorber passive Q-switching and mode-locking have been achieved for a number of neodymium lasers (LiYF₄, YAG, KGd(WO₄)₂) at 1.06 μm and 1.34 μm. The shortest and strongest mode-locked pulses were obtained at 1.06 μm from a Nd-KGd(WO₄)₂ laser and at 1.34 μm Nd-YAlO₃. In both cases the trains of ultrashort pulses were observed with single pulses of about 0.3-0.5 ns in duration and up to 1 mJ in energy.Q-switched laser pulses at 1.06 μm were 20-30 ns in duration and 15-20 mJ in energy at flash-lamp pump energy of 30 J.

V^{3+}:YAG was shown to be also an effective saturable absorber for Pr-YAlO₃ and Ti-Al₂O₃ lasers, generated at 747 and 780 nm, respectively. The $^3A_2$ - $^3T_1$ (^3F) transition of tetrahedral V^{3+} was saturated by the emission of these lasers. A flat high reflector and 92% reflectivity, 1,5 m radius output coupler were used in both these mode-locked lasers. The laser rod of Pr^{3+}:YAlO₃ was 6 mm in diameter and 50 mm in length. The titanium sapphire laser rod was 5x100 mm² in size and had a FOM (figure of merit) of about 200. A pair of TF-10 glass prisms was placed in the titanium sapphire laser cavity to narrow the spectral bandwidth of the laser emission down to 2 nm. The ultrashort pulses observed were 0.8-1.0 ns in duration and up to 0.5 mJ in energy both at 747 nm of Pr-laser ($T_0 = 75\%$) and at 780 nm of a Ti-laser ($T_0 = 60\%$) at a flash-lamp pump energy of about 40 J. The duration of flash-lamp light pulse for these lasers was about 5-7 μs.

Simultaneous Coherent Red, Green and Blue Radiation of a Frequency Converted Nd:YAG-Laser

G. Mann[1], G. Phillipps[2]

[1] : Technische Universität Berlin, Straße des 17. Juni 135, 10623 Berlin

[2] : Laser- und Medizin-Technologie Berlin gGmbH, Festkörper-Laser-Institut
Straße des 17. Juni 135, 10623 Berlin

During the last years, a lot of work has been done on realizing dual wavelength oscillation with Nd-Lasers via the $^4F_{3/2} \rightarrow {}^4I_{11/2}$ and $^4F_{3/2} \rightarrow {}^4I_{13/2}$ fluorescence channels with different operation modes. Q-switching of Nd:YAG-laser[1-3], free running pulsed operation of Nd:YAP-laser[4,5] and continuous wave operation of Nd:YAP-Laser[5] and Nd:YAG-laser[6] have been investigated.

In the steady state regime a stable dual wavelength operation is difficult to achieve, since a) the laserfields compete for the same upper laser level, and b) the laser transitions of Nd-ions are dominantly homogeneous broadened in crystals at room temperature[7]. Therefore laser action is expected only from one transition. However, when the resonator round trip gain minus the losses is nearly the same for laser fields from different transitions, dual wavelength operation is possible due to spatial nonuniform saturation and an incomplete overlap between different modes. The limit of the tolerance was measured by the authors of reference /6/, who used in their continuous wave Nd:YAG-laser (1064 nm + 1319 nm) an acoustooptical control to modulate the losses for the 1064 nm radiation field.

In our system; we used a nonlinear crystal for frequency doubling of the 1319 nm radiation field inside the resonator to enable a stable, dual wavelength operation. The experimental setup is shown in fig.1.

We used a 1.1 at% Nd:YAG crystal with dimensions of 0.25 inch in diameter and 4 inch in length as laser active element which was pumped by two flashlamps. The power supply deliveres pulses with a temporal rectangular shape and a pulse duration up to several milliseconds, which is much longer than the lifetime of Nd-ions in the $^4F_{3/2}$-level with 0.23 ms. Therefore the laser operates in the steady state regime.

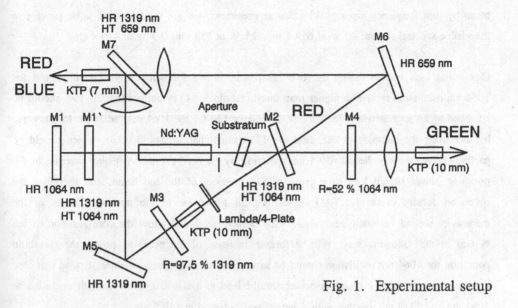

Fig. 1. Experimental setup

Two resonators are formed by the mirrors M1` - M2 - M3 and M1 - M4 for the wavelengths 1319 nm and 1064 nm, respectively. The dichroitic beamsplitter M2 separates the wavelengths. An aperture is located in front of the Nd:YAG-crystal to force a TEM_{00} operation for both wavelengths. In the 1319 nm-resonator arm a nonlinear type II KTP-crystal with 10 mm in length is placed for frequency doubling which generates coherent red radiation with a wavelength of 659.5 nm. Due to the high reflectivity of the mirror M3 at 659.5 nm, the red radiation leaves the resonator via the highly transparent mirror M2. A quarter wave plate is located in front of the KTP-crystal to avoid the "green problem"[8]. A special dielectric coated substratum is used to suppress the competitive oscillation via the 1338 nm-channel, which exhibits nearly the same stimulated cross section as the 1319 nm-channel.

With a fixed repetition rate the system of fig.1 shows a stable multi-wavelength operation, and therefore we use this system to generate the three visible colors red, green and blue. The green radiation is generated by focussing the output of the 1064 nm radiation into a type II KTP-crystal.

Through the mirror M3 a part of the 1319 nm radiation leaves the resonator, which is focussed simultaneously with the red radiation into a 7 mm long type II KTP-crystal to generate the blue beam by sum frequency mixing. With this arrangement, we got the following pulse powers in the visible spectral region: 40 W at 659.5 nm, 25 W at 532 nm, 0.3 W at 440 nm.

Our initial idea was to design the two resonators in the following way: The threshold for 1064 nm oscillation should be higher than the threshold for 1319 nm oscillation. This should be achieved by an appropriate reflectivity of the mirror M4 for the 1064 nm radiation. In this case, by increasing the pumping power, first the 1319 nm radiation should occur, which should be partially converted to coherent 659.5 nm radiation by the KTP-crystal. A further increase of the pumping power should lead to a grow of the intensity of the red beam, and therefore the inversion density in the Nd:YAG crystal should grow due to the nonlinear losses of the intracavity second harmonic generation. This behaviour should allow the configuration of the system in the following way: With a further increase of the pumping power the threshold condition for 1064 nm oscillation should be satisfied. Our rate equations show, that in that case a further increase of the pumping power should lead to a stable dual wavelength operation at 1064 nm and 1319 nm together with coherent red radiation at 659.5 nm.

Our experimental results confirm the expected behaviour, but the system is very difficult to align. The overlap of the different modes inside the active medium depends on the thermal lensing of the Nd:YAG-rod. This leads to the effect, that the ratio of the powers at 1064 nm and 1319 nm could be varied by changing the repetition rate (e.q. the pumping power and the thermal lensing). Therefore, further experiments were done with a ring resonator, which doesn't show such a strong sensitivity against alignment. With this system, the expected behaviour described above could be clearly demonstrated.

References:

1: C.G. Bethea; J. Quantum Electron. 9, 254 (1973)
2: G.A. Henderson; J. Appl. Phys. 68, 5451 (1990)
3: W. Vollmar, M.G. Knights, G.A. Rines, J.C. McCarthy, E.P. Chicklis; CLEO '83, Digest of technical papers, 188 (1983)
4: H.Y. Shen, W.X. Lin, Y.P. Zhou, G.F. Yu, C.H. Huang, Z.D. Zeng, W.J. Zhang, R.F. Wu, Q.J. Ye; Applied Optics 32, 5952 (1993)
5: H.Y. Shen, R.R. Zeng, Y.P. Zhou, G.F. Yu, C.H. Huang, Z.D. Zeng, W.J. Zhang, Q.J. Ye; J. Quantum Electron. 27, 2315 (1991)
6: V.E. Nadtocheev, O.E. Nanii; Sov. J. Quantum Electron. 19, 444 (1989)
7: T. Kushida; Phys. Rev. 185, 500 (1969)
8: M. Oka, S. Kubota; Optics Letters 13, 805 (1988)

Unstable Resonators with Supergaussian Mirror

J.MARCZAK, J. FIRAK and S. SARZYŃSKI
Institute of Optoelectronics, Military University of Technology
2 Kaliskiego Str., 01–489 Warsaw. P.O.Box. 49 ,POLAND

ABSTRACT

In this paper, on the basis of literature and the own works there has been made a review of optical and electrooptical elements, leading to an enlargement of TEM_{00} mode volume and to a smoothing of its transverse distribution of irradiance, both in a near and a far field. The special attention has been paid to an unstable resonator, "p–branch" type as well as to an output mirror with a radially variable reflection coefficient. Moreover design procedures of variable reflectance mirrors fo laser output coupler are proposed. Properly choosing both the geometry of the substrate and the masking system the quasigaussian or supergaussian reflectivity profiles were obtined. Two kinds of experiments for supergaussian mirrors deposited on concave–convex and thin optical vedge have been carried out.

1. INTRODUCTION

The basic, transversal mode of a laser oscillator with a big diameter, smooth transversal distribution of irradiation in a near and far field, plane wave front ensuring the diffraction beam divergence, was and still is the aim for designers of laser resonators [1]. The current design works are concentrated mainly on the unstable resonators as well as on a manufacturing technology of so called apodizers, it means optical or electrooptical elements by means of which one can realize the smoothing of transversal profile of a laser beam, inside or out of the oscillator.

Widespread, stable resonator, in which the laser beam has insignificant diffraction losses and is kept near an optical axis of resonator, produces a basic transversal mode which surface amounts approximately : $\sim \lambda x L$, where λ – is a laser wavelength , L – optical length of a resonator. If the diameter of an active medium is "2a" then the ratio of surface of a cross–section to a cross–section of a mode of the lowest order has the same value as a Fresnel number: $N_F = \pi a^2 / \lambda L$ which characterizes the laser system. Next, if the value of this number is much higher than one, the basic mode will choose only a part of energy, from the available energy in the active medium, equal to $\sim 1/N_F$ or laser will have to work with the much greater Fresnel numbers (multimode work) in order to extract a whole energy accumulated in a medium. The same, in order to produce the diffraction–limited output beam, the Fresnel number should be of the order of unit, it means $N_F = \pi a^2/\lambda L < 1$ and next it limits the diameter of amplifying medium to the value regulated by this number.

For the given type of laser (L = const.), the effective cross–section of a basic mode of a stable resonator in practice can be enlarge by:

– extension of a resonator length "L";
– application of optical telescope – expander inside a laser resonator [2].

However, in practice, a cross–section of TEM_{00} mode is still limited and optical elements, in the area of the narrowest beam, are subjected to damages.

Unlimited extension of a cross–section of a basic mode can be obtained using an unstable resonators which are adequate for an optical system of the focusing of a divergence beam. The basic mode in such a resonator expands in the sequent reflections to total filling of a cross–section of an active material [3]. The active material, used in such a resonator should have suitable high amplification coefficient for a double cycle, namely should fulfil the relationship: 2x G x L > 1.5, where G – amplification coefficient of an active medium per length unit, for a weak signal ; L – length of amplifying medium.

The main purpose of this elaboration is to present two types of unstable resonator configurations, as well as optical element which enable an extension of transversal dimensions of TEM$_{00}$ mode. Moreover, they enable avoiding, completely or partialy any harmful reasons, leading to a disturbance of a transversal distribution of an irradiation intensity in a laser beam, both in a near and far field.

2. OPTICAL AND ELECTROOPTICAL ELEMENTS OF RADIALLY VARIABLE TRANSMISSION.

The biggest disadvantage of a diffraction method of putting out a laser beam out of resonator is an emission of a ring beam which subjects the strong diffraction. This disadvantage can be partly reduced by means of application of polarization or transmission method of putting out a beam from an oscillator. However, the problem of obtaining the smooth transversal distribution of irradiation of TEM$_{00}$ mode, with an adequate volume, still exist. In the other technological solutions which enable obtaining the smooth distribution of irradiation intensity, both in a near and far field, are used so called apodizers, it means optical and electrooptical elements, by means of which one can realize, radially variable transmission or reflection coefficient. These elements can be used both in stable and unstable resonators what is schematicaly presented in Fig.1.

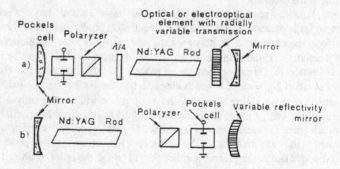

Fig.1. The exemplary schemes of resonators, using elements with the radialy variable coefficient of reflection or transmission.

Moreover, these elements should additionally ensure:
– matching to a profile of amplification coefficient of an active material;
– realization of maximal coefficient of filling an amplifying medium, it means:

$$F_{FIL} = \frac{1}{\pi a^2 I_0} \int_0^a 2\pi I(r) r dr \qquad 1$$

where: I(r) – radiation intensity; I$_0$ – radiation intensity in a beam centre; r – distance from the beam centre and: a–radius of an active material for which I = 1/e²I$_0$ should be equal unity.

– plane profile of irradiation intensity in the output beam;
– resistance at relatively high power;
– significant contrast $I_{MAX}/I_{MIN} > [10^2 - 10^3]$, it means a ratio of a transmission in a centre $I_0 = I_{MAX}$ to a transmission at an edge $I_0 = I_{MIN}$;
– high optical quality;
– possibility of making an aperture with any diameter;
– simple construction, reliability and low price of manufacturing;

Till now, many elaborations and constructions of optical and electrooptical elements with radially variable transmission or reflection coefficients, have been presented which enable to attain a balance between two mutually excluded demands: big volume of the basic mode and smooth distribution of irradiation in a near field.

In general, these elements can be divided into two types: to the first type are included these elements which placed inside the laser resonator realize only the smoothing of the profile of irradiation intensity distribution, e.g. a dish with an absorption liquid with the radially variable absorption coefficient. The second type, these are elements which, besides of realization of the first function, namely smoothing the transversal beam profile, additionally realize a back coupling (feedback) in a resonator (see Fig.1).

In order to diffraction image of diaphragm with "a" radius could undergo smoothing at the a smoothing, at a distance "z" from diaphragm, the change of Fresnel number: $N_F = a^2/\lambda z$ caused by imperfection of broadening of aperture edges should be near unity, it means: $\Delta N_F = 2a\Delta a/\lambda z = (2a\Delta a)N_F = 1$. Therefore, one can find a necessary size of imperfection or broadening of apodizer edges, namely: $\Delta a \sim a/2N_F = \lambda z/2a$.

Below, there are mentioned an existing and elaborated optical and electrooptical elements which enable realization of radially variable transmission or reflection coefficient, preventing harmful diffraction effects, occuring in the processes of generation, amplification as well as proccesing of laser radiation into the higher harmonics.

A. Photographic aperture with a radially variable stage of photo–emulsion blackenning [4].
B. Metalic diaphragmes with a radially variable thickness of a deposited layer [5].
C. Apertures with a radially variable transmission of a liquid absorber [6].
D. Apertures with a radially variable transmission of a solid–state –glass absorber [7].
E. Profiled apertures based on a disturbance of a coefficient of total internal reflection [8].
F. Apertures with radially variable transmission, using electro and magnetooptical effects[9]
G. Profiled aperture using double refracting crystal [10].
H. Apertures with a radially variable transmission, using F–P interferometr [11].
I. Diffraction grating with a variable depth, [1].
J. Passive and active apertures using liquid crystals [12].
K. Dielectric mirrors with a radially variable coefficient of a transmission [13–15].

The most perfect configuration of a resonator, in which can be used a single optical element with a radially variable transmission and can be ensured a back coupling, it seems to be an unstable resonator p–branch type, presented in Fig.2, with the dielectric output mirror with a profiled reflection coefficient [13–15].

If the profile of reflection coefficient of an output mirror is a Gauss curve, it means $R = R_0 \exp[-r^2/w_m^2]$, where: R_0 – value of reflection coefficient in the centre of mirror; w_m – radius of reflecting, dielectric layer for which occures the power decrease of e^{-2} value, the generated beam will be collimated beam, Gauss type, with a radius $w_b =$

$w_m, (M^2 - 1)^{0.5}$; where: $M = R_{WKL}/R_{WYP}$ is an optical magnification of resonator. As it is seen, a mode radius is fully described by a value of magnification M and radius w_m.

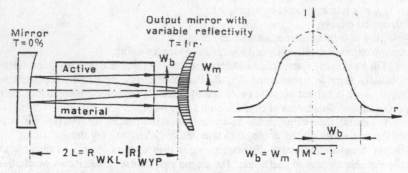

Fig. 2. Unstable resonator p–branch type with a dielectric output mirror with a radially variable coefficient of reflection.

Moreover, an effective back coupling for the basic and sequent mode is: R_0/M^2 and R_0/M^4, what influences on the velocity of discrimination of the higher order modes. The putting out a beam out of resonator follows through the output mirror and a distribution of radiation intensity can bedefined in form:

$$I(r) = I_0 \exp(-r^2/w_b^2)[1 - R_0(r^2/w_b^2)] \qquad 2$$

The distribution of irradiation in an output beam will have a dip in a centre if the below condition is not fulfil: $R_0/M^2 < 1$.

The output beam in a near field will further have the diffraction rings, arrised on the rod aperture. However, in a far field there will be a smooth distribution, even for filling a rod with a beam of diameter $2w_b = 0.7d$; where: d – rod diameter. By means of the mirrors with the reflection profile of a Gauss type, it is not possible to obtain simultaneously the smooth distribution of irradiation intensity in a near and far field. However, a resonator configuration with such a mirror is further very attractive, even for active media with a small amplification coefficient.

3. CONSTRUCTION OF VARIABLE REFLECTANCE DIELECTRIC SUPERGAUSSIAN MIRROR

Optical scheme of the variable reflectance mirror (VRM), which consist of three dielectric lyers is shown in Fig.3.

The construction shown above is made in such a way in order to the layers with refraction index n_2 and n_3 have had so called quarterwave thickness for $\lambda = 1.06\mu m$. These layers have been deposited on the substrate with refraction index n_s so such a design has the lowest coefficient of reflection – antireflection layer. The layer 1, with refraction index n_1 has different thickness versus its own cylindric axis. It means that such a layer has the variable reflectance and analitical formulas to describe such construction are described in [16]. Figure 4, shows the value of reflectivity stack versus the phase thickness of the layers, calculated from the formula (5) from reference [16]. The phase thickness was changed from zero to 90% ($\pi/2$).

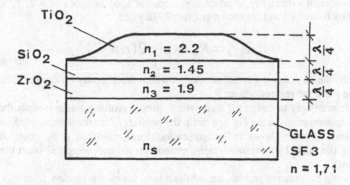

Fig.3. Scheme of the design variable reflectance dielectric mirror.

where: n_1; n_2; and n_3 – refraction index of the layers 1,2, and 3 respectively; n_s refractive index of the substrate.

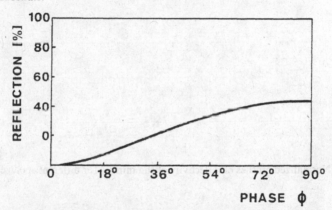

Fig.4. The value of the refractivity coefficient versus phase thickness of the layer 1, where: $n_s = 1{,}71$; $n_2 = 1{,}45$; $n_3 = 1{,}9$.

For design processes of profile mirrors is necessery the layers n_2 and n_3 (with the quarterwave thickness to be created the antireflection coationg on a substrate n_s (reflection R=0%). In this case, on the glass substrate SF$_3$ was deposited the ZrO$_2$ and SiO$_2$ layers which refractive index were 1,9 and 1,45 respectively.The reflectivity of the light from this construction, calculated from formula 5 was R=0,0004%. Of course, in the real mirror these refractive indexes of evaporate layers are differ from theoretical values and thicknesses are different too.

The transition from minimum to maximum of reflectivity,(which in this case is R$_{max}$=43%), is received through a deposition of thin layer TiO$_2$ with the refraction index n_1=2,2. Profiled reflection is received by a profilling of thickness a layer TiO$_2$ (n_1). Minimum and maximum of reflectivity are obtained for the zero and the quarterwave values of thickness of the layer, respectively.

Profiles of reflectivity of profiled mirrors for apodize index N = 2, 5, and 10, are described by formula 3 and are shown graphically in Fig.5.

$$R(r) = R_0 \exp[-2(r/w)^N]$$

the order N, the spot size w (radial distance at which the intensity reflectance falls to $1/e^2$ of peak value) and the real reflectance R_0.

It is necessery to design the mask with a proper profile because it needs the suitable profile of transmission. In the Fig.6, a and b the profiles of masks determinated for apodizes index 2 and 10 were shown in magnification form, diminish it by photo and a next performing the mask by a photolitography method on the coppering foil (with the thickness of 0.05 mm).

During evaporation process, the substrate is in the rotates motion relatively to its own axis, but the mask is in any motion. The symetric position of the mask in respect to the rotating of substrate is very important and difficult (especially for a convex mirror).

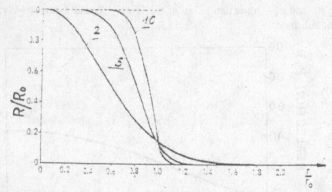

Fig.5. Normalized profiles of reflectivity of the mirrors for different apodize index N.

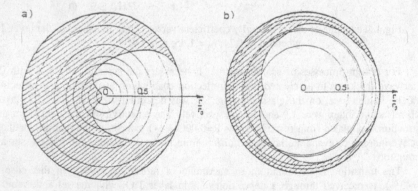

Fig.6. Examples of the masks for the apodize index N = 2, and 10.

4. EXPERIMENT

The experimental investigation of oscillators with the unstable two different resonators: classical and p – branch type were carried out adopting the optical elements used earlier in the other oscillators. The Nd:YAG rod of the diameter $\Phi = 4$mm and lenth 75mm was pumped by a linear flash lamp placed in a diffusion cavity reflector. Fig.7 shows the optical schemes of the two oscillators in which the laser beams were led outside the oscillators by means of the dielectric variable reflectivity mirrors for the same coefficient of apodization, N = 10. As one can seen from Fig. 7 the shapes of substrate that mirrors are different. The first dielectric mirror has been evaporated on the thin optical wedge. The second one has been evaporated on the meniscus. The curvature radius of the output variable reflectivity mirror (meniscus) were: $R_1 = 1033$mm, and $R_2 = -2016$mm, respectivly.

The concave–convex output mirror has been made on the substrate of SF–6 glass in the form of meniscus of thickness 4,5mm. Both, the concave and convex surfaces of that mirror were first coated with antireflective layers and then the convex surface was coated with profiled layers. The transmission of those layers was measured by means of a conventional light source and CCD camera, and is shown in Fig.8. The conventionalsource of light, instend of a coherent one was used in order to avoid heterogenity of the beam intensity, influence of diffraction and interference disturbing the measurements.

The measured transmission profile of the output mirror was approximated by the supergaussian function of the following form:

$$T(r) = \exp[1 - 0,4(r/r_0)^{10}] \qquad 4$$

where: r – the current radius measured from the centre; r_0 – 0,9mm is scalingradius of the mirror.

The best agreement of the measurement results with the approximating curve was obtained for the apodization factor N = 10,6.

Fig.9 shows the output – vs – input energy curves of the free running Nd:YAG oscillator for confocal p–branch type resonator. In the experiment we utilizing two different resonator configurations but with the same supergaussian reflectivity profile mirrors as the output couplers in order to compare the boath oscillators.

The measurements indicate that in the far field more than 80% of the beam energy is contained in the central spot corresponding to a divergence of <0,3mrad. The beam point stability of unstable classical resonator (Fig.7a) was an order less than for the concave resonator from Fig.7b.

In literature, there are analysed theoretically and experimentally the other reflection profiles of dielectric output mirrors of unstable resonators (linear, parabolic and the like). One of such profiles is so called super–Gauss for which a reflection R(r) can be written in an analytical form, namely: $R(r) = R_0 \exp[-2(r/w_m)^N]$, where R_0 – value of reflection coefficient in the centre of a mirror; r – radial coordinate, w_m – radius of a reflection layer of a mirror, for which the reflection coefficient attains value $1/e^2 R_0$, and "N" – coefficient of apodization. Mirrors of a such type are diposited on the glass base, plane [14] or spherical one [15]. It was schematically presented in Fig.7. The profiles ofcurvatures, for various values of apodization exponent, shows Fig.8. If "N" tends to an infinity, the smoothness disapears and such a mirror degenarates to an ordinary, homogeneous reflective mirror with a radius w_m and now we have diffraction method of putting out a beam out of resonator.

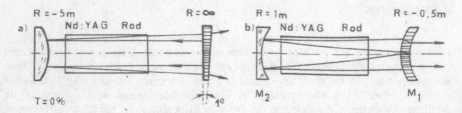

Fig.7. Scheme of the construction of unstable resonators: a) classical, b) confocal p—branch

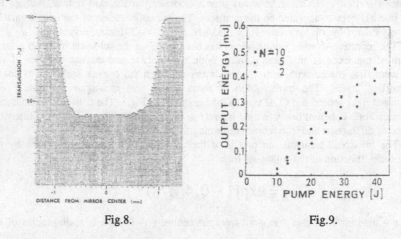

Fig.8.   Fig.9.

The dimension of a beam w_b which incidences on a mirror with a profile of super—Gauss type is given by the following expression: $w_b = w_m(M^N - 1)^{1/N}$, while a reflected beam is also in super—Gauss form with a dimension $w_m = w_b/M$.

In spite of the fact that manufacturing technology of the dielectric mirrors with a radially variable reflection coefficient is for time being very expensive, the obtained gain means that radiance [W/cm^{2}sr] is several dozen times higher than in a stable resonator.

5. CONCLUSIONS

In the paper, there have been presented advantages and disadvantages of resonators, more precisely speaking – ather their configurations which enable the smoothing of transversal profile of a laser beam. In this chapter, there is given more general summarization of advantages and disadvantages of previously described constructions.

The classical unstable resonators have two very significant disadvantages, namely:
– output beam in form of ring in a diffraction method of its putting out out of resonator or heterogeneous transversal distribution with the visible diffraction rings in a polarization method of a beam putting out;
– a possibility of simultaneous coexistence of two modes when a resonator "works" near the integral equivalent Fresnel number;

One from among the methods of overcoming these advantages it is an usage of the soft—edges mirrors [13] which reflections change from zero in the direction of mirror edge and the same it is a possibility of optimal fulfilement of a amplifying medium and a smoothing of transversal distribution of an output beam in a near field. In particular, the unstable resonators with a radially variable reflection coefficient of the output mirrors have the following advantages:

– folding of a transversal distribution of an irradiation intensity in a near field is significantly reduced; this is a consequence of decreased irradiation intensity from so called diffraction edge waves [1];

– a near field is more homogeneous without a typical houl which characterizes the diffraction method of putting out a beam out of resonator and in consequence, the sidebands in a far field are significantly decreased [13–15];

– bigger and faster discrimination of the transversal modes of a higher order [1].

6. BIBLIOGRAFIA

[1] – A.E. Siegman, "Lasers",Oxford U. Press, Oxford, 858–922{1986}.

[2] – D.C. Hanna, C.G. Sawyers and M.A. Yuratich, "Large Volume TEM_{00} Mode of Nd: YAG Lasers", Optics Communic., vol. 37, No 5, p.359,{1981}.

[3] – A.E. Siegman, "Unstable Optical Resonators for Laser Applications", Proceedings of the IEEE, vol.53, p. 277, {1965}.

[4]– E.W.S. Hee, "Fabrication of Apodized Apertures for Laser Beam Attenuation", Optics and Laser Technology, April 1975.

[5]– P. Giacomo, B. Roizen–Dossier et S. Roizen, "Preparation par Evaporation Sous Vide, D'Apodiseurs Circulaires", Journal de Physique, tome25, page 285, {1964}.

[6]– V.R. Costich, and B.C. Jonson,"Apertures to Shape Highpower Beams", Laser Focus, September 1974.

[7] –A.D. Cvetkov i inni, "Steklane Apodizirujuszcze Diafragmy z Supergausovskoj Funkciej Propuskania", Ż.P.S. tom 45, No. 6, s. 1022, {1986}.

[8]– E. Armandillo and G. Giuliani, "Achievment of Large–Sized TEM_{00} Mode from an Excimer Laser by means of a Novel Apoditic Filter", Opt. Lett., vol.10, No.9, {1985}

[9]– W.W. Simons, G.W. Leppelmeier, and B.C, Jonson, "Optical Beam Shaping Devices Using Polarization Effects", Applied Optics, nol.13, No. 7, p. 1629, {1974}.

[10]– G. Giuliani, Y.K. Park, and R.I. Byer, "Radial Birefringent Element and its Application to Laser Resonator Design", Opt. Lett. vol.5, No.11, p.491, {1980}.

[11]– S.DeSilvestri, P. Laporta, V. Magni, and O. Svelto, Radially Variable Reflectivity Output Coupler of Novel Design for Unstable Resonators", Optics Letters, vol.12, No.2, p.84,{1987}.

[12]– S.D. Jacobs, "Liquid Crystal Devices for Laser Systems", J. of Fusion Energy, vol. 5, No. 1, p. 65, {1986}.

[13]– N. McCarty and P. Lavigne, "Large–size Gaussian Mode in Unstable Resonators Using Gaussian Mirrors", Optics Letters, 10(11), 553–555{1985}.

[14]– J. Firak, J. Marczak, and A. Sarzyński, "Unstable Resonator with a Supergaussian Dielectric Mirror for Nd:YAG Q–Switched Laser", SPIE, Vol. 1391, p.42{1990}.

[15]– S.DeSilvestri, P. Laporta, V. Magni, and O. Svelto, and B. Majocchi, "Unstable Laser Resonator with a SuperGaussian Mirror,Optics Letters, vol.13, No.3, 201–203,{1988}.

[16]– J.Firak, and J.Marczak, "Optical Mirrors with VariableReflectance", SPIE, Vol. 2202, p.196{1993}.

Theoretical and Experimental Investigations of Passive Q-Switching of Diode-Pumped Solid-State Lasers

A. Pfeiffer[+], P. Peuser[*], N.P. Schmitt[*], M. Kokta[#]

[+]Ludwig-Maximilians-Universität München, Sektion Physik, Germany
[*]Daimler-Benz AG, Forschung und Technik, Box 800465, D-81663 München, Germany
[#]Union Carbide Corporation, 750 South 32nd Street, Washougal, Washington 98671, USA

Abstract: Starting with the rate equations, the dynamical behavior of passively Q-switched, diode-pumped solid-state lasers was investigated. The importance of the absorption cross section of the saturable absorber material was evaluated, and the relevant parameters of a passively Q-switched laser were studied. The theoretical calculations were compared with results from experiments performed with a longitudinally diode-pumped Nd:YLF laser which was Q-switched by means of Cr^{4+}:YAG as saturable absorber.

The dynamics of a passively Q-switched solid-state laser can well be described on the basis of the rate equations. The results can then be used to optimise a diode-pumped laser, Q-switched, e. g., by the new Cr^{4+}-doped, saturable absorber crystals. Preliminary considerations concerning the theoretical implications on passive Q-switching have already been dicussed in [1].

The ratio σ_s / σ_a (σ_s: absorption cross section of saturable absorber, σ_a: stimulated emission cross section of active material) is a parameter of special importance for passive Q-switching. This can be seen by investigating the relation between the population densities of the active material and the saturable absorber during the laser pulse. The following expression can be derived from the rate equations:

$$n_s / n_{si} = \left(n_a / n_{ai} \right)^{\frac{2 \, \sigma_s}{\sigma_a}}$$

(n_s: inversion population density of saturable absorber; $n_s = n_{s2} - n_{s1}$ (two-level system, equal level degeneracies assumed); n_a: inversion population density of active material, the index "i" means initial conditions, i.e. before onset of lasing).

The term on the left side reveals in how far the absorber is bleached compared to the depletion of the inversion density of the active material. Its value lies between zero and one :

(n_s / n_{si}) = 1 : the absorber is not bleached at all; corresponds to (n_a / n_{ai}) = 1, i.e. the situation before onset of lasing.

(n_s / n_{si}) = 0 : the absorber is perfectly transparent for the lasing wavelength; corresponds to (n_a / n_{ai}) = 0 , i.e. entire depletion of the upper laser level.

Both population densities are plotted simultaneously in Fig. 1 for different values of σ_s / σ_a.

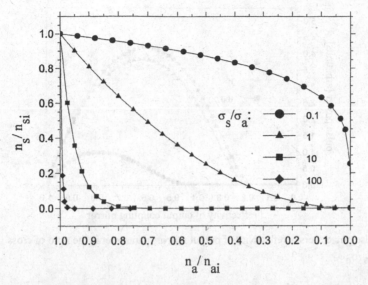

Fig. 1: Relative population density of the saturable absorber versus relative population density of the active material for different ratios of the corresponding cross sections

It is important to notice that the abscissa in this diagram starts with $(n_a/n_{ai}) = 1$ (beginning of laser pulse) and ends with $(n_a/n_{ai}) = 0$ (end of laser pulse). However, the exact time structure of the pulse dynamics cannot be seen by these curves. Furthermore, the population densities do not necessarily reach their lowest possible value (zero) at the end of the laser pulse in every case. Nevertheless, an important conclusion results from this diagram concerning the influence of the ratio σ_s/σ_a :

The larger the ratio σ_s/σ_a, the better the bleaching of the saturable absorber for the same values of n_a/n_{ai}. If the cross section of the saturable absorber exceeds the cross section of the active material by ten to a hundred times, the absorber gets already entirely bleached for minimal depletion of the inversion density of the active material, i.e. only a few photons have to be ´sacrificed´ for the passive Q-switch, whereas the majority of the photons enter into the actual laser pulse.

In order to optimize a passively Q-switched solid-state laser with respect to pulse energy, pulse duration and peak power, one can again refer to the rate equations which give suitable expressions after some calculations. Several of these expressions cannot be evaluated analytically and therefore have to be calculated numerically.

For a given laser system some figures have to be measured, the most important of which are the maximum possible inversion density of the active material (equivalent to the gain coefficient) and the constant (i.e. not saturable) losses. These figures enter directly as constants into the equations, whereas the reflectivity of the output mirror serves as a variable parameter. The saturable losses have to be adapted to the reflectivity according to the first threshold in order to prevent lasing as long as the maximum gain has not yet been reached. Some results of such calculations are plotted in Fig.2.

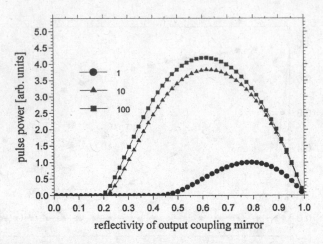

Fig. 2: Pulse power versus reflectivity of output mirror; parameter is the ratio of cross sections σ_s / σ_a .

By concerning the laser threshold conditions, one can determine the initial inversion of the active material n_{ai} as a function of the absorber length. It turns out that n_{ai} increases linearly with increasing absorber length. By using the rate equations, one can also calculate numerically the remaining population density of the active material after the laser pulse, i.e. the part of stored energy that has not been extracted from the active material. It can be seen that the relation between the initial population density of the active material and the final population density at the end of the laser pulse is similar to an exponentially decreasing curve. The stored energy is used more effectively the higher the initial gain, similar to the case of active Q-switching.

The influence of the absorber thickness (i.e. the initial saturable losses) on the dynamics of the laser pulse is demonstrated in Fig.3.

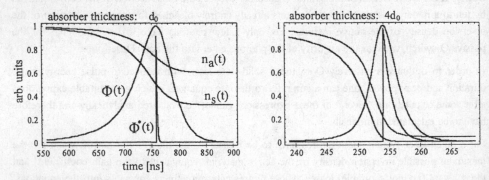

Fig. 3: Time dependence of photon density $\Phi(t)$, population density $n_a(t)$ of active material and population density of saturable absorber $n_s(t)$ for different absorber thicknesses; $\Phi^*(t)$: exponential decay of maximum photon density corresponding to cavity photon lifetime τ_c , i.e. without saturable absorber.

Increasing the initial saturable loss results in an increase of the initial population density of the active material, which means a more efficient extraction of the stored energy by the laser pulse. Moreover, the absorber gets bleached better and faster for higher initial gain. In addition, the duration of the laser pulse is more and more determined by the cavity photon lifetime.

Some of these results were experimentally verified for a small Nd:YLF laser, which was longitudinally pumped by a 3 W cw-diode laser, and which was Q-switched by means of Cr^{4+}:YAG as a saturable absorber crystal [1]. The cylindrical laser crystal was 8 mm long, and the resonator length was 40 mm. By optimising this laser configuration, applying the calculations described above, maximum pulse energies of > 80 μJ and minimum pulse lengths of 7 ns have been achieved (Fig.4). The pulse repetition rate decreased with increasing absorber thickness. A maximum value of 10 kHz was measured at a length of 2 mm of the absorber crystal and 10 % transmission of the output coupling mirror. It is interesting to note that the duration of the pump pulse (until laser pulse emission begins) decreased with increasing pump power and, on the other hand, increased for greater absorber thicknesses.

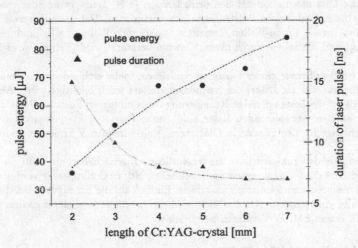

Fig. 4: Experimental values of pulse energy and pulse length plotted as functions of the length of the saturable absorber crystal

References:

[1]   A. Pfeiffer, S. Heinemann, A. Mehnert, N.P. Schmitt, P. Peuser:
      Passive Q-switching of diode-pumped solid-state lasers with Cr^{4+}:YAG crystals;
      in: Laser in der Technik, Laser 93, ed.: W. Waidelich, Springer-Verlag, Berlin, 1994, p. 94 -
      97

Charakterisierung eines frequenzmodulierbaren Nd:YAG-Ringlasers für die Brillouin-Fasersensorik

R. Philipps und J.W. Czarske
Universität Hannover, Institut für Meß- und Regelungstechnik im Maschinenbau,
D–30167 Hannover, Germany

Kurzfassung:
Der Beitrag stellt ein FMCW-Verfahren als neue Möglichkeit vor, einen intrinsischen Glasfasersensor unter Nutzung der stimulierten Brilouin-Streuung zu realisieren. Die Eignung eines diodengepumpten monolithischen Nd:YAG Ringlasers für dieses Verfahren wird experimentell untersucht.

Einführung:

Bei der verteilten Glasfasersensorik unter Nutzung der stimulierten Brillouin-Streuung ist es bisher üblich, die Ortsinformation der Sensormeßgrößen (z.B. Temperatur oder mechanische Spannung) durch die Analyse von Pulslaufzeiten zu bestimmen. Dabei werden zwei Laser als Quelle von Pump- bzw. Probe-Wellen eingesetzt, wobei eine Welle den Brillouin-Gewinn in der Faser erzeugt und die andere durch diesen Gewinn verstärkt wird (Prinzip eines Brillouin-Verstärkers).

Um einen intrinsischen Sensor großer Länge zu realisieren, bietet sich jedoch auch ein FMCW-Verfahren an, für das nur *ein* Laser, der frequenzmoduliert wird, benötigt wird. Als Vorteile dieses Verfahrens sei die kostengünstige Realisierung eines einfachen Sensoraufbaus angeführt. Anforderungen an den einzusetzenden Laser sind: hohe Leistung, Frequenzmodulierbarkeit, geeignete Wellenlänge für den Einsatz in Glasfasern, single-frequency Emission und eine hohe Kohärenzlänge.

In diesem Beitrag werden zunächst kurz die wesentlichen Eigenschaften eines für das FMCW-Verfahren geeigneten dioden-gepumpten monolithischen Nd:YAG Ringlasers vorgestellt. Weiterhin wird auf die Frequenzmodulation und deren Einfluß auf die Erzeugung der stimulierten Brillouin-Streuung eingegangen. Als Ausblick wird die mögliche Realisierung eines verteilten Fasersensors mit einem FMCW-Verfahren beschrieben.

1. Monolithischer Nd:YAG Ringlaser

Als geeigneter Laser für den Einsatz in einem Glasfasersensor bietet sich ein monolithischer Laserdioden-gepumpter Nd:YAG-Ringlaser an. Der verwendete Laser wurde im Laser - Zentrum Hannover e.V. entwickelt, weist eine Wellenlänge von 1319 nm, eine hohe Ausgangsleistung von bis zu 110 mW und eine Möglichkeit der schnellen Frequenzmodulation auf. Dieser Laser wird im weiteren als MISER (monolithic isolated single-frequency end-pumped ring) bezeichnet [FREITAG(a),KANE]. Zu den weiteren Forderungen nach hoher Kohärenzlänge, single-frequency Emission sowie Frequenzmodulation werden im weiteren kurz die experimentellen Ergebnisse vorgestellt. Für nähere Angaben sei auf [CZARSKE] verwiesen.

Kohärenzlänge: Wird auch nach großen Gangunterschieden von zwei Lichtwellen noch Interferenzfähigkeit dieser Wellen verlangt, so wird der maximal nutzbare Gangunterschied durch die Kohärenzlänge des Lasers bestimmt. Es wurde zur direkten Messung der Kohärenzlänge

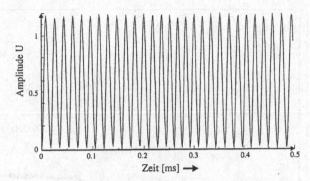

Abbildung 1: Typisches Beatsignal der beiden Interferometerwellen.

ein all-fiber-Mach-Zehnder-Interferometer mit großem Gangunterschied (ca. 57 km) aufgebaut. Die Interferometerstrahlen wurden auf einer Photodiode überlagert und das entstehende Beatsignal, siehe Abb. 1, ausgewertet. Aus der sich ergebenden Visibilität wurde ein Kohärenzgrad bestimmt, aus dem durch Extrapolation (Abfall der Autokorrelierten auf $1/e$) die Mindestkohärenzlänge zu 1500 km bestimmt werden konnnte. Über die Fouriertransformation entspricht dies einer FWHM-Linienbreite von weniger als 60 Hz.

Modenverhalten: Im üblichen, ungestörten Betrieb emittiert der MISER [KANE] singlefrequency Strahlung mit hoher Amplituden- und Frequenzstabilität. In Experimenten mit Einkopplung der Laserleistung in eine Glasfaser hat sich jedoch herausgestellt, daß durch den an der Oberfläche der Faser auftretenden Fresnel-Reflex eine Rückkopplung in die Lasermode auftritt. Als Ergebnis wurde eine instabile Oszillation in bis zu vier longitudinalen Moden beobachtet, siehe Abb. 2. Die Rückkopplung muß demnach durch einen optischen Isolator im Strahlengang genügend geschwächt werden, in unserem Aufbau um > 17 dB.

Im Laufe dieser Experimente stellte sich zudem heraus, daß auch ohne Rückkopplung stets ein Beatsignal bei dem Abstand der longitudinalen Moden (ca. 5.66 GHz) zu beobachten war. Dies ist das Resultat der verstärkten spontanen Emission im Laser. Bei bestimmter Wahl der Laserparameter (Pumpleistung, Kristalltemperatur) war es möglich, eine über mehrere Minuten stabile Oszillation von genau zwei longitudinalen Moden zu erreichen. Die Ursache liegt hier in einem Modenübergang, d.h. durch relative thermische Verschiebung sowohl der Laserfrequenz als auch der Nd:YAG-Gewinnkurve weisen zwei Moden etwa die gleiche Verstärkung auf. Der Modenwettkampf wird vermutlich aufgrund von Anteilen inhomogener Linienverbreiterung in Nd:YAG verringert, so daß beide Moden anschwingen können. Über die Beobachtung der Beatsignale zwischen den longitudinalen Moden war es möglich, u.a. die optische Weglänge eines Resonatorumlaufes sehr genau zu bestimmen: bei einer Kristalltemperatur von 20^0C ergab sich eine Umlauflänge von 52.933 mm, mit einem thermischen Koeffizienten von 0.8 μm/K.

Fazit: Es wurde gezeigt, daß bei geeigneter Kontrolle des Beatspektrums eine hochstabile single-frequency Emission gewährleistet ist. Die sehr große Kohärenzlänge von > 1500 km bedeutet keine Einschränkung für real nutzbare Sensorlängen.

2. Frequenzmodulation des MISER
2.1. Realisierung der Frequenzmodulation
Die Frequenzmodulation bei dem eingesetzten MISER erfolgt durch die Ansteuerung eines piezoelektrischen Kristalles, der auf der Oberseite des monolithischen Laserkristalles befestigt ist

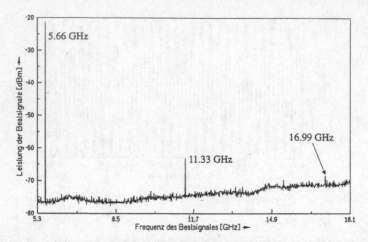

Abbildung 2: Beatsignale, entstanden durch die Überlagerung von mehreren gleichzeitig os-
zillierenden longitudinalen Moden. Der Modenabstand beträgt etwa 5.66 GHz, so daß aus
der Existenz des Signales bei 16.99 GHz sofort auf die Oszillation von mindestens vier Moden
geschlossen werden kann.

[FREITAG(b)]. Durch die Ansteuerung des Piezos ändert sich die Resonatorweglänge und damit auch die Emissionsfrequenz des Lasers.

2.2. Meßergebnisse

Eine Kenngröße bei der Frequenzmodulation ist der Konversionsfaktor, der hier die Umsetzung
der Spannungsamplitude am Piezo in den Frequenzhub der Laserwelle angibt. Er wurde folgendermaßen bestimmt: Durch Überlagerung der frequenzmodulierten Welle des MISER mit
der unmodulierten Welle eines weiteren, unabhängigen MISER (Localwelle) wurde auf einem
Photodetektor ein Beatsignal erzeugt, siehe z.B. die Local/Pump-Beatsignale in Abb. 4. Die
spektrale Breite dieses Beatsignales läßt sich über einen Spektrumanalysator direkt ausmessen
und ist zudem gegeben durch das Produkt aus Konversionsfaktor und Spannungsamplitude am
Piezo. Dabei mußte sichergestellt werden, daß der Frequenzdrift zwischen den beiden Lasern
im Meßzeitraum zu vernachlässigen war. Es stellt sich heraus, daß der Konversionsfaktor ca.
1.5 MHz/V beträgt, jedoch einen nichtlinearen sowie frequenzabhängigen Verlauf aufweist.
Eine Beschränkung der Spannungsamplitude am Piezo und somit des Frequenzhubes wird
durch die Amplitudenmodulation hervorgerufen, die sich parasitär durch die vorgenommene
Frequenzmodulation ergibt. In der Umgebung der Systemresonanzen, d.h. der Nd:YAG-
Relaxationsfrequenz sowie der Piezoeigenfrequenz, steigt die Amplitudenmodulation auf 100%
an. Da die Frequenz der Relaxationsschwingungen in Nd:YAG bei den von uns genutzten Laser-
Ausgangsleistungen kleiner als die Resonanzfrequenz des Piezos ist, ergibt sich daraus konkret
die Beschränkung, die Modulationsfrequenz unterhalb der Nd:YAG Relaxationsschwingungen
zu wählen, d.h. unter ca. 100 kHz.

3. Brillouin-Streuung bei Frequenzmodulation des MISER
3.1. Fragestellung

Bevor der frequenzmodulierte MISER zu Sensorexperimenten unter Nutzung stimulierter Brillouin-Streuung herangezogen werden kann, muß geklärt werden, welchen Einfluß die Frequenz-

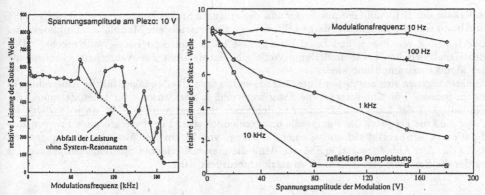

Abbildung 3: Verhalten der erzeugten Stokes-Leistung, wenn der Pumplaser frequenzmoduliert wird. Die Pumpleistung betrug für eine Faserlänge von 19 km in beiden Fällen etwa 23 mW.

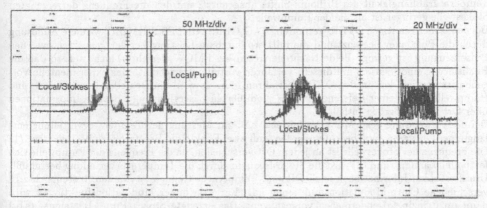

Abbildung 4: Beatsignale zwischen Local-/Stokes-Welle (jeweils links) und Local-/Pumpwelle (jeweils rechts) für rechteckförmige Modulation (linkes Spektrum) und dreieckförmige Modulation (rechtes Spektrum). Die Modulationsfrequenz beträgt 1 kHz, der Spannungsamplitude am Piezo 30 V (Spitze-Spitze).

modulation auf die Erzeugung der stimulierten Brillouin-Streung ausübt. Konkret mußte der Wertebereich von Spannungsamplitude und Modulationsfrequenz bestimmt werden, in dem stimulierte Brillouin-Streuung erzeugt wird.

3.2. Ergebnisse

Man stellt für eine sinusförmige Frequenzmodulation der Laser-Pumpwelle eine Abnahme der erzeugten Stokes-Leistung mit zunehmendem Spannungsamplitude bzw. Modulationsfrequenz fest. Dies ist in Abb. 3 dargestellt. Qualitativ bietet sich für dieses Verhalten folgende Erklärung an: Bei sinusförmiger FM ist das Spektrum gegeben durch die Bessel-funktionen 1.Art und n-ter Ordnung mit dem Argument (Modulationsindex) $\beta = UK/f_m$, wobei U die Spannungsamplitude, K der Konversionsfaktor und f_m die Modulationsfrequenz ist. Durch steigende Modulationsfrequenz f_m wird der Abstand der Seitenbänder vergrößert, woraus sich eine Verringerung der Pumpleistung unter der Brillouin-Verstärkungskurve ergibt. Dies führt zu einer Abnahme der Stokes-Leistung. Bei steigendem Spannungsamplitude U und somit des Modulationsindex β ergibt sich eine Umverteilung der Leistung in höhere Seitenbänder, die ebenfalls außerhalb des Brillouin-Gewinns liegen und die erzeugte Stokes-Leistung verringern [DAVIS]. In Abb. 4 sind die Beatsignale dargestellt, die sich für verschiedene Modulationsfunktionen bei Überlagerung von Pump- und Local-Welle, sowie bei Stokes- und Local-Welle ergeben. Das Beatsignal zwischen Local- und Pump-Welle gibt dabei direkt die Wahrscheinlichkeitsdichte der Modulationsamplitude wieder.

Konkret ergeben sich aus diesen Untersuchungen folgende Überlegungen für die Anwendung in einem Sensor. Für eine vorgegebene Faserlänge muß in der Sensorik für jeden Zeitpunkt die Eindeutigkeit der Laserwellenfrequenz in der Faser gefordert werden, d.h. an jedem Ort der Faser muß die Frequenz der Pumpwelle unterschiedlich sein. Für dreieckförmige Frequenzmodulation (Chirp) ergibt sich so bei einer Faserlänge von 20 km eine Modulationsfrequenz von < 5 kHz. Aus Abb.3 (rechts) ergibt sich dann die maximal erlaubte Spannungsamplitude, bis zu der die Brillouin-Streuung nicht zu stark unterdrückt wird.

4. Ausblick: FMCW-Sensor / Ortsauflösung

Der Sensor nutzt als Meßeffekt die stimulierte Brillouin-Streuung [BOYD]. Dabei wird die Temperaturabhängigkeit des Brillouin-Shifts ausgenutzt, um den die Frequenz der erzeugten Stokeswelle geringer ist als die der Pumpwelle.

Bei den bisher genutzten OTDR-Verfahren [BAO,GERS,NIKLES,SHIMIZU] wird die Auflösung durch die Pulsdauer begrenzt und die mittlere nutzbare Leistung bei der genutzten externen Pulserzeugung ist gering. Beide Nachteile treten bei FMCW-Verfahren nicht auf, so daß die Realisierung eines Sensors mit diesem Verfahren vorteilhaft erscheint. Das experimentelle Vorgehen ist ähnlich dem in den OTDR-Verfahren, d.h. es wird ein Beatsignal von Stokes- und Pumpwelle (bzw. einer Referenzwelle des modulierten Lasers) erzeugt, das die Informationen über Ort und Meßgröße enthält.

Bei der Anwendung des FMCW-Verfahrens, z.B. zur Messung der Temperaturverteilung entlang einer Glasfaser, muß beachtet werden, daß zwei Effekte die Beatfrequenz beeinflussen: (1) Die Frequenz der Stokeswelle ist an jedem Ort verschieden, da die Frequenz der Pumpwelle sich fortlaufend ändert, und (2) wird die Stokesfrequenz durch die Temperaturverteilung beeinflußt. Um die beiden Informationen nach Ort und Temperatur trennen zu können, bieten sich z.B. verschiedene Chirp-Modulationsfunktionen an. Für jede dieser Modulationsfunktionen wird dann im Prinzip folgendermaßen vorgegangen: Das erzeugte Stokes/Pump-Beatsignal (s.o.) wird durch einen Bandpaß gefiltert und mit einem Sample-Oszilloskop ausgewertet. Dies wird für verschiedene Mittenfrequenzen des Bandpaßfilters und anschließend für eine andere Chirp-Modulation durchgeführt. Für die beiden Größen Ort und Temperatur stehen dann mehrere Datenreihen zur Verfügung, die die Separation der Informationsgrößen erlauben.

5. Zusammenfassung

Es wurde ein dioden-gepumpter monolithischer Nd:YAG Ringlaser vorgestellt, der sich aufgrund seiner Charakteristika für den Einsatz in der Glasfasersensorik anbietet. Ein neues Verfahren für einen verteilten Temperatur-Glasfasersensor unter Nutzung der stimulierten Brillouin-Streuung wurde vorgeschlagen: Statt des Pulslaufzeitverfahrens (OTDR) verspricht das FMCW-Verfahren einen einfacheren Sensor mit den prinzipiellen Vorteile von FMCW- gegenüber OTDR-Verfahren.

Danksagung

Diese Arbeit wurde finanziell unterstützt von der DFG (Bo 860/8-2). Die Glasfasern wurden von der Firma Siecor (D-Neustadt) zur Verfügung gestellt. Weiterhin sei dem Laser-Zentrum Hannover e.V., speziell Herrn Dr.rer.nat. I. Freitag, für die breite Unterstützung gedankt.

Literatur

- BAO X., Webb D.J., Jackson D.A., 'Recent progress in experiments on a Brillouin loss based distributed sensoir', Proc. SPIE **2360** (1994), 506.
- BOYD R.W., 'Nonlinear Optics', Academic Press 1992
- CZARSKE J., Philipps R., Freitag I., 'Spectral properties of diode-pumped nonplanar monolithic Nd:YAG ring lasers', Appl.Phys.B, im Druck
- DAVIS M.A., Kersey A.D., 'Scheme for Negating the SBS Power Limit in Remotely Interrogated Interferometric Fiber Sensor Arrays', SPIE **2071** (1993), 112.
- FREITAG I. (a), Kröpke I., Tünnermann A., Welling H., 'Electrooptically fast tunable miniature diode-pumped Nd:YAG ring laser', Opt.Commun. **101** (1993), 371.
- FREITAG I. (b), 'Entwicklung und Charakterisierung einer Laserstrahlquelle für den interferometrischen Nachweis von Gravitationswellen', Dissertation Universität Hannover (1994).
- GERS F., Czarske J.W., 'Untersuchungen zur verteilten Temperatur-Fasersensorik mit stimulierter Brillouin-Streuung' im Tagungsband Kongreß C dieses Jahres.
- KANE T.J.,Byer R.L., 'Monolithic, unidirectional singla-mode Nd:YAG ring laser', Opt.Lett. **10** (1985), 65.
- NIKLES M., Thevenaz L., Robert P.A., 'Simple Distributed Temperature Sensor based on Brillouin Gain Spectrum Analysis', Proc. SPIE **2360** (1994), 138.
- SHIMIZU K., Horiguchi T., Koyamada Y., 'Measurement od distributed strain and temperature in a branched optical fiber network by using Brillouin OTDR', Proc. SPIE **2360** (1994), 138.

Coherent Coupling of Diode-Pumped Solid-State Lasers
Kohärente Kopplung diodengepumpter Festkörperlaser

J. Plorin°, P. Peuser*, H. Weber
Optisches Institut der Technischen Universität Berlin,
Straße des 17. Juni 135, D-10623 Berlin, Germany
*Daimler-Benz AG, Forschung und Technik, Box 800465, D-81663 München, Germany
°present address: Daimler-Benz Aerospace, Box 801149, D-81663 München, Germany

Abstract: An analytical expression for the locking range of two lasers was derived by a theoretical model, describing the coherent coupling in a common cavity by means of an internal lens and a Fourier-filter. The results from the calculations were experimentally proved using two diode-pumped solid-state lasers by thermally detuning one laser and analysing the far field pattern in the locked and in the unlocked state.

Zusammenfassung: Um die kohärente Kopplung zweier Laser zu untersuchen, welche in einem gemeinsamen Resonator mit interner Linse oszillieren, wurde mit Hilfe eines theoretischen Modells ein analytischer Ausdruck für die gemeinsame Kopplungsbandbreite entwickelt. Im Experiment konnte diese Kopplungsbandbreite mittels zweier diodengepumpter Festkörperlaser bestätigt werden, indem die Auswirkungen einer thermisch hervorgerufenen Frequenzverstimmung zwischen den Lasern auf deren Fernfeld analysiert wurden.

Longitudinally diode pumped solid-state lasers exhibit principal thermal problems at low pumping levels of a few Watt already [1]. Coherent coupling of several individual lasers, each of moderate power, might be an alternative to solve these problems [2,3]. One possibility is the coupling of several independent lasers within a common cavity by internal Fourier-filtering [4,5].

A mathematical model has been developed giving an expression for the locking range of two diode pumped solid-state lasers which were coherently coupled by a simple Fourier-filter consisting of two absorbing areas. The electromagnetic field amplitude and the arbitrary phase difference generated in the laser crystal (Fig. 1) propagate through the laser resonator which can be unfolded as shown in Fig. 2. The field and the phase in the focus of an intracavity lens are described by a Fourier-transformation.

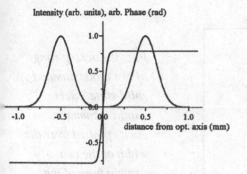

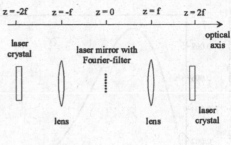

Fig. 1: Amplitude and (arbitrary) phase distribution of two lasers areas

Fig. 2: Confocal "unfolded" laser resonator used for the model calculations

In the focus of the lens the amplitude and phase distributions have to be multiplied by the transmittance function of the Fourier-filter. In this approach the Fourier-filter is assumed as a grating consisting of two perfectly absorbing bars of width g_0, located symmetrically to the optical axis at the minimum intensity of an in-phase locked laser array. After a complete round-trip within the laser resonator (and another Fourier-transformation) the resulting electromagnetic field amplitude and phase distributions are compared with the initial ones. The difference between the two phases as a function of the relative phase difference of the two lasers at the beginning of the round-trip yields the total phase shift per round-trip. Its maximum provides the locking range corresponding to Adler's equation [6].

The locking range as a function of the diameter of the absorbing areas of the Fourier-filter gives an essential correlation for the design of the Fourier-filter and is shown in Fig. 3 for two diode-pumped solid-state lasers with beam waists of 180 μm and distance of 1 mm.

The locking range was determined by detuning one laser which was achieved by varying the pump diode current. The experimental set-up is shown in Fig. 4. Outside the laser cavity four laser spots appear, two from each pumped laser (A and B). The laser oscillation was detected by the photodiode for each laser separately, while the interference pattern was analysed in the focus of an external lens (Fig. 5).
The signal from the photodiode placed in the beam spot of laser B disappeared when the laser system left the locking range.

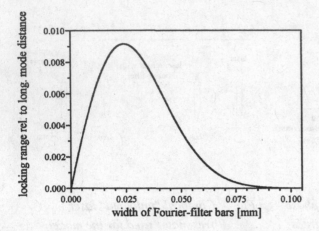

Fig. 3: Locking range
of two diode pumped
solid-state lasers
within a common
cavity plotted over the
width of the two
grating bars of the
Fourier-filter

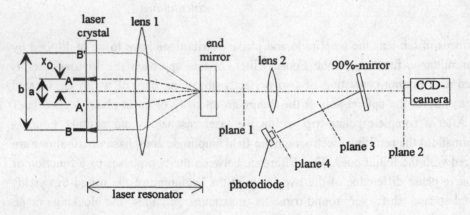

Fig. 4: Experimental set-up to determine the locking range

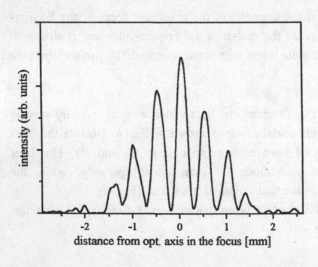

Fig. 5: Interference
pattern of the nearly
in-phase locked laser
array, recorded with
the CCD-camera

The periodical variation of the pump current of laser A and the resulting intensity of laser B measured with the photodiode are shown in Fig. 6.

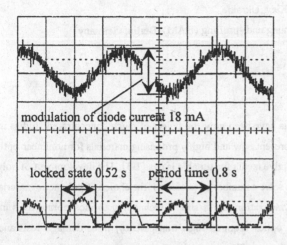

Fig. 6: Variation of the pump current of laser A (upper plot) and corresponding oscillation of laser B within the locking range (lower plot)

The maximum current variation of ± 9 mA resulted in a frequency shift of 7.8 MHz which is 1.3 times the theoretical locking range. Thus the laser system was about 70 % of a full period of 1.6 s in the locked state.

References:

1] M. Oka et al., IEEE J.Quantum Electron. **28,** 1142 (1992)
2] A. Okulov, J. Opt. Soc. B **6,** 1045 (1990)
3] V. Antyukhov et al., JETP Lett. **44,** 79 (1986)
4] R. Rediker et al., Appl. Phys. Lett. **46,** 133 (1985)
5] C. Nabors, LEOS 1992 Conf. Proc., pp. 497
6] R. Adler, Proc. of IEEE **61,** 1380 (1973) (reprint from 1946)

Fast Relaxation Processes in the Excited State of Polymeric Laser Media

O. Przhonska[1], Yu. Slominsky[2], U. Stahl[3], M. Senoner[3] and S. Dähne[3]

[1] Institute of Physics, Kiev, Ukraine,

[2] Institute of Organic Chemistry, Kiev, Ukraine,

[3] Bundesanstalt für Materialforschung und -prüfung (BAM), Berlin, Germany

Introduction

Dye-activated polymeric media have been now the object of much concentrated studies in basic research and applications. They represent new and highly promising materials for nonlinear optics and quantum electronics, especially as active media for tunable lasers [1-3]. The development of polymeric laser elements makes it necessary to solve a number of problems: the choice of polymeric materials and organic dyes, mechanism of polymerization and a simple technology to incorporate dyes into the matrix. The application in laser optics requires high transparency, laser damage resistance and photochemical stability of the polymers. To satisfy these requirements, high elastic polyurethane acrylate (PUA) was chosen as the optical matrix [3].

There is a number of peculiarities in the selection and optimization of dye molecules for lasing in a polymeric matrix. The requirements for the "optimal" dye structure are, in general, known, but the effect of the polymeric medium on the lasing properties of the dyes is ambiguous. For many dyes, lasing well in liquid solution, incorporation into polymeric media decreases the laser efficiency which can be associated with a decrease in the Stokes shift and with the heterogeneity of the polymeric medium. At the same time the use of polymeric matrices makes it possible to obtain the lasing effect at many organic dyes which in liquid solution either do not lase at all or are characterized by a high threshold of lasing. Positive effects of the polymeric medium include increasing fluorescence quantum yields and decreasing losses connected with photoisomerization processes in liquid solutions [4]. Therefore, the main effort of this paper was directed to study the effect of the polymeric matrix on excited-state relaxation processes of polymethine dye molecules using time-resolved spectroscopy.

Results and discussions

All experiments were made using two special selected polymethine dyes PD 1 and PD 2 synthesized in the Institute of Organic Chemistry (Kiev, Ukraine), see structure formula:

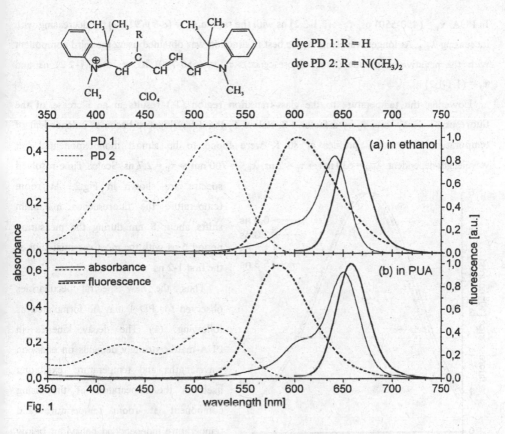

dye PD 1: R = H

dye PD 2: R = N(CH₃)₂

Fig. 1

Fig.1 (a,b) shows the absorption and fluorescence spectra of these dyes at room temperature. The absorption spectra demonstrate only slight shifts in the long wavelength region without any essential changes in the structure and half-width of the band. The fluorescence spectra do not depend on excitation wavelength in ethanol solution and exhibit some peculiarities in PUA-media. The most important are the "red-edge effect" and the "red limit" at the position of anti-Stokes fluorescence.

1. Polymeric medium and rearrangement of the dye environment

For the unsubstituted dye PD 1 the planar trans conformation is the energetically most preferable due to the absence of steric hindrances between chromophore fragments. In the polymeric medium trans-cis isomerization processes are restricted. Therefore, the fluorescence kinetics reflects only the relaxations in spectrally-inhomogeneous media. We found two components in the fluorescence decay of PD 1 at room temperature. In ethanol: $\tau_1 = (60\text{-}90)$ ps and $\tau_2 = (750\text{-}800)$ ps with a ratio of the amplitudes $a_1/a_2 = (3\text{-}5) / (95\text{-}97)$ % independent of excitation (λ_{EX}) and emission (λ_{EM}) wavelengths.

In PUA: $\tau_1 = (450\text{-}550)$ ps, $\tau_2 = (2.1\text{-}2.2)$ ns with the ratio $a_1/a_2 = (8\text{-}9)/(91\text{-}92)$ %. decreasing with increasing λ_{EM}. At longer wavelengths, the best fit function was obtained using the third component with the negative amplitude (rising fluorescence): $\tau_1 = (650\text{-}700)$ ns, $\tau_2 = (2.1\text{-}2.2)$ ns and $\tau_3 = (1.1\text{-}1.2)$ ns.

Lowering the temperature to the glass-transition region (T_g) results in an increase of the fluorescence decay time τ_2 by a factor 1.2. Below T_g the decay kinetics becomes independent of temperature. The decay profiles at 80 K were found to be almost monoexponential but wavelength-dependent: $\lambda_{EM} = 630$ nm $\rightarrow \tau_F = 2$ ns, $\lambda_{EM} = 700$ nm $\rightarrow \tau_F = 2.7$ ns. Scaled Time-resolved

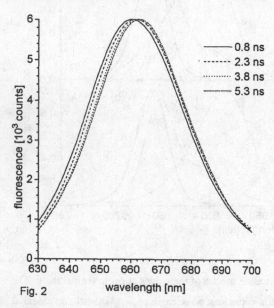

Fig. 2

spectra are shown in Fig.2. At room temperature the fluorescence maximum shifts about 8 nm during the measuring period 6 ns with the most essential shift at the first 1-2 ns.

Thus, the new spectral peculiarities observed for PD 1 may be formulated as following: (1) The decay kinetics in PUA-matrix strongly depends on emission wavelengths and temperature, exhibiting itself in the appearance of the rising component at room temperature and temperature independend behaviour below glassing region. (2) The dynamic Stokes shift is observed in polymeric medium confirming the role of excited-state relaxational processes with a time comparable with fluorescence lifetime.

2. Polymeric medium and large Stokes shift

For the dimethylamino (DMA) substituted dye PD 2 the great steric hindrances between the chromophore fragments strongly effect the probability of trans-cis isomerization processes at the ground state. That leads to a dominant cis conformation at the ground state confirmed by NMR-measurements.

The nature of the large Stokes shift for PD 2 (see Fig.1) has been investigated using time-resolved fluorescence spectroscopy at the solutions of different polarities: acetonitrile, o-dichloro-benzene, ethanol, and PUA. We found at room temperature and at $\lambda_{EX} = 422$ nm in acetonitrile: $\tau_1 \leq 30$ ps,

$\tau_2 = (400\text{-}450)$ ps and $\tau_3 = (3.2\text{-}3.5)$ ns with amplitudes strongly depending on λ_{EM}. At $\lambda_{EM} = 525$ nm the amplitude of the long component and at $\lambda_{EM} = 650$ nm the amplitude of the short component is dominant. In o-dichloro-benzene: $\tau_F = (1.4\text{-}1.5)$ ns, almost monoexponential and independent of λ_{EM}. In ethanol: $\tau_1 = (70\text{-}80)$ ps, $\tau_2 = (110\text{-}120)$ ps, $\tau_3 = (700\text{-}800)$ ps, slightly depending on λ_{EM}. In PUA: $\tau_1 = (4.0\text{-}4.1)$ ns and $\tau_2 = (200\text{-}250)$ ps, almost λ_{EM}-independent.

The analysis of the decay curves allowed to propose the following hypothesis based on the twisted intramolecular charge transfer (TICT) model [5,6]. For polymethine dyes TICT-phenomena seem to be almost uninvestigated. For additional confirmation we studied the fluorescence decay kinetics in liquid solutions by adding acetic acid. In all cases, this leads to a decrease in the TICT-component due to the acceptor properties of the ammonium group partly closing the TICT-channel.

So, from our point of view, all results may be described by the following schemes:

in ethanol: in o-dichloro-benzene:

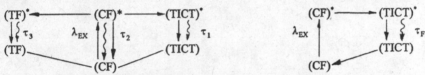 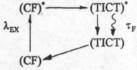

(CF), (TF), (CF)*, and (TF)* are cis- and trans- forms at the ground and excited states, respectively. Experiments show that the main pathways of deactivation of the excited-state energy are: in ethanol TICT-channel (70-80%), (CF)⇌(CF) transitions (20-30%) and about 1% (TF)⇌(TF) processes (at the short wavelength region) and in o-dichloro-benzene TICT-channel fully dominates.

in acetonitrile: in PUA:

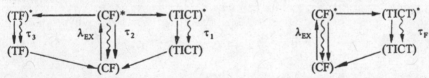 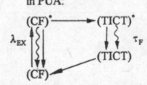

The main pathways of deactivation are: in acetonitrile at the short wavelength region (TF)⇌(TF) transitions and at the long wavelength edge via TICT-channel and in PUA via TICT-state with a small addition of (CF)⇌(CF) transitions about 1%.

Conclusion

(1) The dynamic Stokes shift is observed in elastic PUA matrix confirming that the microenvironment of dye molecules is changing during the fluorescence lifetime. (2) The free rotation of the DMA-group may be realized in microcavities of free volume resulting in a great static Stokes shift. Therefore, this type of dye molecules shows promise for efficient lasing in solid media. Both conclusions are extremely important for applications of the dye-activated polymeric media in lasers.

292

References

1. R.O'Connel, T.Saito, Opt. Engeneer. 22 (1983) 383.
2. C.Caligaris, Photonics spectra 1 (1995) 87.
3. V.Bezrodnyi, M.Bondar, G.Kozak, O.Przhonska, E.Tikhonov, J.Appl.Spectr. (translation of Zhurnal Prikladnoi Spektroskopii) 50 (1989) 441.
4. M.Bondar, O.Przhonska, E.Tikhonov, and N.Fedotkina, J.Appl.Spectr. (translation of Zhurnal Prikladnoi Spektroskopii) 52 (1990) 352.
5. Z.R.Grabovski, K.Rotkiewicz, A.Siemiarczuk, D.J.Cowley, W.Baumann, Nouv.J.Chim. 3 (1979) 443.
6. W.Rettig, Topics in Current Chemistry, 169 (1994) 253.

Temporal- and Spatial Resolved Measurements of Thermal Lensing

M.A. Rohwedder, M.H. Niemz
Universität Heidelberg
Institut für angewandte Physik
Albert-Ueberle-Str. 3-5
D-69120 Heidelberg

1 Introduction

Although the theoretical background of thermal lensing for cylindrical rods is well known, it is difficult to calculate the thermal focal length in a specific setup. The amount of the pump power which is absorbed by the laser material and the cooling conditions are usually not exactly known. In addition, new laser materials are always in development.

In this work, temporal and spatial resolved thermal lensing measurements have been carried out on cw and repetitive Xe-arc lamp pumped cylindrical laser crystals.

2 Theoretical Background

The solution of the heat conduction equation for a long, cylindrical solid with a homogenous source of heat inside and a constant temperature on the surface is a parabolic temperature distribution. The optical distortion of a beam on the axis of the cylinder depends on three effects:

1. Gradient of the diffraction index n_0 because of $dn_0/dT \neq 0$
2. Stress induced effects
3. Deformation of the end surfaces

A lens-like effect results with the focal length f [Koe92]:

$$f = \frac{KA}{P_a} \left(\frac{dn}{2dT} + \alpha C_{r,\phi} n_0^3 + \frac{\alpha r_0 (n_0 - 1)}{L} \right)^{-1} \tag{1}$$

with: K: heat conduction coefficient; A: rod crossection; P_a: absorbed power; α: thermal expansion coeffizient; $C_{r,\phi}$: photoelastic coefficient; n_0: index of refraction; L: length of the crystal. The three terms in the bracket correspond to the mentioned effects. Note that in this model, $1/f$ depends linearly on P_a.

The *time dependent* solution of the heat conduction equation of a cylinder ($l = \infty$) with a constant Temperature T=0 on the surface is given by [Car59]:

$$T(r,t) = \frac{2}{r_0^2} \sum_{n=1}^{\infty} \exp(-\beta_n^2 t/\tau) \frac{\beta_n^2 J_0(\beta_n r/r_0)}{(A^2 + \beta_n^2) J_0^2(\beta_n)} \int_0^{r_0} r T(r,0) J_0(\beta_n r/r_0) dr \tag{2}$$

with the roots β_n of the eigenfuction equation $\beta_n J_1(\beta_n) = A J_0(\beta_n)$, the cooling parameter $A = r_0 h/K$, the surface coefficient h and the time constant $\tau = r_0^2/k$.

In strongly absorbing rods with diffuse surfaces, most of the pump energy is deposited near the surface, causing an initial concave temperature distribution. Calculations ([Koe73]) show, that cooling on the surface causes an inversion of the temperature distribution.

3   Method

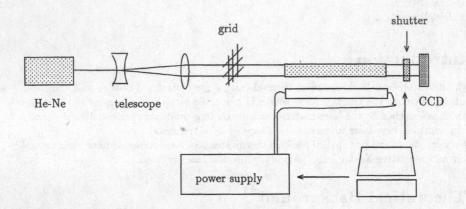

grid

shutter

He-Ne   telescope   CCD

power supply

The applied method [Pau92] uses a bundle of probe beams which is passed through the crystal. The beam of a He-Ne-laser (633 or 543 nm) was extended to crystal diameter (4 mm) and passed through a 400 μm-grid (a quadratic pattern of transparent spots on a dark slide). The probe beams were imaged through the crystal onto a CCD-camera and were recorded on videotape. Later, sequences of up to 10 s were digitized, allowing a resolution of 50 pictures/s with 512x256 Pixels, 8 bit/pixel. Position measurements were carried out by digital image processing methods. The segmentation algorithm used two global thresholds in a neighborhood-oriented search .

By comparing the distances of two beams before (d_0) and after (d_1) pumping the crystal, a thermal focal length f can be calculated: $f = a\frac{d_0}{d_0 - d_1}$ The only length that is actually measured is the distance a between the center of the rod and the CCD-Chip. In the setup for repetitive pumping, a shutter was operated synchronously with the pump flash to protect the CCD-chip from the intense flashlamp light.

4   Results

4.1   Repetitive pumped Er:YAG, Ho:YAG and Nd:YAG

The crystals were pumped in a silver coated single ellipse cavity (no flowtubes) by a Xe-flashlamp with a bore diameter of 4 mm.

By averaging over 1.5-4 seconds with different repetition rates and flash energies, the dependence of the thermal focal length from input power was measured. A system specific parameter c was extracted through linear regression: $1/f = cP_a$.(Tab. 1)

In addition to the fast oscillations, longer term deviations (0.5 s) of up to 25% of the average focal strength $1/f$ were measured. A more homogeneous cooling could possibly reduce these long-term oscillations. The measurement of a transition from a negative thermal lens directly after the pump flash to a positive thermal lens can be visualised with a vector field showing the movements of the beams (Fig. 3).

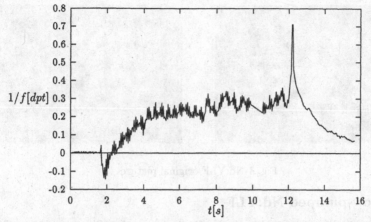

Fig. 2 Temporal development of the thermal lens in the center of a Nd:YAG. Rep.-Rate 15 Hz, pulse energy 65 J. The fast oscillations are caused by the pump flashes.

crystal	$c\ [dpt/kW]$	$\Delta c[dpt/kW]$	illum. length [mm]	pumping
Er:YAG, 50%	1.34	0.1	35	rep
Ho:YAG	3.4	0.4	76	rep
Nd:YAG	0.94	0.1	76	rep
Nd:YLF, $\pi_{\parallel}$	-0.022	0.003	76	cw
Nd:YLF, $\pi_{\perp}$	-0.029	0.003	76	cw
Nd:YLF, $\sigma_{\parallel}$	0.0048	0.0015	76	cw
Nd:YLF, $\sigma_{\perp}$	-0.0053	0.0015	76	cw

Tab. 1 Thermal lensing parameter c, rod diam. 4 mm; explanation for Nd:YLF next section

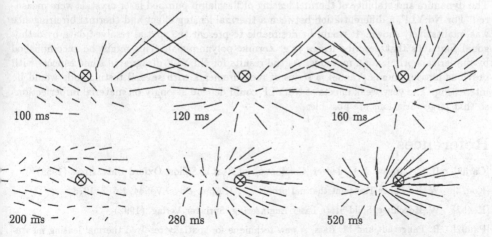

Fig. 3: Nd:YAG, 1Hz, 65 J/flash grid 0.4 mm equiv. 23 mrad (20 times larger). Pump flash at 90 ms. Max. negative thermal lens at 120 ms, intermediate strong anisotropy, decaying positive thermal lens after 280 ms

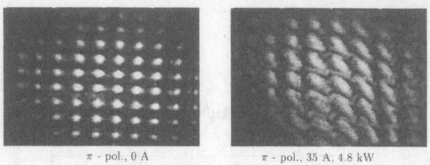

π - pol., 0 A π - pol., 35 A, 4.8 kW

Fig. 4 Nd:YLF original pictures

4.2 4.2 cw-pumped Nd:YLF

The Nd:YLF crystal was pumped in a single ellipse pump cavity (Quantronix Laser Head 116). As Nd:YLF exhibits a strong natural birefringence, polarizers were used in front of the crystal and the CCD. The end faces of the rod were cut parallel to the c-axis of the crystal. The σ-transition at 1.053 μm is linearly polarized $\perp$ c (ordinary ray), the π-transition at 1.047 μm is polarized $\parallel$ c, therefore polarization of the probe beam is labeled by σ and π. $\perp$ and $\parallel$ in Tab. 1 apply to the relation of the spatial orientation and the direction of polarization $\vec{E}$. The weak thermal lens of the ordinary ray (σ) shows different signs for different spatial orientations when measured by the described method. The beamlets are deformed elliptically pependicular to the expected deformation induced by the thermal lens. A distortion of the same orientation is measured on the Nd:YLF-laser beam (A modelocked oscillator with a two head regenerative amplifier at 1053 nm). This deformation is thought to be a thermal birefringence effect, which overrides the thermal lensing. The effects for the π-polarization are much stronger.

5 Conclusion

The dynamics and stability of thermal lensing of flashlamp pumped laser crystals were measured. For Nd:YLF, a differentiation between a thermal lensing effect and thermal birefringence was qualitatively shown. It would be desireable to present the spatial resolved data by orthogonal functions that describe lenses (e.g. Zernike-polynomials). This might be accomplished by numerical optimization methods. As the results fot Nd:YLF disagree in some aspects with previous measurements [Van88], a series of measurements with several testing rods would be interesting. The very weak effects of Nd:YLF could depend strongly on material imperfections so that large variations are possible.

References

[Car59] H. S. Carslaw, J. C. Jaeger, Conduction of Heat in Solids, Oxford Univ. Press (1959)

[Koe73] W. Koechner,Transient thermal profile ... , J. Appl. Phys, Vol.44, No. 7, 1973

[Koe92] W. Koechner, Solid-State Laser Engineering, Springer Verlag, (1992)

[Pau92] R. Paugstadt und M. Bass, A new technique for spatially resolved thermal lensing measurements, Optics and Laser Technology Vol. 24, No. 3 (1992)

[Van88] H. Vanherzeele: Thermal lensing measurement and compensation in a continuous-wave mode-locked Nd:YLF laser, Optics Letters Vol. 13, No. 5, Mai 1988

Dual Q-Switching and Laser Action in Nd^{3+}:YAG and Cr^{4+}:Forsterite Laser System

M.Skórczakowski, Z.Jankiewicz, G.Skripko*, I.Tarazevich*, A.Zając, W.Żendzian
Military University of Technology, Institute of Optoelectronics,
ul. Kaliskiego 2, 01-489 Warszawa, Poland,
*Fotek, 66 Partizanckij Prospekt, Minsk, Belarus.

Introduction

Cr^{4+} doped forsterite crystal is an easily lasing medium. The first laser action in this crystal has been achieved by Petricevic et. al. [1]. Many authors have described their experiments presented Cr:Forsterite laser actions in different regimes (CW, pulsed, mode-locked) [2, 3, 4]. It has been established that stimulated transitions between $^3T_2$ and $^3A_2$ levels of the Cr^{4+} ion are responsible for lasing of this medium in the range of wavelength 1,15 - 1.35 μm [5]. The Cr:Forsterite laser is usually pumped by 1.06 μm Nd:YAG laser radiation. The absorption at 1.06 μm results in population of the $^3T_2$ state (an upper laser level) and depopulation of the ground $^3A_2$ state. The ground $^3A_2$ state may be strongly depopulated till complete blenching of the Cr:Forsterite crystal. This is a characteristic behaviour of a saturable absorber.

The idea of utilisation of Cr:Forsterite crystal as a Q-switching device (saturable absorber) has also been presented earlier [6]. However, the authors of the work did not present temporal dependence of the Q-switched by Cr:Forsterite Nd:YAG laser intensity.

The Cr:Forsterite crystal may be successfully applied as a lasing medium or a saturable absorber. We propose the utilisation of the Cr:Forsterite crystal simultaneously in the double role. The idea of such an employment of the similar Cr^{4+}:YAG crystal has been presented by Spariosu et. al. in [7]. Although the active centre in these both crystals (YAG, Forsterite) is the same (Cr^{4+} ion), the results of laser actions in these ones should be different, because of the difference in their ESA properties.

We have measured the non-linear absorption of Cr:YAG and Cr:Forsterite crystals. The results of the measurements are presented in Fig.1 and Fig.2.

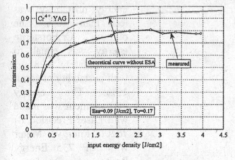

Fig.1. Non-linear Cr^{4+}:YAG transmission vs 1.06 μm energy density, input pulse duration 20 - 50ns.

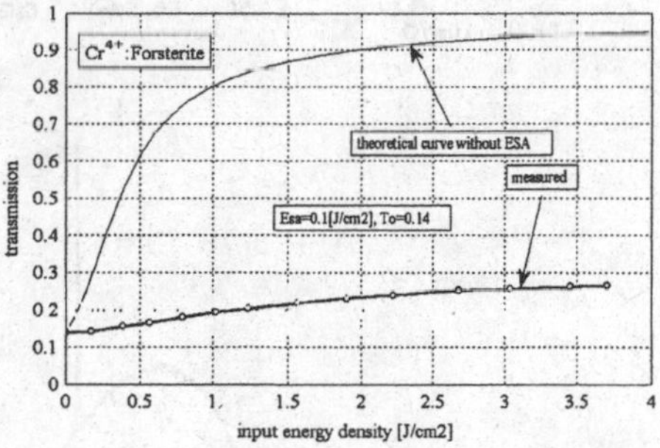

Fig.2. Non-linear Cr^{4+}:Forsterite transmission vs 1.06 μm energy density, input pulse duration 20 - 50ns.

Using a theoretical expression of the transmission dependence on input energy density for a slow absorber with ESA published in [7] we have found the values of the absorption cross sections for the both media. According to our estimations the values are as follows.

For Cr^{4+}:YAG : $\sigma_{1a} = 1.91 \times 10^{-18}$ cm^2, $\sigma_{2a} = 0.098\, \sigma_{1a}$.

For Cr^{4+}:Forsterite : $\sigma_{1a} = 1.88 \times 10^{-18}$ cm^2, $\sigma_{2a} = 0.58\, \sigma_{1a}$.

It is seen, that Q-switched Nd:YAG laser operation with applying the Cr:Forsterite saturable absorber will be strongly depended on ESA phenomena.

Experimental set up

We have carried out our experiment with the laser, which scheme is presented in Fig.3. The Nd:YAG laser resonator consists of two flat mirrors M1 and M2, Nd:YAG rod ϕ 4mm x 3,5''in the

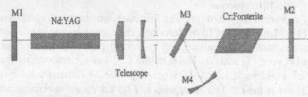

Fig.3. Experimental set up scheme, Nd:YAG resonator length is 50 cm. Cr:Forsterite resonator length is 14.5cm.

LMI 1520 laser cavity, a 2X magnification telescope, a diaphragm ϕ 1.4 mm, a Cr:Forsterite Q-switch. The M2 mirror is a totally reflective for Nd:YAG and its reflectivity equals approximately 85% for Cr:Forsterite. M1 is a 50% transmission mirror for Nd:YAG. The Cr:Forsterite resonator includes three mirrors : M2 mirror, a flat tilted dichroic M3 mirror totally reflecting in the range of 1.15 - 1.35 μm and transmitting 97% at 1.06μm, the concave 152mm M4 mirror totally reflecting for Cr:Forsterite and the Brewster cut, initial transmission 14% Cr:Forsterite crystal 18 mm in length. The Cr:Forsterite resonator is close to hemispherical. The both resonators are coupled by the Cr:Forsterite lasing and Q-switching medium.

Experimental results

Dual operation of the Cr:Forsterite as a Q-switching and as a lasing medium has been obtained. The oscilloscope traces of the temporal dependence of pumping power and laser action are shown in Fig.4. Only one pulse was emitted for pumping energy of 12 J. The output 1.06 μm energy

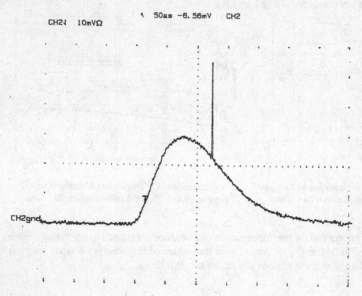

reached 2.6 mJ and energy unstability was less than 10%. The output Cr:Forsterite energy was equal to approximately 0.28 mJ with unstability of about 15%. The energy measurement was made with Laser Precision Rj 7620 Energy Meter with RjP 735 Energy Probe.

When pumping energy was increased to about 14 J, the second pulse appeared after approximately 35μs. The total 1.06 μm energy was in this case about 4 mJ.

Fig.4. Oscilloscope trace of pump power and laser action, the time scale is 50μs/div.

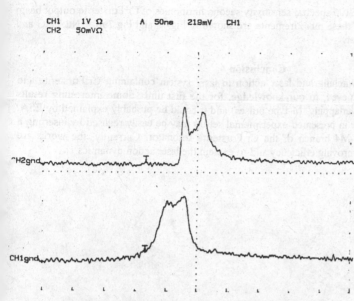

```
CH1   1V Ω        A  50ns   219mV  CH1
CH2   50mVΩ
```

Fig.5. Simultaneous temporal evolution of the Cr:Forsterite output (upper trace) and the Nd:YAG output (bottom trace), time scale 50 ns/div.

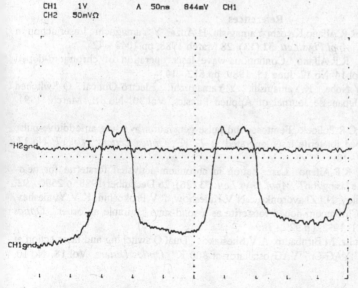

```
CH1   1V         A  50ns   844mV  CH1
CH2   50mVΩ
```

Fig.6. Temporal evolution of the Nd:YAG output when Cr:Forsterite resonator is closed, time scale 50 ns/div.

A temporal dependence of simultaneous laser actions in Nd:YAG and Cr:Forsterite is presented in Fig.5. There is the characteristic overlapping of Nd:YAG and Cr:Forsterite output powers, whereas analogous dual laser actions in Nd^{3+}:YAG and Cr^{4+}:YAG [7] are approximately 400 ns separated. Moreover, the shape of our pulses is a M-type, what may be caused by ESA. The second 1.06μm pulse, similar to the first one, appeares when the Cr:Forsterite resonator is closed. This fact is illustrated in Fig.6.

Dynamics of the lasing processes in Nd:YAG and Cr:Forsterite was measured with 2440 Tektronix Digital Oscilloscope and registered with HC-100 Tektronix Color Plotter. The EGG type C30617E fast photodiode was used as a detector.

It should be pointed out, that only free-running 1.06 μm laser action was obtained when no telescope was applied inside the resonator M1-M2. When the telescope magnification has been increased 3X, the temporal behaviour of Nd:YAG laser action was changed a little and Cr:Forsterite output pulse duration was only 13 ns and had a normal Q-switching (gain switching) shape.

The energy density distribution in the laser beams was measured by Spiricon Laser Beam Analyser Model LBA 100A with Spiricon Laser Beam Sampler Model LBS 100. Becouse of

300

limitations of the SamplerLBS100 spectral sensitivity second harmonics of Cr:Forsterite output beam was measured. The results of these measurements are shown in Fig.7a and Fig.7b for Nd:YAG and Cr:Forsterite output respectively.

Conclusion

The dual passive Q-switching and laser action in laser system containing Cr:Forsterite and Nd:YAG media has been achieved, to our knowledge, for the first time. Some interesting results have been obtained (pulses overlapping, M-type pulses) and it could be probably explained by ESA. The tunability of Cr:Forsterite in presented experimental set up may be easily realised by inserting a birefrigent filter into the M3-M4 branch of the Cr:Forsterite resonator. Currently, the works are being developed toward to improving efficiency and to explain the laser action dynamics.

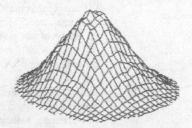

Fig.7a. Spatial distribution of the Nd:YAG output beam.   Fig.7b. Spatial distribution of the Cr:Forsterite output beam.

References

1. V.Petricevic, S.K.Gayen, R.R.Alfano,Kiyoshi Yamagishi, H.Anzai, Y.Yamaguchi „Laser action in chromium - doped forsterite", *Appl.Phys.Lett.* **52** (13), 28 March 1988, pp.1040 - 42.
2. V.Petricevic, S.K.Gayen, R.R.Alfano „Continuous-wave laser operation of chromium-doped forsterite", *Optics Letters*, Vol.14, No.12, June 15, 1989, pp.612 - 14.
3. A.Sugimoto,Y.Segawa, Y.Nobe, K.Yamagishi, Y.Yamaguchi „Electro-Optical Q-Switched Tunable Forsterite Laser", Japanesse Journal of Applied Physics, Vol.30, No.3B, March, 1991, pp.L495 - L496.
4. A.Sennaroglu, T.J.Carrig, C.R.Pollock „Femtosecond pulse generation by using an additive-pulse mode-locked chromium-doped forsterite laser operated at 77 K", *Optics Letters*, Vol.17, No.17, September 1, 1992, pp.1216 - 18.
5. V.Petricevic, S.K.Gayen, R.R.Alfano „Laser action in chromium-activated forsterite for near-infrared excitation: Is Cr^{4+} the lasing ion?", *Appl.Phys.Lett.* **53** (26), 26 December 1988, p.2590 - 92.
6. M.I.Demchuk, V.P.Mikhailov, N.I.Zhavoronkov, N.V.Kuleshow, P.V.Prokoshin, K.V.Yumashev, M.G.Livshits, B.I.Minkov „Chromoum-doped forsterite as a solid-state saturable absorber", *Optics Letters*, Vol.17, No.13, July 1, 1992, pp.929 - 30.
7. K.Spariosu, W.Chen, R.Stultz, M.Birnbaum, A.V.Shestakov „Dual Q switching and laser action at 1.06 and 1.44 μm in a Nd^{3+}:YAG-Cr^{4+}:YAG oscillator at 300 K", *Optics Letters*, Vol.18, No.10, May 15, 1993, pp.814 - 16.

The Theoretical Model of Simultaneous Utilization of a Saturable Absorber as a Q-Switch and as a lasing Medium

Zdzisław Jankiewicz, Marek Skórczakowski,
Military University of Technology, Institute of Optoelectronics
01-489 Warszawa, ul. Kaliskiego 2 , Poland

Abstract

A theoretical model of simultaneous saturable absorber Q-switched laser action in an active laser medium (LM) and laser action in saturable absorber (SA) is proposed. The new theoretical model of a double laser operation when both laser actions are coupled with SA medium is described. Results of numerical calculations are presented for a few different cases. The most important conclusions are the following: 1) simultaneous laser action in a passively Q-switched LM and in a SA is possible, 2) change of the laser beam diameter in the SA is useful in order to obtain to efficient Q-switch or to optimize the SA output, 3) LM and SA laser actions may overlap in a specific case.

1. Introduction

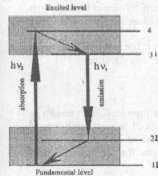

Excited level

$h\nu_2$ $h\nu_1$

absorption   emission

4

3¹

22

11

Fundamental level

Fig.1. Schematic diagram of a saturable absorber energy levels.

The method of passive Q-switching of laser resonator by using a nonlinear saturable absorber has been well known for many years. A strong single pulse similar to the pulses obtained with other (active) Q-switching methods is emitted by such a laser. The main disadvantage of passive Q-switching is impossibility to control the lasing process and Q-switch energy losses. However, the passive Q-switching method is simpler and cheaper in comparison with an active one.

Many media have been used as passive Q-switches. The common feature of all these materials is the existence of broad bands in their energy level structure (fig.1). Nonlinear change of transmission of these media always consists in saturation of absorption of the fundamental level up to the complete blenching. Thus, population inversion of the saturable absorber is a consequence of its utilizing as a Q-switch. It is well known that population inversion is a fundamental condition of lasing process. Therefore, there is a possibility of application of a nonlinear absorption medium as a lasing material unless its stimulated emission probability is equal to zero. The most of nonlinear absorbers, particularly the single Cr^{4+} crystals, accomplish this condition. The other condition of laser action in Q-switching medium is application of its suitable resonator. So that, employment of two coupled resonators is necessary to realize the conception of simultaneous Q-switching and laser action in nonlinear absorber.

302

2. Conception of laser generation with resonators coupled by nonlinear saturable absorber

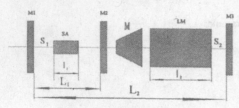

Fig.2. The coupled lasers setup.

According to the above presented suggestions we consider a laser composed of a laser medium (LM) and a saturable absorber (SA). Moreover, the laser contains a multimirror resonator which the simplest version is presented in fig. 2. The resonator mirrors have appropriate spectral characteristics to ensure a coupling between mirrors: M1 and M2 for SA (wavelength λ_1) and M1 and M3 for LM (wavelength λ_2). Furthermore, the laser includes a telescope which magnification is equal to M.

$$M^2 = \frac{S_2}{S_1}$$

where: $S_i = \frac{\pi d_i^2}{4}$

d_i- diameter of LM or SA.

It is obvious that the following conditions concerning the mirrors of the coupled resonators should be fulfilled for the proper laser operation:

M1: transmission $T(\lambda_1)$ - optimal feedback,
transmission $T(\lambda_2)$ - minimal,
M2: transmission $T(\lambda_1)$ - minimal,
reflection $R(\lambda_2) = 0$,
M3: transmission $T(\lambda_2)$ - minimal.

Laser medium (fig. 2) is externally excited which increase the population inversion of medium till to the threshold level. Then, the laser action begins in the resonator M1-M3 and SA starts to blench. An absorption of SA rapidly decreases and a single pulse at wavelength λ_2 is generated inside the cavity M1-M2. Laser action at wavelength λ_1 in SA may appear if its upper laser level is populated enough. When SA is completely blenching the level is excited extremely and SA may efficiently lases.

The conception of a similar generation has been used earlier, [1] but nobody analyzes it theoretically. In the paper, we try to describe this process with the modified rate equations .

3. Rate equations

The problem of description of generation process in the above presented laser consist in finding the solutions of four coupled differential equations. These well known equations [2] characterize the changes of photon density in a laser cavity and population inversion of a laser medium. In our analysis, we apply some modified equations. We describe the problem with four functions which denote respectively [3]:

$J_1(t)$ - averaged power density in the laser cavity M1-M2 [W/cm^2],
$J_2(t)$ - averaged power density in the laser cavity M1-M3 [W/cm^2],
$k_1(t)$ - averaged SA gain coefficient [1/cm],
$k_2(t)$ - averaged LM gain coefficient [1/cm].

Our equations are as the following [4]:

$$\frac{dJ_1(t)}{dt} = V_{R1}[k_1(t) - \rho_1]J_1(t)$$

$$\frac{dk_1(t)}{dt} = \eta_1 M^2 \frac{J_2(t)[\kappa_1 - k_1(t)]}{E_{Sa}} - \frac{k_1(t)}{\tau_1} - \frac{k_1(t)J_1(t)}{E_{S1}}$$

$$\frac{dJ_2(t)}{dt} = V_{R2}\{k_2(t) - \rho_{2s} - \frac{l_1}{l_2}\beta[\kappa_1 - k_1(t)]\} J_2(t)$$

$$\frac{dk_2(t)}{dt} = \eta_2 W_{p2}(t)\kappa_2 - \frac{k_2(t)}{\tau_2} - \frac{k_2(t) J_2(t)}{E_{s2}}$$

with initial conditions:

$$k_1(0) = 0,$$
$$k_2(0) = 0,$$
$$J_1(0) = J_{10},$$
$$J_2(0) = J_{20},$$

where initial noise power density is :

$$J_0 \approx \frac{n_a}{\tau} h \nu \frac{\Omega}{4\pi} l$$

and the following symbols are :

$V_{R} = c\frac{l_i}{Li}$ - light velocity in resonators M1-M2 or M1-M3 [cm/s],

ρ_i - loss coefficient of SA resonator [1/cm],

ρ_{2s} - static loss coefficient of LM resonator [1/cm],

W_{p2} - pumping rate of LM [1/s],

τ_i - upper laser level lifetimes [s],

$E_{si} = \frac{h\nu_i}{\sigma_e}$ - saturation energy densities [J/cm^2],

$E_{se} = \frac{h\nu_1}{\sigma_{el}}$ - saturation energy density of absorption [J/cm^2],

η_i - quantum efficiencies,

$\kappa_i = \sigma_n n_{0i}$ - emission cross section [cm^2] and concentration [1/cm^3] product; theoretically maximal gain coefficient [1/cm],

$\beta = \frac{\sigma_d}{\sigma_e}$ - saturable absorber absorption to emission cross section quotient.

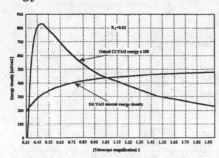

Fig.3. The Cr:YAG output energy per 1cm^2 of the Nd:YAG rod area section and the internal Nd:YAG energy density as the functions of telescope magnification M^2.

We assume that both LM and SA are fourth level laser materials. This assumption is often valid, especially for Nd^{3+}:YAG and Cr^{4+}:YAG or Cr^{4+}:Mg$_2$SiO$_4$ crystals. Moreover, the assumption that excited state absorption (ESA) is negligible was done. This is true for Cr^{4+}:YAG in relation to fundamental level absorption where ESA cross section is only 10% of σ_{a1}' , but for Cr^{4+}:Forsterite this value reaches almost 60% [6]. For Cr:YAG emission of λ_1 wavelength the difference between emission cross section and ESA cross section reaches about 50% [5]. It has been accepted that ESA actually decreases the emission Cr^{4+}:YAG cross section.

4. Numerical solution

The differential equation system describing the presented problem of simultaneous generation in the coupled lasers has not analytical solution because of its non-linearity. A numerical solution with Runge - Kuty method has been performed to obtain temporal characteristics of power densities and gain coefficients. Moreover, rate equation model is also limited by the well-known condition: $2l(k-\rho) < 1$. So, losses in both resonators, even in numerical analysis, should be sufficiently high not to cause an increase in the calculations error.

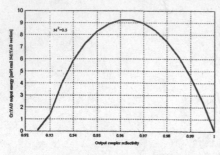

Fig.4. The Cr:YAG output energy per 1 cm^2 of the Nd:YAG rod area section dependence on Cr:YAG output coupler reflectivity.

A constant pumping rate of LM W_{p2} has been taken and only the first pulse characteristics has been found. Calculations are started when LM gain coefficient k_2 reaches the initial losses level : $\dfrac{l_1}{l_2}\beta\kappa_1 + \rho_{2s}$. This level cannot be too high, because of a gain limitation in a real laser material. For instance, the initial Cr^{4+} saturable absorber transmission should be in the range of about 20-30% to Q-switch Nd:YAG efficiently and to make the generation possible. In our calculations 30% of initial transmission has been assumed. Our calculations have been performed for Cr^{4+}:YAG - Nd^{3+}:YAG coupled lasers, as partly adequate to [5], for the following input data :

$\dfrac{l_1}{l_2}\beta\kappa_1$ =0.16 [1/cm] - initial dynamic loss coefficient,

($T_0 = 0.3$),

$\rho_{2s} = 0.1$ [1/cm] ,

$\rho_{As} = 0.096$ [1/cm], (FOM = 25),

$\eta_1 = \eta_2 = 1$,

$l_1 = 0.5$ cm, $l_2 = 7.5$ cm, $L_1 = 10$ cm, $L_2 = 40$ cm,

$\kappa_1 = 0.2$ [1/cm] , $\beta = 12$.

$E_{S1} = 343 mJ\ /\ cm^2$

$E_{S2} = 214 mJ\ /\ cm^2$

Two other parameters were being changed : M^2 in the range from 0.35 to 2 and R_1 in the range from 0.92 to 1. Relatively large value of the static losses of M1-M3 resonator has been put in order to satisfy the rate equation approximation.

Results of calculations of the output Cr^{4+}:YAG energy per 1 cm^2 of Nd:YAG area section versus telescope magnification and output coupler reflectivity are plotted in fig 3 and 4, respectively. Moreover, in fig. 3 there is plotted an internal Nd:YAG energy dependence on telescope magnification. An optimal point can be clearly seen in the both figures. A maximal Cr^{4+}:YAG output energy is extracted for M=0.5 and $R_1 \approx 0.96$. When telescope magnification increases, the output energy decreases, which is caused by decrease of the Cr:YAG active volume. The 1.45μm output energy vanishes for M^2=0.35. It is the result of not completely Cr^{4+}:YAG blenching. The threshold gain coefficient value of k_1=0.147 cm^{-1} is reached

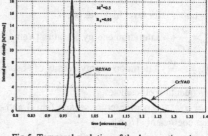

Fig.5. Temporal evolution of the laser actions in coupled Cr:YAG and Nd:YAG resonators for optimal telescope magnification.

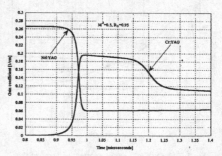

Fig.6. Temporal evolution of Cr:YAG and Nd:YAG gain coefficients for optimal telescope magnification.

for this condition (R$_1$=0.95, M^2=0.35), whereas for M=0.5, Cr^{4+}:YAG absorption is almost completely saturated: k$_1$=κ$_1$=0.2cm^{-1}, as it is shown in fig.6.

It is important, that for threshold conditions (M^2 = 0.35), about 74% of Cr^{4+} ions are excited. This situation is not characteristic for four level laser medium, where usually only several percent of active centers are inverted. The cause of such situation is large value of β and a bottom limitation of initial transmission of saturable absorber $T_o = \exp(-\beta\kappa_1 l_1) \approx 0.3$. For higher values of β it is necessary to decrease Cr^{4+} concentration (or Cr:YAG length) in order to keep transmission constant. It ensures a constant value of gain when laser action starts. Moreover, Cr^{4+}:YAG saturation, during its emission, is not complete what is clearly seen in fig. 6. The Cr^{4+}:YAG gain coefficient decreases from the value of about 0.2 to the value of 0.11 mainly for the reason of laser action and frequently for the reason of luminescence, because the pulse evolution time is very long (more then 150 ns). Thus, less than 50% of the energy, stored in Cr^{4+}:YAG is extracted and the output energy is additionally decreased by the factor : $\frac{A - A_s}{A_s} \approx 0.35$.

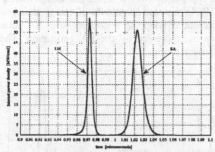

Fig.7. Temporal evolution of laser actions in Nd:YAG laser medium and saturable absorber with β=4 and κ_1=0.3cm⁻¹. The SA output energy is 18.8mJ per 1cm² of LM area section.

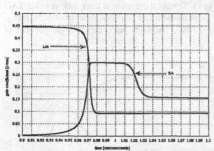

Fig.8. Temporal evolution of Nd:YAG (LM) and SA gain coefficients for β=4 and κ_1=0.3cm⁻¹.

The above mentioned limitations cause that Cr^{4+}:YAG cannot be very effective laser medium in the proposed configuration, although the efficiency exceeds 11% of energy absorbed by Cr^{4+}:YAG in the presented example. This value has been experimentally obtained [5] with use of the external excitation by active Q-switched Nd:YAG laser. However, the both coupled Nd:YAG and Cr:YAG lasers operate in the giant pulse regime, without employment of additional Q-switching devices. In order to obtain high efficiency it is also very important to use a crystal of good quality (FOM>50) and a proper resonator (optimal output coupler).

As it was mentioned above, Cr^{4+}:YAG saturable absorber is not an optimal laser medium for the laser presented in this paper. It seems that the large value of β is the main reason of this fact. It is important to determine the influence of β parameter on characteristics of the coupled resonators, composed of the laser medium and the saturable absorber. Some simple calculations have been done with the following input data :

$\eta_1=\eta_2=1$, $l_1= 1.5$ cm, $L_1 = 10$ cm, $l_2 = 7,5$ cm, $L_2 = 40$ cm ,
$E_{S1} = 1.06$ J/cm² , $E_{S2} = 0.21$ J/cm² ,$\tau_1 = 3.5$ μs, $\tau_2 = 240$μs,
$\rho_1 = 0.05$ cm , $\rho_2 = 0.2$ cm , $R_1 = 0.6$, $M^2 = 9$.

Similarly as for Cr:YAG, the initial transmission of SA has been fixed on the level of 17.4 % for all considered cases. It ensures a constant amount of energy stored in LM at the moment of threshold achieving. The β-parameter was changed which results in change of SA absorption cross section (or E_{Sa}) because the emission cross section (or E_{S1}) was settled. Thus, κ_1 value (or concentration) had to be changed to ensure initial transmission constant.

The results of these calculations are presented in figures from 7 to 12. The various temporal evolution of laser action intensities and gain coefficients are remarkable. This is not only due to the change of β-parameter but also, as it was earlier mentioned, due to the changes of SA active centers concentration.

The following trends may be noticed:

1. Increase in the β-parameter makes the saturable absorber more susceptible to blenching, so it works better as Q-switch. But for high value of β the emission cross section of SA (or κ_1) is rather low. So the laser action runs close to the threshold exceeding. It causes that SA energy extraction is

not complete because this is a basic feature of laser operation in Q-switch or gain-switch regimes for low threshold exceeding.

2. Low values of β cause a weak Q-switch effect and low energy extraction from the laser medium. The absorber excitation is not strong and it works close over threshold. Such a saturable absorber will not lase, if the laser beam is not strongly compressed in it.

3. A low threshold exceed results in relatively long SA pulse evolution time. The pulses of LM and SA are separated what can be seen in figs. 7 and 11.

4. There is an optimal value of β for which the SA generation efficiency is maximal. For this intermediate value, the LM laser is still Q-switched well and saturable absorber concentration is sufficient to store quite a big amount of energy on the upper laser level. It seems that this β-value is close to 1.5 for the presented example (see fig. 9 and 10). In this case about 64% of energy absorbed by SA is extracted out of cavity. Moreover, a characteristic overlapping is noticed when SA pulse evolution time is short enough. Then LM generation is suppressed by the SA generation. The saturable absorber is an intermediary agent in this suppression process.

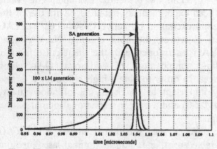

Fig.9. Temporal evolution of laser actions in Nd:YAG laser medium and saturable absorber with β=1.5 and κ_1=0.8cm^{-1}. The SA output energy is 79mJ per 1cm^2 of LM area section.

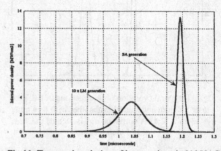

Fig.11. Temporal evolution of laser actions in Nd:YAG laser medium and saturable absorber with β=1and κ_1=1.2cm^{-1}. The SA output energy is 9.5mJ per 1cm^2 of LM area section.

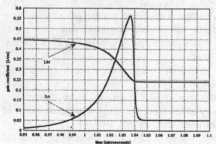

Fig.10. Temporal evolution of Nd:YAG (LM) and SA gain coefficients for β=1.5 and κ_1=0.8cm^{-1}.

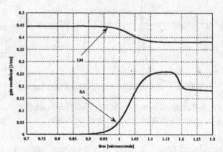

Fig.12. Temporal evolution of Nd:YAG (LM) and SA gain coefficients for β=1 and κ_1=1.2cm^{-1}.

5. Conclusions

A simple theoretical model of simultaneous laser action in a laser medium and in a saturable absorber (acting additionally as a Q-switch) in coupled resonators is presented. This model has been successfully applied to estimate Cr:YAG - Nd:YAG coupled lasers. For such coupled laser media, the change of beam diameter inside a laser cavity is necessary to reach an optimal output efficiency. Particularly, the beam diameter should be increased by the factor of about $\sqrt{2}$ at the Cr:YAG saturable absorber - laser medium. This value of optimal telescope magnification depends on effective quantum cross sections of saturable absorber and may change for other materials e.g. Cr^{4+}:Forsterite. The occurrence of a characteristic overlapping of generated intensities is the most important feature of the proposed generation regime. This phenomenon appears only under the specific conditions.

Excited state absorption has not been taken directly into account in this paper. Currently, the present model is extended to include ESA for both λ_1 and λ_2 transitions.

6. References

1. K. Spariosu, W. Chen, R. Stultz, M. Birnbaum and A. V. Shestakov, "Dual Q switching and laser action at 1.06 and 1.44 μm in a Nd^{3+} :YAG - Cr^{4+} :YAG oscillator at 300 K," *Optics Letters*, Vol. 18, pp. 814 - 816, May 15, 1993.

2. A. E. Siegman, *Lasers*, University Science Books, Mill Valley, California, 1986.

3. Z. Jankiewicz, "Generacja serii impulsów laserowych metodą stopniowego wyłączania strat rezonatora," *Dodatek do biuletynu Nr 8 (336) Wojskiwej Akademii Technicznej*, Warszawa, 1980 (in Polish).

4. M. Skórczakowski, Z. Jankiewicz, "Analiza generatora laserowego z nieliniowym absorberem jako ośrodkiem laserującym," *Biuletyn Wojskowej Akademii Technicznej*, (in Polish), (in printing).

5. N. I. Borodin, V. A Zhitniuk, A. G. Okhrimchuk and A. V. Shestakov, "Generacya v oblasti dlin voln 1.34 -1.6 mkm laziera na osnovie $Y_3Al_5O_{12}$:Cr^{4+}," *Izviestya Akademi Nauk SSSR - Seria Fizicheskaya*, T. 54, No 8 pp.1500-1505, 1990, (in Russian).

6. M.Skórczakowski, Z.Jankiewicz, G.Skripko, I.Tarazevich, A.Zając, W.Żendzian „Dual Q-switching and laser action in Nd^{3+}:YAG and Cr^{4+}:Forsterite laser system" (in this issue).

Efficient Generation of Pulsed Single-Mode Radiation Tunable in the UV-C Region

A. Steiger, K. Grützmacher, M. I. de la Rosa

Physikalisch-Technische Bundesanstalt, Abbestr. 2-12, D 10587 Berlin, Germany

Two-photon laser spectroscopy has become a powerful diagnostic tool in the fields of high-temperature, plasma and combustion physics because it gives access to ground-state atoms of interesting species [1], such as H, C, N, O. These applications can be extended if tunable pulsed UV-C radiation of best spectral and spatial quality is available. Future diagnostics of hydrogen istopes under magnetically confined fusion-plasma conditions need much higher peak power than commercial laser systems can presently provide. To overcome present limitations we developed a new type of pulsed single-longitudinal-mode (SLM) laser system utilizing commercially available solid-state lasers.

Our concept is based on a consequent use of the higher harmonics of injection-seeded Q-switched Nd:YAG lasers to generate tunable UV radiation as efficient as possible [2]. To produce radiation at, e.g., 243 nm, which is required for two-photon excitation of atomic hydrogen, we use the scheme shown in Fig 1. In this case, the second (SH) and third harmonic (TH) of the Nd:YAG laser are generated with approximately the same

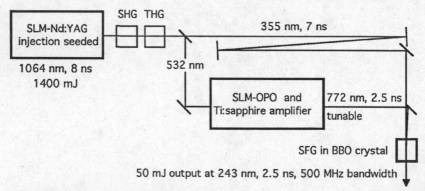

Fig.1: *Schematic layout of the pulsed single-mode laser system. The values given refer to the generation of 243 nm radiation required for two-photon exitation of atomic hydrogen.*

pulse energy. The SH is converted into tunable near-infrared radiation at 772 nm by frequency splitting in a single-mode optical parametric oscillator (OPO) based on a KTP crystal, and the OPO output is subsequently amplified in Ti:sapphire crystals pumped by the SH. Finally, 243 nm radiation is obtained by sum-frequency generation (SFG) of the 772 nm and 355 nm radiation (TH) in a BBO crystal.

By this set-up, 1.4 J pulse energy provided by the Nd:YAG laser is converted into 50 mJ pulse energy at 243 nm with excellent performance at 10 Hz repetition rate. The bandwidth at 243 nm is < 500 MHz, measured by optogalvanic detection of the Doppler-free two-photon absorption when tuning the 243 nm output across the 1S-2S resonance of atomic hydrogen produced by thermal dissociation in a small reference cell [3]. The pulse duration is 2.5 ns, mainly caused by pulse reduction in the OPO. The beam divergence is about 100 μrad and the pointing stability is better than 50 μrad.

With UV pulse energies of 50 mJ at 243 nm, the efficiency is better than 3% for the energy conversion from the 1064 nm fixed-frequency infrared output of the Nd:YAG laser into tunable UV-C radiation. Since the pulse duration of the UV output is reduced to about one third of that of the Nd:YAG laser, the pulse peak-power conversion efficiency is even as high as 10 %.

Although this highly efficient UV generation is achieved with commercially available solid-state laser systems, some details of the set-up are important for successful operation. Special care was taken in choosing a suitable Nd:YAG laser providing about 1.4 J pulse energy which is converted into about 350 mJ pulse energy each for SH and TH. Beam profiles and pointing stability have to be as good as promised by the manufacturer. For better performance, the OPO-Ti:sapphire laser (type: HRL100 Z with three Ti:sapphire crystals from former STI Optronics Inc.) was improved. The first Ti:sapphire amplifier was changed from a four-pass to a six-pass arrangement, and a diverging lens was placed behind the OPO in order to compensate thermal lensing of the strongly pumped Ti:sapphire amplifiers. To optimize the UV generation, the beam profiles of both input beams, i.e., the TH of the Nd:YAG laser and the output from the OPO-Ti:sapphire laser, had to be best adapted to the acceptance of the BBO crystal with its 13 mm clear aperture.

First investigations with a smaller BBO crystal of 7 mm clear aperture exhibited a saturation-limit of about 30 mJ pulse energy at 243 nm. However, using our UV laser system with the small BBO crystal in 1994 for a six month measurement campaign for the first successful demonstration of two-photon induced Lyman-α fluorescence [4] at the plasma generator PSI-1 of the Max-Planck-Institute of Plasma Physics, Section Berlin, revealed that the BBO crystal did not suffer any degradation of its efficiency.

The successful realization of this concept combines several advantages resulting in highly efficient UV generation. The initial SLM quality of the Nd:YAG radiation is preserved in all conversion processes. Generating tunable radiation pulses in an OPO, the time jitter of the output is negligible even for SLM operation. The Nd:YAG-pumped Ti:sapphire amplifiers provide high output power with a smooth, nearly Gaussian radial profile and excellent pointing stability. All this is important for the final the UV generation by SFG in the BBO crystal. Additionally, this SFG yields better over-all efficiency than the corresponding frequency doubling process, because two sufficiently different frequencies, corresponding to 772 nm and 355 nm, are used. As only one of these is tunable, our scheme provides better spectral quality in the UV.

The results obtained and our experience with the efficient generation of 243 nm radiation are of general validity for UV-C generation down to the transmission limit of BBO, as shown by the following examples:

(i) The present configuration of our pulsed SLM UV laser system is limited by the Ti:sapphire amplifier tunability, 720 nm to 920 nm, which corresponds to an UV range from 238 nm to 256 nm. It should be noted that the center of this interval coincides well with the KrF*-excimer laser wavelength at 248 nm.

(ii) Replacing the third harmonic (TH) of the Nd:YAG laser in our scheme by its fourth harmonic (FH) will shift the UV wavelengths to the range of 194 nm to 206 nm. The lower wavelength is close to the ArF*-excimer laser wavelength of 193 nm. There is no doubt that radiation within this range can be generated with similarly high efficiency as demonstrated for 243 nm.

(iii) Replacing the Ti:sapphire amplifiers by an optical parametric amplifier (OPA) based on a BBO crystal pumped by the TH of the Nd:YAG laser will allow UV generation from 205 nm to 280 nm. This correspondes to a tuning range of the OPO/OPA output from 486 nm to 1.3 μm, as already commercially available, when mixed with the TH of the Nd:YAG laser.

In conclusion, our method of generating tunable UV radiation yields peak-power and efficiency which are, to our knowledge, higher than in other processes, e.g., frequency-doubled single-mode OPOs driven by the third harmonic of a Nd:YAG laser or the third harmonic of a single-mode Ti:sapphire lasers.

Nevertheless, the over-all efficiency for UV generation obtained by our system is still limited by imperfections of the beam profile of the Nd:YAG lasers which are available on the commercial market. We are convinced that better Nd:YAG laser beam profiles will result in even better conversion efficiency. But already now, following our concept, just a few hundert mJ pulse energy of a well suited Nd:YAG laser is sufficient to produce several mJ in the whole UV-C region, which is enough for most applications.

Of course, higher pulse energies as demanded by UV lidar, sheet diagnostics of atomic and molecular states in turbulent media, e.g., combustion processes, require more powerful Nd:YAG lasers. Pulse energies of about 100 mJ at 243 nm with highest spectral and spatial beam quality as needed for magnetically confined fusion-plasma diagnostics [5] will be soon generated with our UV laser system. Therefore, we will install a second Nd:YAG laser, one of them can be used exclusively for pumping the OPO-Ti:sapphire system by its SH while the other provides only the TH for converting the output of the OPO-Ti:sapphire system to the 243 nm radiation.

References:

[1] R. Dux, K. Grützmacher, M.I. de la Rosa and B. Wende, Phys. Rev. E **51**, 1416 (1995)

[2] K. Grützmacher and A. Steiger, P19.48 in Verhandl. DPG **27**, 1371 (1992)

[3] A. Steiger, K. Grützmacher, R. Dux and M.I. de la Rosa,
Q3D.1 in Verhandl. DPG **29**, 654 (1994)

[4] K. Grützmacher, M.I. de la Rosa, A. Steiger, W. Bohmeyer, H. Grote and E. Pasch,
P7.4 in Verhandl. DPG **30**, 176 (1995)

[5] K. Grützmacher, J. Seidel, A. Steiger and D. Voslamber,
in Jahresbericht der PTB Kap. 2.9.9 p. 268 (1993),
also in Proc. 21st EPS Conf. Contr. Fusion and Plasma Physics **18B**, 1288 (1994)

Die Bonding and Thermosonic Wire Bonding of Laser Diodes on CVD-Diamond Heatsinks

Stefan Weiß, Elke Zakel*, Herbert Reichl

Technical University of Berlin, Center of Microperipheric Technologies

Sekr. TIB 4/2-1, Gustav-Meyer-Allee 25, 13355 Berlin, Germany

*Fraunhofer-Institute IZM-Berlin

Gebäude 17, Gustav-Meyer-Allee 25, 13355 Berlin, Germany

Abstract: The die bonding process developed for mounting of 1 Watt laser diodes on CVD diamond heatsinks with a AuSn-sandwich type of metallization is presented. And a reliable thermosonic wire bonding process of these devices is described.

Introduction: Today high power GaAlAs/GaAs-laserdiodes emitting at a wavelength of 808 nm have a wide field of application e.g. material processing like drilling and cutting or pumping of solid state lasers. Because of the high mismatch in the coefficient of thermal expansion (CTE) between GaAs (~ 5.8 ppm/K) and Cu (~ 17.8 ppm/K) mounting is only possible applying soft solders. However soft solders like eutectic PbSn- or In-solder lead to migration and whisker growth [1,2] at the bond, which both effect catastrophic failure of the chip. All these effects are not observed when using the hard solder like Au(80wt%)Sn(20wt%) [3] for low power lasers up to 5 mW. Diamond heatsinks with a CTE < 2.3 ppm/K are very interesting for high power application due to their high thermal conductivity (> 1000 W/mK). The high thermal conductivity can lead to a longer lifetime and to a reduced linewidth of the laser diode. A 120 W laser bar on a diamond heatsink with excellent thermal characteristcs was presented by Sakamoto et al.[4].

Today wire bonding is the standard chip interconnection technique for laser diodes. Thermocompression bonding [5] is preferred for delicate semiconductor devices like GaAs. With thermocompression bonding a high capillary bonding temperature of 300 - 350°C is needed for a good contact, this will cause a local pre-damage in the GaAs-bulk material or micro cracking beneath bond pads, which can seriously affect the performance of the laser diode. A temperature reduction is only possible with a thermosonic wire bond. But GaAs is very sensitive to ultrasonic energy, which is the most important bonding parameter for the reliability of the contact in this process. Infrared microscopy is used for semiconductor chip inspection since several years especially in failure analyses of electronic components. The method has been used to prove bond quality [6] and to make visible the bonding stress of die-attach bonds by polarized infrared light [7].

Die-Bonding: First the laser diodes, which were commercial laser diodes from Siemens, were bonded on the commercial diamond heatsinks (Diamonex, Sumitomo) with a Au(80)Sn(20)-layer thickness of 3 µm. This is a typical thickness for p-side-down mounting. These laserdiodes are suffering under high stress. An increase in the threshold (typical 500 mA), a decrease in the slope efficiency η (typical 0.8 -1.0) and a large linewidth (typical 2-3 nm) in the laser diodes could be observed. The metallization layer was changed, shown in Fig.1 to compensate the difference in CTE and to achieve laserdiodes, which are bonded with low stress. A 2 µm thick Au-layer was added under the 3 µm AuSn-layer. The P-I-characteristic and the spectrum of a laser diode bonded on this modified heatsink is shown in Fig.2 and Fig.3. At a heatsink temperature of 25°C a maximal power of 3.4 W at a current of 4.0 A could be achieved. The slope efficiency was η = 1.0. The threshold was in the typical range of 500 mA. The linewidth was 1.5 nm at an output power of 1 Watt (I = 1.5 A). This is smaller than the linewidth of commercial packages with 2-3 nm. This effect is caused by the better thermal management of the diamond heatsink. The results of the first accelerated aging tests in Fig.4 expected a total lifetime in the range of 10000 h. A detailed description of the bonding process can be found in [8].

Wire-Bonding: To achieve a good, reliable and reproducible bond contact a variation of the following bond parameters was performed: bonding force (20 - 110 cN)/ultrasonic energy (1 - 5 a.U. (arbitrary Units)). The adhesion of the bond interface was investigated with destructive analysis methods (shear-tests). Therefore 25 contacts were tested for every investigation. The aim of the optimization was to minimize the percentage part of the delamination as well as to maximize the shear forces without destruction of the GaAs. This is the classical method for the optimization of wire bonds. The optimal bonding window for ball-wedge-bonding is presented in Fig.5 at a given substrate temperature of 120°C and a constant bondtime of 70 ms. The optimal bondparameters in these figures are the parameters, where the required failure criteria are met, i.e. no delaminations (0%), no mechanical destructions or pre-damaging and a minimal shearforce of 40 cN, which corresponds to the ASTM-standard [9] for gold ball-bonds of 70 µm (2.8 mils). The range of destruction was determined using destructive shear tests as well as infrared microscopy. In the window of bad bondability it was not possible to obtain sticking contacts or the contacts

314

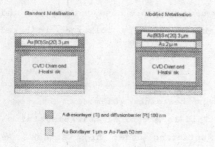

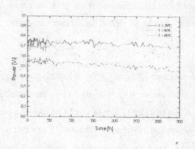

Fig.1:Standard and modified layer
     structure

Fig.4:Accelerated aging tests at
      constant current of 1.5 A and
      different temperatures

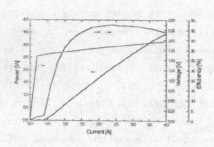

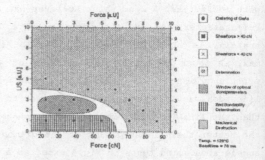

Fig.2:Power-current characteristic
      of a laser diode bonded on the
      modified heatsink

Fig.5:Range of bondparameters for
      ball-bonds at a substrate
      temperature of 120°C

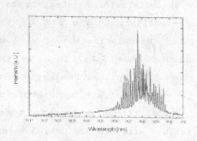

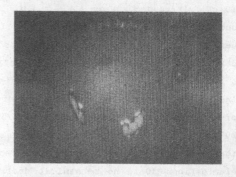

Fig.3:Spectrum of the laser diode of
      Fig.2 at an output power of 1
      W and a current of 1.5 A.

Fig.6:IR-microscope: Pre-damaging by
      a ball-bond, bondparameters(
      120°C, 60 cN, 4 a.U.)

delaminated in the shear tests. The window of optimal bondparameters is the smallest window of these three. The window of optimal bondparameters has a size of 2 - 3 a.U in the ultrasonic energy range and 20 - 50 cN in the bonding force range. There is a well defined limit to the area of mechanical destructions. Accelerated aging tests at a storing temperature of 125°C showed a good reliability after storing times up to 1000 h. A detailed description of the optimization for wedge-wedge bonds can be found in [10].

Infrared-Microscopy: With infrared-microscopy it is possible to analyze the bond quality by the reflection of the infrared beam at the metallized pad interface. With this analyzing method it was possible to decrease the quantity of the bond- and shear-tests for the classical optimizing method described above. Fig.6 shows a pre-damage generated by a ball-bond. A large area with a size of 10 to 20 µm which indicates destructions can be seen. These destructions are placed in the circle form of the ball. The most pre-damaging with a size up to 5 µm detected by infrared microscopy, did not lead to craterings in the shear tests. These damages could cause sudden failures in the metallization/chip interface.

References:

[1] Fukuda, M., Fujita, O., Iwane, G.. "Failure Modes of InGaAsP/InP Lasers Due to Adhesives," IEEE Trans. Comp. Hybrids Manufactur. Technol., vol. CHMT-7, p.202-206, 1984

[2] Mizuishi, K., "Some aspects of bonding-solder deterioration observed in long-lived semiconductor lasers: Solder migration and whisker growth," J. Appl. Phys., vol.55, p.289-295, 1984

[3] Fujiwara, K., Fujiwara, T., Hori, K., Takusagawa, M., "Aging characteristics og $Ga_{1-x}Al_xAs$ double-heterostructure lasers bonded with gold eutectic alloy solder," Appl. Phys, Lett., vol. 34, p.668-670, 1979

[4] Sakamoto, M., Endriz, J.G., Scifres, D.R., " 120 W cw output power from monolithic AlGaAs (800nm) laser diode array mounted on diamond heatsink," Electronics Letters, Vol.28, No.2, p.197-199, 1992

[5] Riches, S.T., White, G.L., "Wire Bonding To GaAs Electronic Devices," Proc. ISHM Europe, 1987, p.143-151

[6] Footner, P.K., Richards, B.P., Stephens, C.E., Amos, C.T., "A study of gold ball bond intermetallic formation in PEDs using infra-red microscopy," Proc. IEEE IRPS, 1986, p.102-108

[7] Cohen, B.G., "Infrared Microscopy For Evaluation Of Silicon Devices And Die-Attach Bonds," SPIE Vol. 104 Multidisciplinary Microscopy 1977, p.125-131

[8] Weiß, S., Zakel, E., Reichl, H., "Mounting of High Power Laser Diodes on Diamond Heatsinks," Proc. 45th ECTC, Las Vegas, 1995

[9] ASTM F1269-89 Test Methods for Destructive Shear Testing of Ball Bonds

[10] Weiß, S., Zakel, E., Reichl, H., "A Reliable Thermosonic Wire Bond of GaAs-Devices Analysed by Infrared Microscopy", Proc. 44th ECTC, Washington, p.929-937, 1994

The Optimization of Output Characteristics of Flash Lamp Pumped Cr:Tm:Ho:YAG Laser

W. Żendzian, Z.Jankiewicz, J.K. Jabczyński, M.Skórczakowski, A.Zając
Institute of Optoelectronics, Military University of Technology,
01-489 Warsaw, ul. Kaliskiego 2

ABSTRACT

The lamp pumped Cr:Tm:Ho:YAG laser, with slope efficiency about 2%, and up to 17 W average power was demonstrated. The output energy, average power, as well as M^2 parameter were investigated in dependence on temperature and cavity configuration for wide rang e of pump power levels. The optimized cavity with concave facets rod and flat mirrors enabled to shift stability range to 2 kW of pump power.

1. INTRODUCTION

At present, solid-state lasers generating within the medium infrared (2÷3 μm) are especially extensively investigated and developed because of the possibilities of their applications for medicine (ophthalmology, surgery, and dentistry). Among the lasers applied in medicine the most attractive are the ones in which the active ions are trivalent erbium ions [Er^{3+}], holmium ions [Ho^{3+}], and thulium ions [Tm^{3+}].

The paper presents the investigation and optimisation of generation conditions of the pulse Cr:Tm:Ho:YAG (CTH:YAG) laser which generates radiation at wavelength of λ=2.09 μm. The efficiency and spatial characteristics of the laser beam depend mainly on thermal phenomena occurring in the laser medium. The main purpose of our work was to optimise the parameters of the CTH:YAG laser beam radiation for high values of pump average power. We have investigated the dependence of generation energy, average generation power as well as of laser beam quality on average pump power and laser rod temperature for several configurations of resonators.

2. INFLUENCE OF TEMPERATURE ON ENERGETIC CHARACTERISTICS OF CTH:YAG LASER

Short distance of the low laser level from the basic level of Ho ions (~ 470 cm⁻¹) causes the increase of generation threshold and the decrease of total generation efficiency together with the growth of laser rod temperature.[1] It is a consequence of density of states increase in low laser level with the temperature growth in the laser rod which causes the decrease of density of states inversion in the generation channel. This phenomenon is llustrated in Fig.1. The stable laser head made by Laser Modules Inc. with CTH:YAG rod of dimensions 4 x 89 mm were used for experiments. The laser was operating at 0.1 Hz frequency to eliminate .an influence of thermal focusing. The cooling system with temperature stabilisation, enabling the changes of temperature of liquid coolant in the range of 5÷50°C was used for laser head cooling.

As was shown in Fig. 2 the increase of temperature of the CTH:YAG rod from 10°C to 30°C results in the generation threshold increase from 20 J up to 30 J. However, the slope efficiency was invariable within this temperature range and amounted to 1.7 %. From the above results it can be seen that it is necessary to stabilise a temperature of the active medium in holmium laser.

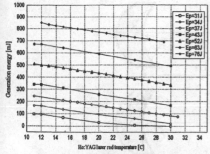
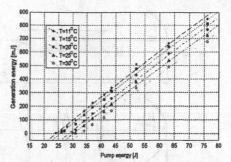

Fig.1. Generation energy of CTH:YAG laser as a function of temperature. Fig.2. Generation energy of CTH:YAG laser as a function of pump energy.

3. THERMAL FOCUSING IN THE CTH:YAG LASER ROD

The performance of continuos work or high repetition pulse solid-state lasers are determined by the heat dissipation in the active medium. The heat sources in the laser rod are generated as a result of (see e.g. [2]):

   a) The quantum defect i.e. the difference between the energy of pumping band and the top laser level energy absorbed by the active medium host during the radiationless transition.

   b) The quantum efficiency of fluorescence is lower than 1.

   c). Not well matching of the spectrum of lamp luminescence with the absorption spectrum of the active dopant.

The last effect is the main purpose, causing approximately 100 % of pump energy changing into heat in the case of lam pumping. This effect can be significantly lowered in the laser diode pumped solid-state lasers due to the excellent matching of diode wavelength to the peak of absorption spectrum of active dopant.

The heat generated in the laser rod, among the others, induces the thermal focusing effect which modifies the properties of laser resonator and, as a result, changes energetic and spatial parameters of output beam. The optical power of thermal lens depends mainly on optical pump power, laser medium properties, laser head parameters, and heat dissipation power in the cooling system. Its value can be expressed as:[2]

$$\frac{1}{f} = \beta P \qquad (1)$$

where f is the focal length of thermal lens of the laser rod; P is the pumping power; β is the constant dependent on the parameters of laser medium describing which part of electric energy of the pumping system is exchanged into heat in the laser rod. The value of β parameter can be determined experimentally.

Fig.3 presents an optical scheme of a real resonator (3a) and its substitute scheme (3b) which was received by replacing the laser rod by the thin lens with the optical power dependent on pump power (1). For a general case, the laser rod faces can have concave facets with curvature radius R', so we have additional lenses with the focal lengths f' described by:

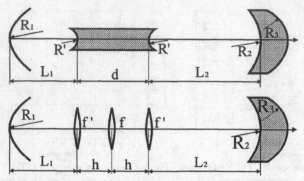

Fig.3. Resonator of the solid-state laser with the cylindrical rod (a); equivalent resonator with the thin lens (b).

$$\frac{1}{f'} = -\frac{(n-1)}{R'} \tag{2}$$

where n is the refraction index of the active medium.

The following notations are used in Fig. 3 : R_1, R_2 are the curvature radii of the totally reflected mirror and the output mirror, respectively; R_3 is the radius of curvature of the output surface of the output mirror (it makes it possible to correct the wave front of the generated laser beam), h=nd/2.

Using the ABCD method, the generator's parameters g_1^*, g_2^* of the resonator presented in Fig.3b can be determined as:[3]

$$g_i^* = 1 - L'_j\left(\frac{1}{f} + \frac{2}{f'} - \frac{h}{ff'}\right) - \frac{L^*}{R_i} \quad ; \quad i,j = 1,2 \tag{3}$$

where:

$$L'_i = L_i + h - \frac{hL_i}{f'} \tag{4}$$

$$L^* = \frac{L'_1 + L'_2 - \frac{2L'_1 L'_2}{h}}{1 - \frac{h}{f'}} - L'_1 L'_2\left(\frac{1}{f} - \frac{2}{h}\right) \tag{5}$$

The g_1^*, g_2^* parameters enable us to calculate the parameters of the basic TEM_{00} mode of a resonator and to determine the range of resonator stability from the conditions ($0 < g_1^* g_2^* < 1$). It results from the relationship (3) that they linearly depend on the optical power $1/f$ of the thermal lens of the laser rod. The points of intersection of straight line $1/f$ with the axes g_1^*, $g_2^* = 0$ and with the hyperbola $g_1^* g_2^* = 1$ determine the values of optical power of the thermal lens of laser rod for which the resonator state is on the limit of the stability range.

$$\frac{1}{f} = \frac{1}{1 - \frac{h}{f}}\left[-\frac{2}{f'} + \frac{1}{h + \varepsilon_1\left(1 - \frac{h}{f'}\right)} + \frac{1}{h + \varepsilon_2\left(1 - \frac{h}{f'}\right)}\right] \tag{6}$$

where:

$$\varepsilon_1 = L_2 \quad ; \quad \varepsilon_2 = L_1 - R_1 \quad \text{for } g_1^* = 0$$
$$\varepsilon_1 = L_1 \quad ; \quad \varepsilon_2 = L_2 - R_2 \quad \text{for } g_2^* = 0$$
$$\varepsilon_1 = L_2 \quad ; \quad \varepsilon_2 = L_2 \quad \text{for } g_1^* g_2^* = 1$$
$$\varepsilon_1 = L_1 - R_1 \quad ; \quad \varepsilon_2 = L_1 - R_1 \quad \text{for } g_1^* g_2^* = 1$$

(7)

Knowing the pump power which correspond to upper limit of stability diagram for various data of cavity parameters the value of the β parameter for a real laser can be estimated. In our work this method was used for the CTH:YAG laser. The configurations of the examined resonators of the holmium laser are presented in Table 1.

Table 1. Geometrical parameters of three examined resonators.

	Classical resonator	Dynamically stable resonator	Plane-plane resonator
R_1 [mm]	2500	1200	∞
R_2 [mm]	∞	600	∞
R_3 [mm]	∞	186	∞
R' [mm]	∞	∞	600
d (Ho:YAG) [mm]	89	89	89
φ (Ho:YAG) [mm]	4	4	4

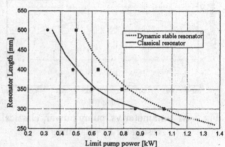

Fig.4. Dependence of an limit pump power P_g which causes the break of CTH:YAG laser generation on length of Lr resonator.

Fig.4 presents the limit pump power of the P_G pump as a function of a resonator length for classical resonator and dynamically stable one. The solid curves were obtained from the theoretical model for $\beta = 0.15$ m^{-1}kW^{-1} whereas the circle and square points correspond to experimental data. Comparing the value of β parameter of the investigated CTH:YAG laser with typical values of these parameters for the cw Nd:YAG lasers one can notice that the thermal focusing in the CTH:YAG rods is almost one order of magnitude higher despite the same crystallite base of the active ions. It is supposed that the reasons of this fact are as follows:

- Cr and Tm sensitizers expand the spectrum of the CTH:YAG crystal absorption adding the bands enabling increase of pumping efficiency as well as the bands which cause direct exchange of pump energy into the heat. The CTH:YAG crystal is opaque, almost in the whole visible range of spectrum.

- Efficiency of energy transfer from Cr and Tm ions to the active Ho ions is lower than 1.[4]

3.1. Influence of thermal focusing phenomenon on the M^2 quality parameter of the holmium laser beam

The ABCD matrix method was used also for determination of the M^2 beam quality parameter as a function of average pump power. Transforming the fundamental mode TEM$_{00}$ through particular optical elements of resonator, the maximum value of diameter of the fundamental mode (2w) in the laser rod was determined and then, according to [5]:

$$M^2 \cong \left(\frac{\phi}{2w}\right)^2$$

$$\theta \cong \sqrt{M^2} \cdot \theta_0 \tag{8}$$

where Φ is the diameter of a laser rod; θ is the divergence angle of laser radiation beam ; θ_0 is the divergence angle of the basic mode beam. Of course, there was assumed that the aperture which limited a laser beam is a diameter of the laser rod. The results of numerical analysis and measurements of the M^2 quality parameter as a function of average values of pump power, for which the resonator is stable, are presented in four consequent figures 5-8.

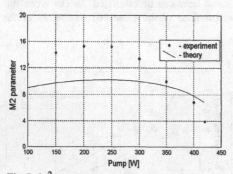

Fig.5. M^2 parameter vs. pump power, classical resonator Lr=400 mm.

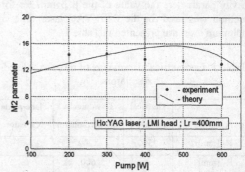

Fig.6. M^2 parameter vs. pump power, dynamically stable resonator with the rod having plane faces.

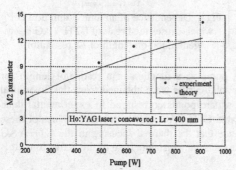

Fig.7. M^2 parameter vs. pump power, dynamically stable resonator with the rod having concave faces.

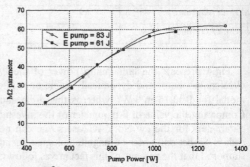

Fig.8. M^2 parameter vs. pump power, plane-plane resonator with the rod having concave faces.

The solid curves in Figs.5,6 and 7 have been obtained from theoretical calculations. It can be seen that for dynamically stable resonator (Fig.6), the beam quality parameter M^2 slightly depends on average pump power what improves the efficiency of laser beam input into optical fiber for wide range of pump power levels. However, an application of this resonator does not ensure a significant increase in the limit pump power ($P_g = 650$ W) in comparison with classical resonator ($P_g = 430$ W), thus, does not ensure a significant increase of the generation power of the holmium laser. The dynamically stable resonator with the rod having concave faces (Fig.7 $P_g = 900$ W) and plane-plane resonator with the similar rod ensure to achieve significantly higher average output power. As it can be seen from Fig.8, the plano-plane resonator starts to generate from the pump power of the order of 450 W. The limit pump power P_g is higher than 1400 W (in experiments the pumping power was not higher than 1400 W). The drawback of this resonator is high value of the M^2 parameter ($M^2 \approx 20 \div 60$, $\theta \approx 10 \div 30$ mrad) as well as its changes as a function of the average pump power. However, it does make possible to efficiently introduce a laser beam into the quartz optical fiber of 400 μm core diameter.

3.2. Influence of thermal focusing phenomenon on energetic characteristics of the CTH-YAG laser

The results of generation energy measurements of the CTH:YAG laser with the flat-flat resonator having concave faces as a function of the pumping energy for repetition rate from 7 to 14 Hz are presented in Figs. 9 and 10.

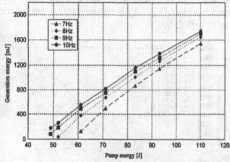 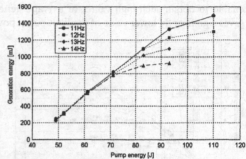

Fig.9. Generation energy of the CTH:YAG laser with the flat-flat resonator as a function of pump energy.

Fig.10. Generation energy of the CTH:YAG laser with the flat-flat resonator as a function of pump energy.

For the repetition rates higher than 9 Hz, the generation threshold was of order of 40 J and practically does not change with the change of repetition rate. For the diminishing repetitions, the generation threshold increases because the average pump power is insignificant to get the resonator into the stability range. Moreover, the thermal state of the rod is not stated in this low repetition range and in principle the operation conditions are not quasi stationary in such a case. It can be an additional reason of the significant differences in energy of the generated pulses (about 19%). The lowest repetition rate at which the generation threshold attains minimum value (about 40 J) is about 9 Hz. It is equivalent to the average pump power of about 350 W. Above this average power the resonator is being stable. In respect of the maximum generation energy of the examined laser the optimum repetition rate is of 10 Hz and it is equivalent to the maximum use of the active medium volume. Above this frequency value, the differences in energies of the generated pulses were of the order of 2%. The slope efficiency of generation for this repetition rate is 2.4 %. The energetic parameters are invariable for repetition rate from 9 Hz to 10 Hz. This range of repetition rates is recommended in order to effectively generate the pulses of maximum energy. The energy of 1650 mJ was obtained as a result of the pulse generation for the pump pulse energy of 110 J. The pump energy exceeding 110 J was not applied in fear that mirrors and rod faces might damage. However, it is supposed that using adequate treatments (dust hoods and air dehumidification in the beam region) there is possible to obtain the output energy of about 2 J. It should be noticed that the measurements were carried out at the liquid coolant temperature of about 19 °C. The above results were obtained for the resonator with the minimum length possible, i.e., Lr = 27 cm. As it can be seen from Fig.10, for repetition of 11÷14 Hz, the decrease in output energy, caused by diminishing the generation region in the rod and because of its temperature growth, is not rapid.

The optimum repetition range in respect of the average output power will not be identical as in respect of maximum generation energy. In Fig. 11 the dependencies of the average output power on the average pump power are shown. It can be seen that optimum repetitions range in respect of the average generation power depends mainly on the pumping pulse energy. The maximum power, obtained for the examined laser was about 17 W of the average output power in case of pump excitation pulse of 110 J (about 1100 W). Taking into account the above mentioned precautions, the laser was not examined in the conditions of extreme emission of the output power. An extrapolation of the obtained results enables to think that there is possible to generate in the examined laser the power of about 20 W. Characteristic is the fact of the curves convergence in the point (350W, 0) what is shown in Fig. 11. It confirms the previous conclusion regarded to the minimum pump power which get the resonator into the stability range.

An application of the dynamically stable resonator with the rod having concave facets does not enable to obtain such high values of average generation power (Fig. 12) because such a resonator gives significantly lower value of the limit pump power $P_g = 900$ W.

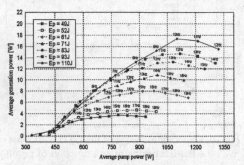

Fig. 11. Dependence of the average generation power of the CTH:YAG laser with the flat-flat resonator on the average pump power (the rod with concave faces).

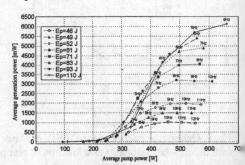

Fig. 12. Dependence of the average generation power of the CTH:YAG laser with the dynamically stable resonator on the average pump power (the rod with concave faces).

4. CONCLUSIONS

It results from the analysis of the carried out investigations that for the holmium laser design the following aspects should be considered:

- stabilisation and decreasing of the laser medium temperature ;

- optimisation of configuration both a resonator and a laser head within the range of the used average powers of the pumping system;

- increase in the average generation power can be obtained by application of the resonator having higher value of the average maximum power P_g which exceeding causes the generation break.

5. ACKNOWLEDGEMENTS

The work has been done within the frame of the grant No. 8 8606 015 06 supported by Polish Committee for Scientific Research.

6. REFERENCES

1. G.J. Quarles, A. Rosenbaum, C.L. Marquardt, L. Esterowitz, 'High- efficiency 2.09 μm flashlamp-pumped laser', *Appl. Phys. Lett.*Vol. 55 ,pp. 1231-1235, (1989).
2. W. Koechner, *Solid-State Laser Engineering*, Springer, New York, (1976).
3. E. Cojocaru, 'Simple relations for different resonators of solid-state lasers', *Appl. Opt.*, Vol.33, pp.3454-3459, (1994)
4. T.S. Kubo, J.Kane,'Diode-pumped lasers at five eye-safe wavelengths', *IEEE J. QE*, Vol.28, pp. 1033 -1039, (1992).
5. A.Siegman, *Lasers*, University Sciences Books, Mill Valley, California, 1986

Efficient Optical Amplifier for Use in Coherent Space Communication

M. Fickenscher

DLR Oberpfaffenhofen, Institut für Optoelektronik, D-82230 Weßling

1. Introduction

For coherent optical communication in space based on Nd:YAG (1064 nm) two alternative concepts exist to reach a transmission power of at least 1 Watt [1]:

i) a high power laser with subsequent bulk phase modulator

ii) a medium power laser followed by an integrated optical modulator (IOM) and (a) laser amplifier(s).

Because of some characteristics of the bulk modulator, e.g. limited bandwith, mechanical resonances and high power consumption, the latter concept has been found to be better if the optical power rating of the IOM and the efficiency of the amplifier(s) are high enough.

For our experiments with the optical transmitter we used an IOM developed at Friedrich-Schiller University Jena [2]. The characteristics at high optical energy densities were investigated [3] to find out realistic values for the input power of the amplifier(s).

2. Optical Amplifier

Demands

Although the IOM is capable to transmit an optical power of more than 100 m W without problems [3], the amplifier should show up a high efficiency even at much lower power levels of the input signal (e.g. 10 mW). This is not to waste more power than necessary in the IOM, which has a total insertion loss of about 4 dB. By restricting the amplification to the small area of the core of a singlemode fiber, the signal intensity exceeds the saturation intensity of the amplification process and gain saturation occurs. This way an efficient transfer of absorbed pump light energy to signal energy can be achieved.

Using a doped fiber for amplification also would allow full fiber coupling within the transmitter, i.e. a compact and robust setup which is necessary for space applications. Furthermore a fiber amplifier would allow redundancy of the pump sources, i.e. to maintain the output power of more than 1 Watt even in case of failures of laser diodes used for pumping.

Because the high power laser diodes used for pumping are broad area emitters with inherent bad beam quality compared to a TEM_{00} laser beam, their radiation can not be coupled efficiently into a singlemode fiber. So a double clad fiber (DCF), manufactured at IPHT (Institut für Physikalische Hochtechnologie) in Jena is adopted.

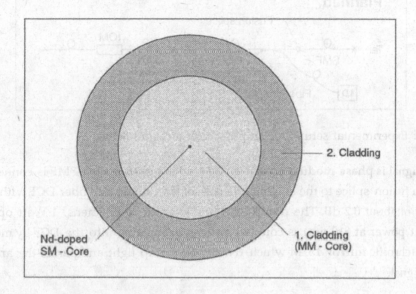

Fig. 1: Nd-doped double-clad fiber for signal amplification

The signal is coupled into the 5 μm singlemode core of the fiber, which is Nd-doped for amplification at the desired wavelength 1064 nm. The first cladding consists of pure silica, giving a numerical apertur NA = 0.16 for the singlemode core. This cladding is 110 μm in diameter and acts simultanuosly as a multi-mode core with NA = 0,38 by means of a second cladding of silicone. The laser diode pump light is launched into the multimode core and partially absorbed each time it crosses the doped singlemode core. So it is nearly fully absorbed over the length of the fiber.

Experimental setup

The experimental setup used for amplification measurements is shown in the upper part of Fig. 2 („In the moment").

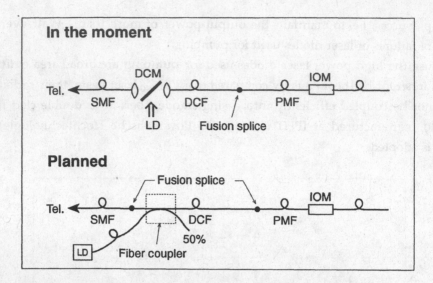

Fig. 2: Experimental setups for amplification measurements.

The signal is phase modulated in the IOM, which output fiber PMF is connected with a fusion splice to the singlemode core of the double clad fiber DCF with low losses of about 0.2 dB. The pump light of a laser diode (Siemens, 1 Watt optical output power at 810 nm) is coupled counterpropagating into the DCF by means of a dichroitic mirror DCM which reflects the pump light and passes the ampli-fied signal.

Experimental results

The experimental results achieved in this first setup are as follows:

The pump light can be launched with an efficiency of about 70% into the DCF. The power for gain saturation is rather low, $P_{sat} \approx 5$ mW. So for an input power $P_{in} \gg P_{sat}$ the amplifier is saturated and the change in output power is propor-tional to the pump power, i.e. $P_{out} = P_{in} + \eta_{slope} \cdot P_{pump}$ with $\eta_{slope} = 0{,}17$. This slope efficiency is the overall efficiency, including for instance the launch effi-cieny. So with a laser diode of 1 Watt optical output and an signal input power of 40 mW an output power of 210 mW were achieved.

A test of the phase modulated and amplified signal in the transmission system of our collegues at DLR Oberpfaffenhofen, Institut für Nachrichtentechnik, sho-wed no degradation of the signal due to amplification, i.e. no broadband phase noise which would cause an error floor in bit error ratio (BER) measurements as a function of received power.

Plans

In the next step we aim at full fiber coupling of the transmitter system according to the lower part of Fig.2 („Planned"). For this purpose a fiber coupler between the multimode fiber of a pigtailed laser diode and the multimode core of the DCF has to be developed. First experiments with couplers made between hard clad silica (HCS) multimode fibers show promising results and have to be transferred to the system HCS - DCF. Because maximum 50% of the power can be coupled, the remaining part should be backreflected with a mirror or used in a second coupler. Using pigtailed laser diodes and fiber couplers for pumping should allow to scale the amplifiers output power to several watts [4] by using a number of laser diodes for pumping.

3. Conclusions

We believe to reach our aim of a full fiber coupled, efficient amplifier for use in coherent optical space communication within next time. This amplifier would be useful for a lot of other purposes, e.g prededection amplification in direct detection communication systems or in lidar.

Acknowlegement

The author is indebted to IPHT Jena for the loan of a sufficient length of double clad fiber. Thanks also to Daimler Benz AG for loan of the grinding / polishing machine and to Laser Zentrum Hannover LZH for helpful discussions.

[1] Pribil, J., Johann, U., Sontag, H.: „SOLACOS: a diode pumped Nd:YAG laser breadboard for coherent space communication system verification", Opt. Space Communication II, J. Franz, Editor, Proc. SPIE 1522 (1991), pp. 36 - 47

[2] Rasch, A., Buß, W., Göring, R., Steinberg, S., Karthe, W.: „Optical carrier modulation by integrated optical devices in lithium niobate", Opt. Space Communication II, J. Franz, Editor, Proc. SPIE 1522 (1991), pp. 83 - 92

[3] Fickenscher, M., Rasch, A.: „Characteristics of an Integrated Optical Phase Modulator at High Optical Energy Densities", Laser in Engineering; Proceedings of the 11th International Congress Laser '93, W. Waidelich, Editor, Springer 1994, pp. 796 - 799

[4] Zellmer, H., Willamowski, U., Tünnermann, A., Welling, H., Unger, S., Reichel, V., Müller, H.-R., Kirchhof, J., Albers, P.: „High-power cw neodymium-doped fiber laser operating at 9.2 W with high beam quality", Opt. Lett. 20 (1995) 578 - 580

Diode-Pumped Multmode Fiber Generating Low-Order Laser Mode

U. Griebner, R. Grunwald, H. Schönnagel

Max-Born-Institut für Nichtlineare Optik und Kurzzeitspektroskopie

Rudower Chaussee 6, D-12489 Berlin, FRG

Introduction

Different types of diode-pumped fiber laser arrangements including fiber array lasers are of increasing interest for compact and efficient systems particularly because of their specific thermal properties. A large improvement to scale single mode fibers to higher power levels has been demonstrated through the use of cladding-pumping schemes[1].

The potential of multimode fibers to generate high output power - up to 100 W (array, flashlamp pumped) - has also been demonstrated, but with restrictions concerning to the brightness[2]. In this case, the participating waveguide modes of each single fiber are coupled within a limited coherence volume due to the relatively high mode conversion coefficient and the mode dispersion of the multimode fiber[3]. By the lateral coupling of the fibers, a supermode can be generated with a random distribution in phase and amplitude but stable over the gain relaxation time. During this time, longitudinal mode locking can be realized[4]. The spectral locking width is limited by the coherence volume.

The bad beam quality is related to the mode conversion in the multimode fibers. We have successfully overcome this problem by length shortening of the multimode fibers down to 10 mm. In this way the external resonator dominates the beam quality of the fiber laser. With the relatively large multimode core and high acceptance angle, this geometry provides an efficient means to capture pump light from a low brightness diode pump source and ensuring high brightness laser output.

In this contribution, we report on a theoretical consideration of the resonator design and experimental results concerning diode-pumping of short and heavily Nd-doped multimode phosphate-glass fiber lasers capable of emitting diffraction-limited output.

Calculation of fiber and resonator design

The aims are the selection of the fundamental mode in multimode fibers and a high stability of the laser operation. For a core diameter $d_c \sim 100\ \mu m$, more than 100 waveguide modes are possible. Therefore, the mode conversion has to be drastically reduced. The number of generated waveguide modes is a function of the mode conversion coefficient D_m [4] and the fiber length l. D_m must be determined experimentally for the specific used fiber - in our case D_m has been measured to be about $11*10^{-4}$ rad^2/m. To prevent the generation of higher-order modes due to mode conversion the fiber length l must be less than

$$l < \frac{\lambda^2}{8 D_m d_c^2} < 14 mm,$$

(1)

where $\lambda = 1053$ nm is the laser wavelength ($d_c = 100$ μm). The propagation of the fundamental mode is described by the Bessel function $J_0(ur)$ in a fiber as waveguide mode and by a Gaussian function $A(r)$ outside of the fiber, respectively. To achieve the best fit, the following condition must be approximately fulfilled within the entrance face of the fiber:

$$A(r) \approx J_0(ur), \text{ with } u = \frac{d_c}{2}\sqrt{n_c^2 k^2 - \beta^2}, \quad \beta = n_c k \sin\theta_1,$$

(2)

where u is the normalized transverse propagation constant, β the propagation constant, n_c the refractive index of the core and k the wave vector. For the fundamental mode in the fiber u yields ~ 2.44. For the Gaussian beam follows a beam waist w_0 adapted to the fundamental waveguide mode depending on the core diameter:

$$A(r) \approx e^{-\frac{r^2}{w_0^2}}, \quad \frac{2w_0}{d_c} \approx 0.65 + \frac{1.615}{V^{1.5}} + \frac{2.879}{V^6} + ... \Rightarrow 3.07 w_0 \approx d_c.$$

(3)

In Eq. (3), V is the fiber parameter. After discussing the conditions of the fiber design we have further to examine the requirements for mode selection by an external resonator.

The Fresnel diffraction loss can be determined approximately for different modes m in dependence on the mirror distance of the resonator mirror from the fiber end face dz.

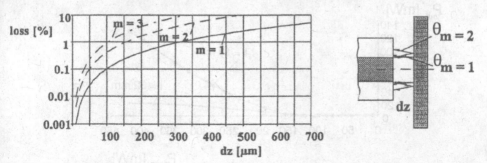

Fig. 1: Fresnel diffraction loss versus distance fiber - resonator mirror dz for different transversal modes m

The results of the calculation and a graphical explanation of the propagation direction θ for different modes m have been depicted in Fig. 1. With the demand of a fundamental mode loss ($m = 1$) much less than 1% and a higher-order mode loss greater than 1%, a selection of the fundamental mode should be possible for a mirror distance dz in the range of 150 μm, as can be seen from Fig. 1.

Experimental set-up

The fiber laser system described in the present paper is based on phosphate glass fibers doped with 2.0 wt% Nd_2O_3 in the core - what is about two orders of magnitude higher than in silica doped fibers - and without doping in the cladding (*Kigre* preforms). The drawn fibers are of multimode structure ($d_{core} = 100$ μm) with a numerical aperture NA of 0.44 (*Fiberware*). The loss at the laser wavelength λ = 1.053 μm has been determined to be ≤ 4 dB/m. The pump source was a diode laser (*Adlas*) with a nominal maximum operating power behind the pigtail (200 μm diameter with a NA of 0.22) of 0.6 W cw at 807 nm. To investigate the laser performance, short fibers (length up to 32 mm) with cleaved ends were placed between two flat dichrioc mirrors ($R_{1;1053\,nm} = 99.9\%$, $R_{1;802\,nm} = 7\%$; $R_{2;1053\,nm} = 96.8\%$, $R_{2;802\,nm} = 99\%$). The fiber surface was only at the side of high reflectivity butted against the mirror. Index matching was used to minimize loss and unwanted Fabry-Perot effects. Passive cooling has been achieved by placing the fiber at a copper heat sink.

Experimental results

The pump light was launched into the fiber by an AR-coated two element lens system and provides a 2 : 1 reduction and reimages the pigtail's output onto the input face of the fiber laser. Approximately 70% of the incident pump power reach the active fiber within the NA of 0.44. The fibers reported here are 13 mm and 32 mm long, which ensure complete pump light absorption through the fiber (absorption coefficient: 11 cm^{-1}).

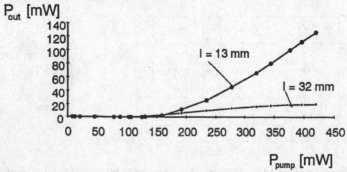

Fig. 2: Fiber laser output power versus absorbed pump power

The gain has been found in good agreement with the calculated values[5]: $g_0 \, l = (0.06 \text{ W}^{-1}) \, P_a$. This corresponds to a small-signal gain-length product of 0.252 for the maximum absorbed pump power P_a. As Fig. 2 shows, a maximum output power of 20 mW could be achieved for the fiber with a length of 32 mm. The over-all loss per pass is therefore 5.4%. By shortening the fiber length to 13 mm the single-pass loss has been reduced to 2.2%. The change of laser output with pump power striking the 13 mm long fiber is essentially linear with a maximum output power of 130 mW and a threshold of about 98 mW. This corresponds to a slope efficiency of 40% and an optical efficiency of 22%.

The transverse mode profile of the laser framed for three different pump powers and depicted in Fig. 3 shows that mostly the lowest-order waveguide modes are generated and the main energy is concentrated in a maximum of 1.5 times the diffraction limit which corresponds to only 5% of the NA of the fiber. The mode discrimation is due to the mode selection by the Fresnel loss and by the increase of waveguide losses for higher-order waveguides modes. Varying the distance between the fiber end face and the mirror surface around 150 μm permits the generation of the fundamental mode.

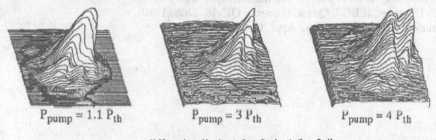

$P_{pump} = 1.1 \, P_{th}$ $P_{pump} = 3 \, P_{th}$ $P_{pump} = 4 \, P_{th}$

diffraction limit: $1.2 \times \theta_1 \perp$, $1.5 \times \theta_1 \|$

Fig. 3: Far-field distribution at different absorbed pump powers of the multimode fiber laser

It would be desirable to achieve a more efficient mode discrimination and to incorporate some optical elements into the resonator. This problem should be solved by a micro external spherical resonator with the help of micro-lenses. The nearly diffraction-limited imaging properties of the micro-lenses make further investigations with this elements for external resonators particularly attractive[7]. First examples with an adapted focus distance of 4 mm have been developed and examined.

Conclusions

Diode-pumping of short Nd-doped multimode phosphate glass fibers has been demonstrated. With a compact longitudinal pump scheme, 70% of the pump power have been absorbed in the active fiber. By using a fiber with a length of 13 mm and a core diameter of 100 μm a maximium output power of 130 mW at a pump power of 420 mW with a slope efficiency of

332

40% has been achieved. The selection of the fundamental mode has been reached mainly by Fresnel diffraction. Laser emission at 1.053 μm close to diffraction-limited performance and nearly independent on the pump power has been obtained.

Recent advances in optical fabrication technologies allow to integrate micro-optical components in diode pumped systems to realize more compact coupling optics as well as micro spherical resonators for single transverse mode selection. We developed micro-lenses and graded micro-mirrors appropriately designed for these requirements.

This work has been supported by the Bundesministerium für Bildung, Wissenschaft, Forschung und Technologie under contract no. 13 N 6356.

References
1. H. Po, J.D. Cao, B.M. Laliberte, R.A. Minns, R.F. Robinson, B.H. Rockney, R.R. Tricca, Y.H. Zhang, Electron. Lett. **29**, 1500 (1993).
2. R. Koch, U. Griebner, R. Grunwald, Appl. Phys. B **58**, 403 (1994).
3. H. Schönnagel, P. Glas, J. Opt. Soc. Am. B, submitted.
4. P. Glas, M. Naumann, A. Schirrmacher, H. Schönnagel, Opt.Comm. **109**, 101 (1994).
5. W.A. Gambling, D.N. Payne, H. Matsumara, Appl. Opt. **14**, 1538 (1975).
6. M.J.F. Digonnet, IEEE J. Quant. Electron. **QE-26**, 1788 (1990).
7. R. Grunwald, U. Griebner, Pure Appl. Opt. **3**, 435 (1994).

Fiber Laser Optimization by Fluorescence Measurements

V. Reichel, A. Schwuchow, H.-R. Müller

Institute for High Technology Jena e.V., Helmholtzweg 4, D-07743 Jena, Germany

H. Zellmer

Lasercentre Hannover e.V., Hollerithallee 8, D-30419 Hannover, Germany

supported by the Federal Ministery of Science and Technology (BMFT; FKZ: 13 N 6143)

Abstract

Silica based, rare-earth doped fiber lasers become strong competitors for other solid state lasers thanks to their compactness, inherent stability and excellent beam quality. For high output power and maximum conversion efficiency, composition and geometric structure of the laser fibers must be optimized. Problems arise especially due to the limited rare-earth solubility in the silica matrix. With Nd-doped fibers as an example, we describe, how fluorescence measurement techniques can be used to find optimal compositions giving high lifetimes of the upper laser level. For that, we not only use measurements at fibers, but also nondestructive spatially resolving methods at fiber preforms.

Such data enable us to consider the non-constant distribution of the Nd ions in the fiber core for the design of efficient lasers. We report experimental details of the measurements as well as attained fiber laser results, optimized fibers delivered ≈ 9 W cw output power from the fiber core of about 12 μm diameter and 2.5 W monomode cw output power of a 5 μm core diameter, respectively.

Introduction

The development of optimized structures of rare-earth doped fibers for the application in high-power fiber lasers demands a very complicated technology (MCVD, drawing technique) and a comprehensive measurement technique for the characterization of the behaviour of different types of active fibers. In addition to the standard measurements (spectral attenuation, refractive index, cut-off wavelength etc.) the investigation of the fluorescence of these fibers plays an important role for the prediction of their laser behaviour.

Fluorescence measurements

Therefore, we developed a measurement set-up which allows the nondestructive investigation of the spectral and the time behaviour of the fluorescence spatially resolved in the rare-earth doped region of the preform (Fig.1).

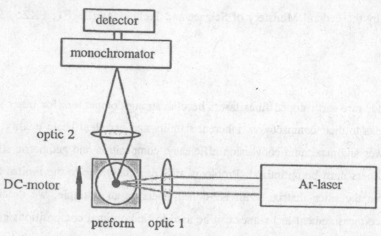

Fig.1: Principle of the set-up of the nondestructive fluorescence measurement at preforms

By means of the spectral fluorescence measurement we get information on the central wavelength and the linewidth and can investigate the influence of different codopants (P, Al, Ge), too.

For the time measurements additionally an optical switch (polarizer, Pockels-cell) is necessary, which allows us to observe the lifetime of the upper laser level online at an oscilloscope.

Fig. 2 shows a typical relation between the fluorescence lifetime and the spatial position at the preform and a fluorescence intensity profil (dashed line) which gives some information on the concentration distribution of the rare earth.

The results of the spectral and lifetime measurements we are used for the optimization of the manufacturing process (rare earth concentration, relation beetween the codopants etc.). The aim is a low-loss, high efficient active fiber with a high fluorescence lifetime and a high amplification, respectively.

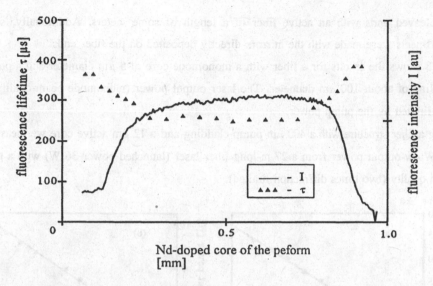

Fig.2: Fluorescence intensity and -lifetime dependent on the spatial position at the Nd-doped preform

For the characterization of the drawn fibers it is necessary to measure the main optical properties as spectral attenuation, refractive index profile, geometry and especially, in the case of active fibers, the fluorescence behaviour. By means of a special measurement set-up we are able to measure the time and spectral fluorescence of the fibers by longitudinal pumping with an argon-ion laser (488/514 nm), respectively.

The comparison of the results at the preform and at the fibers shows the influence of the drawing process. In our experiments we could not observed any significant difference between both measurements.

The results at the fibers give us the possibility to predict the amplifying and lasing behaviour in dependence on the fiber structure and composition. So we can optimize the fibers in a first step without the comprehensive and relative complicated fiber laser experiments. But, the fluorescence measurements are only a possibility to characterize the fiber behaviour quickly not a substitution of the laser experiments.

Fiber laser results

The Nd-doped double-clad fibers were experimentally tested in a fiber laser setup consisting of a diode-laser ($\lambda \approx 810$ nm) and wavelength-selective mirrors which were buttcoupled to the cleaved ends with an active fiber of a length of some meters. Additionally, some experiments were made with the mirrors directly deposited on the fiber ends.

Fig. 3 shows the results for a fiber with a monomode core of 5 μm diameter and a pump-cladding of about 100 μm diameter. The laser output power (monomode beam-quality) is only limited by the pump-power.

Using a fiber structure with a 400 μm pump-cladding and a 12 μm active core we measured 9.2 W cw-output power from a 27 m-long fiber laser (launched power 36 W) with a good beam quality (two times diffraction limited).

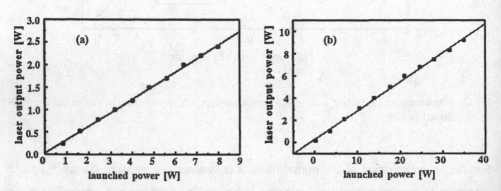

Fig.3: CW-fiber laser output power with Nd-doped double-clad fibers and a high beam quality (a) monomode ; (b) two times diffraction limited in dependence on the lauched pump power

Conclusion

By means of the nondestructive measurement of the spectral, spatial and time behaviour of the fluorescence in Nd-doped fibers and/or preforms we are able to optimize the composition of these active fibers for maximum cw-output power and high pump-efficiency. With so called double-clad structures we demonstrated a fiber laser with about 9 W cw output power (only limited by the available pump power) and an efficiency of 25%.

A Novel Design for High Brightness Fiber Lasers Pumped by High Power Diodes

P. Glas, M. Naumann, A. Schirrmacher, and J. Townsend*
*Max-Born-Institut für Nichtlineare Optik und Kurzzeitspektroskopie,
Rudower Chausse 6, D-12489 Berlin, Germany*
University Southampton, Southampton SO17 1BJ, UK

Abstract

Lasing action is reported in a new type of fiber design. The fiber consists of an undoped core and doped glass cladding surrounded by a plastic coating. The refractive indices obey the relation $n_{clad} > n_{core} \gg n_{coat}$. The refractive index/doping concentration distribution recommends to the designation M-profile fiber (MPF). The brightness and the output power of the MPF laser may at least be as high as that of a cladding pump design. For one of the first properly working MPF lasers the near and far field intensity distributions are reported along with measurements concerning the pump light distribution in the fiber, the output power, lasing spectrum and polarisation of the emitted radiation.

Introduction

This work was motivated by ideas to improve existing fiber laser geometries to accomplish a miniaturized high power, short pulse laser source with special emphasis on the realization of a coupled fiber array.

The requirements to reach this goal are the following:

Effective optical pumping with a high power diode laser (array) necessitates both a high numerical aperture and a large entrance area of the fiber. A high brightness source, however, is characterized by a low numerical aperture. Furthermore, a high as possible saturation power is desirable for the laser. The design of the M-profile fiber is plotted schematically in graph 1 (G-1). In contrast to conventional fibers, the MPF resembles a tube waveguide being firstly described theoretically by STOLEN /1/ in 1975 as a passive light guide. We have modified the structure to get a lasing one by surrounding the undoped low index core with a concentric ring of high index glass, being doped with Nd_2O_3. Such a fiber geometry has not been used to realise a laser up till now, as we believe. The pump light is guided in both the core and the ring and total reflection occurs mainly at the ring-coating interface. In this way, a large fiber diameter is available to couple the pump light efficiently into the fiber. Traversing the doped ring, the pump radiation suffers losses from absorption. The excited ions in the ring area emit radiation into a solid angle of 2π rad. The fraction which is guided within the numerical aperture of the ring contributes to the build-up of the laser oscillation. Optical feedback at the emission wavelength is provided by high reflectivity mirrors attached to the fiber ends (butt-coupling) or by directly coating the fiber ends. To illustrate the potential of the MPF design, we compare in the following its key parameters with those of the standard monomode fiber and the cladding pump design. In the next three graphs the corresponding data are listed concerning the M-profile fiber, the monomode fiber and the cladding pump design for comparison. We define the following quantities: numerical aperture $NA = n_2\sqrt{n_2 - n_1} / \sqrt{n_2}$, $n_{2,1}$ refractive indices of the doped/undoped material; brightness $B = P/A \cdot (2NA)^2$, A-area of the doped material, P-laser power; saturation power $P_s = 4\pi^2 n_2^2 \Delta \upsilon h\upsilon \cdot A/\lambda^2$, $\Delta\upsilon$–fluorescence bandwidth, $h\upsilon$–photon energy, λ-laser wavelength.

G-1 M-profile fiber

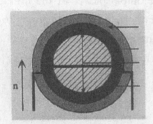

coating diam. - 120µm (silicon rubber) ⎤
cladding thickness - 5µm (SiO₂: Nd) ⎤ NA=0.4
core diam. - 50µm (SiO₂) ⎦ NA=0.12 ⎦

Nd-doped

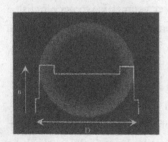

$B = P/3 * 10^{-7}$ W/cm²sr

$P_{damage,\ laser} = 500$ W

$P_{sat} \sim 90$ mW

$B = 3 * 10^5$ W/cm²sr

G-2 Standard monomode fiber

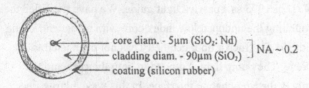

core diam. - 5µm (SiO₂: Nd) ⎤
cladding diam. - 90µm (SiO₂) ⎦ NA ~ 0.2
coating (silicon rubber)

$B \sim P/3 * 10^{-8}$ W/cm²sr

$P_{pump} \sim 150$ mW (limited by diode pump power)

$P_{damage,\ laser} \sim 25$W

$P_{sat} \sim 5$ mW

$B \sim 3 * 10^5$ W/cm²sr

G-3 Cladding pump design

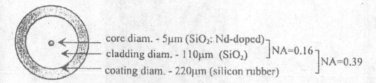

core diam. - 5μm (SiO$_2$: Nd-doped)
cladding diam. - 110μm (SiO$_2$) $\Big]$NA=0.16 $\Big]$NA=0.39
coating diam. - 220μm (silicon rubber)

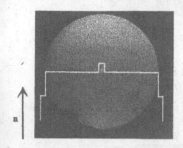

$B \sim P/2*10^{-8}$ W/cm^{2}sr

$P_{damage,laser} = 25W$

$P_{sat} \sim 5$ mW

$B \sim 5*10^6$ W/cm^{2}sr

Experimental results and discussion

The first properly working fiber laser with a M-refractive index/doping profile was made from phosphate glass doped with Nd$_2$O$_3$. Better results were obtained using silica instead of phosphate glass. The silica fiber was manufactured in collaboration with the University of Southampton. The geometrical design of the phosphate glass fiber is given in the 1st graph (G-1). The Nd$_2$O$_3$ doping concentration is about 1000 ppm. The characteristic parameters of the silica fiber have been determined as follows: small signal gain $g_o \sim 0.006$ cm^{-1}; absorption loss of the pump radiation at λ=804nm $\alpha_{l,p} \sim 0.026$ cm^{-1} · fiber loss for the laser radiation guided in the ring $\alpha_{l,l} \sim 2*10^{-4}$ cm^{-1}. The laser emits at a wavelength of λ=1060nm. The spectral bandwidth amounts to about 10nm. The laser set-up is conceivably simple. The mirrors are butt-coupled possessing a reflectivity of R_1=99% and R_2=60%, resp. The output mirror is a dichroic one. The radiation of the pump diode is focused with an optics onto the fiber front surface facing a mirror, that is AR-coated at 804nm and HR-coated at 1054nm. The output beam is analysed with a CCD-camera (near-, far field), a spectrograph and a calibrated power meter. Fig. 1 shows the distribution of the pump light and the ASE within the silica M-profile fiber. Absorption occurs mainly within the doped ring. Obviously, the structure is not optimised, because a large amount of pump light is contained in the second cladding bordering on the doped ring. This cladding will be thinned or completely removed in the next improved design, cf. G-1 .
Optimum results concerning the conversion efficiency were obtained for the case $NA_{pump} \leq NA_{MPF}^{core,coat}$.

The near and far field pattern of the radiation emitted by the MPF laser are shown in Fig. 2a, 2b.

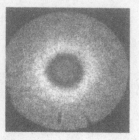

Fig. 1a

Fig.1b

Fig.1 near field intensity distribution of the pump radiation at 804nm (Fig.1a) and the fluorescence at 1060nm (Fig.1b)

28μm

9°

Fig.2a

Fig.2b

Fig.2 near field intensity distribution of the M-fiber laser above threshold (Fig.2a) and the corresponding far field pattern (Fig.2b)

The near field pattern can be defined as a $LP_{l,m}$ mode with l=0,1 and m=2-4. The concentric rings in the far field stem from the contribution of the $LP_{0,m}$ modes, resembling the diffraction pattern at a ring aperture. For an obscuration ratio of 0.75 about 30% of the total intensity should be contained in the central maximum. The experimental value is ~ 10%, which means a considerable amount of fundamental mode radiation.

The output power is shown as a function of diode pump power in Fig. 3.

From a FINDLAY-CLAY /2/ analysis the optimum output coupling was determined to 45%. The slope efficiency σ_s for the example shown is equal to 12 %.

The emitted radiation has a preferred state of polarization, with a ratio of the main axis of 10:1. The beam parameter product amounts to 3 mm mrad.

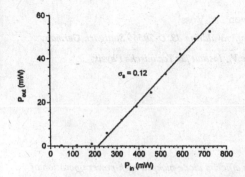

Fig.3 M-fiber laser output power vs. absorbed pump power ($R_1=1$, $R_2=0.6$)

The refinements planned for the next future concern a better fiber endface finish and a improved conversion efficiency. For this, we estimate a realistic output power of 160mW and a slope efficiency of 34% for 600mW pump power.

In summary, we have proposed and tested a new design for a fiber laser which promises comparable performances related to the double clad fiber. The MPF may serve as a potential candidate for a high power, high brightness fiber laser source. MPFs are light guides with a near-surface guiding structure being advantageous related to fiber coupling, modulation and distributed reflection.

This work is sponsored by BMFT under contract 13N 6363.

The phosphate glass fiber has been manufactured by FIBERWARE GmbH, Berlin, the silica glass fiber by Opt.El. Res. Cent., Univ. Southampton.

References

/1/ R.H.Stolen
Appl. Opt. 14(7)1975, 1533

/2/ cf. W.Koechner, Solid State Laser Engineering (Springer, Berlin, 1992)

New Concepts for High Power Diode Pumped Solid State Lasers

A. Giesen[1], U. Brauch[2], M. Karszewski[1], Chr. Stewen[1], A. Voss[1]

[1]Institut für Strahlwerkzeuge, Universität Stuttgart, Pfaffenwaldring 43, D-70569 Stuttgart, Germany

[2]Deutsche Forschungsanstalt für Luft- und Raumfahrt e.V., Institut für Technische Physik,
Pfaffenwaldring 38-40, D-70569 Stuttgart, Germany

Abstract:

The new, scalable thin disc laser concept allows efficient diode-pumped high-power operation of solid state lasers with high beam quality. This concept is very useful to operate quasi-three level systems like Yb:YAG. The use of a very thin crystal disc with one face mounted on a heat sink together with the low heat generation of Yb:YAG (< 11 %) enables high pump power densities (> 5 kW/cm²) required for efficient operation without strong thermal degradation or optical distortion. Compact diode-pumped solid-state lasers in the kilowatt range seem to be achievable by increasing the pump beam diameter and/or by using several crystal discs.

In this contribution the latest results from an Yb:YAG thin disc laser are presented. The scaling laws to high output power levels are presented and discussed in detail. Simulations of the efficiency are also presented. The broad tuning range of Yb:YAG shows the potential of this material for femtosecond pulse generation.

1 Introduction

Ytterbium[3+] - as an ion with a f ^{13} electron configuration - is a simple 2 level system with a $^2F_{7/2}$ ground state and a $^2F_{5/2}$ excited state [1] split in a crystal field into 4 and 3 sublevels, respectively. Compared to other rare earth ions, the electron-phonon coupling is extremely strong which leads to broad zero-phonon transitions and phonon sidebands making it still difficult to assign the correct electronic transitions [2-6].

Reports on laser operation were already published in the sixties [4, 7-13] mainly in garnets using pulsed excitation and liquid nitrogen cooling in order to obtain a nearly ideal 4 level system. Most of the crystals were codoped with Nd^{3+} to impove the absorption efficiency for the black-body-like radiation of the flashlamps. More efficient laser operation was only possible with spectrally narrow light sources: LEDs with their relatively low power density still allowed only pulsed operation at 80 K [14] while dye, Ti:Sapphire, or diode pump lasers made room-temperature operation possible for the first time. Fiber or planar waveguide geometries allow to enhance the confinement of pump and laser beam.

The first room-temperature operated Yb laser was a 7-m single-mode silica fiber doped with 600 ppm Yb. An output power of 4.5 mW at 40 mW absorbed pump power and a tuning range from 1015 to 1140 nm was achieved [15]. With 100-m fiber 660 mW output power were reported with a launch efficiency of 45 %, a slope efficiency of 62 % and a threshold of 270 mW [16]. Using the technique of clad-pumping the launch efficiency can be drastically increased: With a Ti:Sapphire laser (750 mW power incident on the fiber at 975 nm) a maximum output power of 430 mW and a slope efficiency of 80 % was achieved[17]. In planar (epitaxially grown) Yb:YAG waveguides slope efficiencies > 77 % with thresholds < 43 mW and maximum output powers of 270 mW in three lobes were realized [18]. The first room-temperature operation of a bulk Yb:YAG crystal was published by Lacovara et al. [19]. They achieved 71 mW threshold and 46 mW output with an absorbed pump power of 178 mW from a Ti:Sapphire laser in a longitudinal pump configuration. The slope efficiency of 56 % was predicted to increase to 70 % at 3 times above threshold because of the quasi-three-level nature and the Gaussion intensity distribution of pump and laser beam. The corresponding data for diode pumping were 234 mW threshold, 23 mW output, and 345 mW pump power, respectively. The authors pointed out that Yb:YAG should be a very interesting material for both high efficiency and high power InGaAs diode pumped solid-state lasers because of the broad absorption regime around 940 nm, the small heat generation of < 11 % [20] and the absence of concentration quenching.

A survey at the LLNL [21,22] of a number of host crystals for Yb^{3+} showed that the fluorapatite and its homologues have the lowest pump saturation densities (2.0 kW/cm^2 for FAP compared to 28 kW/cm^2 for YAG) and the highest emission cross sections with similar lifetimes but reduced absorption and emission linewidths. A drawback is the extremely poor thermal conductivity of 2 $Wm^{-1}K^{-1}$ (YAG: 10 $Wm^{-1}K^{-1}$) making FAP only useful for low-power or low-duty-cycle high-energy operation. An Yb:S-FAP diode-pumped through an external lensing duct (23 kW/cm^2) delivered 0.55 J in 1-ms long pulses with a repetition rate of 1 Hz [23] and a slope efficiency of 30 %.

The fluorescence lifetime of Yb:YAG of nearly 1 ms and the broad emission linewidth of several tens of nanometers make this system attractive for Q-switched and mode-locked operation. Fan et al. [24] realized an InGaAs-diode-pumped Yb:YAG laser with 72 μJ/pulse with a pulse length of 11 ns. In mode-locking operation an average output power of 390 mW with 1,3 W incident on the laser crystal was demonstrated, but the pulse length of 80 ps was longer than expected from the laser linewidth of 7 GHz [25].

All the experiments show the potential of Yb^{3+} lasers with optical laser efficiency of more than 50 %. Prerequisites are a very high pump power density of > 5 kW/cm^2 in case of Yb^{3+}:YAG to be sufficiently far above threshold and a good beam overlap of pump and laser beam. This has been realized first with a longitudinal pumping scheme, removing the heat radially from the crystal.

Because of the temperature sensitivity of Yb:YAG and the required high pump power density an improved cooling design is necessary. For this geometries with high surface to volume ratio e. g. fiber, slab or disc have been proposed [19, 26-30].

The maximum ouput power of a single-mode fiber laser is limited because of the extremely high power density in the fiber core of some 5 µm diameter. On the other hand, a slab should have a thickness of well below 1 mm. Therefore it is difficult to find a reasonable design for scaling the laser output power with good beam quality simultaneously. The thin-disc geometry used as an active mirror design [26-32] seems to be the best choice. A disc of a few hundred µm thickness of the laser material is connected onto a heat sink. The disc, high-reflection coated on the back side, acts as an active mirror. Therefore the temperature gradient is approximately collinear to the laser radiation which reduces the thermal effects (thermal lensing, stress induced birefringence) drastically. The power-density-handling capability can be increased by further reducing the disc thickness provided the heat transfer into the heat sink is sufficient. The total power can be scaled by increasing the disc diameter and / or putting several discs in series [28-32]. Quasi-longitudinal pumping allows to match the laser-mode and pumped volume which is necessary for an efficient single-mode operation.

2 Thin disc laser design

The laser active element is a thin disc (7 mm diameter, 3 mm thick) of YAG material doped with 11 at % Yb, AR coated on the front and HR coated on the back side for both the pump and laser wavelength. It is mounted on a copper heat sink with an indium foil. Quasi-longitudinal pumping allowed to match the laser-mode and pumped volume necessary for an efficient single-mode operation. The problem of the short absorption length of twice the disc thickness was solved by

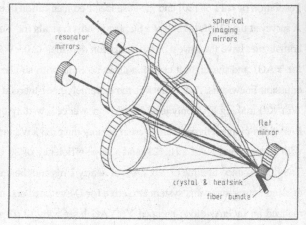

Fig. 1: Three-dimensional view of the pumping scheme and the resonator of the diode-pumped thin-disc laser.

using pump optics that allowed four double passes in the crystal by imaging the transmitted pump light back onto the crystal (Fig. 1) [28-32]. For pumping the crystal InGaAs laser diodes (Siemens) were used each coupled into a silica-silica fiber (core ⌀ 125 µm, cladding ⌀ 140 µm, N.A. = 0.4) giving 1.2 W maximum output power at the end of the fiber. Bundles of 19 and 37 fibers resulted in pumped diameters of 0.7 and 1.0 mm, respectively, corresponding to a power density of nearly 6 kW/cm².

3 Experimental results

3.1 Output power and scalability

To demonstrate the scaling of the output power with the diameter of the pumped area the disc was pumped using fiber bundles with 19 and 37 fibers, respectivly. The experimental results (laser output power versus total pump power for different temperatures) are shown in Fig. 2. The maximum output power of 11 W for the small pumped area has been obtained with a pump power of approximately 24 W. The slope efficiency at a temperature of T_C = -70 °C was η_{sl} = 63 %. The maximum output power using 37 fibers was 21 W with a pump power of 45.5 W. The corresponding slope efficiency η_{sl} was 66 %, and a beam quality of M^2 = 1.70 was measured with a Coherent Modemaster. As expected for a quasi-three-level system the laser threshold is proportional to the pumped volume. Increasing the current of the diodes the central wavelength moves towards longer wavelengths, because of the rising diode temperature. This leads to a reduction of the slope efficiency at higher pump powers.

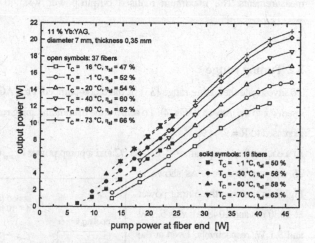

Fig. 2: *Output power versus pump power at various temperaures of the cooling fluid for two pumped areas*

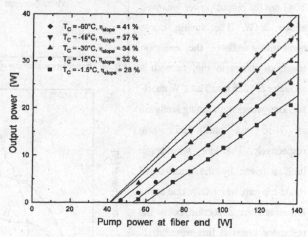

Fig. 3: *Output power versus pump power for various temperatures*

Further increasing the output power of this laser system has been realised by using 17 fiber coupled laser diode bars (Jenoptik Laserdiode) with a nominal output power of 10 W out of a 800 µm core diameter silica fiber (N. A. = 0.20). With a conical fused silica rod (taper) the pump power density was increased by a factor of three giving a maximum pump power of 136 W (pump beam diameter 2.7 mm, power density 2.4 kW/cm², N. A. = 0.35). Fig. 3 shows the experimental results. The discussion of the total efficiency is quite difficult because the degradation of the pump power was 15 % during the measurements. The maximum realised output power was 40 W at a cooling fluid temperature of $T_C = -60$ °C.

3.2 Tuning range

To investigate the tuning range [33] of the diode-pumped Yb:YAG laser, a 3-plate birefringent filter was inserted into the resonator (Fig. 4). To achieve the maximum tuning range the output coupler reflectivity was increased to R = 98.4 %.

At a cooling fluid temperature T_C of -55 °C and a pump power P_{pump} of 26.5 W a tuning range of more than 35 nm was observed. As shown in Fig. 5, in this case the output power at 1030 nm and 1048 nm was 5.3 W and 3.1 W, respectively. Even at the minimum between these peaks near 1043 nm the output power was well above 2 W. The tuning curve qualitatively reflects the emission spectrum shown in Fig. 6, with a shoulder at 1025 nm. The 1 W short- and long-wavelength tuning limits are at 1018 nm and 1053 nm, respectively. The long wavelength limit is given by the small-signal round-trip gain, which falls, due to the low emission cross section, below the resonator losses at this wavelength. When moving the birefringent filter beyond the short wavelength limit, the laser emission wavelength jumps to values near 1030 nm, indicating the insufficient suppression of this

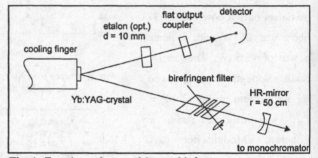

Fig. 4: Experimental setup of the tunable laser

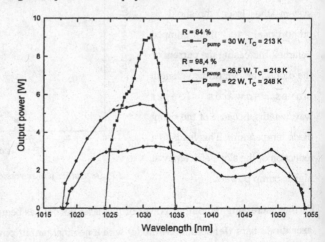

Fig. 5: Tuning curves for different output coupler reflectivities (R) and cooling fluid temperatures T_C

wavelength by the filter. Therefore, the tuning range may be extended further by adding a blocking filter for 1025-1035 nm.

At a higher temperature (-25°C) and a reduced pump power of 22 W a flatter curve with 3.3 W at 1030 nm and 2.2 W at 1048 nm is obtained (see Fig. 5). The tuning range then is slightly reduced and shifted towards longer wavelengths. At a cooling fluid temperature of -60°C and maximum

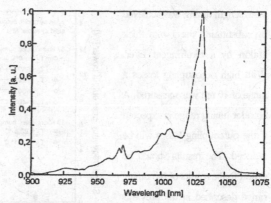

Fig. 6: Room temperature emission spectrum of Yb:YAG excited at 969 nm.

pump power (30 W) the best tuning range obtained with high output coupling (R = 84 %) was 1024 nm to 1034 nm (10 nm) with a maximum output power of 9.1 W (see Fig. 5).

The emission bandwidth of the tunable Yb:YAG laser was measured with a 0.5 m Jobin Yvon monochromator, operated in the second order. The value obtained (0.07 nm) is assumed to be resolution limited.

3.3 Single frequency operation

The best results were achieved using the crystal as one end mirror of the resonator. In this case strong spatial hole burning occurs within the crystal [34]. The crystal itself, even with an anti-reflective coating on the front side, served as a weak etalon. The experimental setup is shown in Fig. 7.

With only two etalons (100 μm thick uncoated, and 1 mm thick, 30 % reflection) single frequency operation has been realised.

The experimental results are shown in Fig. 8. At a pump power of 45.5 W the maximum single frequency output power was 14.5 W with a beam quality of M²

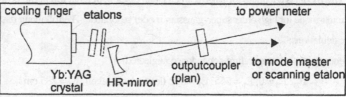

Fig. 7: Experimental setup of the single frequency laser

= 1.02. This has to be compared to Fig. 2 where the output power is 18.5 W (temperature -40 °C) with a beam quality of M² = 1.70. The linewidth measured with the scanning etalon is less than 10 MHz.

The scalability of the output power is realised by increasing the pumped area at constant pump power density as shown above. If the active mirror is used as end mirror, the beam waist is on the crystal and the resonator length has to be scaled as well. The decrease of the slope efficiency shown in Fig. 8 at high pump power is caused by the wavelength shift of the pump diodes as described in chapter 3.1.

Wavelength tuning was achieved by substituting the 1 mm thick etalon by a birefringent filter. With high outcoupling losses a range of 10 nm was measured. A broader tuning range is expected if the outcoupling losses will be reduced. The results should be similar to the results of the tuning range described in the previous chapter.

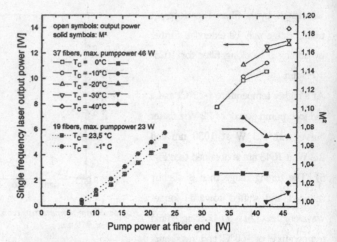

Fig. 8: Single frequency output power and beam quality for different temperature of cooling fluid.

4 Yb:YAG thin disc laser model

In order to describe, predict, and optimize the behaviour of the Yb:YAG thin disc laser system, especially at power levels not available experimentally yet, a selfconsistent model of this quasi-three-level system has been developed which includes the following features:

- 1-dimensional calculation along the disc axis
- 7-level system with thermal redistribution of the population densities
- temperature dependence of absorption/emission cross sections and heat conductivity
- selfconsistent calculation of absorption, resonator photon flux density, and temperature distribution
- optimization of output coupling and doping concentration

This model is adequate for large diameter thin discs with a flat top pump power density profile and a highly multi-mode (or high order super-gaussian mode) resonator. The following data of Yb:YAG are used for the calculations:

energy levels (temperature dependence neglected):

$^2F_{7/2}$: $E_0 = 0$, $E_1 = 565$, $E_2 = 612$ (lower laser level), $E_3 = 785$ cm^{-1}

$^2F_{5/2}$: $E_4 = 10327$ (upper laser level), $E_5 = 10624$ (pump level), $E_6 = 10679$ cm^{-1}

peak stimulated emission cross section:

$\sigma_{em}(T) = 3.25 \cdot 10^{-20}$ cm^2 · 235 K/(T - 60 K) for absolute temperature T > 60 K

(fit to data of Sumida and Fan [34] plus correction for thermal population)

peak absorption cross section:

$\sigma_{abs}(T) = 1.2 \cdot 10^{-20}$ cm^2 / (1 + (T/350 K)^2) (fit to data of Bazin [35])

radiative lifetime (temperature independent, [34]):

$\tau_{sp} = 9.5 \cdot 10^{-4}$ s

heat conductivity:

$\lambda_{th}(T) = 13 \text{ Wm}^{-1}\text{K}^{-1} \cdot 235 \text{ K}/(T - 60 \text{ K})$ for absolute temperature $T > 60$ K

All non-saturable losses (scatter etc.) are summarized in a constant α. The stationary solution of the local rate equations under consideration of the resonator internal photon flux density s_R gives the following equation (n_i denotes the population density of the i-th level, s_p the local effective pump power density):

$$s_R \cdot \sigma_{em} \cdot (n_4 - n_2) + \frac{n_4}{\tau_{sp}} - s_P \cdot \sigma_{abs} \cdot (n_0 - n_5) = 0 \tag{1}$$

with some additional conditions describing the thermal Boltzmann-distribution within the lower ($^2F_{7/2}$) and upper ($^2F_{5/2}$) manifold, respectively. Eq. (1) can be solved for the inversion density $i = n_4 - n_0$ by eliminating n_0 using the additional conditions with the only unknown s_R, giving the *local* equilibrium condition; to obtain s_R the *global* equilibrium condition has to be considered also:

$$G \cdot L = \frac{\int\limits_{-\infty}^{\infty} \int\limits_{-\infty}^{\infty} s_R(x,y) \cdot e^{\int\limits_{0}^{l} i(x,y,z) \cdot \sigma_{em}(x,y,z) dz} \, dxdy}{\int\limits_{-\infty}^{\infty} \int\limits_{-\infty}^{\infty} s_R(x,y) dxdy} \cdot R \cdot (1 - \alpha) = 1 \tag{2}$$

where G denotes the global gain and L the global loss per roundtrip, respectively. R is the output coupler reflectivity, z the direction of the propagation, and l the length of the beam path through the active medium, including multiple passes. To solve the system of coupled equations ((1)+(2)) an iteration has to be executed, starting with a guess for $s_R(x,y)$. Next, the pump power density distribution $s_P(x,y,z)$ is calculated by taking into account the local absorption coefficient $\alpha_{abs}(x,y,z) = (n_0(x,y,z) - n_5(x,y,z)) \cdot \sigma_{abs}(x,y,z)$. In the end-pumped configuration this can be done by propagation of s_P along z, starting with the pump power distribution imaged from the fiber bundle onto the crystal and taking into account the m multiple passes of the pumplight through the crystal of thickness d (nonideal imaging ignored) :

$$s_P(x,y,z) = s_P(x,y,0) \cdot \sum_{j=0}^{\frac{m}{2}-1} \left(e^{-\int\limits_{0}^{(2j-1)d+z} \alpha_{abs}(x,y,z')dz'} + e^{-\int\limits_{0}^{(2j+1)d-z} \alpha_{abs}(x,y,z')dz'} \right) \tag{3}$$

From the distribution of absorbed power the temperature distribution is calculated. In the case of pure 1-dimensional heat conduction this is can be done by double integration of the heat sources:

$$T(x,y,z) = T(x,y,d) + \frac{E_5 - \Delta E_{42}}{\lambda_{th}} \cdot \int\limits_{d}^{z} \int\limits_{0}^{z'} \alpha_{abs}(x,y,z'') \cdot s_P(x,y,z'') dz'' dz' \tag{4}$$

T(x,y,d) denotes the temperature at the cooled face of the crystal. From the temperature distribution the thermal redistribution within the manifolds and temperature dependent constants can be derived. Using these

350

results, the inversion density i(x,y,z) is calculated by using (1); the population of the manifolds has to be updated accordingly. Finally, the amplification of $s_R(x,y)$ is determined using (2). Unless convergence is achieved, these steps are repeated, starting from the calculation of absorption.

The program calculates the pump power density required for a given temperature difference within the crystal, the optimized doping concentration (found by systematic search), the optimized output coupler, the resulting absorption efficiency, all population densities (averaged along z), the effective local pump power density achieved by multiple passes, and the obtained optical-to-optical efficiency. For the calculations the crystal is divided into 100 discs of equal thickness, and to each of them a disrete set of values is assigned.

If not stated otherwise the following calculations were performed for a crystal thickness of 0.3 mm, a temperature difference in the crystal of 40 K, for 8 passes of the pump light and for internal resonator losses of 1 %.

Fig. 9 shows the calculated optical-to-optical efficiency as a function of the pump power density and the heat sink temperature. Doping concentration and output coupler are optimized separately - for each operational condition. At a heat sink temperature of 200 K the threshold pump power density is quite low (approx. 150 W/cm²) and the efficiency approaches 75 % at pump power densities above 5 kW/cm². With increasing heat sink temperature the threshold increases nearly exponentially (due to the thermal population of the lower laser level, note the logarithmic scale) whereas the efficiency drops remarkably (e.g. 58.5 % at 300 K with 6.2 kW/cm²).

In Fig. 10 the optical-to-optical efficiency is plotted versus the crystal thickness with the heat sink temperature as additional parameter. The pump power density is kept constant while varying the thickness. The efficiency drops almost linearly with the thickness. Therefore, and to minimize

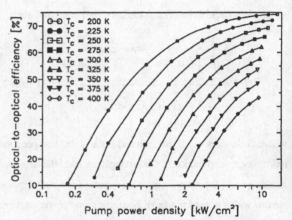

Fig. 9: *Optical-to-optical efficiency of the 1-dim Yb:YAG thin disc laser vs. pump power density for different heat sink temperatures, crystal thickness 0.3 mm.*

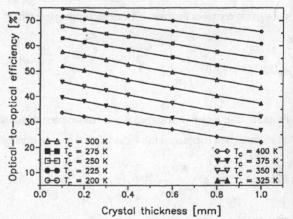

Fig. 10: *Optical-to-optical efficiency vs. thickness of the Yb:YAG disc for different heat sink temperatures*

thermal lensing, the crystal should be kept as thin as possible to get good results. The lower limit for the crystal thickness is given by the required doping concentration, which increases inversely proportional to the thickness and by manufacturing and handling problems.

The influence of the internal losses of the resonator is shown in Fig. 11. The heat sink temperature dependence is also given. The results indicate that even small losses reduce the efficiency strongly. For example, at 200 K 0.01 % loss leads to 84.7 % efficiency, which drops to 81.7 % at 0.1 % and to 72.8 % at 1 % loss. Therefore, extreme efforts should be made to reduce the losses to the lowest possible value. On the other hand, experimental determination of the optimal output coupler can give an estimation of the losses since the calculated optimal ouput coupler strongly depends on the losses.

Finally, Fig. 12 shows the effect of different numbers of passes of the

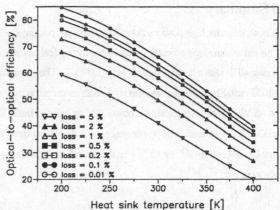

Fig. 11: Optical efficiency of the Yb:YAG thin disc laser vs. heat sink temperature for different internal losses

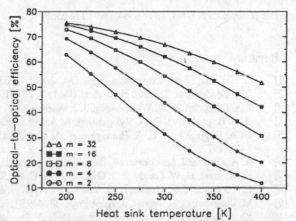

Fig. 12: Optical efficiency of the Yb:YAG thin disc laser vs. temperature for different numbers of multiple passes

pumplight through the crystal on the efficiency as a function of temperature. At 200 K an 8-fold pass gives a near-optimum result (72.8 %), whereas at 300 K a 32-fold pass gives 66.9 %, a 16-fold pass 62.0 %, and an 8-fold pass only 54.5 % efficiency.

In conclusion, the heat sink temperature plays a major role for the obtainable efficiency. The higher the temperature, the more critical the other factors are. The comparison with the experimental results achieved so far indicates some additional effects not mentioned in the model, which reduce the efficiency. Some of the more important are:

-   heat resistance of the HR-coating on the cooled face of the crystal
-   heat resistance of the cooling finger
-   some additional heat generation within the crystal due to upconversion induced by impurities (strongly depending on the population of the upper laser level and on the doping concentration)

5 Summary

A new design for high-power solid-state lasers was presented and successfully demonstrated with Yb:YAG: The maximum slope efficiency was 68 %, the optical to optical efficiency 50 %, and the maximum output power 40 W with a beamquality of $M^2 = 1.0 - 1.7$. The maximum tuning range extended over 35 nm (1018 - 1053 nm). Single-frequency operation with an output power of 14.5 W was achieved with a pump power of 45 W. This concept allows the power scaling into the multi 100 W range with similar efficiency and beam quality. It is also useful for power scaling of other quasi-three level systems like Tm^{3+}, Ho^{3+}, $Nd^{3+}(^4F_{9/2})$ and tunable lasers like Ti^{3+} or Cr^{3+}. For the near future the realisation of a multi-100-W Yb:YAG system is planned.

Acknowledgement

This work was supported by the Bundesministerium für Bildung, Wissenschaft, Forschung und Technologie under contract 13 N 6093, 13 N 6364, and 13 N 6365.

Literature

[1] G. H. Dieke, H. M. Crosswhite, Appl. Opt. 2, 675 (1963)

[2] J. A. Koningstein, Theoret. chim. Acta (Berl.) 3, 271 (1975)

[3] J. A. Buchanan, K. A. Wickersheim, J. J. Pearson, G. F. Herrmann, Phys. Rev. 159, 245 (1967)

[4] G. A. Bogomolova, D. N. Vylegzhanin, A. A. Kaminskii, Sov. Phys.-JETP 42, 440 (1976)

[5] G. A. Bogomolova, L. A. Bumagina, A. A. Kaminskii, B. Z. Malkin, Sov. Phys. Solid State 19, 1428 (1977)

[6] A. A. Kaminskii, Laser crystals, 2nd ed., Springer-Verlag, Berlin, Heidelberg 1990

[7] H. W. Etzel, H. W. Gandy, R. J. Ginther, Appl. Opt. 1, 534 (1962)

[8] L. F. Johnson, J. E. Geusic, L. G. Van Uitert, Appl. Phys. Lett. 7, 127 (1965)

[9] M. Robinson, C. K. Asawa, J. Appl. Phys. 38, 4495 (1967)

[10] Kh. S. Bagdasarov, G. A. Bogomolova, D. N. Vylegzhanin, A. A. Kaminskii; A. M. Kevorkov, A. G. Petrosyan, A. M. Prohkorov, Dok. Akad. Nauk SSSR 216, 1247 (1974)

[11] Kh. S. Bagdasarov, A. A. Kaminskii; A. M. Kevorkov, A. M. Prokhorov, S. E. Sarkisov, T. A. Tevosyan, Sov. Phys. Dokl. 19, 592 (1975)

[12] Kh. S. Bagdasarov, A. A. Kaminskii; A. M. Kevorkov, A. M. Prokhorov, Dok. Akad. Nauk SSSR 218, 810 (1974)

[13] A. A. Kaminskii, T. I. Butaeva, A. M. Kevorkov, V. A. Fedorov, A. G. Petrosyan, M. M. Gritsenko, Inorg. Mater. (USSR) 12, 1238 (1976)

[14] A. R. Reinberg, L. A. Riseberg, R. M. Brown, R. W. Wacker, W. C. Holton, Appl. Phys. Lett. 10, 11 (1971)

[15] D. C. Hanna, R. M. Percival, I. R. Perry, R. G. Smart, P. J. Suni, J. E. Townsend, A. C. Tropper, Electron. Lett. 24, 1111 (1988)

[16] C. J. Mackechnie, W. L. Barnes, R. J. Carman, J. E. Townsend, D. C. Hanna, OSA Proceedings on Advanced Solid-State Lasers (Optical Society of America, Washington, DC 1993), Vol. 15, p. 192

[17] D. C. Hanna, H. M. Pask, J. E. Townsend, J. L. Archambault, L. Reekie, A. C. Tropper, in Advanced Solid-State Lasers, Technical Digest, 1994 (Optical Society of America, Washington, DC 1994), p. 192

[18]   D. Pelenc, B. Chambaz, I. Chartier, B. Ferrand, C. Wyon, D. P. Shepherd, D. C. Hanna, A. C. Large, A. C. Tropper, Opt. Commun. **115**, 491 (1995)

[19]   P. Lacovara, H. K. Choi, C. A. Wang, R. L. Aggarwal, T. Y. Fan, Opt. Lett. **16**, 1089 (1991)

[20]   T. Y. Fan, IEEE J. Quant. Electron. **QE-29**, 1457 (1993)

[21]   L. D. DeLoach, S. A. Payne, L. L. Chase, L. K. Smith, W. L. Kway, W. F. Krupke, IEEE J. Quant. Electron. **QE-29**, 1179 (1993)

[22]   S. A. Payne, L. K. Smith, L. D. DeLoach, W. L. Kway, J. B. Tassano, W. F. Krupke, IEEE J. Quant. Electron. **QE-30**, 170 (1994)

[23]   C. D. Marshall, S. A. Payne, L. K. Smith, R. J. Beach, M. A. Emanuel, H. T. Powell, W. F. Krupke, *Advanced Solid-State Lasers,* Technical Digest, 1995 (Optical Society of America, Washington DC 1995), p. 218

[24]   T. Y. Fan, S. Klunk, G. Henein, Opt. Lett. **18**, 423 (1993)

[25]   S. R. Henion, P. A. Schulz, Conference on Lasers and Electro-Optics, OSA Technical Digest Series (Optical Society of America, Washington, DC 1992), Vol. 12, p. 540

[26]   K. Ueda, N. Uehara, Proceedings SPIE Vol. 1837 (1992), p. 336

[27]   P. J. Morris, W. Lüthy, H. P. Weber, Yu. D. Zavartsev, P. A. Studenikin, I. Shcherbakov, A. I. Zagumenyi, Opt. Commun. **111** (1994) 493

[28]   A. Giesen, H. Hügel, A. Voss, K. Wittig, U. Brauch, H. Opower, *Advanced Solid-State Lasers*, Technical Digest, 1994 (Optical Society of America, Washington, DC 1994), Postdeadline Paper

[29]   A. Giesen, H. Hügel, A. Voss, K. Wittig, U. Brauch, H. Opower, Appl. Phys. **B 58**, 365 (1994)

[30]   A. Voss, U. Brauch, K. Wittig, A. Giesen, *SPIE Proceedings of the 9th Meeting on Optical Engineering*, Tel Aviv 1994

[31]   A. Giesen, L. Berger, U. Brauch, M. Karszewski, Chr. Stewen, A. Voss, *Advanced Solid-State Lasers,* Technical Digest, 1995 (Optical Society of America, Washington DC 1995), p. 227

[32]   U. Brauch, A. Giesen, M. Karszewski, Chr. Stewen, A. Voss, Opt. Lett. **20**, 713 (1995)

[33]   G. J. Kintz, T. Baer, IEEE J. Quant. Electron. **26**, 1457 (1990)

[34]   D.S. Sumida, T.Y. Fan, OSA Processings on Advanced Solid-State Lasers **20** 100 (1994).

[35]   N. Bazin, Univ. Stuttgart, II. Phys. Inst., private communications

High Effects High Power Diode-Pumped Nd:YAG Laser

U. J. Greiner, H. H. Klingenberg
Institut für Technische Physik, Deutsche Forschungsanstalt für Luft- und Raumfahrt
D-70503 Stuttgart, Germany

D. R. Walker, C, J. Flood, H. M. van Driel
University of Toronto
Toronto, Canada

Laser diode arrays are an efficient light source for optically pumping solid state laser crystals. With the availability of ever larger laser diode arrays and stacks of arrays, there is an increasing effort in using laser diodes for pumping high power lasers. In this paper we report on the results of scaling a diode pumped Nd:YAG laser. Starting out with a pump power of 60 W an output power of 12 W TEM_{00} mode was achieved. Applying pump powers of 90 W yielded 14.5 W in a TEM_{00} mode 1.1 times diffraction limited.

The experimental setup of the pump module was described in detail by GREINER[1] et al. and WALKER[2] et al. Therefore only a brief description will be given here. In the module the pump light is focussed into the laser rod with elliptically shaped mirrors. Three such units of laser diodes and mirrors are arranged symmetrically around the laser rod, separated by 120 degrees. The laser rod itself is mounted inside a flow tube to ensure efficient cooling, required at high pump powers. The flow tube was coated with gold reflector stripes on the side opposite to the elliptical mirrors to reflect the pump light not absorbed on the first pass.

The laser diodes arrays, were placed on copper cooling blocks to keep the temperatur at approximately 25 ^{0}C. At this temperatur the emission wavelength of the diodes matched the absorption band of the Nd:YAG crystal. Furthermore, the diodes and mirrors were placed on translation stages to adjust the pump light distribution inside the laser rod.

In the first module two 10 W laser diode arrays (Siemens Heimann) were placed next to each other on each unit. Therefore, a total pump power of 60 W was available. The laser rod used in this modules was a 1.1% doped Nd:YAG crystal with a diameter of 2 mm and a length of 40 mm.

In the second, scaled version we added another 10 W laser diode array to each unit. A total of 90 W of pump power was now available and the laser rod had to be at least 10 mm longer. In this module two different laser rods wer tested: first, a 1.1 % doped Nd:YAG rod with a diameter of 3 mm and a length of 50 mm and second, a laser rod with a diameter of 2 mm and a length of 60 mm.

The heat generated inside laser crystals while being pumped results in thermal distortions[3], in the best case a spherical lens. For an easy method to compensate the thermal lens in the laser rod we used an adjustable curvature mirror (ACM) described in earlier[4]. The results of the two systems using the ACM for the 2 mm diameter laser rods and a fixed resonator for the 3 mm diameter rod are listed in Table 1. In figure 1 the input vs. output curves for the two pump modules using a laser rod with a diameter of 2 mm and the ACM is shown.

The experiments show the following: adding an additional row of laser diodes to scale the system to higher powers decreases the optical to optical efficiency from

Table 1. Summary of the experimental results of the two different modules and the different laser crystals.

Pump Power [W]	Laser Rod Dia.[mm] × L [mm]	Laser Power [W]	Efficiency %
60	2 × 40	12, TEM_{00}	20
90	2 × 60	14.5, TEM_{00}	16
90	3 × 50	10, TEM_{00}	11

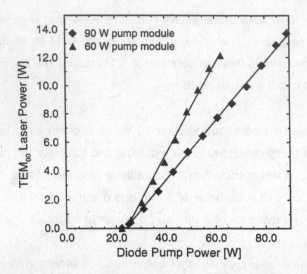

Fig. 1 *The input vs. output power with the ACM in the case of the 60 W pump module and the 90 W pump module, both using a laser rod with a diameter of 2mm.*

20 % down to 16 % for the case of the 2 mm laser rod. For a 3 mm laser rod the drop in efficiency was even larger. These effects can be understood if one considers the thermal effects in the laser crystal. The dioptric power of the laser rod is dependent on several factors: the total pump power, the pump light distribution, and in general the diameter of the laser rod, and also the length of the laser rod. Adding another row of laser diodes, thereby lengthening the crystal, but keeping the same pump light distribution, will produce a stronger thermal lens in the crystal. Furthermore, phase front errors not correctable with a spherical lens become more relevant. This explains the slightly lower efficiency of the 90 W module with the 2 mm diameter laser rod.

Changing the diameter of the laser rod to 3 mm, while keeping the pump light distribution and providing a similar pump density, results in a larger temperature gradient. This will increase thermal distortions in the laser crystal. When the pump light distribution is changed towards a larger area in the crystal, the thermal distortions will be smaller. On the other hand, the pump density will also decrease, lowering the output power. This effect can be seen in the case of the 90 W pump module with the 3 mm diameter laser rod.

The experiments show, that in principal a scaling is possible, however, a decrease in efficiency will result due to thermal effects in the laser rod. Nevertheless, with the 90 W pump module we achieved 14.5 W in a TEM_{00} mode 1.1 times diffraction limited which results in a optical to optical efficiency of 16 %.

This research project was supported by the Ministerium für Wirtschaft, Mittelstand und Technologie, Baden- Württemberg, Germany and the Ontario Ministry of Trade and Economic Development, Canada.

References

1. U. J. Greiner, H. H. Klingenberg, D. R. Walker, C. J. Flood, H. M. van Driel, Appl. Phys. B **58** 393 (1994)
2. D. R, Walker, C. J. Flood, H. M. van Driel, U. J. Greiner, H. H. Klingenberg, Opt. Lett. **19,** 1055 (1994)
3. W. Koechner, *Solid-State Laser Engineering*, 3rd Ed. (Springer-Verlag, Berlin 1992), p. 381
4. U. J. Greiner, H. H. Klingenberg, Opt. Lett. **19**, 1207 (1994)

Lateral Gain Optimization of a Diode Pumped Nd:YAG Laser

Keming Du, Yutian Lu*, R. Wester, M. Quade and P. Loosen

FhG-ILT, Aachen, Germany; *Institute for Optics and Fine Mechanics CSA, Shanghai, China

Performances of a diode side pumped Nd:YAG laser have been studied. Au-coated elliptical mirrors with high energy transfer efficiency (up to 98%) were used to focus the diode emission on the Nd:YAG rod. With respect to efficiency, beam quality and the depolarisation loss the lateral gain profile was optimized by changing the distance from the secondary focus lines of elliptical mirrors to the rod axis. 60W multimode power was obtained with diode power of 190W, which corresponds to a slope efficiency of 37.5%. The pump power threshold in TEM_{00} mode operation was 50W. 22W TEM_{00} mode power was achieved for a pump power of 160W. The corresponding slope efficiency is 20%.

1 Introduction

There are two principal different diode pumping configurations. These are end-pumping and side-pumping [1]. In general the side-pumping configuration is used to scale the output power. To obtain highly efficient laser operation with high beam quality it is essential to choose focusing optics with high energy transfer efficiency and to match the lateral pump power distribution and the laser mode. The use of diamond turned Cu elliptical mirrors permits energy transfer efficiencies of >95% [2]. In our side pumped design, Au-coated elliptical mirrors with an energy transfer efficiency of up to 98% are used. In this paper we will discuss the theoretical and experimental results obtained with the side-pumped Nd:YAG-laser.

2 Diode pumping module

As shown in fig. 1, the pumping module consists of six 1cm diode bars and six elliptical cylinder mirrors, which are arranged symmetrically around the rod. Each diode bar is located on the primary focus line of the mirror and its emission is focused on the secondary focus line. The lateral gain profile can be adjusted by changing Δx, the distance from the secondary focus lines of elliptical mirrors to the rod axis. Two such pumping modules are used in our laser. In order to minimize the azimuthal gain modulations and the heat-induced inhomogeneous inside the rod, the two pumping modules are rotated 30 degrees towards each other.

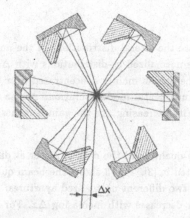

Fig. 1 A diode pumping module with 6 diode laser bars and elliptical cylinder mirrors as focusing optics.

x = 5mm   x = 10mm   x = 30mm

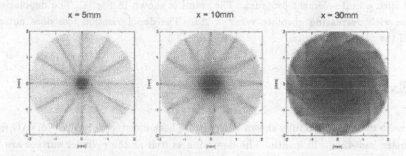

Fig. 2 The lateral pump power distributions for different Δx.

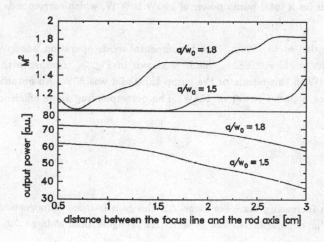

Fig. 3 The output power and the beam quality as a function of the distance Δx.

3 Theory

A ray-tracing program was developed to calculate the lateral distributions of the pump power and the small signal gain (g_0). Fig. 2 shows the normalized g_0-distributions with $\Delta x = 5$mm, 10mm and 30mm, respectively. For small Δx the gain is mainly concentrated near to the rod axis. The high axial gain is advantageous for fundamental mode operation. At the same time there is a strong azimuthal gain modulation. With increasing Δx the gain becomes more and more homogeneously distributed.

With the computed gain distributions the beam quality and the output power at different Δx are obtained using the multimode model presented in [3]. Fig. 3 shows the beam qulaity (M^2) and the output power in dependence on Δx for two different normalized apertures. In general the value of M^2 increases and the output power decreases with increasing Δx. For $a/w = 1.8$ there is an optimal Δx regarding the beam quality.

Starting from the pump power distributions the dependence of depolarisation losses on Δx is studied with a finite-element-program. The result is shown in Fig. 4. The depolarisation loss increases with decreasing absolute value of Δx. The depolarisation loss does not change so much, if the $abs(\Delta x) > 5$mm.

4 Experiments

A laser with two pump modules shown in Fig. 1 has been constructed. The Nd:YAG rod is 4mm in diameter and 42mm in length. The flow tube as well as the cylinder surface are uncoated. To check the pump efficiency a plane/plane resonator with a length of 12cm was used. The transmission of the output coupler was 9%. With this resonator the pumping threshold is 30W. The multimode output power for a total pump power of 190W is 60W, which corresponds to a slope efficiency of 37.5%.

The resonator was then lengthened to 150cm. The fundamental mode operation was investigated. The far-field intensity profile of TEM_{00} mode is shown in Fig. 3, together with the multimode intensity profile. With this resonator the pump threshold was 50W. An output power of 22W was obtained for a pump power of 158W. The corresponding slope efficiency is 20%.

5 Conclusions

The axial gain reduces with increasing Δx. For large Δx the gain is more homogeneousely distributed inside the rod. The depolarisation loss at small Δx is higher than at large Δx.

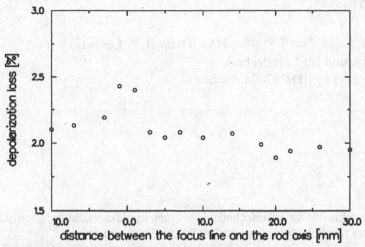

Fig. 4 The depolarisation loss as a function of the distance Δx.

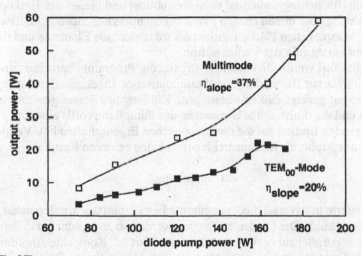

Fig. 5 The output power in fundamental and multimode operation as a function of the distance Δx.

References

[1] FAN T. Y. AND BYER R. L. Diode laser-pumped solid state lasers *IEEE J. of Quantum Electronics* 24 6(1988) 895-912

[2] WALKER D. R. ET AL: Efficient continuous-wave TEM$_{00}$ operation of a transversely diode-pumped Nd:YAG laser *Optics Letter* 19 14(1994) 1055-57

[3] DU KEMING, HERZIGER G., LOOSEN P., RÜHL F.: Computation of the statistical properties of laser light *Optical and Quantum Electronics* 24 (1992) 1095-1108

Neues Konzept für einen seitengepumpten kontaktgekühlten Festkörperlaser

D. Mühlhoff, K.-M. Du, R.Wester, H.G. Treusch, P. Loosen
Fraunhofer-Institut für Lasertechnik
Steinbachstraße 15   D - 52074 Aachen

Abstract

Das Fraunhofer Institut für Lasertechnik plant einen kontaktgekühlten, gepulsten Festkörperlaser mit Pulsleistungen von 300 W zur Feinbearbeitung.
Der Laserstab ist hier in einen zylindrischen Kühlkörper aus undotiertem YAG mit sechseckigem Querschnitt eingebettet, welcher die Pumpstrahlung führt und formt. Drei der Mantelflächensegmente sind eben ausgebildet und dienen zur Einkopplung der Pumpstrahlung. Die diesen ebenen Flächen gegenüberliegenden zylindrisch gekrümmten, verspiegelten Flächen wirken als fokussierende Elemente und dienen als Kontaktfächen zur Abfuhr der Verlustwärme.
Das Pumplichtprofil wurde mit einem 2D-ray tracing Programm berechnet und damit eine FEM-Analyse zur Berechnung der thermooptischen Eigenschaften durchgeführt. Diese Analyse hat gezeigt, daß eine homogene Kühlung des Stabes gewährleistet werden kann und die dreistrahlige Symmetrie des Pumplichtprofils einen zu vernachlässigenden Einfluß auf die thermooptischen Eigenschaften hat. Vergleichende Rechnungen mit Saphir als Kühlmantel ergeben keine besseren Resultate.

1. Einleitung

Zur Zeit realisierte transversal diodengepumpte Festkörperlaser werden meist durch eine Flowtube gekühlt. Die Diodenlaserbarren werden so angeordnet, daß deren Emissionskante parallel zur optischen Achse orientiert ist. Kompakte Anordnungen werden durch direkte Kopplung erreicht welche für eine optimale Gainverteilung sehr geringe Fertigungstoleranzen erfordern. Flexiblere Anordnungen arbeiten mit Einkoppeloptiken (Linsen oder Spiegel). Der gegenüber der direkten Kopplung erhöhte Raumbedarf wird durch die verwendeten Fokuslängen bestimmt. Die erreichbare Pumpleistungsdichte ist in beiden Fällen durch die geometrische Anordnung der Dioden begrenzt.
Alternativ zur Flowtube - Kühlung wurden Laserkonzepte mit Kontaktkühlung theoretisch und experimentell untersucht /1/. Die erprobten Geometrien zeichnen sich durch ihre effiziente Pumplichtführung bei minimaler Justierempfindlichkeit aus. Trotz der Kompaktheit der Systeme kann eine gute Kühlung gewährleistet werden.

2. Neues Konzept für einen seitengepumpten, kontaktgekühlten Festkörperlaser

Mit dem Ziel, die einfache und effiziente Kontaktkühlung mit einer hohen Pumpleistungsdichte zu verbinden, wurde vom Fraunhofer Institut für Lasertechnik eine neue Geometrie entwickelt /2/. Der Laserstab ist hier in einen zylindrischen Kühlkörper mit sechseckigem Querschnitt eingebettet (Abb. 1). Drei der Mantelflächensegmente sind eben ausgebildet und zur Einkopplung der Pumpstrahlung aus Diodenstacks antireflexbeschichtet. Die Diodenstacks sind vor diesen Einkoppelflächen so angeordnet, daß die Emissionskanten der einzelnen Arrays senkrecht zur Stabachse verlaufen. Der starken Divergenz der Pumpstrahlung in Richtung der Stabachse wird dadurch begegnet, daß sie nach Eintritt in den Kühlkörper an den Stirnflächen Totalreflexion erleidet. So wird die Pumpstrahlung bis zu den zylindrisch gekrümmten, metallisch verspiegelten Mantelflächensegmenten geführt, die den Einkoppelflächen gegenüberliegen und von dort in den Stab fokussiert. Gleichzeitig dienen die gekrümmten Flächen als Kühlkontakt zur Abfuhr der Verlustwärme.

Bei einem Durchmesser des Gesamtkristalls von nur 26 mm und einer Ausdehnung in Strahlrichtung von 12 mm können drei Diodenstacks mit je 8 Dioden angeordnet werden. Die Anzahl der Dioden pro Stablänge liegt damit dreifach höher als bei Anordnungen die mit anderer Orientierung der Dioden arbeiten.

Wesentliches Merkmal des Aufbaus ist die Grenzfläche zwischen aktivem Medium und Kühlkörper. Durch ein neuartiges Bonding-Verfahren ist es möglich geworden auf Klebemittel zu verzichten. Die entstehende Verbindungsfläche zeichnet sich aus durch verschwindenden Wärmeübergangswiderstand und durch eine Reflektivität, die allein durch die Brechungsindizes der verwendeten Materialien bestimmt ist.

Da der Laserstab vollständig durch den Kühlkörper gehalten wird, treten keine ungepumpten Endzonen auf. Diese Eigenschaft macht die Anordnung auch für Drei-Niveau-Systeme interessant.

3. Theoretische Ergebnisse

Die geometrischen Parameter der Anordnung müssen im Hinblick auf hohe Kopplungseffizienz und ein gaussähnliches Gainprofil optimiert werden. Die Abmaße der Mantelflächensegmente sind hierbei durch die Breite der Diodenstacks und durch die Divergenz der Diodenstrahlung in Slow-Richtung bestimmt.

Das Gainprofil hängt empfindlich vom Krümmungsradius des verspiegelten Mantelflächensegments ab. Bei zu kleinem Krümmungsradius treten starke Randüberhöhungen auf und der Zentralbereich des Stabes wird nur schwach gepumpt. Bei zu großem Krümmungsradius wird zwar die Mittenüberhöhung des Gainprofils erreicht, es treten jedoch starke azimutale Schwankungen auf. Der zu schließende Kompromiß muß vertretbare azimutale Schwankungen mit möglichst gaussähnlichem Profil und minimierten Randüberhöhungen verbinden (vgl. Abb. 2). Die Kopplungseffizienzen in diesem Bereich liegen mit 68 % nahe am theoretischen Maximum von 70 % (Stabdurchmesser = 3,0 mm, Absorptionslänge = 2,5 mm, einfacher Durchgang) und schwanken bei Variation des Krümmungsradius nur schwach.

364

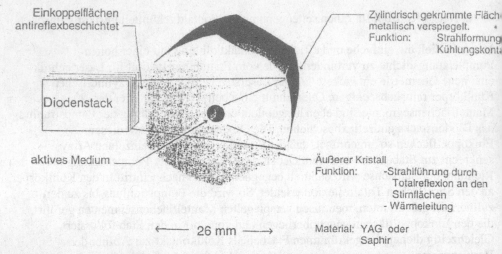

Einkoppelflächen
antireflexbeschichtet

Zylindrisch gekrümmte Fläch
metallisch verspiegelt.
Funktion: Strahlformung
                Kühlungskont.

Diodenstack

aktives Medium

Äußerer Kristall
Funktion: -Strahlführung durch
             Totalreflexion an den
             Stirnflächen
          - Wärmeleitung

←——— 26 mm ———→

Material: YAG oder
          Saphir

Abb. 1 :   Prinzipskizze

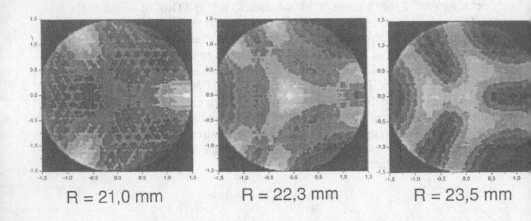

R = 21,0 mm   R = 22,3 mm   R = 23,5 mm

Abb. 2 :   Pumplichtverteilungen für verschiedene Krümmungsradien R
             Stabdurchmesser = 3,0 mm

Mit den unter diesen Aspekten bestimmten geometrischen Parametern wurde eine FE -
Analyse des Gesamtkristalls durchgeführt. Wegen der dreiachsigen Symmetrie des
Aufbaus kann die Analyse auf ein Sechstelausschnitt beschränkt werden.
Das resultierende Temperaturprofil zeigt eine im Inneren kreisförmig verlaufende
Isotherme. D.h. obwohl nur die Hälfte des Umfanges des Kühlkörpers als
Kühlkontaktflächen ausgebildet sind, kann eine homogene Kühlung des Laserstabes
sichergestellt werden.

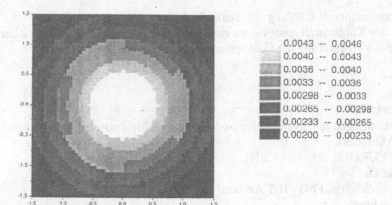

	0.0043 -- 0.0046
	0.0040 -- 0.0043
	0.0036 -- 0.0040
	0.0033 -- 0.0036
	0.00298 -- 0.0033
	0.00265 -- 0.00298
	0.00233 -- 0.00265
	0.00200 -- 0.00233

Abb. 3 :   Änderung der optischen Weglänge über den Bereich des aktiven
Mediums [mm]

Die maximale Temperaturerhöhung im Stab liegt für eine zeitlich gemittelte
Wärmeleistung von 60 Watt und für eine Stablänge von 20 mm bei 33K (67K) für
Saphir (YAG) als Kühlmantel.
Die Untersuchung der optischen Weglängenänderung für den Stab, welche durch die
Temperaturverteilung bestimmt wird zeigt eine fast perfekte Radialsymmetrie. Die
Dreiersymmetrie der Pumplichtverteilung ist nur noch in Ansätzen vorhanden. Die
azimutalen Schwankungen der optischen Weglänge sind geringer als 0,1 µm und damit
im Bereich der durch den Fertigungsprozeß bedingten Oberflächentoleranzen (Abb. 3).

Der Vergleich von undotiertem YAG und Saphir als Kühlmaterialien liefert folgende
Aussagen:
a) Durch die dreifach höhere Wärmeleitfähigkeit des Saphir wird die Kerntemperatur
des YAG-Stabes abgesenkt ohne das für die thermooptischen Eigenschaften
bestimmende Temperaturprofil zu ändern.
b) Wegen des kleineren thermischen Ausdehnungskoeffizienten des Saphir wird der
YAG-Stab bei Umhüllung mit Saphir komprimiert. Durch die Radialsymmetrie des
Temperaturprofils ist diese Kompression homogen und führt nicht zu erhöhten
Depolarisationsverlusten.
c) Durch den geringeren Brechungsindex des Saphir wirkt die Grenzfläche des Stabes
gleichzeitig strahlbegrenzend und damit nachteilig für die Strahlqualität.

Die Bewertung dieser Aspekte führt zu dem Ergebnis, daß YAG als Kühlmaterial zu
bevorzugen ist.

Die für die Beurteilung des Systems erforderlichen Berechnungen sind abgeschlossen;
das System befindet sich zur Zeit im Aufbau. Mit experimentellen Ergebnissen ist bis
Ende 1995 zu rechnen.

Da es für die ausreichende Kühlung des Stabes nicht notwendig ist die ganze Mantelfläche des Kühlkristall auszubilden sind auch andere Formgebungen denkbar um eine weitere Verbesserung der Gainverteilung zu erreichen.

Literatur:
/1/   S.D. Jackson
        Efficient thermal management for pasively cooled high power diode pumped
        Nd:YAG Lasers
        CLEO EUROPE `94   CFH1
/2/   Keming Du
        Patentapplication, FhG - ILT Aachen
/3/   W. Koechner
        Solid state laser engineering
        Springer Verlag 1992
/4/   Lü
        Dissertation, TH Berlin 1992

Diode-Pumped Slab and Rod Lasers with High cw Output Power

T. Brand, F. Hollinger

Festkoerper-Laser Institut Berlin, a division of

Laser- und Medizin-Technologie gGmbH, Berlin (LMTB)

Straße des 17. Juni 135, 10623 Berlin, Germany

1. Introduction

With the availability of reliable high power laser diode modules, diode-pumped solid-state lasers attract increasing attention in the field of material processing. In soldering and hardening applications, with its low demands on beam quality and power density, the laser diodes themselves can be used [Bey]. However, for marking and drilling of metals a peak power density beyond 10^8 W/cm² is required. When this has to be combined with a beam-quality in the range of 10-20 mm*mrad, the aid of a solid-state laser is necessary. Stacked arrays of microchannel-cooled cw bars, emitting up to 200W/cm² pumping power density, enable the realization of compact and stable high-average power solid-state lasers. Optical efficiencies of 30% are routinely obtained in high power cw and high repetition rate q-switched operation.

2. Laser-diode pumped cw slab-laser

Diode-pumped slab lasers with a cw output power of more than 1kW are under intense development now. The large side-faces of a slab crystal are well adapted

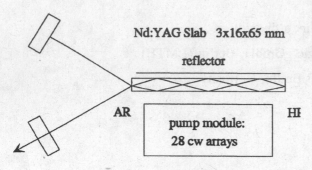

Fig. 1 Resonator scheme of the 140W rectangular-endface Nd:YAG slab laser.

to the twodimensional emitting surfaces of stacked arrays. In a first experiment, the slab crystal was pumped from one side with a microchannel cooled 500W cw stacked array as shown in fig.1. A folded stable resonator was used with two external mirrors and one high-reflecting endface of the slab. With this geometry, the stored energy is restricted by the onset of parasitic oscillations. To avoid thermooptical distortions it is crucial that the vertical dimension of the slab is illuminated homogeneously. Even slight differences in pumping wavelength and output power from the different laser-bars have to be mixed before striking the slab crystal. Limited by the available pumping power, 140W cw Nd:YAG output power have been obtained with an optical efficiency of 30% [Bra1]. To improve the beam quality, folded and unstable resonators are under investigation.

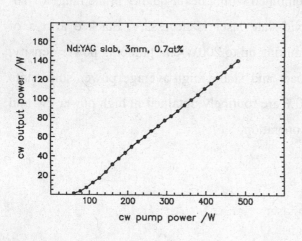

Fig. 2 Output power characteristic.

3. Side-pumped Nd:YAG rod laser

Circular pumping schemes: Usually, medium power side-pumped Nd:YAG rod lasers make use of a close-coupled circular symmetric pumping arrangement. Scaling the output power means to increase the number of diode bars and to increase their distance to the rod surface [Gol]. Coated, large aperture optics

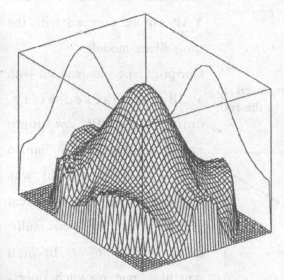

Fig. 3 Calculated absorbed power distribution in a rod with 3fold pumping symmetry.

have to be used to transfer the pumplight to the laser rod. Generally, the circular pumping scheme produces a deposition of power that has a distinct maximum in the centre (Fig. 3). It is obvious that at high average power, a concentration of the absorbed pumping power near the rod axis leads to a higher on-axis temperature and focusing and a larger stress in the crystal. This results in strong higher order aberrations of the

thermal lens and prevents the operation of high power Nd:YAG rod lasers with a near diffraction limited beam quality.

Side-pumping with nonimaging reflectors: An alternative approach to the scaling problem, is the nonimaging pump cavity (Fig. 4) [Bra2]. It permits the application of large area stacked arrays to pump small diameter laser rods. The cavity is closed by the gold-coated front of the stack to avoid losses of pumping power. Because multiple passes of the pumplight through the laser rod are possible, the absorbed power profile can be formed smooth. The influence of the wavelength-shift of the laser diodes with increasing power and of a higher

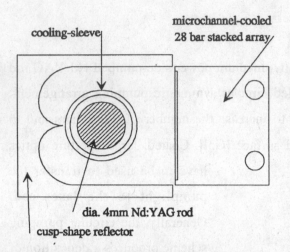

Fig. 4 Nonimaging concentrator, used to couple a stacked diode laser array to the rod.

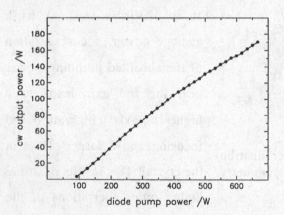

Fig. 5 Output power characteristic of the nonimaging-cavity 170W cw rod laser.

bandwidth of the emission is reduced significantly. This makes the wavelength selection less critical and reduces the costs of the pump modules. Also different Nd doped laser materials like YAG, YLF and YAP can be pumped with the same diode modules.

Our prototype was realized with a 28 bar stacked array, delivering 660W cw pump power, shifting from 805nm to 812nm. The pumplight was transferred with 26% optical efficiency to 170W cw multi-mode output power. In high repetition rate q-switch operation, 110W average power at 580W pump power are obtained.

References:
[Bey] E. Beyer, P. Loosen, S. Pflüger, Proc. Conf. "Hochleistungslaser im Maschinenbau", 25.-27. April 1995, ISL, Saint-Louis (Haut Rhin) France, Société Française des Mécaniciens.
[Bra1] T. Brand, F. Hollinger, Proc. SPIE Vol. 2382, 22 (1995).
[Gol] D. Golla, S. Knoke, W. Schöne, G. Ernst, M. Bode, A. Tünnermann, H. Welling, *300-W cw diode-laser side-pumped Nd:YAG rod laser,* Opt. Lett. 20, 1148 (1995).
[Hod] N. Hodgson, H. Weber, IEEE J. QE. 29, 2497 (1993).
[Bra2] T. Brand, *Compact, 170 W cw diode-pumped Nd:YAG rod laser, using a cusp-shape reflector,* accepted for publication in Optics Letters.

Supported by the Federal Ministry of Education, Science, Research and Technology, Contract no. 13 N 6358

Neue Konzepte für transversal diodengepumpte Slab-Laser

G. Schmidt, K.-M.Du, H. D. Hoffmann, H. G. Treusch, P. Loosen
Fraunhofer Institut für Lasertechnik
Steinbachstr.15
D - 52074 Aachen

Abstract : Ein gepulster, diodengepumpter Slab-Laser für das Feinstschneiden von Metallen wurde entwickelt, der im Gegensatz zu bisher üblichen Systemen durch die zwei schmalen Seiten gepumpt wird. Ziel der Entwicklung ist die Minimierung der thermischen Störungen. Durch Einführung eines Sandwich-Slab-Laser ist eine Reduzierung der thermischen Brechkraft um einen Faktor 5 gelungen. Die theoretischen Ergebnisse werden präsentiert.

Beschreibung des Lasers

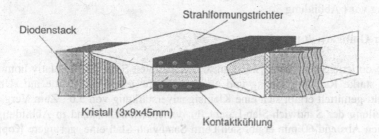

Abbildung 1: Prinzipskizze des Lasers

Die Diodenbarren stehen senkrecht zur optischen Achse des Slabs, d.h. der Strahlformungstrichter muß die Diodenstrahlung nur in Richtung der kleineren Divergenz zusammenführen und in die schmale Seite des Slabs (1.35 cm²) einkoppeln. Die breiten Seiten des Slabs sind metallisch verspiegelt und werden für eine effiziente Kontaktkühlung genutzt. Durch Anordnung der Dioden in einem Diodenstack ist eine hohe Packungsdichte gewährleistet. Bei einer Stapelhöhe der Dioden von 1.5 mm können auf jeder Seite des Nd:YAG Kristalls (3x9x45 mm) je 30 Dioden angeordnet werden. Damit steht eine Pumpleistung im Pulsbetrieb der Dioden von 3000 W auf einer Fläche von 5 cm² zur Verfügung. Der Aufbau ist unempfindlich gegen Dejustierung der Pumpquellen und mechanisch einfach zu realisieren.

Es werden zwei verschiedene Kristalle eingesetzt :

        a) homogen mit 1 % at. Nd dotierter YAG Kristall
        b) Sandwich-Slab (Abbildung 2), aufgebaut aus Schichten unterschiedlicher
           Dotierung

372

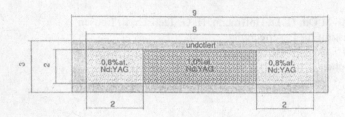

Abbildung 2: Sandwich-Slab

Optimierung

Einziger zu optimierender Parameter ist der Abstand des Diodenarray vom Slab. Mittels eines 2D ray-tracing Programms wird der Abstand und damit der Winkel des Strahlformungstrichter gegen die Horizontale unter den Aspekten einer hohen Kopplungseffizienz sowie einer möglichst homogenen Pumplichtverteilung bestimmt. Bei einem Abstand von 60 mm liegt für den homogen dotierten Kristall eine Kopplungseffizienz von 95 % und eine relativ homogene Gainverteilung vor (Abbildung 3).

Vergleich der Gainverteilungen

Der homogen dotierte Slab-Kristall weist an den gepumpten Seiten trotz relativ homogenem Verlauf eine starke Randüberhöhung und eine gering ausgeleuchtete Mitte auf. Über den Slabquerschnitt gemittelt ergibt sich eine Kleinsignalverstärkung von 5.6 . Zum Vergleich ist die Gainverteilung des Sandwich-Slab-Lasers für den gleichen Abstand in Abbildung 4 gezeigt. Für einen Abstand 60 mm ergibt sich beim Sandwich-Slab eine geringere Kopplungseffizienz von 83 %. Der Sandwich-Slab ist wesentlich homogener gepumpt, die Randüberhöhung ist durch die geringere Dotierung ebenfalls abgebaut. Hier ergibt sich eine gemittelte Kleinsignalverstärkung von 4.5 .

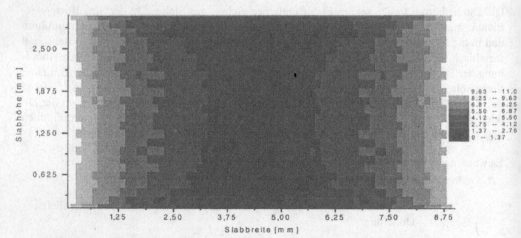

Abbildung 3 : Verlauf der Kleinsignalverstärkung (g $_0$ l) des homogen dotierten Slabs

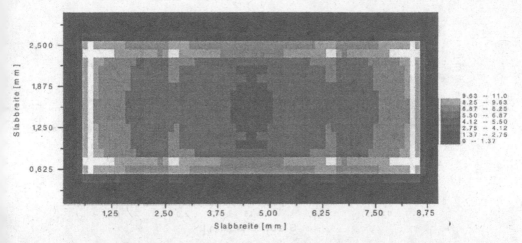

Abbildung 4 : Verlauf der Kleinsignalverstärkung (g₀ l) des Sandwich - Slab

Thermooptische Effekte

Mittels eines FEM-Programms werden die thermischen Effekte unter Berücksichtigung der Endeffekte in den gepumpten Slabs berechnet. Die Temperaturverteilungen werden mit einer mittleren Pumpleistung von 300 Watt und einem Heizwirkungsgrad von 50 % berechnet, sie stellen also eine Abschätzung der thermischen Effekte nach oben dar. Bei der Änderung der optischen Weglänge werden die thermisch induzierte Variation des Brechungsindex sowie die Deformation der Endflächen berücksichtigt.

In Abbildung 5 ist die Änderung der optischen Weglänge für den homogen dotierten Slab dargestellt. Sie zeigt eine starken Gradienten in beiden Richtungen. In Y-Richtung,d.h. in Richtung des bevorzugten Wärmeflusses, liegt eine positive thermische Linse [PEN88] vor mit einer Brennweite von f_{therm} (Y) = 383 mm. In X-Richtung liegt eine negative thermische Linse vor mit einer Brennweite von f_{therm}(X) = -2400 mm.

Die thermische Linsenwirkung in Y- Richtung ist im wesentlichen durch die Pumpleistung und die Kühlanordnung vorgegeben. Die thermische Brechkraft in X-Richtung kann durch geschickte Wahl der Dotierungen beim Sandwich-Slab jedoch erheblich verkleinert werden.

Die Änderung der optischen Weglänge für den Sandwich-Slab (Abbildung 5) zeigt einen quasi eindimensionalen Verlauf. Die thermische Linsenwirkung ist in beiden Richtungen positiv. Die resultierende Brennweite in Y-Richtung liegt mit einer Brennweite von 220 mm in der Größenordnung des homogen dotierten Slabs. Die Fokuslänge wird nur aufgrund der kleineren Ausdehnung des Slabs in Y-Richtung kürzer.
In X-Richtung ist die thermische Linsenwirkung **positiv** mit einer Brennweite von f $_{therm}$ (X) = **12000** mm, d.h. eine Reduktion der Brechkraft um einen Faktor 5 ist gelungen.

374

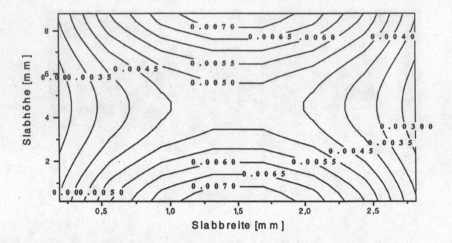

Abbildung 5: Änderung der optischen Weglänge in Millimeter für homogen dotierten Slab

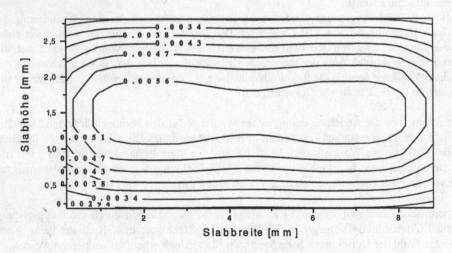

Abbildung 6: Änderung der optischen Weglänge in Millimeter für Sandwich-Slab

Zusammenfassung:

Das vorgestellte Konzept bietet die Möglichkeit, die thermooptischen Störungen bei diodengepumpten, kontaktgekühlten Slab-Lasern durch einen Schicht-Aufbau zu minimieren. Eine Skalierung zu hohen mittleren Leistungen ist möglich. Die theoretische Analyse ist abgeschlossen, der Laser befindet sich im Aufbau.

Literatur :

[PEN88] A.Penzkofer, Prog. Quant.Elect.12 (1988) 291

Power Tunable Nd-Oscillators with Diffraction Limited Beams via SBS Phase Conjugation

Axel Heuer, Ralf Menzel, Martin Ostermeyer
Lehrstuhl Photonik, Institut für Experimentalphysik, Universität Potsdam
Am Neuen Palais 10
14469 Potsdam, Germany

Abstract

Phase distortions such as thermal lensing in laser rods can be compensated by optical phase conjugation. The phase conjugating mirror (PCM) can be easily realized by selfpumped stimulated Brillouin scattering (SBS) in liquids and gases. Pulsed Nd:YAG master-oscillator-double-pass-amplifier-systems (MOPAs) with PCMs were frequently built and are commercially available, now. Oscillators with PCMs are much simpler in their construction but more difficult in their design. Nevertheless, Q-switched Nd-rod oscillators with PCM can be realized with diffraction limited beam quality. Even with a not optimized set-up using a phase conjugating SBS-cell as highly reflecting resonator mirror in a flash lamp pumped Nd:YALO rod laser we achieved tunable average output power between 3 and 20 Watts. Thermal lensing of the rod with focal lengths of less than 30 cm was compensated.

Introduction

Phase conjugating mirrors based on stimulated Brillouin scattering (SBS) can be used to improve the beam quality of high power solid state lasers up to the diffraction limit [1]. These selfpumped SBS mirrors are easy to realize. Many systems applying phase conjugating SBS-mirrors in master-oscillator-double-pass-amplifiers were reported. [e.g. 2-8]. Average output powers between 100 and 150 Watts and almost diffraction limited beam quality have been achieved [e.g. 2-6]. With a diode pumped slab laser system even 800 Watts were reached [7]. Up to now only one system with PCM is commercially available. It has an average output power of 40 Watts [8]. In these systems the SBS mirror is used to compensate phase distortions caused in the highly pumped amplifier material during the second pass of the light. Thermal lensing effects with focal lengths of less than 30 cm could be compensated in this manner [5].

Only few publications using SBS cells directly as a resonator mirror in laser oscillators are known [9 - 15]. 10 Watts average output power from a single rod Nd:YAG laser with very good beam quality were reported [12].

For many applications, e.g. marking, micro material processing, in scientific research and medicine, solid state laser radiation with average output powers in the range between 10 and 100 Watts and very good beam quality are desired. Therefore we investigate the possibilities in designing simple

Funded by the Bundesministerium für Bildung und Forschung (BMBF) und the Verein Deutscher Ingenieure (VDI).

solid state rod lasers with output powers in that range and good beam quality. For compensating phase distortions, especially thermal lensing in the highly pumped rods, we used a gas cell as SBS mirror.

The phase conjugating mirror has to replace the high reflector of the resonator. The properties of the phase conjugating mirror are determined by the parameters of the laser itself. Therefore, it is an adaptive mirror with varying curvature and reflectivity. Thus, these lasers with phase conjugating mirror show completely new properties.

If the PCM is based on stimulated Brillouin scattering it will show a non-linear reflectivity as a function of incident light power and energy, and therefore a „threshold"-power has to be exceeded to obtain sufficient reflectivities for practical applications.

In laser oscillators this initial light power can be provided by setting up an additional start resonator (see fig. 1).

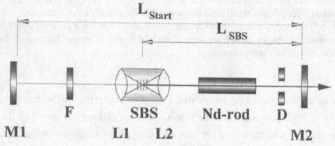

Fig. 1: Scheme of a laser oscillator with a phase conjugating SBS resonator of the length L_{SBS} and start resonator of the length L_{start}

This start resonator consists of the auxiliary mirror M1 and the output coupler M2. The net reflectivity of the mirror M1 can be varied by the filter F. It has to be smaller than the actual reflectivity of the phase conjugating sound wave grating mirror between L1 and L2. With its variation the transition of the laser mode from the start resonator to the SBS resonator is determined, and thus the Q-switch conditions.

For designing such lasers, the following problems have to be studied:
- The start resonator has to be stable. This is much less a stringent requirement than the demand that the start resonator would have to provide a good beam quality itself [13].
- The length of the two resonators have to be tuned to the Brillouin shift of the material [15].
- The transversal mode structure of the start and the SBS resonators have to be investigated. For smooth transition they have to be matched to each other [14].
- The whole system has to be tuned for high average output power with best beam quality [12].

Detailed investigations about these problems are in progress in our group. Even with a not yet optimized single rod Nd laser oscillator we were able to obtain 20 Watts of average output power. We expect a further increase of the average output power in the near future.

After successful optimization of this laser (as shown in fig. 1), the elements of the phase conjugating SBS mirror M1, F, L1, and L2 can be combined into one component (see fig. 2).

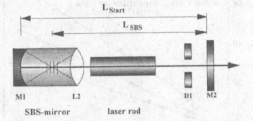

Fig. 2: Scheme of a laser oscillator with phase conjugating SBS mirror and integrated start resonator mirror M1

In this case the reflectivity of M1 has to equal the net reflectivity of M1 in figure 1 in combination with twice the transmission of F. A laser designed according to the scheme of figure 2 has the same construction principles as a conventional resonator, but the high reflecting mirror is replaced by a compact new component that has to be optimized for the given laser head and power parameters of the system.

Experimental

For the present study we chose the laser oscillator configuration of figure 1. A conventional outcoupling mirror M2 (R = 10 %) and the SF_6 cell under a pressure of 20 bar, replacing the standard high reflector, formed the SBS resonator with an optical length L_{SBS} of about 120 cm. The length of the start resonator determined by the optical distance of M1 and M2 was approximately 190 cm. The standing light wave of this resonator generates a hypersonic wave, which is further increased by the feedback inside the SBS resonator and finally leads to a Q-switched laser operation.

The SBS cell consisted of a stainless steel tube and BK-7 glass windows. A telescope realized by the two lenses L1 and L2 was mounted inside the cell. Their focal lengths of about 50 mm were well suited to produce the required intensity as well as the stability of the auxiliary start resonator. All optical surfaces were provided with antireflecting coatings.

As laser rod we used a neodymium doped YALO (YAP) crystal, with a diameter of 4 mm and 79 mm long. Compared with YAG the YALO crystal has only a negligible thermally induced birefringence. Thus, no thermally induced depolarization limits the beam quality of this oscillator with phase conjugating SBS-mirror. In addition the efficiency of Nd:YALO is about 40 % higher as compared to Nd:YAG (see figure 3). Replacement of the standard reflector of the commercial laser by a specular europium doped quartz glass cavity increased the efficiency by further 47 %. This cavity converts the otherwise harmful UV-radiation into the visible spectrum. This transformation prevents the occurrence of color centers in YALO crystals and raises the effective pump pulse energy.

The disadvantage of the roughly two times more pronounced thermal lensing of YALO (see figure 4) is expected to be fully compensated by phase conjugation.

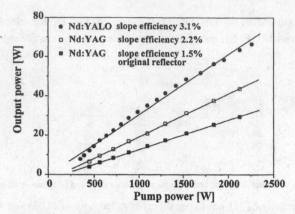

Fig. 3: Efficiency of the used specular cavities with Nd:YAG and Nd:YALO crystals

Thermal lensing of the laser rod was detected using a HeNe-laser probe beam. Thereby the change of the focal length z_f was measured as a function of the electrical input power P_{in}, and is shown in figure 4.

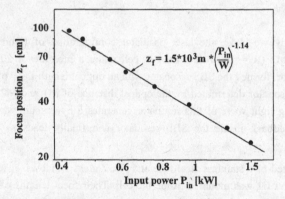

Fig. 4: Focus position of the HeNe-laser beam as a function of the pump power of the flash lamp

The rod was pumped by a flashlamp with an electrical input energy of up to 60 J/pulse and a repetition rate up to 25 Hz. The pulse length of the flashlamp was 1 ms.

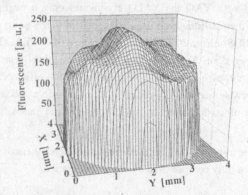

Fig. 5: Spatial flourescence distribution across the Nd:YALO laser rod

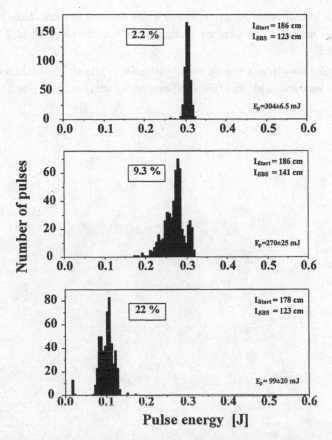

Fig. 6: Histograms of the pulse energy for optimal lengths (top), detuned SBS-resonator (middle) and detuned start resonator (bottom).

Figure 5 shows the pump light distribution across the laser rod characterized by the fluorescence. The fluorescence light was imaged by a lens on a CCD-camera connected to a personal computer via a grabber interface. As can be seen the laser rod is largely homogeneously illuminated.

The laser pulse energy was measured after reflection from a beam splitter and additional attenuation with neutral density filters by a pyroelectrical detector. Calibration of the whole arrangement was performed at small light powers by a direct measurement of the laser output. The time profile of the pulses was detected employing a PIN photodiode and a fast digital oscilloscope (2 G samples/sec)

Results

As mentioned above the resonator lengths of both the start and the SBS based resonators have to be tuned to the Brillouin frequency of the used material [15]. The detuning will decrease the output energy per flash and thus the average output power and it will increase fluctuations from shot to shot (see figure 6).

From this figure it is obvious that the start resonator length seems to be more crucial than the length of the SBS based resonator. In optimized configurations fluctuations of about 2 % or less are commonly reached.

Although no detailed knowledge about the transversal mode profile of SBS-based resonators is yet available [14], we were able to obtain a Gaussian beam profile with our set-up (see figure 7).

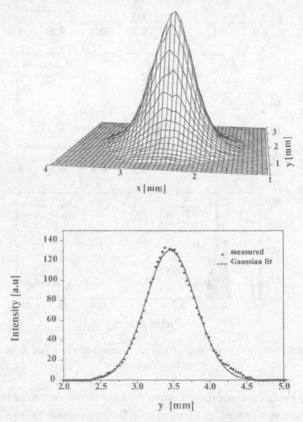

Fig. 7: Beam profile of the Q-switched SBS-based Nd:laser

Although the beam quality has not yet been determined, the stable Gaussian shape indicated in earlier observations with SBS-based resonators an almost diffraction limited value [12]. As expected the strong thermal lensing in the highly pumped Nd:YALO rod with focal lengths of less than 25 cm was almost perfectly compensated by the phase conjugating mirror.

The nonlinear reflectivity of the SBS-mirror produces a Q-switching of the SBS based resonator. The positive feedback between increasing reflectivity in the cell and the growing amplification in the laser rod up to saturation leads to pulses with ns widths as in conventional passive Q-switching. The pulse width is determined by the stored energy in the rod and other laser parameters, as known. This energy deposition can be varied by changing the transmission of the used filter F in figure 1. With lower transmission more energy will be stored before the Q-switch is activated.

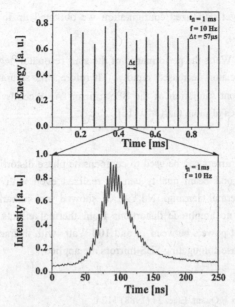

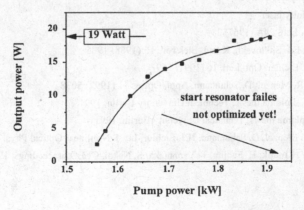

Fig. 8: Time profile of the Q-switched laser radiation.

We chose a transmission of about 50 %. With this value a train of 14 Q-switch pulses occurred (see top of figure 8).

A train of 14 pulses with a spacing of 60 μs was generated with flashlamp repetition rates of 10 to 25 Hz. Within each Q-switch pulse a modulation with the Brillouin frequency could be observed. The overall pulse width was about 40 ns. With a pulse train structure as in figure 8 the peak power of the radiation could be reduced compared to the case of a single pulse per flash. This time structure was shown to be useful in material processing [6]. Although some energy fluctuations were observed between the single Q-switch pulses the fluctuations for the whole train was rather small (see figure 6).

Fig. 9: Average output power of the Nd:YALO oscillator with phase conjugating SBS mirror

With the presently not yet optimized configuration we obtained up to 19 Watts average output power (see figure 9).

Above approximately 10 Watts the performance of the start resonator degraded and thus the power for the SBS based resonator decreased rapidly. Therefore, the saturation effect observable in figure 9 does not represent saturation of the SBS mirror. As recently shown, SBS mirrors can operate with pulse energies of more than 14 J [16].

Conclusion

Phase conjugating SBS-mirrors can be used to compensate phase distortions within solid state rod laser oscillators. Very good beam quality can be realized even with highly pumped rods and consequently strong thermal lensing. Nd:YALO showed no remarkable thermally induced depolarization and thus no amplitude distortions from thermal effects. With simple single rod oscillators average output powers between 10 and 100 Watts with diffraction limited beam quality seem to be possible if phase conjugating SBS-mirrors are applied.

References

[1]   D.A. Rockwell, IEEE J. Quant. Elec. 24 (1988) 1124

[2]   H. J. Eichler, A. Haase, R. Menzel, J. Schwartz, Pure Appl. Opt. 3 (1994) 585

[3]   C. B. Dane, L. E. Zapata, W. A. Neumann, M. A. Norton, and L. A. Hackel, IEEE J. Quant. Electr. 31 (1995) 148

[4]   C. B. Dane, M. A. Norton, J. D. Wintemute, W. L. Manning, B. Bhachu, L. A. Hackel, CLEO proceedings, 1995, p. 80

[5]   H. J. Eichler, A. Haase, R. Menzel, 100-Watt Average Output Power 1.2 Diffraction Limited Beam from Pulsed Neodymium Single-Rod Amplifier with SBS Phase Conjugation, IEEE J. Quantum Electron., vol. 31, 1995, to be published

[6]   H. J. Eichler, A. Haase, R. Menzel, CLEO proceedings, 1995, p. 61

[7]   R. J. St. Pierre, H. Injeyan, J. Berg, R. C. Hilyard, M. E. Weber, M. G. Wickham, R. Senn, G. Harpole, C. Florentino, F. Groark, J. Endriz, J. M. Haden, D. C. Mundinger, CLEO proceedings 1995, postdeadline papers, CPD30

[8]   Coherent, Infinity laser

[9]   D. Pohl, Phys. Lett. 24A (1967) 239

[10] P.P. Pashinin, E.I. Shklovskii, Kvant. Elektron. 15 (1988) 1905

[11] H. Meng, H.J. Eichler, Opt. Lett. 16 (1991) 569

[12] H. J. Eichler, R. Menzel, D. Schumann, Appl. Optics 31 (1992) 5038

[13] D. Schumann, diploma thesis, Technical University Berlin, 1991

[14] E. Geinitz, diploma thesis, Technical University Berlin, 1991

[15] Kummrow, R. Menzel, D. Schumann, H.J. Eichler, Int. J. Nonlinear Optical Phys. 2 (1993) 261

[16] H. Yoshida, V. Kmetik, H. Fudjita, T. Yamanaka, S. Nakai, CLEO proceedings, 1995, p. 1

Diode-Pumped High-Power cw Nd:YAG Lasers

W. Schöne, D. Golla, S. Knoke, G. Ernst, M. Bode, A. Tünnermann and H. Welling

Laser Zentrum Hannover e.V.

Hollerithallee 8, D-30419 Hannover, Germany

ABSTRACT

Diode laser side-pumped, cw Nd:YAG rod lasers operating at pump powers up to 1.1 kW will be reported on. In multimode operation at 1064 nm, output powers of more than 300 W cw are observed. Higher pump powers up to several 100 W/cm can be achieved by using fiber-coupled diode lasers as pump sources.

1. INTRODUCTION

Diode-pumped solid-state lasers operating at high cw power levels are attractive sources for various applications in materials processing and laser based metrology, because of their reliable and highly efficient operation and lifetimes in the range of several 10.000 hours. Low power end-pumped solid-state lasers have demonstrated their superiority compared to lamp-pumped lasers for many years [1,2]. But power scaling possibilities of end-pumped configurations are limited because of the thermally-induced stress fracture of the laser materials [3]. Therefore, the transverse pump geometry has to be used for output powers beyond 100 W cw [4].

This contribution reports on the realization of high power diode laser side-pumped Nd:YAG lasers in rod geometry. As pump sources linear diode laser arrays as well as fiber-coupled diode lasers are used. The laser performance and power scaling of the side-pumped Nd:YAG rod lasers at an emission wavelength of 1064 nm will be discussed.

2. ND:YAG ROD LASERS SIDE-PUMPED WITH LINEAR DIODE LASER ARRAYS

Our first approach to high power systems was based on the application of linear diode laser arrays. A detailed description of the experimental setup has been published elsewhere [4]; hence, we only mention some essential features here. A sketch of the pumping arrangement is shown in Fig. 1. The laser rod (diameter: 5 mm, length: 220 mm, Nd-doping: 0.9 at.%) is mounted inside a flow tube, which is antireflection coated at 808 nm, for direct water cooling.

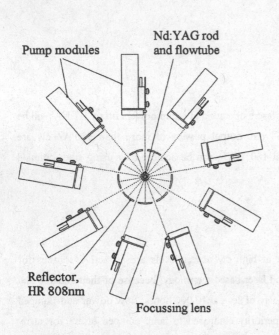

Pump modules

Nd:YAG rod and flowtube

Reflector, HR 808nm

Focussing lens

Fig. 1. High power Nd:YAG rod laser, side-pumped by 108 diode lasers.

The optical pump source consists of 108 linear diode arrays (Jenoptik Laserdiode JO LD 10L) with a nominal output power of 10 W each at 808 nm. For each pump module, 6 diode lasers with similar emission wavelengths are selected. Each diode laser is attached to a copper heatsink. The radiation of the diode lasers of each module is imaged into the laser rod by an antireflection coated cylindrical lens (diameter: 3 mm, length: 80 mm).

Eighteen pump modules are arranged in a nine-fold symmetry around the laser rod yielding a total pump power of 1100W at the rod. Pump light reflectors are mounted around the rod opposite each pump module in order to reflect back transmitted diode laser radiation into the active medium.

The output characteristics of the side-pumped rod lasers at the laser wavelength of 1064 nm are investigated in a short, linear planar-concave resonator with a highly reflecting mirror at a radius of curvature of 5000 mm. A maximum output power of 300 W is achieved at the maximum pump power corresponding to a pump power at laser threshold of 190 W, and an optical slope efficiency of 32 %. The wallplug efficiency is determined to about 7 %, the beam parameter product to approximately 30 mm·mrad. In an 800 mm long, symmetrical resonator with flat mirrors, a beam parameter product of 5 mm·mrad is obtained at an output power of 200 W cw. This radiation can be efficiently coupled into an optical fiber with a diameter of 100 μm and 0.1 numerical aperture (NA).

3. ND:YAG ROD LASERS SIDE-PUMPED WITH FIBER-COUPLED DIODE LASERS

Significant power scaling beyond 300W within the simple single rod scheme requires - because of the limited rod lengths - an increase of the linear pump power density. But due to the large-sized dimensions of linear diode laser arrays, pump power densities of more then 100W/cm are difficult to realize with these devices in a reliable and simple setup.

In contrast, pump power densities of several hundred watts per cm are easily achieved with fiber-coupled diode lasers. Moreover, the concept of fiber coupling with its intrinsic separation of laser head and pump source has clear advantages as compared to the direct use of diode laser arrays. The diode lasers with electrical supply, mechanical holders, thermal control and coolant are mounted in an enlarged version of a power supply, where they can be sealed off from environmental influences. Spatial problems at the laser head and those of coolant handling are almost completely avoided. There is only a bundle of optical fibers that is attached to the laser head. This bundle can be easily shaped to match the geometry of the active volume. Any failure of individual diode lasers means only having to plug in the corresponding fiber connector to a new diode laser without touching the laser head, possibly without interrupting laser operation. We believe that these are crucial features to turn diode-pumped high power lasers into reliable products. Another advantage is the possibility to easily shape the pump light distribution, allowing a reduction of thermally-induced effects.

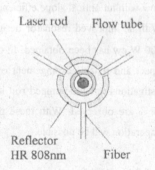

Fig. 2. Threefold pumping scheme with fiber-coupled diode lasers.

Fig. 2 shows a schematic drawing of a pump arrangement with a pump power of 150 W/cm based on fiber-coupled diode lasers. As pump sources, fiber-coupled diode lasers (Jenoptik Laserdiode) are applied. The output power at the end of the fiber is in the range of 10 W cw. The optical quartz fiber has a diameter of 800 μm and 0.22 NA. The Nd:YAG rod (length: 56 mm, diameter: 4 mm) is pumped from three sides by linear fiber arrays with a lateral distance of 2 mm between the centers of adjacent fibers. The diode laser radiation directly irradiates the laser rod without any additional focussing optics. Pump light reflectors are mounted around the rod at a distance of approximately 4 mm from the flow tube.

In our setup, the distance between fiber ends and laser rod can easily be varied in order to modify the pump light distribution within the rod. Fig. 3 depicts the distributions over a rod cross-section for different distances. With 170 W of pump power, a maximum multimode output power of 62 W at a fiber-to-flow tube spacing of 5 mm was obtained. The pump power at threshold was 27 W, the

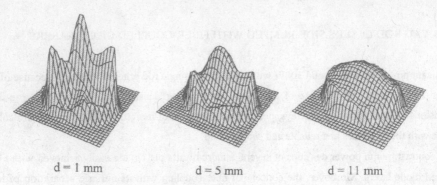

$d = 1 \text{ mm}$ $d = 5 \text{ mm}$ $d = 11 \text{ mm}$

Fig. 3. Pump light distributions for different distances d between fibers and flow tube.

slope efficiency 45 %. At a spacing of 11 mm, the pump light distribution is nearly homogeneous in the central section of the rod, while the obtained output power and slope efficiency (59 W and 43%, respectively) are only slightly lower. Additionally, the observed focal length of the thermally-induced lens increases with the fiber-to-flow tube spacing from 25 cm to 37 cm, respectively. Hence larger spacings are preferrable for high beam quality and TEM_{00} mode operation.

4. CONCLUSION

In conclusion, we have reported on side-pumped Nd:YAG rod lasers using linear diode laser arrays as pump sources. This laser produces more than 300 W cw with an optical slope efficiency of 32 % and an overall electrical efficiency of more than 7 %. With an improved resonator design, a beam quality in the range of 5 mm·mrad at output powers of 200 W cw has been obtained. In order to achieve higher pump power densities and to realize a compact and reliable arrangement of the laser head, fiber-coupled diode lasers are applied. In first investigations on side-pumped rod lasers using these devices as pump sources, slope efficiencies of 45 % are observed. With these pump sources, power scaling in the range of 800 W cw in multimode operation will be possible.

This research has been supported by the German Ministry of Education, Science, Research and Technology under contract 05 5MQ 11I, 13 N 5723 and 13 N 6361.

5. REFERENCES

1. R. L. Byer, Science 239, pp. 742-747, 1988.
2. I. Freitag, I. Kröpke, A. Tünnermann and H. Welling, Opt. Commun. 101, pp. 371-376, 1993.
3. S. C. Tidwell, J. F. Seamans and M. S. Bowers, Opt. Lett. 18, pp. 116-118, 1993.
4. D. Golla, S. Knoke, W.Schöne, G. Ernst, M. Bode, A. Tünnermann and H. Welling,
   Opt. Lett. 20, 1995 (To be published)

Microcrystal Lasers for the Generation of Tunable Synthetic Wavelength in the Microwave Region

N. P. Schmitt[1], W. Waidelich[2], K.-D. Salewski[3], J. Schalk[1], P. Peuser[1]

[1] Daimler Benz AG, Forschung und Technik, 81663 München /D
[2] Ludwig-Maximilians-Universität, 80799 München /D
[3] Ernst-Moritz-Arndt-Universität, 17489 Greifswald /D

The characteristics of diode-pumped microcrystal lasers have been investigated intensively for the last 5 years. The frequency separation of longitudinal resonator modes is inversely proportional to the resonator length. If the mode separation exceeds at least half of the gain bandwidth of the laser transition, only one single longitudinal resonator mode is able to achieve gain. The laser will emit on a single longitudinal mode only for a suitable position of the resonator mode frequency near the maximum gain frequency (or generally, independent of the resonator frequency, if the mode separation exceeds the whole gain linewidth). Such lasers are called "microcrystal lasers" further on. The geometrical length of these resonators, which can be made of a polished monolithically coated piece of a laser crystal, typically is lessthan 1mm, which in this case is identical with the crystal thickness itself. Microcrystal lasers can be miniaturized in a very high grade by integrating them together with a pumping laser diode in a single hybrid chassis of about 25 mm square [1, 2].

If the mode separation is clearly above half of the gain bandwidth, the laser frequency can be tuned by changing the resonator frequency while keeping the single frequency emission characteristic. Since no mode beating occurs, spatially undepleted gain regions can now allow resonator modes coinciding with other gain curves of further laser transitions to lase, and so microcrystal lasers are able to operate on several laser transitions simultaneously. Heterodyning tunable single frequency emission with laser radiation of another laser or with simultaneously emitted other modes (corresponding to the same or different laser transition) allows the generation of "synthetic" microwaves, which is a modulation of the signal amplitude (beat) with a frequency that is equal to the frequency difference of the heterodyned laser beams.

Detecting a superimposed beam of two frequencies with a nonlinear element such as a photo diode will give a signal modulated with the frequency difference at the output. Another way of interpreting this phenomenon is to say that the fringe pattern of not perfectly collinear beams will dynamically change with the frequency of the beat, which will be detected by the photo diode as an amplitude modulation. As the detectable frequency range is limited by the speed of the photodiode, phase changes of higher frequency beats may be detected by modulating the laser beams with a RF frequency (e.g. by AOMs) and demodulating the photodetector signal in superheterodyne technology (analogous to radio frequency demodulation). This can be done by the quadrature of all beat frequency components and filtering afterwards, which gives a dc signal proportional to the relative phase change of the carrier frequencies (interferometric method).

Using those synthetic microwaves generated by frequency beating is advantageous for several measuring methods: the carrier frequency is an optical wave which proposes a high brightness and therefore reduces the necessary ´antenna´ diameter. The synthetic wavelength itself is in the microwave region (or longer) which eliminates the ambiguity in absolute distance measurements for large distances for example. Furthermore, the influence of perturbations such as reflector vibrations is reduced, since in heterodyne measuring arrangements the perturbation would lead to a modulation of the phase of the carrier only, not of the synthetic wavelength. High precision perturbation-insensitive optical heterodyne measuring systems where the measuring signal is proportional to the phase change of a synthetic wave become therefore more and more popular.

Microcrystal lasers allow multiple ranges of synthetic microwaves to be generated by quite different schemes of generation, whereas the inherently small linewidth yields a large coherence length of the carrier which is important for most kind of interferometric systems.

Tuning a microcrystal laser with mode separation clearly above half of the gain bandwidth allows tuning about 100-200 GHz, depending on the resonator design and the pump power[3]. If this radiation is superimposed with the radiation of a second microcrystal laser, which is advantageously tuned to an edge of the gain curve, a tuneable synthetic microwave can be generated with a frequency range up to 100-200 GHz (Fig. 1a). If the laser frequencies meet, the difference frequency will disappear. Thus the synthetic wavelength $\Lambda = c/\Delta\nu$ can be tuned continuously between ∞ and 3 mm, respectively 1.5 mm. The beams are separate and can be further modulated separately or may be combined in one collinear beam. Using Nd:YAG-microcrystals of 300 µm thickness single frequency tuning ranges of > 130 GHz have been demonstrated at output powers of 21 mW (up to 245 GHz could be reached by decreasing the pump power down to an output of 2 mW) [3, 4].

A synthetic wavelength in the longer wavelength range can be generated with one single laser, utilizing the birefringence induced by the photoelastic effect [3, 5]. If the pump spot and therefore the laser mode is not totally concentric to the isotherms inside the laser crystal, this induced birefringence splits the laser frequency of the single longitudinal mode into two components with orthogonal polarization (Fig. 1b). The output of a single microcrystal laser contains two frequency components which differ in the range of 0 to 200 MHz and can be tuned by lateral displacement of the pump focus or modulation of the pump power [4]. The beams can easily be separated by a polarizing beam splitter and can be fed into different optical paths.

Another way of generating two frequency components is to design the microcrystal laser in a way that two longitudinal modes may coincide with one gain curve (Fig. 1c). Since the mode separation in microcrystal lasers is quite large (typically 50 to 150 GHz), synthetic microwaves in a range of 6 mm to 2 mm can be generated. The output is collinear and is not separable if isotropic crystals are used. Changes of the resonator length affect both lasing modes, the frequency difference stays nearly constant and the synthetic wavelength can be very frequency stable. For two independent lasers, the instability of the synthetic wavelength is $\dfrac{\delta\Lambda}{\Lambda} = \dfrac{\Lambda}{\lambda}\sqrt{2}\cdot\dfrac{\partial l}{l}$, whereas for two modes of the same resonator it is only $\dfrac{\delta\Lambda}{\Lambda} = \dfrac{\partial l}{l}$. Using microcrystal lasers of 300 µm crystal thickness, lasing at the 1.064 µm-transition (A-Line) of Nd:YAG for example, a temperature variation of 1 K leads to a change of the synthetic wavelength Λ=3mm of 148 µm (or in other words the beat frequency changes about 5 GHz) for two independent lasers and only about 37 nm (or 1.2 MHz for the beat frequency) for two modes of the same resonator, which is a factor of $4.2\cdot10^3$ better. Synthetic waves generated in this way are inherently stable and may be tuned by a small amount as calculated above. Using ferroelectric crystals with overlapping gain curves for both polarizations would make it possible to tune the beat frequency up to 10 GHz and the beams could be separated by polarizers.

Experimentally, using a 800 µm Nd:YAG microcrystal laser, two resonator modes lasing on the A-Line have been observed with a frequency difference of 100 GHz [5].

As mentioned in the second paragraph, microcrystal lasers are in general able to emit on two longitudinal modes corresponding to two different laser transitions simultaneously [1]. In this way,

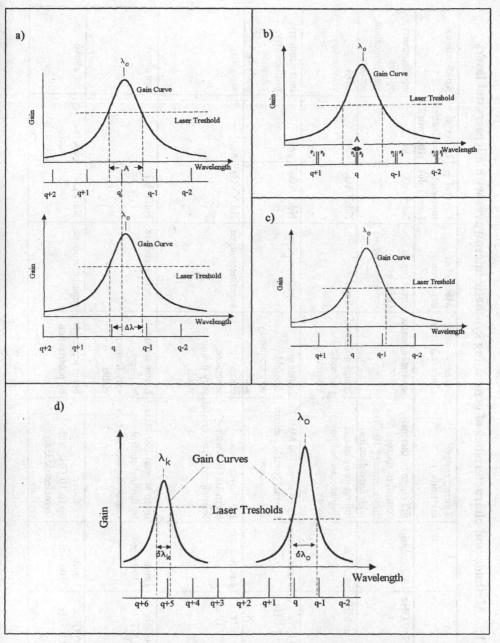

Fig. 1. Schemes for the generation of synthetic microwaves with microcrystal lasers:
a) Two independent lasers, one of them tuned
b) Frequency splitting of longitudinal modes due to the photoelastic effect
c) Two longitudinal modes corresponding to the same laser transition
d) Two longitudinal modes corresponding to different laser transitions, simultaneously oscillating

Table 1: Methods and characteristics of generating synthetic microwaves with microcrystal lasers

Principle of generation	$\Delta\nu$	Λ	Tuning range / methods	other characteristics	Method of detection	Crystal material (example)	Experim. verification (this work)	Applications
two independent lasers	0..240 GHz	∞..1.2 mm	full range by thermal control or mirror displacement (10 GHz with ferroelectrica)	separated beams	direct (up to 60 GHz) or superheterodyne	Nd:YAG	yes	interferometry, optical communication, sensors
birefingence due to photoelastic effect	0-120 MHz	∞..2.5 m	full range by lateral displacement of pump focus	beams polarized perpendicular, separable	direct	Nd:YAG	yes	LDA velocimetry
two longitud. modes of same laser transition	50 GHz ..150 GHz	6 mm..2 mm	up to some 100 MHz by thermal control / mirror displacement	beams not separable, very frequency stable	superheterodyne	Nd:YAG	yes	interferometry, profilometry, vibration analysis
as above, perpendic. polarized in naturally birefringent media	50 GHz ..150 GHz	6 mm..2 mm	see above; up to 10 GHz with ferroelectric laser crystal	beams separable, very frequency stable	superheterodyne	Nd:YVO$_4$	no	see above
simultaneous emission on different laser transitions	280GHz ..2.8 THz	1 mm ..100 µm	up to several 100 MHz by thermal control / mirror displacement	beams separable by filters, very frequency stable	superheterodyne	Nd:YAG	yes	see above
see above, birefringent laser crystal	280 GHz ..2.8 THz	1 mm ..100 µm	as above up to 10 GHz with ferroelectric laser crystals	beam separable by polarization	superheterodyne	Nd:YLF	yes (not micro-crystals)	see above

synthetic microwaves with nearly the same stability as mentioned above can be generated with wavelength in the range of 600 μm to 100 μm (depending on the laser transitions) (Fig. 1d). For example, simultaneous emission on the Nd:YAG A- and B-line (1064 nm resp. 1061 nm) has been observed which gives a wavelength difference of 3.13 nm, corresponding to a synthetic wavelength of 360 μm, and also on the A- and C-line (1052 nm) which shows a difference of 12 nm corresponding to a synthetic wave of 95 μm length, respectively. The laser beams are inherently collinear, but the frequency separation might be large enough that they can be separated by dichroitic dielectric mirrors or gratings. Using birefringent crystals such as Nd:YVO$_4$ or Nd:YLF, polarization separation is possible (with the latter simultaneous emission was observed on the 1047 nm and the 1053 nm transition, but in a resonator containing two different crystals). Ferroelectric crystals would again allow a tuning of the beat frequency of about 10 GHz.

Table 1 summarizes these six methods for generating synthetic microwaves with microcrystal lasers and lists their properties as well as possible application fields.

Three typical applications have been verified with different partners, using hybrid integrated, diode-pumped microcrystal lasers (Nd:YAG, λ=1064 nm) as light sources for optical heterodyne measuring systems.

In a first application, realized together with Carl Zeiss, Jena, two microcrystal lasers, one of them being frequency tuneable, were used in an absolute distance measuring interferometer setup for measuring distances up to 50 m. Tuning the beat frequency from 0 to 70 GHz (λ=∞ to 4.3 mm) a precision of 7 μm on a distance of 7 m could be verified which corresponds to a phase detection resolution in the order of 10^{-3} [6].

A second application makes use of the induced birefringence due to the photoelastic effect. The frequency split radiation of a single microcrystal laser was used at the PTB in Braunschweig in a LDA system to measure the speed of aerosole particles with sensitivity to the velocity vector [7].

A third application was in the field of velocimetry, which was realized together with the University of Hannover using the same laser arrangement as above, which is published in [8].

In conclusion, due to their large mode spacing microcrystal lasers where found to be very suitable to generate synthetic microwaves in a range from GHz up to some THz. Six fundamental methods for generating synthetic microwaves have been identified, five of them experimentally verified (see Table 1). Dependent on the scheme of generation (Fig. 1), the synthetic wavelength can either be tuned from infinity to the millimeter or micrometer range, or can be generated very frequency stable with a typical stability of 37 nm / K for Λ = 3 mm (Nd:YAG, l=300 μm) which equals Δν = 1.2 MHz/K. Three typical applications of optical heterodyne measuring systems have been verified with different partners and so microcrystal lasers seem to have demonstrated their potential as simple but unique sources for optical heterodyne techniques, which will expand in future.

Literature:
[1]   N.P. Schmitt, S. Heinemann, A. Mehnert, P. Peuser, Laser 91 Optoelektronik (W. Waidelich edt.), Springer-Verlag 1992, p. 599
[2]   P. Peuser, N.P. Schmitt: "Diodengepumpte Festkörperlaser", Springer-Verlag 1995
[3]   N. P. Schmitt: "Abstimmbare Mikrokristall-Laser", submitted as dissertation (Ludwig-Maximilians-Universität München) 1994
[4]   N.P. Schmitt, S. Heinemann, A. Mehnert, P. Peuser, Laser 93 Optoelektronik (W. Waidelich edt.), Springer-Verlag 1994, p. 8
[5]   N.P. Schmitt, P. Peuser, S. Heinemann, A. Mehnert, Opt. and Quant. Electr. 25 (1993) p. 527
[6]   N.P. Schmitt, K.-H. Bechstein, W. Fuchs, K.-D. Salewski, CLEO Europe '94, Techn. Dig. p. 56
[7]   R. Kramer, H. Müller, D. Dopheide, J. Czarske, N.P. Schmitt, Seventh intern. Symposium on Application of Laser Techniques to Fluid Mechanics, Lisbon, July 11-14, 1994, p. 14.4.1-14.4.4
[8]   J. Czarske, H. Müller, A. Wüste, X. Liu, N.P. Schmitt, 96. Tagung der Deutschen Gesellschaft für angewandte Optik, Rügen, 6.-10. Juni 1995

Design and Performance of High Average Power Diode Array Pumped Frequency Doubled YAG Laser

B. Le Garrec, G. Razé.
CEA/DCC/DPE/SPL, Centre d'études de Saclay
F-91191 Gif sur Yvette, FRANCE

We report the demonstration of a transversally diode array pumped Nd:Yag laser using thirty 20 watts CW linear diode arrays. At a 18 kHz repetition rate the laser produces 40 watts average power at 532 nm when intracavity doubled with a KTP crystal, leading to a 2 % optical/electrical efficiency.

1. INTRODUCTION.

In order to obtain high average power in the green, a diode pumped laser program is under development. High average power (tens of watts) and high repetition rate (tens of kilohertz) require: cw diode array pumping, high reprate Q-switching and intra-cavity second harmonic generation.

We report the demonstration of a transversely diode array pumped frequency doubled Nd:Yag. The diode laser module is based on a 30 bars structure made of 5 water cooled submounts. Each submount incude 6 bars and 2 collimating cylindrical lens. The 30 linear diode arrays are arranged radially around a single Nd:Yag rod providing up to 635 watts of pumping power around 808 nm.

2. DIODE PUMPED LASER MODULE.

The design of the diode pumped laser module follows very simple and specific rules: one single water cooling design, one single electrical supply.

Thirty 20 watts cw diode laser arrays supplied by Spectra Diode Labs (SDL-3470-S) are optically coupled to a Nd:YAG rod (6.2mm diameter, 105 mm long). It is not possible to mount each 1 cm long diode array on a water cooled brass submount like we did in our early experiments [1] .

The diode laser module is based on a 30 bars structure made of 5 water cooled submounts arranged in a fivefold symmetry around the laser rod. Each submount incudes 6 bars and 2 collimating cylindrical lens (anti-reflection coated cylindrical lens, diameter 6mm, length 90 mm). The collimating cylindrical lens holder is part of the water cooled submount. The thirty diode arrays are positioned at angles of 36° with respect to each other and arranged alternatively to pump the rod along 9 cm (see fig.1). To avoid suppliing the bars in parallel, we use a non conductive material for the cooling submount (Delrin). The cooling submount is provided with 6 square apertures (three on each side) . Each diode bar is mounted with a square flat joint and the water flows along the anode case of each bar throughout the cooling module (see fig.2).

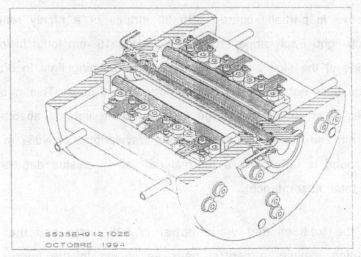

S535EH9 12 102E
OCTOBRE 1994

Fig.1. Schematic overview of the diode pumped laser head.

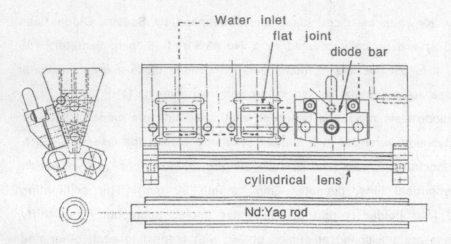

Fig. 2. Cross sectional view of the diode pumped laser module showing the water cooled submount, the linear diode array and the collimating cylindrical lens.

The Nd:Yag rod (diameter 6.2 mm, length 105 mm, 1% Nd doping) is polished and is mounted inside a glass sleeve for water cooling. The outer part of the sleeve is partially coated with 30 stripes of a highly reflective coating for pump light. Each stripe (1.5 mm wide and 15 mm long) is located on the other side of the sleeve with respect to the incoming light in order to reflect the diode light not absorbed by the rod at first pass. This geometry offers a number of advantages: a large fraction of the light is absorbed at both the 808.5 nm and the 812.5 nm absorption wavelengths of Nd3+ in YAG; the absorption band is larger (i.e a thermal drift of 1° celsius degree does not affect the total absorptance).

At 30 A, the total emitted diode power is 525 watts and the pump density distribution exhibits a central peak as shown in the fluorescence profile (figure 3) at 1064 nm when recorded with a low aperture CCD camera [2]. This distribution has been verified through gain and threshold measurements in different 1064 nm CW cavity configurations. The output caracteristics of the side-pumped rod laser at 1064 nm are investigated in a linear plano-concave resonator (300 mm long) with a highly reflecting mirror at a radius of curvature of 2000 mm. A maximum output power of 105 W in multimode operation is achieved with a 10 % transmittance output coupler. The focal length of the rod is equal to 250 mm when pumped with 525 watts.

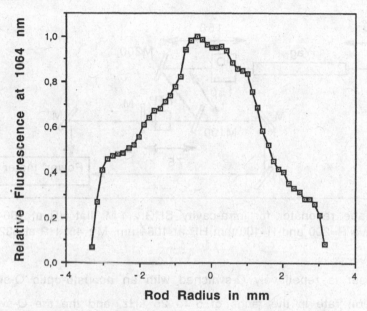

Fig. 3. Fluorescence profile at 1064 nm along the rod diameter.

3. LASER RESONATOR FOR INTRA-CAVITY SHG.

In order to achieve Q-switched operation of the laser at 532 nm, a Z-shape cavity has been designed. The principle of the cavity has been proposed by Kuizenga and coworkers [3, 4]. In classical linear or L-shape cavities, the spotsize in the non-linear crystal decreases as the pump power is increased, and this can cause damage in the non-linear crystal. In the Z-shape resonator, an optical relay is formed by two concave mirrors between the laser medium and the non-linear crystal. The optical relay images the laser rod aperture into the non-linear crystal. Figure 4 illustrates this optical relay in the case of a magnification of 2 (mirrors curvatures: 200 and 100 mm) and a 4*4*6 mm3 KTP crystal. The curved mirrors are tilted with an angle of 12° and the four cavity mirrors are high reflectors at 1064 nm.

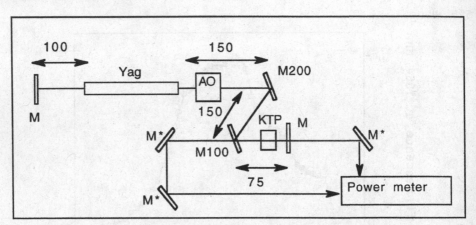

Fig.4. Z shape resonator for intra-cavity SHG: M: flat HR at 1064 nm, M: M100 concave R=200 and R=100 mm HR at 1064 nm, M*: 45° HR at 532 nm.

The laser is repetitively Q-switched with an acousto-optic Q-switch at high repetition rate in the range of 5 to 25 kHz, and the the Q-switch off time and RF power where adjusted for maximum output power.

The laser resonator enables SHG along two opposite directions into the KTP crystal and the two main laser beams at 532 nm are extracted through the mirrors closest to the KTP crystal.

At 30 amps diode drive current, (total emitted diode power: 525 watts, electric supply: 1800 watts) the average green power we have obtained is 38 watts at 15 kHz in a 275 ns pulse (FWHM) leading to an optical to electrical efficiency of 2.1 %. At a maximum diode drive current of 32 A, we have obtained 40.2 Watts at 18 kHz in a 280 ns pulse (FWHM) leading to an optical to electrical efficiency of 2 %.

4. CONCLUSION.

At 30 A diode drive current (525 watts pumping power), we have been able to obtain 38 watts average power at 532 nm in a Z shape cavity Q-switched at 15 kHz. The maximum output power is 40 Watts at 32 A diode drive current. In both cases the efficiency is greater than 2%. The laser beam was intra-cavity doubled in a 6 mm long KTP crystal. The pulse to pulse stability is better than 1% over several hours.

5. REFERENCES.

[1] B. Le Garrec, Ph. Féru: "High power diode-array-pumped frequency doubled Nd:Yag laser" Conference on lasers and electro-optics CLEO'94 Anaheim May 8-13, 1994 paper CTHC5 Volume 8 CLEO'94 Technical Digest p283-284.

[2] B. Le Garrec, P. Féru : " Laser Nd:Yag pompé transversalement par diodes lasers ". ESI Publications Opto 92, 1992, p433-437.

[3]   M. Ortiz, J. Fair, D. Kuizenga : " High average power second harmonic generation with KTiOPO4 ". OSA Proceedings on Advanced Solid State Lasers Vol.13, 1992, p361-365.

[4] D.J. Kuizenga: US Patent 4907235, 3-6-1990.

Towards a 1 khz Diode-Pumped Narrow-Linewidth Tunable Light Source

U. Stamm, W. Zschocke, N. Deutsch, D. Basting

Lambda Physik GmbH, Hans-Böckler-Str. 12,

D-37079 Göttingen, Germany

Using an OPO 0.6 W tunable output (VIS + IR) have been generated from a nanosecond Q-switched diode pumped Nd:YAG laser. The overall conversion efficiency 1064 nm to tunable output is 10%.

A schematic of the experimental setup is shown in Fig. 1.

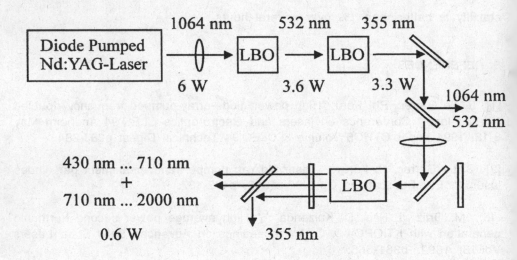

Fig. 1: Schematic of the OPO based high repetition rate widely tunable all-solid-state laser source.

The heart of the system was an electro-optic Q-switched Nd:YAG laser with stable resonator geometry. The laser was pumped by 20% duty cycle laser diodes in a side pumping arrangement. In maximum we obtained up to 10 mJ of TEM_{00} mode output with 1 kHz pulse repetition rate. The pulse duration at 1064 nm was 15 ns. Measurements of the beam divergence proved the beam quality to be 1.1 ± 0.1 x diffraction limited.

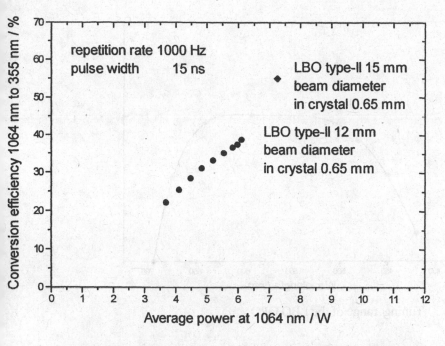

Fig. 2:   Frequency tripling conversion efficiency in dependence on average
           power at 1064 nm.

requency doubling / tripling of the output pulses from the pump laser was performed in
vo subsequent LBO type-I / LBO type-II crystals. Maximum conversion efficiency
064 nm to 532 nm was 61 %. To achieve optimum frequency mixing efficiencies in the
econd LBO crystal the conversion into 532 nm had to be sicnificantly reduced. This was
erformed by detuning the phase matching angle of the doubling crystal. Maximum
onversion efficiency 1064 nm to 355 nm of 55 % have been achieved corresponding to
ore than 5 W UV output. Fig 2. gives an example of the dependence of the conversion
fficiency on fundamental average power for two different mixing crystal lengths. Both
ie 532 nm and the 355 nm beam retained the diffraction limited (divergence
ieasurement yielded 1.1 ± 0.1 x diffraction limited) characteristic of the fundamental.

he 355 nm output was used to pump a LBO based optical parametric oscillator. The
)PO employed a linear standing wave cavity singly resonant for the signal wave with
5 mm cavity length. A single set of broad band reflecting mirrors covered the entire
ignal wavelength range (430 nm - 710 nm). To prevent damage of the optical
omponents (in particular the OPO outcoupling mirror and the frequency tripling crystal

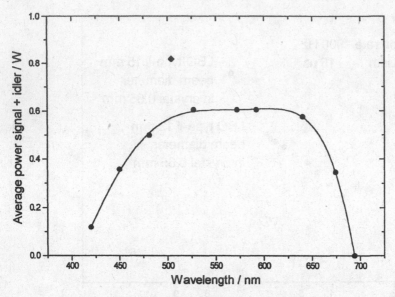

Fig. 3: Tuning range of the LBO OPO.

of the pump laser), the maximum UV pump power has been limited to levels below 4 W. The 355 nm beam was focused into the 15 mm type-I LBO crystal of the OPO. With 15 % outcoupling the total OPO output was above 0.6 W (signal wavelength 560 nm, 0.21 W ; idler wavelength 970 nm, 0.41 W). The pulse duration of the OPO output was with 7 ns significantly shorter than the pump pulse duration (approximately 15 ns). The short cavity length in combination with the proper focusing into the OPO crystal gave rise to stable TEM_{00} oscillation of the OPO. With detuning of the cavity mirrors higher order transverse modes could be observed. Under optimum alignment the beam quality of the OPO output was 1.3 ± 0.1 x diffraction limited.

The tuning curve of the OPO is depicted in Fig. 3. While the output power was rather flat over a wide range, changes in the mirror reflectivities as well as larger Fresnel losses at the LBO crystal caused a drop in output energy to both the blue and the red end of the signal tuning range. Optimized mirror reflectivities as well as AR coatings of the crystal faces could increase the output energy in thoses ranges.

Spectrally resolved measurements of the OPO output revealed that the linewidth is significantly narrower than reported earlier /1,2/. The minimum measured linewidth of 1.5 cm^{-1} is limited by the resolution of the used grating spectrometer.

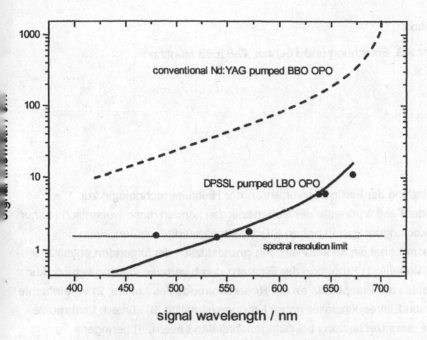

Fig. 4: Linewidth of the LBO OPO signal output in dependence on wavelength. For comparison the linewidth of a conventional BBO based device is given (dashed curve). The resolution of the used spectrometer was 1.5 cm^{-1}.

In conclusion, we have demonstrated a narrow linewidth 1 kHz nanosecond pulsed tunable all-solid-state light source based on a LBO OPO The maximum achieved conversion efficiency (355 nm to tunable output) of 18 % is limited by the onset of damage to optical components which is consistent with earlier observations /1/ . Power levels of several hundred mW from UV to IR mark an encouraging step towards the development of highly efficient compact spectroscopic systems for widespread applications.

This research is supported by the Niedersächsische Minister für Wirtschaft, Technologie und Verkehr.

1. Y. Cui, D. E. Withers, C. F. Rae, C. J. Norrie, Y. Tang, B. D. Sinclair, W. Sibbett, M. H. Dunn, Opt. Lett. 18, 122 (1993).
2. A. Fix, T. Schröder, R. Wallenstein, Laser und Optoelektronik 23, 106 (1991).

Miniaturisierter diodengepumpter Festkörperlaser (µ-Las)

Th. Halldórsson

Daimler Benz AG, Forschung und Technik, D-81663 München

Übersicht

Mit der Einbindung der Fertigungsverfahren der Halbleitertechnologie zur Herstellung der Festkörperlaser der Seltenenerden können diese wesentlich kleiner gebaut und kostengünstiger in hohen Stückzahlen gefertigt werden. Beim Festkörperlaser bringt die Miniaturisierung grundsätzlich die folgenden physikalisch-technischen Vorteile: 1) Erhöhung der Effizienz durch verbesserte Anpassung des Strahlungsfeldes der Pumpdiode an die Resonatormode des Lasers, 2) vereinfachte Herstellung von Einfrequenzlaser hoher Frequenzstabilität, 3) kürzere Laserpulse und höhere Pulsspitzenleistung bei gütegeschalteten Lasern, 4) geringere Empfindlichkeit gegenüber Umwelteinflüssen wie mechanischer Belastung, Beschleunigungen und Temperaturänderungen. Die wirtschaftlichen Vorteile sind dagegen 1) Kostenreduzierung durch verringertes Materialvolumen und 2) Möglichkeit der kostengünstigen Massenanfertigung durch automatisierte Fertigungstechniken. Mit diesen Vorteilen eröffnen sich neue Anwendungsfelder der Lasertechnik vor allem in der Meßtechnik, Medizin und Displaytechnik die vorher wegen der hohen Kosten und aufwendiger Technik verschlossen blieben.

Die Aufgabe des von dem Bundesministerium für Bildung, Wissenschaft, Forschung und Technologie (BMBF) geförderten Verbundprojektes µ-Las mit 4 Hochschulinstituten, 3 mittelständischen und kleinen Unternehmen und Daimler Benz AG als Partner ist zu erforschen mit welchen technologischen Ansätzen und Mitteln die Miniaturisierung und Massenherstellung von diodengepumpten Festkörperlasern mit dem heutigen Stand der Mikorelektronik, Mikrooptik und Mikrosystechnik ermöglicht werden kann. Zum Erreichen der festgelegten Projektziele wurde eine stufenweise Entwicklung von Festkörperlasern mit einem zunehmenden Grad der Miniaturisierung und erhöhter technischer Komplexität durchgeführt. Aus der Vielfalt möglicher Festkörperlaser mußte für die Demonstration der Endergebnisse eine gewisse Auswahl getroffen werden. Es sollten bedeutende Eigenschaften des Festkörperlaser und vor allem Eigenschaften, die für die Anwendung von besonderem Interesse sind, realisiert werden. Diese sind:

- Leistungslaser mit Ausgangsleistung bis 1 W bei $\lambda = 1,06$ µm und 200 mW bei $\lambda = 0,53$ µm.

- Gepulste Laser im Leistungsbereich über 1 kW mit Pulswiederholfrequenzen bis in den kHz-Bereich, Pulsbreiten im ns-Bereich bei $\lambda = 1,06$ µm

- Effiziente Dauerstrichlaser bei der Wellenlänge um 2,8 µm

- Grundsatzversuche zu Lasertätigkeit auf verschiedenen anderen Wellenlängen wie 1,44 µm, 1,6 µm und 2,1 µm.

- Einmoden-Dauerstrichlaser mit Frequenzdurchstimmung über einen Frequenzbereich bis 100 GHz mit Ausgangsleistungen von einigen zehn Milliwatt mit Hilfe von Mikro-Aktoren.

Die Arbeiten an dem Vorhaben das Mitte 1995 abgeschlossen wird, waren in folgende Teilaufgaben der Partner gegliedert:

Simulation und Modellierung

Die Erfahrungen aus der Mikroelektronik haben gezeigt, daß die Simulation von Komponenten und Schaltungen große Einsparungen an experimentellen Testreihen und Prozeßdurchläufen bringt. Bis jetzt war bei der Herstellung von Lasern die Simulation auf Strahlausbreitung- und Resonatorberechnung, meist analytische Methoden beschränkt.

An der Technischen Universität Berlin, Forschungsschwerpunkt Technologien der Mikroperipherik wurden Kühlerstrukturen durch Simulation mit dem Einsatz von Fast Fourier Transformationen (FFT) optimiert. Bei Micro Systems Design Dr. Altmann GmbH wurde mit Hilfe von Finite Element Methoden (FEM) die Temperaturverteilung und die thermische Verformung der Laserkristalle bei unterschiedlichen Pumpgeometrien und Kühlgeometrien berechnet /1/. Mit Hilfe er Beam Propagation Methode (BPM) wurde dann die Modenstruktur nach dem Iterationsprinzip von Fox und Li berechnet, wobei die Ergebnisse der thermomechanischen Rechnungen als Eingabedaten dienen. Die Berechnung der Eigenlösungen der transversalen Wellengleichung mit Hilfe eines äquivalenten mechanischen Modells ermöglichte die Modellierung der Rückwirkung der Stahlauskopplung auf die Modenstruktur und somit komplette FEM-Simulation des dynamischen Laserverhaltens /1/.

Prozeßtechnologien

Mikrokühler für Laserdioden und Mikro-Aktorik für Laserspiegel wurden von der Firma Daimler Benz in Siliziumwafern prozessiert. Am Institut für Laserphysik der Universität Hamburg wurden spezielle Laserkristalle für Mikrolaser gezogen. Ihre Massenherstellbarkeit und Bearbeitung wurden von der Firma HAM, Kristalltechnologie, Andreas Maier GmbH untersucht.

Aubau- und Verbindungstechniken

Im Mikrolaser sind eine Reihe heterogener Komponenten, die Laserdiode, ihre Kühlstruktur und Teile der Netzversorgung auf einem Siliziumwafer integriert. Mit Hilfe von Photolithographie und anisotropem Ätzen sind Mikrokühlkanäle zur Flüssigkeitskühlung der Laserdiode und einiger elektronischer Bauteile strukturiert. Durch anodisches Bonden des Siliziums mit Pyrex-Glas werden die Kühlkanäle abgedichtet. Das Netzteil besteht aus Dickfilmkomponenten, die auf einer Keramikplatte prozessiert sind.

Der Laserrsonator besteht aus einer Anzahl gleichartiger Haltelemente, die um mehrere Achsen gegneinander justierbar sind. Sie erlauben einen Aufbau des Lasers in drei Dimensionen und werden nach der Justage miteinander durch Laser geschweißt (s. Abb. 1). Die Gestaltung der Laserresonatorstruktur sieht eine automatisierbare Bestückung, Justage und Fixierung vor.

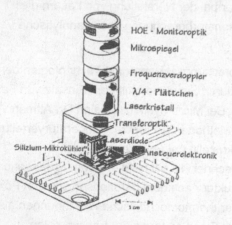

HOE - Monitoroptik
Mikrospiegel
Frequenzverdoppler
λ/4 - Plättchen
Laserkristall
Transferoptik
Laserdiode
Silizium-Mikrokühler
Ansteuerelektronik

Abb. 1

Pumpmodul mit Resonator

Komponentenherstellung

Die Laserdioden werden als Halbleiterchips im Entwicklungsauftrag extern gekauft. Die Ausgangsleistung der Dioden wurde auf 2 W beschränkt, die eine

Ausgangsleistung der Festkörperlaser bis 1 W ermöglichte. Es kommen nur AlGaAs und InGaAs-Dioden zum Einsatz. Eine besondere Herausforderung für die Projektpartener ist das Löten der Dioden auf die Mikro-Kühler /2/.

Eine weitere Aufgabe in dem Projekt war es geeignete Lasermaterialen und Beschichtungen für miniaturisierte Laser zu identifizieren und zu untersuchen /3/. Die Arbeiten konzentrierten sich vor allem auf Nd-dotiertes $GdVO_4$ und $LaSc_3(BO_3)_4$ und Er-dotiertes YAG, YSGG und GGG. Das Nd:LSB erwies sich aufgrund der hohen Dotierbarkeit und geringen Resonatorverluste als besonders geeignet um effiziente Neodymium Mikrolaser zu realisieren /4/. Es konnten mehrere single-frequeny Erbium laser bei 1,55 µm und 1.64 µm aufgebaut werden /5/ und ein effizienter 2,8 µm Er:Laser für endoskopische Anwendungen in der Medizin aufgebaut werden /6/.

Zur Strahlungstransfer von Pumpdiode zum Festkörperlaser als auch für des Monitoren der Laserstrahlung wurden diffraktive Optiken und Hybride aus diffraktiven, gradienten und refraktiver Elementen entwickelt /7/. Diese Arbeiten wurden an der Uni Stuttgart, TH Darmstadt und bei der Firma HSM in München durchgeführt

Eine besondere Herausforderung bei der Komponentenentwicklung ist die Entwicklung eines miniaturisierten Laserauskoppelspiegels mit Hilfe de Silizium-Prozesstechnologie, der mit einem elektro-statischen Aktor für Laserfrequenzstabilisierung bzw -durchstimmung und Güteschalten schnell in axialer Richtung schnell bewegt werden kann.

Zusammenfassung

Es konnten verschiedene technologische Verfahren der Mikroelektronik und Mikromechanik auf den Bau von Komponenten und die Gesamtstruktur eines diodengepumpten Festkörperlasers übertragen werden. Der endgültige Durchbruch zur verbilligten Festkörperlasern ist jedoch erst durch die Massenfertigung für eine sehr breite Anwendung solcher Laser möglich.

/1/ K. Altmann, Full FEM simulation of a miniaturized end-pumped laser rod, Proc. of the 13th International Congress, LASER 95, Munich, Springer Verlag, Berlin (1996)

/2/ S. Weiss, H. Reichl, Die bonding of laser diodes on CVD-diamond and thermosonci wire bonding of this device, Proc. of the International Congress, LASER 95, Munich, Springer Verlag, Berlin (1996)

406

/3/   G. Huber, Development and applications of new laser crystals, Proc. of the 13th International Congress, LASER 95, Munich, Springer Verlag, Berlin (1996)

/4/   J.-P. Meyn, G. Huber, Laser diode-pumped, intracavity frequency doubled neodymium laser, Proc. of the 13th International Congress, LASER 95, Munich, Springer Verlag, Berlin (1996)

/5/   S. Nikolov , L.Wetenkamp, Diode-pumped single frequency erbium lasers at 1.55 μm and 1,64 μm, Proc. of the 13th International Congress, LASER 95, Munich, Springer Verlag, Berlin (1996).

/6/   S. Nikolov, N.P. Schmitt, G.Reithmeier, Th. Halldorsson, Fiber coupled diode pumped Er:YSGG-Laser at 2,8 μm, Proc. of the 13th International Congress LASER 95, Munich, Springer Verlag, Berlin (1996)

/7/   Ch. Haupt, M. Daffner, A.Rothe, H. Tiziani, Herstellung und Test hybrider Mikrooptiken zur Korrektur von Hochleistungslaserdiodenstrahlung, Proc. of the 13th International Congress, LASER 95, Munich, Springer Verlag, Berlin (1996).

Development and Applications of New Laser Crystals

Günter Huber

Institut für Laser-Physik   Universität Hamburg
Jungiusstr. 11, D-20355 Hamburg   Germany

This paper reviews the properties of laser active centers in crystals as well as recent advances in crystalline solid-state lasers achieved with new rare-earth and transition metal ion doped crystals in the spectral region 500 nm to 3000 nm. The progress in this field is mainly based on the growth of new laser crystals, the application of diode laser pumping, and the realization of visible upconversion lasers.

In the near infrared spectral region high efficiencies have been achieved with diode pumped oxide and fluoride crystals doped with Nd^{3+}, Tm^{3+}, Ho^{3+}, Er^{3+}, and Yb^{3+}. Transition metal doped crystals based on the ions Ti^{3+}, Cr^{3+}, and Cr^{4+} emit tunable radiation in the wavelength range 0.68 μm to 1.63 μm. In the visible spectral region Er^{3+}, Tm^{3+}, and Pr^{3+}-lasers operate at several red, green, and blue transitions. Especially, Er^{3+}-doped crystals can create cw-551 nm laser light under upconversion excitation near 810 nm or 910 nm.

Diode Pumped CW Tm:YAG Laser Operating at Room Temperature

Rolf Heilmann

Deutsche Forschungsanstalt für Luft- und Raumfahrt (DLR)

Institut für Optoelektronik

D-82230 Oberpfaffenhofen

Efficient eye safe lasers ($\lambda > 1.4$ µm) can be used in laser radar (lidar) systems. Thulium doped YAG crystals ($\lambda \approx 2$ µm) have a relatively high absorption coefficient at 785 nm (~ 5 cm^{-1}) and a long flourescent lifetime of ~10 ms [1]. In addition to that, there is a Tm-Tm cross relaxiation energy transfer process which increases pump quantum efficiency [2]. Thus, Tm:YAG is a promising candidate for diode pumped continous wave (CW) and Q-switch operation, respectively.

However, atmospheric transmission near 2 µm is dominated by strong absorption lines of H_2O and CO_2 (Fig. 1). For application in lidar systems the lasers have to be tunable to a transmission window (for example at 2021.8 nm). Tm:YAG lasers allow for this.

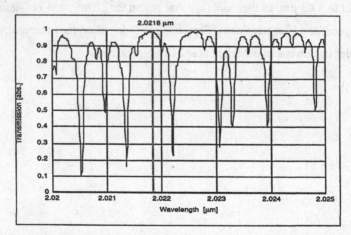

Fig. 1: Atmospheric transmission spectrum near 2.02 µm, calculated on the basis of the HITRAN database [3] for sea-level, path length 1 km

The significant disadvantage of Tm:YAG relative to Nd:YAG is the partially three-level-nature which leads to a high threshold at room temperature. Thus, Tm:YAG lasers with output powers > 100 mW (pump power ~1 W) operate normally with a crystal temperature of approximately - 40 ... -20 °C [4, 5].

In this paper it is shown that efficient CW operation at room temperature is possible. The configuration of the laser is figured in Fig. 2. The Tm:YAG rod (3mm x 6mm, concentration 4 %) is pumped with a 785 nm diode (1.25 W) via a multiple longitudinal mode fiber (core diameter 100 µm, numerical apertur 0.4) and a gradient index (GRIN) lens. A thermoelectric (TE) cooler is used to compensate the heat generated by pump light absorption. The radius of the 1.5 % output coupler is 100 mm.

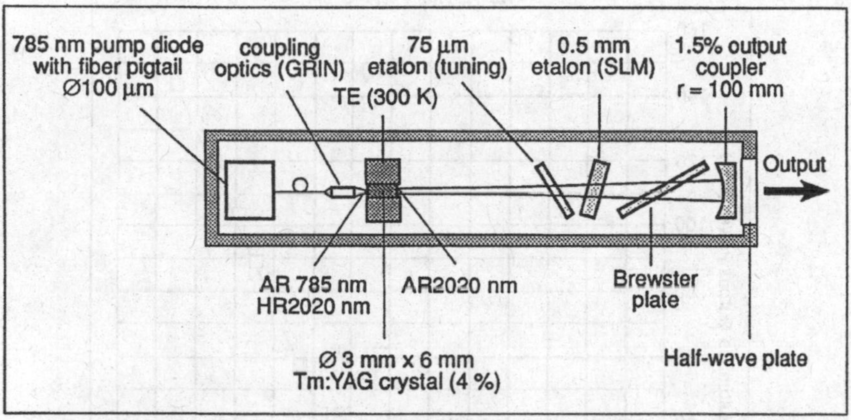

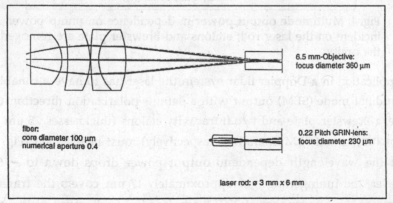

Fig. 2: Cavity configuration of a CW SLM Tm:YAG laser operating at room temperature

Room temperature operation is possible by optimization of the pump light coupling optics (see Fig. 3). An optimal matching of pump and laser mode

Fig. 3: Ray tracing calculations for diode beam shaping optics: 6.5 mm objective and 0.22 P GRIN lens

volume was realized within the 6 mm-crystal with an AR coated 0.22 Pitch GRIN lens. (With "classical" complex optics and/or other crystals and a pump power ≤ 1.2 W, CW operation could not be achieved at 20 ° C!)

Fig. 4 shows the input-output-characteristic of the optimized laser. A multimode output power of 145 mW has been reached with a pump power of 1.25 W. The threshold was ~ 0.7 W. The slope efficiency was estimated to be 0.26.

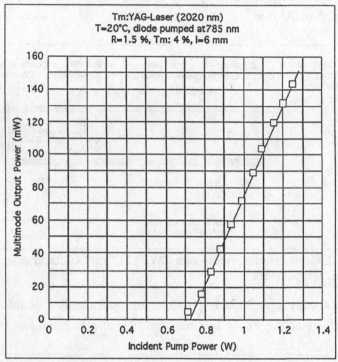

Fig. 4: Multimode output power in dependence on pump power incident on the laser rod; etalons and Brewster plate are removed from the cavity

For application in a Doppler lidar system the laser has to have a tunable single longitudinal mode (SLM) output with a definite polarization direction (> 98 %). Hence, a Brewster plate and two intracavity etalons (thicknesses 75 µm and 500 µm for tuning and SLM operation, respectively) must be installed (Fig. 2). As a result the wavelength dependend output power drops down to ~ 60 mW. However, the tuning range of approximately 17 nm covers the transmission window of the atmosphere at 2021.8 nm (see Fig. 5). Because of the achieved parameters the laser can be used succesfully as master oscillator in an eye safe coherent Doppler lidar system.

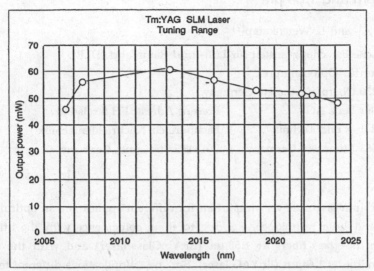

Fig. 5: SLM output power in denpendence on the wavelength; the transmission window of the atmosphere at 2021.8 nm is indicated (cf. Fig. 1)

The author would like to thank Mrs. S. Nikolov, DLR, for ray tracing calculations and F. Heine, Universität Hamburg, for providing Tm-doped crystals for test measurements.

This project is supported by the *Bayerische Forschungsstiftung e. V. München*, Germany.

References

[1] T. S. Kubo and J. Kane, IEEE J. Quantum Electron. **28**. 1033 (1992)

[2] G. Huber, E. W. Duczynski, and K. Petermann, IEEE J. Quantum Electron. **24**, 920 (1988)

[3] L. S. Rothman, R. R. Gamache, A. Goldman, L. R. Brown, R. A. Toth, H. M. Pickett, R. L. Pointer, J.-M. Flaud, C. Camy-Peyret, A. Barbe, N. Husson, C. P. Rinsland, and M. A. H. Smith, Appl. Opt. **26**, 4058 (1987)

[4] P. J. M. Suni and S. W. Henderson, Opt. Lett. **16**, 817 (1991)

[5] J. D. Kmetec, T. S. Kubo, T. J. Kane, and C. J. Grund, Opt. Lett. **19**, 186 (1994)

Diode Pumped Single Frequency Erbium Lasers at 1.55 μm and 1.64 μm

S. Nikolov[1] and L. Wetenkamp[2]

(1) Deutsche Forschungsanstalt für Luft- und Raumfahrt (DLR),
Institut für Optoelektronik
D-82230 Oberpfaffenhofen, Germany

(2) Daimler Benz AG   Present Adress: FH Stralsund
Forschung und Technik   Fachbereich Nachrichtentechnik
D-81663 München, Germany   D-18435 Stralsund, Germany

Eye safe erbium lasers are very important for different applications in optical communication and lidar techniques, e.g. due to their coincidence with the attenuation minimum of glass fibers at 1.55μm (Er:Yb:Glass laser) and with the methane absorption line at 1.64μm (Er:YAG laser). For these applications different techniques for obtaining single frequency operation have been investigated.

Single frequency operation of a diode pumped Er:YAG laser at 1.64μm was achieved first time with halfmonolithic laser crystals in C3-configuration. This special three mirror technique, where the uncoated internal surfaces of the two laser rods act as an air etalon, was formerly only applied to diode lasers. The C3-design offers especially for laser materials with low absorption coefficients (e.g. in Er:YAG: $\alpha=0.7$ cm^{-1} at 964 nm) an alternative solution for single frequency operation, where a short monolithic cavity cannot be applied. Single frequency operation with short monolithic resonators have been demonstrated in [1] with high Yb^{3+}-doped Er:Yb:Glass.

The most interesting C3-configuration is a sandwich combination from Er:YAG and Er:Yb:Glass, where the laser lines at 1.64 μm and 1.55 μm emitting simultanously, both pumped by the 964 nm laser diode. To our knowledge this was achieved first time. Until now only intracavity pumping of Er:YAG by an flashlamp pumped Er:Yb:Glass laser has been demonstrated [2].

Fig. 1 shows the experimentel setup for the C3-lasers. The input coupler (with radii of R=80 or 100 mm) was evaporated on the front surface of the first laser rod and the plane (pl) output coupler on the rear surface of the second laser rod. Both laser rods were temperature stabilised at 20°C with Peltier coolers.

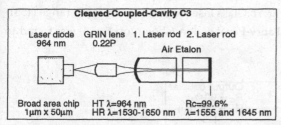

Fig. 1: Experimental Setup

Several laser rods, with lengths between 1 and 3 mm, made from Er:YAG (1.1 at. % Er^{3+}) and Er:Yb:Glass (Kigre QE-7S), have been investigated in different C3-configurations. Single frequency operation was achieved in laser cavities consisting of two Er:YAG rods as well as two Er:Yb:Glass rods. The air gap between both laser rods was approx. 100-200 μm, with a corresponding free spectral range of $\Delta\lambda$=12-6 nm, respectively. For an Er:YAG-C3-Laser (2mm R=80/pl+1mm pl/pl) the single frequency output power at 1646 nm was approx. 1 mW at P_p=900 mW diode pump power, with a laser threshold of 580 mW.

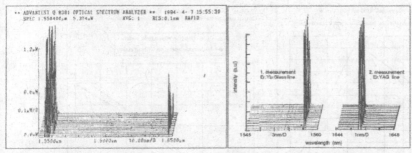

Fig. 2: Multimode Spectra of   Fig. 3: Er:Glass-YAG-Laser (sep. meas.)
        Er:Glass-YAG sandwich laser   Single Mode Spectra for each laser line

In a sandwich laser, containing an Er:YAG rod and an Er:Yb:Glass rod, the pump light has to be focussed on the Er:YAG crystal due to the high threshold of the Er:YAG laser at P_0=650 mW and the thermal damage threshold of the Er:Glass laser at P_0>700 mW. Fig. 2 shows the multimode spectrum of the Er:Glass-YAG Sandwich laser 2mm R80/pl + 1mm pl/pl, recorded by an optical spectrum analyser.

The output power at P_p= 900 mW for the Er:YAG line was approx. 1mW and for the Er:Glass line 5.5 mW. For each line of this laser it was also possible to achieve single frequency operation (Fig. 3).

For single frequency operation of an Er:Yb:Glass laser with a broad tuning range (1531-1939 nm) the technique with a tilted Fabry-Perot etalon (shown on Fig. 4) was found to be the best.

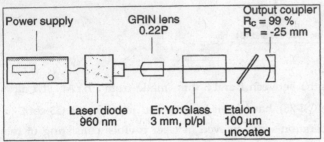

Fig. 4: Experimental setup for SLM operation with etalon

A halfmonolithic laser rod (plane/plane, 3mm length) and separate output coupler (R=-25 mm and R_c=99% at 1535 nm) was tuned with a 100µm-quartz etalon (FSR=104 GHz or 8.2 nm).

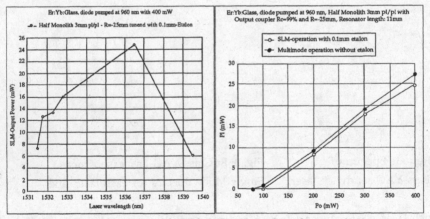

Fig. 5: SLM-Tuning range with Etalon Fig. 6: In-Out-characteristic

Fig. 5 shows the SLM-output power for a laser diode power of 400 mW with respect to laser wavelength. The laser was tuned by tilt the etalon. At the fluorescence maximum of Er:Yb:Glass at 1536 nm the SLM output power is only 10% lower than the multimode output power (see Fig. 6).

An important influence on the tuning range has the mirror coating as well. It can be extended if the reflection coefficient for 1555 nm is aprox. 0.05% higher than for 1535 nm. For these experiments the halfmonolithic glass rods for the sandwich laser (shown on Fig. 1) were used, because these rods were coated together with the Er:YAG crystals. At the same pump power the laser can be tuned within in the range 1552-1560 nm and around 1545 nm as well. For this tuning range the FSR of the 100μm-etalon is not sufficient and 3 laser modes oscillate.

A second etalon (this can be an uncoated internal resonator surface as well) is needed for SLM operation. Two etalons causes higher losses, therefore at P_p=400 mW an output power of 15 mW for 1536 nm and 4 mW for 1556 nm has been obtained.

Summary:
It has been demonstrated, that single frequency operation with the C3-technique is also applicable to diode pumped solid state lasers. The C3-technique offers new opportunities in miniaturisation of diode pumped solid state lasers for single frequency operation. For using the wide tuning range of Er:Yb:Glass a tilted Fabry-Perot etalon is the best mode selective element. This tuning range can be extended up to 1560 nm at the same pump power by a special mirror coating.

Acknowledgement:
This work has been supported by Daimler Benz AG Munich within the BMBF-Project "μ-Las".

References:
[1] P. Laporta, S. Longhi, S. Taccheo, O. Svelto, G. Sacchi: "Diode-pumped micro chip Er:Yb:glass laser", Optics Lett., Vol. 18, No. 15, Aug. 1993, pp. 1232-1234

[2] K. Spariosu, M. Birnbaum: "Intracavity 1.549μm Pumped 1.634 μm Er:YAG-Lasers at 300 K", IEEE J. of Quantum Electronics, Vol. 30, No. 4, April 1994, pp. 1044-49

Emission Properties of SrLaGa$_3$O$_7$:Pr^{3+} Crystals

W. Woliński, M. Malinowski, I. Pracka[1]) and R. Wolski

Institute of Microelectronics and Optoelectronics IMiO, Warsaw University of Technology, ul.Koszykowa 75, 00-662 Warsaw, Poland
[1]) Institute of Electronic Materials Technology ITME, ul. Wólczyńska 133, 01-910 Warsaw, Poland

Recently, laser action in trivalent praseodymium (Pr^{3+}) activated solids has been extensively studied and reported in many systems [1,2,3].

Here we report on the emission properties of Pr^{3+} ion in disordered melilite structure SrLaGa$_3$O$_7$ (abbreviated SLG) crystals. The principal interest in studying rare earth doped melilite type materials is due to their structural disorder and resulting inhomogeneous broadening of the optical transitions [4,5]. Broadened absorption and emission spectra are suitable for diode laser pumping and tunable laser generation respectively. SLG offers also good thermal conductivity and possibility of high doping levels. The crystals used in this study were prepared at the Institute of Electronic Materials Technology (ITME) laboratory using methods described earlier [6]. Two crystals of praseodymium concentrations of 0.5 and 1 at. % have been investigated. After excitation by the 476 nm blue line of an argon laser, or pulsed dye laser pumping, emission transitions from the $^3P_0$ (20648 cm^{-1}) and $^1D_2$ (16536 cm^{-1}) excited states to several lower lying multiplets have been observed in the visible and near infrared (IR) region, see Fig1.

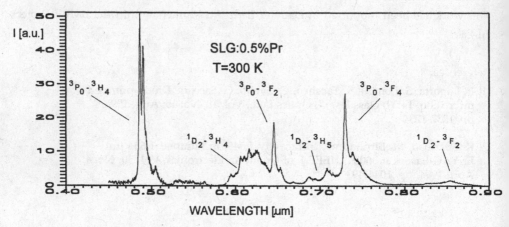

Fig.1. Room temperature emission spectrum of SrLaGa$_3$O$_7$:Pr^{3+} crystal.

It could be noticed that SLG:Pr^{3+} fluorescence is dominated by the transitions from the excited $^3P_0$ state, weak emission originating from the $^1D_2$ multiplet, populated by the non-radiative relaxation from the $^3P_0$ state, could be observed in the 700 nm and 850 nm range. The strongest observed fluorescence was in the 490, 644 and 732 nm bands and was assigned to the $^3P_0 \rightarrow {}^3H_4$, $^3F_2$, $^3F_4$ transitions respectively. The determination of individual Stark levels of the Pr^{3+} ions was made by analysing low temperature absorption, excitation and emission spectra which showed sharper lines. However, it should be noticed that even at 10 K the optical transitions of Pr^{3+} ion in SLG are broader than in other crystals. Also, low temperature emission linewidths and lineshapes were found to be excitation wavelength dependent. Thus, some of the individual Stark components are not well resolved and reported here energy levels are good within 5 cm^{-1}. A summary of the energy levels derived from our measurements is given in Table 1.

Table I. Energy levels of Pr^{3+} in SLG measured at 10 K.

[S'L'J']	E [1/cm]
$^3P_2$	22331, 22702
$^1I_6$	
$^3P_1$	21130, 21355, 21510
$^3P_0$	20648
$^1D_2$	16536, 16853, 17003, 17088, 17219
$^1G_4$	9964*, 10037*
$^3F_4$	6903, 6956*, 7017*, 7099, 7450
$^3F_3$	6514, 6629,
$^3F_2$	5148, 5159, 5208, 5310, 5377
$^3H_5 + ^3H_6$	2118, 2228, 2337, 2448*, 2580*, 2682, 2810, 2887*, 2861, 2844, 2918, 2958, 3109, 3500, 3554, 4249
$^3H_4$	0, 126, 163, 228, 348, 445, 718

* - denotes less accurate results

Figures 2 and 3 show the decay profiles of the $^3P_0$ and $^1D_2$ levels. In 0.5 % Pr^{3+} doped sample at room temperature these directly excited decays are exponential with time constants of 28 μs and 245 μs respectively, at higher Pr^{3+} concentration fluorescence decays were faster and non exponential. Following excitation of the $^3P_0$ level a rise time of 15 μs is observed in the decay profile of the $^1D_2$ emission. This rise time can be attributed to the weak nonradiative decay from the $^3P_0$ level to the $^1D_2$ level. Room temperature absorption spectrum allowed to determine oscillator strengths for several transitions and to perform calculations in the framework of Judd-Ofelt theory [7,8]. From the least-square fit of measured and calculated oscillator strengths the three intensity parameters are found to be; $\Omega_2 = 0.4 \times 10^{-20}$ cm^2, $\Omega_4 = 3.6 \times 10^{-20}$ cm^2 and $\Omega_6 = 2.3 \times 10^{-20}$ cm^2 [9]. In performing the calculation the reduced matrix elements were taken from Pappalardo [10]. The RMS deviation of 1.6×10^{-6} is comparable to values found by applying Judd-Ofelt theory to Pr^{3+} ion in other systems. From the calculated set of Ω_t intensity parameters the electric dipole transition probabilities from the excited $^3P_0$ state together with the resulting branching ratios were calculated. The highest calculated value of the branching ratio, $\beta = 0.70$, is for the $^3P_0 \rightarrow ^3H_4$ transition. This is found to be in agreement with the experimentally determined from the room temperature emission spectrum, showing a very intense blue emission, value of $\beta = 0.56$. One of the most important spectroscopic parameters of an active medium is the effective emission cross section σ_{em}. For the strongest blue fluorescence peak at 484.3 nm, σ_{em} has been determined to be 1.9×10^{-20} cm^2, see Fig.4, which is close to the peak σ_{em} value of 1.8×10^{-20} cm^2 reported for blue emission in Pr^{3+} in ZBLAN. Since in SLG:Pr^{3+} the second Stark level is 126 cm^{-1} above the ground state and is only slightly thermally populated at room temperature, the intense emission transition from $^3P_0$ to $^3H_4(2)$ at 487.3 nm is a quasi four level system and is of practical interest for blue laser.

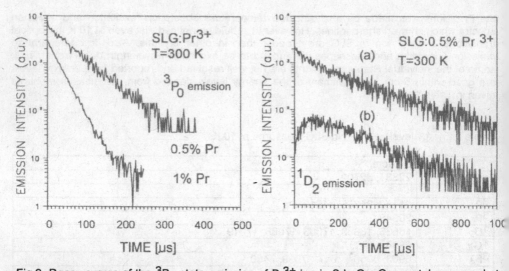

Fig.2. Decay curves of the $^3P_0$ state emission of Pr^{3+} ion in $SrLaGa_3O_7$ crystal measured at room temperature

Fig.3. Decay curves of the $^1D_2$ emission of Pr^{3+} ion in $SrLaGa_3O_7$ after: a) direct, orange excitation of the $^1D_2$ state, b) after blue excitation of the $^3P_0$ state, at room temperature.

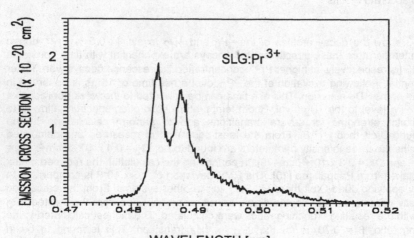

Fig.4 Blue emission cross section spectrum corresponding to the $^3P_0 \rightarrow {}^3H_4$ transition of Pr^{3+} ions in SLG crystal.

IV. Conclusions.

The spectroscopic properties of Pr^{3+} ion in $SrLaGa_3O_7$ crystals were studied and analysed. Absorption spectra have been obtained leading to the determination of the energy level diagram and Judd-Ofelt intensity parameters. Because SLG is a partially disordered host, the observed linewidths are correspondingly broader and emission cross sections are lower than in ordered crystals. SLG:Pr^{3+} presents longer lifetimes than most of praseodymium doped oxide crystals. This, together with the high value of the branching ratio for the $^3P_0 \rightarrow ^3H_4$ transition and large splitting of the ground state, makes this material useful in developing efficient blue laser. In Table II comparison of the spectroscopic and laser properties of Pr^{3+} doped SLG and YAG [11] crystals has been presented.

Table II. Comparison of the structural and spectroscopic properties of Pr^{3+} doped SLG and YAG crystals measured in our laboratory. (N - active ion concentration, ΔE - transition bandwidth, τ- fluorescence lifetime, E- energy)

	$SrLaGa_3O_7$:Pr^{3+} (SLG)	$Y_3Al_5O_{12}$:Pr^{3+} (YAG) [11]
structure	melilite	garnet
symmetry, point group	tetragonal, C_2	cubic, D_2
optical class	isotropic	uniaxial
refractive index (633 nm)	1.82	1.84
N(1 at% Pr) [cm^{-3}]	0.56×10^{20}	1.385×10^{20}
E($^3P_0$) [cm^{-1}]	20648	20534
$\Delta E(^3P_0)$ [cm^{-1}]	68	1.5
$\tau(^3P_0)$ [μs]	28	12
E($^1D_2(1)$) [cm^{-1}]	16534	16400
$\tau(^1D_2)$ [μs]	320	245
laser emission λ [nm] (after pulsed excitation of $^3P_1$ state at 10 K) $^3P_0 \rightarrow ^3H_4$ $^3P_0 \rightarrow ^3F_2$	484 645	487.9 616
max. emission cross section σ ($^3P_0 \rightarrow ^3H_4$) [10^{-20} cm^2]	1.9	7.3

References.

[1] A.A. Kaminskii OSA Proc. on Advanced Solid-State Lasers, 1993, Vol.15, eds. A.A. Pinto and Tso Yee Fan, 1993 Optical Society of America, p.266.
[2] T. Danger, A. Bleckmann and G. Huber, Appl.Phys. B **58** (1994) 413
[3] M. Malinowski, M.F. Joubert, R. Mahiou, and B. Jacquier, J. de Physique IV (1994) C4-541
[4] A.A. Kaminskii, E.L. Belokoneva, B.V. Mill, S.E. Sarkisov, and K. Kurbanov, Phys.Stat.Sol. (a) **97** (1986) 279
[5] W. Ryba-Romanowski, S. Gołąb, G. Dominiak-Dzik, and M. Berkowski, Mat.Science and Engineering B **15** (1992) 217
[6] I. Pracka, W. Giersz, M. Świrkowicz, A. Pajączkowska, S. Kaczmarek, Z. Mierczyk and K. Kopczyński, Mat.Science and Engineering B **26** (1994) 201
[7] B.R. Judd, Phys.Rev. **127** (1962) 750
[8] G.S. Ofelt, J.Chem.Phys. **37** (1962) 511
[9] M. Malinowski, I. Pracka, B. Surma, T. Łukasiewicz, W. Wolinski,and R. Wolski, submitted to Chem.Phys.Lett.
[10] R. Pappalardo, J.Lumin. **14** (1976) 159
[11] M.Malinowski, R.Wolski and W.Woliński, Solid State Comm. **74** (1990) 17

Saturable Absorbers for Ruby Laser Based on Cr^{4+}-Doped Silicates

V.P.Mikhailov, N.V.Kuleshov, N.I.Zhavoronkov, A.S.Avtukh, V.G.Shcherbitsky, B.I.Minkov
International Laser Center, Kurchatov str.7, Minsk 220064, Belarus

Cr^4-doped Y$_2$SiO$_5$, Gd$_2$SiO$_5$ and Mg$_2$SiO$_4$ crystals have been investigated as passive Q-switches for ruby laser. The saturation of absorption at 694 nm was observed under laser pumping, and ground state and excited state absorption cross sections were estimated.

Passive Q-switching in a ruby laser was demonstrated using Cr^4-doped Mg$_2$SiO$_4$ and Y$_2$SiO$_5$ as intracavity saturable absorbers. The hemispherical laser cavity was formed by 1.0 m radius high reflector and 50% reflectivity flat output coupler. Laser rod was 10 mm in diameter and 120 mm in length. The polished samples of saturable absorber crystals were 2-10 mm in thickness and were not antireflection coated. In the free running operation (without saturable absorber Q-switch) laser pulses obtained were 0.5 J in energy and about 1 ms in duration. At the same pumping conditions single pulses of about 80-100 ns in duration and up to 240 mJ in energy were generated from Q-switched laser using intracavity Cr:forsterite and Cr:YSO saturable absorbers. Low optical damage threshold of Cr:Gd$_2$SiO$_5$ crystal did not allow to use this crystal as a passive laser shutter.

Optische Mess- und Prüftechnik
Optical Measuring Technique
H. J. Tiziani

Einleitung/Introduction

H.J. Tiziani, Institut für Technische Optik
Universität Stuttgart, D - 70569 Stuttgart, Germany

Optical measuring techniques are becoming useful tools for industrial applications. Compact and reliable coherent light sources such as small and rubust semi - conductor laser as well as the information processing with powerful computer together with solid state detector arrays have led to new applications. The acceptance of optical contactless measuring systems in manufacturing and process controll systems as well as for environmental - and medical applications is increasing.

For the distance measurement improvements in accuracy for time of flight as well as phase measurements were obtained. To improve the accuracy but especially to extend the range further, new interferometric principles based on heterodyne together with multiple wavelength were introduced in order to obtain high precision absolut distance measuring devices. Futhermore, interferometry ist applied for the measurement of technical surfaces (optically rough surfaces). It should be noted, that the interference at the microstructure leads to the well know speckle phenomen on.

Speckling can be used however as a useful tool in metrology and applied in methods based on speckle interferometry or speckle photography. More often however, speckling is limiting the accuracy of coherent optical measuring techniques.

Optical probes have many advantages, especially for high speed, applications. They are non contacting and high precision measurements can be obtained.

For industrial applications, calibration and standards become important. Especially when using optical measuring techniques.

In the congress of optical measurement 67 papers together with posters were presented. The main emphasis was on Interferometry, Holography and Speckle applications together with 3-D-measurement as well as surface analysis and spectroscopical methods. The knowledge of the 3-dimensional shape of an object as well as the 3-dimensional deformation is important for a wide range of industrial, as well as for medical applications.

A variety of procedures have been proposed for 3-D-range and data acquisition based on optical triangulation stereo image processing or Moiré techniques. The projected fringe and coded light approach leads to robust measuring principles. Interferometry, multiple wavelength or short coherence interferometrie were developed for technical surface measurements.

The application of intelligent sensor systems can lead to very efficient sensor devices. Image sensors have taken advantage from the advances in semiconductor technology. Image sensors carying out image processing algorithms on the sensor chip has led to powerful "seeing chips".

Quality assurance of products for instance is an example for increasing need for optical high precision measurement techniques. Scanning of models in connection with CAD systems or the recording of the shape of human organs for plastic surgery is very useful. Recently, a lot of research was carried out in the field of 3-D-data aquisition. However, so far only a limited number of applications have been implemented. It seems that the maturity and efficiency of the developed systems has advanced to such a degree that a vast breakthrougt is near. I hope that. The congess will help to spread the knowledge.

Multiple-Wavelength and White-Light Interferometry

R. Dändliker, E. Zimmermann, U. Schnell, Y. Salvadé
Institute of Microtechnology, University of Neuchatel,
CH-2000 Neuchatel, Switzerland

Abstract: Absolute distance measurement by optical interferometry is possible using multiple-wavelength or white light sources. Multiple-wavelength interferometry enables to increase the range of unambiguity and to reduce the sensitivity of the measurement. It can also be operated on rough surfaces. The accuracy depends on the properties of the source. White-light interferometry is most convenient for short distances. We have extended this method to include effects of thin transparent films deposited on the target, as it is often the case in profilometry on micro-electronic circuits or for auto focusing in photolithography.

1. Introduction

The interferometric measurement at different, well-known optical wavelengths enables the generation of new synthetic wavelengths, which are longer than the optical wavelengths [1,2,3]. For two wavelengths λ_1 and λ_2, one obtains the synthetic wavelength $\Lambda = \lambda_1\lambda_2/|\lambda_1-\lambda_2|$. This method thus makes it possible to increase the range of unambiguity for interferometry and to reduce the sensitivity of the measurement. In order to obtain this new, synthetic wavelength, the different optical wavelengths must be interrelated. Real-time electronic signal processing is mandatory for practical applications.

White-light interferometry is commonly used to measure absolute lengths by path length compensation. By adding a dispersive element (grating) the white-light interferometer becomes a multiple-wavelength interferometer. However, the maximum working distance is limited by the spectral resolution of the dispersive element.

Examples for both types of interferometry for absolute distance measurement will be given in the following.

2. Distance Measuring with Multiple-wavelength Interferometry

Two-wavelength heterodyne interferometry was reported by Fercher et al. [4]. By simultaneous phase measurement at both wavelengths, the interference phase at the synthetic wavelength Λ can be directly determined. This method provides fast measurement and works also for rough surfaces. However, because the two wavelengths must be optically separated (prism, grating) before detection, the technique can be used only for relatively large wavelength differences and thus small synthetic wavelengths Λ. Superheterodyne detection, introduced by Dändliker et al. [5,6], enables high-resolution measurements at arbitrary synthetic wavelengths Λ without the need for interferometric stability at the optical wavelengths λ_1 and λ_2 or separation of these wavelengths optically. This is of great importance for range finding and industrial distance measuring with sub-millimeter resolution.

Modern laser sources provide a great variety of optical frequencies or wavelengths to get interesting synthetic wavelengths $\Lambda = c/\Delta\nu$. A selection is given in Table 1. The resolution δL for the distance measurement is calculated for an assumed phase interpolation of $2\pi/100$ at the synthetic wavelength Λ. For highly accurate distance measurements ($\delta L/L \leq 10^{-5}$), the synthetic wavelength has to be known with at least the same accuracy. Therefore the two laser sources must be stabilized with respect to each other and the synthetic wavelength must be calibrated [7].

Laser sources			δL $(2\pi/100)$
HeNe laser, 632.8 nm, stabilized:			
2 longitudinal modes	$\Delta\nu \approx 1$ GHz	$\Lambda \approx 30$ cm	1.5 mm
HeNe laser lines:			
632.8 nm, 635.2 nm		$\Lambda = 167.5$ μm	850 nm
629.4 nm, 632.8 nm		$\Lambda = 117.1$ μm	600 nm
629.4 nm, 632.8 nm, 635.2 nm		$\Delta\Lambda = 389.2$ μm	
GaAlAs diode lasers, tunable:			
tuning by current	$\Delta\nu < 30$ GHz	$\Lambda > 10$ mm	50 μm
tuning by temperature	$\Delta\nu < 600$ GHz	$\Lambda > 0.5$ mm	2.5 μm
Nd:YAG laser, tunable	$\Delta\nu < 60$ GHz	$\Lambda > 5$ mm	25 μm

Table 1   Selection of laser sources for multiple wavelength interferometry. The resolution δL is given for a phase interpolation accuracy of $\delta\phi = 2\pi/100 = 3.6°$.

3. Signal processing

Superheterodyne detection [5,6]: Two laser sources of different optical frequencies ν_1 and ν_2, corresponding to the wavelengths λ_1 and λ_2, are used to illuminate simultaneously a Michelson-type heterodyne interferometer. For this purpose, each source is followed by a device which creates two orthogonal polarizations of slightly different frequency. These frequency differences can be produced by acousto-optical modulators and are typically $f_1 = 40.0$ MHz and $f_2 = 40.1$ MHz. Two photodetectors behind appropriate polarizers produce reference and interferometer signals $I_r(t)$ and $I(t)$ of the form

$$I(t) = a_0 + a_1 \cos(2\pi f_1 t + \phi_1) + a_2 \cos(2\pi f_2 t + \phi_2), \tag{1}$$

which is the sum of the two heterodyne signals for the wavelengths λ_1 and λ_2, with the corresponding interferometric phases ϕ_1 and ϕ_2. For an interferometric path difference L, the phases ϕ_1 and ϕ_2 are given by

$$\phi_1 = 4\pi L/\lambda_1 = 4\pi \nu_1 L/c \quad \text{and} \quad \phi_2 = 4\pi L/\lambda_2 = 4\pi \nu_2 L/c. \tag{2}$$

They are sensitive to path changes ΔL of the order of the optical wavelengths. Therefore, to determine the phase difference $\phi = \phi_1 - \phi_2$, the phases ϕ_1 and ϕ_2 must be measured accurately, which requires interferometric stability of the setup. However, if the two heterodyne signals are fed to a quadratic detector (mixer), the phase difference $\phi = \phi_1 - \phi_2$ can be measured directly. Because $f_1 - f_2$ is chosen small compared with f_1 and f_2, the detector output [Eq. (1)] has the form of a carrier-suppressed amplitude-modulated signal with carrier $(f_1+f_2)/2$ and modulation frequency $(f_1- f_2)/2$. After amplitude demodulation, one gets therefore

$$I_{dem}(t) = a_{12} \cos[2\pi (f_1 - f_2)t + (\phi_1 - \phi_2)]. \tag{3}$$

This signal at $f = f_1 - f_2$ makes it possible to measure directly the phase difference

$$\phi = \phi_1 - \phi_2 = 4\pi (v_1 - v_2) L/c = 4\pi L/\Lambda, \tag{4}$$

which is now only sensitive to the effective wavelength $\Lambda = c/(v_1 - v_2) = c/\Delta v$.

Successful application of superheterodyne detection has been reported for multiple-wavelength interferometry with different types of sources, namely two detuned single frequency Ar lasers ($\Lambda = 60$ mm) [6], diode laser and acousto-optic modulator for a 500 MHz frequency shift ($\Lambda = 0.6$ m) [8], two-wavelength HeNe laser ($\Lambda = 55.5$ μm) [9], tunable Nd:YAG lasers ($\Lambda = 0.12 \ldots 1.5$ m) [10].

Synthetic fringe detection: In some cases, a simpler detection method might be of interest. If the two heterodyne frequencies f_1 and f_2 are chosen equal ($f_1 = f_2$), Eq. (1) becomes

$$I(t) = a_0 + a_1 \cos(2\pi f_1 t + \phi_1) + a_2 \cos(2\pi f_1 t + \phi_2). \tag{5}$$

The interference fringe function for the synthetic wavelength Λ is then obtained by detecting the electrical power of the ac part of this signal, which is

$$P_{ac} = \frac{1}{2} [a_1^2 + a_2^2 + 2a_1a_2 \cos(\phi_1 - \phi_2)]. \tag{6}$$

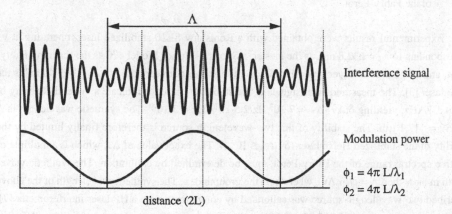

Interference signal

Modulation power

$\phi_1 = 4\pi L/\lambda_1$
$\phi_2 = 4\pi L/\lambda_2$

distance (2L)

Fig. 1   Interference signal for two wavelengths λ_1 and λ_2 as a function of the distance L (Eq. (5) with $f_1 =0$) and the corresponding modulation power of this interference signal (Eq. (6)).

The interference phase $\phi = \phi_1 - \phi_2 = 4\pi\, L/\Lambda$ of the synthetic wavelength can now be determined by techniques similar to those used for phase interpolation in interferometry. In practice, the frequency shift f_1 can be obtained by an acousto-optical modulator or by path length modulation in the interferometer. Figure 1 shows the detected interference signal for two wavelengths λ_1 and λ_2 as a function of the distance L (Eq. (5) with $f_1 = 0$) and the corresponding modulation power of this interference signal (Eq. (6)).

4. Stabilized laser wavelengths

For highly accurate distance measurements ($\delta L/L \le 10^{-5}$), the synthetic wavelength $\Lambda = c/\Delta v$ has to be known with at least the same accuracy. Therefore the two laser sources must be stabilized with respect to each other. This can be done with the help of a common reference length in the form of a Fabry-Pérot resonator, as shown in Fig. 2. Absolute accuracy can be obtained, if the scanning Fabry-Pérot is stabilized with respect to a frequency stabilized master laser. The slave laser is then tracked by means of the Fabry-Pérot, which is stabilized with respect to the master laser.

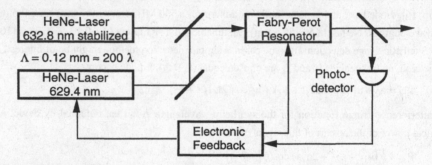

Fig. 2   Stabilized two-wavelength source. The Fabry-Pérot is stabilized with respect to the frequency stabilized laser (632.8 nm). The slave laser (629.4 nm) is then tracked by means of the Fabry-Pérot.

Experimental results were obtained with a Renishaw SL10 stabilized laser, operating at v_1 corresponding to $\lambda_1 = 632.8$ nm, as the master, a Spectra-Physics Model 470 as the scanning Fabry-Pérot, and a Spindler + Hoyer type ML500, operating at v_2 corresponding to $\lambda_2 = 629.4$ nm, as the slave laser [3]. The measured error signals of the stabilization loops indicate an overall stability of about $\pm\, 7$ kHz, yielding $\delta(\Delta v)/\Delta v = 7\cdot10^3$ Hz$/2.5\cdot10^{12}$ Hz $\approx 3\cdot10^{-9}$. The synthetic wavelength is $\Lambda = c/\Delta v = 117.1\ \mu$m. The stability of this two-wavelength source is therefore finally limited by the stability of the stabilized HeNe laser ($\delta v_1/v_1 \le 10^{-7}$). The exact value of Δv, which is a multiple of the free spectral range of the Fabry-Pérot, must be determined by calibration. The synthetic wavelength in air is then $\Lambda = c/(n_g\Delta v)$, where n_g is the group index. The synthetic wavelength of the above described two-wavelength source was calibrated by comparison with a HP laser interferometer [7]. As a result we got $\Lambda = 115{,}715.89 \pm 0.08$ nm. This gives a relative accuracy of $\delta\Lambda/\Lambda = 0.7\cdot10^{-6}$, which is consistent with an interpolation accuracy of $2\pi/100$ (or 0.6 μm) for a distance of $L \approx 1$ m.

An even more versatile multiple-wavelength source is obtained by locking several diode lasers with appropriately chosen wavelengths to a common Fabry-Pérot resonator, which is stabilized with respect to a HeNe master laser. In this case, however, the coherence length of the diode lasers may become a limiting factor for the multiple-wavelength interferometry.

We have also investigated a source with two diode-lasers (Sharp LT027MD), operating at $\lambda = 780$ nm [11]. The difference between them was about 6 nm. The measured synthetic wavelength was $\Lambda = 0.12$ mm. The wavelength was adjusted via the junction temperature and stabilization was achieved by controlling the injection current. One of the diode-lasers was locked to the resonance of a Rubidium vapor cell at 780 nm to give the absolute stability. This laser serves as the master to track a Fabry-Pérot to which, in turn, the second diode-laser is tracked as a slave. Generation of the error signal can be achieved by modulation of the injection current rather than the Fabry-Pérot length. From the measured error signals in the stabilization loops we estimated an overall stability of the laser frequencies of about $\pm$ 150 kHz, which corresponds to $\delta\Lambda/\Lambda \approx 5\cdot10^{-8}$. The linewidth $(\Delta\nu)_L$ of these lasers was measured to be about 20 MHz, which gives a coherence length of $c/\pi(\Delta\nu)_L = 4.8$ m.

5. Calibrated multiple-wavelength source

The absolute accuracy of distance measurement by multiple-wavelength interferometry depends essentially on the stability and the calibration of the different frequencies (or wavelengths). The accuracy δL for the measured distance L is given by

$$\delta L/L = \delta(\Delta\nu)/\Delta\nu , \tag{7}$$

where $\Delta\nu$ is the frequency difference corresponding to the synthetic wavelength $\Lambda = c/\Delta\nu$. Figure 3 shows the concept of a multiple-wavelength source with absolute calibration by electronic beat frequency measurement [7]. This three-wavelength source consists of three diode lasers operating at the frequencies ν_1, ν_2 and ν_3. Two of them (ν_1 and ν_2) are stabilized on two consecutive resonances of a common stable Fabry-Pérot resonator. In our experiment, the Fabry-Pérot resonator has a free spectral range of 0.75 GHz, as shown in the bottom part of Fig. 3. The corresponding beat frequency $\nu_{21} = \nu_2 - \nu_1 = 0.75$ GHz is detected and measured by a frequency counter with electronic accuracy. The third laser is tuned (without mode hopping) over N resonances of the Fabry-Pérot. The frequency difference $\nu_{31} = \nu_3 - \nu_1$ is then known with the same relative accuracy as the electronically calibrated beat frequency ν_{21}. For N = 100 we obtain $\nu_{31} = N\times\nu_{21} = 75$ GHz ($\Lambda_{31} = 4$ mm). To get an accuracy of $\delta L/L = 10^{-6}$, the free spectral range of the Fabry-Pérot must be calibrated with an accuracy of $\delta\nu_{21}/\nu_{21} = 10^{-6}$, or in our case $\delta\nu_{21} = 0.75$ kHz, which is feasible for a stable Fabry-Pérot by long time averaging. The required short time stability of the laser diodes for the high resolution distance measurement with Λ_{31} is obtained from $\delta\nu_{31}/\nu_{31} = 10^{-6}$, which is in our case $\delta\nu_{31} = 75$ kHz. With such a multiple-wavelength source it would be possible to measure distances without ambiguity within 200 mm ($\Lambda_{21} = 400$ mm) and with a resultion of 20 μm ($\Lambda_{31} = 4$ mm, $2\pi/100$ interpolation).

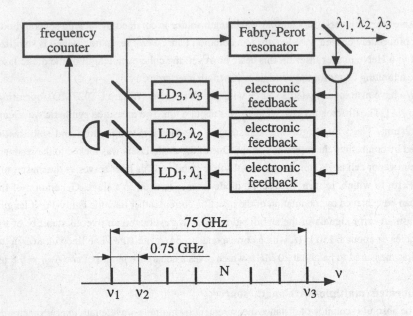

Fig. 3   Multiple wavelength source using laser diodes with absolute calibration by electronic beat frequency measurement.

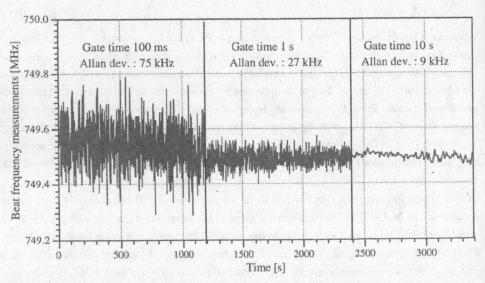

Fig. 4   Beat frequency measurements for different gate times T of the frequency counter.

Experimental investigations were performed with commercial GaAlAs laser diodes (Sharp LT027) emitting at 780 nm with about 5 mW optical power. The measured closed loop stability (square root of the Allan variance) of the diode lasers was better than 10 kHz. The detected beat

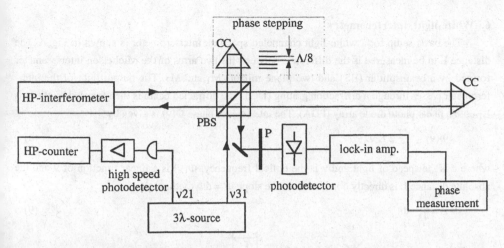

Fig. 5   Optical setup used for the calibration of the synthetic wavelength by comparison with an HP laser interferometer.

frequency signal at 0.75 GHz was 35 MHz wide, corresponding to the sum of the spectral widths of the two lasers. The exact value of the beat frequency was determined with a calibrated frequency counter (HP 53131). The results of the measurements for different integration times of the counter are shown in Fig. 4. The ratio of the Allan deviation for the gate times of 10 s and 100 ms is approximately 10, as expected for a noise limited system. It was verified, that the fluctuations of the beat frequency are mainly caused by the low frequency noise of the laser diodes. For an integration time of 10 s the difference frequency Δv_{21} is therefore calibrated with an accuracy of $\delta v_{21}/v_{21} = 1.3 \cdot 10^{-5}$, and the same accuracy can be expected for Δv_{31}, which gives for N = 100 a synthetic wavelength of about $\Lambda_{31} = c/\Delta v_{31} = c/N\Delta v_{31} \approx 4$ mm. In order to prove these concept, we compared the calibration of the synthetic wavelength Λ_{31} obtained from the beat frequency measurement at Δv_{21} with a calibrated HP laser interferometer as reference.

Figure 5 shows the calibration setup with the HP laser interferometer. We used v_1 and v_3 to illuminate the Michelson interferometer. A number of N = 100 Fabry-Pérot resonances were counted while tuning the third laser diode from v_1 to v_3. The detection at the output of the two-wavelength interferometer is done by a photodiode and a lock-in amplifier to obtain the power of the heterodyne signal, as described by Eqs. (5) and (6). To measure the phase of the synthetic wavelength, we moved the reference mirror in steps of 0.5 mm to get five 90° phase steps. From the corresponding measured values of the heterodyne signal power the phase is calculated with a standard phase stepping algorithm. By measuring over a common path difference of about 1 m we got for the synthetic wavelength Λ_{31} = 4.00014 ± 0.00006 mm (in air with TCN = 0.9997288), corresponding to an interpolation accuracy of $2\pi/100$. Then we determined the beat frequency from ten measurements obtained with an integration time of 10 s, yielding v_{21} = 749.25 ± 0.01 MHz. We found with this value a synthetic wavelength of 4.00014 ± 0.00005 mm (in air as above). This result proves that a calibration of the synthetic wavelength with an accuracy of at least $1.25 \cdot 10^{-5}$ is achieved.

6. White-light interferometry

The basic setup for a white-light channeled spectrum interferometer is shown in Fig. 6. The distance L to be measured is the difference between the two arms of the Michelson interferometer formed by a beamsplitter (BS) and two plane mirrors M_1 and M_2. The output light of the interferometer passes through a diffraction grating (DG). The diffracted beam is then focused by the lens L_1 onto a linear photo diode array (PDA). The interference phase $\Phi(\nu)$ is given by

$$\Phi(\nu) = \frac{2\pi}{c} 2 L \nu , \tag{8}$$

where c is the speed of light and ν is the optical frequency. $\Phi(\nu)$ is a linear function of ν and the absolute distance L is directly obtained from the slope $\Phi' = d\Phi/d\nu$ through

$$L = \frac{c}{4\pi} \Phi' . \tag{9}$$

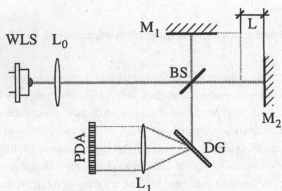

Fig. 6.
White-light channeled spectrum interferometer for absolute distance measurement using a diffraction grating (DG) and a photo diode array (PDA) for the signal processing.

In our case the broadband source was a light-emitting-diode (LED ABB 1A225) emitting at $\lambda \cong$ 890 nm with a spectral bandwidth of $\Delta\lambda \cong 50$ nm. After the lens L_0 the beam is collimated with a diameter of approximately 4 mm. The mirror M_2 of the Michelson interferometer is mounted on a piezo translator in series with a stepper motor. The output light of the interferometer passes through a diffraction grating (DG) with 1732 lines/mm. The diffracted beam is focused by the lens L_1 of 60 mm focal length onto a linear photodetector array (PDA) with 512 pixels of 25 µm separation. In this case we get a spectral resolution of 60 GHz/pixel, which corresponds to a coherence length of 5 mm.

For a working distance of 125 µm we get a channeled spectrum as shown in Fig. 7. The fringe spacing is 20 pixels, as expected from the calculated spectral resolution. To make the phase evaluation simple, we have adopted a synchronized sampling with four samples per period [12,13]. The corresponding values are also show in Fig. 7. A five-frame algorithm is than used to calculate the local phases Φ_n of the spectral fringe pattern. This algorithm is quite insensitive to phase shift errors and gives also correct results when the samples are not spaced exactly by $\pi/2$. The absolute distance as then directly obtained from the slope of the unwrapped phases Φ_n [13].

The maximum operating distance is limited by the spectral resolution and the minimum distance by the spectral width of the source. Using an LED source ($\Delta\lambda \approx 80$ nm) and a simple diffraction

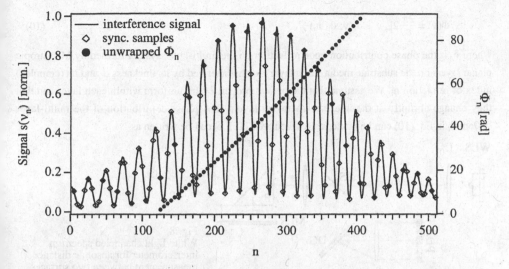

Fig. 7 Processing of the channeled spectrum: the observed interference fringes in the spectrum are synchronously detected with four samples per fringe period to calculate the fringe phases. Every fourth sample is graphically emphasized to demonstrate that the sampling is (nearly) synchronous to the fringe pattern.

grating (14000 lines), the range of operation is from 8 μm to 1.9 mm. It has been shown, that the minimum sampling concept for the fringe evaluation, based on four samples per period, works for 2.5 to 10 samples per period with a phase accuracy of better than $\Delta\Phi = 0.1°$. The method has been experimentally demonstrated for a working distance of 125 μm with an operating range from 50 μm to 150 μm. A statistical analysis of the experimental results showed, that for 30 or more samples the reproducibility of the distance measurement was always better than 0.1 μm, which is better than $\lambda/8$. Therefore, one of the measured local phases Φ_n can now be used to determine without ambiguity the exact distance with interferometric accuracy. The maximum variation of an individual local phase Φ_n was $\delta\Phi = 25.12$ mrad and the standard deviation was $\sigma\Phi = 5.55$ mrad, which gives an accuracy of $\delta L = 1.78$ nm ($\sigma_L = 0.39$ nm) for the distance L.

7. Effects of thin film layers

The schematic drawing of a white-light dispersive interferometer for auto-focusing in proximity printing or profilometry is shown in Fig. 8. The beam of the white-light source (WLS) is collimated by the lens L_0. The lens L_1 focuses the beam onto the test surface of the Fizeau interferometer. The interferometer is formed by one surface of the prism P and the multi-layer system MLS. The distance L to be measured is the air gap between the reference surface of the prism and the surface of the multi-layer system on the substrate. An optical profilometer, which uses a similar approach of white-light interferometry with channeled spectrum detection, has recently been published by Schwider et al [14]. In the case of thin film layers on the substrate, Eq. (8) for the interference phase $\Phi(\nu)$ is modified to

$$\Phi(v) = \frac{2\pi}{c} 2L\, v + \delta_r(v;\, d_i,\, n_i)\,, \tag{10}$$

where δ_r is the phase contribution upon reflection on the multi-layer system formed by M isotropic planar layers on the substrate media. Each layer i is characterized by its thickness d_i and its (complex) index of refraction n_i. We assume that the optical properties are uniform within each layer and that they change abruptly at the interfaces between layers. The phase contribution of the multi-layer system δ_r in Eq. (10) can be numerically calculated [15]. δ_r can be written as

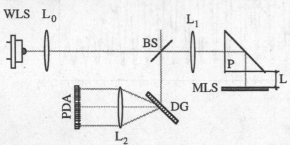

Fig. 8
White-light channeled spectrum interferometer for absolute distance measurement between two surfaces.

$$\delta_r(v) = \frac{2\pi}{c} \left(\sum_{i=1}^{M} 2\, d_i\, n_i \right) v + f_{nl}(v;\, d_i,\, n_i), \tag{11}$$

where the linear term on the right side of Eq. (11) is due to the optical thickness of the multi-layer stack and f_{nl} is a non linear function which describes the effects of multiple reflections within the stack. If the number of measured phase values $\Phi_n(v)$ is sufficiently large, the parameters L, d_i and n_i can be obtained by an appropriate least squares algorithm to fit the calculated phase function $\Phi(v)$ from Eqs. (10) and (11) to the measured values $\Phi_n(v)$.

A series of experimental results for Si substrates with one layer of photo-resist or two layers of SiO_2 and photo-resist have been performed with the experimental arrangement shown in Fig. 8. The light source was now a halogen lamp to get a wider spectrum than with the LED, the channeled spectrum was detected with a resolution of 150 GHz/pixel on the photo detector array and averaged over two adjacent pixels for the determination of phases $\Phi_n(v)$. It was assumed that the refraction indices of the Si substrate ($n_{Si} = 3.85 - 0.02 \cdot i$), the photo-resist ($n_r = 1.644$) and the silica ($n_{SiO2} = 1.46$) are known and independent of the optical frequency. Figure 9 shows experimental results obtained by fitting with a Levenberg-Marquardt least squares algorithm [16]. The results of the fit, f_{opt}, after deduction of the linear parts of $\Phi(v)$ in Eqs. (10) and (11), are compared with the measured phase values. In order to verify the method, the thicknesses d_r of the photo-resist and d_{SiO2} of the SiO_2 have been measured with a mechanical profilometer (TENCOR Instruments, α-step 250). The corresponding calculated phases, f_{mec}, are also shown in Fig 9. More experimental results are listed in Table 2. A comparison of the optically and the mechanically determined thicknesses shows excellent agreement, for most samples better than 10 nm.

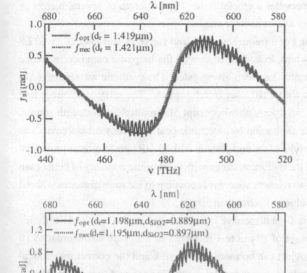

Fig. 9a.
Comparison of measured and calculated non linear phase contribution f_{nl} vs. optical frequency ν (wavelength λ) for photo-resist (d_r) on Si substrate.

Fig. 9b.
Comparison of measured and calculated non linear phase contribution f_{nl} vs. optical frequency ν (wavelength λ) for photo-resist (d_r) on SiO_2 (d_{SiO2}) on Si substrate.

Photo-resist on Si substrate					Photo-resist on SiO_2 on Si substrate			
d_r [μm]					d_r [μm]		d_{SiO2} [μm]	
mech.	opt.	L_{uc}	L_c	ΔL	mech.	opt.	mech.	opt.
1.145	1.142	126.37	123.74	2.63	1.295	1.289	0.298	0.299
1.198	1.202	125.27	123.53	1.74	1.490	1.483	0.298	0.303
1.266	1.259	125.14	122.68	2.46	1.195	1.198	0.897	0.889
1.338	1.342	126.63	124.48	2.15	1.290	1.295	0.897	0.890
1.421	1.419	124.09	122.80	1.29	1.455	1.463	0.897	0.884

Table 2 Comparison of mechanical and optical measurements of layer thickness (d_r: photo-resist, d_{SiO2}: SiO_2) and measured distance L without (L_{uc}) and with correction (L_c) for effects of the thin layers ($\Delta L = L_{uc} - L_c$).

8. Conclusions

Multiple-wavelength interferometry can be adapted for distance measurements from millimeter to micrometer accuracy by appropriate choice of wavelengths (stabilized HeNe-lasers, diode lasers) without requiring interferometric stability. Range, resolution and accuracy depend very much on the source, on the detection and on the signal processing. For high accuracy measurements, the stabi-

436

lization of the wavelength differences becomes a crucial issue. For distances of several meters or more, the coherence length of diode lasers will be another one.

A two-wavelength source consisting of a frequency stabilized HeNe master laser ($\lambda_1 = 632.8$ nm) and a HeNe slave laser ($\lambda_2 = 629.4$ nm), locked together with the help of a common reference length in the form of a Fabry-Pérot resonator, has been investigated. The synthetic wavelength ($\Lambda = 116$ μm) was calibrated by comparison with a HP laser interferometer. The relative accuracy and stability was found to better than $\delta\Lambda/\Lambda = 10^{-6}$. A novel concept of a multiple-wavelength source, using tunable diode lasers, with absolute calibration by electronic beat frequency measurement has been proposed and tested. Preliminary results of the comparison with an HP interferometer prove that a calibration of synthetic wavelengths in the millimeter range with an absolute accuracy of better than 10^{-5} is feasible. With the reported three-wavelength source it is possible to measure distances without ambiguity within 200 mm and with a resultion of better than 20 μm.

It has been demonstrated that white-light dispersive interferometry can be used efficiently for absolute distance measurement in the range of 10 μm to 1 mm with an accuracy of the order of 10 nm. Effects of thin films on the target object can be measured 'in-situ' and the correct mechanical distance to the top surface can be determined. Experimental results for one and two layers (photoresist and SiO2 on Si) show that the thickness of these layers can also be determined with an accuracy of the order of 10 nm (assuming that the refractive indices are known). Extension of the method to more than two layers and to layers with unknown refractive index seems possible. The presented method can be readily integrated in profilometers and auto focusing systems based on white-light interferometry.

References

[1] see, e.g., J.C. Wyant, "Holographic and moiré techniques", in Optical Shop Testing, D. Malacara, ed., (John Wiley, New York, 1978), pp. 397-402.

[2] C.R. Tilford, "Analytical procedure for determining lengths from fractional fringes", Appl. Opt. **16**, 1857-1860 (1977).

[3] R. Dändliker, "Distance measurements with multiple wavelength techniques", 2nd Internat. Workshop on High Precision Navigation, ed. K. Linkwitz, U. Hangleiter, (Ferd. Dümmlers Verlag, Bonn, 1992), p. 159-170.

[4] A.F. Fercher, H.Z. Hu, and U. Vry, "Rough surface interferometry with a two-wavelength heterodyne speckle interferometer", Appl. Opt. **24**, 2181-2188 (1985).

[5] R. Dändliker, R. Thalmann, D. Prongué, "Two-wavelength laser interferometry using super-heterodyne detection", Proc. SPIE 813, 9-10 (1987).

[6] R. Dändliker, R. Thalmann, D. Prongué, "Two-wavelength laser interferometry using superheterodyne detection", Opt. Lett. **13**, 39-341 (1988).

[7] R. Dändliker, K. Hug, J. Politch, E. Zimmermann, "High accuracy distance measurement with multiple-wavelength interferometry", to be published in Opt. Eng. **34** (1995).

[8] S. Manhart, R. Maurer, "Diode laser and fiber optics for dual-wavelength heterodyne interferometry", Proc. SPIE 1319, 214-216 (1990).

[9] Z. Sodnik, E. Fischer, T. Ittner, H. J. Tiziani, "Two-wavelength double heterodyne interferometry using a matched grating technique", Appl. Opt. **30**, 3139-3144 (1991).

[10] E. Gelmini, U. Minoni, F. Docchio, "Tunable, double-wavelength heterodyne detection interferometer for absolute-distance measurement", Opt. Lett. **19**, 213-215 (1994).

[11] R. Dändliker, "High accuracy distance measurements with multiple wavelength techniques", EOS Annual Meeting Digest Series, (European Optical Society, Paris, 1993), p. 243-246

[12] R. Dändliker, E. Zimmermann and G. Frosio, "Electronically scanned white-light interferometry, a novel noise–resistant signal processing", Opt. Lett. **17**, pp. 679–681 (1992).

[13] U. Schnell, E. Zimmermann, R. Dändliker, "Absolute distance measurement with synchronously sampled white-light channeled spectrum interferometry", to be published in Pure Appl. Opt. (1995)

[14] J. Schwider, L. Zhou, "Dispersive Interferometric Profilometer", Opt. Lett., **19**, 995–997 (1994).

[15] R.M.A. Azzam, N.M. Bashara, "Ellipsometry and Polarized Light", (North-Holland, Amsterdam, 1987) pp. 283-286 and 332-340.

[16] Levenberg-Marquardt algorithm implemented in 'Optimization Toolbox for Use with MATLAB', The MathWorks, Inc., Natick, Mass. 01760-1500, 1994.

Anforderungen der Interferometrie an die nächste Generation von Lasern

J. Thiel, T. Pfeifer

Fraunhofer-Institut für Produktionstechnik IPT

Steinbachstr. 17, D-52074 Aachen

1. Einleitung

Neuere Entwicklungen im Bereich der Interferometrie, so z.B. in der Distanz-, Formprüf- sowie Speckle-Interferometrie, bleiben aufgrund der begrenzten Eigenschaften herkömmlicher Lasersysteme in starkem Maße in Ihrem Entwicklungspotential eingeschränkt. Obwohl in der Vergangenheit zunehmend Halbleiterlaser oder Festkörperlaser anstelle des mittlerweile als klassisch zu bezeichnenden HeNe-Lasers eingesetzt wurden, geraten auch diese Systeme mehr und mehr an ihre Grenzen. Sehr deutlich wird dieser Umstand insbesondere dadurch, daß die Meßtechnik stärker denn je von Lasersystemen lebt, die ursprünglich für völlig andere Applikationen entwickelt wurden. So werden etwa 95% der jährlich gefertigten Halbleiterlaser von insgesamt ca. 80 Millionen Stück in optischen Speichermedien (CD, CD-ROM, etc.) eingesetzt. Nur ein verschwindender Bruchteil findet einen Einsatz in der Meßtechnik oder darüberhinaus in der Interferometrie. In diesem Beitrag werden zum einen die derzeit in der interferometrischen Meßtechnik geforderten Eigenschaften systematisch zusammengefaßt, zum anderen werden Lasersysteme vorgestellt, die durch gezielte Modifikationen gänzlich neue, sowie interessante Möglichkeiten für die Interferometrie bieten.

2. Interferometrische Applikationen

Die wichtigsten Applikationen der Interferometrie lassen sich in eindimensionale sowie in 3-dimensionale Verfahren unterteilen (siehe Tabelle 1). In der eindimensionalen Interferometrie sind es vornehmlich hochgenaue Längen- und Abstandsmessungen in der Luft- und Raumfahrtindustrie, im Schiff- und Automobilbau, im Werkzeugmaschinenbau, in der Vibrationsmessung und neuerdings in der Mikropositionierung in der Halbleiterindustrie sowie für der Vermessung von Robotern. Aber auch Größen, die auf Längenmessungen rückführbar sind, wie z.B. Winkel, Rechtwinkligkeit, Geradheit sowie Parallelität von Achsen werden - sofern die Präzision gefordert ist - mit Interferometern gemessen. In der 3-dimensionalen Interferometrie liegen die Applikationen einerseits in der Optikprüfung planer sowie sphärischer Flächen, der Asphärenprüfung sowie der Prüfung diamantgedrehter Spiegel und andererseits in der Verformungs-, Form- sowie Schwingungsmessung, wie beispielsweise in der Speckle- und holographischen Interferometrie.

1D-Interferometrie	3D-Interferometrie
Klassische Längeninterferometrie	Formprüf-Interferometrie
Absolute Distanz-Interferometrie	Speckle-Interferometrie
Tracking Interferometrie	Absolute Speckle-Interferometrie
Vibrometrie	Holographische Interferometrie
Interferometrische Wellenlängenmessung	Weißlicht-Interferometrie
Gravitationswellen-Interferometrie	

Tabelle 1. *Anwendungsgebiete der Interferometrie*

3. Problemstellungen in der Interferometrie

Sämtlichen interferometrischen Applikationen ist gemeinsam, daß die Wellenlänge des Laserlichts die Maßverkörperung darstellt. Aufgrund des doppelten optischen Strahlweges entspricht die halbe Wellenlänge $\lambda/2$ gerade einem Skalenteil, und eine Messung ist lediglich in diesem Intervall von $\lambda/2$ eindeutig. Diese Problematik läßt sich durch Hinzunahme einer weiteren Wellenlänge λ_2, die unterschiedlich zur ersten Wellenlänge λ_1 ist, lösen. Hierdurch wird eine synthetische Wellenlänge Λ_S erzeugt (siehe Bild 1), die sich folgendermaßen berechnen läßt: $\Lambda_S = (\lambda_1\lambda_2)/|\lambda_1-\lambda_2|$. Durch geeignete Wahl der beiden Wellenlängen lassen sich synthetische Wellenlängen vom µm-Bereich bis in den m-Bereich generieren. Dieses Verfahren ist an sich nicht neu und wurde schon im Jahre 1979 zur absoluten Distanzmessung mit unterschiedlichen Linien eines CO_2-Lasers durchgeführt [1]. Nach demselben Prinzip arbeiten Endmaßmeßgeräte, wobei die Unsicherheit in der Vorkenntnis der jeweiligen Länge geringer als die synthetische Wellenlänge sein muß [2]. Dieses Verfahren, daß sowohl in der Längen- als auch in der Formprüfinterferometrie Verwendung findet, wird in der Literatur auch als Mehrwellenlängen-Interferometrie beschrieben [3], [4].

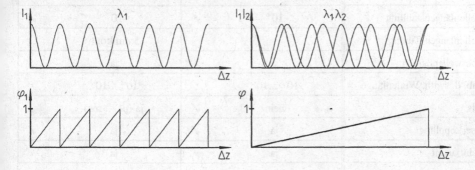

Bild 1. *Vergrößerung des Eindeutigkeitsbereichs durch Verwendung einer zweiten Wellenlänge*

4. Anforderungen der Interferometrie

In Tabelle 2 und Tabelle 3 sind die Anforderungen der Interferometrie an die Laserquellen aufgelistet (erreichte Anforder. mit Haken), wie sie sich für typische Applikationen der klassischen Längen- sowie absoluten Distanz-Interferometrie und der 3D-Interferometrie ergeben. Die Daten sind verständlicherweise nur als Richtwerte zu verstehen und können keinen Anspruch auf Vollständigkeit erheben.

	Klassische Längen-Interferometrie	Absolute Distanz-Interfer. (0...5m)	Absolute Distanz-Interfer. (0...50m)
Leistung	1 mW ✓	2 mW ✓	5 mW ✓
Kohärenzlänge	50 m ✓	20 m	200 m ✓
Wellenlängenstabilität	10^{-8} ✓	10^{-8}	10^{-8}
Wellenlängen-Tuning	-	5 nm kont.	20 nm kont.
Tuning-Geschwindigkeit	-	1... 100 Hz ✓	1... 10 Hz ✓
Mehrwellenlängen	nein	ja	zus. Laser ✓
Stab. d. synth. Wellenlänge	-	10^{-6}	10^{-6}
Faserkopplung	ja ✓	ja ✓	ja ✓
Sichtbarkeit	ja ✓	ja	ja
Kosten	< DM 10.000,- ✓	< DM 10.000,-	< DM 25.000,-

Tabelle 2. *Anforderungen der 1-dimensionalen Interferometrie an die Laserquellen*

	Formprüf-Interferometrie	Speckle-Interferometrie Holographie
Leistung	3 mW ✓	100 ... 200 mW
Kohärenzlänge	1 m ✓	1 ... 20 m
Wellenlängenstabilität	10^{-6} ✓	10^{-6} ✓
Wellenlängen-Tuning	-	5 nm kont.
Mehrwellenlängen	ja	ja
Stab. d. synth. Wellenlänge	10^{-2} ...10^{-3}	10^{-2} ...10^{-3}
Puls	nein ✓	ja (µs ... ms)
Faserkopplung	ja ✓	ja ✓
Sichtbarkeit	ja ✓	ja
Kosten	< DM 10.000,- ✓	< DM 25.000,-

Tabelle 3. *Anforderungen der 3-dimensionalen Interferometrie an die Laserquellen*

5. Laserquellen für die Interferometrie

Neben Halbleiterlasern besitzen diodengepumpte Nd:YAG-Laser bzw. Mikrokristall-Festkörperlaser sehr gute Eigenschaften für die Interferometrie, erfüllen jedoch bei weitem nicht alle geforderten Kriterien. Im Halbleiterlaserbereich existieren spezielle Strukturen, die günstige Eigenschaften aufweisen (siehe Bild 2). Hierzu gehören insbesondere Short External Cavity-Laser (SEC), Distributed Bragg Reflector-Laser (DBR), Distributed Feedback-Laser (DFB) sowie Mehrsegment-Laser. Letztere besitzen aufgrund der unterschiedlichen Ansteuerbarkeit der einzelnen Sektionen weitaus günstigere Durchstimmbereiche der Wellenlänge und zudem eine geringere Linienbreite und damit größere Kohärenzlänge, so daß sie für die Interferometrie besonders interessant sind.

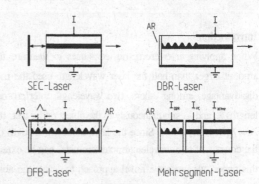

Bild 2. *Halbleiterlaserstrukturen für Anwendungen in der Interferometrie*

Weiteres Augenmerk verdienen sogenannte External Cavity Laser (EC), bei denen der Halbleiterchip einseitig antireflexvergütet ist und so die Strahlung auf ein externes Resonatorgitter trifft. Durch schwenken des Gitters werden modensprungfreie Durchstimmbereiche der Wellenlänge von ca. 10 nm bei einer Kohärenzlänge von mehreren hundert Metern erreicht. Durch ihren komplexen mechanischen Aufbau sind derartige Laser leider noch extrem teuer.

6. Zusammenfassung

Die ideale Lichtquelle für die moderne Interferometrie existiert derzeit leider noch nicht. Obwohl bereits erfolgversprechende Ansätze im Bereich der diodengepumpten Festkörperlaser sowie der Halbleiterlaser zu erkennen sind, ist hier dennoch Forschungs- und Entwicklungsaufwand dringend notwendig.

Literatur

[1]   Bourdet, G. L.; Orszag, A. G.; Absolute distance measurement by CO_2 laser multiwavelength interferometry. Applied Optics, Vol. 18, No. 2 (1979), S. 225-227.

[2]   Bimberg, D.; Laser in Industrie und Technik. expert-Verlag (1985), Sindelfingen.

[3]   Dändliker, R.; Thalmann, R.; Prongué, D.; Two-wavelength laser interferometry using superheterodyne detection. Optics Letters, Vol. 13, No. 5 (1988), S. 339-341.

[4]   Thiel, J.; Haas, C.; Pfeifer, T; Absolutinterferometrie mit wellenlängenstabilisierten Halbleiterlasern. Optik, Vol. 98, No. 4, (1995), Wissenschaftliche Verlagsgesellschaft mbH Stuttgart, S. 163-168.

Absolute Interferometric Distance Measurement Using a FM-Demodulation Technique

Edgar Fischer, Ernst Dalhoff, Silke Heim, Ulrich Hofbauer, Hans J. Tiziani

Institut für Technische Optik, Universität Stuttgart, Pfaffenwaldring 9,

D-70569 Stuttgart, Germany

Introduction

When applying interferometric techniques to measure the distance between two points separated by an amount larger than half the laser wavelength used the measurement become ambigous. To overcome this disadvantage, among others, two wavelength interferometry [1-4] was developed. By applying two wavelengths λ_1 and λ_2 simultaneously to the object under test, the sensitivity is reduced to an effective wavelength $\Lambda = \lambda_1 \lambda_2 / |\lambda_1 - \lambda_2|$. Since the laser diode wavelength can easily be tuned via the laser injection current, the concept of two-wavelength interferometry can be extended to multiple-wavelength interferometry [5-8]. In this paper, we present a novel approach to derive the absolute distance information from the phase modulated detector signal by using a PLL (phase locked loop) based FM-demodulator. In this case, there is no need to generate a reference signal in the interferometer for the purpose of demodulation. By using a FM-demodulator, no unambiguity jump occurs if the phase deviation exceeds 2π. In this case, we are able to determine absolute distances with a noise limited resolution of about 1 μm, even if the distance is much larger than the synthetic wavelength.

Theory of operation

In a two beam interference set-up as shown in figure 1, the injection current of a laser diode is modulated sinusoidally around a certain DC level. This results in a wavelength- as well as an amplitude-modulation of the emitted radiation, namely

$$E(z,t) = E_0(t) \; \exp\left[i \left(2\pi \, \nu(t)\, t - \frac{2\pi}{c} \, \nu(t) z \right) \right] \tag{1}$$

where $E_0(t)$ and $\nu(t)$ are defined by: $E_0(t) = A\,[\,1 + m\cos\,(2\pi f_m\, t + \phi)\,]$; $\nu(t) = \nu_0 + \Delta\nu \cos\,(2\pi f_m\, t + \phi)$
z is the optical path, f_m the modulation frequency, A the electrical field amplitude, m the amplitude modulation index, c speed of light, ν_0 the average frequency of the emitted light, $\Delta\nu$ is the frequency deviation and ϕ the initial phase of the modulation signal. To introduce a high frequency carrier in the detector signal in one arm of the interferometer, an acousto-optical modulator (AOM) is used to shift the frequency of the light by an amount f_H, i. e. the AOM driver frequency.

After passing through the particular interferometer arms, the two laser beams are recombined by means of a beam splitter and interfere. If $2\pi f_m \tau \ll 1$ is fulfilled and defining the new variable $\Delta z = z_{Ref} - z_{Obj}$ the high pass filtered detector signal can be approximated by

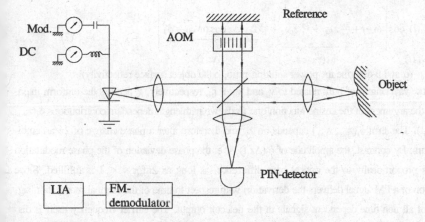

Figure 1: Basic set-up for FM-demodulation interferometry. The phasemodulated detector signal, which has a high frequency carrier, is demodulated with a PLL FM-demodulator. The amplitude of the demodulated signal is proportional to the optical path difference in the interferometer.

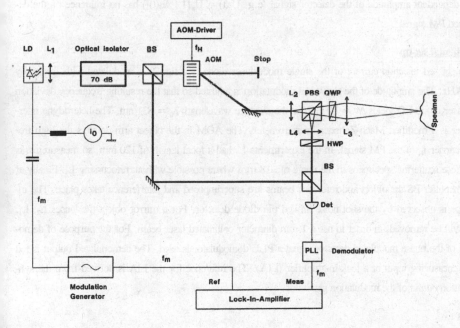

Figure 2: Experimental set-up for measurements onto diffuse scattering objects.

$$u(\Delta z, t) = \underbrace{2\,\gamma\,I_0(t)}_{U_a(t)}\,\cos\left[\omega_H\,t + \underbrace{\frac{2\pi}{\lambda_0}\,z_{Ref} - \frac{2\pi}{\lambda_0'}\,z_{Obj}}_{\phi_1(z_{Ref},\,z_{Obj})} + \underbrace{\frac{2\pi}{c}\,\Delta v\,\Delta z\cos(\omega_m\,t + \phi)}_{\phi_2(\Delta z,\,t)}\right] \qquad (2)$$

where $\gamma = \alpha\,\beta\,\rho$ (α and β describe the power splitting ratio, ρ the object surface reflectivity) λ_0 and λ_0' are the wavelengths of light related to v_0 and $v_0 + f_H$, respectively, I_0 denotes the uniform intensity. Analysing the argument of the cosine function one finds two pathlength dependent contributions $\phi_1(z_{Ref}, z_{Obj})$ and $\phi_2(\Delta z, t)$. The term $\phi_1(z_{Ref}, z_{Obj})$ depends on λ_0 and therefore after a phase change of 2π an ambiguity jump occurs. By contrast, the amplitude of $\phi_2(\Delta z, t)$, i. e. the phase deviation of the phase modulation (PM), increases proportionally to the optical path difference, as long as $2\pi f_m\tau << 1$ is fulfilled. Since FM-demodulation of a PM signal delivers the derivation with respect to time of the original modulation signal we get rid of all non time dependent signals in the detector output. The carrier frequency itself is discriminated by the demodulation process. So we get $\partial\phi_2(\Delta z, t)/\partial t$ after demodulation. The term $\partial\phi_2(\Delta z, t)/\partial t$, which is in principle ambiguous, can be made unambiguous if $2\pi f_m\tau << 1$ is fulfilled at any time. In the following it is assumed that the demodulating PLL is locked and we consider the output signal of the PLL lowpass filter. After some calculation the lowpass filter output can be described by [9]

$$U_S(\Delta z, t) = \frac{2\pi}{c}\,\Delta v\,\Delta z\,\frac{f_m}{k_{VCO}}\,\sin(\omega_m\,t + \phi) \qquad (3)$$

A time dependent amplitude of the detector signal (e.g. $U_a(t) = U_a[1 + \Delta u(t)]$) has no influence on the demodulated PM-signal [9].

Experimental set-up

The d.c. biased injection current of the single mode laser diode (see fig. 2) is modulated with frequency $f_m = 3$ kHz. The amplitude of the sinusoidal modulation is adjusted so that the resulting frequency deviation of the laser diode LD yields $\Delta v = 60$ GHz at the average wavelength $\lambda_0 = 831$ nm. The heterodyne interferometer is a modified Mach-Zehnder interferometer. The AOM in the object arm delivers the high frequency carrier f_H of the PM signal. In the experiments L_3 had a focal length of 120 mm, so measurements onto diffuse scattering specimens in the range of ±18 mm where possible without refocussing L_3. Finally at the beamsplitter BS the object and reference beams are superimposed and interference takes place. The interference is observed by the shot noise limited pin diode detector. For a mirror object the lenses L_2, L_3, and L_4 will be removed, in order to use a 1 mm diameter collimated laser beam. For the purpose of demodulation of the phase modulated detector signal a PLL demodulator is used. The demodulated output is fed into the measuring input of a lock-in-amplifier (LIA). The reference for the LIA is derived from the synchronisation output of the modulation generator.

Results

- **Long duration measurement:** The resolution of the phase deviation measurement can be derived from the standard deviation of the lock-in signal of a long duration measurement. An RMS amplitude of 0.16 mV was gained from a measurement over a time period of 400 s this corresponds to a long term stability of ±16 μm and the noise limited resolution was found to be 1.6 μm.

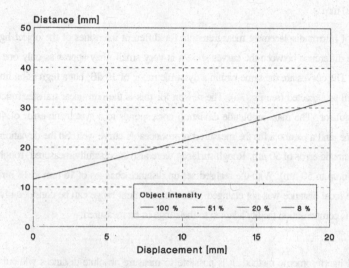

Figure 3a: Displacement measurements for different target reflectivities; object: mirror with variable reflectivity (for more details see fig.3b and text).

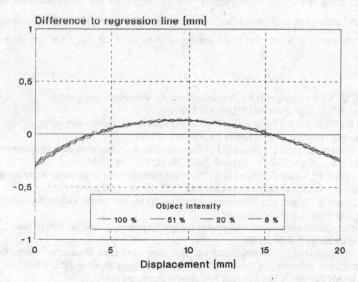

Figure 3b: Deviation from expected linear relation in fig. 3a.

- Measurement of displacement: In the displacement measurements, distances up to +25 mm were realised (limited by the maximum travel range of the specimen translation stage used). The amplitude of the lock-in signal is equal for both shift directions, i. e. +25 mm and -25 mm, however the phase changes by 180° if the path difference changes from positive to negative values. Therefore the measurement range in our experiment is 50 mm .

In fig. 3a, curves of mirror displacement measurements for different intensities of the object light are depicted. Because the difference between the curves shown is very small, they appear as only one (for more details see fig. 3b). The curves are the same within a dynamic range of 10 dB, but a regression line does not fit the values as well as expected (see fig. 3b). The reason for this is the non linear transfer function of the detector and demodulator. The max. amplitude deviation corresponds to a systematic error of 400 μm. A combination of a line and a parabola fits the measured reproduceable curve well. So the deviation decreases to a remaining systematic error of 50 μm. Rough surfaces were also successfully measured (roughness standards, $R_z = 1.6$ μm up to 50 μm). With the realised set-up distance changes of 16 mm to 18 mm could be measured while the focal distance was not changed. The measurement range can be considerably increased by a additional focus compensation facility adapted to the range to be measured.

Conclusion

With the presented interferometric method, it is possible to measure absolute distances without ambiguity problems. With our experimental set-up, measurements on reflecting surfaces yielded a resolution of 1.6 μm by the measuring range of 50 mm. Repeatability of the displacement measurement was 50 μm. The maximum dynamic range of the object light intensity was 10 dB without any compensation of the decreasing signal. Measurements on optically rough surfaces were also carried out. The measuring range of rough surfaces was slightly less than 50 mm due to limited depth of focus of the image forming device used for rough surfaces.

References:

1. Williams C.C, Wickramasinghe H.K.: "Optical ranging by wavelength multiplexed interferometry", J. Appl. Phys. **60**, 1900-1903, (1986)
2. Dändliker R., Thalmann R., Prongué D.: "Two-wavelength laser interferometry using superheterodyne detection", Optics Letters **13**, 339-341, (1988)
3. Sodnik Z., Fischer E., Ittner T., Tiziani H.J.: "Two-wavelength double heterodyne interferometry using a matched grating technique", Applied Optics **30**, 3139 - 3144, (1991)
4. Fischer E., Ittner T., Sodnik Z., Tiziani H.J.: "Heterodynverfahren für hochgenaue Vermessung im Nahbereich", Zeitschrift für Vermessungswesen **117**, 46-54, (1992)
5. Kikuta H., Iwata K., Nagata R.: "Distance measurement by the wavelength shift of laser diode light", Applied Optics **25**, 2976-2980, (1986)
6. Boef A.J. den: "Interferometric laser rangefinder using a frequency modulated diode laser", Applied Optics **26**, 4545-4550, (1987)
7. Suematsu M., Takeda M.: "Wavelength-shift interferometry for distance measurements using the Fourier transform technique for fringe analysis", Applied Optics **30**, 4046-4055, (1991)
8. Groot P. de, Mc Garvey: "Chirp Synthetic Wavelength Interferometry", Optics Letters **17**, 1626-1628, (1992)
10. Mäusl R.: "Analoge Modulationsverfahren", Hüthig-Verlag, Heidelberg, 2. Auflage, (1992)

Interferometric Measurement of Thin Soap Films

V. Greco and G. Molesini

Istituto Nazionale di Ottica

Largo E. Fermi 6, Firenze 50125, Italy

Surface acting agents, or surfactants, are generally studied by looking at the very thin films they can produce. The films can be drawn out of a soap solution by means of proper frames, and observed during the thinning process that occurs due to draining of the water layer sandwiched between two external layers of surfactant. At the ultimate stage of thinning, Newton black films are produced, in which the action of basic forces such as van der Waals attraction and double layer repulsion results in equilibrium states. The end thickness just before the film rupture can be of the order of a few nanometers. As an alternative to X-ray investigation, phase shifting interferometry has been recently proposed to monitor the 2-D evolution of the thickness map of soap films.[1] A time resolved approach by homodyne interferometry was also demonstrated.[2] These two techniques are here briefly reviewed, and experimental results on soap films with sodium dodecyl sulphate (SDS) surfactant are presented.

For build-up and housing of the soap film under traceable environmental conditions, a precision optical cell has been designed. The cell's main frame is a glass wedge (3° angle) with an average thickness of 30 mm; when in place for measurement, the wedge gives spurious fringes a high spatial frequency that the detector cannot resolve. Windows are made of two glass plates, optically polished to a quarter fringe. A movable frame is included, which allows the soap film to be drawn out of the reservoir solution. Such inner frame holds an optical fibre 50 μm in diameter, tightened horizontally. Openings in the main frame are available for accommodating a micrometer gauge and for service (introducing and removing the solution, probing the temperature). The cell is sealed, so that non-evaporation conditions of the solution can be reached. In operation, the fibre is initially just below the surface of the solution. The inner frame is then raised by the micrometer, so that the fibre comes out of the solution and lifts the soap film. The fibre also defines a clear boundary between air and film, which is taken as a geometrical baseline for optics alignment.

The system being studied is an aqueous SDS solution in the presence of salt (NaCl) at the temperature of 22 ± 0.5 °C. Composition is 1 g l^{-1} SDS in 0.4 M NaCl aqueous solution. Such system has been extensively studied in the literature.[3]

Phase shifting interferometry is implemented by modifying the measuring procedure of a standard system in use for optical testing. The cell is placed between a transmission flat and a reference flat, and is tested in double-pass transmission. Operation of the interferometer is driven by an external computer, where the data of optical thickness are made available in ASCII format. Two software masks are defined, one of the space just above the optical fibre and the other also including the soap film. The first mask serves to define the reference floor for the optical path difference (OPD) through the whole data of the second mask. The interference pattern is sampled at 259x235 locations by the detector array; 10 bit digitization is used.

The water drainage and the onset of Newton black film are monitored by acquiring phase maps at regular time intervals of 20 s. From just after the film is drawn, a sample of the OPD profile along a direction normal to the fibre is given in Fig. 1 a). The Newton black film is not yet formed, and the OPD map exhibits a still swollen section. The perturbation introduced by the fibre is barely visible. In Fig. 1 b) the final Newton black film produced by draining is presented. In the vicinity of the fibre the pattern appears affected by relatively strong irregularities caused by diffraction. Nevertheless, the step in the OPD profile is clearly visible; its height range is about 2.0 nm.

Time resolved monitoring of the optical thickness of soap films is of interest for studying the fluid dynamics of such systems. Limiting the inspection to a single small area of the film, interferometric approaches other than phase shifting have been devised, providing information on the local optical thickness with comparable precision but at considerably higher sampling rate. As a development of already reported optical configurations,[4,5] a polarization interferometer has been set up, working on four quadrature signals and accurately measuring optical path differences.[2,6] Such a technique complements phase shifting interferometry, allowing for locally observing the thinning process of soap films in close real time.

The optical configuration used is basically a Twyman-Green interferometer where polarization of the laser beam and retardation plates are properly handled to produce four quadrature signals. The data is collected with a 16-bit A/D converter interfaced to a desktop computer where processing takes place. The cell is inserted in one of the arms of the interferometer, in such a way that when the cell's inner frame is lifted up the laser beam passes

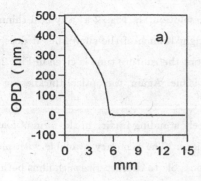
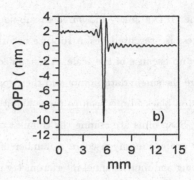

Fig. 1. *Thinning of a soap film, monitored by phase shifting interferometry. a) Optical thickness profile of the soap film just after drawing. b) Optical thickness profile of the Newton black film.*

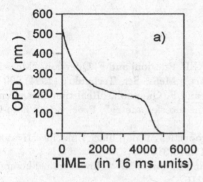
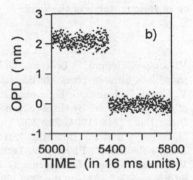

Fig. 2. *Thinning of a soap film, monitored by multichannel homodyne interferometry. a) Time evolution of local optical thickness from film drawing to past rupture. b) Detail of the rupture event.*

through the soap film and the OPD between the two arms is increased. Decreasing of the OPD is then monitored at the sampling rate allowed by the electronics and selected by the computer.

Data reduction includes correction for unbalancing of the signals and for laser power drifts. This is done by means of proper algorithms, which combine the four signal readings and work out the phase value. A calibration procedure provides the system parameters accounting for the actual working conditions; the sources of noise are properly treated, so that the end uncertainty of the measurement is clearly identified.[7] As a result, the time evolution of the OPD is finely measured to sub-nanometre resolution. In Fig. 2 a) a typical thinning process is presented. Film rupture occurs in the right branch of the curve, but it does not show up because of the scale. A magnified view about the rupture time is given in Fig. 2 b), where the single data points and their spread are visible. Again, the optical thickness of the Newton black film is seen to range about 2.0 nm.

Soap films are among the thinnest objects freely standing in air. In the form of foams, they are commonly used in a number of applications. The accuracy available with phase shifting and multichannel interferometry makes it possible to characterize such films both for research and for development purposes. Examples of measurements on SDS surfactant have been given; optical thickness profiles and time evolution curves with unprecedented space and time resolution have been presented.

The authors wish to thank C. Iemmi, S. Ledesma, P. Poggi and A. Tenani for their collaboration in various stages of the work.

1. V. Greco, C. Iemmi, S. Ledesma, G. Molesini, G.P. Puccioni and F. Quercioli, "Measuring soap black films by phase shifting interferometry", Meas. Sci. Technol. 5 (1994) 900-903.
2. V. Greco, C. Iemmi, S. Ledesma, G. Molesini and F. Quercioli, "Real-time measurement of optical thickness by three- channel homodyne interferometry", Proc. Soc. Photo-Opt. Instr. Eng. 2248 (1994) 262-266.
3. O. Bélorgey and J.J. Benattar, "Structural properties of soap black films investigated by X-ray reflectivity", Phys. Rev. Lett. 66 (1991) 313-316.
4. V. Greco, C. Iemmi, S. Ledesma, G. Molesini and F. Quercioli, "Three-channel homodyne interferometer", Appl. Opt. 33 (1994) 8115-8116.
5. V. Greco, C. Iemmi, S. Ledesma, A. Mannoni, G. Molesini and F. Quercioli, "Multiphase homodyne displacement sensor", Optik 97 (1994) 15-18.
6. V. Greco, G. Molesini and F. Quercioli, "Accurate polarization interferometer", Rev. Sci. Instrum. (to appear).
7. V. Greco, C. Iemmi, S. Ledesma, G. Molesini and F. Quercioli, "Multiphase homodyne interferometry: analysis of some error sources", Appl. Opt. (to appear).

Verfahren zur Qualitätsprüfung von Interferometern

J. Miesner, H. Kreitlow
Fachhochschule Ostfriesland
Fachbereich Naturwissenschaftliche Technik
Institut für Lasertechnik
Constantiaplatz 4, 26723 Emden, Bundesrepublik Deutschland

1. Einleitung

Interferometrische Verfahren sind geeignet, um das mechanische Verhalten von Bauteilen ganzflächig und zerstörungsfrei nicht nur qualitativ sondern auch quantitativ zu analysieren.

Der erfolgreiche Einsatz dieser Verfahren, wie die holografische Interferometrie, die elektronische Specklemuster-Interferometrie (ESPI) oder auch die Moiré-Meßtechnik erfordert eine Anpassung und Optimierung des interferometrischen Aufbaus und der Auswerteprogramme für das jeweilige Meßproblem.

In diesem Beitrag wird ein Verfahren vorgestellt, mit dem unter Verwendung eines Rechnerprogramms zur Simulation von Interferenzmustern, systematisch die Qualität interferometrischer Aufbauten und der zugehörigen Auswertesysteme untersucht, bewertet und zum Einsatz für praktische Meßprobleme optimiert werden können.

2. Qualität einer interferometrischen Messung

In Bild 1 ist ein typischer automatisierter Meßprozeß in der Interferometrie schematisch dargestellt. Dabei wird ein Objekt einer Objektänderung unterzogen und mit einem interferometrischen Aufbau bezüglich eines Referenzzustands verglichen. Aus dem resultierenden Interferogramm wird anschließend die Phasenverteilung bestimmt. Diese Phasenverteilung muß noch einem Unwrap-Algorithmus unterzogen werden, um die Phasenwerte zu berechnen, aus denen dann das Objektveränderungsfeld ermittelt werden kann. Das Auswerteergebnis wird abschließend als vollständiger Datensatz des Objektänderungsfelds ausgegeben, in Form einer Grafik dargestellt oder als Kennzahl protokolliert.

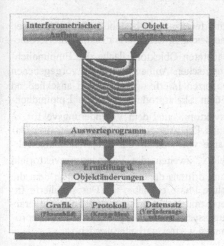

Eine allgemeine Schwierigkeit besteht darin, daß dem Anwender kommerzieller Interferometer die genauen Eigenschaften der einzelnen Hard- und Softwarekomponenten nicht oder nicht genau genug bekannt sind, so daß eine sichere Aussage über die Qualität der aus den Interferogrammen gewonnenen Meßergebnisse nicht möglich ist. Im folgenden wird dieser Sachverhalt durch eine Liste von typischen Problemen bezüglich des interferometrischen Aufbaus sowie des Steuer- und Auswerteprogramms verdeutlicht.

Es wird daraus der im weiteren verwendete Begriff Qualität einer interferometrischen Messung abgeleitet und darauf aufbauend ein Verfahren zur Qualitätsprüfung von Interferometern mit Hilfe eines Interferenzmuster-Simulationsprogramms entwickelt und untersucht.

Bild 1: Meßprozeß in der Interferometrie

452

- **Interferometrischer Aufbau**

  ⇒ das tatsächliche Empfindlichkeitsvektorfeld ist unbekannt, nicht richtig berücksichtigt, bzw. die Näherungsannahmen hierzu sind falsch

  ⇒ die Objektgeometrie wird nicht in das Empfindlichkeitsvektorfeld einbezogen

  ⇒ die Objektbelastung ist nicht exakt genug ausgelegt oder/und realisiert

  ⇒ die Randbedingungen sind schwierig festzulegen, z.B. ist die absolute Phasenordnung unbekannt

- **Steuer- und Auswerteprogramm**

  ⇒ die Auswerteprogramme berücksichtigen die zuvor genannten Sachverhalte nur unzureichend

  ⇒ die Kopplung mit CAD- oder FEM-Programmen wird nicht unterstützt

  ⇒ die Steuerung der Aufnahme der für unterschiedliche Auswerteverfahren benötigten Interferenzmusterfolgen wird nicht sorgfältig genug ausgeführt

  ⇒ die Darstellung der Ergebnisse ist nicht aussagekräftig

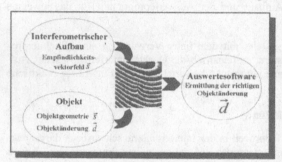

Bild 2: Qualität von interferometrischen Systemen

Der Qualitätsbegriff wird anhand von Bild 2 erläutert. Die Interferenzstreifenmuster, die aus einer interferometrischen Messung resultieren, enthalten außer der Information über die Objektänderung auch wesentliche Beiträge, die aus dem interferometrischen Aufbau und der Geometrie des Objekts selbst herrühren. Beide beeinflussen das Empfindlichkeitsvektorfeld. Die Qualität interferometrischer Systeme ergibt sich nun aus der Kenntnis des genauen Empfindlichkeitsvektorfeldes, bzw. aus dessen Berücksichtigung in den Auswerteprogrammen.

Diese Kriterien werden in 4. mit Hilfe eines Interferenzmuster-Simulationsprogramms untersucht.

3. Das Interferenzmuster-Simulationsprogramm

Der Kern des verwendeten Simulationsprogramms für Interferenzmuster besteht in der punktweisen Berechnung von Interferenzmustern mit Hilfe der Grundgleichung der holografischen Interferometrie: $n\lambda = \vec{d} \cdot \vec{s}$. Dabei wird für jeden Punkt der beobachteten Objektoberfläche der Empfindlichkeitsvektor $\vec{s}$ zu einem vorher festgelegten interferometrischen Aufbau und einer vorgegebenen Objektform berechnet. Die Phasenordnung $n\lambda$ des simulierten Interferogramms wird anschließend aus dem Skalarprodukt zwischen Empfindlichkeitsvektor $\vec{s}$ und dem Veränderungsvektor $\vec{d}$ gebildet [2]. Als Ausgabedaten liegen drei Datensätze vor: erstens das Empfindlichkeitsvektorfeld $\vec{S}$, zweitens das Veränderungsvektorfeld $\vec{D}$ und drittens das Interferogramm, in dem die absolute Phase enthalten ist. Das simulierte Interferenzmuster kann als Grafik mit Bildverarbeitungssystemen oder speziellen Auswerteprogrammen weiterverarbeitet werden.

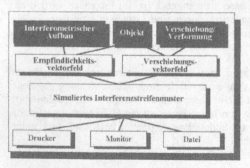

Bild 3: Struktur des Programms zur Interferenzmustersimulation

4. Untersuchung von Qualitätskriterien

4.1 Das Empfindlichkeitsvektorfeld

Der in 2. beschriebene Einfluß der Aufbaugeometrie des Interferometers und der Objektfom auf das Empfindlichkeitsvektorfeld $\vec{S}$ und damit auf die Qualität der Bestimmung des Veränderngsvektorfeldes $\vec{D}$ wird im folgenden untersucht.

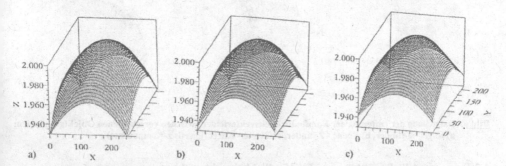

Bild 4: Darstellung der z-Komponente von Empfindlichkeitsvektorfeldern einer a) ebenen Platte, b) einer Zylinderoberfläche und c) einer Kugeloberfläche

Der Empfindlichkeitsvektor $\vec{s}$ wird gebildet aus der Differenz von Beobachtungseinheitsvektor $\vec{c}_2$ und dem Beleuchtungseinheitsvektor $\vec{c}_1$ eines gegebenen interferometrischen Aufbaus. Mit dem Simulationsprogramm (siehe 3.) werden die Empfindlichkeitsvektorfelder eines interferometrischen

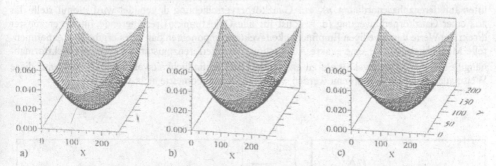

Bild 5: Darstellung der Differenz der Empfindlichkeitsvektorfelder zwischen paralleler und divergenter Beleuchtungsgeometrie: a) ebenen Platte, b) Zylinderoberfläche, c) Kugeloberfläche (z-Komponente)

Aufbaus für verschiedene Objektgeometrien berechnet. Im Bild 4 sind die z-Komponenten dieser Empfindlichkeitsvektorfelder für drei Objektgeometrien dargestellt. Im Bild 5 ist die Differenz zwischen den Werten dieser Empfindlichkeitsvektorfelder und der Konstante 2 dargestellt. Der Empfindlichkeitsvektor hat nämlich im Fall eines parallelen Beleuchtungsstrahlenbündels den Wert 2 und ist ortsunabhängig [1]. Es zeigt sich, daß der relative Fehler in der Größenordnung von 3,5% liegt, wenn das Auswerteprogramm diesen speziellen Objektbeleuchtungsfall annimmt und die tatsächliche Geometrie des interferometrischen Aufbaus nicht berücktsichtigt.

454

Im Bild 6 ist die Differenz zwischen Empfindlichkeitsvektorfeldern verschiedener Objektgeometrien dargestellt. Der relative Fehler beträgt bei der für diese Simulation gewählten Aufbaugeometrie bis zu 0,6%, wenn das Auswerteprogramm die Objektgeometrie nicht berücksichtigt.

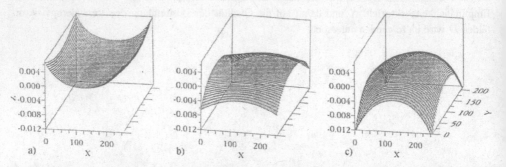

Bild 6: Darstellung der Differenz der Empfindlichkeitsvektorfelder zwischen verschiedenen Objektgeometrien: a) Kugel - Zylinder, b) Ebene - Zylinder, c) Ebene - Kugel (jeweils z-Komponente)

4.2 Experimentelle Ermittlung von Empfindlichkeitsvektorfeldern

Die Grundgleichung der holografischen Interferometrie kann in drei Komponenten zerlegt werden:

$$n\lambda = \vec{d} \cdot \vec{s} = \sum_{i=x,y,z} d_i s_i \Rightarrow s_i = \frac{n\lambda}{d_i} \tag{1}$$

Daraus folgt, daß der Wert der Empfindlichkeitsvektorkomponenten s_i über das Verhältnis der absoluten Interferenzphasenordnung $n\lambda$ zur Ganzkörperverschiebung d_i gebildet wird. Somit stellt das aus einer Ganzkörperbewegung (d_i = const für alle Objektpunkte) resultierende Interferenzmuster direkt die Werte der jeweiligen Empfindlichkeitsvektorkomponenten dar. Dies ergibt eine experimentelle Möglichkeit, das für eine exakte Auswertung der Interferenzmuster notwendige objektformabhängige Empfindlichkeitsvektorfeld zu erzeugen. Das Empfindlichkeitsvektorfeld $\vec{S}$ kann auf diese Weise auch dann richtig bestimmt werden, wenn die Objektgeometrie nicht vorliegt.

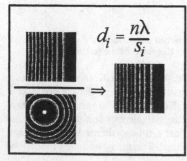

Bild 7: Ermittlung der exakten Vektorkomponenten der Objektänderung

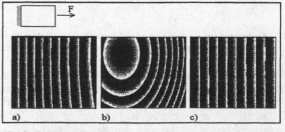

Bild 8: Belastete Zugplatte, a) Interferogramm einer Zugprüfung mittels simuliertem In-Plane ESPI aufgenommen, b) Interferogramm einer zusätzlichen Ganzkörperbewegung in x-Richtung, c) Korrektur der Phasenverteilung in Interferogramm a) mit Hilfe von Interferogramm b)

4.3 Ermittlung der exakten Objektänderungen

Mit Hilfe von experimentell ermittelten Empfindlichkeitsvektorfeldern (siehe 4.2) können die exakten Objektänderungskomponenten d_i bestimmt werden. Dazu muß die Grundgleichung (1) nach d_i aufgelöst werden (siehe Bild 7). Der zugehörige Algorithmus wurde in einem vorhandenen kommerziellen Auswerteprogramm implementiert und liefert die im folgenden Beispiel gezeigten Ergebnisse (siehe Bild 8): Eine ebene Platte wird einseitig eingespannt und am gegenüberliegenden Ende mit einer konstanten horizontalen Streckenlast auf Zug belastet. Die Linien gleicher Dehnung verlaufen bei dieser Randbedingung senkrecht und äquidistant. Das aufgenommene Interferogramm zeigt jedoch gekrümmte Interferenzstreifen, deren Abstand voneinander mit aufsteigender Interferenzstreifenordnung größer wird (siehe Bild 8a). Ein zuvor aufgenommenes Interferogramm einer Ganzkörperverschiebung in x-Richtung (siehe Bild 8b) wird zur Ermittlung des Empfindlichkeitsvektorfelds verwendet. Das korrigierte Interferenzmuster (siehe Bild 8c) gibt den richtigen Dehnungsverlauf wieder.

5. Verfahrenskonzept

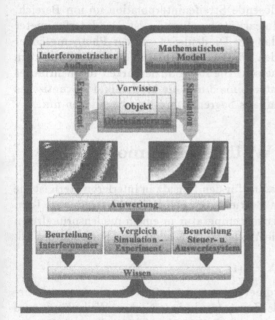

Bild 9: Qualitätsanalysekonzept

Wesentlicher Bestandteil des Verfahrens zur Qualitätsprüfung (siehe Bild 9) von Interferometern ist der Einsatz eines Programms zur Simulation von Interferenzstreifenmustern. Damit können sowohl der interferometrische Aufbau als auch Standardexperimente modelliert werden, um die Qualität der optischen und mechanischen Komponenten des Meßaufbaus, der Halterung des Objekts und der Belastungseinrichtung zu analysieren. Die Qualität des Auswertesystems kann anhand vorgegebener Aufbau- und Objektgeometrien sowie Objektbelastungen hinsichtlich der Ermittlung der exakten Objektänderungen überprüft werden.

6. Zusammenfassung

Es wurde mit Hilfe eines Interferenzmuster-Simulationsprogramms untersucht, welchen Einfluß die Geometrie des interferometrischen Aufbaus und die Geometrie des Objekts auf das zugehörige Empfindlichkeitsvektorfeld haben. Da die Genauigkeit der mit Interferometern zu bestimmenden Objektänderungsfelder von der Berücksichtigung dieses Einflusses auf das Empfindlichkeitsvektorfeld abhängt, kann das Interferenzmuster-Simulationsprogramm auch zur Beurteilung der Qualität von Interferometern herangezogen werden. Durch Einsatz einer Variante dieses Simulationsprogramms bei der experimentellen Ermittlung von Empfindlichkeitsvektorfeldern braucht weder die Aufbaugeometrie des Interferometers, noch die Objektgeometrie bekannt zu sein.

7. Literaturverzeichnis

[1] Jones, R.; Wykes, C.; „Holographic and Speckle Interferometry", Cambridge Univ. Press 1983
[2] Miesner, J.; Kreitlow, H.; „Simulation of interferometric fringe patterns by means of a flexible computer program and the finite element method", SPIE Conference, „Interferometry '94", Warschau 1994

Signalauswertung eines wellenlängenmodulierten Diodenlaserinterferometers

A. Abou–Zeid, P. Wiese
Physikalisch–Technische Bundesanstalt, Labor 5.23
Bundesallee 100, D–38116 Braunschweig

Der Einsatz von Diodenlasern als Lichtquelle in Interferometern ermöglicht aufgrund der Wellenlängendurchstimmbarkeit die Realisierung von neuartigen interferometrischen Methoden. Es wird ein Verfahren vorgestellt, bei dem in einem Michelson Interferometer über eine dreieckförmige Durchstimmung des Diodenstroms ein periodisches Interferenzsignal erzeugt wird. Die Signalauswertung mit einem Signalprozessor (DSP) ermöglicht eine hochauflösende Streifeninterpolation im nm Bereich, wobei jedoch die Korrektur von Nichtlinearitäten notwendig ist. Das Interferometer benötigt keine Polarisationsoptik, so daß der Diodenlaser und der Photoempfänger leicht über Glasfaser angekoppelt werden können. Als Anwendung wurden mit dem System Oberflächentopographien vermessen. Die Reproduzierbarkeit für mehrfaches Aufnehmen einer Profillinie wird in erster Linie durch die Mechanik des Scantisches und thermische Drift des gesamten Aufbaus begrenzt und liegt bei einigen nm.

1 Interferometer mit Wellenlängenmodulation

Eine wichtige Eigenschaft von Diodenlasern für den Einsatz in Interferometern ist die Abhängigkeit der Wellenlänge von den Betriebsparametern Diodenstrom und Temperatur. Stabilisiert man die Wärmesenkentemperatur in einem modensprungfreien Bereich des Kennlinienfeldes, so ist die Wellenlänge λ allein eine Funktion des Diodenstroms i:

$$\lambda = f(i)$$

In einem nicht abgeglichenen Interferometer kann die Interferenzphase gezielt beeinflußt werden, wenn der funktionale Zusammenhang zwischen Wellenlänge und Diodenstrom bekannt ist. Im quasistatischen Fall ergibt sich ein linearer Zusammenhang, für die verwendeten Dioden liegt der Koeffizient $d\lambda/di$ typischerweise bei 0,007 nm/mA. Mit steigender Frequenz nimmt $d\lambda/di$ stark ab [1], da der Einfluß des Diodenstroms auf die Wellenlänge bis zu Frequenzen von 1 MHz in erster Linie auf Temperatureffekte zurückzuführen ist [2].

Ein einfacher Zusammenhang zwischen Diodenstrom und Wellenlänge ist daher nur gegeben, wenn der Diodenstrom sehr langsam oder aber periodisch mit fester Frequenz durchgestimmt wird.

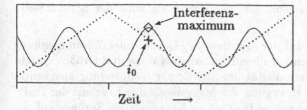

Abbildung 1: Diodenstrom
(dreieckförmig moduliert)
und Interferenzsignal

Bei dem hier vorgestellten Verfahren wird in einem Michelson Interferometer der Diodenstrom dreieckförmig moduliert, um eine Streifeninterpolation und Richtungserkennung zu ermöglichen. Die Abbildung 1 zeigt den zeitlichen Verlauf des Diodenstroms und das zugehörige Interferenzsignal. Die Modulationsamplitude des Diodenstroms wird so gewählt, daß auf einer Flanke des Dreiecksignals mindestens ein Interferenzmaximum liegt. Im Anfangszustand liegt das Interferenzmaximum beim Strom i_0 bzw. der Wellenlänge λ_0. Wird der Meßspiegel des Interferometers ausgehend von der Armlängendifferenz s_0 um die Strecke ds verschoben, so verschiebt sich die Lage des Interferenzmaximum um di bzw. $d\lambda$. Es gilt:

$$ds = \frac{s_0}{\lambda_0}\, d\lambda$$

Man kann also Verschiebungen des Meßspiegels messen, indem man die Lage des Interferenzmaximums relativ zum dreieckförmigen Modulationssignal bestimmt.

2 Signalauswertung mit einer DSP-Karte

Zur Auswertung der oben beschriebenen Signale wurde eine Elektronik zum Anschluß an eine kommerzielle Signalprozessorkarte (Motorola DSP56001, 20 MHz) aufgebaut. Die Elektronik besteht in der Hauptsache aus zwei A/D-Wandlern und Dual Ported RAM Bausteinen und ist in Kombination mit der DSP-Karte als PC-Einsteckkarte konzipiert.

Auf einen externen Triggerimpuls hin werden 1 1/2 Signalperioden (3000 Werte) des dreieckförmigen Modulationssignals (f = 10 kHz) und das zugehörige Interferenzsignal von den zwei A/D-Wandlern (8 Bit, 20 MHz) digitalisiert und in die RAM Bausteine geschrieben.

Der Signalprozessor kann über seinen externen Bus direkt auf die RAM Bausteine zugreifen und die Werte auslesen. Die Auswertesoftware sucht zunächst eine komplette fallende Flanke des Dreiecksignals, nur dieser Teil des Datensatzes wird für die weitere Auswertung betrachtet. Im nächsten Schritt wird die Lage des Maximums des Interferenzsignals relativ zum Dreiecksignal bestimmt. Zur Rauschunterdrückung und Erhöhung der Auflösung werden Digitalfilter eingesetzt.

Der PC liest über den Host-Port die Ergebnisse der Auswertung und speichert die Werte. Die Meßrate beträgt zur Zeit 1 kHz, wobei eine Steigerung um eine Größenordnung durch Verbesserungen sowohl auf der Hard– als auch der Softwareseite möglich ist.

Zur Kalibrierung des Systems wird der Quotient s_0/λ_0 sowie der Zusammenhang zwischen Diodenstrom und Wellenlänge benötigt. Anstatt diese drei Größen einzeln zu messen, kann auch die $\lambda/2$–Periodizität der Signale zur Kalibrierung ausgenutzt werden. Wird bei gleichförmiger Bewegung des Meßspiegels das Ergebnis der DSP–Auswertung über der Zeit aufgetragen, so tauchen im Kurvenverlauf Sprünge auf, die einer Verschiebung von $\lambda/2$ entsprechen. So erhält man eine Zuordnung zwischen den vom DSP gelieferten Zahlenwerten und dem vom Meßspiegel zurückgelegten Weg.

Die Änderung des Kalibrierfaktors mit dem Wegunterschied s_0 muß berücksichtigt werden, wenn das Verhältnis vom Meßbereich zum Anfangswegunterschied in der Größenordnung der angestrebten Auflösung liegt. Sinnvolle Werte für s_0 liegen im Bereich von einem mm bis zu einigen cm. Die Langzeitstabilität der Wellenlänge des parameterstabilisierten Diodenlasers beträgt etwa 10^{-5} für ein Jahr [3]. Das System ist also in erster Linie zur Messung von kleinen Verschiebewegen (< 1 mm) mit Auflösungen von 10^{-4} bis 10^{-5} geeignet.

3 Meßaufbau und Ergebnisse

In einer ersten Anwendung des Systems wurden Oberflächentopographien vermessen. Der optische Aufbau dafür ist relativ einfach (siehe Abbildung 2). Das Licht des Diodenlasers wird über eine Monomodefaser in das Interferometer eingekoppelt. Das Interferometer besteht aus einem Neutralteilerwürfel und einer Spiegel–Linse Kombination als Referenzreflektor. Im Meßarm des Interferometers wird das Licht über ein Mikroobjektiv auf die Oberfläche fokussiert. Da es sich um ein Planspiegelinterferometer handelt, muß der Diodenlaser durch einen optischen Isolator vor Rücklicht geschützt werden. Die Oberfläche wird mit einem x–y–Scantisch abgerastert. Der Photoempfänger am Interferometerausgang kann auch über Multimodefaser gekoppelt werden.

Ein PC als Host–Rechner dient zum Laden der Auswerteprogramme für den DSP, zur Programmierung des Kontrollers für den x–y–Scantisch und zum Speichern und Weiterverarbeiten der Meßwerte.

Als Test für die Reproduzierbarkeit des Systems wurde eine Scanlinie der Länge 0,18 mm von der Oberfläche eines Planspiegels mehrfach (180 fach) direkt nachein-

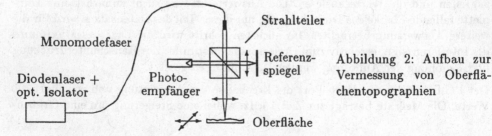

Monomodefaser

Strahlteiler

Diodenlaser +
opt. Isolator

Photo-
empfänger

Referenz-
spiegel

Oberfläche

Abbildung 2: Aufbau zur Vermessung von Oberflächentopographien

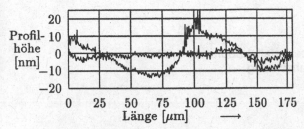

Profil-höhe [nm]

Länge [μm] ⟶

Abbildung 3: Oberflächen-profil eines Planspiegels

ander aufgenommen. Der Sample–Abstand betrug 0,5 μm, die Zeit für das Aufnehmen einer Linie betrug 2 Sekunden. Die vom DSP gelieferten Rohdaten werden im PC skaliert, λ/2 Phasensprünge werden korrigiert und schließlich wird die Differenz zu einer Ausgleichsgeraden gebildet.

Die Abweichungen zwischen den einzelnen Scanlinien sind sehr gering und liegen im Bereich einiger nm. Die Abbildung 3 zeigt exemplarisch eine Scanlinie, wobei die Meßdaten nicht gefiltert sind. Die Abweichungen von der Nullinie liegen im Bereich von $^{+}/$- 20 nm und sind zum größten Teil auf eine nichtlineare Abhängigkeit der Wellenlänge vom Diodenstrom zurückzuführen. Da es sich hier um einen systematischen Fehler handelt, ist eine Korrektur möglich. Das Ergebnis ist ebenfalls in Abbildung 3 zu sehen, die Abweichungen von der Nullinie liegen nun im Bereich von $^{+}/$- 5 nm.

Das vorgestellte System zeigt eine Möglichkeit auf, die Wellendurchstimmbarkeit von Diodenlasern zur Streifeninterpolation in einem Interferometer zu nutzen. Es besteht eine Verwandtschaft zu Phase–Shift–Verfahren [4], wobei der Einsatz eines Signalprozessors neue Ansätze zur Signalauswertung bietet. Die Messungen zeigen, daß mit einem Aufbau zur Oberflächenvermessung Auflösungen im nm–Bereich realisierbar sind, wenn die systematischen Nichtlinearitäten korrigiert werden.

Diese Arbeit wurde von der Deutschen Forschungs Gemeinschaft (DFG) gefördert.

Literatur

[1] *Abou–Zeid, A.; Wiese, P.:* A Surface Topography Sensor Based on a Non-Counting Diode Laser Interferometer: Non Linearity and Scanning Frequency. 3rd Int. Imeko–Symposium on laser metrology for precision measurement and inspection in industry, March 21.–22. 1994, Heidelberg, VDI Berichte 118

[2] *Petermann, K.:* Laser Diode Modulation and Noise. Advances in Optoelectronics (ADOP), Kluwer Academic Publishers, 1988.

[3] *Abou–Zeid, A.:* Spektrale Alterung kommerzieller GaAlAs–Diodenlaser. 9. Int. Kongreß für Laser–Optoelektronik und Mikrowellen, München, 5.–9. Juni 1989

[4] *Packroß, B.:* Interferometrie mit Laserdioden, Berichte aus dem Institut für Technische Optik der Universität Stuttgart, September 1992

A Transportable Lambdameter for Calibration of Laser Wavelength

A. Abou-Zeid, N.Bader
Physikalisch-Technische Bundesanstalt, Laboratory 5.23
Bundesallee 100, D-38116 Braunschweig

Lambdameters can be used to precisely determine the wavelength of a laser source /1,2/. Such a lambdameter essentially consists of a Michelson interferometer fed by two laser sources, namely the reference laser with the vacuum wavelength λ_R and the comparison laser with an unknown vacuum wavelength λ_U which should be determined. By moving the measuring reflector for a defined length d, then the following is valid:

$$d = Z_R * \lambda_R / 8 * n_R = Z_U * \lambda_U / 8 * n_U \qquad 1$$

$4Z_R$ and $4Z_U$ are the number of the interference fringes having passed through the reference and the comparison systems, respectively. n_R and n_U are their refractive indices. From eq. 1, the desired relation for λ_U is obtained:

$$\lambda_U = \lambda_R * (Z_R / Z_U) * (n_U / n_R) \qquad 2$$

As shown in eq. 2 the value of λ_U can be determined by the values of λ_R, Z_R, Z_U, n_R and n_U. If the total differential of eq. 2 is formed, a relation is obtained for the relative wave-length measuring uncertainty $d\lambda_U/\lambda_U$, i.e. the quality of the measurement:

$$d\lambda_U/\lambda_U = dZ_R/Z_R - dZ_U/Z_U - dn_R/n_R + dn_U/ n_U + d\lambda_R/\lambda_R \qquad 3$$

The first and the second two terms on the right side are the relative measurement uncertainties of the interferometer and the refractometer, respectively. The last term represents the relative stability of the reference laser. Eq. 3 shows that the relative uncertainty $d\lambda_R/\lambda_R$ can be reduced by increasing the values of Z_R and Z_U. In the simplest case, this can be done by just increasing the displacement length d. There are, however, limits to this, due to design dimensions of the transportable lambdameter to be developed (d ≤ 20 cm) and the coherence length of the unknown laser.

The aim of this work is to develop and set up a transportable lambdameter
(48 * 25 * 15 cm^3) with a relative uncertainty in the order of 10^{-8} for calibrating
of wavelengths of laser sources in the spectral range between 500 nm and
1000 nm, which can be used in precision interferometry and spectroscopy.

As shown in eq. 3, a value of $\approx 1*10^{-6}$ follows for the relative measurement
uncertainty $d\lambda_U/\lambda_U$, when the following is put in eq. 3, using eq.1: d = 20 cm,
λ_R = 780,24 nm, λ_U = 785nm, $n_U \approx n_R$, $dZ_R = dZ_U = 1$ which corresponds to a
digit error of $\lambda/8$ and $d\lambda_R/\lambda_R \approx 1*10^{-8}$. This means that, in order to achieve a
value of $d\lambda_U/\lambda_U$ of $1*10^{-8}$, the digit error for both lasers must be reduced at least
by a factor 100, meaning a counter resolution of $\lambda/800$ instead of $\lambda/8$, to take
other error sources of the lambdameter into account, which also affect its
uncertainty.

Methods used to reduce the measurement uncertainty

Three methods are developed to reduce the measuring uncertainty, i.e. to
decrease the counter resolution /3/. The results show that the new developed
method of multiple readings is the method best suited to estimate the relative
uncertainty of a lambdameter for the determination of an unknown wavelength
of a cw laser, and this owing to the minor effort involved, relativ high gain
factor, the insensitivity to feedback light and laser modulation etc.. The phase
interpolation method is the second best one and then the concidence method
ranks third. For both last methods, one must be sure that the diode laser is free
from feedback light. To achieve the desired uncertainty with the aid of the
coincidence method, this can be combined with that of multiple readings.

Rb-stabilized diode laser as a reference laser

To determine the absolute frequency shift of the reference laser, the beat
frequency of two diode lasers stabilized to the same absorption line of Rb was
measured. Whereas the stabilization parameters of one of both lasers are kept
constant, those of the other laser tested are varied. For both reference lasers
tested, a long-term wavelength stability better than 10^{-8} is achieved after taking

into account the conditions due to the beam intensity through the absorption cell (< 60 µW/cm^2) and the reference measurement of the optical beam power /3/. The wavelength traceability measurements which are repeated within one year show that the vacuum wavelength of both reference lasers tested which are locked to the strongest $D_2/3$ absorption line of Rb amounts to 780.243830nm ± 0.000006nm.

Error sources of the lambdameter

The relative measurement uncertainty of the lambdameter is affected by some error sources such as adjustment error, beam diameter and dispersion of air for both lasers. To achieve a relative measurment uncertainty of 1 * 10^{-8} for every error source the following conditions have to be considered:

*** the angle of misalignment of reference and unknown laser beams must be smaller than 28 "

*** the beam diameter of both lasers must exceed 4 mm

*** an optical isolation (< 40 dB) for both lasers is necessary

Due to the dispersion effect, i.e. the dependence of the refractive index n on the laser wavelength for the same parameters of the surrounding atmosphere /4/. The correction term of the measured dispersion factor n_U / n_{780} which depends in second order on the air parameters has been implemented in the software of the lambdameter. Before and after every wavelength measurement, the surrounding parameters of air temperature and pressure are measured with the aid of a weather station developed at PTB to determine the correction term due to the dispersion.

Performance of the transportable lambdameter

The developed transportable lambdameter is set up to determine unknown wavelengths of lasers as well as their stabilities. The efficiency of the lambdameter, is tested by the application of the method of multiple readings for the measurements of the following wavelengths and stabilities:

*** the relative wavelength stability of a parameter-stabilized diode laser
(2 µA, 1 mK), $1 * 10^{-7}$

*** the air wavelength of a Fabry-Perot-stabilized diode laser and its relative stability (air parameters: temperature 20°C, pressure 1013 hPa, CO_2 content $4 * 10^{-4}$, rel. humidity 50 %): 778.68399 nm ± 0.00008 nm, $1 * 10^{-7}$

*** the vacuum wavelength of a two-frequency stabilized He-Ne laser and its relative stability: 632.99142 nm ± 0.000006 nm, $1 * 10^{-8}$

*** the vacuum wavelength of a Rb/D_2-stabilized laser and its relative stability: 780.243830nm ± 0.000006nm, $1 * 10^{-8}$

The results show good correspondence with the traceable wavelengths and known stability values of the laser tested. To sum up, unknown laser wavelengths between 500 nm and 1000 nm can be calibrated with the developed transportable lambdameter with a relative measurement uncertainty of better than $4 * 10^{-8}$.

Acknowledgment

The Authors would like to thank the Commission of the European Communities (SMT) for the financial support of this work.

Literature

/1/ Cachenaut, J.; Man, C.; Cerez, P.; Brillet, A.; Stöckel, F.; Jourdan, A., Hartmann, F.: Description and accuracy tests of an improved lambdameter, Rev., Physique Appl. **14**, (1979), p. 685

/2/ Hall, J.L., Lee, S.A.: Interferometric real-time display of cw dye laser wavelength with sub-Doppler accuracy, Appl. Phys. Lett., **29**, (1976), p.367

/3/ Abou-Zeid, A.; Bechstein, H., Enghave, C.: Real-time position and form measurements on CMMs by means of multi-function laser interferometry based on laser diodes, BCR Report EUR 15939 EN, Brussels, 1995

/4/ Edlén, B.: The refractive index of air, Metrologia 2 , (1966), p. 71

Vollständige Verschiebungsbestimmung an mikroelektronischen Bauelementen

O. Kruschke, G. Wernicke
Humboldt Universität zu Berlin   Institut für Physik, Labor für Kohärenzoptik
D-10115 Berlin

Eine Methode zur vollständigen Bestimmung von Verschiebungen, die durch erhöhte Betriebstemperaturen an mikroelektronischen Bauelementen auftreten, wird vorgestellt. Interferometrische Meßgenauigkeit und die Leistungsfähigkeit eines korrigierten Mikroskopes werden von dem dafür entwickelten hologramminterferometrischen Mikroskop verbunden. Durch konjugierte Rekonstruktion kann die Beobachtung der Interferogramme mit von der holographischen Aufzeichnung unabhängiger, variabler Vergrößerung und Objektentfernung erfolgen.

Der Trend in der Mikroelektronik zu immer höheren Packungsdichten herkömmlicher und hochintegrierter Bauelemente auf Leiterplatten ist ungebrochen. Damit verbunden ist die Notwendigkeit, exakte Kenntnis über mikroskopische Veschiebungen aufgrund mechanischer und thermischer Belastungen zu erlangen, um daraus Schlüsse über günstigere Bestückungstechniken oder minimale Abstände zwischen elektrischen Kontaktstellen und Leiterbahnen ziehen zu können. Zur Bestimmung von kleinen Verschiebungen an rauhen, reflektierenden Festkörperoberflächen hat sich die holographische Interferometrie als wirkungsvolles Verfahren erwiesen. Derartige Messungen an Leiterplatten zur Bestimmung von Verschiebungen bei mechanischen Spannungen oder starken Änderungen der Umgebungstemperatur wurden von PRYPUTNIEWICZ durchgeführt[1].

Zusätzliche thermische Belastungen treten infolge des ohmschen Widerstandes einzelner Elemente auf. Die interferometrische Messung der daraus resultierenden Verschiebungen erfordert wegen der geringen Größe mikroelektronischer Bauelemente die Beobachtung der Interferogramme mit mehrfacher Vergrößerung. Für derartige Messungen werden derzeit Interferenzmikroskope eingesetzt, die jedoch außschließlich die Bestimmung nur einer, meist senkrecht zur Leiterplatte stehenden Komponente des Verschiebungsvektors erlauben. Rückschlüsse auf laterale Verschiebungen sind daraus nur bedingt zu ziehen.

In dem im folgenden vorzustellenden hologramminterferometrische Mikroskop (HIM) wird die Projektion des Verschiebungsvektors auf den Sensitivitätsvektor, der Winkelhalbierenden zwischen Beobachtungs- und Beleuchtungsrichtung, gemessen. Für drei linear unabhängige Sensitivitätsvektoren läßt sich der Verschiebungsvektor $\mathbf{d}$ vollständig durch Lösen der Gleichung $N\lambda = \hat{\mathbf{G}}\mathbf{d}$ bestimmen. $\hat{\mathbf{G}}$ ist die Geometriematrix, die die Lage der Sensitivitätsvektoren im gewählten Koordinaten-

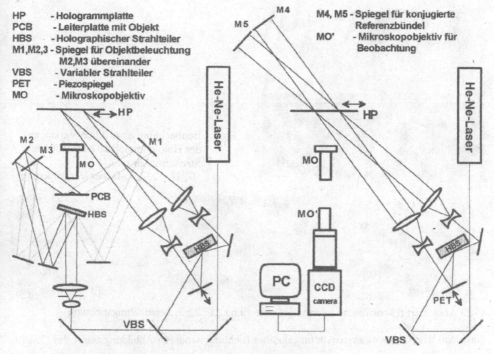

HP   - Hologrammplatte
PCB   - Leiterplatte mit Objekt
HBS   - Holographischer Strahlteiler
M1,M2,3 - Spiegel für Objektbeleuchtung
          M2,M3 übereinander
VBS   - Variabler Strahlteiler
PET   - Piezospiegel
MO   - Mikroskopobjektiv

M4, M5 - Spiegel für konjugierte
          Referenzbündel
MO'   - Mikroskopobjektiv für
          Beobachtung

Fig.1: Aufbau des HIM; links: Aufnahme; rechts: Beobachtung

system beschreibt, λ ist die Wellenlänge und N der Ordnungszahlvektor, der die Verschiebung in Richtung eines jeden Sensitivitätsvektors angibt.

Das optische Layout des HIM ist in Figur 1 skizziert. Die verschiedenen Sensitivitätsvektoren werden durch 3 Beleuchtungsrichtungen realisiert (Spiegel M1,M2,M3). Um eine Phase-Shift-Auswertung der Interferogramme zu ermöglichen und gleichzeitig mehrmalige Zustandswechsel des Objektes bei der Aufnahme von Interferogrammen verschiedener Beleuchtungsrichtungen zu vermeiden, werden drei Doppelbelichtungshologramme mit Zwei-Referenzstrahl-Technik nebeneinander auf einer Hologrammplatte aufgenommen. Für die dafür notwendige Verschiebung der Hologrammplatte zwischen den einzelnen Belichtungen wurde ein kinematischer Plattenhalter mit einer Dreifach-Dreipunktauflage konstruiert[2].

Beleuchtung des Objektes und der Einfall der Referenzwelle unter nicht zu flachem Winkel auf die Hologrammplatte bedingen gewisse minimale Entfernungen von Objekt, Mikroskopobjektiv und Hologrammplatte, die eine abbildungsgenaue Beobachtung des vergrößerten Objektes unmöglich machen. Die von SMITH und WILLIAMS erstmals zur Korrektur von Abbildungsfehlern vorgeschlagene Methode der konjugierten Rekonstruktion[3] wird zur Separation zwischen holographischer Aufnahme und mikroskopischer Beobachtung genutzt. Die konjugiert rekonstruierte Objektwelle

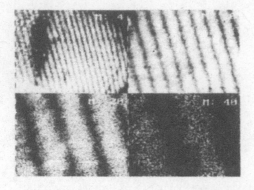

Fig.2: Beobachtung eines durch Verkippen der Hologrammplatte entstandenen Streifensystems mit 4-, 10-, 20-, 40-facher Vergrößerung

Fig.3: konjugiert rekonstruierte Interferogramme v.l.n.r.: 1., 2., 3. Beleuchtungsrichtung

durchläuft das Abbildungssystem in umgekehrter Richtung, wodurch Abbildungsfehler des Objektives MO kompensiert werden. Am Ort des Objektes entsteht ein reelles Bild, welches von seiner Rückseite mit einem zweiten Mikroskopobjektiv (MO') beobachtet werden kann. Die Wahl des Abbildungssystems für die holographische Aufzeichnung kann somit unabhängig von der späteren Beobachtung und entsprechend der für die Aufnahme günstigsten Intensitätsverhältnisse und Distanzen gewählt werden. Demgegenüber erfolgt die Beobachtung in genau der Objektentfernung, für die MO' korrigiert ist, wobei die Vergrößerung beliebig der Streifendichte und Objektgröße angepaßt werden kann (Fig.2). Zusätzlich lassen sich unter den Bedingungen der konjugierten Rekonstruktion die Sensitivitätsvektoren orthogonalisieren. Dadurch werden Fehlerfortpflanzungsfaktoren minimiert und gleiche Sensitivitäten der Interferometergeometrie für alle Verschiebungskomponenten erreicht[4].

Die konjugierten Referenzwellen werden durch Reflexion der parallelen Referenzstrahlen bei senkrechter Inzidenz an den Spiegeln M5 und M6 gebildet. Die beim Durchgang durch die entwickelte Hologrammplatte auftretende Verzerrung der Wellenfronten konnte durch interferometrische Messung als vernachlässigbar klein eingeschätzt werden.

An Messungen des Ausdehnungsverhalten eines oberflächenmontierten Widerstandes (1kΩ) soll die Leistungsfähigkeit des vorgestellten Aufbaus demonstriert werden. Für eine Belastung von 7,5mA sind in Fig.3 die mit 10-facher Vergrößerung aufgenommenen Interferogramme dargestellt. Die im linken Bild eingezeichnete Linie entspricht einer Länge von 0,5mm.

Die mod2π-Phasenverteilung wurde mit einem gegen die Nichtlinearitäten der Piezokennlinie unempfindlichen 5-Schritt phase-step-Algortithmus ermittelt. Die anschließende Filterung erfolgte separat für die sin- und cos-Verteilung, um die Modulation und die Verschiebungsinformation zu erhalten. Für die weitere Auswertung entlang der ausgewählten Strecke wurde der Anstieg einer zur Kompensation von Repositionierungsfehlern subtrahierten Ausgleichsgerade aus dem Mittelwert der Abstände der Phasensprünge berechnet. Die letztendlich berechneten kartesischen Verschiebungskomponenten sind für den Widerstand und vergleichsweise für eine seiner Lötstellen (Belastung: 10 mA) in Fig.4 dargestellt. Die Leiterplatten- bzw. Objektebene wird durch die x- und y-Komponenten aufgespannt. Die z-Komponente entspricht der Normalkomponente und ist der Beobachtungsrichtung entgegengerichtet.

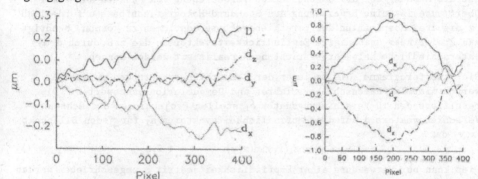

Fig4: kartesische Verschiebungskomponenten: Widerstand(links); Lötstelle(rechts)

In beiden Fällen liegt die Größe der gemessenen Verschiebungen im Bereich der theoretisch abgeschätzten Werte. Deutlich ist erkennbar, daß die Verschiebung des Widerstandes entscheidend durch die laterale x-Komponente bestimmt wird, wogegen die Lötstelle hauptsächlich normale Verschiebungen aufweist. Im Übergangsbereich Widerstand-Lötstelle (ab Pixel 350 im linken Diagramm) ist der wachsende Einfluß der Lötstelle an einer verminderten Verschiebung in x-Richtung bei gleichzeitigem Anwachsen der z-Komponente zu erkennen. Im Übergangsbereich Lötstelle-Leiterplattensubstrat (ab Pixel 350 im rechten Diagramm) kommt es auf Grund großer Differenzen der Ausdehnungskoeffizienten und der Temperaturleitfähigkeit zu starken Spannungen, erkennbar am abrupten Monotoniewechsel der Verschiebungskomponenten.

Für die weitere Arbeit sind die genaue Bestimmung der maximalen Auflösung und der relativen Fehler mittels eichbarer Objekte, Verbesserungen bezüglich des Aufnahmeverfahrens, der Bildung der konjugierten Referenzwellen und der automatisierten Auswertung vorgesehen.

Die Autoren danken der Deutschen Forschungsgemeinschaft für die Unterstützung dieser Arbeit.

1. R.J.Pryputniewicz, "Heterodyne holography applications in studies of small components," *Opt. Eng.* **24**(5), 849-854(1985).
2. O.Kruschke, G. Wernicke,"3D displacement measurement by a holographic interferometric microscope," wird veröffentlich in Proc. zur SPIE-Konf. 'Praktische Holographie IX'
3. R. W. Smith, T. H. Williams, "A depth encoding high resolution holographic microscope," *Optik* **39**,150-155(1972).
4. O.Kruschke,G. Wernicke,H.Gruber"Holographic interferometric microscope with conjugate reconstruction for measurement of 3D displacements," Opt. Eng.**34**(4), 1122-1127(1995).

Holografische Dehnungs- und Spannungsanalyse

H. Steinbichler, H. Klingele und B. Mähner
Steinbichler Optotechnik GmbH
D-83115 Neubeuern

1 Holografische Messung von Verschiebevektoren

Für die Messung der dreidimensionalen Verschiebung von Punkten auf einer
Oberfläche ist eine Erweiterung des Standard-Holografieaufbaus erforderlich.
Um die drei Komponenten des Verschiebevektors bestimmen zu können, benötigt
man drei linear unabhängige Empfindlichkeitsvektoren, die z.B. durch drei
unterschiedliche Beleuchtungsrichtungen realisiert werden können.

Die Interferogramme, die in jeder der Beleuchtungsrichtungen entstehen,
werden mittels Phasenschiebeverfahren und Demodulation ausgewertet. Die
resultierenden 1D-Verschiebungsdaten δ_i stellen Projektionen des echten
Verschiebevektors $\vec{d}$ auf die Empfindlichkeitsvektoren $\vec{s}_i$ für jeden Bildpunkt
(x, y) dar.

$$\delta_i(x,y) = \vec{s}_i(x,y)\,\vec{d}(x,y) \quad , \quad i = 1,2,3 \tag{1}$$

Dies kann bei Verwendung einer Empfindlichkeitsmatrix $\underline{S}$ umgeschrieben werden

$$\vec{\delta}(x,y) = \underline{S}(x,y)\,\vec{d}(x,y) \tag{2}$$

und dann nach dem Verschiebevektor aufgelöst werden

$$\vec{d}(x,y) = \underline{S}^{-1}(x,y)\,\vec{\delta}(x,y) \tag{3}$$

Diese Berechnung ist nur für diejenigen Punkte der Objektoberfläche möglich,
die von allen drei Beleuchtungsstrahlen getroffen und von der Kamera erfaßt
werden.

2 Vermessung der Kontur und Modellierung

Zur Berechnung der Spannungsverteilung benötigt man ein geometrisches Modell
des Meßobjektes, bei dem die Oberfläche durch Knotenpunkte und
Flächenelemente beschrieben wird. Bei flachen Objekten kann hierzu einfach
ein Rechteckgitter verwendet werden. Wenn aber eine gekrümmte oder
verwinkelte Oberfläche vorliegt, dann muß die Kontur zur Erstellung der
Flächenelemente berücksichtigt werden.

Es gibt verschiedene optische Methoden zur Konturvermessung mittels
projizierter Linien. Inzwischen werden flächenhafte Projektionsverfahren
mit Triangulation kombiniert, so daß die absoluten Koordinaten (x,y,z) aller
abgebildeter Objektpunkte berechnet werden können (COMET, TRICOLITE).

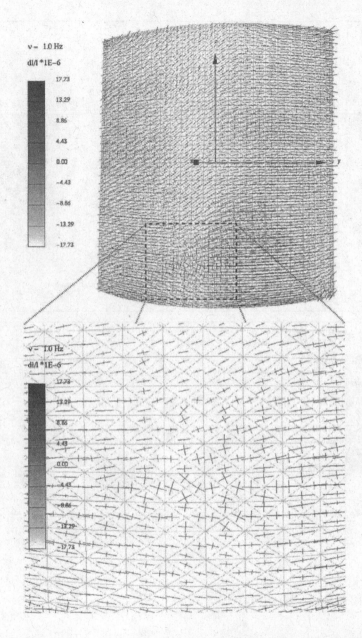

Abbildung 1: Hauptdehnungsverteilung durch punktförmige Belastung eines gekrümmten Bleches.

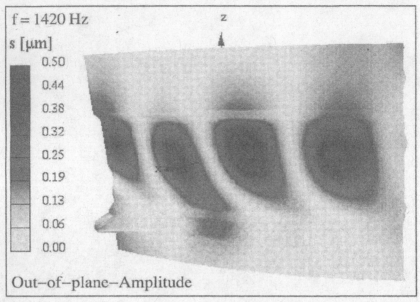

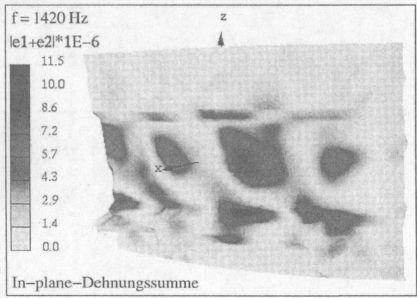

Abbildung 2: Oberflächennormale Verschiebung (oben) und Hauptdehnungssumme (unten) eines Druckbehälters.

Wenn zur Kontur- und Verformungsmessung die gleiche Kameraposition verwendet wird, können beide Datensätze mit hoher Genauigkeit (Pixelauflösung) überlagert werden. Es erfolgt dazu eine Netzgenerierung auf dem Bild der Kontur, bei der Knotenpunkte und Dreieckselemente definiert werden. Der dreidimensionale Verschiebevektor wird dann an den Knotenpunkten aus den ausgewerteten Interferogrammen eingefügt.

3 Berechnung der Dehnungsverteilung

Zur Berechnung der Dehnungsverteilung wird auf jedem der Oberflächenelemente ein linearer Verlauf für die Verschiebevektoren angenommen. Da bei der Dehnungsberechnung im Prinzip eine numerische Ableitung der Verschiebevektoren durchgeführt wird, führt das zu einem über das gesamte Element konstanten Spannungszustand.

Die Hauptdehnungsrichtungen (die im Falle eines isotropen Materials den Hauptspannungsrichtungen entsprechen) und die Werte ϵ_1 und ϵ_2 für die Hauptdehnungen werden jeweils für das lokale Koordinatensystem auf dem Dreieck berechnet.

4 Statische Verformung

Bei statischen Verformungen kann holografisch die relative Dehnung ermittelt werden, die zwischen einem beliebigen Referenzzustand und einem diesem gegenüber verformten Zustand auftritt.Ω Abbildung 1 zeigt das Ergebnis einer derartige Dehnungsbestimmung an einem gekrümmten Blech, das in einer geeigneten Vorrichtung punktförmig belastet wurde. Die Verteilung der Hauptdehnungen zeigt einerseits die durch die verwendete Stahlkugel hervorgerufene Belastung in Form isotroper Ausdehnung, und andererseits das von der Belastungsstelle ausgehende, gerichtete Spannungsfeld (Abbildung 1).

5 Dynamische Belastung

Unter Verwendung von Pulsholografie oder von Dauerstrichlasern mit Stroboskoptechnik kann für sinusförmige Schwingungen die Berechnungsmethode für die Hauptdehnungsverteilung direkt übernommen werden. Man erhält dann die Momentandehnung z.B. im Maximum der Schwingung bezogen auf den Ruhezustand.

Für einen Druckbehälter ist in Abbildung 2 oben die oberflächennormale Verformung durch eine Eigenschwingung dargestellt, unten ist die Summe der Hauptdehnungen gegenübergestellt, die sich aus der Dehnungsberechnung ergibt.

Literatur

[1] H.Marwitz (Ed.): Praxis der Holografie, expert-Verlag (1990).

Automated Moiré Interferometry for Residual Stress Determination in Experimental Mechanics Applications

M. Kujawińska
Institute for Design Precise and Optical Instruments
Warsaw University of Technology, 8 Chodkiewicza St., 02-525 Warsaw, Poland

1. INTRODUCTION

Manufacturing processes, the action of load and thermal stresses create residual stresses in engineering elements [1]. These stresses; acting in conjunction with those produced by live loads, can threaten the safety of an engineering structure operation by increasing the hazard of creation and propagation of fatigue cracks. Difficulty in measuring them nondestructively; the unpredictability of their magnitude, sense, and direction; their adverse ability to combine with stress corrosion, environmental and fatigue situations; the difficulty in removing them can render residual stresses to be extremely troublesome. Residual stresses should be evaluated under operating conditions to assess their effect on the intended service capability of the component. The problem of the experimental analysis of the residual state in elements has been investigated for many years is search of an inexpensive and reliable method of determination of those stresses.

In order to carry out experimental examination of residual stresses one has to answer two questions: the first one is *how to determine* residual stresses from mechanical point of view i.e. how to reach into inside phenomena of a 3D body, the second one is *how to measure* those stresses from experimental point of view, i.e. which experimental technique to apply. Nondestructive techniques that can reach and measure inside phenomena of a 3D body are for instance: rentgenography, neutronography and ultrasonography. They have capability; to an extend, to penetrate the interior of the investigated body and measure the stress state in it. However the results obtained are of different kind and values, depend on the material of the element, and often one of qualitative character only.

Another way of determination of residual stresses is relaying on comparative measurements of displacement/strain values before and after stress relieving. The measurements are performed by strain gauges, photoelastic coatings, moiré or interferometric techniques. Stresses may be relieved by hole drilling (teparation), sectioning of a body into small pieces or by annealing.

Here the automated moiré (grating) interferometry, GI, method has been selected as an experimental technique leading to strain determination on the surface of the sample. This technique in conjunction with numerical analysis based on an theoretical model of stresses enables solving the problem of determination of the original 3D residual stresses.

The paper describes the methodology and instrumentation of measurement for two cases: big engineering element, namely railway rail and fracture mechanics fregue type specimen with a notch.

2. DETERMINATION OF RESIDUAL STRESS IN RAILROAD RAILS

One of the best known and successful technique for 3D residual stress determination in rails, that has been applied, is so called Battlle technique [2,3]. It is based on the general concepts of destructive stress relief and consists of two main sectioning steps for determination of the in-plane residual stresses:

1. Yasojima - Machii (YM) slicing (for δ_{xx}, δ_{yy}), which is done by cutting out of a rail slab a thin transverse slice (Fig. 1) and dicing it;
2. Meier sectioning (for δ_{zz}), which is done by cutting a rail into longitudinal rods (4 x 4 x 500 mm size).

Here the in-plane displacement fields at YM slices are determined using the moiré interferometry with temperature resistant grid [4]. Residual stress in the samples was released by the annealing process, which was performed according to the temperature scheme given in Fig 2.

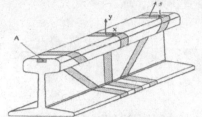

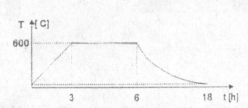

Fig. 1. Orientation of slices cutout of the railroad rail.

Fig. 2. Temperature changes versus time during the annealing process of YM slice.

The interferometric system for in-plane displacement measurement is based on the two beam interference. Two mutually coherent light-beams illuminate symmetrically the reflection type cross-pattern diffraction, grating fixed to the sample. The first diffraction orders of two illuminating beams interfere and produce a fringe pattern with information about in-plane displacement. Residual stresses cause the state of deformation which, when released by thermal process, causes in turn the opposite deformation. The interferogram analysis gives the field of displacements (u, v), which by differentiation may be converted into strains $(\varepsilon_x, \varepsilon_y, \gamma_{xy})$. The system presented (Fig 3) was designed to be used for big components testing (measuring area 75 x 75 mm^2) [5]. In the arrangement, one-mirror head and the rotary mounts of the specimen and CCD camera enable the sequential measurement of u and v displacements. Interferograms are collected by CCD camera using imaging objectives O1 and O2 and video signals are digitised using VFG512 frame grabber. The automatic analysis of fringe patterns was performed by the temporal phase shifting [6] method. Phase shifting was introduced by tilting the plane parallel plate PP. The specimen mount enable the accurate replacement of the object and therefore the systematic errors connected with the sample grating, aberration of the system and nontotal residual stress relieving can be eliminated by sample subtraction of the displacement maps.

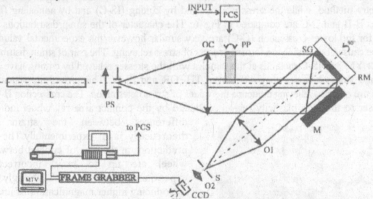

Fig 3. Scheme of the grating interferometry system: He-Ne laser. PS - pinhole system, OC -collimator objective, SG - specimen grating, RM, M - mirrors, PCS - phase-shifting control.

The interferogram obtained after annealing are shown in Fig. 4 together with the displacement fields (u, v) which were obtained by subtraction of the final and initial displacement fields $(u'=u_1-u_0, v'=v_1-v_0)$. The experimental strain fields calculated from (u', v') displacements are shown in Fig. 5 a-c and are compared with their numerical predictions (Fig. 5 d-f). The formulation of analytical model [7] relies on the motion of a shakedown state established by the highest load to which the rail was subjected.

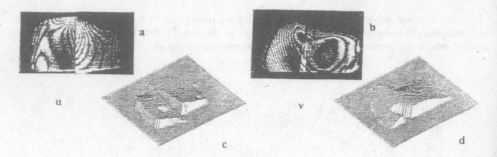

Fig. 4. Interferograms representing in-plane u_1 (a) and v_1 (b) displacement fields in YM slice of rail after the residual stress relieving by annealing and 3-D plots of the calibrated displacment fields $u=u_1-u_0$ (c) and $v=v_1-v_0$ (d).

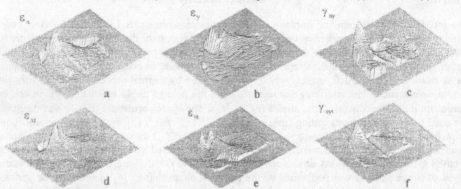

Fig. 5. 3D plots of experimental strain distribution calculated from u and v displacement fields: ε_x (a), ε_y (b), γ_{xy} (c) and numerically calculated strains ε_{xt} (d), ε_{yt} (e), γ_{xyt} (f) within the same area as experimental data.

The theoretical and experimental strain distribution determined by strain gauges (YM) and grating interferometry method, while the stress was relieved by cutting (IS-C) and by annealing (IS-W) in the crossections B-B and C-C are compared in Fig. 6. The character of the strain distributions for THEOR and IS-W for the lower crossection (C-C) are very similar however the experimental values are lower. This may be caused by nonfully performed process of stress relieving. The pair of strain distribution in C-C obtained for YM and IS-C methods at the samples with the stresses relieved by cutting have again similar character and values however they differ from THEOR and IS-W. This proves the statement that the physical nature of stress relieving influence the values of stress determined. The crossection B-B, which is placed closer to the high plasticity region produced by the contact area rail/wheel indicates big or differences between the strain determined theoretically and experimentally.The theoretical predictions in the area of contact between rail and wheel are up to now, incorrect and the experimental data are strongly required. Introducing higher magnification of imaging optics in the system presented in Fig. 3 enables the analysis within the area A (see Fig. 1). The interferogram analysis within this region is shown in Fig. 8. This experimental date help proper modeling of the rail behavior in the contact area.

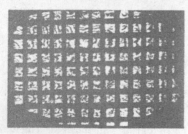

Fig. 6. The interferograms obtained at the rail slice with the stress relieved by cutting. B-B and C-C: the localization of crossections analysed in Fig. 7

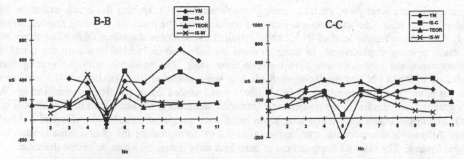

Fig. 7. The strains ε_x distributions obtained due to stress relieving by cutting and by: strain gauges (YM) and grating interferometry (IS-C) measurements, by annealing and grating interferometry measur (SI-W) compared with numerical calculations (THEOR). The results in crossection B-B (a) and C-C (b) by shown in Fig. 6.

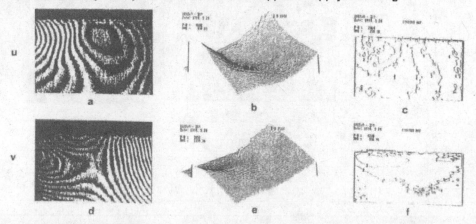

Fig. 8. Interferograms representing in-plane displacement fields u (a) and v (d) at the area A (see Fig. 1), 3-D plots of u (b) and v (e) displacements and contour maps of strain ε_x (c) , ε_y (f) in the contact area rail/wheel.

The combination of grating interferometry and stress relieving by annealing process was proved to be very efficient method for in-plane displacement/strain distribution determination, however at the early stage it replace only strain gauge technique at YM slices for δ_{xx} and δ_{yy} determination. Recently the procedure using the displacement data obtained from obliquely cutted slices (see Fig. 1, co-ordinates (t, s)) was proved to be able to successfully replace Meier sectioning to restore δ_{zz} stress component [3]. In this way a new experimental-numerical procedure restores a full 3D stress state that was originally in the rail.

3. THE IMPORTANCE OF RESIDUAL STRESSES IN FRACTURE MECHANICS
Stress intensity factor or strain energy release rate are both used to predict crack growth following conventional fracture mechanics. For isotropic materials, stress intensity factor, K_I, has been used to predict fatigue crack growth and fracture. For composite materials, strain energy release rate G_I, has been found to be more convenient to predict cracks or delamination [8]. The determination of both values, K_I and G_I, depends on the displacements values which here are measured by moiré interferometry method combined with a spatial-carrier phase shifting method.
In this investigation a titanium alloy specimen with a single edge crack for Mode I fracture (Fig. 9) has been used to obtain the values K_I and G_I [9]. The crack was extended by fatigue loading using sinusoidally varying loads (5 Hz, F=0÷3,2 kN, N=4500 cycles). Then a cross-line diffraction grating (1200 l/mm) was produce on the specimen, the sample was placed again at the loading machine

together with the workshop, portable grating interferometer [10]. During the crack extension by fatigue loading plastic deformation was expected to occur near the crack tip, creating residual stress field. A maximum fatigue load of P=3,2 kN is required to overcome the effect of residual stress on the crack opening displacement. In order to show the influence of residual stress on the u and v displacements, the measurements were made in three steps. The specimen with the grating was loaded up to P=0,83 N and displacement fields u, v were measured (Fig. 10a). Then the specimen was put in an oven at temperature of 600 ^{0}C for 1 h and cooled down to release residual stress. A new grating was bounded to the specimen to obtain displacement at the same load of P=0,83 N (Fig. 10b). The displacement before and after stress relief were compared to show the effect of residual stress. Afterwards the specimen was loaded up to 3,2 kN to reproduce the plastic zone created by fatigue loading. The residual displacement at zero load were measured again to reveal the extent of plastic deformation created during the loading process (Fig. 10c).

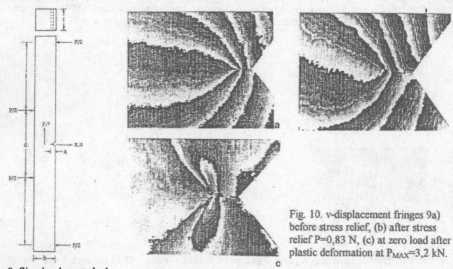

Fig. 10. v-displacement fringes 9a) before stress relief, (b) after stress relief P=0,83 N, (c) at zero load after plastic deformation at P_{MAX}=3,2 kN.

Fig. 9. Simple edge cracked specimen under four point bending (a=0,5 mm, t=8 mm, l=d=25 mm).

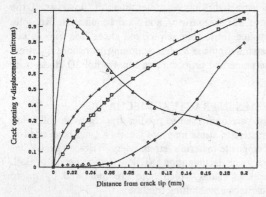

Fig. 11. Crack opening v displacement along the crack live. (◇) Measured before stress relief; (□) measured after stress relief; (△) residual displacement after P=3,2 kN; (+) finite element prediction [9].

The effect of residual stress on the crack opening displacement is show in Fig. 11. Before stress relief, the displacement near the crack tip is very small and increases away from the crack tip. After stress relief, there is significant increase in displacement, particularly near the crack tip. The difference of displacement before and after stress relief can be explained by the residual stress created during fatigue loading in which a plastic zone was created near the crack tip. The difference of displacement before and after stress relief can be explained by the residual stress created during fatigue loading in which a plastic zone was created near the crack tip. The size of the plastic zone depends on the magnitude of the load applied and may be additionally monitored by out-of-plane displacement testing. The residual displacement at zero load after applying

P_{MAX}=3,2 kN is shown in Fig. 11. It represents the plastic deformation which is not recoverableduring unloading. It can be seen that a significant amount of plastic deformation occurs near the crack tip and decreases away from the crack tip. The plastic deformation was produced by tensile loading and therefore a compressive residual stress was created and additional work is required to open the crack, explaining why the displacement before the stress relief was lower then after the stress relief. The value of K_I and G_I of the titanium cracked specimen have been determined experimentally and numerically. For the experimental results, when displacements close to the crack tip were used the values of K_I and G_I were much lower then the reference value. The difference can be explained by the presence of residual stresses near the crack tip, which tends to maintain the crack faces in contact.

The experiments show that due to variation in the crack-tip plastic zone size and accompanying residual stresses, the state of stress cannot be approximated by the elastic stress with $1/\sqrt{r}$ singularity. Grating interferometry can be used to measure the history of the stress state in materials and can be applied to study the crack growth process in various loading system.

4. CONCLUSIONS

The widespread occurrence of residual stresses and their importance in determining component reliability is obvious especially now with the increasing requirements for both big and miniature engineering elements.

The moiré interferometry combined with the annealing method of residual stress releasing was shown to be very efficient experimental method to analyse displacement fields in the elastic, plastic-elastic and plastic zones of the specimen. The data provided by this method are well suited to be used in the hybrid experimental-numerical methodology to determine full 3D residual stress distribution in engineering elements (eg. railroad rails, materials formed by explosive method or welded elements) as well as to predict fracture mechanics figures of merit including stress intensity factor, strain energy release rate or time relaxation constants of elements used e.g. as electronic microconnectors.

Automated grating interferometry is the practical alternative to the expensive and often qualitative nondestructive methods of residual stress determination i.e. neutronography or rentgenography. It is definitely excellent laboratory tool for modeling the behavior of materials and elements however it is difficult at the moment to determine the industrial methodology for on-line residual stress testing.

5. REFERENCE

1. R.E. Rowlands, "Residual stress" in SEM *Handbook of Experimental Mechanics*, A.S. Kobayasahi, Ed., Prentice-Hall, Englewood Cliffs, New York, 1987.
2. J.J. Groom, "Determination of residual stresses in rails", Final Report for US DOT No DOT/FRA/ORD-83/05, May 1983.
3. J. Magiera, J. Orkisz, "Experimental-numerical analysis of 3D residual stress state in railroad rails by means of oblique slicing technique", Proc. SPIE, v. 2342, 314-325, 1994.
4. C. Forno, "High temperature resistant grating for moiré interferometry", Proc. SPIE, v. 2342, 238-243, 1994.
5. M. Kujawińska, L. Sałbut, "Grating interferometry for analysis of residual stresses relieved by annealing", Proc. SPIE, v. 2342, 1994.
6. M. Kujawińska, "The architecture of a multipurpose fringe pattern analysis system", Optics and Lasers in Eng., v. 19, 261-268, 1993.
7. J. Orkisz et al., "Discrete analysis of actual residual stress resulting from cyclic loadings", Computers & Structures, v. 35, 397-412, 1990.
8. C.W. Smith and A.S. Kobayashi, "Experimental fracture mechanics", in SEM Handbook on Experimental Mechanics, A.K. Kobayashi, ed, Prentice-Hall Englewood Cliffs, New York, 1987.
9. C.Y. Poon, C. Ruiz, "Hybrid experimental-numerical approach for determining strain energy release rates", Theoret. Appl. Fracture Mechanics, v. 20, 123-131, 1994.
10. L. Sałbut, M. Kujawińska, G. Dymny, "Portable automatic grating interferometer for laboratory and field studies of mechanical elements", Proc. SPIE, v. 2342, 58-65, 1994.

Auswertung holografischer Interferenzmuster mit künstlichen neuronalen Netzen

Th. Kreis, R. Biedermann, W. Jüptner

BIAS - Bremer Institut für angewandte Strahltechnik

D-28359 Bremen

Einleitung

Künstliche neuronale Netze sind ein Versuch, die Funktion eines biologischen Hirns auf einem Computer nachzuvollziehen, indem das Wissen in einem Lernprozeß gewonnen wird und als synaptische Stärke der Verbindungen zwischen den einzelnen Neuronen gespeichert wird /1/. Für eine Reihe von Anwendungen wurde bereits gezeigt, daß künstliche neuronale Netze verallgemeinern, abstrahieren, klassifizieren, voraussagen und filtern, sowie aus unvollständigen Daten assoziieren können. Der neuronale Ansatz eignet sich besonders zur Lösung von Problemen, die zu vielfältig und kompliziert sind, um vollständig durch sequentielle Algorithmen implementiert werden zu können.

Um die vielfältigen Möglichkeiten künstlicher neuronaler Netze auch in der Auswertung von Interferenzmustern zu zeigen, werden im folgenden beispielhaft drei Anwendungen vorgestellt: (1) Die Demodulation der Interferenzphase modulo 2π durch ein zeitdiskretes rekurrentes Hopfield-Netzwerk, (2) die Skelettierung eines Streifenmusters durch eine selbstorganisierende Merkmalskarte nach Kohonen und (3) die automatische Erkennung von charakteristischen Teilmustern in holografischen Interferenzmustern durch ein lernendes mehrlagiges Netzwerk zur Erkennung von Defekten in technischen Bauteilen.

Künstliche neuronale Netze

Die informationsverarbeitenden Einheiten eines neuronalen Netzes sind die Neuronen. Jedes Neuron hat n Eingangspfade, deren Eingangsdaten x_i, $i = 1, ..., n$, mit den Gewichten w_{ji} multipliziert werden, wobei j das Neuron numeriert, Bild 1. Die aufsummierten gewichteten Eingänge repräsentieren das interne Aktivitätsniveau $I_j = \sum_i w_{ji}x_i$ des Neurons j. Diese Aktivität wird durch eine steigende, nach oben und unten beschränkte sogenannte Transfer-Funktion f modifiziert. Das Ergebnis $y_j = f(\sum_i w_{ji} x_i)$ wird auf den Ausgangspfad des Neurons gegeben, der das Ergebnis entweder der Außenwelt zur Verfügung stellt oder es an die Eingangspfade weiterer Neuronen verteilt.

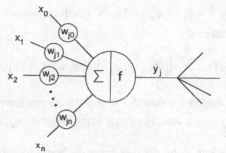

Bild 1. Künstliches Neuron

Die Neuronen eines künstlichen neuronalen Netzes sind im allgemeinen in sogenannten Lagen organisiert. Ein typisches Netzwerk besteht aus einer Eingangslage, einer oder mehrerer sogenannter versteckter Lagen und einer Ausgabelage. Während die grundlegende Operation eines Neurons immer die gleiche ist, unterscheiden sich die verschiedenen Typen von neuronalen Netzen in ihrer Architektur, d. h. Anzahl von Eingängen und Ausgängen, Anzahl der Lagen, Anzahl der Neuronen in jeder Lage, Anzahl der Gewichte jedes Neurons, Verbindungen zwischen Neuronen innerhalb der Lagen und des gesamten Netzes, Korrektursignale und Transferfunktionen f können variieren. Charakteristisch für die verschiedenen Konzepte ist auch die Art, wie die Information eingespeichert wird, das Training des Netzwerks. Obwohl ein einzelnes Neuron keinerlei relevante Informationen verarbeiten kann, ist das Netzwerk aus vielen verbundenen Neuronen und mit den in den synaptischen Gewichten gespeicherten Informationen in der Lage, komplexe Probleme zu lösen.

Interferenzphasendemodulation durch Hopfield-Netze

In einem Hopfield-Netz nehmen die Neuronen nur zwei Zustände an, i. a. $\{1, 0\}$ oder $\{1, -1\}$. Jedes Neuron x_j ist mit allen anderen außer sich selbst verbunden, d. h. die Gewichtsmatrix ist $w_{jj} = 0$ für alle j. Sie sei weiter symmetrisch: $w_{ji} = w_{ij}$ für alle i, j. Hopfield konnte zeigen, daß unter diesen Voraussetzungen die Neuronen ihre Zustände so ändern, daß die Energiefunktion

$$E = -\frac{1}{2}\sum_{j=1}^{N}\sum_{i=1}^{N} w_{ji} x_i x_j - \sum_{j=1}^{N} I_j x_j,$$

minimiert wird. Hierbei ist I_j ein zum Neuron x_j gehörender Ausgleichs-Eingang.

Um eine Interferenzphasenverteilung Φ_i modulo 2π zu demodulieren, ist eine Stufenfunktion $2\pi f_i$ zu addieren, wobei f_i nur ganzzahlige Werte annimmt /2/. Die demodulierte Interferenzphase ist

$\Phi_i^{(u)} = \Phi_i + 2\pi f_i$, $i = 1, ..., N$. Um das optimale $f = \{f_i, i = 1, ..., N\}$ zu finden, wird eine Bewertungsfunktion

$$E(f) = \sum_{i=1}^{N} \left\{ \left[\left(\Phi_{i-1} + 2\pi f_{i-1}\right) - \left(\Phi_i + 2\pi f_i\right)\right]^2 + \left[\left(\Phi_{i+1} + 2\pi f_{i+1}\right) - \left(\Phi_i + 2\pi f_i\right)\right]^2 \right\}$$

eingeführt, welche die Varianz der demodulierten Phase für jede Wahl von f im Sinne der kleinsten Quadrate mißt. Um $E(f)$ mit Hilfe eines Hopfield-Netzes zu minimieren, werden die f_i in $f_i = \sum_m x_{i,m}$ zerlegt, wobei die $x_{i,m}$, m nur die Werte 0 oder 1 annehmen. Hiermit wird die Bewertungsfunktion auf die Form der Energiefunktion gebracht und man erhält

$$E(x) = -\frac{1}{2} \sum_i \sum_m \sum_{i'} \sum_{m'} \left\{ -4\pi \delta_{i,i'}\left(1 - \delta_{m,m'}\right) + 2\pi \delta_{i-1,i'} + 2\pi \delta_{i+1,i'} \right\} x_{i,m} x_{i',m}$$
$$- \sum_i \sum_m \left(-2\Phi_i + \Phi_{i-1} + \Phi_{i+1} - 2\pi\right) x_{i,m}$$

wobei $\delta_{ij} = 1$ für $i = j$ und $\delta_{ij} = 0$ für $i \neq j$ ist. Die Interferenzphasenverteilung Φ_i wird also demoduliert durch ein zeitdiskretes Hopfield-Netz, dessen Neurodynamik durch

$$u_{i,m}(t) = \sum_{i'} \sum_{m'} w_{i\,m\,i'\,m'} \, x_{i',m'}(t) + I_{i,m}$$

und $x_{i,m}(t + 1) = f\left(u_{i,m}(t)\right)$ definiert ist. Hierbei bezeichnen t die diskrete Zeit, $w_{imi'm'} = -4\pi\delta_{i,i'}$ $\times (1 - \delta_{m,m'}) + 2\pi\delta_{i-1,i'} + 2\pi\delta_{i+1,i'}$ die synaptischen Stärken, $I_{i,m} = 2\Phi_i + \Phi_{i-1} + \Phi_{i+1} - 2\pi$ die Ausgleichs-Eingänge und

$$f(x) = \begin{cases} 0 & x < -1 \\ x + 1 & -1 \leq x \leq 0 \\ 1 & x > 0 \end{cases}$$

die Transfer-Funktion. Die entsprechende Neurodynamik in zwei Dimensionen mit 4-Pixel-Nachbarschaften ist

$$E(x) = -\frac{1}{2} \sum_i \sum_j \sum_m \sum_{i'} \sum_{j'} \sum_{m'} w_{i,j,m,i',j',m'} x_{i,j,m} x_{i',j',m'} - \sum_i \sum_j \sum_m I_{i,j,m} x_{i,j,m}$$

mit

$$w_{ijmi'j'm'} = -8\pi\delta_{i,i'}\delta_{j,j'}\left(1 - \delta_{m,m'}\right) + 2\pi\delta_{i,i'}\left(\delta_{j-1,j'} + \delta_{j+1,j'}\right) + 2\pi\delta_{j,j'}\left(\delta_{i-1,i'} + \delta_{i+1,i'}\right)$$

$$I_{i,j,m} = \left(\Phi_{i,j-1} + \Phi_{i,j+1} + \Phi_{i-1,j} + \Phi_{i+1,j} - 4\Phi_{i,j}\right) - 2\pi$$

Eine Anwendung in einer Dimension zeigt Bild 2.

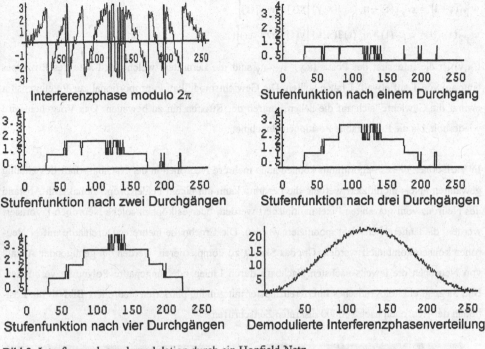

Bild 2. Interferenzphasendemodulation durch ein Hopfield-Netz

Skelettierung durch selbstorganisierende Merkmalskarten

Selbstorganisierende Karten sind Modelle, bei denen beliebig-dimensionale Signale in niedrig-dimensionale Neuronen transformiert werden, so daß Signalähnlichkeit in Lagenachbarschaft der erregten Neuronen umgesetzt wird.

Bei der Skelettierung von Interferenzmustern werden Intensitätsmaxima oder -minima identifiziert und markiert, d. h. die Interferenzstreifen auf Linien reduziert. Um dies mit einer selbstorganisieren-den Karte zu erreichen, werden die Pixel-Koordinaten (x,y) auf das Einheitsquadrat $0 \leq x \leq 1$, $0 \leq y \leq 1$ reduziert /3/. Eine Anzahl N von Neuronen werden definiert, denen die zweidimensionalen Ge-wichtsvektoren $\vec{w}_j = \left(w_{j1}, w_{i2} \right)$, j = 1,2, ..., N zugeordnet sind. Diese synaptischen Gewichte kön-nen als Koordinaten im Einheitsquadrat interpretiert werden. Zu Beginn des Prozesses sind die Neu-ronen zufällig verteilt. Während des Lernprozesses werden zu jedem Zeitpunkt t die zufälligen Pixel-koordinaten (x(t), y(t)) gewählt und das Neuron j*, dessen Gewichtsvektor $\vec{w}_{j*}$ im euklidischen Sinn am nächsten an (x(t), y(t)) liegt, gesucht. Die Gewichte dieses Neurons werden modifiziert durch

$$w_{j*1}(t+1) = w_{j*1}(t) + \eta_x(t)\, I(x,y) \left[x(t) - w_{j*1}(t)\right]$$

$$w_{j*2}(t+1) = w_{j*2}(t) + \eta_y(t)\, I(x,y) \left[y(t) - w_{j*2}(t)\right]$$

$I(x,y)$ ist die Intensität des Pixels (x,y), η_x, η_y sind die Lernraten, welche im Laufe des Prozesses langsam von 1 auf etwa 0.1 fallen sollten. Die Gewichtsmodifikation proportional zur Pixelintensität zwingt die Gewichte, sich auf die hellen Zentren der Streifen hin zu bewegen. Das Vorgehen wird wiederholt, bis die Pixel stabile Positionen einnehmen.

In Variationen dieses Algorithmus können auch mehrere Neuronen in der Nachbarschaft des zufällig gewählten Pixels modifiziert werden, die Lernrate kann mit einem Fehler proportional zum Abstand des Neurons vom gewählten Pixel multipliziert werden, oder es können solche Neuronen favorisiert werden, die längere Zeit nicht modifiziert wurden. Die Ergebnisse mehrerer Durchläufe mit N Neuronen können kombiniert werden. Um das Skelett zu komplettieren, werden bei genügender Anzahl von Neuronen die jeweils nächsten Nachbarn durch Linien oder angepaßte Polynome verbunden. Bild 3a zeigt ein holografisches Interferenzmuster mit zufällig plazierten Neuronen, Bild 3b die Positionen der Neuronen nach 100000 diskreten Zeitschritten.

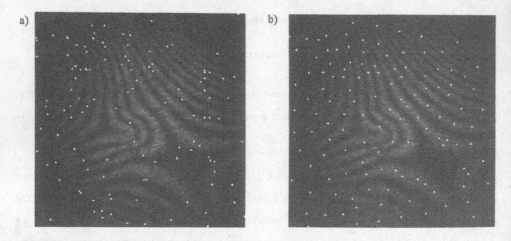

Bild 3. Interferenzstreifenskelettierung durch selbstorganisierende Merkmalskartei

Fehlererkennung durch mehrlagige Netzwerke

Bauteilfehler manifestieren sich in holografischen Interferenzmustern belasteter technischer Bauteile durch lokale Teilmuster, die vom globalen Muster auf charakteristische Weise abweichen. Um eine automatische Erkennung solcher Teilmuster mit mehrlagigen neuronalen Netzen zu erreichen, werden die Muster auf wenige, hier 16, translations- und rotationsinvariante Parameter reduziert /4/. Dazu wird an jedem Pixel aus den Grauwerten seiner Nachbarschaft nach dem Verfahren der kleinsten Quadrate die lokale Steigung berechnet. Nach einer Einteilung des gesamten Musters in gleich große parkettierende Rechtecke wird jedem Rechteck die maximale Steigung aller seiner Pixel als Kennwert zugeordnet. Die Abweichung dieser Kennwerte von denen benachbarter Rechtecke wird mit einem Laplace-Filter gemessen. In Vergleichen hat es sich als vorteilhaft erwiesen, die jeweils 4 größten Ergebnisse der Laplace-Filterung für Einteilungen des Gesamtmusters von 512 x 512 Pixel in 8 x 8, 16 x 16, 32 x 32 und 64 x 64 Rechtecke als charakteristische Parameter für ein Interferenzmuster zu benutzen. Das Netzwerk, dessen im Bild 4 gezeigte Struktur sich als besonders geeignet erwiesen hat, wird nach dem Backpropagation-Algorithmus trainiert. Die Lernstichprobe von 500 fehlerfreien und 500 fehlerbehafteten holografischen Interferenzmustern wird hierbei nicht experimentell erzeugt, sondern rechnerisch simuliert.

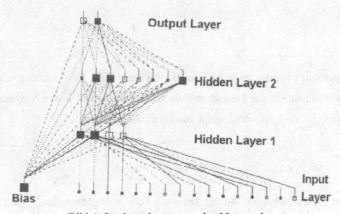

Bild 4. Struktur des neuronalen Netzwerks

Von den in Bild 5 gezeigten holografischen Interferenzmustern wurden die in 5a und 5b als fehlerbehaftet, das in 5c als fehlerfrei erkannt.

484

a)   b)   c)

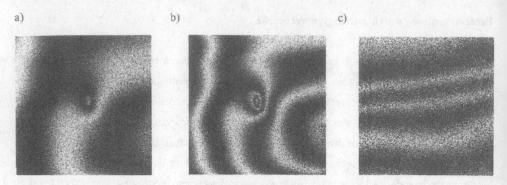

Bild 5. Holografische Interferenzmuster mit (a und b) und ohne (c) Bauteilfehlern

Zusammenfassung

Künstliche neuronale Netze sind ein Konzept zur Datenverarbeitung, welches auch zur Auswertung holografischer Interferenzmuster herangezogen werden kann. Dies wird beispielhaft an den Problemen der Interferenzphasendemodulation, der Streifenskelettierung und der automatischen Fehlstellen detektion vorgeführt. Der Einsatz neuronaler Netze steht erst am Anfang, so sind die gezeigten Lösungsansätze noch weiter zu optimieren, weitere Einsatzmöglichkeiten sind vorstellbar.

Danksagung

Die Forschungsarbeiten wurden teilweise vom Bundesministerium für Forschung und Technologie, Vorhaben 13N5870 und von der Deutschen Forschungsgemeinschaft in den Vorhaben Kr 953/5-1, Kr 953/5-2 und Kr 953/5-3 gefördert, wofür ausdrücklich gedankt sei.

Literatur

/1/   S. Haykin, Neural Networks: A Comprehensive Foundation, Macmillan, New York, 1994

/2/   M. Takeda, K. Nagatome, Y. Watanabe, Phase Unwrapping by Neural Network, in W. Jüptner und W. Osten (Eds.), Fringe'93: 2nd International Workshop on Automatic Processing of Fringe Patterns, Akademie-Verlag Berlin, 136-141, 1993

/3/   V. Srinivasan, S.-T. Yeo, P. Chaturvedi, Fringe Processing and analysis with a neural network, Opt. Eng., Vol. 33, No. 4, 1166-1171, 1994

/4/   Th. Kreis, W. Jüptner, R. Biedermann, Neural network approach to holographic nondestructive testing, Appl. Opt., Vol. 34, No. 8, 1407-1415, 1995

Speckle- and Digital Holographic Interferometry (A Comparison)

G. Pedrini, Y. Zou and H. J. Tiziani

Institut für Technische Optik,

Pfaffenwaldring 9, D-70569 Stuttgart

1. Introduction

There exists many types of optical systems which allow deformation and shape measurements which are based on interferometry. Different names are used: holographic interferometry, digital holographic interferometry, direct holography, speckle correlation interferometry, electronic speckle interferometry (ESPI), TV-holography, shearing interferometry. The difference between all this systems is subtle, in general we speak about "speckle interferometry" when a medium with low resolution (usually a CCD) is used to record the interference between the one wave front coming from one object and a reference beam. Since the size of the speckle should be greater than the medium resolution, when we use a CCD sensor we need speckle larger than the CCD-pixel size. For the holography a high resolution recording medium (usually a photographic film) is used to record holograms. The resolution of CCD cameras as well as computer capacities are constantly increasing. It is now possible to record an off-axis hologram using a CCD-camera and evaluate it using a computer. The electronic recording of one hologram enables to compare two wave fields without recourse to any form of photographic processing and plate relocation. This allows a fast evaluation and is thus better suited to be used in an industrial environment. In this paper we will pay our attention to the comparison of different kind of electronic off-axis holograms.

2. The spatial carrier method

A hologram can be seen as a carrier wave whose spatial frequency is modulated by the object information. If we denote the object wave with $u(x,y)$ and an off-axis plane reference wave with $r(x,y)$, the intensity of the interference pattern can mathematically be described by the relation

$$I(x,y) = |r(x,y)|^2 + |u(x,y)|^2 + 2|r(x,y)||u(x,y)|\{\cos[\phi(x,y) + 2\pi f_0 x]\}$$

which describes a set of carrier interference fringes of spatial frequencies f_0 which are modulated in phase by $\phi(x,y)$. In order to determine the phase $\phi(x,y)$ in Eq. 1, the sinusoidal-fitting method can be used. For the application of this method, the phase must change much more slowly than $2\pi f_0 x$. (This condition is satisfied if the spatial extension of the speckles is large). If we use a CCD-sensor as a recording device, the phase changes, from one camera pixel to the adjacent one will be $\alpha = 2\pi\Delta x f_0$, where Δx is the pixel size. This means that three (or more) adjacent pixels can be used to calculate the phase $\phi(x,y)$ according to the standard phase shifting algorithm. Another method to determine the phase is to use the Fourier method [1].

3. Image plane hologram

If a lens is used to image the object on the recording device as shown in Fig. 1.a), we have an image plane hologram. In order to have large speckles, we can insert an aperture which is located in the front focal plane of the lens.

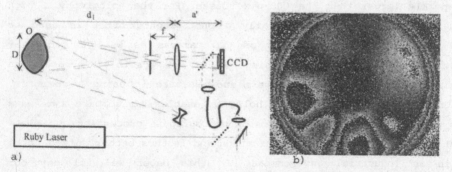

a)

b)

Fig. 1 a) Experimental set-up for the recording of an image plane hologram. b) Phase-map of a vibrating loudspeaker. Frequency 992 Hz. Pulse separation 100 µsec.

Figure 1.b) shows one experimental result obtained using a vibrating loudspeaker. The source was a pulsed ruby laser which can emit two high energy pulses separated by few microseconds. The object vibrates and the two different state of the object are recorded using the two pulses [2]. The phase of the two recorded interferograms is then calculated using the sinusoid fitting method. By phase subtraction we obtain a map which contains the information about the deformation in the interval between the two pulses. We evaluated the same pattern by using the Fourier method. No difference was observed in the obtained results, but it should be noticed that the evaluation using the Fourier method takes more time since it involves the calculation of Fourier tranforms using the FFT algorithm.

4. Fresnel hologram

Figure 2.a) shows an arrangement for the recording of an off-axis Fresnel hologram on a CCD-sensor. To record a hologram of the entire object, the resolution of the CCD must be sufficient to record the fringes formed by the reference wave and the wave from the object point farthest from the reference point. This can be obtained in two ways: a) by increasing the distance d between the object and the camera; b) by recording an hologram of the demagnified image of the object (see Fig. 2.b)).

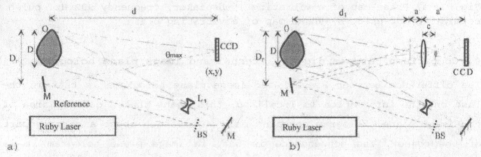

Fig. 2 Experimental set-up for the recording of a Fresnel hologram. a) without lens. b) with lens.

The reconstruction of the hologram is carried out by simulation of the reference wave which illuminates the hologram [3,4]. By Fresnel diffraction the complex amplitude in a given plane at a distance z from the hologram can be calculated. This calculation can be carried out using a Fast Fourier Transform (FFT) algorithm. Since we reconstruct the complex amplitude of the wave front, we can directly calculate its phase. Figure 3 shows two experimental result obtained using two vibrating objects. The result shown in Fig.3a) was obtained using the arrangement shown in Fig. 2.a), the test object was small (a loudspeaker with diameter 6 cm located at a distance d=150 cm from the CCD sensor). Figure 3b) shows the results obtained with a vibrating metal plate having a diameter of 20 cm, for the recording we used the arrangement shown in Fig. 2.b). Notice that by digital reconstruction of the hologram we reconstruct even the conjugate wavefront(see Fig. 3). This means that if

we record the hologram with NxN points in the reconstructed wave front
the image of the object will occupy in the most favourable case NxN/2
pixels.

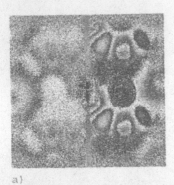
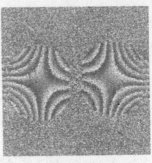

a)   b)

Fig. 3. a) Phase-map of a vibrating loudspeaker, frequency 992 Hz. pulse
separation 100 μsec. b) Phase-map of a vibrating plate.

4. Comparison between digital Fresnel and image plane holograms

The difference between Fresnel and image plane holograms is that in the
last one the information is localised, this means that a certain area of
the image plane hologram contains only information about a certain part
of the object. The consequence is that in image plane hologram it is
possible to obtain the phase simply by using a sinusoidal fitting. In our
experiment with 512x512 pixels this operation was carried out in about
one second. For the evaluation of the Fresnel hologram we need a digital
reconstruction which is carried out using a FFT algorithm, this is time
consuming. Our PC (486 DX 33) can carry out this calculation in about 20
seconds. In the image plane hologram we have a reconstruction which fills
more pixels but with reduced resolution. In the Fresnel hologram we have
less image points (compare the results reported in Fig. 1.b and thus
reported in Fig.3) but the speckle size is smaller. Thus the spatial
resolution in the two cases is the same. Experimentally it was possible
to verify it be analysing fringes obtained using the two methods.

References

1. Takeda, Hideki, Kobayashi, *J. Opt. Soc. Am. A* **72**, 156-160, (1982).
2. G. Pedrini, H. J. Tiziani, in Appl. Opt., **33**, 7857- 7863 (1994).
3. U. Schnars, *J.   Opt. Soc. Am. A   11**, 2011-2015, (1994)
4   G. Pedrini, Y. Zou, H. J. Tiziani, J. M. Opt. **42**, 367-374 (1995)

Numerische Rekonstruktion von Hologrammen in der interferometrischen Meßtechnik

U. Schnars, Th. Kreis, W. Jüptner

BIAS - Bremer Institut für angewandte Strahltechnik

Klagenfurter Str. 2

28359 Bremen

1. Einleitung

Die holografische Interferometrie (HI) ist ein optisches Meßverfahren zur berührungslosen Erfassung von Oberflächenformen oder Formveränderungen opaker Körper sowie von Brechzahländerungen in transparenten Medien. Die HI kann als Zweistufenprozeß verstanden werden: Im ersten Schritt werden die Wellenfelder der zu untersuchenden Oberflächen holografisch gespeichert. Nach Entwicklung und Repositionierung des Hologrammes werden die gespeicherten Wellenfelder in einem zweiten Schritt rekonstruiert und miteinander zur Interferenz gebracht. Die Analyse der so entstehenden Interferenzmuster erlaubt z. B. Rückschlüsse auf Form- oder Formveränderungen der Oberfläche. In der konventionellen holografischen Meßtechnik auf der Basis von fotografisch gespeicherten Hologrammen werden beide Schritte, d. h. sowohl die Hologrammaufzeichnung als auch die Erzeugung von Interferenzstreifen, optisch ausgeführt.

Die Digitale Holografie ist ein neues Verfahren der holografischen Meßtechnik [1]: Die Hologramme der zu untersuchenden Objekte werden direkt mit einem hochauflösenden CCD-Sensor aufgezeichnet und digital gespeichert. Die Rekonstruktion der Objektwelle, die optisch durch Beleuchtung einer entwickelten Fotoplatte vorgenommen wird, kann somit numerisch durchgeführt werden. Im Gegensatz zur optischen Rekonstruktion kann im mathematischen Rekonstruktionsprozeß die Phasenverteilung der gespeicherten Wellenfront direkt aus dem digitalen Hologramm errechnet werden. Dadurch eröffnen sich neue Möglichkeiten für die kohärent-optische Meßtechnik: In jedem Objektzustand wird ein Hologramm elektronisch aufgezeichnet und mathematisch rekonstruiert. Durch einen rechnerischen Vergleich der rekonstruierten Phasen des Grund- und des Lastzustandes kann die Interferenzphase direkt aus den Hologrammen bestimmt werden, ohne den Zwischenschritt über ein Interferenzmuster.

2. Digitale Holografie

Anfang der 70er Jahre wurde von mehreren Forschergruppen die Möglichkeit untersucht, die fotografische Registrierung der Hologramme durch eine elektronische Aufzeichnung mit Fersehkameras zu ersetzen. Das Ergebnis dieser Untersuchungen ist die Elektronische Specklemuster-Interferometrie, ESPI [2]. In der Specklemuster-Interferometrie werden die zu untersuchenden Objekte auf die lichtempfindliche Fläche einer Kamera abgebildet und hier mit einer Referenzwelle überlagert. Es werden also Hologramme fokussierter Bilder elektronisch gespeichert. Durch Subtraktion von zwei in unterschiedlichen Objektzuständen aufgezeichneten Speckle-Interferogrammen entstehen Korrelationsstreifen, die ähnliche Eigenschaften wie optisch erzeugte Interferenzstreifen aufweisen. Allerdings geht bei den ESPI-Verfahren die Phasenverteilung der gespeicherten Wellenfronten verloren. Die Phaseninformation kann nur indirekt aus mehreren phasengeschobenen Interferogrammen erechnet werden [3] (Elektrooptische Holografie, EOH). Dies setzt jedoch einen höheren technischen Aufwand voraus, da für jeden Lastzustand mindestens drei phasengeschobene Speckle-Interferogramme erzeugt und ausgewertet werden müssen.

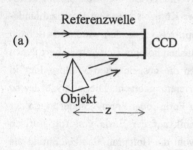

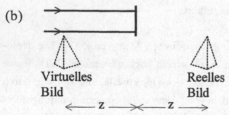

Bild 1: Prinzip der digitalen Holografie
  (a) Digitale Hologrammaufzeichnung
  (b) Numerische Rekonstruktion

Demgegenüber werden in der Digitalen Holografie keine Bildfeldhologramme, sondern echte Fresnelhologramme aufgezeichnet, Bild 1. Zur Rekonstruktion wird die Beugung der Rekonstruktionswelle an der holografischen Mikrostruktur modelliert und numerisch nachgebildet. Im Unterschied zur optischen Hologrammrekonstruktion kann im numerischen Rekonstruktionsvorgang die Phaseninformation direkt aus den digitalen Hologrammen berechnet werden und muß nicht, wie bei den konventionellen Verfahren der holografischen Interferometrie und wie bei den ESPI-Verfahren, indirekt aus mehreren Intensitätsmessungen errechnet werden. Meßtechnisch interessante Phasenänderungen infolge von Objektdeformationen (Interferenzphase) können durch Vergleich der Phasenverteilungen zweier Lastzustände sehr einfach gemessen werden. Die Einzelheiten dieser numerischen Rekonstruktion sind in der Literatur dargestellt [1,4] und sollen hier nicht ausführlich behandelt werden.

3. Anwendungsbeispiel

Der experimentelle Aufbau zur digitalen Aufzeichnung von Hologrammen ist in Bild 2 dargestellt. Der Strahl eines Argon-Ionen Lasers wird mit einem Strahlteiler in die Referenzwelle und in die Objektbeleuchtungswellen aufgeteilt. Die Referenzwelle wird mit einem Teleskop aufgeweitet und beleuchtet den CCD-Sensor. Der hier eingesetzte Sensor besteht aus einer Matrix von 2048 × 2048 lichtempfindlichen Pixeln mit einer lichtempfindlichen Fläche von jeweils 9 μm × 9 μm. Die Pixel sind ohne Zwischenraum aneinander gereiht, so daß der Mittenabstand benachbarter Pixel ebenfalls 9 μm beträgt. Die Hologramme werden mit 256 Graustufen pro Pixel aufgezeichnet und in einem digitalen Bildverarbeitungssystem gespeichert.

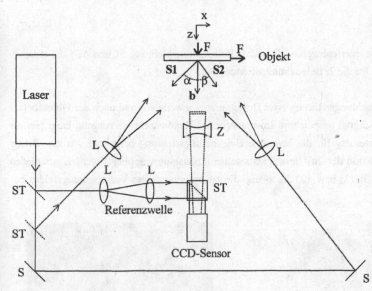

Die maximale Ortsfrequenz der Mikrostruktur eines Hologrammes wird durch den Winkel zwischen den interferierenden Lichtwellen bestimmt. Da CCD-Sensoren z. Z. noch ein relativ geringes räumliches Auflösungsvermögen (etwa 100 Linienpaare/mm gegenüber einigen 1000 Linienpaaren/mm bei hochauflösenden Fotoplatten) haben, muß man sich bei der Hologrammaufnahme auf kleine Winkel zwi-

Bild 2: Experimenteller Aufbau (Z: Zerstreungslinse; L: Linse; ST: Strahlteiler; S: Spiegel; **S1, S2**: Beleuchtungsvektoren; **b**: Beobachtungsvektor)

schen der Referenz- und Objektwelle und auf kleine Objektwinkel beschränken. Bei vorgegebenem Abstand zwischen Objekt und Sensor können nur relativ kleine Objekte holografisch gespeichert werden. Um dennoch "große" Objekte aufnehmen zu können, wird der Objektwinkel mit einer Zerstreuungslinse verkleinert [4], Bild 2. Die Zerstreuungslinse erzeugt ein verkleinertes virtuelles Bild des Objekts. Es wird also nicht das Objektwellenfeld selbst aufgezeichnet, sondern das vom verkleinerten virtuellen Bild ausgehende Wellenfeld. Bei der Überlagerung dieses Wellenfeldes mit der ebenen Referenzwelle entstehen daher geringere Ortsfrequenzen als bei der Interferenz zwischen der orginalen Objektwelle und der Referenzwelle. Anders als in der Speckle-Interferometrie (ESPI) dient die Linse hier nicht zur Abbildung des Objekts in die Sensorebene, sondern lediglich zur Winkelverkleinerung.

Zur Demonstration des Meßverfahrens wird ein Metallstreifen verwendet, der in z-Richtung ("out-of-plane") gebogen und in x-Richtung ("in-plane") auf Zug belastet werden kann. Eine Verformung in y-Richtung tritt nicht auf. Weiterhin wird angenommen, daß der Empfindlichkeitsvektor über dem ganzen Objekt konstant ist. Daher reichen zur Berechnung des Verformungsfeldes zwei Phasenmessungen mit unterschiedlichen Empfindlichkeitsvektoren aus. Dies wird hier durch eine Variation der Beleuchtungsrichtung realisiert, Bild 2. Aufgrund des symmetrischen Aufbaus ($\alpha=\beta$ =30°) ergibt sich die Verschiebungskomponente in z-Richtung als Summe und die x-Komponente als Differenz der Interferenzphasen:

$$d_x = \frac{\lambda}{4\pi} \frac{\Delta\varphi_2 - \Delta\varphi_1}{\sin\alpha} \qquad (1)$$

$$d_z = \frac{\lambda}{4\pi} \frac{\Delta\varphi_1 + \Delta\varphi_2}{1 + \cos\alpha} \qquad (2)$$

$\Delta\varphi_1$ bezeichnet dabei die Interferenzphase für die 1. Beleuchtungsrichtung S1 und $\Delta\varphi_2$ die entsprechende Interferenzphase für die 2. Beleuchtungsrichtung S2, vgl. Bild 2.

Es werden für jede Beleuchtungsrichtung zwei Hologramme, jeweils vor und nach der Objektbelastung, aufgezeichnet und digital gespeichert. Im numerischen Rekonstruktionsvorgang kann hieraus die Interferenzphasenverteilung für die jeweilige Beleuchtungsrichtung berechnet werden, Bilder 3 und 4. Nach Demodulation der 2π-Unstetigkeitsstellen ("phase-unwrapping") werden aus diesen Phasenverteilungen mit Gl. (1) bzw. (2) die x- und die z-Komponente des Verschiebungsfeldes berechnet, Bilder 5 und 6.

Bild 3: Interferenzphasenverteilung $\Delta\varphi_1$ Bild 4: Interferenzphasenverteilung $\Delta\varphi_2$

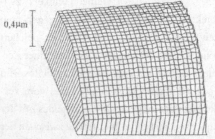

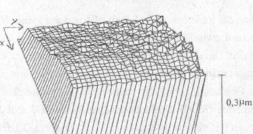

Bild 5: z-Komponente der Verschiebungsfeldes   Bild 6: x-Komponente des Verschiebungsfeldes
("out-of-plane"-Verschiebung)   ("in-plane"-Verschiebung)

4. Zusammenfassung

Die holografische Interferometrie wird gewöhnlich als Zweistufenprozeß verstanden: Im ersten Schritt, der primären Interferenz, werden Wellenfelder der zu untersuchenden Objektoberfläche holografisch gespeichert. Hierfür wurden bisher zumeist fotografische Aufzeichnungsmaterialien verwendet. Im zweiten Schritt, der sekundären Interferenz, werden die holografisch gespeicherten Wellenfelder optisch rekonstruiert und überlagert. Die Auswertung des dabei entstehenden Interferenzmusters erlaubt Rückschlüsse auf die Oberflächendeformation des Objektes.

In der hier vorgestellten Methode der digitalen Holografie werden die Hologramme mit einem elektronischen Sensor aufgezeichnet und numerisch rekonstruiert. Der zweite Schritt in der konventionellen holografischen Meßtechnik wird durch einen rechnerischen Vergleich der Wellenfelder ersetzt. Alle für die Beurteilung der Objektveränderungen wesentlichen Informationen, insbesondere Phasenänderungen infolge von Objektdeformationen, können direkt aus den digitalen Hologrammen berechnet werden.

5. Literatur

[1] U. Schnars, "Direct phase determination in hologram interferometry with use of digitally recorded holograms", Journ. Opt. Soc. Am. A, vol. 11, no. 7, 2011-2015 (1994)

[2] O. J. Lokberg und G. A. Slettemoen, "Basic Electronic Speckle Pattern Interferometry", Applied Optics and Optical Engineering, vol. 10, 455-505 (1987)

[3] K. Creath, "Phase shifting speckle-interferometry", Appl. Opt., vol. 24, no. 18, 3053-3058 (1985)

[4] U. Schnars, Th. Kreis und W. Jüptner, "CCD-recording and numerical reconstruction of holograms and holographic interferograms," erscheint in Proc. Interferometry VII: Techniques and Analysis, SPIE Vol. 2544

Untersuchungen zur Genauigkeit des scherografischen Prüfverfahrens

S. Naasner, K.-P. Gründer[1], G. Wernicke
Humboldt-Universität Berlin/D; Institut für Physik
[1]Bundesanstalt für Materialforschung und -prüfung, Berlin/D

Die Scherografie ist ein kohärent-optisches Ganzfeldverfahren zur Registrierung von Ableitungen der Oberflächenverschiebungen. Zwei grundlegende Merkmale des Verfahrens sind hervorzuheben.

- Der Referenzstrahl zur Erzeugung von Interferenzen wird durch ein Scherelement im Objektstrahlengang realisiert. Die Abbildung des Objekts interferiert leicht verschoben mit sich selbst [1]. Deshalb zeichnet sich das Verfahren durch einen relativ einfachen und damit störunanfälligeren Aufbau aus. Innerhalb der Scherweite können keine Verformungsdifferenzen bestimmt werden. Dies führt zu einem begrenzten Auflösungsvermögen des Verfahrens.

- Das in einem Scherogramm gespeicherte Streifensystem entspricht dem relativen Verschiebungsgradienten (Verformung) $\partial u/\partial x$, $\partial v/\partial x$ und $\partial w/\partial x$ (erste Ableitung des Verschiebungsvektors $\vec{d}=\vec{d}(u,v,w)$) auf der Objektoberfläche. Diese stehen über die Materialgesetze mit den Beanspruchungen von Bauteilen in Verbindung.

Das Ziel der vorgenommenen Untersuchungen ist die Überprüfung des Verfahrens auf quantitative Eignung zur Bestimmung von Dehnungsdifferenzen und Gestaltsänderungen.

1. Systematik

Die Grundgleichung der Scherografie für senkrechte Beobachtung lautet:

$$\Delta(x,y) = \frac{2\pi}{\lambda}\left(\sin(\varphi)\,\frac{\partial u}{\partial x}(x,y) + \left(1+\cos(\varphi)\right)\frac{\partial w}{\partial x}(x,y)\right)\delta x \qquad (1)$$

Δ ist eine dem Phasenwinkel im Scherogramm entsprechende Größe, (x,y,z) die Objektkoordinaten, λ die Wellenlänge, δx die Scherweite und φ der Beleuchtungswinkel in der x-z-Ebene. Diese Gleichung enthält folgende Näherungen: Linearer Abbruch der Taylorentwicklung [2], der Verschiebungsvektor $\vec{d}=\vec{d}(u,v,w)$ ist vernachlässigbar gegen die Beleuchtungs- und Beobachtungsentfernung. Vorausgesetzt werden außerdem parallele Strahlenbündel.

Von einer Objektverformung sind durch die Variation von Scherrichtung δx und δy, Beleuchtungsrichtung x und y sowie jeweils zwei Beleuchtungswinkel acht (2^3) linear unabhängige Scherogramme darstellbar. Die Kombinationen von Scherogrammen ermöglicht die Erfassung einzelner Verformungskomponenten :
die Neigungen $\partial w/\partial x$ und $\partial w/\partial y$, die Dehnungen $\partial u/\partial x$ und $\partial v/\partial y$ sowie die Komponenten der Verzerrung $\partial u/\partial y$ und $\partial v/\partial x$.

Aus sechs unterschiedlichen Scherogrammen einer Verformung können die Differenzen dieser sechs Verformungskomponenten über die Objektoberfläche bestimmt werden. Aus experimentellen Gründen wurde eine Auswahl von sechs Scherogrammen vorgenommen: $\Delta_{x,1}$, $\Delta_{x,2}$, $\Delta_{x,3}$, $\Delta_{y,1}$, $\Delta_{y,2}$ und $\Delta_{y,3}$. Der erste Index steht für die Scherrichtung (Verschiebung der Objektbilder zueinander) und der zweite für die Beleuchtungsrichtung. Diese erfolgt für 1und 2 aus der x-z-Ebene mit einem möglichst kleinen bzw. großen Winkel $0° < \varphi < 90°$ und für 3 aus der y-z-Ebene mit möglichst großem Winkel bezüglich der Oberflächennormalen.

Zur Auswertung eignet sich die Zusammenfassung des entstehenden Gleichungssystems in einer Matrix [3].

$$
\begin{pmatrix} \Delta_{x,1} \\ \Delta_{x,2} \\ \Delta_{x,3} \\ \Delta_{y,1} \\ \Delta_{y,2} \\ \Delta_{y,3} \end{pmatrix} = \begin{pmatrix} S_{x,1} & 0 & 0 & 0 & K_{x,1} & 0 \\ S_{x,2} & 0 & 0 & 0 & K_{x,2} & 0 \\ 0 & 0 & S_{x,3} & 0 & K_{x,3} & 0 \\ 0 & S_{y,1} & 0 & 0 & 0 & K_{y,1} \\ 0 & S_{y,2} & 0 & 0 & 0 & K_{y,2} \\ 0 & 0 & 0 & S_{y,3} & 0 & K_{y,3} \end{pmatrix} \cdot \begin{pmatrix} \partial u / \partial x \\ \partial u / \partial y \\ \partial v / \partial x \\ \partial v / \partial y \\ \partial w / \partial x \\ \partial w / \partial y \end{pmatrix}
\tag{2}
$$

Die $\Delta_{i,j}$ bilden einen Vektor, der die Meßergebnisse enthält (Meßvektor). Die Matrix beinhaltet die von der geometrischen Anordnung abhängigen Parameter. In die Konstante $S_{i,j}$ geht der Vorfaktor $2\pi / \lambda$, die Scherung (δx bzw. δy) und der Sinus des Beleuchtungswinkels ein. $K_{i,j}$ enthält statt dem Sinus des Beleuchtungeswinkels dessen Kosinus plus eins. Der Lösungsvektor (rechte Spalte) faßt die Verformungskomponenten zusammen.
Die Umstellung des Gleichungssystems nach dem Lösungsvektor ermöglicht eine einfache Handhabung zur Bestimmung einzelner Verformungskomponenten.

$$
\begin{pmatrix} \dfrac{\partial u}{\partial x} \\[2mm] \dfrac{\partial u}{\partial y} \\[2mm] \dfrac{\partial v}{\partial x} \\[2mm] \dfrac{\partial v}{\partial y} \\[2mm] \dfrac{\partial w}{\partial x} \\[2mm] \dfrac{\partial w}{\partial y} \end{pmatrix} = \frac{1}{Z} \begin{pmatrix} K_{x,2} & -K_{x,1} & 0 & 0 & 0 & 0 \\ 0 & 0 & 0 & K_{y,2} & -K_{y,1} & 0 \\ \dfrac{S_{x,2}K_{x,3}}{S_{x,3}} & \dfrac{-S_{x,1}K_{x,3}}{S_{x,3}} & \dfrac{Z}{S_{x,3}} & 0 & 0 & 0 \\ 0 & 0 & 0 & \dfrac{S_{y,2}K_{y,3}}{S_{y,3}} & \dfrac{-S_{y,1}K_{y,3}}{S_{y,3}} & \dfrac{Z}{S_{y,3}} \\ -S_{x,2} & S_{x,1} & 0 & 0 & 0 & 0 \\ 0 & 0 & 0 & -S_{y,2} & S_{y,1} & 0 \end{pmatrix} \cdot \begin{pmatrix} \Delta_{x,1} \\ \Delta_{x,2} \\ \Delta_{x,3} \\ \Delta_{y,1} \\ \Delta_{y,2} \\ \Delta_{y,3} \end{pmatrix}
$$

$$
Z = S_{x,1}K_{x,2} - S_{x,2}K_{x,1}
\tag{3}
$$

2. Beispiele

Die obige Auswertmethode wurde anhand einer Dehnungsmessung an einer Zugprobe erprobt.
Die Dehnungsdifferenz zwischen den Bereichen großer und kleiner Dehnungen wurde anhand der Phasensprünge entlang eines Schnittes im Scherogramm ermittelt. Als Referenzverfahren wurden Dehnmeßstreifen (DMS) verwendet.

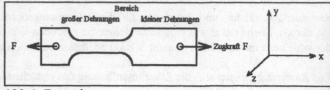

Abb.1 Zugprobe

Aus der obigen Gleichung ergibt sich für eine definierte Position im Scherogramm:

$$\frac{\partial u}{\partial x} = \frac{K_{x,2} \cdot \Delta_{x,1} - K_{x,1} \cdot \Delta_{x,2}}{S_{x,1} K_{x,2} - S_{x2} K_{x,1}} \tag{4}$$

Die Dehnungsdifferenz zwischen breiter und schmaler Stelle der Zugprobe läßt sich folgendermaßen bestimmen:

$$\left(\frac{\partial u}{\partial x}\right)_{[br]} - \left(\frac{\partial u}{\partial x}\right)_{[schm]} = \frac{\lambda}{2\pi\delta x} \cdot \frac{(1+\cos\varphi_2)(\Delta_{x,1[br]} - \Delta_{x,1[schm]}) - (1+\cos\varphi_1)(\Delta_{x,2[br]} - \Delta_{x,2[schm]})}{\sin\varphi_1(1+\cos\varphi_2) - \sin\varphi_2(1+\cos\varphi_1)} \tag{5}$$

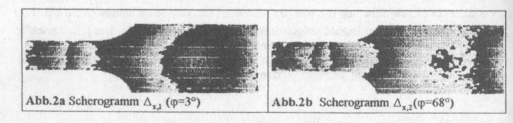

Abb.2a Scherogramm $\Delta_{x,1}$ ($\varphi=3°$)   Abb.2b Scherogramm $\Delta_{x,2}$ ($\varphi=68°$)

Für die Experimente wurde ein HeNe-Laser verwendet. Die auf die 5mm starke, 200mm lange und 30mm (bzw. 10mm) breite Kunststoffprobe wirkende Kraft lag zwischen 100N und 500N. Bei Dehnungsdifferenzen von 100-250 µm/m traten Meßfehler im Vergleich zur DMS-Messung zwischen 5 und 14% auf. Die Genauigkeit der Ergebnisse ist unter der Berücksichtigung der bei Fulten [2] gemachten Fehleranalyse der Gleichung (1) und der begrenzten Genauigkeit der Phasenwertbestimmung im Scherogramm im allgemeinen tolerierbar.

Als zweites Beispiel dient die Gleichung der Beschreibung der Verformung einer auf Druck belasteten Kreisscheibe aus Kunststoff (∅ 128mm, Dicke 10mm). Entlang der Hauptachsen können die In-Plane-Komponenten der Verformung $\partial u/\partial x$, $\partial u/\partial y$, $\partial v/\partial x$ und $\partial v/\partial y$ bestimmt werden. Dazu sind alle sechs Scherogramme notwendig, die im folgenden dargestellt sind. Zur Überprüfung der Ergebnisse wurde als Referenzmethode die mathematische Beschreibung einer belasteten Kreisscheibe herangezogen [4]. Die Auswertung der in Abb.3 dargestellten Scherogramme erfolgt in der folgenden Tabelle von der Mitte der Kreisscheibe hin zum Rand.

Beleuch-tungswinkel	$\dot{\varphi}_1$ (4° in x-Richtung)	φ_2 (76° in x-Richtung)	φ_3 (78° in y-Richtung)
Scherrich-tung: δy			
Scherrich-tung: δx			

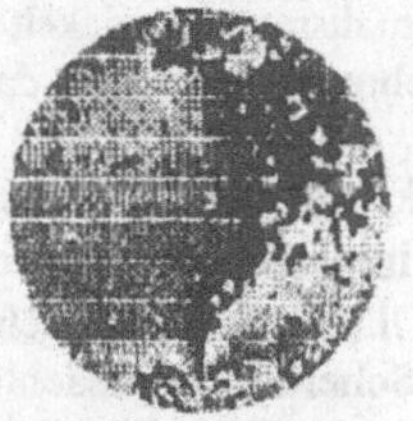

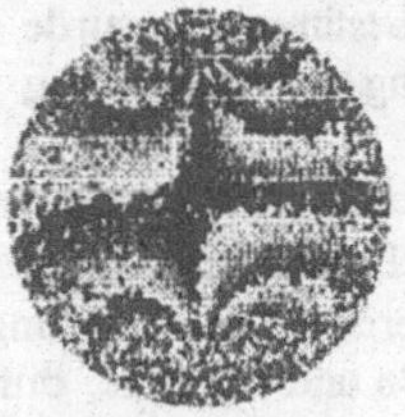

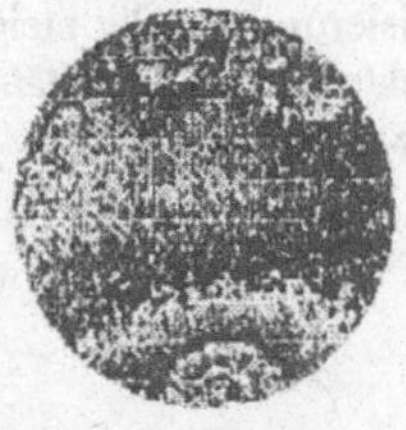

Abb 3 Scherogramme einer Kreisscheibe

vertikale Achsc	Scherogr. best. Vcrf.	$\Delta\left(\dfrac{\partial u}{\delta x}\right)=120\,\dfrac{\mu m}{m}$	$\Delta\left(\dfrac{\partial v}{\delta y}\right)=-41\,\dfrac{\mu m}{m}$	$\Delta\left(\dfrac{\partial u}{\partial y}+\dfrac{\partial v}{\delta x}\right)=-5\,\dfrac{\mu m}{m}$
	Math. best. Verformung	$\Delta\left(\dfrac{\partial u}{\partial x}\right)=169\,\dfrac{\mu m}{m}$	$\Delta\left(\dfrac{\partial v}{\partial y}\right)=-60\,\dfrac{\mu m}{m}$	$\Delta\left(\dfrac{\partial u}{\partial y}+\dfrac{\partial v}{\delta x}\right)=0\,\dfrac{\mu m}{m}$
horizon-tale Achse	Scherogr. best. Verf.	$\Delta\left(\dfrac{\partial u}{\partial x}\right)=-116\,\dfrac{\mu m}{m}$	$\Delta\left(\dfrac{\partial v}{\partial y}\right)=65\,\dfrac{\mu m}{m}$	$\Delta\left(\dfrac{\partial u}{\partial y}+\dfrac{\partial v}{\delta x}\right)=-14\,\dfrac{\mu m}{m}$
	Math. best. Verformung	$\Delta\left(\dfrac{\partial u}{\delta x}\right)=-89\,\dfrac{\mu m}{m}$	$\Delta\left(\dfrac{\partial v}{\partial y}\right)=57\,\dfrac{\mu m}{m}$	$\Delta\left(\dfrac{\partial u}{\partial y}+\dfrac{\partial v}{\delta x}\right)=0\,\dfrac{\mu m}{m}$

Tabelle 1 Vergleich der Ergenisse der Verformungsbestimmung an eine Kreisscheibe

Die Abweichungen zwischen den theoretischen Ergebnissen der Referenzmethode und denen der Scherografie sind beachtlich (bis zu 40%). Sie setzen sich aus den Unsicherheiten der experimentellen Ausführung, insbesondere der schwer zu realisierenden symetrischen Krafteinleitung, und den scherografischen Meßunsicherheiten zusammen.

3. Zusammenfassung

Mit der Scherografie können dreidimensionale Oberflächenverformungen bestimmt werden. Für die In-Plane-Verformungsbestimmung wurde ein Formalismus entwickelt und beispielhaft angewandt, der die Bestimmung der Verzerrung und Dehnung aus sechs dazu notwendigen Scherogrammen beschreibt.

Die gemessenen Dehnungsdifferenzen liegen je nach Stärke der Scherung zwischen 10 μm / m und 300 μm / m. Bei stärkeren Dehnungsdifferenzen verschlechtert sich der Kontrast durch die Ganzkörperverschiebung gravierend. Die Näherungen in Gl.(1) führen je nach Anwendung zu einer Unsicherheit zwischen 4% und 6%. Die, durch die Scherung verursachte eingeschränkte Ortsauflösung im Scherogramm verringert die Genauigkeit der Ergebisse um ca.10%. Damit liegt die zu erwartende Unsicherheit bei 15%.

Der Vorteil flächenhafter Verformungserfassung für Realisierung ständig steigender Anforderungen der Industrie auf den Gebieten der Bauteilentwicklung und Qualitätsicherung ist ein besonderes Merkmal der Scherografie. Quantitative Abweichungen in der oben genannten Größe treten dagegen in den Hintergrund.

4. Literatur

[1] Y.Y.Hung *A speckle-shearing interferometer: A tool for measuring derivatives of surface displacements* Opt. Comm. 11 (1974) No 2 ,132

[2] J.P. Fulton, M.Namkung, L.D.Melvin: *Practical estimates of the errors associated with the governing shearography equation;* Review of Progress in Quantitativ Nondestructive Evaluation Vol. 12 (1993) S.427

[3] St. Naasner: *Untersuchungen zur Entwicklung des scherografischen Prüfverfahrens zum Meßverfahren und dessen Anwendung;* Diplomarbeit Humboldt Universität zu Berlin 1994

[4] M.M.Frocht: *Photoelasticity;* New York John Wily & Sons 1948

Kombiniertes Simulationsverfahren zur Optimierung holografisch-interferometrischer Prüfprozesse

Petra Aswendt

Fraunhofer Institut für Umformtechnik und Werkzeugmaschinen IUW
Reichenhainer Str. 88 D-09126 Chemnitz

1. Einleitung

Jüngste Entwicklungen zur gerätetechnischen Umsetzung holografisch-interferometrischer Prüfverfahren [1-3] sowie zur automatischen Streifenauswertung [4-6] lassen ihren verstärkten Einsatz im Rahmen der Qualitätssicherung erwarten. Doch jeder neue Anwendungsfall wirft eine Reihe von Fragen auf:

- Welche Fehler werden erkannt?
- Wie groß ist der Schädigungsgrad, insbesondere bei den nicht detektierbaren Fehlern?
- Welches der Meßverfahren liefert die besten Fehlersignale?
- Sind Verfahrenskombinationen sinnvoll?

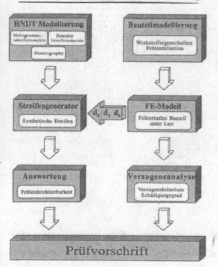

Bild 1. Prinzipielles Vorgehen

Die Antworten darauf wurden bisher überwiegend empirisch gefunden und sind nicht frei von subjektiven Einflüssen. Besonders schwer fällt die Entscheidung, wenn Bauteile zu prüfen sind, die aus komplexen Verbundwerkstoffen bestehen. Der vorliegende Beitrag beschreibt, wie die Entscheidungsfindung im Stadium der Planung erleichtert und objektiviert werden kann. Vorgestellt wird ein umfassendes Simulationsverfahren, das sich zusammensetzt aus einer FEM-Modellierung des fehlerbehafteten Bauteils und der Simulation der verschiedenen interferometrischen Anordnungen. Am Beispiel eines kohlefaserverstärkten Epoxidharzrohres (CFK-Rohr) wird gezeigt, wie die Kenntnis vom Werkstoff und seinen Fehlern mit dem Wissen über den optischen Prüfprozeß kombinierbar und welche Ergebnisse daraus ableitbar sind.

2. Simulationsverfahren

Der Rahmen einer Prüfaufgabe ist fest abgesteckt: auf der einen Seite steht das fehlerbehaftete Bauteil und auf der anderen die aus der Gruppe der holografisch-interferometrischen Verfahren (HNDT) auszuwählende Prüfmethode. Dafür eine durchgängige Simulation bereitzustellen, ist das Hauptanliegen dieser Arbeit. Das entwickelte Vorgehen ist im Bild 1 schematisch dargestellt. Die rechentechnische Umsetzung erfolgte erstmals durchgängig auf einer HP Workstation und bietet damit eine komfortable, nutzerfreundliche Lösung.

2.1. Simulation des Bauteilverhaltens

Die Ausgangssituation für den Prüfer ist kompliziert: für ein Bauteil mit unbekannten Fehlern soll bei einer Testbelastung anhand der experimentellen Streifenbilder eine Aussage zum Schädigungsgrad getroffen werden. Wie kann er diese Aufgabe in ihrer Komplexität lösen? Am Beispiel eines CFK-Rohres unter Innendruckbelastung wird das Vorgehen demonstriert.

Die Basis bildet ein Finite-Element Modell des Bauteils, welches die Berücksichtigung einer Vielzahl relevanter Fehler erlaubt. Bei einem solchen laminierten Rohr können das sein: Risse, Kratzer, ungetränkte Lagen etc. Diese Fehlstellen variieren hinsichtlich ihrer Lage, Orientierung und Stärke. Hier ist eine systematische Zusammenstellung erforderlich, deren Umsetzung in das FE-Modell dann keine Schwierigkeit darstellt. Bild 2 zeigt zum einen die vernetzte Struktur des Rohres und zum anderen, wie ein Fehler im Laminataufbau modelliert wird.

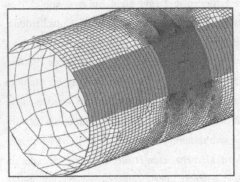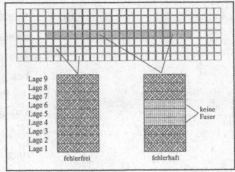

Bild 2. FE-Modell eines fehlerbehafteten CFK-Rohres

Im vorliegenden Beispiel wurden 19 Fehlertypen simuliert. Als Ergebnis stehen für jeden Fehlerfall neben den berechneten Verschiebungsdaten auch die jeweiligen Spannungswerte im Laminat zur Verfügung. Erstere werden für die weitere Simulation des Prüfprozesses benötigt und die Spannungsfelder bilden die Grundlage für eine weiterführende Versagensanalyse, auf die im Pkt. 3 eingegangen wird.

2.2. Modellierung des Prüfprozesses

Die holografisch-interferometrische Prüftechnik beruht in vielen Fällen darauf, daß sich Werkstoff-
fehler in Unregelmäßigkeiten eines Interferenzstreifenmusters widerspiegeln. Um entscheiden zu
können, welche Fehler mit den unterschiedlichen Verfahren detektierbar sind, werden die aus der
FEM-Simulation resultierenden Verschiebungsdaten herangezogen und verfahrensspezifische,
wirklichkeitsnahe Streifenbilder generiert. Gegenstand der Untersuchung sind dabei die Hologramminterferometrie (HI), die digitale Specklemusterinterferometrie (DSPI) mit ihren unterschiedlichen
Sensitivitätsrichtungen sowie die Shearography (SHEAR) mit zwei Varianten des Bildversatzes. Die
Grauwerte I(x,y) der synthetischen Interferenzstreifenbilder werden nach der, allen Meßverfahren
gemeinsamen Gleichung

$$I(x,y) = I_0(x,y) \cdot R_{sp}(x,y) \cdot \left(1 \pm v(x,y) \cdot \cos\left(\frac{2\pi}{\lambda} \, \underline{S} \cdot \underline{d}(x,y) \right) \right) \tag{1}$$

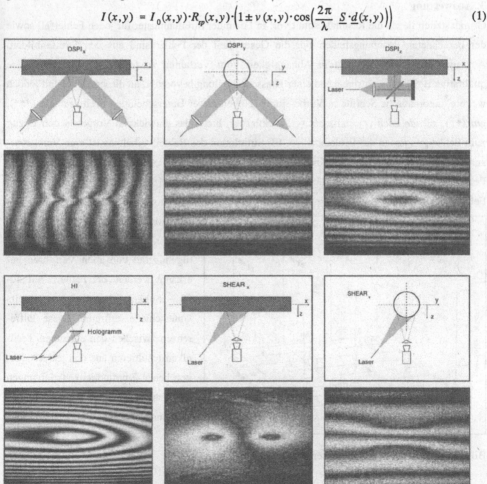

Bild 3. Simulierte Streifenmuster für unterschiedliche Interferometeranordnungen

ermittelt, wobei folgende Einflußgrößen in die Rechnung eingehen:

I_0 Grundhelligkeit (Laserbeleuchtung, Reflektivität)

R_{sp} Specklerauschen (Blendenöffnung, Wellenlänge)

v   Streifenkontrast (Specklegröße, Starrkörperbewegung)

S   Sensitivität (Meßverfahren, Interferometergeometrie).

Die verfahrensspezifischen Unterschiede äußern sich in erster Linie im Sensitivitätsvektor, der bei DSPI und annähernd auch bei der Shearography für alle Bildpunkte konstant ist, bei der Hologramminterferometrie aber typischerweise geometrieabhängig variiert. Die simulierten optischen Anordnungen sind zusammen mit einem entsprechenden Streifenmuster im Bild 3 angegeben.

3. Auswertung

Grundsätzlich liegen die Resultate somit vor: in Form der Streifenmuster für jeden Fehlerfall sowie der berechneten Spannungsfelder. Für die Gesamtheit der Fehler sind aus den Streifenbildern Aussagen zur Detektierbarkeit in Abhängigkeit vom Verfahren zu treffen. Derzeit wird eine qualitative Bewertung entsprechend einer Klasseneinteilung bevorzugt, an dieser Stelle sind jedoch weitere, automatische Schritte in Vorbereitung. Neben dieser Unterscheidung nach *sehr gut* (***), *gut* (**), *gerade noch* (*) und *nicht* (-) detektierbar, bietet das entwickelte Vorgehen den Bezug zum Schädigungsgrad des Fehlers. In die FE-Simulation des Bauteilverhaltens ist eine Versagensanalyse auf Grundlage des Tsai-Kriteriums [7] integriert, die Aussagen zur Schadensrelevanz der Fehler ermöglicht. Eine ausführliche Beschreibung dazu ist in [8] zu finden. Im Bild 5 sind Fehlerdetektierbarkeit und Schädigungsgrad für vier Prüftechniken gegenübergestellt.

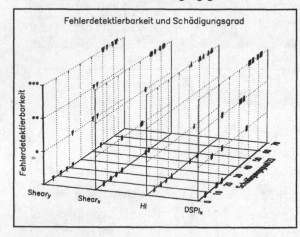

Als allgemeine Tendenz läßt sich ablesen, daß Fehler mit hohem Schädigungsgrad von allen Verfahren gut erkannt werden, bei Fehlern mit 40-50% Schädigung hingegen bereits Unterschiede auftreten. Diese Differenzen zwischen den einzelnen Techniken resultieren aus den jeweils vorgegebenen Empfindlichkeitsrichtungen sowie aus der starken Anisotropie des Verbundrohres.

Bild 5. Vergleich der Simulationsergebnisse

Die Vorteile von $DSPI_x$ - beste Nachweisquote und deutliche Fehlersignale - werden beeinträchtigt durch die Notwendigkeit sehr hoher Testdrücke. Das Gegenteil ist bei der gewählten holografischen

Anordnung der Fall, bei ihr reichen kleine Testdrücke, um Streifenmuster zu erhalten. In diesem Druckbereich zeichnen sich aber insbesondere die innen liegenden Fehler noch nicht ab, und eine Erhöhung der Belastung hat dann eine zu große Streifendichte zur Folge. Bei der Shearography ist die Ausgangssituation besser, durch die variable Empfindlichkeit liefern mittlere Testdrücke gute Ergebnisse und die Kombination von horizontalem und vertikalem Bildversatz erhöht die Nachweisquote, so daß für eine zerstörungsfreie Prüfung des CFK-Rohres unter Produktionsbedingungen die Shearography zu bevorzugen ist.

Diese Schlußfolgerung beinhaltet das im Bild 1 ausgewiesene Ziel, die Formulierung einer Testvorschrift auf Grundlage des kombinierten Simulationsverfahrens. Die aufgeführten Ergebnisse sprechen für das Vorgehen und seine Anwendung auf eine breite Palette von Prüfaufgaben.

4. Literatur

[1] Ettemeyer, A; Honlet, M.: Neuer optischer Aufbau für kombinierte TV-Holografie und Shearography. Proc. LASER 93 in W.Waidelich (Ed.) Laser in der Technik, Berlin: Springer-Verlag 1994, S.180-183

[2] Pedrini, G; Tiziani, H.: Double Puls-Electronic Speckle Interferometry. Proc. LASER 93 in W.Waidelich (Ed.) Laser in der Technik, Berlin: Springer-Verlag 1994, S.162-165

[3] Aswendt, P; Höfling, R.: Interferometrische Dehnungsmessung - Aufbau und Anwendung eines DSPI-Meßsystems. Proc. LASER 93 in W.Waidelich (Ed.) Laser in der Technik, Berlin: Springer-Verlag 1994, S.172-175

[4] Jüptner, W.: Nondestructive Testing with Interferometry. Proc. Fringe'93 in W.Jüptner, W.Osten (Eds.) Physical Research Vol.19, Berlin: Akademie Verlag 1993, S.315-321.

[5] Höfling, R.; Priber, U.: Automatic Fringe Pattern Recognition using Invariant Moments: a Feasibility Study. Proc. Fringe'93 in W.Jüptner, W.Osten (Eds.) Physical Research Vol.19, Berlin: Akademie Verlag 1993, S.374-381.

[6] Kreis, T., Geldmacher, J., Jüptner, W.: Phasenschiebe-Verfahren in der interferometrischen Meßtechnik: Ein Vergleich. Proc. LASER 93 in W.Waidelich (Ed.) Laser in der Technik, Berlin: Springer-Verlag 1994, S.119-126

[7] Chawla, K.K.: Composite Materials Science and Engineering. New York: Springer-Verlag 1987.

[8] Höfling, R., Aswendt, P.: Werkstoffkundliche Grundlagen für die holografisch-interferometrische Meßtechnik. in: Holografisch-interferometrische Meßtechnik, Abschlußbericht zum gleichnamigen BMBF-Verbundprojekt, VDI-TZ Physikalische Technologien, Düsseldorf, 1995, S. 3-18

Danksagung
Die Ergebnisse, die in der vorliegenden Arbeit vorgestellt wurden, konnten in einem vom BMBF unter dem Kurzzeichen HOLOMETEC FKZ 13N 59098 geförderten Vorhaben gewonnen werden.

All-Optical Correlation in Speckle Photography for Deformation Mapping in Monuments

E.Gärtner[1], G.Gülker[2], H.Hinrichs[2], K.Hinsch[2], P.Meinlschmidt[2], F.Reichel[1], K.Wolff[2]

[1]Jenoptik Technologie GmbH, 07739 Jena

[2]Angewandte Optik, FB Physik, Carl von Ossietzky Universität Oldenburg, 26111 Oldenburg

1. Introduction

Monuments of our cultural heritage -such as historical buildings from stone or wood- are suffering from environmental influences like climatic changes and pollution. This leads to deformations and cracks on the valuable historical objects. To prevent such tremendous damages it is necessary to detect decay processes in the very beginning, before they are visible to the unaided eye. Due to the delicacy of the object, a remote and nondestructive sensing technique is desirable.

The Oldenburg group is involved in such tasks mainly with two methods. Electronic speckle pattern interferometry (ordinary ESPI and shear-ESPI) is applied for deformations in the range of a few microns, laser speckle photography (LSP) [1] for lateral displacements in the range from 20 μm up to about 1 mm. Application of the latter method will be reported here. It is a well established and simple technique for the mapping of transverse deformation fields. The photographic recording material allows the interrogation of large fields with high resolution. Recent developments of optically addressable spatial light modulators (OASLM) promise an essential improvement concerning the processing time of such recordings. This has been demonstrated in similar applications in the field of particle image velocimetry (PIV) [2], and will be applied to LSP in the present paper.

2. Speckle double exposure recording

When a rough surface is illuminated with laser light a speckle pattern is generated due to the surface microstructure. This pattern is photographed on holographic material before and after the object is loaded mechanically or thermally. Usually, both records are superimposed on the same plate. A lateral deformation results in two shifted almost identical speckle patterns. To eliminate ambiguity in the displacement, i.e. to determine its sign, a bias shift has to be applied between the exposures. This should point approximately in the direction of the unknown displacement and exceed it in magnitude. To overcome difficulties due to the apriori uncertainty about the displacement field a sandwich technique known from holographic interferometry [3] using two separate single plates for the exposures can be chosen. This allows to produce the necessary bias shift upon evaluation and to compare records taken at different times.

3. Evaluation of the photographic speckle recordings

The quantitative analysis of a double-exposure transparency (either superimposed or sandwiched exposures) calls for the determination of local displacements and can be achieved by 2-D autocorrelation of respective small interrogation areas in the transparency. Separation of the correlation peaks yields the displacement. The complete displacement field is obtained by scanning the input record. 2-D autocorrelation is performed employing two Fourier transformations (FTs), followed by peak detection in the autocorrelation domain. The FTs can be performed digitally, still a time consuming task unless a very powerful processor is employed. Time can be gained by optical performance relying on the FT property of diffraction. A conventional all-optical technique, however, requires photographic recording and processing of the first Fourier transform (the familiar Young's fringes); thus needing even more time.

To avoid this disadvantage the evaluation system utilizes an optically addressable liquid crystal SLM (type 300/01 of Jenoptik, analog working, active area $\varnothing$ 32 mm) [4]. In this LC-SLM (Fig.1) an alternating voltage applied to transparent electrodes is divided according to the impedance of a batch of different layers. The write light intensity pattern changes the conductivity of the photoconductor material, which leads to a change of the electric field in the LC-layer and hence to a variation of the orientation of the molecules. The nematic untwisted liquid crystal is an uniaxial birefringent medium, so that the change of the orientation of the molecules induces a change of the refractive index. Thus the write intensity modulation is converted to a phase modulation. The linear polarisation of the readout beam has to be parallel to the extraordinary axis of the crystal to achieve maximum variation.

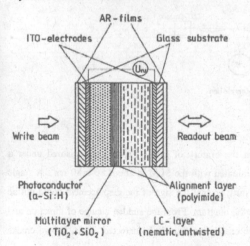

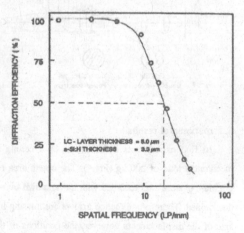

Fig.1: Principal buildup of optically addressable liquid spatial light modulator (LC-SLM)

Fig.2: Modulation transfer of LC-SLM

The magnitude of the phase shift is determined by the driving parameters and spectral sensitivities and assumes a maximum of about 2π. The dynamic behavior of the LC-SLM is determined by the speed of voltage change between the photoconductor- and LC-layers as well as by the dynamics of the arrangement of the LC molecules. The switching time is about 20-30 ms [4] [5], so that the system is capable of frame rates in the order of video frequency. The device is easy to handle with operating voltage of some Volts at 50 - 5000 Hz.

The resolution of the SLM is defined by a 50% decrease of the diffraction efficiency of a rectangular 1:1 grating, which is projected onto the SLM. It is found to be about 20 lp/mm (Fig.2).

The experimental realization performing the two necessary optical FTs combines both steps in one setup (Fig.3). Collimated light from the HeNe-laser is focused as write beam onto the left side of the SLM by a FT lens. The double-exposure record is placed in the converging part of the beam. This arrangement provides for an adjustable interrogation spot size on the recording and for scaling of the spectrum. This spectrum -the Young's fringe pattern- is thus written onto the light modulator. A second HeNe-laser provides the collimated readout beam from the right of the SLM. This light is phase modulated by the device according to the write intensity distribution. Due to a dielectric mirror inside the SLM the readout beam is reflected back and passes the LC-layer twice. Another FT lens performs the second FT, thus yielding the 2-D spectrum of the Young's fringe pattern, i.e. the autocorrelation function (ACF) of the speckle double-exposure transparency. The ACF is recorded by a video camera and a computer program detects the location of the correlation peaks and produces a map of displacement vectors. The computer also controls the stepper motors to scan the record. The processing speed is about 5 spots per second. Previously, such a system has been employed successfully in the interrogation of PIV records [5]. The present study extends the range of application to speckle images.

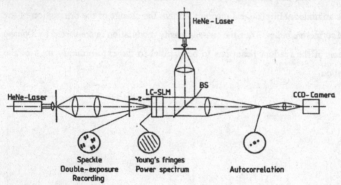

Fig.3:

All-optical autocorrelator with OASLM for the evaluation of photographic speckle records.

5. Experimental results

In first experiments deformations of a ceiling in the oratory of Brühl Castle are measured under a mechanical load of 200 kg (Fig.4). The object area evaluated with the SLM is about 80×80 cm^2. A single double-exposure was recorded with an unknown bias shift, so that the sign of the displacement could not be determined. Therefore, direction arrows are missing in the diagram. From the sudden change of direction and size of the displacements however, the positions of the cracks can already be detected. Thus, known cracks (full lines) could be verified, but also hitherto unknown cracks (dashed lines) could be discovered.

Another object under investigation was an experimental half-timbered wall in a large climatic chamber at the Wilhelm-Klauditz-Institut in Braunschweig. This was an extremely adverse object of greatly varying backscattering with dark timbers surrounded by bright plaster or brick. The different thermal response at the bonds of wood and the different building materials was of great interest. The inspected region ($\varnothing$ about 40 cm) was heated to about 80 $^\circ$C using a radiator in front of the wall. Two separate records were made at about 70 $^\circ$C and 50 $^\circ$C during the cooling phase. The results of the evaluation of the sandwiched records are presented in Fig.5, the vectors indicate the relative displacement. Only some regions could be evaluated, since the records are of poor quality because of decorrelation by the drastic heating process and uneven exposure due to the locally varying backscattering. Problematic areas are discerned by rather chaotic distributions of displacement vectors. Nevertheless, transitions at the boundaries of the wooden parts can be recognized.

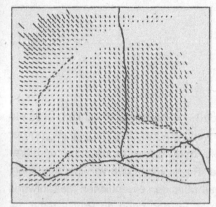 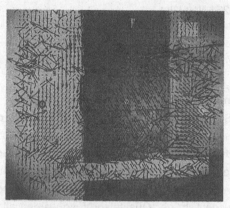

Fig. 4: Displacement field of an 80 x 80 cm² region at the ceiling of the oratory of Brühl Castle, induced by a load of 200 kg.

Fig 5: Thermally generated deformations in a half-timbered wall. Deformation field due to cooling from 70 °C to 50 °C.

6. Conclusions

The successful application of an OASLM for the evaluation of speckle double-exposure records provides an improvement concerning processing time and comfort. Further refinement would be the simultaneous parallel evaluation by an array of small lenses, each serving as a FT lens and thus producing a set of Young's fringe patterns on the input side of the SLM [6]. Readout with a respective second lenslet array provides a set of ACFs for peak evaluation. An array of 10 x 10 small lenses, for instance, could increase the frame rate by a factor of 100. Preliminary investigations with the SLM have shown that resolution is sufficient to support such a concept. Further work is in progress.

References

[1] Erf, R.K.: Speckle Metrology. Academic Press, New York (1978).

[2] Vogt, A.; Raffel, M.; Kompenhans, J.: A comparison of optical and digital evaluation of photographic PIV recordings. Conference Proceedings of the Sixth International Symposium on Application of Laser Techniques to Fluid Mechanics, Lisbon (1992), pp.24.3.1-24.3.6.

[3] Abramson, N.: Sandwich hologram interferometry: a new dimension in holographic comparison. Appl.Opt. 13 (1974), 2019-2025.

[4] Gärtner, E.; Reichel, F.: An optical autocorrelator for evaluation of PIV and LSP transparencies. To be published in: Measurement Science and Technology.

[5] Vogt, A.; Reichel, F.; Kompenhans, J.: A compact and simple all optical evaluation method for PIV recordings. Proc. Seventh International Symposium on Application of Laser Techniques to Fluid Mechanics, Lisbon (1994), pp.35.2.1-35.2.8.

[6] Hinsch, K.; Arnold, W.: Spatially multiplexed optical correlation for displacement mapping in speckle metrology. Proc. SPIE 1136 (1989), 327-334.

Miniaturisierte Speckle-Sensoren für die Qualitätsprüfung

E. ETTEMEYER
Ettemeyer Qualitätssicherung
Memminger Str. 72/207
D-89231 Neu-Ulm

1. Einleitung und Aufgabenstellung

Die Speckle-Interferometrie zeichnet sich aus durch hohe Meßempfindlichkeit, bildhafte und somit flächenhafte, berührungslose Messung, weitgehende Unabhängigkeit von der Beschaffenheit der Objektoberfläche, Schnelligkeit, quantitative Auswertung und grafische Darstellung des Ergebnisses. Sie liefert in der Regel eine Fülle von Informationen, deren ingenieurmäßige Auswertung und Analyse weit über den eigentlichen Meßaufwand hinausgeht. In der Qualitätsprüfung werden allerdings Forderungen an Prüfsysteme gestellt, die nur auf einzelne, definierte Eigenschaften der Speckle-Meßverfahren abzielen. Sie können allgemein wie folgt charakterisiert werden:

- Die Prüfung sollte in den Produktionsprozeß integriert sein.
- Das Prüfsystem sollte einfachst bedienbar sein bzw. besser vollautomatisch arbeiten.
- Das Prüfsystem sollte wartungs- und justagefrei sein.
- Die Prüfergebnisse müssen eindeutig und zuverlässig sein.

Unser Konzept ist es daher, nur die für die jeweilige Prüfaufgabe wesentlichen Verfahrensmerkmale in ein Prüfsystem zu integrieren und das System dafür so einfach, leicht und klein zu machen wie möglich.

Dieses Konzept soll am Beispiel eines Optischen Dehnungssensors vorgestellt werden.

2. Optischer Dehnungssensor

In der Materialprüfung werden entweder berührende Wegmeßsysteme oder mit Marken arbeitende optische Systeme eingesetzt. In jedem Fall erhält man nur eine Information über die Materialdehnung zwischen 2 Punkten auf der Objektoberfläche. In einigen Fällen (z.B. bei inhomogenen Werkstoffen) ist es wichtig, die ganzflächige Information über das Dehnungsverhalten zu gewinnen. Dazu kann die Speckle-Meßtechnik vorteilhaft eingesetzt werden.

Folgende Forderungen standen am Anfang der Gemeinschaftsentwicklung mit einem Hersteller von Prüfmaschinen:

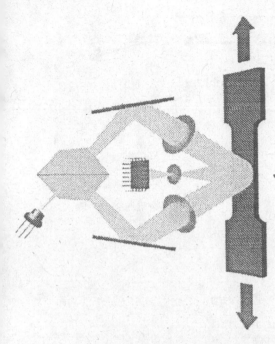

Abb. 1: Prinzip der Inplane-Speckle-Interferometrie mit dualer Beleuchtung

- Messung von Verformungs- und Längsdehnungsfeldern auf Probenoberflächen
- Berührungslose Messung
- Markenfreie Messung
- Hohe Empfindlichkeit
- Einfachste Bedienung, keine optischen Einstellarbeiten.

Die Speckle-Interferometrie erlaubt die Messung von Verschiebungsfeldern durch symmetrische Beleuchtung mit Laserlicht, **Abb. 1**. Dieses Prinzip wurde auch für den vorliegenden Sensor eingesetzt. Das Licht eines Lasers wird in zwei Strahlenbündel aufgeteilt und beleuchtet die Objektoberfläche. Das reflektierte Licht wird von einer Videokamera aufgenommen und mit einem Rechner verarbeitet.

Die Meßrichtung dieses Verfahrens wird durch die beiden Beleuchtungsstrahlen definiert. Sie liegt auf der Senkrechten zur Winkelhalbierenden zwischen den beiden Beleuchtungsstrahlen in der durch diese beiden Strahlen aufgespannten Ebene, **Abb. 2**. Dies bedeutet, daß sich die Meßrichtung bei divergenter Beleuchtung über der Probenoberfläche ändert. Die interferometrische Meßtechnik ist ein Relativmeßverfahren, d.h. man bekommt nur Relativbewegungen zwischen den betrachteten Punkten im Meßfeld und keine Absolutbewegung. Daher kann dieser Effekt der Meßrichtungsänderung ohne weitere Information (z.B. aus einem weiteren, absolut messenden Punktmeßverfahren) nicht

510

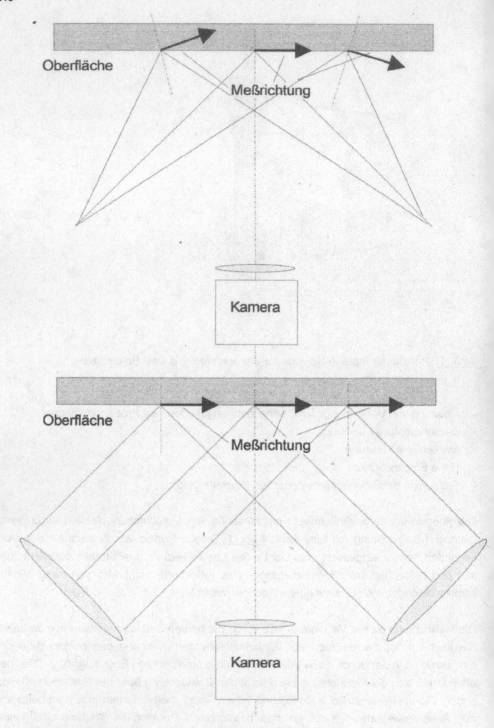

Abb. 2: Meßrichtung bei der Inplane-Speckle-Interferometrie: nur bei parallelen Beleuchtungsbündeln ist die Meßrichtung konstant

Abb. 3: Optischer Dehnungssensor im Einsatz an einer Materialprüfmaschine

numerisch kompensiert werden. Bei der Dehnungsberechnung eines auf diese Weise gemessenen Verschiebungsfeldes würden sich zum Rand hin unzulässige Fehler ergeben, deren Größe zudem aus dem Meßergebnis nicht abschätzbar ist.

Daher wurde der Aufbau des Sensors so ausgeführt, daß zwei parallele Lichtbündel auf die Objektoberfläche treffen. Die Meßrichtung ist an allen Meßpunkten konstant und es tritt kein optisches Übersprechen anderer Bewegungskomponenten auf das Meßergebnis auf.

Das komplette optische System wurde für einen festen Meßabstand (240mm, Vorgabe des Maschinenherstellers) und ein definiertes Meßfeld (ca. 30 x 40 mm^2) ausgelegt, **Abb. 3**. Dadurch zeichnet sich das System durch sehr einfache Bedienung und justagefreies Arbeiten aus.

Abb. 4 zeigt ein typisches Meßergebnis beim Einsatz des Sensors beim Zugversuch einer Stahlprobe. Im Übergangsbereich zwischen elastischem und plastischen Verhalten treten die sog. Lüdersbänder (Scherbänder) auf.
Der Meßablauf wird vollautomatisch durch die Prüfmaschine gesteuert, so daß auch Serienmessungen durchgeführt werden können und der Meßbereich in den Millimeterbereich erweitert werden kann.

512

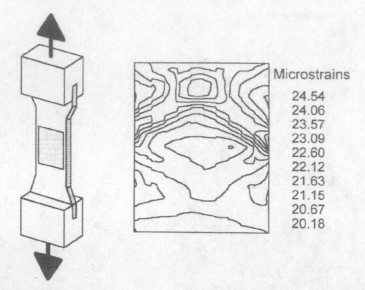

Microstrains

24.54
24.06
23.57
23.09
22.60
22.12
21.63
21.15
20.67
20.18

Abb. 4: Lüdersbänder einer Stahlprobe. Das Dehnungsfeld ist durch Linien gleicher Dehnung dargestellt.

Abb. 5: Miniaturisierter Shearing Sensor für die Werkstoffprüfung

3. Weitere Sensoren

Dieses Prinzip der Vereinfachung und Reduktion auf die geforderten Prüfeigenschaften kann natürlich für jedes Verfahren der Speckle-Meßtechnik durchgeführt werden, sobald die relevanten Eigenschaften klar definiert sind. So führten diese Arbeiten zu miniaturisierten Shearing-Sensoren für die zerstörungsfreie Werkstoffprüfung, **Abb. 5** und ESPi-

Sensoren zur Verformungs-/Dehnungsanalyse. Beide zeichnen sich durch geringes Gewicht und kleinste Baugröße aus, so daß sie sehr leicht in Qualitätsprüfsysteme integriert werden können. Außerdem sind sie leicht transportabel und können direkt an die zu untersuchenden Komponenten angesetzt werden. Dadurch werden viele der sonst problematischen Randbedingungen für die Speckle-Meßtechnik (Umgebungseinflüsse wie Schwingungen, Erschütterungen, etc.) sehr einfach eliminiert.

4. Resumé

Die Miniaturisierung und Vereinfachung von Speckle-Meßgeräten führt zu neuen Einsatzmöglichkeiten dieser Verfahren in der Qualitätsprüfung. Derartige Systeme werden künftig überall dort verbreiteten Einsatz finden, wo industrielle Fragestellungen nach Produkt- und Prozeßqualität beantwortet werden sollen.

The Phase-Shifting Electronic Shearography for Vibration Analysis

**Sung-Hoon Baik, Jai-Wan Cho, Cheol-Jung Kim, Jang-Seob Choi*,
Young-June Kang*,**
Quantum Optics Laboratory, Korea Atomic Energy Research Institute,
*Dept. of Mechanical Design, Chon-Buk National University,

1.Introduction

There are many optical interferometric methods for measuring vibration, such as holography, speckle and moire interferometer. The electronic shearography technique has a lot of advantages in practical use[1] : low sensitivity for mechanical unstability, low limit in laser coherence, simple setup, etc.. The fringe patterns obtained by time-average shearography represent derivatives of amplitude of vibration[2]. Therefore, through a proper integration technique, the useful information of vibration modes can be obtained from shearograms.

A Michelson interferometer type phase-shifting electronic shearography was developed for the study of vibrating objects in noisy environments. With electro-optic holography method, the speckle images were processed and the numerical evaluation of vibration mode was achieved by integration of the shearing phase map. Experimental results of this method, applied to measure vibration of objects, cantilever and boiler, are described and discussed.

2. Electronic Shearography system

The experimental setup of electronic shearography is shown in Fig.1. For the reduction of speckle noise of shearogram, 4-frame phase shifting method and speckle averaging technique were used. ITI VFG frame grabber was used for the data acquisition and processing. Due to the limit in arithmetic capability, the time for processing one shearogram was about 4 sec.'s. The fringe visibility was dependent on the $F\#$ of the camera. The obtained shearographic fringes with various $F\#$ of CCD camera are shown in Fig.2. These figures show that the best fringe visibility was obtained at $F/5.6$. When the $F\#$ of the camera was either larger than 8 or smaller than 5.6, the fringe visibility decreased rapidly. This is not the same results as the characteristics of ESPI, in which fringe visibility was good when F# is larger than 11. Because A 5W Ar+ laser without etalon and large aperture of CCD camera (F#=5.6) were used, the measuring area could be larger than 2x2 m^2 .

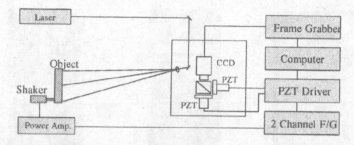

Fig.1. Experimental setup of electronic Shearography

Fig.2. Shearographic Fringes with various F# of CCD camera

3. Measurement of Vibration modes of Vibrating Cantilever

The electro-optic holography method[3] was used for the evaluation of the fringe patterns. The adjustment of vibration phases between object and phase modulating PZT was controlled by a 2-channel function generator. The adjustment process for phase and bias,we used, was the method described by Pryputniewics[3]. But, It was difficult to adjust the phases and amplitude. So it seems that there should be some errors in the phases and the bias modulation.

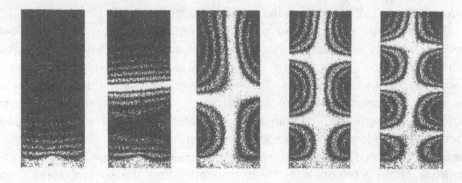

Fig.3. The Longitudinally Sheared Vibration Fringes of Cantilever

a) Vibration mode   b) Lateral Shearing   c) Longitudinal Shearing

Fig.4. Shearing Fringes of Vibration modes of Cantilever (f = 747 Hz)

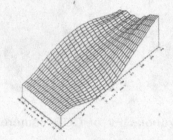 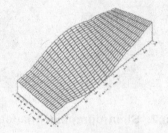

a) From lateral shearogram   b) from longitudinal shearogram

Fig.5. Vibration modes calculated from shearogram

The shearographic fringe patterns measured are shown in Fig. 3. and Fig.4. And the vibration modes calculated from shearographic fringes are shown in Fig.5.. There are some differences between two results. The errors are seems to be due to the phase and magnitude errors of the bias modulation.

4. Applications : Vibration modes of Gas Boiler

The electronic shearographic method for vibration measurement, described previous section, was applied to vibration study of gas boiler in practical mounting condition. When running condition, there are no time average fringes beause of the irregular vibration. To investigate the vibration mode of boiler at the same mounting condition as practical use, the vibration modes of boiler, excited by shaker, were measured by the same method. The shearographic module was mounted at a photographic tripod. Fig.6 show the time-average shearographic fringe patterns at several frequencies. Fig.7 are the slope and amplitude of the vibration at f=115Hz, calculated from the shearographic fringe pattern with phase modulation. This result shows that this shearographic technique can be applied to the study of vibration in noisy environment.

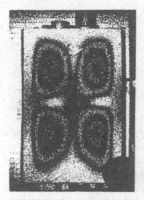

Fig.6. Shearographic Fringes Measured with several resonance frequency

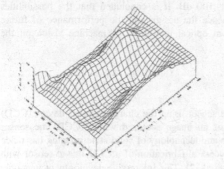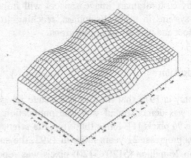

a) 3-D plot of slope of vibration   b) 3-D plot of Vibration mode

Fig.7. Calculated Vibration mode pattern of Boiler

5. Discussions

The experimental results show the feasibility of applications of electronic shearography for vibration analysis of large object that can be hardly mounted on the optical table. Despite of the errors, the results show that the shearographic method can be applied to measure the vibration mode shapes. The studies about the effects of F# of CCD camera to fringe visibility, the phase errors due to the vibration phase between object and phase modulating PZT and magnitude of bias modulation, are needed.

References

1. Y.Y. Hung, Proceedings of SPIE, Vol.1554B, 29-45(1991)
2. S.L. Toh, et al., Optics & Laser Technology, Vol.27, 51-55(1995)
3. Ryszard J. Pryputniewicz, Karl A. Stetson, Proceedings of SPIE, Vol.1162, 456-467(1989)

Custom Smart Image Sensors for Optical Metrology and Machine Vision

Peter Seitz
Paul Scherrer Institute Zürich
Badenerstrasse 569
CH - 8048 Zürich, Switzerland

Abstract : Thanks to the advances in semiconductor technology it is possible today to fabricate solid state image sensors with 5k×5k and more pixels, while reducing the geometry of the pixels at the same time to 5×5µm^2 and less. This enables one to integrate electronic circuitry with each pixel, without compromising the fill factor substantially. Such pixels with added functionality are the basis of custom photo-ASICs, application-specific integrated circuits containing photosensitive elements, and so-called smart image sensors : Single-chip cameras with on-chip analog-to-digital converters for less than $10 are advertised; image sensors have been developed including novel functionality such as real-time selectable pixel size and shape, the capability of performing arbitrary convolutions simultaneously with the exposure, as well as variable, programmable sensitivity of the pixels leading to image sensors with a dynamic range exceeding 100 dB. It is concluded that the possibilities offered by custom smart image sensors will influence the design and the performance of future imaging systems in many disciplines, reaching from optical metrology to machine vision on the factory floor and in robotics applications.

1. Introduction

It was only in 1971 that the first solid-state image sensor using the charge-coupled device (CCD) principle was demonstrated with the publication of an image acquired with a CCD line sensor exhibiting 96 pixels [1]. Since that time, the science and technology of solid state imaging has made enormous progress: 21 years later, in 1992, the successful fabrication of a CCD image sensor with more than 26 million (5120×5120) pixels was reported [2]. This impressive development was only possible because of the amazing progress made by silicon semiconductor technology, largely driven by the ever-increasing demand for digital computer memories with higher and higher storage capacity. In fact, it is not only with larger number of pixels that image sensors have profited from the advances in semiconductor technology. On-chip functionality has increased as well, and first successes with image sensors carrying out image processing algorithms on the sensor chip has led researchers to declare the imminence of powerful "seeing chips" [3].

In Section 2 recent developments in semiconductor technology are reviewed, in particular the reduction in minimum feature size. Image sensors can profit from this development not only with sensors offering larger numbers of pixels, the pixels exhibit also smaller and smaller geometry. It is becoming increasingy possible, therefore, to add functionality with electronic circuitry to each pixel without compromising their photosensitive area substantially. In Section 3 the physical limits of photodetection are discussed, pointing out that today's technology is close to where photon quantum-limited performance can be achieved over a wide illumination range without cooling. Examples of smart pixels offering photodetection with integrated functionality is discussed in Section 4. Such smart image sensors form the basis of novel optical measurement techniques and metrology systems with interesting properties and capabilities, a few of which are discussed in Section 5. There is an on-going discussion, briefly reviewed in Section 6, whether smart image sensors are on their way to becoming "seeing chips" with applications in automatic manufacturing, transportation, robots or even house-hold applications where visual recognition tasks might be desirable. In the concluding

Section 7 the major facts and findings of this paper are summarized, and conclusions for further developments and expected influences on optical metrology are given.

2. Semiconductor technology for larger numbers of smaller pixels with more functionality

Semiconductor technology has made significant and surprisingly predictable progress over the past decades. The most obvious measure for this development is the minimum feature size which is available in a certain technology, also called design rules. The evolution of the minimum feature size is illustrated in Fig. 1, from which an exponential trend is easily derived : The design rules are very predictably reduced by an average of about 10% per year. In the early 70's, a minimum feature size of 6-8 μm was employed, while today's most advanced DRAM (dynamic memory) chips are fabricated with 0.35 μm design rules. Using the above stated empirical rule, it can be predicted that the minimum feature size will have dropped to below 0.2 μm by the year 2000 [4].

At the same time that design rules were continuously shrinking, the diameter of the silicon wafers has increased, making 20 cm (8 inches) the standard for modern silicon foundries. The technology employed for image sensors is less advanced, making use today of 15 cm (6 inches) diameter silicon wafers, and often employing larger design rules. These two developments, firstly, the increase of the wafer diameter, and, secondly, the reduction of the design rules, demand much more expensive fabrication equipment. The investment is worthwhile, however, because more electronic circuitry can be placed on the same chip area, and

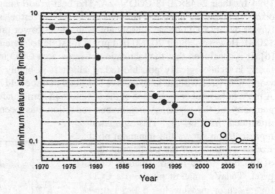

Fig. 1 : Evolution of the minimum features size (design rules) in semiconductor manufacturing technology. Values are taken from published data (solid dots) and the predictions made in Ref. [4] (open dots).

more wafer area can be processed at a time in the batch processes of semiconductor technology. As it turns out, advances in equipment manufacture just about compensate the increasing production capability of this more expensive equipment: The cost per area of processed silicon has remained approximately constant over the years, on the order of $50/cm^2, while offering increased functionality on the same chip area.

One of the most visible consequences of these impressive advances of semiconductor technology is the marked increase in resolution of image sensors, illustrated in Fig. 2. This graph shows the evolution of the record number of pixels on a CCD image sensor, as a function of the date when the work was published. There is a marked lack of progress in the number

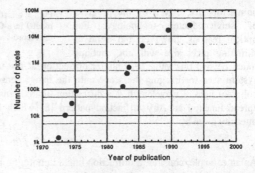

Fig. 2 : Evolution of the number of pixels on a CCD image sensor as a function of the year when the record-breaking number was published.

of pixels in the years 1975-1983. Our interpretation of this phenomenon is the following: The number of pixels in image sensors was increased rapidly by different research groups, until enough pixels on an image sensor were available for the realization of solid-state video cameras. After that initial period of research activity, it took significant time and effort to develop the semiconductor technology, necessary for the mass-fabrication of these devices with high yield. It was only then, after 1983, that the technology was pushed again, and image sensors with increasingly large numbers of pixels were fabricated. The largest image sensor available commercially today is a frame-transfer CCD with 5120×5120 pixels, covering an area of 61.4×61.4 cm^2 [2]. This sensor is fabricated using silicon wafers with 100 mm diameter. Today's CCD fabrication technology has recently changed to silicon wafers with 150 mm diameter, while at the same time offering reduced design rules. It should be technologically feasible, therefore, to fabricate CCD image sensors with 15,000×15,000 pixels or more [5], one large-area CCD image sensor covering a whole wafer. A major reason why such image sensors have not been offered commercially is the yield problem. An indication of this problem is the pricing policy of the CCD manufacturers, for example the one published by Loral Fairchild in mid-1994 for their 2048×2048 CCD 442A. The best-grade imagers can still suffer from up to 200 bad pixels and typically three bad columns per CCD. Nevertheless, such an image sensor costs $15,000. Lower quality CCD image sensors are available at lower prices. A grade 4 image sensor costs just $2000. However, about 20,000 bad pixels must be tolerated, and typically 100 columns will be bad [6]. It is concluded that the fabrication of large-area CCD image sensor with more than 100 million pixels is technologically possible today, albeit it is highly improbable that a sufficient number of working devices without an excessive bad pixel count could be manufactured.

A more influential consequence of the advances in semiconductor manufacture is the shrinking of the geometry of the pixels. A continuous reduction has been observed over the past 25 years, as illustrated in Fig. 3. This graphical representation shows the minimum pixel size as a function of the publication date of the then world-record smallest CCD pixel. Today's smallest pixels in a CCD image sensor measure just 5.1×5.1 µm^2 [7]. Similar to the minimum feature size shown in Fig. 1, an exponential trend in the decrease of the minimum pixel size is clearly visible in Fig. 3, with an average reduction rate of also about 10% per year. By the year 2000, a minimum pixel size of around 3×3 µm^2 will be reached, assuming this trend continues uninterruptedly, driven by the progress of semiconductor technology. For optical reasons, however, the trend of shrinking pixel size will not continue

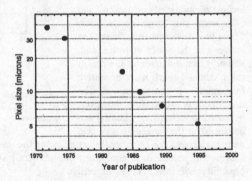

Fig. 3 : Evolution of the minimum pixel period (of square pixels) in a CCD image sensor, as a function of the date when the record-breaking work was published.

much beyond this value. Fraunhofer diffraction at the usually circular aperture of the optical lens system used for imaging the scene onto the image sensor's surface blurs the images. For spatially coherent light of wavelength λ, a diffraction-limited lens system with f-number f/# has as its point-spread-function an Airy diffraction pattern [8] which exhibits a diameter d of the first zero of intensity given by

$$d = 2.44 \, \lambda \, f/\# \tag{1}$$

As an example, consider a diffraction-limited photographic lens with f/2.8, for which a diameter of the Airy disk of d=3.8µm is obtained for green light (λ=555 nm). Actual, low-cost photographic lenses - for example in video applications or consumer electronic photography - will not be of the highest quality, and they will not reach diffraction limited imaging performance. It quite obvious, therefore, that the pixel size will not drop substantially below about 2-3 μm, which is expected to be reached shortly after the year 2000.

This situation, where minimum feature size is continuously shrinking, while a lower limit to the minimum dimension of a pixel exists, has an obvious consequence: It will be possible to supply the individual pixels with added functionality in the form of electronic circuitry, without substantially increasing the geometrical dimensions of the pixels. Image sensors with such pixels offering added functionality are called Active Pixel Sensors (APS) [9]. It is the main purpose of the present paper to investigate the possibilities offered by APS and related technologies, and to discuss the consequences for optical measurement techniques, optical metrology, electronic photography and machine vision.

3. Physical and technological limits of photodetection

The original motiviation for the development of APS technology was the competition between image sensors based on CCD technology and photodiode array image sensors fabricated with standard CMOS technology [9]. In the case of the CCD, a MOS (Metal-Oxide-Semiconductor) capacitor is used for the separation and the storage of the photogenerated electron-hole pairs, see Fig. 4a.

In a photodiode array, a reverse-biased pn-junction is employed for the same purpose, see Fig. 4b. The large difference between the two types of image sensors is caused by the method with which the accumulated photo-charge is read out. In CCDs, the individual

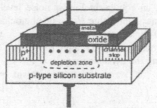

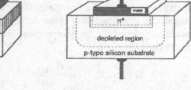

Fig. 4a : Schematic representation of a MOS capacitor, the basic building block of CCDs

Fig. 4b : Schematic representation of a pn-junction, a photodiode.

photo-generated charge packets are shifted out laterally, by using appropriate clocking pulses at the insulated MOS capacitor electrodes. In this way, the charge packets are transported, one by one, to the input of a charge detector circuit, which transduces this charge signal into a measurable voltage. In conventional photodiode array image sensors, the individual pixels are connected sequentially to a main signal line via individual MOS-FET switches, one for each pixel. This signal line usually consists of long, thin strips of metal, to which the pixel MOS-FET switches are connected. The signal line ends at the input of a charge detector circuit. In CCDs the photocharges themselves are physically transported, therefore, and in photodiode arrays, the electrical signal information is transported on the signal line. What is the big difference? To answer this question, consider the basic building block of a charge detector, consisting of a MOS-FET used in a (multiple) source-follower circuit as illustrated in Fig. 5. As discussed in Ref. [10], the main contribution to the electronic noise (distinguished from the photon quantum noise) stems from the Johnson noise of the transistor's channel region. One obtains the following equation for the equivalent statistical uncertainty ΔN of the number of charge carriers on the transistor's gate :

$$\Delta N = \frac{C}{e} \sqrt{\frac{4kT\,B\,\alpha}{g_m}} \qquad (2)$$

with C denoting the transistor's total gate capacitance, the elementary charge $e = 1.602 \times 10^{-19}$ As, Boltzmann's constant $k=1.3806 \times 10^{-23}$ J/K, the absolute temperature T, the bandwidth B, the transistor's transconductance g_m, and a factor α which depends on the operating characteristics of the transistor. For typical operation conditions of MOS transistors $\alpha=2/3$ can be assumed [11].

522

It is now clear from equation (2) why CCDs with their very small effective gate capacitances of a few 10 fF offer much reduced noise of typially a few tens of electrons at video bandwidth and room temperature. This noise level is at least an order of magnitude lower compared to the one of conventional photodiode array image sensors. They suffer from the substantial capacitances introduced by the switches and the long signal line to the amplifier. It is obvious, therefore, how APS technology can reduce the noise level of photodiode image sensors to levels previously only available in CCD sensors: Each pixel is supplied with its own source follower, row selection and reset transistor, as illustrated in Fig. 6. As an

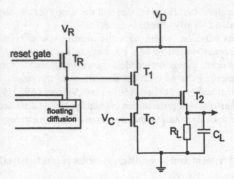

Fig. 5 : Schematic diagram of a typical charge detection circuit consisting of a double-stage source follower (T_1 and T_2) with active load T_C and reset transistor T_R

example of the noise performance of such an APS photodiode array image sensor at room temperature and for video bandwidth, an equivalent lowest noise level of about 20 electrons is reported in Ref. [12]. Since a typical maximum value N_{max} of photocharges (in CCDs called "full well charge capacity") of a few 100,000 photocharges can be obtained, the described low noise levels make very high dynamic ranges possible. In photosensing the dynamic range D/R is defined by

$$D/R = 20 \, ^{10}\log \frac{N_{max}}{\Delta N} \qquad (3)$$

The image sensor described in Ref. [12], for example, exhibits a maximum charge value of N_{max}=370,000 electrons, leading to a dynamic range of D/R=85 dB. In this way, very high D/R values are obtained that were previously only accessible to cooled image sensors operated with slow-speed readout [13]. Combined with the naturally high quantum efficiency of silicon in the visible and the near infrared spectral wavelength region of 50-95% [14], today's image sensors offer a performance that is limited over a very wide illumination range by the ubiquitous photon noise and not by the electronic detection noise. As an example, we would like to mention the utilization of a modified CCD process for the development of a novel charge detection circuit with an effective detection capacitance of less than 1

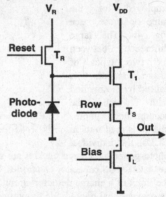

Fig. 6 : Schematic diagram of a simple APS architecture example. The APS pixel consists of a photodiode, a reset transistor T_R, a source follower ("amplifier") transistor T_1 and a row select T_S. The active load T_L is required only once per column.

fF. A surprisingly low photocharge detection noise level of less than one electron r.m.s. is reported for this image sensor at room temperature and for video bandwitdh [15].

APS technology is by no means restricted to the realization of photodiode array image sensors with pixels having their own "pre-amplifier". Much more functionality is possible - for photodiode arrays as well as CCD image sensors, of course at the price of increased circuit complexity, but at lower and lower real-estate cost on the silicon.

4. Smart pixels : photodetection with integrated functionality

Traditionally, the detection of light in an image sensor has been restricted to the simplified signal chain illustrated in Fig. 7. A semiconductor device with an associated electric field (usually a photodiode or a precharged MOS capacitor) is employed for the separation of photo-generated charge pairs, resulting in a photocurrent that is highly linear with the incident light intensity over nine orders of magnitude or more [16]. This photocurrent is integrated over a certain time, the so-called exposure time, and the photocharges are retained on a suitable storage device. The individual pixels are then sequentially scanned with a suitable switching mechanism. The pixels' charge signals are read out, and they are amplified, one by one, to complete the detection process. At the same time, the storage device is cleared (reset), so that it is ready for the next exposure cycle.

Fig. 7 : Simplified signal chain in traditional solid-state image sensors. Incident light generates a photocurrent in each pixel. The photocurrent is integrated and stored. During sequential scanning photocharge is amplified and read out.

Modern semiconductor processes and the reduced feature sizes for electronic circuits are the basis for functionality in the individual pixels that is much increased above what is illustrated in Fig. 7. Some of the possibilities and novel functionalities offered at the different stages of the image acquisition chain are symbolized in Fig. 8. This forms the basis of "smart pixels" that can be tailored to a specific application in optical metrology or other fields in which the system task is to acquire spatio-temporal light distributions and process the contained information. Before discussing the consequences of the availability of such smart pixels, the symbolic representations of the capabilities in Fig. 8 are presented in a little more detail:

The generation of a photocurrent proportional to the incident light is not restricted to rectangular pixel geometries as employed traditionally. Applications exist, where a suitable choice of geometry serves as a nonlinear transformation of the incident light, making a digital processor obsolete in an optical distance measurement setup [17]. It is possible to "calculate" the (complex) Fourier transform of a one-dimensional light distribution with a suitable sensor shape, as used in an optical position encoder [18]. Examples have even been described of making the effective shape of the photosensors programmable, i.e. electrically adaptable to changed conditions in real-time [19]. It is also possible to change the spectral sensitivty of a detector

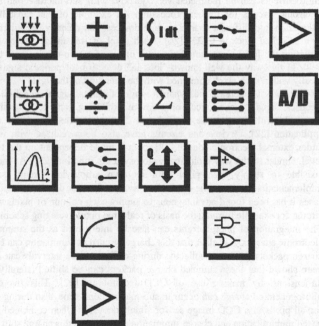

Fig. 8 : Enhanced image acquisition and processing chain in modern solid state image sensors, making use of APS (active pixel sensor) and related technologies for the realization of smart image sensors. This picture is a symbolic and incomplete representation of the possibilities offered by image sensors with smart pixels for novel optical measurement techniques and integrated optical metrology.

with an electrical signal [20], or to stack photosensors with different spectral sensitivity on top of each other for a solid-state color pixel without filters [19].

To increase the dynamic range of image sensors, offset currents can be added to or substracted from the photocurrents, as used for example for non-uniformity or background cancellation [21]. It is possible to supply each APS pixel with its own, real-time programmable offset, so that dynamic image acquisition becomes feasible, optimized for given image sensing applications. Typical saturation currents of single MOS-FETs, used as current sources in such offset cancellation circuits, are doubled with each increase of gate voltage by 30-40 mV. A very wide range of offset currents can be realized in this way, ranging from fA to µA [22], and covering up to nine orders of magnitude. There are several ways in which multiplications and divisions can be implemented: With CCDs, split gates divide charge packets according to (fixed) geometrical capacitance ratios, which has been used, for example, for the realization of FIR filters [23]. The easiest way to change the sensitivity multiplicatively is to vary the exposure time. This technique is used in most commercial video cameras for adapting the image sensor's sensitivity to varying light levels. Recently, a technique has been published, with which the exposure time - and therefore the effective sensitivity of each pixel - can be programmed individually for each pixel and in real-time [24]. Another proposal for adapting the gain of an image sensor relies on a three-transistor circuit, used for multiplying the photocurrent with a factor that it selectable over many orders of magnitude [25]. This solution does not lend itself well to integration into each pixel, but competing proposals for pixel-individual programmable gain are currently under investigation. Two- and four-quadrant multiplier electronic circuits requiring only a moderate number of transitors have been known for quite some time [22].

Photocurrents can be redirected very quickly, with sub-microsecond switching times, to different electronic circuits for further processing. An example of this capability, combined with repeated integration into different charge storage devices, has recently been demonstrated for the realization of a so-called "lock-in CCD" [26]. Each pixel of this image sensor is capable of synchronously detecting the local phase, amplitude and offset of a two-dimensional temporally modulated wave-field. In this way the well-known "lock-in" detection of periodic signals [27] can be implemented locally within each pixel, combined with the detection of the light.

Another way to increase the dynamic range of optical signal detection is to process the converted electrical signals with a circuit or a device exhibiting non-linear transfer characteristics. The well-known logarithimc response of diodes or diode-connected transistors are preferrably used for this application [22]. Devices and circuits have also been realized with which such a non-linearity is under external electronic control [28], or it is made dependent on the locally detected average light level, similar to the way biological vision systems function [29]. As mentioned, it is of course also possible to supply each pixel with its own amplifying circuit (see Ref. [30] for a recent implementation), as was the original motivation for the development of APS technology. In some cases it has been found advantageous to supply each column of pixels with its own charge amplifier circuit, for example forming the basis of real-time pixel averaging schemes [31]

The integration of photocurrents can also be interpreted as the summation of the photogenerated electronic charges. It is clear that this charge or current summation can be temporally interrupted, i.e. charge packets that were collected during different time intervals can be added in this way. It has been shown that these summed charge packets can be shifted laterally, bi-directionally and in two dimensions, by making use of CCD technology [32]. This two-dimensional lateral shift of photogenerated charge can occur in sub-microsecond time, also during the exposure, and it happens for all pixels in a CCD image sensor simultaneously. When combined with suitable, exposure-time based multiplication and charge summation capabilities, this charge shift makes it possibel to realize a CCD image sensor which can acquire and convolve an optical scene simultaneously, with a freely programmable convolution kernel of any extent [33].

While traditional image sensors have relied on sequential scanning of the individual pixels for readout with a single output amplifier, many types of CCD image sensors have been developed in the past few years, offering several output amplifiers working in parallel, see for example Ref. [34]. These multi-tap image sensors offer a much increased frame rate, albeit at the cost of increased complexity of the external image data acquisition circuitry. It is also possible to pre-process the image data on-chip, by making use of other parallelisms. Analog circuitry, such as comparators, differentiators, maximum finders, etc. can be combined in each pixel or for each column with digital circuitry for controlling the data acquisition and preprocessing functions. Analog-to-digital converters of various precision can be integrated with each pixel [35] or - with improved performance - are integrated for each column [21].

In summary, the individual pixels of a modern, custom-designed image sensors can contain a wide variety of analog and digital circuitry, giving the pixel astonishing levels of functionality. Such "smart pixels" profit directly from the on-going development in semiconductor technology, because the shrinkage of design rules translates directly into more functionality per area in a pixel of a given size.

5. Custom smart image sensors (photo-ASICs) for integrated optical measurement techniques

The monolithic integration of photosensitive, analog and digitial electronic devices has led to the realization of a wide range of integrated circuits — also called photo-ASICs or opto-ASICs — that are of direct interest in many optical measurement techniques. Thanks to multi-project wafer services, where several users share the cost of a complete process run, even small numbers of 10-100 application-specific photo-ASICs can be fabricated at a total cost of about $4,000-$10,000. This capability opens up the wide field of optical measurement techniques for highly integrated, miniature solutions that can be very cost-effective. At the same time superior performance can be offered, compared with the traditional hybrid approaches, or it becomes feasible to implement novel types of optical measurement techniques that were not practical before.

Early work in the mid '80s, for example the single-chip optical position encoder described in Ref. [36], demonstrated the practical usefulness of such an integrated technology. Another obvious use of photo-ASICs is the integration of all components of a video camera on one single chip, as described in [37]. Such single-chip video cameras are commercially available now [38]. Recently, an improved version of a single-chip digital camera has been advertised, combining a 160×160 photodiode pixel array, autoexposing circuitry, all necessary analog and digital control/timing electronics, as well as an on-chip A/D converter with interface to processor-compatible serial and parallel ports. Volume price for such a single-chip digital camera is around $10, making it very attractive for many practical applications such as surveillance, automatic manufacturing, process control, picture telephony, etc. It is not difficult to imagine that the next step can be taken as well, i.c. the co-integration of such an electronic camera with a general-purpose digital processor, capable of evaluating the acquired imagery directly on chip. Such camera-processor products, either based on line or area cameras, are already commercially available now [39], with first successful industrial applications, primarily in automatic manufacturing and process control.

Various types of photo-ASICs for different approaches to range imaging have been realized. For variations of the well-established triangulation distance-measuring technique, photo-ASICs exhibiting the following properties have been described: A prescribed one-dimensional spatial response can be obtained with suitably shaped photosensors [17]. A range imager working at video speed, producing live depth imagery according to the CCIR video standard, is based on a linear array of pairs of triangular photosites [40]. A similar technique makes use of a linear array of PSDs, position-sensitive devices [41]. A two-dimensional array of "time-to-maximum-light pixels" is the basis of another triangulation setup with swept sheets of light [42]. Stereo depth vision can also be considered to be a (passive) triangulation technique, for which special high-speed stereo depth vision chips have been proposed, see for example [43].

Another type of range imaging, offering much longer measuring distances at reduced resolution, is based on time-of-flight measurements. Due to the lack of two-dimensional phase-sensitive detectors, time-of-flight measurements are traditionally carried out point-by-point, using mechanically scanned laser beams. The introduction of the lock-in CCD, as described for example in Ref. [33], makes it possible to perform time-of-flight range imaging without moving parts.

For highest distance accuracy, interferometric methods are commonly employed. Absolute distance measurements are based on multiple-wavelength interferometry, as described for example in Ref.

[44]. Such techniques also work for rough, diffusely scattering surfaces, requiring photodetectors with an increased dynamic range. If combined with heterodyne modulation and detection, such a technique requires lock-in signal detection as described in Ref. [18], making it possible to realize two-dimensional, absolute interferometric range imaging without any moving parts. Another type of interferometry, known under the names of white-light interferometry, low-coherence reflectometry (LCR), or coherence radar, see for example Ref. [45], could also profit enormously from such developments. Two-dimensional arrays of smart pixels would be required, offering extremely high dynamic range because of the high background and low signal modulation, where the time-to-maximum could be detected and stored. Modifications of the smart image sensor described in [42] might be an excellent solution for real-time, two-dimensional coherence radar range imaging without rastering the objects in two dimensions.

The few examples given above should serve as an indication that optical measurement techniques can profit enormously from the developments in the field of smart image sensing. They make it possible to miniaturize, improve or extend known measurement techniques, while at the same time often reducing the cost and increasing the performance. It can also be expected that these novel capabilities offered by custom image sensors might form the basis for the development of new types of optical measurement techniques. This combination of conventional optics with optoelectronics and smart image sensing in innovative ways will lead to the integration of optical microsystems with unique functionality and capabilities, relevant for many practical applications.

6. From smart image sensors to seeing chips

The rapid development of image sensors with more and more integrated functionality led a promiment researcher in the field to proclaim the imminence of "seeing chips" [3]. A few examples of image sensors with complete, integrated image processing hardware have been reported for certain tasks, such as the fingerprint recognition and identification chip described in Ref. [46]. Various successful smart image sensors have been demonstrated that are capable of carrying out certain important, but still only basic functions for the vision process on a single chip, see for example [47]. The suspicion that "vision is difficult [3]" has been fully verified, and it has become obvious that the early expectations of monolithically integrated single-chip vision systems were too high. As demonstrated for example by the fingerprint verification chip [46] it is possible today to co-integrate an image sensor and all the necessary processing circuitry on a single chip for the solution of a given - still not too complex - machine vision problem. However, this would be far removed from the original idea of a seeing chip which visually perceives some aspects of its surroundings, and in most cases it would make no economical sense.

The basic philosophy behind the seeing chip, as formulated very clearly for example in References [22] and [29], is to distribute the processing power over the photosensitve part. This strategy is inspired by the biological concept of highly parallel, low-speed and low power distributed analog computing, which is the basis of nature's marvellous visual perceptive systems, such as our own highly-developed sense of vision. In contrast to the planar, essentially two-dimensional semiconductor fabrication technology, nature realizes fully three-dimensional processing systems, in which each "pixel" is backed by a tremendous number of nerve cells (more than 10^5 in the human visual system [48]) performing the necessary calculation for the sense of vision. In the near future, it is unrealistic to expect that each pixel on a solid-state image sensor will contain more than a few ten transistors, while maintaining a useful pixel size of the order of 30x30 mm^2 and a reasonable fraction of at least 10% of the pixel area being photosensitive.

As a consequence, recent developments in the area of integrated machine vision also consider architectures based on different planes: an image acquisition plane might be followed by several (analog) preprocessing planes, an (essentially digital) classification plane and an output plane, all connected using suitable high-bandwidth bus schemes with an appropriate software protocol. This guarantees a maximum fill factor for the image sensing part and allows to use optimal architectures and technologies for the different parts of the complete system. Such an approach does not necessarily mean that every plane resides on its own chip; different planes can be integrated on the same chip, as envisaged for example in the feature extractor imager described in [49]. The technology for stacking and interconnecting silicon chips, so called 3D or z-plane technology, has been developed [50], but the appealing idea of a low-cost single-chip vision system, a seeing chip, becomes seriously compromised.

The conclusion is that smart image sensors - offering additional on-chip functionality - and integrated vision systems are certainly trends that will lead to a wide range of practical products, albeit rarely in the form of single, self-contained seeing chips. It can rather be expected that smart image sensors with extended capabilities for the dynamic acquisition of images will be part of an integrated vision system. This will consist of an economically sensible combination of imager, analog and digital processing parts. Special properties built into such smart image sensors include lower noise, higher D/R, programmable sensitivity, on-chip non-uniformity and shading correction, variable exposure and timing control, region-of-interest capability, dynamic pixel size and shape, on-chip image pre-processing which can be carried out for all pixels in parallel, etc. It might well be that "seeing chip" is a misnomer, and that the silicon retina [29], raising less exaggerated expectations and suggesting more of a front-end image acquisition/pre-processing module, is a much more appropriate name for the current and future development directions in the field of integrated image acquisition and processing systems.

7. Conclusions and summary

It was only about 10 years ago that a few researchers started to exploit one the exciting capabilities offered by silicon technology: The monolithic integration of photosensitive and analog/digital processing circuits. The present paper has described some of the results of these efforts. Many other examples of this young but quickly evolving field of research and development exist, and they all underline the point to be made: Today's image sensors are not restricted anymore simply to the acquisition of optical image signals; modern image sensors can be supplied with monolithically co-integrated analog and digital electronic circuitry, giving them increased functionality for the solution of certain ranges of optical measurement problems. As we have seen, it is not always optimal to add this required custom functionality in the form of smart pixels; an increase in functionality is often coupled with a larger proportion of a pixel's surface being used for electronic circuitry, leaving only a small part for the photosensitive areas. Each new optical measurement problem has to be inspected, therefore, what the optimum partitioning is, regarding technical and economical issues. One has to consider not only smart pixels but also on-chip functionality in a monlithically co-integrated processor part, adjacent to an image sensing area; this might be complemented with functionality that is supplied with off-chip, perhaps commercially available electronic components.

In summary, 25 years after the demonstration of the first CCD image sensor [1], image sensing has come of age, not only concerning the ever-increasing size and dropping price of image sensors, but primarily because today image sensing can be regarded as being part of solving an optical measurement problem. A wide range of integrated building blocks exists or can be custom-developed, with which an optimum solution with just the needed functionality can be realized. This functionality can now be introduced exactly at the place in the image signal acquisition and processing chain where it makes the most sense in terms of technical or economical requirements.

8. References

[1] M.F. Tompsett, G.F. Amelio, W.J. Bertram, R.R. Buckley, W.J. McNamara, J.C. Mikkelsen jr. and D.A. Sealer,"Charge-Coupled Imaging Devices: Experimental Results", *IEEE Trans. Electr. Dev.*, Vol. 18, 992-996 (1971).

[2] P. Suni, "CCD Technology at Orbit Semiconductor Inc.", *International Conference on.Scientific Optical Imaging*, Georgetown, Grand Cayman Islands, Dec. 2-6, 1992.

[3] C. Koch, "Seeing Chips: Analog VLSI Circuits for Computer Vision", *Neural Computation*, Vol. 1, 184-200 (1989).

[4] "Technology Roadmap for Products and Systems", BPA Technology Management Ltd., BPA House, 250-256 High Street, Dorking, UK - Surrey RH4-1QT. Tel. +44-1306-875-500, FAX +44-1306-888-179.

[5] A.J.P. Theuwissen, personal communications, Philips Imaging Technology, Nederlandse Philips Bedrijven B.V., Prof. Holstlaan 4, WAG14, NL - 5656 AA Eindhoven, The Netherlands. Tel. +31-4074-2734, FAX +31-4074-3390.

[6] Loral Fairchild Imaging Sensors, 1801 McCarthy Blvd., Milpitas, CA 95035, USA; Tel. (408) 433-2500, FAX (408) 433-2508.

[7] B. Bosiers et.al., "A True Progressive Scan 640×480 FT-CCD For Multimedia Applications", *Proc. IEDM'94*, Dec. 11-15, 1994.

[8] E. Hecht, "Optics", 2nd edition, Addison-Wesley (1987).

[9] E.R. Fossum: "Active pixel sensors (APS) - Are CCDs dinosaurs ?", *Proc. SPIE*, Vol. 1900, 2-14 (1992).

[10] K. Knop and P. Seitz, "Image Sensors", in *Sensors Update*, (Eds. H. Baltes, W. Göpel and J. Hesse), VCH-Verlag, Weinheim, *in print* (1995).

[11] F.M. Klaassen and J. Prins, "Thermal noise of MOS transistors", *Philips Research Reports*, Vol. 22, 505-514 (1967).

[12] R.H. Nixon, S.E. Kemeny, C.O. Staller and E.R. Fossum, "128×128 CMOS Photodiode-Type Active Pixel Sensor with On-Chip Timing, Control and Signal Chain Electronics", *Proc. SPIE*, Vol. 2415, 117-123 (1995).

[13] J. Janesick, T. Elliott, A. Dingizian, R. Bredthauer, C. Chandler, J. Westphal and J. Gunn, "New advancements in charge-coupled device technology - sub-electron noise and 4096x4096 pixel CCDs", *Proc. SPIE*, Vol. 1242, 223-237 (1990).

[14] J. Kramer, "Photo-ASICs : Integrated Optical Metrology Systems with Industrial CMOS technology", Ph.D. thesis No. 10186, Federal Institute of Technology (ETH), Zurich (1993).

[15] Y. Matsunaga, H. Yamashita and S. Ohsawa, "A Highly Sensitive On-Chip Charge Detector for CCD Area Image Sensor", *IEEE J. Solid State Circ.*, Vol. 26, 652-656 (1991).

[16] W. Budde, "Multidecade linearity measurements on Si photodiodes", *Applied Optics*, Vol. 18, 1555-1558 (1979).

[17] J. Kramer, P. Seitz and H. Baltes, IEEE Quantum Electronics, "Planar Distance and Velocity Sensor", *IEEE J. Quantum Electr.*, Vol. 30, 2726-2730 (1994).

[18] P. Seitz, T. Spirig, O. Vietze and K. Engelhardt, "Smart sensing using custom photo-ASICs and CCD technology", *Optical Engineering*, to appear in August 1995.

[19] P. Seitz, D. Leipold, J. Kramer and J.M. Raynor, "Smart optical and image sensors fabricated with industrial CMOS/CCD semiconductor processes", *Proc. SPIE*, Vol. 1900, 21-30 (1993).

[20] Q. Zhu, H. Stiebig, P. Rieve, H. Fischer and M. Böhm, "A Novel *a*-Si(C):H Color Sensor Array", *Proc. MRS Spring Meeting*, Symposium A - Amorphous Silicon Technology, San Francisco, April 4-8, 1994.

[21] B. Pain, S.K. Mendis, R.C. Schober, R.H. Nixon and E.R. Fossum, "Low-power low-noise analog circuits for on-focal-plane signal processing of infrared sensors", *Proc. SPIE*, Vol. 1946, 365-374 (1993).

[22] C.A. Mead, "Analog VLSI and Neural Systems", Addison-Wesley, Reading (1989).

[23] J.D.E. Beynon and D.R. Lamg, "Charge-coupled devices and their applications", McGraw Hill, London (1980).

[24] S. Chen and R. Ginosar, "Adaptive Sensitivity CCD Image Sensor", *Proc. SPIE*, Vol. 2415, 303-310 (1995).

[25] E.R. Fossum and B. Pain, "Infrared Readout Electronics for Space Science Sensors: State of the Art and Future Directions", *Proc. SPIE*, Vol. 2020, 262-283 (1993).

[26] T. Spirig, P. Seitz, O. Vietze and F. Heitger, "Lock-in CCD", *IEEE J. Quantum Electronics*, in print (1995).

[27] M.L. Meade, "Lock-in amplifiers : principles and applications", Peregrinus, London (1983).

[28] B.C. Doody, S.G. Chamberlain and W.D. Washurak, "A high photosensitivity wid dynamic range silicon linear image sensor array", *Advance Printing of Paper Summaries, Electronic Imaging '87*, Institute for Graphic Communication Inc. (pub.), 254-259 (1987).

[29] M. Mahowald and C.A. Mead, "The Silicon Retina", *Scientific American*, May '91, 40-46 (1991).

[30] E-S. Eid, A.G. Dickinson, D.A. Inglis, B.D. Ackland and E.R. Fossum, "A 256x256 CMOS Active Pixel Sensor", *Proc. SPIE*, Vol. 2415, 265-275 (1995).

[31] C. Claeys, I. Debusschere, N. Ricquier, P. Seitz, M. Stalder, J.M. Raynor, G.K. Lang, G. Cilia, C. Cavanna, U. Müssigmann and A. Abele, "An active machine vision system for surface quality inspection", *Proc. SPIE*, Vol. 2183, 205-213 (1994).

[32] C. Séquin and M.F. Tompsett, "Charge Transfer Devices", Academic Press, New York (1975).

[33] P. Seitz, T. Spirig, O. Vietze and P. Metzler, "Lock-in CCD and the convolver CCD : Applications of exposure-concurrent photocharge transfer in optical metrology and machine vision", *Proc. SPIE*, Vol. 2415, 276-284 (1995).

[34] D.J. Sauer, F.L. Hsueh, F.V. Shallcross, G.M. Meray, P.A. Levine, G.W. Hughes and J. Pellegrino, "High Fill-Factor CCD Imager with Highr Frame-Rate Readout", *Proc. SPIE*, Vol. 1291, 174-184 (1990).

[35] E. R. Fossum, "Architectures for focal plane image processing", *Optical Engineering*, Vol. 28, 865-871 (1989).

[36] P. Aubert, H.J. Oguey and R. Vuillemier, "Monolithic optical position encoder with on-chip photodiodes", *IEEE J. Solid State Circ.*, Vol. 23, 465-473 (1988).

[37] D. Renshaw, P.B. Denyer, G. Wang and M. Lu, "ASIC Vision", *Proc. of the IEEE 1990 Custom Integrated Circuits Conf.*, 7.3.1-7.3.4 (1990).

[38] VLSI Vision Limited, Aviation House, 31 Pinkhill, UK - Edinburg EH12 7BF. Tel. (031) 539-7111, FAX (031) 539-7141.

[39] Integrated Vision Products AB, Teknikringen 1, S-58330 Linköping, Tel. +46-13-21-4160, FAX +46-13-21-3724.

[40] J. Kramer, P. Seitz and H. Baltes, "Inexpensive range camera operating at video speed", *Applied Optics*, Vol. 32, 2323-2330 (1993).

[41] A. Kawasaki, M. Goto, H. Yashiro and H. Ozaki, "An array-type PSD (position-sensitive detector) for light pattern measurement", *Sensors and Actuators A*, Vol. A21-23, 529-533 (1990).

[42] A. Gruss, L.R. Carley and T. Kanade, "Integrated Sensor and Range-Finding Analog Signal Processor", *IEEE J. Solid State Circ.*, Vol. 26, 184-192 (1991).

[43] J.M. Hakkarainen, J.J. Little, H. Lee and J.L. Wyatt, "Interaction of algorithm and implementation for analog VLSI stereo vision", *Proc. SPIE*, Vol. 1473, 173-184 (1991).

[44] R. Dändliker, K. Hug, J. Politch and E. Zimmermann, "High Accuracy Distance Measurements with Multiple Wavelength Interferometry", *Optical Engineering*, to appear in August 1995.

[45] J.A. Izatt, M.R. Hee, D. Huang, E.A. Swanson, C.P. Lin, J.S. Schuman, C.A. Puliafito and J.G. Fujimoto, "Micron-Resolution Biomedical Imaging with Optical Coherence Tomography", *Optics & Photonics News*, October 1993, 14-19 (1993).

[46] P.B. Denyer, D. Renshaw and S.G. Smith, "Intelligent CMOS Imaging", *Proc. SPIE*, Vol. 2415, 285-291 (1995).

[47] C. Koch, "Implementing early vision algorithms in analog hardware - An overview", *Proc. SPIE*, Vol. 1473, 2-15 (1991).

[48] D.H. Hubel, "Eye, Brain and Vision", Scientific American Library, New York (1988).

[49] J. Tanner and J. Luo, "A single chip imager and feature extractor", *Proc. SPIE*, Vol. 1473, 76-87 (1991).

[50] J.E. Carson (Ed.), "Materials, Devices, Techniques and Applications for Z-Plane Focal Plane Array Technology", *Proc. SPIE*, Vol. 1097 (1989).

Holographic Modal Analysis

Dr. R. Freymann[1], W. Honsberg[1], F. Winter[1], Dr. H. Steinbichler[2])

[1] BMW AG, München

[2] Labor Dr. Steinbichler, Neubeuern

Summary

State of the art experimental modal analysis investigations in the structural area require the determination of the dynamic response of the structural system to an external excitation. Normally this information is generated from the frequency response of accelerometers distributed all over the structure.

A newly developed technique now allows the experimental determination of the dynamic structural frequency response by making use of the holographic interferometry. The main advantages of this technique are the realization of a non-contact (optical) measuring system as well as the nearly unlimited resolution with regard to the density of the measuring points.

A description of this newly developed modal analysis technique will be given. Results from practical applications will be shown.

1. Introduction

Nowadays structural analysis and optimization procedures are basic tools applied in the development process of vehicle constructions as well as in many other engineering areas. Thereby experimental techniques, such as modal analysis [1] and holography [2], are widely used for determination of the vibrational behavior of structural systems.

Referring to automotive applications, modal analysis techniques are primarily used for identification of the modal parameters of beam-shaped structural systems (e.g. car body frame) in the lower frequency range, whereas holographic measurements are mostly performed to investigate the dynamic behavior of flat shaped structural parts (e.g. floor panel structure of a vehicle) in the higher frequency range.

Apart from the differences with regard to the area of application of these measuring techniques to differently shaped structural systems and to different frequency ranges, there exists a significant difference in the information obtained from both measurements. State of the art modal analysis investigations, based on the experimental determination of the frequency response of a structural system by means of accelerometers and the subsequent processing of these data in a computer system [3, 4], allow the determination of the so-called modal parameters in a defined (lower) frequency range. The complete set of modal parameters is defined by the eigenfrequencies, the normal mode shapes (eigenmodes), the generalized masses and the (modal) damping coefficients of the structural system being investigated. These parameters are fundamental with regard to the performance of numerical structural investigations such as finite-element-model updating [5], structural optimization [6], dynamic response [7] and structural-acoustic coupling calculations [8].

It is a well-known fact that these very relevant modal parameters cannot be provided at the present moment when applying holographic measuring techniques. All we can identify from such investigations is the dynamic response of the structure to an external excitation, depicted in form of interferograms of the structural displacements. The lack for providing the structural modal data must be regarded as a large disadvantage of holographic investigations in the area of structural dynamic applications.

The scope of this paper is to describe a newly elaborated experimental-numerical approach allowing the determination of the structural modal parameters when making use of holographic measuring techniques. In the following we will focus on the basics of this new approach, its practical application to a real structural system and the results obtained therefrom.

2. Normal mode shape or dynamic structural response?

Exciting a structural system at a defined circular frequency ω we will excite all of its normal mode shapes. Thus, we can formulate, according to [1], the dynamic response vector x of the structural deformations as follows:

$$(1) \qquad x \cdot e^{i\omega t} = \sum_{r=1}^{N} \frac{Q_r}{K_r} \cdot \frac{\left(1-\eta_r^2\right)-i \cdot 2\vartheta_r \cdot \eta_r}{\left(1-\eta_r^2\right)^2+4\vartheta_r^2\eta_r^2} \cdot \phi_r \cdot e^{i\omega t} \;,$$

with
ϕ_r as the vector of the displacements in the r-th eigenmode,
Q_r as the generalized excitation force with regard to the r-th eigenmode,
K_r as the generalized stiffness related to the r-th eigenmode,
ϑ_r as the modal damping coefficient of eigenmode r,
$\eta_r = \omega/\omega_r$ as a dimensionless frequency factor, ω_r being the circular eigenfrequency of the r-th eigenmode.

Eq. (1) underlines that it is normally impossible to excite just one single of the N eigenmodes under consideration, even if the external excitation occurs at the eigenfrequency of a defined eigenmode. This especially applies for more complex structures featuring a high modal density and thus a significant modal coupling [9]. This statement entails that when applying holographic measuring techniques there can only be a determination of the dynamic response of the structural system, say of a state of structural vibrations which must be considered as the superposition of an undefined number of normal mode shapes (eigenmodes).

532

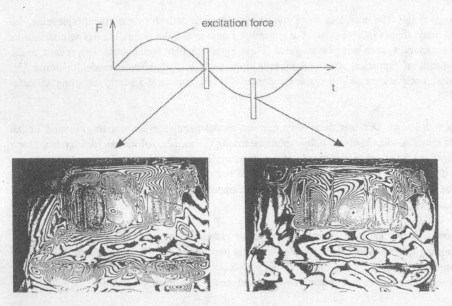

Fig. 1: Interferograms obtained at different timing points

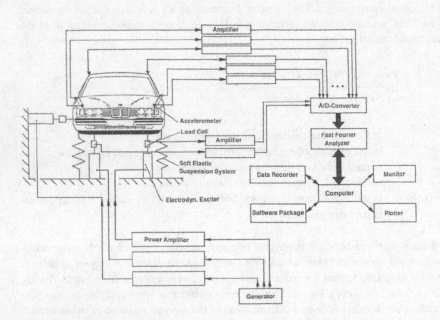

Fig. 2: Conventional modal analysis test setup configuration

Since, according to Eq. (1), this superposition is achieved in a complex form, the dynamic response obtained from the experiment is depending on the instant the "snapshot" is taken in case of a harmonic excitation of the structural system. Fig. 1 clearly depicts this situation. The two interferograms were produced at different instants, say with different phase shifts compared to the harmonic external structural excitation force as a reference signal.

Looking at the two interferograms it is difficult to interprete which one is depicting "the correct state of deformations". From the point of view of a modal consideration, none of these snapshots is delivering the thus required information.

3. Determination of the frequency response behavior of the structural system

The foregoing reflections indicate that a detailed structural analysis does require more than the questionable interpretation of snapshot-interferograms taken at arbitrary timing points and frequencies of the excitation force. In order to perform a thorough modal analysis investigation it is required to determine the complete frequency response of the structural system to a defined external excitation. This is in agreement with conventional modal analysis procedures where the frequency response behavior is determined by means of accelerometers distributed all over the structural system (Fig. 2). As will be shown in the following, the holographic measuring technique can also provide this information.

For the performance of the testing, we use the setup, as depicted in Fig. 3. Let's assume that the object (the structural system) is excited by a harmonic external force F with circular frequency ω. In order to generate the frequency response with regard to the amplitude and the phase of the different object points, holographic snapshots are taken at two timing points t_1 and t_2 corresponding to two different states of deformation of the structure (Fig. 4). Making use of an ESPI or TV holography setup [10, 11] allows an easy determination of the interferograms related to the timing points t_1 and t_2, respectively. From these interferograms, featuring the structural deformations $x(t_1)$ and $x(t_2)$ at normally 512 x 512 grid points, the real and imaginary components of the deformations at a given point P with reference to the excitation force can be formulated as follows [12]:

$$(2) \quad \mathrm{Re}(x_P(\omega)) = \frac{x_P(t_2)\sin(\omega t_1) - x_P(t_1)\sin(\omega t_2)}{\sin(\omega t_1)\cos(\omega t_2) - \sin(\omega t_2)\cos(\omega t_1)}$$

$$(3) \quad \mathrm{Im}(x_P(\omega)) = \frac{x_P(t_1)\cos(\omega t_2) - x_P(t_2)\cos(\omega t_1)}{\sin(\omega t_1)\cos(\omega t_2) - \sin(\omega t_1)\cos(\omega t_1)}.$$

Performing the above mentioned investigations at a number of different frequencies of the excitation force within a defined frequency range we are interested in, allows to determine the frequency response of the structural deformations. Fig. 5 depicts the procedure of generating the frequency response data of a structural point P from interferograms produced at different frequency steps. It is well understood that the frequency response data are determined for all of

534

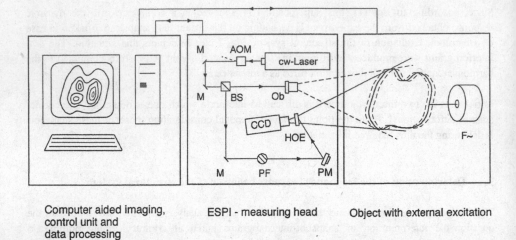

Computer aided imaging,
control unit and
data processing

ESPI - measuring head

Object with external excitation

Fig. 3: Holographic test setup

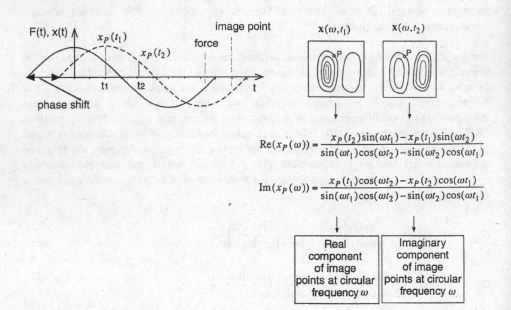

Fig. 4: Determination of the real and imaginary components of structural deformations

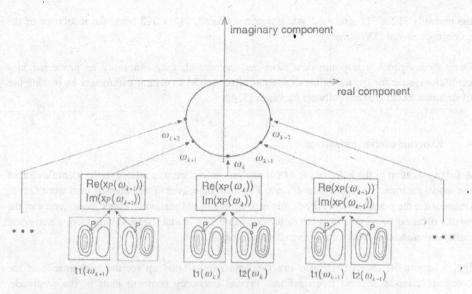

Fig. 5: Determination of the structural frequency response

Fig. 6: Flat plate system test setup

the normally 512 x 512 structural grid points considered, 512 x 512 being the resolution of the recording material (TV-camera).

Once these frequency response data have been generated, they can easily be processed in a computer system for the numerical evaluation of the modal structural parameters by making use of common modal analysis software packages [3, 4].

4. Experimental investigations

A first application of the holographic modal analysis procedure was related to the identification of the modal parameters of a flat plate system. The main objectives of this investigation were first to demonstrate the practical feasibility of this new approach and second to check the accuracy of the results obtained by comparing them to the corresponding modal data resulting from a numerical finite-element analysis and a conventional modal analysis test.

Fig. 6 depicts the object (flat plate structural system) as build up for the performance of the holographic modal analysis investigations. Typical frequency response plots of the amplitude, related to one structural point P, are plotted in Fig. 7, one plot being related to the experimental result generated by the numerical evaluation of interferograms according to Eqs. (2) and (3), the other one resulting from the numerical (least square error) approximation achieved by the modal analysis software. In this context it has to be mentioned that the numerical determination of the modal data is achieved on basis of the approximated frequency response curves.

Typical results obtained from the modal holographic investigations are outlined in Tab. 1 and depicted in Figs. 8 and 9. The plots of Fig. 9 indicate the "almost unlimited" resolution with regard to the display of normal mode shapes, due to the possibility of considering up to 512 x 512 = 262144 structural grid points.

Another test was performed on the same structure by applying conventional modal analysis techniques. For determination of the structural frequency response, 25 accelerometers were fixed to the plate system (Fig. 10). Typical results obtained from this investigation are plotted in Fig. 11. There is no need to point out that, due to the very limited number of grid points which can be considered, the interpretation of normal mode shapes with numbers higher than 4 must be considered as "insufficient".

Moreover a rather large deviation between the results obtained from the holographic and conventional modal analysis investigations was notified (Tab. 1). Further more detailed experiments revealed that the dynamic behavior of the structural system was significantly influenced by the sensoric, say by the mass of the accelerometers installed as well as by the mass/stiffness properties of their wiring. This effect points out another advantage of the holographic measuring equipment consisting in the non-contact determination of the structural deformations, thus not effecting the dynamics of the elastic structure.

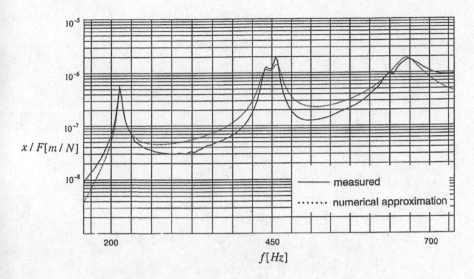

Fig. 7: Frequency response plots

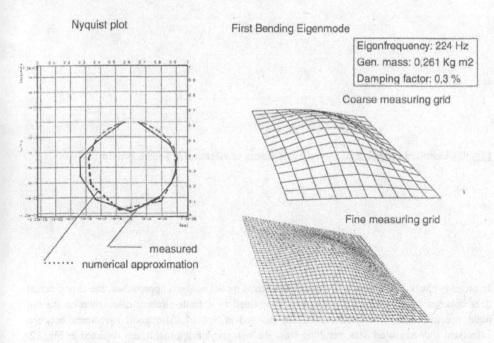

Nyquist plot

First Bending Eigenmode

Eigenfrequency: 224 Hz
Gen. mass: 0,261 Kg m2
Damping factor: 0,3 %

Coarse measuring grid

Fine measuring grid

—— measured
······· numerical approximation

Fig. 8: Typical results of a holographic modal analysis

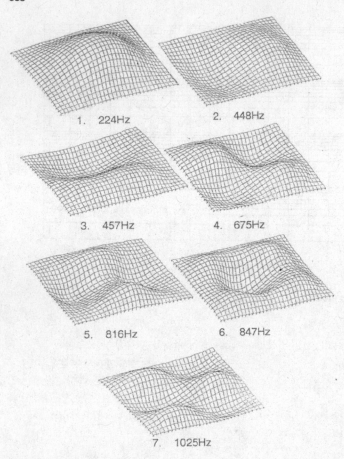

1. 224Hz
2. 448Hz
3. 457Hz
4. 675Hz
5. 816Hz
6. 847Hz
7. 1025Hz

Fig. 9: Identified fundamental natural mode shapes of vibration of a plate system

In order to check the accuracy of both experimental modal analysis approaches, the experimental data obtained were compared to the results provided by a finite-element calculation on the flat plate structure. The calculated results are outlined in Tab. 1. The good agreement between calculated and measured data, resulting from the holographic approach, are depicted in Fig. 12. The graph indicates an average deviation of 2 % between calculated and "holographically measured" eigenfrequencies whereas the deviation between these values is - for each value - larger than 15 % when comparing numerically obtained and "conventional modal analysis" results to one another.

Fig 10: Conventional modal analysis test setup

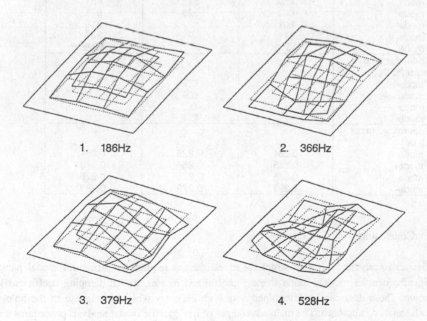

1.   186Hz

2.   366Hz

3.   379Hz

4.   528Hz

Fig. 11: Fundamental eigenmodes of the plate system

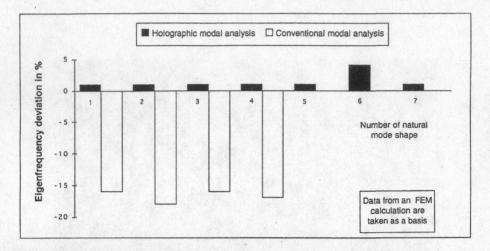

Fig. 12: Accuracy of results

Tab. 1: Modal data of the plate system

	Finite - element calculation	Conventional modal analysis	Holographic modal analysis
Eigenfrequency [Hz]			
1 mode	221	186	224
2 mode	450	366	448
3 mode	450	379	457
4 mode	663	528	675
5 mode	807	-	816
6 mode	811	-	847
7 mode	1012	-	1025
Modal damping factor [% D_{krit}]			
1 mode	-	0,7	0,3
2 mode	-	0,9	0,3
3 mode	-	0,7	0,4
4 mode	-	2,4	0,5
Generalized mass [kg m^2]			
1 mode	0,284	0,229	0,261
2 mode	0,285	0,254	0,273
3 mode	0,285	0,275	0,276
4 mode	0,293	0,279	0,283

5. Conclusion

Advanced numerical dynamic structural investigations require the structural modal parameters (eigenfrequencies, normal mode shapes, generalized masses, modal damping coefficients) to be known. These data can be determined with high accuracy when making use of the holographic modal analysis approach. The main advantages of holographic modal analysis procedures are

- the non-contact measuring technique not effecting the dynamic behavior of (light-weight) structural systems,
- the detailed representation of normal mode shapes, with up to 250000 grid points, due to the high resolution of the recording material.

Moreover, the practical application of holographic modal analysis procedures is simplified if the measured data are produced in a form which allows their further processing by standard modal analysis software packages.

6. References

[1] Freymann, R.
Die experimentelle Modalanalyse, ein Mittel zum Verifizieren von strukturdynamischen Rechenmodellen
in: Hrsg. H. P. Willumeit: Computergestützte Berechnungsverfahren in der Fahrzeugdynamik, VDI-Verlag, Düsseldorf, 1991, S. 305 - 339

[2] Roesener, K.; Wagner, S.; Jüptner, D.
Holografische Untersuchungen an großflächigen CFK-Bauteilen
in: Hrsg. W. Waidelich: Laser/Optoelektronik in der Technik. Springer Verlag, Berlin/ Heidelberg/New York/London/Paris/Tokyo/Hong Kong, 1990, S. 327-331

[3] LMS CADA-X User Manual
Revision 3.3, LMS International, Leuven, 1994

[4] I-DEAS Master Series Test and Measurement Package User Manual
V. 1.3, Structural Dynamics Research Corp. (SDRC), Milford (Ohio), 1994

[5] Niedbal, N.; Klusowski, E.
Die Verknüpfung strukturdynamischer Rechenmodelle mit gemessenen Eigenschwingungs-Kenngrößen.
Zeitschrift für Flugwiss. und Weltraumf. (ZFW), Band 12, Heft 2, 1988, S. 99-110

[6] Freymann, R.; Stryczek, R.
Dynamische strukturelle Optimierung von Fahrzeugstrukturen
VDI-Berichte Nr. 816, 1990, S. 745-755

[7] Försching, H.
Grundlagen der Aeroelastik
Springer-Verlag, Berlin/Heidelberg/New York, 1974

[8] Freymann, R.
An Energetic Approach for Derivation of the Generalized Equations of Motion of Coupled Structural-Acoustic Systems
American Inst. of Aeron. and Astron., AIAA-CP-942 (1994), Part 4, S. 1868-1880

[9] Zaveri, K.
Modal Analysis of Large Structures
Brüel & Kjaer Doc. (1985), Naerum

[10] Creath, K.
Speckle: Signal or Noise?
in: Hrsg. W. Jüptner, W. Osten: Physical Research, Akademie Verlag, Berlin, 1993, S. 97-102

[11] Applikationslabor für opto-elektronische Meß- und Prüftechnik
Labor Dr. Steinbichler, Neubeuern, 1993

[12] Winter, F.
Verknüpfung von holographischer Meßtechnik und Modalanalyse in der Strukturdynamik
Diplomarbeit an der Fakultät für Physik der Technischen Hochschule München, 1994

Topometrie mittels schneller Datenaufnahme

R. Windecker, H.J. Tiziani
Universität Stuttgart, Institut für Technische Optik
Pfaffenwaldring 9
D-70559 Stuttgart

Kurzfassung

Es wird ein Verfahren vorgestellt, bei dem einige Probleme statischer Streifenauswert-metho-den vermieden werden. Für die Phasenauswertung werden nur zwei Bilder benötigt, die mit herkömmlichen CCD-Kameras innerhalb von 80ms erfaßt werden. Dadurch wird es möglich dieses Verfahren ebenfalls an Meßobjekten anzuwenden, die eine schnelle, robuste aber gleichzeitig auch eine genaue Analyse erfordern. Als Beispiel wird eine Anwendung der Streifenprojektion zur Messung der Hornhauttopometrie am menschlichen Auge vorgestellt.

Einleitung

Die Analyse von kodierten Intensitätsbildern spielt in der modernen Oberflächenmeßtechnik eine wichtige Rolle [1]. Viele Anwendungsbeispiele im industriellen Umfeld aber auch in Medizin verlangen möglichst kurze Datenerfassungszeiten, insbesondere wenn es sich um bewegte Strukturen handelt. In den vergangenen Jahren wurde eine Vielzahl an Meßverfahren und unterschiedlicher Auswertalgorithmen entwickelt und zur Topografiemessung vorge-schlagen. Es kristallisierten sich hierbei bei der Streifenanalyse zwei Richtungen heraus [2]. Eine Methode extrahiert die Höheninformation aus einer Serie von Intensitätsbildern pixel-weise (**dynamische Algorithmen**), die andere Methode gewinnt die Informationen aus einer einzigen Aufnahme (**statische Algorithmen**). Gerade die zweite Methode, die immer dann eingesetzt wird, wenn es sich um zeitkritische Anwendungsfälle handelt, besitzt an schwieri-gen Oberflächen erhebliche Nachteile. Sie führt zu Phasenfehlern, wenn die Probe z.B. eine inhomogene lokale Reflektivität oder große Oberflächensteigungen aufweist [3]. In solchen Fällen ermöglicht u.a. eine Kombination aus beiden Verfahren eine vorteilhafte Synthese, die positive Eigenschaften beider Methoden in sich vereinigt.

Theorie

Im Folgenden wird ein Trägerfrequenz-Streifenmuster angenommen, wie es zum Beispiel bei der Streifenprojektion auftritt. Die lokale Intensitätsverteilung kann mathematisch in der folgenden Form allgemein beschrieben:

$$I(x,y) = I_0(x,y)\{1 + V(x,y)F[f_x, x, \Phi(x,y)]\}.$$

(1)

Die Funktion $F[f_x, x, \Phi(x,y)]$ ist für $\Phi(x,y) = $ konst. eine periodische Funktion, die das Profil der Intensitätsstreifen beschreibt. Sie läßt sich für diesen Fall i.a. in Form einer Fourierreihe

$$F[f_x, x, \Phi(x,y)] = \sum_{i=0}^{\infty} a_i \cdot \cos[i \cdot [2\pi f_x x + \Phi(x,y)] + \varphi_i] \qquad (2)$$

schreiben. Die Terme $I_0(x,y)$ und $V(x,y)$ geben die lokale mittlere Intensität und den Modulationsgrad wider. Aufgrund der optischen Übertragungsfunktion der Optiken und der CCD-Kamera werden jedoch die höheren Ordnungen der Reihe nicht oder nur teilweise übertragen.

Durch eine geeignete Binärkodierung (Bild 1) in Verbindung mit den Übertragungseigenschaften einer dafür ausgelegten Optik, der CCD-Kamera und der in diesem Artikel beschriebenen Datenaufnahme kann bereits das Spektrum ab der dritten Oberwelle bei projizierten Streifen unterdrückt werden.

Bild 1: Binär kodiertes Sinusmuster für Streifenprojektionsverfahren

Eine statische Methode zur Auswertung solcher Intensitätsbilder findet sich in einer lokalen Fourier-Analyse der Muster [4]. Bei diesem Algorithmus wird nur ein Bild benötigt. Allerdings werden an die Intensitätsverteilung Voraussetzungen gestellt, die in der Praxis nur schwer einzuhalten sind. Die Terme $\Phi(x,y)$, $I_0(x,y)$ und $V(x,y)$ dürfen sich innerhalb einer Streifenperiode kaum ändern, andernfalls ergeben sich systematische Fehler. Gerade an stark gekrümmten Flächen (wie bei der Hornhaut des Auges) ist das allerdings kaum einzuhalten, da hier die Phase $\Phi(x,y)$ sich rasch verändern kann.

Durch Einführung weiterer Intensitätswerte pro Pixel läßt sich das Verfahren deutlich verbessern, indem die additiven und multiplikativen Terme durch pixelweise Subtraktion bzw. Normierung mit Hilfe des Streifenkontrastes herausgefiltert werden [3], was zu einem erweiterten Fourier-Analyse-Verfahren führt, das sehr tolerant auf statistisches Rauschen (von der Kamera bzw. Frame Grabber), Phasenschiebefehler und höhere Ordnungen der Streifenmusters reagiert.

Datenaufnahme

Um eine effektive Datenaufnahme zu ermöglichen, bietet es sich an, die Phase des Intensitätsmusters kontinuierlich mit einer konstanten Geschwindigkeit zu schieben. Viele CCD-Kameras arbeitet im Interlaced-Modus. Hierbei werden innerhalb von 40ms zwei Halbbilder aufgenommen. Ein Halbbild wird durch die geraden und eins durch die ungeraden Zeilen gebildet. Ist die Geschwindigkeit der Phasenschiebung gerade so gewählt, daß innerhalb eines Kamerabildes ($t = \dfrac{1}{f_{CCD}}$) das Streifenmuster um eine halbe Periode p verschoben wird

$$\frac{\partial \varphi}{\partial t} = \frac{p \cdot f_{CCD}}{2},$$

so beträgt die Phasenschiebung zwischen den Halbbildern 90°. Innerhalb von zwei Bildern bekommen wir also eine Phasenschiebung über eine volle Streifenperiode, wobei bei jedem Halbbild über eine Phase von annähernd 90° integriert wird. Bild 2 zeigt ein Streifenmuster, das an einer ebenen Fläche aufgenommen und aus zwei Bildern zusammengesetzt wurde. Jeweils zwei Zeilen gehören abwechselnd zur ersten oder zur zweiten Aufnahme. Unter Halbierung der lateralen Auflösung in einer Koordinate läßt sich somit die Datenaufnahmezeit ebenfalls halbieren.

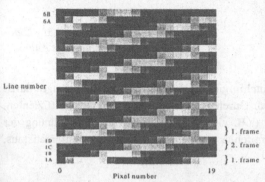

Bild 2: Darstellung des Streifenmusters, das sich ergibt, wenn jeweils zwei Zeilen zweier aufeinanderfolgender Aufnahmen zusammengefaßt werden. Man erkennt deutlich die lineare Phasenschiebung von jeweils 90°

In Bild 3 ist eine Serie von zwei aufeinanderfolgenden Kamerabildern gezeigt, die *in vivo* an einer Hornhaut eines Auges aufgenommen wurden. In der Darstellung wurden jeweils die Halbbilder getrennt nebeneinander angeordnet. Bei diesem Anwendungsbeispiel ist eine schnelle Datenerfassung aufgrund der permanenten Augenbewegungen essentiell notwendig.

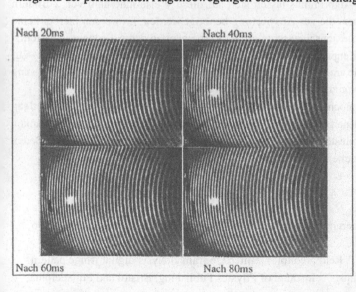

Bild 3: Intensitätsbilder am Auge. Die oberen beiden Bilder sind das erste und das zweite Halbbild im InterlacedModus und die unteren Bilder zeigen das dritte und das vierte Halbbild. Zwischen jedem Bild wurde die Phase um 90° verschoben.

Meßbeispiele

In Bild 4 ist die Topografie einer *in vivo*-Messung an einer menschlichen Hornhaut gezeigt. Das Objektfeld beträgt 8mm×8mm. Durch den mittleren Krümmungsradius von 7,9mm ergeben sich gerade im Randbereich große Oberflächensteigungen von etwa 30°. Die longitudinale Auflösung, die mit dem erweiterten Fourier-Algorithmus erzielt werden, liegt bei 1μm.

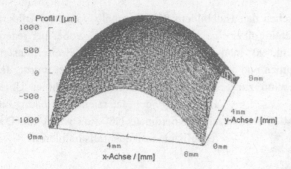

Bild 4: In vivo gemessene 3D-Topografie an der Hornhaut des Auges

Eine Anwendung findet sich in der Hornhautchirugie, bei der die Brechkraft des Auges durch Formänderung der Hornhaut korrigiert wird. Durch Berechnung von Zernike-Koeffizienten, dem Vergleich mit einem Referenzkörper (z.B. einer Sphäre) oder durch Berechnung der lokalen Krümmungsradien lassen sich Formfehler der Hornhaut, wie z.B. Astigmatismus, visualisieren.

Zusammenfassung

Vorgestellt wurde eine Methode, mit der sehr schnell Meßdaten aufgenommen werden können. Unter Verwendung eines 4 Phasen-Algorithmuses wie zum Beispiel dem modifizierten statischen Fourier-Analyse-Algorithmus lassen sich auf diese Weise innerhalb von 2 Bildern die Informationen aufnehmen, die zur Auswertung benötigt werden. Es werden hierbei keine Voraussetzungen an die Reflexionseigenschaften der Probe oder an deren Topografie gestellt.

Unter Einsatz eines Streifenprojektionsaufbaus konnte der Einsatz des Verfahrens zur hochaufgelösten Topometrie am menschlichen Auge demonstriert werden. Die rms-Auflösung liegt bei einem Objektfeld von mehr als 8mm×8mm bei etwa 1µm.

In diesem Zusammenhang möchten wir sehr herzlich Prof. Dr. H. Thiel und Prof. Dr. B. Jean für die intensive Zusammenarbeit bei der Hornhautvermessung danken. Ebenfalls danken möchten wir dem Bundesministerium für Forschung und Technologie sowie dem Land Baden Württemberg für die finanzielle Unterstützung.

Literatur

[1]   H.J. Tiziani, "Rechnerunterstützte Laser-Meßtechnik", Technisches Messen 54, No. 6, 221-230 (1985).

[2]   D.W. Robinson, G.T. Reid , (editors) „ Interferogram analysis-digital fringe pattern measurement techniques", Institute of Physics Publishing, Bristol and Philadelphia, (1993).

[3]   R. Windecker, H.J. Tiziani, „A new semi spatial, robust and accurate phase evaluation algorithm", eingereicht in Applied Optics.

[4]   B. Dörband, „Analyse optischer Systeme", Dissertation, Universität Stuttgart, 1986, Seite 39.

Measurement of Absolute Rotation Angle by Combining Fiber-Optics Distance Sensors with Gravity Deformed Cantilevers

H. Kreitlow, U. Samuels, L. Tiase, K. Boer, M. Schlaf
Fachhochschule Ostfriesland
Fachbereich Naturwissenschaftliche Technik
Institut für Lasertechnik
Constantiaplatz 4, D-26723 Emden

ABSTRACT

The principle of a fiber optical sensor (FOS) for sensitive measurements of absolute angles has been investigated. The sensor principle is based on the well defined angular dependence of the bending of a cantilever under the influence of its own weight. The cantilever bending is measured via a non-contact and high resolution method using two fiber optical sensors. These sensors are oriented perpendicular to each other and fixed opposite to the cantilever so that they are rotated synchronously during every angular movement. The current values of the rotation angle are determined in an analysing and data processing unit and visualised on the computer monitor.

This high precision absolute rotation angle sensor has the advantages of being compact, insensitive to external influences such as humidity, atmospheric densitiy fluctuations, electromagnetic fields, and nuclear radiation and is therefore applicable in nearly every kind of angular measurement problem and in unfavorable environmental conditions.

1. CANTILEVER

The cantilever with length l, width b, and height h is clamped at one end for bending under its own weight (see figure 1).

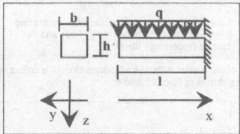

Figure 1: Cantilever model

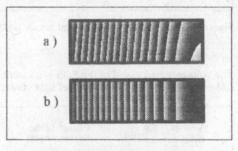

Figure 2: Fringe pattern simulation for cantilever optimization by holographic interferometry: a.) torsion; b.) ideal adjustment

The angular dependent deflection of a cantilever with square cross section (b=h) and isotropic material along the two principle axes at the point of maximum deflection is given by equations (1) and (2) where ρ is the density of the cantilever material, g is the acceleration due to gravity at the Earth's surface, and E is the module of elasticity.

$$fy(\alpha) = \frac{3\rho g l^4}{2Eh^2}\sin(\alpha) \qquad (1)$$

$$fz(\alpha) = \frac{3\rho g l^4}{2Eh^2}\cos(\alpha) \qquad (2)$$

Using this result, the cantilever as one part of the whole sensor can be optimized theoretically with respect to the FOS as the second important part of the rotation angle sensor. Experimantal optimization of the cantilever was performed by holographic interferometry supported by a computer programme for holographic fringe pattern simulation (see figure 2).

2. FIBER OPTICAL SENSOR (FOS)

By means of a FOS, the mechanical bending of a cantilever as depends on its orientation compared to the vertical can be measured, to give the absolute rotation angle. For measuring the cantilever deflection with highest possible sensitivity, fiber optical sensors based on the principle of external light intensity modulation is employed as described: the light from a light source (guided to the cantilever surface by the sending fiber) is reflected and scattered back from the cantilever surface into the coaxial receiving fibers for opto-electronic conversion by a photodiode [1,2], see figures 3, 4, 5.

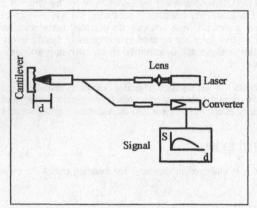

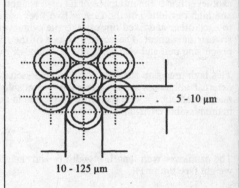

Figure 3: Principle of the fiber optical sensor (FOS)

Figure 4: Cross-section of the FOS measuring head with the central (sending) fiber and the coaxial (receiving) fibers

The voltage output signal of the FOS is a function of the distance from and orientation of a reflecting and scattering surface at the points of light contact as shown in figure 5 and 6.

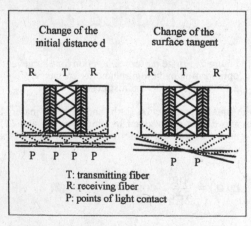

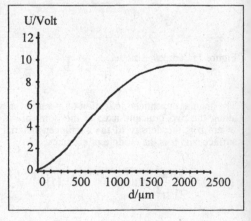

Figure 5: Input of the reflected light in the fiber optical sensor depending on the distance and the orientation of the reflecting and scattering surface

Figure 6: Output U of the FOS as a function of the nominal displacement d

3. COMBINATION OF FOS AND CANTILEVER

With regards to sensor construction, the smallest measureable parameter changes and the measurement range are the major demands. In this development the rotation angle sensor has to work over the full-angle (360 degrees) with a resolution as high as possible. For this reason, the cantilever has to have a maximum angular-dependent bending amplitude. According to equations (1) and (2), the deflection of the cantilever is nearly linear at some ranges of both deflection curves with the additional advantage that these are also the high slope parts required for sensitive measurements. Using a cantilever with square cross section, the deflection curves fy(α) and fz(α) have equal amplitudes. Since the sine function bending characteristic is nearly linear in a range from 0 degree up to 45 degrees while the cosine function showing the bending of the orthogonal component delivers a nearly linear bending in the range from 45 degrees up to 90 degrees, these two curves should be used alternatively for linear rotation angle determination. Following from the behaviour of the sine and cosine function, this method of angle determination is also applicable for angles between 90 and 360 degrees. Since the above mentioned ranges have also the characteristic of the highest slope, that is, within the deflection range between 0 and $1/\sqrt{2}$ of the maximum signal value, the angle measurement is also performed with highest sensitivity. In order to eliminate the ambiguity in the angle determination (each amplitude of both signals can be obtained by two different angles) an additional piece of information is required.

Through use of equations (1) and (2), various combinations of parameters may be found, allowing optimal design of the cantilever using the full linear working range determined above. The following parameters were chosen:

Table 1: Cantilever parameters material: Ceramtec ®
	E:	$360kN/mm^2$
	ρ:	$3.8Kg/dm^3$
	l:	50mm
	h:	0.1mm
	b:	5mm

4. SENSOR DESIGN

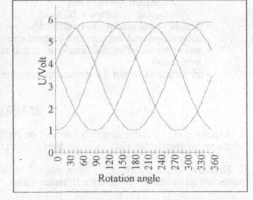

Figure 7: Design of the full angle sensor   Figure 8: Output signal of the full angle sensor

For the measurement of absolute rotation angles with a measurement range of 360 degrees, a sensor was developed which is based on two cantilevers which are fixed perpendicular to each other in combination with four FOS which are rotated synchronously with the cantilever. Figure 7 shows a picture of the full-angle sensor. The output signal of the four FOS as a function of the rotation angle is sketched in figure 8.

For special applications a miniaturized sensor for a measurement range of 180 degrees based on one FOS has been developed (see figure 9). The output signal of this sensor is shown in figure 10.

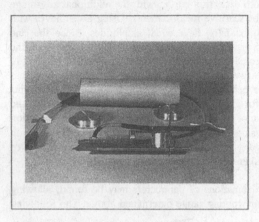

Figure 9: Miniaturized sensor design

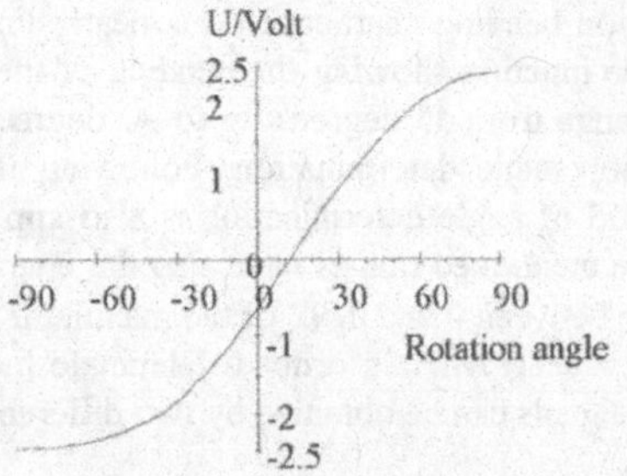

Figure 10: Output signal of the miniaturized sensor

5. RESULTS

The characteristics of the sensor developements are sumerized in the following table

- measurement range: 360 degrees
- resolution :1/1000 degree
- no measurement influence due to:
    - humidity
    - high temperature
    - electromagnetic fields
    - nuclear radiation < 100krad
- flexible signal guiding by modular fiber system
- noise minimized electronics for signal conversion and analysing
- measurement of angular acceleration and angular velocity by computer signal evaluation
- miniaturized sensor for measurement range: 180 degrees

REFERENCES

1. Kreitlow, H., "Faseroptische Sensoren" in Pfeifer, T., et. al., "Optoelektronische Verfahren zur Messung geometrischer Größen in der Fertigungstechnik", Ehningen, expert-verlag, 1993

2. Kreitlow, H., Samuels, U., "Fiber optical sensor for absolute measurement of rotation angles", presented on the SPIE Conference "Interferometry '94", Warsaw, 1994

Präzisions-Kalibrierung von topometrischen 3D-Sensoren

B. Breuckmann, E. Klaas, F. Halbauer
Breuckmann GmbH
Torenstr. 14, D-88709 Meersburg

1. Einleitung

Topometrische Meßverfahren auf der Basis von bildgebender Triangulation mit strukturierter Beleuchtung haben in den vergangenen Jahren Eingang in die industrielle Meß- und Prüftechnik gefunden. Die Oberflächeninspektion an Blechteilen im Preßwerk oder das optische Digitalisieren von Modellen und Werkzeugen für das Rapid Prototyping sind typische Anwendungsbeispiele.

Die Meßgenauigkeit topometrischer 3D-Sensoren hängt dabei nicht nur vom jeweiligen Verfahrensprinzip und optischen Aufbau ab; sie wird auch entscheidend von der Systemkalibrierung bestimmt, mit der insbesondere auch Fehler der strukturierten Beleuchtung und Abbildungsfehler der Optiken erfaßt und korrigiert werden müssen. Neben der Grundkalibrierung durch den Hersteller ist dabei eine schnelle und einfache Systemüberprüfung und Präzisionskalibrierung durch den Anwender von ausschlaggebender Bedeutung.

Für das topometrische 3D-Meßsystem optoCAM wurde ein Kalibrierverfahren entwickelt, welches eine hohe Meßgenauigkeit gewährleistet und gleichzeitig den genannten Praxisanforderungen gerecht wird.

2. Topometrische 3D-Koordinatenmeßtechnik

Das der optoCAM-Sensorik zugrunde liegende Funktionsprinzip der bildhaften Triangulation ist einfach am Beispiel der Projected Fringe Technik zu erklären: Über eine miniaturisierte Projektionseinheit werden rechnergesteuert Streifensysteme mit unterschiedlicher Gitterperiode auf das Objekt projiziert. Die Streifensysteme werden von einer CCD-Kamera erfaßt, die unter einem Winkel ϑ zur Beleuchtungsrichtung angeordnet ist (siehe Abb. 1).

Die Koordinaten eines beliebigen Punktes P der Meßszene berechnen sich nach den bekannten Triangulationsgesetzen anhand der Geometrieparameter des optischen Aufbaus (b, α_0, β_0, $\Delta\alpha$, $\Delta\beta$) sowie der Beleuchtungs- und Beobachtungswinkel α und β. Die Parameter des optischen Aufbaus müssen anhand von Kalibriermessungen ermittelt werden. Die Triangulationswinkel α und β ergeben sich aus den Meßgrößen φ_n und (Px, Py). Dabei sind φ_n die Phase und Ordnung der auf den Meßpunkt P projizierten Streifenlinie, die mittels eines kombinierten Graycode- und Phasenshiftverfahrens bestimmt werden, und (Px, Py) die Pixelkoordinaten des Bildpunktes von P auf dem Kameratarget.

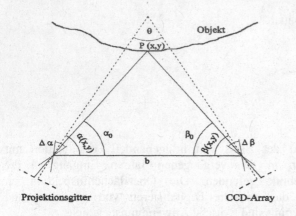

b - Basislänge
α_0 - Beleuchtungswinkel
β_0 - Beobachtungswinkel
$\Delta\alpha$ - Divergenz der Beleuchtungsoptik
$\Delta\beta$ - Divergenz der Beobachtungsoptik

Abb. 1 : Geometrieparameter in der topometrischen Meßtechnik

3. Vorgehensweise bei der Präzisionskalibrierung

Für den praktischen Einsatz der optoCAM-Sensoren wurde ein Kalibrierverfahren entwickelt, das sich an den folgenden Anforderungen orientiert :

❏ einfache Bestimmung der Sensor-Geometrie durch den Hersteller (*Grundkalibrierung*)
❏ 3-dimensionale Erfassung von Gitter- und Abbildungsfehlern in Projektions- und Beobachtungssystem mit hoher Auflösung im gesamten Meßvolumen (*Präzisionskalibrierung*)
❏ einmaliges Teachen des Präzisions-Kalibriervorganges
❏ schnelle und einfache Präzisionskalibrierung vor Ort
❏ automatische Korrektur der Meßdaten
❏ ortsaufgelöste Darstellung der Sensorspezifikationen im Meßvolumen
❏ Berücksichtigung der lokalen Güte der Meßdaten bei der Weiterverarbeitung

Die Notwendigkeit zur Erfassung und Korrektur von Gitter- und Abbildungsfehlern zeigt die nebenstehende Grafik (Abb. 2), in der die Meßfehler für die Z-Koordinate bei fehlender Präzisions-Kalibrierung dargestellt sind. Zwar kann die hier auftretende Verkippung durch eine einfache Ebenenkorrektur deutlich reduziert werden. Die verbleibenden Restfehler liegen aber immer noch im Bereich von ca. 0.5 mm; damit ist die Meßgenauigkeit eine Größenordnung schlechter als mit Präzisions-Kalibrierung (siehe auch Abb. 5 und 6).
Um die Erfassung der Gitter- und Abbildungsfehler mit hoher Ortsauflösung zu gewährleisten, ist es notwendig, beim Kalibrieren das Meßvolumen an einer entsprechend großen Anzahl von Stützstellen zu vermessen.

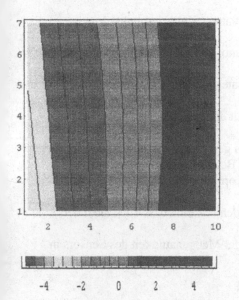

Es hat sich gezeigt, daß dazu je nach Güte der verwendeten Optiken bis zu 1000 im Meßvolumen annähernd gleichmäßig verteilte Meßpunkte notwendig sind. Dieser 3-dimensionale Vorgang kann in der Praxis am einfachsten dadurch realisiert werden, daß eine ebene Kalibrierplatte mit bekannten Kalibrierpunkten (Paßmarken) in mehreren Ebenen im Raumvolumen rechnergesteuert positioniert wird. Abb. 3 zeigt ein Beispiel für den schematischen Aufbau einer solchen Kalibrierplatte.

Meßfehler [mm]

Abb. 2 : Meßfehler in der Z-Koordinate ohne Präzisionskalibrierung

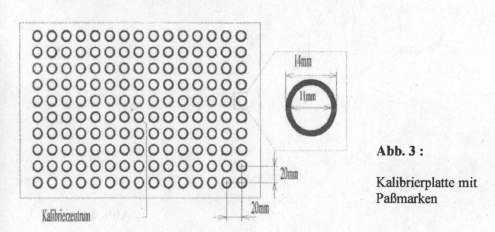

Abb. 3 :

Kalibrierplatte mit Paßmarken

Eine Prinzipskizze von Sensor, Kalibriervorrichtung, Bildverarbeitungssystem und Steuerung ist in Abb. 4 darstellt. Die feste Anordnung von Sensor und Kalibriersystem ermöglicht insbesondere, den gesamten Kalibriervorgang einmalig zu teachen, sodaß der Anwender jederzeit die Sensorspezifikationen überprüfen und gegebenenfalls innerhalb von nur 5 - 10 Minuten eine automatische Präzisionskalibrierung durchführen kann.

Der automatische Kalibriervorgang verläuft in mehreren Stufen, wobei die 3 letzten der Darstellung der Sensorspezifikationen dienen und somit nur optional durchgeführt werden :

554

- Vermessung der Paßpunkte der Kalibrierplatte (Kalibrierpunkte) in unterschiedlichen Z - Ebenen
- Bestimmung der Pixelkoordinaten, Streifenordnungen und Phasen der Kalibrierpunkte
- Berechnung der zugehörigen 3D-Koordinaten anhand der Grundkalibrierung (*unkorrigierte Meßdaten*)
- Vergleich der unkorrigierten Meßdaten mit den bekannten SOLL-Daten der Kalibrierpunkte
- Berechnung einer Korrekturmatrix für das gesamte Meßvolumen des Sensors
- erneute Vermessung der Kalibrierpunkte, Berechnung der zugehörigen 3D-Koordinaten unter Verwendung der Korrekturmatrix (*korrigierte Meßdaten*)
- Vergleich der korrigierten Meßdaten mit den bekannten SOLL-Daten der Kalibrierpunkte
- Berechnung und graphische Darstellung der Meßgenauigkeit des Sensors in Abhängigkeit vom Meßvolumen

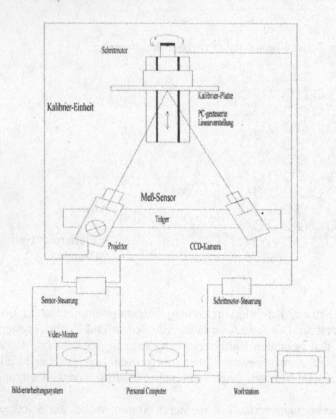

Abb. 4 :

Aufbau der automatischen Kalibriervorrichtung mit Sensor, Kalibrierplatte, Bildverarbeitungssystem und Steuerung

4. Ergebnisse

Mit präzisionskalibrierten optoCAM-Sensoren lassen sich Merkmalsgenauigkeiten von bis zu 1/4.000 der Meßfeldgröße erzielen. Dies ist eine Verbesserung gegenüber der Grobkalibrierung mit Ebenenkorrektur um ca. eine Größenordnung (siehe Abb 5). So werden beispielsweise mit einem typischen optoCAM-Sensor folgende Sensorspezifikationen erreicht:

Meßfeld : 120 * 150 mm, Meßtiefe : 100 mm, Meßabstand : 400 mm
Merkmalsgenauigkeit : +/- 50 µm, Meßunsicherheit (U95) : +/- 50 µm

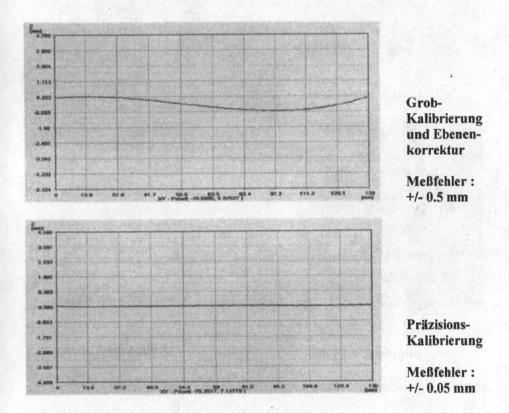

Grob-
Kalibrierung
und Ebenen-
korrektur

Meßfehler :
+/- 0.5 mm

Präzisions-
Kalibrierung

Meßfehler :
+/- 0.05 mm

Abb. 5 : Vergleich der Meßgenauigkeit von Profilschnitten

Dem Anwender steht außerdem eine graphische Darstellung zur Verfügung, in der die verbleibenden Restfehler im Meßvolumen des Sensors ortsaufgelöst und quantitativ abzulesen sind (siehe Abb 6).

556

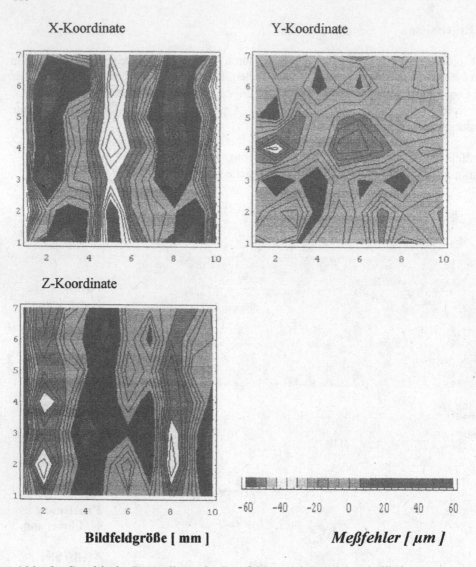

X-Koordinate

Y-Koordinate

Z-Koordinate

Bildfeldgröße [mm]

-60   -40   -20   0   20   40   60

Meßfehler [µm]

Abb. 6 : Graphische Darstellung der Restfehler nach Präzisionskalibrierung

5. Zusammenfassung

Die Meßgenauigkeit von topometrischen 3D-Sensoren läßt sich durch eine Präzisions-Kalibrierung, die Gitter- und Abbildungsfehler erfaßt und korrigiert, deutlich steigern. Für optoCAM-Sensoren wurde hierzu eine Vorgehensweise entwickelt, welche dem Anwender nach einmaligem Teachen des Kalibriervorganges jederzeit eine schnelle und automatische Präzisionskalibrierung ermöglicht.

3D-Verschiebungsmessungen an komplexen 3D-Objekten
3D-Displacement Measurement on Complex 3D-Surfaces

P. Andrä[*], W. Jüptner[*], W. Nadeborn[*], W. Osten[*] und M.-A. Beeck[**]

[*] BIAS - Bremer Institut für angewandte Strahltechnik, Klagenfurter Str. 2, D-28359 Bremen
[**] Volkswagen AG, Forschung und Entwicklung, D-38436 Wolfsburg

Abstract

In der holographischen Interferometrie sind für die Messung von drei Verschiebungskomponenten mindestens drei unabhängige Messungen notwendig, die aus unterschiedlichen Beobachtungs- oder Beleuchtungsrichtungen erfolgen können. Jedem Meßpunkt werden als primäre Meßgrößen mindestens drei Interferenzphasen- bzw. -phasendifferenzwerte zugeordnet, die mit bewährten Standardverfahren der Phasenmeßtechnik [1] erfaßt werden. In dieser Arbeit soll ein praktikables Meßverfahren für die experimentelle Verformungsanalyse vorgestellt werden, das insbesondere für komplexe Oberflächen die Raumpunkt- und damit Objektform- und Objektlageabhängigkeit des Sensitivitätsvektors berücksichtigt. Dafür müssen jedem Meßpunkt seine 3D-Objektkoordinaten zugeordnet werden, die mit Methoden der optischen Formerfassung bestimmt werden [2, 3]. Liegen die gemessenen Objektkoordinaten, die Interferenzphasen und die Parameter des Meßaufbaus in einem gemeinsamen Koordinatensystems des Interferometers vor, können nun die drei Verschiebungskomponenten berechnet und bewertet werden. Die Vorteile dieses Verfahrens und einige Ergebnisse werden an einem einfachen Testobjekt sowie an einem komplexeren Automobilteil demonstriert.

According to the basic equation of holographic interferometry for the measurement of the three displacement components at least 3 independent equations are necessary. Usually 3 or more observation and illumination directions, respectively, are chosen. The quantities to be measured are the interference phases at each measuring point and its three Cartesian coordinates in a predefined coordinate system within the interferometer. In practice the acquisition of both kind of data is reduced more and more to a phase measurement. Modern topometric methods based on structured object illumination by projected fringes are used for the measurement of absolute coordinates. Because the investigation of complex surfaces makes it necessary to calculate the sensitivity vectors for each measuring point these coordinates have to be transformed into the new coordinate system of the interferometric set-up for displacement measurement. Using the corresponding phase values measured in the relevant object points the 3 displacement components can be calculated and evaluated with respect to their accuracy. This procedure and its results will be demonstrated on an automotive component.

1   Einleitung

Im Zusammenhang mit der experimentellen Verformungsanalyse besitzt die optische 3D-Formerfassung folgende Bedeutung:

Zum einen soll sie erforderliche Eingangsinformationen liefern, um den ortsvariablen Sensitivitätsvektor für 3D-Verschiebungsmessungen präzise bestimmen zu können. Zum anderen erleichtert es die Ergebnisinterpretation wesentlich, wenn experimentell ermittelte Verformungsverteilungen grafisch über der Grundform des Meßobjektes dargestellt werden können. Schließlich ist die Kenntnis der Objektgeometrie unerläßlich, wenn es um eine Verbindung zwischen experimenteller und numerischer Verformungsanalyse geht, sei es rein zu Vergleichszwecken oder um z. B. realitätsnahe Randbedingungen für letztere bereitstellen zu können.

Die Objektform wird im gegebenen Fall mit Hilfe inkohärenter Streifenprojektion und räumlicher Triangulation bestimmt. Dies bietet im Vergleich zu kohärenten Verfahren, wie z. B dem holografischen Contouring, den Vorteil der größeren Robustheit.

Bild 1 veranschaulicht den Datenfluß und die Abhängigkeiten zwischen den einzelnen Meß- und Auswertekomponenten des Gesamtverfahrens. Bild 1 verdeutlicht ebenfalls das hohe Maß an Komplexität der Methode, deren praktische Beherrschung, sowohl in experimenteller als auch in algorithmischer und softwaretechnischer Hinsicht mitentscheidend ist für die nicht nur qualitative, sondern auch quantitative Zuverlässlichkeit von Form- und Verformungsmeßdaten.

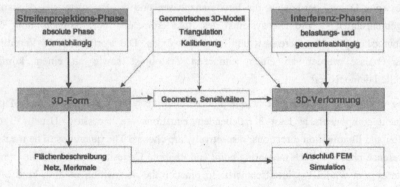

Bild 1: Datenfluß und Abhängigkeiten

Um ein verifizierbar genaues und gleichzeitig mit vertretbarem Aufwand praktikables Meßverfahren zu realisieren sind eine ganze Reihe von Einzelproblemen zu lösen, die hier nur kurz erwähnt werden.

- Die Phase des auf das Objekt projizierten Streifenmusters ist absolut zu bestimmen in dem Sinne, daß die Streifenordnung in jedem Objektpunkt unabhängig von seinen Nachbarpunkten ermittelt werden kann und daß kein willkürlicher Offset verbleibt [4]. Dabei ist die von den Phasenschiebetechniken her gewohnte Relativmeßgenauigkeit absolut zu bewahren.
- Zur Umrechnung von Phasendaten in 3D-Koordinaten ist ein Geometrie- bzw. Triangulationsmodell aufzustellen, das einen in allen Komponenten flexiblen Meßaufbau zuläßt, gleichzeitig jedoch kompakt und effektiv handhabbar bleibt. Es ist eine Kalibrierprozedur zu

entwickeln, die relativ schnell, mit geringem Aufwand und möglichst ohne zusätzliche, externe Meß- oder Justiermittel durchführbar ist und eine systematische Koordinatenverzerrung weitgehend ausschließt, d. h. auf das bereits durch das Phasenrauschen gegebene Maß von ca. D / 3000 (D - laterale Objektausdehnung) beschränkt [3].

- Um eine möglichst vollständige Objekterfassung zu gewährleisten, sind im allgemeinen mehrere Ansichten miteinander zu vereinigen, wiederum möglichst ohne wesentliche Genauigkeitsverluste. Im Hinblick auf eine substantielle Reduzierung des Datenvolumens sowie die Anbindung an den Bereich von CAD und numerischer Simulation sind Mittel für eine Flächenrückführung aus der 3D-Punktewolke bereitzustellen. Dabei sind sowohl Regelflächen als solche zu erkennen und zu bewahren als auch Freiformflächen adäquat zu beschreiben.

- Da Form und Verformung im allgemeinen in unterschiedlichen Aufbauten und Bezugssystemen gemessen werden, muß vor der Berechnung der ortsvariablen Sensitivität eine entsprechende Koordinatentransformation gefunden und ausgeführt werden.

- Um schließlich 3D-Verschiebungsvektoren mit einer korrekten Verteilung auf die einzelnen Komponenten zu erhalten ist ein lineares Gleichungssystem zu lösen, in welchem für jeden Meßpunkt mindestens 3 - z. B. aus unterschiedlichen Beleuchtungsrichtungen resultierende - Interferenzphasenwerte und die zugehörigen Sensitivitäten miteinander verknüpft werden.

In den folgenden Abschnitten wird ein praktikables Verfahren für die beiden letzt genannten Probleme vorgestellt und untersucht.

2 Beschreibung des Meßverfahrens

2.1 3D-Verschiebungsmessung mit Holografischer Interferometrie

Um das Verschiebungsfeld eines belasteten Objekts zu berechnen, ist es notwendig die drei Vektorkomponenten der Verschiebung für jeden Punkt P auf der Oberfläche eindeutig zu bestimmen. Ausgangspunkt ist die eindimensionale Grundgleichung der Holografischen Interferometrie:

$$(\bar{e}_B + \bar{e}_Q) \cdot \bar{d}(P) = \bar{S}(P) \cdot \bar{d}(P) = \lambda \, N(P) \tag{1}$$

in der $\bar{S}(P)$ den Sensitivitätsvektor darstellt, $\bar{d}(P)$ den Verschiebungsvektor und $N(P)$ die Interferenzordnung, die für den Punkt P gemessen wird. Der Sensitivitätsvektor besteht bei der verwendeten statischen Meßmethode aus der Summe der Einheits-Richtungsvektoren vom Objektpunkt P zum Beobachtungspunkt B, dem Vektor $\bar{e}_B$, und dem Richtungsvektor $\bar{e}_Q$ von P zum Beleuchtungspunkt Q. $\bar{S}(P)$ zeigt also in Richtung der Winkelhalbierenden zwischen Beobachtungs- und Beleuchtungsrichtung. Als Beobachter wird eine CCD-Kamera verwendet.

Da die Grundgleichung aufgrund der Eigenschaften des Skalarprodukts die Verschiebung nur als Projektion auf die Raumrichtung von $\bar{S}(P)$ liefert, benötigt man für die 3D-Verschiebungsmessung mindestens drei Gleichungen mit unterschiedlichen Sensitivitätsvektoren. Bei variabler Beleuchtungsrichtung besteht der Aufbau zur Verschiebungsmessung aus mindestens drei unterschiedlichen Beleuchtungspunkten Q1, Q2 und Q3, woraus sich auch drei unterschiedliche Sensitivitätsvektoren ergeben. Rechnerisch ergibt sich, ausgehend von der Grundgleichung (1), $\bar{d}(P)$ als die Lösung des linearen Gleichungssystems

$$G(P) \cdot \bar{d}(P) = \lambda \cdot \bar{N}(P), \tag{2}$$

wobei die Geometriematrix **G** aus den einzelnen Sensitivitätsvektoren in der Form

$$G(P) = \begin{bmatrix} \bar{S}_1(P) \\ \bar{S}_2(P) \\ \bar{S}_3(P) \\ \vdots \end{bmatrix} = \begin{bmatrix} \bar{e}_B + \bar{e}_{Q_1} \\ \bar{e}_B + \bar{e}_{Q_2} \\ \bar{e}_B + \bar{e}_{Q_3} \\ \vdots \end{bmatrix} \tag{3}$$

besteht und der Ordnungsvektor $\bar{N}(P)$ sich aus den entsprechenden Interferenzordnungen ergibt. Da sowohl $\bar{S}(P)$ als auch $G(P)$ von der Lage des Punktes P abhängig ist, müssen beide Größen für jeden Oberflächenpunkt auf dem Objekt neu berechnet werden.

Um den ortsvariablen Sensitivitätsvektor für die 3D-Verschiebungsmessung (Gl. 2, 3) berechnen zu können, müssen die durch die Formerfassung gewonnenen Objektkoordinaten $\bar{r}$, die Koordinaten der Beleuchtungspunkte Q_i und die des Beobachtungspunktes B in einem einheitlichen Koordinatensystem vorliegen. Zweckmäßigerweise wird die Verschiebungsmessung geschlossen im Objektkoordinatensystem OKS durchgeführt, das aus der Formmessung im Streifenprojektionsaufbau resultiert. Da die Beleuchtungspunkte und der Beobachtungspunkt im Interferometeraufbau vermessen bzw. kalibriert werden, muß mit Hilfe von wiedererkennbaren Objektmarkierungen, deren Punktkoordinaten $\bar{r}$ von einer separaten Formerfassung her bekannt sind, eine Lageerkennung des Objekts im Interferometeraufbau durchgeführt werden. Aus dieser Lageerkennung läßt sich eine Vorschrift für die Transformation zwischen den beiden Aufbauten, d.h. zwischen dem Objektkoordinatensystem OKS und dem Interferometerkoordinatensystem IKS, gewinnen (vgl. Bild 2).

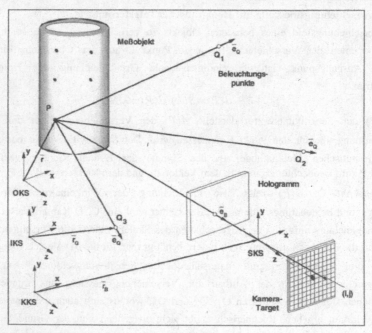

Bild 2: Interferometeraufbau für Verformungsmessung

2.2 Bestimmung der Koordinatentransformationen

Ausgehend vom geometrischen Modell der Beobachtung, das die Abbildung von Objektpunkten $\bar{r}_{IKS}$ in Bildpunkte (i,j) beschreibt,

$$\bar{r}_{IKS} = \bar{r}_B + \lambda\left[\bar{c}_0 + \bar{c}_i(i - i_0) + \bar{c}_j(j - j_0)\right] \tag{4}$$

wobei die Vektorkoeffizienten $\bar{c}$ Funktionen der internen und externen Kameraparameter und λ einen freien Abstandsparameter darstellen [3], beruhen die benötigten Transformationen auf folgender Vorgehensweise:

• Durch eine Kalibrierung der CCD-Kamera im Interferometeraufbau werden die Kameraparameter in Gl. (4) identifiziert. Diese Parameter werden als zeitlich konstant angenommen. Die benötigte Kalibrierplatte legt gleichzeitig das Interferometerkoordinatensystem IKS fest, in dem zusätzlich die Beleuchtungspunkte definiert werden.

• Die Koordinaten $\bar{r}_Q$ der Beleuchtungspunkte Q_i werden bezüglich der Kamerakalibrierebene durch Längenmessungen bestimmt. Die Beobachtungspunktkoordinaten $\bar{r}_B$ liegen ebenfalls im IKS vor und müssen für die weitere Auswertung ins OKS transformiert werden.

• Mit Hilfe der Markierungen am Objekt und den zugehörigen Bildkoordinaten (i,j) kann dessen Lage und somit die Translations- und Rotationsparameter für die Transformation zwischen OKS und IKS bestimmt werden:

$$\bar{r} = R_1 \cdot \bar{r}_{IKS} + \bar{l}_1 \tag{5}$$

Mit der Transformation der relevanten Aufbauparameter (Lage der Beobachtungs- und Beleuchtungspunkte) in das Objektkoordinatensystem werden für jeden interessierenden Objektpunkt mit $\bar{r}(P)$ der Sensitivitätsvektor $\bar{S}(P)$ nach Gl. (3) berechnet. Im folgenden muß noch die Interferenzordnung N(P) der Verschiebung dieser Objektpunkte bestimmt werden. Mit den ermittelten Kamera- und Transformationsparametern können die zugehörigen Bildpunkte (i,j) berechnet und die gemessenen Interferenzordnungen N(i,j) zugeordnet werden. Folglich liegen alle Größen, die in das zu lösende Gleichungssytem Gl. (2) eingehen, in einem gemeinsamen Koordinatensystem OKS vor.

Im folgenden soll der Einfluß der objektformabhängigen Sensitivität auf die Genauigkeit der Verschiebungsmessung untersucht werden. Außerdem wird der Einfluß anderer Meßparameter auf die Verschiebungsmessung diskutiert. Dazu werden sowohl experimentelle Daten als auch FEM-Simulations-Ergebnisse an einem Testobjekt und einem Automobilteil herangezogen.

3 Untersuchungen zum Einfluß des Sensitivitätsvektors

3.1 Versuchsdurchführung und -auswertung

Bei der holografischen Verschiebungsmessung wird, wie auch bei nachfolgenden Berechnungen, ein beidseitig geschlossener und einseitig fixierter Aluminiumzylinder mit einem Durchmesser von 100 mm und einer Länge von 200 mm, wie in Bild 2 dargestellt, durch Anlegen eines veränderlichen Innendrucks belastet.

Die 3D-Objektkoordinaten des Zylinders (Bild 4a) und der aufgebrachten Markierungen werden mit Hilfe eines im BIAS realisierten Streifenprojektions-Aufbaus bestimmt. Für die holografischen Untersuchungen wird ein Interferometer mit vier Beleuchtungsrichtungen (Bild 2) aufgebaut. Die

Position aller Beleuchtungs- und Beobachtungpunkte werden im Interferometerkoordinatensystem IKS ermittelt. Mit der vermessenen Objektform und der Kenntnis über den interferometrischen Aufbau werden sowohl die ortsvariablen Sensitivitätsvektoren bzw. Geometriematrizen nach Gl. (3) erstellt als auch zu Vergleichszwecken allen Meßpunkten ein konstanter Sensitivitätsvektor aus der Objektmitte zugeordnet.

Für jede Beleuchtungsrichtung wird nacheinander das zugehörige Interferogramm unter Benutzung von insgesamt nur einem Hologramm in Doppelbelichtungstechnik aufgezeichnet. Bild 3 zeigt die korrespondierenden Interferogramme des Zylinders bei einer Druckdifferenz von etwa 1.2 bar.

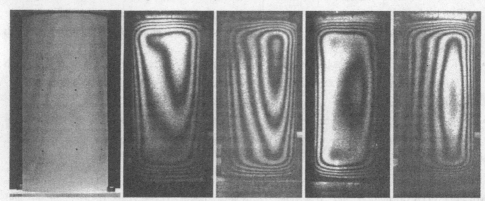

Bild 3: Zylinder mit Markierungen, Interferogramme für die Beleuchtungsrichtungen Q_1, bis Q_4

Für die Auswertung der Interferogramme sind das bekannte Skelettierungs- bzw. Phase-Schiebe-Verfahren geeignet. Die Interferenzordnungen $N(P)$ können für jedes Interferogramm bezüglich eines gemeinsamen Refenzpunktes (Phase 0) ermittelt werden. Mit den berechneten Geometriematrizen und den ermittelten Interferenzordnungen wird nunmehr das überbestimmte Gleichungssystem (2) an jedem Meßpunkt P gelöst. Das Ergebnis ist ein Feld, das für eine Druckbelastung des Objekts jedem Punkt P die zu ihm gehörende räumliche Verschiebung $\vec{d}(P)$ angibt (Bild 4b).

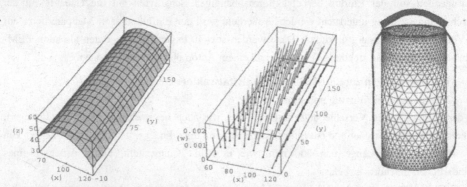

Bild 4: a) Geometriedaten des Zylinders, b) Verschiebungsfeld aus der Messung mit ortsvariabler Sensitivität über der Zylinderoberfläche (Ausschnitt) aufgetragen, $|\hat{d}| \approx 1.5 \mu m$ (alle Angaben in mm), c) Verformung des Zylinders aus der Simulation

3.2 Einfluß des Sensitivitätsvektors auf die 3D-Verschiebungsmessung

Im Vergleich zu den experimentell ermittelten Verschiebungskomponenten mit ortsvariabler Sensitivität zeigt sich bei der Annahme konstanter Sensitivität (Bild 5) in der radialen Komponente entlang des Zylinderquerschnitts ein nur wenig veränderter Verlauf. Jedoch ergibt nun die Auswertung eine hohe und sogar das Vorzeichen wechselnde Tangentialkomponente, was der zu erwartenden Rotationssymmetrie völlig zuwiderläuft und auch nicht durch eine eventuelle Starrkörperverschiebung zu erklären wäre. Auch in der Axialkomponente in Richtung der Längsachse sind deutlich Auswirkungen feststellbar. Die Gründe dafür sind in den Verfahrenseigenschaften der holografischen Verschiebungsmessung zu suchen. Demnach kann die out-off-plane-Komponente robust gemessen werden. Dagegen ist die Messung der in-plane-Komponenten sehr sensibel gegen Fehler aller Art, so auch der Geometriefehler.

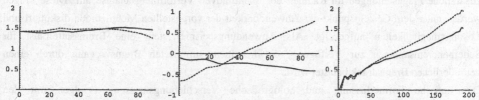

Bild 5: Radiale, tangentiale und axiale Komponenten für Verschiebungsmessung [µm] mit ortskonstanter – – – und ortsvariablen Sensitivität ——— (Abszisse: Zylinderkoordinaten φ [°] bzw. z [mm])

Diese globale systematische Fehlercharakteristik kann ebenfalls durch Auswertungen der simulierten Verschiebungsdaten nachgewiesen werden.

3.3 Vergleich mit FEM-Berechnungen

Als Randbedingung für die Modellierung wird lediglich die Einspannung am unteren Rand (keine Verschiebung) und die Druckbelastung vorgegeben. Bild 4c stellt die durch den Innendruck deformierte Wandung als Schnitt über dem FEM-Netz dar.

In der Simulation ist die radiale Verschiebung aufgrund der Rotationssymmetrie entlang des Zylinderquerschnitts konstant während die tangentiale Komponente verschwindet (Bild 6). Die axiale Verschiebung steigt in Richtung der Längsachse annähernd linear an.

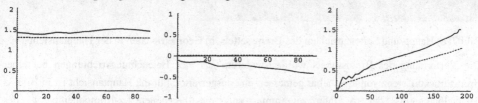

Bild 6: Berechnete – – – und experimentell ——— ermittelte radiale, tangentiale und axiale Verschiebungskomponenten (Abszisse: Zylinderkoordinaten φ [°] , z [mm])

Im Vergleich zur Simulation zeigen sich bei den experimentell ermittelten Verschiebungskomponenten mit objektvariabler Sensitivität (Bild 6) in der radialen Komponente geringfügige Abweichungen vom konstanten Verlauf sowie eine negative, leicht abfallende tangentiale Komponente.

564

Als Ursache für die verbleibenden Differenzen kommen in Frage: eine gewisse Starrkörperverschiebung des Zylinders bei der Innendruckvariation und mögliche Phasenoffsets durch ungenaue Lokalisierung der nullten Streifenordnung bei der Messung sowie vereinfachende Annahmen bzgl. Geometrie und Randbedingungen bei der Simulation.

Ungeachtet dessen kann jedoch eindeutig festgehalten werden, daß eine Berücksichtigung der Ortsvariabilität der Sensitivität insbesondere für die in-plane-Komponente zu einer wesentlich verbesserten Annäherung der experimentell ermittelten Verschiebungen an den theoretisch zu erwartenden Verlauf mit sich bringt.

4   Verschiebungsmessung an einem Industrieteil

Im Vergleich zum Testobjekt Zylinder werfen industrielle Bauteile bei praxisnaher Belastung zusätzliche Fragestellungen im Rahmen der quantitativen Verformungsanalyse auf. Diese Fragen werden unter dem Gesichtspunkt der Anwendbarkeit der vorgestellten Meßmethode diskutiert und Lösungsmöglichkeiten aufgezeigt. Als Anwendungsbeispiel steht eine Bremssattelfaust für Scheibenbremsanlagen zur Verfügung, die in einem simulierten Bremsvorgang durch einen veränderlichen Bremsdruck belastet wird.

Die optische Formerfassung und holografische Verschiebungsmessung erfolgen mit den beschriebenen Meßaufbauten. Die nachstehenden Abbildungen zeigen verschiedene Ansichten des Bremssattels (Größe ca. 100 mm x 100 mm x 80 mm, s. Bild 7a) und die vermessene Objektform (s. Bild 7b) in der Hauptansicht. Bereits hier zeichnet sich ab, daß für eine vollständige Formerfassung einer Partie mehrere Ansichten zu verknüpfen sind.

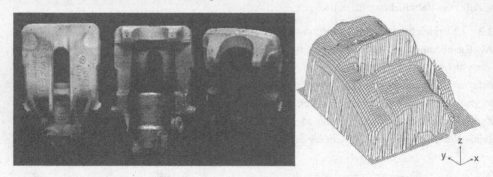

Bild 7: a) Haupt- und Seitenansichten des Bremssattels, b) Geometriedaten in der Hauptansicht

Die Verformung des Bremssattels wurde wiederum aus vier Beleuchtungsrichtungen bei einer Bremsdruckdifferenz von etwa 5 bar gemessen und ausgewertet. Für die Hauptansicht ist in Bild 8a beispielhaft eins der 4 Interferogramme und die berechnete z-Komponente w des Verschiebungsvektorfeldes $\bar{d}$ (s. Bild 8b) dargestellt.

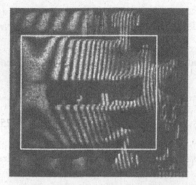 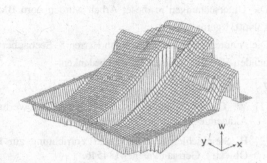

Bild 8: a) Interferogramm, b) z Komponente w der Verschiebung für dic Hauptansicht

Die vom Druckzylinder aufgebrachte Bremskraft wirkt beidseitig über die Bremsklötze auf die Bremssscheibe und führt zu der deutlich sichtbaren Durchbiegung des Faustsattels im Bereich von 4 μm der z-Verschiebungskomponente.

Neben der Schwierigkeit einer korrekten Zuordnung von Geometrie- und Verschiebungsdaten treten weitere Schwierigkeiten für eine quantitative Verformungsanalyse ganzer Bauteile auf:

• Korrekte Messung der Interferenzordnungen ohne systematischen Phasenoffset
• Durch das Konstruktionsprinzip bedingte Starrkörperverschiebungen, die unter der Randbedingung mehrerer Beleuchtungsrichtungen optisch kompensiert oder aus den gemessenen Verformungsdaten numerisch extrahiert werden müßten.
• Verschiebungsmessung über den gesamten Arbeitsbereich oder Detektion lokaler Unstetigkeiten bei größeren Belastungen.
• Vereinigung von Form- und Verformungsmeßdaten aus mehreren Ansichten unter gleichen Belastungsverhältnissen.

Folgende Lösungswege werden hier bevorzugt:

• Anwendung von Verfahren zur absoluten Phasenmessung für die holografische Verschiebungsmessung [5].
• Einsatz adaptierbarer optischer Techniken zur Kompensation von Starrkörperbewegungen oder globalen Verformungen (z.B. aktive Phasenkompensation über geeignete addressierbare Bauelemente).
• Messung von Form und Verformung in einem Aufbau und deren Fusion.

5   Zusammenfassung

Es wurde ein Gesamtverfahren für die experimentelle Verformungsanalyse vorgestellt, deren praktische Beherrschung, sowohl in experimenteller als auch in algorithmischer und softwaretechnischer Hinsicht mitentscheidend ist für die nicht nur qualitative, sondern auch quantitative Zuverlässlichkeit von Form- und Verformungsmeßdaten. Dazu sind eine ganze Reihe von Einzelproblemen gelöst worden. Insbesondere wurde für die holografische Verschiebungsmessung eine Methode dargestellt und verifiziert, die für komplexe Oberflächen die Raumpunkt- und damit Objektform- und Objektlageabhängigkeit des Sensitivitätsvektors berücksichtigt.

6 Danksagung

Die Untersuchungen in dieser Arbeit wurden vom BMFT im Rahmen des Verbundprojektes Nr. 13N6031 gefördert.

Die Autoren möchten sich bei den Herren S. Seebacher und A. Stephens für Ihr hohes Engagement bei den Verschiebungsmessungen bedanken.

7 Literatur

[1] W. Osten: "Digitale Verarbeitung und Auswertung von Interferenzbildern", pp. 39 - 75, Akademie-Verlag Berlin, 1991

[2] H. Steinbichler: Verfahren und Vorrichtung zur Bestimmung der Absolutkoordinaten eines Objektes. German Patent 4134546.

[3] W. Nadeborn, P. Andrä, W. Osten: "Model based identification of system parameters in optical shape measurement", In: Jüptner, W.; Osten, W.(Eds.): Proc. Fringe'93. 2nd International Workshop on Automatic Processing of Fringe Patterns. Akademie Verlag Berlin 1993, pp.214-222.

[4] W. Nadeborn, P. Andrä, W. Osten: A Robust Procedure for Absolute Phase Measurement. (to be published Optics & Lasers Eng.)

[5] W. Osten, P. Andrä, W. Nadeborn, W. Jüptner: Modern Approaches for Absolute Phase Measurement. (to be published SPIE)

Ein neuer Ansatz zur 3D-Objektdigitalisierung nach dem codierten Lichtansatz

Dirk Bergmann
Experimentelle Mechanik
TU Braunschweig
38106 Braunschweig

1 Einleitung

Aufgrund gestiegener Qualitätsanforderungen in der industriellen Fertigung stellt die automatische Konturmessung von Objekten eine wichtige Kontrollfunktion dar. Besondere Vorteile bieten hier optische Meß- und Prüftechniken durch eine berührungslose und zerstörungsfreie Prüfung. Eine Vielzahl von Methoden ist in den letzten 10 - 20 Jahren entwickelt worden (Lichtschnittverfahren, Codierter Lichtansatz [1], Projected Fringes, Moiré [2]).

Die bislang auf diesen Methoden beruhenden, verfügbaren Geräte basieren überwiegend auf dem Prinzip der aktiven Triangulation. Hier werden, wie in der Photogrammetrie, Strahlenbündel im Meßvolumen zum Schnitt gebracht. Dabei ersetzt man eine der Kameras durch eine Projektionseinrichtung, die in den Objektraum eine definiert strukturierte Beleuchtung ausstrahlt. Projektor, Oberflächenpunkt und Kamera bilden ein Dreieck mit Projektor und Kamera als Basis. Ist die Basislänge und der Winkel zwischen Lichtstrahl und Basis bekannt, so kann der Ort des Oberflächenpunktes bestimmt werden, Abbildung 1.

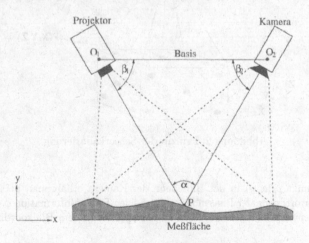

Abbildung 1: Prinzip der aktiven Triangulation

Als strukturierte Beleuchtung wird in modernen Sensoren vielfach die Phasenshiftmethode in

Kombination mit dem codierten Lichtansatz verwendet. Zur Erzeugung der Linienstrukturen dient z.B. ein LCD-Display mit einem vorgeschalteten Weißlichtprojektor. Es stehen LCD-Displays zur Verfügung, die orthogonal zueinander orientierte Linienstrukturen erzeugen. Eine zentrale Aufgabe beim Einsatz der auf der aktiven Triangulation basierenden Meßverfahren besteht häufig in der Kalibrierung des Meßaufbaus.

Im folgenden wird ein neuer Ansatz vorgestellt, der Projektor und Kamera in einem gemeinsamen geometrischen Modell beschreibt. Er beruht auf der Idee, die Abhängigkeit zwischen Bildkoordinaten, Phasen und Objektkoordinaten in einem einheitlichen Modell zu erfassen. Voraussetzung hierfür ist die Projektion zweier orthogonal zueinander orientierter liegenden Phaseninformationen.

Im folgenden wird das Meßprinzip und die Kalibrierung des Systemes beschrieben sowie durch ein Beispiel belegt.

2   Meßprinzip

Der grundlegende Gedanke besteht darin, das Koordinatensystem des Projektors direkt über die projizierte Phaseninformation festzulegen. Abbildung 2 verdeutlicht diesen Zusammenhang.

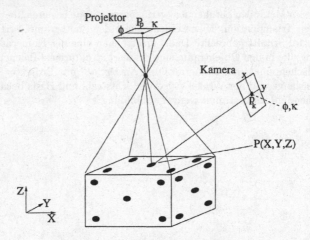

Abbildung 2: Prinzip der Sensorkalibrierung

Für jeden Bildpunkt $p_k(x, y)$ in der Bildebene der Kamera erhält man bei der Verwendung zweier zueinander orthogonaler Phasenbilder die beiden Phaseninformation ϕ und κ an diesem Ort. Diese Phasenwerte beschreiben einen Punkt $p_p(\phi, \kappa)$ in dem Bildkoordinatensystem des Projektors.

Sind die Modellparameter (innere und äußere Orientierung von Kamera und Projektor) im Rahmen einer Kalibrierung ermittelt worden, so kann der zu den homologen Bildpunkten $p_k(x, y)$ und $p_p(\phi, \kappa)$ gehörende Objektpunkt $P(X, Y, Z)$ z.B. über einen räumlichen Vorwärtsschnitt berechnet werden.

3 Kalibrierung

Wie bereits erwähnt, muß für eine Berechnung der Objektkoordinaten P(X,Y,Z) ein geometrisches Modell vorgegeben werden. Die freien Parameter dieses Modelles ermittelt man üblicherweise im Rahmen einer Kalibrierung. Als Kameramodell dient hier beispielhaft die in der Photogrammetrie verwendete Zentralprojektion:

Dort gilt für die Kamera:

$$\begin{pmatrix} x - x_0 - d_x \\ y - y_0 - d_y \\ c \end{pmatrix} = s_i * \mathbf{R} * \begin{pmatrix} X - X_0 \\ Y - Y_0 \\ Z - Z_0 \end{pmatrix} \tag{1}$$

x, y - Bildkoordinaten

x_0, y_0 - Bildhauptpunkt

d_x, d_y - Verzeichnungsparameter

c - Kammerkonstante

$\mathbf{R}$ - Rotationsmatrix

X_0, Y_0, Z_0 - Projektionszentrum

$X.Y, Z$ - Objektpunkt

und für den Projektor:

$$\begin{pmatrix} \phi - \phi_0 - d_\phi \\ \kappa - \kappa_0 - d_\kappa \\ c \end{pmatrix} = s_i * \mathbf{R} * \begin{pmatrix} X - X_0 \\ Y - Y_0 \\ Z - Z_0 \end{pmatrix} \tag{2}$$

ϕ_x, κ_y - Bildkoordinaten

ψ_0, κ_0 - Bildhauptpunkt

ϕ_x, κ_y - Verzeichnungsparameter

c - Kammerkonstante

$\mathbf{R}$ - Rotationsmatrix

X_0, Y_0, Z_0 - Projektionszentrum

$X.Y, Z$ - Objektpunkt

Die unterschiedlichen Beobachtungen, Marken und anderen Oberflächenpunkte werden in einem Modell mittels der Ausgleichsrechnung gemeinsam bestimmt. Der Ausgleich erfolgt anhand der grundlegenden Fehlergleichung:

$$\mathbf{v} = \mathbf{A} * \mathbf{dx} - \mathbf{l} \tag{3}$$

$\mathbf{v}$ Verbesserungsvektor

$\mathbf{A}$ Funktionalmatrix

$\mathbf{dx}$ Vektor der verkürzten Unbekannten

$\mathbf{l}$ Vektor der verkürzten Beobachtungen,

wobei der Widerspruch zwischen Bildkoordinaten und Projektorphasen zu minimieren ist.

Für die Kalibrierung des Systems müssen auf dem Objekt an verschiedenen Orten Marken aufgebracht worden sein. Sind die Objektkoordinaten dieser Marken hochgenau bekannt, so kann eine Simultankalibrierung bei der Messung realisiert werden. Andernfalls sind Kalibrierung und Messung nacheinander durchzuführen.

4 Messung und Beispiel

Für eine Messung in einem kalibrierten System wird das Meßobjekt in Form von zwei Phasenbildern als primäre Meßinformation erfaßt. Für jeden Bildpunkt der Kamera können nun in Verbindung mit den an dieser Position vorhandenen Phasenwerten aus den Meßbildern mit Hilfe des räumlichen Vorwärtsschnittes die zugehörigen Objektkoordinaten bestimmt werden.

Abbildung 3: Meßobjekt

Abbildung 4: 3D-Punkte

Als Beispiel wurde ein Ventilator vermessen. Bild 3 zeigt das zu vermessende Objekt, Bild 4 die dazugehörigen ermittelten Objektkoordinaten. Die Meßdauer ist abhängig von der verwendeten Bildverarbeitungshardware; für das Beispiel lag sie bei etwa 10 Sekunden. Die Bestimmung der 3D-Objektpunkte wiederum ist abhängig von Rechenleistung des verwendeten Auswerterechners.

5 Ausblick

Es wurde ein Ansatz zur Konturvermessung auf der Basis einer aktiven Triangulation vorgestellt, der eine einfache Kalibrierung des Projektors ermöglicht. Hierzu wird ein Projektor benötigt, welcher zueinander orthogonale Streifensysteme projizieren kann.

Das mathematische Modell ist flexibel. Es kann eine beliebige Anzahl von Kameras sowie Projektoren eingesetzt werden.

Literaturverzeichnis

[1] Thomas Stahs und Friedrich Wahl. *Oberflächenvermessung mit einem 3D-Robotersensor.* Zeitschrift für Photogrammetrie und Fernerkundung, (6), 190–202, 1990.

[2] G. Seib und H. Höffler. *Überblick über die verschiedenen Moirétechniken.* Vision & Voice Magazine, 4(2), 1990.

Konfokale Anordnung mit Interferometrischen Nachweis in der optischen Tomographie

J. Rosperich-Palm, M. Fernandes, R. Engelhardt

Medizinisches Laserzentrum Lübeck GmbH

23562 Lübeck

Abstract:

Bei der konfokalen Mikroskopie können Strukturen innerhalb von stark streuenden Medien nur auf einem hohen Untergrund nachgewiesen werden. Dieser Untergrund kann reduziert werden, wenn das aus einer über die konfokale Anordnung definierten Tiefe gesammelte Licht mit einer phasenmodulierten Referenzwelle überlagert wird, und anstatt der Gesamtintensität die Amplitude der sich ergebenden Interferenzmodulation detektiert wird. Die physikalische Ursache für diese Reduktion kann mit zwei Effekten erklärt werden, die bisher nur in der Theorie getrennt werden können.

Auf der experimentellen Seite verwendeten wir ein Michelson Interferometer mit einem adaptierten konfokalen System. Wir werden zeigen, daß es möglich ist, auf der Basis eines interferometerischen Nachweissystems in einem konfokalen Aufbau Strukturübergänge innerhalb streuenden Medien bei einer deutlichen Reduktion des Streulichtuntergrundes nachzuweisen.

Grundlagen

Mikroskopierverfahren in stark streuenden Medien sind extrem schwierig, da durch Viel fachstreuung ein sehr hoher Untergrund vorhanden ist. Durch diesen Umstand lassen sich etwa bei einer rein konfokalen Anordnung keine kleinen Signale nachweisen (s. Bild 1).

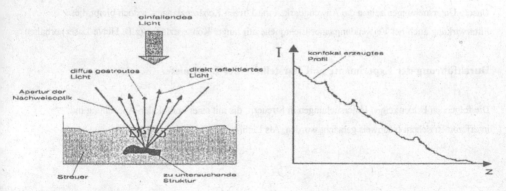

Bild 1: konfokaler Nachweis

572

Unsere Untersuchungen zeigen, daß der durch Vielfachstreuung hervorgerufene Untergrund durch geeignete Filter effektiv unterdrückt werden kann.. Diese Filter lassen sich durch einen interferometrischen Nachweis realisieren, der mit einer geeigneten Anordnung zur Ortsauflösung gekoppelt ist (z. B. konfokal).

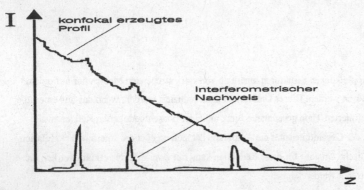

Bild 2: Reduktion des Untergrundes durch Interferometrischen Nachweis

Die physikalische Ursache dieser Filterwirkung bei einem interferometrischen Nachweis kann mit zwei Effekten erklärt werden, die sich experimentell zur Zeit nicht trennen lassen:

a) die Phasenbeziehung für einen Punkt in der Weglängenverteilung ist zeitlich nicht konstant, dadurch ist nur der Nachweis von einfach gestreuten Photonen möglich; bei einem interferometrischen Nachweis werden vielfach gestreute Photonen unterdrückt.

b) die Phasenbeziehung ist konstant, der Nachweis ist dann auf Gradienten in der Weglängenverteilung beschränkt, die signifikant innerhalb eines Bereiches sind, der kleiner als eine Wellenlänge ist. Dieses Verhalten läßt sich mathematisch über eine Fouriertransformation der Modulationstransferfunktion zeigen. Unsere Untersuchungen zeigen die Anwendbarkeit mit kurzen Kohärenzlängen, jedoch bleibt diese Filterwirkung auch bei Verwendung einer Lichtquelle mit langer Kohärenzlänge (z.B. HeNe-Laser) erhalten.

Durchführung der Experimente und Darstellung der Ergebnisse

Die folgenden Bilder zeigen Untersuchungen in Streuern, die mit einer konfokalen Anordnung mit interferometrischem Nachweis gemacht wurden. Als Lichtquelle diente ein HeNe-Laser.

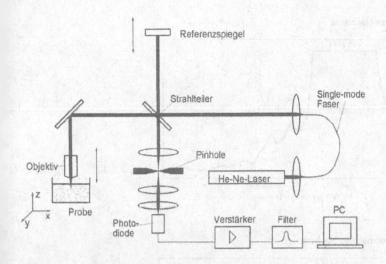

Die Phasenmodulation der Referenzwelle wurde durch wobbeln des Referenzspiegels erzeugt. Prinzipiell ist dies auch mittels eines elektrooptischen Phasenmodulators denkbar.

Untersucht wurde folgende Strukturen in einem streuenden Medium (Intralipid) mit einem Streukoeffizienten von $\mu_s=50\,\mathrm{cm}^{-1}$:

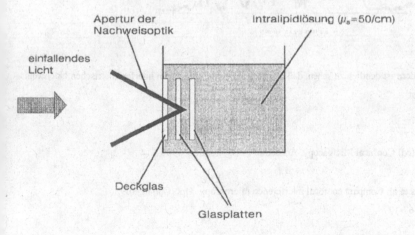

Die nächsten beiden Bilder demonstrieren die prinzipielle Machbarkeit des Verfahrens.

574

1. Messung mit konfokalem Nachweis:

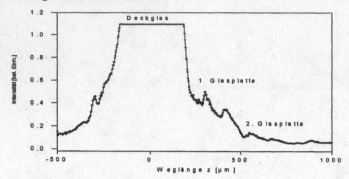

2. Messung mit interferometrischem Nachweis

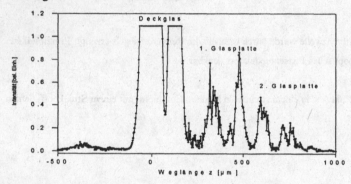

Auf den beiden Bildern ist deutlich zu sehen, daß der Streulichtanteil bei einem interferometrischen Nachweis deutlich reduziert ist.

Literatur:

[1]:   T. Wilson (ed) Confocal Microscopy, Academic Press, London, 1990

[2]:   R. Juskaitis et al, Compact confocal interference microscopy, Opt. Comm. 109, 1994, p. 167-177

[3]:   J. M. Schmitt et al, Confocal microscopy in turbid media, J. Opt. Soc. Am., Vol 11, No. 8, 1994, p. 2226-2235

[4]:   Y. Pan et al, Low coherence optical tomography in turbid tissue, Applied Optics, in press

Speckle-Interferometrische Ansätze zur Messung der Oberflächentopographie von Kurvenscheiben

T.Pfeifer, J.Evertz, C.Faber
Fraunhofer-Institut für Produktionstechnologie
52074 Aachen

Speckle-Interferometrische Verfahren werden bereits seit längerem zur berührungslosen meßtechnischen Erfassung von Oberflächenverformungen eingesetzt. Insbesondere zur raschen Lokalisierung punktueller Materialfehlstellen unter verschiedenartigen Belastungen haben sich die sogenannten *shearographischen Verfahren* etabliert, welche die direkte Bestimmung der partiellen Ableitung einer Oberflächendeformation in beliebig vorgegebener Richtung erlauben.

Abbildung 1 zeigt einen typischen speckle-shearographischen Meßaufbau. Für ein prinzipielles Verständnis des Meßprinzips sei festgestellt, daß bei Betrachtung des Prüfobjektes durch die Abbildungslinse L aufgrund des diesem vorgeschalteten Michelson-Interferometers *zwei* Bilder der zu untersuchenden Oberfläche erzeugt und in der Bildebene I kohärent überlagert werden. Wird einer der beiden in Abbildung 1 dargestellten Spiegel *M1*, *M2* um eine vorgegebene Achse verkippt, so bewirkt dies in linearer Näherung eine senkrecht zur Kippachse orientierte laterale Verschiebung der beiden Objektbilder zueinander. Der in die Gegenstandsebene zurückprojizierte Verschiebevektor wird als *Shearvektor* $\vec{\Delta S}$ bezeichnet. Er ergibt sich unmittelbar aus den vorliegenden Geometrieparametern der Abbildung und dem Kippwinkel der beiden Spiegel.

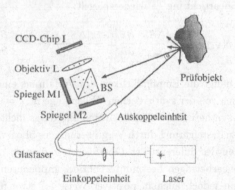

Abbildung 1: *Grundlegender speckle–shearographischer Aufbau*

Die Phase eines solchen speckle-shearographischen Signals (wie sie etwa durch Verwendung eines konventionellen zeitlichen oder räumlichen Phasenschiebe–Algorithmus gewonnen werden kann) ist für sich genommen im allgemeinen noch nicht sinnvoll interpretierbar, da durch die hier angenommene optische Rauhigkeit der Objektoberfläche dem eigentlichen Nutzsignal durch den Speckle-Effekt ein statistisches, in der Phase im wesentlichen gleichverteiltes Störsignal überlagert ist. Aus diesem Grund beschränkte sich die klassische Speckle-Shearographie auf die Bestimmung der Gradienten von Objekt*deformationen*, bei der sich nach getrennter Phasenauswertung vor und nach Einsetzen der Verformung durch anschließende Differenzphasenbildung der statistische Störterm — unter Einhaltung gewisser Bedingungen — gerade heraushebt. Das auf diese Weise erhaltene Differenzphasensignal $\Delta\phi$ kann mathematisch wie folgt beschrieben werden:

$$\Delta\phi(x,y) = \frac{2\pi|\vec{\Delta S}|}{\lambda} \cdot \frac{\Delta z(x+\Delta s_x, y+\Delta s_y) - \Delta z(x,y)}{|\vec{\Delta S}|} \approx \frac{2\pi|\vec{\Delta S}|}{\lambda} \cdot \frac{\partial}{\partial\vec{n}}(\Delta z) \qquad (1)$$

wobei λ und Δz die eingesetzte Laserwellenlänge respektive die Objektverformung bezeichnen und $\frac{\partial}{\partial\vec{n}}$ für die partielle Ableitung in Shearrichtung steht. Bei der Interpretation dieser Gleichung ist jedoch zu beachten, daß der dort auftretende Differenzenquotient nur als Näherung für die nachgestellte partielle Ableitung aufzufassen ist.

Ist man nun anstelle von Objektdeformationen eher an der Erfassung von Oberflächentopographien bzw. (wie etwa Falle von Kurvenscheiben) deren Ableitungen interessiert, so muß dafür Sorge getragen werden, daß der speckle-bedingte statistische Phasenstörterm auf andere Weise eliminiert wird. Eine Möglichkeit hierzu besteht in der Verwendung zweier diskreter Laserwellenlängen λ_1 und λ_2. Bei ansonsten analoger Vorgehensweise zu oben, jedoch nun unter Vermeidung einer Objektdeformation (diese wird hier ja gerade durch eine Wellenlängenänderung ersetzt), erhält man ein Differenzphasensignal, welches unmittelbar den Verlauf der lokalen Oberflächensteigung in Shearrichtung $\frac{\partial z}{\partial\vec{n}}$ wiederspiegelt:

$$\Delta\phi(x,y) \approx \frac{2\pi|\vec{\Delta S}|}{\Lambda_{syn}} \cdot \frac{\partial z}{\partial\vec{n}} = \frac{2\pi|\vec{\Delta S}|(\lambda_1 - \lambda_2)}{\lambda_1\lambda_2} \cdot \frac{\partial z}{\partial\vec{n}} \qquad (2)$$

Wie im klassischen Fall hängt die Empfindlichkeit des Verfahrens einerseits vom Shearbetrag ΔS, andererseits aber auch von der synthetischen Wellenlänge $\Lambda_{syn} = \frac{\lambda_1\lambda_2}{\lambda_1-\lambda_2}$ ab. Die beiden Parameter Sheargröße und synthetische Wellenlänge sind allerdings nicht als gleichwertig anzusehen, da eine Empfindlichkeitssteigerung durch Vergrößerung des Shearvektors eine entsprechend schlechtere Approximation des differentiellen Oberflächengradienten nach sich zieht. Die Tatsache, daß bei großen Shearbeträgen bestimmte Fourierkomponenten des Signals vollständig unterdrückt werden, kann jedoch auch in positiver Weise — etwa für eine Fehlererkennung bei periodischen Strukturen — ausgenutzt werden. Sind solche Tiefpaßeffekte hingegen unerwünscht, so ist der Shearbetrag möglichst klein zu halten und die erforderliche Empfindlichkeit durch Wahl einer entsprechend kleinen synthetischen Wellenlänge zu gewährleisten. Aufgrund des Mehrwellenlängen–Ansatzes wird der Einstellbereich des Shearbetrags in keiner Weise durch die vorliegende mittlere Specklegröße begrenzt.

Ein interessantes Einsatzfeld für die Zweiwellenlängen-Speckle-Shearographie bilden all diejenigen Meßaufgaben, bei denen die zu bestimmende Meßgröße in unmittelbarem Zusammenhang mit der ersten oder höheren Ableitungen eines gegebenen Oberflächenverlaufes steht. Naheliegende Beispiele finden sich bei der Rundheits- und Ebenheitsprüfung, vor allem aber bei der Untersuchung von dynamisch beanspruchten Kurvenscheiben wie etwa Nockenwellen. Hier hängt die Beschleunigung des Abtriebstößels unmittelbar von der Krümmung der Lauffläche ab, so daß diskontinuierliche Übergänge im Krümmungsradius zu unerwünschtem mechanischem Aufschwingverhalten führen. Da eine solche sprunghafte Krümmungsradiusänderung mit einer entsprechend unstetigen Änderung der zweiten Ableitung der Oberfläche verknüpft ist, was gleichbedeutend mit einem Knick in der ersten Ableitung ist und somit zu einer stetig differenzierbaren Objektoberfläche führt, lassen sich Schwachstellen im dynamischen Verhalten von Kurvenscheiben durch Messung der Prüflingsoberfläche selbst nur schwer erfassen.

Abhilfe kann hier die Zwei-Wellenlängen-Shearographie leisten, da bei dieser unmittelbar die erste Ableitung der Prüflingstopographie bestimmt wird. Ein entsprechendes Meßergebnis ist Abbildung 2 zu entnehmen. Das für die weitere Diskussion gewählte Beispiel ist eine Nockenwelle, deren einzelne Nocken aus mehreren Kreis- und Geradenstücken bestehen. Indem der in Abbildung 1 dargestellte Laser durch eine Halbleiter–Laserdiode ersetzt und deren Wellenlänge durch Temperatur- und Stromverstimmung über einen Modensprung hinweg um $\Delta\lambda = 0.4$ nm verschoben wurde, konnte der Oberflächengradient einer einzelnen Nocke in einem Übergangsbereich der Krümmungsradien meßtechnisch erfaßt werden. Oberflächenbereiche, welche bei dynamischer Beanspruchung zu unerwünschtem Schwingverhalten führen werden, treten im Meßsignal deutlich als scharfe Knicke hervor (s. Bild 2 rechts).

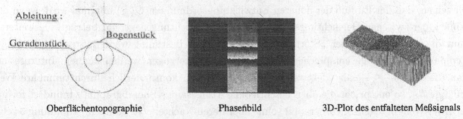

Oberflächentopographie Phasenbild 3D-Plot des entfalteten Meßsignals

Abbildung 2: *Erkennung dynamisch kritischer Bereiche bei Kurvenscheiben am Beispiel einer Nockenwelle*

Als gewichtiger Nachteil dieses Verfahrens verbleibt der Umstand, daß zur Meßdatenerfassung zwei bekannte, frequenzstabile Laserwellenlängen verwendet werden müssen. Dies führt zum einen zu einem erhöhten experimentellen Aufwand, zum anderen zu einer Einschränkung der durch die Wahl der synthetischen Wellenlänge vorgegebenen Meßdynamik und -auflösung. Selbst bei Einsatz durchstimmbarer Laserquellen wie etwa Halbleiter-Laserdioden hat dies aufgrund der derzeit noch recht starken Begrenzung des verfügbaren kontinuierlichen Durchstimmbereiches eine spürbare Limitierung der Systemflexiblität zur Folge. Desweiteren ist insbesondere bei der Verwendung von kleinen synthetischen Wellenlängen auf das Problem der meßwertverfälschenden chromatischen Aberration hinzuweisen.

Einen Ausweg aus diesem Problemkreis eröffnet ein Ansatz, der eine von der Form des Prüfobjektes abhängige Änderung des Gangunterschiedes durch kontrollierte *Verschiebung des Quellpunktes* der eingesetzten Laserquelle erzeugt. Um konkrete Aussagen darüber treffen zu können, wie sich eine kleine Verschiebung δl des Laserquellpunktes auf eine Änderung des optischen Gangunterschiedes bei gegebenem Shearvektor auswirkt, sind aufwendigere Rechnungen und Näherungen durchzuführen, die den hier bereitgestellten Rahmen sprengen würden. Soviel sei angemerkt: Wählt man den experimentellen Aufbau derart, daß die Verschiebung δl des Laserquellpunktes klein gegen den Abstand l der Laseraustrittsfläche von der Prüfoberfläche ist, und untersucht die Spezialfälle einer Lasertranslation — in der durch die Beleuchtungs- und Beobachtungsrichtung vorgegebenen Ebene — zum einen in, zum anderen senkrecht zur Richtung der Beleuchtungachse, so läßt sich die durch die Laserverschiebung induzierte Änderung des Gangunterschieds in einer Form darstellen, die bei Reihenentwicklung starke Analogien zur Multipolentwicklung in der Elektrostatik aufweist. Es zeigt sich, daß im zuerst genannten Fall nur eine geringe Änderung des Gangunterschiedes bei Translation des Quellpunktes hervorgerufen wird, was auch anschaulich zu erwarten war. Im zweiten Fall ergibt sich hingegen eine starke Änderung des Gangunterschiedes sowie eine dadurch bedingte Änderung der Interferenzphase $\Delta\phi$ der folgenden Form:

$$\Delta\phi(x,y) \approx \frac{2\pi}{\lambda} \cdot \frac{\delta l}{l} \cdot |\vec{\Delta S}| \sin\alpha \cdot \frac{\partial z}{\partial \vec{n}} + const + \mathcal{O}(\tfrac{1}{l^2}) \tag{3}$$

wobei α den Winkel zwischen Beleuchtungs- und Beobachtungsrichtung angibt. Nähere Analysen zeigen, daß der Einfluß der höheren Entwicklungsordnungen $\mathcal{O}(\tfrac{1}{l^2})$ minimal wird, wenn der Winkel, der zwischen Beleuchtungs- und Beobachtungsrichtung liegt, 45° beträgt. Desweiteren kann die genaue Form der „Störterme" $\mathcal{O}(\tfrac{1}{l^2})$ explizit bestimmt werden, so daß die aufgenommenen Meßsignale entsprechenden Korrekturen unterzogen werden können. Interpretiert man die Größe $\frac{\lambda \cdot l}{\sin\alpha \cdot \delta l}$ als effektive, in weiten Bereichen kontinuierlich durchstimmbare Wellenlänge $\Lambda_{\delta l}$, so entspricht die oben beschriebene Translations-Shearographie zumindest formal der Zwei-Wellenlängen-Shearographie, ohne aber deren Nachteile zu besitzen. Abbildung 3 zeigt Phasendatensätze des Oberflächengradienten eines Nockenabschnitts, links gemessen mittels Zwei-Wellenlängen-Speckle-Shearographie, rechts mittels Verschiebung des Laserquellpunktes.

Phasenbild mit Zwei-Wellenlängen-
Shearographie
　　　　　　　　Phasenbild mit Translations-
Shearographie

Abbildung 3: *Vergleich der vorgestellten Topographieerfassungstechniken*

Literatur

R. Jones and C. Wykes, *Holographic and Speckle Interferometry*, Cambridge Studies in Modern Optics 6, Cambgridge University Press (1983)

A. Ettemeyer and M. Honlet, *Nondestructive Testing with TV-Holography and Shearography*, **VDI**-Berichte 1118, VDI-Verlag (1994)

Y. Iwahashi, K. Iwata and R. Nagata, *Single-Aperture Speckle Shearing Interferometry with a Single Grating*, **Appl. Op.** Vol.23 No.2 (1984)

Grundlagen zur Dreidimensionalen Modalanalyse mit einem Mehrkanal-Laservibrometer

M. Sellhorst, S. Seiwert, R. Noll

Fraunhofer-Institut für Lasertechnik, Steinbachstraße 15, D-52074 Aachen

1 Einleitung

Bei Neuentwicklungen im Werkzeugmaschinen- und Automobilbau ist die Kenntnis und entsprechende Berücksichtigung des dynamischen Verhaltens komplexer mechanischer Baugruppen häufig ausschlaggebend für die Funktion und Qualität des neuen Produktes. Die Analyse an Hand von theoretischen Modellen ist in vielen Fällen unzureichend. Eine direkte Messung des dynamischen Verhaltens ist erforderlich.

Das in der Praxis dafür bewährte Meßverfahren ist die Experimentelle Modalanalyse (EMA) [1]. Dabei wird der Zeitverlauf der räumlichen Bewegung mehrerer ausgewählter Strukturpunkte in Abhängigkeit von einem eingeleiteten Kraftsignal bestimmt. Die Meßdaten liefern die interessierenden dynamischen Eigenschaften des Meßobjekts. Dieses Meßverfahren ist sehr zeitaufwendig, und beeinflußt durch das Aufbringen von elektromechanischen Beschleunigungssensoren die dynamischen Eigenschaften des Meßobjekts.

Seit einiger Zeit werden als Ergänzung zu den Standardmeßaufnehmern Laservibrometer eingesetzt. Mit den Vibrometern wird der Zeitverlauf der Schwingbewegung eines Objekts in einer Dimension berührungslos gemessen.

Für die Modalanalyse des dynamischen Objektverhaltens muß der Zeitverlauf der Verlagerungs- bzw. Geschwindigkeitsvektoren an mehreren diskreten Objektpunkten bekannt sein. Da die Laservibrometer nur die Information über eine Komponente des Geschwindigkeitsvektors liefern, muß das Verfahren erweitert werden, um als Meßverfahren für die Modalanalyse eingesetzt werden zu können.

2 Experimentelle Modalanalyse

Bei der experimentellen Modalanalyse werden die komplexen Übertragungsfunktionen für Nachgiebigkeiten zwischen diskreten Strukturpunkten bestimmt. Die Nachgiebigkeit ist als das Verhältnis zwischen einer Komponente der Verlagerung und einer Komponente der eingeleiteten Kraft definiert. Die Nachgiebigkeiten aller Komponenten ergeben den Nachgiebigkeitstensor. Ist dieser für zwei Strukturpunkte bekannt, so ist es möglich, den zeitlichen Verlauf der Verlagerung an dem einen Strukturpunkt für ein an dem anderen Punkt eingeleitetes Kraftsignal zu bestimmen.

Bei der experimentellen Modalanalyse (EMA) werden die Nachgiebigkeitsfrequenzgänge für eine diskrete Zahl von Strukturpunkten in einem festgelegten Frequenzbereich gemessen. Das gemessene dynamische Verhalten wird durch ein System von Einmassenschwingern simuliert, so daß die Nachgiebigkeitsfrequenzgänge von Objekt und Modell in einem bestimmten Frequenzbereich übereinstimmen. Die Parameter des Modellsystems werden auch als modale Parameter bezeichnet. Sind die modalen Parameter für eine bestimmte Anregungsart bekannt, so kann das Objekt durch ein Gittermodell modelliert (Bild 1a) und die Schwingungsform dreidimensional animiert werden (Bild 1b).

Aus den Analysedaten werden oft wichtige Hinweise für Verbesserungen der Konstruktion von Maschinen- oder Fahrzeugteilen abgeleitet.

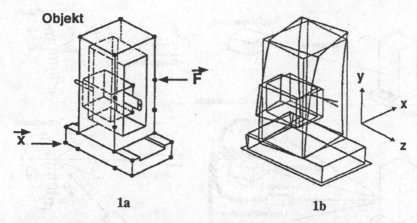

Objekt

$\vec{F}$

$\vec{x}$

y
x
z

1a

1b

Bild 1a: Strukturpunkte für die Krafteinleitung und die Verlagerungsmessung
1b: Analyse der Schwingungsform aus den Nachgiebigkeitsfrequenzgängen

3 Eindimensionale (1D) - und dreidimensionale (3D) - Laserdopplervibrometrie

In vielen Fällen, etwa bei heißen oder rotierenden Meßobjekten ist die Messung mit herkömmlichen, berührenden Bewegungssensoren nicht mehr möglich. Ebenso bei Objekten deren schwingende Masse in der Größenordnung der Sensormasse liegt muß auf berührende Meßverfahren verzichtet werden, um die dynamischen Eigenschaften des Meßobjekts nicht zu beeinflussen. Ein Meßverfahren, das die herkömmlichen Bewegungssensoren ersetzt, ist die Laserdopplervibrometrie. Bei der Laserdopplervibrometrie wird die Dopplerverschiebung des von einer bewegten Oberfläche gestreuten Lichts zur Schwingungsmessung ausgenutzt [2]. Der Zeitverlauf der Dopplerverschiebung ist dabei direkt proportional zum Zeitverlauf der Objektgeschwindigkeit.

Bei der bekannten 1D-Vibrometrie wird nur die zur Einstrahlrichtung kollineare Komponente des Geschwindigkeitsvektors erfaßt. Diese Information allein ist für die Modalanalyse nicht ausreichend. Aus diesem Grund wird versucht das Meßverfahren der Laserdopplervibrometrie so zu erweitern, daß der Zeitverlauf aller kartesischen Komponenten des Geschwindigkeitsvektors am Meßort kontinuierlich und phasenrichtig erfaßt wird.

Hierzu wird die Tatsache ausgenutzt, daß die Dopplerverschiebung des Streulichts von der Streurichtung abhängt. Bild 2 zeigt die Meßanordnung bei der 1D- und bei der 3D-Vibrometrie.

Bei der 3D-Vibrometrie wird das Streulicht aus drei Richtungen beobachtet, die nicht in einer Ebene liegen. Die zugehörigen Empfindlichkeitsvektoren spannen in diesem Fall ein schiefwinkeliges Koordinatensystem auf, in welches die Komponenten des Geschwindigkeitsvektors projiziert werden. Auf diese Weise kann durch Rücktransformation der gemessenen Komponenten in ein kartesisches Koordinatensystem der Zeitverlauf des gesamten Geschwindigkeitsvektors ermittelt werden.

Bei der beschriebenen Anordnung wird die Dopplerverschiebung des Streulichts ebenfalls mit einem Heterodyn-Interferometer bestimmt. Hierbei sind die Intensität, Kohärenz und der Polarisationsgrad des Objektstreulichts von ausschlaggebender Bedeutung für den optischen Aufbau und die Auslegung der Signalverarbeitungselektronik. Die Diagramme in Bild 3 zeigen die Abhängigkeit des Polarisationsgrads des Streulichts und der Reflektivität vom Beobachtungswinkel für verschiedene Beispieloberflächen. Um den systematischen Fehler aufgrund der Projektion des Ge-

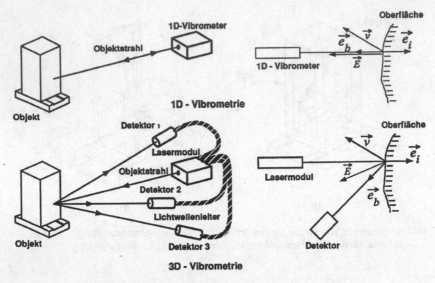

Bild 2: Skizze zur dreidimensionalen Laserdopplervibrometrie

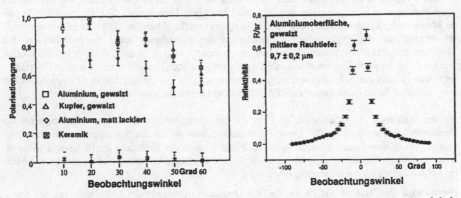

Bild 3: Abhängigkeit des Polarisationsgrades und der Reflektivität vom Beobachtungswinkel bei senkrechter Einstrahlung auf die Objektoberfläche

schwindigkeitsvektors in ein schiefwinkeliges Koordinatensystem für die Durchführung einer Modalanalyse hinreichend klein zu halten, muß der Beobachtungswinkel größer als 40° sein. Wie die Messungen zeigen, fällt der Polarisationsgrad in diesem Fall für einige Proben auf ca. 60%. Die Reflektivität pro Raumwinkeleinheit liegt bei der Aluminiumprobe für diesen Winkel bei etwa 0,05/sr und ist damit um etwa eine Größenordnung niedriger als bei zur Einstrahlrichtung kollinearer Beobachtungsrichtung.

Ein entscheidender Unterschied der 3D-Vibrometrie zur 1D-Vibrometrie ist die Bewegung des abgebildeten Objektleuchtpunkts auf der Detektoroberfläche. Hierdurch wird eine Frequenzmodulation des Interferenzsignals verursacht, die nicht proportional zum Zeitverlauf der Objektbewegung ist.

Für die Auswertung der Streulichtsignale wurde eine Vibrometeranordnung mit Verstärker und Signalverarbeitungselektronik aufgebaut, welche zur Zeit erprobt und verbessert wird. Bild 4 zeigt eine Vergleichsmessung zwischen der Vibrometeranordnung und einem berührenden, induktivem Wegaufnehmer bei der Grenzfrequenz des induktiven Wegaufnehmers von 90 Hz. Bis auf eine leichte Amplitudenmodulation des Wegaufnehmersignals, die von Netzeinstreuungen herrührt, stimmen die beiden Signale in Amplitude, Frequenz und Phasenlage innerhalb der Meßunsicherheit überein.

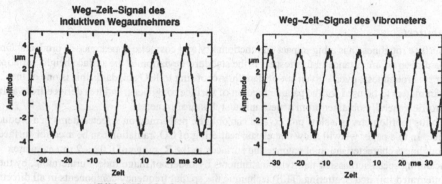

Bild 4: Vergleichsmessung zwischen einem konventionellem
Wegaufnehmer und dem Laservibrometer

4 Zusammenfassung

Durch die Detektion des Streulichts von einem Meßobjekt aus drei nicht koplanaren Raumrichtungen ist es möglich, das Meßverfahren der Laserdopplervibrometrie so zu erweitern, daß der Zeitverlauf des Geschwindigkeitsvektors am Meßort vollständig erfaßt wird. Die so gewonnenen Signale sind für eine Modalanalyse des dynamischen Objektverhaltens geeignet. Gegenüber dem herkömmlichen Meßverfahren mit berührenden Meßaufnehmern ist die Messung mit einem Laservibrometer berührungs- und trägheitslos. Auf diese Weise wird das dynamische Verhalten des Meßobjekts nicht beeinflußt und der Meßort kann beliebig gewechselt werden.

Die vorgestellten Arbeiten werden teilweise durch das BMBF, Förderkennzeichen 13N5961, gefördert.

5 Literatur

[1] M. Weck,   Werkzeugmaschinen, Fertigungssysteme, Band 4, 4. Auflage, VDI-Verlag, Band 4, 4. Auflage (1992) Kap. 6,

[2] R. Dändliker, J.F. Willemin, Measuring Microvibrations by Heterodyne Speckle Interferometry, Opt. Lett. 6 (1981) 165-167

CO$_2$ Laser Scattering for Surface Characterisation of Engineering Surfaces

Lars Mattsson
Surface Evaluation Laboratory
Institute of Optical Research
S-100 44 Stockholm, Sweden

Abstract

Surface roughness has a big impact on functionality and cosmetic appearance of products. One parameter of importance in this respect is the standard deviation of the surface height variations, R_q or rms roughness, over a given cut-off length. According to ISO standards this is one of the parameters to be used for the parametrization of surface roughness, and it is derived from a line profile generally obtained from a mechanical stylus profilometer.

Stylus profilometry is a slow process and can only be performed on selected items in a production line. In this paper we will show that elastic scattering of CO$_2$ radiation can be used for surface roughness characterisation of engineering surfaces in the R_q range of 0.02 - 2 μm at an area scanning speed surpassing the stylus instruments by orders of magnitude. Furthermore, by the Total Integrated Infrared Scattering (TIIS) technique the spatial frequency components in all directions of the surface are covered in a single measurement. This technique which averages the roughness over an area of about 1 mm^2 per measurement point has allowed us for the first time to map the roughness variation across large ground surfaces. On a firm basis of more than 1200 measurements on different surfaces we can also claim that the agreement between stylus data and roughness values derived from laser scatter data is excellent.

Introduction

Surface finishing operations are critical issues in industrial production. A fault at this stage of the process means large economic losses. The purpose of the finishing process can be cosmetic, to make a surface appear attractive to the eye, or it can be functional, i.e. the surface needs a specified surface roughness to fulfil its requirements.

In the cosmetic case it is the interaction between the surface roughness and visible light that must be well established for automatic production control. This is typically done by gloss meters, i.e. the measurement of specular reflectance relative to a known standard surface.

In the functional case a specified surface roughness e.g. a value of the average roughness[1] R_a or root mean square (rms) roughness[1] R_q is assigned to the surface already at the drawing board. These parameters represent averages of the amplitude of the roughness measured along a given cut-off length on the surface. Their values will have a direct or indirect influence on e.g. oil distribution capabilities between two contacting surfaces, the adhesion properties or the light scattering properties of a matte surface. The observation one can make in the engineering industries of today, is the demand for a specific surface roughness, rather than the previously common "smoother than" demand.

These requirements on surface finish and texture also means harder constraints on measurement procedures and in many cases also calls for new types of measurements of roughness over areas rather than along a line. In other words, three dimensional characterisation of roughness is becoming essential.[2]

A symbolic representation of inspection techniques for surface roughness is illustrated in Fig 1. First there is the subjective way of using visual perception for comparing a reference surface with the item to be evaluated. This is particularly good for finding defects and getting a general view of the appearance of surfaces, as the eye is very sensitive to deviations in texture. A similar human related inspection technique is the scraping of the surface with the nail and compare it with a reference surface. However, these qualitative ways of surface roughness characterisation are far from sufficient when it comes to specific tolerances in roughness specifications.

The quantitative techniques are represented in Fig. 1 by light scattering, interferometry and stylus techniques. This paper will concentrate on the results we have obtained using CO_2 laser scattering on engineering surfaces, but some introductory comments about the other techniques are given below.

Interferometric techniques provide 3-D topographic information over small or large areas depending on magnification. By image capturing and computer analysis of interference or moiré fringe patterns the technique provide a faster analysis technique than mechanical stylus techniques do. The WYKO Rough Surface Tester[3] is a typical representative of today's white light interferometers for rough surfaces, but as for all interferometric techniques it requires a vibration stable environment.

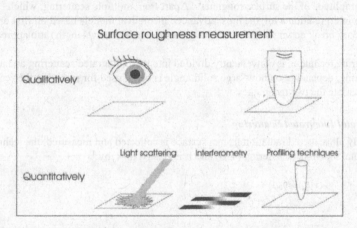

Fig. 1. Several measurement principles exist for surface roughness characterisation.

Environmental sensitivity of optical techniques is one reason for the complete dominance of mechanical stylus instruments in mechanical workshops. Another is the prescribed standardisation of roughness measurements and its close connection to mechanical stylus profiling. An advantage of the mechanical stylus technique is that it is quite easy to understand that a sufficiently sharp tip will reproduce a profile of the surface, while the optical techniques are sensitive to both optical properties of the surface and the surface topography. The latter has been shown to be a serious problem in some cases for optical stylus instruments.[4]

Nevertheless, none of the existing optical or mechanical instruments are suitable for on line analysis of machined surfaces passing by at high speeds and having roughness in the µm - sub-µm range. This was one of the reasons for starting a joint European project on Non-contact surface metrology.[5] The particular objective at the Surface Evaluation Laboratory was to develop and test infra-red (IR) scattering at λ 10.6 µm as a possible tool for rapid assessment of the roughness of engineering surfaces. This paper gives a review of the results we have achieved throughout this project, and after an introduction about the scattering technique, some features of the instrumentation is given, before ending with the results and comparisons between conventional stylus measurements and total integrated infrared scattering.

Surface roughness and light scattering

The use of CO_2 laser scattering as a tool for determining surface roughness is just an extension of previous applications of the interaction between electromagnetic radiation and rough surfaces.[1] The scattering technique was first explored for radar wavelengths, but in the sixties the scattering at visible wavelengths was applied by Bennett and Porteus[6] as a tool for surface roughness measurements of high quality optical surfaces. In 1987 Total Integrated Scattering (TIS) at the HeNe wavelength of λ 0.6328 μm became an ASTM standard for effective surface roughness measurement of optical surfaces.[7] This standard is applicable for surfaces having root mean square roughness $(R_q) < 35$ nm. With the most sensitive TIS instrument, surfaces can be measured at atomic height levels with R_q values smaller than 0.1 nm.[8]

Earlier studies have been made, using visible scattering as a tool for roughness characterisation of machined surfaces, e.g. Peters.[9] However, one major obstacle is, the short wavelength of the light in relation to the amplitude of the surface roughness. Apart from multiple scattering, which complicates the interpretation a lot, the most advanced theoretical models based on first order perturbation theory break down when the rms roughness / incident wavelength ratio approaches 10 % - 20 %.

The light scattering technique is conveniently divided into total integrated scattering and angle resolved scattering, depending on how large solid angle is being used for receiving the scattered light. Fig. 2 illustrate the two techniques.

Principles of Total Integrated Scattering

When practically all scattered radiation from a surface is collected and measured, the technique is referred to as total integrated scattering (TIS) and it is related to R_q by:[1]

$$TIS = \frac{R_d}{R_d + R_s} = 1 - e^{-(4\pi R_q \cos\theta_0/\lambda)^2} \tag{1}$$

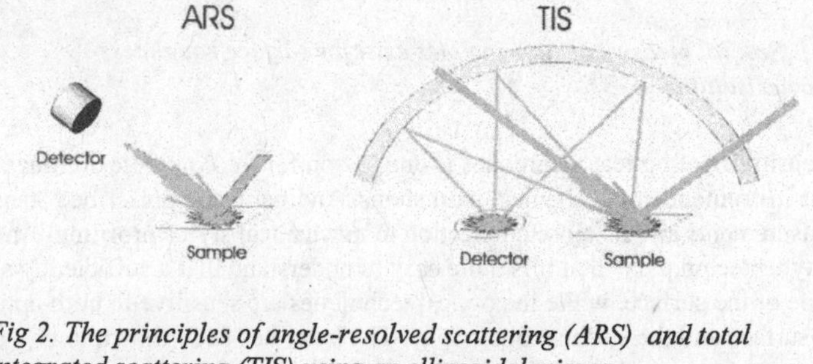

Fig 2. The principles of angle-resolved scattering (ARS) and total integrated scattering (TIS) using an ellipsoidal mirror.

where λ is the wavelength of the radiation, θ_0 the angle of incidence, and R_s and R_d are the specular and diffuse reflectance of the surface. The sample must be opaque for the wavelength of the radiation, otherwise bulk scattering and back surface reflections may interfere with the measurement.

Expression (1), which is based on scalar scattering theory,[1] is valid under the conditions that

$$R_q \cos \theta_0 \ll \lambda \qquad (2)$$

From (1) it is seen that by measuring the diffusely and specularly reflected light, the R_q value can be immediately revealed. The R_q value will be determined within a certain surface spatial frequency bandwidth, (the reciprocal of the surface spatial wavelengths) determined by the collection angles of the diffusely scattered light. By performing the collection within a spherical geometry R_q values will be obtained for the true 3-dimensional topography. The technique also averages the R_q over the entire spot being irradiated. Hence, by working with different spot diameters, a trade off can be made between lateral sampling area and the measurement speed. High speed scanning of surface roughness variations over large areas can thus be made with this technique.

Principles of Angle-Resolved Scattering

The principle of angle resolved scattering measurements is shown in Fig. 2. It resembles the function of the eye when looking at a surface. By moving the eye relative to the incident light, different intensity levels will be observed, representative of the amplitude of the surface roughness and the correlation length and direction of the surface lay. In contrast to the TIS technique which yields an integrated single parameter value, R_q, the angle-resolved technique requires multiple recording at several scatters angles and post processing to provide the surface roughness information.

Parameters involved in the angle resolved scattering process are the wavelength of the radiation λ the angle of incidence θ_0 and the angle of scattering θ_s, detection solid angle $d\Omega$, the optical factor Q involving polarisation state and optical constants of the surface, and the topography of the surface described by the power spectral density (PSD) function $\mathbf{g} (f_x, f_y)$, where f_x and f_y represent the spatial frequencies in the x and y directions of the surface. The officially and ASTM accepted abbreviation for the output of reflected angle-resolved scattering is BRDF (Bi-directional Reflectance Distribution Function)[10] and its relation to the above mentioned parameters is[11]

$$BRDF = \frac{dP_s/d\Omega}{P_0 \cos \theta_s} = \frac{16\pi^2}{\lambda^4} \cos \theta_0 \cos \theta_s Q \mathbf{g} (f_x, f_y) \qquad (3)$$

where P_0 is the incident power, dP_s is the scattered power received by the detector within the given detection solid angle. If the surface is opaque and the optical constants are known, the PSD function is calculated from the BRDF values by properly scaling the data with the optical and the geometrical factors given in (3). The PSD in its general 2-dimensional form holds information about both height and lateral dimensions in all directions on the surface, and is the fundamental basis for comparison between different surface roughness measurement techniques. The R_q values are obtained by integrating the PSD function within the spatial frequency limits given by the scatter measurements, provided eq. 2 still holds, i.e. the roughness must be considerably smaller than the wavelength of the incident radiation. However, in contrast to the TIS technique which covers spatial frequencies in all directions on the surface, only a limited spatial frequency band in the 2-dimensional power spectrum will be covered by the angle resolved measurement. This is due to the fact that the detector normally scans in a single scattering plane at a time. The results obtained with angle resolved scattering will thus be critically dependent of the lay of the surface, and in which direction the scattering is measured.

Implementing CO_2 laser radiation for scatter measurements

From equation 1 and 3 it is seen that the TIS and BRDF values are simply related to the wavelength, λ. The larger the wavelength the smaller the scattering will be for a given surface roughness. The upper limit of the useful R_q range, before breakdown of the roughness-scattering models, is about 0.035 µm for visible wavelengths at normal incidence. By operating at large angles of incidence and by changing the laser wavelength from the HeNe 0.6328 µm to the CO_2 laser-wavelength of 10.6 µm, the useful R_q limit can be extended up to 2 µm. The spatial frequency bandwidth will also be shifted towards lower frequencies, making it possible to cover the standard cut-off length of 0.8 mm for the R_q measurements.

Based on this possibility we built a combined Angle Resolved Infrared Scattering (ARIS) and Total Integrated Infrared Scattering (TIIS) system. The performance of the system has been optimised for roughness measurements of machined surfaces in the R_q range of 0.02 - 2 µm, and it operates by irradiating the sample with a ~1 mm diameter spot of s-polarised radiation. By laterally scanning the sample at a speed of 1 mm²/s we can now for the first time rapidly quantify the roughness variations on large ground surfaces, over areas up to 100 x 100 mm².

Results

In total, more than 1200 scattering measurements on different metal and glass surfaces have been made within the project. Some of them as a parts of the three round robin tests that were performed. The results presented here will just exemplify the technique over the engineering roughness range. R_q roughness values have been calculated using the scalar scattering theory for the TIIS measurements (eq. 1) while first order perturbation theory[11] (eq. 3) was applied for the angle-resolved scatter measurements. The results of the scatter measurements have been compared with the roughness values obtained by the other techniques within the project, taking care of always using the same spatial frequency bandwidth for the roughness intercomparisons. Particularly good agreement was obtained between TIIS, confocal microscopy and conventional stylus profiling techniques.[5]

In addition to the project specific surfaces, a series of measurements were made on ground Rubert roughness standards.[12] Figure 3 shows the logarithm of the BRDF measured with the ARIS system at λ 10.6 µm and at an angle of incidence of 60°, for two of the Rubert roughness standards labelled R_a 0.05 µm and 0.8 µm respectively. The corresponding R_q values obtained by the Talystep[14] were 0.09 µm and 1.1 µm respectively over 0.8 mm cut-off length, using a 1 µm radius stylus. Note the pronounced specular reflection peaks showing up at 60° for both the smooth and the rough surface. This is a good indicator that the roughness is $\ll \lambda$ and that the first order perturbation theory can be applied for extracting R_q values from the PSD-data. In figure 4, the corresponding BRDF result is given for the same surfaces but measured with HeNe-laser light at λ 0.6328 µm. The loss of the specular reflex for the R_a 0.8 µm surface reveals that the condition of equation 2 is not fulfilled, and there is no way the scattering theory can be used for evaluating the roughness. The graphs of figure 4 show very clearly why attempts of using visible wavelength scattering techniques for roughness characterisation fail. The wavefront of the visible, reflected light from the 0.8 µm rough surface is too much disturbed to be treated as just a perturbation, and no simple relationship exists between the scattering and the roughness.

The six ground Rubert standard surfaces having R_a values of 0.05, 0.1, 0.2, 0.4, 0.8 and 1.6 µm were also measured with the TIIS instrument and with the Talystep surface profiler. The data is summarised in Table I for the TIIS measurements made at 75° angle of incidence and in Fig 5 the R_q values of TIIS are plotted vs the R_q values obtained from the Talystep. Thanks to the grinding process, these surfaces have almost a uniform lay direction, and the rms roughness is totally dominated by the grinding grooves. This is why the measured TIIS data and Talystep data fits so

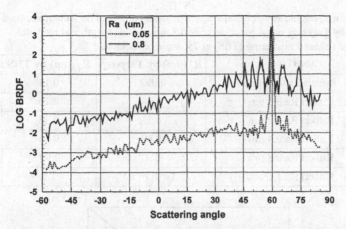

Fig 3. Angle-resolved infrared scattering (s-polarised) at λ 10.6 mm on two Rubert standard surfaces, having R_a values of 0.05 μm and 0.8 μm. Note the distinct and equally intense specular reflex for both surfaces at 60° scattering angle indicating that the first order perturbation theory for scattering can be applied for roughness determination.

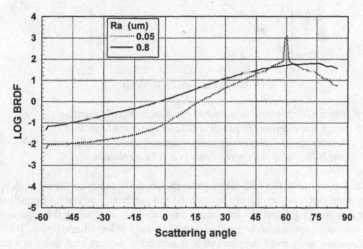

Fig 4. Angle-resolved visible scattering (s-polarised) at λ 0.6328 mm and 60° angle of incidence on two Rubert standard surfaces, having R_a values of 0.05 μm and 0.8 μm. Note the complete loss of the specular reflex for the rougher surface, and the shift of the maximum scatter towards larger scatter angles.

well. If a considerable cross lay should have existed, the R_q values of the TIIS are expected to give the more correct and higher roughness value than what would be obtained by the line profile measurements.

Table I. Rms roughness values (average of 5 traces) obtained on Rubert standard surfaces by Talystep stylus profiling over 0.8 mm cut-off length and by Total Integrated Infrared Scattering (TIIS) at 75° angle of incidence

Surface	R_q (µm) by Talystep	R_q (µm) by TIIS
Rubert Ra 0.05 µm	0.092	0.11
Rubert Ra 0.1 µm	0.14	0.16
Rubert Ra 0.2 µm	0.31	0.33
Rubert Ra 0.4 µm	0.49	0.54
Rubert Ra 0.8 µm	1.1	0.99
Rubert Ra 1.6 µm	1.7	1.7

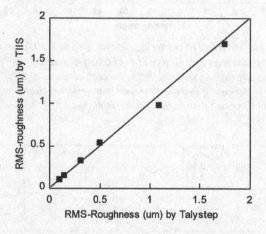

Fig 5. Comparison between rms roughness (R_q) obtained by mechanical stylus (Talystep) and Total Integrated Infrared Scattering for Rubert roughness standards made out of ground steel. The solid line corresponds to a 1:1 correlation.

The greatest achievement with the TIIS technique is its speed. By scanning the surfaces at 1 mm²/s, large areas of ground surfaces can quickly be assessed and patches of deviating roughness, caused by e.g. a defect in a grinding wheel can immediately revealed. By optimising the technique we expect that measurement speeds can be boosted by two to three orders of magnitude. It is not sensitive to vibrations, and curved samples can be measured as well, at least down to a radius of 20 mm with the present system. The drawback compared to the conventional stylus technique is the requirement of a clean surface, and a relatively clean operating environment. But that is a general requirement for all optical surface roughness measurement techniques, and can be handled by proper cleaning and careful design.

Conclusion

The application of CO_2 laser scattering as a tool for rapid assessment of surface roughness of engineering surfaces has been demonstrated. Both angle-resolved infrared scattering (ARIS) and total integrated infrared scattering (TIIS) have been tested. The TIIS technique provides a rapid parametric technique for measurement of the true area rms roughness up to 2 μm over large sample areas, at a speed of 1 mm²/s. It shows an exceedingly good correlation with stylus profiling techniques for ground surfaces with a uniform lay. The ARIS technique is particularly useful for evaluating lay directions and rms roughness in different directions of a surface. Both techniques have the potential of being operated for on-line inspection in an industrial environment.

Acknowledgement

This work has been performed as a joint European project supported by the European Community under the BCR Programme (now the SMT Programme) and by the National Swedish Board for Technical Development. Måns Bjuggren who designed and set up most of the IR-scatter equipment in the Surface Evaluation Laboratory and also analysed the results, and Laurent Krummenacher who did most of the measurements are greatly acknowledged.

References

1. J. M. Bennett and L. Mattsson, *Introduction to Surface Roughness and Scattering*, Optical Society of America, Washington, 1989.
2. K.J. Stout, P.J. Sullivan, W.P. Dong, E. Mainsah, N. Luo, T. Mathia and H. Zahouani, "The development of methods for the characterization of roughness in three dimensions," Publication EUR 15178 EN of the Commission of the European Communities Dissemination of Scientific and Technical Knowledge Unit, Directorate-General Information Technologies and Industries and Telecommunications, Luxembourg
3. WYKO Rough Surface Tester Plus, manufactured by WYKO Corporation, 2650 E. Elvira Road, Tucson, Ariz. 85706.
4. L. Mattsson and P. Wågberg, "Assessment of surface finish on bulk scattering materials. A comparison between optical laser stylus and mechanical stylus profilometers," Precision Engineering, **15**, (1993), 141 - 149
5. W.H. Simmonds, R.J. Smith, R.E. Renton, G. Gregoriou, J. Tripp, C.H.F. Velzel, L. Mattsson, M. Bjuggren, H.J. Tiziani, H.J. Jordan, "Optical Non-Contact Techniques for Engineering Surface Metrology", Final report of the project "Non-Contact Surface Metrology," funded by the European Community under the BCR Programme. Report EUR 16161EN, European Commission, Directorate - General XIII, Telecommunications, Information Market and Exploitation of Research, L-2920 Luxembourg
6. H. E. Bennett and J. O. Porteus, "Relation between surface roughness and specular reflectance at normal incidence," J. Opt. Soc. Am. **51**, 123-129 (1961).
7. Standard Test Method for Measuring Effective Surface Roughness of Optical Components by Total Integrated Scattering, ASTM F 1048-87, American Society for Testing and Materials, 1916 Race Street, Philadelphia, PA 19103, USA.
8. L. Mattsson, "Characterization of supersmooth surfaces by light scattering techniques," in *Surface Measurement and Characterization*, J. M. Bennett, ed., Proc. Soc. Photo.-Opt. Instrum. Eng. **1009**, 165171 (1989).
9. J. Péters, "Messung des Mittenrauhwertes zylindrischer Teile während des Schleifens," VDI-Berichte Nr 90, 27-31 (1965)
10. "Standard practice for angle resolved optical scatter measurements on specular or diffuse surfaces. ASTM E1392-90, American Society for Testing and Materials, 1916 Race Street, Philadelphia, PA 19103, USA
11. J. M. Elson and J. M. Bennett, "Vector scattering theory," Opt. Eng. **18**, 116124 (1979).
12. M. Bjuggren, L. Krummenacher and L. Mattsson, "Quality asessment of engineering surfaces by infrared scattering," in Proc. Soc. Photo-Opt. Instrum. Eng. **2536**, 327-336 (1995).
13. Talystep is a trademark of Rank Taylor Hobson Ltd, P.O. Box 36, Leicester, England

Gleichzeitige Prüfung von Form und RMS-Rauheit mittels Laser-Triangulation

H. Rothe und A. Kasper
Universität der Bundeswehr Hamburg
FB Maschinenbau/Institut für Automatisierungstechnik
Lehrstuhl für Meß– und Informationstechnik
D–22039 Hamburg

Zusammenfassung Mit der hier vorgeschlagenen Variante des Lasertriangulationsverfahrens i
es möglich, ein einfaches und genaues Meßgerät für die Profilometrie zu realisieren, welches gleic
zeitig eine 3D–Formerfassung in einem Meßvolumen von $(50\times50\times5)mm^3$ mit einer Standardabwe
chung der Einzelmeßpunkte von ca. 1.5 μm und eine Rauheitsklassifikation im Bereich von 0
bis 2.5 μm ermöglicht. Folglich kann ein Verhältnis von Meßbereich zu Standardabweichung d
Koordinatenmessung von $> 3000 : 1$ erreicht werden. Die rms–Rauheitswerte der Oberfläche we
den aus dem für die Triangulation gemessenen Strahlprofil durch ein statistisches Verfahren d
Mustererkennung gewonnen.

Grundidee Die Meßgenauigkeit der Lasertriangulation wird wesentlich durch Speckles begrenzt
Es wurde vorgeschlagen, Speckles vom Nutzsignal zu dekorrelieren.[2,3] Das vorgestellte Verfahre
basiert hauptsächlich auf der Auswertung der Korrelationsmatrix von mittels einer CCD–Matri
zeilenweise verfügbar gemachten Profilen des Laserspots. Es wird davon ausgegangen, daß die Va
rianz des Meßsignals hauptsächlich vom Gauß–Profil des Laserspots beeinflußt wird. Stellt man sic
an Stelle des Meßobjekts einen Spiegel vor, dann erfährt das Gauß–Profil keinerlei Deformation. I
diesem Fall sind 100% der Information des Meßsignals (=Signalvarianz) im 1. Eigenwert der Kor
relationsmatrix konzentriert. Alle anderen Eigenwerte der Korrelationsmatrix sind Null. Die rea
immer vorhandene Rauhigkeit der Oberfläche führt zur Streuung des einfallenden Strahls und zu
Speckle–Bildung. Das Gauß–Profil wird – je nach Mikrotopographie der angetasteten Oberfläch
und den Daten des Meßaufbaus (Apertur) – mehr oder weniger stark gestört. Bei geeigneter Aus
legung des Sensorsystems bleiben 70%–90% der Signalinformation im 1. Eigenwert konzentriert
Diese Information wird zur Signalrekonstruktion für die Abstandsmessung herangezogen, d.h. di
Subpixelverfahren werden auf ein mit den signifikanten Eigenwerten und Eigenvektoren rekonstru
iertes Gauß–Profil angewendet.
In diesem Sinne wären die dekorrelierten Signalanteile "Informationsabfall". Daß sie dennoch nutz
bringend zur Bestimmung von Oberflächeneigenschaften herangezogen werden können, zeigten LEI
u.a.[4] durch Auswertung von Grauwerthistogrammen des Laserspots.

Durch frühere Untersuchungen konnte nachgewiesen werden, daß die für die rms–Rauheitsbestim
mung erforderliche Oberflächeninformation, die in der Streulichtverteilung kodiert ist, mittels empi
rischer statistischer Momente erfaßt werden kann.[5] Weiterhin ergab sich, daß die Hauptkomponen
tentransformation systematische und stochastische Anteile der Oberflächeninformation trennt,[6]
und somit zur von systematischen Meßfehlern bereinigten Schätzung der rms–Rauheit bei profilo
metrischen Messungen herangezogen werden kann. In diesem Sinne kann das Gaußprofil des La
serspots als "systematischer" Signalanteil, die Speckles und das Streulicht als der die rms–Rauhei
beschreibende "stochastische" Anteil betrachtet werden. Im Gegensatz zur profilometrischen Mes
sung gibt es jedoch i.a. keinen direkten Zusammenhang zwischen Specklemuster, Streulicht und
rms–Rauheit (RAYLEIGH–Limit).

Deshalb war es notwendig, mit lernfähigen Verfahren der Mustererkennung zur Klassifizierung de

xperimente[8–11] Für die Triangulationsmessung wurden verschiedene Scanmodi verwendet. Zur alibrierung und Ermittlung der Antastgenauigkeit und Linearität in z–Richtung (Meßrichtung, ▸tische Achse) wurden eindimensionale Scans mit verschiedenen ebenen Probekörpern ausgeführt. m eine präzise Kalibrierung zu ermöglichen, wird ein z–Scan durch den vollen Meßbereich ge-hren und der Abbildungsmaßstab als Funktion der z–Koordinate durch eine Ausglcichsparabel ▸proximiert. In Tabelle 1 sind die Meßergebnisse für die eingesetzten Subpixelverfahren bei un-rschiedlichen Versuchsbedingungen zu sehen. Die Werte sind in Pixeln angegeben und stellen die ;andardabweichung einer Geradenregression von jeweils 100 Einzelmeßpunkten dar. Die Versuchs-edingungen waren:

- telezentrisches Objektiv (TZO) $\beta = 1 : 1$, numerische Apertur (NA) 0.01;
  Dokumar $\beta = 1 : 1.6$, NA 0.1

- z–Scan, 100 Meßpunkte, 5 μm Schrittweite, Triangulationswinkel: 30 Grad

- 8–Bit–Analogkamera (XC 77), 12–Bit–Digitalkamera (DigiCam63x12, Eigenbau)

		Einzelprofil			Gemitteltes Profil			PCA–Profil		
	P	Med.	Par.	COG	Med.	Par.	COG	Med.	Par.	COG
TZO	P1	1.26	1.45	1.65	0.49	**0.49**	1.32	0.55	0.55	1.46
NA 0.01	P2	0.68	0.84	1.49	0.54	0.73	1.23	**0.50**	0.74	1.26
8Bit	P3	1.22	1.29	1.81	0.72	0.89	1.45	**0.67**	0.82	1.44
	P4	1.66	2.17	2.44	0.85	0.97	1.83	**0.77**	0.89	1.75
	P5	0.77	0.87	1.28	0.36	0.47	0.96	**0.32**	0.40	0.94
	P6	2.13	2.18	2.92	0.67	0.70	1.35	**0.56**	0.62	1.39
	P7	0.38	0.41	0.95	**0.23**	0.24	0.78	**0.23**	0.24	0.78
	P8	1.15	1.22	1.80	0.63	0.75	1.23	**0.57**	0.67	1.16
Dokumar	P1	0.57	0.60	1.37	**0.17**	0.20	0.60	**0.17**	0.20	0.62
NA 0.1	P2	0.55	0.77	1.47	0.24	0.30	0.66	**0.23**	0.28	0.65
8Bit	P3	0.62	0.63	1.21	0.24	0.24	0.65	0.24	**0.23**	0.63
	P4	0.33	0.41	0.86	0.18	0.29	0.46	**0.17**	0.29	0.48
	P5	0.57	0.59	1.50	0.25	0.32	0.71	**0.24**	0.30	0.69
	P6	0.64	0.73	1.17	0.18	0.25	0.58	**0.17**	0.24	0.56
	P7	0.14	0.17	0.49	**0.11**	0.13	0.30	**0.11**	0.13	0.31
	P8	0.68	0.59	1.71	0.33	0.24	0.77	**0.32**	0.24	0.73
TZO	P1	0.40	0.54	0.86	**0.14**	0.33	0.56	**0.14**	0.33	0.57
NA 0.01	P2	0.57	0.50	0.81	0.26	0.35	0.59	**0.21**	0.34	0.58
12Bit	P3	0.43	0.42	0.62	0.32	0.42	0.60	**0.28**	0.39	0.59
	P4	0.39	0.56	0.87	0.25	0.34	0.53	**0.24**	0.33	0.52
	P5	0.38	0.46	0.75	0.37	0.51	0.79	**0.31**	0.48	0.78
	P6	0.17	0.35	0.62	**0.12**	0.29	0.50	**0.12**	0.28	0.49
	P7	0.09	0.24	0.50	**0.06**	0.16	0.32	**0.06**	0.16	0.32
	P8	0.38	0.51	0.80	**0.21**	0.37	0.62	**0.21**	0.39	0.65
Dokumar	P1	0.07	0.21	0.37	**0.03**	0.16	0.30	**0.03**	0.16	0.29
NA 0.1	P2	0.19	0.34	0.55	**0.07**	0.16	0.30	**0.07**	0.16	0.30
12Bit	P3	0.11	0.20	0.34	**0.10**	0.20	0.34	**0.10**	0.20	0.34
	P4	0.10	0.24	0.42	**0.02**	0.15	0.28	**0.02**	0.15	0.28
	P5	0.08	0.21	0.37	**0.05**	0.18	0.32	**0.05**	0.17	0.32
	P6	0.08	0.23	0.44	**0.03**	0.15	0.29	**0.03**	0.15	0.29
	P7	0.03	0.15	0.29	**0.02**	0.15	0.29	**0.02**	0.15	0.29
	P8	0.09	0.22	0.38	**0.03**	0.16	0.29	**0.03**	0.16	0.29

Tabelle 1: Standardabweichungen in Pixeln für z–Scans in Abhängigkeit von NA, Grau-wertauflösung, Probe und Subpixelverfahren (Med.: Median, Par.: Parabelfit, COG: Center of Gravity, PCA: Pricipal Component Analysis – Hauptkomponentenanalyse). Die jeweils besten Werte sind mit fetter Schrift hervorgehoben.

Für die zur Klassifizierung verwendete Methode der Diskriminanzanalyse sind Daten zum Anlern
und natürlich auch zur eigentlichen Klassifizierung notwendig. Bei den Triangulationsmessung
wurde neben dem Positionswert ein Eigenschaftsvektor pro Meßpunkt gespeichert. Zum Anl
nen wurden 30 Eigenschaftsvektoren und zum Klassifizieren 20–30 Eigenschaftsvektoren mit je
Variablen (statistischen Momenten) verwendet. In Tabelle 2 sind die Klassifizierungsresultate
Abhängigkeit von der Probe und den Versuchsbedingungen dargestellt.

	Material	Stahl			Messing		
	Probe	P1	P2	P3	P4	P5	P6
NA 0.01	Anlernen	29/30	30/30	29/30	30/30	30/30	30/30
XC77	Klassifizieren	20/20	19/20	20/20	20/20	13/20	20/20
NA 0.1	Anlernen	30/30	30/30	30/30	30/30	30/30	30/30
XC77	Klassifizieren	20/20	20/20	20/20	20/20	20/20	20/20
NA 0.01	Anlernen	28/30	26/30	21/30	23/30	25/30	23/30
DigiCam63x12	Klassifizieren	17/20	15/20	18/20	20/20	11/20	18/20
NA 0.1	Anlernen	30/30	30/30	30/30	30/30	30/30	30/30
DigiCam63x12	Klassifizieren	20/20	20/20	18/20	19/20	20/20	20/20

Tabelle 2: Ergebnisse der Klassifizierung von verschiedenen Rauheiten für Stahl und Messing

Die Klassifizierung wurde für die Proben P1–P3 und P4–P6 (d.h. für Messing und Stahl) getren
vorgenommen, wobei die erste Zahl die Anzahl der richtigen Klassifizierungen und die zweite Za
die Gesamtanzahl der Klassifizierungen darstellt. Für jede Klassifizierung wurden 11 Einzelprof
mit jeweils 51 Pixeln verwendet. In Tabelle 3 wurden die Proben unabhängig vom Material
Rauheitsklassen eingeordnet und klassifiziert. Die Zahlen stellen dabei wieder das Verhältnis d
Anzahl der richtigen Klassifizierungen zur Gesamtanzahl der Klassifizierungen dar.

	rms/μm	0.1	0.3–0.4	0.7	1.0	1.4	1.9–2.1
	Probe	P7	P1,P4	P2	P5	P3	P6,P8
NA 0.1	Anlernen	30/30	29/30	30/30	30/30	30/30	30/30
XC77	Klassifizieren	20/20	20/20	20/20	20/20	20/20	20/20
NA 0.1	Anlernen	30/30	20/30	30/30	28/30	29/30	27/30
DigiCam63x12	Klassifizieren	20/20	25/30	20/20	19/20	19/20	25/30

Tabelle 3: Ergebnisse der Klassifizierung von verschiedenen Rauhheiten unabhängig vom Materi

Schlußfolgerungen

- Die genauesten und robustesten Triangulationsergebnisse wurden bei der Anwendung d
  Medianverfahrens auf ein hauptkomponentengefiltertes Profil erreicht. Dabei erwies sich d
  Auswertung von etwa 10 Profilen und die ausschließliche Verwendung des 1. Eigenwertes u
  1. Eigenvektors als sinnvoll.

- Durch Erhöhung der Grauwertauflösung und der Verwendung von Sensoren mit einem hoh
  Dynamikbereich war es möglich, die Meßgenauigkeit der Triangulation in bezug auf die Sta
  dardabweichung der Einzelmeßpunkte zu verdoppeln. Als sinnvolle Einzelempfängergröß
  stellten sich Pixelgrößen heraus, die gerade das Abtasttheorem für die mittlere Specklegrö
  erfüllen. Eine Überabtastung hat sich für die Triangulation als nicht zweckmäßig gezeigt.

- Die besten Ergebnisse in bezug auf die Triangulation als auch für die Charakterisierung d
  Oberfläche wurden bei NA = 0.1 erreicht.

- Die empirischen statistischen Momente können als robuste Schätzer für die Klassifizierung von Oberflächeneigenschaften angesehen werden.

- Zur Klassifizierung von Rauheiten ist eine Überabtastung sinnvoll, um möglichst viel Information bezüglich der Mikrotopographie zu erhalten.

- Als günstige Kombination für einen z–Scanbereich von ca. 5mm werden folgende Parameter vorgeschlagen: NA 0.1, Abbildungsmaßstab 1:1...2, Pixelgröße 10...25 μm (s. Abb. 1).

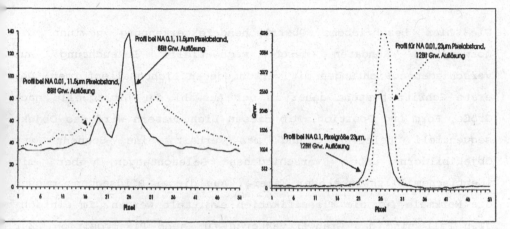

Abbildung 1: Typische Strahlprofile für unterschiedliche Aperturen und Sensoren

Literatur

[1] J.C. Dainty, "Laser Speckle and Related Phenomena", Berlin, Springer–Verlag, 1990.

[2] H. Rothe u.a., "Improved Accuracy in Laser Triangulation by Variance Stabilizing Transformations", Optical Engineering, Vol. 31, No. 7, pp. 1538–1545.

[3] Rothe, H.; Truckenbrodt, H., "Laser Triangulation near the Diffraction Limit", International Journal of Optoelectronics, Vol. 8, Nos. 5/6, pp. 655 – 667, (1993).

[4] C.S. Lee u.a., "An In–Process Measurement Technique Using Laser for Non–Contact Monitoring of Surface Roughness and Form Accuracy of Ground Surfaces", Annals of the CIRP, Vol. 36/1/1987, pp. 425–428.

[5] Rothe, H.; Truckenbrodt, H., "Discrimination of Surface Properties Using BRDF–Variance Estimators as Feature Variables", International EUROPTO/SPIE Symposium on Optical Systems Design, Proc. SPIE No. 1781 (1992), ed. L. Baker, pp. 152–162.

[6] Rothe, H.; Duparré, A., "The Principal Surface Approach as Means of Precise rms–Roughness Estimation in Noncontact Profilometry", Proc. SPIE No. 2263 (1994), in press.

[7] Rothe, H.; Duparré, A.; Jacobs, S., "Generic Detrending of Surface Profiles", Optical Engineering, Vol. 33, No. 9, pp. 3023 – 3030, (1994).

[8] O. Specht, "Entwurf, Aufbau und Erprobung von sensornaher Hard– und Software einer metrischen CCD–Kamera mit 12 Bit Digitalisierungstiefe", Diplomarbeit, TU Dresden 1993.

[9] A. Kasper, "Entwurf, Implementierung und Analyse von Algorithmen zur subpixelgenauen Konturfindung", Diplomarbeit, Friedrich–Schiller–Universität Jena 1994.

[10] Rothe, H.; Riedel, P.; Specht, O., "BRDF–Sensing with Fiber Optics, Programmable Laser Diodes and High Resolution CCD Arrays", Proc. SPIE No. 2260 (1994), in press.

Anwendung von Klassifikatoren für die Bildsequenzverarbeitung zur Oberflächendefekterkennung

P. Lehle, H. Gärtner und H. J. Tiziani
Institut für Technische Optik, Universität Stuttgart
D-70569 Stuttgart

Die hier beschriebene Oberflächendefekterkennung gewinnt die notwendigen Rohdaten durch sequentielle Beleuchtung aus verschiedenen Richtungen mit verschiedenen Lichtquellenformen. Der erste Schritt besteht daher in der Auswahl der Lichtquelle nach Größe, Form und Position. Mit diesen Lichtmustern wird das Objekt sequentiell beleuchtet und man erhält eine Sequenz von Objektbildern unter verschiedenen Beleuchtungen, aber mit identischem Blickwinkel der Kamera. Aus dieser Bildsequenz müssen die Merkmale für die Klassifikation ermittelt werden. Im einfachsten Fall sind das Grauwertvektoren für jede Pixelposition. Zur Gewinnung von rotationsinvarianten Merkmalen läßt sich hier noch eine FFT durchführen. Entsprechend den Aufnahmebedingungen bzw. dem Objekt kann auch eine Kontrastanpassung realisiert werden. Der Klassifikator ordnet dann jedem Merkmalsvektor eine Klasse zu. Anschließend kann aus dem Bild mit den Klassenzugehörigkeiten die für den speziellen Anwendungsfall relevante Information wie Flächenanteile bzw. Überschreiten eines Grenzwertes für den Flächenanteil ermittelt werden.

Klassifikation

Klassifikatoren unterscheiden sich vor allem im Hinblick auf die Art und Komplexität der Speicherung ihres 'Wissens'. Die Beschreibung komplexer Merkmalsräume mit wenigen, aber vor allem nur relevante Informationen tragenden Parametern, also eine Wissenskonzentration, erfordert aufwendiges Anlernen.

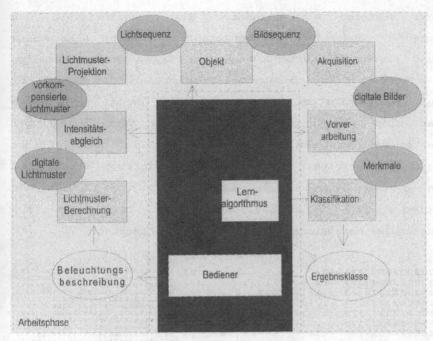

Gesamtprinzip der Oberflächendefekterkennung

Die von den Klassifikatoren vorgenommene Zuordnung geschieht über mathematische Funktionen, deren Art und Aufbau von der Struktur des Klassifikators abhängt. Die Parameter dieser Funktionen und teilweise auch die Struktur werden beim Lernvorgang bestimmt und enthalten somit das Wissen über die Zuordnung.

Durch diese approximierenden Funktionen wird der Merkmalsraum implizit in Klassen eingeteilt, d.h. es werden Trennflächen definiert, die die einzelnen Klassen voneinander separieren. Die zugrundeliegenden Funktionen müssen diese Trennflächen beschreiben. Daher hängt es von der Form der Trennflächen ab, wie kompliziert die einzelnen Funktionen sein müssen, um ausreichende Genauigkeit zu erzielen.

Nächster-Nachbar-Klassifikator

Der Nächster-Nachbar-Klassifikator verwendet keine geschlossene mathematische Funktion, sondern sucht sich aus einer gespeicherten

Menge von Vektoren mit bekannter Klassenzugehörigkeit den nächstliegenden aus. Die Probenvektoren können hier ohne Bearbeitung übernommen werden. Unstetige Klassengrenzen stellen bei diesem Verfahren kein besonderes Problem dar. Die Anzahl der notwendigen gespeicherten Proben und die damit verbundenen Suchzeiten sind jedoch relativ groß. Eine Verringerung der gespeicherten Vektoren kann durch Überprüfung der Proben auf ihre Wichtigkeit hin erfolgen. Entfernt man unwichtige Proben, so entstehen an den Klassengrenzen doppelschichtartige Anordnungen der Proben. Zur Veranschaulichung sei hier ein nur zweidimensionaler Merkmalsraum mit 5 Klassen betrachtet.

```
12222222222222222222223   12................23    12222222222222222222222223
11222222222222222222233   .12...............23.   11222222222222222222222233
11122222222222222222333   ..12.....22......23..   11222222222222222222222333
11112222222222222222333   ...12...2552....23...   11122222222222222222222333
11111222555522223333333   ....12225..522223....   11111122222555222223333333
11114445555444333333      ...144445..54443.....   11111112222555555222223333333
11144444455444443333      ..14....4554...43....   11111114444455555544433333333
11444444444444443333      .14......44.....43...   11114444444455554444433333333
14444444444444444333      14.............43..     11144444444444444444333333
44444444444444444433      4................43.    44444444444444444444443333
44444444444444444443      .................43     44444444444444444444444443
```

links: exemplarischer Merkmalsraum, mitte: verwendete
Stützstellen, rechts: Klassifikationsergebnis

Polynomklassifikator

Der Polynomklassifikator approximiert die Wahrscheinlichkeit der Zugehörigkeit eines Probenvektors zu einer Klasse durch ein Polynom, dessen maximaler Grad als Strukturmerkmal festzulegen ist.

```
12222222222222222223    12222222222222222222222223    12222222222222222222222223
11222222222222222233    11222222222222222222222233    11222222222222222222222233
11122222222222222333    11122222222222222222222333    11122222222222222222222333
11112222552222223333    11112222222222222222223333    11112222222222222222223333
11111222555522223333    11111222222222222222223333    11111222222222222222223333
11114444555544433333    11111112222222222222233333    11111111222555555222233333
11144444455444443333    11111144444444442222333333    11111114444455555553333333
11444444444444443333    11114444444444444443333333    11111444444455554444433333
14444444444444443333    11144444444444444443333333    11144444444444444443333333
44444444444444444433    14444444444444444443333       11444444444444444444443333
44444444444444444443    44444444444444444443           44444444444444444444444443
```

links: exemplarischer Merkmalsraum, mitte: Ergebnis Klassifikation
mit Grad 3, rechts: Grad 4

Neuronales Netz

Neuronale Netze tragen als Funktionen über Gewichte gekoppelte nichtlineare Funktionen in mehreren Schichten hintereinander. Die Struktur der resultierenden Funktionen ist daher schwer überschaubar. Die Ermittlung der Gewichte geschieht über rekursive Näherungsverfahren, um die Parameter in hintereinandergeschalteten Schichten zu optimieren.

Einsatz

Je nach Einsatzgebiet treten verschiedene Eigenschaften der Klassifikatoren in den Vordergrund. Im anwendungsorientierten Einsatz liegt der Schwerpunkt auf der Rechenzeit und der Klassifikationsgüte. Aufwenige Anlernprozeduren können dabei in Kauf genommen werden. Im Labor zur Untersuchung von Machbarkeit, Veränderungen und Verbesserungen wird die Klassifikator-charakteristik häufig verändert und anderen Bedingungen angepaßt. Langwierige Einlernprozesse rechtfertigen hier die beschleunigte Auswertung kaum, da meist nach wenigen Auswertungen bereits ein neuer Lernprozeß erfolgt. In diesem Zusammenhang kann auch die Fähigkeit zu inkrementellem Anlernen große Bedeutung erlangen, wenn die Stichprobenmenge häufig, aber jeweils nur geringfügig modifiziert wird.

Schürmann, J. (1977). Polynomklassifikatoren. Oldenbourg, München
Rumelhart D. E., McClelland, J. L. (1986). Parallel Distributed Processing. Volume 1. MIT Press, Cambridge
Malz, R. (1991). Erkennung und Analyse von Mikrodefekten auf technischen Oberflächen mit dem POLARIS-Beleuchtungsprozessor. 2. Symposium Bildverarbeitung, TAE Eigenverlag
Langer, M. S.; Zucker, S. W. (1994). Spatially Varying Illumination: A Computational Model of Converging and Diverging Sources. Lecture Notes in Computer Science, Vol. 801, Jan-Olof Eklundh, Computer Vision - ECCV '94, Springer Verlag

Konturfehlererkennung an Profilbrettern mittels Mehrfachlichtschnitt-Triangulation

M. Becker und E. Schubert
Zentrum für Sensorsysteme (ZESS)
Universität - GH Siegen
Paul-Bonatz-Str. 9-11
57068 Siegen

1   Einleitung

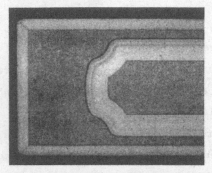

Bild 1: Ausschnitt MDF mit eingefrästem Profil

Bei der Herstellung von Möbeln u.a. werden beschichtete oder furnierte Profilbretter aus sog. mittel-dichten Faserplatten (MDF) verwendet. Konturfehler im Bereich der gefrästen Profile stellen eine Qualitätseinbuße für das spätere Endprodukt dar; vielfach ist eine Beschichtung oder Weiterverarbeitung der fehlerhaften Profilbretter sogar gänzlich unmöglich. Es ist daher notwendig, im Produktionsprozeß durch Erfassung der 3D-Geometrie, die Qualität der Profile und den Zustand des Werkzeuges kontrollieren zu können, um Ausschuß und Maschinenschäden zu vermeiden.

Ein Fehler macht sich durch eine Kerbe oder einen Grat im Profil bemerkbar. Um die obigen Forderungen zu erfüllen, müssen Abweichungen die größer als 100 μm sind erkannt werden. Dies liegt nur knapp oberhalb der Struktur, die das Material selbst aufweist.

2   Der Prüfaufbau

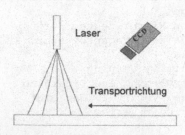

Bild 2: Prinzip des Meßaufbaus

Für die Erfassung dieser Konturfehler wurde ein Lichtschnittsystem, bestehend aus einem Laser mit Mehrfach-Linienoptik und einem PC-basierten Bildverarbeitungssystem mit Standard-CCD-Kamera aufgebaut.

Das Mehrfach-Lichtschnittverfahren wird eingesetzt, um Fehlmessungen zu eliminieren, welche durch die statistisch verteilten Materialstrukturen hervorgerufen werden.

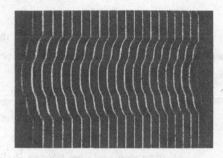

Bild 3: Beispielhaftes Kamerabild

Diese heben sich bei der Mittelung über alle Lichtschnitte auf. Ein Fehler, der durch einen beschädigten Fräser entsteht, macht sich dagegen entlang des gesamten Profils, d.h. in jedem Lichtschnitt bemerkbar. Dadurch wird die Erkennungssicherheit für Profilfehler erheblich gesteigert. Mit diesem Verfahren können Konturfehler in der Größenordnung von ca. 100 μm sicher detektiert werden.

3 Ablauf eines Meßzyklus

Ein Produktionszyklus verläuft nach folgendem Schema:

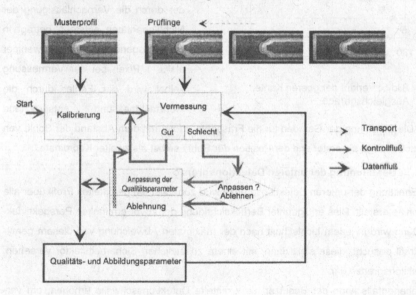

Bild 4: Ablauf des Produktionszyklus

Zuerst muß das Meßsystem auf ein Gutprofil kalibriert werden. Dabei werden Perspektiv-Korrekturfaktoren und die minimale Detektionsschwelle errechnet.

Danach kann der Benutzer das System in 2 Modi betreiben:

1. Automatische Ablehnung eines als schlecht erkannten Profils
2. Nachfrage, ob das als schlecht erkannte Teil noch innerhalb der Produktionstoleranzen liegt und entsprechende Anpassung der Detektionsschwelle.

3.1 Kalibrierung

Durch die Vermessung eines Gutprofils ermittelt die Software eine untere Grenze für die Fehlerdetektion, indem die Struktur des Profils durch Ermittlung von Abweichungen der einzelnen Linien gegenüber der gemittelten Profilkontur analysiert wird.

3.1.1 Bestimmung des Perspektivfaktors

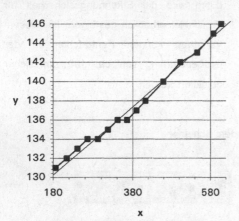

Bild 5: Verlauf der oberen Kante, Ausgleichsgerade.

Es werden für jeden Lichtschnitt jeweils von einer oder mehreren Profilkanten und vom tiefsten Punkt der Fräsung die Koordinaten extrahiert. Diese Koordinaten werden jeweils durch eine Ausgleichsgerade angenähert; der maximale Fehler, der durch die Vernachlässigung der Nichtlinearitäten entsteht, beträgt in der vorliegenden Anordnung weniger als ± 1 Pixel; bei der Vermessung selbst wird der Fehler durch die Mehrfachmessung entsprechend kleiner. Die Ermittlung der Geraden für die Frässohle erfolgt mit dem Abstand der Sohle von der Brettoberkante als erster und der Position der Sohle selbst als zweiter Koordinate.

3.1.2 Bestimmung der unteren Detektionsgrenze

Zur Ermittlung der unteren Detektionsgrenze wird zunächst ein gemitteltes Profil über alle Lichtschnitte erstellt. Dies erfolgt unter Berücksichtigung der zuvor ermittelten Perspektivfaktoren. Dann wird in jedem Lichtschnitt nach der maximalen Abweichung von diesem gemittelten Profil gesucht, diese stellt dann, mit einem zusätzlichen Sicherheitsfaktor versehen, die Detektionsgrenze dar.

Gegebenenfalls kann der Benutzer die ermittelte Detektionsschwelle erhöhen, um verschiedene Qualitätsstufen zu erreichen.

3.2 Vermessung

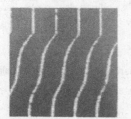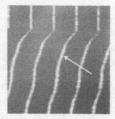

In der nebenstehenden Abbildung ist ein korrekt gefrästes Profil (links) und ein Profil mit Fehler (Pfeil, rechts) als Ausschnittvergrößerung abgebildet. Die Vermessung eines Profils erfolgt jeweils mit einem einzigen

Kamerabild, da durch das Mehrfach-Lichtschnitt-Verfahren bereits mehrere Messungen gleichzeitig ausgeführt werden. Zu diesem Zweck wird ebenfalls ein gemitteltes Profil errechnet. Diese Mittelung berücksichtigt die bei der Kalibrierung errechneten Perspektive-Faktoren. Das Ergebnis der Mittelung ist ein Querschnitt durch das vermessene Profil, das nicht in reale Koordinaten umgerechnet werden muß, sondern direkt im Pixelkoordinatensystem ausgewertet werden kann. Dazu wird es mit dem Referenzprofil verglichen, um eine Entscheidung über die Qualität des gefrästen Profils zu treffen. Dabei ist zu beachten, daß das eingefräste Profil eine leichte seitliche Verschiebung gegenüber dem Referenzprofil aufweisen kann; diese wird durch eine Korrelation mit dem Kalibrierprofil ermittelt und kompensiert. Überschreitet sie eine vom Benutzer einstellbare Grenze, wird das Profil ebenfalls als "schlecht" klassifiziert und das betreffende Brett verworfen bzw. die Erkennungsschwelle entsprechend angepaßt.

Es ist zur Vermessung nicht notwendig, das Brett zu stoppen, d.h. die Meßvorrichtung kann durch eine automatische Handhabung mit Brettern im Durchlauf versorgt werden. Die Bewegung der Bretter hat dabei keinen negativen Einfluß auf die Detektionsschwelle, da dadurch das Speckle-Rauschen der Laser durch zeitliche Integration unterdrückt wird. Auf die Profilfehler wirkt sich die Integration nicht aus, da diese in der Bewegungsrichtung des Brettes verlaufen und sich daher zeitlich nicht ändern.

4   Zusammenfassung

Das Multilichtschnitt-Triangulationssystem ist ein flexibles, leicht einzurichtendes System zur Qualitätskontrolle das nicht nur in der Holzverarbeitung verwendet werden kann und sehr einfach zu handhaben ist. In der vorliegenden Anordnung werden folgende Spezifikationen erreicht:

Meßbereich:	22 x 22 mm
Meßabstand:	ca. 100 mm
Meßrate:	2/s
Auflösung:	0.05 mm
Genauigkeit:	0.1 - 0.2 mm

Das System läßt sich leicht auf andere Meßbereiche, Qualitätsansprüche und Genauigkeiten anzupassen, ohne daß eine aufwendige Kalibrierung durchzuführen ist.

Literatur

K. Engelhardt, Optische 3D-Meßtechnik, Technische Rundschau, Heft 31, 1993

Kalibrierung eines Lichtschnittsensors zur optischen 3D-Konturerfassung

W. Heckel und M. Geiger

Lehrstuhl für Fertigungstechnologie, Universität Erlangen-Nürnberg,
Egerlandstr. 11, D-91058 Erlangen

Einleitung

Optische 3D-Konturerfassung nach dem Lichtschnittprinzip bietet sich für prozeßnahe Messungen in der Produktionstechnik an. Als absolut messendes Verfahren liefert es 3-D Koordinaten eines Profils der Werkstückoberfläche entlang der beleuchteten Linie. Mit einer einzigen Messung kann man die absoluten 3D-Koordinaten von mehreren hundert (hier: 750) Punkten der Objektoberfläche berechnen. Hochgenaue 3D-Messungen erfordern eine Kalibrierung des Meßsystems, die nach jeder Veränderung des Detektionssystems erneut durchgeführt werden muß. Eine Anwendung des Lichtschnittverfahrens zur on-line Winkelmessung beim freien Biegen wurde bereits früher vorgestellt [1], [2]. Der Schwerpunkt dieses Beitrags liegt auf den mathematischen Grundlagen zur Berechnung der 3D-Koordinaten aus dem Kamerabild und zur Kalibrierung des Meßsystems. Das verwendete photogrammetrische Modell der "Direkten Linearen Transformation" erlaubt die Berechnung der 3D-Koordinaten für jede mögliche Orientierung der Lichtebene, wodurch sowohl eine flächenhafte 3D-Konturerfassung des Meßobjekts mit mehreren Aufnahmen als auch eine Anpassung der strukturierten Beleuchtung an die jeweilige Meßaufgabe ermöglicht werden. Bei Einsatz eines positionsgeregelten Galvanoscanners kann eine automatische Kalibrierung des Meßsystems erfolgen. Die erzielte Absolutgenauigkeit liegt dabei im Subpixelbereich.

Gewinnung der 3D-Koordinaten

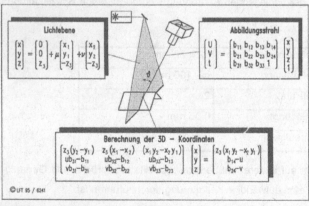

Bild 1:   *Berechnung der 3D-Koordinaten*

Bild 1 erläutert die prinzipielle Vorgehensweise bei der Gewinnung der 3D-Koordinaten. Das Objekt wird mit einer Lichtebene beleuchtet, die z.B. durch Scannen eines fokussierten Laserstrahls über das Meßobjekt während einer Integrationsperiode des Detektors erzeugt werden kann. Die Schnittlinie der Lichtebene mit der Werkstückoberfläche, das sogenannte "Profil", wird mit einer CCD-Kamera unter dem Triangulationswinkel ϑ detektiert. Die durch den Scanner aufgespannte Lichtebene legt den ersten geometrischen Ort zur Gewinnung der 3D-Koordinaten fest und kann direkt aus Startpunkt (x_1, y_1, z_1=0) und Endpunkt (x_2, y_2, z_2=0) der programmierten Spiegelbewegung sowie dem Spiegelmittelpunkt (x_3=0, y_3=0, z_3) berechnet werden, siehe *Bild 1*. Der zweite geometrische Ort wird durch den sogenannten Abbildungsstrahl definiert, also den Hauptstrahl der geometrisch-optischen Abbildung. Dieser wird in der Gleichung nach *Bild 1* durch die auf *ABDEL-AZIZ* und *KARARA* [3] zurückgehende "Direkte Lineare Transformation" dargestellt. Dabei werden alle Transformationsschritte zwischen Bild- und Weltkoordinaten durch eine einzige lineare Transformation ausgedrückt. Dies erleichtert die Kalibrierung des Meßsystems. Der Schnittpunkt von Abbildungsstrahl und Lichtebene ergibt die 3D-Koordinaten der Werkstückoberfläche. Nach entsprechender Umformung liefern die Gleichungen für Lichtebene und Abbildungsstrahl drei unabhängige Gleichungen für die drei Weltkoordinaten (x, y, z), siehe *Bild 1*. Durch Lösen dieses linearen Gleichungssystems z.B. mit der Cramer'schen Regel erhält man die einem Bildpunkt (u, v) zugeordneten 3D-Weltkoordinaten (x, y, z). Das Gleichungssystem gilt für jede beliebige Lichtebene, solange diese nicht parallel zur optischen Achse der CCD-Kamera ist. Damit ist auch eine flächenhafte 3D-Konturerfassung durch sequentielle Antastung mit verschiedenen Lichtebenen möglich.

Die Direkte Lineare Transformation

Die Gleichung für den Abbildungsstrahl in *Bild 1* ist die normierte Matrixdarstellung der Direkten Linearen Transformation zwischen homogenen Bild- (U, V, t) und Weltkoordinaten (x, y, z, 1) und entspricht einer allgemeinen Darstellung der Zentralprojektion eines dreidimensionalen Objektraums auf einen zweidimensionalen Bildraum [4]. Die kartesischen Bild- und Weltkoordinaten (u, v) bzw. (x, y, z) ergeben sich durch Normierung der homogenen Koordinaten (U, V, t) und (X, Y, Z, w) auf die letzte Komponente. Die Direkte Lineare Transformation beschreibt eine lineare Beziehung zwischen den Punkträumen der Welt- und der Bildkoordinaten, die der in der Photogrammetrie üblichen Kollinearitätsbeziehung entspricht [4]. Die Zentralprojektion ist eine singuläre Kollineation und damit nicht umkehrbar eindeutig. Erst die strukturierte Beleuchtung des Objekts mit der Lichtebene ergibt die zweite Bedingung zur Auflösung dieser Vieldeutigkeit.

Zur Gewinnung der 3D-Koordinaten benötigt man die Koeffizienten b_{ij} der Direkten Linearen Transformation. Diese werden bei der photogrammetrischen Kalibrierung des Meßsystems durch Vermessen von Paßpunkten, also signifikanten Punkten mit bekannten Weltkoordinaten, ermittelt. Die zusammengehörigen Bild- und Weltkoordinaten der Paßpunkte formen ein überbestimmtes lineares Gleichungssystem für die Transformationsparameter. Der lineare Modellansatz erlaubt die Lösung des Gleichungssystems in einem Schritt, während die Einbeziehung nichtlinearer Terme (z.B. für die Linsenverzeichnung) aufwendige, iterative Lösungsverfahren erfordern würde.

Die Bestimmung der 11 Parameter b_{ij} der Direkten Linearen Transformation erfordert die Lösung eines überbestimmten linearen Gleichungssystems, dessen Struktur in Glg. (1) dargestellt ist. Jeder der n zur Kalibrierung verwendeten Paßpunkte legt ein Paar von zugehörigen Weltkoordinaten (x_i, y_i, z_i) und Bildkoordinaten (u_i, v_i) fest, wobei der Index i = 1...n die Nummer des ausgewerteten Paßpunkts bezeichnet. Die Koeffizientenmatrix dieses Gleichungssystems besitzt 11 Spalten (Zahl der Unbekannten) und 2 Zeilen pro ausgewertetem Paßpunkt. Das Gleichungssystem wird durch

606

Auswertung von mindestens 6 (geeignet gewählten) Paßpunkten lösbar. In der Regel wird man 50-100 Paßpunkte auswerten, um statistische Fehler bei der Aufnahme der Paßpunkte ausgleichen zu können. Die Lösung umfangreicher überbestimmter Gleichungssysteme erfolgt vorteilhaft mit der Methode der Singulärwertzerlegung [5], die bei hoher Effizienz und numerischer Stabilität die im Sinne der kleinsten Fehlerquadratsumme optimale Lösung liefert.

$$
\begin{pmatrix}
\dfrac{x_1}{u_1} & \dfrac{y_1}{u_1} & \dfrac{z_1}{u_1} & \dfrac{1}{u_1} & 0 & 0 & 0 & 0 & -x_1 & -y_1 & -z_1 \\
0 & 0 & 0 & 0 & \dfrac{x_1}{v_1} & \dfrac{y_1}{v_1} & \dfrac{z_1}{v_1} & \dfrac{1}{v_1} & -x_1 & -y_1 & -z_1 \\
 & \cdot & & & & \cdot & & & & \cdot & \\
 & \cdot & & & & \cdot & & & & \cdot & \\
 & \cdot & & & & \cdot & & & & \cdot & \\
\dfrac{x_n}{u_n} & \dfrac{y_n}{u_n} & \dfrac{z_n}{u_n} & \dfrac{1}{u_n} & 0 & 0 & 0 & 0 & -x_n & -y_n & -z_n \\
0 & 0 & 0 & 0 & \dfrac{x_n}{v_n} & \dfrac{y_n}{v_n} & \dfrac{z_n}{v_n} & \dfrac{1}{v_n} & -x_n & -y_n & -z_n
\end{pmatrix}
\begin{pmatrix}
b_{11} \\ b_{12} \\ b_{13} \\ b_{14} \\ b_{21} \\ b_{22} \\ b_{23} \\ b_{24} \\ b_{31} \\ b_{32} \\ b_{33}
\end{pmatrix}
=
\begin{pmatrix}
1 \\ 1 \\ \cdot \\ \cdot \\ \cdot \\ 1 \\ 1
\end{pmatrix}
\qquad (1)
$$

Kalibrierung des 3D-Konturmeßsystems

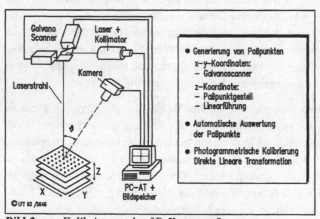

Bild 2: *Kalibrierung des 3D-Konturmeßsystems*

Eine Kalibrierung des 3D-Konturmeßsystems erfordert die Erzeugung von Paßpunkten im Objektraum, deren Weltkoordinaten hochgenau bekannt sind und deren Bildkoordinaten gemessen werden. Dazu müssen diese Paßpunkte vom Detektionssystem eindeutig und mit hoher Genauigkeit erkannt werden können. Im Vergleich zu den bisher bekannten Verfahren der Paßpunktaufnahme (z.B. [6], [7] bietet das 3D-Kontur-meßsystem entscheidende Vorteile aufgrund der aktiven, strukturierten Beleuchtung mittels Laserdiode und Galvanoscanner. Eine unabhängige Kalibrierung des Galvanoscanners steigert dessen absolute Positioniergenauigkeit um eine Größenordnung [8]. Die Strategie der automatisierten Kalibrierung des 3D-Konturmeßsystems ist in *Bild 2* dargestellt. Der Galvanoscanner erlaubt das flexible Setzen von Paßpunkten innerhalb einer x-y-Ebene. Die z-Koordinate der Paßpunkte kann z.B. durch eine rechnergesteuerte Linearführung definiert werden. Das in die Kamera eingebaute Interferenzfilter blendet das Umgebungslicht aus, so daß eine automatische Auswertung der Paßpunktkoordinaten problemlos möglich ist. Nach der Paßpunktaufnahme in mehreren x-y-Ebenen mit verschiedenen z-Koordinaten kann die Kalibrierung des Detektionssystems mit dem vorgestellten Formalismus der Direkten Linearen Transformation erfolgen. Eine Kalibrierung muß nach jeder Veränderung der Kameraorientierung oder Wechsel des Kameraobjektivs bzw. Veränderung der Brennweite bei Zoom-Objektiven erneut durchgeführt werden. Bei höchsten Genauigkeiten sollte selbst nach jeder Veränderung der Fokussierung des Kameraobjektivs neu kalibriert werden. Unter diesem Gesichtspunkt wird die Bedeutung einer vollautomatischen Kalibrierung des 3D-Konturmeßsystems deutlich.

Absolute Meßgenauigkeit des 3D-Konturmeßsystems

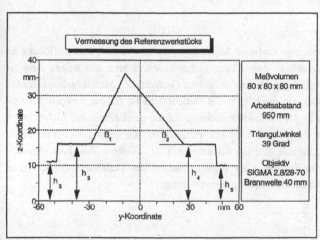

Bild 3: *Vermessung des Referenzwerkstücks*

Zur Ermittlung der absoluten Meßgenauigkeit des 3D-Konturmeßsystems wurde ein Referenzwerkstück mit 5 definierten Höhenstufen h_1, ..., h_5 und zwei im Winkel von $ß_1$ und $ß_2$ geneigten Flächen gefertigt. Die wahren Werte für diese Formmerkmale wurden mit einem Koordinatenmeßgerät ermittelt. *Bild 3* zeigt die vom Lichtschnittsensor gelieferten Rohdaten in der y-z-Ebene. *Tab. 1* zeigt die mit dem 3D-Konturmeßsystem und dem Koordinatenmeßgerät gewonnenen Ergebnisse für die Formmerkmale des Referenzwerkstücks im Vergleich. Nicht alle Formmerkmale können mit einer Messung erfaßt werden, da das Referenzwerkstück größer als das Bildfeld der Kamera ist.

	Koo.meßgerät	Lichtschnitt	Abweichung	rel. Fehler
Höhe h_2	10,999 mm	11,062 mm	0,063 mm	0,57%
Höhe h_3	15,996 mm	16,085 mm	0,089 mm	0,56%
Winkel $ß_1$	45,005°	44,957°	- 0,048°	- 0,11%
Winkel $ß_2$	30,004°	29,911°	- 0,093°	- 0,31%
Höhe h_4	15,992 mm	16,048 mm	0,056 mm	0,35%
Höhe h_5	10,001 mm	10,053 mm	0,052 mm	0,52%

Tab. 1: *Formmerkmale des Referenzwerkstücks*

Die in *Tab. 1* aufgeführten Werte sind das Ergebnis einer Regressionsanalyse bzw. Mittelwertsbildung über alle Meßpunkte des Formmerkmals. Die einzelnen Meßpunkte streuen in einer Spanne von 0,2 mm um diesen Mittelwert. Diese Schwankungen sind bei Wiederholmessungen zum Teil reproduzierbar, werden also auch durch die Oberflächenstruktur des Meßobjekts hervorgerufen. Die Werte in *Tab. 1* belegen einen relativen Fehler der mit dem 3D-Konturmeßsystem gemessenen Absolutwerte von unter 1%. Bei dieser Konfiguration des 3D-Konturmeßsystems entspricht die Größe eines CCD-Sensorpixels im Meßvolumen einem Bereich von rund 0,2 mm in y- und 0,4 mm in z-Richtung. Die Fehlergrenze des 3D-Konturmeßsystems ist mit 0,15 mm bzw. 0,4 Px deutlich besser als ein Pixelabstand. Die automatische Kalibrierung ermöglicht also eine Absolutgenauigkeit im Subpixelbereich. Für andere geometrische Konfigurationen kann von einer Fehlergrenze von 0,2% des Meßbereichs ausgegangen werden. Die Meßzeit beträgt 0,5 s pro Bild und damit pro Meßpunkt unter 1 ms.

Ausblick

Das geometrische Meßprinzip erlaubt eine einfache Anpassung von Meßbereich und Auflösung an verschiedene Meßaufgaben und erschließt damit dem Lichtschnittverfahren ein hohes Anwendungspotential. Meßbereich und Auflösung können in weiten Grenzen der Meßaufgabe angepaßt werden, die erreichbare Fehlergrenze von rund 0,2% des Meßbereichs ist für die meisten Anwendungen ausreichend. Meßaufgaben in der Produktionstechnik, für die der Einsatz des Lichtschnittprinzips vorgesehen ist, sind zum Beispiel eine on-line Prozeßkontrolle beim Laserstrahlbiegen [9] oder beim thermischen Richten von Karosseriebauteilen [10]. Weitere denkbare Anwendungsfelder sind die Detektion von Reißern oder Einschnürungen bei Tiefziehteilen oder eine Teile- und Lageerkennung für verschiedenste Fertigungsverfahren.

Literatur

[1] Heckel, W.; Geiger, M.: 3D-Lichtschnittsensor zur Biegewinkelerfassung. In: Waidelich, W. (Hrsg.): Laser in der Technik. Vorträge des 11. Internationalen Kongresses LASER 93. Berlin: Springer, 1994, S. 306-309.

[2] Heckel, W.; Geiger, M.: In-process measurement of bending angles. In: International Archives of Photogrammetry and Remote Sensing, **30** (1994) 5, S. 163-170.

[3] Abdel-Aziz, Y.I.; Karara, H.M.: Direct linear transformation into object space coordinates in close-range photogrammetry. In: Proc. Symp. on Close-Range Photogrammetry, University of Illinois, Urbana, Illinois, 26.-29. Jan. 1971, S. 1-18.

[4] Krauß, H.: Das Bild-n-Tupel. Ein Verfahren für photogrammetrische Ingenieurvermessungen im Nahbereich. Deutsche Geodätische Kommission, Reihe C: Dissertationen, Heft Nr. 276. München: Bayer. Akademie der Wissenschaften, 1983.

[5] Press, W.H.; Flannery, B.P.; Teukolsky, S.A.; Vetterling, W.T.: Numerical recipes in C. Cambridge, Mass.: Cambridge University Press, 1988.

[6] Föhr, R.: Photogrammetrische Erfassung räumlicher Informationen aus Videobildern. Braunschweig: Vieweg, 1990. Zugl.: Aachen, T.H., Diss., 1990.

[7] Beyer, H.A.: Advances in characterisation and calibration of digital imaging systems. In: International Archives of Photogrammetry and Remote Sensing **29** (1992) B5, S. 545-555.

[8] Heckel, W.: Optische 3D-Konturerfassung und on-line Biegewinkelmessung mit dem Lichtschnittverfahren. Reihe Fertigungstechnik Erlangen, Nr. 43. München: Hanser, 1995. Zugl.: Erlangen, Nürnberg, Univ., Diss., 1994.

[9] Nakagawa, T.: Neue Entwicklungen bei der Fertigung von Karosserieteilen. In: Blech Rohre Profile **41** (1994) 2, S. 100-103.

[10] Geiger, M.; Vollertsen, F.; Deinzer, G.: Flexible straightening of car body shells by laser forming. In: Proceedings of the NADDRG/SAE Sheet Metal Forum, Detroit, Michigan, 1.-5. März 1993, SAE Paper 930279, S. 30-33.

Hochfrequenzkinematographische Untersuchungen an laserinduzierten Oberflächenplasmen

M. Althaus, M. Hugenschmidt

Deutsch-Französisches Forschungsinstitut Saint-Louis (ISL)

F-68301 SAINT LOUIS CEDEX, 5 RUE DU GÉNÉRAL CASSAGNOU

Frankreich

ABSTRACT

Mit einem repetierend gepulsten CO_2-Laser der Wellenlänge $\lambda = 10,6$ µm (Leistungsdichte bis zu 100 MW/cm^2, variable Repetitionsraten) wurden auf verschiedenen Targets Oberflächenplasmen erzeugt. Versuchsparameter waren dabei Umgebungsdruck und verschiedene Schutzgase. Untersucht wurden verschiedene Metalle, Keramiken sowie PMMA.

Zur Erfassung von Ausbreitungsgeschwindigkeiten, Elektronentemperaturen und -dichten dienten zeitlich und räumlich auflösende kurzzeitphysikalische Verfahren wie Interferometrie und Spektroskopie. Als Diagnostiklaser wurde ein frequenzverdoppelter Nd:YLF-Laser mit einer Halbwertsbreite von 10 ns eingesetzt.

Im folgenden werden die Elektronentemperaturen und -dichten, welche mit den oben genannten Verfahren ermittelt wurden, vorgestellt und diskutiert.

EINLEITUNG

Bei der Materialbearbeitung von Werkstücken mit Lasern werden je nach Leistungsdichte des bearbeitenden Laserstrahls Plasmen auf oder sogar vor dem Werkstück gezündet. Diese spielen eine wesentliche Rolle bei der Materialbearbeitung, da sie, abhängig von Elektronendichte und -temperatur, zur Energieeinkopplung in das Zielmaterial beitragen können, aber auch das zu bearbeitende Werkstück so abschirmen können, daß eine weitere Energieeinkopplung nicht möglich ist.

Um den Energieeintrag zu optimieren, ist eine genaue Kenntnis der plasmabestimmenden Parameter wie Elektronendichte, Elektronentemperatur, Ausdehnung und Ausbreitungsgeschwindigkeit nötig. Daher wurden im folgenden diese Größen mittels unterschiedlicher Diagnostikmethoden bestimmt.

EXPERIMENTELLER AUFBAU

Bei den hier beschriebenen Versuchen wurde ein TEA-CO_2-Laser mit einer Energie von 150 mJ/Puls und einer Gesamtpulsdauer von 2 µs eingesetzt, mit dem bei geeigneter Fokussierung Leistungsdichten von bis zu 100 MW/cm^2 erreicht werden können. Der Laserstrahl wurde über Spiegel durch ein ZnSe-Fenster in eine Vakuumkammer gelenkt und dort mittels einer ZnSe-Linse (f = 19 cm) auf die Targetoberfläche fokussiert. Das entstehende Plasma konnte mit Hilfe verschiedener Diagnostiktechniken untersucht werden.

INTERFEROMETRIE

Zur Erzeugung von Interferogrammen wurde ein frequenzverdoppelter Nd:YLF-Laser (t_{FWHM} = 10 ns) aufgeweitet, so daß das Plasma von parallelem Licht durchleuchtet wird. Die Targetebene wurde mit einer zweiten Linse über ein Wollaston-Prisma und einen Polarisator auf eine Mattscheibe abgebildet. Mit Hilfe des Prismas und des Polarisators konnten so gleichzeitig Schattenbilder und Interferogramme erzeugt werden [HUG]. Den schematischen Aufbau zeigt Abb. 1.

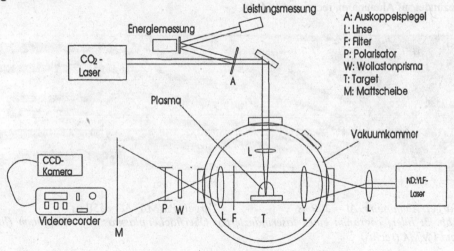

Abb. 1: Schematischer Aufbau zur Interferometrie

SPEKTROSKOPIE

Zur direkten Bestimmung von zeitlich aufgelösten Elektronentemperaturen und -dichten diente die Spektroskopie. Ein definierter Bereich des Plasmas wurde auf einen Lichtwellenleiter abgebildet. Dieser führte das Licht zu einem Monochromator, dessen Ausgang fest mit einem Photomultiplier verbunden war. Das Ausgangssignal konnte dann mit einem Oszillographen registriert werden. Durch Intensitätsmessungen zweier Spektrallinien der gleichen Ionisationsstufe lassen sich Elektronentemperaturen, bzw. bei Aufzeichnung zweier Linien unterschiedlicher Ionisationsstufen Elektronendichten bestimmen [GRIEM; KONJEVIC](Abb. 2).

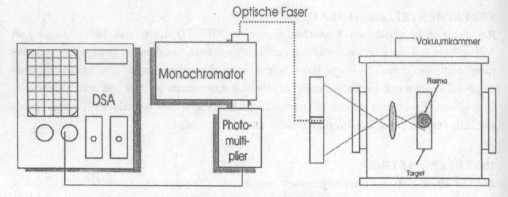

Abb. 2: Zeitaufgelöste Spektroskopie

ERGEBNISSE

Die Abb. 3 zeigt Beispiele von Interferogrammen laserinduzierter Oberflächenplasmen, links gezündet auf Aluminium, rechts auf PMMA.

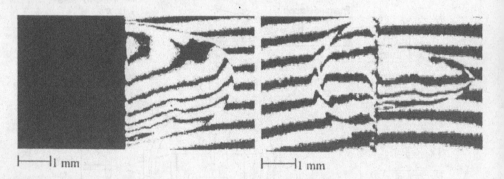

Target: Aluminium, $\Delta t = 300$ ns   *Target: PMMA, $\Delta t = 550$ ns*
Abb. 3: Interferogramm eines laserinduzierten Oberflächenplasmas auf Aluminium (links) und PMMA (rechts)

Aus den Streifenverschiebungen lassen sich u. a. die Ausbreitungsgeschwindigkeiten berechnen. Diese liegen für Aluminium in Luft bei 13,4 km/s in den ersten 200 ns, nehmen dann jedoch rasch ab, bis sie nach einigen Mikrosekunden Schallgeschwindigkeit erreicht haben. Reduziert man den Umgebungsdruck, so erhöht sich die Ausbreitungsgeschwindigkeit in den ersten 200 ns bis auf 15,4 km/s bei 300 mbar. Ein ähnliches Ergebnis erhält man für Kupfer und Messing. Bei PMMA konnte auch die Ausbreitung einer Stoßfront im Innern des Materials beobachtet werden. Hier wurde sowohl bei Atmosphärendruck wie auch bei Drücken unter 1 mbar eine Geschwindigkeit von 2,74 km/s gemessen, was im Rahmen der Meßungenauigkeit sehr gut dem Wert der Schallgeschwindigkeit in PMMA von 2,72 km/s entspricht.

Aus den Interferogrammen bestimmen sich die maximalen Elektronendichten im Plasmazentrum zu einigen 10^{18}cm^{-3}.

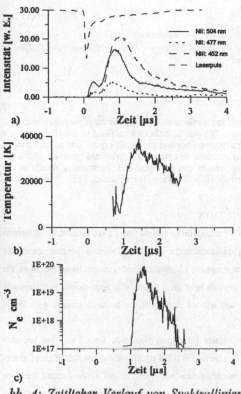

a)

b)

c)

*bb. 4: Zeitlicher Verlauf von Spektrallinien
a), Elektronentemperatur (b) und
lektronendichte (c)*

Die Spektroskopiedaten liefern höhere Werte für die Elektronentemperaturen und -dichten. In Abb. 4a ist der zeitliche Verlauf dreier Spektrallinien von Stickstoff dargestellt. Man erkennt einen kleinen Peak zu Beginn des Laserpulses und einen zweiten nach 500 ns. Bei Druckverminderung des Umgebungsgases ändert sich das relative Verhältnis der beiden Amplituden zueinander, der erste Peak wächst, der zweite wird kleiner und der zeitliche Abstand zwischen ihnen vermindert sich ebenfalls. Dies läßt sich durch die Zündung erst einer LSD-Welle und dann einer nachfolgenden LSC-Welle erklären. Wie in Abb. 4b dargestellt werden Elektronentemperaturen von bis zu 38000 K und Elektronendichten bis zu einigen 10^{19} cm^{-3} erreicht.

ZUSAMMENFASSUNG

Mit Hilfe der Spektroskopie war es möglich, direkt Elektronentemperaturen und -dichten zu ermitteln. Letztere lagen je nach Druckbereich zwischen einigen 10^{17}cm^{-3} und einigen 10^{19}cm^{-3}. Bei PMMA gelang es, mit interferometrischen Methoden gleichzeitig die Schockwellenausbreitung innerhalb als auch außerhalb des Materials sichtbar zu machen, außerdem wurden Ausbreitungsgeschwindigkeiten der Stoßfronten bei verschiedenen Drücken gemessen.

LITERATUR

- Griem, *Plasma Spectroscopy*, McGraw-Hill Book Company, 1964
- Konjevic, *A critical review of the Stark Widths and Shifts of spectral lines from non-hydrogenic Atoms*, J. Phy. Chem. Ref. Data, Vol. 5, No 2, 1976
- M. Hugenschmidt, *Recent developments in high-resolution optical diagnostics of repetitively pulsed laser-target effects*, Proc. of the 21st Int. Congress on High Speed Photography and Photonics, Korea, 1994 SPIE

Stress Effects in Highly Birefringent Optical Fibers with Liquid Crystalline Side Holes

T. R. Woliński, W. J. Bock*, R. Dąbrowski**, A. Jarmolik, J. Parka**, A. Zackiewicz

*Institute of Physics, Warsaw Univ. of Technology, Koszykowa 75, 00-662 Warszawa, Poland, *University of Quebec at Hull, Canada; **Military University of Technology, Poland*

Abstract - This paper discusses the influence of hydrostatic stress effects on lowest-order mode propagation in new HiBi fibers with holes located on both sides of an elliptical core. The side holes have been filled with a chiral nematic liquid crystal. The whole system composed of the highly birefringent fiber and the liquid crystal acts as an optically anisotropic medium. Since stress effects occurring in the system due to external perturbations generate additional birefringence, a new class of fiber-optic hydrostatic pressure sensors can be introduced. Preliminary results of the fiber-optic liquid crystalline pressure sensor utilizing the side-hole HiBi fibers are also presented.

1. INTRODUCTION

Over the last decade highly birefringent (HiBi) polarization maintaining optical fibers have focused great interest both from theoretical and practical points of view. In sensing applications, the influence of various external parameters on mode propagation in HiBi fibers is of special interest, since a number of physical quantities can be measured on the basis of the fibers. Also, very recently it appeared that liquid crystals hold great potential in applications to hydrostatic pressure metrology utilizing fiber optic sensing techniques and a number of different pressure sensors with liquid crystals as an active element has been recently proposed [1-6].

When high linear birefringence is introduced into a regular (isotropic) two-mode fiber, the fiber becomes anisotropic and two effects come into being. First, the mode degeneracy is lifted which means that the different modes will have different propagation constants and the greater the birefringence the greater the difference. Second, the even and the odd second-order LP_{11} modes will have different cut-off wavelengths that can reduce a number of propagating modes in the HiBi fiber. Since highly birefringent fibers have a pair of preferred orthogonal axes of symmetry, the two orthogonal quasilinear polarized field components $HE_{11}{}^x$ and $HE_{11}{}^y$ of the fundamental mode HE_{11} (LP_{01}) which propagate for all values of frequency (wavelength), have electric fields that are polarized along one of these two axes. Hence, light polarized in a plane parallel to either axis will propagate without any change in its polarization, but with different velocities. For wavelengths slightly shorter than a critical value (cutoff wavelength) the next higher-order mode with greater propagation velocity compared with that of the fundamental mode is guided. The relevant feature of HiBi fiber is that only two second-order modes ($LP_{11}{}^{even}$) propagate instead of four. Over a large region of the optical spectrum the fiber guides only four modes: two orthogonal linearly polarized eigenmodes for each of the fundamental (LP_{01}) and the second-order ($LP_{11}{}^{even}$) spatial modes. The separation between the cutoff wavelengths for the $LP_{11}{}^{even}$ and the $LP_{11}{}^{odd}$ modes will increase with increasing birefringence.

2. STRESS EFFECTS IN ELLIPTICAL-CORE SIDE-HOLE FIBERS

The modal behavior of lowest-order mode HiBi fibers under various stresses is of special interest for sensors and device applications. A number of physical quantities can be measured on the basis of HiBi fibers: hydrostatic pressure, strain, vibration, temperature, acoustic wave, etc.

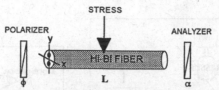

Fig. 1   Stress effects in a HiBi side-hole fiber modulate light intensity after the analyzer:
$I(z=L) = 1/2 [1 + cos2\alpha \cdot cos2\phi + sin\alpha \cdot sin2\phi \cdot cos\Delta\Phi_i]$, where $\Delta\Phi_i$ - the differential phase of the output light.

A symmetric stress effect influences propagation constant β_i in every mode due to the changes in fiber length (L) and the refractive indices of the core and the cladding. This leads to changes in the phase $\Phi_i = \Delta\beta_i L$ along the fiber:

$$\delta(\Delta\Phi_i) = \delta(\Delta\beta_i) L + \Delta\beta_i \delta L \qquad 1)$$

where: $i = 0,1,x,y$; $\Delta\Phi_x$, $\Delta\Phi_y$ are responsible for changes in the output light intensity distribution, and $\Delta\Phi_0$, $\Delta\Phi_1$ correspond to the output polarization state changes. First case is the intermodal interference between X- or Y-polarized LP_{01} and LP_{11} spatial modes; whereas the second is in fact polarization interference between polarization modes $LP_{01}{}^x$ and $LP_{01}{}^y$ (or $LP_{11}{}^x$ and $LP_{11}{}^y$) and requires use of an analyzer, see Fig. 1. In a two-mode HiBi fiber, the variation of the phase difference under e.g. hydrostatic stress (pressure, δp) may be written as [7]:

$$\frac{\delta(\Delta\Phi_i)}{\delta p} = \Delta\beta_i \frac{\partial L}{\partial p} + L\frac{\partial(\Delta\beta_i)}{\partial p} \qquad (2)$$

3. HiBi FIBERS WITH LIQUID CRYSTALLINE SIDE HOLES

We consider a dielectric waveguide with a chiral liquid crystal side holes. It consists of a cylindrical isotropic region of with electrical permeability ϵ, two side holes occupied by a chiral nematic liquid crystal surrounded by an isotropic cladding with dielectric constant ϵ_{cl}. The dielectric properties of the liquid crystal side-hole single domain are described by the dielectric tensor written in laboratory frame as [8]

$$\epsilon(x,y,z) = \begin{bmatrix} \epsilon + \delta\cos2\beta z & \delta\sin2\beta z & 0 \\ \delta\sin2\beta z & \epsilon - \delta\cos2\beta z & 0 \\ 0 & 0 & \epsilon_3 \end{bmatrix}$$

where

$$\beta = 2\pi/Z, \quad \epsilon = \frac{\epsilon_1 + \epsilon_2}{2}, \quad \delta = \frac{\epsilon_1 - \epsilon_2}{2} \qquad (4)$$

and Z is pitch length. The dielectric ellipsoid representing this tensor has two principal axes ϵ_1, ϵ_2 normal to z axis and ϵ_3 along z. The axis of the length $\epsilon_1 = \epsilon + \delta$ is in the x direction at $z = 0$, and it spirals with a pitch $Z = 2\pi/\beta$ as a value of z is increases.

Since a very low energy is required to induce a helical twist within a mesophase, the molecular arrangement of a cholesteric liquid crystal is expected to be affected by external pressure. We expect, a decrease of a pitch Z since the centers of mass of the molecules come closer together as a result of the compressibility of liquid crystals. On the other hand, there may be an additional structural change with pressure due to variation of the relative arrangement of the molecules [9].

4. EXPERIMENTAL RESULTS

The experimental set-up is shown in Fig 2. A HiBi fiber with liquid crystal side holes was attached to the input/output single-mode fibers which were used to transport optical signals in and out of the pressure region.

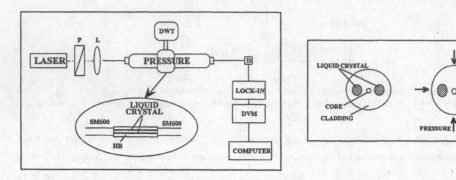

Fig. 2 The experimental set-up (to the left): P - polarizer, L - lens, D - detector, DWT - deadweight tester, LOCK-IN - lock-in amplifier, DVM - digital multivoltmeter, SM600- single-mode leading fibers, along with the cross section of the HiBi fiber with liquid crystal side holes

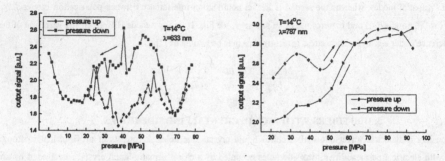

Fig. 3 Pressure characteristics of HiBi fibers with liquid crystalline side holes obtained for 632 and 787 nm lasers

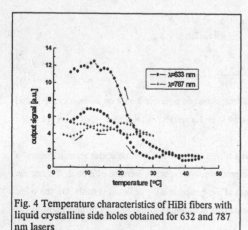

Fig. 4 Temperature characteristics of HiBi fibers with liquid crystalline side holes obtained for 632 and 787 nm lasers

The liquid crystalline material was a special mixture of a nematic liquid crystal characterized by refractive indices n_o= 1.517, n_e= 1.732 doped with an optically active chiral compound and was introduced to the side holes by capillary action. The sensor assembly was placed inside a standard thermally stabilized pressure chamber. Pressure generation and calibration up to 100 MPa was performed using a Harwood DWT-35 deadweight tester. As a light source a He-Ne laser operated at 633 nm wavelength and a laser diode at 787 nm were used, modulated using standard techniques. The experimental details were described elsewhere [10]. Fig. 3 presents, for the first time to our best knowledge, pressure characteristics of HiBi fibers with liquid crystalline side holes obtained for two laser sources. It is evident that hydrostatic stress influences the optical signal propagating along the fiber. A similar effect is caused by thermal stresses as it is presented in Fig. 4.

5. CONCLUSIONS

The influence of hydrostatic and thermal stress effects in HiBi fibers with liquid crystal side holes can have several origins. Since compressibility of liquids splits into geometrical and structural terms, a structural effect can be expected. According to existing theories [11] helical pitch depends on pressure

$$Z = C_1 / T(1 - C_2\eta) \tag{5}$$

where C_1 and C_2 are molecular constants and viscosity η strongly depends of pressure

$$\eta_p = \eta_1 \exp(\alpha p) \tag{6}$$

where η_1 is a viscosity at $p = 0.1$MPa, and $\alpha = (1/\eta_p)d\eta_p/dp$. An increase of Z with pressure is expected which exceeds the pitch decrease caused by geometrical compressibility. Since cholesteric mesophases exhibit a selective reflection the maximum reflection λ_R is connected to the pitch Z by the following relation [12]:

$$\lambda_R = \bar{n}Z = \bar{n}C_1 / T(1 - C_2\eta(p)) \tag{7}$$

where $\bar{n}$ is the average refractive index.

To summarize the observed hydrostatic stress effects include changes in fiber birefringence, the selective Bragg refection in chiral nematic liquid crystals and its changes under hydrostatic pressure and effects of light scattering of liquid crystal medium. Since liquid crystal are highly scattering medium. the rate of scattering from liquid crystal side holes will also change with pressure asymmetrically. This creates a new inroads in future constructions of fiber optic sensors with liquid crystals for pressure measurement.

ACKNOWLEDGMENT

The work was supported by the Polish Committee for Scientific Research (KBN) under the grant no. 8/T11D/006/08 and by the University of Québec at Hull, Canada.

REFERENCES

[1] T. R. Woliński, W. J. Bock, "Cholesteric Liquid Crystal Sensing of High Hydrostatic Pressure Utilizing Optical Fibers", *Mol. Cryst. Liq. Cryst.*, Vol. 199, 7-17 (1991).

[2] T. R. Woliński, W. J. Bock, "Fiber-Optic Liquid Crystal High Pressure Sensor", in *"Fiber-Optic Sensors: Engineering and Applications"*, 14-15 March, The Hague, The Netherlands, *Proc. SPIE (USA)*, vol. 1511, 281-288 (1991).

[3] T. R. Woliński, W. J. Bock, R. Dąbrowski, "Fiber-Optic Measurement of High Hydrostatic Pressure with Cholesteric Liquid Crystals," *SPIE* Vol.1845 *Liquid and Solid State Crystals*, 558-562 (1992).

[4] T. R. Woliński, R. Dąbrowski, W. J. Bock, A. Bogumil, S. Klosowicz, "Liquid Crystalline Films for Fiber Optic Sensing of High Hydrostatic Pressure", *Thin Solid Films*, Vol. 247, 252-257 (1994).

[5] W. J. Bock, T. R. Woliński, "Method for Measurement of Pressure Applied to a Sensing Element Comprising a Cholesteric Liquid Crystal", *US Patent No. 5 128 535*, issued on July 7, 1992.

[6] T. R. Woliński, R. Dąbrowski, W. J. Bock, A. Bogumil, S. Klosowicz, "Liquid Crystalline Films for Fiber Optic Sensing of High Hydrostatic Pressure", *Thin Solid Films*, 1994 (in press).

[7] W. J. Bock, T. R. Woliński, "Hydrostatic pressure effects on mode propagation in highly birefringent two-mode bow-tie fibers", *Optics Letters*, vol. 15, 1434-1436, 1990.

[8] D.W. Berreman,"Optics in stratified and Anisotropic Media: 4x4 Matrix Formulation,"*J.Opt.Soc.Am.*62, 502-510 (1972).

[9] P. Pollman, H.Stegemeyer, "Pressure dependence of helical strukture of cholesteric mesophase", *Chem. Phys. Letters*,V.20, No.1 (1973) 87.

[10]T. R. Woliński, R. Dąbrowski, A. Bogumil, Z. Stolarz, "Liquid Crystalline Cells for Fiber Optic Sensing of Low Hydrostatic Pressure", *Mol. Cryst. Liq. Crys.* 1995 (in press).

[11]P.N.Keating, *Mol. Cryst. Liq. Crys.* 8, p.315 (1969), B.Boettcher, *Chem. Z.* 96 (1972) 214.

[12]Hl. de Vries,"Rotatory power and another properties of certain liquid crystals," *Acta. Cryst.* 4, 219-226 (1951).

Mehrkomponentiges Laser-Doppler-Velocimeter mit erstmaliger Nutzung eines frequenzmodulierten Nd:YAG Lasers und Glasfaserumwegstrecken

J.W. Czarske und H. Müller*)

Universität Hannover, Institut für Meß- und Regelungstechnik im Maschinenbau,
D–30167 Hannover, Germany *)Physikalisch Technische Bundesanstalt, Laboratorium für Strömungsmeßtechnik, D–38116 Braunschweig, Germany

Kurzfassung:

Es wird ein neuartiges Laser–Doppler–Velocimeter (LDV) vorgestellt, das eine richtungsempfindliche und zweikomponentige Geschwindigkeitsmessung von Strömungen erlaubt. Das LDV-System hat durch die Nutzung von Faseroptik einen vereinfachten Aufbau und mit dem Einsatz eines diodengepumpten Nd:YAG Lasers wird eine hohe Leistung im Meßvolumen ermöglicht.

This contribution presents a novel directional heterodyne laser Doppler velocimeter (LDV) for two dimensional fluid velocity measurements. The heterodyne technique is realized without the use of an additional frequency shifter. Hence, together with fiber optics the realization of a compact and alignment insensitve LDV system is possible. The used Nd:YAG ring laser enables the realization of a LDV–system with a high optical power.

1. Einleitung

Das in der Laser–Doppler–Velocimetrie (LDV) übliche Heterodyn–Verfahren ermöglicht bekanntermaßen sowohl eine richtungsempfindliche als auch eine mehrkomponentige Geschwindigkeitsmessung von Fluiden. Zur Realisierung des Heterodyn–Verfahrens werden bisher Frequenzshift–Elemente, z.B. Bragg–Zellen eingesetzt. Die Verwendung von Bragg–Zellen bedingt jedoch unvermeidbare Lichtverluste, einen hohen Justageaufwand und behindert die Miniaturisierung des LDV–Systems.

In diesem Beitrag wird ein Heterodyn–LDV vorgestellt, das keine externen Frequenzshift–Elemente benötigt und daher als besonders kompaktes, justageunaufwendiges LDV–System realisiert werden kann.

2. Das Grundprinzip

Die unmittelbare Realisierung des Heterodyn–Verfahrens kann durch Chirp–Frequenzmodulation eines Lasers in Verbindung mit einer Verzögerungsstrecke (Gangunterschied) erreicht werden [JONES, SORIN]: Die Frequenzdifferenz Δf der interferierenden Strahlen ergibt aus der Laserfrequenzmodulation $f_L(t)$, dem Gangunterschied Δl und der Lichtgeschwindigkeit c zu:

$$\Delta f = \frac{l}{c}\frac{df_L(t)}{dt}. \tag{1}$$

Dieses Verfahren wurde bisher lediglich durch die Nutzung von strommodulierten Laserdioden vorgenommen [JONES, SCHRÖDER], die eine geringe Leistung, eine parasitäre Leistungsmodulation bei durchgeführter Frequenzmodulation und eine geringe Kohärenzlänge aufweisen.

In diesem Beitrag soll zur Realisierung des Heterodyn–Verfahrens ein frequenzmodulierbarer Nd:YAG Laser in Verbindung mit Glasfaserumwegstrecken eingesetzt werden. Der Nd:YAG Laser ermöglicht bei hoher single–frequency Leistung eine definierte Frequenzmodulation [FREITAG] und weist eine hohe Kohärenzlänge auf [CZARSKE (a)], die die Realisierung einer Verzögerungsstrecke großer Länge erlaubt.

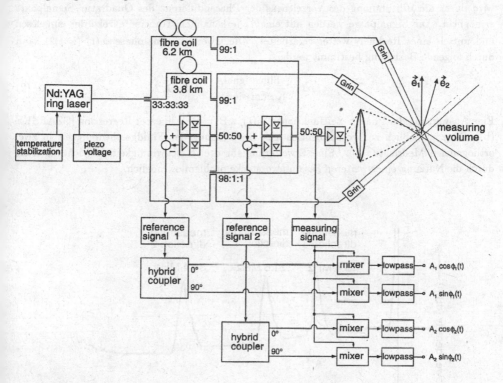

Abbildung 1: Schematische Darstellung des zweikomponentigen Heterodyn–Laser–Doppler–Velocimeters.

3. Zweidimensionales Laser–Doppler–Velocimeter

3.1 Aufbau

In Abb. 1 ist der Aufbau des LDV–System dargelegt: Ein fiberpigtailed Nd:YAG Ringlaser mit einer Leistung von 160 mW bei 1319 nm Wellenlänge wird mittels einer dreieckförmigen Piezospannung frequenzmoduliert. Die fasergeführte Laserwelle wird mit Richtkopplern geteilt und unterschiedlich verzögert. Mittels Gadient–Index–Linsen (Grin) werden die Laserwellen ins Meßvolumen fokussiert und das Meßsignal erzeugt und durch Auskopplung eines geringen Leistungsanteils der Laserwellen werden die Referenzsignale erzeugt, s. Abb. 1.

Durch die Nutzung verschiedener Verzögerungslängen werden unterschiedliche Differenzfrequenzen der Laserwellen bzw. Trägerfrequenzen der elektrischen Interferenzsignale erhalten, s. Abb.2. Daher kann ein mehrkomponentiges LDV–System durch Nutzung von Frequenzdemultiplex–Techniken realisiert werden, d.h. es werden unterschiedliche Frequenzbereiche zu den verschiedenen Meßkomponenten zugewiesen.

620

Als Signalverarbeitungsmethode wurde die Quadratur–Demodulation–Technik genutzt, s. Abb.1 (unten): Das Meßsignal wird mit beiden Referenzsignalen (Meßrichtung 1: Trägerfrequenz 21.6 MHz, Meßrichtung 2: Trägerfrequenz 34.4 MHz, s. Abb. 2) gemischt und tiefpaßgefiltert, wobei mittels eines breitbandigen 90^0 Phasenshifters jeweils ein Quadratur–Signalpaar (Kosinus–/Sinussignal) im Basisband entsteht. Die Richtungserkennung der Fluidströmung wird durch die Bestimmung des Vorzeichens der Phasendifferenz des Quadratur–Signalpaars ermöglicht. Die Signalpaare werden mit einem vierkanaligen Transientenrekorder eingelesen und mittels eines Rechners weiter verarbeitet. Die jeweilige Signalphase $\phi_i(t)$ (i=1,2) kann durch folgende Beziehung bestimmt werden:

$$\phi_i(t) = \arctan \frac{\text{Sinussignal}_i}{\text{Kosinussignal}_i}, \quad i = 1, 2. \tag{2}$$

Eine Auswertung der Phasen–Zeitfunktionen $\phi_i(t)$, z.B. mittels linearer Regression [CZARSKE (b)], führt schließlich zur vorzeichenrichtigen Bestimmung der Fluidgeschwindigkeit in zwei orthogonalen Meßrichtungen. Eine Erweiterung für drei Geschwindigkeitsmeßrichtungen ist durch die Nutzung einer weiteren Faserumwegstecke problemlos möglich.

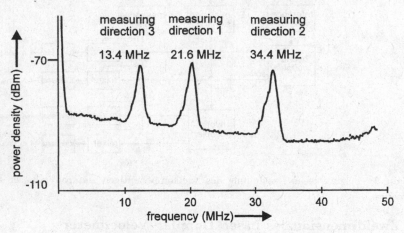

Abbildung 2: Trägerfrequenz–Spektrum des Meßsignals.

3.2 Meßergebnisse

Es wurde eine dreieckförmige Frequenzmodulation des Ringlasers mit 15 V Spannungsamplitude und 4 kHz Modulationsfrequenz durchgeführt. Nach einer Faserumwegstecke von 3.8 km bzw. 6.2 km ergaben sich die Meßrichtung 1 bzw. 2 folgende Trägerfrequenzen: 21.6 MHz bzw. 34.4 MHz, siehe Gl.(1) und Abb. 2. Die ersichtliche Frequenzverbreiterung der Träger wird durch die oben beschriebene Korrelation zwischen Meß– und Referenzsignal eliminiert.

Als typisches Ergebnis der Signalverarbeitungseinheit ist in Abb. 3 ein Quadratur–Signalpaar dargestellt. Mittels Gl.(2) kann die laufende Signalphase bestimmt werden, die als Basisdatensatz für die Bestimmung der Fluidgeschwindigkeit dient (s.o.).

4. Diskussion

Das kurz beschriebene zweidimensionale LDV–System mit Richtungserkennung weist folgende Besonderheiten auf:

- Der genutzte Nd:YAG–Laser ermöglicht sowohl eine hohe Leistung als auch eine schnelle, definierte Frequenzmodulation.

- Die eingesetzte Faseroptik ermöglicht eine einfache Realisierung der verschiedenen Verzögerungsstrecken zur Erzeugung der Trägerfrequenzen sowie eine leichte Traversierung des LDV–Meßkopfes.

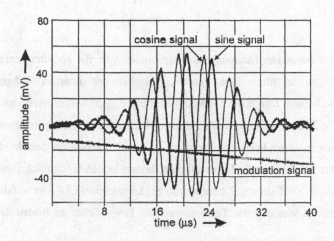

Abbildung 3: Burst–Signalpaar der Meßrichtung 1.

- Die Quadratur–Demodulations–Technik ermöglicht die Vermeidung von parasitären Trägerfrequenzschwankungen im auszuwertenden Quadratur–Signalpaar. Daher kann die Dopplerfrequenz bzw. die Fluidgeschwindigkeit mit einer hohen Präzision bestimmt werden.

Literatur:

CZARSKE (a), J., Philipps, R., Freitag, I.: 'Spectral properties of diode–pumped monolithic Nd:YAG ring lasers', Applied Physics B (in press).

CZARSKE (b), J., Hock, F., Müller, H.: 'Minimierung der Meßunsicherheit von Laser–Doppler–Anemometern', Technisches Messen, 1994, S.168–182.

FREITAG, I., Kröpke, I., Tünnermann, A., Welling, H.: 'Electrooptically fast tunable miniature diode–pumped Nd:YAG ring laser', Optics Communications, 1993, 101, S.371–376.

JONES, J., Corke, M., Kersey, A., Jackson, D.: 'Miniature solid–state directional laser Doppler velocimeter', Electronics Letters, 1982, 18, S.968–969.

SCHRÖDER, J., Nielsen, C.: 'Wide–range directional LDV employing modulated semiconductor laser source', Electronics Letters, 1986, 22, S.469–470.

SORIN, W., Donald, D., Newton, S., Nazarathy, M.: 'Coherent FMCW reflectometry using a temperature tuned Nd:YAG ring laser', IE³ Photon. Techn. Lett., 1990, 2, S.902–904.

Charakterisierung von Leistungsdioden durch Messung der internen Infrarot-Laser-Deflektion

G. Deboy, H. Brunner, W. Keilitz*, U. Müller*, R. Thalhammer* und G. Wachutka*
Siemens AG, Zentrale Forschung und Entwicklung,
Otto-Hahn-Ring 6, D-81739 München
* Technische Universität München, Lehrstuhl für Technische Elektrophysik,
  Arcisstr. 21, D-80290 München

1. Einleitung

Die Entwicklung moderner Leistungshalbleiterbauelemente für hochfrequente Pulswechsel-
richteranwendungen erfordern eine genaue Kenntnis der internen Trägerdichten- und
Temperaturverteilungen. Durch die Marktanforderungen nach höhersperrenden Bauelementen
und verbesserter Stromtragfähigkeit wird dessen thermische Belastung insbesondere während
der Schaltflanken zu einem kritischen Problem. Die elektrothermische Simulation des internen
Verhaltens liefert dabei wertvolle Informationen, bedarf jedoch hinsichtlich ihrer Aussagekraft
der experimentellen Verifizierung. Daher wurde ein kontaktloses IR-Laserverfahren entwickelt,
das die gleichzeitige Messung der Temperatur- und Trägerdichte im Innern des Bauelements
ermöglicht.

2. Meßprinzip

Das Verfahren beruht auf der Detektion von Brechungsindexänderungen in Silizium, die durch
eine Variation der Temperatur oder der Trägerdichte hervorgerufen werden. Ein Gradient
dieser Größen verursacht eine Winkelablenkung des Laserstrahls längs seiner Bahn durch das
Bauelement. Das Prinzip ist schematisch aus Abb. 1 ersichtlich.

Der Strahl wird über die Seitenflächen in das Bauelement fokusiert; eine Abrasterung in
vertikaler und horizontaler Richtung ist möglich. Die Tiefeninformation längs der optischen
Achse kann durch Drehung des Bauelements um 90° und eine horizontale Abrasterung
gewonnen werden.

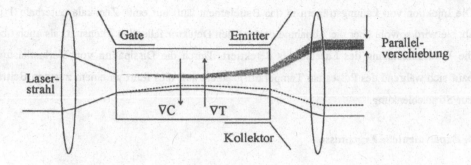

Fig. 1: Prinzip des IR-Laser-Ablenkungsverfahrens.

Die Winkelablenkung des Laserstrahls wird durch ein Auskoppelobjektiv in eine Parallelverschiebung umgesetzt, die über einen positionsempfindlichen Detektor direkt gemessen wird. Dabei lassen sich gleichzeitig die horizontale und vertikale Strahlverschiebung, sowie die Intensitätsabnahme des Strahls infolge Absorption an freien Ladungsträgern bestimmen; durch **eine** Messung gewinnt man somit über den Realanteil des Brechungsindex die Gradienten der Trägerdichte und der Temperatur in zwei Raumrichtungen und über den Imaginäranteil den Absolutwert der Trägerdichte. Die Anteile der Temperatur bzw. der Ladungsträgerkonzentration zur lokalen Änderung des Brechungsindex lassen sich über die zeitliche Auflösung des Verfahrens voneinander trennen. Abb. 2 zeigt ein Meßbeispiel an einer Leistungsdiode, die durch einen 70 µs breiten Puls mit einer Stromdichte von 150 A/cm² belastet wird.

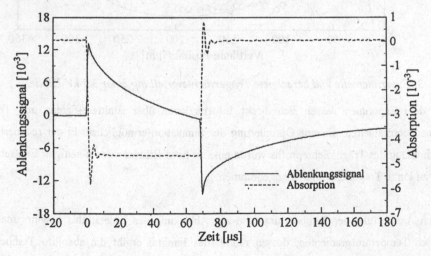

Abb. 2: Absorptions- und Ablenkungssignal einer pulsförmig belasteten Leistungsdiode (Oszillationen auf Absorptionssignal durch elektronische Verstärkung verursacht).

Die Injektion von Ladungsträgern in das Bauelement läuft auf einer Zeitskala unterhalb 1 µs ab; sie wird sowohl über die Abnahme der auf den Detektor fallenden Intensität, als auch über die Winkelablenkung des Laserstrahls detektiert. Durch die Dissipation von Verlustleistung baut sich während des Pulses ein Temperaturgradient auf; dies führt zu einem zweiten Beitrag zur Strahlablenkung.

3. Experimentelle Ergebnisse

Das Verfahren wurde auf eine neuartige Hochleistungsdiode mit einem Sperrvermögen bis 3500 V angewandt. Abb. 3 zeigt einen Vergleich zwischen der experimentell bestimmten Trägerdichteverteilung und einer Simulation mit Hilfe des Programm DESSIS (200 A/cm²).

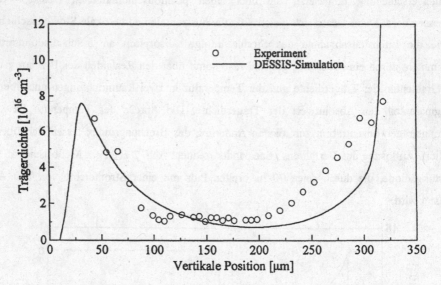

Abb. 3: Experimentelle und berechnete Trägerdichteverteilung einer 3.5 kV-Diode.

Aus dem Experiment lassen sich direkt Informationen über Emittereffizienz und Trägerlebensdauern ableiten, die eine Optimierung der Simulation ermöglichen. In der rechnerischen Bestimmung des Trägerdichteprofils wurde eine zu hohe Effizienz des n-seitigen Emitters und eine zu kurze Trägerlebensdauer angenommen.

Durch Auswertung der Temperaturanteile zur Ablenkung des Laserstrahls gewinnt man den lokalen Temperaturgradienten; dessen räumliches Integral ergibt die absolute Temperaturdifferenz im Bauelement bezogen auf die Rückseite. Abb. 4 zeigt das Ergebnis.

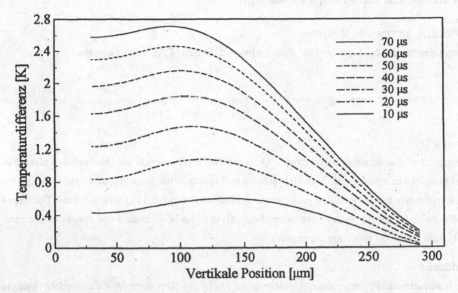

Abb. 4: Experimentelle Temperaturverteilung im Substrat einer 3.5 kV-Diode (Strompuls von 150 A/cm², 70 μs).

Das Temperaturprofil spiegelt im wesentlichen die Aufheizung durch Joulesche Wärme wider. Die Temperaturabsenkung zur Oberfläche ist z.Zt. noch nicht völlig verstanden. Eine lokale Wärmeableitung über den Bonddraht ist für sehr kurze Pulse vorstellbar.

Ausblick

Das vorgestellte Verfahren wird routinemäßig auf verschiedene Bauelemente angewandt. Dabei soll sowohl die Leistungsfähigkeit neuer Bauelemente untersucht werden, als auch die Modellvorstellungen und Parametersätze der Simulation verbessert werden.

Multi-Element-Analyse von Stahllegierungen mit Laser-Emissionsspektroskopie

R. NOLL, R. SATTMANN, V. STURM

Fraunhofer-Institut für Lasertechnik, Steinbachstr. 15, D-52074 Aachen , Germany

Abstract

The effects of double pulse bursts from a Q-switch Nd:YAG laser on the laser-induced plasma of steel samples are investigated. Material removal and spectral line intensity are found to be increased significantly by using double pulse bursts in comparison to single Q-switch pulses. The use of double pulse bursts for quantitative microchemical analysis is discussed and results of chemical analysis of low alloy steels are presented.

Einleitung

Die laserinduzierte Plasmaemissionsspektroskopie [1] ist ein Verfahren für die chemische Analyse von festen, flüssigen oder gasförmigen Stoffen. Es nutzt die elementspezifische Linienstrahlung eines Plasmas, das entsteht, wenn ein gebündelter Laserstrahl hoher Intensität auf das zu analysierende Proben- oder Werkstück gerichtet wird. Aufgrund der Leistungsmerkmale

- berührungslose Analyse
- Meßrate $\approx$ 10 - 500 Hz,
- Nachweisempfindlichkeit $\approx$ 10 - 1000 ppm,
- Meßgenauigkeit $\approx$ 2 - 15 %

ist das Verfahren prädestiniert für die on-line Analyse zur Prozeßkontrolle, die Qualitätssicherung oder andere industrielle Anwendungen [2, 3].

Diese Arbeit beschreibt, wie sich durch eine gezielte Steuerung der Laserleistung die Wechselwirkung zwischen der Laserstrahlung eines Nd:YAG-Lasers und Metallproben beeinflussen läßt. In Hinblick auf eine quantitative chemische Analyse können dadurch vorteilhafte Plasmabedingungen erzeugt werden. Weiterhin werden ein Aufbau zur Multi-Element-Analyse, der für industrielle Anwendungen geeignet ist, und damit erzielte Ergebnisse vorgestellt.

Experimentelle Grundlagen

Abbildung 1 zeigt den Aufbau für die laserinduzierte Plasmaemissionsspektroskopie. Der Laserstrahl wird mit Spiegeln umgelenkt und mit einer Linse auf die Probe fokussiert. Ein Lichtleiterbündel führt die emittierte Plasmastrahlung zu einem Spektrometer.

Der Laser ist ein mit 10-30 Hz gepulst gepumpter, gütegeschalteter Nd:YAG-Laser. Die Spannung der Pockelszelle wird in mehreren Stufen erhöht, so daß bei jeder Stufe ein Laserpuls emittiert wird. Damit lassen sich innerhalb eines Blitzlampenpulses Pulsgruppen von 1 bis 6 Einzelpulsen mit einstellbarem Einzelpulsabstand und Einzelpulsenergien erzeugen. Die maximale Energie in einem Sechsfachpuls mit 15 μs Einzelpulsabstand beträgt ≈ 1 J. Im folgenden vergleichen wir Ergebnisse, die mit einem Standard Q-switch-Puls von 80 mJ und einem Doppelpuls mit einer Energie von 80 mJ und 6 μs Einzelpulsabstand erzielt werden. Die Pulsdauern der Pulse im Doppelpuls betragen ≈ 40 ns (FWHM) und im Einfachpuls ≈ 20 ns.

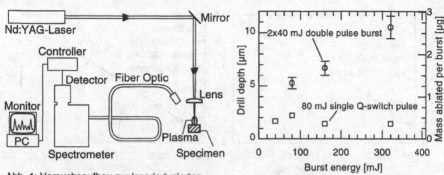

Abb. 1: Versuchsaufbau zur laserinduzierten Plasmaemissionsspektroskopie

Abb. 2: Materialabtrag mit 1x80 mJ-Einfach- und 2x40 mJ-Doppelpulsen

Ergebnisse und Diskussion

Abbildung 2 zeigt den Materialabtrag, der von Einfach- und Doppelpulsen unter atmosphärischen Bedingungen bei 1013 mbar in Abhängigkeit von der Laserenergie hervorgerufen wird. Die Bohrrate von Einfachpulsen bleibt im wesentlichen aufgrund der abschirmenden Wirkung des laserinduzierten Plasmas etwa konstant [4]. Mit Doppelpulsen wird die Sättigung des Materialabtrags im untersuchten Energiebereich vermieden, der Materialabtrag ist bei gleicher Laserenergie bis etwa einen Faktor 7 größer.

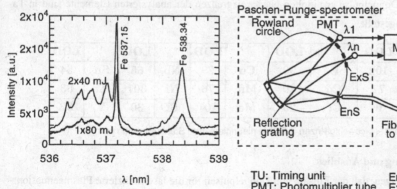

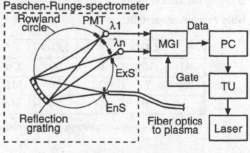

Abb. 3: Spektrum angeregt durch einen 1x80 mJ-Einfach und einen 2x40 mJ-Doppelpuls

TU: Timing unit   EnS: Entrance slit
PMT: Photomultiplier tube   ExS: Exit slit
MGI: Multichannel gateable integrators

Abb. 4: Versuchsaufbau mit Paschen-Runge-Spektrometer

Abbildung 3 zeigt, daß durch Doppelpulse gleicher Gesamtenergie auch die vom Plasma emittierte Linienstrahlung deutlich erhöht wird. Dargestellt ist die mit einem intensivierten Photodiodenarray detektierte Emission im Bereich von 536 bis 539 nm. Die Belichtungszeit beträgt für beide Spektren 2 μs, beginnend 0,8 μs nach dem Einfachpuls bzw. dem zweiten Puls vom Doppelpuls. Die Linienintensität der markierten Fe-Spektrallinien ist nach dem Doppelpuls etwa einen Faktor 2 größer als nach dem Einfachpuls.

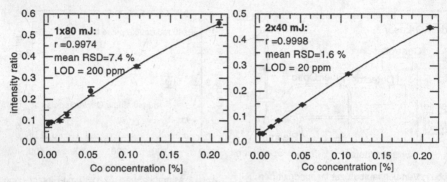

Abb. 5: Kalibrierkurven für Kobalt für 1x80 mJ-Einfach- und 2x40 mJ-Doppelpulse

Eine Multi-Element-Analyse niedriglegierter Stahlproben wurde mit einem in Abbildung 4 dargestellten Paschen-Runge-Polychromator durchgeführt. Er ist mit mehreren Photomultipliern ausgerüstet, so daß Kohlenstoff, Silizium und über 10 metallische Legierungsbestandteile simultan analysiert werden können. Kalibrierkurven für Kobalt bei Einfach- und Doppelpulsen mit 80 mJ sind in Abbildung 5 dargestellt. Die Standardabweichung der Meßpunkte verringert sich von etwa 7.4 % für Einfach- auf 1.6 % für Doppelpulse. Der Korrelationskoeffizient r, der angibt, inwieweit die Meßpunkte um die Ausgleichskurve eines Polynoms 2. Grades streuen, verbessert sich von 0.9974 auf 0.9998 für Doppelpulse. Die Nachweisgrenze verbessert sich von etwa 200 auf 20 ppm. Die für Doppelpulse ermittelten Nachweisgrenzen der analysierten Elemente sind in Tabelle 1 zusammengestellt.

	LOD		LOD		LOD		LOD		LOD
Al	102	C	90	Cu	29	Nb	65	Sn	54
As	77	Co	20	Mn	76	Ni	301	Ti	83
B	6	Cr	25	Mo	101	Si	30	V	32

Tab. 1: Nachweisgrenzen (LOD) in niedriglegierten Stahlproben (in ppm)

Zusammenfassung und Ausblick

Die Ergebnisse zeigen, daß der Einsatz von Doppelpulsen für die laserinduzierte Plasmaemissionsspektroskopie die analytische Leistungsfähigkeit verbessern kann. In Verbindung mit einem Paschen-Runge-Polychromator für die Simultananalyse und einer an die Laseranalyse angepaßten Auswerteelektronik ergeben sich vielfältige Einsatzmöglichkeiten für die laserinduzierte Plasmaemissionsspektroskopie in der industriellen Prozeß- oder Qualitätskontrolle. Beispiele sind die on-line Analyse flüssigen Stahls in der Pfanne (Abbildung 6) oder die Materialprüfung mit anschlie-ßender Beschriftung von Metallhalbzeugen in der Qualitätskontrolle (Abbildung 7).

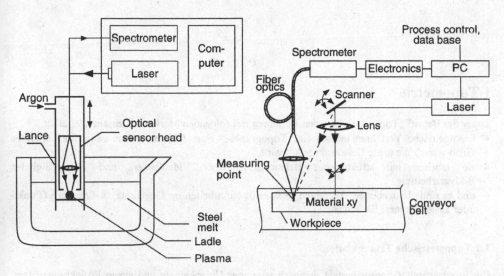

Abb. 6: Analyse von Stahlschmelzen

Abb. 7: Materialprüfung und Beschriftung von metallischen Halbzeugen

Danksagung

Diese Arbeiten wurden teilweise gefördert von der Europäischen Kommission für Kohle und Stahl (Nr. 7210-GD/109 und PHIN 940015) und von der Thyssen Stahl AG.

Literatur

[1] Runge E. F., Minck R. W. und Bryan F. R. *Spectrochim Acta* **20**, 733 (1964)

[2] Noll R., Sattmann R. und Sturm V. in *Optical Measurements and Sensors for the Process Industries*, C. Gorecki, R.W.T. Preater, Editors, Proc. SPIE 2248, 50 (1994)

[3] Sattmann R., Sturm V. und Noll R., submitted to *Journal of Physics D: Applied Physics*

[4] Piepmeier E. H. und Osten D. E. *Appl. Spectrosc.* **25**, 642 (1971)

Die topometrische 3D-Meßtechnik mit miniaturisierter Projektionstechnik erschließt neue Anwendungsgebiete

E. Klaas, B. Breuckmann
Breuckmann GmbH, D - 88709 Meersburg

1 Topometrie

Unter der Begriff „Topometrie" werden Verfahren mit folgenden Merkmalen zusammengefaßt:
- topometrische Verfahren messen die „Topographie" eines Objekts, indem neben den lateralen Größen auch die dritte Dimension erfaßt wird,
- sie arbeiten mit strukturierter Beleuchtung, einer Videokamera und einer digitalen Bildverarbeitung
- und es sind „bildgebende" Verfahren, daß heißt sie arbeiten im Gegensatz zu Scannern (Punkt- oder Linienscanner) flächenhaft.

1.1 Topometrische Triangulation

Ein topometrischer Sensor besteht demnach aus einer Videokamera und einem Projektionssystem, deren optische Achsen sich unter einem Winkel schneiden. Ist dieser Triangulationswinkel und einige anderen geometrischen Parameter des Aufbaus bekannt, müssen bei der Meßaufnahme nur die Pixelkoordinaten (p_x, p_y) eines Punktes und der Winkel (α) der auf diesen Punkt projizierten Linie bestimmt werden, um die Punktkoordinaten zu berechnen.

1.2 Kombination von Graycode und Phasenshift

Neben der Bestimmung der Kamerapixelkoordinaten (p_x, p_y), gestaltet sich jedoch die Bestimmung des Projektionswinkels (α) als aufwendiger:
In der Praxis hat sich heute die Kombination des sogenannten „Graycode-" und des „Phasenshiftverfahrens" durchgesetzt, um einen guten Kompromiß zwischen Auflösung und Zuverlässigkeit zu erreichen. Dabei wird die „Grobauflösung" mit dem digitalen Graycode und anschließend die „Feinauflösung" mit dem analogen Phasenshiftverfahren realisiert.

2 Bisher: LCD-Projektoren

In praktisch eingesetzten Projektoren für Graycode-Phasenshift-Sequenzen wurden bisher hauptsächlich Liquid-Christal-Displays eingesetzt, deren einzelne Linien rechnergesteuert transparent bzw. undurchlässig geschaltet werden. Aus dieser technischen Realisierung ergeben sich für einen LCD-Sensor folgende charakteristische Merkmale:

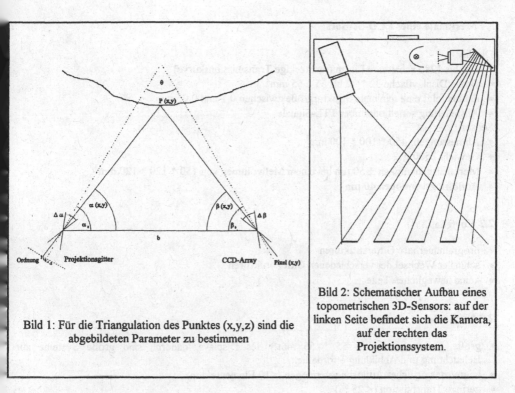

Bild 1: Für die Triangulation des Punktes (x,y,z) sind die abgebildeten Parameter zu bestimmen

Bild 2: Schematischer Aufbau eines topometrischen 3D-Sensors: auf der linken Seite befindet sich die Kamera, auf der rechten das Projektionssystem.

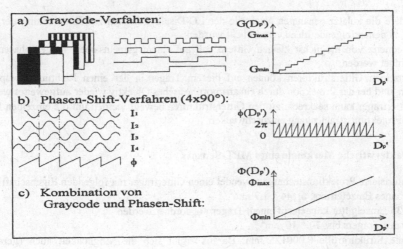

Bild 3: Verfahrensüberblick

2.1 Merkmale eines LCD-Sensors

Display:
- 320 bis 1280 schaltbare Linien (rechteckige Transmissionskurve)
- aktive Displayfläche 32 * 32 bis 55 * 55 mm^2
- daraus folgt eine minimale Strukturgröße zwischen 0.1 und 0.05 mm
- Ansteuerung seriell oder über TTL-Signale

Sensor:
- Abmessungen: 450 * 100 * 100 mm^3
- Gewicht 2500 g
- Merkmalsgenauigkeit ± 50 μm bei einem Meßvolumen von 150 * 120 * 120 mm^3
- Auflösungsvermögen 40 μm

2.2 Vorteile

- programmierbare Gitterstrukturen
- schneller Wechsel der verschiedenen Gitterstrukturen
- keine beweglichen Teile

2.3 Nachteile

- große Abmessungen (bis 55 * 55 mm^2) des Displays, dadurch sind große Systeme zur Beleuchtung und Abbildung erforderlich
- begrenztes örtliches Auflösungsvermögen (<10 Elemente/mm)
- geringe Transmission (< 25 %)

3 Neu: MPT, die miniaturisierte Projektionstechnik

Insbesondere die zuletzt genannten Nachteile der LC-Displays führten zur Entwicklung der MPT-Sensoren, in denen folgende Ideen verwirklicht wurden:
- eine Sequenz von möglichst kleinen Gittern soll auf einem gemeinsamen Träger phasenrichtig angeordnet werden,
- die einzelnen Gitterstrukturen können auf diesem Träger in der einen Richtung komprimiert werden, und bei der Projektion durch ein anamorphotisches Objektiv wieder aufgeweitet werden,
- der Gitterträger kann senkrecht zu den Gitterstrukturen bewegt werden um die einzelnen Muster in den Projektionsstrahlengang einzuschwenken.

3.1 Charakteristische Merkmale eines MPT-Sensors

Die miniaturisierte Projektionstechnik verwendet einen Gitterträger mit folgenden Eigenschaften:
- Größe eines Einzelgitters: 6,144 * 1,7 mm^2
- bis zu 35 Einzelgitter können auf einem Träger angeordnet werden
- Größe des Trägers bis: 70 * 10 mm^2
- minimale Strukurgröße 0,000125 mm. Daraus ergibt sich die Möglichkeit auch (gerasterte) sinusförmige Strukturen aufzubringen.

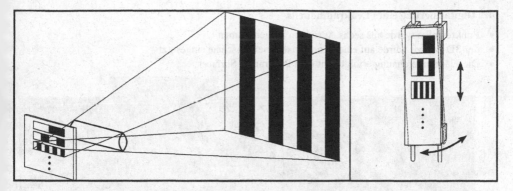

Bild 4: Schema des optischen Aufbaufs mit dem beweglichen und justierbaren Gitterträger sowie der anamorphotischen Abbildung

Mit diesem Gittertyp gebaute Sensoren, können derzeit folgende Daten erreicht werden:

- Abmessungen: 380 * 86 * 100 mm^3
- Gewicht 1450 g
- Merkmalsgenauigkeit ± 50 μm bei einem Meßvolumen von 230 * 160 * 150 mm^3
- Auflösungsvermögen 40 μm

3.2 Vorteile

- kleinere Sensorabmessungen
- kleineres Gewicht, dadurch wird der Einsatz auch auf kleinen Koordinatenmeßmaschinen und einfachen Dreh- Schwenkköpfen ermöglicht.
- größere Schärfentiefe durch kleinere Brennweite beim Projektionsobjektiv
- höhere Auflösung durch die Projektion sinusförmiger Streifen

3.3 Nachteile

- bewegte Teile im Sensor

4 Wofür?

Folgende Beispiele zeigen sowohl die unterschiedlichen Einsatzgebiete topometrischer 3D-Sensoren, als auch die Möglichkeit das Meßsystem bezüglich Meßvolumen und Genauigkeitsanforderungen diesem Einsatzgebiet anzupassen.

4.1 Digitalisierung eines Lenkradmodells

- Punktewolke wurde aus sechs Ansichten aufgenommen
- der 3D-Sensor wurde auf einer Koordinatenmeßmaschine eingesetzt
- Die Flächenberchnung erfolgte mit dem Programm Surfacer

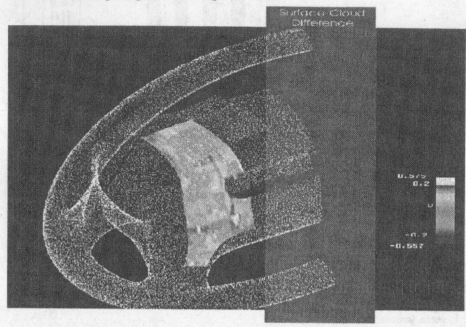

4.2 Vermessung eines Stoßdämpfertellers

- Flächenrückführung mit topometrisch bestimmten 3D-Koordinaten
- auf einme Rotationstisch wurde das das Teil um jeweils 60° gedreht und aufgenommen

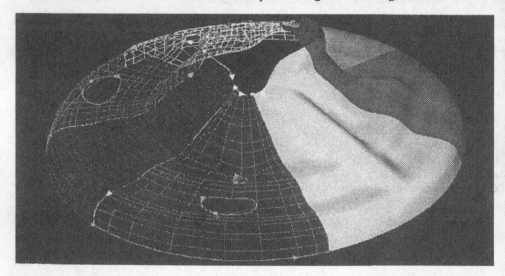

The Measurement of Local Transient Deformation with a fast-shuttered Shearing ESPI

Cheol-Jung Kim, Sung-Hoon Baik, Jae-Wan Cho
Korea Atomic Energy Research Institute, POB 105, Yu-Sung, Taejon, Korea

Summary

A local transient deformation of metal plates induced by a pulsed Nd:YAG laser was measured with a shearing ESPI. To measure the transient deformation, the frame grabber of ESPI was synchronized to the laser pulse with a trigger generator and a fast-shuttered CCD camera was used. Also, to increase the shutter speed, the imaging optics was adjusted to see only 8 x 6 mm target area and the illuminating laser beam was condensed over this small target area. The dependence of the fringe visibility on the shutter-speed was investigated. Furthermore, the characteristics of local deformations of Al and stainless steel plates were measured with a fast-shuttered shearing ESPI and compared with the measurements from a vibrometer. Both results showed the same time-dependence of the deformation. Finally, this shearing ESPI was applied to the inspection of defects in metal plates and could detect the defect of 1.5 mm depth in Al plate of 4 mm thickness.

1. Experimental Setup

The experimental setup is shown in Fig.1. A traditional Michelson type shearing ESPI was used, but a imaging lens of 50 mm focal length was adjusted to 1 : 1 magnification to reduce the target area up to 8 x 6 mm. By reducing the target size, the shutter speed could be increased up to 1/4,000 sec with 5 mW He-Ne laser. However, the Gaussian distribution of the laser beam intensity generated quite non-uniform image for a shearing ESPI. So, a 100 mW 2nd harmonic Nd:YAG laser was used and a spatial filter was introduced for uniform illumination. Furthermore, only small central portion of the laser beam was used in these experiments.

DT 2867 frame grabber was used for data acquisition. Due to the limit in memory size, it could store only three frames. So, one frame was designed to store the image just before the laser pulse and the other two frames were supposed to store the images after the laser pulse with different delays. Then, subtraction was used to obtain the fringes. On the other hand, the laser pulse from RSY-1000P Nd:YAG laser has a jitter of about 50 msec in delay after the external trigger pulse due to the slow response of mechanical shutter. To synchronize these three frames to the laser pulse as designed, a special trigger circuit was implemented as shown in Fig.2. The CCD camera was triggered externally with a 60 Hz vertical sync from the trigger circuit. After the manual start signal input, the trigger circuit produces a trigger pulse for laser after pre-entered adjustable delay. Also, the trigger circuit produce three external trigger signals for

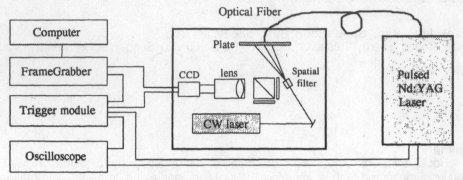

Fig.1: Experimental Setup for Measurement of Local Transient Deformation

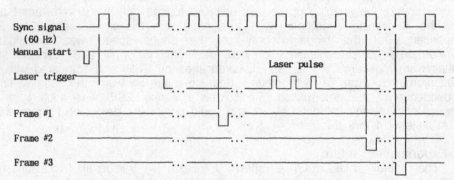

Fig.2: Timing Diagrams of Trigger Pulses

grabbing three frames after pre-entered adjustable delays. The delay of laser trigger pulse can be adjusted in msec and the delays of three frames can be adjusted in number of frames.

2. Dependence of Visibility on Shutter Speed

The deformation was found to be induced in less than 20 msec after the laser pulse of 5 msec width (as shown in Fig.4). The visibility was compared during this fast responding region. The fringes obtained from 4 mm thick Al plate in less than 10 msec after laser pulse (5msec, 15J) at shutter speed of 1/60, 1/250 and 1/1000 sec respectively are shown in Fig.3. As shown in the figure, the fringe was clear enough even at shutter speed of 1/250 sec. The shutter speed was maintained at 1/250 sec during the remaining experiments.

3. Temporal Variation of Deformations on Al and Stainless Steel Plates

To see the difference in temporal variation of deformation on Al and stainless steel plates, the deformations induced on a 4 mm thick Al plate and a 0.5 mm thick stainless steel plate by laser pulses were measured with a VS-1000

(a) 1/60 sec   (b) 1/250 sec   (c) 1/1000 sec

Fig.3: Dependence of Visibility of Fringes on Shutter speed

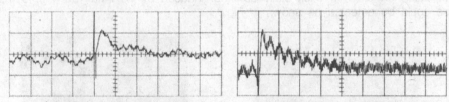

(a) Al (Div/ 50 msec)   (b) SUS304 (Div/200 msec)

Fig.4: Temporal Variation of Deformation on Al and SUS304 plate measured with Vibrometer

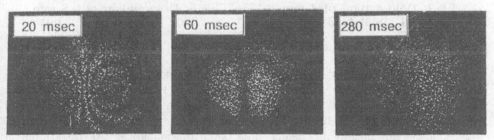

Fig.5: Temporal Variation of Deformation on Al plate measured with ESPI

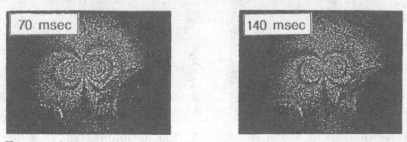

Fig.6:Temporal Variation of Deformation on SUS plate measured with ESPI

vibrometer with displacement measuring option. The temporal variation of deformation is shown in Fig.4. As shown in Fig.4, the deformation induced on Al plate is recovered after about 100 msec, but that of stainless steel remains at half-maximum even after 200 msec. The difference in decay time can be explained with the difference in thermal conductivity of 229 $Jm^{-1}sec^{-1}K^{-1}$ in Al and 25 $Jm^{-1}sec^{-1}K^{-1}$ in stainless steel.

Fig.5 shows the temporal variation of deformation on 4 mm thick Al plate induced by a 15 J, 5 msec pulse. It shows the deformation was recovered much even after 60 msec. Fig.6 shows the same variation on 0.5 mm stainless steel induced by a 2 J, 5 msec pulse. It shows the deformation was still maintained at high level even after 150 msec. These results show the same tendency as those from the vibrometer. The number of fringes can be adjusted by changing the shearing distance and Fig.7 shows the fringes from the deformation induced by the same laser pulse as in Fig.5 but at twice larger shear. Furthermore, the deformation on 0.5 mm thick stainless steel induced by as small as 1 J, 5 msec pulse could be detected as shown in Fig.8.

Fig.7: Deformation of 4 mm Al at shear of 3 mm

Fig.8: Deformation of 0.5 mm thick SUS304 by 1 J pulse

4. Inspection of Defects

The artificial defects of 1mm width but with 1.5 and 3.5 mm depth each were engraved on the 4 mm thickness Al plate. The 15 J, 5 msec pulse was illuminated just next to the engraved defects from the backside. Fig.9 shows the temporal variation of the deformation with defect of 3.5 mm depth. The abrupt change in fringe pattern at defect area was noticeable at 20 msec, but it disappeared at 120 msec. Fig.10 shows the same variation with shear direction

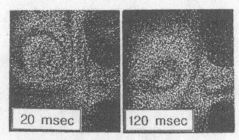

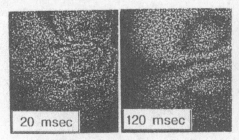

Fig.9:Deformation of 4 mm Al with 3.5 mm defect(X-shear)

Fig.10:Deformation of 4 mm Al with 3.5 mm Defect (Y-shear)

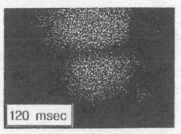

Fig.11: Deformation of 4 mm thick Al with 1.5 mm defect (Y-shear)

rotated by 90 degree. The non-symmetrical fringe pattern shows the defect area at 20 msec, but it becomes more symmetrical at 120 msec. Fig.11 shows the temporal variation of deformation from 1.5 mm defect and shows the same tendency.

5. Conclusion

A fast-shuttered shearing ESPI was implemented and was applied to the inspection of defects in metal plates. The defect as small as 1.5 mm depth in 4 mm thick Al plate could be detected. However, to detect defects, the deformation has to be measured just after the laser pulse and before the deformation becomes uniform. Therefore, if a high-speed camera and a fast frame grabber were used, the sensitivity of detection of defects could be improved. Furthermore, if front illumination of pulsed laser is implemented, the usefulness can be increased for industrial application.

Speckle Interferometer with very Simple Optical Setup

B. Lau, W. Göggel, J. Hoffmann
Fachhochschule Ulm
Institut für Angewandte Forschung (IAF) Automatisierungssysteme
Prittwitzstr. 10
D-89075 Ulm/Donau

Abstract

We present a speckle pattern interferometer for out-of-plane ESPI deformation measurements with a very simple optical setup of Fizeau type. A speckled reference wave is generated by means of a spherically curved transmitting / reflective scattering element placed directly in front of the object. It is slightly roughened and diffusely scatters the divergent beam of a semiconductor laser into the objective lens of the video camera which simultaneously receives the object wave. The transmitting / reflective scattering element is the only optical component needed, thus the device is inexpensive and very easy to align and to operate. The reference wave intensity easily can be matched to that of the object wave. As the common path configuration ensures minimum sensitivity to environmental vibration and air turbulences, this device is very well suited for industrial nondestructive testing.

Introduction

Electronic speckle pattern interferometry (ESPI)[1-4] (or TV holography) is a very sensitive and powerful tool for nondestructive testing of materials and products. Like holographic interferometry it allows to measure deformations in the micrometer and submicrometer range which are caused by very small forces applied to the specimen under test. As the interferograms are recorded with a video camera and evaluated with a computer, no material costs for the measurements arise and results are obtained very quickly. Thus this method principially is very suited for large-scale industrial quality inspections. However, up to now electronic speckle pattern interferometers are not wide-spread in industry due to the complexity of the optical setup and therefore their high price and difficulties in alignment and operation which require skilled personnel.

Another requirement for an interferometer to be used in an industrial environment is that it must be insensitive to mechanical vibrations and air turbulences. This can be achieved by sophisticated insulation, special numerical evaluation techniques[5] or by making the object and reference waves to travel nearly the same path. In such a "quasi-equal-path" or "common path" arrangement[6,7] mechanical vibrations and air turbulences have approximately the same influences on both object and reference wave and hence their effects are considerably decreased.

Speckle pattern interferometers can be subdivided into two types. One of them uses a "smooth" reference wave[1] which is a part of the wave emitted by the laser. This wave is guided to the camera sensor (CCD array) either through the objective aperture or behind the objective lens onto a beam combiner. It can produce very high speckle contrast, but alignment is critical and the camera objective cannot be easily exchanged. The other type uses a "speckled" reference wave[8,9] with randomly varying phase generated by means of a diffusor screen. The video camera "sees" both the specimen under test and the diffusor whose images are superimposed in the optical setup. Alignment of the optical setup is much easier in this type which is preferable for industrial applications. In both types the reference wave intensity must be fitted to the object wave intensity in order to get a speckle contrast sufficiently high for quantitative evaluation.

Principle and optical setup

We describe a very compact and simple speckle interferometer with a speckled reference wave for the measurement of out-of-plane deformations. A 10 mW semiconductor laser (Toshiba TOLD 9125, 670 nm or Hitachi HL 6741 G, 675 nm) is placed beneath the video camera. Its elliptical radiance profile with half divergence angles of about 28° and 8° allow it to illuminate objects having a size up to about 10 cm without using any beam expanding optics.

The speckled reference wave is generated by means of a spherically curved transparent element placed directly in front of the object (Fig. 1). Similar devices - both of them producing a smooth reference wave by directed reflection - were described by Virdee et al[10] who use a specially constructed camera lens and housing, and by Joen-

athan and Khorana[11]. The beam paths of both of them are rather complicated and require careful alignment. However, in our device we do not use the directly reflected light. Instead, our transparent element is slightly roughened and thus diffusely reflects the laser radiation into the objective lens of the video camera which simultaneously receives the object wave. The direct reflection which would produce a very bright spot in the image is trapped by the objective aperture by slightly tilting the transmitting / reflectice scattering element which also allows to match the intensity of the diffusely scattered part. To achieve this, the laser diode (regarded as a point source) must be approximately imaged into the objective aperture plane of the video camera. This determines the radius of curvature of the transparent element.

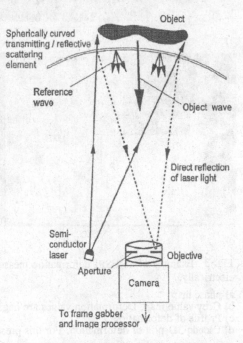

Fig. 1 Principal scheme of ESPI setup.

Fig. 2 (left)
Photograph of ESPI setup.

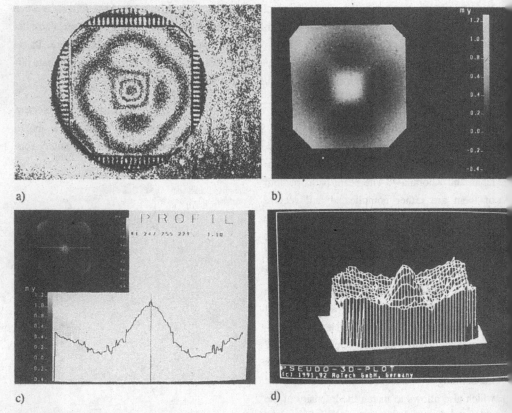

a)

b)

c)

d)

Fig. 3 Example of an ESPI deformation measurement. Object is an integrated circuit heated electrically.

a) phase map
b) Grey value map. Deformation values are negative in direction to the camera
c) Profile of deformation
d) Pseudo-3D-plot of deformation. For this presentation a smoothing filter was applied.

The optical setup is mounted on a 30x30 cm honeycomb plate supported by four rubber pads. It can be operated on a usual laboratory table without special vibration insulation. The transparent element is mounted on a translational stage which can be shifted piezoelectrically. The speckle interferograms are evaluated quantitatively using the phase stepping method[12] with a commercial holographic image processing system (Fringe Expert, Rotech Gesellschaft für Prüftechnik, Amerang) implemented on a PC 486. A photograph of the optical setup is given in Fig. 2.

Experimental results

As an example, the deformation of an integrated circuit (22 x 22 mm) electrically heated with a power dissipation of 2 W is measured. Its surface oriented to the camera is supported at its four corners which form the reference system for deformation measurement. The silicon chip inside the IC bends on its support. Thus its edges press against the housing, the out-of-plane deformation of which is displayed. Fig. 3a shows the phase map, Fig. 3b the grey value map obtained after phase unwrapping, Fig. 3c a profile section and Fig. 3d a pseudo-3D-plot.

Conclusion

A new electron speckle pattern interferometer is presented which is extremely friendly for industrial automated inspection. The specimen under test is illuminated with a laser point source. A speckled reference wave is generated by means of a spherically curved transparent element which is slightly roughened. As this transmitting / reflective scattering element itself acts as beam splitter and as it is the only optical component, our speckle interferometer hardly can be surpassed in economy and simplicity of its optical setup and in ease of alignment and operation. Optically skilled personnel is not required for operation. Any camera objective can be used and easily exchanged. Moreover, being a Fizeau type interferometer and a "common path" arrangement it is rather insensitive to mechanical vibrations and air turbulences. It can be operated under adverse environmental conditions often present in industry. Owing to the low cost and extreme simplicity of the system new application fields in industrial nondestructive testing are arising for ESPI.

Acknowledgements

Thanks are due to Dipl.-Ing. (FH) E. Mattes for many helpful discussions and Dipl.-Ing. (FH) E. Vetter for help in preparing the figures. The IC was provided by Siemens AG, München. This work was supported by the "Schwerpunktprogramm zur Förderung der Forschung an Fachhochschulen" (programme for support of reseach at polytechnics) of the state Baden-Württemberg.

References

1. J. N. BUTTERS, J. A. LEENDERTZ, "Holographic and video techniques applied to engineering measurement," Transactions of the Institute of Measurement and Control 4, 349-354 (1971).
2. J. C. DAINTY, Laser Speckle and Related Phenomena, (Topics in Applied Physics, 9). Springer-Verlag, Berlin (1975/1984).
3. R. K. ERF (ed.), Speckle Metrology, Academic Press, New York (1978).
4. R. JONES, C. WYKES, Holographic and Speckle Interferometry, (Cambridge Studies in Modern Optics, 6), Cambridge University Press, London (1983/1989).
5. B. F. POUET, S. KRISHNASWAMY, "Additive subtractive decorrelated electronic speckle pattern interferometry," Optical Engineering 32, 1360-1369 (1993).

644

6. S. PENG, C. JOENATHAN, B. M. KHORANA, "Quasi-equal-path electronic speckle pattern interferometric system," Optics Letters 17, 1040-1042 (1992).
7. C. JOENATHAN, B. M. KHORANA, "Quasi-equal-path electronic speckle pattern interferometric
   system," Applied Optics 32, 5724-5726 (1993).
8. J. A. LEENDERTZ, "Interferometric displacement measurement on scattering surfaces utilising
   speckle effects", Journal of Scientific Instruments 3, 214-218 (1970).
9. G. Å. SLETTEMOEN, "Electronic speckle patt   rn interferometric system based on a speckled
   reference beam," Applied Optics 19, 616-623 (1980).
10. M. S. VIRDEE, D. C. WILLIAMS, J. E. BANYARD, N. S. NASSAR, "A simplified system for
    digital speckle interferometry," Optics and Laser Technology 22, 311-314 (1990).
11. C. JOENATHAN, B. M. KHORANA, "A simple and modified ESPI System," Optik 88, 169 (1991).
12. K. CREATH, "Phase-shifting speckle interferometry," Applied Optics 24, 3053-3058 (1985).

PC-basierte Rekonstruktion von Phasenverteilungen mit automatischer konsistenter Markierung von Interferenzstreifen-Skeletten

F. Elandaloussi, U. Mieth, W. Osten
Bremer Institut für Angewandte Strahltechnik
Klagenfurter Str. 2
28359 Bremen

Abstract

Optical-interferometrical methods are gaining more and more importance in quality control because of their high sensitivity and noncontact working principle. The basic principle of these methods consists in the modification of the object illumination by coherent superposition of light waves representing different states of the object under test. As the result fringe patterns can be observed. These patterns contain the information about the displacement or the shape of the object. For their evaluation several methods are used. The procedure described here is based on the so-called fringe tracking or skeleton method which can be used without restriction for the evaluation of any kind of fringe patterns. As manually evaluation of fringes takes several hours, it is convenient to use computers for the skeletonization and phase reconstruction. The new procedure described here is implemented on an usual PC and makes it possible to evaluate any kind of interferograms in a few minutes with minor a-priori knowledge and only few user interactions. The basis of that semi-automatic phase evaluation is a new algorithm for fringe numbering. Using a graph formulation of the problem, it is possible to automate the consistent labelling of the fringes in the case of a continuous phase field. The remaining ambiguous points can then be determined in an arbitrary or interactive way.

Einleitung

Zur Rekonstruktion der Phasenwerte eines einzelnen Interferogramms kann beispielsweise die Skelett-Methode [1] herangezogen werden. Wesentliches Merkmal dieser Methode ist, die Reduktion der Information eines Streifenbildes auf ein Skelettbild. In dem binären Skelett sind nur noch Orte gleicher Phasendifferenz δ enthalten, die durch Extremwerte (helle oder dunkle Streifen) im Interferogramm repräsentiert wurden. Durch einen abschließenden Interpolationsprozeß zwischen den Skelettlinien erhält man die gwünschte Phasenverteilung. Dieser Prozeß stellt eine genaue Kopie der manuelle Streifenauswertung dar. Ein Problem bei der Automatisierung dieser Methode ist die konsistente Markierung eines Skelettbildes, da durch die Geradheit der Cos-Funktion Aussagen über die Monotonie des Phasenfeldes begrenzt sind. Abbildung 1 verdeutlicht die Problematik. Während der Schluß von Abbildung a nach Abbildung b eindeutig ist, gibt es für den umgekehrten Schluß theoretisch unendlich viele Lösungen. Beschränkt man sich jedoch auf jene, in denen die Objektdeformation nicht nur stetig, sondern auch so einfach wie möglich ist, so erhält man die 4 in Abbildung c dargestellten Fälle. In diesem Beitrag wird eine Möglichkeit vorgestellt, die das Problem der Mehrdeutigkeit durch minimale Interaktion eines Benutzers konsistent löst.

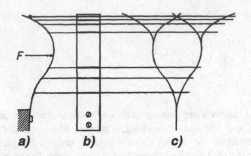

Abbildung 1: Problem der Mehrdeutigkeit der Monotonie

Skelettierung

Die Intensität eines Interferogramms $I(i,j,t)$ in dem Pixel (i,j) für den Zeitpunkt t kann wie folgt beschrieben werden: $I(i,j,t) = a(i,j,t) + b(i,j,t) \cdot \cos \delta(i,j)$. Der Term $a(i,j,t)$ beschreibt die additiven Komponenten wie zum Beispiel elektronisches Rauschen oder Hintergrundintensität, $b(i,j,t)$ repräsentiert die multiplikativen Komponenten wie Speckle-Rauschen. Die Phasendifferenz $\delta(i,j)$ beinhaltet Information über die Verschiebung der Oberfläche zwischen den beiden Objektzuständen.

Das Skelett wird nach der in [2] beschrieben Methode erzeugt. In dem ersten Verarbeitungsschritt wird das Signal-Rausch-Verhältnis durch Filtern und einer Shading-Korrektur verbessert. Anschließend wird das verbesserte Interferogramm durch eine Segmentierung in Regionen heller (Berg) und dunkler (Tal) Streifen zerlegt. Zusätzlich wird Information über die Richtungen der Streifen in einem Richtungsbild gesammelt. Dieses Bild wird für das anisotrope Filtern des segmentierten Interferogramms herangezogen. Kleinere Lücken im anisotrope gefilterten Bild werden durch Gebietswachstum geschlossen. Im letzten Verarbeitungsschritt wird das nun verbesserte segmentierte Bild durch Verdünnung in ein Skelettbild gewandelt.

Konsistente Markierung eines Skelettes mit Hilfe eines Graphen

Ein Graph besteht aus Knoten und Kanten, die die Beziehung zwischen den Knoten repräsentieren. Ist die Beziehung zwischen den Knoten nur in einer Richtung gültig, so spricht man von einem orientierten Graph. Der Weg in einem Graph über verschiedene Kanten wird Pfad genannt. Diese Zusammenhänge können am Beispiel eines Straßenkarte veranschaulicht werden. Die Städte auf der

Karte stellen die Knoten dar, während die Straßen die Kanten repräsentieren. Ein Pfad ist die Verbindung von Stadt A über Stadt B nach Stadt C.

Soll ein Skelett in einen Graph transformiert werden, so stellt jede Skelettlinie einen Knoten dar. Zwei Knoten werden nur dann durch eine Kante verbunden, wenn diese benachbart sind. Der aus dieser Transformation resultierende Graph ist nicht orientiert.

Zur Generierung eines Graphen können verschiedene Methoden angewendet werden. So zum Beispiel ein Füllalgorithmus für geschlossene Gebiete, der jedoch die wichtige Voraussetzung stellt, daß die Skelette entweder geschlossen sind oder den Rand des Bildes berühren. Durch die Skelettierung ist diese Voraussetzung nicht immer erfüllt, da durch Filtern der Rand des Interferogramms in den meisten Fällen für Berechnungen ausgespart wird, und somit keine Skelettlinien enthält, bzw. diese dort abrupt beginnen oder enden. Für den Fall, daß das Skelett unerlaubt verzweigte Linien enthält, werden diese an den Kreuzungspunkten aufgebrochen, und als neue Linie in den Graphen transformiert. Abbildung 2 und Abbildung 3 veranschaulichen anhand eines Beispiels diese Transformation.

Abbildung 2: Skelett

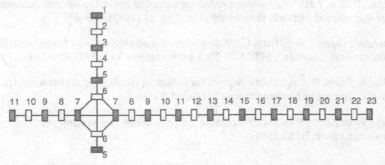

Abbildung 3: Aus der Transformation resultierender Graph

Für die konsistente Markierung des Skelettes sind unter den oben gemachten Annahmen wenige Regeln erforderlich [3]:

1. Zwei benachbarte Knoten des gleichen Grauwertes erhalten die selbe Ordnung.

2. Zwei benachbarte Knoten mit verschiedenen Grauwerten erhalten eine Ordnungsdifferenz von $\pm\pi$.

3. Ein nicht-verzweigter Pfad mit Knoten abwechselnder Grauwerte erhält eine monotone Numerierung.

4. Zwei benachbarte Knoten des gleichen Grauwertes in einem sich nicht-verzweigenden Pfades bewirken einen Wechsel der Monotonie.

Neben diesen wenigen Regeln müssen lediglich die absolute Phase und Monotonie im Startpunkt bekannt sein, und der in [3] hergeleitete Algorithmus ist in der Lage eine konsistente Markierung vorzunehmen. Sind verzweigte Linien in dem Skelett enthalten, so muß der Benutzer aus seinem a-priori Wissen die Monotonie definieren. In diesem Fall stoppt der Automatismus und erwartet vom Benutzer eine Interaktion.

Zusammenfassung und Ausblick

In diesem Beitrag wurde eine Methode zur konsistenten Markierung eines Skelettes beschrieben. Für einfache Skelette stellt das System ein robustes Verfahren zur automatisierten Auswertung von Interferogrammen dar. Für verzweigte Skelette wird das Problem, eine Aussage über den Verlauf der Monotonie zu machen, durch minimale Interaktion des Benutzers gelöst, da nur dieser aufgrund seines a-priori Wissens in der Lage ist, das korrekte Ergebnis zu definieren. Durch Transformation in einen Graphen und Anwendung weniger Regeln ist das beschriebene Verfahren auf jedem handelsüblichen PC einsetzbar und ist im Fringe Processor™ [4] implementiert.

Literatur

[1]   Becker, F.; Yu, Y.H.:"Digital fringe reduction technique appield to the measurement of three-dimensional transonic flow fields", Opt. Eng. 24 (1985), 429-434

[2]   Osten, W.; Jüptner, W.; Mieth, U.:"Knowledge assisted evaluation of fringe patterns for automatic fault detection", SPIE Vol. 2004 Interferometry VI (1993), 256-268

[3]   Colin, A; Osten, W.:"Automatic support for consistent labelling of skeletonized fringe patterns", Journal of Modern Optics 42 (1995) 5, 945-954

[4]   Fringe Processor™: Windows™-Software zur pc-basierten Auswertung von Interferogrammen, BIAS 1995

Out-of-Plane and In-Plane Strain Measured by Shearography

W. Steinchen, M. Schuth, L. X. Yang, G. Kupfer
University of Kassel, Laboratory of Photoelasticity, Holography and Shearography,
Dept. of Mechanical Engineering, FRG

ABSTRACT

The shearographic interferometry is employed as a nondestructive full field, optical testing and measuring method without contact. Fringes of constant strain (so called isotase, tasis (Greek)=strain) can be observed in real-time on the surface of the investigated machine parts and structures of any material and are represented by the shearogram. Using shearography two states of deformation are recorded by doubly exposing a Holotest film in an ordinary camera or stored by an electronic image processing system. In the objective of the camera a shearing element is integrated or the lateral Michelson shearing interferometer is used. Rigid body motions of the object are not recorded. Local deformation irregularities caused by a defect under or on the surface of the specimen create strain concentrations; the homogeneous surrounding is poorly superimposed by an interference pattern. The shearogram shows dark and bright fringes which are the functions of the displacement derivative. The holographic interferometry measures the out of plane deformations directly. Terms of the out-of-plane strain can be determined by the shearographic method as well as the in-plane strain fringes which are described in the following.

1. INTRODUCTION

Shearography is a coherent optical method which allows the whole-field derivatives of surface displacement to be measured directly. Therefore, it is a tool well suited for either nondestructive testing or for strain analysis. The basic principle of shearography is shown in Fig. 1. The tested object is illuminated by an expanded laser beam. The light reflected from the surface of the object is focused on the image plane of an image shearing camera. In this camera a shearing element is

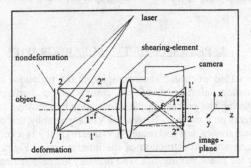

Fig. 1 Basic principle of shearography[1,4,5]

implemented in front of its lens so that a pair of laterally sheared images of the object is generated. The two laterally sheared images interfere with each other producing a random interference pattern commonly known as a speckle pattern. When measuring, usually a comparison is made of the speckle patterns created by two states of deformation, first when the object is undeformed and second when the object is loaded. In the conventional technique of shearography, i.e. photographic shearography, the photographic emulsion is employed as recording media. The images are normally recorded on a high-resolving Holotest-film.

TV-shearography is a new technical development of the conventional shearography. Two speckle patterns created by two states of deformation are registered with a CCD-camera and processed with a computer. This leads to rapidly increased testing speed and enables automatic evaluation of the test results. Furthermore, this technique can perform not only double exposure technique, but also real-time technique, which are suitable for the static and dynamic investigations and therefore it is rapidly gaining acceptance by industry. In this paper the basic principle of TV-shearography and its application for measuring the out-of-plane and in-plane strains exactly in conjunction with the phase shifting technique are presented.

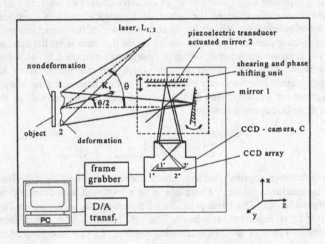

Fig. 2 Principle setup of the TV-shearography including the phase shifting unit

2. PRINCIPLE OF TV - SHEAROGRAPHY

TV-shearography, also called electronic shearography, is the technique using laser speckle as the carrier of the deformation information of the testing object surface, recording the speckle interference field, comparing and processing the information by elecronic methods and displaying the interference fringe patterns on a monitor. Between TV-shearography and conventional shearography is no difference in optical theory, but technically TV-shearography is a computerized process which eliminates wet processing and reconstruction of the interferogram. Fig. 2 shows the setup of TV-shearography including the phase shifting unit. The ordinary image-shearing camera is replaced by an image-shearing CCD-camera in which a Michelson interferometer is implemented in front of its lens so that a pair of laterally sheared images of the object are generated. The advantage of the lateral Michelson shearing interferometer is that the shearing direction and the shearing amount can be simply altered by rotating the mirror 1 from normal position to the corresponding position. As mentioned above, the two shearing images interfere with each other producing a speckle pattern where the CCD-array is positioned. This speckle pattern is registered by the CCD-camera and processed with a computer. When the object is deformed, the intensity distribution of the speckle pattern is slightly altered and recorded by the CCD-camera again. The subtraction operation between two digitized informations is displayed on the monitor and yields the so called "TV-shearogram". The fringes of the TV-shearogram describe the relative phase change Δ which is induced by the relative displacement between the two points whose distance is the shearing amount due to the object deformation between the two exposures.

It has been shown[2] that the relative phase change Δ due to the relative displacement of the object surface is related to the deformation derivatives by

$$\Delta_x = \delta x \left(\frac{\partial u}{\partial x} \, k_s \bullet i + \frac{\partial v}{\partial x} \, k_s \bullet j + \frac{\partial w}{\partial x} \, k_s \bullet l \right), \tag{1}$$

where Δ_x is the relative phase change when the shearing direction is in the x-direction, δx is the shearing amount in the object surface. u, v, and w are the components of deformation vector in the x, y and z-direction, i, j and l are the unit vectors in the x, y and z-direction, k_s is the sensitivity vector of shearography[2] expressing the dimensions of the shearographic setup. If the shearing is in the y-direction, the relative phase change Δ becomes

$$\Delta_y = \delta y \left(\frac{\partial u}{\partial y} \, k_s \bullet i + \frac{\partial v}{\partial y} \, k_s \bullet j + \frac{\partial w}{\partial y} \, k_s \bullet l \right) \tag{2};$$

where the relative phase change Δ_y describes the deformation derivative with respect to y, δy is the shearing amount in the y-direction.

The equations (1) and (2) show that shearographic fringes are whole-field representations of the loci of the first derivative of deformation. Therefore, shearography is very useful for nondestructive testing and strain analysis. As a nondestructive testing tool, shearography is more practical than other nondestructive testing methods, because the defects of the object induce usually strain concentrations. Shearography reveals defects directly created by strain anomalies rather than displacement anomalies[3]. Moreover, rigid body motions do not produce strain, thus shearography is insensitive against such motion and does not require any particular device for vibration isolation.

In the simple TV-shearography the CCD-camera registers only one speckle pattern before and after stressing respectively. The subtraction between the two digitized speckle pattern yields the TV-shearogram[2]. This technique is very simple, it can perform both real-time and double exposure method.

3. BASIC PRINCIPLE OF QUANTITATIVE FRINGE EVALUATION

The quantitative evaluation of fringes needs to use the phase-shifting interferometry. When this technique is used in TV-shearography, a fringe pattern of phase modulo 2π (i.e. the phase-shifted TV-shearogram) not only is obtained, but also the shape of a surface deformation and the values of deformation derivative can be determined. In order to calculate the phase before and after deformation state, three, four or more interferograms (speckle patterns) are required to be recorded. The phase shift is usually introduced with a piezoelectric transducer. In this case the ordinary mirror

2 in Fig. 2 is replaced by the piezoelectric transducer actuated mirror. For each data frame, the intensity is integrated over the time so it takes time to move the reference mirror 2 linearly shifted by the piezoelectric transducer through $\pm 120°$ change in phase. These three frames of intensity data are recorded in this way:

$$I_1 (x,y) = I_0 \{1 + \gamma \cos [\phi(x,y)]\},$$

$$I_2 (x,y) = I_0 \{1 + \gamma \cos [\phi(x,y) + 120°]\}, \tag{3}$$

$$I_3 (x,y) = I_0 \{1 + \gamma \cos [\phi(x,y) - 120°]\},$$

where I_0 is the average intensity and γ is the modulation of the interference term. There are only three unknowns in three equations above, therefore the phase calculated at each detected point (x,y) in the speckle pattern is

$$\phi_1 = \arctan \frac{\sqrt{3} \ (I_3 - I_2)}{2 \ I_1 - I_2 - I_3}, \tag{4}$$

The object is then deformed, and three more frames of intensity data are taken while shifting the phase for the same amount as for the first set of data. This is analogous to the second exposure of the double exposure method. Hence, the phase ϕ_2 after deformation can also be calculated like ϕ_1. Once these data are stored, the relative phase change can be calculated simply by subtracting ϕ_2 from ϕ_1 for each corresponding speckle point:

$$\Delta = \phi_2 - \phi_1 \tag{5}$$

This calculation yields a phase-shifted shearogram expressing precise fringes full in contrast.

4. OUT-OF-PLANE SHEAROGRAM

Now rewriting equations (1) and (2) by considering the relative phase change of the doubly exposing shearography and the dimensions of the experimental arrangement[5]

$$\Delta_x = -\frac{2 \pi}{\lambda} \left[\sin \theta \frac{\partial u}{\partial x} + (1 + \cos \theta) \ \frac{\partial w}{\partial x} \right] \delta x$$

$$\Delta_y = -\frac{2 \pi}{\lambda} \left[\sin \theta \frac{\partial u}{\partial y} + (1 + \cos \theta) \ \frac{\partial w}{\partial y} \right] \delta y \tag{6}$$

where Θ is the angle between the direction of the illumination and the observation.

The angle Θ can be varied approximately between $0°$ and $90°$, Fig. 2; when $\Theta \approx 0°$ then $\sin \Theta \approx 0$ and $\cos \Theta \approx 1$. Hence equations (6) may be simplified to read,

$$\Delta_x = -\frac{4\pi\delta x}{\lambda} \ \frac{\partial w}{\partial x}, \qquad \Delta_y = -\frac{4\pi\delta y}{\lambda} \ \frac{\partial w}{\partial y}, \tag{7}$$

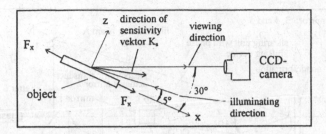

Fig. 3 Shearographic setup for measuring the in-plane strain components of a small
angle of viewing and a small angle of incidence (ESPSI)

5. IN-PLANE STRAIN BY MEANS OF SHEAROGRAPHY

When the light source L_1 and the camera C are situated in the x,z-plane as shown in Fig. 2 causing
the angle Θ_{xz} it results in the relative phase change Δ_{xx} indicating the shear direction x and Δ_{yx} the
shear direction y (equations (8)). The light source L_2 and the camera C form the angle Θ_{yz} in the y,z-
plane yielding the equation Δ_{xy} considering the shear direction x, and Δ_{yy} the shear direction y in the
following equations (8) where Θ_{xz} is perpendicular to Θ_{yz}.

The approach of the angle Θ between the illuminating and the observing directions is approximately
90° i.e., $\sin \Theta \approx 1$ and $\cos \Theta \approx 0$. We will get the following equations[6,7].

$$\Delta_{xx} = -\frac{2\pi}{\lambda}\left[\frac{\partial u}{\partial x} + \frac{\partial w}{\partial x}\right]\delta x \quad \text{(x shear direction, } \Theta \text{ in the x,z-plane } (\Theta_{xz})) \tag{8.1}$$

$$\Delta_{yx} = -\frac{2\pi}{\lambda}\left[\frac{\partial u}{\partial y} + \frac{\partial w}{\partial y}\right]\delta y \quad \text{(y shear direction, } \Theta \text{ in the x,z-plane } (\Theta_{xz})) \tag{8.2}$$

$$\Delta_{xy} = -\frac{2\pi}{\lambda}\left[\frac{\partial v}{\partial x} + \frac{\partial w}{\partial x}\right]\delta x \quad \text{(x shear direction, } \Theta \text{ in the y,z-plane } (\Theta_{yz})) \tag{8.3}$$

$$\Delta_{yy} = -\frac{2\pi}{\lambda}\left[\frac{\partial v}{\partial y} + \frac{\partial w}{\partial y}\right]\delta y \quad \text{(y shear direction, } \Theta \text{ in the y, z-plane } (\Theta_{yz})) \tag{8.4}$$

If measuring the in-plane strains the terms $\partial w/\partial x$ resp. $\partial w/\partial y$ are disturbance factors in eqs. (8).

The knowledge to measure the in-plane deformations $\partial u/\partial x$, $\partial u/\partial y$, $\partial v/\partial x$ and $\partial v/\partial y$ are sufficient for
describing the two-dimensional state of strain. Two ways are possible to measure the strain
components $\varepsilon_{xx} = \partial u / \partial x$ and $\varepsilon_{yy} = \partial v / \partial y$ resp. the partial derivatives

$\partial u/\partial y$ and $\partial v/\partial x$: Either by using the one beam-shearography for approximate or by the dual
illumination method for exact determination. The one beam-shearography considers the sensitivity
vector k_s in the x- or the y-shearing direction (Fig. 6). The calculation how to direct the illumination
and the observation beam shall be adjusted with respect to the sensitivity vector k_s for minimizing
the out-of-plane component[8]. It can be shown that the maximum relative error for the proposed in-
plane measurement will be less than 10 %.

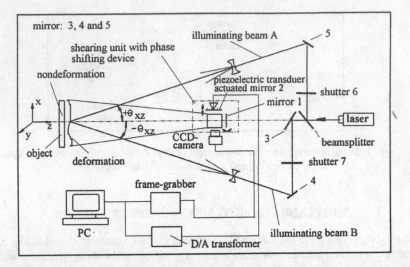

Fig. 4 Schematic arrangement for distinguishing the phase distributions at the illumination angles $+ \Theta_{xz}$ and $- \Theta_{xz}$ determining the relative phase changes of the exact in-plane and out-of-plane strains

The dual illumination device was developed for measuring the in-plane as well as the out-of-plane components exactly by means of the digital shearography using the phase-shifting unit which will be represented in the following (Fig. 4).

5.1. ONE BEAM-SHEAROGRAPHY CONSIDERING THE SENSITIVITY VECTOR

It seems that $\partial u / \partial x$ and $\partial u / \partial y$ could be measured from the equations (1) and (2) when the sensitivity vector k_s lies in the x-coordinate axis and $\partial v / \partial x$ and $\partial v / \partial y$ could be measured when the sensitivity vector k_s lies in the y-coordinate axis. But technically it is not possible to realize that, if the illuminating and the viewing direction lie both in the x-axis or the y-axis, the object can not be observed by the CCD-camera. Therefore, the measurement of the in-plane strain components compared with the measurement of out-of-plane strain components becomes difficult.

In order to measure the in-plane strain components the illuminating beam therefore should form a small angle to the object surface, that is near the grating incidence. Considering the prerequisite where the object can be observed suitably, the viewing direction should be at an angle as small as possible to the object surface too. A suggestion for shearographic measuring the in-plane strains is proposed and shown in Fig. 6. The illuminating direction is in the xz-plane and at an angle of 5° to the object surface and the viewing direction is in the xz-plane too and at an angle of 30° to the object surface. Thus the angle θ composed of the illuminating and the viewing direction is 25°. The sensitivity vector lies in xz-plane too and along the bisector of the angle θ. The angle between the sensitivity vector and the x-axis is $5° + θ / 2 = 17.5°$ (Fig. 6). The term $k_s \bullet i$ can be calculated as following

$$k_s \bullet i = |k_s| \, |i| \, \cos 17.5° = 2 \, (2\pi/\lambda) \cos (25°/2) \times \cos 17.5° = 1.86 \times (2\pi / \lambda) \qquad (9)$$

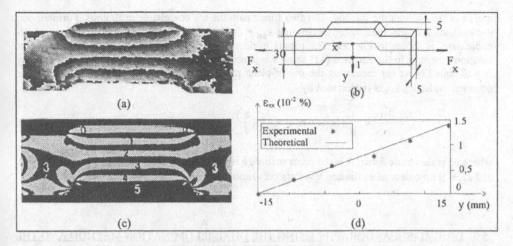

Fig. 5 Shearogram depicting the in-plane strain component $\partial u/\partial x$ for a cranked tensile bar (ESPSI),
(b) Dimensions of the cranked tensile bar, (c) Isochromatic fringes $\sigma_1-\sigma_2$ of the cranked tensile bar,
(d) Evaluation of the shearogram (ε_{xx} vs. y-coordinate) for the cranked tensile bar as $F_x = 35$ N.

Under these circumstances it is considered that $k_s \bullet l$ is very small compared with $k_s \bullet i$ and
$k_s \bullet j$, moreover, because of the state of plane stress, $\partial w/\partial x$ and $\partial w/\partial y$ compared with the terms $\partial u/\partial x$, $\partial u/\partial y$, $\partial v/\partial x$ and $\partial v/\partial y$ are small too. Therefore, the terms of the out-of-plane strain components
in the equations (8.1-4) can be neglected, and the equations (8.1) to (8.4) become

$$\Delta_{xx} \approx \delta x \frac{\partial u}{\partial x} \, k_s \bullet i = 1.86 \frac{2\pi}{\lambda} \delta x \frac{\partial u}{\partial x},\tag{10.1}$$

$$\Delta_{yx} \approx \delta y \frac{\partial u}{\partial y} \, k_s \bullet i = 1.86 \frac{2\pi}{\lambda} \delta y \frac{\partial u}{\partial y}\tag{10.2}$$

$$\Delta_{xy} \approx \delta x \frac{\partial v}{\partial x} \, k_s \bullet j = 1.86 \frac{2\pi}{\lambda} \delta x \frac{\partial v}{\partial x}\tag{10.3}$$

$$\Delta_{yy} \approx \delta y \frac{\partial v}{\partial y} \, k_s \bullet j = 1.86 \frac{2\pi}{\lambda} \delta y \frac{\partial v}{\partial y}\tag{10.4}$$

where Δ_{xx} corresponds to the situation that the shearing is in the x-direction and the illuminating and
the viewing direction lie both in the xz-plane, i.e. the angle θ composed of the illuminating and the
viewing direction lie in the xz-plane. The relative phase changes Δ_{yx}, Δ_{xy} and Δ_{yy} correspond to the
situations of the y-shearing direction and the angle θ in the xz-plane, the x-shearing direction and the
angle θ in the yz-plane, and the y-shearing direction and the angle θ in the yz-plane respectively.

An example is the shearographic evaluation of the in-plane strain components for a cranked tensile
bar loaded at both ends (material: Araldite B) using the sensitivity vector. Fig. 5a shows the
shearogram depicting the in-plane strain component $\partial u/\partial x$ created by the force $F_x = 35$ N. Fig. 5c
shows the isochromatic fringes i.e. the difference of the principal stresses $\sigma_1-\sigma_2$ of the tensile bar by

means of the photoelastic method. The two fringe patterns are completely analogous. Furthermore, the evaluation of the shearogram illustrating $\varepsilon_{xx} = \partial u/\partial x$ in y-direction, i.e. the graph ε_{xx} vs. y-coordinate, is shown in Fig. 5d. The results evaluated by the shearogram and by the analytical solution according to the equation (11) show a high correspondence. It proves that the shearography is well suited either for measuring the out-of-plane strain components or the in-plane strains. The equation evaluating ε_{xx} is represented by

$$\varepsilon_{xx} = \frac{\partial u}{\partial x} = \frac{1}{E}(\sigma_x - \nu\sigma_y) = \frac{F_x}{E}\left(\frac{a\,y}{J_{zz}} + \frac{1}{A}\right) \tag{11}$$

where F_x is the tensile force, A is the cross section, a is the cranking distance (it is 5 mm in Fig. 7b), and J_{zz} is the moment of inertia and ν is Poisson's ratio.

5.2. DIGITAL-SHEAROGRAPHY USING THE DUAL-ILLUMINATION METHOD AND THE PHASE-SHIFTING TECHNIQUE

The optical arrangement illustrated in Fig. 4 is slightly changed compared with Fig. 2[9]. The dual illumination configuration required for in-plane sensitive ESPSI (Electronic Speckle Pattern Shearing Interferometry) is schematically similar to TV-holography. Mutually, but in sequence coherent laser beams a and b lie in the xz-plane and are incident on the flat test surface at equal angles, $+\Theta_{xz}$ and $-\Theta_{xz}$ to the z-axis. Additionally two shutters 6 and 7 are included in the experimental setup.

When the shutter 7 is closed first and the shutter 6 is opened (Fig. 4), then the object is illuminated at the angle $+\Theta_{xz}$ in the original undeformed state. If the speckle-interference pattern reflecting the undeformed test surface by the illumination beam a is stored in the electronic memory, the phase distribution Φ_{+1} of the speckle interference pattern will be calculated by using the different intensities, eq. (4). After that the shutter 6 is closed and the shutter 7 is opened. Afterwards the object is illuminated at the angle $-\Theta_{xz}$ and the phase distribution Φ_{-1} of the speckle interference pattern generated by the laser beam b is calculated in the same way.

Then the object is deformed and the phase distribution Φ_{+2} created by the illumination beam a and this one called Φ_{-2} caused by illumination beam b are determined analogously to the phase calculation Φ_{+1} and Φ_{-1} before. Using the equation for calculating Δ analogously the relative phase change is given by the relationship

$$\Delta_{+\theta} = \Phi_{+2} - \Phi_{+1} \tag{12.1}$$

at the illumination angle $+\Theta_{xz}$ and as well

$$\Delta_{-\theta} = \Phi_{-2} - \Phi_{-1} \tag{12.2}$$

of the illumination angle $-\Theta_{xz}$.

The relative phase change $\Delta_{+\theta}$ (acc. to $\angle +\Theta_{xz}$) and $\Delta_{-\theta}$ (acc. to $\angle -\Theta_{xz}$) using the dual illumination configuration in the x,z-plane can be expressed as follows

$$\Delta_{+\theta} = -\frac{2\pi}{\lambda}\left\{\sin(\Theta_{xz})\frac{\partial u}{\partial x} + [1+\cos(\Theta_{xz})]\frac{\partial w}{\partial x}\right\}\delta x \tag{13.1}$$

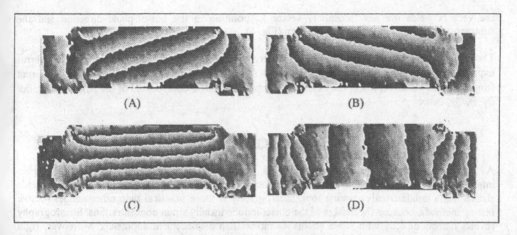

Fig. 6 Shearograms of a cranked tensile bar showing (a) the relative phase changes $\Delta_{+\theta}$
(acc. to $\angle +\Theta_{xz}$), (b) $\Delta_{-\theta}$ (acc. to $\angle -\Theta_{xz}$), (c) the exact in-plane strain $\partial u/\partial x$
and (d) the exact out-of-plane component $\partial w/\partial x$

$$\Delta_{-\theta} = -\frac{2\pi}{\lambda}\left\{\sin(-\Theta_{xz})\frac{\partial u}{\partial x} + \left[1+\cos(-\Theta_{xz})\right]\frac{\partial w}{\partial x}\right\}\delta x \qquad (13.2)$$

Fig. 6 a and b show the relative phase change $\Delta_{+\theta}$ and $\Delta_{-\theta}$ for the cranked tensile bar according to eqs. (13).

The oblique fringe pattern on the tensile bar contains the in-plane and out-of-plane strain components. The bar is subjected beside tension and bending additionally to distortion (Fig. 6 a and b). The relative phase changes $\Delta_{+\theta}$ and $\Delta_{-\theta}$ were calculated by the digital-shearography using the phase shift technique and therefore the exact in-plane strain $\partial u/\partial x$ can be determined as a fringe pattern by digital subtracting $\Delta_{+\theta}$ and $\Delta_{-\theta}$ considering $\sin(-\theta) = -\sin\theta$ and $\cos(-\theta) = \cos\theta$:

$$\Delta_{S1} = \Delta_{+\theta xz} - \Delta_{-\theta xz} = -\frac{4\pi\delta x \sin(\Theta_{xz})}{\lambda}\frac{\partial u}{\partial x} \quad \text{(x,z-illumination plane, x-shearing direction)} \qquad (14)$$

Beside measuring the in-plane strain the experimental arrangement can be used for determining the exact out-of-plane strain component $\partial w/\partial x$ furthermore by digital adding $\Delta_{+\theta}$ and $\Delta_{-\theta}$

$$\Delta_{A1} = \Delta_{+\theta xz} + \Delta_{-\theta xz} = -\left[1+\cos(\theta_{xz})\right]\frac{4\pi\delta x}{\lambda}\frac{\partial w}{\partial x} \quad \text{(x,z-illumination plane, x-shearing direction)} \qquad (15)$$

Fig. 6 c and d represent the results by digital subtracting and adding $\Delta_{+\theta}$ and $\Delta_{-\theta}$ showing the exact in-plane strain $\partial u/\partial x$ and the exact out-of-plane term $\partial w/\partial x$ for the tensile bar.

As mentioned above the direction of the sensitivity vector k_s according to the illumination direction $+\theta$ lies in the course of the bisecting line $+\theta/2$ resp. to the illumination direction $-\theta$ in the course of the bisector $-\theta/2$. The amount of the two formed sensitivity vectors is the same. The superposition of the vectors yields the new sensitivity vector k_s pointing in the out-of-plane direction and the subtraction indicates in the in-plane direction (Fig. 4).

The different shearing directions and the different planes of illumination create seperate terms expressing the relative phase changes $\Delta_{+\theta}$ and $\Delta_{-\theta}$[9]. The digital subtraction originates different fringe patterns which represent the exact and the mixed in-plane components of the strain tensor: $\partial u/\partial y$, $\partial v/\partial x$, $\partial v/\partial y$

6. CONCLUSIONS

As a strain measuring tool, shearography is simpler than speckle pattern and holographic interferometry since it does not require differentiating the measured displacements and hence it yields strains. As a nondestructive testing tool, shearography is more practical than other nondestructive testing methods, because the defects of the object induce usually strain concentrations. Shearography reveals defects directly with strain anomalies rather than displacement anomalies. Moreover, rigid body motions do not produce strain, thus shearography is insensitive against such motion and does not need any particular device for vibration isolation.

The electronic shearography further develops and perfects the technique. It realizes a filmless, fast, real-time and no vibration isolation requiring optical testing, and therefore, the optical method is ready for leaving the laboratory to be applied in the industry. This technique is very well suited for either strain measurement or nondestructive testing, and for either static or dynamic investigations. It measures not only the out-of-plane strain components but also the in-plane strain components. In measuring the out-of-plane strain this technique has a pure out-of-plane sensitivity. The proposed shearographic in-plane method has almost the in-plane sensitivity if the optical setup is adjusted as described in this paper.

The digital-shearography using the dual-illumination method and the phase-shifting technique determines the in-plane strain and the out-of-plane strain terms exactly by measuring two different phase distributions and calculating the relative phase changes.

7. REFERENCES

[1] W. Steinchen, M. Schuth, L.X. Yang
    Strain measurement on the surface of plates and discs by the means of shearography
    Journal Strain, July (105-108) and Nov. (139-141), 1994, Febr. 1995 (25-29)

[2] W. Steinchen, L.X. Yang, M. Schuth, G. Kupfer
    Electronic Shearography (ESPSI) for direct measurement of strains, European Symposium on Optics for Productivity in Manufacturing, EOS und SPIE (EUROPTO), 20./24.06.1994, Frankfurt/Main, 210-221

[3] W. Steinchen
    Quality control of fiberreinforced composites by means of the holographic and shearographic methods, SPIE, vols. 1756-25, 240-251, 1992, San Diego, California, USA

[4] P. Carrè, Installation et utilisation du compateur photoelectrique et interferential du Bureau International des Poids et Mesures, Metrologia 2, 13 (1966)

[5] Y.Y. Hung
Shearography: A new optical method for strain measurement and nondestructive testing,
Optical Engineering, 21(1982) 391-395

[6] W. Steinchen, L.X. Yang, M. Schuth und G. Kupfer
Precision Measurement and Nondestructive Testing by means of the Speckle and Speckle-Shearing Interferometry, VDI-Berichte Nr. 1118, 1994, 27-32, Germany

[7] Steinchen, W., Yang, L.X., Schuth, M., Kupfer, G.
Electronic speckle pattern shearing interferometry (ESPSI) and its application,
2nd Internat. Conf. on Optoelectronic Science and Eng. 15.-18.08.1994,
Peking, China, 249-252

[8] M. Schuth, W. Steinchen, L.X. Yang
Application of shearography to measure strain and strain concentration,
SEM, Spring Conference, July 1994, Baltimore, USA

[9] Steinchen, W., Yang, L.X., Schuth, M. und Kupfer, G.
Dehnungsmessung mit digitaler Shearografie,
Technisches Messen (tm), 9 (1995)

Miniaturisiertes holografisches Interferometer für die Kurzzeitanalyse mit Oberflächen-Biegewellen

S.Wild, H.Kreitlow, U.Samuels
Fachhochschule Ostfriesland
Fachbereich Naturwissenschaftliche Technik
Institut für Lasertechnik
Constantiaplatz 4, D-26723 Emden

Einleitung

Die Kenntnis des dynamischen Verhaltens technischer Bauteile ist notwendige Voraussetzung zu Beurteilung ihrer Beanspruchung und zur optimierten werkstofftechnischen sowie konstruktive Auslegung. Neben der stochastischen und periodischen· Belastung treten an einer Vielzahl vo Bauteilen impulsartige Anregungen ʼauf, z.B. bei Umformmaschinen oder bei Fahrzeugen, wi Flugzeugen beim Aufsetzen auf die Landebahn.

Impulsartige Belastungen durch Querkräfte können zu örtlichen Spannungsüberhöhungen führer Eine Berechnung der Störungsausbreitung ist bereits bei einfachen Objekten mit großem Aufwan verbunden. Andererseits ist eine meßtechnische Erfassung der gesamten Ausbreitungsform eine Störung durch Querkräfte mit herkömmlichen Meßverfahren am Originalbauteil nicht möglic Mit Hilfe der holografischen Meßtechnik können vollständige dreidimensionale Verformungs felder selbst bei kleinen Auslenkungen von weniger als 1µm berührungslos am realen Baute sichtbar gemacht und gemessen werden.

Kurzeitanalyse von Oberflächenbiegewellen

Für die Kurzzeitanalyse von Biegewellenvorgängen wird das holografische Doppelbelichtungs verfahren eingesetzt. Mit zwei zeitlich getrennten sehr kurzen Laserriesenimpulsen wird mit de ersten Belichtung ein unbelasteter Objektzustand sowie zeitlich dazu verzögert, nach de stoßartigen Biegewellenanregung, ein zweiter Objektzustand mit einer zweiten Belichtun holografisch gespeichert (siehe Bild 1).

Die durch die Doppelbelichtung entstehenden holografischen Interferenzstrukturen ermögliche es, das räumliche und zeitliche Ausbreitungsverhalten der Oberflächenbiegewelle zu analysiere und daraus z.B. richtungsabhängige mechanische Kennwerte wie den E-Modul in anisotrope Werkstoffen zu ermitteln oder Materialfehler zu erkennen (siehe Bilder 2 und 3).

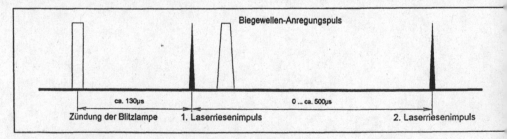

Bild 1.: Ablaufdiagramm zur holografischen Aufnahme von Biegewellen

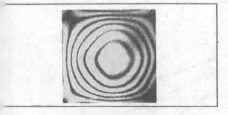

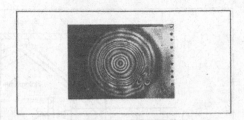

ild 2.: Biegewelle auf einer quadratischen
Platte nach Reflexionen an Rändern.

Bild 3.: Fehlerdetektion mit Biegewellen.

)ie Anregung der Biegewellenvorgänge erfolgte bisher größtenteils mit mechanischen
Anregungsverfahren. Der Nachteil hierbei ist, daß die minimale Anregungszeit auf etwa 50μs
egrenzt ist. Hierdurch wird die kleinste erreichbare Biegewellenlänge und damit der minimal
uflösbare Materialfehler festgelegt.

Biegewellenanalyse mit einem miniaturisierten holografischen Interferometer

m folgenden wird ein Verfahren vorgestellt, bei dem mit Hilfe eines Laserriesenimpulses, der
lurch einen Lichtwellenleiter geführt wird, Anregungsimpulse erzeugt werden können, die kleiner
als 1μs sind. Gleichzeitig wird durch die Lichtwellenleiterlänge eine definiert reproduzierbare
Verzögerungszeit zwischen erster Belichtung und der Biegewellenanregung festgelegt, wodurch
ei der ersten Belichtung eine Aufnahme im Ruhezustand des Objektes garantiert wird. Der für
lie Biegewellenanregung benötigte Laserimpuls wird durch Mehrfach-Strahlaufteilung mit Hilfe
sines diffraktiven optischen Elements (DOE) in der nullten Beugungsordnung erzeugt. Die ersten
Beugungsordnungen des DOE liefern die Objektbeleuchtungs- und Referenzwelle des
nolografischen Interferometers.

Typen:
- Gitter
- Off-axis-Linse

Herstellungsverfahren:
- interferometrisch durch holografischen
  Aufbau
- parallel-lithografisch durch Maskenprojektion
- seriell-lithografisch durch Laser Pattern
  Generator

Parameter für die DOE
Beugungswinkel, Brennweite
Intensitätsverteilung für λ=694nm
I≈76% ⇒ Biegewellenanregung
I≈6%   ⇒ Objektbeleuchtung
I≈5%   ⇒ Referenzwelle

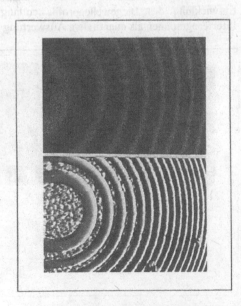

Tabelle 1: Herstellung von diffraktiven
Mehrfachstrahlteilern

Bild 4: interferometrisch und durch
Maskenprojektion hergestelle DOE

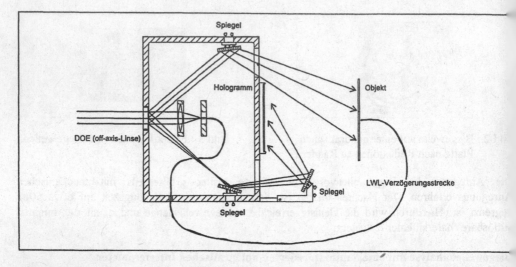

Bild 5.: Miniaturisiertes holografisches Interferometer mit LWL-Verzögerungsstrecke und DOE

Mit Hilfe unterschiedlicher Herstellungsverfahren (siehe Tabelle 1) wurden Mehrfachstrahlteiler angefertigt (siehe Bild 4), die die Entwicklung eines miniaturisierten holografischen Interferometers zur Biegewellenanalyse ermöglichten (siehe Bild 5).

Für die Aufnahme der Biegewellenfolge an einem 0,1mm dicken Titanblech (siehe Bild 6) wurde ein miniaturisiertes Interferometer mit einer off-axis-Linse als DOE-Mehrfachstrahlteiler und einer LWL-Verzögerungsstrecke von 10m (Verzögerungszeit ca. 50ns) eingesetzt. Die zeitliche Entwicklung des Biegewellenprofils entlang eines Biegewellendurchmessers ist unter den Interferogrammen als quantitative Auswertung zu sehen.

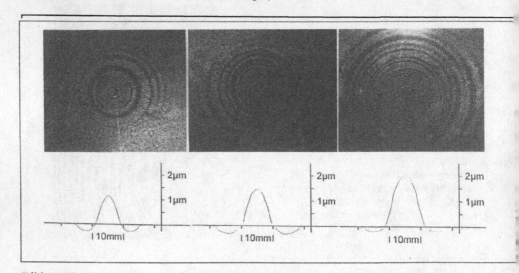

Bild 6.: Biegewellenfolge an Titanblech; 10, 40 und 60µs nach Biegewellenanregung

Zusammenfassung

Laserriesenimpulse zur Biegewellenanregung ermöglichen sehr kurze Anregungszeiten, woraus auch kurze Biegewellenlängen und eine damit verbundene hohe Ortsauflösung ergeben, wie sie für die Materialfehleranalyse benötigt wird. Der Einsatz von Lichtwellenleitern zur Führung und Verzögerung des Anregungsimpulses gegenüber der ersten Belichtung ermöglicht eine einfache und reproduzierbare Biegewellenanregung. Durch einen diffraktiv optischen Mehrfachstrahlteiler kann sowohl der Biegewellen-Anregungsimpuls als auch die Objekt- und Referenzwelle erzeugt und das gesamte holografische Interferometer miniaturisiert werden.

Faseroptisches modulares Speckle-Interferometer

B. Ströbel
Fachbereich Mathematik und Naturwissenschaften der Fachhochschule Darmstadt
Haardtring 100, D-64295 Darmstadt

1. Einleitung

Die computergestützte Speckle-Interferometrie (bekannt als ESPI, DSPI oder TV-Holografie) sowie die verwandten Verfahren (z. B. Speckle-Shearografie) haben sich als robuste Alternative zur holografischen Interferometrie für die Darstellung und Vermessung von geringfügigen Oberflächenverformungen bewährt. Dabei wird der durch die Speckle-Statistik gegebene Nachteil der geringeren lateralen Auflösung aufgewogen durch die vielfältigen Möglichkeiten, welche durch die direkte Rechnersteuerung zur Verfügung stehen: So lassen sich z. B. Echtzeit-Streifenbeobachtung, Phasenschiebetechniken [1], Mittelungen über wiederholte Messungen zur Rauschunterdrückung [2] oder Auswertungen der Specklestatistik zur Optimierung der Messbedingungen problemlos in den Meßablauf einfügen. Leider steht einer weiteren Verbreitung der Speckle-Interferometrie in Ausbildung, Entwicklung und Anwendung häufig der Preis der kommerziell erhältlichen Geräte und Software entgegen. Meist sind die Geräte in aufwendiger Technik gebaut und für Routineanwendungen ausgelegt; für die Realisierung eigener Entwicklungen ist der Anwender auf die Kooperationsbereitschaft des Herstellers angewiesen. Aufgrund dieser Situation entstand an der Fachhochschule Darmstadt der Plan, ein kostengünstiges „low end" Speckle-Interferometer samt der benötigten Software selbst zu entwickeln.

2. Gerätekonzept

Herzstück des Geräts ist ein 2 x 2 Singlemode-Faserschmelzkoppler, bei dem in einem Arm ein piezomechanischer Phasenschieber integriert wurde. Vorteile dieses faseroptischen Konzepts [3] sind:

- einfacher, kostengünstiger und platzsparender Aufbau
- leichte Justierbarkeit
- geringe Anfälligkeit für äußere Störungen, z. B. Erschütterungen oder Zugluft
- hohe Flexibilität beim Umbau auf andere Meßverfahren und Probengeometrieen

Als Lichtquellen wurde ein HeNe-Laser sowie eine kollimierte Laserdiode (λ = 690 nm) erfolgreich eingesetzt. Der unbenützte vierte Arm des Schmelzkopplers gibt die Möglichkeit, eine zweite Lichtquelle einzukoppeln (z. B. zur Erweiterung des Meßbereichs), oder kann für andere Aufgaben (z. B. zur Kalibrierung des Phasenschiebers) Verwendung finden.

Der modulare Aufbau des Geräts erlaubt die Realisierung verschiedener Meßverfahren durch einfachen Umbau (vgl. **Abb. 1**):

- **Speckle-Dekorrelation** zur Beobachtung von Veränderungen der Probenoberfläche
- **ESPI out-of-plane** zur Messung von Verschiebungen senkrecht zur Bildebene
- **ESPI in-plane** zur Messung von Verschiebungen in der Bildebene
- **Speckle-Topometrie** nach der Zweistrahl- [4] oder der Zweiwellenlängenmethode [5] zur Vermessung der Oberflächenform einer Probe
- **Speckle-Shearografie** zur Messung des Gradienten der Verschiebung (in diesem Fall wird vor der Kamera ein Scherelement eingefügt, an dem auch die Phasenverschiebung ansetzt)

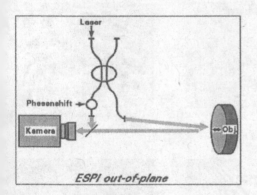

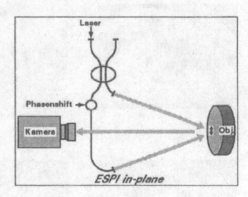

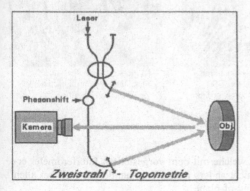

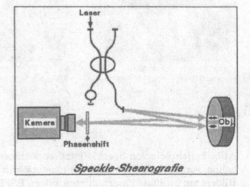

Abb. 1: Schematische Darstellung einiger Meßverfahren der Speckle-Interferometrie

666

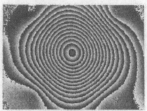

a) Eingespannte Blechplatte, in der Mitte punktförmig belastet, ESPI out-of plane

b) Blech mit zwei Schlitzen, in der Mitte punktförmig belastet, ESPI out-of-plane

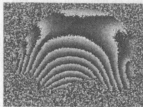

c) Fahrradhelm unter Kompression in Längsrichtung, Bildfeld ca. 32 cm x 24 cm

d) Eingespannte Blechplatte, wie oben a), jedoch ESPI in-plane

e) Blech mit zwei Schlitzen, wie oben b), jedoch ESPI in-plane

f) Blumentopf aus rotem Ton mit zwei Rissen, unter geringfügiger Erwärmung

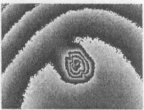

g) Kunststoff-Verbundplatte mit Impact-Schaden, unter geringfügiger Erwärmung

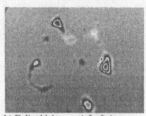

h) Folienklebung mit Lufteinschlüssen, unter geringfügiger Erwärmung

i) Häuschen aus Kunststoff-Bausteinen, von oben belastet

j) Glühlampe mit geringer Stromstärke betrieben, Phasenverschiebung in Transmission

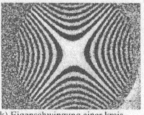

k) Eigenschwingung einer kreisförmigen Metallscheibe, ESPI-Streifenbild im Zeitmittelverfahren

l) Sparschwein aus Ton, Speckle-Topometrie nach dem Zweistrahlverfahren

Abb. 2: Beispiele von Speckle-Interferogrammen, welche mit dem vorgestellten Interferometer erhalten worden sind. Falls in der Bildunterschrift nicht anders angegeben, handelt es sich bei allen Bildern um gefilterte Phasendifferenzbilder, ESPI out-of-plane.

Für eine spätere Erweiterung zur 3D-Meßgerät bietet sich wegen der günstigen Kosten eine dreifache Auslegung von Laser und Schmelzkoppler mit rechnergesteuerten Verschlüssen an.

Das Gerät wurde aus Standardkomponenten (Mikrobank, Viedokamera, PC) aufgebaut. Als Schnittstelle zwischen Rechner und Kamera wird eine handelsübliche Bildverarbeitungskarte eingesetzt; aufgrund der Weiterentwicklung der Rechnertechnik soll diese in Zukunft durch eine einfache Framegrabber-Karte ersetzt und die Bildverarbeitung vollständig in den Rechner verlegt werden. Die Demodulation der Phasendifferenzbilder zwischen Grund- und Lastzustand geschieht derzeit nach einem modifizierten Pixel-Queue [6] Verfahren. Die in **Abb. 2** gezeigten Bildbeispiele (gefilterte Phasendifferenzbilder) geben einen Eindruck von den bisher realisierten Untersuchungen und der erzielten Bildqualität.

3. Schlußbemerkung

Ein leistungsfähiges Speckle-Interferometer braucht nicht aufwendig und teuer zu sein. Für das vorgestellte „low end" Gerät bieten sich vielfältige und neue Einsatzmöglichkeiten in Ausbildung und Entwicklung sowie in kleineren Firmen und Prüflabors an.

Danksagung: Ich danke Herrn Thomas Leitner für seine engagierte Mitwirkung bei der Realisierung des vorgestellten Projekts, sowie dem Hessischen Ministerium für Wissenschaft und Kunst für finanzielle Unterstützung.

Literatur:

[1] Th. Kreis, J. Geldmacher, W. Jüptner: *Phasenschiebeverfahren in der interferometrischen Meßtechnik: ein Vergleich*, Laser in der Technik, Hrsg. von W. Waidelich, Springer Verl. 1994, S. 119 - 126
[2] Th. Floureux: *Improvement of electrical speckle fringes by addition of incremental images*, Opt. Laser Technol. vol. 25, p. 255 - 258 (1993)
[3] P. Aswendt, R. Höfling: *Interferometrische Dehnungsmessung - Aufbau und Anwendung eines DSPI-Meßsystems*, Laser in der Technik, Hrsg. von W. Waidelich, Springer Verl. 1994, S. 172 - 175
[4] Y. Zou, H. Diao, X. Peng, H. Tiziani: *Geometry for contouring by electronic speckle pattern interferometry based on shifting illumination beams*, Appl. Opt. vol. 31, 31, p. 6616 - 6621 (1992)
[5] X. Peng, Y. L. Zou, H. Y. Diao, H. J. Tiziani: *A simplified multi-wavelength ESPI contouring technique based on a diode laser system*, Optik Bd. 91, 2, S. 81 - 85 (1992)
[6] H. A. Vrooman, A. M. Maas: *Image processing algorithms for the analysis of phase-shifted speckle interference patterns*, Appl. Opt. vol. 30, 13, p. 1636 - 1641 (1991)

Herstellung und Test hybrider Mikrooptiken zur Korrektur von Hochleistungslaserdiodenstrahlung

Ch. Haupt, M. Daffner, A. Rothe und H.J. Tiziani
Institut für Technische Optik
Universität Stuttgart
Pfaffenwaldring 9
D-70569 Stuttgart

Einleitung:

Für viele Anwendungen sind Hochleistungslaserdioden von großem Interesse. Besonders geeignet sind sie wegen ihres hohen Wirkungsgrades und ihrer Schmalbandigkeit zum Pumpen von Festkörperlaserkristallen. Hochleistungslaserdioden emittieren jedoch unter verschiedenen Divergenzwinkeln und besitzen wegen der auftretenden transversalen Moden eine schlechte Strahlqualität und Fokussierbarkeit. Ziel ist es geeignete Transferoptiken zum effektiven Pumpen von zwei Nd:YAG-Festkörperlasern mit einem monolitischen bzw. einem halbmonolithischen Resonator zu finden. Da das gesamte Lasersystem ein max. Volumen von $10\,\mathrm{cm}^3$ besitzt, dürfen die Transferoptiken eine Baulänge von 6mm nicht überschreiten und sollten aus Kostengründen möglichst einfach aufgebaut sein.

Bei den untersuchten Laserdioden handelt es sich um Breitstreifendioden der Firma Siemens (Typ 480201) bei einer Streifenbreite von 200µm und einer max. Ausgangsleistung von 1W.

Besonders geeignet zum Pumpen von Festkörperlasern sind anamorphotische Systeme, bestehend aus einer Grinlinse zum Einfangen der Laserdiodenstrahlung in Kombination mit weiteren optischen Elementen mit asphärischer Wirkung.

Im Zuge einer immer weiter fortschreitenden Miniaturisierung von Lasern besitzen insbesondere hybride Mikrooptiken ein großes Potential zum Einsatz als Pumpoptik. Hybride Mikrooptiken bestehen aus der Kombination von refraktiven und diffraktiven Elementen. Mit den refraktiven Elementen werden die divergenten Strahlen der Laserdiode gesammelt und gebündelt. Mit dem nachfolgenden diffraktiven Element wird dann eine Strahlkorrektur und Fokussierung durchgeführt. Die berechneten und untersuchten hybriden Mikrooptiken bestehen aus einer Grinlinse und einem computergenerierten Hologramm, welches direkt auf die Endfläche der Grinlinse in Photoresist belichtet wird. Das Hologramm besitzt eine zonenplattenartige Struktur und wirkt fokussierend. Besonders vorteilhaft bei dieser Aufteilung der optischen Kräfte ist die Reduktion der Pumpoptik auf ein einteiliges System. Durch das Aufbringen des planen Hologramms auf der Grinlinse erreicht man eine Miniaturisierung der Pumpoptik und gleichzeitig eine Korrektur der Laserdiodenstrahlung, die mit rein refraktiven Elementen nicht möglich ist. Zum Vergleich wurden auch refraktive Mikrooptiken bestehend aus einer Grinlinse mit eingeschliffener Zylinderlinse oder eingeschliffener Sphäre und einer weiteren Zylinderlinse untersucht.

Transferoptik:

Ausgangspunkt der im folgenden dargestellten Beurteilung der Leistungsfähigkeit der entwickelten Systeme war die bisher zum Pumpen eingesetzte NSG-Kataloglinse mit eingeschliffener Sphäre. Die Hauptprobleme ,die bei Einsatz der einzelnen Grinlinse auftreten, sind die sehr asymmetrischen Focii, der schlechte Modenüberlapp und dadurch bedingte hohe Laserschwellen sowie das Auftreten höherer Transversalmoden.

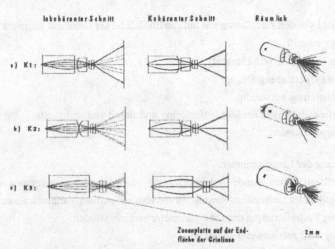

Bild 1: Strahldurchrechnung für verschiedene Realisierungen der Transferoptiken. K1: In die Grinlinse eingeschliffene Zylinderfläche und weitere Zylinderlinse. K2 In die Grinlinse eingeschliffene Sphäre und zusätzliche Zylinderlinse. K3: Hybride Variante mit auf die Endfläche der Grinlinse aufgebrachter Zonenplatte.

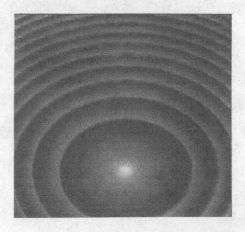

Bild 2: Draufsicht auf die Zonenplatte in Variante K3 aus Bild 1. Geschrieben mit 256 Stufen und Samplingabstand 0.8μm.

Mit den in Bild 1 dargestellten Transferoptiken konnten um ein Drittel kleinere Focii erreicht werden, die trotz der starken Asymmetrie der verwendeten Laserdiode nahezu symmetrisch sind. Die kohärente Schnittebene ist dabei die durch den großen Divergenzwinkel gekennzeichnete senkrecht zur aktiven Schicht ausgerichtetete Ebene . Die inkohärente Schnittebene ist die parallel zur aktiven Schicht ausgerichtete Ebene, in der bedingt durch die hohe Streifenbreite der Breitstreifendiode verschiedenc Transversalmoden höherer Ordnung auftreten.

Vorteile der hybriden Ausführung mit auf die Endfläche der Grinlinse aufgebrachter 256-stufiger Zonenplatte sind:
- einfacherer Aufbau (ein Element),
- verkürzte Übertragungslänge,
- keine Justierung notwendig,
- sowie die Möglichkeit der Massenfertigung und damit verbunden eine
- Senkung der Kosten.

Verbesserung der Laserparameter:
Durch die damit einhergehende Verbesserung der Modenüberlappung konnten nach Einsetzen der Transferoptiken im Laseraufbau sowohl für die rein refraktiven Varianten als auch für die hybride Ausführung Verbesserungen der Laserparameter erreicht werden:
- Senkung der Laserschwelle um 60%
- TEM$_{00}$- Betrieb bis 20mW
- Erhöhung der slope - efficiency (Bild 4)

Bild 3: Gefaßte, rein refraktive, Variante und einzelne Zylinderlinse

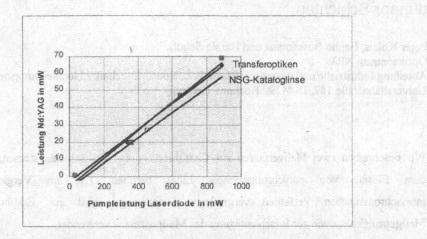

Bild 4: Slope - efficiency für Nd:YAG-Laserkristall (monolithischer Resonator)

Polychromatische on-line Meßverfahren zur Charakterisierung dünner Schichten

Peter Kohns, Benno Buschmann und Harald Schulz
Optikzentrum NRW
Abteilung Industrielle und wissenschaftliche Präzisionsmeßtechnik / Hochleistungsoptik
Universitätsstraße 142, D-44799 Bochum, Germany

Wir beschreiben zwei Meßverfahren zur Charakterisierung von Schichtsystemen, die auf dem Einsatz von polychromatischem Licht beruhen. Die im Vergleich zu monochromatischen Verfahren vergrößerte Datenmenge wird zur Erhöhung der Meßgenauigkeit sowie zur Beschleunigung des Meßprozesses verwendet.

1. Einleitung

Die industrielle Bedeutung der Oberflächen- und Beschichtungstechnik hat in den letzten Jahren enorm zugenommen. Ein wichtiger Grund hierfür ist die ständige Weiterentwicklung und Verbesserung der unterschiedlichsten Beschichtungstechnologien.

Parallel dazu verlangen die neuen Beschichtungsverfahren auch eine ständige Weiterentwicklung der Meßtechnik zur Schichtcharakterisierung. Dieser Artikel beschreibt zwei Meßverfahren, die im Gegensatz zu den heute meist verwendeten monochromatischen Schichtdickenmeßverfahren mit polychromatischem Licht arbeiten und die Schichtdicke on-line im Fertigungsprozeß messen.

2. Meßverfahren zur on-line-Kontrolle des Auftrags eines Schichtsystems

Zur Kontrolle des Schichtwachstums während des Bedampfungsprozesses werden heute standardmäßig interferometrische Verfahren verwendet. Dabei wird die Änderung der reflektierten Lichtintensität bei einer Wellenlänge λ als Funktion der Beschichtungsdauer beobachtet. Zur Auswertung muß der Brechungsindex der aufgebrachten Schicht genau bekannt sein. Hierbei besteht allerdings das Problem, daß der Brechungsindex eines Beschichtungsmaterial von den Betriebsparametern der Beschichtungsanlage abhängt. Ein weiterer Nachteil ist die Verwendbarkeit des Verfahrens nur bei Einfachschichten. Zur Steuerung des Auftrags eines Vielfach-Schichtsystems wird daher für jede einzelne Schicht ein neues, unbeschichtetes Probenglas mitbeschichtet und vermessen. Dabei führt das

Wechseln des Probenglases zu unerwünschten, aber bisher unvermeidlichen Verzögerungen und Meßfehlern bei der Messung der spektralen Eigenschaften des gesamten Schichtsystems. Ein neuartiges Verfahren ("Typ I") dient zur Beseitigung der beschriebenen Nachteile. Dabei wird simultan die gesamte spektrale Abhängigkeit der Reflexion $R_i(\lambda_j)$ und der Transmission $T_i(\lambda_j)$ der Probe jeweils nach dem Aufbringen der i-ten Schicht (i = 1,2,...,N) bei M verschiedenen Wellenlängen λ_j (j = 1, ...,M) bestimmt. Diese Werte werden in einem Steuerrechner mit den theoretisch vorgegebenen Charakteristiken $T_i'(\lambda_j)$ und $R_i'(\lambda_j)$ verglichen, die mit einem Matrixalgorithmus bestimmt werden [1]. Dazu wird jede einzelne Schicht eines Mehrfachschichtsystems durch eine komplexe Matrix M_i beschrieben, die als Parameter die gewünschte Schichtdicke d_i sowie den Realteil $n_{R,i}$ und den Imaginärteil $n_{I,i}$ des Brechungsindex der Schicht i enthält. Nach Vorgabe der Brechungsindices und der Schichtdicken aller N Schichten ermittelt der Steuerrechner die zu jeder Schicht gehörenden 2 x 2-Matrizen $M_i(\lambda_j)$ und deren Produkte $M_i^*(\lambda_j) = M_i(\lambda_j) \cdot \ ... \ \cdot M_1(\lambda_j)$, aus denen $R_i'(\lambda_j)$ und $T_i'(\lambda_j)$ folgen.

Der Vergleich mit den gemessenen Charakteristiken $R_i(\lambda_j)$ und $T_i(\lambda_j)$ ergibt die im Beschichtungsprozeß vorliegende Schichtdicke und den Brechungsindex der Schicht i. Zur Ermittlung der drei Parameter d_i, $n_{R,i}$ und $n_{I,i}$ wird verwendet, daß sehr viele Wellenlängen λ_j in einem großen Spektralbereich (UV bis NIR) verwendet werden und genügend Meßwerte zur Inversion des Matrixgleichungssystems zur Verfügung stehen. Insbesondere wird die Schichtdicke dabei in den Wellenlängenbereichen bestimmt, in denen der Brechungsindex eine sehr geringe Wellenlängenabhängigkeit (Dispersion) aufweist.

Im Gegensatz zu der bekannten interferometrischen Schichtdickenmessung von Einzelschichten kann die Wellenlängenabhängigkeit von Reflexion und Transmission eines sequentiell mit i Schichten beschichteten Probeglases gemessen und mit den theoretisch berechneten Daten des i-Schichtsystems verglichen werden, so daß das Probeglas nicht gewechselt werden muß. Die Beschichtungsanlage wird so gesteuert, daß die Abweichung der gemessenen Charakteristik von der vorgegebenen im gesamten untersuchten Spektralbereich (relevante Werte λ_j) minimal wird. Außerdem ergibt sich abhängig vom Schichtsystemaufbau die Möglichkeit, die Abweichungen des bis zur i-ten Schicht aufgebrachten Schichtsystems von den theoretischen Vorgaben durch Änderung der Schichtdicken der noch aufzubringenden N-i Schichten so zu korrigieren, daß die Zielcharakteristik des gesamten Schichtsystems bestehend aus N Schichten erreicht wird.

3) Verfahren zur Bestimmung der Dicke absorbierender Schichten auf einem transparenten Substrat

Ein weiteres Verfahren ("Typ II") dient zur Bestimmung der Dicke absorbierender Schichten auf transparenten Substraten. Dabei wird die schichtdickenabhängige Transmission von polychromatischem Licht durch die Probe bestimmt. Die im Vergleich zu monochromatischen Verfahren gesteigerte Datenzahl kann zur Reduzierung von Fehlern z.B. durch Schmutz- oder Staubablagerungen verwendet werden. Das Verfahren wird hier für den Fall eines Systems von N Schichten beschrieben.

Zur Messung der N Schichtdicken d_1 ....d_N wird die Transmission der Probe bei N + 1 Wellenlängen λ_1 ....λ_{N+1} bestimmt, die von den Schichten unterschiedlich absorbiert werden. Dazu können die Strahlen von N+1 monochromatischen Lichtquellen vor der Probe durch geeignete dichroitische Strahlteiler kombiniert werden. Die Lichtquellen werden - von einem Rechner gesteuert - abwechselnd eingestrahlt. Als Detektor kann in diesem Fall eine Photodiode verwendet werden. Alternativ ist auch die Einstrahlung von weißem Licht möglich, das nach Durchlaufen der Probe spektral zerlegt wird. Ein Steuerrechner erhält die Ausgangsspannungen der den einzelnen Photodioden nachgeschalteten Verstärkern. Die Verstärkerausgangsspannung bei der Wellenlänge λ_j beträgt bei durchstrahlter Probe

$$U_{j,mit} = U_{j,\,ohne} \cdot (1 - V_{Substrat}) \cdot (1 - V_{Schmutz}) \cdot \exp\left[-d_1/\mu_1\,(\lambda_j) - d_2/\mu_2\,(\lambda_j) - .... - d_N/\mu_N(\lambda_j)\right]$$

Dabei gibt $V_{substrat}$ die Lichtverluste durch das Substrat an. Mit $V_{Schmutz}$ seien die Lichtverluste durch Schmutz und Staub gekennzeichnet. $\mu_i(\lambda)$ sei die 1/e-Absorptionslänge der Schicht i bei der Wellenlänge λ. $U_{j,\,ohne}$ ist die Ausgangsspannung, die bei der Wellenlänge λ_j ohne Probe gemessen wird. Diese Spannungen sollten in bestimmten Abständen (z.B. beim Probenwechsel) neu ermittelt und gespeichert werden. Aus den Meßwerten bildet der Rechner Werte Q_K (K=1,...,N):

$$Q_K = \ln\left[(U_{N+1,\,ohne} \cdot U_{K,\,mit}) / (U_{N+1,\,mit} \cdot U_{K,\,ohne})\right]$$

$$= d_1 [1/\mu_1(\lambda_{N+1}) - 1/\mu_1(\lambda_K)] + d_2 [1/\mu_2(\lambda_{N+1}) - 1/\mu_2(\lambda_K)]+ .... + d_N [(1/\mu_N(\lambda_{N+1}) - 1/\mu_N(\lambda_K)]$$

Offensichtlich spielen bei den Werten Q_K die o.a. Fehlerquellen keine Rolle mehr. Die Koeffizienten in den eckigen Klammern sind schichtdickenunabhängige Materialkonstanten. Bei der Verwendung geeigneter Wellenlängen ist das erhaltene System aus N Gleichungen lösbar, so daß die Schichtdicken d_1, ..., d_N bestimmt werden können.

Anwenderspezifische Schichtdickenmeßgeräte, in denen die oben genannten Meßverfahren realisiert sind, bieten wir kommerziell an. Beispiele für die Geräteausführung und Anwendung der polychromatischen Schichtdickenmessung sind in den Abbildungen 1 bis 3 angegeben.

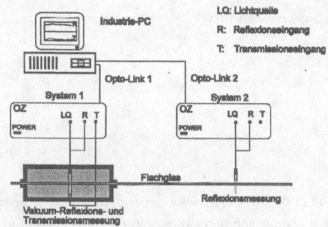

Abb.1: Einsatz des Meßgerätes Typ I bei der Beschichtungskontrolle von Flachglas

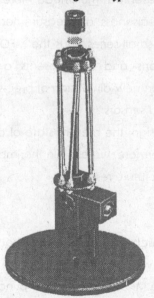

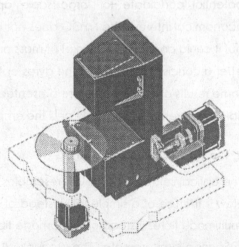

Abb. 2:Ausführungsbeispiel des Meßkopfes des Gerätes Typ I. Der Meßstrahl wird von unten zugeführt. Das durch die Probe transmittierte Licht wird oben von einer Faser aufgenommen. Der Reflex der Probe wird durch eine rechts anzuschließende Faser gesammelt.

Abb. 3: Meßgerät Typ II zur Schichtdickenbestimmung bei optischen Speichermedien. Der Meßstrahl durchläuft die Probe (in diesem Fall eine CD) von oben nach unten. Der untere Teil des Gerätes enthält den Detektor.

[1] M. Born, E. Wolf. "Principles of Optics". Pergamon Press Oxford 1980

Multimode Fibre Optic Gyroscope, Recent Progress and Limits

M. Eckerle, Peter Krippner, M. Bouamra, A. Chakari, P. Meyrueis

Laboratoire des Systèmes Photoniques,

ENSPS, Université Louis Pasteur,

Boulevard Sébastien Brandt,

F-67400 Illkirch-Graffenstaden, France

1 Introduction

The widespread application of a sensor depends mainly on its performance, on cheap standard components and low assembly costs. In contrast to single mode fibre optic gyroscopes (SFOG) [1]-[16] the presented multimode fibre optic gyroscope (MFOG) meets both a sufficient preciseness of measurement and already low manufacturing costs, if produced in small series. Thus the MFOG is a potential candidate for large scale applications and it represents a great economical interest. The MFOG does not reach an new dimension of preciseness, but it could conquer the market of mass produced sensors.

After a concise discussion of the gyros optical design, the present state of art and some results of our research are presented. Furthermore, we explain the impact of fibre parameters on sensitivity and the emphasis in future research.

2 The Optical Setup of the MFOG

The optical setup of the MFOG resembles very much the one of a classical SFOG. New is the use of a simple LED instead of expensive SLD's or Lasers and the use of multimode fibres instead of monomode fibres. The MFOG bases on the principle of a Sagnac interferometer: The Y-coupler II Fig. 1 splits the light with the intensity I_0 into two light beams counterpropagating in the fibre coil. If the gyro rotates with the angular velocity $\bar\omega$, the Y-coupler II reunifies the light beams with the Sagnac phase shift $\Delta\Phi_R \neq 0$, whereas $\Delta\Phi_R \sim \omega$. The light interferes and the photo diode Fig. 1 captures this interference as a variation of intensity

$$I = \frac{I_0}{2}\left(1 - \cos\Delta\Phi_R\right).$$
(1)

The piezo electric modulator and the lock in amplifier are introduced due to the low sensitivity for small angular velocities $\bar\omega$ and because of the $1/f$ proportional noise.

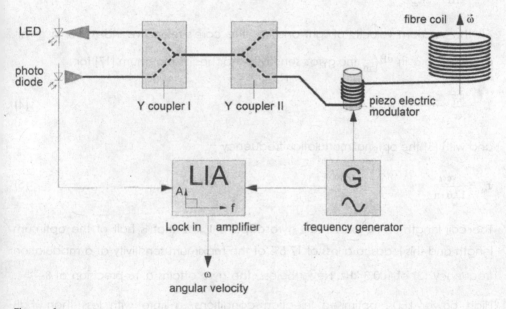

Figure 1
The optical setup of the multimode fibre optic gyro

3 Aspects of Optimisation

Supposing a sinusoidal phase modulation with an amplitude ϕ_0 and a frequency f_m one can express (1) as a sum of Besselfunctions J_n. One of the summands, the so called first harmonic

$$S_1(t) \approx 2vS_0 J_1\left[2\phi_0 \sin(\pi f_m \Delta\tau)\right] \cdot \sin\left[2\pi f_m\left(t - \frac{\Delta\tau}{2}\right)\right] \sin\Delta\phi_R$$
(2)

is demodulated in a lock in amplifier (Fig. 1) whose output $S_1(t)$ is proportional to the rotation rate $\bar{\bar{\omega}}$. $\Delta\tau$ is the group delay time of light in the fibre coil, S_0 is the bias and υ is the visibility factor of the interference fringes. The fibre length of the coil L_c and the modulation frequency f_m are calculated to minimise the electric power consumption of the piezoelectric modulator while the Besselfunction J_1 remains maximal. Thus side effects of the modulator like phase distortion and amplitude modulation reach their minimum, when

$$f_m = \frac{1}{2\Delta\tau} = \frac{c}{2n_{co}L_c} . \tag{3}$$

c is the vacuum velocity of light and n_{co} the core's refractive index. For a given attenuation α in $\frac{dB}{km}$ the gyro's sensitivity reaches it's maximum [17] for

$$L_c = \frac{8.7}{\alpha} \tag{4}$$

and with (3) the optimal modulation frequency

$$f_m = \frac{c\alpha}{17.6 \cdot n_{co}} \tag{5}$$

The coil length in the presented gyroscope is 1 km, that is half of the optimum length and this leads to a loss of 17.5% of the maximum sensitivity at a modulation frequency (3) of 100.3 kHz. Nevertheless, the gyro attains a 1σ-precision of $8\frac{deg}{h\sqrt{Hz}}$.

High power LEDs, optimised injection conditions, a fibre with less than 4 dB attenuation as well as more carefully chosen standard fibre components will lead to further improvements.

The mode-coupling analyser [18] and the resulting knowledge of mode coupling properties of the gyro's optical components enables us to forecast the visibility factor and the gyro's performance. The difference between the measured visibility factor the forecasted results is small.

4 Summary

To summarise, the unique combination of a highly dispersive, multimode fibre with a broadband light source leads to a cheap gyro with a surprisingly good performance. Its cheapness is based on the simple optical design, without any need to monitor polarisation and to master backscattering like in SFOGs. Furthermore, the manufacturing tolerances of MFOGs are 10 times higher than the ones of SFOGs.

5 References

[1] Ezekiel, S., and Udd, E. eds.
Fibre Optic Gyros: 15th Anniversary Conference
SPIE Proceedings, Vol. 1585, 1991

[2] Liu, R. Y., El-Wailly T. F. and Dankwort, R. C.
Test Results of Honeywell's First Generation High Performance Interferometric Fibre-Optic Gyroscope
Spie Proceedings, Vol 1585, 1991, pages 262-275

[3] Pavlath, G. A.
Production of Fibre Gyros at Litton Guidance and Control Systems
Spie Proceedings, Vol 1585, 1991, pages 2-6

[4] Page, J. L.
Multiplexed Approach for the Fibre Optic Gyro Inertial Measurement Unit
Spie Proceedings, Vol. 1367, 1990, pages 93-102

[5] Sakuma, K.
Fibre Optic Gyro Production at JAE
Spie Proceedings, Vol. 1585, 1991, pages 8-16

[6] Hayakawa, Y., Kurokawa, A.
Fibre Optic Gyro Production at Mitsubishi Precision Co.
Spie Proceedings, Vol. 1585, 1991, pages 30-39

[7] Malvern, A. R.
Progress Towards Fibre Optic Gyro Production
Spie Proceedings, Vol. 1585, 1991, pages 48-64

[8] Böschelberger, H. J. and Kemmler M.
Closed Loop Fibre Optic Gyro Triad
Spie Proceedings, Vol. 1585, 1991, pages 89-97

[9] Lefèvre, H. C., Martin, P., Morisse, J., Simonpiétri P. Vivnot, P., and Arditty H. J.
Fibre Optic Gyro Productization at Photonetics
Spie Proceedings, Vol. 1585, 1991, pages 42-47

[10] Auch, W., Oswald, M., Regener, R.
Fibre Optic Gyro Production at Alcatel-SEL
Spie Proceedings, Vol. 1585, 1991, pages 65-79

[11] Blake, J., Feth, J., Cox. J., and Goettsche R.
Design and Test of a Production Open Loop All-Fibre Gyroscope
Spie Proceedings, Vol. 1169, 1989, pages 337-346

[12] Kajoka, H. T., Kumagai, H., Nakai, H., Tizuka, H., Nakamura, M., and Yamada, K.
Fibre Optic Gyro Productization at Hitachi
Spie Proceedings, Vol. 1585, 1991, pages 17-29

[13] Hartl, E., Trommer, G., Möller, R., Poisel, H.
Low Cost Passive Fibre Optic Gyroscope
Spie Proceedings, Vol. 1585, 1991, pages 405-416

[14] Weed, G., Sanders, G. A., Adams, W. and Johnson, A. P.
Fibre Optic Gyro Productization at Honeywell, Inc.
Fibre Optic Gyro 15th Anniversary Conference, Paper 1585-01, SPIE, 1991, oral presentation only

[15] Fibre Optics News
16 December 1991, pages 3-4

[16] Military and Commercial Fibre Business
10 January 1992, page 3

[17] Lefèvre, Hervè
The Fibre Optic Gyroscope
Artech House Inc., 1993, page 21, ISBN 0-89006-537-3

[18] Eckerle, M.
Mode coupling analysis in multimode optical waveguides and their application
VDI Verlag, Düsseldorf 1994, ISBN 3-18-331910-1

Measurement of Diameter and Velocity of Fibers Using a DualPDA

H. Mignon, G. Gréhan, F. Schöne[1], M. Stieglmeier[1], C. Tropea[2], T.-H. Xu[2]
CORIA, University Rouen/F, [1] Fa. Dantec/Invent, Erlangen/D,
[2] LSTM, University Erlangen-Nürnberg/D

ABSTRACT

Extension of the Phase Doppler technique to the measurement of cylindrical particles is explored. Theoretical predictions as well as experimental results are reported.

1. INTRODUCTION

The Phase Doppler Anemometry is a technique used for the measurement of spherical particles. However, in a lot of cases, particles are cylindrical (wires, fibers, threads, glass fibers). In the present paper, the application of PDA to the sizing of cylindrical particles is investigated. In the following section, some numerical simulations are presented, based on geometric optic, Mie theory and Generalized Lorenz-Mie theory. Experimental results follow in section 3, confirming the sensitivity of the particle orientation, as shown by simulation. Finally, some remarks about applications and future works are made in section 4.

2. NUMERICAL SIMULATIONS

- Computation with a single beam

The main difference between the scattering of a beam by a sphere or by a cylinder is that the sphere scatters the light in all directions, whereas the cylinder scatters the light in a cone surface, with a half-angle équal to the angle between the axis of the cylinder and the incident beam. This property is shown in figure 1, where the shape of the light cone scattered by a glass fiber (d= 2 μm, refractive index= 1.54 + 0.0i) for four incidence angles (90, 60, 30 and 2 degrees) of the laser beam are described. The lobe number, location and intensity are functions of the incident wave length, the cylinder diameter, the refractive index and the incidence angle. The results suggest that a planar Phase Doppler geometry is the most appropriate for sizing of cylindrical particles orientated parallel to the X-axis.

Figure 2 presents the intensity scattered by a cylindrical particle (d= 2 μm, refractive index= 1.333 + 0.0i), located at different positions along the Y-Axis and illuminated by a Gaussian beam (the wavelength is 0.6328 μm and the beam waist diameter is 10 μm) at perpendicular incidence.

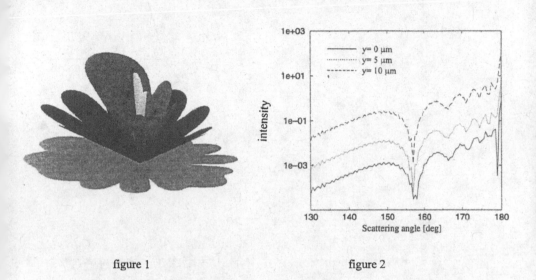

figure 1   figure 2

- Computation with 2 beams

Figure 3 displays the phase difference computed using geometric optic for an arbitrary orientated cylinder (d= 30 μm, refractive index 1.5+0.0i), for a planar PDA geometry defined by figure 4. Two detectors are located in the Y-Z plane as the two incident beams, with a scattering angle of 30 degrees. The orientation of the fiber is given by two angles (γ and τ). The results indicate that if the inclination angle γ is smaller than 2 degrees, the rotation angle τ has no influence. When the rotation angle is equal to 0, 90, 180, 270 degrees, the inclination angle has no influence.

Figure 5 displays the position of the scattered light from the two beams on a detector. As shown in figure 1, the scattered light forms a thin cone. In case the cylindrical particle is tilted, the two scattered light cones from the two incident beams, may have a separation on a detector. If this separation is large, there would be no interference, thus no Doppler signal. The same parameters as for figure 3 are used for this simulation.

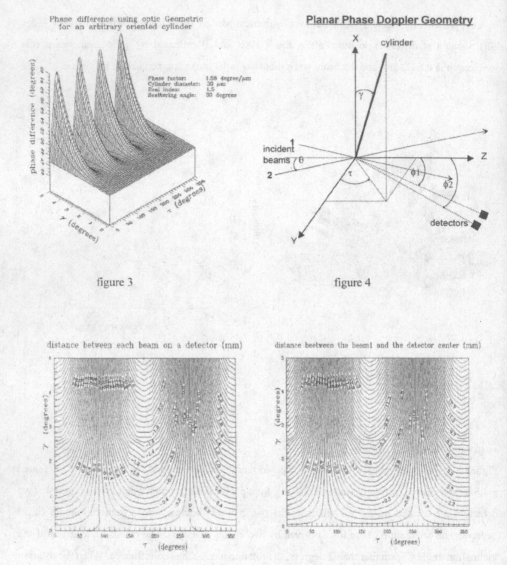

figure 3

figure 4

figure 5

figure 6

3. EXPERIMENTAL RESULTS

Figure 6 and 7 display experimental results using a DualPDA system. The parameters are the same as those described previously.

Figure 6 shows the measured diameter for different fiber positions on the Y-Axis, under the condition τ=0 and γ=0 degree.

In Figure 7, the influence of the fiber tilt on the measured diameter is investigated. In the particulary case of $\tau=0$ degree, the fiber diameter can be measured with a variation of 2 degrees for γ.

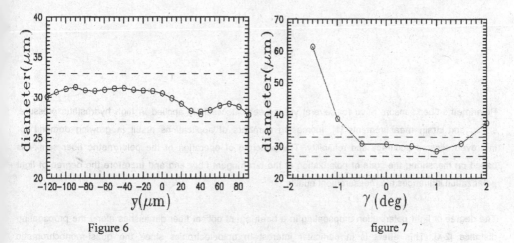

Figure 6   figure 7

4. CONCLUSION AND FUTURE WORK

Light scattering of one and two laser beams by a circular cylinder was simulated by means of different theories. Tilting of cylinder causes movement of the scattered light cone. The location of scattered light cones of two incident beams might be separated if the cylinder is tilted. Computed phase difference of a planar Phase Doppler Anemometer shows a narrow region for possible application of the PDA. Experimental results regarding diameter of a fiber and possible tilt angles are in good agreement with theoretical results. Further computational work will focus on reducing the restrictions concerning application of PDA for measurement of fibers.

RELATED LITERATURE:

E. Zimmermann, R. Dändliker, N.Souli, B. Kratinger: *Scattering of an off-axis Gaussian beam by a dielectric cylinder compared with a rigorous electromagnetic approach*; J.O.S.A. A, 12(2):398–403, 1995.

G. Gousbet, G. Gréhan: *Interaction between shaped beams and an infinite cylinder, including a discussion of Gaussian beams*; Part. Part. System Charac., 11: 299–308.

G. Gousbet, G. Gréhan: *On the interaction between a Gaussian beam and an infinite cylinder, using non sigma-separable potentials*; J.O.S.A. A, 11(12): 3261–3273, 1994.

Polarimetric Fiber Sensors for Partially Polarized Laser Source

Andrzej W. Domański, Mirosław A. Karpierz, Adam Kujawski, Tomasz R. Woliński

Institute of Physics, Warsaw University of Technology, Koszykowa 75, 00-662 Warszawa, Poland

Polarimetric fiber sensors have for several years been successfully applied in high hydrostatic pressure, stress and strain measurements [1]. Increasing numbers of applications result in growing demand for improved sensors accuracy and reliability. The principle of operation of the polarimetric fiber sensor is based on measuring the state of polarization at the birefringent fiber end and therefore the degree of light polarization influences the measurement quality.

The degree of light polarization propagating in a birefringent optical fiber diminishes along the propagating distance [2-4]. This effect is in particular interest in optoelectronics since the quasi-monochromatic semiconductor sources are not perfectly coherent. Therefore, measurements of the degree of polarization of light propagating along the fiber has been applied to determine the coherence characteristics of laser diodes [4]. In this paper the problem of polarization's degree fading in polarimetric fiber optic sensors is presented.

The degree of light polarization is defined by the ratio $P = I_{pol}/I_{tot}$, (where I_{pol} is a polarized light intensity and I_{tot} is a total intensity of light) and it can be expressed with the aid of the coherency matrix as follows:

$$P = \sqrt{1 - 4\det[\mathbf{J}]/\operatorname{tr}[\mathbf{J}]^2} \qquad (1)$$

where $\det[\mathbf{J}]$ is the determinant and $\operatorname{tr}[\mathbf{J}]$ is the trace of the coherency matrix $\mathbf{J}$ defined as an averaged product $J_{ij} = \langle u_i u_j \rangle$ (for $i,j=1,2$) where u_i is a linear eigenpolarization component of the electrical field. Averaging is strictly connected with the mutual coherence function γ and since the degree of polarization is calculated from the formula:

$$P = \sqrt{1 - 4\frac{1-\gamma^2}{(|u_1/u_2| + |u_2/u_1|)^2}} \qquad (2)$$

Note, that the degree of polarization and consequently the sensor sensitivity depends on the splitting power between two eigenpolarizations. For $|u_1|=|u_2|$ which can be realised by launching to the fiber the linearly polarized light at the angle 45^0 to the birefringent fiber principal axis or by launching the circurarly polarized

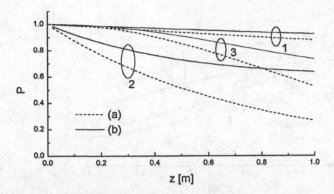

Fig.1. The degree of polarization P changes along the propagation distance z in the birefringent fiber with the beat length L_B=1mm for the lasers sources: (1) laser diode with the line width $\Delta\lambda$=0.1nm, (2) laser diode with $\Delta\lambda$=1nm, (3) gas laser with $\Delta\lambda$=1nm and for two different initial power splitting: (a) $|u_1|=|u_2|$ (then $P=\gamma$) and (b) $|u_1|=2|u_2|$.

light, the degree of polarization is equal to $P=\gamma$. On the other hand, for the limit $|u_j|=0$ (j=1 or j=2) the linear polarization with P=1 is obtained.

The mutual coherence function γ at the fiber end is connected with the light source spectral intensity and at the distance z is given by

$$\gamma = \exp\left[-\left(\frac{\Delta\nu\Delta N}{2\sqrt{2}c}z\right)^2\right] \tag{3}$$

for a gaussian spectrum whose profile has been observed in gas lasers, and

$$\gamma = \exp\left[-\frac{\Delta\nu\Delta N}{c}z\right] \tag{4}$$

for a Lorentzian spectrum whose line shape has been observed in semiconductor lasers. Here $\Delta\nu$ is a spectral width and ΔN is a effective indices difference of fiber polarization modes which is connected with a birefringence beat length L_B as follows: $\Delta N=\lambda/L_B$. The fig.1 presents the degree of polarization changes along the propagation distance for typical values of the fiber birefringence and the spectral width of laser diode sources. Note, that for unequal power splitting between polarization eigenmodes the degree of polarization is higher that for $|u_1|=|u_2|$ case.

In the analyzed sensors at the end of the sensing fiber it is placed the analyzer at the 45° to the eigenmode polarization axes. The dynamic η of the sensors depends on the maximal and minimal light intensities (I_{max} and I_{min} respectively) passing through the analyzer: $\eta=(I_{max}-I_{min})/(I_{max}+I_{min})$. For partially polarized light and perfect analyzer the sensitivity is equal to:

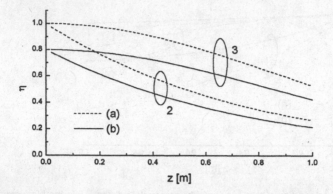

Fig.2. The dynamic of fiber sensors η changes vs the length z of the measuring birefringent fiber for parameters defined in fig.1.

$$\eta = \gamma \frac{2}{|u_1/u_2|+|u_2/u_1|} \qquad (5)$$

The dependence of the dynamic h on the measuring fiber length and the initial power splitting is presented in fig.2. Note, that the power splitting is influenced on the dynamics η significantly and therefore the problem of initial eigenmodes excitation is important for fiber polarization sensors [6].

The equation (5) can be rewritten in the terms of the degree of polarization in the form:

$$\eta = \sqrt{\left(\frac{2}{|u_1/u_2|+|u_2/u_1|}\right)^2 - 1 + P^2} \qquad (6)$$

The dependence of the dynamic η on the polarization degree for various power splitting (calculated from the eq.(6)) is presented in fig.3.

In conclusions it should be pointed that decreasing the dynamic of a polarimetric sensors cases also decreasing their sensitivity and accuracy due to the influence of noise. Therefore the degree of polarization and the light source spectrum is very important for proper operation of the polarimetric stress, strain and high pressure sensors.

This work was partially supported by the Polish Committee of Scientific Research under the grant no. 2P03B172 08.

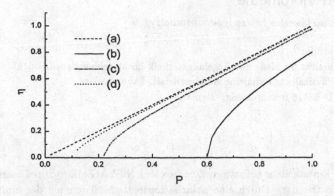

Fig.3. The dynamic of fiber sensors η changes vs the degree of polarization P for different power splittings: (a) $|u_1|=|u_2|$, (b) $|u_1|=0.5|u_2|$, (c) $|u_1|=0.8|u_2|$, and (d) $|u_1|=0.9|u_2|$

References

[1] A.W.Domański, T.R.Woliński, and W.J.Bock, Proc. SPIE 2341 (1994) 21-28.
[2] J.Sakai and T.Kimura, IEEE J.Quantum Electron. QE-18 (1982) 59-65; 488-495.
[3] W.K.Burns and R.P.Moeller, J.Lightwave Technol. LT-1 (1983) 44-49.
[4] A.W.Domański and D.Malinowski, Laser in Engineering, Proc. of the 10 Int.Congress Laser 91, ed. W.Waidelich (Springer Verlag 1991) 88-91.
[5] J.W.Goodman, Statistical Optics, (John Wiley & Sons 1985).
[6] A.W.Domański, T.R.Woliński, and W.J.Bock, IEEE Trans. on Instr. and Measur. 41 (1992) 1050-2

Neuartiger Zweifrequenz-Mikrochip-Laser für die Heterodyn-Vibrometrie

Two–frequency microchip laser for heterodyne vibrometry

J.W. Czarske und H. Müller*)
Universität Hannover, Institut für Meß- und Regelungstechnik im Maschinenbau, D–30167
Hannover, Germany, *)Physikalisch Technische Bundesanstalt, Laboratorium für
Strömungsmeßtechnik, D–38116 Braunschweig, Germany

Kurzfassung:
Der Beitrag stellt den Einsatz eines neuartigen Zweifrequenz Nd:YAG Mikrochip–Lasers in der
Laser–Doppler–Vibrometrie dar. Durch eine polarisationsoptische Trennung der emittierten
zwei Moden wird eine direkte Erzeugung der Heterodyn–Trägerfrequenz ermöglicht, so daß keine
zusätzlichen Frequenzshift–Elemente für die Realisierung des Heterodyn–Vibrometers benötigt
werden.

This contribution presents the application of a novel diode–pumped monolithic Nd:YAG
microchip laser in laser Doppler vibrometry. The heterodyne carrier frequency is generated by
the separation of the different polarized modes of the microchip laser. The realized compact
heterodyne vibrometer can be used for the measurement of out–of–plane vibration components.

1. Einleitung

Die Laser–Doppler–Vibrometrie ist ein bewährtes Verfahren zur berührungslosen Erfassung von
schwingenden Oberflächen, z.B. an Werkzeugmaschinen. Bisher werden für Vibrometer überwiegend HeNe–Laser eingesetzt, die u.a. hinsichtlich der geforderten Miniaturisierung als problematisch zu bewerten sind. Daneben wird bisher zur Realisierung des Heterodyn–Verfahren,
mit dem die notwendige Richtungserkennung der Objektbewegung vorgenommen wird, eine
justageaufwendige Bragg–Zelle verwendet. Eine erhebliche Vereinfachung des Vibrometers läßt
sich bekanntermaßen durch die Verwendung von Zweifrequenz–Lasern erreichen. Geeignete
Zweifrequenz–Laser mit hoher Differenzfrequenz und kompaktem Aufbau standen bisher jedoch nicht zur Verfügung. Diese knapp umrissene Situation hat sich durch das Aufkommen
von diodengepumpten Festkörperlasern grundlegend geändert: Es stehen zuverlässige miniaturisierte Laser mit hoher Leistung und sehr guter Frequenzstabilität zur Verfügung [HEINE-
MANN(a)]. In diesem Beitrag wird ein neuartiger Zweifrequenz–Laser, basierend auf einem
diodengepumpten Nd:YAG Mikrochip–Laser, vorgestellt und dessen Einsatz für ein kompaktes
Vibrometer–System beschrieben.

2. Nd:YAG Mikrochip–Laser

In Abb. 1 ist der schematische Aufbau des Nd:YAG Lasers dargestellt. Mit einem Laserdioden-
array von 250 mW wird über eine Kopplungsoptik der monolithische Nd:YAG Laser gepumpt.
Aufgrund des kurzen Resonators von 300 μm Länge ergibt sich ein Modenabstand, der größer als
die Nd:YAG Verstärkungsbandbreite ist, so daß nur ein Mode oszillieren kann [HEINEMANN
(a,b)]. Wird in den Nd:YAG Kristall eine Doppelbrechung induziert, z.B. aufgrund mechani-

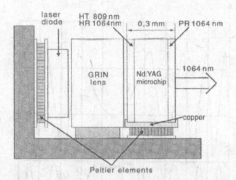

Abbildung 1: Schematische Anordnung des Zweifrequenz Nd:YAG Mikrochip–Lasers.

scher Spannung, dann kann dieser Mode in zwei unterschiedlich polarisierte Moden verschiedener Frequenz, aber gleicher Modenzahl aufspalten [SCHMITT, YOSHINO]. Bemerkung: Das genutzte Nd:YAG Kristall zeichnet sich im Vergleich zu anderen Lasermaterialen, z.B.Nd:YVO$_4$, daß einen höheren Pumplicht–Absorptionskoeffizienten aufweist, u.a. durch einen polarisationsunabhängigen Emission–Wirkungsquerschnitt aus [BOWKETT], so daß Nd:YAG Mikrochip–Laser zwei Moden mit gleicher Leistung emittieren können.

Es stand für unsere Experimente vom Hersteller, der Daimler Benz AG, ein Nd:YAG Mikrochip–Laser der Emissionswellenlänge 1064 nm zur Verfügung, der eine zweimodige Emission mit einer temperaturabstimmbaren Differenzfrequenz von 20 MHz bis zu 43 MHz und eine näherungsweise orthogonale zirkulare Polarisation aufwies [Czarske (a)]. Bei der eingestellten Nd:YAG Temperatur von 35^{0}C wurde eine stabile Moden–Differenzfrequenz von 22.8 MHz erreicht.

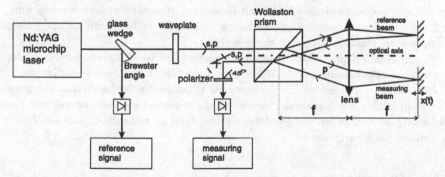

Abbildung 2: Schematische Anordnung des richtungsempfindlichen Heterodyn–Vibrometers

3. Heterodyn–Laser–Doppler–Vibrometer
3.1. Aufbau

Der in Abb. 2 dargestellte optische Aufbau des LDV–Systems basiert auf einem Michelson–Interferometer nach dem Heterodyn–Verfahren [CZARSKE (b)]: Die verschieden polarisierten Moden (s, p) Mikrochip–Lasers werden polarisationsoptisch getrennt, so daß direkt die notwen-

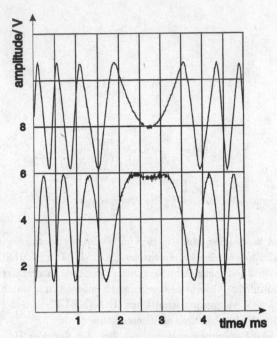

Abbildung 3: Quadratur–Signalpaar bei Bewegungsrichtungsumkehr des Meßobjektes (Quadratur: konstante +/-90⁰ Phasenverschiebung zwischen den Signalen)

dige Differenzfrequenz zwischen Meß– und Referenzinterferometerstrahlen erreicht wird. Das durch Zweistrahlinterferenz erzeugte Meßsignal enthält die richtungsempfindliche Bewegungsinformation des Meßobjektes (hier: Meßspiegel).

Die eingesetzte Signalverarbeitungseinheit besteht aus einem Quadratur–Korrelator und einer Phasenwinkelmessung [CZARSKE (b,c)]: Es resultiert ein Quadratur–Signalpaar, dessen Differenzphase von +/-90⁰ die Richtung der Objektbewegung bestimmt. Die gemessene Phase–Zeitfunktion des Signalpaars ermöglicht die zeitaufgelöste Bestimmung von Position, Geschwindigkeit und weiteren abgeleiteten Größen; dabei ist zu beachten, daß die Meßposition proportional zur Signalphase ist.

3.2. Meßergebnisse

In Abb. 3 ist ein Quadratur–Signalpaar bei einer Richtungsumkehr des Meßspiegels dargestellt: Vor dem Umkehrzeitpunkt bei ≃2.7 ms folgt das untere Signal dem oberen, nach dem Umkehrzeitpunkt eilt das untere Signal dem oberen voraus. Die Auswertung der Differenzphase des Signalpaars ermöglicht es daher eine Richtungserkennung durchzuführen.

In Abb. 4 ist für eine sinusförmige Anregung eines elektrodynamischen Schwingers, auf dem der Meßspiegel befestigt wurde, das erzeugte Signalpaar sowie die demodulierte Signalphase dargestellt.

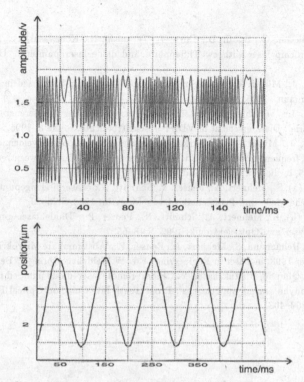

Abbildung 4: Sinusförmige Bewegung des Meßspiegels: (OBEN) Quadratur-Signalpaar, (UNTEN) Position-Zeitkurve des Meßspiegels

4. Diskussion

Es wurde in knapper Form die Verwendung eines Zweifrequenz Nd:YAG Mikrochip-Lasers für die Heterodyn-Vibrometrie dargelegt. Der Mikrochip-Laser emittiert in zwei Moden verschiedener Polarisation und mit einer Differenzfrequenz von 23 MHz. Der Laser hat einen kompakten Aufbau (Gehäuseabmessungen: 25mmx25mmx10mm), eine geringe Frequenzjitter-Linienbreite und eine für die Vibrometrie ausreichende Leistung von ca.25 mW bei 1064 nm Wellenlänge. Daher ist dieser neuartige Zweifrequenz-Laser prädestiniert für den Einsatz in der Laser-Doppler-Vibrometrie.

Das realisierte Vibrometer ermöglicht die präzise Vermessung der out-of-plane Schwingungskomponenten. Dabei wird mit der Quadratur-Korrelationstechnik (entspricht einem Lock-in-Verstärker) der Einfluß von Moden-Differenzfrequenzfluktuationen des Lasers eliminiert. In den ersten Verfikationsexperimenten wurde eine Phasenwinkeldigitalisierung mit einer Auflösung von 8 Werten pro Periode erreicht, die i.allg. für die Vibrationsmessung ausreichend ist.

Danksagung: Die Autoren danken Herrn N.P. Schmitt (Daimler Benz, Technologieforschung, München) für die Bereitstellung des Nd:YAG Mikrochip-Lasers. Für hilfreiche Diskussionen sei Herrn Prof.em. F. Hock gedankt.

Literatur:

BOWKETT, G., Baxter, G., Booth, D., Taira, T., Teranishi, H., Kobayashi, T.: 'Single–mode 1.34-μm Nd:YVO$_4$ microchip laser with cw Ti:Sapphire and diode–laser pumping', Optics Lett., 1994, S.957–959.

CZARSKE (a), J., Müller, H.: 'Birefringent Nd:YAG microchip laser used in heterodyne vibrometry', Optics Commun., 1995, S.223–229.

CZARSKE (b), J.: 'Verfahren zur Messung und Auswertung der Interferenzphase in der Laser–Doppler–Velocimetrie', Diss., erscheint in VDI–Fortschrittberichte, Reihe 8, 1995.

CZARSKE (c), J., Müller, H.: 'Mehrkomponentiges Laser–Doppler–Velocimeter mit erstmaliger Nutzung eines frequenzmodulierten Nd:YAG Lasers und Glasfaserumwegstrecken', Proc. 12th Int.Conf.Laser 1995, in diesem Tagungsband.

HEINEMANN (a), S., Mehnert, A., Peuser, P., Schmitt, N.: 'Laserdiodengepumpte Mikrokristall–Laser für die optische Meß– und Prüftechnik', Laser Magazin, 1992, S.26–29.

HEINEMANN (b), S., Mehnert, A., Schmitt, N., Peuser, P.: 'Diodenlaser–gepumpte Miniatur–Festkörperlaser', Laser und Optoelektronik, 1992, S.48–55.

SCHMITT, N., Heinemann, S., Mehnert, A., Peuser, P.: 'Abstimmbare Mikrokristall–Laser', Proc. 11th Congress Laser 1993, in Laser in Engineering, Ed. W.Waidelich, Springer, Berlin, 1994, S.8–12.

YOSHINO, T., Qlmude, B., Takahashi, Y.: 'Fiber–remote laser–diode pumped microchip Nd:YAG laser and its heterodyne sensing applications', Proc. Tenth Int. Conf. on Optical Fibre Sensors SPIE Vol.2360, 1994, S.400–403.

Zweikomponenten-LDA auf Basis von ND:YAG-Ringlasern mit differierenden Emissionsfrequenzen

R. Kramer, H. Müller und D. Dopheide

Physikalisch-Technische Bundesanstalt

Laboratorium für Strömungsmeßtechnik

Bundesallee 100

D-38116 Braunschweig

Einleitung

Ein Laser-Doppler-Anemometer gestattet es, im Schnittbereich sich kreuzender kohärenter Beleuchtungsstrahlen (Kreuzstrahlanordnung) eine Geschwindigkeitskomponente von lichtstreuenden Teilchen zu messen. Passieren die Teilchen das Meßvolumen, erzeugen sie dopplerverschobene Streuwellen, die mit Hilfe der Empfangsoptik auf einem Photoempfänger überlagert werden. Die Signalfrequenz f_S der detektierten Signalbursts ist als Schwebungssignal mit der Dopplerdifferenzfrequenz f_D proportional zum Betrag der Teilchengeschwindigkeit. Eine vorzeichenbehaftete Messung der Dopplerdifferenzfrequenz ist möglich, werden im Meßvolumen Laserstrahlen mit einer optischen Differenzfrequenz zum Schnitt gebracht. Die Differenzfrequenz ist als eine Trägerfrequenz zu betrachten und wird bei bekannten LDA-Systemen meist mit Hilfe von akustooptischen Modulatoren durch eine Frequenzverschiebung (Frequenzshift) der Laserstrahlen realisiert. Die Signalfrequenz f_S der detektierten Streulichtbursts folgt dann aus der Summe der vorzeichenbehafteten Dopplerdifferenzfrequenz f_D und der Trägerfrequenz f_{SH}. Bei mehrkomponentigen LDA können durch Generierung von Strahlenpaaren mit unterschiedlichen optischen Frequenzverschiebungen, die erzeugten Trägerfrequenzen sowohl für die Vorzeichenbestimmung als auch für die Signalkomponententrennung genutzt werden /1/. Alle bisher eingesetzten LDA-Anordnungen basieren jedoch auf einer Strahlaufspaltung eines Laserstrahls und damit einer Verteilung der Laserleistung auf mehrere Strahlen. Durch den Einsatz jeweils eines leistungsstarken, abstimmbaren Lasers für die Beleuchtungsstrahlen läßt sich hingegen

- eine Verdreifachung der Lichtleistung im Meßvolumen und
- eine beliebige Einstellung der Trägerfrequenzen der Kreuzstrahlsysteme

erreichen.

Nd:YAG-Ringlaser

Für das Zweikomponenten-LDA eignen sich diodengepumpte Nd:YAG-Ringlaser /2/, die kleine Abmessungen, Ausgangsleistungen bis über 1W bei annähernd gaußförmigem Strahlprofil, Linienbreiten von 10 kHz sowie eine geringe Empfindlichkeit gegen Rückreflexe besitzen. Durch Änderung der Kristalltemperatur ist eine Durchstimmbarkeit der Emissionsfrequenzen innerhalb der etwa 10 GHz breiten, regelmäßigen, gut reproduzierbare Plateau's in den Modenkarten möglich, so daß ein Betrieb der Laser mit optoelektronisch detektierbaren Differenzfrequenzen durch Wahl geeigneter Betriebstemperaturen erreicht werden kann.

Aufbau des Zweikomponenten-LDA

Bei dem vorgestellten LDA werden die drei Ausgangsstrahlen der Nd:YAG-Ringlaser nach einer Strahlaufweitung durch die Frontlinse im Meßvolumen fokussiert und zum Schnitt gebracht (Bild 1).

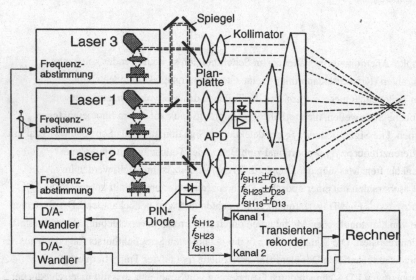

Bild 1 Zweikomponenten-LDA mit Frequenzshift auf Basis von drei Nd:YAG-Ringlasern

Entsprechend den Differenzfrequenzen der drei sich im Meßvolumen kreuzenden Meßstrahlenpaare ergeben sich aus den drei Emissionsfrequenzen f_1, f_2 und f_3 der Laser die Trägerfrequenzen zu:

$$f_{SH12} = f_2 - f_1 \qquad f_{SH23} = f_2 - f_3 \qquad f_{SH13} = f_1 - f_3 \tag{1}$$

Mit Hilfe der Empfangsoptik werden die bei einem Partikeldurchgang durch das Meßvolumen gestreuten, dopplerverschobenen Wellen auf einer Avalanchediode überlagert und als Schwebungssignal detektiert.

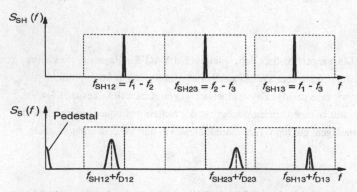

Bild 2 Referenzsignalspektrum S_{SH} und Burstspektrum S_S des Zweikomponenten-LDA

Die Trennung bzw. Zuordnung der Signalanteile zu den einzelnen Kreuzstrahlsystemen erfolgt trägerfrequent, in dem, wie in Bild 2 gezeigt, für jede Komponente unter Berücksichtigung der maximal zu erwartenden Dopplerdifferenzfrequenzen angepaßte Trägerfrequenzen und damit Signalfrequenzbereiche festgelegt werden, die sich nicht überschneiden. Mit dem LDA können drei Geschwindigkeitskomponenten koinzident gemessen werden von denen jedoch nur zwei Komponenten linear unabhängig sind. Die dritte Geschwindigkeitskomponente kann für eine Signalvalidierung genutzt werden.

Um einen Einfluß von Emissionsfrequenzfluktuationen der Laser auf die Meßgenauigkeit der Dopplerdifferenzfrequenzen zu vermeiden, ist es notwendig, die Trägerfrequenzen der Kreuzstrahlsysteme mit Hilfe von Referenzsignalen zu detektieren und mit Regelkreisen zu stabilisieren.

Zur Erzeugung der Referenzsignale wird mit Hilfe von Planplatten ein Teil der Laserstrahlen ausgekoppelt und auf PIN-Dioden überlagert. Durch paarweise Überlagerung der Teilstrahlen können für die einzelnen Trägerfrequenzen separate Referenzsignale gebildet werden, die für die Realisierung von schnellen Hardwareregelkreisen zur Trägerfrequenz-stabilisierung geeignet sind. Werden hingegen, wie in Bild 1 gezeigt, die Teilstrahlen aller Laser auf einem Referenzempfänger überlagert kann mit einem Transientenrekorderkanal das Referenzsignal zeitgleich zu Signalbursts detektiert und die drei Trägerfrequenzen aus dem Spektrum bestimmt werden. Die Regelabweichungen der Trägerfrequenzen werden mit einem Softwareregelkreis zur Ansteuerung

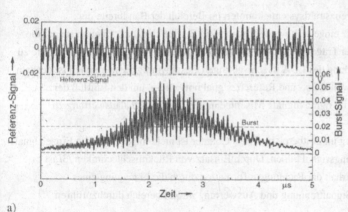

a)

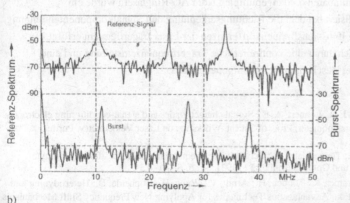

b)

Bild 3 Meßsignale des Zweikomponenten-LDA im Zeit- (a) und im Frequenzbereich (b)

von D/A-Wandlern zur Beeinflussung der Laserkristalltemperaturen von 2 Lasern genutzt. Die zeitgleiche Erfassung von Burst und Referenzsignal bietet den Vorteil, daß sich Regelabweichungen der Trägerfrequenzen nicht auf Meßgenauigkeit die Dopplerdifferenzfrequenzen auswirken.

Ergebnisse

Das vorgestellte LDA-System wurde in einer Luftströmung erprobt, der Wassertröpfchen in einem Durchmesserbereich von ca. 1- 5 μm zugesetzt wurden. Typische Signale zeigt Bild 3a, die bei einem Partikeldurchgang mit Hilfe eines zweikanaligen Transientenrekorders im Zeitbereich aufgenommenen wurden. Die aus dem Referenz- bzw. Burst-Signal mit Hilfe eines FFT-Algorithmus berechneten Spektren sind in Bild 3b dargestellt. Es sind sowohl im Referenz- als auch im Signalspektrum drei Signalanteile detektierbar, deren Mittenfrequenzen durch Interpolation der Frequenzanteile im Bereich der Peaks bestimmt werden. Die Dopplerdifferenzfrequenzen der einzelnen Geschwindigkeitskomponenten folgen aus den Mittenfrequenzen der Signalpeaks und den zugehörigen Referenzpeaks.

Die Trägerfrequenzen der Kreuzstrahlsysteme konnten im Bereich der Bandbreite der Photoempfänger bis 800 MHz eingestellt werden, wobei ein Einfluß der Laserlinienbreite auf die Breite der Signalpeaks nur bei Trägerfrequenzen kleiner oder vergleichbar der Laserlinienbreite zu beobachten war. Bei Anwendung des LDA in sehr geringen Geschwindigkeitsbereichen ist deshalb eine gleichzeitige Erfassung von Burst- und Referenzsignal notwendig, um den Einfluß der Laserlinienbreite beispielsweise durch digitale Mischtechniken bei der Signalauswertung zu berücksichtigen.

Eine direkte Signalauswertung im Zeitbereich mittels Counter ist nicht möglich, da die Einzelsignale ein unregelmäßiges Schwebungssignal bilden. Durch Einsatz von Rückmischtechniken /3/ ist es möglich bei gleichzeitigem Erhalt der Richtungsinformation durch die Erzeugung eines Quadratursignalpaares eine Signaltrennung und Auswertung im Zeitbereich durchzuführen.

Zusammenfassung

Durch Einsatz von drei abstimmbaren diodengepumpten Nd:YAG-Ringlasern wurde ein Zweikomponenten-LDA realisiert bei dem die Richtungserkennung und Signalkomponententrennung durch Nutzung der optischen Emissionsfrequenzdifferenzen der Laser trägerfrequent erfolgt. Das LDA-System zeichnet sich durch eine kontinuierliche Trägerfrequenzeinstellung und eine einkanalige Erfassung der drei Geschwindigkeitskomponenten aus.

Literatur
/1/ Crosswy, F. L.; Hornkohl, J. O.; Lennert, A. E.: Signal characteristics and signal-conditioning electronics for a vector velocity laser anemometer. Proc. of 1st Int. Workshop on Laser Velocimetry. Purdue University, 1972, 396
/2/ Freitag, I.; Rottengatter, P.; Tünnermann, A.; Schmidt, H.: Frequenzabstimmbare, diodengepumpte Miniatur-Ringlaser. Laser und Optoelektronik **25** (1993), Nr. 5, S. 70 - 75
/3/ Müller, H.; Czarske, J.; Kramer, R.; Többen, H.; Arndt, V.; Wang, H.; Dopheide, D.: Heterodyning and Quadrature Signal Generation: Advantageous Techniques for Applying New Frequency Shift Mechanisms in Laser Doppler Velocimetry. Seventh international symposium on applications of laser techniques to fluid mechanics, Lisbon, Portugal, July 11-14, 1994, Conference proceedings Volume II, 23.3.1-23.3.8

Ausgewählte Beispiele für den Einsatz der optischen Meßtechnik in der Industrie

J. Kieckhäfer, H. Schulz, H. Wiesel
Optikzentrum NRW, Abt. Ind. und Wiss. Präzisionsmeßtechnik / Hochleistungsoptik
Universitätsstr. 142, D-44799 Bochum, Deutschland

1. Vollständigkeitskontrolle von Bauteilen in der Automobilindustrie

In der Automobilherstellung kommt es immer wieder vor, daß unvollständige oder Fehler behaftete Bauteile eingebaut werden. Dies führt zu Qualitätsmängeln und hohen Produktionskosten.

Vorgestellt wird hier beispielhaft ein Meßsystem, welches das Vorhandensein von Stoßdämpferhalterungen optisch überprüft. Dieses Meßsystem ist in einem Handlingroboter integriert und sortiert automatisch unvollständige Bauteile in einem frühen Stadium der Produktionskette aus.

Durch frühzeitige Qualitätsüberwachung werden unnötige Weiterverarbeitungsschritte von fehlerhaften Teilen vermieden.

Das Meßsystem liefert seit einem Jahr eine 100% Qualitätskontrolle und hat sich durch Verringerung der Produktionskosten innerhalb eines Monats amortisiert.

2. Qualitätskontrolle von großflächigen Verbundglasscheiben

Getrübte Aussichten haben Autofahrer und Brillenträger nicht nur, wenn Scheiben und Brillengläser verschmiert oder beschlagen sind. Auch kleinste Mängel, wie Luftbläschen, Einschlüsse und Materialfehler, können die Sicht stark beeinflussen. Um solche Produktionsfehler zu vermeiden, werden die Gläser vor dem Verkauf nochmals von einem Kontrolleur in Augenschein genommen.

Die Mitarbeiter des Optikzentrums haben speziell für dieses Meßproblem ein Beleuchtungsverfahren entwickelt.

Mittels einer optimierten Lichtquelle und einer angepaßten Optik werden die Materialfehler wie Einschlüsse, Blasen und Kratzer vergrößert und mit starkem Kontrast sichtbar gemacht. Fingerabdrücke oder Staubpartikel haben keinen Einfluß auf die Fehlererkennung.

Die Qualität der Scheiben wird hierdurch gewährleistet, kostenintensive Reklamationen werden vermieden. Automobil- und Brillenglashersteller haben bereits ihr Intesse an der zum Patent eingereichten Scheibeninspektion bekundet.

3. Optisches Meßsystem zur Führung eines Bühnenwagens im Theater

An vielen Theatern Deutschlands, auf denen Bühnenwagen unterschiedlicher Größe verfahren werden, taucht immer wieder das Problem auf, daß diese Wagen beim Verfahren verkanten und festfahren. Besonders unangenehm wirkt sich dies während einer

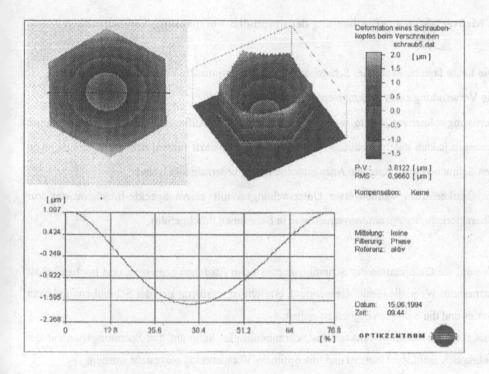

Theatervorstellung aus. Es kommt zu Verzögerungen des Ablaufs, im Extremfall können bei zu schneller Fahrt Bühnenaufbauten in Mitleidenschaft gezogen werden.

Im Optikzentrum NRW wurde ein Meßsystem entwickelt, welches den Abstand des zu verfahrenden Bühnenwagens zu den seitlichen Bühnensegmenten mißt und eine Regelung der Bühnenwagen-Antriebsmotoren erlaubt.

Für eine genaue Positionierung des Bühnenwagens längs des Fahrweges bietet das Optikzentrum ein Entfernungsmeßgerät mit einer Genauigkeit von ± 1 mm auf eine Entfernung bis zu 100 m an. Beide Systeme zusammen erlauben einen reibungslosen Vorstellungsablauf mit beliebiger Positionierung des Bühnenwagens.

4. Meßverfahren zur Bestimmung der optimalen Vorspannung von Schrauben

Die heute übliche Methode, Schrauben mit einer bestimmten Vorspannung anzuziehen, ist die Verwendung eines Drehmomentschlüssels.

Fertigungstoleranzen, Grate und Ölrückstände an der Auflagefläche der Schraubenköpfe erzeugen jedoch unterschiedliche Reibungseffekte, die dazu führen, daß die Spannungen in den Schrauben trotz gleichem Anzugsmoment sehr unterschiedlich sind.

Im Optikzentrum wurden erste Untersuchungen mit einem Speckle-Interferometer zur Charakteristik der Spannungsverhältnisse in Schrauben durchgeführt.

Es wird die Deformation des Schraubenkopfes beim Anziehen gemessen, und hochaufgelöst dargestellt. Wenn die ideale Vorspannung erreicht ist, verformt sich der Schraubenkopf nicht weiter und die Schraube beginnt zu reißen.

Aus dem Verformungsverhalten des Schraubenkopfes kann auf den Spannungszustand der Schraube geschlossen werden und die optimale Vorspannung eingestellt werden.

Sensor zur berührungslosen optischen Profilmessung von Innengewinden bis < M1

H. Kreitlow, V. Prott[*]
Fachhochschule Ostfriesland
Fachbereich Naturwissenschaftliche Technik
Constantiaplatz 4, D-26723 Emden

1   Einleitung

Schraubverbindungen sollten bei automatisierten Einschraubverfahren eine Paßgenauigkeit von nahezu 100 % aufweisen [1]. Dies bedingt sehr hohe Anforderungen an die Qualitätsprüfungen von Gewinden vor dem Einschraubvorgang. Eine übliche Praxis ist es, stichprobenartige Prüfungen mittels Handlehren durchzuführen [2]. Dieses Verfahren der mechanischen Gewindeprüfung ist zeitaufwendig und wegen der kurzen Taktzeiten bei einem automatisierten Herstellungsverfahren problematisch [1]. Einige Gewindefehler lassen sich darüberhinaus mit Handlehren nicht detektieren, wie z. B. nach Bohrerabbrüchen entstandene Profilformen oder abgebrochene Gewindespitzen.

Demgegenüber stehen die optoelektronischen Meßverfahren, die Vorteile wie die berührungslose und zerstörungsfreie Abtastung mit dem Vorteil einer sehr kurzen Prüfzeit vereinigen. Die Prüfzeit ist im wesentlichen abhängig von der Gewindelänge und beträgt z.B. bei einem 30 mm langen Gewinde etwa 1,5 s [1].

Faseroptische Sensoren (FOS) sind für optoelektronische Meßverfahren hervorragend geeignet, da ihre Handhabung i.a. unproblematisch ist und die Bauform des Sensors sehr klein gehalten werden kann. Bei FOS werden am Sensor selbst keine Leitungen zur Strom- und Spannungsversorgung benötigt, da bei dieser Art von Sensoren Licht die Rolle der Energieversorgung des Sensors und die Signalübertragung übernimmt. Die Vorteile von FOS sind nach [3] u.a. die Einsatzmöglichkeit an schwer zugänglichen Meßpunkten und das berührungslose und rückwirkungsfreie Messen. Außerdem sind sie unempfindlich gegenüber allen Arten von elektromagnetischen Störfeldern und weitestgehend wartungsfrei.

Für die optoelektronische Gewindeprüfung ist bisher ein Verfahren entwickelt worden, welches auf der Basis eines faseroptischen Sensors nach dem Rückstreuprinzip arbeitet [4]: Ein Lichtwellenleiterbündel, bestehend aus einer Monomodefaser als Sender und zwei umliegenden Multimodefasern als Empfänger, wird in das zu prüfende Innengewinde eingefahren. Damit das aus der Faser austretende Lichtbündel auf das Gewindeprofil trifft, wurde das Faserbündel um 90° mit einem kleinen Radius gekrümmt und im Sensorkopf fixiert. Nach diesem Prinzip sind allerdings nur Prüfungen von Gewinden möglich, die eine Größe von M12 nicht unterschreiten [5].

Im vorliegenden Bericht wird ein Sensor vorgestellt, der es ermöglicht, Gewinde bis herunter zu einer Größe von M1 zu vermessen. Dies wird durch eine Verkleinerung des Sensorkopfes unter Verwendung eines einzigen, nicht gekrümmten Lichtwellenleiters und eine Veränderung des Meßdatenerfassungsprinzips erreicht.

---

[*] jetzt bei Fa. LASOR Laser Sorter GmbH, Rudolf-Diesel-Str. 24, D-33813 Oerlinghausen

2 Prinzip des Sensors für Innengewinde bis M1

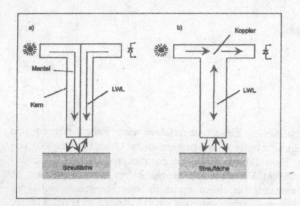

Bild 1: Abstandmessung mittels Rückstreuung durch
(a) zwei getrennte LWL und durch
(b) einen gemeinsamen LWL als Sender und
Empfänger, nach [6]

Der Sensor arbeitet ebenfalls nach dem Prinzip des Rückstreuverfahrens. Bei der Abstandsmessung mittels Rückstreuung unter Verwendung von LWL kann nach folgenden zwei Prinzipien gearbeitet werden [6]:

Mit einem LWL wird das Licht zur messenden Oberfläche gesendet und mit einem zweiten LWL das reflektierte Licht empfangen und einem Detektor zugeleitet (Bild 1(a)).

Statt getrennter LWL für Sende- und Empfangsfaser wird nur ein LWL und ein nachgeschalteter Koppler verwandt (Bild 1 (b)).

Als Lichtquelle des Sensors dient eine Laserdiode, deren abgestrahltes Licht über die Sendefaser auf die zu messende Oberfläche geleitet wird. Das während des Abtastvorganges vom Gewinde rückgestreute Licht wird mit Hilfe derselben Faser auf eine Photodiode geleitet, anschließend verstärkt und ausgewertet. Durch Aufnahme einer Abstandskennlinie, die den Zusammenhang zwischen dem Abstand der Lichtaus- bzw. Eintrittsfläche des Sensors gegenüber der rückstreuenden Oberfläche und dem Meßsignal darstellt, kann der Sensor kalibriert werden und liefert bei Wiederholung der Messung reproduzierbare Abtastergebnisse.

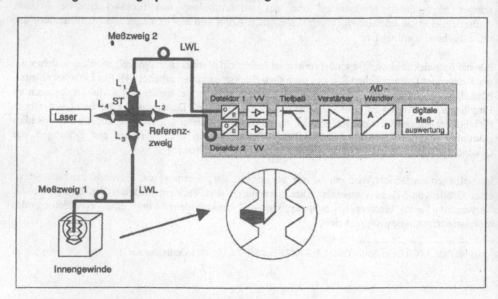

Bild 2: Prinzipieller Meßaufbau des entwickelten Sensors

Der prinzipielle Aufbau des entwickelten Sensors läßt sich aus Bild 2 entnehmen. Durch die Aufteilung des Lichtes einer Halbleiterdiode in Meßzweig 1 und Referenzzweig mit Hilfe eines Strahlteilers ST kann das Meßsignal normiert werden, so daß eventuelle Schwankungen der Lichtquelle ausgeglichen werden: Der Referenzzweig wird direkt auf einen Detektor 2 geführt, während das Licht aus dem Meßzweig 1 zur Abtastung des zu vermessenden Innengewindes verwendet wird. Das am Gewinde reflektierte bzw. gestreute Lichtbündel gelangt dann über den Strahlteiler in den Meßzweig 2, von wo aus die erhaltene Meßinformation, die von der Entfernung vom Sensor zur reflektierenden Gewindeoberfläche abhängt, auf einen weiteren angeschlossenen Detektor 1 geleitet wird. Das normierte Signal wird verstärkt und nach einer A/D-Wandlung einem Rechner zur Auswertung zugeführt.

2.1 Signalformung durch ein Gewindeprofil

Trifft ein divergentes Lichtbündel auf ein isometrisches Gewindeprofil, so hängt es vom Öffnungswinkel der Faser, der Entfernung der Lichtaustrittsfläche zum Gewinde und der Gewindegröße ab, ob die beleuchtete Fläche sich über einen oder mehrere Gewindegänge erstreckt.

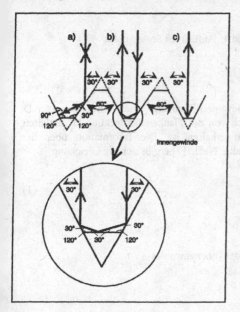

Betrachtet man axial aus der Faser austretende Lichtbündel, so lassen sich drei Fälle unterscheiden:

(a) Das Lichtbündel trifft im oberen Bereich der Flanke auf das Gewinde und wird von der gegenüberliegenden Flanke in die Faser reflektiert.

(b) Das Lichtbündel trifft im unteren Teil der Flanke auf und wird parallel versetzt reflektiert.

(c) Das Lichtbündel trifft senkrecht auf den Gewindegrund auf und wird direkt auf gleichem Wege reflektiert.

Wegen der besonderen Form eines isometrischen Gewindes (60°-Flankenwinkel) ergeben sich keine Probleme bei der Erfassung des Gewindeprofils. Es ist allerdings für Fall (b) darauf zu achten, daß der Faserdurchmesser im Verhältnis zum Abstand zweier Gewindespitzen nicht zu klein ist, da sonst in diesem Abtastbereich keine Lichtsignale in die Faser zurücklaufen, und deshalb entstehende Meßsignale nicht ausgewertet werden dürfen.

Bild 3: Ablenkung von Lichtstrahlen, die auf ein isometrisches Gewinde treffen

3 Aufbau des Sensorkopfes

Ein Problem bei der Konfiguration eines Sensorkopfes für einen Innengewindesensor zur Vermessung sehr kleiner Innengewinde ist die Umlenkung des aus der Faser austretenden Lichtes in Richtung des Gewindeprofils. Um das Licht von der Faser auf das Gewindeprofil zu leiten, wurde der Faserkopf so bearbeitet, daß sich durch die Faserform eine Lichtumlenkung ergibt.

704

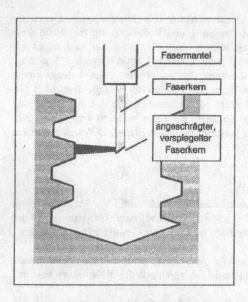

Fasermantel

Faserkern

angeschrägter,
verspiegelter
Faserkern

Dazu wurde die Faser abisoliert und auf 45° angeschrägt. Eine anschließende Bedampfung der Faserendfläche mit Aluminium ermöglicht eine 90°-Lichtumlenkung in Richtung des Gewindeprofils. Die Reflexion des Lichts am Gewinde ist ausreichend, um ein Meßsignal auf dem gleichen Wege zurück in die Faser einzukoppeln und zur Auswertung zum Detektor zu führen.

In Bild 4 ist der prinzipielle Aufbau des Sensorkopfes dargestellt.

Bild 4: Aufbau des Sensorkopfes

4 Gewindeparameter

Die für eine Qualitätskontrolle von Innengewinden relevanten Parameter, wie Kerndurchmesser D_1 und Flankendurchmesser D_2 , lassen sich mathematisch von der Flankenüberdeckung H_1 herleiten, die direkt aus dem abgetasteten Gewindeprofil zu erhalten ist. Die Information über den Durchmesser des Sensorkopfes ist hierbei nicht notwendig. Nach [7] ergibt sich die Gleichung

$$H_1 = \frac{D - D_1}{2} \tag{1}$$

mit

H_1 :=	Flankenüberdeckung
D :=	Außendurchmesser (Nenndurchmesser) des Innengewindes
D_1 :=	Kerndurchmesser des Innengewindes

Nach D_1 aufgelöst ist

$$D_1 = D - 2H_1 \tag{2}$$

Somit läßt sich der Kerndurchmesser D_1 direkt aus der Flankenüberdeckung H_1 bestimmen. Mit den Gleichungen aus [8]

$$D_2 = D - 0,64952 * P \tag{3}$$

und

$$H_1 = 0,54127 * P \tag{4}$$

folgt dann

$$D_2 = D - 1,1999926 * H_1 \tag{5}$$

Damit erhält man auch den Flankendurchmesser D_2 in Abhängigkeit von der Flankenüberdeckung H_1. Mit einer geeigneten Auswertesoftware können nun die gewonnenen Werte mit den Kleinst- und Größtmaßen aus DIN 13, Teil 13 verglichen werden, so daß zwischen Gut und Ausschuß bzw. Nachbearbeitung in einem automatisierten Fertigungsprozeß entschieden werden kann.

5 Meßergebnisse

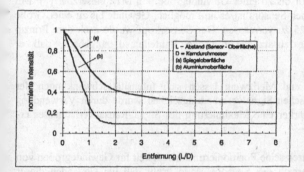

Bild 5: Charakteristische Sensorkennlinien

Die Sensorkennlinie in Bild 5 stellt den Zusammenhang zwischen dem Abstand des Sensors zur reflektierenden Oberfläche und der detektierten Intensität dar.

Der Graph enthält einen linearen Bereich bei L/D zwischen 0 und 1. Die Meßempfindlichkeit des Sensors für ein zu vermessendes Gewinde kann durch die Wahl des Faserdurchmessers optimiert werden.

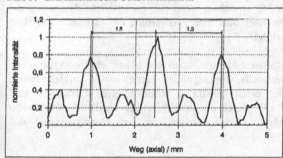

Bild 6: Profilvermessung an einem Innengewinde der
Größe M10 mit einem Kunststoffasersensor
(Durchmesser der Faser: 1 mm)

Nach dem Aufnehmen der charakteristischen Kennlinie, kann für die Innengewindeprofilvermessung ein Weg-Längen Diagramm erstellt werden. Bild 6 zeigt ein aufgenommenes Gewindeprofil der Größe M10 und der Steigung P = 1,5. Deutlich sind die angesprochenen lokalen Minima zwischen Gewindesspitze und Gewindegrund zu sehen (s. 2.1(b)).

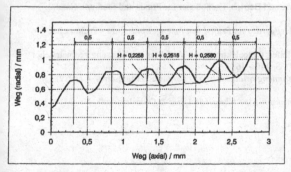

Bild 7: Profilerfassung eines Innengewindes der Größe M3
mit der Steigung P = 0,5

In Bild 7 ist das Profil eines Innengewindes der Größe M3 mit der Steigung P = 0,5 zu sehen, in das die gemessenen Werte der Flankenüberdeckung H_1 und die Steigung P eingetragen ist. Mit diesen Werten können relevante Parameter wie Kerndurchmesser D_1 und Flankendurchmesser D_2 anhand der Gleichungen 2 und 5 bestimmt werden. Das vermessene Gewinde wurde nach DIN 13 geprüft. Die relevanten Parameter liegen hier alle innerhalb der Toleranzgrenze.

6 Zusammenfassung

Mechanische Prüfverfahren zur Vermessung von Innengewinden mit Lehrdornen lassen aufgrund des relativ langsamen Prüfprozesses nur niedrige Taktzeiten zu und erfordern bei der Vermessung unterschiedlicher Gewinde einen Lehrdornwechsel. Ein optoelektronischer Sensor ermöglicht dagegen eine schnelle Prüfung und ist ohne Sensorkopfwechsel für alle Gewindegrößen verwendbar.

Bisherige optische Prüfverfahren sind nur bis zu einer Gewindegröße von M12 einsetzbar [5]. Mit der Weiterentwicklung eines faseroptischen Sensors ist es nun möglich, Gewinde bis zu einer Größe von M1 mit einer Steigung von P = 0,25 zu vermessen. Der Sensorkopf wurde nach dem Prinzip eines Reflektors konzipiert und ohne zusätzliche optische Bauteile so aufgebaut, daß er kostengünstig gefertigt werden kann.

Durch die Anordnung von Referenz- und Meßzweig lassen sich Intensitätsschwankungen der Lichtquelle eliminieren. In der elektronischen Signalverarbeitungseinheit wurde durch Verwendung einer Photodiode mit integriertem Verstärker eine Maximierung des Signal/Rauschverhältnisses erreicht.

Insbesondere zeichnet sich der Sensor durch seine Positionierunabhängigkeit für Gewindegrößen von M3 bis M1 aus. Der lineare Arbeitsbereich des Sensors ist so groß, daß bei Gewinden dieser Größenordnung keine Justierung in Bezug auf die Mantellinie notwendig ist. Diese Eigenschaft läßt sich grundsätzlich auch für Gewinde größer M3 erreichen, wenn eine maßstäbliche Anpassung der Sensorgeometrie an den Gewindekerndurchmesser oder eine Auswertung der Meßdaten mit Hilfe des nichtlinearen Teils der Sensorkennlinie erfolgt.

Vorteile wie berührungsloses, rückwirkungsfreies Messen, auch bei hohen elektromagnetischen Störfeldern, einfache Handhabbarkeit, größtenteils wartungsfreier Betrieb sowie vergleichsweise kurze Prüfzeiten machen diesen Sensor für einen Einsatz in automatisierten Fertigungs- und Prüfprozessen besonders geeignet, so daß z.B. durch die kontinuierliche Qualitätskontrolle von Gewindeprofilen Bohrerstandzeiten verlängert werden können, da der Abnutzungsgrad des Schneidgerätes verfolgt werden kann.

7 Literatur

[1] G. Kuhn: *Berührungsloses automatisiertes Prüfen von Innengewinden*, Industrieanzeiger, 12 (1986), S. 34 - 35

[2] W. Langsdorf: *Messen von Gewinden*, Fertigung und Betrieb, Fachbücher für Praxis und Studium, Band 2, Springer Verlag, Berlin, 1974

[4] H. Kampa et al.: *Flexibles rechnergesteuertes Meßgerät zum Prüfen von Innengewinden*, tm - Technisches Messen, 52 (1985) 12, S. 465 - 470

[3] H. Kreitlow: *Faseroptische Sensoren*, aus: T. Pfeifer, Optoelektronische Verfahren zur Messung geometrischer Größen in der Fertigung , Expert-Verlag (1993), S. 140 - 16

[5] T. Pfeifer, M. Molitor: *Berührungslose optische Innengewindemessung*, Technische Rundschau, 4 (1989), S. 53 - 55,57

[6] K. Spenner et al.: *Faseroptische Multimode-Sensoren: Eine Übersicht*, Laser & Optoelektronik, Nr. 3 (1983). S. 226 - 234

[7] DIN 13, Teil 13

[8] DIN 13, Teil 1

Die Nutzung abstimmbarer Monomode-Laserdioden für Heterodyntechniken, dargestellt am Beispiel eines LDA-Strömungssensors

H. Müller, D. Dopheide
Physikalisch-Technische Bundesanstalt
Laboratorium für Strömungsmeßtechnik
Bundesallee 100
D-38116 Braunschweig

Kurzfassung

Im Gegensatz zu konventionellen Heterodyninterferometersystemen, die für die Erzeugung zweier in ihrer Lichtfrequenz sich unterscheidender Interferometerteilstrahlen im allgemeinen Zeeman-Laser oder akustooptische Modulatoren einsetzen, wird ein Konzept zur direkten Nutzung der optischen Frequenzdifferenz zweier Monomode-Laserdioden vorgestellt.
Die Rückmischung der resultierenden trägerfrequenten Signale ins Basisband und die Auswertung von Quadratursignalen erlaubt eine von den Emissionslinienbreiten und -schwankungen der Monomode-Laserdioden unabhängige Auswertung der Meßinformation.
Das Konzept wird anhand eines leistungsfähigen LDA-(Laser-Doppler-Anemometer)-Strömungssensors erläutert.

Einleitung und Grundprinzip des vorgestellten Verfahrens

Heterodyntechniken werden in interferometrischen Meßsystemen zur Weglängen- oder Geschwindigkeitsmessung für die Bestimmung des Vorzeichens eingesetzt. Die Grundvoraussetzung für den Einsatz dieser Techniken in der Zweistrahl- oder Mehrstrahlinterferometrie ist die Erzeugung einer optischen Frequenzdifferenz der zur Schwebung gebrachten Laserstrahlungen.

Mit Ausnahme von Interferometersystemen, die im Bereich kleiner Laserleistungen die Frequenz-aufspaltung von Zeeman-Lasern oder den Modenabstand zweier benachbarter longitudinaler Moden eines HeNe-Lasers, bzw. im Bereich einiger 10 mW neuerdings auch die Polarisations-aufspaltung longitudinal singlemodiger Mikrokristall-Laser nutzen, werden in konventionellen Systemen stets zusätzliche optoelektronische Komponenten für die Erzeugung einer optischen Frequenzdifferenz benötigt.

Neuartig hingegen ist der Einsatz von jeweils einem abstimmbaren Monomode-Laser pro Interferometerteilstrahl und die direkte Nutzung der optischen Differenzfrequenz als Trägerfrequenz. Auf diese Weise lassen sich durch die Einsparung von Strahlteilern und akustooptischer Modulatoren kompakte Meßsysteme realisieren, die sich durch die Verdopplung der Laserleistung insbesondere für Meßaufgaben zur Auswertung von Streulichtsignalen eignen.

Dabei lassen sich Einflüsse von Emissionslinienbreiten und - schwankungen der einzelnen Monomode-Laser auf die Meßinformation eliminieren, wenn man das resultierende trägerfrequente Meßsignal (Trägerfrequenz und Meßinformation) mit einem Quadratur-Referenzsignalpaar (Trägerfrequenz , 0 Grad - 90 Grad Phase) korreliert. Bei diesem Verfahren wird die Trägerfrequenz lediglich als Hilfsträger für die Erzeugung eines Quadratursignalpaares im Basisband genutzt, das die vollständige Meßinformation (Betrag und Vorzeichen) beinhaltet /1/.

Die Korrelation von Meß- und Referenzsignal kann mit low-cost Komponenten aus der HF-Technik (Leistungsteilern Breitband-Phasenschiebern (Hybrid-Kopplern), Mischern, Tiefpässen) durch die Mischung des Meßsignals mit einem Referenzsignalpaar realisiert werden (s. Bild 1). Das Referenzsignalpaar läßt sich aus dem Schwebungssignal von Teilintensitäten beider Laser (vgl. Bild 3) durch Aufteilung mit Hilfe eines breitbandigen Phaseschiebers erzeugen.

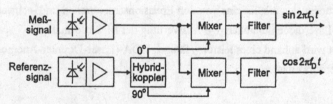

Bild 1: Rückmischanordnung zur Korrelation von Meß- und Referenzsignal
→ Eliminieren der Trägerfrequenz
→ Erhalt der Vorzeicheninformation durch Erzeugen eines Quadratursignalpaares

In dem im Basisband resultierenden Quadratur-Meßsignalpaar ist die Meßinformation in der durch das Signalpaar definierten momentanen Phase $\phi(t)$ vollständig enthalten. Im vorgestellten Anwendungsbeispiel eines LDA-Strömungssensors ist $\phi(t) = 2\pi f_D t$, wobei die Dopplerfrequenz f_D direkt proportional zu der zu messenden Geschwindigkeit ist.

Auswahl geeigneter abstimmbarer Monomode-Laserdioden

Eine notwendige Voraussetzung für die betriebssichere Realisierung des neuen Konzeptes ist die Schwebungsfähigkeit von Monomode-Laserdiodenpaaren, die eine kontinuierliche Abstimmbarkeit der Emissionswellenlängen über den Betriebsstrom oder die Betriebstemperatur der Dioden in einem, wenn auch geringen, Wellenlängenüberlappungsbereich erfordert.

ei konventionellen Monomode-Laserdioden mit Fabry-Perot-Resonator -Struktur-gestaltet sich
e Selektion geeigneter Laserdioden aufgrund des selbst innerhalb einer Fertigungsserie recht
nterschiedlichen Mode-hopping- und Hysterese-Verhaltens mehr oder weniger aufwendig.
iese Verhältnisse ändern sich jedoch grundlegend, wenn man Laserdioden mit Bragg-Reflektor -
esonatorstrukturen nach dem DFB- (distributed feed back) oder DBR- (distributed bragg reflector)
rinzip verwendet /2/, die neben einem erweiterten kontinuierlichen Abstimmbereich, im Gegen-
tz zu konventionellen Monomode-Laserdioden mit Emissionslinienbreiten bis zu 100 MHz,
missionslinienbreiten von nur wenigen MHz aufweisen.

ei den seit kurzem auch für Wellenlängen im 850 nm-Bereich verfügbaren leistungsstarken
)BR-Laserdioden konnte ohne vorherige Selektion über weite Parameterbereiche eine sichere
chwebungssignalbildung mit einer Bandbreite von unter 10 MHz erzielt werden.
ild 2 zeigt die Wellenlängenabhängigkeit zweier SDL 5712 100mW-Dioden in Abhängigkeit von
em Betriebsstrom für verschiedene Betriebstemperaturen und Bild 3 ein typisches Schwebungs-
ignal im 500 MHz-Bereich mit geringer Bandbreite.

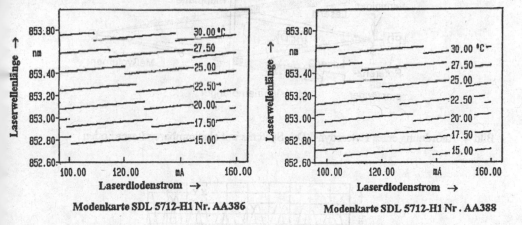

Bild 2: Abstimmverhalten (Modenkarten) zweier DBR-Laserdioden vom Typ SDL 5712

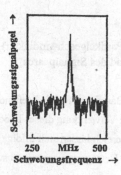

Bild 3: schmalbandiges Schwebungssignal der DBR-
Laserdioden nach Bild 2 im 500MHz-Bereich

710

Nutzung der optischen Frequenzdifferenz zweier Laserdioden in einem LDA-Sensor

Bild 4 zeigt die optische Anordnung eines LDA-Systems nach dem Doppler-Differenz-Vefahren.
Die Ausgangsstrahlen zweier Monomode Laserdioden werden direkt in das Meßvolumen
fokussiert und das aus dem Kreuzungsbereich von beiden Teilstrahlen an einem in der Strömung
mitgeführten Partikel gestreute Licht auf einem in Rückwärtsrichtung angeordneten Avalanche-
Meßsignalempfänger überlagert, dessen Ausgang das trägerfrequente Meßsignal liefert.
Die Referenzsignalerzeugung erfolgt durch Schwebungssignalbildung der an einer Planplatte in
Rückwärtsrichtung reflektierten und mit Hilfe eines Gitters auf den Referenzsignalempfänger
vereingten Teilintensitäten beider Laserdiodenausgangsstrahlen.
Die Wahl der Betriebswellenlängen und die Stabilisierung der optischen Frequenzdifferenz als
Trägerfrequenz des Systems geschieht über die Temperaturstabilisierung beider Laserdioden und
die Stromregelung einer der beiden Monomode-Laserdioden.

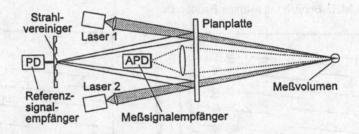

Bild 4: Anordnung eines Heterodyn-LDA-Systems mit abstimmbaren Laserdioden

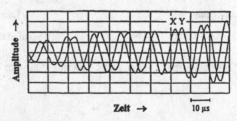

Bild 5: Ausschnit eines LDA-Quadratursignalpaares im Basisband
der Betrag der Meßinformation liegt in der Frequenz (f_D ~ Partikelgeschwindigkeit),
die Vorzeicheninformation in der Phasenbeziehung ± 90 Grad des Signalpaares

Bild 5 zeigt den Ausschnitt eines typischen LDA-Burstsignalpaares im Basisband für die Meßanordnung
nach Bild 4 mit selektierten 40 mW HL 8318 Laserdioden. Trotz der bei Verwendung dieses Diodentyps
resultierenden relativ großen Trägerfrequenzbandbreiten im 100 MHz-Bereich lassen sich infolge der
Rückmischtechnik geringe Dopplerfrequenzen und selbst quasistatische Bewegungsvorgänge auswerten.

Zusammenfassung

Durch den Einsatz von Rückmischtechniken zur Erzeugung und Auswertung eines Quadratur-signalpaares im Basisband läßt sich die optische Frequenzdifferenz von Monomode-Laserdioden zur Heterodyntechniken nutzen. Hierfür müssen lediglich die resultierenden Trägerfrequenzen und Trägerfrequenzbandbreiten in der Photoempfängerbandbreite liegen.

Die kommerzielle Verfügbarkeit leistungsstarker über weite Wellenlängenbereiche kontinuierlich abstimmbarer Monomode-Laserdioden mit Bragg-Resonator-Strukturen erlaubt bereits ohne Selektionsaufwand die Realisierung interessanter Heterodyn-Meßkonzepte.

Literatur

1/   Müller, H.; Czarske, J.; Kramer, R.; Többen, H.; Arndt, V.; Wang, H.; Dopheide, D.: Heterodyning and
     Quadrature Signal Generation: Advantageous Techniques for Applying New Frequency Shift Mechanisms
     in Laser Doppler Velocimetry. 7th Int. Symp. on Applications of Laser Techniques to Fluid Mechanics,
     Lisbon, Portugal, July 11-14, 1994 , Conference proceedings Volume II, 23.3.1-23.3.8

2/   Amann, M.-C.: Wavelength tunable single mode laser diodes Advances in Solid State Physics, Vol. 31,
     p. 201-218, 1991

Untersuchungen zur verteilten Temperatur-Fasersensorik mit stimulierter Brillouin-Streuung

F.A. Gers, J.W. Czarske
Institut für Meß- und Regelungstechnik im Maschinenbau
der Universität Hannover, Nienburger Str. 17, D - 30167 Hannover

Kurzfassung:
Es wird die Zielsetzung verfolgt, einen kostengünstigen Temperatursensor über eine Glasfaserstrecke größer 20 km Länge zu realisieren. Als Meßeffekt wird die Temperaturabhängigkeit des Stokesfrequenzshifts bei stimulierter Brillouin–Streuung genutzt. Es wurde eine präzise Kalibierung der Frequenzshift–Temperatur–Kurve vorgenommen. Es wurde eine ortsauflösende Temperaturmessung mit einer Glasfasermeßstrecke von 25 km durchgeführt. Die Ortsauflösung wurde mit einem Pulslaufzeit–Meßverfahren (OTDR) realisiert. Zur Senkung der notwendigen Pulsleistung wird ein neues Sensorprinzip mit einer vierfach gefalteten Faser vorgeschlagen. Unter Verwendung eines Multimodelasers ist es möglich, mit einem statt mit zwei Lasern, einen ortsauflösenden Glasfaser-Temperatursensor aufzubauen.

Einführung:
Verteilte Temperatursensorik wird bisher üblicherweise mit extrinsischen Sensoren vorgenommen. Die Realisierung eines ortsauflösenden Temperatursensors, als intrinsischer Sensor in einer Glasfaser, bietet gegenüber den herkömmlichen extrinsischen Sensoren entscheidende Vorteile: Im Vergleich zu einem extrinsischen Sensor entfällt die kostenträchtige Verwendung vieler Einzelsensoren. Durch diesen Vorteil bietet sich die Nutzung eines Glasfasertemperatursensors zur Überwachung von Strecken größerer Länge, z.B Pipelines und Starkstromkabeln, an.
Unter Nutzung von stimulierter Brillouin-Streuung (SBS) ist es möglich, die Frequenz und nicht, wie z.B. beim Ramansensor, die Amplitude des Meßsignals als Meßinformation zu nutzen und damit einen streckenneutralen Sensor zu realisieren. In dieser Arbeit wird die Temperaturabhängigkeit der SBS als Meßeffekt genutzt.
Bisher wird zur Realisierung eines ortsauflösenden Brillouin-Temperatursensors das Pump-Probe-Verfahren eingesetzt, d.h. eine Pumplaserwelle erzeugt einen Brillouingewinn, der durch eine entgegengesetzt laufende Probelaserwelle abgetastet wird, wodurch sich die Brillouinfrequenz und damit die Fasertemperatur bestimmen läßt. Nachteile dieses Verfahrens sind erstens die Notwendigkeit von zwei sehr gut frequenzstabilisierten Lasern, deren Differenzfrequenz im 13 GHz Bereich zu messen ist, und zweitens, daß beide Enden der Glasfaserstrecke zur Messung zugänglich gemacht werden müssen.
In dieser Arbeit wird gezeigt, daß es möglich ist, einen ortsauflösenden Brillouin–Glasfaser-Temperatursensor mit niederfrequenter Meßsignalauswertung und nur einem Laser zu realisieren. Dazu werden zwei unterschiedliche Moden eines Multimodelasers gleichzeitig als Pump-bzw. Referenzwelle genutzt.

1. Meßprinzip
SBS ist ein Wechselwirkungsprozeß von drei Wellen. Die Lichtwelle des Pumplasers wechselwirkt in der Glasfaser mit thermisch angeregten Schallwellen in der Weise, daß eine zweite Lichtwelle erzeugt wird. Diese, als Stokeswelle bezeichnete Welle, läuft entgegen der Pumpwelle und besitzt eine um den Brillouinshift kleinere Frequenz.
Der Meßeffekt beruht auf der linearen Temperaturabhängigkeit des Brillouinshifts, also dem linearen Zusammenhang zwischen Stokeswellenfrequenz und Fasertemperatur.

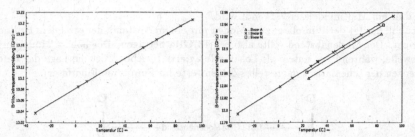

Abbildung 1: Temperaturabhängigkeit der Brillouinverschiebung mit linearer Regression von vier unterschiedlich dotierten Fasertypen (links: (Sumitomo) 6km Fluor, links: (Siecor)(N, A, B) 19 km Germanium) sind.

Der Brillouinfrequenzshift und –temperaturkoeffizient ist abhängig von der Dotierung der verwandten Glasfaser.
Die Information des Ortes, an dem die Meßgröße, der Brillouinshift, die gemessene Brillouinfrequenz annimmt, wird durch ein Laufzeitmeßverfahren (OTDR: Optical Time Domain Reflectometry) bestimmt.

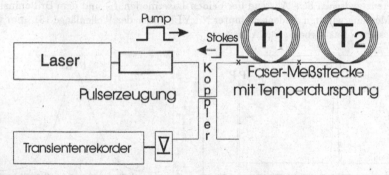

Abbildung 2: Schematische Anordnung für eine verteilte Glasfaser-Temperaturmessung

Abb. 2 zeigt den Versuchsaufbau schematisch. Ein Pumppuls läuft über einen Richtkoppler in die Glasfasermeßstrecke, die die zu vermessende Temperaturverteilung aufweist. Die dort generierte rücklaufende Stokeswelle wird mit einer Referenzwelle (Heterodynverfahren) gemischt und im Zeitbereich aufgezeichnet. Kurzzeit FFT's über verschiedene Abschnitte dieses Signals liefern die an den entsprechenden Meßfaserorten erzeugten Brillouinfrequenzverschiebungen und somit die Temperaturen an diesen Orten.

2. Einsatz eines Multimodelasers

Um die Brillouinfrequenzverschiebung mit einer elektronischen Signalauswertung zu bestimmen, muß die Stokeswelle mit einer Referenzwelle überlagert werden. Im einfachsten Fall verwendet man als Referenzwelle die Pumplaserwelle und erhält so ein Beatsignal, dessen Frequenz gleich dem Brillouinshift ist. Der Brillouinshift liegt aber im Bereich von 13 GHz, so daß sich ein Mikrowellen-Meßsignal ergibt. Wünschenswert ist ein Meßsignal im Hochfrequenzbereich unter 1 GHz, um konventionelle Detektoren und Signalverarbeitungstechniken verwenden zu können. Es besteht die Möglichkeit, ein geeigneteres Referenzsignal mit einem zweiten Laser oder unter Nutzung externer Frequenzshifttechniken zu erzeugen. Dieser kostenträchtige Aufwand kann vermieden werden, wenn als Referenzwelle ein zweiter Lasermode zur Verfügung

714

steht, d.h. ein Multimodelaser verwandt wird.

Im Idealfall besitzt der Pumplaser zwei Moden mit einem Abstand, der etwa dem Brillouinshift entspricht. Der Modenabstand sollte also ca. 13 GHz betragen. Der höhere Mode pumpt die Stokeswelle, während der untere als Lokalwelle agiert. In Abb. 3 ist die Lage der beteiligten Frequenzen der schematisch dargestellt (Zahlenwerte für Sumitomo Fluorfaser).

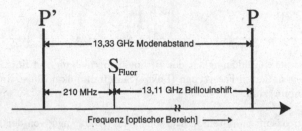

Abbildung 3: Schematische Darstellung der optischen Signale von Lasermoden und Stokeswelle.

Durch Mischen von Stokes- und Laserlicht ergibt sich zwischen dem unteren Lasermode und der Stokeswelle das niederfrequente Beatsignal (SP'). Die Frequenz $f_{SP'}$ dieses Signals entspricht der Differenz zwischen dem Abstand der beiden Lasermoden f_M und dem Brillouinshift f_B.

Für die Messung wird ein diodengepumpter Nd:YLF Laser der Wellenlänge 1313 nm verwandt. Die gemessenen Beatsignale zeigt Abb. 4.

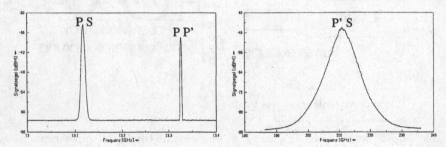

Abbildung 4: Beatsignale von Lasermoden und Stokeswelle als Mittelung über 100 Spektren.

Der Modenabstand entspricht der Frequenz des Beatsignals der Lasermoden PP' von 13,33 GHz. Der Brillouinshift der verwandten Faserspule beträgt, wie das Beatsignal zwischen Pumpmode und Stokeswelle PS zeigt, bei Zimmertemperatur $f_{PS} \approx 13,11$ GHz. Erwartungsgemäß ergibt sich ein niederfrequentes Beatsignal zwischen dem zweiten Lasermode und der Stokeswelle $P'S$ bei $f_{P'S} = 210$ MHz $\approx 13,33$ GHz $- 13,11$ GHz.

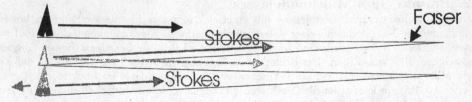

Abbildung 5: Ortsauflösendes Sensorverfahren mit einer vierfach gefalteten Glasfaser

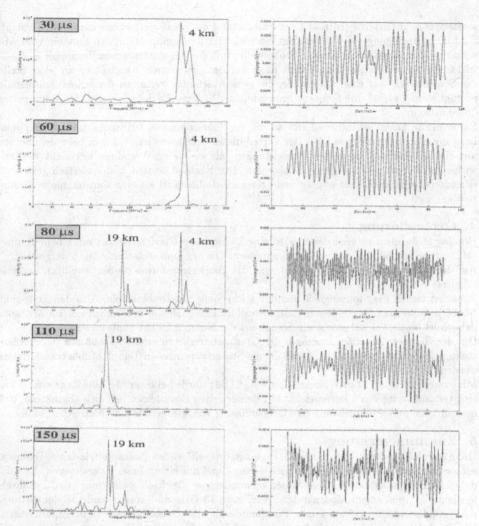

Abbildung 6: Rechts: Stokes-Lokal-Beatsignale auf dem Transientenrekorder, links: Die Fouriertransformierte (FFT) des Signals. Die Signale gehören zu den jeweils links-oben angegebenen Zeitpunkten der rücklaufenden Stokeswelle ($t = 0$: Eintritt des Pumppulses in die Faser).

3. Sensor mit vierfach gefalteter Faser

Da bei dem angestrebten Glasfaser-Temperatursensor mit nur einem Laser größere Pumpleistungen erforderlich sind, als bei den bisher üblichen Pump-Probe Verfahren, soll zur Verminderung der benötigten Pumppuls-Leistung ein ortsauflösendes Sensorverfahren mit einer vierfach gefalteten Glasfaser vorgestellt werden. Das Verfahren beruht darauf, daß ein vorlaufender Pumppuls spontane Brillouin-Streuung erzeugt, die in einem nachlaufenden Pumppuls verstärkt wird, so daß dort ein stimulierter Brillouin- Streuungs-Prozeß entsteht.

Um die Wechselwirkung zu synchronisieren, muß die Glasfaser vierfach gefaltet, d.h. die Sensorfaserlänge geviertelt werden und der Pulsabstand der Glasfaserstrecke entsprechen (Abb.5). Sind diese Bedingungen erfüllt, erzeugt ein vorlaufender Pumppuls auf den hinteren zwei Abschnitten der Glasfaserstrecke eine Stokeswelle, mit der der nachfolgende Pumppuls wechselwirkt. Dieser nachfolgende Puls trifft dann auf den ersten zwei Abschnitten an jeder Stelle der Sensorstrecke auf eine Stokeswelle, die zwei Abschnitte weiter an der selben Sensorstelle erzeugt wurde, so daß ihre Frequenz immer mit der Stokeswelle des nachfolgenden Pumppulses übereinstimmt.

Experimente haben gezeigt, daß sich so die zur Erzeugung von SBS notwendige Pumppuls-Leistung erheblich verringen läßt. Der Vorteil dieses Verfahrens ist, daß man bei einem Sensor mit einem Laser mit geringeren Pulsleistungen, als sie für ein Verfahren mit nicht wechselwirkenden Pulsen notwendig sind, auskommt. Der Nachteil besteht in der vierfach gefalteten Glasfaserstrecke, die teurer und aufwendiger zu installieren ist als eine konventionelle Sensorstrecke.

4. Ortsauflösung

Um das Meßprinzip zu verifizieren, wird eine 25 km lange Glasfaserstrecke mit einem Temperatursprung von $59,0^\circ$C nach 4,4 km vermessen. Die Ergebnisse sind in Abb. 6 dargestellt.

Auf der linken Seite von Abb. 6 sieht man die Fouriertransformierte des jeweiligen Zeitausschnitts.

Zunächst treffen die Stokessignale aus der 4 km Spule am Detektor ein. Um den Zeitpunkt $t \approx 80$ μs sieht man für ein kleines Zeitintervall Stokessignale aus beiden Spulen. Im Zeitraum danach ist bis zu $t = 200$ μs nur ein Signal aus der hinteren 19 km Spule zu sehen.

Um den Temperaturverlauf innerhalb der Glasfaserstrecke zu erhalten, muß nur die Zeitdarstellung in eine Längendarstellung und die die Stokesfrequenzen in Temperaturen transformiert werden.

Mit extern erzeugten 13 mW Rechteckpulsen (EOM) wurde bei einer Meßfaserlänge von 25 km eine Ortsauflösung von 3 km erreicht. Mit den erzielten Ergebnissen läßt sich abschätzen, daß man mit 600 mW Pumpleistung eine Ortsauflösung von weniger als 100 m erreichen kann.

5. Zusammenfassung

Diese Arbeit hat gezeigt, daß es möglich ist, einen ortsauflösenden Brillouin- Glasfaser–Temperatur sensor mit niederfrequenter Meßsignalauswertung und nur einem Laser zu realisieren. Es wird dabei als Pumplaser ein Multimodelaser verwendet, der gleichzeitig als Pump- und Lokalwelle dient. Er muß einem Modenabstand von etwa 13 GHz und ausreichende Modenleistung, d.h. mindesten 600 mW auf einem Pumpmode für eine Ortsauflösung < 100 m, aufweisen. Damit wird, im Vergleich zum bisher realisierten Pump-Probe-Verfahren, das zwei frequenzstabilisierte Laser und den Zugang zu beiden Faserenden benötigt, ein Verfahrn vorgestellt, das diese Nachteile nicht aufweist und damit eine wesentlich einfachere Realisierung eines Glasfaser-Temperatursensors mit nur einem Laser ermöglicht.

Literatur

–X. Bao, D.J. Webb, D.A. Jackson, '*32-km distributed temperature sensor based on Brillouin loss in an optical fiber*', Opt.Lett. **18** (1993), 1561.

–M. Tateda, T. Horiguchi, T. Kurashima, K. Ishihara, '*First Measurement of Strain Distribution Along Field-Installed Optical Fibers Using Brillouin Spectroscopy*', J.Lightw.Techn. **8** (1990), 1269.

–R. Philipps, J.W. Czarske, '*Charakterisierung eines frequenzmodulierbaren Nd:YAG - Ringlasers für die Brillouin-Fasersensorik*', Laser'95 (Kongreß C) Tagungsband.

–J.W. Czarske, H. Müller, '*Heterodyne detection technique using stimulated Brillouin scattering and a multimode laser*', Opt.Lett. **19** (1994), 1589.

2D-Brillouin-Strömungssensor

H. Többen, H. Müller, D. Dopheide
Physikalisch-Technische Bundesanstalt
Laboratorium für Strömungsmeßtechnik
Bundesallee 100, 38116 Braunschweig

Einleitung

In der Laser-Doppler-Anemometrie werden Frequenzshifttechniken eingesetzt, um neben dem Betrag auch das Vorzeichen einzelner Geschwindigkeitskomponenten in Strömungen bestimmen zu können. Hierbei wird das Meßvolumen durch sich kreuzende Laserstrahlen unterschiedlicher Lichtfrequenz gebildet. Die von in der Strömung mitgeführten Partikeln beim Durchgang durch das Meßvolumen gestreute Strahlung erfährt aufgrund des Doppler-Effektes eine zur Strömungsgeschwindigkeit proportionale Frequenzverschiebung. Ein Photoempfänger detektiert die Streustrahlung und liefert ein Ausgangssignal, das als Meßsignal für die Geschwindigkeitsbestimmung dient. Für eine zweidimensionale Geschwindigkeitsmessung wird das Meßvolumen von drei Laserstrahlen gebildet. Dabei läßt sich jeder Geschwindigkeitskomponente ein trägerfrequentes Meßsignal zuordnen, dessen Trägerfrequenz direkt durch die optische Frequenzdifferenz der Laserstrahlen, auch Shiftfrequenz genannt, gegeben ist.

Bei dem hier vorgestellten Verfahren wird die Shiftfrequenzerzeugung zur Vorzeichenbestimmung und Mehrkomponentenmessung mit nur einer Laserquelle durch Nutzung der stimulierten Brillouin-Streuung (SBS) in drei Einmodenfasern vorgenommen. Dabei erfährt die in den Einmodenfasern in Rückwärtsrichtung gestreute sogenannte Stokes-Strahlung eine brechzahlabhängige Frequenzverschiebung im GHz-Bereich. Die Frequenzdifferenzen der von der Pumpquelle in den drei unterschiedlichen Fasern generierten Stokes-Strahlen liegen im MHz-Bereich und lassen sich als Frequenzshift nutzen.

Stimulierte Brillouin-Streuung

Um mit faseroptischen Komponenten durch stimulierte Brillouin-Streuung Stokes-Strahlung generieren und nutzen zu können, wird eine Laserquelle, ein Faserkoppler

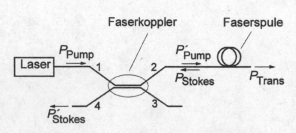

Bild 1 Faseroptischer Aufbau zur Generierung und
Nutzung der stimulierten Brillouin-Streuung

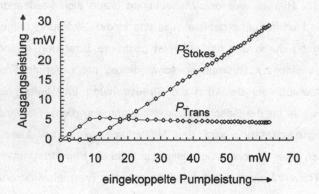

Bild 2   Leistungskennlinien

(besser: ein optischer Zirkulator) und eine Faserspule benötigt, vgl. Bild 1. Hierbei wird die im Kern der Faserspule geführte Pumpstrahlung P'_{Pump} an akustischen Phononen, die in der Faser durch thermische Fluktuationen ständig angeregt sind, gestreut. Überschreitet die Pumpstrahlung einen bestimmten Wert (Brillouin-Schwelle) bildet sich durch Elektrostriktion eine Schallwelle aus, und der Streuprozeß findet stimuliert statt [1]. Aufgrund der Bragg-Bedingung und der wellenführenden Eigenschaft der Faser wird die Pumpstrahlung an der Schallwelle

in Rückwärtsrichtung, als sogenannte Stokes-Strahlung P_{Stokes}, gestreut. Die Leistung der Pumpstrahlung wird dabei nahezu vollständig in Stokes-Strahlung konvertiert [2], vgl. Bild 2. Der nicht konvertierte Pumpstrahlungsanteil bleibt annähernd konstant und kann als transmittierte Pumpleistung P_{Trans} am Faserspulenende abgegriffen werden.

Die Stokes-Strahlung wird hierbei gegenüber der Pumpstrahlung um den Betrag der Schallwellenfrequenz zu kleineren Frequenzen hin verschoben. Diese Frequenzverschiebung ist u.a. materialabhängig und beträgt in Standard-Telekommunikationsfasern bei einer Pumpwellenlänge von 1,3 µm ca. 13 GHz. Durch gleichzeitiges Pumpen von Ein-modenfasern unterschiedlicher Brechzahl lassen sich somit frequenzverschiedene Stokes-Wellen erzeugen, die dann als Sendestrahlen für die Meßvolumenbildung eines Strömungssensors zur mehrkomponentigen Geschwindigkeitsmessung genutzt werden.

Experimenteller Aufbau

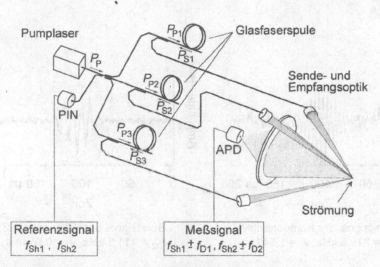

Bild 3   Experimenteller Aufbau des 2D-Brillouin-Strömungssensors

Den experimentellen Aufbau des realisierten faseroptischen Strömungssensors für zwei-dimensionale Geschwindigkeitsmessungen zeigt Bild 3. Ein 1,32 µm-Nd:YAG Ringlaser mit einer maximalen Ausgangsleistung von 150 mW wird als Pumplaser eingesetzt. Die Pumpstrahlung (P_P) wird über Faserkoppler zu gleichen Anteilen (P_{P1}, P_{P2} und P_{P3}) auf drei Faserspulen (Durchmesser: 23 cm; Höhe: 11 cm) mit Faserlängen von 6 km, 6,4 km und 19 km verteilt. Die drei in den Spulen mit unterschiedlichen Brechzahlen generierten frequenzverschiedenen Stokes-Wellen (P_{S1}, P_{S2} und P_{S3}) gelangen über die Sendeoptik als fokussierte Strahlen ins Meßvolumen. Ein Bruchteil der Stokes-Wellen gelangt über die Faserkoppler zum PIN-Detektor. Das Referenzsignal entsteht durch die Schwebungssignalbildung der Stokes-Wellen, wobei von den resultierenden Träger-frequenzen die Frequenzen f_{Sh1} und f_{Sh2} als Shiftfrequenzen zur Signalkomponenten-

720

trennung und Vorzeichenbestimmung der Geschwindigkeitskomponenten genutzt werden.

Die trägerfrequenten Signalkomponenten des Meßsignales, $f_{Sh1} \pm f_{D1}$ und $f_{Sh2} \pm f_{D2}$, unterscheiden sich von denen des Referenzsignales lediglich um die geschwindigkeitsproportionalen Doppler-Frequenzanteile $\pm f_{D1}$ bzw. $\pm f_{D2}$. Das Vorzeichen ist abhängig von der Strömungsrichtung. Die gleichzeitige Auswertung von Meß- und Referenzsignal liefert dann die Aussage über Betrag und Vorzeichen der beiden Geschwindigkeitskomponenten [3]. Bild 4 zeigt die Burstsignale der beiden simultan gemessenen Geschwindigkeitskomponenten.

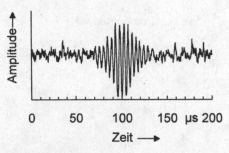 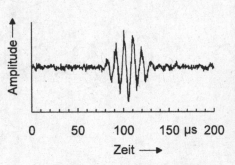

Bild 4   Burstsignal der Komponente 1   Burstsignal der Komponente 2
         $f_{D1} = 219,8$ kHz, $v_1 = 1,54$ m/s $f_{D2} = 111,1$ kHz, $v_2 = 0,95$ m/s

Zusammenfassung

Die Nutzung von Einmodenfasern für die Lichtführung und für die Shiftfrequenzerzeugung in einem Laser-Doppler-Anemometer ermöglicht die Realisierung eines 2D-Strömungssensors mit folgenden Vorteilen:

- Einsatz von faseroptischen Standardkomponenten
- Justageunempfindlichkeit des Meßsystems
- Frequenzshifterzeugung ohne zusätzliche optoelektronische Bauteile
- Vorzeichenbestimmung bei einer zweidimensionalen Geschwindigkeitsmessung

Das Forschungsvorhaben wird von der Deutschen Forschungsgemeinschaft, DFG,

(AZ.: Do292/3-1) gefördert.

Literaturverzeichnis

[1]   Agrawal, G.P.: *Stimulated Brillouin Scattering*, in Nonlinear Fiber Optics, Academic
        Press, Boston-London-Toronto, 1989, S. 263-288

[2]   Többen H.; Müller H.; Dopheide D.: *Brillouin Frequency Shift LDA*, 7th Int. Sym. on
        Applications of Laser Techniques to Fluid Mechanics, 11-14 Juli 1994, Lisabon,
        Portugal, Conference proceedings paper 14.2

[3]   Többen H.; Müller H.; Dopheide D.: *Geschwindigkeitsbestimmung über den
        Phasenwinkelzeitverlauf von Quadratursignalen*, 4. Fachtagung: Lasermethoden in
        der Strömungsmeßtechnik, 12.-14. Sept. 1995, Universität Rostock, Tagungsband
        vom Verlag Shaker, Aachen

Metrological Instruments for Energy Measurements of Laser Working with Q-Modulation

J. Owsik
Institute of Optoelectronics, Military University of Technology
2 Kaliski Str., 01-489 Warsaw, Poland

ABSTRACT

The paper presents metrological instruments for radiation energy measurement of pulse lasers working at wavelengths of 1.06 and 0.53 μm, within energy range of 30 ÷200 mJ for pulse durations of 10 ns to 1 s with an error not higher than 1 %. There has been performed the analysis of component errors of measuring converter of calorimetric type which influence on accuracy of energy measurement of pulse laser radiation.

1. INTRODUCTION

An accuracy of energy measurements of the Q-switching lasers requires an application of measuring converters (MC) which have the following metrological and technical characteristics:
- stability of the conversion coefficient for the MC within a wide range of laser pulse durations;
- nonselectivity of the MC within a wide spectral range;
- stability of the conversion coefficient of the MC for the power densities of the order of 10^8 W/cm^2.

The converter of thermoelectric or calorimetric type meets the above mentioned requirements. In such convertes a beam energy which is incident on an absorber is absorbed and converted into heat. The temperature of absorber will increase and by measuring this growth the absorbed energy can be determined. The change of absorber temperature can be measured using sensitive thermocouple. The calorimeter sensitivity will be the higher the higher is be temperatures difference causeds by the incident radiation. An important advantage of calorimeters as radiation receivers is linear dependence of temperature increase on incident energy, as well as the fact that calorimeter similarly responds to radiation energy of different wavelengths [1].

The absorber in Li and Sims calorimeter [2] was made in form of a graphite cone. Three thermistors connected in series were attached to it. Another version of conical calorimeter has been worked out by Edwards [3]. Temperature of a carbonic cone was measured by means of eight-constantan thermocouples.

The calorimeter worked out by Koozekanani et al [4] was made as a cavity, simulated absolute black body. The conical absorber was made of metal. Internal surface of the cone is blacking.

The methods of power and energy measurements as well as description of conical calorimeters for laser energy measurement are described [5,6,7,8].

In [9] the measuring instruments with colorimetric receiver used in standards are presented.

The scheme of calorimetric type MC is shown in Fig.1. The MC comprises two identical absorbing receiving elements: the working and compensating ones, in order to compensate the influence of ambient temperature changes. Only the working receiving element is illuminated by laser radiation.

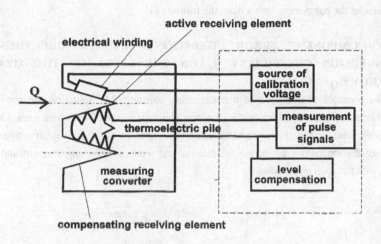

Fig.1. Scheme of the measuring converter (MC).

At the working element there is electrical winding for transmission the known values of electrical energy to the MC.

The error of energy measurement of Q-switched lasers by means of such MC, consists of the instrument error (S_{MC}) and constituent errors which characterize the lasers beam parameters.

Only the S_{MC} error will be presented below. The expression describing a self-error is of the following form:

$$S_{MC} = \sqrt{\left(\Theta_Q^2 + \Theta_\phi^2 + \Theta_\lambda^2 + \Theta_{XY}^2\right)/3}$$

(1)

where Θ_Q - is the component error dependent on the relation of MC conversion coefficient on a level of the measured energy; Θ_ϕ - is the component error dependent on the relation of the conversion coefficient of the MC on the laser beam diameter; Θ_λ - is the component error dependent on the relation of the MC conversion coefficient on a laser radiation wavelength;

Θ_{xy}- is the component error dependent on the relation of the MC conversion coefficient on the laser beam adjustment;

The formula (1) is valid only when the following conditions are fulfilled:
- the number of component errors is not higher than three;
- values of component errors are of the same order;
- distribution laws for error components (S_{MC}), regarded as systematic errors are not known and therefore one can assume a uniform distribution;
- confidence interval is 0.95.

Let us consider the component errors from the formula (1).

2. THE COMPONENT ERROR CONDITIONED BY THE DEPENDENCE OF CONVERSION ÇOEFFICIENT K_c ON THE LEVEL OF THE MEASURED ENERGY Θ_Q

The principle of operation of the measuring converter MC relies on the conversion of energy of laser radiation, incident on the receiving element, into the heat which causes heating of the receiving element and then, produces a thermo EMF at the thermoelectric pile. The hot junctions of thermocouples are in the thermal contact with the receiving element and the cold junctions with the receiver housing.

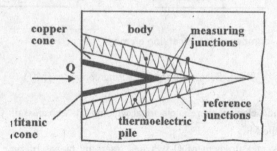

Fig.2. Receiving element of measuring converter MC.

In the ideal case (i.e. when the temperature of working element is exactly proportional to the incident energy) the coversion coefficient K_c (being the ratio of voltage at the thermoelectric pile vs pulse energy) is constant and the considered error does not exist.

The ideal case can be assumed for the considered construction of converter when the maximum temperature arising at the hot junctions of thermoelectric pile does not exceed the tenth parts of a degree with the energy 100 mJ. One should expect that for durations of $1 \div 2 \cdot 10^{-8}$ s and for energies of the order of $0.1 \div 0.5$ J, the receiving element will be heated up to relatively high temperature (of the order of $1000°$) what causes the non-linear phenomena in the heated place an then a disturbance of K_c constancy within the required range. The rate of thermal process spreading is so high that the temperature is constant during the period $1 \cdot 10^{-6} \div 1 \cdot 10^{-5}$ s. As it is seen from the analysis, the part of energy used for the new conversion into radiation as a result of the heating of the receiving element does not exceed 10^{-2} % for this case.

Let us consider the pulse energy recording and the errors related to this process. The process of incidence of single laser pulse on the MC can be decribed by means of two exponents which characterize heating up duration and cooling down of active element absorber. According to above, the output voltage tracking (following) the curve of temperature changes for the receiving element can be defined as the formula [10]:

$$U(t) = kQ(exp\,t\,/\,\tau_1 - exp\,t\,/\,\tau_2)\,/\,(\tau_1 - \tau_2) \tag{2}$$

where:τ_1 and τ_2 are the heating up and cooling down constants, respectively; k is the proportionality coefficient; Q is the pulse energy.

The maximum of this relation is:

$$t_{max} = \left[\tau_1\tau_2\,/\,(\tau_2 - \tau_1)\right]ln\,\tau_2\,/\,\tau_1 \tag{3}$$

The output sinal of the MC includes three componetns, openly described: signal increase up to 0.99 level of its maximum value, during the period of $4\div6$ s; signal stabilization at the level of 0.99, during $0.8\div1.5$ s; signal decrease, to the initial level, during the period of $3\div5$ min.

Two methods describing energy of pulse radiation on the basis of this curve are known. The first method it is a definition of surfaces under the curve; it corresponds to the total amount of heat received by the thermoelectric pile. This method is very time-consuming and not enough accurate and also it does not make possible to define the integration range. The second method is a definition of the output voltage amplitude, of the output voltage, which is exactly proportional to the value of the pulse energy as it results from (2) and (3) formulae.

In practice, these dependencies are not always preserved. Within the energy range of $30\div200$ mJ the K_c is non-linear what causes the necessity of the Θ_Q error consideration.

3. COMPONENT ERROR CONDITIONED BY THE DEPENDENCE OF CONVERSION COEFFICIENT K_c ON THE LASER RADIATION WAVELENGTH Q_λ.

Material of the receiving element cone is chosen in respect of the power densities which it should withstand [6]. Such a power density is $5 \cdot 10^7$ W/cm^2 for the pulse duration of the order of $1 \cdot 10^{-8}$ s, energy 0.1 J, beam diameter $4\div5$ mm, and the cone top angle 15°. The known black coatings do not withstand such densities. Due to this reason, the materials with the high melting and boiling points should be selected.

These materials have the spectral absorption coefficient decreasing with the wavelength changes. The receiving elements should operate within the wavelength range of $0.48\div1.06$ µm. An estimation Q_λ error within this range of wavelength is necessary. Definition of the K_{cs} value for the given wavelength is obtained by the MC verification with the standard. The K_c value is defined by the formula (4) for any second wavelength within the given spectral range:

$$K_{c\lambda} = K_{cs}\alpha_\lambda / \alpha_s \qquad (4)$$

where α_s is the absorption coefficient of the MC receiving element for the standard wavelength, α_λ is the absorption coefficient of the MC receiving element for the given wavelength within the range of $0.48 \div 1.06$ µm.

Experimental investigations of the receiving elements, made in the form of titanic cones with the cone top angle $15°$ and the internal copper coatings, were carried out at the wavelength of $\lambda = 0.5$ µm. The results of experiments show that a typical value of $\alpha_{0.5}$ is 0.9981. The coefficients for wavelengths of 0.53 µm; 0.63 µm; 0.69 µm, and 1.06 µm are $\alpha_{0.53} = 0.9980$; $\alpha_{0.63} = 0.9978$; $\alpha_{0.69} = 0.9976$ and $\alpha_{1.06} = 0.9941$, respectively. According to the formula (4), regarding the values $\alpha_{0.5}$ and $\alpha_{0.53}$, one can assume:

$$K_{c_{0.53}} = K_{c_{0.5}} = K_{c_{0.48}}$$

For the other wavelengths we obtain:

$$K_{c_{0.63}} = 0.9975 \, K_{c_{0.5}}$$
$$K_{c_{0.69}} = 0.9970 \, K_{c_{0.5}}$$
$$K_{c_{1.06}} = 0.9930 \, K_{c_{0.5}}$$

The analysis of $K_{c0.48} \div K_{c1.06}$ values shows that a typical change of K_c is 0.7% within the spectral range of $0.48 \div 1.06$ µm. Then, we can define a component of the error $Q_\lambda = 0.35\%$, assuming $K_c = K_{c0.5}$ for the spectral range interesting for us.

4. COMPONENT ERRORS CONDITIONED BY THE DEPENDENCE OF K_c ON THE LASER BEAM DIAMETER Θ_ϕ AND IT'S ALIGNMENT Θ_{xy}.

The component errors occur in the case when the radiation pulse with the exactly the same energy produces the output signal of the various values if the pulse reaches various parts of the receiving element surface. The conversion coefficient can be different if the same energies, the beams of various diameters are applied.

Such errors are caused by several evident reasons. The special conditions of the receiving element costruction are one of them. The receiving element should have the low thermal capacity and should be made of the thin materials in order to obtain enough sensitivity and possibility of high accuracy of the low energy beams recording. The heat flow to the thermopile should not be high (in the contrary the decrease of an output signal occurs) because it causes significant limitations for the surfaces of termocouples which are placed directly under the receiving element. On the basis of these conditions it results that for the termocouples which are not uniformly placed at the external surface of the receiving element, the incident radius with the small diameter causes a various heating of the termocouples, especially under the heated part of the surface.

The second reason causing the occurence of the Θ_ϕ and Θ_{xy} errors is a difference of the thermoelectric power of the termocouples used in the receiving element. The thermo EMF power depends on a thickness of the copper-nickel coating. It is impossible to obtain the same coatings for several hundred of thermocouples by means of the electrolitic process. The radiation incident on the various parts of receiving elements heats up various thermocouples what results in the changes of the output signal for the same energy.

The next reason of the Θ_ϕ and Θ_{xy} errors ocurrence is the "optical zone characteristics". The essence of this reason is the fact that the part of non-absorbed radiation should depend on the place of incident radiation. This part of radiation is relatively large at both the edge and the center of the receiving element where the constructional nonuniformity exist and it is small for the intermediate parts of the receiving element.

Small deviations in the receiving element location in respect of the optical axis causes the occurence of Θ_ϕ and Θ_{xy} errors.

The above mentioned reasons characterize the essential design and technological conditions which cannot be exactly regarded. The best method of these errors reduction is to make the models of the receiving element and next to precise examine of the zone characteristic for each model.

The error Θ_{xy} is considered as a systematic one and it is described by the formula:

$$\Theta_{xy} = \left| K_c[max, min] - \overline{K}_c \right| / \overline{K}_c \qquad (5)$$

where $\overline{K}_c$ - is an average value of the MC conversion coefficient, constant for the whole scanning zone; ane $K_c[max, min]$ is the conversion coefficients which value is the most different from $\overline{K}_c$.

The change of the K_c of the MC as a function of a beam diameter, i.e., Θ_ϕ error gives is lower than Θ_{xy} error. It is evident that for central alignment and next inreasing (decreasing) its diameter the radiation interaction is averaged with receiving element of the MC what results in compensation of non-unifprmity of K_c of the MC.
Experimentally described that Θ_ϕ is

$$\Theta_\phi \leq (0.2 \div 0.5)\Theta_{xy}$$

5. CONCLUSIONS

The above mentioned component errors were considered during the metrological certification of the MC. The values of these errors are presented in Table 1.

Table 1

Values of the MC component errors

No	Component error	Error symbol	Values [%]
1	Conditioned by the K_c dependence on the energy level	Θ_Q	0.8
2	Conditioned by the K_c dependence on the laser radiation wavelength	Θ_λ	0.35
3	Conditioned by the K_c dependence on the beam diameter of laser radiation	Θ_ϕ	0.5
4	Conditioned by the K_c dependence on the alignment of the laser beam	Θ_{xy}	1.1

Then, the error calculated according to the equation (1) amounts:

$$S_{MC} = 0{,}86\ \%$$

The basic technical characteristics of the previously described MC presents Table 2.

Table 2

Basic technical characteristics of the MC

No	Characteristic	Units	Values
1	Range of energy reproduction	mJ	30 - 200
2	Spectral range	μm	0.48 - 1.06
3	Pulse durations range	s	$10^{-8} - 1$
4	Maximum of power density	W/cm^2	10^8
5	Conversion coefficient	mV/J	5 - 8
6	Instrument error	%	0.9
7	Diameter of initial aperture	mm	10

REFERENCES

[1] Higgins T.V., *Laser Focus World*, 1994, vol 30, 11, 65-73.

[2] Li.T.,Sims S.D., *Appl. Optics*, 1962, vol 1, 3,325-328.

[3] Edwards J.G., *J.Sci. Instrum.*, 1967, vol 44, 835-838.

[4] Koozekanani S., Deybe P.P., Krutchkoff A., Ciftan M., *Proc.IEEE*, 1962, vol 50, 207-209.

[5] Calviello J.A., *Proc. IEEE*, 1963, vol 51, 611.

[6] Killick D.E., Batmen D.A., Brown D.R., Moss T.S., DeLa Perrelle E.T., *Infrared Physics*, 1966, vol 6, 85-109.

[7] Scott B.F., *J.Sci. Instrum.*, 1966, vol 43, 685-687.

[8] Brinbaum G., Brinbaum M., *Proc.IEEE*, 1967, vol. 55, 6, 1026-1031.

[9] Smathers S.E., Maksymonko G., *IEEE Transactions on Instrumentation and Measurement*, 1972 vol .IM-21, 4, 430-433.

[10] Hadson R., *Infrared system* (translated from English into Russian by Wasilczenko N.W., edited by Mir), 1977, 255.

Kalibrierung von Leistungsmeßgeräten für Hochleistungslaser

F. Brandt, K. Möstl
Physikalisch-Technische Bundesanstalt (PTB), Abt. Optik (4.11)
Postfach 3345   D-38023 Braunschweig

Abstract: The paper describes standard radiometers developed for the calibration of laser power meters in the power range from some mW to 1 kW. The calibration chain is linked to a cryogenic radiometer the PTB primary standard for radiant power. The measurement uncertainties (2σ-values) are in between 0.1 % and 1 %.

Ein wichtiges Glied in der Kette zur Darstellung von Laserstrahlungsleistungen Φ ist das **Normal für kleine und mittlere Leistungen** (LM7), das für Leistungen von 5 mW bis 10 W einsetzbar ist [1, 2]. Es arbeitet nach dem Prinzip eines elektrisch kalibrierbaren Radiometers. Als Absorber hat es einen schwarz vernickelten Hohlkegel mit einer Öffnung von 10 mm Durchmesser, der einfallende Laserstrahlung der Wellenlänge λ im Bereich von 405 nm $\leq \lambda \leq$ 1064 nm zu 99,83 % absorbiert; bei $\lambda = 10,6$ μm werden noch 96,61 % absorbiert. Der spektrale Absorptionsgrad $\alpha(\lambda)$ wurde aus dem bei den verschiedenen Laserwellenlängen gemessenen spektralen Reflexionsgrad $\rho(\lambda)$ errechnet: $\alpha(\lambda) = 1 - \rho(\lambda)$. Die Absorption der Laserstrahlung führt zu einer Erwärmung des Kegels gegenüber der Wärmesenke. Die Erwärmung wird mit einer Thermosäule als Thermospannung U_{th} gemessen.

Zur elektrischen Kalibrierung ist eine Heizwicklung außen am Kegel angebracht, mit der eine Erwärmung des Absorbers mit elektrischer Leistung möglich ist. Durch genaue Messung von Heizstrom I_H und Heizspannung U_H läßt sich die elektrische Empfindlichkeit s_{el} bestimmen:

$$s_{el} = U_{th}/(U_H \cdot I_H) \tag{1}$$

Die zur Messung der Laserstrahlungsleistung $\Phi = U_{th}/s_{Str}$ notwendige Strahlungsempfindlichkeit s_{Str} ergibt sich aus der raumtemperatur- und leistungsabhängigen elektrischen Empfindlichkeit s_{el}, dem wellenlängenabhängigen Absorptionsgrad $\alpha(\lambda)$ und einem Korrekturfaktor f_K, der in guter Näherung nicht von Parametern wie Temperatur, Leistung und Wellenlänge abhängt:

$$s_{Str} = s_{el} \cdot \alpha(\lambda) \cdot f_K \tag{2}$$

Der Korrekturfaktor f_K wurde jetzt mit einem Kryoradiometer als dem **Primärnormal der PTB für die Strahlungsleistung** neu bestimmt [3]. Dazu wurde das Normal LM7 über einen Zwischenschritt mit dem Kryoradiometer bei $\lambda = 632,8$ nm kalibriert. Da das Kryoradiometer bei $\Phi = 1$ mW arbeitet und für LM7 die Leistung $\Phi \geq 5$ mW sein muß, erfolgte diese Kalibrierung über einen Trapempfänger als **Transfernormal** [4]. Vom Trapempfänger ist auf Grund der verwendeten Si-Photodioden aus eigenen Messungen bekannt, daß für Strahlungsleistungen $\Phi \leq 10$ mW keine Nichtlinearität auftritt.

Durch elektrische Kalibrierungen von LM7 wurden der Temperaturkoeffizient β_t und der Leistungskoeffizient β_P bestimmt ($\beta_t = 0,088$ %/°C; $\beta_P = 0,02$ %/W) und können somit zur Korrektur der Meßergebnisse benutzt werden.

Damit steht mit LM 7 ein Normal zu Verfügung, mit dem Laserstrahlung im sichtbaren und nahen infraroten Spektralbereich sowie bei 10,6 µm (CO_2-Laser) im genannten Leistungsbereich mit einer Meßunsicherheit < 0,1 % gemessen werden kann.

Für den Leistungsbereich bis 120 W wurde ein **Normal für mittlere Leistungen** (HLR300) gebaut und vermessen [5]. Es besteht aus einer kommerziellen ebenen Absorberscheibe (strahlungsempfindlicher Durchmesser 18 mm), bei der auf der Rückseite neben der Thermosäule auch eine elektrische Heizvorrichtung angebracht ist. Diese Absorberscheibe ist in ein wassergekühltes Gehäuse eingebaut worden. Der Empfänger wurde mit Strahlungsleistungen von 5 W $\leq \Phi \leq$ 10 W eines Nd:YAG-Lasers (λ = 1064 nm) und eines CO_2-Lasers (λ = 10,6 µm) mit LM7 kalibriert, d.h. es wurde das Produkt $\alpha(\lambda) \cdot f_K$ aus Gl. (2) bestimmt. Dabei wurde der Empfänger mit auf 20°C temperiertem Wasser bei einer Durchflußrate von 1 l/min gekühlt. Der Leistungskoeffizient konnte durch elektrische Kalibrierung mit Hilfe der Heizvorrichtung bis $\Phi \leq$ 120 W gemessen werden (β_P = -0,015 %/W). Damit steht mit HLR300 ein Normal zur Verfügung, mit dem Kalibrierungen von Laserleistungsmessern im Leistungsbereich bis 120 W mit einer Meßunsicherheit von 0,5 % möglich sind.

Für den anschließenden Leistungsbereich bis 1 kW wurde ein kommerzieller Durchflußempfänger so modifiziert und vermessen, daß er als **Normal für hohe Leistungen** (SCT391) eingesetzt werden kann. Die zu messende Strahlung fällt durch eine Eintrittsöffnung mit einem Durchmesser von 55 mm in ein 580 mm langes Absorberrohr. Außen ist um das Absorberrohr eine Kühlschlange gewickelt, die, von Wasser durchströmt, die absorbierte Leistung abführt. Ein zwischen den Windungen der Kühlschlange angebrachtes Heizband ermöglicht eine elektrischen Kalibrierung. Die Erwärmung des Kühlwassers dient als Meßsignal für die Leistung. Sie wurde bei dem unmodifizierten Empfänger mit einem thermoelektrischen Sensor gemessen, der zwischen Vorlauf- und Rücklauf-Kühlleitung befestigt war.

Da die Empfindlichkeit des Durchflußempfängers wesentlich von der Durchflußrate des Kühlwassers abhängt, sollte nach Herstellerempfehlung vor jeder Strahlungsmessung eine elektrische Kalibrierung mit Netzspannung (230 V~) bei konstant gehaltener Durchflußrate durchgeführt werden. Diese Art der elektrischen Kalibrierung mit Wechselspannung und bei nur einer Leistung (1,5 kW) ist für ein Normal, das im Bereich von 100 W bis 1 kW eingesetzt werden soll, zu ungenau. Deshalb wurde die elektrische Kalibrierung auf Betrieb mit einem Gleichspannungsnetzgerät umgestellt, mit dem eine rechnergesteuerte Kalibrierung über den ganzen Einsatzbereich möglich ist. Die Durchflußrate des Kühlwassers wurde bei diesen elektrischen Kalibrierungen konstant auf $V_{S,0}$ = 1,5 kg/min gehalten und zusätzlich mit einem Massedurchflußmesser bestimmt. Die auf Normbedingungen ($V_{S,0}$ = 1,5 kg/min; $P_{el,0}$ = 100 W) bezogene Empfindlichkeit $s_{el,0}$ eignet sich zur Langzeitüberwachung der Konstanz des Normals. Für andere Durchflußraten V_S und Leistungen $P_{el} = U_H \cdot I_H$ ergibt sich dann die reale Empfindlichkeit zu:

$$s_{el,real} = s_{el,0} \cdot (V_{S,0}/V_S) \cdot [1 + \beta_P \cdot (P_{el} - P_{el,0})] \tag{3}$$

Die Strahlungsempfindlichkeit s_{Str} wurde durch Kalibrierung mit HLR300 bei Φ = 100 W und λ = 10,6 µm bestimmt. Hieraus ergibt sich wieder das Produkt $\alpha(\lambda) \cdot f_K$, so daß zusammen mit der im Bereich 100 W $\leq \Phi \leq$ 1000 W zu bestimmenden elektrischen Empfindlichkeit auch die Strahlungsempfindlichkeit bekannt sein sollte.

Es zeigte sich jedoch, daß die Reproduzierbarkeit der elektrischen Empfindlichkeit wie auch der

Strahlungsempfindlichkeit relativ schlecht war. Zusätzlich ergab sich bei Verwendung von Leitungswasser zum Kühlen die Schwierigkeit, daß dessen Temperatur jahreszeitlich schwankte und im Sommer unterhalb des Taupunktes lag. Mit dem Einsatz eines Thermostaten konnte zwar die Temperatur auf 20°C eingestellt werden, jedoch reicht die Temperaturkonstanz von ±0,1°C des einlaufenden Kühlwassers offensichtlich nicht aus, da die von 100 W Heizleistung erzeugte Temperaturerhöhung nur ca. 1°C beträgt. Zur Verbesserung der Reproduzierbarkeit wurde die Vorlauf- und Rücklauftemperatur getrennt gemessen, indem der vorhandene thermoelektrischen Sensor durch je einen Pt100-Widerstand in der Vorlauf- bzw. Rücklaufleitung ersetzt wurde. Damit kann bei der computergestützten Auswertung berücksichtigt werden, daß sich eine Temperaturänderung am Vorlauf zeitverzögert auf die Rücklauftemperatur auswirkt.

Zur Analyse dieser Zeitverzögerung wurde in die Vorlaufleitung eine Heizpatrone eingebaut, die für die Dauer von 1 s mit 250 W beaufschlagt wurde, während die eigentliche Heizvorrichtung stromlos blieb. Die Kurve (1) in Bild 1 zeigt, wie sich daraufhin kurzzeitig die Vorlauftemperatur um ca. 1,3°C erhöht. Etwa 6 s später tritt eine wesentlich niedrigere, aber länger andauernde Temperaturerhöhung am Rücklauf auf (Kurve 2). Die Verzögerung rührt daher, daß das Kühlwasser so lange für das Durchfließen der Kühlschlange benötigt. Die Verbreiterung wird hauptsächlich durch die Pufferung der Wärme in der Kühlschlange und im Absorberrohr bewirkt. Mathematisch formuliert, ergibt sich der zeitliche Verlauf der Rücklauftemperatur $\vartheta_R(t)$ aus der Faltung der Vorlauftemperatur $\vartheta_V(t)$ mit der Antwortfunktion $A(t)$. Letztere erhält man aus der gemessenen Antwortfunktion in Bild 1, Kurve (2) (für eine Anregungsfunktion endlicher Länge) durch iterative Entfaltung und anschließende Normierung, derart daß das Integral $\int A(t)\mathrm{d}t = 1$ wird. Diese normierte Antwortfunktion für eine δ-förmige Anregung bei $t = 0$ ist ebenfalls in Bild 1, Kurve (3) dargestellt.

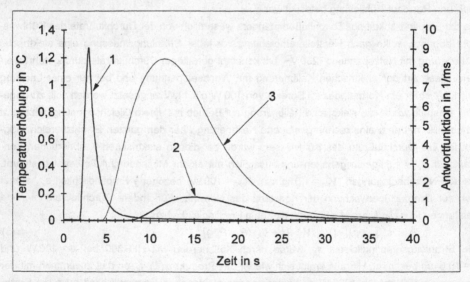

Bild 1: Die Kurve (1) zeigt die zur Untersuchung des Durchflußempfängers SCT391 benutzte Vorlauftemperaturerhöhung durch einen kurzen äußeren Heizimpuls, während Kurve (2) die daraus resultierende Rücklauftemperaturerhöhung zeigt. Kurve (3) zeigt die normierte Antwortfunktion.

Die korrigierte Temperaturdifferenz bei Bestrahlung (oder elektrischer Kalibrierung) ist dann:

$$\vartheta_{Diff}(t) = \vartheta_R - \int_0^t \vartheta_V(t') \cdot A(t-t')dt' \tag{4}$$

Die elektrische Empfindlichkeit wird dementsprechend neu definiert:

$$s_{el} = \vartheta_{Diff}/P_{el} \tag{5}$$

Zur Prüfung·des Modells wurde der elektrische Heizer 90 s nach Beginn einer Messung mit einer Leistung von ca. 100 W beaufschlagt, die nach weiteren 90 s wieder ausgeschaltet wurde. Außerdem wurde die Temperatur am Vorlauf durch zwei künstlich erzeugte Störungen kurzzeitig stark erhöht. Die Kurve (1) in Bild 2 zeigt das Ausgangssignal, wie es der unmodifizierte Empfänger liefern würde: Die Störungen am Eingang führen zu einer starken Verfälschung des Ausgangssignals. Kurve (2) zeigt, daß durch eine rechnergestützte realtime-Faltung gemäß Gl. (4) die Temperaturdifferenz vollständig von den Störungen befreit werden kann.

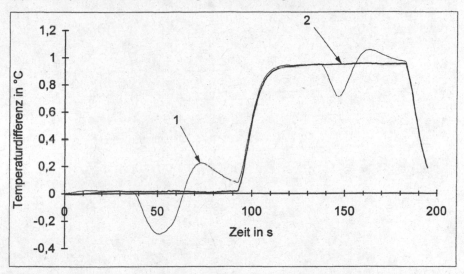

Bild 2: Zeitlicher Verlauf der Temperaturdifferenz als Ausgangssignal des Durchflußempfängers SCT391 für einen Meßzyklus mit 100 W Heizung und zwei zusätzlichen Störungen im Vorlauf (bei ca. 40 s bzw. 140 s). Kurve (1) zeigt die Temperaturdifferenz ohne und Kurve (2) mit Korrekturrechnung nach Gl.(4).

Mit dem so verbesserten Meß/Auswerteverfahren konnte die Reproduzierbarkeit der Messungen erheblich gesteigert werden. Die Messung des Leistungskoeffizienten im Bereich von 100 W bis 1 kW ergab einen nur kleinen Wert von $\beta_P = -4 \cdot 10^{-6}$ W^{-1}.

Es ist damit möglich, den modifizierten Empfänger SCT391 mit der Korrekturrechnung als Normal für hohe Leistungen einzusetzten. Die Kalibrierunsicherheit (2σ-Werte) von Laserleistungsmessern durch Kalibrierung mit SCT391 wird vorläufig mit 1 % abgeschätzt.

Literatur:

1 Möstl, K. in *Laser/Optoelektronik in der Technik*. Vorträge des 7. Internationalen Kongresses, Laser 85 Optoelektronik, 254-258
2 Möstl, K. in *New Developments and Applications in Optical Radiometry*. Inst. Phys. Conf. Ser. **92**, Institute of Physics, Bristol and Philadelphia, 1989, 11-18
3 Stock, K.D., Hofer, H., *Metrologia*, 1993, **30**, 291-296
4 Stock, K.D. et al. :*"Trapempfänger und Thermosäulen als Bindeglied zwischen Kryoradiometer und Gebrauchsempfänger"* Laser 95, Kongress K9, P11
5 Brandt, F., Möstl, K. in *PTB-Jahresbericht 1994*, 209

Optische Komponenten und Systeme
Optical Components and Systems

H. Opower, H. Hügel

Einleitung

Im Kongress F werden zwei unterschiedliche Aspekte der Lasertechnik behandelt,
nämlich einmal Gesamtsysteme, d.h. Laser, die durch Einführung spezieller, meist
komplexer Konfigurationen besondere Leistungsmerkmale erhalten, die sie für besonders
anspruchsvolle Anwendungen geeignet machen, und zum anderen neue Komponenten,
die der modernen Entwicklung der Optik, insbesondere in Richtung auf Integration und
Miniaturisierung, Rechnung tragen.

Zur ersten Gruppe gehören beispielsweise Ultrakurzpulslaser, Halbleiterlaseranordnungen
und die Nutzung der Phasenkonjugation zur Verbesserung der Strahlqualität bei
Hochleistungssystemen; zur zweiten Gruppe zählen insbesondere mikrooptische
Komponenten, Lichtleitfasertechnik und diffraktive Elemente.

Die genannten Themen sind nur im globalen Maßstab mit Einbeziehung des weltweit
vorhandenen Wissens kompetent zu behandeln. Es ist deshalb außerordentlich positiv zu
bewerten, daß nicht nur für die eingeladenen Vorträge international renommierte
Wissenschaftler und Ingenieure gewonnen werden konnten, sondern auch die Fach-
beiträge ein sehr hohes Maß an Qualität aufwiesen. Die Diskussionen und der lebhafte
Gedankenaustausch unter den Teilnehmern werden sicherlich für alle eine Bereicherung
und eine Vertiefung ihrer Kenntnisse gebracht haben. In diesem Sinne soll die
nachfolgende Zusammenstellung der Einzelbeiträge zur Auffrischung der während des
Kongresses gebotenen Präsentationen dienen.

Hans Opower

Introduction

The congress F deals with two different topics of laser technology, namely complete laser systems characterized by configurations of high complexity and special adaption to sophisticated applications on the one hand and new components especially with respect to devices of integrated optics on the other hand.

Examples for systems are ultrashort pulse lasers, semiconductor laser devices and the implementation of phase conjugated mirrors in order to improve the beam quality of high power lasers. Examples for new components are microoptic lenses, fiber technology and diffractive elements.

The areas of research mentioned above need to be treated by comprising the experiences available all over the world. In this respect we highly appreciate it that not only outstanding scientists and engineers came to present their invited papers, but also all other contributions showed a very high degree of quality triggering animated discussions. In this sense the following pages may serve as a recollection of what has been presented during the congress.

Hans Opower

Optical Components and Systems for Industrial Application. Future Demands and Present Approaches

Bernd Wilhelmi

JENOPTIK Technologie GmbH

D-07739 Jena

INTRODUCTION

The technological pace has been quickening during transition from the mobility age towards the information age. On both types of highways - the one for persons and freight and the one for information - as well as in many other fields photonics is needed in increasing quality and quantity, where the term photonics comprizes all branches of modern optics - microoptics, integrated optics, fiber optics and nonlinear optics included -, lasers, electrooptics and optoelectronics. With higher and higher speed photonics is penetrating into new fields of industry, agriculture, mining, forestry, transport, service, environmental monitoring and every day life, and it is playing the dominant part in all kind of tasks connected with sensorics and monitoring. Already now photonic systems, devices and components are met at almost all manufacturing shop floors, in supermarkets and at home. However, in a rapidly rising number of cases optical components and subsystems are hidden in overall systems which contain additionally electrical and mechanical subsystems and mostly an informatic part with specific hardware and software in the very centre, see Fig. 1.

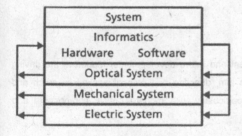

Fig.1
Structure schema of typical systems for industrial use

It is particularly this link between photonics on the one hand side with electronics and informatics on the other hand which spark future fast growth in key applications. In connections with these facts, the role and importance of the "classical" optical and precision-mechanical industry experienced dramatic changes, see S. BÜTTNER 1993. Whereas in the past most companies of this branch manufactured optical components and assembled systems for final users, nowadays many of them supply components and subsystems to system-integrating companies in other branches. The positive part of this message is the following: the comparatively small turnover with optical components triggers the progress in large branches, e.g., semiconductor lasers, where the turnover is about $ 500 Mio per year, became decisive components in the much larger communication and information market.

Thus, demands for optical components and optical systems do no longer arise from only pure optical device concepts, but more and more from general technological, customer-oriented approaches. And the final customer is in general interested mostly in features of the total system which he intends to purchase, in its performance, functionality, reliability, smartness and, of course, costs. Whether methods and components applied in such systems originate from optics, optoelectronics, electronics or mechanics is only of importance as far as this fact influences performance and costs of the overall system. The user need not understand or even notice components and internal procedures. On the other hand, these components and procedures - in particular optical ones - set the margin within which innovative systems can be created for the market.

Thus, when discussing the future of opticals components and systems, we have to search for their role in systems for customers' applications and customers' benefit. We have to ask where optical methods determine the measuring or manufacturing strategy and where optical components are required for support of other methods. Where do we need first of all improved, cheaper or more reliable "classical" optical components, and where are demands for novel optical procedures, components and systems? Where can we make use of the potential of present optical knowledge and technology, and where do we need novel R&D results? And which innovative results do we need most urgently?

In the following we will restrict our considerations mainly to optical systems containing lasers and to components applied in connection with lasers because our congress is embedded in LASER 95 and to industrial requirements because most other application fields are tackled at other congresses proceeding here simultaneously.

1. MARKET AND INNOVATIONS

Science and technology provide the potential for present and future developments. Market is the driving force, provides expenditures required and determines aim, direction and path of applied research and development. These more general statements hold particularly with modern optical components and systems.

1.1 Needs of the Industrial Market

First, increasing requirements of industrial customers for equipment containing optical components and systems go along with increasing efforts to make all industrial processes leaner and more efficient and to manufacture more and more high-quality products. Recently "hundred per cent quality" and "product liability" became key words.

On-line monitoring, measuring procedures and fast signal processing are required for evaluating decisive product features and for controlling manufacturing processes in real time, almost without delay. Often more than one hundred pieces have to be dealt with per second, or broad materials, e.g., fabric, paper or polymer foils moving between coils are to be checked at speeds of some hundred meter per second. Geometrical parameters, surface properties and quality, material composition, mechanical, thermal, electrical and optical properties and parameters of samples have to be compared, measured and valued. In addition, similar methods and devices are required for highly automated quality assurance before delivering the product to the customer. Quality has to be thoroughly checked and certified.

Second, novel manufacturing technologies are needed for increasing accuracy, precision and efficiency of industrial processes, for replacing "classical" mechanical machinery with its typical problems of tool wear, limited production speed and dead time, for processing special materials and for improving material parameters and particularly surface properties.

1.2 History and Expectations

Optics has played a great role in monitoring and measuring in manufacturing processes, where various of the extraordinarily useful properties of light have been employed for a long time. Coming from Jena, I would like to mention at this point the foundation of an efficient optical workshop by Carl Zeiss 150 years ago which became later on one of the most famous companies for high-precision optical instrumentation, which manufactured optical systems for final users, optical components and measuring devices for quality assurance in industrial processes. Many companies followed that way, and at the end of the last century, optical industry had already become a well established industrial branch.

At the beginning of the second half of our century, everything changed in high-tech industry, and this process started with the marriage between optics and precision mechanics on the one hand side and electronics and informatics on the other hand.

Some time later, the advent of lasers opened new doors for employing light. Particularly high carrier frequency, potential for fast modulation, short pulses, monochromaticity, small divergency, high power and intensity have been under consideration since that time, see Fig. 2.

Great expectations have existed from the very beginning in the sixties. First of all, these new expectations concerned communication, optical interconnects, information storage, information processing, high-precision measuring technology, spectroscopy, material processing and medicine.

1.3 Story of Failures and Successes Since the Advent of Lasers

Inspite of the great potential of laser light for novel applications, the early expectations mentioned and many ingenious attempts, this early period was a story of many failures and rare successes in practical applications. Industry had to wait for successes in using laser light for a rather long time, and even now not all of these early expectations have been set into industrial reality. Which are main reasons of such delays and which were and are measures against?

EXPECTED SUCCESSES OF LASERS (OF LIGHT)	REASONS FOR EXPECTATIONS FEATURES OF LIGHT	NEEDS
communication optical interconnects	high carrier frequency fast modulation superposition	reliable source low-loss transmission system approach
information storage information processing	small focus short pulses inform. content of image	reliable source storage materials nonlinear system
measuring technology(1D-3D)	monochromaticity frequency stability short pulses	reliable source nonlinear control interferometry radar device
spectroscopy	monochromaticity tunability short pulses	spectral range VUV-FIR nonlinear optical modulators scanners system approach
material processing	high power small divergence small focus	reliable sources modulators scanners beam shaping system approach

Fig. 2 Expected applications of laser light

I see mainly three reasons and they are all connected with the topic of this talk, see Fig. 3.

First, for a rather long period almost all lasers had not been reliable. Thus lasers had been rarely available as "certified" optical components integrable into complex systems without risk like ordinary lamps, lenses and filters.

Only after tremendous efforts in R&D and manufacturing technology a few of the huge variety of possible (and in many cases already existing in research labs) lasers reached the necessary level of performance and reliability for industrial applicaton. This process of maturing could only be driven by the market. And thus, not the researcher, but the customer decided which lasers became most important.

Second, existing optical components were not matched to lasers and to the expected new applications. Some decisive components (e.g., fibers, wave guides, fast photo detectors, modulators and scanners) did not exist at all. Again it needed high expenditure in R&D to match existing components to new applications and to

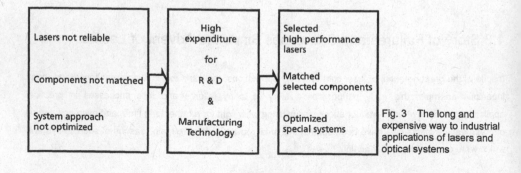

Fig. 3 The long and expensive way to industrial applications of lasers and optical systems

Component	Driver	Other Users
Laser Diodes	Communication Information Storage Material Processing (Pump sources)	Measuring Technology Spectroscopy Multi-Purpose Sensorics
Autofocus Unit	Information Storage	Measuring Technology
Fibers	Communication	Measuring Technology Spectroscopy Sensors Nonlinear Optics
Liquid Crystals	Displays	Modulation Image Processing

Fig. 4   Examples for market-driven development of optical components

develop and produce novel ones. The success of light in communication was, for instance, only possible after inventing and developing low-loss fibers and matched light sources, modulators and detectors.

Third, the integration of new components in existing system approaches failed in many cases. Novel system concepts had to be developed which take into account this specific advantages and drawbacks of all components and subsystems composing the total system. Information storage and material processsing may serve as examples.

Typically, not the market as a whole initiated and payed these developments of specific lasers, components and systems, but important and fast developing single branches, which urgently needed new approaches, served as driving forces, see Fig. 4. Among them there have been first of allcommunication, information storage and material processing, where the first two took responsibility for reliable semiconductor lasers, components and systems matched to them, and the third pushed forward high power CO_2 lasers and solid-state lasers and in conjunction scanners, modulators and mirrors with high damage threshold.

Then, under the heading "diversification", other branches have advantageously made use of performance and reliability of such mature components, and furthermore of low prices in case of mass production for the driving branch. E.g., semiconductor lasers and matched optical components developed for communication, information storage and bar code scanners have been extensively employed in measuring technology.

2.   THE INNOVATIVE POTENTIAL OF OPTICAL COMPONENTS AND SYSTEMS.
CONTRIBUTIONS OF OUR CONFERENCE

2.1 General Aspects

The innovative potential of optics and the other branches of photonics is on demand for improving the parameters of existing components and systems and for inventing and developing completely novel ones. The general function of optical systems and components can be easily described.

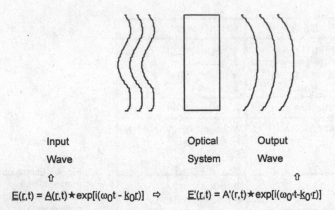

Input | Optical | Output
Wave | System | Wave

⇧ | | ⇧

$E(\underline{r},t) = \underline{A}(\underline{r},t) \star \exp[i(\omega_0 t - \underline{k}_0\underline{r})]$ ⇨ $\underline{E}'(\underline{r},t) = A'(r,t) \star \exp[i(\omega_0't - \underline{k}_0'\underline{r})]$

Fig. 5   Optical system for transforming light waves

A light beam is acting in the measuring device or in the processing unit via its electric field vector $\underline{E}'(\underline{r},t)$ which depends in a specific way on space and time coordinates and is composed of polarization direction vector, complex amplitude and propagation phase factor. For optimum interaction in the device under consideration, the electric field should exhibit appropriate center frequency, space-time dependence of the complex amplitude and polarization direction vector.

First, one optical system is needed for tranforming the light wave with field strenghts $\underline{E}(\underline{r},t)$ emitted by the source, e.g., a laser, into the desired field $\underline{E}'(\underline{r},t)$, see Fig. 5.

This transformation can be of linear or nonlinear type. Linear optical systems may consist of refractive, reflective and diffractive components, linear spatial and spectral filters, and polarizers, and they are capable of changing wave vector, complex amplitude, and polarization direction vector. And - via the complex amplitude - pulse shape, spatial profile and divergency can be affected. Here the frequency of a monochromatic wave can never be changed. By use of nonlinear optical systems the frequency of such entrance waves can be modified, too, and new frequency components can be generated. Moreover nonlinear optical filters may be used for nonlinear amplification, attenuation and stabilization as well as for optically controlled parameter changes. Such control of light parameters can also be achieved by employing electrooptical or optomechanical loops.

Second, another optical system is needed for transforming the interaction field $\underline{E}'(\underline{r},t)$ into another field $\underline{E}''(\underline{r},t)$ which fits to the features of the detector unit.

2.2 Selected topics of the innovative potential

In this section we will treat some selected topics of the innovative potential of optics which are of specific importance with respect to application, and in this context we shall preferentially refer to papers of our conference, i.e. more or less, this section intends to be a "commented table of content" of this conference.

Refractive and reflective optics

Highly corrected optical systems have become available for various practical applications by improved computer aided simulation and design and novel, more efficient manufacturing methods. Solving the inverse problem for optical systems has become a useful tool in design. High-quality aspheric and non-rotation-symmetric lenses and mirrors have been manufactured , and their use allow the reduction of the overall number of optical surfaces in optical systems; see M. Weck et al. 1995. Manufacturing cheap optical systems for low-end optics and even medium-level optics has experienced dramatic changes by employing pressed lenses made of glass and plastics.

The wave length range for which reliable optical systems with appropriate cost-performance ratio are available has been extended on both sides of the spectrum, which is of particular interest when employing ultraviolet and infrared radiation in material processing and medicine, see CH. DEUERLING et al. 1995.

Sophisticated coatings increase the performance of optical systems in all spectral ranges. Also here the inverse problem has been solved for complex spectral and spatial requirements. New technologies for manufacturing low-loss, high-performance components have been introduced, see D. RISTAU et al. 1995.

Let me draw your attention to some sophisticated optical subsystems which determine main features of devices from JENOPTIK Technologie.

First, in the laser direct printer DP 40, which writes, e.g., directly on printed circuit boards or flat panels up to 600mm by 600mm size with high resolution and extrem accuracy, the optical subsystem is responsible for focussing UV laser radiation to $10\mu m$ spots over the entire field. This is achieved by using a highly corrected 600mm optical system composed of "slice"-shaped lenses, where the optical parameters of the coating layers vary appropriately with distance from the axis.

Second, JENOPTIK Technologie presents at LASER 95 its novel high-performance glass cutting device where the laser in conjunction with a beam of cooling agent produces invisible micro distortions in extremely narrow and exactly predetermined regions which serve as sources for the final breaking process. Extraordinary quality of the edges is achieved. Here the specifically designed optical head transforms the laser radiation into a beam of desired shape and intensity profile, the parameters of which can be varied in accordance with glass type, thickness and processing speed.

Third, our company developed customer-specified laser materials processing equipment, which cuts, perforates or trenches compound materials at high speed and with high accuracy of perforations and trenches. This is achieved by rapid control of laser pulse parameters and focal spot width. Particularly, high-perfomance optical systems have been designed for the spectral range near $10\mu m$. See Fig. 6.

Diffractive Optics

Recently diffractive optics has experienced extraordinarily fast progress in R&D, and it is starting now to enter application, either in pure diffractive optical systems or in conjunction with reflective and refractive optical components. By combining advanced simulation, modelling and design of binary and multilevel phase structures with very efficient new manufacturing procedures - partially "borrowed" from microelectronics -, see, H. ZARSCHITZKY 1995, high-performance, low-cost components for various functions have become available. Applications concern for example mode control in lasers, beam shaping, beam splitting and power

746

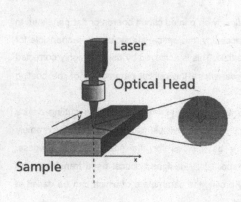

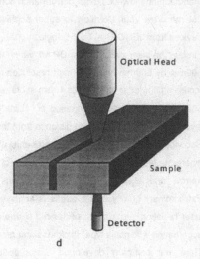

Fig. 6   Laser materials processing, JENOPTIK Technologie GmbH
  a) Robotic laser materials processing
  b) System with 3 lasers in parallel for processing materials rapidly moving   between coils
  c) Optical head for shaping and profiling the laser beam
  d) Detection and control of process parameters

attenuating, see H. U. GRUBERT et al. 1995, CH. BUDZINSKI et al. 1995, a, b, D. ASSELIN et al. 1995, J. LEGER 1995 and M. PALKE et al. 1995.

In future, 2D electronically controllable phase plates will enter application when solving sophisticated real-time problems in modifying optical waves, image processing and synthetic holography, Th. Beth et al. 1995.

Optomechanics and Electrooptics

For switching, scanning, deflecting, modulating and stabilizing light beams, fast controllable components have been constructed where electrooptic, micromechanic and piezoelectric principles and elements are in use, see, Y. DANILEIKO 1995 and I. HINKOV et al. 1995.

JENOPTIK Technologie presents at LASER 95 LCD components capable of switching and modulating light patterns. In our company such components are used in several devices, e.g., in purely optical image processors where sample features can be compared and valuated. In particular, there are novel systems for characterizing substantial features of stochastic-structured materials, e.g., sugar, coffee, cement and sand as well as certain biological objects, via incoherent-optical space frequency analysis, see Fig. 7 and E. GÄRTNER et al. 1995.

2D electronically controllable components (miniaturized arrays of deflectors, scanners, switches etc.) will gain increasing importance in connection with image display and image processing, where micro-system technology will provide efficient manufacturing concepts and processes.

Particle size determination

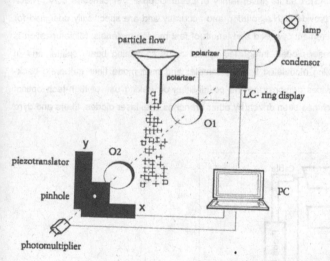

Fig. 7 Incoherent-optical space frequency analysis

Microoptics, Integrated Optics, Fiber Optics and Optical Wave Guides

For achieving reliability, compactness and comparatively low production costs, optics has been following the general lines of miniaturization and integration in recent years. Additionally, optical fibers, fiber optics as well as optical wave guides provide unique opportunities for transmitting light at low losses, which is of great importance not only for weak signals in optical communication but also in material processing and other laser applications where high power has to be transmitted from source to interaction volume , see V. UHL et al., 1995, J. Y. ZHOU et al. 1995, M. DIEKMANN et al. 1995, G. HILLRICHS et al. 1995.

Meanwhile, microoptics is capable of connecting semiconductor lasers directly and at low losses to fibers as well as fibers to optoelectronic units and radiation detectors. The coupling of laser diode arrays to fibers has also been dealt with, see K. J. EBELING 1995.

Integrated optics may moreover contain all "classical" optical functional elements, e.g., lenses, gratings, interferometers, filters, modulators etc., see G. SZARVAS et al. 1995.

However, until now, miniaturized systems composed of lumped components dominate in most practical applications because of the greater experience in production and application. For instance, integrated optical read & write heads for optical data storage application, which contain as central part the focussing grating coupler providing output- and input-coupling as well as controlled focussing to the surface of the optical disk, were developed already about one decade ago, see, S.URA et al.1987 and D. VUKOBRATOVICH et al. 1992 and have even now problems to compete with "classical" devices on the market.

Fiber sensors and integrated optical sensors promise great advantages in measuring technology, see K. PRIBIL et al. 1995. Fiber sensors with distributed Bragg gratings or SBS reflectors provide furthermore sensors with spatial resolution along the fiber. Fiber amplifiers pumped by laser diodes and coupled to fibers, optoelectronic or integrated optical circuits, see J. J. PAN et al. 1995 a, b, have successfully been developed for communication lines and are now going to be used in measuring technology, too.

JENOPTIK Technologie exhibits at LASER 95 its novel family of Laser Doppler Velocimeters LDV. These small, compact and reliable devices provide high resolution and accuracy and are specifically designed for industrial use, e. g., for on-line measurement of speed and length of fast moving materials. Miniaturization is achieved by applying high-stability laser diodes, integrated optical chips containing beam splitter and hf modulation, single-mode fibers coupling modulated light to sample, and multi-mode fiber gathering back-scattered signal radiation. These devices demonstrate the possibilities of making use of high-tech optical components, the development of which had been driven by other branches, i. e. laser diodes, fibers and gyro chips. See Fig. 8.

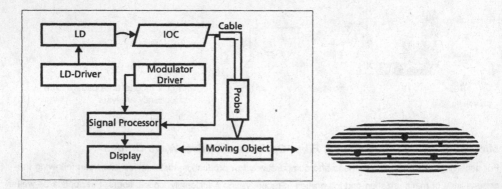

Fig. 8 Laser Doppler Velocimeter LDV, JENOPTIK Technologie GmbH

Nonlinear optics

Nonlinear optics has the greatest potential for future applications. However, at present the various nonlinear processes play only a minor part in application, where the intrinsic nonlinearities in lasers from which the laser-oscillator effect arises, the formation of light pulses by using saturable absorbers, see V. P. MIKHAILOV 1995, and the generation of harmonics are exeptions.

Let me introduce here as an example JENOPTIK's new family of laser-diode pumped solid state lasers presented at LASER 95, where the fundamental wave is converted to higher harmonics. For instance, green 3W cw radiation is now available with small, reliable and efficient Nd:YAG devices.

The lack of broad application of other nonlinear optical processes results from the high fields and interaction lengths required for achieving sufficiently intense phenomena because of low nonlinearities in most materials. Recent progress has been achieved in constructing and employing novel materials with high nonlinear susceptibilities and in using nonlinear guided waves in fibers, integrated optical devices and in specific resonant structures where the effective interaction length is increased.

In addition to the generation of harmonics, sum and difference frequency generation, recently the parametric optical oscillator has found its way from the research lab to real applications, and it provides now rather broad and reliable wavelength tuning in the visible and infrared spectral range, see U. STAMM 1995. Nonlinear optical wave mixing can furthermore be applied in measuring devices. The temporal shape of ultrashort light pulses can, e.g., be measured by using nonlinear optical auto correlators or cross correlators, see P. ZHOU 1995.

Self focussing and self defocussing are optical phenomena being of great importance in most high-intensity applications. However, the result is a negative one in most situations as, with increasing intensity, there appear beam instabilities in space and time which lead to irregularities in the light signal and eventually to beam filamentation and damage of optical components, see O. HESS 1995. On the other hand, beam narrowing resulting from self focussing in nonlinear optical samples inside laser resonators has been advantageously used for generating short light pulses because the stronger focussed high intensity part of the laser pulse is preferred when the light beam passes a narrow aperture. Self-phase modulation and cross-phase modulation can be applied for generating ultrashort light pulses and for fast optical-optical switching and scanning, see B. WILHELMI et al. 1990, W. RUDOLPH et al. 1989 and H. S. ALBRECHT et al. 1993.

In principle phase conjugation should yield ideal imaging procedures, and it can serve as a phase corrector which compensates for even complex phase distortions experienced by the light beam in a sample when the sample is transmitted before and after the pulse is reflected at the phase conjugator. Phase conjugation has successfully been applied with high-power laser amplifiers where the main amplifier stages are passed by the laser pulses on their way to the phase conjugator and back, see H. J. EICHLER 1995, and H.J. EICHLER et al. 1995.

In long single mode fibers and wave guides even weak-power laser pulses lead to interesting nonlinear phenomena because of the strong optical confinement where, e.g., solitons and soliton-like wave packets can be built up by the simultaneous action of nonlinear optical self-phase modulation and linear optical group velocity dispersion, see Z. LI et al. 1995. Moreover strong optical instabilities resulting from nonlinearities can be observed in such fibers.

A variety of nonlinear phenomena appear when nonlinear optical samples are placed into optical resonators. Bistable optical devices and optical transistors may serve as examples with great future potential.

Adaptive Optics

Adaptive optics can either be realized by employing only linear optical components in an additional control loop or by use of nonlinear optical components where self control is possible.

Recently the principles of adaptive optics, which have been already successfully applied for many years in astronomic telescopes, is now being used in small devices where, for instance, the curvature of a thin reflecting membrane is varied via an electric voltage in the control loop, see also A. SAFRONOV 1995.

Phase conjugators can be employed for nonlinear optical adaptive devices.

CONCLUSIONS

Optics in conjunction with all other branches of photonics provides an almost unlimited and rapidly increasing potential for important future applications, particularly in communication and information processing, manufacturing technology and quality assurance.

The employment of this potential requires novel high-perfomance lasers, optical components and systems which are to be designed specifically for main applicators and customers. Here special attention is to be paid to high reliability, small deviations from specified parameters, long mean time between failures high life time and low manufacturing costs. The needs of other braches, e. g., with smaller number of pieces, are to be served, first of all, by employing and modifying well developed concepts, systems and components.

Summarizing, joint and simultaneous efforts in R&D, manufacturing technology and application are required for making rapid and efficient use of this potential.

REFERENCES

H. S. ALbrecht, P. Heist, D. V. Lap, T. Schröder, B. Wilhelmi, **Ultrafast modulation of light by light,** Intern. J. Optoelectronics 8, 605-616, 1993

D. Asselin, P. Galarneau, J. M. Lacroix, P. Langlois, Y. Painchaud, M. Poirier, S. Barsetti, **Power attenuators for industrial lasers; The HVA (holographic variable attenuator) solution**, LASER 95, CONGRESS F, MÜNCHEN 1995, P1

Th. Beth, H. Aagedal, St. Taiwes, H. Schwarzer, Modern Concepts for computer-aided design in diffractive optics, OSA Proceedings International Optical Design Conference, Rochster, USA, 1994, pp. 257-260

Ch. Budzinski, H. J. Tiziani, P. Lippert, G. Lensch, G. Bostangolo, **Diffraktiver Teiler zur Strahldiagnose für Hochleistungs-Festkörperlaser**, LASER 95, CONGRESS F, MÜNCHEN 1995 a

Ch. Budzinski, H. J. Tiziani, **Strahlfeste diffraktive optische Elemente für CO$_2$-Hochleistungslaser**, LASER 95, CONGRESS F, MÜNCHEN 1995 b

S. Büttner, **Die Wettbewerbsituation in der feinmechanischen und optischen Industrie**, Verlag P. C. O., Bayreuth 1993

Y. Danileiko, B. Denker, I. Kertesz, A. Korchagin, N. Kroo, V. Osiko, A. Prokorov, S. Sverchkov, **Piezoelektrische periodische Güteschaltung für Laser mit hoher mittlerer Leistung**, LASER 95, CONGRESS F, MÜNCHEN 1995

M. Dieckmann , U. Willamowski, H. Zellmer, D. Ristau, **Dielectric coatings on fiber end faces for high power laser applications and fiber lasers**, LASER 95, CONGRESS F, MÜNCHEN 1995, P 12

Ch. Deuerling , M. Nübler-Moritz, P. Hering, H. Niederdellmann, W. Falkenstein, B. Bückle, W. Prettl, **Transmissionssysteme für den Er:YAG-Laser (2,94 µm) - eine Standortbestimmung**, LASER 95, CONGRESS F, MÜNCHEN 1995

K. J. Ebeling, **Vertikal emittierende Didenlaserarrays für effiziente Strahleinkopplung in optische Fasern**, LASER 95, CONGRESS F, MÜNCHEN 1995

H. J. Eichler, **Phasenkonjugierende Spiegel für Hochleistungsverstärker mit beugungsbegrenzter Strahlqualität**, LASER 95, CONGRESS F, MÜNCHEN 1995

H. J. Eichler, R. Menzel, R. Sander, M. Schulzke, J. Schwartz, **Phase conjugation SBS mirrors for pulsed dye lasers**, LASER 95, CONGRESS F, MÜNCHEN 1995, P10

R. Erdmann, K. Teegarden, **A self modelocked erbium doped fiber laser for field use**, LASER 95, CONGRESS F, MÜNCHEN 1995

E. Gärtner, G. Gülker, H. Hinrichs, K. Hinsch, P. Meinlschmidt, F. Reichel, K. Wolff, **All-optical correlation in speckle photography for deformation mapping in monuments**, Laser 95, Congress C, München 1995

H.-U. Grubert, J. Willner, H.-D. Hartmann, **Holographisch-optische Verbindungen für ein Multichip-Modul auf Silizium**, LASER 95, CONGRESS F, MÜNCHEN 1995

O. Hess, **Dynamik der Selbstfokussierung und Filamentierung in Hochleistungshalbleitern**, LASER 95, CONGRESS F, MÜNCHEN 1995

G. Hillrichs, K.-F. Klein, J. Langlitz, P. Schließmann, **Laserstrahlformung mit optischen Fasern**, LASER 95, CONGRESS F, MÜNCHEN 1995

I. Hinkov, V. Hinkov, E. Wagner, **Hochauflösende elektrooptische Strahlablenker**, LASER 95, CONGRESS F, MÜNCHEN 1995

V. I. Igoshin, V. A. Katulin, S. Yu. Pichugin, **Physical conceptions of developing high-power pulsed chemical lasers on photon-branching chain reactions**, LASER 95, CONGRESS F, MÜNCHEN 1995, P 16

V. S. Kazakevich, K. V. Morozov, A. L. Petrov, G. N. Popkov, **Cryogenically cooled repititively pulsed electron-beam controlled CO Laser**, LASER 95, CONGRESS F, MÜNCHEN 1995 a, P18

V. S. Kazakevich, K. V. Morozov, A. L. Petrov, G. N. Popkov, **Investigation of a pulsed EBCD CO-laser supplied from an LC transmission line**, LASER 95, CONGRESS F, MÜNCHEN 1995 b, P17

W. Koechner, **Neuere Festkörperlasersysteme für Produktionsverfahren**, LASER 95, CONGRESS F, MÜNCHEN 1995

J. Leger, **Neuartige Laserstrahlformung und Modenkontrolle mit diffraktiven optischen Elementen**, LASER 95, CONGRESS F, MÜNCHEN 1995

Z. Li, G. Zhou, **Optical solitary wave solutions in dispersion linear waveguides**, LASER 95, CONGRESS F, MÜNCHEN 1995, P3

J. W. Loughboro, **Thermal regulation of cathodes in industrial laser flashlamps**, LASER 95, CONGRESS F, MÜNCHEN 1995

752

K. R. Mann, B. Wolff-Rottke, **Cleaning of metal mirror surfaces by eximer laser radation**, LASER 95, CONGRESS F, MÜNCHEN 1995, P4

V. P. Mikhailov, K. V. Yumashev, P. V. Prokoshin, I. V. Bodnar, **CuInSSe-doped glass saturable absorbe for the Nd lasers**, LASER 95, CONGRESS F, MÜNCHEN 1995, P14

A. V. Michtchenko, A. D. Margolin, V. M. Shmelyov, **High power electrical discharge CO and CO_2 lasers utilizing vortex gas flows**, LASER 95, CONGRESS F, MÜNCHEN 1995, P19

M. Pahlke, C. Haupt, L. Niessen, H. J. Tiziani, **Diffractive optics for CO_2 laser up to 5kW**, LASER 95, CONGRESS F, MÜNCHEN 1995

J. J. Pan, Y. L. Huang, **Variable polarization beam splitter and combiner with 30 dB extinction ratio an low polarization-dependent loss**, LASER 95, CONGRESS F, MÜNCHEN 1995, P6, a

J. J. Pan, M. Shih, P. Jiang, J. Chen, J. Y. Xu and S. Cao, **Integrierter Isolator mit WDM-Modul und optischem Monitor für kompakte Faserverstärker,** LASER 95, CONGRESS F, MÜNCHEN 1995, b

K. Pribil, U. Brandenburg, **Präzisionssensor auf der Basis einer rotierenden optischen Faser**, LASER 9 CONGRESS F, MÜNCHEN 1995

G. Rabczuk, P. Kukietto, G. Sliwinski, **Output beam properties of a transverse flow CO_2 laser with DC excitation**, LASER 95, CONGRESS F, MÜNCHEN 1995, P15

U. Rebhan, K. Brunwinkel, P. Zacharias. R. Pätzel, D. Basting, **1 kHz Excimerlaser mit hoher Pulsenergiestabilität**, LASER 95, CONGRESS F, MÜNCHEN 1995

D. Ristau, R. Henking, **High-performance optical coatings produced by ion beam sputtering**, LASER 9 CONGRESS F, MÜNCHEN 1995, P2

W. Rudolph, B. Wilhelmi, **Light pulse compression**, Harwood, Chur, 1989

A. Safranov, **One-channel laser adaptive mirrors**, LASER 95, CONGRESS F, MÜNCHEN 1995, P7

J. Schirmer, P. Kohns, H. Schulz, **Achromatisches Phasenverzögerungselement auf Flüssigkristallbasis**, LASER 95; CONGRESS F, MÜNCHEN 1995, P9

M. Sierakowski, A. W. Domanski, M. Swillo, **Power-flow stabilisation of lasers by liquid crystalline optical attenuator**, LASER 95, CONGRESS F, MÜNCHEN 1995, P8

U. Stamm, **OPOs advance in Europa, but challenges remain**, Photonics Spectra 29, Iss. 3, 110-116, 199

K. D. Stock, H. Hofer, M. Pawlak, H.-Ch. Holstenberg, J. Metzdorf, **Trapempfänger und Thermosäulen als Bindeglied zwischen Kryoradiometer und Gebrauchsempfänger**, LASER 95, CONGRESS F, MÜNCHE 1995, P11

G. Szarvas, M. Barabas, G. Erdei, **Auslegung anisoptroper Wellenleiterlinsen mit einem kommerziellen 3D Konstruktionsprogramm**, LASER 95, CONGRESS F, MÜNCHEN 1995,

V. Uhl, K. O. Greulich, J. Wolfrum, H. Albrecht, D. Müller, T. Schröder, W. Triebel, **Quarzglas-Wellenleiter für die Transmission von Hochleistungslaserimpulsen,** LASER 95, CONGRESS F, MÜNCHEN 1995,

S. Ura, T. Suhara, H. Mishihara, N. J. Koyama, **Focussing grating for integrated optical disk drive,** Electron. and commun. 70, 1-10, 1987

D. Vukobratovich, **Principles of optomechanical design**, Applied Optics and Optical Engineering, Vol. XI, editors: R. R. Shannon, J. C. Wyant, Academic, Boston 1992

M. Weck, J. Everts, T. Henning, R. Lebert, H. Özmeral, S. Zamel, **Nicht-rotations-symmetrische Optiken: Herstellung und Anwendung in der Lasertechnik,** LASER 95, CONGRESS F, MÜNCHEN 1995, P5

B. Wilhelmi, M. Kaschke, W. Rudolph, **Ultrafast wavelength shift of light induced by light,** in: Optics in complex systems, SPIE VOL. 1319, pp. 81-82, Washington 1990

H. Zarschizky, **Binäre und mehrstufige Elemente mit Submikrometerstrukturen,** LASER 95, CONGRESS F, MÜNCHEN 1995

J.-Y. Zhou, G.-N. Lu, Z.-X. Li, **Ein metallischer Hohlleiter für nichtlineare Frequenz-konversion und FIR-Wellenleitung,** LASER 95, CONGRESS F, MÜNCHEN 1995

P. Zhou, H. Schulz, P. Kohns, **Korrelationsmeßsysteme für ultrakurze Laserimpulse,** LASER 95, CONGRESS F, MÜNCHEN 1995

Transmission Systems for the Er:YAG-Laser (2,94 μm) – State of the Art

C. Deuerling*, W. Prettl*, M. Nuebler-Moritz[+], H. Niederdellmann[+], P. Hering[‡], W. Falkenstein[#], B. Rückle[#]

*Institute of Applied Physics, University of Regensburg, 93040 Regensburg, Germany;
[+]Department of Oral and Maxillofacial Surgery, University of Regensburg, 93042 Regensburg, Germany;
[‡]Institute for Laser Medicine, University of Düsseldorf, 40225 Düsseldorf and Max-Planck-Institute for Quantum Optics, Garching, Germany;
[#]Baasel Lasertechnik GmbH, 82319 Starnberg, Germany

INTRODUCTION

Various facilities to transmit Er:YAG laser radiation at a wavelength of 2,94 μm exist or are under development. Because the Er:YAG laser was primarily designed for medical applications, a proper transmission system for this laser has to meet the following requirements for medical as well as for physical and optical demands[1,2,3,4]:

- high transmission of the specific laser wavelength
- high damage threshold
- high flexibility
- low bending losses (to minimize them the numerical aperture should be at least 0.2, but not much larger than 0.5, to keep the beam divergence at the distal fiber end in an acceptable range)
- sterilizability
- low, or better no toxicity and/or hermetic sealing of the transmitting materials
- inertness to organic fluids such as blood
- durability against ablation debrides
- the material should not be hygroscopic
- acceptable costs and/or long lifetime

DISCUSSION

In 1965, KAPANY and SIMMS reported about experiments with an As-S-chalcogenide glass fiber[5]. Thus chalcogenide fibers were the first optical fibers investigated in the infrared field[6]. They have sufficient stability against moisture and acids. The Er:YAG laser wavelength of 2,94 μm is in the transmission range of various chalcogenide glasses with an attenuation of about 1 dB/m[6,7,8]. Nevertheless, because of a generally high dn/dT (n: refractive index; T: temperature) and the resulting selffocussing, power transmittance is strongly limited[3]. The mechanical and, above all, the toxic drawbacks of these materials, made from elements such as As, S or Te, brought these fibers to an outsider position among the Er:YAG transmission systems meanwhile[2,3].

Zirconiumfluoride glasses were developed by Marcel and Michel POULAIN in 1974 and patented as an optical fiber in 1977 [9]. Corresponding to the very low intrinsic losses of heavy metal fluoride glasses (HMFG) an attenuation of 0,04 dB/m at 2,94 µm was reported for a ZBLAN-fiber[3] (ZrF_4-BaF_2-LaF_3-AlF_3-NaF-Na_4, IFS). According to the IFS company, the minimum bending radius of a 300 µm glass cladded IFS-fiber currently is 14 mm. The production costs of glass fibers are on an acceptable level because of the simpler production methods compared to crystalline fibers. Like the chalcogenide fibers, HMFG-fibers are brittle, easily soluble and hygroscopic. Hoya Corp. managed to improve both hardness and inertness with an AlF_3-ZrF_4-YF_3-CaF_2-NaCl - HMFG-fiber, but this improvement led to a greater minimum bending radius and to an increased attenuation of about 1 dB/m [3]. The most important problem of HMFG-fibers, their toxicity, prohibits direct use of the fiber end onto tissue.

However, the high transmittance of HMFG-fibers enables a multicomponent solution of light transmittance: Aesculap-Meditec Corp. has already commercialized an European-licensed Er:YAG laser system. After being transmitted through a zirconium fluoride fiber over the long distance, the laser beam is coupled into a short fused silica fibertip of 400 µm core diameter. According to the company, the maximum distal pulse energy is up to 500 mJ at a pulse duration of about 200 µs. These data allow the use of this transmission system even in medical non-contact or shaping treatments[10]. Due to the necessity of a coupling optic which is integrated into a handpiece, only the final fibertip is flexible. Though an endoscopic application has not yet been reached with this concept, the Er:YAG-laser was introduced at least in medical surface disciplines like ophthalmology, dermatology or as a dental laser.

The single-crystal sapphire fiber was first realized as an optical quality fiber in 1989 at Stanford University by JUNDT, FEYER and BAYER[11] and is distributed by Saphikon since 1994[12]. Despite of the high price, high fragility and poor flexibility of the fiber, it is currently favored for Er:YAG-transmission systems. According to the company, the effective numerical aperture is 0.2 and the minimal attenuation is approximately 0.8 dB/m (R_{core} = 800 µm). Commercially available samples usually have 3 dB/m [11]. The advantages of this fiber are it's inertness, high damage threshold and non-toxicity. In our own investigations the damage threshold surpassed 1 kJ/cm² and therefore permits most medically desired energy densities at the distal end, assuming that a fiber length of 3 meters is not exceeded. The high additional loss caused by the polyester cladding restricts the cladded area to the final 10 cm at the distal end, just to protect the fiber end from humidity. The slow and sensitive edge-defined film-fed growth (EFG)[12] production method leads to high prices and certain deviations in transmission properties. The main disadvantage of this fiber system is the missing cladding and flexibility for various invasive medical applications. Nevertheless, the utility of the sapphire fiber has been demonstrated in successful pre-clinical in vitro Er:YAG-sclerectomy and trabecular ablation[12,13].

Baasel Lasertech Corp. has achieved a non-toxic output of Er:YAG radiation at the application region via the new concept of an Er:YAG-handpiece-laser: a handpiece weighing 70 g with an outer diameter of 25 mm and a length of 140 mm includes the whole Er:YAG-laser source. It is water-cooled and power supplied externally. The radiation is directly coupled into a 35 mm fiber tip, at the moment being low-OH fused silica with core diameter of 400 µm. The loss of the actual transmission device is in the range of 3 dB. The distal pulse energy is currently limited by the company to 40 mJ at a pulse duration of 200 µs and repetition rates of 4 Hz. A distal energy density of 30 J/cm² is obtained and has already been applied successfully in contact to porcine meniscus in a preliminary in vitro study[14] as well as for Er:YAG scerectomy[15].

	Chalco-genide	Halo-genide	Halogenide + handpiece	Sap-phire	handpiece laser	hollow waveguide	Liquid core fiber	Silicon
Transmission	+	++	o	o	o	--	o	?
Damage threshold	--	+	o	++	-	o	+	?
Flexibility	o	+	-	-	-	-	++	+
Numerical Aperture	o	+	+	+	+	-	+	?
Sterilizability	o	o	+	++	+	o	o	++
Biocompatibility	--	-	+	++	++	++	o	++
Inertness	o	o	+	++	+	+	+	+
Surface hardness	-	o	+	++	+	o	++	+
not hygroscopic	o	-	+	+	+	+	-	+
Costs	o	o	o	-	+	+	++	?

table 1: Capability of the different Er:YAG transmission systems according to the demanded requirements.

Hollow waveguides development towards 3 µm transmission is observed with increasing interest. CROITORU's dielectric layered, plastic hollow waveguides have shown an attenuation of 12 - 16 dB/m at the Er:YAG wavelength, respectively, depending also on the thickness of the dielectric AgI-layer[16]. In 1992, HARRINGTON mentioned that the damage threshold of metallic hollow waveguides is just as low for 2,94 µm because little work had been done in this field[3]. However, the necessity of gas cooling and the strong dependence of transmission on bending radius[17], which may overheat and eventually damage the waveguide and/or the touched tissue, hinders this transmission type from being used endoscopically. Furthermore, the contamination or even occlusion of the bare distal end by ablation debrides can affect the transmission in an uncontrollable manner[18].

Experiments with liquid core waveguides are carried out by HERING and coworkers at the Institute of Laser Medicine in Düsseldorf and the Max-Planck-Institute for Quantum Optics in Garching. The flexibility of this waveguide is very high and the damage threshold of the used fluid itself is higher than 700 J/cm^2. The attenuation of a complete fiber is about 3 dB/m. The current configuration uses tetrachloromethane for the core, a special teflon-FEP tube as cladding and fused silica as windows. Because the fluid is toxic and has a low boiling point of 77°C, efforts have been made to avoid the risk of leakage. The principal ability to transmit Er:YAG laser radiation with high pulse energy has been proven repeatedly. Present investigations are described in a recent publication[19].

Since 1989 Amorphous Materials Corp. is working on high purity float zone silicon. Semiconductors are IR-transparent insofar as the bandgap is larger than the photon energy, so silicon could be fabricated into fibers which transmit 2,94 µm[1]. The stage of development of these fibers, including the research of F. Falk's group in Jena, Germany, do not yet allow conclusions of the utility of these fibers. A comparison of the available or future transmission systems are given in table 1, however with preliminary personal subjective judgements.

CONCLUSION

Transmission systems for the Er:YAG-laser comparable to fused silica fibers with respect to handling and biocompatibility are not yet available. At the moment, the decision for the most suitable transmission system has to be fitted to the main demands with compromises for each single application. Though, none of the discussed transmission systems is fully developed, but each of them has potentials for further improvements.

Literature:

1. S.M. Fry: SPIE Vol.1067, Opt. Fib. Med. IV, pp. 242-247, 1989
2. V. Artjushenko, N. Ivchenko, V. Krupchitsky et al.: SPIE Vol. 1420, Opt. Fib. Med. III, pp.157-168, 1991
3. J.A. Harrington: SPIE Vol. 1649, Opt. Fib. Med. VII, pp. 14-22, 1992
4. G. Müller, K. Dörschel, J. Helfmann et al.: SPIE Vol. 1649, Opt. Fib. Med. VII, pp. 148-161, 1992
5. N.S. Kapany and R.J. Simms: Infrared Phys., Vol. 5, pp. 69-80, 1965
6. T. Miyashita and T. Manabe: IEEE J. Quant. Electron., Vol. QE-18(10), pp. 1432-1450, 1982
7. M. Saito and M. Takizawa: J. Appl. Phys., Vol. 59(5), pp. 1450-1452, 1986
8. P.A. Tick and D.A. Thompson: Photonics Spectra, pp. 65-68, July 1985
9. M. Poulain: Photonics Spectra, pp. 68-69, July 1985
10. M. Moritz, H. Niederdellmann, C. Deuerling et al.: J. Clin. Las. Med. Surg. 13, pp. 23-26, 1995
11. D.H. Jundt, M.M. Fejer and R.L. Byer: Appl. Phys. Lett. 55 (21), pp. 2170-2172, Nov 1989
12. J.J. Fitzgibbon, H.E. Bates, A.P. Pryshlak et al.: Proc SPIE, Biomed. Opt., Feb 1995
13. C.T. Troy: Photonics Spectra, pp. 135-140, Sept 1994
14. C. Deuerling: Diplomarbeit, 1995
15. W. Wetzel, R. Otto, W. Falkenstein et al.: German J. Opth. 4, pp. 242ff, 1995
16. N. Croitoru, M. Dror, M. Alaluf et al.: SPIE Vol. 1649, Opt. Fib. Med. VII, pp. 24-33, 1992
17. C.C. Gregory, J.A. Harrington, R.I. Altkorn et al.: SPIE Vol. 1420, Opt. Fib. Med. VI, pp. 169-175, 1991
18. P.H. Cossmann, V. Romano, S. Spörri, et al.: Las. Surg. Med. 16:66-75, 1995
19. S. Diemer, W. Fuß, M. Haisch, J. Meister and P. Hering: Proc. SPIE 2396, Biomed. Fib. Opt., 1995.

Laser Beam Shaping with Optical Fibers

G. Hillrichs[1]*, K. Mann[1], K.-F. Klein[2], J. Langlitz[2], P. Schließmann[2]
[1] Laser-Laboratorium Göttingen, Hans-Adolf-Krebs-Weg 1, D-37077 Göttingen, [2] Fachhochschule Gießen-Friedberg, * Present Address : Fachhochschule Merseburg

Abstract
The spatial intensity profile of a laser beam is of crucial importance for many laser applications. Homogeneous illumination or ablation of a sample by laser radiation requires a "flat top" intensity distribution. As an alternative to complex optical systems the use of optical fibers has been tested for beam shaping of XeCl excimer lasers. By controled bending of a small fiber segment flat top intensity distributions can be generated. First applications are demonstrated.

Introduction
The spatial intensity profile of a laser beam is one of the most important properties of a laser beam. For many laser types (i.e Nd:YAG and excimer lasers) this profile deviates strongly from a "flat top"-distribution[1]. However, homogeneous irradiation or ablation at a well defined fluence level is important for many industrial[2] and medical laser applications[3]. Especially for excimer lasers several sophisticated optical systems[1,4,5] have been developed to transform the laser beam profile into a flat-top profile without hot spots. These methods are based on splitting and mixing of different parts of the laser beam. As a possible alternative to complex bulk optical systems we studied the use of large core multimode optical fibers for homogenization of a XeCl excimer laser beam. Averaging over different parts of the laser beam is achieved by coupling of laser radiation into many fiber modes[6]. Controled bending of the fiber leads to mode mixing and hence to a distribution of the laser intensity over the whole fiber core. Then a "flat top" intensity distribution should be achievable at the output endface of the fiber.

Experimental
A setup similar to that in Ref. 7 is used. The laser beam is coupled into the fiber core unter carefully controled coupling conditions by a video microscope. The fiber is guided without bendings or mechanical stress to a pair of two fiber holders with a fixed distance of typically 3 cm. Rotating around the midpoint of the distance between the two fiber holders enables controled choice of a single fiber bending with defined bending radius. The near field intensity at the fiber output endface is observed by a beam profiling system (Laser-Laboratorium Göttingen).
Fused silica fibers (Heraeus Quarzglas) with core diameters from 300 μm to 1000 μm were used. We investigated the influence of different bending radii and different coupling conditions on the profile and the total intensity at the fiber output endface. The stability of the intensity profile against further bendings of the fiber has been studied. This is important in laser applications, where the fiber output end is not fixed in space but is moved to various positions on the sample surface.

As a first application example the ablation of polyimide with fiber guided and homogenized 308 nm excimer laser radiation was tested. Holes are drilled by imaging the intensity distribution at the fiber output endface onto the sample surface.

Results

In Fig. 1a a cut through the laser beam profile at the fiber front surface is shown. The laser beam has been coupled centrally into the fiber core with a spot size diameter (FWHM) of about 0.8 times the fiber core diameter (600 μm). A cut through the intensity profile on the fiber output endface is shown in Fig. 1b for a straight fiber without bending. The laser energy is highest near to the fiber axis and the outer profile diameter is given by the fiber core dimenension. Under this conditions homogenization is not observed. A corresponding result for a fiber with one bending with radius 60 mm is shown in Fig. 1c. The output intensity profile approaches a nearly ideal flat top profile for this conditions. The full two-dimensional profile is shown in Fig. 1d. When decreasing the bending radius we observed a nearly unchanged output profile for a wide rande of radii. Decreasing the radius below a certain limit leads to a change of the profile to a nearly ideal flat top profile. A measurement of the fiber transmission as a function of bending angle (or bending radius) is shown in Fig. 2. Transmission is constant over a wide range and there is no drop of transmission at the bending radii which are appropriate for homogenization. Only at very small bending radii fiber transmission falls off. A ray tracing picture (Fig. 3) shows how the laser beam tends to fill the whole fiber core after passing the bending region.

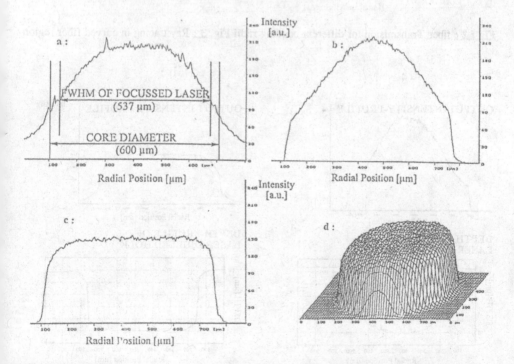

Fig 1: Laser intensity profiles. a : coupling surface, b : "linear" fiber, c, d : curved fiber

760

We checked this result for misaligned coupling geometries and found that the homogenization effect is not disturbed for a wide range of coupling conditions.

The stability against further bendings at different locations of the fiber has been tested, too. After homogenization of the output profile by a first controlled bending only a very strong second bending has a significant influence on the output profile.

As an application example we choose hole drilling into a polyimide sample, a material which is widely used in microelectronics. In a non-contact way the fiber output intensity distribution has been imaged (1:1) onto the sample by a fused silica lens system. Fig. 4a shows a cut through the intensity profile and the contour of the laser drilled hole for the straight fiber. There is a close

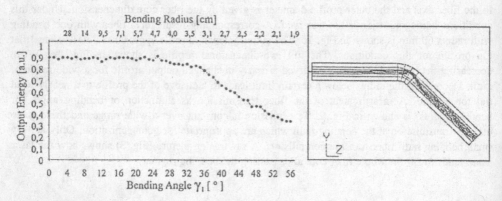

Fig : 2 : fiber Transmision for different bending radii **Fig. 3 :** Ray tracing in curved fiber region

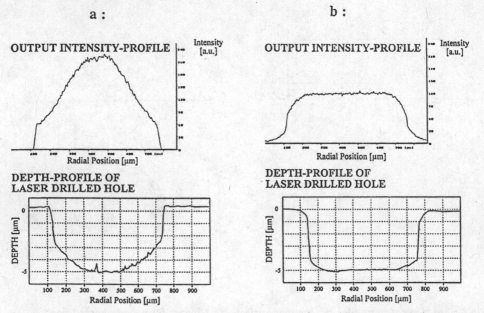

Fig. 4: Laser intensity profiles and polyimide hole contours. a: "Linear" fiber ., b: Curved fiber

correspondence between laser intensity profile and hole shape. The slighly smaller ratio of central to outer hole depth is due to a certain saturation of the polyimide ablation at higher fluences[8]. The same result for a laser beam profile homogenized by fiber bending is shown in Fig. 4b. The flat top profile of the laser intensity is completely reproduced in the hole contour.

Qualitatively similar results have been obtained for different fibers and also for different laser types.

Discussion
The beam profile studies and the application example show that fiber optic homogenization of 308 nm excimer laser radiation is possible. For this laser type and wavelength fiber optic or waveguide homogenization may become an alternative to complex optical systems for certain applications. The upper limit for the pulse energy is given by the surface damage threshold of the fiber, so that for high energy transmission large core fibers or bars should be used. Also the UV induced photodegradation of the waveguiding fused silica glass material would be less important for thicker fibers and hence lower energy densities in the material.

For industrial applications it is even more interesting to look on the 248 nm KrF excimer laser. Unfortunately, the transmission properties of commercial fuesd silica fibers are very poor at this wavelength. First results with a new modified fused silica material optimized for short wavelength transmission show very encouraging results for 248 nm transmission[9]. The formation of color centers induced by the laser light can be strongly reduced. A detailed study of the performance of the new fibers and their application to beam homogenization is cuurently under way.

In summary, beam profile shaping by waveguiding optical devices appears to be a reasonable approach. This is especially true for such applications in which a fiber optic beam guiding system can replace further homogenization and beam guiding optics. With further improvements of fiber optic materials the useful wavelength range can be extended.

References
1. K. Mann, A. Hopfmüller, H. Gerhardt : Laser und Optoelektronik 24, 42 (1992)
2. J.R. Lankard Sr., G. Wolbold: Appl. Phys. A 54, 363 (1992)
3. T. Seiler, T. Bende, J. Wollensak : Klin. Mbl. Augenheilk. 191, 179 (1987)
4. Y.E. Kawamura, Y. Itagaki, K. Toyoda, S. Namba : Opt. Comm. 48, 44 (1983)
5. M.R. Latta, K. Jain : Opt. Comm. 49, 435 (1984)
6. R.E. Grojean, D. Feldmann, J.F. Roach : Rev. Sci. Inst. 51, 375 (1980)
7. G. Hillrichs, M. Dressel, H. Hack, R. Kunstmann, W. Neu :
   Appl. Phys. B 54, 208 (1992)
8. H. Schmidt, Ph.D. Thesis, Göttingen 1994
9. K.-F. Klein, G. Hillrichs, P. Karlitschek, U. Grzesik : accepted for publication in
   "Laser in Medicine", Laser'95, Springer Verlag 1995

Advanced Solid-State Laser Systems for Manufacturing Processes

W. Koechner, R. Burnham, M. Rhoades, J. Kasinski, P. Bournes, D. DiBiase, K. Le and G. Moul
Fibertek, Inc. Herndon, Virginia

J. Unternahrer, M.J. Kukla, and M. McLaughlin
General Electric CRD

ABSTRACT

Several advanced diode-pumped Nd:YAG lasers are described which range from a picosecond system for laser-induced molecular beam epitaxy, to high average power long pulse and Q-switched systems for precision machining.

1. INTRODUCTION

An important goal of the work described here is the demonstration of excellent beam quality from high-power diode-pumped laser systems. Diode-pumped lasers have efficiencies of 3 to 4 times those of lamp-pumped versions, and the heat loading of the laser medium is reduced proportionally. This lower thermal load leads to smaller thermal aberration in the active medium of the laser. Higher overall brightness (W/cm^2-sr) will therefore be achievable from a diode-pumped laser. In industrial applications, the brightness of a laser determines process throughput and is therefore a good measure of the utility of the laser system.

Because the repetition rates for the lasers described in this paper are lower than the reciprocal of the storage lifetime for Nd:YAG, the efficiencies of the systems are optimized using pulsed diode-array pumping. The cost of the pump source is also minimized since the cost per watt of pump is lower for pulsed arrays. High average power diode-pumped lasers have been made possible by the advent of high power, high duty-cycle diode pump arrays with good efficiency and reliability [1,2]. At the present time, diode arrays operating at 20 % duty cycle with peak powers of 50 W/cm and efficiencies of over 40 % are available commercially. Typically, high duty-cycle arrays are packaged on intensely cooled heat sinks using micro-channels or impingement type fluid cooling schemes.

763

2. PICOSECOND, HIGH AVERAGE POWER Nd:YAG LASER

Epitaxial layers of GaN grown on SiC wafers are of interest for the fabrication of blue emitting laser diodes. One approach of creating such a structure is laser-induced molecular beam epitaxy (LIMBE), a technique pursued at DLR, Institut für Technische Physik, Stuttgart. In this technique, a thin layer of Ga is ablated with a laser, tiny Ga particles react with N_2 and form an epitaxial layer of GaN on the SiC substrate. Very short pulses on the order of tens of picoseconds are required to avoid heat conduction and melting of the Ga surface, because this leads to droplet formation. A high average power is desirable to achieve a reasonable epitaxial growth rate.

Figure 1 shows an optical block diagram of a high average power, modelocked Nd:YAG laser designed for LIMBE applications.

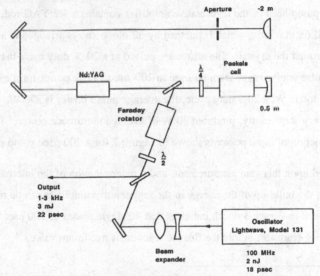

Figure 1. Optical schematic of regenerative amplifier

The system is comprised of a regenerative amplifier which is seeded with pulses derived from an end-pumped, acousto-optic modelocked oscillator (Lightwave Model 131). The Nd:YAG oscillator produces 18 nsec long pulses, with pulse energies of 2 nJ at a repetition rate of 100 MHz. The beam from the oscillator is expanded to match the mode size in the regenerative amplifier. A Faraday rotator and half wave plate provide separation of input and output at the regenerative amplifier.

The resonator of the amplifier, having a length of 1.5 m, consists of a concave-convex mirror combination and two turning mirrors to reduce the physical length of the system. The resonator mode is largest close to the concave mirror, therefore all components are located as close as possible to this mirror. Side-pumping of laser rods creates a relatively large pump cross-section which needs to be matched by a large TEM_{oo} resonator mode for good energy extraction efficiency. Furthermore, a large beam waist reduces the possibility of damage in the $LiNbO_3$ crystal.

An individual pulse from the pulse train generated by the oscillator is trapped in the resonator by switching the Pockels cell to $\lambda/4$ voltage. Once the pulse has been amplified, it is ejected by switching the Pockels cell to $\lambda/2$ voltage. The Pockels cell which has a rise time of 4 nsec, uses an acoustically damped $LiNbO_3$ crystal [MEDOX, Model 700].

The pump head of the regenerative amplifier contains a Nd:YAG rod, 4 mm in diameter and 68 mm long, which is pumped by 32 diode arrays arranged in a four-fold symmetry around the crystal. The arrays are pulsed at a 20 % duty cycle between 1 and 3 KHz, i.e. pulse widths range from 66 μsec to 200 μsec. Peak power form each array is 50 W, or 1.6 KW, total. At a 20% duty cycle, the average pump power is 320 W. The pump head, tested in a very short cavity, produced 80 W of average multimode power. The single pass gain as a function of input power is shown in Figure 2 for a 200 μsec pump pulse.

Based upon this gain measurement, and a determination of the internal losses which are about 15 %, the build-up of the energy in the regenerative amplifier can be modeled [3]. The result is plotted in Figure 3 which indicates that 40 single passes or 200 nsec are required in the 1.5 m long resonator before the fluence reaches its maximum value.

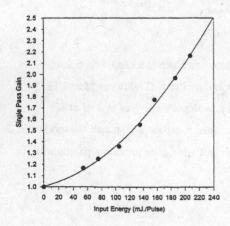

Figure 2.
Single pass gain as a
function of input power.

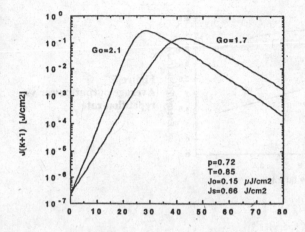

Figure 3
Fluence in the resonator vs
number of passes through
the gain medium

A trace of the output pulse taken with an autocorrelator is shown in Figure 4. The seed pulses which have a pulse width of 18 psec are broadened by the regenerative amplifier to 22 psec. Due to the high gain, the amplifier operates over a wide bandwidth, and therefore relatively little pulse broadening occurs.

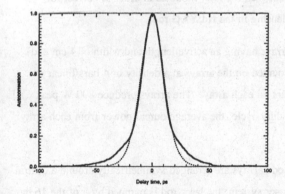

Figure 4
Auto correlation of
output pulse

The maximum output energy is 3 mJ per pulse, which represents a 63 dB gain over the seed pulse. Average output vs. repetition rate is shown in Figure 5. Because the amplifier is pulsed rather than cw pumped, the average power stays fairly constant at around 7W over the measured range of repetition rates.

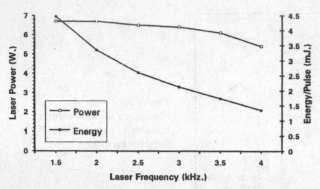

Figure 5
Average output power vs repetition rate

3. MULTI-HUNDRED WATT AVERAGE POWER
LASER FOR PRECISION HOLE DRILLING

Fibertek, Inc. has recently developed the first diode pump module for industrial laser applications using impingement-cooled diode arrays. The diode-pumped high-power laser used two identical pump modules mounted in either a stable or unstable linear resonator cavity. A 90° quartz optical rotator was mounted between the laser modules in order to correct for thermally-induced birefringence and bi-lensing in the Nd:YAG rods.

Each head is pumped by diode arrays having an active length and width of 4 cm and 0.5 cm, respectively. Diode bars are mounted on the arrays at a density of 4 bars/linear cm, at 1.2 mm transverse spacing with 16 bars on each array. The arrays produce 800 W peak power output at 806 ± 3 nm. At 20 % duty cycle, the average output power from each array is 160 W.

In the laser pump module, the diode arrays are arranged symmetrically around a central liquid-cooled laser rod. In the present laser system, the laser rod is pumped by 5 of the 16 bar arrays. The maximum optical pump input to the rod is therefore 4000 W peak, or 800 W average power over a 4 cm active length. The pump distribution in the laser rod is designed to have a maximum on-axis intensity with a center-to-edge contrast ratio of ~ 1.5:1. This centrally-peaked pump distribution helps to improve the overlap between the energy stored in the rod and the mode volume of the laser. This overlap contributes directly to the efficiency of the diode-pumped laser.

**Figure 6
Photograph of
prototype diode
pump module**

The present pump module is capable of operating with laser rod diameters of between 0.8 and 1.0 cm. The laser rod and diode arrays in the pump head are cooled through common coolant manifolds located at the ends of the pump module. De-ionized water to cool the diode arrays and laser rod enters the pump head through face seals located at the ends of the module. The pump module has the shape of a cube with 6.5 cm sides. A photograph of a prototype diode-pumped rod laser module is shown in Figure 6.

The laser was first operated with a close-coupled stable resonator using 8.5 mm diameter rods to determine the base line performance of the system. As shown in Figure 7, at 15 % duty cycle and 540 W output power, the laser reached a maximum electrical efficiency of 16 %, based on electrical input from the DC power supply.

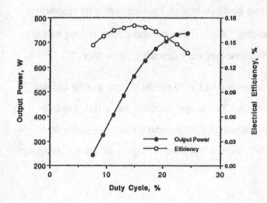

**Figure 7
Output power and efficiency
vs. duty cycle**

The electrical slope efficiency was 24 %. A maximum output of 750 W was reached at a duty cycle of 25%. The nonlinear increase in output power for duty cycles above 15 % indicates that the diode array junction temperature is rising significantly. This leads to a spectral mismatch between the emitted pump light and absorption in the Nd:YAG crystal.

Subsequent testing and optimization of the laser was performed at a 15 % duty cycle. Although the laser was capable of repetition rates in the KHz regime, the application dictated operation with 750 μsec pulses at 200 pps or 1000 μs pulses at 150 pps. At the 15 % duty cycle, the performance of the laser with a stable resonator is shown in Figure 8. The beam quality with the stable resonator was 60 mm-mrad (DΘ/4), or approximately 94 times the diffraction limit of a uniformly filled circular aperture (0.64 mm-mr). Both the near field and far field radiation patterns from the laser were uniformly-filled circular spots.

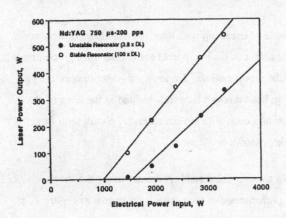

Figure 8
Laser output at 15 % cycle with a stable and unstable resonator

Several unstable resonators were investigated, and a rod imaging unstable resonator was chosen in the final design [4]. The negative-branch rod-imaging unstable resonator is much less sensitive to mirror misalignment than positive-branch confocal unstable resonators commonly used for high-power laser applications. Also, it is only about half as long as the near-concentric unstable resonator, which otherwise has very desirable properties [5].

Our particular rod-imaging unstable resonator had a 30 cm ROC rear mirror located 30 cm from the midpoint between the two laser rods. The output coupler was a flat, graded-reflectivity super-Gaussian mirror with a spot radius of 3.0 mm and a Gaussian order of 6. The central reflectivity of the mirror was 85%. Figure 9 shows the layout of the rod imaging unstable resonator.

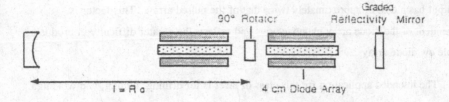

Graded
90° Rotator   Reflectivity Mirror

l = R o   4 cm Diode Array

gure 9   Schematic diagram of the rod-imaging unstable resonator used for the
high-power diode-pumped laser

The performance of the diode-pumped laser using the rod-imaging unstable resonator is
also shown in Figure 8. An average output power of 335 W was extracted at 15 % duty
cycle, with an electrical input of 3300 W. The overall electrical efficiency of the laser was 10
%, and the electrical slope efficiency was 19 %. The far-field beam parameter (spot size-
divergence product) from the laser was measured to be 3.8 mm-mrad, corresponding to 5.8
times the plane-wave diffraction limit. Therefore, a 156-fold improvement in brightness was
achieved.

4. DESIGN OF A MULTI-KILOWATT DIODE-PUMPED LASER

As part of the White House's Technology Reinvestment Program (TRP), Fibertek,
Inc., and General Electric are developing a multi-kilowatt, Q-switched Nd:YAG laser with a
repetition rate in the KHz regime.

The most recent significant development in pump array technology has been the
dramatic increase in duty cycle made possible through the use of intense impingement or
micro-channel cooled heat exchangers mounted intimately to the diode array bars. These
coolers allow the duty cycle of the arrays to be raised from a few percent to over 20 % at the
present time. With these cooling techniques, the same number of diode array bars can produce
over 10 times the average power density without a significant increase in cost.

For interpulse periods longer than the storage lifetime of Nd:YAG (i.e., repetition rates
below 4 KHz), system efficiency will be significantly higher with pulsed diodes, as compared
to cw laser diodes. In addition to efficiency considerations, there is a cost advantage in favor
of pulsed pumping. The proposed diode arrays have a peak power capability of 50 W per bar.
At 20 % duty cycle, this corresponds to 10 W average power. A comparable cw bar with 10

W output has a cost of approximately twice that of the pulsed array. This factor is independent of the diode array manufacturer, and reflects the greater difficulty in producing reliable cw diode arrays.

The intended application for this class of laser is for drilling, cutting, and welding. The generation of a near diffraction limited beam is an important goal of this program. The architecture chosen for the system is a master oscillator-power amplifier configuration as shown in Figure 10. The major components of the laser are the two diode-pumped Nd:YAG slab gain modules. Each gain module is composed of the solid-state laser slab and its mounting and cooling assemblies, and the diode pump arrays with their mounting and cooling hardware.

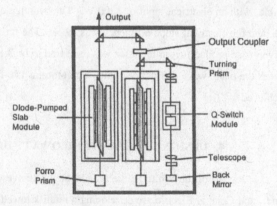

**Figure 10
Optical layout of 2.5 KW
oscillator-amplifier system**

The dimensions of the two identical Nd:YAG slabs are 21 cm x 2 cm x 0.7 cm. The slabs are used in a double pass zig-zag arrangement. The porro prisms shown in Figure 10 return the beam for a second zig-zag pass through each slab. The oscillator employs an unstable resonator with variable reflectivity output coupling. Several acousto-optic Q-switch assemblies are required to hold off lasing action. The diode arrays are designed to be mounted on a common cooling manifold. The large-area planar array thus formed will be optically coupled to the laser slab with high efficiency since no intermediate optics are required. The diodes match the planar pump face of the zig-zag slab. The laser slab will be pumped on both sides to provide uniform energy deposition and heating.

In general, it has been determined that micro-channel and impingement coolers have comparable thermal impedances. The choice of cooling technique must be based on practical considerations. Micro-channel coolers depend on coolant passages with openings of 50 to 100

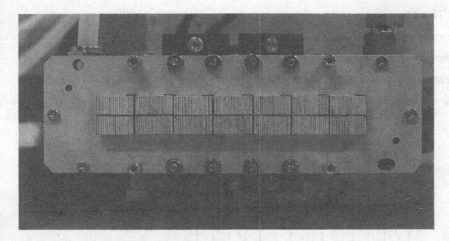

Figure 11   Diode array pump module

μm to reduce the coolant boundary-layer temperature gradient. In practice, the coolant flowing through these channels must be kept extremely clean to prevent channel clogging which causes catastrophic heating of the adjacent diode array junction. Typically, micron-sized filters and biological inhibiters must be employed to maintain coolant purity. These components reduce the coolant flow efficiency and require maintenance.

Impingement coolers, on the other hand, utilize mm-scale passages, and therefore are an order of magnitude less susceptible to clogging and wear. Typically, coolant supplies for impingement coolers require no in-line filters, and can be protected against corrosion using passive chemical inhibiters. Impingement-cooled packages feature brazed hard seals, and have extremely low failure rates from leaks and mechanical failure. Impingement-cooled diode array packages are available in formats that are highly compatible with pumping planar slab-type laser media.

Figure 11 shows a photograph of an impingement-cooled diode array module produced by Spectra Diode Laboratories which contains 14 arrays with 16 bars each. At an output of 5 mJ per bar in a 200 μsec pulse, total optical pump power form this module is 1.1 Joules per pulse. A total of four of these modules are used to pump the two laser heads of the zig-zag slab oscillator-amplifier configuration.

The optical design of the resonator, as well as performance modeling was greatly facilitated by operating the zig-zag slab in a flashlamp-pumped test bed. Diode pump loading simulation was performed under equivalent thermal loading and stored energy conditions. At an average output power of 500 W, this flashlamp laser produced a near diffraction limited beam as shown in Figure 12.

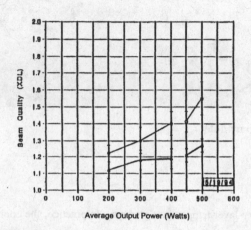

Figure 12
Beam quality vs output power M=2.0 unstable resonator

Based upon the simulation experiments, and the measured subsystem performance of the diode-pumped laser, it is expected that the system will have an average output power of 2.5 KW at a repetition rate of 2.5 KHz in a 3 x diffraction limited beam. Overall electrical efficiency is expected to be around 10 %. The laser has been assembled and is currently undergoing final subsystem check-out. Performance characterization is scheduled to begin in July 1995.

5. ACKNOWLEDGEMENT

The development of the modelocked laser was performed under a contract with DLR/Institut für Technische Physik/Stuttgart. The work on the high average power laser was carried out under the sponsorship of the Advanced Research Projects Agency, Defense Sciences Office. The authors wish to thank the sponsors for their continuing support and encouragement.

6. REFERENCES

1.   J.G. Endriz, M. Vakili, G.S. Browder, M. DeVito, J.M. Haden, G.L. Harnagel, W.E. Plano, M. Sakamoto, D.F. Welch, S. Willing, D.P. Worland, H.C. Yao, IEEE J. Quant. Electr., 29, 952 (1992).

2.   R. Beach, W.J. Benett, B.L. Freitas, D. Mundinger, B.J. Comaskey, R.W. Solarz, M.A. Emanuel, IEEE J. Quant. Electr., 28, 966 (1992).

3.   W.H. Lowdermilk, J.E. Murray, J. Appl. Phys., 51, 2436 (1980).

4.   V. Magni, S. DeSilvestri, J.J. Qian, and O. Svelto, "Rod-imaging unstable resonator for high power solid-sate lasers," Opt.Comm., 94, 87 (1992).

5.   N. Hodgson, G. Bostanjoglo and H. Weber, "Multirod unstable resonators for high-power solid-state lasers," Appl. Opt., 32, 5902 (1993).

1 kHz Excimer Laser with Excellent Pulse Energy Stability

U. Rebhan, I. Bragin, K. Brunwinkel, R. Pätzel, and D. Basting

Lambda Physik GmbH

D-37039 Göttingen

1. Motivation

Equipment development for optical microlithography is now aiming to Gigabit DRAM chips. The required feature size is shrinking below quarter micron. A major challange for optical lithography has been while the minimum features were reduced, the device dimensions and the necessary field size of the projection lenses were significantly growing with every equipment generation (WITTEKOEK). The increase in costs for lithographic lenses is becoming problematic from an economical point of view. A solution for this problem is the scanning technique. The wafer and the reticle are synchronously scanned through a limited field size of an economical projection in order to project a large chip. The scanning approach requires new equipment development and demands an advanced laser source with special features like kHz repetition rate for high throughput and excellent pulse energy stability for precise exposure dose control on the wafer.

2. KrF Laser Development

We are developing a laser source for the scanning tool generation. The basic requirements for a DUV 248nm laser are well defined:

repetition rate	1kHz minimum
average power	up to 30W broad band operation
pulse energy stability	sigma = 0.7%
gas lifetime	100 million pulses

The starting point for our 1kHz KrF laser development is the metal-ceramic lasertube technology NOVATUBETM. In order to meet the requested specifications we first optimized the gas circulation system. The electrode gap was reduced from 19mm to 16mm. A side effect is the low charging voltage of less than 16kV. The gas velocity was improved by blower optimization. Our modified preionization allows for more efficient gas exchange between consecutive laser pulses. We obtained in our test laser a clearing ratio at 1kHz operation bewteen 4 and 5.

The optimization of the gas circulation system resulted in a shock wave free operation up to 1.2kHz. The average output power is increasing nearly linear with repetition rate; the deviation is less than 3% at 1kHz. Excellent pulse energy stability demands for a high voltage power supply with an output voltage regulation precision of 0.1% because an uncontrolled voltage change of 100V results in a 1% output energy variation. The electrical excitation circuit must also be designed for low pulse energy fluctuations and high repetition rate operation. When magnetic pulse compression methods are applied - this is done to reduce the load of the main circuit switch and to extend its lifetime - unstable resetting of the saturable inductors will negatively affect the output energy stability.

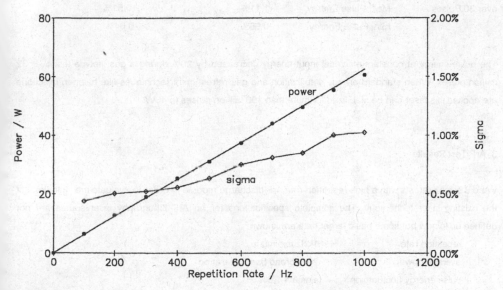

Output power and pulse energy fluctuation (1 sigma correspond to about 68% of all pulses) of our 1kHz KrF test module.

An important factor for constant laser beam parameters is the preionization of the gas discharge itself. A powerful and spatially uniform preionization is a precondition for a reproducible discharge and therefore for beam profile homogeneity and pulse energy stability. The electron density, generated by our conventional uv-spark preionization, is slightly modulated with a period of 5mm along the main electrodes. This nonuniformity is reduced in our 1kHz discharge module which is equipped with a modified preionization. There is no measureable preionization modulation along the electrode and therefore the spatial uniformity is significantly improved. The result is an excellent pulse energy stability up to 1kHz operation.

The total chip area must be exposed uniform. The width of the exposure slit and the scanning speed define the number of pulses which are superimposed at every point. Therefore, the stability of a moving average over typically 30 consecutive laser pulses is more important for scanning applications than the single pulse fluctuation. The corresponding specification is a fluctuation of ±0.7% peak to peak which is fulfilled by the modified preionization even at 1kHz. The following table shows a comparison of our standard uv-spark and the modified preionization:

Average Condition	Measured parameter	Standard Preionization	Modified Preionization
Single Pulse	Sigma	1.75%	0.74%
	Max. Pulse Energy	+5.51%	+1.97%
	Min. Pulse Energy	-5.19%	-2.64%
Moving Average	Sigma	0.40%	0.23%
over 30 Pulses	Max. Pulse Energy	+1.11%	+0.51%
	Min. Pulse Energy	-0.96%	-0.64%

The pulse energy at constant electrical input energy decreased by 20% during a gas lifetime test over 32 million pulses. When standard energy stabilization and gas refreshment techniques like halogen injections are applied this laser can be stabilized for more than 100 million pulses to 40W.

3. ArF Test Results

We did some pretests with a high repetition rate ArF discharge module in order to evaluate the limitations of the existing laser technology. The complete specifications for an ArF lithography laser source are not defined up to now but some basic target data are known:

repetition rate	1kHz minimum
average power	20W broad band operation
pulse energy fluctuatuion	sigma = 1.5%

The present status of 193nm laser technology is a maximum repetition rate of only 400Hz and typical pulse energy fluctuations are in the order of 3%. Additionally, the lifetime of the laser gas and the laser tube window is well below the target data.

We were able to demonstrate ArF operation up to 600Hz with a 20W laser module. The power is increasing nearly linear with repetition rate; the deviation is less than 2% at 600Hz. The pulse energy fluctuations are below 2% but they still exceed the target value. At very low repetition rate the ArF pulse energy is extremly stable and the measured value of sigma = 0.5% is comparable with the KrF data. But the pulse energy stability decreases with increasing repetition rate by typically 0.2% for 100Hz. The reason for this high rate is unknown and will be investigated in detail.

A first gas lifetime test at constant electrical input energy resulted in an energy drop of 22% after 12 million pulses. When standard energy stabilization and gas refreshment techniques are applied a gas lifetime of more than 25 million pulses can be expected at an energy level of 20mJ.

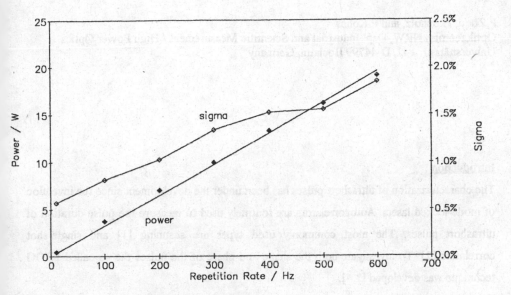

Output power and pulse energy fluctuation (1 sigma correspond to about 68% of all pulses) of our ArF test module.

4. Conclusions

Stable 1kHz operation at a power level of 60W is achieved. The laser pulse energy is extremly stable up to 1kHz due to the improvements of the high voltage power supply, the gas circulation system and the preionization. The key specification for scanning applications, the moving average over 30 consecutive pulses, is fulfilled by this KrF discharge module. Our high reprate pretest demonstrated a 600Hz discharge pumped ArF laser for the first time. The linear increase in power indicates that 600Hz operation is not the limit and a repetition rate of about 800Hz for ArF is within the scope of our gas circulation concept. The pulse energy stability improved when compared with commerially available ArF excimer lasers. The significant increase with repetition rate indicates that technical improvements will be possible.Therefore, our new high repetition rate discharge modules can be a successful basis for the next lithography laser generation.

5. Acknowledgement

This work was performed within the JESSI project E232. Research and development was supported by the German Ministery for Research and Technology (BMBF).

S. Wittekoek, Microcircuit Engineering Conference, Maastricht, 1993

Correlator-Measuring-System for Ultrashot Laser Pulses

P.Zhou, H.Schulz, and P.Kohns
Optikzentrum NRW, Dep. Industrial and Scientific Measurement / High Power Optics
Universitätsstr. 142, D-44799 Bochum, Germany

Introduction

The characterization of ultrashort pulses has been under the development since the invention of mode-locked lasers. Autocorrelators are routinely used to measure the pulse duration of ultrashort pulses. The most commonly used types are scanning [1] and single-shot correlators [2]. To investigate both the shape and phase of the pulses the so called FROG technique was developed [3,4].

We present a microprocessor controlled modular correlator system which contains all types of correlators mentioned above. In contrast to [3] a second harmonic generation (SHG) process is used to characterize the single pulse simultaneously with temporal and spectral resolution.

Operation modes of correlator system

1) Scanning and stepwise autocorrelator

The correlator system consists of several modules. The basic instrument is a scanning autocorrelator (fig. 1). The pulse is split into two parts which are delayed differently by a glass plate because the optical path length depends on the angle of incidence. The delay time τ is scanned by rotating the glass plate. With beamstairing retro reflectors and mirrors (M_1 - M_4) the two pulses are focussed into an LBO crystal for background free sum frequency generation (SHG). The intensity of SHG is recorded as a function of τ by a microcontroller and displayed on a liquid-crystal display. Using glass plates with different thicknesses we are able to cover a measurement range from several fs up to 15 ps. The rotation frequency ($\leq$ 1 Hz - 25 Hz) is controlled and stabilized to $\Delta\omega/\omega \leq 10^{-3}$ for all frequencies resulting in a very good measurement accuracy. Low rotating frequencies are of special interest for low repetition rate laser systems.

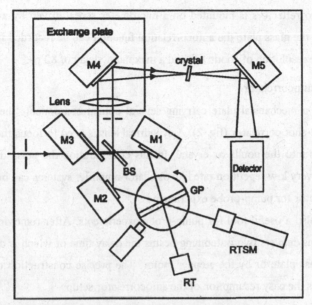

Fig. 1: Scanning and stepwise autocorrelator (BS: beam splitter, M1-M5: mirrors, RT: retro reflector,GP: glass plate, RTSM: retro reflector with stepper motor)

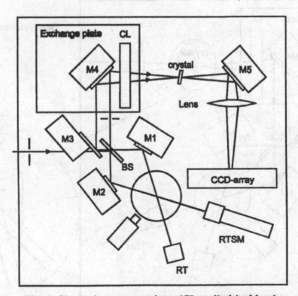

Fig. 2. Single shot autocorrelator (CL: cylindrical lens)

One of the retro reflectors is mounted on a micrometer screw driven by a stepper motor. After removing the glass plate the autocorrelation function can be recorded in a very precise way with a time resolution of about 1 fs and a maximum delay of 80 ps.

2) Single shot autocorrelator

By exchanging a mechanical plate carrying several optical components the system can be used as a single-shot correlator (fig. 2). A cylindrical lens is used to focus the two replica of the laser pulses into the nonlinear crystal. In this configuration the pulse duration of laser systems with a very low repetition rate like oscillator-amplifier systems can be measured.

3) Delay generator for pump-probe experiments

The system is also a useful tool for pump-probe experiments. After removing the exchange plate the systems delivers two outcoming beams the delay time of which is defined either by the rotating glass plate or by the stepper motor. The precise construction of the exchange plate guarantees the easy resumption of the autocorrelator setup.

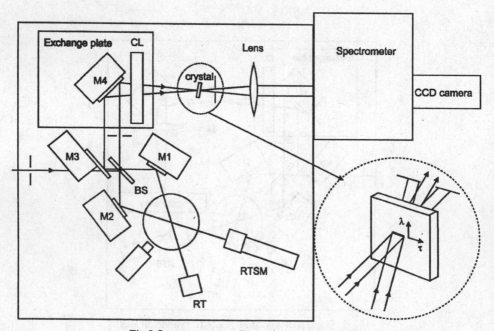

Fig.3 System as spectrally resolving autocorrelator

4) Spectrally resolved autocorrelator

By adding a small grating spectrometer the system can be expanded to measure both the pulse shape and phase (fig. 3). This measuring instrument uses a SHG process in contrast to already known FROG correlators[3]. Therefore in our case much lower pulse energy is sufficient making this technique available for high repetition rate fs lasers with low single pulse energy (0.2 nJ).The SHG is produced by an LBO crystal at the focus line of the cylindrical lens. The length of the SHG emitting area is proportional to the pulswidth. This line is spectrally diffracted by a grating spectrometer and detected by a CCD-camera. Two-dimensional information about the spectra at different delay times is simultaneously recorded.

The pulse form and phase information can be obtained from this records using a phase-retrieval algorithm as described by PAYE [4]. In contrast to [4] our measurement is done in real time, not stepwise. The spectrally resolved single shot correlation can also be used as a pulse monitor to optimize the laser system.

The arrangement of [4], where the system is configurated as scanning autocorrelator with a spectrometer can also be realized. The spectrally resolved autocorrelation can be recorded step by step with a micrometer-screw driven by a stepper motor.

Fig. 4 shows a typical record of a single shot spectrally resolved autocorrelation from a self mode-locked Ti:Sapphire laser (λ=780 nm, τ=120 fs, pulse energy 0.2 nJ). The inclination of the elliptical area indicates a linear chirp of the pulses.

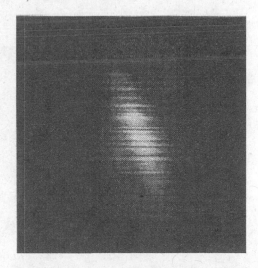

Fig. 4 A typical single shot record of a spectrally resolved autocorrelation from a self mode-locked Ti:Sapphire laser

Data acquisition and processing

The different operation modes of the system are controlled by a microprocessor. The advantage is the simultaneous recording and calculation of the data. Changes of the laser power can be measured by an additional channel in the microprocessor in order to normalize the output signal.

To reduce the noise the processor can simply be programmed as a fast scanning measuring system or lock-in amplifier for pulses with high repetition rates and as a boxcar integrating system for low repetition rates. The dynamic range is only limited by the resolution of the A/D converter. The system is equipped with 12 bit A/D converter and therefore has a dynamic range up to 4000:1.

[1] A. WATANABE at.al., Rev. Sci. Instrum. 58 (10), p1852, (10.1987)
[2] F.SALIN, et.al., Applied Optics, vol. 26, no. 21, p4528, (11.1987)
[3] D. J.KANE und R. TREBINO, Optics Letters, vol 18, no.10, p823 (05.1993)
[4] J.PAYE et al., Optics Letters vol.18, no.22, p1946 (11.1993)

Vertical Cavity Laser Diodes and Arrays for Efficient Fiber Coupling and High Speed Data Transmission in Optical Interconnects

K.J. Ebeling, B. Möller, G. Reiner, U. Fiedler, R. Michalzik,
B. Weigl, C. Jung, E. Zeeb

Department of Optoelectronics, University of Ulm
D–89069 Ulm, Germany

1 Introduction

Vertical cavity laser diodes (VCSELs) form a new class of semiconductor lasers extremely prospective for a wide range of applications, e.g. in parallel fiber links, optical interconnects and cross connects, bar code scanners, compact optical disk systems and laser printers. Parallel fiber busses using VCSEL transmitters have just entered the market. Compared to conventional edge emitting laser diodes VCSELs have distinct advantages like low divergence, astigmatic-free output beams for efficient fiber coupling or dynamic single-mode emission up to 40 Gbit/s data rates under direct modulation. Two-dimensional independently addressable laser diode arrays are easily produced and may be applied for display or monitoring purposes or coupling to fiber bundles. Like light emitting diodes VCSELs are manufactured on wafer scale and testing is also performed directly on-wafer. This is in contrast to edge emitting lasers where cleaving is required and testing is performed after dicing and facet coating which leads to comparatively high fabrication costs. Moreover, mounting and packaging requirements are expected to be much less for VCSELs than for edge emitters.

Recently, very efficient VCSELs have been presented for the 980 and 850 nm emission wavelength regime [1-3]. VCSELs for 650 nm red emission have also improved performance steadily [4] whereas VCSELs for 1.3 or 1.55μm wavelength emission are still being far from practical applications in optical communications due to immense difficulties in producing low threshold continuous wave operating devices [5]. Quite generally, VCSELs are considered attractive for fiber optics since 90 % coupling efficiency into single-mode fibers is easily achieved by simple butt-coupling. Similarly, VCSELs are ideal light sources for integrated optics since their beam profiles can be adjusted to the waveguide mode and VCSELs are also good for free space optics since two-dimensional array sources are easily formed.

In this contribution we describe basic characteristics of planar proton implanted VCSELs, discuss fiber coupling, modulation behavior, high bit rate fiber transmission and two-dimensional array performance. In more detail, we present VCSELs with very low laser threshold current, above 10mW maximum output power, 17.6 % electrical-to-optical power conversion efficiency [2] and record low emission linewidth of 30 MHz [6]. We demonstrate above 90% single-mode fiber coupling at $6\mu m$ lateral alignment tolerance and demonstrate 1 and 3 Gbit/s data transmission [7] over single-mode and multi-mode fiber at widely reduced feedback sensitivity. Finally, we discuss properties of 10 x 10 independently addressable VCSEL arrays [8], its coupling to multi-mode fiber bundles and output characteristics under high speed current modulation.

2 Proton implanted VCSELs

In edge emitting laser diodes light propagates parallel to the plane of the pn-junction and emission occurs perpendicularly from the cleaving planes of the chip. Vertical cavity lasers, on the other hand, radiate perpendicularly from the wafer surface like in a light emitting diode. The resonator axis is perpendicular to the plane of the pn-junction. The active length of the resonator is extremely short ($< 1\mu m$) and highly reflective mirrors are necessary in order to achieve reasonably low threshold currents. Fig. 1 shows a planar proton implanted vertical cavity laser structure for 980nm wavelength emission. The active layer contains three typical 8 nm thick strained layer InGaAs quantum wells embedded in GaAs barriers and AlGaAs claddings. Upper and lower mirrors consist of AlAs-GaAs Bragg reflector stacks with quarter wavelength thick epitaxially grown layers. The emission wavelength of the InGaAs quantum well is larger than the bandgap wavelength of GaAs and therefore the surrounding layers and the substrate are transmissive for the lasing light. The reflectivity of the Bragg mirrors has to be above 99% requiring about 20 layer pairs in each of the p-doped top and n-doped bottom reflector. Current is supplied through both reflectors and modulation and δ-doping is applied in the Bragg stacks to reduce the series resistance for electron and hole transport across the heterojunctions. Proton implantation serves for current confinement. From the calculated longitudinal field distribution in Fig. 2 it is seen that the effective length of the vertical cavity laser is in the order of just one micrometer. It is important to place the quantum wells exactly in an antinode of the standing wave pattern to attain a good coupling of carriers and photons. Fig. 2 also indicates that step grading of the AlAs-GaAs interfaces in the mirrors is applied to further reduce the series resistance. Typical output characteristics in Fig. 3 for VCSELs

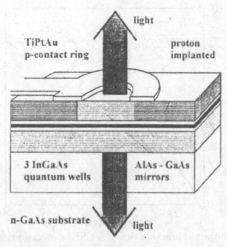

Fig. 1: Planar proton implanted vertical cavity surface emitting laser diode (VCSEL)

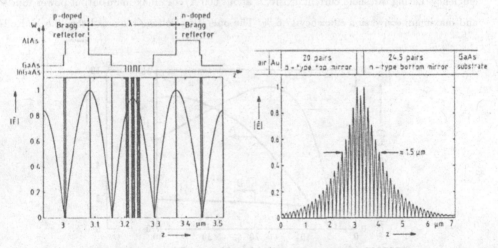

Fig. 2: Electric field distribution and band structure of VCSEL

of various diameters indicate that maximum output power increases and threshold current decreases with increasing size of the device. The threshold current of the 13µm VCSEL is 1.1mA corresponding to a threshold current density of 830 A/cm^2. The related figures for the 50µm diameter VCSEL are 8mA threshold current and 400 A/cm^2 threshold current density. Further improvement in these figures is generally observed when Bragg reflectors of still higher reflectivity are applied. In this case, however, output power tends to decrease. Fig. 4 shows light output and current voltage characteristics of a highly efficient planar proton implanted vertical

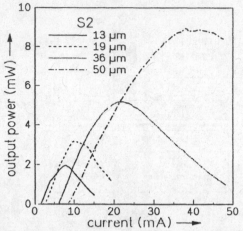

Fig. 3: Output characteristics of VCSELs of various diameters

cavity laser of 25μm active diameter. Also included is the electrical to optical power conversion efficiency. Lasing threshold current density is about 600 A/cm^2, maximum output power 15mW and maximum conversion efficiency 17.6 %. The operating voltage remains below 2.5 V. Roll-

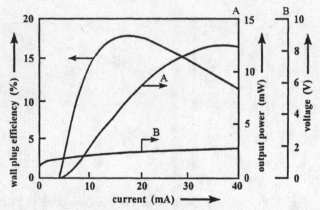

Fig. 4: Output power, driving voltage, and electrical to optical conversion efficiency as a function of the driving current of a 25μm diameter vertical cavity laser diode

over of the light output characteristics is due to device heating. Due to the short cavity length only modes of lowest longitudinal order can oscillate. At threshold the fundamental transverse mode starts lasing and depending on the diameter of the device higher order transverse modes can start to oscillate at higher current levels. For the device of Fig. 4 emission remains single-mode up to 2.5mW output power.

3   Transverse mode behavior

Due to current heating the refractive index of planar proton implanted vertical cavity lasers is increased on the symmetry axis. The lateral dependence of the refractive index can be well approximated by the parabolic profile

$$\bar{n}(r) = \bar{n}_0 - \bar{n}_2 r^2/2 \quad . \tag{1}$$

Assuming that the index profile is parabolic throughout the cavity the transverse eigenmodes are Laguerre-Gaussian functions. For laser diameter in the order of about $10\mu m$ emission predominantly occurs in the lowest order transverse mode. The Gaussian intensity distribution

$$I(r, z) = I_0 \exp\left\{-\pi\sqrt{\bar{n}_2\bar{n}_0}r^2/\lambda\right\} \tag{2}$$

with wavelength λ is observed as near field pattern on the laser endface. The corresponding far field pattern is also circularly symmetric and of Gaussian shape. Fig. 5 shows that Gaussian far field patterns are observed experimentally.

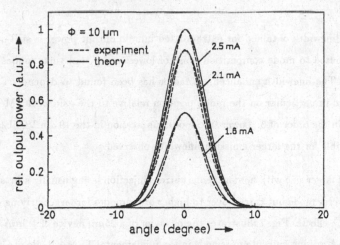

Fig. 5: Gaussian far field pattern of a $10\mu m$ single-mode VCSEL for various driving currents

For $10\mu m$ diameter laser diodes the full width half-maximum divergence angle is about five degrees. This is very much in contrast to edge emitting lasers where the radiation cone is elliptical with typical angles of $10°$ in the plane of the pn-junction and $40°$ normal to it resulting in an astigmatic beam.

The emission linewidth of a single-transverse mode VCSEL decreases linearly with decreasing inverse output power as indicated in Fig. 6. We have observed a record low linewidth of about 30 MHz in VCSELs of 10μm diameter. The slope of the curves in Fig. 6 defining a linewidth-power-product of 2.2 MHz·mW is less than in distributed feedback laser diodes.

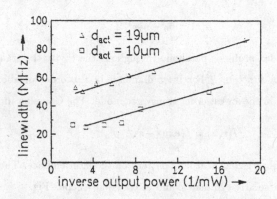

Fig. 6: Emission linewidths of 10μm and 19μm single-mode VCSEL as functions of the inverse output power

The residual linewidth obtained for extrapolated infinite output power is still quite large and may be attributed to mode competition noise of lowest order transverse modes of orthogonal polarization. The linewidth enhancement factor has been found to depend on the operating conditions and in particular on the mode position relative to the gain peak. Measured figures are typically in the order of 3. Lower side mode suppression in the 19 μm VCSEL is considered to be responsible for the larger emission linewidth observed.

In VCSELs of larger size with homogeneous current injection lasing usually starts on the fundamental transverse mode but carries over to higher order modes for larger driving current due to gain saturation effects. Fig. 7 illustrates the behavior of a 25μm device of 4.9mA threshold current. At 5.5mA driving current emission is in the fundamental TEM_{00} mode of Gaussian shape. At 6.7mA the TEM_{01}^* so-called dounut mode dominates and at 9.1mA still higher modes come up and produce multi-mode oscillation. Note that higher order modes have shorter emission wavelengths than the fundamental mode.

VCSELs of still larger size usually show simultaneous oscillation on several tranverse modes. Due to the narrow spacing individual modal emission lines overlap and produce a relatively wide spectral emission well suited far multi–mode fiber transmission at reduced modal noise.

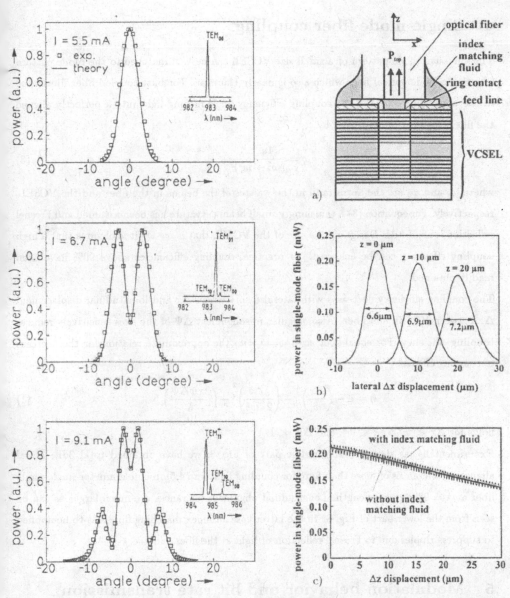

Fig. 7: Far field characteristics of $25\mu m$ VC-SEL (I_{th}=4.9 mA) for various driving currents I=5.5 mA, 6.7 mA, and 9.1 mA displaying fundamental mode, higher transverse mode and multi-mode oscillation, respectively

Fig. 8: Alignment tolerances of VCSEL-single-mode fiber coupling
a) Schematic of butt-coupling arrangement,
b) Dependence of coupled power on lateral VCSEL-fiber displacement for various distances Z=0, 10, 20μm between VCSEL and fiber endface,
c) Dependence of coupled power on longitudinal VCSEL-fiber displacement

4 Single-mode fiber coupling

The Gaussian mode pattern of a small size VCSEL can be well matched to the mode pattern of a single-mode optical fiber which also is nearly Gaussian. For butt-coupled fiber illustrated in the upper part of Fig. 8 the coupling efficiency for launching light into a perfectly aligned the fiber is approximately given by

$$\eta = \frac{4w_v^2 w_f^2}{(w_f^2 + w_v^2)^2},$$ (3)

where w_f and w_v are the beam radii in the waistes of the beams in the fiber and the VCSEL, respectively. For equation (3) a vanishingly small beam curvature has been assumed and Fresnel reflection is neglected. Designing the size of the VCSEL that $w_v = w_f$ it is obvious that a high coupling efficiency can be achieved. In practice, coupling efficiencies above 90% have been readily obtained.

The coupling efficiency decreases with lateral displacement Δx and longitudinal displacement Δz between VCSEL and fiber. Also, angular misalignment $\Delta \Psi$ of the axes sensitively reduces coupling efficiency. For equal spot sizes $w=w_v=w_f$ the approximate relation for the decrease of coupling efficiency

$$\eta = 1 - \left(\frac{\Delta x}{w}\right)^2 - \left(\frac{\lambda \Delta z}{2\pi w^2}\right)^2 - \left(\frac{2\pi w \Delta \Psi}{\lambda}\right)^2$$ (4)

holds for $\Delta x \ll w$, $\Delta z \lambda \ll w^2$, and $\Delta \Psi \ll 1$.

Experimentally, as shown in the middle part of Fig. 8, we have obtained total 3dB lateral alignment tolerances of more than 6μm for coupling light into 4.5μm core diameter single-mode fiber at $\lambda = 980$nm wavelength. Longitudinal alignment tolerances are much larger as can be seen from the lower part of Fig. 8. In the latter case an index matching fluid has to be applied to suppress ripples due to Fresnel reflection of light at the fiber endface.

5 Modulation behavior and bit rate transmission

The output power of a VCSELs like any other laser diode can be simply modulated by varying the driving current. A small signal analysis of the laser diode rate equations can already provide insight in the general dynamic behavior. In particular, this analysis gives the frequency transfer characteristics

$$\frac{P(\nu)}{P(\nu = 0)} = \frac{\nu_r^2}{\nu_r^2 - \nu^2 + i\gamma\nu/2\pi}$$ (5)

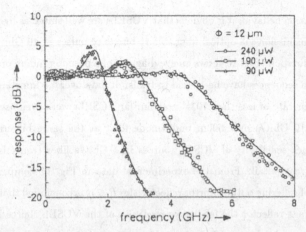

Fig. 9: Modulation response of a single-mode VCSEL for various average output powers

which describes the relative output power fluctuations at Fourier component $P(\nu)$ when the driving current is sinusoidally modulated at frequency ν about the constant bias current $\hat{j}$, i.e. $j(t) = \hat{j} + \Delta j \exp \{2\pi i \nu t\}$. The resonance frequency ν_r and the damping constant γ both increase with increasing bias current $\hat{j}$, i.e. increasing average output power. The predicted behavior is in accordance with experimental data presented in Fig. 9, where transfer characteristics of a single-mode VCSEL are plotted for various bias conditions. The 3dB frequency limit is pushed to above 5 GHz at comparatively low bias levels. Higher frequency limits of 12 GHz have been observed with improved microwave packaging of the devices. However, in this case the agreement with the theoretical transfer characteristics (5) is less convincing most probably due to electrical parasitics not included in the simple analysis.

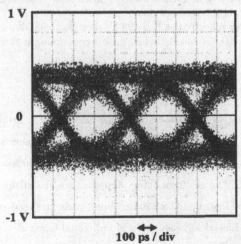

Fig. 10: Eye-diagram of 3 Gbit/s pseudo random excitation of 10 μm VCSEL

From the modulation behavior it is obvious that VCSELs are well suited as light sources for high bit rate data transmission via optical fiber. Short bursts of pulses at 40 Gbit/s were produced under direct modulation. Fig. 10 shows an eye diagram of a pseudo random optical bit sequence at 3 Gbit/s. Such sequences have been used to transmit data over 4.8 km standard single-mode fiber at a bit error rate of less than 10^{-11} and similar VCSELs were even used to demonstrate transmission at 10 Gbit/s over 500m multi-mode fiber at the same bit error rate of 10^{-11}. Moreover, feedback sensitivity of VCSEL sources in 1 Gbit/s fiber transmission experiments was studied in great detail. From the experimental data in Fig. 11 comparing two different VCSEL sources of differing output mirror reflectivities R_{top} it is concluded that by proper design of the output Bragg reflector the feedback sensitivity of the VCSEL source is largely reduced and can be made much less than that of standard edge emitting laser diodes.

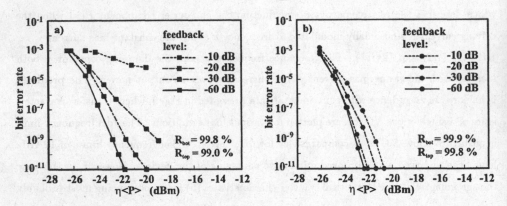

Fig. 11: Feedback sensitivity in 1 Gbit/s single-mode fiber transmission. Plotted is the bit error rate as a functions of the received power.
a)   VCSEL with mirror reflectivities R_{bot}=99.8% and R_{top}=98.5%,
b)   VCSEL with R_{bot}=99.9% and R_{top}=99.5%. Fiber coupling is through the top mirror.

6   Two-dimensional VCSEL arrays

In contrast to edge emitting laser diodes VCSELs can be easily arranged in densely packed two-dimensional arrays to be used for instance for parallel transmission in fiber bundles or to produce high output power. Fig. 12 shows the layout of an independently addressable 10 x 10 VCSEL array. VCSELs of 12μm active diameter are spaced by 250μm and electrically connected by 30μm wide lines to 120 x $120\mu m^2$ bond pads arranged at the edges of the array of 5 x 5mm^2 total size. Electrically insulation of individual lasing elements is achieved by deep proton implantation. Light output is through the bottom of the substrate.

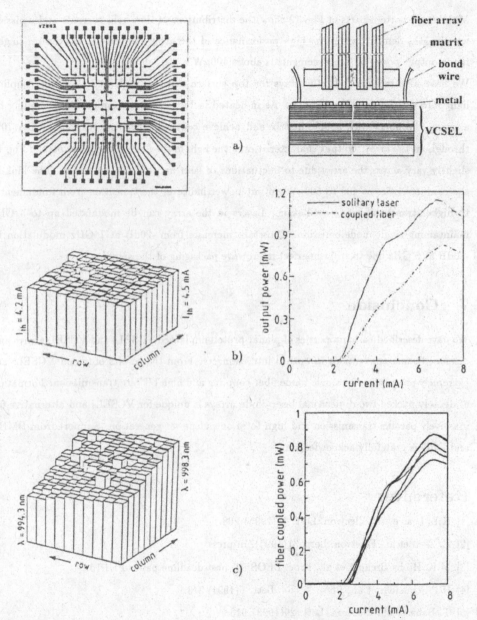

Fig. 12: Two-dimensional 10x10 VCSEL array.
a) Layout,
b) Threshold current distribution of $12\mu m$ diameter lasing elements,
c) Distribution of emission wavelengths

Fig. 13: Coupling of VCSEL array to multi-mode fiber bundle.
a) Schematic of butt-coupling,
b) Output characteristics of solitary laser and light in fiber,
c) Comparison of light power in various fibers

Middle and bottom parts of Fig. 12 show the distributions of threshold currents and emission wavelengths demonstrating the high performance of the array fabricated. Maximum single-mode output power of lasing elements is about 400μW.

We have also fabricated 10x10 arrays for top surface emission and used them for coupling light into multi-mode fiber bundles. As indicated in Fig. 13 we use simple butt-coupling by aligning the fibers with a hole matrix and achieve coupling efficiencies of larger than 70% throughout the array. Output characteristics of the light in the fiber in the lower part of Fig. 13 slightly vary across the array due to fluctuations of laser and coupling behavior. The kink in the characteristics at 400 to 500μW output power indicates the transition from fundamental to higher transverse mode oscillation. Lasers in the array can be modulated up to 5 GHz maintaining single-mode emission. Crosstalk increases from -20dB at 1 GHz modulation to -10dB at 5 GHz due to still imperfect microwave packaging of the sample.

7 Conclusion

We have described basic properties of planar proton implanted VCSELs and VCSEL arrays and discussed some applications in optical interconnects. From the results obtained VCSELs are extremely prospective for single-mode fiber coupling and high bit rate transmission. Formation of densely packed two-dimensonal laser diode arrays is unique for VCSELs and alternative for massively parallel transmission and high total output power generation. Support from BMBF and DFG is gratefully acknowledged.

References

[1] K.L. Lear et al., Electron. Lett. 31(1995) 208.

[2] E. Zeeb et al., Electron. Lett. 31(1995), in press.

[3] M.K. Hibbs-Brenner et al., Proc. LEOS'94, post-deadline paper PDI.15.

[4] R.P Schneider et al., Phot. Techn. Lett. 6(1994) 313.

[5] T. Baba et al., Electron. Lett. 29(1993) 913.

[6] G. Reiner et al., Phot. Techn. Lett. 7(1995), in press.

[7] U. Fiedler et al., Proc. ECOC'95, in press.

[8] B. Möller et al., Phot. Techn. Lett. 6(1994) 1056.

Microscopic Dynamic Aspects of Self-Focusing and Filamentation in High-Power Semiconductor Lasers

ORTWIN HESS

Institut für Technische Physik, DLR, Pfaffenwaldring 38 - 40, D – 70569 Stuttgart, Germany

Introduction

Today's users of semiconductor lasers still have to face the choice between the coherent and stable operation of relatively low-power semiconductor lasers or to profit from ample light power of high-power devices which emit light with several Watts, however, at rather poor emission quality. In the high-power semicoductor lasers, seveve instabilities are frequently responsible for the generally much inferior beam quality and spectral characteristics when compared to common semiconductor lasers with moderate output power. The transverse width of their active zone being about an order of magnitude larger than that of conventional stripe-geometry semiconductor laser diodes promotes filamentation effects [1]. In spite of considerable initiative being invested to find laser designs and configurations to aviod these effects, even in the steady state the profile of the temporally averaged optical near-field at the output facet of most broad-area laser devices appears spatially incoherent [2]. Allthemore, the dynamical variation of the intensity distribution of the broad-area semiconductor lasers shows bewilderingly complex spatio-temporal behavior [3, 4, 5].

In this presentation, results of extensive numerical simulations on the basis of the microscopic semiconductor- Maxwell-Bloch model equations [3, 5] are presented, showing the influence of the macroscopic resonator geometry on the microscopic physical processes and their relevance to the macroscopic behavior of high-power broad-area lasers. In this theoretical approach [4, 5] the polarization dynamics is fully included in the numerical modelling of the dynamics of various semiconductor lasers, while simultaneously taking into account the microscopic properties of the active semiconductor laser medium and the macroscopic characteristics of the laser device. This allows the identification of the relevant physical effects which lead to the formation of the spatio-temporal patterns and which are involved in the mutual interactions between the light field and the active semiconductor medium, on *microscopic* scales manifesting themselves e.g. in the simultaneous relevance of spectral and spatial hole-burning effects in the charge carrier distributions of the broad-area laser [4, 5]. Conventional and tapered resonator geometries are considered for broad-area lasers and the formation and propagation of the optical filaments is shown. Those recent numerical results of the spatio-temporal dynamics of broad-area semiconductor lasers are in striking agreement with new experimental streak-camera measurements [6].

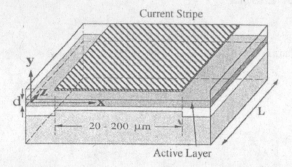

Figure 1: Schematic of a broad-area semiconductor laser. The active $GaAs$ layer (shaded dark) is sandwiched between 2 cladding layers of $Al_xGa_{1-x}As$ (white). Charge carriers which are injected through the contact stripe at the top of the device (hatched) recombine in the active zone. Light propagates in the longitudinal (z) direction. The transverse ribbed cladding structure at the bottom serves as a passive waveguide.

Theory: Microscopic Model Equations For Broad-Area Semiconductor Lasers

In a semiconductor laser, the active laser medium is the source of the optical radiation which is absorbed and amplified in the laser. In it the fundamental radiative processes, i.e. absorption, stimulated and spontaneous emission, occur simultaneously and mutually influence each other. Thus, in the theoretical description of the dynamics of the semiconductor laser, the optical properties of the laser device have to be considered self-consistently with those of the active laser medium. In the broad-area lasers, schematically shown in Fig.1 with a simple conventional resonator geometry, it is vital to carefully consider the spatial dependence of the material and optical properties. For the description of the spatially dependent dynamics of the interactions between the optical fields and the active laser medium of the broad-area laser, the recent theoretical approach [4, 5] is followed. Therby, typical shortcomings of more phenomenological models [8], such as the assumption of a specific functional dependence of the gain on the density of charge carriers or the introduction of the linewidth enhancement factor α to model the influence of the spectrally asymmetric gain, are no longer necessary. For a detailed discussion of the theoretical model and the numerical schemes, please see [3, 7].

Spatio-Temporal Dynamics

In the broad-area laser the width of the transverse stripe contact on the top of the laser is about one order of magnitude wider than the typical single-stripe laser. The lower cladding layer is frequently realized without any transverse structure. Lacking optical confinement and the confinement of the diffusing injected charge carriers in the transverse x-direction, the possibility of self-focusing effects and multi-transverse mode operation is thus given. The dynamic filamentation process in the near-field output intensity of conventional broad-area lasers process has recently been theoretically analysed [5] and experimentally observed[6]: the near-field intensity displays a bewilderingly complex

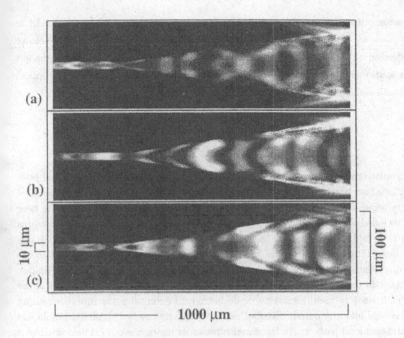

10 μm

100 μm

(a)

(b)

(c)

1000 μm

Figure 2: Propagation of optical filaments in a flared broad-area semiconductor laser. The individual plots display snapshots the intra-cavity intensity $I(x, z, t_{(i)}) \sim \left(\mid E^+(x, z, t_{(i)}) \mid^2 + \mid E^-(x, z, t_{(i)}) \mid^2 \right)$, with $i = a, b, c$ at 1.700 ns (a), 1.714 ns (b), and 1.728 ns (c) after the start of the laser. Dark shading corresonds to low intensity and bright colours to areas of high intensity. The out-coupling facet is located at the right side of the laser. Its transverse stripe width is $w_2 = 100 \mu m$. The stripe width at the left side of the laser is $w_1 = 100 \mu m$. The longitudinal extension corresponds to 1000 μm. Both facets are uncoated, i.e. $R_1 = R_2 = 0.33$.

sequence of bright and dark transverse patches and stripes which seem to migrate transversely accross the laser facet as time passes. The complex spatio-temporal patterns with wide, randomly changing channels of light and "darkness" are a result of the perpetually varying interactions between various optical filaments. The computed spatio-temporal change of the current density shows high values of the density at the edge of the laser stripe indicate that the optical field has created by the process of gain-guiding its own optical dynamically changing waveguide [5]. In the flared semiconductor laser, the active stripe width increases from one end of the laser cavity to the other one. Typically, the lasers have a length of 1 to 2 mm, a narrow stripe width of about 10 μm at one side and at the signal-out-coupling facet a transverse width of 100 to 200 μm. Thereby it is hoped to increase the output power of the laser and still suppress most of the salient spatial [2] and spatio-temporal filamentation effects which are so typical for broad-area lasers with conventional geometries. As Fig. 2 illustrates, which displays snapshots of the intra-cavity intensity $I(x, z, t_{(i)}) \sim \left(\mid E^+(x, z, t_{(i)}) \mid^2 + \mid E^-(x, z, t_{(i)}) \mid^2 \right)$, with $i = a, b, c$ at the times (a) $t_{(a)} = 1.700$ ns, (b) $t_{(b)} = 1.714$ ns, and (c) $t_{(c)} = 1.728$ ns after the start of the laser, the formation of optical filaments indeed is prevented if the transverse tripe width is norrow enough (left facet, $w_1 = 10\mu m$). However, it cannot be avoided at the wide side of the cavity ($w_2 = 100\mu m$). Reflection of the filaments at this cavity end and back-propagation towards the left side causes complex spatio-temporal intensity distributions and thus dynamically incoherent optical fields in the flared broad-area semiconductor laser.

Conclusions

Recent results of extensive numerical simulations on the basis of space-dependent semiconductor laser Maxwell-Bloch equations have been presented which show the spatio-temporal behavior of conventional and tapered-geometry broad-area lasers. Important relevant microscopic properties of the active laser medium have been included. In broad-area semiconductor lasers, complex spatio-temporal patterns are observed in the near-field output intensity as a result of self-focusing, filamentation and transverse modulational instabilities. The numerical modelling of the spatio-temporal dynamics of the broad-area laser by direct integration of the partial differential equations reveals that the combination of self-focusing effects with diffusion of charge carriers and spatial and spectral nonlinearities in the distribution of the gain leads to multi-transverse mode behavior. Soon after the initial relaxation oscillation in the near-field intensity output distribution, the transverse modes become unstable and lead to complex spatio-temporal patterns. In the flared broad-area lasers, the optical field initially is spatially uniform in the narrow end of the device. At its wide edge however, filamentation sets in – similar as in the laser with a conventional resonator geometry. Due to reflections at the outcoupling facet, the filaments are back-coupled into the resonator and thus strongly influence its behavior. In spite of some improvements in spatial coherence as compared to the conventional broad-area laser, the consequences of such a monolithic integration of "low power" coherent and "high-power" incoherent laser sections into one device still lead to dynamically rapidly changing optical filaments at the outcoupling facet and, in particular, to strong phase fluctuations. It is those fast spatio-temporal variations of the optical phase which is hard to control but vital for the realization of coherent high-power semiconductor laser sources. Clearly, the fundamental microscopic physical effects determine the behavior of the whole device. Increasing our knowledge about the mechanisms and underlying processes which lead to the self-organized, spontaneous formation of dissipative structures in the high-power semiconductor laser devices may thus be at the same time of fundamental significance and relevant for the design of new devices in which such effects may be avoided.

References

[1] G. H. B. Thompson. *Physics of Semiconductor Laser Devices*. Wiley, New York, 1980.

[2] R. J. Lang, A. Hardy, R. Parke, D. Mehuys, S. O'Brien, J. Major, and D. Welch. Spontaneous filamentation in broad-area diode laser amplifiers. *IEEE J. Quant. Electr.*, QE-30:685–694, 1994.

[3] Ortwin Hess. *Spatio- Temporal Dynamics of Semiconductor Lasers*. Wissenschaft und Technik Verlag, Berlin, 1993.

[4] Ortwin Hess. Spatio- temporal complexity in multi-stripe and broad-area semiconductor lasers. *Chaos, Solitons & Fractals*, 4:1597–1618, 1994.

[5] O. Hess, S. W. Koch, and J. V. Moloney. Filamentation and beam propagation in broad-area semiconductor lasers. *IEEE J. Quant. Electr.*, QE-31:35–43, 1995.

[6] I. Fischer, O. Hess, W. Elsäßer, and E. Göbel. Complex spatio-temporal dynamics in the nearfield of a broad-area semiconductor laser. *(submitted)*, 1995.

[7] O. Hess and T. Kuhn. Spatio-temporal dynamics of semiconductor lasers: Theory, modeling and analysis. *Prog. Quant. Electr.*, (in press), 1995.

[8] H. Adachihara, O. Hess, E. Abraham, P. Ru, and J. V. Moloney. Spatio-temporal chaos in broad-area semiconductor lasers. *J. Opt. Soc. Am. B*, 10:658, 1993.

Phase-Conjugating SBS Mirrors for Realisation of High-Power Laser Systems with Diffractation-Limited Beam Quality

HANS J. EICHLER, ANDREAS HAASE, BAINING LIU and JÖRG SCHWARTZ

Technische Universität Berlin, Optisches Institut PI1, Straße des 17. Juni 135, D-10623 Berlin, FRG

1. Introduction

Lasers with near diffraction-limited beam quality at high average output power are essential for many applications, such as material processing, nonlinear frequency conversion and plasma generation. In the case of material processing, small beam waists become possible for fabrication of micrometer structures. The Rayleigh length increases also with the beam quality, so that deep cuts with high aspect ratio can be produced. The large focal length for a given beam waist results in an increased distance between the target and the focusing optics.

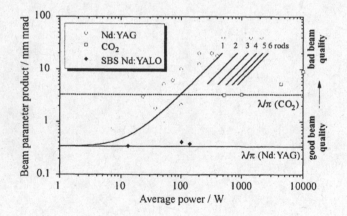

Fig.1: Actual situation of the achieved beam quality using different industrial laser systems [1]

Fig. 1 shows the achieved beam quality for two kinds of industrial laser systems for different average output powers. The CO_2 laser delivers a constant beam quality up to powers of several thousand Watts. Lower beam parameter products are possible for solid-state lasers because of the shorter wavelength. However, solid-state laser systems usually show a rapid deterioration of the beam quality with increased output power. Typically, the beam quality is reduced to ten or hundred times the diffraction-limit at average powers above 100 Watts. The beam distortions are generated by thermally induced lensing and stress birefringence in the active material.

Using phase conjugating mirrors, based on stimulated Brillouin scattering (SBS), phase distortions can be perfectly compensated and the beam quality can be improved to the theoretical diffraction-limit. We have applied this technique to different laser systems.

2. Phase conjugation by stimulated Brillouin scattering

Stimulated Brillouin scattering (SBS) is the major process used for the realisation of self-pumped phase conjugating devices. Liquid and gaseous media of high purity as well as solid materials like glass fibres can be used with high efficiency. SBS means interaction between the incident laser with hypersonic sound waves. This process can be described in the following steps:

1. A pump beam is scattered spontaneously in all directions from sound waves of the thermal noise.

2. A backscattered, slightly frequency-shifted wave interferes with the incident wave.

3. Electrostriction leads to strong density variations with the period of the interference structure.

4. The induced density variations have the frequency of the initial sound wave, which is amplified therefore.

5. The amplified sound wave reinforces the backscattering, leading to an exponential rise of the reflected signal.

Acoustic waves are pressure and density changes, they are correlated with a refractive index modulation. Therefore backscattering from sound waves can be assumed as reflection from a multi-layer system, comparable to a dielectric mirror. The most important feature of SBS mirrors is that they are self-induced and any disturbation of the incident wavefront will result in a self-adapted mirror curvature with response times in the ns range. As a result the reflected signal is phase conjugated and will be completely corrected after a second pass through stationary phase aberrations. Furthermore, SBS mirrors are self-adjusting, this property is very useful e.g. for resonator mirror applications.

For the construction of SBS mirrors the properties of the used laser have to be considered carefully. In most cases the mirror consists of a cell filled with liquids of gases and a focusing lens to increase the intensity. Focal length and scattering material have to be chosen suitable to achieve a high SBS reflectivity and a good reproduction of the transversal mode. Unwanted side-effects in the material like absorption, optical breakdown or other scattering processes have to be avoided.

3. Oscillator-amplifier systems

Master oscillators with power amplifiers (MOPAs) are an effective combination to achieve high average power. Phase distortions in highly pumped amplifiers can be compensated using SBS mirrors. Such SBS-MOPAs with output powers of more than 100 Watts have been realised [2].

Near-infrared solid state MOPAs

We developed a double-pass MOPA with a simple Nd:YAlO single rod amplifier (see Fig. 2) [3]. Carbon disulfide (CS_2) is used as the SBS medium because this liquid shows a low SBS-threshold of about 20 kW and high SBS-reflectivity up to 80%.

The TEM$_{00}$ oscillator delivers an average output power up to 6 Watt with diffraction-limited beam quality. The system works with a repetition rate of 100 Hz and emits per shot a burst of Q-switched pulses generated by a Cr^{++}-doped YAG crystal. Additional etalons increase the coherence length to achieve a high SBS reflectivity.

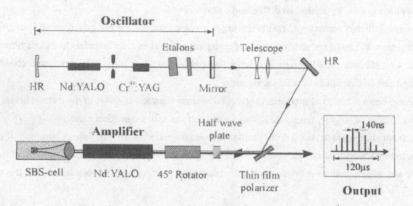

Fig. 2: Single rod Nd:YAlO double-pass amplifier with SBS-mirror

The beam of the oscillator is coupled into the amplifier rod by a thin film polariser. After the first pass, the transversal beam profile is disturbed by phase distortions and focused by the thermal lensing. After the second pass, thermal lensing and other phase distortions are compensated due to effective SBS-phase conjugation, so that the transversal beam profile is reproduced and a 1.1 times diffraction-limited beam quality is obtained (see Fig. 3). Because thermal lensing is compensated, the output power can be tuned from 4 W to 140 Watts without changes of the beam profile and the beam quality.

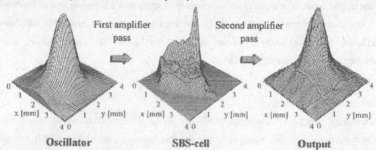

Fig. 3: Reproduction of the transverse mode profile

In contrast to commonly used Nd:YAG, the Nd:YAlO-crystals are optically anisotropic and show no remarkable stress birefringence. Therefore, depolarisation losses in the pumped laser crystals are negligible. Losses in the output power and a decrease of the beam quality due to depolarisation can be avoided. For adjusting the polarisation of the incident beam to the polarisation of maximum amplification in the YAlO crystal we used a half wave plate.

To achieve an average output power in the range of several 100 Watt serial and parallel arrangements of two such amplifier systems are under development [4].

SBS phase conjugation for pulsed dye and excimer lasers

SBS mirrors have also many perspectives in the visible and ultraviolet spectral range. Widely used pulse lasers in this range are tuneable dye lasers and excimer lasers. Also flash-lamp pumped Ti:sapphire lasers or frequency doubled or tripled Nd-lasers are of interest. The application of SBS mirrors to them has not reached the same level as for 1 μm wavelength lasers. Nevertheless, the SBS process is well characterised already. The SBS-threshold decreases from the IR to the UV so that phase-conjugating SBS-mirrors are more conveniently realised [5].

Excimer lasers are the most commonly used powerful UV laser sources. Their output characteristics are not ideal for generation of small focal spots because of oscillation in a mixture of high order transverse modes. Therefore, SBS phase conjugation has been considered for improvement of their beam characteristics similar as for solid-state lasers. However, the poor temporal coherence and beam quality of excimer lasers are possible limitations for efficient SBS.

It is essential to find suitable materials for the design of SBS mirrors for excimer lasers. Gases have been found to be less useful, because they have a comparably large "threshold" so that large pulse energies are necessary to obtain considerable reflectivity. Therefore 26 liquids with low UV absorption have been

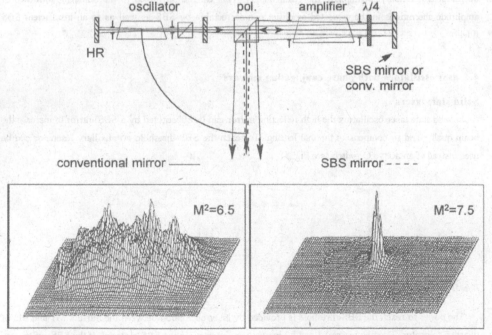

Fig. 4: XeCl excimer laser oscillator with double pass amplifier. The output characteristics are compared using a conventional mirror and SBS mirror. The given times-diffraction-limited-factors M^2 refer to the use of 2 mm intra-cavity apertures [8]

compared [6], n-hexane, cyclohexane, carbon tetrachloride and methanol have been found to be most suitable. They all have to be used with high purity grades because chemical impurities as well as solid dust particles strongly disturb the SBS process [7].

Most disadvantageous for SBS is the small temporal coherence of excimer lasers, e.g. an unmodified XeCl laser has a bandwidth of more than 1000 GHz. As result, strong focusing of the beam into the SBS liquid is necessary in order to get a low threshold [9]. But this increases the probability of other high-intensity effects decreasing the SBS fidelity. Separation of single lines from the structured XeCl laser spectrum e.g. by using a intra-cavity grating is a relatively simple solution for this problem. This is related to a bandwidth reduction to 100 GHz and leads to a output pulse energy decrease by about 40 %. On the other hand, the maximum reflectivity of the SBS mirror is roughly doubled and also the fidelity of phase conjugation increases up to about 80 %.

SBS mirrors have been applied to a XeCl laser oscillator-double-pass amplifier configuration (Fig. 4). Similar to solid state lasers, it is expected that the SBS mirror can compensate phase distortions in the highly pumped amplifier. In fact, it can be observed that the application of the SBS mirror leads to a convergent outcoupled beam with uniform shape, i.e. a reproduction of the initial oscillator beam waist. Use of a conventional mirror leads to a divergent beam with inhomogeneous profile. But if spatial beam quality is measured using the M^2 theory [10], i.e. the beam parameter product is compared, the application of the SBS mirror does not really an improve the beam quality. This can be attributed to the dominant influence of amplitude aberrations in the saturated amplifier non-correctable by SBS as well as to an insufficient SBS fidelity.

4. Laser oscillators with phase conjugating mirrors

Solid state lasers

In solid state laser oscillators the high reflecting mirror can be substituted by a SBS-mirror to increase the beam quality and to compensate thermal lensing. To reach the SBS-threshold an auxiliary resonator can be used instead of an active Q-switch (see Fig. 5).

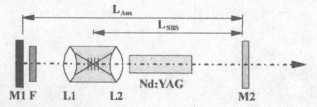

Fig. 5: SBS resonator set up

The power to reach the SBS-threshold is provided by laser oscillation between the mirrors M1 and M2 of the start resonator. The two lenses L1 and L2 increase the intensity in the SBS-material. If the SBS-reflectivity is sufficient, the laser oscillates between M2 and the SBS-mirror, leading to Q-switching. In this scheme the longitudinal and transversal mode structure of the two coupled resonators has to be matched to achieve a stable operation of the laser.

Experimentally, a system with an average output power up to 17 Watts (Fig. 6) has been realised in a near diffraction limited beam. Because thermal lensing is compensated, the laser power can be tuned from 3 to 17 Watts with no remarkable changes in the transversale beam profile. The temporal emission of the oscillator is a train of 10 Q-switched pulses per shot. The number of the pulses can be adjusted by using different attenuation filters F. To reduce the complexity of such a system the lens L1 and the mirror M1 can be integrated in the SBS-cell.

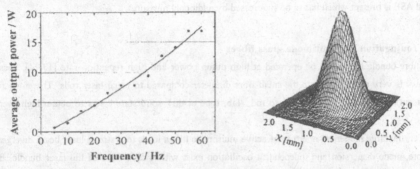

Fig. 6: *Average output power vs. frequency and spatial beam profile of a Nd:YAG-SBS-oscillator at 17 Watts output power [11]*

Excimer lasers

The most important difference of excimer SBS oscillators is that they do not need an auxiliary resonator like solid state lasers. Due to the large gain of the laser medium the initial amplified spontaneous emission (ASE) is sufficient for starting the SBS reflection. The highly reflecting mirror of the resonator can be simply replaced by the SBS mirror. The output characteristics of such resonator are shown in Fig. 7 compared with a conventional plane-plane resonator. From the temporal behaviour (a) it can be observed that pulses from the SBS resonator are about 3 times shorter than from the conventional resonator. A similar pulse shortening effect has been realised recently by a nonlinear absorber inside a XeCl resonator [12]. In contrast to that, the application of self-adjusting SBS mirrors is much simpler.

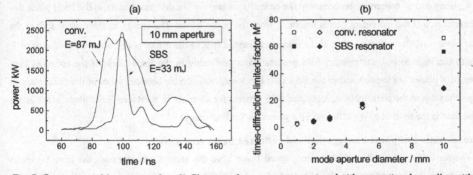

Fig. 7: *Comparison of the emission from XeCl excimer laser resonator equipped with conventional as well as with SBS resonator: (a) temporal behaviour, (b) spatial beam quality expressed by M^2 as a function of the intracavity mode aperture diameter*

Fig. 7 (b) shows the spatial beam quality as a function of the diameter of two mode apertures placed inside the resonator. When the beam diameter is limited to 1 mm, both resonators types deliver a near diffraction-limited beam. Also at larger apertures both resonator types have a comparable spatial beam quality. With nearly open apertures (10 mm diameter), i.e. near to the unmodified excimer laser output, the times-diffraction-limited factor M^2 of the SBS resonator will become better than that of the conventional one. The corresponding pulse energies E are given in Fig. 7 (a). It must be mentioned that a background of non-phase-conjugated ASE is present which has to be suppressed by additional measures.

5. Phase conjugation for multimode glass fibres

Glass fibre bundle lasers can be operated at high pump power and high repetition rate [13,14], because heat removal is very effective due to the small fibre diameter compared to usual laser rods. Therefore, fibre bundle lasers offer a potential alternative to rod, slab, tube or disk solid state lasers to obtain high average power.

However, the large numerical aperture of active multimode fibres leads to a strong laser beam divergence. Furthermore, mode dispersion and independent oscillation exist within each fibre of the laser bundle. Both factors seriously reduce the beam quality of fibre bundle lasers.

As a first step we investigated an ordered passive fibre bundle of 0.07 numerical aperture (NA) in a double-pass set-up using SBS phase conjugating mirrors (PCMs) [15]. We used also CS_2 as SBS medium. The used ordered fibre bundle consisted of about 1000 single quartz fibres melted together in a compact array with a diameter of 4 mm and a length of about 15 mm. The diameter of the fibre cores was about 100 mm and of the cladding 120 μm.

A pulsed 1.06 μm laser delivered the TEM_{00} input beam with pulse duration of 60 ns and pulse energy up to 100 mJ. The coherence length is about 1 m. About 30 fibres can be illuminated at same time.

The radiation passing the fibre bundle is collected by a lens. The SBS cell could be placed into two characteristic planes (see Fig. 8). One is the focal plane, where one observes the Fourier transformed intensity distribution of the exit surface of the fibre bundle (Fig. 8a). Another one is the image plane, where one gets the image of the exit surface (Fig. 8b).

A strong energy concentration occurs in the central spot and the six diffracted spots in the focal plane due to interference of the individual beams emerging from the single fibres. The energy of the central spot in the focal plane is much higher than that of the maximum imaged spot in the image plane, almost by a factor of 7. Due to this high energy, the primary SBS process starts preferably in the focal plane of the collected lens, when both planes are located within the SBS cell. However, the distance between spots in the focal plane is larger than that in the picture plane. A coupled SBS process for all spots is, therefore, more likely in the image plane than in the focal one. We achieved SBS processes for both cases.

Demonstration of phase conjugation and reflected beam structure

The far-field of the reflected phase conjugated beam after the second pass through the fibre bundle is shown in Fig. 9. This pattern was also observed, if the Nd:YAG laser beam passed through fibre bundle areas with bad optical quality or if the adjustment of the fibre bundle was not optimised. However, the SBS-threshold increased due to the decreased intensity.

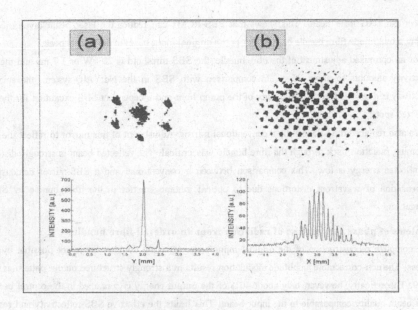

Fig. 8: Nd:YAG laser beam behind an ordered fibre bundle:

(a) far-field distribution in the focal plane and (b) image of the exit surface of the fibre bundle

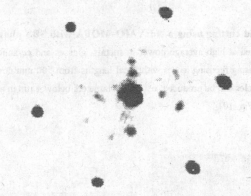

Fig. 9: Two dimensional far-field distribution of the reflected phase conjugated Nd:YAG-laser beam after second pass
through the fibre bundle

The output beam in Fig. 9 is regularly structured due to non-correctable amplitude distortions. The output far-field distribution shows therefore the typical diffraction pattern of a coherently radiating, two-dimensional aperture array. The main output energy appears in the central and diffracted beams within the diffraction angle of a single fibre core. The divergence of the central spot corresponds almost to the divergence angle of the incident Nd:YAG laser beam. This shows that SBS-PCMs can reduce the laser beam divergence double-passing a multimode fibre bundle back to that of the original input beam in the central peak.

For an optimised adjustment of the fibre bundle, the SBS-threshold is 28 kW or 1.7 mJ and the maximal reflectivity amounts to about 25 %. In comparison with SBS in the Nd:YAlO-system, the smaller SBS reflectivity is due to reduced beam quality of the pump light and independent SBS excitation for the different diffracted spots.

We also replaced the SBS-cell by a conventional mirror. Adjustment of this mirror to reflect the divergent outcoming radiation back through the fibre bundle was critical. The reflected beam is strongly distorted and the reflected energy is low. This comparison between a conventional and a SBS-mirror demonstrates the compensation of wavefront distortions due to optical inhomogeneities in the fibre bundle by SBS-phase conjugation.

Problems of phase conjugation of radiation from an ordered fibre bundle

A complete reconstruction of the original input beam after the double-pass is not possible by the SBS process. The non-correctable amplitude modulation results in a strongly structured output pattern as shown in Fig. 9. Theoretically, however, only about 40% of the output energy is contained in the central beam which has a beam quality comparable to the input beam. This limits the effective SBS-reflectivity and restricts the final application of SBS phase conjugation to fibre bundle amplifiers or oscillators. A possible solution for this is the application of an appropriate diffractive optics similar to fan-in plates as shown in Ref. 16 to combine all output spots coherently again.

6. Precision drilling and cutting using a Nd:YAlO-MOPA with SBS phase conjugation

Drilling was investigated at high average power in metals, glasses and ceramics with different thickness from 50 µm up to 1 cm, using focusing lenses with focal lengths from 500 mm down to 100 mm. Due to the high beam quality small holes can be produced, e.g. hole diameters below 4 µm in aluminium with a thickness of 100 µm were realised (Fig. 10).

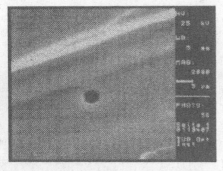 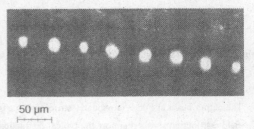

Fig. 10: Left: Precise hole in aluminium (thickness 100 µm, bore diameter 4 µm). Right: Array of holes in aluminium (bore diameter 20 µm, bore distance 45 µm)

The used laser repetition rate of 100 Hz allows fast drilling of hole arrays (one hole can be drilled with one shot). Fig. 10 (right) shows an example photograph of a hole array in aluminium. The bore diameter is 20 µm and the bore distance 45 µm. The translation stage was moved with a speed of about 0.45 cm/s. The drilling time per hole is in the range of 100 µs, so that the geometry of the holes cannot be deformed by the used translation speed. The different diameters are due to material inhomogeneities.

Because of the high Rayleigh length cuts and holes can be realised with a high aspect ratio and dimensions in the µm-range. Figure 11 shows a cutting kerf in ceramics for different positions of the focal point relative to the target. For the best configuration a kerf width of 30 µm was realised with an aspect ratio of 1:37. At high average output power, holes with high aspect ratio in thick material are possible, e.g. hole diameters of 60 µm in aluminium with a thickness of 0.5 cm and a drilling time of 0.3 seconds were demonstrated. These results show the advantages of the used high beam quality of 1.1 times the diffraction limit.

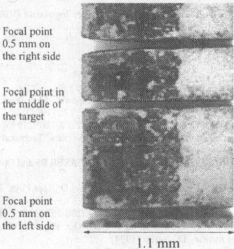

Focal point
0.5 mm on
the right side

Focal point in
the middle of
the target

Kerf width 30 µm

Aspect ratio 1:37

Focal point
0.5 mm on
the left side

1.1 mm

Fig. 11: Cutting kerf in ceramics

7. Summary

Phase conjugation by SBS is a powerful tool to increase the beam quality of high power solid state lasers up to the diffraction limit. Master oscillator amplifier systems based on Nd:YAlO as active material were realised with an average output power up to 140 Watts in a near diffraction limited beam (1.1xDL). Thermal lensing was compensated resulting in a constant beam diameter independent of the output power. Also Nd:YAG-oscillators with phase conjugating mirrors were realised with an average output power up to 17 Watt and high beam quality.

With these systems material processing experiments were carried out. Due to the high beam quality structures of 4 µm in aluminium and cuts in ceramics with a kerf width of 30 µm and an aspect ratio of 1:37 are demonstrated.

In the UV range, excimer oscillators with SBS mirrors deliver the same or a somewhat better beam quality than conventional plane-plane resonators in 3 times shorter pulses. Excimer oscillator-amplifier arrangements have been constructed with SBS mirrors. Improvement of the beam quality has still to be demonstrated in this case using excimer laser discharges without amplitude distortions. SBS mirrors for the visible and ultraviolet spectral range are well characterised and can be applied also to other laser types, e.g. Titanium:sapphire or Praseodymium lasers.

A fibre bundle distorts both amplitude and phase of a transmitted beam. Also in this case the phase distortions can be healed by a SBS phase-conjugating mirror.

Acknowledgements

The authors wish to thank the following co-workers for experimental help and stimulating discussions: D. Berger, M. Duelk, E. Geinitz, S. Heinrich, S. Jobst, O. Mehl and R. Menzel. Financial support from the Bundesministerium für Bildung, Wissenschaft, Forschung und Technologie (BMBF), the Deutsche Forschungsanstalt für Luft- und Raumfahrt (DLR) and the Verein Deutscher Ingenieure (VDI) is greatfully acknowledged.

References

1   N. Hodgson, H. Weber: "Optische Resonatoren", Springer-Verlag (1992)
2   C.B. Dane, L. da Silva, L.A. Hackel, J. Harder, D.L. Matthews, S. Mrowka, M.A. Norton, Conference on Lasers and Electro-optics (1994), Technical Digest Series, Vol. 8, p. 175 (1994)
3   H.J. Eichler, A. Haase, R. Menzel, IEEE J. Quantum Electron. 31, p. 1265-1269 (1995)
4   H.J. Eichler, A. Haase, R. Menzel, Conference on Lasers and Electro-optics, Technical Digest Series, Vol. 15, 61-62 (1995)
5   H. J. Eichler, R. Menzel, R. Sander, M. Schulzke, J. Schwartz: Proc. LASER 95 and Optics Comm., to be published
6   H.J. Eichler, R. König, R. Menzel, H.-J. Pätzold, J. Schwartz: J.Phys.D: Appl.Phys. 25, 1161-1168 (1992)
7   H.J. Eichler, R. Menzel, R. Sander, B. Smandek: Optics Comm. 89, 260-262 (1992); H.J. Eichler, R. König, R. Menzel, R. Sander, J. Schwartz, H.-J. Pätzold: Int. J. Nonl. Opt. Phys. 2, 267-270 (1993)
8   D. Berger: Diploma thesis, Optisches Institut, TU Berlin (1994)
9   H.J. Eichler, R. König, H.-J. Pätzold, J. Schwartz: Appl. Phys. B, in print
10   A.E. Siegman: Proc. SPIE Vol. 1224, 2-14 (1990)
11   R. Menzel, H.J. Eichler, E. Geinitz, A. Haase, Conference on Lasers and Electro-optics, Technical Digest Series, Vol. 15, p. 79-80 (1995)
12   M.R. Perrone, Y.B. Yao: Opt. Commun. 110, 187-191 (1994)
13   U. Griebner, R. Grunwald and R. Koch: Proc. Int. Conf. on Lasers, 319-325 (1993)
14   L.E. Zapeta: J. Appl. Phys. 62, 3110-3115 (1987)
15   H.J. Eichler, B. Liu, M. Duelk, Z. Lu, J. Chen: submitted to Optics Comm. (1995)
16   J. Morel, A. Woodtli and R. Dändliker: Opt. Lett. 18, 1520-1522 (1993)

Piezoelectric Driven Devices for Periodical Q-switching of High Average Power Lasers

Danileiko Yu., Denker B., Kertesz I*., Korchagin A., Kroo N.*,
Osiko V., Prokhorov A., Sverchkov S.
General Physics Institute, Russia Academy of Sciences
117333 Vavilova 38, Moscow, Russia. Fax (095) 1350270.
* Research Institute for Solid State Physics, Budapest, Hungary.

Introduction.

The most essential specific demand for a Q-switching device of a high average power continuously pumped laser operating at 1.06 μm is proper operation in a very high optical flux without overheating or any kind of damage.

The search of an appropriate shutter for periodically Q-switched continuously pumped laser is somewhat simplified by the fact that the gain and thus the giant pulse forming rate in continuously-pumped lasers are much lower than those typical for pulsed pumping. That is the reason why relatively slow piezoelectric driven optomechanical devices (namely Frustrated Total Internal Reflection (FTIR) shutter and Variable Base Fabri-Perot (VBFP) interferometer) may be appropriate Q-switchers for powerful continuously pumped lasers. Another potential advantages of these devices are: low (~100 V) driving voltage, high optical and thermal damage thresholds, low losses and also absolute or practical insensitivity to polarisation of the modulated laser beam. The purpose of the present paper is the study of Q-switching possibilities of FTIR and VBFP devices in powerful continuously pumped YAG:Nd lasers.

Laser description.

The laser was based on two Nd:YAG rods each of 6.3 mm in diameter and 130 mm long. Each rod was mounted in a separate water-cooled gilded pump cavity and pumped by a single Kr arc lamp. The laser power supply was able to produce both DC and several milliseconds long rectangular current pulses with variable duration-to-period ratio. The laser supplied with a 60 % reflectivity flat output mirror had output power reaching 300 W and the efficiency of about 3.5%.

FTIR shutter operation.

The principle of Q-switching with the help of FTIR shutters has been known for decades. But only recently the design and technology of these devices has advanced so far they have become reliable and reproducible enough to compete with other types of Q-switches (see, for example, [1]). The FTIR shutter used in our experiments consisted of a pair of truncated triangle fused silica prisms. The prisms have mechanically rigid connection with each other, and a calibrated gap of about

0.5 micrometers between them. The gap can be rapidly varied from zero to approximately 1 micrometer due to elastic deformations of the prisms with the help of piezoelectric drivers mounted on the truncated facets of each prism. The optical arrangement of the FTIR shutter controlled laser we used in our experiments is shown in Fig. 1 A.

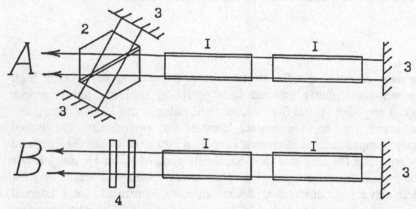

Figure 1. Laser cavity schemes.
A - FTIR shutter controlled laser.
B - VBFP interferometer controlled laser.
1 -   laser rods; 2 -FTIR shutter; 3 - highly reflecting mirrors;
4 -   VBFP interferometer

In this arrangement the FTIR shutter together with a pair of highly reflecting mirrors plays the role of variable reflectivity outcouple of the laser cavity. The equivalent reflectivity coefficient R is connected with the shutter gap transparency T by the formula : $R=(1-T)/(1+T)$.

In the Q-switched regime the FTIR shutter showed several obtuse resonant frequencies of mechanical vibrations (approx. 38 and 66 kHz in our case). At these frequencies the driving AC voltage needed to produce Q-switching did not exceed 10 - 15 V. In order to make the shutter less sensitive to the driving frequency we have found it convenient to damp the resonances with the help of lead cylinders mounted to the outer surfaces of piezoelectric driving tablets. With such damping AC voltage of 40 - 50 V was found enough to produce Q-switching of the laser at any frequency in the range from 5 to 100 kHz. At any driving frequency the needed AC voltage has to be fitted to it and to pump power of the laser in order to produce accurate pulse trains. At the same time DC voltage (or initial mechanical compression of the shutter) can be used for optimisation of time   average outcoupling. In the frequency range from 10 to 40 kHz the laser average power was very close to that in the free-run mode (see Fig. 2), and the typical giant pulse duration was about 1 -2 μs.

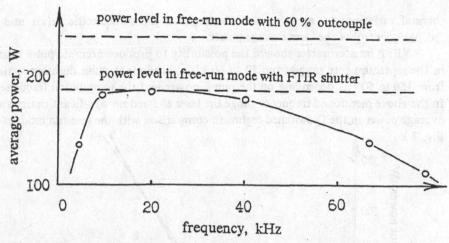

Figure 2. FTIR shutter Q-switched laser average power versus modulation frequency.

At higher frequencies both pump power and modulation amplitude (at 60 V AC shutter driving voltage) were not enough to produce accurate pulse trains. The pulse amplitude varied too much or even some of the pulses skipped. At frequencies lower than 5 kHz the period between pulses is comparable with neodymium metastable level lifetime. This resulted in some energy losses. At the same time the Q-switching rate of the shutter driven by nearly-sinusoidal AC voltage is not enough to produce single-spike giant pulses (they may consist of 2 - 3 spikes).

VBFP interferometer operation.

Q-switching with the help of VBFP interferometer is also based on the dependence of its reflectivity on its base (the distance between two mirrors), which is varied by piezoelectric drivers [2]. The device also serves as a variable reflectivity outcouple of the laser cavity (Figure 1 B). In the case when both semitransparent mirrors of the interferometer have the same reflectivity R the resulting reflectivity of the device can be varied from zero to $4R/(1+R)^2$. Since any laser media has a finite spectral width of the gain band Δv the condition of low interferometer reflectivity between the giant pulses should be fulfilled in the whole band. This circumstance limits the possible length L of the interferometer base: $L<<L^*=(1-R)/(2\pi\Delta v\sqrt{R})$

In case of YAG:Nd gain media $\Delta v = 6$ cm^{-1}

Both its mirrors have reflectivity of 23 %, providing maximum reflectivity of the device equal to 61 %. The initial base length is about 15 - 20 micrometers, which is much less than the critical value $L^* = 400$ micrometers. The base distance is varied by a piezoelectric ring providing translations of about 7 nm per 1 Volt of the driving voltage. The driving circuit of the VBFP interferometer was able to provide automatic optimisation of time average reflectivity and compensation of thermal variations by applying static voltage on the piezoelectric driver and a feedback control of the laser average power.

VBFP interferometer showed the possibility to produce accurate pulse trains in the repetition rate range from 10 to 60 kHz. The giant pulse duration varied from 150 to 500 ns depending on the pump power and the modulation frequency. In the above mentioned frequency range the laser showed no significant changes of average power in the Q-switched regime in comparison with the free-run mode (see Fig. 3).

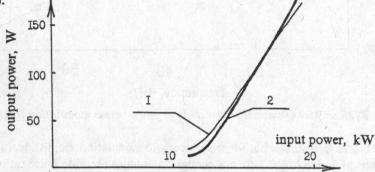

Figure 3. Power characteristics of the laser:
1 - free-running, equipped with 60 % reflectivity output mirror.
2 - Q-switched by VBFP interferometer at 50 kHz repetition rate.

Conclusions.

Both FTIR and VBFP devices showed themselves to be appropriate Q-switches for 300 W average power continuously pumped Nd:YAG laser. We hope they still can operate in lasers with average power in the range of 1 - 2 kW too. The further perfection of both shutters is associated with the improvements of the driving system. The system should provide automatic optimisation and thermal stabilisation of the time average equivalent reflectivity for both types of modulators. At low (5 kHz or less) repetition rates the system should drive the piezoelectric cells with separate voltage pulses and should provide fast electric dumping of mechanical vibrations of the Q-switching device. This should provide the possibility to produce accurate separate giant pulses at low repetition rates without parasitic lasing between pulses caused by mechanical vibrations of the shutters.

References.
1. B.I.Denker, V.V.Osiko, S.E.Sverchkov, Yu.E.Sverchkov, A.P.Fefelov, S.I.Khomenko. Hihgly efficient erbium glass lasers with Q-switching based on frustrated total internal reflection, Sov. J. Quantum Electron. v.22, p.500 (1992).
2. A.I.Ritus. Fabri-Perot interferometer as a laser Q-switcher . Sov. J. Quantum Electron., v.20, N 2, p.198 (1993), in Russian.

Digital Electro-Optical Laser Beam Deflector with Domain-Inverted Prism-Array

V. Hinkov, I. Hinkov and E. Wagner
Fraunhofer Institute of Physical Measurement Techniques
Heidenhofstr. 8, D-79110 Freiburg i. Br.

1. Introduction Laser-beam deflectors are important optical elements with several applications in the communication and measurement techniques. The commonly used mechanical or acousto-optical devices have limitations in regard of operation speed, are of complex design and quite expensive in system applications. An alternative offer the electro-optical deflectors, various forms of which have been investigated. The basic form is usually an electro-optically controllable prism, which refracts the laser beam at different angles according to the applied electric voltage [1]. Single [2] or several successively aligned prisms [1] have been used, but they have limited resolution and need a high driving voltage. Prism-arrays aligned perpendicular to the beam propagation direction (see Fig.1) have also been proposed and tested in bulk [3] or integrated optical [4] form. They principally allow to achieve a high resolution at low drive voltages, but the devices being developed produced strong phase distortions of the deflected beam and were not suitable for practical applications.

We have developed and investigated an integrated electro-optical prism-array deflector with improved characteristics and simple fabrication technology. The deflecting prisms were fabricated using the technology of domain inversion in LiNbO$_3$ substrates. The device was driven in digital mode and deflection over ten orders was observed, the beam quality being good preserved.

2. Deflector fabrication A schematical diagram of the electro-optical laser-beam deflector is shown in **Fig.1**. It consists of nine deflecting prism-pairs with 600 μm aperture and 2 cm length, incorporated in a monolithical LiNbO$_3$ crystal sample by means of domain inversion. +Z-cut LiNbO$_3$ substrates 10x25x0.5 mm^3 were used in the experiments. To fabricate the domain-inverted prisms first 5 to 15 nm thick Ti-film was evaporated on the sample surface and photolithographically patterned. The Ti-film was indiffused at 1100°C for 15 to 30 min in oxygen atmosphere, the samples being closed in a platin box together with LiNbO$_3$ pieces to suppress the LiO$_2$ outdiffusion. In preliminary experiments it was found that this technique is suitable for fabricating large domain-inverted structures. Domain inversion takes place in the Ti-doped regions that was revealed by etching the crystal surface. The domain inverted parts are etched much faster than the rest of the surface. **Fig.2** shows the surface profile measured with a profilometer, domain-inverted regions of approx. 3 μm depth are clearly seen.

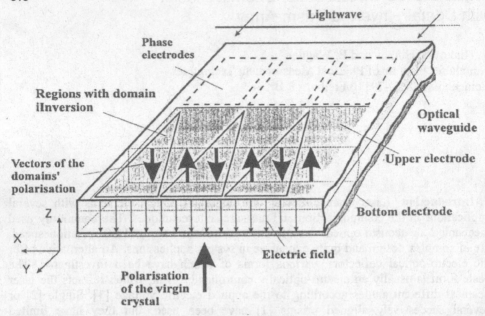

Fig.1 Schematic diagram of the electro-optical deflector

A single mode slab waveguide was fabricated on the sample surface by first performing proton exchange in 1%-buffered (with lithium benzoate) benzoic acid for 2 to 3 hours at 180°C. After the exchange the samples were annealed for some hours at 350°C in oxygen atmosphere. A 200 nm thick SiO_2 buffer layer was deposited on the waveguide surface to avoid the light absorption from the electrodes. In a last fabrication step the metal electrodes were evaporated on both sides of the sample overlapping the prism-array.

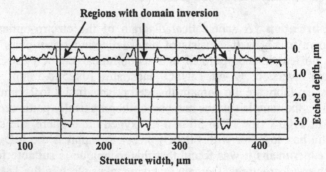

Fig2. Etch-test of the domain inverted structure

3. Measurements of the deflector The experimental set-up used to measure the electro-optical deflector is schematically shown in **Fig.3**. The deflector chip was mounted on a movable stage and electrically connected to the output of a voltage supply. As a light source a 670 nm laser diode was used, the output beam of which was first shaped in a form of a thin stripe by means of a cylindrical and a spherical lenses. It was then coupled into the waveguide through the polished input face of the chip. The deflected beam was coupled out of the waveguide again with a combination of a cylindrical and a spherical lenses. The deflection was monitored on a video camera and was also observed visually. The output spot was scanned through a 100 µm slit with a photodiode mounted on a motorised stage to measure its intensity distribution.

DC-voltage was applied to the deflector to perform the digital deflection of the laser beam. About ±80 V were necessary to move the outcoupled spot to the first deflection maxima. For intermediate voltages part of the light intensity was coupled also to the next (higher or lower) maxima. A good quality of the optical beam was preserved for deflections up to the +5 and -4 maxima. In **Fig.4** the intensity distribution of the output spot for different applied driving voltages is presented as measured with the above described experimental set-up.

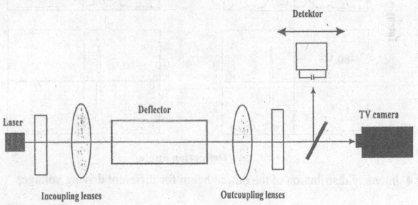

Fig.3 Measurement set-up

4. Discussion The prism-array deflector developed allows a high resolution to be achieved at low drive voltages. Only ±80 V were necessary to switch the light between the deflection orders. This corresponds to a deflection sensitivity of about 2 mrad/100 V and is factor three better than the corresponding value reported in [2]. Another important advantage of the presented deflector is its simple fabrication technology and the improved electric field distribution inside the $LiNbO_3$ sample. In its digital form the deflector can find several applications as a high-speed switch or intensity modulator in the optical communication systems, but also in optical measurement, data processing or memory systems. A continuous deflection can also be achieved by using phase-controlling electrodes in front of the prism pairs (Fig.1). This will also lead to a large reduction of the driving voltage by a factor equal to the number of the prism-pairs used (in this case nine). We have already fabricated such deflectors and the results of their experimental investigations will be published in a future paper.

818

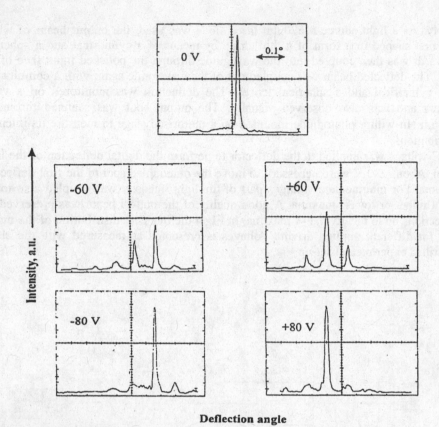

Deflection angle

Fig.4 Intensity distribution of the output beam for different driving voltages

5. Conclusions   We have presented a digital electro-optical laser-beam deflector for use in the optical communication, measurement, memory, etc. systems. High deflection sensitivity was achieved at low drive voltage. A continuous type deflector is also in development.

References

1. J. F. Lotspeich, IEEE Spectrum, February 1968, pp. 45-52
2. Q. Chen, Y. Chiu, D. N. Lambeth, T.E. Schlesinger and D. D. Stancil, J. Ligthw. Technol., 12 (1994) pp. 1401-1404
3. Y. Ninomia, IEEE J. Quantum. Electr., QE-9 (1973) pp. 791-795
4. C. H. Bulmer, W. K. Burns and T. G. Giallorenzi, Appl. Opt., 18 (1979) pp. 3282-3295

Binary and Multilevel Diffractive Elements with Submicron Feature Sizes

H.Zarschizky, A.Stemmer, F.Mayerhofer and G.Springer[*]
Siemens Corporate Research and Development
[*]Technical University Munich
D-81739 Munich

1. Introduction

Diffractive elements may be identified as miniaturised or micro optical components which affect an incoming beam of light by the laws of diffraction. In contrast to other micro optical components like refractive ball lenses and micro calottes or graded index lenses diffractive optics can be attributed as flat and in general aspheric optical elements which can perform multiple optical functions simultaneously.

The diffractive optical elements (DOEs) discussed throughout this paper are synthetic elements. This means that the diffractive pattern of the element is computer generated and the pattern data are coded in certain data formats suited to be processed from common lithographic facilities such as electron beam writers. The beginning of computer generated optical elements started nearly 30 years ago when LOHMANN and PARIS [1] constructed the first synthetic holograms, which could generate artificial images. These early elements were of binary amplitude type, which means that light can pass through small and especially arranged apertures in an opaque screen or is blocked otherwise. The poor brightness of the images called for purely transmissive holograms which influence the phase of light instead of performing a spatial amplitude modulation. A specific result is the "kinoform" devoped by LESEM, HIRSCH and JORDAN [2] at the end of the sixtees. The problems for practical applications of these early elements were twofold: on the one hand these "digital holograms" need long computing times and a large amount of data. On the other hand the restricted resolution power of the lithographic equipment was fairly above the wavelength of light which only allowed for small deflection angles and long focal lengths, respectively. They therefore served mainly for demonstrators but not for usefull compact set-ups.

During the last decades, however, the fast development of microelectronics has lead to dramatically enhanced possibilities for diffractive microoptics. Firstly, computation time and data handling is no more a significant problem. Secondly, the achievable minimum feature size of diffractive patterns is nowadays well below one micrometer and therefore rapid prototyping as well as small volume production of ambitious DOEs is possible.

Since the last years the applications of synthetic optical elements have changed from displaying artificial images to optical interconnection. Therefore the historical denotation "holographic optical element" (HOE) should be replaced by "diffractive optical element" DOE especially if the element don't make use of a multiplex diffractive pattern. The very simplest type of a DOE is the linear diffraction grating which allows for beam splitting and deflection, next is the chirped grating which may represent one half of a cylindrical lens. The curved chirped grating will finally lead to zone lenses. DOEs made of interleaved zone lenses of different focal lengths will have multifunctional purposes for simultaneous beam splitting, beam deflection and individual beam focussing. An example of a multifunctional synthetic element for optical clock signal distribution may be found in [3].

Certainly the benefit of synthetic DOEs is the large freedom in design to match the shape and phase of the incoming light. The main problem is the efficiency of the DOEs. DOEs are basically gratings and will therefore exhibit several diffraction orders. For many applications DOE optimisation means to shift almost all light into the desired diffraction order. This is intended by an accurate and sophisticated profiling of the local grating structure of the diffractive pattern. The simplest DOE profile is binary, which means there are two profile levels to create the required phase shift of light. Sophisticated profiles have to be constructed from multiple levels to give a more accurate phase control within one grating period. This is often a staircase approach which requires steps of considerably small widths and heights. If the multilevel DOEs should operate with laser light in the near infra red or even in the visible regime they therefore must exhibit features of submicron size.

This paper will give a brief overview of our current methods for DOE design and calculation based on scalar diffraction theory and the use of a commercially CAD-program. We will describe the current and future fabrication steps utilising direct write electron beam lithography and direct print grey tone mask photolithography, respectively. The pattern transfer from resist into the substrate is done with reactive ion etching. We will give some impressions of the measured performance of our elements in terms of imaging characteristic and efficiency of various multilevel lenses. The problem of multiple diffraction orders with respect to crosstalk will be sketched and the high chromatic sensitivity of DOEs will be pointed out. An example for a facetted multifunctional element will be shown and we will end with the conclusion.

2. Design and Calculation of Diffractive Lenses

Our current methods for the calculation of diffractive patterns are based on scalar diffraction theory. One method simulates the interferometric recording of a hologram [3], the other uses Fermat´s principle to construct kinoform lenses as proposed by BURALLI, MORRIS and ROGERS [4]. The first method has the advantage to calculate the pattern of multifunctional elements and gives a multiplex pattern. The result is a two dimensional transmission function of sinusoidal type which has to be converted to binary or multilevel type. In contrast the second method doesn´t need any assumptions about the exact phase profile and the binary or multilevel data are derived by quantisation of the kinoform profile. But, only the calculation of diffractive lenses or mirrors is possible. Multifunctionality has to be achieved by facetted types of diffractive elements. Examples are given in chapter 4.

Algorithms of the two methods are implemented in the CAD-Layout editor SIGRAPH-Optik® [5] which allows for the lens data calculation and the generation of appropriate GDSII-data files for further postprocessing by lithography facilities. More details about the CAD-program may be found in [6].

Fig.1 shows three different imaging geometries and examples of the corresponding diffractive lens patterns as calculated with method 1 and represented as GDSII-data.

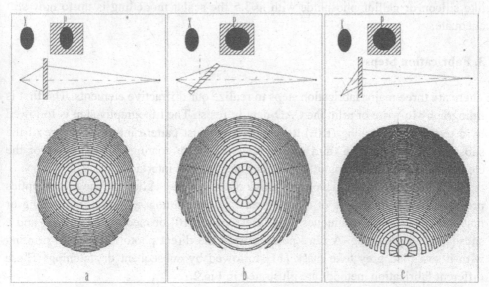

Fig.1: Examples for on-axis and off-axis imaging geometries (upper row) and corresponding GDSII-data of the binary diffraction patterns (bottom row).
(a) On-axis geometry, elliptical beam cross section X, elliptical DOE aperture P.
(b) Off-axis geometry, elliptical beam cross section X, circular DOE aperture P.
(c) Off-axis geometry, elliptical beam cross section X, circular DOE aperture P.

Due to the assumption that the light source has an elliptical beam cross section and according to the type of set-up the lens apertures get an elliptical or circular shape for maximum light capture. The corresponding diffractive patterns may consist of concentric circular fringes around the aperture center as in Fig.1(a), of decentered elliptical fringes as shown in Fig.1(b) or the fringe system may be shifted away from the aperture center as depicted in Fig.1(c).

The construction methods based on scalar theory are appropriate for diffractive optical elements with relatively large local grating periods Λ. Investigations, for example, carried out by NOPONEN, TURUNEN and VASARA [7] show a considerable drop in diffraction efficiency if Λ becomes smaller than 10 times the wavelength λ of the operation light. For practical elements with numerical apertures of N.A.$\approx$0.5 the minimum grating period is in the range of $\Lambda=2\lambda$ and the efficiency from the respective part of the diffractive element drops to zero, even for multilevel elements. Complicated and time consuming modelling based on the exact electromagnetic theory helps to rise the efficiency. However, the limits of scalar diffraction theory strongly depend on the refractive index **n** of the substrate the element. is made of. Highly refractive substrates show a much more relaxed dependence of efficiency on the grating period as demonstrated by POMMET, MOHARAM and GRANN [8]. Because our substrates are semiconductor materials like silicon or galliumphosphide with n$\approx$3.5 the scalar modelling is up to now still adequate.

3. Fabrication Steps

There are three major fabrication steps to realize our diffractive elements. The first is lithography to write or print the CAD-data in resist. The lithography step is followed by a reactive ion etching (RIE) to transfer the resist pattern into the surface of the substrate of choice. The third step is a very precise coating of both sides of the element to reduce reflection of light at the substrate-air interfaces.

The fabrication of multilevel diffractive elements may either use the binary optics process [9] which consists of repeated runs of photolithography and dry etching or may use direct write techniques with laser beams [10] or electron beams [6] and a subsequent dry etching. A third possibility is the direct photolithographic printing with a so-called grey tone mask [11] followed by subsequent dry etching. These different fabrication methods are illustrated in Fig.2.

The binary optics procedure as shown in the left part of Fig.2 is the complete adaption of semiconductor fabrication techniques. To realize a diffractive element with a number of 2^m levels the run of photolithography and subsequent dry etching has to be repeated m times. The example of Fig.2 needs m=2 runs and requires m=2 different masks to realize a 4-level element. After the first run the second mask has to be very precisely aligned to the already patterned and again resist coated substrate.

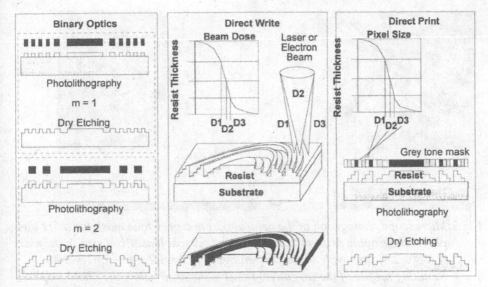

Fig.2: Three fabrication methods for multilevel diffractive elements (from left):
binary optics procedure using m=2 runs of photolithography and dry etching,
direct write lithography and subsequent dry etching,
direct print lithography and subsequent dry etching.

The direct write procedure with laser or electron beams creates the multilevel resist profile within one single cycle. If the resist height in dependence on beam dose is known (see characteristics on top of the center part of Fig.2) the 3-D resist pattern can be achieved by adjusting the beam dose during the exposure step. A 4-level element needs three dose values D1, D2 and D3. After the direct write step the resist profile is transfered into the substrate surface by the final RIE process.

The third method is based on the direct print photolithography as illustrated in the right part of Fig.2. The three different light intensities D1, D2 and D3 can be achieved by a special kind of photomask denoted as "grey tone mask". Although this mask is still made of transparent and opaque regions it allows for a variable exposure intensity via respective "pixel densities" within the mask. A pixel is a small area of the mask with a fixed size which contains a centered opening of certain dimensions. The size of this opening must be small enough to be not resolved by photolithography. Indeed, the projected light intensity depends on the size of this opening. The dependence of resist thickness and pixel size has to be evaluated as it is the case for the direct write procedure. Knowing the characteristics (as sketchted on top of the right part in Fig.2) the 3-D profile for the multilevel pattern can be coded in a mask pattern with completely "black", "shaded" or fully transparent regions with pixel ensembles made of sizes D1, D2 or D3, respectively. The photolithography step therefore leaves a three dimensional patterned resist which is transfered into the substrate with reactive ion etching.

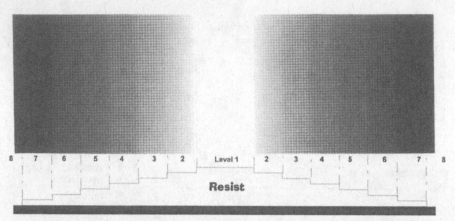

Fig.3: Microscope photograph of the central part of a grey tone mask for a 5:1 direct photolithographic printing of a 8-level cylindrical lens. Pixel size in the mask is (2.4 x 2.4) μm² with minimal openings of (1.2 x 1.2) μm²for level #8. Resulting profile of the (negative) resist is sketched below the photograph.

Fig.3 shows a microscope photograph of the central part of a grey tone mask pattern fabricated for step-and-repeat photolithography with a wafer stepper of 0.7 μm specified resolution. The mask produces a 8-level cylindrical lens in a negative photo- resist. Due to the stepper resolution the widest opening in the pixel (2.4 μm x 2.4 μm) ensemble for level #2 is (2 x 2) μm² going down to (1.2 x 1.2) μm² for level #8 (compare to bottom schematic of Fig.3).

For fast prototyping the direct write methods show the best flexibility. Direct writing with electron beams, however, is much faster than with laser beams and allows for a very high pattern resolution down to 0.2 μm or less. This is necessary for multilevel diffractive lenses with high numerical apertures of N.A.≈0.5 as, for example, required for beam shaping of edge emitting semiconductor laser diodes. Typical writing times for electron-beam lithography is about one minute for a 4-level lens. Nevertheless electron-beam lithography is by no means a production technique for high volume fabrication and is not low cost.

Photolithography as used by the binary optics method and the direct printing with grey tone masks allows for high throughput and typical printing times are in the range of a second inclusively wafer step. In contrast to the binary optics procedure the direct printing avoids the highly accurate alignment procedure required for repeated process runs. The problem, however, is shifted to the fabrication of the complicated grey level mask. According to the resolution power of todays photolithographic equipment diffractive lenses (of first order) with an upper limit in the numerical aperture of about N.A.≈0.35 are possible. Higher numerical apertures might be achievable with higher order diffractive lenses [12].

4. Experimental Results

A big effort has been made to fabricate and investigate binary and multilevel diffractive lenses for optical telecommunication purposes in the infra red wavelength regime of $\lambda=1.3$ μm and $\lambda=1.55$ μm. The lens substrate has been silicon which has the advantages to be transparent at these wavelengths, to be available in excellent quality substrates and is technologically very well experienced.

The lenses were designed for semiconductor laser diode to single mode fibre (SMF) coupling with a 1:6 imaging geometry. Due to the highly divergent laser diode beam the numerical aperture of the lenses was chosen to a N.A.=0.5. The lenses had a diameter of $\varnothing=600$ μm, a focal length of f=400 μm and a total number of grating periods of 80 with minimum period width of 1.8 μm. This means that the binary version of the lens needs a minimum fringe width of 0.9 μm at the outer edge, the 4-level version needs a minimum step width of 0.45 μm and for the 8-level version a minimum step width of 0.225 μm is necessary. The step heights depend not only on the wavelength of light but also on the refractive index of the lens substrate. Because of the high refractive index of silicon (n=3.5) the step heights were 0.26 μm for the binary version and 0.065 μm for each step of the 8-level lens.

Fig.4 shows a schematic of different lens profiles, the respective maximum diffraction efficiencies and three scanning electron micrographs (SEM) of the central parts of the lenses under discussion. The lenses were realised with direct write electron

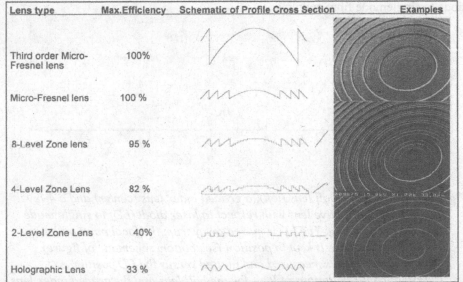

Lens type	Max.Efficiency	Schematic of Profile Cross Section	Examples
Third order Micro-Fresnel lens	100%		
Micro-Fresnel lens	100 %		
8-Level Zone lens	95 %		
4-Level Zone Lens	82 %		
2-Level Zone Lens	40%		
Holographic Lens	33 %		

Fig.4. Lens profiles and maximum theoretical diffraction efficiencies predicted from scalar theory as well as examples of fabricated lenses. The SEM-micrographs show the centers of 2-level, 4-level and 8-level lenses realized in silicon.

beam lithography followed by a subsequent reactive ion etching (RIE) with SF_6, H_2 and Ar. Finally the lenses were double sided anti reflection coated with siliconnitride Si_3N_4 using plasma enhanced chemical vapour deposition (PECVD). Going from bottom to the top of Fig.4 the diffraction efficieny increases in dependence of the specific kind of lens profile. The sinusoidal profile as created by an interferometric recording in photoresist can direct 33% of the incoming beam to the desired focus. The 2-level (binary) lens allows about 40% efficiency. The 4-level and 8-level version promise efficiency values of 82% and 95% while the continuous profiles of the micro Fresnel lens and the third order lens may allow for the maximum diffraction efficiency of 100%. As pointed out in chapter 2 these efficiencies η_{theo} are maximum values predicted by the scalar theory for diffractive lenses with considerable small numerical apertures. However, lens substrates of high refractive index should help to increase the apertures and diffraction efficiencies η_{meas} and this is confirmed by our measurements:

The 2-level lens shows a measured efficiency η_{meas} =30%, η_{theo}=40%. The 4-level lens shows η_{meas} =64%, η_{theo}=82% and the 8-level lens shows η_{meas} =72%, η_{theo}=95%. The experimental results are about a factor of 0.75 lower than the maximum values and this is with respect to the high numerical lens aperture of N.A.=0.5 a quite satisfying result. Further measurements show that the desired imaging geometry is precisely fulfilled and that the diffractive lenses are free from spherical aberrations. This is demonstrated in Fig.5 where a miniature ball lense and graded index lens are compared to a 4-level diffractive lens with respect to laser diode to singlemode fibre coupling.

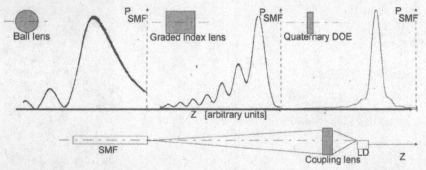

Fig.5: Comparison of a ball lens (left), a graded index lens (center) and a 4-level (quaternary) diffractive lens with respect to laser diode(LD) to single mode fibre (SMF) coupling. The LD is moved away from the focal point of the lenses while the fibre is kept in position (see bottom schematic of figure). The fibre light P_{SMF} is measured and plotted versus the LD position z. The significant oscillations of P_{SMF} for the ball lens and the graded index lens are attributed to spherical aberration (SA). The 4-level diffractive lens show a smooth response and is free from SA.

The measurements illustrated in Fig.5 show significant oscillations of P_{SMF} for the ball lens and the graded index lens which can be attributed to spherical aberration (SA). In contrast the 4-level diffractive lens shows a smooth response and is free from SA due to a proper design and precise fabrication.

However, there is an intrinsic mismatch of the shapes of the image point of the LD and the fibre spot and therefore the coupling efficiency η_c differs from the diffraction efficiency η_d. For example, even if the diffraction effiency for the 8-level lens is η_d =72% only η_c=45% could be coupled into the single mode fibre.

As a result of the experiments on laser to fibre coupling with on-axis diffractive lenses of multilevel type it should be derived that diffractive elements show excellent imaging characteristics but high efficiency is difficult to achieve. The superiority of diffractive elements relies on functionality but not in efficiency. From this point of view the replacement of single micro optic lenses for on-axis imaging by diffractive lenses is surely not the best application for diffractive elements. The use of diffractive on-axis lenses is much more justified if arrays of lenses are of interest because there is a large freedom for lens shape and focal length. Diffractive off-axis lenses are good examples for superior function and multifunctionality within one single element.

The examples of Fig.6 and Fig.7 show such multifunctional diffractive elements which were designed with the CAD-tools mentioned in chapter 2 and realized by methods described in chapter 3. The pictures in Fig.6 and Fig.7 are SEM micrographs of 4-level resist profiles generated with direct write electron beam lithography. Both structures are composed of parts of an on-axis lens and an off-axis lens. The pattern of Fig.6 is based on a rectangular mesh and the two lenses are interleaved in a chestboard like manner. The pattern of Fig.7 is based on a hexagonal mesh and the two lenses are interleaved in concentric fringes built up from single hexagons.

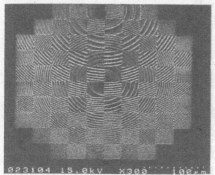 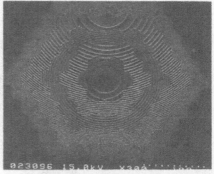

Fig.6: Multifunctional diffractive element composed of an on-axis and an off-axis lens arranged in chestboard like manner. 4-level element.

Fig.7: Multifunctional diffractive element composed of a on-axis and an off-arranged in an hexagonal mesh. 4-level element.

5. Critical aspects for use of diffractive elements

One basic problem of diffractive optics is the restricted efficiency as already been emphasized throughout this paper. Besides the fact that the desired image may be not intense enough one has to consider the question what problems the remaining part of the input light might cause. To illustrate this question Fig.8 gives an example of an on-axis and an off-axis lens designed to focus an incoming parallel beam of light. The upper part of Fig.8 shows the on-axis case. The desired image point is the focus which is imaged by the +1. diffraction order. In the denotation of holography this is the direct "real" image. Due to a restricted efficiency there are at least the -1. order and the zeroth order. The -1.order is denoted as the conjugate "virtual" image which causes a diverging cone of light which superimposes the real images wave. The undiffracted 0.order will additionaly lead to interference and the desired +1.order image will be appear in a more or less modulated background. The brightness of the background dependes strongly on the element efficiency.

The off-axis arrangement shown in the bottom part of Fig.8 can help to reduce significantly the problem of overlapping diffraction orders. While the wave of the undiffracted zeroth order moves straight ahead the desired +1.order is deflected. Additionally, the undesired -1.order is deflected to the opposite side of the +1.order and so the principal diffraction orders can be completely separated. In fact, this is well known in holography for a long time and nearly every display hologram is recorded and reconstructed in such an off-axis set-up as invented by LEITH and UPATNIEKS in the early sixtees [13]. For more complicated set-ups as depicted in Fig.8 the locations of the respective orders for specific arrangements can be calculated with help of the holographic imaging equations given by CHAMPAGNE [14].

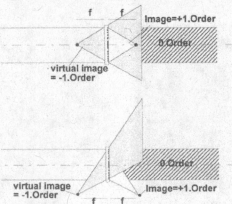

Fig.8: Principal diffraction orders of a diffractive lens.
Top schematic: on-axis focussing and overlapping orders.
Bottom schematics: off-axis focussing and separated diffraction orders.

It is obvious that interconnections using arrays of diffractive on-axis lenses may suffer from considerable optical cross talk which arises from the extended background light according to orders 0. and -1.

The other major problem of diffractive optics is the extreme chromatic sensitivity. Especially off-axis set-ups behave like spectrometers which resolve wavelength shifts in the nanometer range. Fig.9 sketches an example of a laser diode to single mode fibre coupling with an off-axis arrangement using a lens of $20°$ tilt angle.

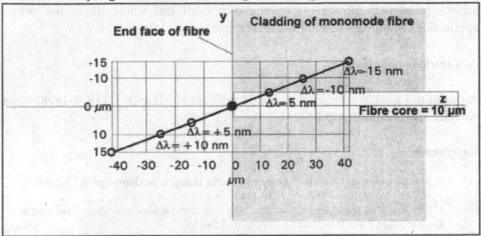

Fig.9: Displacement of image point due to wavelength changes $\Delta\lambda$ for an off-axis set-up for laser diode to single mode fibre coupling using a lens with 20 degree tilt angle. For $\Delta\lambda=0$ the image point is centered at the fibre end face (black dot) but moves in the y-z plane for $\Delta\lambda\neq0$ (open dots).

If the wavelength shift is in the range of $\Delta\lambda= \pm15$ nm the image point (black dot) moves $(\Delta z,\Delta y) = \pm(40, 15)$ µm away from the fibre end face (open dots). Fibre coupling is practically impossible if the absolute value of $\Delta\lambda>5$ nm. Such wavelength shifts are of practical concern if uncooled transceiver modules have to work under extreme environmental conditions.

6. Conclusions

We gave a brief description of our design, calculation and data generation for various synthetic computer generated diffrative lenses using an extended CAD layout editor. Although we modell lenses with high numerical apertures with methods based on scalar diffraction theory the measured results show that these methods are still adequate in terms of imaging quality and satisfying diffraction efficiencies up to about 70%. These results are related to the highly refractive substrates of use which on the one hand relax the design problem of the necessarily small grating periods and on the other hand simplify the fabrication because of low grating depths and accordingly small aspect ratios. The fabrication is up to now carried out with direct write electron-beam lithography and followed by a dry etching step. Future fabrication will focus on direct print photolithography with grey level masks. It was pointed out that diffractive optics is superior in functionality but not in efficiency. Examples of multifunctional elements are shown. Critical aspects for the application of diffractive elements are on the one hand the restricted efficiency which can give rise for image interference and crosstalk. On the other hand the extreme chromatic sensitivity of the diffractive elements is pointed out which might not permit applications whith very tight dimensional tolerances.

Acknowledgements

A part of this work is sponsored by BMBF project PHOTONIK II (1994-1998) under contract number 01 BP 459/0.

References

[1] A.W.Lohmann and D.P.Paris, "Binary Fraunhofer Holograms, Generated by Computer", Appl.Opt., Vol.6(10), (1967), pp.1739-1748.
[2] L.D.Lesem, P.M.Hirsch and J.A.Jordan, "The Kinoform: A New Wavefront Reconstruction Device", IBM J.Res.Dev., Vol.13, (1969), pp.150-155.
[3] H.Zarschizky, Ch.Gerndt, A.Stemmer and H.W.Schneider, "A Multifacet Diffractive Mirror for Optical Clock Signal Distribution", SPIE, Vol.1732, (1992), pp.297-306.
[4] D.A.Buralli, G.M.Morris and J.R.Rogers, "Optical Performance of Holographic Kinoforms", Appl.Opt., Vol.28(5), (1989), pp.976-983.
[5] SIGRAPH-Optik is a registered trademark of Siemens-Nixdorf Informationssysteme
[6] H.Zarschizky, A.Stemmer, F.Mayerhofer, G.Lefranc and W.Gramann, "Binary and Multi-level Diffractive Lenses with Submicron Feature Sizes", Opt.Eng., Vol.33(11), (1994), pp.3527-3536.
[7] E.Noponen, J.Turunen and A.Vasara, "Electromagnetic Theory and Design of Diffractive Lens Arrays", J.Opt.Soc.Am.A, Vol.10(3), (1993), pp.434-443.
[8] D.A.Pommet, M.G.Moharam and E.B.Grann, "Limits of Scalar Diffraction Theory for Diffractive Phase Elements", J.Opt.Soc.Am.A, Vol.11(6), (1994), pp.1827-1834.
[9] W.B.Veldkamp and T.J.McHugh, "Binary Optics", Scientific American, (May 1992), pp.50-55.
[10] M.T.Gale, M.Rossi, R.E.Kunz and G.L.Bona, "Laser Writing and Replication of Continuous Relief Fresnel Microlenses", in Diffractive Optics, Vol.11, Technical Digest Series OSA, (1994), pp.306-309.
[11] Y.Oppliger, P.Sixt, J.M.Stauffer, J.M.Mayer, P.Regnault and G.Voirin, "One-Step 3D Shaping Using a Gray-Tone Mask for Optical and Microelectronic Applications", ME 23, (1994), pp.449-454.
[12] J.C.Marron, D.K.Angell and A.M.Tai, "Higher Order Kinoforms", SPIE, Vol.1211, (1990), pp.62-66.
[13] E.N.Leith and J.Upatnieks, J.Opt:Soc.Am., Vol.52, (1962), pp.1123.
[14] E.B.Champagne, "Nonparaxial Imaging, Magnification, and Aberration Properties in Holography", J.Opt.Soc.Am., Vol.57(1), (1967), pp.51-55.

Diffraktiver Teiler zur Strahldiagnose für Hochleistungs-Festkörperlaser

Ch. Budzinski, H.J.Tiziani, Institut für Technische Optik, Universität Stuttgart, FRG
P. Lippert, G. Lensch, NU-TECH GmbH, Neumünster, FRG
G. Bostanjoglo, Festkörper-Laser-Institut Berlin, FRG

Abstract

Wir stellen eine konzeptionelle Lösung zur Leistungsauskopplung von ca. 0.1% für Festkörperlaser mit Leistungen bis 2kW vor. Für die Auskopplung wurde ein diffraktiver Strahlteiler aus Quarz entwickelt. Die diffraktiven Strukturen für die Strahlteilung wurden nach einem Modell der rigorosen Beugungstheorie berechnet, holografisch in Resist hergestellt und durch Ionenätzen in Quarz übertragen. Bei normaler Inzidenz tritt die erste Ordnung als Diagnosewelle seitlich aus dem Strahlteiler aus. Es wurden verschiedene Exemplare von Strahlteilern hergestellt und deren Wirkung in einem vorhandenen Festkörperlasersystem geprüft. Die Strahlungsfestigkeit der Teilerstrukturen wurde getestet.

Für die Diagnose eines Hochleistungslaserstrahles mit einer Strahlungsleistung im kW-Bereich wird ein diffraktiver Strahlteiler mit einem Teilungsverhältnis von 1:1000 bis 1:10 000 entwickelt. Aus der örtlichen Verteilung der ersten Ordnung und ihrer zeitlichen Funktion werden die Strahlparameter abgeleitet. Zur Strahlformung und Anpassung an den Sensor werden auf die Seitenflächen computergenerierte diffraktive Strukturen durch Replikation aufgebracht. Die Auskopplung der Diagnosestrahlen erfolgt durch eine diffraktive Struktur auf einem Quarzglas-Substrat.

Die Abhängigkeit des Diagnosestrahles von der Ortsfrequenz, Profilform und dem Substratmaterial ist durch Computersimulation untersucht worden. Ausgehend von der "Start"-Geometrie (siehe Abb. 1) wurden für verschiedenen Geometrien die Strahlengänge, die Auswertemöglichkeiten sowie die polarisationsabhängige Teilung untersucht. Die Abb. 2 zeigt exemplarisch die profiltiefenabhängige Auskopplung durch

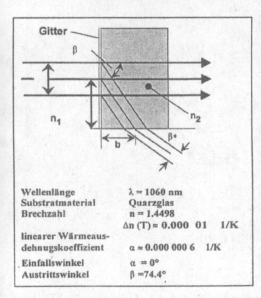

Abb.: 1 Geometrie "Start" (Modell 1)

Modell 1

θ =0°, λ = 1060nm, d = 759nm

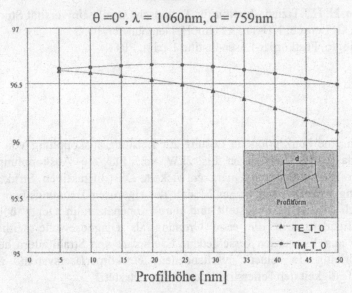

Profilhöhe [nm]

Modell 1

θ =0°, λ = 1060nm, d = 759nm

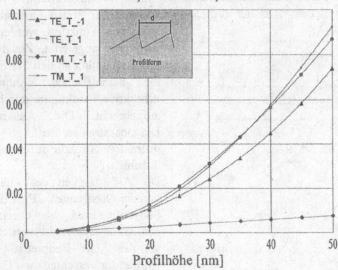

Profilhöhe [nm]

Abb. 2: Beugungswirkungsgrad

ein Blazeprofil. Die prozentual ausgekoppelte Energie der 1. Ordnung für TE- bzw. TM-Polarisation sind von gleicher Größenordnung. Damit ist ein polarisationsunabhängiger Diagnosestrahlteiler mit den oben genannten Teilungverhältnissen herstellbar.

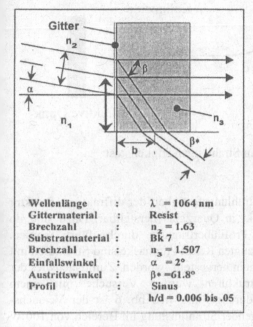

Wellenlänge : $\lambda = 1064$ nm
Gittermaterial : Resist
Brechzahl : $n_z = 1.63$
Substratmaterial : Bk 7
Brechzahl : $n_3 = 1.507$
Einfallswinkel : $\alpha = 2°$
Austrittswinkel : $\beta* = 61.8°$
Profil : Sinus
$h/d = 0.006$ bis .05

Abb. 3: Funktionsmodell

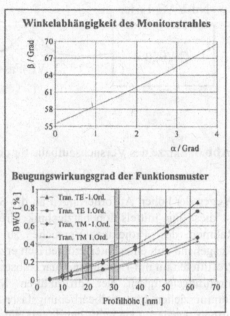

Abb. 4: a)Winkelübertragung
b)Ausgekoppelte Energie

Die Funktion wurde an einem Funktionsmuster in Resisttechnik überprüft. Die Abb. 3 zeigt den Strahlengang und die zugehörigen Daten. In der Abb. 4 ist die berechnete Winkelübertragung und die ausgekoppelte Energie für drei gefertigte Profiltiefen (schraffierter Bereich) dargestellt. Die Untersuchung hinsichtlich der optischen Eigenschaften der Funktionsmuster erfolgte in einem Laboraufbau mit einer Nd:YAG-Laserstrahlquelle. Es wurden die Auskoppelverhältnisse bei kleiner Leistung unter Berücksichtigung der Polarisation gemessen. Die Teilungsverhältnisse betrugen 0.1% für das Muster 1111-04 bis 0.01% für das Muster 1111-05. Die Abbildung 5 zeigt der Einfluß des Einfallswinkel auf den ausgekoppelten Strahl für das „Muster111-04". Der Teiler ist für einen Einfallswinkel von 2° ausgelegt. Der Leistungsabfall bei 0° bzw. 4° ist bedingt durch die Apertur der Auskoppelfläche.

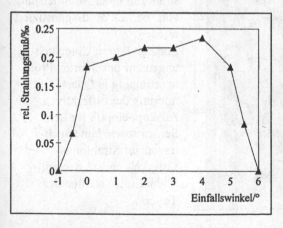

Abb. 5: Winkelabhängigkeit der ausgekoppelten Strahlungsleistung

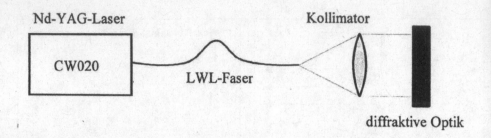

Abb.6: Skizze des Versuchsaufbaus für den Strahlungsfestigkeitstest

Wegen der hohen Anforderungen an die Strahlungsfestigkeit der diffraktiven Struktur ist diese nur mittels Ionenätztechnik direkt in Quarzglas herstellbar. Dazu sind im Rahmen des Projektes Versuche zu Profilübertragung durchgeführt worden. Ausgehend von einer interferometrisch erzeugten Resist-Ätzmaske sind Strukturen mit Profiltiefen im nm-Bereich in Quarzsubstraten hergestellt worden. Zum Nachwies der Strahlungsfestigkeit der diffraktiven Strukturen wurden Versuche mit einem kommerziellen Materialbearbeitungslaser durchgeführt. In Abb. 6 ist der Versuchsaufbau skizziert. Das DOE wurde mit steigender Strahlleistung im Bereich von 100 W bis 1000 W bei einem Strahldurchmesser von 7,5 mm belastet. Der Laser wurden im CW-Betrieb gefahren. Die Einwirkdauer betrug pro Leistungswert 30 Sekunden. In Abb. 7 wird eine Mikrofotografie des getesteten DOE gezeigt. An dem DOE sind keine Veränderungen erkennbar. Unter Voraussetzung der maximalen Bestrahlungsstärke auf dem DOE und einem Strahldurchmesser von 7,5 mm während des Versuches kann bei einem Strahldurchmesser von ca. 12 mm bis 15 mm ein Laserstrahl mit einer Strahlleistung von ca. 2kW diagnostiziert werden

Zur Zeit laufen Untersuchungen zur optimierten Profilübertragung in Quarz, Optimierung der diffraktiven Auskoppeloptik für den Sensor sowie Untersuchungen zur Strahlungsfestigkeit und zu den optischen Eigenschaften des Teilers.

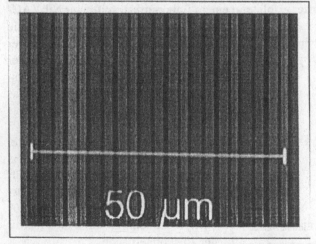

Abb. 7: DOE nach Strahlungsfestigkeitstest Position 1

A Compact Modelocked Erbium Doped Fiber Laser for Field Use

Reinhard Erdmann

Rome Laboratory Photonics Center

25 Electronics Pkwy, Griffiss Air Force Base, NY 13441

Kenneth Teegarden

University of Rochester

Institute of Optics, Rochester, NY

ABSTRACT

A compact, modelocked erbium doped fiber laser is constructed with fiber Bragg gratings. Passive modelocking is achieved with a MQW saturable absorber micro-assembled onto the fiber tip. The pulses produced can be used as a source for an OTDR.

CONFIGURATION

The cavity shown in Fig 1 gives the initial test layout using bulk optics to focus on the saturable absorber. In the *final* version the fiber tip is butt coupled to the SA surface, which was cut to mm dimensions permitting a micro bond. This all fiber cavity requires no coupling to waveguide or bulk optics, nor polarization control or isolation. The erbium fiber is tightly wound on a 1.5 inch coil: it and all other components including a 100 mw/980 nm pump laser with WDM, can be fit into a module about 1x3x5 inches. A fiber Bragg grating formed directly in SM fiber is fusion spliced and serves as output coupler and spectral filter. This application requires gratings with a spectral bandpass in the nm range to support short pulses. The other end of the cavity is formed by a 50 layer MQW saturable absorber on InP, which generates passive modelocking.

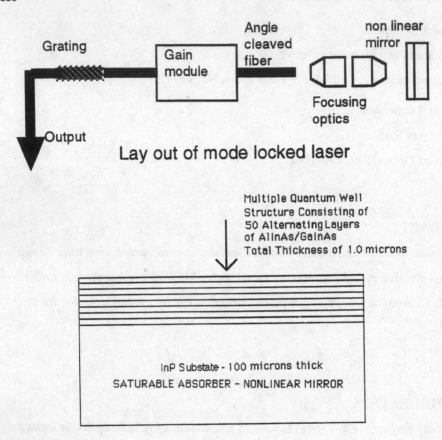

Lay out of mode locked laser

Multiple Quantum Well
Structure Consisting of
50 Alternating Layers
of AlInAs/GaInAs
Total Thickness of 1.0 microns

InP Substate - 100 microns thick
SATURABLE ABSORBER - NONLINEAR MIRROR

DATA

The output pulse train was stable and self-starting at a low threshold of 20 mw; at 50 mw pump power, the pulse energy was between 1 and 2 nj at 1.55 µm.

	Fiber Grating #1	Fiber Grating #2
Tx Bandpass	0.3 nm	1.5 nm
Pulse Spectral Width	0.15 nm	0.3 nm
Pulse Temporal Width	16.5 ps	8.0 ps
Time Bandwidth Product	.314	.305

SPECTRAL BANDPASS OF FIBER GRATING

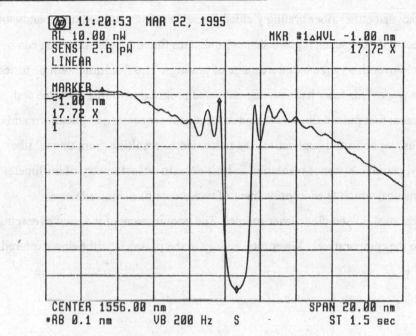

SPECTRUM OF LASER OUTPUT

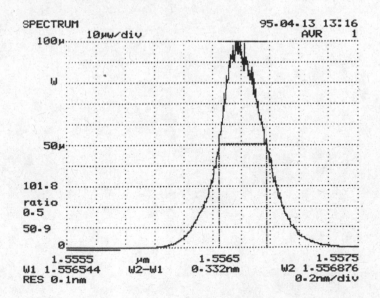

CONCLUSIONS

The fact that either fiber grating yielded a near transform limited time bandwidth product {.31 for $(\text{sech}^{-1})^2$ pulse shapes} indicates that the pulse duration can be selected over this range by interchange of gratings. The full limits will be tested in future work. This laser has also demonstrated progress toward a portable and alignment free pulsed fiber laser source for signal processing applications and field use. as an initial application, the pulse widths resulting from this all fiber cavity, permit a distance resolution in time domain reflectometry of millimeter dimensions, which is an improvement of two orders of magnitude over conventional pulsed diode laser sources. The requirements for signal averaging due to the comparatively lower total energy in the pulses is still being explored.

Power Attenuators for Industrial Lasers: The HVA (Holographic Variable Attenuator) Solution

D. Asselin, P. Galarneau, J.M. Lacroix, P. Langlois, Y. Painchaud, M. Poirier and S. Barsetti

National Optics Institute, Québec, Canada and Gentec, Québec, Canada

Summary

Due to the dependence of the Nd:YAG laser beam characteristics on pumping conditions and passive loss level, it seems logical to operate the laser at a constant power and use an external attenuator to adjust the power at the workpiece for industrial applications requiring variation of the incident laser power such as featured cutting and high resolution trimming and marking. We have successfully demonstrated the feasibility of Holographic Variable Attenuator (HVA) made of fused silica substrates. The HVA fulfils the following requirements: simple design, polarisation insensitive, good transmission range, absence of apodization and high power handling at 1.06 μm. The demonstrators realized were 40 mm long and 13 mm wide. For the AR-coated HVA set, the transmission ranges from 44% to 97% and the spatial uniformity of the transmission was better than 1.4% for a surface of 5 mm in diameter.

1. Introduction

The Eureka project *EU-226: High Power Solid-State Lasers* was aimed at developing new solid-state laser tools for industry, comprising high power Nd:YAG lasers and associated optics and delivery systems, adapted to industrial needs. Within this framework, we had the mandate to develop methods and means to perform laser beam monitoring and control to improve both the system effectiveness and friendliness.

Due to the dependence of the Nd:YAG laser beam characteristics on pumping conditions and passive loss level[1], there is a strong belief that external modulator are needed for material processing such as featured cutting and high resolution trimming and marking. This paper presents an Optical Power Modulator based on Holographic Variable Attenuators (HVA) etched directly onto the surface of fused silica substrates.

2. The Holographic Variable Attenuator

It is well-known that a phase-relief diffraction grating separates an incident beam into many diffracted orders. Amoung these, the zero order is simply an attenuated replica of the incident beam since some energy flows in higher orders. As higher orders propagate away from the optical axis, it then becomes easy to separate them from the zero order. The zero order diffraction efficiency (η) of a sinusoidal phase-relief grating is given by

$$\eta = J_0^2(\pi \, \Delta n \, M / \lambda)$$

where J_0 is the Bessel function of the first kind of order 0, Δn is the refractive indices difference between the diffractive element and the surrounding media, λ is the vacuum wavelength and M is the modulation depth (peak to peak). It is possible to vary the transmission of the HVA by adjusting the modulation depth with respect to the transverse position on the diffractive optical element.

If the transmission of the HVA varies exponentially with respect to the transverse position, the use of two symetric HVAs will attenuate an incident laser beam uniformaly over its cross section, a mandatory property of the attenuator to prevent from beam apodization. By varying the relative position of the HVAs, one will control the transmission of the system. The characteristic of such a variable attenuator system will be mainly determined by the relief profile of a single HVA element, the accuracy on the positioning system, the length of the HVA and the desired clear aperture of the system.

Recently, such diffractive optical elements were realised for the first time at NOI[2-4]. The desired modulation profiles were obtained by adjusting the recording time using a moving slit in front of the photoresist layer. The profiled photoresist reliefs were subsequently transferred onto the fused silica substrates using CAIBE (Chemically Assisted Ion Beam Etching) techniques.

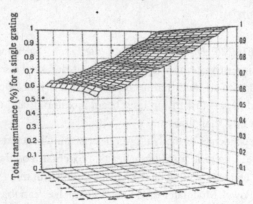

Figure 1: Transmission mapping of a single HVA

3. The Optical Power Modulator

Based on the HVA technology, Gentec has developed and now commercialise a system to externally attenuate high-power lasers without affecting the laser beam propagation characteristics: the Optical Power Modulator (OPM). Such a system is made of two symetrical HVAs mounted on two reverse worm gears. The enclosure has two water-cooled beam dumps for compatibitlity with high power lasers. The electronic control may be address both digitally and analogically. Moreover, a feedback input is also available for power stabilisation and process control.

The OPM is designed to have a transmission between 44% and 97% at 1.064 μm. The HVAs are 40 mm long and 13 mm wide and the spatial uniformity of the transmission is better than 1.4% for a surface of 5 mm in diameter which is typical for that kind of optical elements. As already observed with the HBS, the polarisation sensitivity is dependent on the modulation depth but still remains below the 2% level even for strong attenuation.

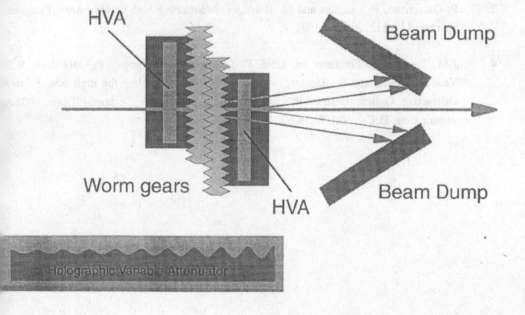

Figure 2: Schematic of the Optical Power Modulator

4. Conclusion

The OPM fulfils the following requirements: simple design, polarisation insensitive, good transmission range, absence of apodization and high power handling at 1.064 μm. The OPM not limited to the actual design. It is possible to achieve larger dimensions, attenuation at other wavelengths and different attenuation ranges. It thus opens the possibility to improve the process reliability, effectiveness and friendliness for several type of applications.

5. References

1. N. Hodgson, C. Rahlff and H. Weber, *Dependence of the refractive power of Nd:YAG rods on the intracavity intensity*, Optics & Laser Tech., **25**, 179-185, 1993.

2. S. Barsetti and P. Galarneau, *Making industrial laser systems smarter with on-line beam monitoring*, Photonics Spectra, 153-160, March 1992.

3. P. Galarneau, P. Langlois and M. Bélanger, *Monitoring high-power lasers*, Photonics Spectra, 112-113, January 1991.

4. J.M. Trudeau, J. Fréchette, M. Côté, P. Langlois, M. Bélanger, P. Galarneau, R.E Vander Haeghe and S. Barsetti, *Holographic beam samplers for high power laser* Diffractive Optics: Design, Fabrication, and Applications Technical Digest (OSA Washington, D.C.), Vol. **9**, 58-60, 1992.

High Quality Coatings Produced by Ion Beam Sputtering

R. Henking, D. Ristau, H. Welling

Laser Zentrum Hannover e.V.
Hollerithallee 8
30419 Hannover, Germany

Introduction

In the last years, the considerable progress of laser technology was made possible in a high degree by the simultaneous development of high quality coatings. Conventional deposition techniques like E-Beam- and boat evaporation are well established and are used for the production of standard all-dielectric mirrors, filters and anti reflective coatings. This components are characterised by low losses, high damage thresholds and low costs. But there are certain disadvantages of these thermal evaporation techniques, which have to be mastered for the realisation of special coatings. One drawback is the need of heating the substrates which restricts the production of high quality coatings on temperature sensitive substrates. Another point is the great expenditure for the realisation of coatings with sophisticated spectral specifications, not to mention an automation of the deposition process. But the most important point is the micro structure of the layers, which is completely different from the bulk structure. This polycrystalline columnar structure causes high scatter losses and an incorporation of contaminants (especially water) into the layer.

There are different methods to enhance the quality of the coatings with respect to the problems mentioned problems. Most of these methods are based on ion- or plasma assisted techniques. The prominent ways are ion assisted deposition, ion plating, magnetron or dc sputtering and plasma activation. For the production of extremely low loss components, the ion beam sputtering (IBS) deposition is the method of choice because of its unique film characteristics [1]. The micro structure of the IBS-coatings is similar to the bulk and therefore they show no water incorporation, low scatter losses and a high thermal and chemical stability. Due to a constant and reproducible sputter rate, the IBS-process is well suited for the production of coatings with defined spectral characteristics over a wide wavelength range and also for a complete automation of the process.

Deposition Technique

The IBS-plant is based on a Balzers BAK 640 vacuum chamber, which is equipped with an cryogenic pump to achieve a clean oil-free vacuum of 10^{-7}mbar. The vacuum system and the deposition process are automatically controlled by a computer unit. The ion source consists of a hollow cathode and a discharge chamber with an accelerating grid of 5cm in diameter. This kaufman source [2] emits an ion current in the range of 40-70mA with an energy of 1.2keV towards the target. This might be the dielectric oxide e.g. SiO_2 or the corresponding metal e.g. silicon. In both cases a proper amount of oxygen has to be added to achieve a stoichiometric oxide layer. A second hollow cathode (the neutraliser) injects electrons into the ion beam to avoid charging of the target, which would cause a broadening of the ion beam and arcing during the deposition process. The film thickness is controlled by an optical monitor and - because of the constant deposition rate - by the time. Typical deposition times are 24 hours for a high reflecting mirror for the wavelength of 1064nm.

Applications

Low-Loss Mirrors

At the Laser Zentrum Hannover, mirrors for the visible and near infrared spectral range produced by ion beam sputtering exhibit low optical losses due to absorption and scattering. Scatter losses are mainly induced by the surface roughness of the substrates. To achieve scatter losses of less than 6ppm, super polished substrates are necessary with a rms-roughness of less than 0.1nm. The additional portion due to inhomogenities in the micro structure of the coatings is less than 1ppm. The absorption losses are essentially caused by stoichiometric defects and residual impurities like iron, chromium etc.. With an improved deposition process these defects can be avoided and the absorption can be decreased significantly. Therefore, mirrors for the wavelength of 514nm and 1064nm with losses of less than 10ppm can be produced as a matter of routine.

In addition, the IBS-mirrors for the Nd:YAG-Laser resist also high laser intensities. The damage threshold for these components is in the range of 60J/cm^2 and comparable to the best values of advanced e-beam coatings.

Coatings for Frequency Conversion

The frequency doubling and tripling of solid state lasers, especially the Nd:YAG-Laser is an efficient method to produce coherent light in the visible and near ultraviolet spectral regime. Coatings used in this application have to meet sophisticated specifications at different wavelengths. Ion beam sputtering is a well suited method for the realisation of these coatings because of the reproducible and stable deposition rate and of the high quality of the deposited films. Figure 1 shows an example of a double high reflector produced with IBS compared with the calculated target. A very good agreement is obvious, which shows the high performance of the IBS-process.

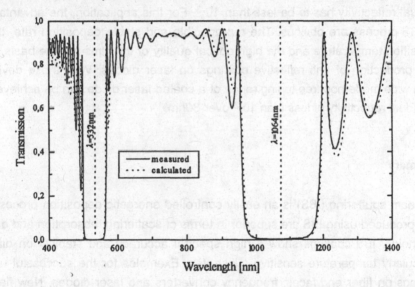

Fig. 1: Calculated and realised spectral behaviour of an IBS-double-high reflector

Fiber coatings

Optical fibers are an important tool in today's laser applications. They are used in material processing, communication technology and to transfer the light of laser diodes for efficient pumping of solid state lasers. With anti reflective coatings on the end faces the throughput of this devices can be increased by 7%. One difficulty here is that the coating has to be made on a ready to use fiber to achieve a process on a profitable basis. Ready to use fibres involve the fibre with cladding, plastic cover mechanical shielding and adapters. Therefore, these parts could not be heated during the deposition process. IBS coatings are made at temperatures of less than

50°C, so no thermal influencing takes place and coatings with a high optical quality can be made on fiber end faces. Typical lots contain 10 fibres with a length of 3m.

Anti reflective coatings on laser diodes

Compact tuneable laser sources are of a great interest for the detection of gaseous species in the atmosphere. Laser diodes are available for a variety of wavelengths from 600nm to the infrared, show broad fluorescence spectra and are well suited for the detection systems mentioned above. Tuning can be achieved with a grating in Littrow- or Littman-arrangement. Therefore an anti reflective coating on one end face of the laser diode is necessary. To realise a wide mode hop free tuning range, the residual reflectivity has to be less than 10^{-3}. For this application, the advantages of the IBS-process are obvious. The reproducible and stable deposition rate, the low deposition temperature and the high optical quality of the films are the basis for the easy production of anti reflective coatings on laser diodes. With those devices, a 30nm wide mode hop free tuning range of a coated laser diode can be achieved with a residual reflectivity of less than 10^{-3} over 30nm.

Summary

Ion beam sputtering (IBS) is an easily controlled energetic deposition process. The films produced using IBS are superior in terms of scattering, absorption and damage thresholds. IBS-coatings show a high spectral accuracy and stability on different, particularly temperature sensitive substrates. Examples for the successful use are coatings on fiber end faces, frequency converters and laser diodes. New fields for applications are the production of water-free coatings for the 2µm-3µm spectral range, high quality coatings on miniaturised devices and coatings for the VUV and XUV-spectral range.

1.   T.W. Jolly, R. Lalezari, "Ion beam sputter deposition techniques for the production of optical coatings of the highest quality", in Thin Films for Optical Systems, Proc. SPIE 1782, pp 250-254, (1990).
2.   H.R. Kaufman, J. J. Cuomo, J. M. Harper, "Technology and application of broad-beam ion sources used in sputtering, Parts 1 and 2", J. Vac. Sci. Technol., 21(3), pp725-756, (1982).

Optical Solitary Wave Solutions in Dispersive Linear Waveguides

Zhonghao Li, Guosheng Zhou
Department of Electronic and Information Technology,
Shanxi University, Taiyuan, Shanxi, 030006, P.R. China

Optical solitons, and particularly solitary waves, which preserve their shape and energy as they propagate through a medium, have attracted considerable attention because of their potential applications in ultra-high bit-rate optical communication and ultrafast signal-routing systems [1]. So far the most of the researches on the optical solitons are focused on the nonlinear region. However, an envelope optical solitary wave solution has been obtained in a dispersive linear medium recently [2]. This may open a new way of realizing very high bit-rate optical communications. In this paper, we present a novel type of exact optical (envelope) solitary wave solution in cylindrical linear optical waveguides. Unlike the conventional bright solitary wave in cubic nonlinear media, the present bright solitary wave solution can be obtained in the normal (positive) GVD regime. In addition, a peculiar feature of the solution is that it may achiéve a condition of "zero-dispersion" to the media so that a solitary wave of arbitrarily small amplitude may be propagated with no dependence on its pulse width.

In the development that follows, let us consider the propagation of an optical pulse in a bulk dispersive homogeneous linear medium. After removing the terms describing inhomogeneity and nonlinearity of the medium in Ref.[3], we can obtain the governing equation for the complex envelope function $A(\mathbf{r},t)$ of the optical pulses under cylindrical coordinate in the form:

$$[\frac{\partial^2}{\partial r^2}+\frac{1}{r}\frac{\partial}{\partial r}+\frac{1}{r^2}\frac{\partial^2}{\partial\varphi^2}+\frac{\partial^2}{\partial z^2}+2i(q\frac{\partial}{\partial z}+k_0 k_0'\frac{\partial}{\partial t})$$

$$-(k_0'^2+k_0 k_0'')\frac{\partial^2}{\partial t^2}+k_0^2-q^2]A(r,\varphi,z,t)=0 , \tag{1}$$

where q is the reference constant of propagation along z direction and $k\equiv\omega n/c$ the wave number, r is the radial axis, φ the azimuthal axis, the primes indicate the derivatives with respect to ω, and the subscript 0 indicates evaluation at the carrier center frequency ω_0. k_0'' describes the GVD effect.

Equation (1) is a (3+1)-dimensional (three spatial and one temporal) space-time coupled linear wave equation, we can treat it by separation of variables. It is easy to verify that the solution of Eq.(1), which satisfies the following natural boundary conditions:

$$F(r,\varphi)\,|_{r\to\infty}=0\ ,\tag{2}$$

$$F(r,\varphi)\,|_{r\to 0}\to finite\ ,\tag{3}$$

$$F(r,\varphi+2\pi)=F(r,\varphi)\ ,\tag{4}$$

can be written by

$$A(r,\varphi,z,t)=F(r,\varphi)\,G(z,t)\ ,\tag{5}$$

with

$$F(r,\varphi)=J_m(\beta r)\begin{matrix}\cos(m\varphi)\\ \sin(m\varphi)\end{matrix}\ ,\tag{6}$$

$$G(z,t)=A_0 sech[\,(t-z/V_g)/\tau)\exp[i(\beta_3 z-\Delta\omega t)]\ ,\tag{7}$$

$$V_g=\pm(k_0'^2+k_0 k_0'')^{-1/2}\ ,\tag{8}$$

$$\beta_3+q=k_0 k_0' V_g+\Delta\omega/V_g\ ,\tag{9}$$

$$\beta^2=k_0^3 k_0'' V_g^2\ ,\tag{10}$$

where V_g is group velocity in z direction, parameter τ is the pulse half-width, β_3 is the propagation constant change in z direction, $\Delta\omega$ is the frequency shift and A_0 is the maximum amplitude of pulse.

If all of the parameters are reasonably chosen, we can expect to obtain the solitary wave solutions described by Eq.(6) and (7). In order to fulfill the physical conditions in transverse directions, the parameter β^2 must be positive. From Eq.(10) one can see that there exist solitary wave solutions only in the case of normal (positive) GVD region. This is contrary to that of (1+1)-dimensional NLSE, in which the sech-like solitary wave solutions exist only in the anomalous (negative) GVD region. The existence of present solitary wave solution indicates that the physical effects of transverse confinement seems to counteract the effect of normal GVD. Therefore, it is possible to realize undistorted transmission of optical pulses even in the bulk cylindrical linear media of normal GVD if the physical effects of transverse confinement balance the normal GVD effect under appropriate conditions. This may be a new way for the realization of ultra-high bit-rate optical soliton communication systems. Comparing with the nonlinear method of utilizing the nonlinear dependence of refractive on pulse intensity suggested by Hasegawa and Tappert [4], the present one has three features as follows: For the first, it may achieve a condition of "zero-dispersion", in which a solitary wave of arbitrarily small amplitude may be propagated with no dependence on its pulse width. While the pulse amplitude A_0 is proportional to the inverse of pulse half-width τ for the nonlinear refractive index case. This implies that the pulse intensity will increases rapidly with the decrease of pulse half-width (to the second order). Therefore, in realizing ultra-high bit-rate optical soliton communications, it will finally meet the limit set by the damage threshold of optical guide materials and other nonlinear effects. This difficulty may be overcome easily in our case as one may achieve ultra-high bit-rate transmission of pulses in optical soliton communication systems, in which the pulse half-width is narrow enough while the intensity still keeps at a low level. For the second, it may utilize all of the advantages of linear techniques (e.g. wavelength division multiplex)in the future optical soliton communication systems. For the third, the optical bright solitary wave can be achieved in the normal (positive) GVD region. This feature can greatly extend the range of optical wavelength for realizing transmission of the bright solitary wave. It is unnecessary to search for special light source, of which the wavelength lies in the range of anomalous GVD for optical guide materials.

From a practical standpoint, we should also consider whether there also exist optical solitary wave solutions in a realistic case (i.e. the radial variable r is finite). In order to verify the existence of the optical solitary wave solutions in such a medium, let us consider a simple case, in which the optical pulses propagate in an optical fiber with a perfect metal-cladding (i.e., its conductivity is infinitely large). Comparing with the case of infinite optical waveguides in transverse directions, the only difference between them is the boundary condition (4). In the present case, the boundary condition (4) should be replaced by the following condition:

$$F(r,\varphi)\big|_{r=R}=0 \ . \tag{11}$$

As before, we can treat the Eq.(1) by separating the variables and obtain the solitary wave solution described by Eq.(5)-(10), in which only $F(r,\varphi)$ should be subjected to the conditions:

$$J_m(\beta R) = 0 \ . \tag{12}$$

This implies that the undistorted transmission of optical pulses (i.e. optical solitary wave) may be achieved in a practical linear regime in the presence of normal GVD if we choose appropriate core radius of the optical fiber with perfect metal-cladding. It should be noted that there is a substantial difference from the case of infinite media in transverse directions, not all the integer-order Bessel functions fulfill simultaneously the condition (12). For an appropriate core radius of a special optical fiber there could be only one integer-order Bessel function $J_m(\text{ß}r)$ that satisfies the above mentioned condition. That is, only certain single mode optical pulse is permitted to propagate in such a special optical fiber no matter of its size of core radius. Of course, one can choose the core radius to fit the transmission of optical pulse of any desired transverse mode if necessary. In practice, the most interested case is the transmission of optical pulse of the fundamental mode (i.e. zeroth-order Bessel function $J_0(\text{ß}r)$). One can obtain no node optical wave field if the first zero for zero-order Bessel function $J_0(\text{ß}r)$ is chosen. The minimum of corresponding core radius, which lies at the first zero for zero-order Bessel function, can be given by

$$R = 2.405 \left[(k_0'^2 + k_0 k_0'') / (k_0 k_0'') \right]^{1/2} / k_0 \ . \tag{13}$$

In conclusion, we have found a novel type of exact optical (envelope) solitary wave solution by taking into account the interplay between the physical effects of transverse confinement and the GVD in cylindrical linear media. It may open a new way for the realization of ultra-high bit-rate optical soliton communication systems. Unlike the conventional bright solitary wave in cubic nonlinear media, the present solitary wave solution can be obtained in the normal GVD regime. In addition, a peculiar feature of the solution is that it may achieve a condition of "zero-dispersion" to the media so that a solitary wave of arbitrarily small amplitude may be propagated with no dependence on its pulse width. Finally, it is shown that the transmission of optical solitary wave in practical optical fiber with metal-cladding could also be realized under the same idea.

[1] A. Hasegawa, Optical Solitons in Fibers (Springer-Verlag, Berlin, 1989).
[2] Zhonghao Li and Guosheng Zhou, Chinese Science Bulletin **40**, 119 (1995).
[3] M. Jain and N. Tzoar, J. Appl. Phys. **49**, 4649 (1978).
[4] A. Hasegawa and F. Tappert, Appl. Phys. Lett. **23**, 142 (1973).

Cleaning of Metal Mirror Surfaces by Excimer Laser Radiation

K. Mann, B. Wolff-Rottke

Laser-Laboratorium Göttingen
Hans-Adolf-Krebs-Weg 1
D-37077 Göttingen

Particle contamination is known to be a general problem for high quality optical components. This is particularly the case for astronomical mirror surfaces, since adsorbed dust particles cause losses in reflectivity and a high image background due to light scattering, resulting in prolonged observation times. Hence, especially for future large telescope systems, improved non-contact cleaning procedures have to be developed. Since the effect of particle removal from solid state surfaces by pulsed UV laser irradiation has been successfully applied already for the restoration of icones[1] and the cleaning of Si wafers[2], laser cleaning is a promising technique also for optical surfaces.

The work presented in this paper was initiated by demands of astronomers from ESO (European Southern Observatory/Garching, FRG). ESO plans to build "Very Large Telescope (VLT)" systems with 8.2m diameter primary mirrors on Mount Paranal in Chile. We report on results of a feasibility study, in which the influence of the various relevant laser parameters like wavelength, pulse duration, energy density, pulse repetition rate and number of pulses per site on the particle removal efficiency and corresponding reflectivity enhancement was investigated systematically.[3]

The investigations were performed on Al coated BK7 and Zerodur substrates provided by ESO, which had been exposed to dust contamination at Mount Paranal in Chile, the projected site of the VLT, for various times from 3 to 12 months. The main contamination is quartz sand with particle sizes from several μm up to a few hundred μm.

The efficiency of laser cleaning on the mirror samples is monitored by on-line reflectiv
measurements using a He-Ne laser, as well as by video microscopy. In combination w
digital image processing techniques, the latter is used to evaluate the particle density on t
tested site. Fig. 1 shows the number of particles on a given sample area irradiated
successive excimer laser pulses (248nm, 30ns) as a function of the pulse number. Obvious
particle detachment is most efficient during the first few laser pulses. A saturation value
observed after 5-10 pulses, indicating that only about 3-5 pulses per site have to be applied t
efficient cleaning.

In Fig. 2 the reflectivity increase is shown as a function of the energy density for a strong
polluted Al coating on BK7 glass. The reflectivity is found to increase with increasing ener;
density, being limited however by laser induced damage of the coating, which is responsit
for the decrease at high energy densities. Therefore, parameters influencing the dama
threshold of metal coatings like wavelength, pulse width and number of pulses have be
studied in detail.

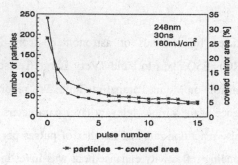

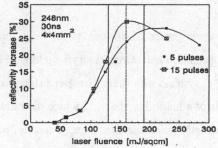

Fig. 1: Particle density of contaminated Al mirror as a function of number of laser pulses

Fig. 2: Relative increase in reflectivity of a contaminated Al sample as a function of energy density

Taking into account laser induced damage and degradation effects, the process parameters fe
optimum cleaning performance on Al coatings can be defined as follows:

wavelength λ = 248nm

pulse length τ = 30ns

fluence H = 160 ± 30mJ/cm^2, *i.e. homogeneous beam profile*

pulse number per site n = 3-5

effective pulse repetition rate on each site f_p ≤ 5Hz

sing this "process window", cleaning is possible not only for typical contaminants (dust, 1artz sand), but also for other types of pollutants like water marks and finger prints.

'ith respect to the mechanisms responsible for particle desorption the results indicate that the riving force is the *absorption of UV radiation by the Al coating* and not by the particles 1emselves, since the process works even for quartz sand as contaminant with a very low UV osorptivity. Moreover, it is observed that *humidity* is of great importance for cleaning. It is onceivable that moisture, which is always present in atmosphere (e.g. dew), builds a very thin quid film beneath the dust particles due to capillary forces. This film is explosively vaporated upon the laser pulse by heat transfer from the irradiated metal surface, causing the jection of the dust particles.[2]

)ata indicate that on Al coated BK7 and Zerodur samples KrF laser radiation yields the ptimum cleaning result, with cleaning efficiencies comparable to polymer film stripping. The nitial reflectivity of the clean coating can nearly be reinstalled, in particular when an additional olvent film on the sample surface is applied. Hence, laser desorption seems to be a viable nethod also for cleaning large Al mirror surfaces.

References:

1. E. Hontzopoulos, C. Fotakis, M. Doulgeridis: SPIE Vol. 1810, 784 (1992)

2. A.C. Tam, W.P. Leung, W. Zapka, W. Ziemlich: J. Appl. Phys. 71 (7), 3515 (1992)

3. K. Mann, F. Müller, B. Wolff-Rottke: "UV laser cleaning of Very Large Telescope (VLT) mirrors", feasibility study for ESO, May 1994

One-Channel Laser Adaptive Mirrors

Andrei G. Safronov.
TURN Ltd, P.O.Box 19, Center, Moscow,103104, Russia.

An adaptive optics for low order correction, the so-called "small" adaptive optics, is of particular actuality and interest because of its simplicity and low cost. The present paper is dedicated to one-channel deformable mirrors designed for use in above mentioned systems including industrial ones.

The construction of the mirrors is a realization of a well-known bimorph structure [1, 2]. The body is made of molybdenum, piezoceramics used is of ЦТС-19 (PZT) type. Total number of deformable mirrors of a given construction produced and tested is 5, three of the mirrors have flat initial shape, outer diameter of the body is 60 mm and inner diameter of 50 mm (mirrors № 2, 3, 4); one mirror is analogous to the previous one but has spherical concave initial optical surface, R = - 40 m (mirror № 1), and the last mirror is flat, 70 mm in outer diameter , 54 mm in inner diameter (mirror № 5).

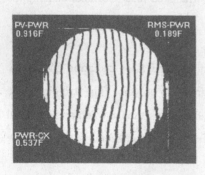

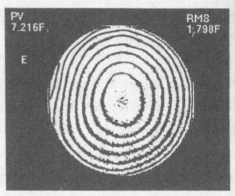

Fig. 1. Interferogram of the initial shape of surface for deformable mirror №3 over the full light diameter (42 mm). F = λ/2.

Fig. 2. Interferogram of reflecting surface for deformable mirror № 4 under control voltage v = 20 V over the full light diameter. F=λ/2.

All the mirrors of the family under consideration have copper reflecting coating and protective coating made of sapphire (Al₂O₃). Regular reflectances of all the mirrors are no than 98.5%. Control voltage ranges from - 200 V to + 300 V. Capacitance of the control ?trode is about 150 nF. Weight of a mirror is about 250 g.

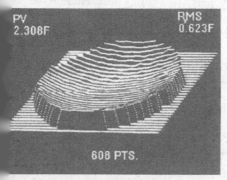

;. 3. Response function of the 'ormable mirror № 4 at v=10V. :λ/2.

Measurements of the initial shapes of the mirrors were made on the automated complex based on "MARK-3" interferometer. A typical interferogram of the initial shape of one of the deformable mirrors is shown in fig. 1. Fig. 2 shows an interferogram of the reflecting surface of the mirror under control voltage. The response function reconstructed upon the experimentally measured data for one of the deformable mirrors is shown in fig. 3. As a result of experimental measurements and further statistical data processing the following

ues of sensitivity (PV) of the adaptive mirrors were obtained: deformable mirror № 1 - 5 µm/kV, deformable mirror № 2 - 47.5 µm/kV, deformable mirror № 3 - 53.8 µm/kV, ormable mirror № 4 - 74.7 µm/kV, deformable mirror № 5 - 79.4 µm/kV..

Therefore the empirical formula for the response function w(r) of a mirror under itrol voltage v can be represented as following (without considering the effect of teresis):

$$w(r) = -K(r/r_1)^2 v ,$$

ere r_1 - radius of the reflecting surface of the mirror (here r_1=21 mm); K - sensitivity (PV).

is known that piezoelectric deformable mirrors are characterized by the presence of :tromechanical hysteresis. Numerical values of hysteresis for the mirrors under estigation are: 10.3% (№1), 13.4% (№2), no more 13% (№3), 12.8% (№4), no more 14% :5).

cause of employment of various materials in the mirror construction, its reflecting surface il be deformed at the temperature changes and therefore under the action of a laser beam. the mirror during its exploitation in the optical system. The computer simulation results the thermal deformations are given in table, numerical data are given for deformable rror № 5 with 70 mm outer diameter. Shape of thermal deformations under the all nsidered conditions coincides with the obtained earlier for cooled deformable mirrors [4] d practically matches the defocusing $Z_4 = \sqrt{3} \cdot \left(2r^2 - 1\right)$. So the results obtained show that the presence of a water outer cooling thermal deformations of the described adaptive irrors under the action of a laser beam with power 1 kW are small compared to the intrinsic ntrollable displacements of the reflecting surface. Considering the above sensitivity values the mirrors, it is easy to verify that from 6% to 10% of the range of control voltage is :cessary to compensate their intrinsic thermal deformations.

Conditions of thermal loading	Thermal deformations, μm		Increase of the mean temperature, °C, of	
	PV	RMS	reflecting plate	piezo-ceramics
1.Change of environmental temperature by 1°C	0.20	0.06	1	1
2.Action of laser beam of 40 mm in diameter and integral power 1 kW, Gaussian distribution. Air cooling. Air rate 10 m/sec.	6.1	1.7	27.6	27.3
3. Action of laser beam of 40 mm in diameter and integral power 1 kW, Gaussian distribution. Water cooling. Water rate 5 m/sec:				
a) Heat removal through the cylindrical surface;	2.2	0.6	9.7	9.3
б) Heat removal through the rear end.	2.3	0.7	10.0	9.6

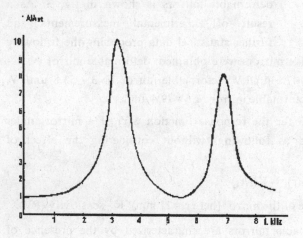

Fig.4. Amplitude of reflecting surfa displacements of deformable mirr № 5 (A/A$_{st}$) as a function of contr voltage frequency (f). A$_{st}$-sta amplitude of mirror deformations (f=0).

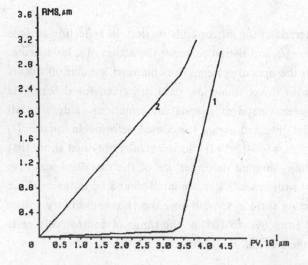

Fig.5. Minimal root-mean squa compensation error (RMS) of axi symmetric Zeidel aberrations as function of span of the aberratio for deformable mirror № 5:
    1 - defocusing,
    2 - spherical aberration.

Frequency response of the mirrors were determined by means of a computer ilation. The reflecting surface displacements as a function of control voltage frequency mirror № 5 are shown in fig.4. As one can see, first resonance frequency of this mirror als to 3.2.kHz. The shape of the amplitude-frequency response for deformable mirrors № is similar to the shown in fig. 4. main resonance frequency for these mirrors is 3.8 kHz.

Efficiency of the described mirrors has been estimated by their ability to compensate axial symmetric wavefront distortions. Calculational results for deformable mirror № 5 represented in fig.5. The case of normal radiation incidence on the mirror surface was imed. It is seen from the represented results that the described adaptive mirrors are very ctive for compensation of the defocusing of wavefront. For mirror № 5 the satisfactory npensation is observed up to about 32 microns, fig.5. Because the calculated response ctions of the adaptive mirrors were used at the computer simulation, the actual efficiency he mirrors are higher, by 25% for mirror № 1 and by about 40% for mirrors № 4 and 5.

Results of the investigations have shown that the presented one-channel adaptive rors compensate effectively large-scale axial symmetric wavefront distortions in laser tics with power level up to 1 kW. Operational range of the mirrors is about 20 microns in plitude of the optical surface displacements and up to 1 kHz in frequency.

ferences:

E. Steinhaus, S.G. Lipson. J. Opt. Soc. Am., 6, 478-481 (1979).

A.V. Ikramov, I.M. Roshchupkin, A.G. Safronov. Quantum electronics, 24, 613-617 (1994).

Power-Swing Stabilisation of Lasers by Liquid-Crystalline Optical Attenuator

M. Sierakowski , A.W.Domański, M.Świłło

Institute of Physics, Warsaw University of Technology,

Koszykowa 75, 00-662 Warsaw, Poland

ABSTRACT

Previously[1] we have demonstrated an optical attenuator for fibre-optic applications based on a new scattering-mode liquid crystalline modulator[2]. Now we present an optical control unit operating with our liquid crystal light attenuator. The device is destined for stabilisation of long-term power swing of lasers used to feed optical sensing, telecommunication and control systems. This paper presents the concept and results of measurements of principal optical parameters of the device.

INTRODUCTION

Simple semiconductor diode lasers are widely used in optoelectronics as a cheap, small-size and easy-to-handle coherent light sources. However in certain applications, especially in sensing and control systems the light intensity must be maintained at a stable level. Disadvantageous changes of signal level in an optical network are caused by power fluctuations of lasers feeding the system and by ageing of the system itself. The changes which exceed over the limits of acceptance of the network have to be compensated. Of course, the stabilisation of the average light intensity should be accomplished without suppressing working signal modulation.

THE IDEA OF LASER POWER STABILISATION

Stabilisation of diode lasers and compensation of light changes in an optical network can be realised by electrical or optical methods. The first are based on a control of laser supply current. This way is simple and effective, but it is accompanied by simultaneous changes in emitting mode polarisation, generating spectrum, light coherence, and has disadvantageous consequences essential in such applications as polarimetric sensing, coherent transmission or optical logic.

In the simplest optical method light power can be stabilised by using electrically controllable optical attenuators. They operate with light modulators which are driven electrically according to voltage signal from monitor element. Generally, electrically controllable optical modulators can be divided onto two groups: interference devices and dispersion devices. Among them very convenient in optoelectronics seem to be liquid crystalline (lc) devices, which comply with most of requirements and demands of optoelectronic systems. The interference devices utilise controlled birefringence of an active element (here - lc layer). As applied to light intensity control, they have to be provided with two crossed polarisers in addition to the birefringent layer. This requirement causes that they rotates the polarisation plane of the controlled component of light. Moreover, the complementary component of the beam is erased and is irretrievably lost.

Therefore, more convenient for this purpose are dispersive attenuators. Liquid crystalline dispersive attenuators are made as PDLC[3] devices. The PDLC modulators which work on light scattering effect control the transmitting light independently on its polarisation, but they give an output beam highly depolarised.

The SLC modulator which we have invented [2] works selectively with respect to the incoming light polarisation and maintain its plane . It means that only one (extraordinary) component of the transmitted beam is voltage - controlled while the other is unaffected (see Fig. 2).

CONSTRUCTION OF THE OPTICAL POWER STABILISATOR

The construction of the light stabilising unit is shown in Fig. 1. It consist of SLC attenuator mentioned above , two optical fibres for input and output connections, and electrooptical feedback. For testing purposes the arrangement shown in Fig. 1 is provided with fluctuated laser source and controlling detector.

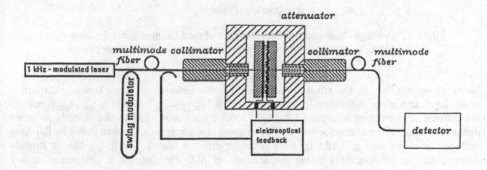

Fig.1. Construction of the optical power stabilisator and testing circuit.

EXPERIMENTAL EXAMINATION OF THE DEVICE

Performance of the SLC attenuator was measured for light of laser diodes of 670 nm and 1300 nm. All parameters were taken in room temperature. We have examined operating range, attenuation ratio, light losses, coherence and polarisation noises of the device. Finally stabilisator circuit was completed and stabilisation efficiency was measured.

The voltage - transmission characteristics of the attenuator are shown in Fig 2. Operating range of the device is within 0 ÷(-8) dB for driving voltages 0 ÷ 8 V (depending on wavelength). The total light losses are less than 1 dB.

In order to determine coherence noise introduced by the device we have compared coherence of the light outgoing from the attenuator to that of direct laser beam. As a degree of coherence we assumed - according to Zernike[4] - Michelson´ s visibility of interference patterns, $\gamma = (I_{max} - I_{min}) / (I_{max} - I_{min})$. The measurements were made by means of Michelson interferometer for optical path difference of the both equally divided rays close to zero. One could expect that γ will decrease in high-attenuation region in regard of increasing light scattering.

However, results of our measurements, as presented in Fig. 3, show that the coherence, within the measurement error is not perturbed by attenuation. It probably will worsen with temperature because of enhancing thermal orientational fluctuations in l.c. layer. Possible explanation is that the attenuator at the output prevailingly collect axial rays less scattered.

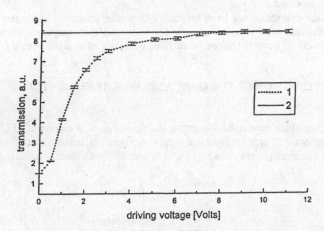

Fig. 2. The voltage - transmission characteristics of the SLC attenuator for laser light 1300nm. Transmission of an : 1 - extraordinary beam, 2 - ordinary beam

More apparently the device affects polarisation of the transmitted light. Linear polarisation coefficient is in analogy defined as $lp = (I_{max} - I_{min}) / (I_{max} - I_{min})$, where I_{max}, I_{min} refer to the extreme intensities of light passed through rotated polarisator. The lp for linearly polarised light at the input of the attenuator was 0.998, whereas at the output dropped to 0.988 by full drive voltage , as shown in Fig. 3 (the lp value is for an ordinary beam). The lp coefficient strongly depends on lc cell treatment during preparation of SLC element, for it determines surface boundary conditions and establishes optic axis alignment in lc cell. We have prepared our lc modulator in rather crude processes, therefore it can be in many points still markedly improved.

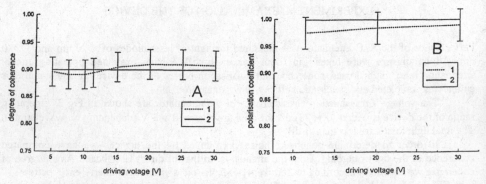

Fig. 3. Degree of coherence (A) and linear polarisation coefficient (B) for : 1 - light beam transmitted through the SLC attenuator and 2 - for direct laser beam.

The main parameter of our optical stabilisator , i.e. efficiency of stabilisation was investigated in the test circuit shown in Fig. 1. The 1 kHz - modulated signal laser was equipped with an additional slow modulation to simulate output power swing. The imposed fluctuations of beam intensity ranged from 25% to 40% of the average signal level. Then the beam was put through examined device and the power fluctuations at the output was measured, as it is indicated in Fig. 1. They remain below 2.5% ÷ 4% of the average output intensity, i.e. the input power swing was suppressed by factor of 10, whereas working modulation (1 kHz) was unaffected.

SUMMARY

We have constructed an optical power stabilisator for lasers in fibre optic systems. The stabilisator is based on our new liquid crystalline scattering mode (SLC) optical attenuator. In contrast to other devices of this type, the SLC attenuator maintain polarisation of the controlled light, does not perturb its coherence, and shows polarisation-selective operation. Those properties are essential in such an optoelectronic applications as polarisation coded optical sensing and information processing or coherent transmission. The principal parameters of our stabilisator are as follows:
- stabilisation range up to 40% power swing,
- stabilisation ratio (input/output swing amplitude) 10,
- total light losses less than 1 dB (depending on wavelength),
- polarisation-selective mode of operation,
- undetectable coherence noises (no perturbance of controlled light coherence),
- small polarisation noises (maintenance of polarisation plane).

ACKNOWLEDGEMENTS

This work was supported by the KBN Contract No 8 S 504 022 66.

REFERENCES

1. M.W.Sierakowski, A.W.Domański :"Liquid Crystal Light Attenuator for Fiber Optic Network", accepted for publication in Mol.Cryst.and Liquid Cryst.,
2. M.Sierakowski, A.Domański :"A novel liquid crystal light modulator", Liquid Crystals, vol 14, no 2 , 1993.
3. I.C.Khoo, S.T.Wu, "Optics and nonlinear optics of liquid crystals", World Scientific, London 1993.
4. F.Zernike, "The concept of degree of coherence and its application to optical problems" in Selected Papers on Coherence and Fluctuations of Light, Mandel and Wolf Edition, Dover Publication, N.York,

Achromatic Phase Retarder Using Liquid Crystals

Jörg Schirmer, Peter Kohns, and Harald Schulz
Optikzentrum NRW, Abt. Industrielle u. Wissenschaftliche Präzisionsmeßtechnik/Hochleistungsoptik
Universitätsstraße 142, D-44799 Bochum, Germany

1. Introduction

Liquid crystals (LC) are a low cost alternative to birefringent crystals commonly used in polarization optics, especially if large apertures are needed. One possible application are achromatic phase retarders (e.g. $\lambda/2$-plates) to modulate the polarization of a white light continuum. Such devices can be assembled from at least two layers of different birefringent materials [1]. In contrast to crystals LCs offer a large variety in range and dispersion of refractive indices and allow tuning of birefringence by low electrical fields. Since spectral data of refractive indices of commercial LC mixtures are normally not available we developed a technique to determine both birefringence and refractive indices of liquid crystalline compounds. The refractive indices of mixtures can then be found from that of their compounds.

2. Determination of birefringence and refractive indices

The ellipsometric method proposed in [2] using a white light source and a diode line spectrometer (resolution 10 nm) allows the fast and accurate determination of birefringence. Fig.1 shows the experimental setup and the construction of our LC cells.

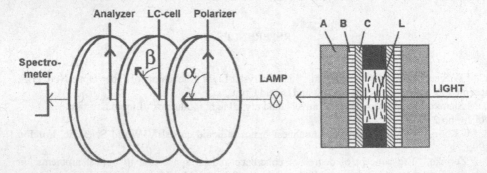

Fig.1: Schematic diagram of experimental setup (left) and LC cell (right).
α : angle between orientations of polarizer and analyzer
β : angle between orientation of polarizer and fast axes of liquid crystal
A : glass substrate (thickness 1.1 mm), B : indium-tin-oxide electrode (ITO, 30-50 nm)
C : alignment layer (polyimide, 10-50 nm), L : liquid crystal layer (3 - 10 µm)

With the fast axes of the LC at β = 45° the birefringent LC splits the incoming light into ordinary and extraordinary ray which suffer different phase retardations and interfere when leaving the LC layer. From the measured transmitted intensities at two polarizer positions I(α=0°) and I(α=90°) the phase difference δ(λ) and birefringence Δn(λ) are calculated [2] :

$$\tan\frac{\delta}{2} = \sqrt{\frac{I(\alpha = 90°)}{I(\alpha = 0°)}} \qquad ; \qquad \Delta n = \frac{2\pi\lambda}{d}\delta \qquad (1,2)$$

The LC layer thickness d is measured before filling the cell using an interferometric method [2,3]. Measurements with polarizer and analyzer aligned parallelly and β = 0° or β = 90° reveal interferences caused mainly by reflections at the indium-tin-oxide (ITO) electrodes of the LC cell. They give additional information on refractive indices, but are not mentioned in [2]. Fig.2 shows an example of the measured spectral characteristic of the four normalized intensities.

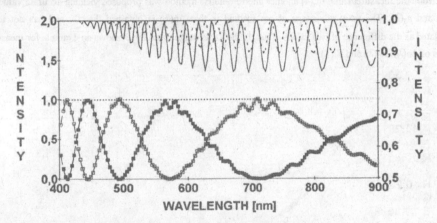

<u>Fig.2:</u> Measured normalized intensities. Upper curves: α=0° (right scale; straight line is for ordinary (β=90°), dotted for extraordinary (β=0°) ray). Lower curves: β=45° (left scale; open squares are for α=90°, filled for α=0°).

To take the interferences into account we consider effective reflection coefficients $R_{p,s}$ for extraordinary and ordinary ray because the thickness of the LC layer is large compared to those of the ITO and the polyimide layers. For both polarizations the cell acts as an etalon transmitting electrical fields E and intensities I (p and s refer to extraordinary and ordinary ray, I_0 is the incoming intensity, τ the transmission coefficient at the interference maxima) :

$$E_{p,s} = E_{p0,s0}\, e^{-i\delta_{p,s}}\, e^{i\omega t}\, \frac{1 - R_{p,s}}{1 - R_{p,s}\, e^{-2i\delta_{p,s}}} \quad ; \quad I_{p,s} = \tau_{p,s} I_0 \frac{1}{1 + \dfrac{4R_{p,s}}{(1 - R_{p,s})^2}\sin^2\delta_{p,s}} \qquad (3,4)$$

With β=45° the electrical fields of p- and s-polarized components add and the intensities for both polarizer positions are found to be :

$$I\begin{pmatrix}\alpha=0°\\\alpha=90°\end{pmatrix} = \tfrac{1}{4}(I_p + I_s) \pm \tfrac{1}{2}\frac{I_p I_s}{\sqrt{\tau_p \tau_s}\, I_0}\left[\cos\delta + \frac{2(R_p + R_s)}{(1 - R_p)(1 - R_s)}\sin\delta_p \sin\delta_s\right] \qquad (5)$$

R_p, R_s, $\sin \delta_p$ and $\sin \delta_s$ are directly obtained from the measured intensities I_p and I_s. Solving eq. 5 for $\delta(\lambda)$ we get corrected values to calculate $\Delta n(\lambda)$. From the interferences in I_p and I_s ordinary ($n_s(\lambda)$) and extraordinary ($n_p(\lambda)$) refractive indices are found. At the maxima $(\lambda_{p,s})_k$ we have (k is an integer value) :

$$k \cdot (\lambda_{p,s})_k = 2 \, n_{p,s} \, d \qquad (6)$$

At the minima we have to replace k by k+½.

Since n does not vary strongly between two subsequent maxima we can find k for at least one maximum by solving eq. 6 for this and the following maximum assuming n to be constant. Once this starting point is found $n(\lambda)$ can be computed for all other extrema. Systematic errors due to phase shifts by the polyimide and ITO layers and to inaccuracy in cell thickness determination are in the order of 1%. The latter can be reduced by at least an order of magnitude if accurate refractive indices for one wavelength are known for example from refractrometric measurements. In [4] another interferometric method was proposed yielding accurate values of $n_p(\lambda)$ and $n_s(\lambda)$. The main advantage of our method is that $\Delta n(\lambda)$ is obtained directly and has not to be calculated as the difference of two values of nearly the same magnitude. Fig.3 shows an example for measured values of $\Delta n(\lambda)$, $n_p(\lambda)$ and $n_s(\lambda)$.

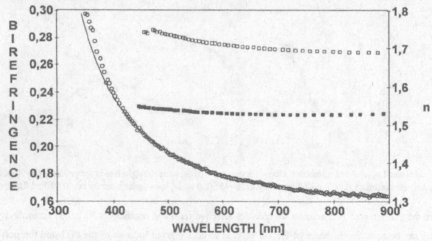

Fig.3: Birefringence (circles, left scale) and refractive indices (squares, right scale) of a liquid crystal. The straight line is a fit to birefingence based on VUK's equation [5] yielding a wavelength dependence of $1/\lambda^2$.

3. Achromatic wave plate

Ideal achromatism is achieved if $\Delta n(\lambda)$ is proportional to λ (eq. 2), which can be approximated by subtracting the retardation of one material with low from another with high wavelength dependence of birefringence [1]. Fig.4 shows the retardation of a combination of two LC cells filled with different LCs. Thicknesses were chosen so that halfwave retardation (180°) was reached near 350 nm and 550 nm. The deviation from the predicted curve can be attributed to thicknesses slightly other than calculated. This could be reduced by using LCs with larger differences in the dispersion of birefringence because these could be used as real zero order waveplates. The ripples near λ=700 nm are due to interferences which could not exactly be canceled. They can be decreased by antireflection coatings on the electrodes. On the other hand interferences may be used to achieve achromatism in small wavelength intervals (here seen at 830 nm). An interesting feature is the achromatic interval arround 600 nm due to deviations from the usual $1/\lambda^2$ wavelength dependence of birefringence of the LCs in that region.

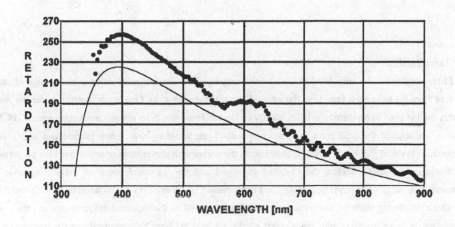

Fig.4: Retardation in [degree] of the combination of two LC cells. The straight line was calculated from fits to birefringence measurements, the points represent measured data.

4. Conclusion

A technique to determine the spectral properties of LCs with an accuracy of 10^{-2} or better in the entire visible spectrum was shown. As an application a double layer waveplate with an achromatic behaviour within a spectral range of 50 nm was assembled. Careful investigation and prediction of interference effects in the multilayer setup of LC cells are necessary for the construction of commercial devices.

Acknowledgements

We would like to thank Prof. R. DABROWSKI, Military University of Technology, Warsaw, Poland, and Prof. P. ADOMENAS, Vilnius University, Vilnius, Lithuania for synthesis of LCs and Dr. S. YAKOVENKO, firm EKMO, Minsk, Belarus for LC cell fabrication.
This work was supported by the European Community under grant No. CIPA-CT94-0157.

References

[1] D. CLARKE, Optia Acta 14, 343-350 (1967)
[2] S. T. WU, U. EFRON, and L. D. HESS, Appl. Opt. 23, 3911-3915 (1984)
[3] D. KINZER, Mol. Cryst. Liq. Cryst. Lett. 5, 147 (1985)
[4] M. WARENGHEM, G. JOLY, Mol. Cryst. Liq. Cryst. 207, 205-218 (1991)
[5] S. T. WU, Phys. Rev. A 33, 1270-1274 (1986)

Phase Conjugating SBS-Mirrors for Pulsed Dye Lasers

HANS J. EICHLER, RALF MENZEL[‡], ROLF SANDER,
MARKUS SCHULZKE[†] and JÖRG SCHWARTZ

Technische Universität Berlin, Optisches Institut P11, Strasse des 17. Juni 135, D-10623 Berlin, FRG

† Lambda Physik GmbH, Hans-Böckler-Strasse 12, D-37079 Göttingen, FRG

‡ Universität Potsdam, Institut für Experimentalphysik, Am Neuen Palais 10, D-14469 Potsdam, FRG

1. Introduction

Phase conjugation by stimulated Brillouin scattering has been demonstrated using a wide range of laser sources from the ultraviolet [e.g. 1] to the infrared spectral range [e.g. 2]. One major motivation for the work in this field is the improvement the beam quality of lasers. Presently, SBS phase conjugating mirror (PCM) devices are applied for solid state laser systems with 1 µm wavelength in many laboratories and a few commercial systems. For use in other types of lasers, the wavelength dependence of important SBS properties, as the phase conjugate fidelity, the so-called threshold and the maximum value of reflectivity have to characterised. As is well known from theory and experiment [3,4], these SBS characteristics are influenced by the chosen scattering material, the pump laser wavelength as well as the experimental geometry and the beam properties. In many investigations, the two last conditions have not been characterised exactly, therefore it is difficult to extract comparable information from numerous previous experimental reports. The intention of the present work is therefore to investigate the SBS performance of liquid and gaseous materials at variable wavelength under constant geometrical and temporal beam parameters.

2. SBS reflection of visible dye laser radiation

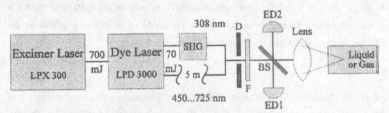

Fig. 1   Experimental set up for dye laser pumped SBS experiments

The experimental set up for the backscattering experiments is depicted in figure 1. Between the laser system (left) and the backscattering experiment (right) an optical isolation is required in order to prevent damage of the dye laser grating by amplified phase conjugated radiation. It was realised by a 5 meter delay distance between the laser and the experiment. No additional isolation was necessary when using the KDP-SHG crystal for generation of UV radiation.

The SBS reflectivity was measured in a usual manner. Incident and backscattered radiation were coupled out by a beam splitter BS and detected by energy detectors ED (Laser Precision, Rj series). The diaphragm D was used to separate the background radiation. Filters F (dielectric mirrors or neutral density filters) controlled the laser beam power. Well-known SBS media with large Brillouin gain coefficient have been used: Liquid n-hexane (C_6H_{14}), carbon disulphide (CS_2) and carbon tetrachloride (CCl_4) as well as gaseous sulphur hexaflouride (SF_6). High purity grades were chosen in order to have a small influence of contamination, observed earlier [4,5].

The measured energy reflectivity vs. pump energy is shown in figure 2 for the largest investigated wavelength of 725 nm. The sample materials n-hexane, CS_2 and SF_6 show a different behaviour. Firstly, they have a very different threshold energy E_{th} (corresponding to an incident energy reaching 2 % reflectivity). The E_{th} value of SF_6 is several times larger than those of the liquids.

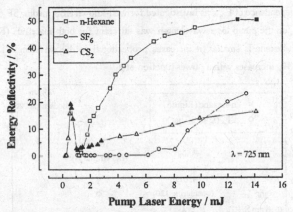

Fig. 2   Energy reflectivity of an SBS mirror using different media as a function of the pump energy. The pump laser was focused by a f = 200 mm lens into the SBS cell. Filled symbols denote the observation of optical breakdown in the focus. (Losses by focusing lens and cell window reflections are not considered, i.e. the intrinsic SBS reflectivity is about 1.4 times larger.)

Secondly, the occurrence of optical breakdown is different. It is correlated to sparks in the focal range and indicated in Fig. 2 by filled data points. As obvious, this effect can strongly disturb the SBS reflection. It was found in all investigated liquids, however at different pump intensities.

In figure 2, the SBS reflectivity of n-hexane shows the usual dependence on the pump energy. In contrast, the behaviour of CS_2 can be subdivided into 4 regions: 1) the SBS threshold is reached, a backscattered signal appears. 2) Optical breakdown starts with increased pump energy and sparks in the centre of the focus can be observed easily. As a result, the reflectivity drops. 3) Above 3 mJ the strongly scattering region moves towards the cell entrance window. As a result, breakdown and sparking disappears. 4) The reflectivity increases now only slowly, being produced in front of the focal region [6]. These observations are in very good agreement with predictions from a recent theoretical study [7] of optical breakdown and SBS interaction. The behaviour of CS_2 in Fig. 2 is calculated in [7] for a situation, where breakdown develops faster than backscattering. The influence of optical breakdown on SBS becomes more important at smaller focal lengths and shorter wavelengths. The reflected signal disappeared completely using CS_2 at $\lambda = 610$ nm in our set up. Also using n-hexane, the onset of breakdown (above 12 mJ in Fig. 2) decreases at shorter wavelengths. As a result, the maximum obtainable energy reflectivity R_{max} was reduced in this material: $R_{max}(725 \text{ nm}) = 50\%$ (see Fig. 2), $R_{max}(610 \text{ nm}) = 46\%$, $R_{max}(450 \text{ nm}) = 40\%$, $(308 \text{ nm}) = 33\%$.

Without breakdown, the reflectivity of gaseous SF_6 at $\lambda = 725$ nm is limited to 22 % by the available laser energy, related to the large threshold (Fig. 2). If the SBS threshold energy of SF_6 could be decreased, it is expected that the maximum reflectivity increases. In fact, this was observed at shorter wavelength, e.g. $R_{max}(SF_6) = 46$ % have been measured at 450 nm.

It should be noted that all R_{max} values given in this section refer to the complete SBS reflector system (cell and focusing lens, without anti-reflection coating), i.e. the intrinsic SBS reflectivity is about 1.4 times larger.

3. Wavelength dependence of the SBS threshold

As mentioned above, the obtainable SBS reflectivity is related to the threshold energy E_{th} which has to be compared to the available pulse energy. Therefore a small threshold energy is desirable for many applications and thus the wavelength dependence of E_{th} was investigated for well-known SBS media, SF_6 and n-hexane.

The dependence of E_{th} on the pump laser wavelength was different for both materials (Fig. 3). In general, the threshold energy of n-hexane is smaller, it increases proportional to the wavelength in the investigated range. In contrast, E_{th} of SF_6 increases with a power function of λ.

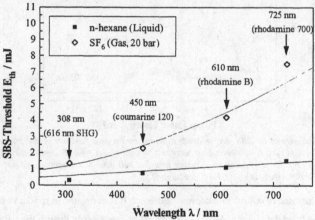

Fig. 3   *SBS threshold energy as a function of the pump laser wavelength. It should be noted that the experimental conditions at $\lambda = 308$ nm are slightly different to all the others. The frequency-doubled laser requires no delay line between laser SBS, incident beam diameter and SBS lens focal length were chosen smaller.*

The established simple SBS model can be used for an interpretation of this observations. Starting from the well-known coupled differential equations for the incident laser electric field amplitude, the backscattered Stokes electric field and the sound wave amplitude, a reflectivity

$$R = r_0 \exp[gI_L L] \tag{1}$$

can be obtained easily for stationary conditions [8], where r_0 is a starting reflectivity initiated by spontaneous scattering. g is the steady-state Brillouin gain coefficient, I_L denotes the pump intensity and L describes an interaction range length.

In contrast to that, the analytical transient solution is more complicated. It can be found e.g. by using Riemann's integration method [9]. Alternative methods are given in Ref. 10 and 11, all of them are leading to a result of the form

$$R = \frac{r_0}{\sqrt{4 g I_L L \dfrac{t_P}{\tau_B}}} \exp\left[-\frac{t_P}{\tau_B} + \sqrt{2 g I_L L \frac{t_P}{\tau_B}} \right] \tag{2}$$

In contrast to (1) the transient reflectivity depends on the ratio t_p/τ_B. This has considerable influence on the threshold behaviour. In both cases, steady-state and transient, the stimulated scattering exceeds the noise if the exponential term in (1) and (2) reaches a magnitude of about 30. This condition represents the threshold of SBS.

Therefore the threshold intensity I_{th} or energy E_{th} is obtained as:

$$E_{th} = t_p \pi w_0^2 I_{th} = t_p \lambda \frac{30}{mg} \qquad \text{(steady-state)} \tag{3}$$

$$E_{th} = \lambda \frac{30}{2mg}\left\{ 7.5\tau_B + t_p + \frac{t_p^2}{30\tau_B} \right\} \qquad \text{(transient)} \tag{4}$$

where m is the interaction length in relation to the Rayleigh range $\pi w_0^2/\lambda$ and w_0 is the beam diameter in the focus. Equation (3) describes a threshold energy proportional to the wavelength for steady-state conditions. This has been found experimentally for liquid n-hexane but not for gaseous SF_6 (Fig. 3).

In the transient case of Eq. (4), the phonon lifetime τ_B leads to a stronger wavelength dependence. It can be expressed in the visible spectral range by [12]

$$\tau_B = 1.99 \cdot 10^{-9}\lambda - 3.83 \cdot 10^{-9} \tag{5}$$

with τ_B in seconds and λ in microns. Substituting this for τ_B in Eq. (4), the curve for SF_6 in Fig. 3 is obtained, using literature data [12] for the SBS gain coefficient $g = 0.020$ cm/MW and $m = 0.9$. For n-hexane the same m value and the corresponding $g = 0.027$ cm/MW [8] has been used to evaluate Eq. (3) for Fig. 3.

In conclusion, the wavelength dependence of the SBS threshold of liquid n-hexane and gaseous SF_6 is different. Due to the small phonon lifetime in the liquid, SBS is stationary and the threshold energy is proportional to the wavelength and remains relatively small over the complete visible range. So liquids are advantageous for the construction of SBS mirrors for widely tuneable lasers if optical breakdown can be avoided. Gases, e.g. SF_6, are less susceptible to breakdown and at small wavelengths in the UV the threshold may become as low as for liquids.

1   M.C. Gower, Opt. Lett. 8, 70-72 (1983)
2   E.K. Gorton, A.M. Richmond, Opt. Commun. 86, 341-350 (1991)
3   Y.-S. Kuo, K. Choi, J.K. McIver, Opt. Commun. 80, 233-238 (1991)
4   H.J. Eichler, R. König, R. Menzel, H.-J. Pätzold, J. Schwartz:, J.Phys.D: Appl.Phys. 25, 1161-1168 (1992)
5   H.J. Eichler, R. Menzel, R. Sander, B. Smandek, Opt. Commun. 89, 260-262 (1992); H.J. Eichler, R. König, R. Menzel, R. Sander, J. Schwartz, H.-J. Pätzold, Int. J. Nonl. Opt. Phys. 2, 267-270 (1993)
6   R. Menzel, H.J. Eichler, Phys. Rev. A 46, 7139-7149 (1992)
7   A. Kummrow, J. Opt. Soc. Am. B, to be published
8   W. Kaiser, M. Maier: "Stimulated Rayleigh, Brillouin and Raman scattering", in Laser Handbook, Vol. 2, F.T. Arecchi, E.O. Schulz-Dubois, eds., (North-Holland, Amsterdam 1972)
9   A. Kummrow, H. Meng, Opt. Commun. 83, 342-348 (1991)
10   N.M. Kroll, J. Appl. Phys. 36, 34-43 (1965); R.G. Brewer, Phys. Rev. 3, A800-805 (1965)
11   B. Ya. Zel'dovich, N.F. Pilipetkii, V.V. Shkunov: in Principles of Phase Conjugation p. 48 (Springer, Berlin 1985)
12   M.J. Damzen, M.H.R. Hutchinson, W.A. Schroeder, IEEE J. Quantum Electron. QE-23, 328-334 (1987)

Trap-Empfänger und Dünnschicht-Thermosäulen als Bindeglieder zwischen Kryoradiometer und Gebrauchsempfänger

K. D. Stock, H. Hofer, M. Pawlak, H.-C. Holstenberg und J. Metzdorf
Physikalisch-Technische Bundesanstalt (PTB)
Bundesallee 100
D-38116 Braunschweig

Kurzfassung
Die nach ISO 9000 geforderte Rückverfolgbarkeit radiometrischer Meßgrößen erfordert geeignete Bindeglieder zwischen dem nationalen Primärnormal und den Gebrauchsempfängern. In diesem Beitrag werden Trap-Empfänger und Dünnschicht-Thermosäulen gegenübergestellt und ihre Verwendbarkeit hinsichtlich ihrer radiometrischen Eigenschaften im Wellenlängenbereich von 200 nm bis 1100 nm diskutiert.

Trap Detectors and Thin-Film Thermopiles as a Link between Cryogenic Radiometer and Working Standard
The traceability of radiometric quantities needed according to ISO 9000 requires suitable transfer standards connecting the national primary standard and the working standards. This paper compares trap detectors and thin-film thermopiles discussing their suitability with regard to their radiometric performance within the wavelengths range from 200 nm to 1100 nm.

1. Einleitung

Der Einsatz von Lasern in der industriellen Meßtechnik erfordert den Nachweis der Rückverfolgbarkeit[1] der verwendeten radiometrischen Meßgrößen auf die nationalen Primärnormale[1]. Für die Meßgröße Strahlungsleistung[2] ϕ (gemessen in W) ist das Laser-Kryoradiometer[3] der Physikalisch-Technische Bundesanstalt (PTB) in Braunschweig das nationale Primärnormal. Im Laser-Kryoradiometer wird die zu messende Laser-Strahlungsleistung in einem geschwärzten Hohlraumabsorber nahezu vollständig absorbiert und durch Substitution elektrischer Heizleistung absolut bestimmt. Die Arbeitstemperatur des Absorbers von etwa 6 K nutzt die gegenüber Raumtemperatur wesentlich günstigeren thermischen Materialeigenschaften aus, um einen hohen Absorptionsgrad (0,99988) bei gleichzeitig relativ kleiner Zeitkonstanten (30 s) zu erreichen: das Absorbermaterial Cu weist bei 6 K eine etwa um den Faktor 1000 geringere spezifische Wärme, sowie eine etwa um den Faktor 10 höhere Wärmeleitfähigkeit auf. Darüber hinaus wirken sich weitere Tieftemperatureffekte günstig auf die Verringerung der Meßunsicherheit aus, wie z. B. die Minimierung von Strahlungsverlusten und das Verschwinden der Heizleistungsverluste in den supraleitenden Zuleitungsdrähten. Die relative Meßunsicherheit[4] des Laser-Kryoradiometers liegt unter 10^{-4} (k=2)[5]. Der Laser-Kryoradiometer-Meßplatz im Reinraumzentrum der PTB[5] ist für den Transfer der Kalibrierungen auf Sekundär- bzw. Transfernormale optimiert und mit Lasern im Wellenlängenbereich von 360 nm bis 1090 nm ausgerüstet. Die geringe Eintrittsapertur und das Brewster-Fenster des Laser-Kryoradiometers läßt nur schwach divergente, monochromatische Laserstrahlen mit einem Bündeldurchmesser < 1 mm zu. Der optimale Arbeitspunkt des Laser-Kryoradiometers liegt bei etwa 1 mW.

2. Transfer-Empfänger

Transfer-Empfänger[1] sind nicht-stationäre Bindeglieder zwischen den i. allg. ortsfesten Primärnormalen und den Normalen geringerer Kategorie. Eine Minimierung des Transferunsicherheit beim Einsatz kalibrierter Transfer-Empfänger an anderen Meßplätzen ist durch sorgfältige Reproduktion der am Laser-Kryoradiometer verwendeten Meßbedingungen möglich. Dieses wird in der PTB bei der Übertragung auf weitere Empfängernormale an einigen Meßplätzen näherungsweise realisiert. Üblicherweise liegen jedoch am Einsatzmeßplatz (z. B. an einem Monochromator) stark veränderte Einflußgrößen wie Divergenz, Leistung, Wellenlänge, spektrale Bandbreite usw. vor, für die die verwendeten Transfer-Empfänger ebenfalls geeignet sein müssen. Nur mit ganz unterschiedlichen Arten von Transfer-Empfängern kann man diesen unterschiedlichen Anforderungen gerecht werden.

2.1 Trap-Empfänger

Trap-Empfänger[6] bestehen aus großflächigen, fensterlosen Photodioden, die in der Art einer Strahlenfalle angeordnet sind: erst nach mehrfacher spiegelnder Reflexion kann ein einfallender Strahl (geringer Divergenz) den Empfänger wieder verlassen (siehe **Bild 1**). Innerhalb eines weiten Wellenlängenbereiches hat ein Trap-Empfänger daher einen Gesamtabsorptionsgrad > 0,99. Der hier beschriebene PTB-Trap-Empfänger hat eine effektive empfindliche Fläche von etwa 7 mm im Durchmesser (Empfängerfläche der einzelnen Si-Photodioden 10×10 mm^2). Zur Überwachung der Empfängertemperatur enthält der Korpus des Trap-Empfängers einen Platinwiderstandssensor vom Typ Pt 100 (nicht eingezeichnet).

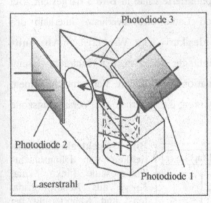

Bild 1: Trap-Empfänger, PTB-Design, Korpus mit Photodioden (Zeichnung maßstabsgerecht). Der Laserstrahl (Pfeil) trifft nacheinander auf die Photodioden 1, 2, 3, 2 und 1 (insgesamt 5 Reflexionen). Die Photodioden 1 und 2 (abgehoben gezeichnet) werden unter einem Einfallswinkel von 45°, Photodiode 3 unter 90° getroffen. Aus- und einfallender Strahl überlagern sich.

2.2 Dünnschicht-Thermosäule

Dünnschicht-Thermosäulen sind massearme, großflächige Thermosäulen mit relativ kleiner Zeitkonstanten. Die in Zusammenarbeit mit der PTB optimierte Thermosäule Typ TS 76[7,8,9] (siehe **Bild 2**) wird in Aufdampf- und anisotroper Ätztechnik hergestellt. Der eigentliche Sensor befindet sich auf einer dünnen Membran aus Si$_3$N$_4$ und SiO$_2$-Schichten. Die Thermosäule ist in ein evakuiertes Glasgehäuse mit einem (zur Vermeidung störender Interferenzen) keilförmigem Quarzfenster (Keilwinkel 1,5°) eingebaut. Ein zusätzlicher Dünnfilm-Heizer unter der Absorberschicht eignet sich zur elektrischen Nachkalibrierung oder auch zur elektrischen Substitution der absorbierten Strahlung.

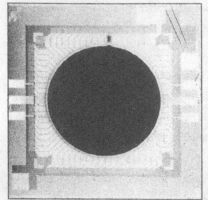

Bild 2: Dünnschicht-Thermosäule TS 76 in Aufsicht. Schwarze Kreisfläche: Absorberfläche aus Ag-Schwarz, Durchmesser 7 mm. Die BiSb/Sb-Thermoelementleiterbahnen führen radial nach außen auf den Rand der Si-Wärmesenke. Der gesamte quadratische Innenbereich mit einer Gesamtdicke von etwa 10 µm wird von einer Membran von etwa 0,8 µm Dicke getragen. Verdeckt unter der Schwarzschicht liegt zusätzlich ein mäanderförmiger Dünnfilm-Heizer.

3. Spektrale Empfindlichkeit

Ein wesentlicher Unterschied der genannten Transfer-Empfänger liegt in ihrem spektralen Verhalten. Die Thermosäule TS 76 ist nahezu unselektiv (siehe **Bild 3**). Der Absorber aus Ag-Schwarz[10] ändert seinen Absorptionsgrad im gezeigten Bereich um weniger als 0,012. Der spektrale Einsatzbereich der Thermosäule wird lediglich durch den endlichen spektralen Transmissionsbereich des Quarzfensters auf 200 nm bis 2500 nm begrenzt (andere Fenster sind optional). Die gemessenen Empfindlichkeiten[11,8] der Thermosäule liegen zwischen 10 V/W und 14 V/W.

Auf Grund der Bandstruktur sind Si-Photodioden nur unterhalb von etwa 1100 nm empfindlich. Die Empfindlichkeitskurve eines idealen Quantenempfängers mit Quantenwirkungsgrad 1 (ein Photon erzeugt genau ein meßbares Elektron im Lastkreis) ist als punktierte Linie in **Bild 3** dargestellt. Bei einer Einzel-Photodiode sorgen Reflexionsverluste und Ladungsträgerverluste innerhalb des Halbleiters dafür, daß die Empfindlichkeit i. allg. unter der Idealkurve liegt. Wie bereits in **Abschnitt 2.1** erläutert, sind die Reflexionsverluste beim Trap-Empfänger durch dessen Hohlraumwirkung weitgehend ausgeschaltet. Es verbleiben interne Rekombinations- und andere Verlustmechanismen, die sich im wesentlichen auf diejenigen Ladungsträger auswirken, die außerhalb der Verarmungszone generiert werden.

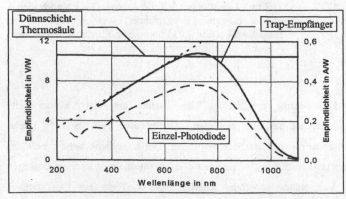

Bild 3: Spektrale Empfindlichkeit von Dünnschicht-Thermosäule (linke Skala, Kurve aus Absorptionsdaten[10] und Kalibrierung bei 633 nm) und Trap-Empfänger (rechte Skala). Im Vergleich dazu sind die spektralen Empfindlichkeiten einer einzelnen Photodiode des im Trap-Empfänger verwendeten Typs und die eines idealen Quantenempfängers (punktierte Linie) gezeigt.

Bild 3 zeigt, wie sich die internen Verluste bei diesem Photodiodentyp bereits ab etwa 700 nm deutlich auswirken. Innerhalb des Bereiches von 400 nm bis etwa 650 nm ist die Empfindlichkeit des Trap-Empfänger in erster Näherung eine Parallele zur Idealkurve, die sich zur Interpolation zwischen den bei einzelnen Laserlinien gewonnenen Empfindlichkeitswerten nutzen läßt.

4. Weitere radiometrische Eigenschaften

	Thermosäule TS 76	Trap-Empfänger
NEP	$< 5 \cdot 10^{-9}$ W/Hz$^{1/2}$	$5 \cdot 10^{-14}$ W/Hz$^{1/2}$
Innenwiderstand	4-5 kΩ	150 MΩ
Zeitkonstante	1 s	2 μs
Temperaturkoeffizient der Empfindlichkeit	$5{,}5 \cdot 10^{-3}$ /K	$< 1 \cdot 10^{-3}$ /K
Nichtlinearität der Empfindlichkeit	$< 3 \cdot 10^{-3}$ für $\phi < 0{,}7$ mW	$< 1 \cdot 10^{-4}$ für $\phi < 10$ mW
Alterung der Empfindlichkeit	0,4 % /a	$< 0{,}1$ % /a
Inhomogenität innerhalb der zentralen Empfängerfläche von 5 mm Durchmesser; 0,5 mm Spot	$\pm$ 1 %	$\pm$ 0,2 %
Änderung der Empfindlichkeit gegenüber Verschiebungen des Strahls aus dem Zentrum	0,2 % /mm	$< 0{,}025$ % /mm

Für den Trap-Empfänger gilt: die elektrischen Kenndaten ergeben sich aus der elektrischen Parallelschaltung seiner Einzel-Photodioden. Temperaturkoeffizient, Nichtlinearität, Alterung und Inhomogenität sind abhängig von der Wellenlänge; die angegebenen günstigen Werte gelten nur für den optimalen Bereich zwischen 400 nm und 650 nm (Photonenabsorption innerhalb der Verarmungszone).

5. Zusammenfassung

Die Gegenüberstellung in **Abschnitt 4** zeigt, daß innerhalb seines optimalen Spektralbereiches der Trap-Empfänger der Dünnschicht-Thermosäule weit überlegen ist. In diesem Bereich eignet er sich in idealer Weise als Transferempfänger zum Anschluß an das Laser-Kryoradiometer. Sind beim weiteren Kalibrier-Transfer Strahlungsleistungen auf $\phi < 1$ mW beschränkt, so ist auch ein anderer Trap-Empfängertyp (mit anderen Photodioden) mit einem erweiterten Optimalbereich bis 800 nm verfügbar. Dünnschicht-Thermosäulen sind weitgehend frei von spektralen Einschränkungen und sind daher (mit etwas größerer Meßunsicherheit) zum Kalibrier-Transfer (auch breitbandig) an Monochromatormeßplätzen gut geeignet.

[1] DIN 'Internationales Wörterbuch der Metrologie', Beuth Verlag Berlin, Wien Zürich 1994.
[2] DIN 5031, Teil 1.
[3] K. D. Stock, H. Hofer, Metrologia **30** (1993) S. 291-296.
[4] DIN 1319, Teil 3.
[5] K. D. Stock, H. Hofer, Metrologia (1995) im Druck.
[6] E. Zalewski, C. R. Duda, Appl. Opt. **22** (1983) S. 2867-2873.
[7] Hersteller: Institut für Physikalische Hochtechnologie e.V., Postfach 100239, D-07702 Jena.
[8] E. Keßler in 'Innovationen in der Mikrosystemtechnik' 26 (1975), S. 107-114. VDI/VDE-Technologiezentrum Informationstechnik, Teltow.
[9] H.-C. Holstenberg, M. Pawlak, K. Möstl, J. Metzdorf in 'Innovationen in der Mikrosystemtechnik' 26 (1975), S. 115-128. VDI/VDE-Technologiezentrum Informationstechnik, Teltow.
[10] J. Vogel, Proc. 9th Int. Symp. Photon Detectors IMEKO (1981) S. 325-335.
[11] DIN 5031. Teil 2.

Dielectric Coatings on Fiber and Faces for High Power Laser Applications and Fiber Lasers

U.Willamowski, H.Zellmer, R.Henking, M.Dieckmann, F.v.Alvensleben

Laser Zentrum Hannover e.V.,
Hollerithallee 8
30419 Hannover, Germany

Abstract

We report on anti-reflective and high-reflective dielectric coatings on the end faces of different types of fibers, using the processes of E-Beam evaporation, Ion Beam Sputtering and Ion Assisted Deposition. The coatings were characterized with respect to their spectral performance, absorption and damage threshold.

Introduction

There are different applications for fibers in laser systems. They can be used as beam guides for the pump light, as a laser by themselves (with a doted core as laser medium) or as beam guides for the laser radiation. Dielectric coatings on the fiber end faces can improve the performance of the laser system: The efficiency of the fibers as beam guides can be increased by anti-reflective coatings, while in the case of fiber lasers, resonator mirrors deposited directly on the end faces render smaller and more compact systems with less optical elements, thus minimizing adjustment difficulties. For the different fiber types and their applications as stated above, different preparation and deposition techniques were applied and tested.

Fiber end face coatings

1. Anti-reflective coatings for fibers used as beam guides for high power Nd:YAG lasers (AR 1064nm, fiber diameter 1000µm) were performed using a conventional PVD process with electron beam evaporation and in comparism, the ion beam sputtering process. The most important requirement was a high damage threshold, as several hundred watts of laser power had to be transmitted without destroying the fiber.

2. In the case of anti-reflective coatings for fibers used as beam guides for laser diode pump light (AR 810nm, fiber diameter 800µm), the question of damage threshold was less important due to the rather low laser power (< 10 W). Because of sit good reliability and accuracy, the technique of ion beam sputtering, which has the general advantage that no heating of the substrates is necessary, was used for this application. (IBS coating plant: Balzers BAK 640, Iontec 5cm ion source with hollow cathode, working gas: Xe, ion current: 45 mA, pump: cryo pump RKP 501 Z (10,000 l/s), target materials: silicon / tantalum, reactive gas: O_2).

3. For Nd:YAG monomode fiber lasers (pump core 150µm, laser core 5µm), thick multi-layer stacks for high reflectivity had to be deposited on the fiber end faces, which made the e-beam evaporation process preferable because of its higher deposition rate than IBS. In order to achieve good adhesion and to overcome the problem of inner stress, the deposition was enhanced by an ion beam, supplying additional

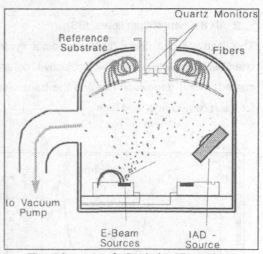

Fig.1 E-Beam plant for Ion Assisted Deposition

energy to the deposited molecules. (Coating Plant: Balzers BAK 640, process unit: BPU 420, pump: diffusion pump with Meissner trap (3360 l/s), E-Beam sources: Balzers ESQ 110, ion source: Denton CC102, film thickness control: quartz monitoring, coating materials: Cerac's SiO_2, Merck's H4, evaporation rates: 1.5 nm/s (SiO_2) / 0.5 nm/s (H4), partial pressure of reactive gas O_2: 1.5×10^{-4} mbar). As laser powers up to several watts were to be out-coupled through an area of a few square micrometers (fiber core), low absorption and a high cw damage threshold were crucial.

For all fiber coatings, reference substrates were placed near the fiber end faces during the deposition process for extended analysis of the coating quality. The

800µm and 150µm fibers were coated without dismantling them from their SMA connectors.

Results

1. AR 1064nm (1000µm fibers; E-Beam / IBS):

Absorption of reference substrates:	5. . . 8 ppm / 15. . .20 ppm
R-on-1 damage threshold of fibers:	15. . .20 J/cm^2

(1064nm, 300 µm spot, 14ns pulse)

Practicability tests with high power lasers:

pulsed system (830 kW/cm^2, 2 min):	undamaged / undamaged
cw system (1.1 kW, 2 min):	undamaged / damaged

2. AR 810nm (800µm fibers, IBS):

The ready to use fibers (SMA-connector system) were mounted into the vacuum chamber complete with their protective outer plastic shielding. Ten fibers were coated in one deposition run and the transmission @810nm was increased by 3% for each coated fiber end face.

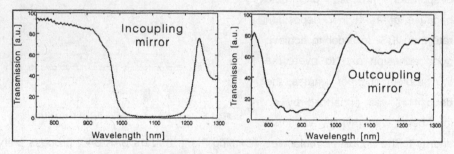

Fig. 2 Spectral transmission of fiber end mirrors

3. Fiber lasers (1060...1075nm, pump light 810nm):

Absorption of reference coatings (laser calorimeter @1064nm):

incoupling mirror: 35 ± 4 ppm,

outcoupling mirror: 48 ± 5 ppm

Damage threshold (1-on-1 @1064nm, 300µm spot, 14ns)

incoupling mirror :   44 ± 10 J/cm^2 (reference coating)

Spectral performance: The spectral transmission of the fiber end faces are given in fig.3. The distortion of the curves is due to incoupling effects and the non-linear spectral response of the photo detector.

Laser performance (see fig.3):

output power   (1065..1075nm):   110mW

input power (810nm):   870mW

max. power density (calculated): > 500 kW/cm^2

(Efficiency similar to the performance of butt-coupled systems.)

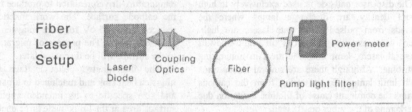

Fig. 3 Setup for testing the fiber laser performance

Summary

Using different deposition technologies, dielectric coatings on fiber end faces have successfully been performed. For the efficient and reliable reduction of reflectivity of fibers used in the medium laser power range, the Ion Beam Sputtering method proved most effective. For high power laser beam guides, IBS and conventional E-beam coatings lead to comparable results. Practicability tests with high power lasers used for laser cutting were successfully performed. Using the technique of Ion Assisted Deposition, incoupling and outcoupling mirror designs were developed and deposited on fiber laser end faces. The coatings' performance as fiber laser resonators was found comparable to the butt-coupled systems.

For further optimization of the fiber coatings, our work is focused on the improvement of the surface preparation techniques (cleaving, polishing). Presently, a method for fiber end face characterization with high spatial resolution is developed, which will contribute to the task of depositing optical high quality coatings on fiber end faces for the use in laser applications.

Thermal Regulation of Cathodes in Industrial Laser Flashlamps

J. W. Loughboro
ILC Technology, Inc.
399 Java Drive, Sunnyvale, California 94089
USA

Introduction

The dispenser cathode is used exclusively in high current density arc discharge lamps where the demands from pulsed solid state lasers are high. When used as pump sources in multi-kilowatt industrial lasers, lamp life is critical in minimizing down-time. Although there are several mechanisms which limit lamp life, sputtering from the cathode remains the dominant cause of lifetime limitation due to darkening of the inner quartz wall.

Until only the past few years, relatively little effort had been made to advance the performance of the cathode in lamps used for laser welding, drilling and cutting applications. Since then, lamp life has been increased by factors of two to five. Benefits to the OEM and end-user are clearly favorable.

Reported here are methods taken to extend lamp life through aggressive restructuring of the dispenser cathode. Several operating conditions are discussed in detail. Only a summary of the behavior of electron emission from a cathode in a high-current arc discharge is presented; extensive treatment of the chemistry and physics is found in the literature.[1-5]

Dispenser Cathode Fabrication

A dispenser cathode is manufactured from a porous pellet of tungsten grains compressed and sintered in hydrogen or vacuum to a density of approximately 80%, with an average grain diameter on the order of 1-10μm. The pellet is then impregnated with a ternary oxide of BaO, CaO and Al_2O_3. Well known mole ratio compositions are 5:3:2 and 4:1:1, $(BaO:CaO:Al_2O_3)$. Other proprietary compounds and ratios are used which give different results specific to the operating conditions in the lamp. During lamp operation, the ternary oxide reacts with the tungsten in a temperature driven reaction to produce free barium at the cathode surface. The work function is thereby reduced from 4.5 eV for pure tungsten to about 2.0 eV for barium. The process of migration of barium to the surface will be discussed later.

The impregnated pellet is then brazed to a tungsten heat sink and machined to a particular shape and size specific to its intended application. The surface area at the tip is made large enough to handle the peak current density during the pulse while positioning the arc during simmer, and is designed to accommodate a uniform distribution of the cathode spot. Anodes are solid materials of tungsten, usually an alloy to aid in machinability. After sealing both the cathode and anode inside the quartz body, the lamp is filled with xenon or krypton gas to a sub-atmospheric pressure.

The cathode spot is that region where arc attachment takes place as shown in Fig. 1. All electron emission at the cathode occurs in this region (except for those electrons emitted by the photoelectric effect outside the cathode spot, which may be neglected). Typical current densities at the cathode spot range from $4\text{-}100 \times 10^3$ A/cm^2 in most industrial laser applications and result from a combination of field-emission and thermionic emission. The relatively low value of work function provided by free barium at the surface allows more electrons to be evaporated at a given temperature than with pure tungsten. When barium has been depleted, or if the cathode pores become "clogged", preventing the migration of barium to the surface, sputtering is accelerated as the higher work function of bare tungsten leads to increased surface temperature. Tungsten grains may separate under such conditions, fusing themselves to the quartz wall yielding localized stress points. Cracking of the quartz at locations of fused tungsten grains is often encountered in a relatively short period of time.

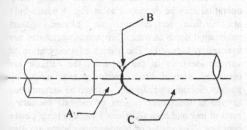

Fig. 1.   Arc attachment: A=cathode. B=cathode spot. C=plasma, (contained by quartz vessel).

Precisely how free barium is generated and the mechanisms of migration to the surface are not completely understood but are described by Dudley.[6] The process of barium production and migration may be described by the temperature driven reaction at 1100°C in the following:[5]

$$4[3BaO \cdot CaO \cdot Al_2O_3] +2W \Rightarrow$$
$$4BaAl_2O_4 + Ca_2BaWO_6 + Ca + 7Ba \qquad (1)$$

The migration of free barium from the bulk to the surface is strongly dependent on both temperature and porosity of the matrix. However, the reaction in (1) does not take place at the surface of the cathode, rather it operates in the bulk of the slug, in regions behind the surface and is depleted slowly during lamp life to depths on the order of a mm, (Fig. 2). The tungstate and aluminate byproducts do not migrate to the surface but remain immobile. Upon reaching the surface of the cathode, free barium reacts with the tungsten to form a monolayer which evaporates during the high current discharge, Fig 3 (reproduced by permission of the author[3]). Therefore, barium is continually generated in the bulk, transported to the surface and evaporated over the life of the cathode, the entire process being temperature driven. Optimum temperature in the bulk where (1) is taking place is thought to be 1000 ± 100°C for many of the ternary oxides in common use. However, as previously stated, different proprietary compounds are used where optimum bulk temperatures may differ from 1000 ± 100°C.

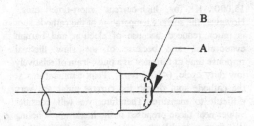

Fig. 2.   Cathode tip: A=cathode spot. B=bulk region where barium is generated.

By controlling both the optimum impregnation mix and bulk temperature for a given set of lamp parameters, lamp life may be maximized. Controlling the temperature profile in the bulk is the topic of the next section.

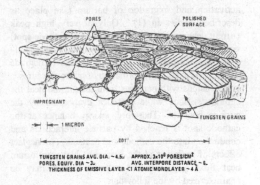

Fig. 3.   Grain structure of the cathode surface showing impregnation.

Thermal Analysis

Typical industrial laser operating conditions require the lamp to handle peak currents in the range of 125 to 1200 amps per pulse; 0.2 to 20 ms pulse duration; 1 to 300 Hz repetition rate and 0.15 to 10 amps simmer current. For example, current density within the column may be 1300 A/cm^2 where electrons are entirely Maxwellian. This translates to a cathode spot current density of 15,000 A/cm^2 including the contributions between thermionic and field emission electrons and positive ions. Gas temperature in the column may range from 6000 to

15,000 . K for high-current short-time arcs.[7] However, the surface temperature at the cathode spot is much reduced because of electron and barium evaporation, and because of the slow thermal response time of tungsten in a pulse-train of relatively low duty cycle, (<10% typ.). Peak temperatures at the cathode spot during the current pulse are very difficult to measure. Therefore, we will consider values near those proposed in the literature,[7-9] noting that for a pulsed lamp, average surface temperatures will be less than that of the peak current discharge. We assume a typical mean surface temperature to be in the range of 1000 to 2500 K.

Thermal cycling which occurs between the on and off states of the current pulse is only a surface effect, and scarcely penetrates into the cathode's bulk because the thermal response time of tungsten is long compared to the period between current pulses. Therefore, we neglect the effects of thermal cycling while examining the bulk where a constant and unvarying temperature is expected, and where activation and migration of barium take place as described earlier in (1). Only at very high peak current densities (eg. ~10^6 A/cm^2 at the cathode spot) with sufficient duration at low repetition rates, eg. 1-2 Hz, has thermal cycling been observed to degrade the cathode surface by formation of micro-fractures.

Collisions with particles from the plasma heat the cathode surface while particles evaporated by the surface cool it. Therefore, energy balance at the cathode spot is reached by particle interactions and conduction into the cathode and heatsink, (we neglect effects of radiation at the surface). Net equilibrium heat flux, q_{net}, into the cathode surface may be characterized by the following,

$$q_{net} = q_c + q_e - q_t - q_n - q_m, \qquad (2)$$

where q_c is heat added to the surface from ion bombardment, q_e is heat added from plasma electrons, q_t is heat removed by evaporation from thermionic and field emission electrons, q_n is heat removed by neutral ions in recombination with electrons and q_m is heat removed by evaporation of barium atoms.

Similarly, we may write the equation for current density at the cathode,[7]

$$J_c V_c = J_t \varphi + P, \qquad (3)$$

where J_c is the positive ion current density, V_c is the cathode drop voltage, J_t is cathode spot current

density, φ is the work function of the emissive surface (~2 eV) and P is the total evaporation power from the cathode spot.

A thermal model for the cathode during lamp operation may be described as in Fig. 4 where bulk and surface temperatures are plotted against penetration depth in mm. (Surface temperatures are average, not peak.) The values of temperature at various locations in the depth of the cathode are estimates derived from measurements made by Richter, Schuda and Degnan[5], as well as assumptions arrived at from observations of cathode sputtering rates at low and high (KW/cm^2) average lamp power densities. The values of temperature chosen are also consistent with data by Cobine and Berger.[7]

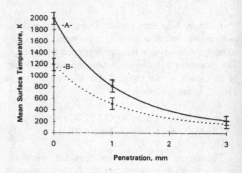

Fig. 4. Temperature profile in the cathode. Curve -A- is optimum; cathode of curve -B- is too "cold".

In cases where bulk temperatures are below the optimum for barium activation and migration (as in curve -B- of Fig. 4), heat choking may be employed, (see Fig. 5, D & E). Heat choking has been employed in lamps at ILC Technology, Inc. for over two decades and is utilized by most lamp manufacturers.

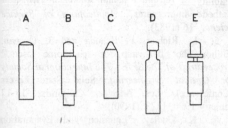

A   B   C   D   E

Fig. 5.   Various types of cathode tip geometries used for controlling bulk temperatures. If bulk temperature is higher than optimum, types A and B may be employed, whereas types D and E may be used if bulk temperatures are low.

Using different styles of heat choking helps the designer optimize the heat flow through the cathode. For example, using tip E from Fig. 5 would cause curve -B- in Fig. 4 to shift upward and "flatten", bringing the bulk temperature more in line with its optimum. Using these techniques for optimizing the cathode bulk temperature often allows us to extend lamp life by two to five times over that which is customary for lamps not incorporating this technology.

Results and Discussion
In the following, we report on advances made in lamp life for given sets of lamp parameters employed in state-of-the-art industrial laser systems as a result of advancements made in cathode performance. Life data noted is that reported by users of the laser systems from which the lamps were in operation. Values of peak current are that of the column and not of the cathode spot.

I)
Prior to cathode improvements, lamps operating under the following conditions routinely achieved 78 million pulses before laser output dropped below the acceptable level of 20% due to sputtering. Following cathode optimization, laser output dropped only 5% at 78 million shots, indicating a four-fold increase in lamp life.

Operating Conditions:
Peak current per pulse = 200 Amps
Pulse width   = 1.0 ms

Pulse repetition rate = 120 Hz
Energy per pulse   = 54 Joules
Simmer current   = 4 Amps

II)
In the following conditions, aggressive cathode development led to an increase in lamp life by a factor of three. The laser system in use operates with a single or multiple mesh pulse forming network, whereby complete capacitor discharge allows for very high values of peak current for applications such as drilling. Therefore, this condition is one of the most challenging for the dispenser cathode.

Operating Conditions:
Peak current per pulse = 800 Amps
Pulse width   = 2.7 ms
Pulse repetition rate   = 3.7 Hz
Energy per pulse   = 400 Joules
Simmer current   = 0.10 Amps

III)
Lastly, we report on another rigorous condition where the application employed by the laser is that of cutting and drilling. As is the case in the previous set of conditions, this laser system utilizes single or multiple mesh pulse forming networks for the current discharge. Improvements in cathode technology have extended lamp operating lifetimes by a factor of five.

Operating Conditions:
Peak current per pulse = 400 Amps
Pulse width   = 1.5 ms
Pulse repetition rate   = 6 Hz
Energy per pulse   = ~150 Joules
Simmer current   = 0.10 Amps

Conclusions
With a thorough understanding of electron emission from a dispenser cathode in a gas and its thermal characteristics, one may make advancements in performance and fabrication. Cathode sputtering may be reduced dramatically.

(1)   Cathode spot current density may be an order of magnitude greater than the column current density. However, because of the evaporation of electrons and barium atoms, and because duty cycles are typically less than ten percent, cathode spot temperature is less than that of the column temperature.

(2)   For best performance, optimum bulk temperature in the cathode must be achieved for a given set of operating conditions required by the laser system. Temperatures may be tailored by utilizing various cathode tip geometries including the use of a heat choke.

(3) Proper activation and migration of barium atoms from the bulk to the surface is a temperature driven reaction. If a cathode is too hot or too cold, sputtering results, diminishing useful light output and reducing lamp life.

(4) By optimizing the temperature within the bulk of the cathode for a given set of operating conditions, we have demonstrated increases in lamp life by factors of two to five.

References

1.   D. N. Hill, R. E. Hann and P. R. Suitch, "THERMOCHEMISTRY OF DISPENSER CATHODE IMPREGNANT MATERIALS: Phase Equilibria in The BaO-CaO-Al2O3 System," Final Report, RADC-TR-87-54, August 1987.

2.   R. E. Taylor, "Thermal Diffusivity of Tungsten Composit Cathodes," a Thermophysical Properties Research Laboratory Report to ILC Technology, Inc., TPRL 833, April 1989.

3.   Green, M., "Dispenser Cathode Physics," Final Report, RADC-TR-81-211, July 1981.

4.   R. A. Lipeles and H. K. A. Kan, "Chemical Stability of Barium Calcium Aluminate Dispenser Cathode Impregnants," *Applications of Surface Science*, 16 (1983).

5.   L. Richter, F. Schuda and J. Degnan, "Billion shot flashlamp for spaceborne lasers," *Proceedings of The SPIE - The International Society for Optical Engineering*, Solid State Lasers (Conference), Los Angeles, California, 15-17 January 1990, 1223 (1990).

6.   K. Dudley, "Emission and Evaporation Properties of a Barium Calcium Aluminate Impregnated Cathode as a Function of its Composition," *Vacuum*, 11 [2] (1961).

7.   J. D. Cobine and E. E. Burger, "Analysis of Electrode Phenomena in the High-Current Arc," *Journal of Applied Physics*, 26 [7] (1955).

8.   K. D. Goodfellow and J. E. Polk, "High Current Cathode Thermal Behavior, Part I: Theory," Final Report, IEPC-93-030, September 1993.

9.   I. S. Marshak. *Pulsed Light Sources*. Consultants Bureau, New York, 1984.

CuInSSe- Doped Glass Saturable Absorbers for the Nd Lasers

K.V.Yumashev, V.P.Mikhailov, P.V.Prokoshin, I.V.Bodnar[*]
International Laser Center, Belarus State Polytechnical Academy,
65 Skaryna av.,   Minsk 220027, Belarus
[*]Chemistry Department, Belarus State University of Informatics and
Radioelectronics, 16 Brovki Str., Minsk 220600, Belarus

By using a $CuInS_{2x}Se_{2(1-x)}$-doped glass as an intracavity saturable absorber, passive mode-locking of a Nd:YAG and $Nd:YAlO_3$ lasers is obtained. Ultrashort pulses of 16 ps duration with 20 mJ of energy (for $YAlO_3$ at 1.08 mm), and 36 ps duration with 45 mJ of energy (for YAG at 1.064 mm) are achieved, respectively. The bleaching relaxation time of CuInSSe-doped glass saturable absorber is found to be ~ 11 ps. The optical damage threshold for CuInSSe-doped glasses estimated from the intensity dependent transmittance measurement with the use of laser pulses of 15 ps duration, was ~ 20 GW/cm^2.

The two $CuInS_{2x}Se_{2(1-x)}$ SA samples used were silicate glass plates containing semiconductor microcrystallites of $CuInS_{1.2}Se_{0.8}$ (sample X-06) and $CuInS_{1.0}Se_{1.0}$ (sample X-05). Fig 1 demon-

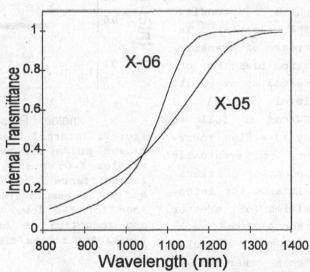

Fig. 1. Linear internal transmittance spectra of (sample X-06) $CuInS_{1.2} S_{0.8}$ and (sample X-05) $CuInS_{1.0}Se_{1.0}$-doped glass saturable absorber samples.

strates linear absorption spectra of both samples. The overage diameter of microcrystallites in both samples was ~ 200 A⁰. The two samples had slightly different CuInSSe band-gap wavelengths; 1.088 mm for sample X-06 and 1.168 mm for sample X-05.

The single-pulse intensity-dependent transmittance and the transient absorption change measurements were performed at the wavelength of 1.08 mm with the totally automated picosecond transient absorption spectrometer based on laser pump-probe technique [1]. The pump (probe) pulse width was ~ 15 ps.

Figure 2 shows the internal transmittance as a function of incident pulse intensity. The initial increase of transmittance bleaching observed at intensity levels of $0.01 - 1$ GW/cm^2 is followed by bleaching recovery and eventually reduction of transmittance for intensities of greater than 4 GW/cm^2.

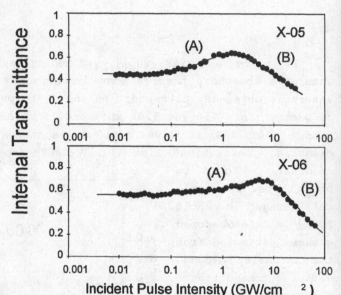

Fig. 2. Internal transmittance versus incident pulse intensity at 1.08 μm for samples X-05 and X-06. The small-signal transmittance is 0.56 and 0.45 for X-06 and X-05, respectively. Bleaching (A) is seen for the lower incident intensities and reduction of transmittance (B) for the higher intensities.

The pulse train generated by the laser oscillator was detected with the use of a detection system with 1.5-ns response time. The trains of pulses separated by the ~ 9.4 ns cavity round-trip time were obtained with a FWHM envelope duration of 150 ns and an output energy of 0.3 mJ. The spatial mode was TEM00 and the beam cross section was close by circular with a diameter of 1 mm. The single-pulse energy measured in the peak of the train was ~ 40 mJ. The asymmetric envelope of pulse train was clearly seen. The trailing part of pulse train envelope is broader than the leading part of one.

The samples used operate as fast saturable absorber for lower intracavity intensities when the increase of intensity leads to the reduction of losses introduced by sample (Fig. 2(A).) This intensity level corresponds to the step of formation of a single ultrashort pulse in laser. The situation is completely different for the higher intensities when the increase of pulse intensity leads to the reduction of sample transmittance (Fig. 2(B)). For the samples used to observe the limiting effect a higher intensity level compared to the that at the beginning of the pulse train is required. So, though the intracavity maximum intensity is limited the high energy of single ultrashort pulse inside the cavity is achieved. The high-energy pulses quickly saturate the amplifier giving rise to the longer trailing part of pulse train and the triangulation of the shape of pulse train envelo

We observed the successful application of the CuInSSe-doped glass samples as saturable absorber in passive mode-locked Nd:YAG laser with active negative feedback without a cooling technique. Introduction of electrooptic feedback loop in the laser

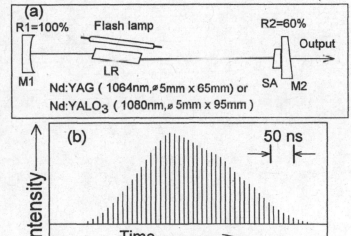

Fig. 3 (a) Schematic of the passively mode-locked laser cavity with a CuInSSe-doped glass saturable absorber. M1 and M2 are resonator dielectric mirrors with reflectivities R1 and R2, respectively. LR, laser rod. SA, CuInSSe-doped glass saturable absorber. (b) Typical pulse train from a passively mode-locked Nd:YAG laser with the CuInSSe-doped glass saturable absorber.

cavity resulted in a generation of pulse train with approximately constant pulse energy for ~ 2 ms, corresponding to a train of 200 pulses. The single ultrashort pulses had an energy of ~ 1 mJ, a FWHM pulse width of ~ 20 ps, and a stability of energy in the steady-state part of a pulse train of ~ 5%. The CuInSSe-doped glass sample operate as only saturable absorber because negative feedback action limits the amplitude of the laser pulse at an intensity level [Fig.3 (A)] which is low to cause the limiting effect with the same CuInSSe-doped glass sample [Fig.3 (B)] in this case with negative feedback on.

1. V.P.Mikhailov, N.V.Kuleshov, N.I.Zhavoronkov, P.V.Prokoshin, et al. Optical Materials. 2, 267 (1993).
2. M.I.Demchuk, V.S.Konevskii. N.V.Kuleshov, V.P.Mikhailov, et al. Optics Commun. 82, 273 (1991).

Investigation of a Pulsed EBCD CO-Laser Supplied from an LC Transmission Line

V.Kazakevich, K.Morozov, A.Petrov, G.Popkov
Lebedev Physics Institute, Samara Branch.
Novo-Sadovaya str.221, Samara 443011, Russia

Summary

The output parameters of a free-running pulse electron-beam-controlled discharge CO-laser (EBCD CO-laser) has been studied experimentally for the first time when collisions of the second kind (superelastic collisions) were of no importance in the active medium of the laser.

1. Introduction

It is known, that among the other gas discharge lasers, the pulse high-pressure electron-beam controlled discharge carbon monoxide lasers possess the very high output characteristics (efficiency η ~45%, $Q_{out \cdot sp}$ ~200-450J/l Amagot). The results of studies of $Q_{out \cdot sp}$ these lasers had been published in numerous papers [1- 3]. These results allow us to consider this type of laser as perspective for technology applications. On the other hand, it is clear, that only practical achievement of high output characteristics makes this technically complicated type of laser (at an average radiation power 15-20kW) attractive for use in industry.

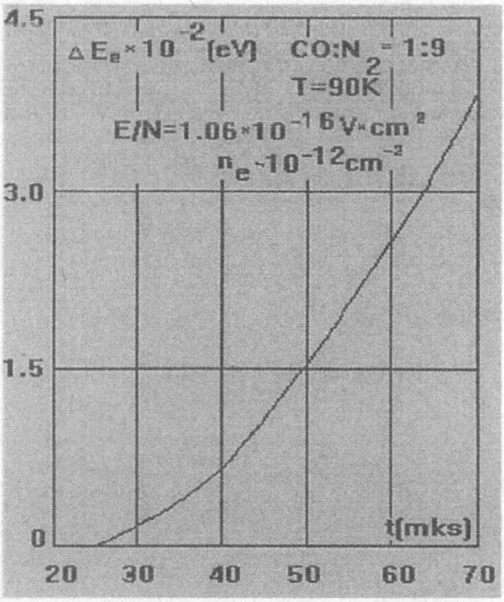

As we suppose one of the possible reasons which may decrease output parameters of a pulsed EBCD CO-laser is collisions of the second kind, i.e. the deactivation of excited vibrational levels of a CO molecule by electrons of discharge, when the vibrational temperature T_v of the

Fig.1: *Time dependence of* ΔE_e

laser levels is above the electronic that T_e. It must be emphasized that this condition is realized in cryogenically cooled EBCD CO laser easily if a capacitor bank is used for the main discharge(C EBCD CO laser). Analysis of obtained results had shown that the superelastic collisions decrease the efficiency of the Q_{out} laser and affect on the the form of output pulse. Analyzing a form of output pulse of a C EBCD CO laser, we had measured the excess of electronic emperature and electronic energy over the mean values, which are determined by the value E/N. The greatest excess of electronic energy over the mean was equal to ΔE_e =0.04eV for our experimental conditions(Fig.1) In order to avoid the negative effect of the superelastic collisions on the output laser parameters we suggested to pump EBCD CO-laser by the discharge of pulse-forming LC line at constant value E/N (LC EBCD CO-laser)[4].

2 Experimental results

The experiments were carried out on a pulse-periodic LC EBCD CO-laser developed and constructed at the Samara branch of the Lebedev Physics Institute. Our experiments shown that the discharge of an LC transmission line across a matched load at constant E/N allows increases output spectrum of the LC EBCD CO-laser in shortwave region more than 1.5 times (Fig.2A) and considerably increases the number of vibrational-rotational transitions in spectrum of oscillation in comparison with pumping from a capacitor bank(Fig.2B). We noted the differences of temporal characteristics

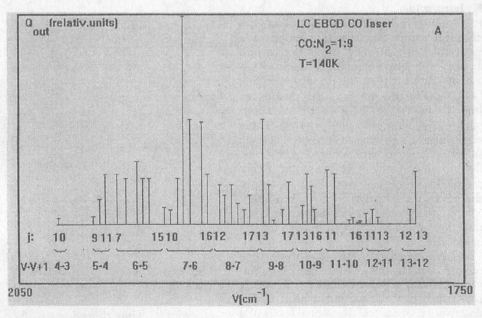

Fig.2A: Output spectrum of the LC EBCD CO laser.

of the radiation emitted for LC EBCD CO-laser and EBCD CO laser with a capacitor bank. The output power from the LC EBCD CO-laser reached its maximum and then remained constant throughout the pump pulse (Fig.3). A small

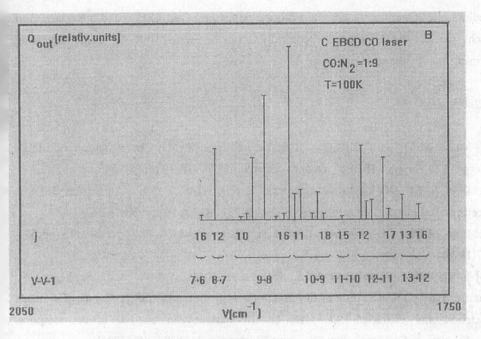

Fig.2B: Output spectrum of the C EBCD CO laser [5]

5%) fall in the radiation power was observed for only the longest (~90μ s) pump pulses. This was qualitatively different from the EBCD CO-lasers with a capacitor bank when a strong (~100%) dip in the radiation power was observed at the end of a pump pulse for a wide range of pump-pulse durations.Under the same conditions the efficiency of the LC EBCD CO-laser (η=22% at T=140K for a CO:N_2=1:19) was higher than the efficiency a capacitor bank(η= 15%).

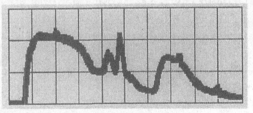

Fig.3.: Pulse profile of radiation from the LC EBCD CO laser. Pump pulse duration τ_{in}=90μs. *CO:N_2=1:9,   T=140K, P_{in}=2.6   kW*cm^{-1}*Amagot^{-1} .   Time scale 50μs/division.*

3 Conclusions

In order to raise output parameters of a pulse high-pressure electron-beam controlled discharge carbon monoxide laser it is necessary to excite its active medium in conditions when superelastic collisions of the CO excited molecules and discharge electrons are absent. Technically it can be carried out by using the pulse-forming line for the main discharge.

4 References

[1]. Mann M.M, Rice D.K.,Eguchi R.G. An Experimental Investigation of High Energy CO lasers. IEEE,J, Quant. Electron,1974, QE-10,pp.682-685.

[2].Mann M.M. CO Electrical discharge Lasers.-AIAA 13th Aerospace Sciences Meeting (Pasadena Calif./ January 20-22, 1975): AIAA Paper, No75-34.

[3].Mann M.M. CO Electric Discharge Lasers/ AIAA J., vol.14, N5.-pp.549-567(1976)

[4] Igoshin V.I., Kazakevich V.S., Morozov K.V., Petrov A.L., Popkov G.N., Chernovaya V.B.:Calculation of the electron energy in the discharge plasma of a pulsed electron-beam controlled CO laser on the basis of an analysis of the temporal characteristics of its [radiation. Quantum Electronics 24(5)395-398 (1994)

[5] Basov N.G. , Kazakevich V.S. Kovsh I.B. *Sov. J. Quantum Electron.* 9,N4,769 (1982)

Cryogenically Cooled Repetitively Pulsed Electron-Beam-Controlled Discharge CO Laser

V.Kazakevich, K.Morozov, A.Petrov, G.Popkov
Lebedev Physics Institute, Samara Branch.
Novo-Sadovaya str. 221, Samara 443011, Russia

Summary

The first description is given of an repetitively pulsed electron-beam-controlled discharge CO laser (RP EBCD CO laser) with the main discharge supplied from an LC transmission line.

1 Introduction

A cryogenically cooled pulsed EBCD CO laser has high quantum efficiency of nearly 90%, so operation with very high conversion efficiency can be obtain. Its output vavelength of 5µm is within the atmospheric transmission window, and therefore applications such as metalworking can be anticipated.

Owing to these excellent characteristics, numerous researches have been done on EBCD CO lasers[1-3]. However, as we know, because of many technical problems, there are no reports on repetitively pulsed EBCD CO laser.

One of the most important problems of a RP EBCD CO laser is effectively cooling the laser mixture. Achievement of a dependable operation for a long time of high voltage feed sourceses of the laser is the second important problem. But the most important problem of RP EBCD CO laser is the achievement of a dependable operation of an electron gun foil unit. These problems we had to decide when we made the RP EBCD CO laser (Fig.1).

2 Experimental setup

We calculated laser's parameters and planed our RP EBCD CO lasers that for perforation metal sheets for aviation industry. The laser could operate at a maximum pump-pulse duration of 90 µs and pulse repetition frequencies up to 10Hz. The lack of a cryogenic heat exchanger with the necessary parameters limited the average output radiation power to 1kW in the pulse-periodic regime. The main discharge of the laser was supplied from an LC transmission line. The ladder-type LC line with a wave impedance of ρ =17.5 Ohm was based on ИК-100-0.4 capacitors (C=0.4 µF) and cons with an inductance of L=120 µH. The maximum reduced electric field E/N was 4kV*cm^{-1}*Amagot^{-1} and was constant during a pump pulse. The use of the LC transmission line as the pump made it possible to weaken considerably superclastic collisions of discharge electrons with the vibrationally excited CO molecules. An unstable resonator with a M=1.4 magnification was used in these experiments. The front and rear copper mirrors had diameters of 65 and 90mm, respectively,

892

and their radii of curvature were R=7 and 19m. The temperature and the density of the active medium(CO:N_2 =1:9 or 1:9) were 140K and 0.5Amagot. We used liquid nitrogen as

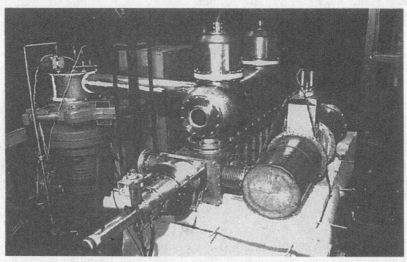

Fig.1 Photograph of the repetitively pulsed
electron-beam-controlled discharge CO laser(without thermo-insulation)

coolant. The largest axial gas flow rate at the discharge region was 30m/s. The resonator was thermally insulated by double CaF_2 windows at the resonator exit and air was pumped out from the gap between these windows. The laser resonator was aligned with the aid of He-Ne laser radiation which passes through a 1mm diameter aperture in the rear mirror.The maximum output pulse energy reached in our system was 360 J and the efficiency was $\eta=Q_{out}/Q_{in}$ =22% (Fig.2). As the discharge-gap volume (9* 9* 100cm³) was 27% higher than the resonator volume the physical efficiency was $\eta_{phys}=Q_{out.sp}/Q_{in.sp}$=28%. It is worth noting that this efficiency is the highest of those known for a pulsed EBCD CO laser at the selected working temperature. This is of

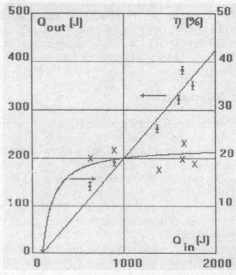

Fig.2 Dependences of the output-pulse energy Q_{out} from an EBCD CO laser and of the lasing efficiency η on the pump-pulse energy Q_{in}. CO:N_2 =1:19; T=140K, Pump pulse duration τ_p =63μs.

practical importance because in some cases there is no need to used liquid nitrogen as a cooling agent. Less expensive Freon can be used in RP EBCD CO lasers without a significant loss in the lasing efficiency. This high energy characteristics of the radiation at a relatively high gas temperature can be explained by weak influences of superelastic collisions in the active medium of the EBCD CO laser which was supplied from the LC transmission line.

The superelastic collisions during excitation also affect the output energy characteristics of the radiation emitted by an EBCD CO laser. Since under otherwise constant conditions the probability of superelastic interactions increases with the serial number of the vibrational level, an artificial transmission line used as an energy storage system ensures that the population of the high vibrational levels at the end of a pump pulse is higher than

the population that can be achieved by the use of capacitor banks. The total energy stored in the system of vibrational levels when an artificial transmission line is employed is therefore higher than for capacitor banks, which must influence the radiation energy of an EBCD CO laser. This is supported by an analysis of the output energy characteristics of our laser. The dependence of the specific energy of the laser radiation pulses on the duration of the pump pulses delivered by the LC transmission line is plotted in Fig.3. When the lasing threshold was reached, the output energy rose linearly with the pump pulse duration up to a certain moment and then fell steeply. This moment was different for different

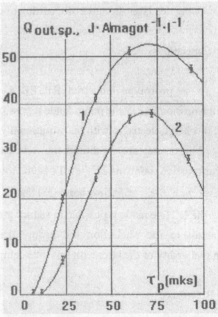

Fig.3 Dependences of the output-pulse specific energy Qout.sp from an EBCD CO laser on the pump-pulses duration τ_p. Initial voltage applied to an LC transmission line $U_0 = 31kV(1)$ and $27kV(2)$, $CO:N_2=1:9$, $T=140K$

specific pump powers and it was obviously governed by the beginning of the increasing influence of VT relaxation and was associated with overheating of the laser mixture. The maximum output radiation energy was governed by the voltage across the discharge gap, i.e. by the pump power, and under our conditions this energy increased when the discharge-gap voltage and pump power were increased.

A major shortcoming of capacitor banks is the high probability of break down of the discharge gap of an EBCD CO laser after a pump pulse and consequent damage to the separation foil. An LC transmission line supplying the discharge in a pulsed EBCD CO laser avoids practically completely this negative influence of breakdown. It is found that discharge of an LC transmission line across a matched load considerably reduced (by a factor of up to 10^5) the probability of damage to a titanium foil which separated the electron gun from the discharge chamber of the laser, compared with the corresponding probability for the laser with a capacitor bank. In fact, the foil was damaged not so much by breakdown of the discharge gap but because of inattention of the operator who permitted the operation of the electron gun under conditions more severe than those assumed in the design of the laser.

Using this laser the successful cutting of carbon composite materials have been made.

3 Conclusions

A prototype industrial RP EBCD CO laser was built and the use of the LC transmission line as the pump made it possible to weaken considerably superelastic collisions of discharge electrons with the vibrationally excited CO molecules;

Under constant conditions the efficiency of the EBCD CO laser pumped by an LC transmission line ($\eta=22\%$ at T=140K for $CO:N_2 =1:19$ mixture) exceed the efficiency of a laser with the capacitor bankη ~15%);

the LC line made it possible to reduce greatly (by a factor of up to 10^5) the probability of damage to the separation foil during breakdown of the discharge gap and thus increase the reliability of the laser with this line, compared with the laser with a capacitor bank.

4 References

[1] Mann M.M., Rice D.K., Eguchy R.G. IEEE J. Quantum Electron. QE-10, 682(1974)
[2] Boness M.J.W. Center R.E. J. Appl. Phys.48 2705 (1977)
[3] Basov N.G., Danilychev V.A., Ionin A.A., et al. Kvantovaya Elektron.(Moscow) 6, 1215(1979)

Novel Laser Beam Shaping and Mode Control with Diffractive Optical Elements

J. R. Leger, D. Chen, G. Mowry, Z. Wang
Department of Electrical Engineering, University of Minnesota,
Minneapolis, Minnesota 55455/USA

Abstract

Diffractive optical elements are shown in this review paper to provide additional flexibility in laser resonator design. Diffractive end mirrors allow the designer to tailor the mode profile of the laser. Additional intracavity diffractive elements increase the discrimination against higher-order modes. Experimental results are shown for Nd:YAG lasers and wide-stripe diode lasers.

1 Introduction

Recently, there have been several attempts to improve the performance of laser resonators by using unconventional optics. Variable reflectivity spherical mirrors improve the performance of unstable resonators by reducing the diffraction from the mirror edges and increasing the uniformity of the near-field intensity[1, 2, 3, 4, 5, 6]. This idea has been extended to variable phase mirrors to provide very flat near-field patterns[7, 8, 9]. Finally, a spatial filter in the form of a grid of opaque strips has been used to flatten the beam profile from a Nd:YAG laser[10].

Diffractive optical elements have also been used inside laser resonators[11, 12, 13, 14]. The phase function of a diffractive element is determined by subtracting integer multiples of a wavelength from a computed continuous phase function, and fabricating this discontinuous

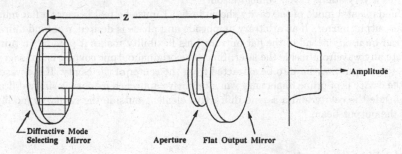

Figure 1: Laser resonator with diffractive end mirror.

surface relief profile on a flat substrate using microfabrication techniques. Since many microelectronic etching techniques are inherently binary in nature, it is convenient to fabricate this surface as a quantized version of the ideal profile. If a sufficiently large number of quantization levels are used, the optical characteristics of this surface can approach the desired response very closely. Since the microfabrication technology can generate very complex shapes, the surface figure of these elements is very flexible. This flexibility makes these elements ideal for producing unusual laser resonators. In this review paper, we describe laser resonators that utilize diffractive elements to improve laser performance and make it possible to tailor the shape, intensity profile, and phase of the fundamental mode.

2 Laser Resonators with Diffractive End Mirrors

The simplest diffractive laser resonator is shown in fig. 1. It consists of a finite aperture planar output mirror and a single diffractive end mirror. The end mirror can be used to tailor the intensity of the fundamental mode by the following method. The designer first specifies the light field at the flat output mirror to be any real two-dimensional function. The field amplitude and phase at the end mirror is calculated by a computer using scalar diffraction theory. The phase of the end mirror is then chosen to return the complex conjugate of the incident light. This complex conjugate distribution propagates back to the output mirror and reproduces itself, thereby establishing itself as a mode of the cavity. If the mirrors are sufficiently large, this mode suffers very little diffraction loss, and becomes the fundamental cavity mode.

One useful choice for the fundamental mode is a high-order super-Gaussian. As the order of a super-Gaussian approaches infinity, the mode distribution approaches an ideal uniform intensity profile with uniform phase. It has been shown[9, 12] that a cavity length of approximately one Rayleigh range of the super-Gaussian will provide the best discrimination against higher-order modes. The finite aperture of the planar mirror can be adjusted to control the losses to the cavity modes. For a high-order super-Gaussian, the mode is very uniform at the flat mirror in a specific area, and dies away very quickly outside of this area. An aperture that is adjusted to just pass the uniform area of this mode will introduce very little loss to the fundamental mode. However, the loss to the second-order mode can be large, making this a very stable cavity configuration.

The fundamental mode of the cavity shown in fig. 1 must be real because a flat mirror is used as an output mirror. If an arbitrary amplitude and phase is desired, a second diffractive element can be used instead of the flat mirror. This flexibility makes it possible in principle to generate any wavefront inside the laser. Since the original output coupler is no longer flat, however, the output may have to be extracted from the center of the cavity. If the wavefront outside the cavity is of prime importance, an alternative approach is to use a single diffractive element inside the cavity and a second diffractive element outside the cavity to modify the phase of the output beam.

2.1 Nd:YAG experiments

An experiment was performed using a flashlamp-pumped Nd:YAG rod as the laser medium. The fundamental mode was chosen to be a 20th-order super-Gaussian of square shape with a full beam width of 1.2 mm. The cavity length was chosen to be 1.1 meters, or approximately one Rayleigh range of the super-Gaussian. The diffractive element was fabricated as a sixteen-phase-level approximation to the calculated surface relief shape. The performance of the resonator was measured by observing the near- and far-field profiles, and by measuring the

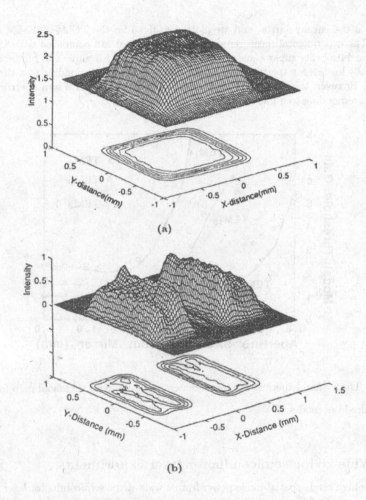

(a)

(b)

Figure 2: Experimental measurements of near-field profile from Nd:YAG laser. a) Fundamental TEM$_{00}$ mode. b) TEM$_{01}$ mode.

lasing threshold. The higher-order modes were excited by placing thin wires in the cavity to introduce loss to the TEM$_{00}$ mode without increasing the loss to the TEM$_{01}$, TEM$_{10}$, and TEM$_{11}$ modes. The measured near-field profiles of the TEM$_{00}$ and TEM$_{01}$ modes are shown in fig. 2. The fundamental mode was fit to a general super-Gaussian and was found to be closest to a 14th-order super-Gaussian with a full width of 1.3 mm. The slight increase in mode size and edge width was attributed to fabrication limitations of our micromachining process.

The discrimination between spatial modes was determined by measuring the relative pump power required to achieve lasing of the various modes. The discrimination was expressed in terms of the modal threshold gain, defined as the round-trip gain required to overcome the diffractive cavity losses to a specific mode. These losses are a function of the size of the aperture at the output mirror, and are plotted in fig. 3 for the TEM_{00}, TEM_{01}, and TEM_{11} modes. The experimental measurements (discrete points) are compared with theory (solid lines) for a 14th-order super-Gaussian with a full width of 1.3 mm. The fundamental mode is essentially lossless for aperture sizes larger that 1.3 mm (corresponding to a threshold gain of unity). However, the threshold gain of the TEM_{01} mode for a 1.3 mm aperture is seen to be 2.5, corresponding to a loss of 60 %.

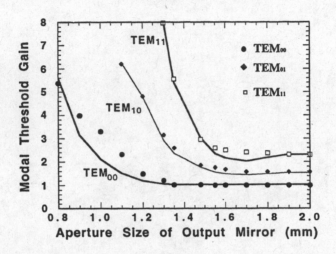

Figure 3: Theory and experimental measurements of the modal threshold gain for the three lowest-order laser modes.

2.2 Wide-stripe semiconductor laser experiments

Obtaining high single-spatial-mode power from a wide-stripe semiconductor laser is challenging because of the difficulty of discriminating against higher-order modes and the tendency of the lasing stripe to filament. The high stability of diffractive laser resonators makes them attractive candidates for external semiconductor laser cavities. In one experiment, a diffractive mirror was designed to establish an 8th-order super-Gaussian across a 500 μm-wide aperture. The laser fabricated by SDL, Inc. contained a 600 μm-wide aperture and was antireflection-coated on both ends. The diffractive mirror was fabricated with an eight-level phase quantization using photolithography and wet chemical etching. Anamorphic optics were used to couple the light from the wide-stripe waveguide to free space.

The laser was operated in a pulsed mode to reduce thermal effects, and measurements of the near- and far-field were made. Fig. 4 shows the results of far-field measurements obtained both with the diffractive mirror and a plane mirror. The far-field pattern in fig. 4a obtained with the diffractive mirror has a divergence corresponding to a uniform 500 μm-wide aperture, indicating single-spatial-mode performance. The amount of power obtained from this laser was 2.8 watts, and no degradation of the far-field pattern was observed at these high power levels. The divergence of the light obtained from the plane mirror (fig. 4b) is approximately 30 times greater, indicating that the laser was operating in many higher-order modes.

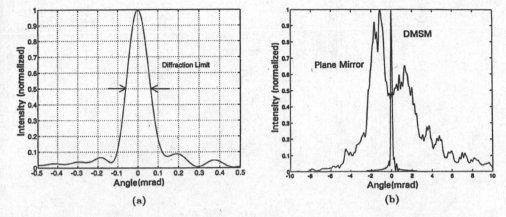

Figure 4: Far-field profile of wide-stripe laser with external cavity. a) Diffractive mirror used as an end mirror. b) Planar mirror used as an end mirror. Note that the scale in figure (b) is increased by a factor of 20. The divergence in figure (a) is replotted in figure (b) for comparison.

3   Intracavity Diffractive Elements

A modification of the simple diffractive laser cavity can be made by introducing intracavity diffractive elements. One possible configuration is shown in fig. 5 consisting of a sinusoidal diffraction grating placed directly after the gain medium. The design of the diffractive end mirror is similar to before, except that the calculated diffraction pattern passes through a diffraction grating on its way to the end mirror. The end mirror then returns the complex conjugate of this more complex distribution. The light retraces its path and reconstructs the original desired fundamental mode at the output mirror. If this mode is chosen to be a high-order super-Gaussian, the field inside the crystal is always approximately uniform. However, the discrimination against higher-order modes is greatly increased by the presence of the grating. This makes it possible to use shorter laser cavities and larger diameter beams while still maintaining single-spatial-mode operation.

The previous Nd:YAG laser cavity was modified to include a two-dimensional sinusoidal grating, and a new diffractive end mirror was fabricated to establish a 20th-order super-Gaussian fundamental mode. Both elements were made by photolithography and wet chemical etching, with the diffraction grating containing 32 phase levels and the end mirror containing 16 phase levels. The size of the mode at the output mirror was 2 mm, and the cavity length was 25 cm. Without the sinusoidal grating, the round-trip loss to the second-order mode in this cavity would only be 0.6 %. By choosing a grating with the optimum spatial frequency (6.0 mm^{-1}), this loss is increased to 60 %. This dramatic increase makes it possible to operate the laser in a single spatial mode. Fig. 5b shows the far-field performance of this laser. The first zero of the diffraction pattern corresponded to a spatial frequency of 0.5 mm^{-1}. This is expected from a diffraction-limited 2 mm fundamental mode.

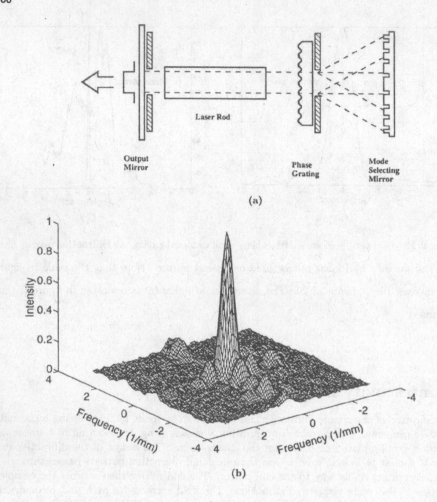

Figure 5: Diffractive laser resonator with intracavity element. a) Optical set-up. b) Experimental result from a Nd:YAG laser.

4 Conclusions

We have reviewed several new laser cavities that contain diffractive optical elements. A single diffractive end mirror can be used to tailor the fundamental mode to have any real field profile. Super-Gaussian profiles were shown to have high modal discrimination when the cavity length is approximately one Rayleigh range. Diffractive mirrors on both ends of the resonator were seen to allow an arbitrary selection of the mode amplitude and phase. Additional intracavity elements were seen to increase the modal discrimination further and permit a reduction of the cavity length. Experimental results were shown using Nd:YAG and semiconductor lasers.

References

[1] P. Lavigne, N. McCarthy, and J.-G. Demers, "Design and characterization of complementary Gaussian reflectivity mirrors," *Appl. Opt.* **24**, 2581-2586 (1985).

[2] D.J. Harter and J.C. Walling, "Low-magnification unstable resonators used with ruby and alexandrite lasers," *Opt. Lett.* **11**, 706-708 (1986).

[3] S. De Silvestri, P. Laporta, V. Magni, O. Svelto, and B. Majocchi, "Radially variable reflectivity output coupler of novel design for unstable resonators," *Opt. Lett.* **12**, 84-86 (1987).

[4] K.J. Snell, N. McCarthy, M. Piché, and P. Lavigne, "Single transverse mode oscillation from an unstable resonator Nd:YAG laser using a variable reflectivity mirror," *Opt. Lett.* **65**, 377-382(1988).

[5] S. De Silvestri, P. Laporta, V. Magni, G. Valentini, and G. Cerullo, "Comparative analysis of Nd:YAG unstable resonators with super-Gaussian variable reflectivity mirrors," *Opt. Commun.* **77**, 179-184 (1990).

[6] A. Parent and P. Lavigne, "Variable reflectivity unstable resonators for coherent laser radar emitters," *Appl. Opt.* **28**, 901-903 (1989).

[7] P.A. Bélanger and C. Paré, "Optical resonators using graded-phase mirrors," *Opt. Lett.* **16**, 1057-1059 (1991).

[8] P.A. Bélanger, R.L. Lachance, and C. Paré, "Super-Gaussian output from a CO_2 laser by using a graded-phase mirror resonator," *Opt. Lett.* **17**, 739-741 (1992).

[9] C. Paré and Pierre-André Bélanger, "Custom laser resonators using graded-phase mirrors," *IEEE J. Quantum Electron.* **QE-28**, 355-362 (1992).

[10] V. Kermene, A. Saviot, M. Vampouille, B. Colombeau, and C. Froehly, "Flattening of the spatial laser beam profile with low losses and minimal beam divergence," *Opt. Lett.* **17**, 859-861 (1992).

[11] J.R. Leger, M.L. Scott, and W.B. Veldkamp, "Coherent addition of AlGaAs lasers using microlenses and diffractive coupling," *Appl. Phys. Lett.* **52**, 1771-1773 (1988).

[12] J.R. Leger, D. Chen, and Z. Wang, "Diffractive optical element for mode shaping of a Nd:YAG laser," *Opt. Lett.* **19**, 108-110 (1994).

[13] J.R. Leger, D. Chen, and K. Dai, "High modal discrimination in a Nd:YAG laser resonator using internal phase gratings," *Opt. Lett* **19**, 1976-1978 (1994).

[14] G. Mowry and J.R. Leger, "Large-area, single transverse-mode semiconductor laser with diffraction-limited super-Gaussian output," *Appl. Phys. Lett.* **66**, 1614-1616 (1995).

Quarzglas-Wellenleiter für die Transmission von Hochleistungslaserimpulsen

V. Uhl (1), K. O. Greulich (2), J. Wolfrum (1), H. Albrecht (3), D. Müller (3), T. Schröder (3), W. Triebel (3)

(1) Physikalisch-Chemisches Institut der Universität Heidelberg
    D-69120 Heidelberg
(2) Institut für Molekulare Biotechnologie
    D-07708 Jena
(3) Institut für Physikalische Hochtechnologie e. V.
    D-07702 Jena

Quarzglas Lichtleiter transmittieren mittlerweile UV-Strahlung mit Wellenlängen deutlich unter 200 nm. Beim Einsatz von Quarzglas zur Transmission von Hochleistungs-UV-Laserpulsen treten jedoch folgende Probleme auf:

- spontane Zerstörung
- geringe Transmission bei hohen Leistungsdichten
- langsam abnehmende Transmission bei längerer Bestrahlung.

Bei Lichtleitern tritt spontane strahlungsinduzierte Zerstörung vor allem an der Einkoppelfläche auf und kann durch Reduzieren der dort herrschenden Leistungsdichte verhindert werden. Hier haben sich getaperte Quarzfasern bewährt, bei denen durch die aufgeweitete Einkoppelfläche bei gleichbleibender Pulsleistung die Leistungsdichte bei der Einkopplung reduziert wird [GREULICH 1988].

Die geringe Transmission bei hohen UV-Leistungsdichten ist auf nicht-lineare Effekte wie zum Beispiel stimulierte Brillouin-Streuung in den Quarzfasern zurückzuführen, die nicht unterdrückt werden können.

Hinzu tritt die langsam zunehmende UV-Absorption im Laufe der Bestrahlung mit leistungsstarken Pulsen, wie beispielsweise typischen Excimerlaserpulsen zur Materialbearbeitung. Diese ist auf die Bildung von absorbierenden Defektzentren im Quarzglas durch die Laserstrahlung zurückzuführen. Im Folgenden werden Effekte für die KrF-Excimerlaserwellenlänge 248 nm dargestellt. Die zusätzliche Absorption durch die strahlungsinduzierte Defektbildung wird als laserinduzierte Absorption bezeichnet.

Durch leistungsstarke KrF-Excimerlaserpulse wird in synthetischem Quarzglas im Laufe der Zeit eine Absorptionsbande mit Maximum bei 210 nm erzeugt, die allgemein dem Defekt E'-Zentrum zugeordnet wird, ein dreifach koordiniertes Siliziumatom mit paramagnetischem Elektron. Der Ausläufer zum längerwelligen Spektralbereich dieser Bande ist Ursache für die zunehmende Absorption der 248 nm-Excimerlaserpulse [LECLERC 1991]. In Quarzglas mit einer OH-Konzentration von 300 ppm/w oder weniger bleibt diese Absorptionsbande auch nach Abschalten der Laserstrahlung bestehen, in Quarzglas mit höherem OH-Gehalt, typisch 800 ppm/w bis 1000 ppm/w, heilt die Absorption in der Zeit von etwa 20 Minuten vollständig aus. Bei erneuter Bestrahlung des selben Probenvolumens erreicht die 210 nm-Absorptionsbande jedoch auch in OH-reichem Quarzglas wieder ihr ursprüngliches Niveau. Dies kann dadurch erklärt werden, daß sich die die Absorptionsbande auslösenden E'-Zentren nach Abschalten der Laserstrahlung in nicht-absorbierende Defekte, sogenannte Vorläuferdefekte, umwandeln. Durch erneute Bestrahlung können aus diesen Vorläuferdefekten jedoch durch Einphotoneneffekt schnell wieder E'-Zentren erzeugt werden [LECLERC 1992]. Als nicht-absorbierender Vorläufer des E'-Zentrums wird in diesem Fall das SiH-Zentrum vorgeschlagen [PFLEIDERER 1993].

Analoge Beobachtungen können auch für die Intensität der Lumineszenzbande bei 650 nm gemacht werden. Diese wird dem NBOH-Zentrum zugeschrieben, ein einfach an ein Siliziumatom gebundenes Sauerstoffatom mit freiem Elektron. In OH-armen Quarzgläsern (300 ppm/w und weniger) konnte in vorgeschädigten Probenvolumen mit einem HeNe-Laser die 650 nm-Lumineszenzbande deutlich sichtbar angeregt werden, während dies in OH-reichen Proben (ca. 800 ppm/w) nicht möglich war. Dies zeigt, daß auch die NBOH-Zentren in OH-reichen synthetischen Quarzgläsern ausheilen können. Als Vorläufer wird von uns ein Defekt SiOH vorgeschlagen.

Achtstündiges Erhitzen von OH-reichem synthetischen Quarzglas auf Temperaturen von bis zu 1100°C führte zu einem Absinken des OH-Gehalts nach Überschreiten einer Temperaturschwelle zwischen 550°C und 700°C. Gleichzeitig reduzierte sich auch die Zahl der Vorläuferdefekte für E'- und NBOH-Zentrum, so daß die Bildung der entsprechenden Absorptionsbande bei 210 nm und der Lumineszenzbande bei 650 nm bei anschließender Excimerlaserstrahlung deutlich geringer war. Bestrahlung mit bis zu 400.000 KrF-Excimerlaserpulsen zeigte jedoch, daß der zunächst vorhandene Vorteil der geringen Absorption und Lumineszenz bei längerer Bestrahlung wieder durch langsame Bildung der Absorptionsbande bei 210 nm bzw. der 650 nm-Lumineszenzbande zunichte gemacht wird (Abbildung 1). Diese langsame Defektbildung erfolgt über Mehrphotonenprozesse. Da die getemperten Proben einen geringen OH-Gehalt aufwiesen, heilte die 210 nm-Absorptionsbande unvollständig oder nicht aus. Die Defektbildung und das Ausheilen sind schematisch in Abbildung 2 dargestellt.

Diese Ergebnisse zeigen, daß die Anwendung von Quarzglas als Werkstoff für Lichtleiter zur Transmission von leistungsstarken 248 nm-Excimerlaserpulsen durch die Bildung der E'-Zentren durch Mehrphotoneneffekt prinzipiell begrenzt ist. Eine erfolgversprechende Möglichkeit für die Transmission von UV-Laserpulsen durch Quarzfasern ist jedoch ein Ausweichen auf die Excimerlaserwellenlänge 308 nm, bei der die laserinduzierte Defektbildung geringer ist [HILLRICHS 1992], und oder Verminderung der Mehrphotonen-Defektbildungsrate durch Reduktion der im Material herrschenden Leistungsdichte, zum Beispiel durch Verteilen der Strahlung auf mehrere Lichtleiter in einem Lichtleiterbündel [DE WITH 1993].

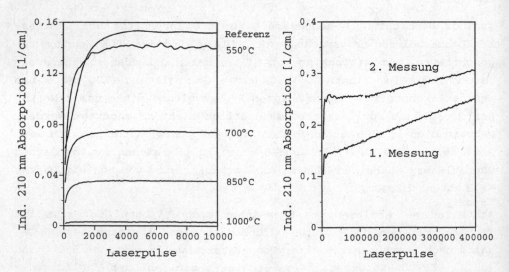

Abbildung 1: Absinken der anfänglichen laserinduzierten Absorption bei 210 nm in zuvor getemperten Proben (links). Angegeben ist die Ofentemperatur. Anstieg der laserinduzierten Absorption bei längerer Bestrahlung (rechts). Bei der 2. Messung im selben Probenvolumen stellt sich durch schnelle Umwandlung der bereits erzeugten Vorläuferdefekte sofort ein höheres Niveau der laserinduzierten Absorption ein.

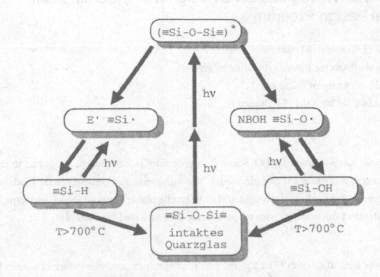

Abbildung 2: Schematische Darstellung der Bildung und des Ausheilens der Defekte E' und NBOH durch KrF-Excimerlaserpulse (248 nm) in OH-reichem (ca. 800 ppm/w) Quarzglas. Durch Zweiphotoneneffekt entsteht aus dem intakten Quarzglas ein angeregter Zustand, der in die Defekte E' und NBOH zerfällt. Wenn Wasserstoff anwesend ist, heilen diese Defekte in Vorläufer SiH und SiOH aus. Aus diesen können durch Einphotoneneffekt leicht wieder E' bzw. NBOH-Zentren entstehen. Durch Tempern der Quarzproben kann ein Ausheilen der Vorläuferdefekte erreicht werden.

Literatur

K. O. Greulich, H. Hitzler, N. Leclerc, J. Wolfrum, K. F. Klein: Transport von leistungsstarken UV-Laserpulsen durch Lichtleiter mit variablem Querschnitt. *Laser und Optoelektronik* 20(4) (1988) 58-60.

G. Hillrichs, M. Dressel, H. Hack, R. Kunstmann, W. Neu: Transmission of XeCl excimer laser pulses through optical fibers: dependence on fiber and laser parameters. *Appl. Phys.* B 54 (1992) 208-215.

N. Leclerc, C. Pfleiderer, H. Hitzler, J. Wolfrum, K. O. Greulich und S. Thomas, H. Fabian, R. Takke, W. Englisch: Transient 210-nm absorption in fused silica induced by high-power UV laser irradiation. *Optics Letters* 16(12) (1991) 940-942.

N. Leclerc, C. Pfleiderer, H. Hitzler, J. Wolfrum K. O. Greulich, S. Thomas, W. Englisch: Luminescence and transient absorption bands in fused SiO_2 induced by KrF laser radiation at various temperatures. *J. Non.-Cryst. Sol.* 149 (1992) 115-121.

C. Pfleiderer, V. Uhl, A. de With, K. O. Greulich, J. Wolfrum: Fiber optics for High power excimer lasers. *EUREKA EU205 International Workshop, Erlangen 1993*, Proc.1 (1993) 61-64.

A. de With, V. Uhl, C. Pfleiderer, S. Rosiwal, H.-W. Bergmann, J. Wolfrum, K. O. Greulich: Oberflächenbearbeitung durch XeCl-Excimerlaser-Pulse mit Hilfe eines Hochleistungs-UV-Taperbündels. *Laser und Optoelektronik* 25(5) (1993) 61-69.

Anisotropic Waveguide Lens Design with a Commercial Optical Design Program

G. Erdei, G. Szarvas, M. Barabás
Department of Atomic Physics, Institute of Physics
Technical University of Budapest
8 Budafoki út, Budapest, 1111 Hungary

Abstract: We have enabled OSLO Series 2, a professional lens design program, to treat integrated lens systems that contain anisotropic media. We summarise the computational model of anisotropic planar lenses and show how we taught OSLO to handle planar uniaxial lenses and gaps. The necessity and usefulness of the new program extensions are demonstrated by examples.

Introduction

In integrated optical devices (like e.g. the real time spectrum analyser) there is a need for diffraction limited, wide field angle lenses. The evaluation and optimisation of these systems require the use of suitable computer software.

Existing lens design programs proved to be appropriate for the calculations concerning isotropic integrated optical (IO) lenses because they are similar to telescope objectives, which already have refined design methods.

To benefit from the acousto-optical characteristics of e.g. $LiNbO_3$, anisotropic waveguide materials are frequently used. The aberrations of waveguide lenses caused by anisotropy can be so large [1, 2] that they must be taken into account in the design process. That is why we need tools for the simulation and optimisation of anisotropic, planar, homogeneous refracting lenses.

Since commercial lens design packages do not permit anisotropic lens materials, we have two ways to obtain the necessary tools. The first one is to develop new software which will be fast but – unless an unacceptable amount of programming is invested – will offer less services than professional codes. An example for this approach is TRACE1D, a program written at the Department of Atomic Physics of the Technical University of Budapest [2]. The other way is to modify an off-the-shelf lens design program. Some of these programs are able to use surfaces with special "user defined properties" which allow for unusual shapes and even unconventional laws of refraction. That OSLO Series 2 (Optics Software for Layout and Optimisation, PC version) does provide such hooks [3] was one of our reasons for choosing this program to work with.

The effect of anisotropy on planar lens systems

For purposes of ray tracing, a planar waveguide is described by the effective refractive index of its fundamental mode. If the waveguide material is anisotropic, this effective index is direction dependent which affects ray tracing in two respects. First, though the law of refraction for the wavefront normal (wave vector) is formally the same as Snell's law, the direction dependent

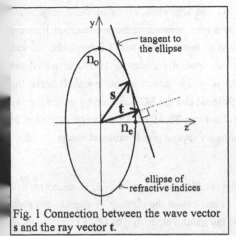

Fig. 1 Connection between the wave vector **s** and the ray vector **t**.

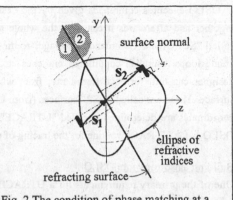

Fig. 2 The condition of phase matching at a line separating uniaxial regions (s_1=incident wave vector, s_2=refracted wave vector).

refractive indices must be used. The second effect is the deviation of the direction of the wave vector from that of the ray vector (which is parallel to the Poynting vector); in the geometric optical approximation the intensity distribution of a beam is determined by the latter. Therefore, when tracing rays, we must use the "non-Snellian" refraction law of the rays instead of the one relating to the wavefront normals.

There is a third phenomenon too: wavefronts emitted by a point source in a uniaxial material are of an elliptical shape. Consequently the algorithm, based on isotropic Fourier optics (and on the Fast Fourier Transformation), by which OSLO computes the diffraction pattern of the focal spot, should also be altered. However, as the change of the intensity distribution due to wavefront ellipticity – and to uniaxial anisotropy in general – was shown [4] to be negligible in most cases of practical interest, the original algorithm can be used.

Figs. 1 and 2 show the condition of phase matching for waves in two adjacent anisotropic regions and the connection between the directions of the wave vector and the ray vector. The curves in these figures are ellipses because we consider only uniaxial media with their optic axis lying in the plane of the waveguide.

Due to the above effects a lens system optimised for an isotropic waveguide but realised on an anisotropic one must be evaluated from the point of view of anisotropic aberrations as well to see whether its characteristics have changed significantly. If so, the system must be re-optimised using the more precise anisotropic law of refraction.

Using OSLO for the evaluation of planar integrated optical lenses

Since OSLO was written with three-dimensional optical elements in mind, it is not quite obvious how planar lenses should be modelled by the program: For purely ray optical simulations we must trace rays only in one plane. For a diffraction analysis, however, a planar lens must be replaced by a cylindrical one because the wavefront leaving the exit pupil is always computed from ray tracing. The other difficulty arises from the fact that the built-in ray tracing routine of OSLO can use only a single refractive index per medium, so we must also adapt it for handling anisotropic media (see the next section for some details).

OSLO executes commands given by the designer, but can also store series of these command together and afterwards it can treat the whole group as a new command. Such command lists are called "SCL" (Star Command Language) routines in which, besides the OSLO commands, we can create loops, establish conditional branches etc. There are special surfaces in OSLO at which the designer can tell, for each traced ray, from where and in which direction the ray shall leave the surface. These are called "USR" surfaces (from USeR defined) and the SCL routine by which the ray coordinates are determined is called "UTRACE" (from User ray TRACE). Our crucial extension of OSLO is the UTRACE routine for the tracing of extraordinary rays in planar uniaxial media.

Building anisotropy into OSLO

One of the primary requirements for a UTRACE routine is efficiency because SCL is an interpreted language and consequently quite slow. For this reason we minimised the number of steps in our code by relying partly on the results computed by the original ray tracing algorithm of OSLO.

When tracing a ray, OSLO starts it from a given point into a given direction. First the program

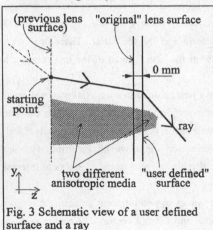

Fig. 3 Schematic view of a user defined surface and a ray

determines the intersection point of the ray with the next lens surface, then it computes the surface normal and the new ray direction from the law of refraction. With an anisotropic medium on either side of the surface, only the intersection point and the surface normal are found accurately. To correct the direction of the refracted ray, we define a USR surface right after the surface under consideration so that they coincide (see Fig. 3). This way, after incorrectly refracting the ray at the original surface, OSLO immediately transfers control to our UTRACE routine at the USR surface. At this point the necessary correction is completed by UTRACE using the anisotropic law of refraction instead of Snell's law.

We economise on computation time by letting the fast built-in OSLO routine determine the surface normal and the intersection of the ray and the surface. Thus we have to correct only the optical path measured from the previous lens surface and the direction of the refracted ray. ·

It is clear that the original isotropic lens system, using the extraordinary indices as refractive indices, must be altered to behave like an anisotropic one. To add the necessary USR surfaces, and store the data needed while tracing rays (e.g. the ordinary indices) we have written an auxiliary routine which performs these operations automatically, alleviating the designer's task.

Examples of anisotropic lenses computed with the aid of the new UTRACE routine

To show that anisotropy must indeed be taken into account during the design of integrated optical systems, we present some results obtained by OSLO and our UTRACE routine.

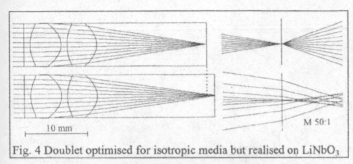

Fig. 4 shows rays passing through a doublet optimised for isotropic media. The focus shift and focal spot deterioration due to anisotropy can well be observed. (If the lenses are isotropic, this system is diffraction limited; the

10 mm   M 50:1

Fig. 4 Doublet optimised for isotropic media but realised on LiNbO₃

unaberrated rays of the isotropic system are not shown.)

Fig. 5 displays the layout of an IO quadruplet optimised for an anisotropic waveguide. The focal plane diffraction patterns of two pencils are plotted in Fig. 6.

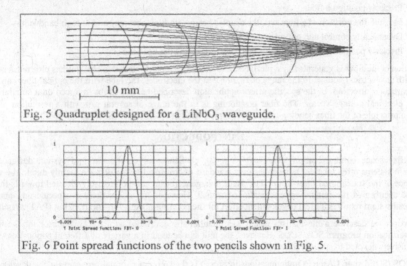

10 mm

Fig. 5 Quadruplet designed for a LiNbO₃ waveguide.

Fig. 6 Point spread functions of the two pencils shown in Fig. 5.

Summary and future plans

We extended OSLO Series 2 with routines for the evaluation of planar IO lenses by providing a set of SCL files to help designers in entering anisotropic lenses and to take over refraction calculations from the built-in OSLO procedure that can treat only isotropic media. We plan to soon re-write the routines for OSLO SIX (the new and improved version of OSLO). Since SCL is replaced in OSLO SIX by a compiled C language called CCL, the migration to CCL will undoubtedly result in considerably faster code.

REFERENCES

1. W. Jiang, and V. M. Ristic, "Study of anisotropy effect in planar lenses for integrated optics", J. Mod. Optics **35**, 849-862 (1988)
2. G. Szarvas, M. Barabás, P. Richter and L. Jakab, "Design of multielement acircular waveguide lens systems in anisotropic media", Opt. Eng. **32**, No.10, 2510-2516 (1993)
3. OSLO Series 2, © 1992 Sinclair Optics, Inc.
4. M. Barabás, G. Szarvas, "Fourier description of the propagation and focusing of an extraordinary beam in a planar uniaxial medium", Appl. Opt. **34**, No. 1, 11-21 (1995)

High-Precision Sensor Based on a Rotating Optical Fiber

K. Pribil, U. Brandenburg, J. Krieger
Dornier GmbH, D-88039 Friedrichshafen

1. ABSTRACT

This paper presents a new sensor for fiber optics applications. The sensor principle is that a free optical beam is coupled into an rotating optical fiber. This produces a modulation of the light, which depends on the relative position of the beam center to the center of the fiber rotation. By a suitable demodulation the information of the beam position can be obtained. This element can be used to:

- Precisely position a fiber
- Measure the position of a fiber wrt. the coupled beam with high accuracy and at high bandwidths
- Dynamically control and optimize the fiber coupling
- Precisely position a fiber end for illumination or imaging applications

The motion of the fiber is generated by a piezo ceramic element. The piezo ceramic element has electrodes at the outer side. With an electrical voltage in the range of 1...100 V at the electrodes, the tube with the optical fiber can be moved in any particular direction. In the specific sensor application described here, the tube is forced on a circular path by a rotating electrical voltage vector. The fiber positioning is in the range of several μm with a resolution of nm. The actuation principle of the fiber bending can be used for any kind of optical fiber (SMF, PPF, MMF).

2. INTRODUCTION

Optical free space communication technology is a big challenge for component and system designers. Future lasercomm systems offer inherent advantages over standard microwave links which are mainly due to the very small divergence of the transmitted laser-beam. The small divergence beam can be precisely directed towards the receive-telescope aperture of the remote communications-terminal. However, to utilize these prospected advantages in an operational system requires a great development effort for the pointing-acquisition and tracking (PAT) system.

In this paper we describe a mechanism which was intentionally developed as a tracking-sensor for the lasercomm system verification program SOLACOS. However, the unit can be used in a variety of different applications which are not exclusively devoted to space technology.

SOLACOS (SOlid state LAser COmmunications in Space) is the German national program for the development of a laser communication system breadboard. The breadboard will comprise two lasercomm terminals and some experimental infrastructure to simulate the orbital behavior of two host satellites. Activities within the program include system engineering as well as key technology development.

The paper briefly describes the basic requirements for the high bandwidth tracking-sensor of SOLACOS and presents the concept of the newly developed element. We present some details of the first prototype units, including results from the hardware tests. Additionally, we give some ideas of other applications in which the component could be advantageously used.

3. LASERCOMM SYSTEM TRACKING SENSOR REQUIREMENTS

The reference scenario for SOLACOS foresees a data transmission of a 650 MBit/s net-data-rate stream over a distance of 45000 km between a LEO and a GEO satellite. The main problem for the pointing-acquisition and tracking system is that a satellite is not a stable platform but induces a broad band of mechanical distortions to the lasercomm terminal. These distortions stem from quite different sources such as solar panel drives, pumps, positioning motors and many more. Additional distortions are produced by the telescope drive. All these distortions sum up to a certain amount of pointing-jitter which results in a loss of receiver signal power at the remote terminal.

After the lasercomm link is established during several acquisition-phases, the terminals enter the stationary communications-mode. In the communications-mode a high bandwidth beam control loop suppresses the terminal distortions. A key element for this tracking-loop is the tracking-sensor. The SOLACOS requirements for this position sensor are:

- High bandwidth (up to 1 kHz)
- High resolution (sub-microradians)
- Small size, mass, high reliability, space qualifyability (materials, voltages, ...), ...

However, the main issue for the terminal is to couple the received free space beam from the telescope into the receiver to regenerate the transmitted data information. So, the tracking function is vital but is only a support function for the system in communications-mode.

4. REALIZATIONS OF FAST TRACKING SENSORS (SOME EXAMPLES)

Several concepts for the data-receiver and for the tracking sensor are well known. The most relevant examples with their main advantages (+) and disadvantages (-) are:

- Quadrant-and multi segment (diode-) sensors (hybrid or monolithic),
  these devices are standard position sensors in simpler applications, many different devices are available.
  + They offer proven technology, simple to build electronics and a simple optical system.
  - Disadvantages are a larger number of electronics hardware and difficult implementation in coherent systems
- Spatial beam splitters onto multi detectors [1],
  + Uses proven technology and simple to build electronics,
  - Requires a difficult optical system and complicated alignment, and multifold electronics for each channel
- Nutating sensor
  This is a well known principle in RF applications and has already been applied to optical systems. The operating principle of a nutating sensor with a fiber is shown and described in fig. 1. Some examples of lasercomm systems implementations are:
    - Electro/acusto-optical beam deflectors
      + No moving parts, the beam is circularly deflected onto a stationary detector
      - High deflection voltages required / additional optical losses induced by the deflection crystal
    - Mechanically nutating fiber [2]; a fiber pigtail is move as a spatial detector
      - Limited lifetime
    - Electrostatically nutating fiber [3], same principle as the mechanically nutating fiber
      + High bandwidth achievable, fiber pigtail is the only moving mass
      - Small electrode gaps required, therefore difficult alignment, sensitive to environmental contamination

5. THE PIEZO-FIBERNUTATOR-SENSOR

We propose a new sensor which is based on a fiber which is forced to nutation by a piezo-mechanism. Figure 2 shows a photograph of a laboratory prototype of a piezo-fibernutator-sensor: The mechanism comprises a piezo-element which deflects a fiber in 2 axes. The figure shows the piezo-tube with 4 outer electrodes and an inner electrode. The outer electrodes are sectioned in 90 deg segments. The segments are driven by 4 sinusoidal deflection voltages. Adjacent drive voltages have 90 deg phase difference thus producing an rotating sum-voltage vector.

The pick-up fiber enters the tube from the rear side through a strain reliever and reaches towards the front surface of the tube where it is rigidly mounted in the center. Most of the present fibers (single-mode-fibers, multi-mode-fibers and polarization-preserving-fibers) can be used without significant modification of the device.

The piezo-tube is fixed in a holder cylinder which also carries the connectors for the electrodes. This cylinder can be integrated into a suitable holder for the optical system. In SOLACOS we use a polarization-preserving-fiber for the receiver path. Therefore we have to mount the piezo-tube with the fiber in a 6 degree of freedom holder.

The main advantages of this new type of sensor are:

- Dynamic and static deflection capability
- Nearly all fibers can be used without significant modification of the device.
- All operating parameters are independently and good adjustable, mainly by the drive electronics

912

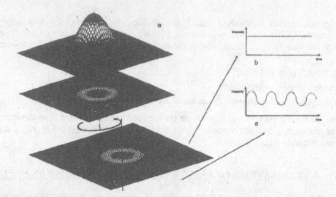

Fig. 1: Operating principle of a nutating sensor with a fiber: an Airy profile (a) from the receiver telescope is coupled into a Gaussian profile (b, c) of a fiber. If the fiber is circularly moved with the axis of the received beam exactly aligned with the axis of the cone (b), no amplitude modulation is generated onto the received signal. If the fiber cone and the received beam are misaligned (c), an amplitude modulation is produced (e).

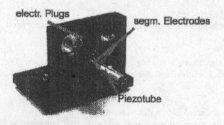

Fig. 2: Photograph of a laboratory prototype of a piezo-fibernutator-sensor.

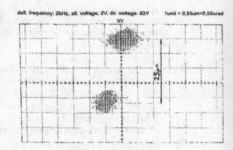

Fig. 5: Measured static deflection performance.

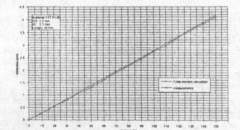

Fig. 3: Calculated and measured deflection values for a standard piezo material.

Parameter	Value	Remarks
Operating frequency	30 kHz	
Field of view	770 urad	4.5 mm lens
Piezo tube diameters	1.1/1.5 mm	
Piezo tube length	3 mm	
Piezo material	PZT PI 151	
Deflection (AC)	80 V/um	30 kHz non resonant

Fig. 6: Technical target characteristics for the SOLACOS fibernutator-tracking-sensor.

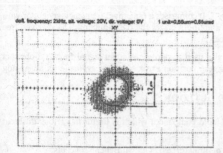

Fig. 4: Measured dynamic deflection performance.

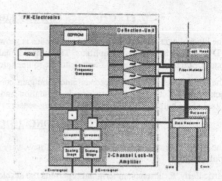

Fig. 7: Block diagram of the fibernutator-electronics

- The device is small in size and lightweight
- The mechanism shows good long term stability, we expect no wear out and a long life time
- The device is insensitive to environmental impacts, the only critical element is the fiber surface
- Depending on the holder of the optical system relaxed manufacturing tolerances are permissible as the device does not operates in resonant mode. Optical tolerances are depending on the construction of the interface.
- The device is easy and cheap to produce. Fabrication can be done in short time.
- The drive electronics are already developed, parameters can be adjusted via RS 232 and stored in an EEPROM.

The performance of the piezo-nutator strongly depends on the geometry and the electrical properties of the piezo element. Figure 3 shows the calculated deflection for a standard piezo material. Figure 4 shows the measured dynamic deflection characteristics, fig. 5 shows the static deflection characteristics.

The required component parameters for the SOLACOS fibernutator-tracking-sensor are listed in fig. 6.

6. APPLICATION AREAS

The piezo-fiber positioning device is very versatile. Besides the tracking-sensor application it can be utilized in several more applications. The underlying idea of statistically and dynamically moving an arbitrary fiber can be applied for four main areas:

1. Precise moving and positioning of fibers (e.g. into the center of a free space beam)
   This application utilizes the mechanism as a (quasi-) static positioning device. The electronics for this application mainly comprises a (4-channel) high voltage DC amplifier.
   → positioning function "in"

2. Produce a 2-dimension position signal which represents the offset between the center of the nutation cone of the fiber and the center of the beam
   In this application the piezo mechanism deflects the fiber along a circular path giving am amplitude modulation of the light intensity in the fiber. The necessary electronics include a drive generator, a (4-channel) high voltage AC-coupled amplifier, an AC detector for the AM on the optical signal in the fiber and a (2-channel) lock-in amplifier for the position signal.
   → sensor function

3. Dynamically optimizing fiber coupling device
   If the sensor function and the positioning function are coupled via a control amplifier, a dynamically optimized fiber-coupling device can be built. This application makes use of the dynamic and the static deflection capability of the piezo-mechanism. By using such a device, a fiber-coupling device with active long term compensation can be built. Additional losses into introduced by the nutation are comparably small. The electronics necessary for this application are the same electronics as in application 2 with the exception that a DC coupled high voltage amplifier is required.
   → dynamic fiber coupled

4. Precise positioning of fibers for illumination other applications.
   → positioning function "out"

914

7. ELECTRONICS FOR THE FIBERNUTATOR-TRACKING-SENSOR

The block diagram of the fibernutator-electronics is shown in fig. 7. It comprises a 6-channel frequency generator with 4 DC coupled high voltage amplifiers. The output signal of the amplifiers drives the piezo-nutator. The received free space beam is coupled into the fibernutator an routed into the receiver.

The control signal from the receiver automatic gain control is also used to derive the input signal for the 2-channel lock-in amplifier. After synchronous demodulation and scaling, the position error signal is available as an analog signal at the output of the fibernutator-tracking-sensor.

8. CONCLUSION AND OUTLOOK

The piezo fibernutator (tracking-sensor) is a new, very easy to build component which is useful in various different applications.

Particularly, for lasercomm terminal it permits the development of miniaturized "Optical Bench" concepts.

Among the next steps with which we are currently planned to pursue the development are the reduction of size, the use of new piezo materials and the optimization of the manufacturing process. In parallel we start extensive lifetime testing.

9. ACKNOWLEDGMENTS

This work is partly supported under DARA contract 50 YH 9201/1. A patent for the mechanism is pending.

10. REFERENCES

[1] Hueber M.F., Scholtz A.L., Leeb W.R., "Heterodyne acquisition and tracking in a free-space diode laser link", SPIE 1417-21
[2] Swanson E.A., Bondourant R.S., "Using fiber optics to simplify free-space lasercomm systems", SPIE 1218-07
[3] U. Johann, K. Pribil, H. Sontag, "A novel optical fiber based conical scan tracking device", SPIE 1522-09

Strahlungsfeste diffraktive optische Elemente für CO_2-Hochleistungslaser

Ch. Budzinski, H. J. Tiziani
Institut für Technische Optik, Universität Stuttgart

Einleitung

Mit diffraktiven Optiken läßt sich im Gegensatz zu refraktiven Optiken im Prinzip jede gewünschte Wellenfront generieren. Deshalb stellen die diffraktiven Optiken eine neue Qualität der optischen Bauelemente dar, die auch in der Lasermaterialbearbeitung zur Strahlformung und -führung von Hochleistungslasern neue Perspektiven öffnet, z. B. zum Härten, Markieren oder Abtragen. Effektivere Laserwerkzeuge können entwickelt werden und durch bessere Einkopplung in die Bearbeitungsfläche kann ein höherer Wirkungsgrad erreicht werden. Allerdings müssen die diffraktiven Optiken, die in Resist vorliegen, in strahlungsfeste Substrate überführt werden. Dazu wurde im Institut für Technische Optik ein preisgünstiges galvanisches Verfahren entwickelt. Mit dieser Methode wurden ein- und mehrstufige Gitter zur Strahldiagnose und zur Strahlteilung sowie ein 8-stufiges Hologramm zur Erzeugung eines Härteprofils hergestellt.

Herstellungsprozeß

Der Herstellungsprozeß besteht i.w. aus den Schritten: Berechnung der lateralen und vertikalen diffraktiven Strukturen, Belichtung und Entwicklung der Resistschichten, evtl. Zwischenreplikation, mikrogalvanische Abformung in Kupfer. Die Systembedingungen sind: eine Strahlleistung von 5 kW cw für CO_2-Laser mit einem Strahldurchmesser von 25-60 mm; die Absorption soll unter 1 % in Reflexion liegen, die Beugungseffektivität wird bei 90 % erwartet.

Nach Berechnung der lateralen und vertikalen Strukturen in der Hologrammebene mit Ray-tracing- Technik oder numerischen Verfahren liegen die Datensätze zur Steuerung des x,y-Tisches eines Laserscanners vor. Im Institut für Technische Optik wurde ein rechnergesteuerter xy- Tisch als Objekthalter zusammen mit einem fokussierten Argonlaser ($\lambda = 454$ nm, NA = 0,9) als Laserscanner aufgebaut [1]. Diese Anlage dient zur Herstellung der Masken und wird auch zum Direktschreiben von computergenerierten Hologrammen benutzt. Erreicht werden Strukturbreiten unter 1μm mit einer Positioniergenauigkeit von 0.2 μm. Die Oberflächenreliefs können durch Kopierprozesse preisgünstig vervielfältigt werden. Die Berechnung der Beugungseffizienz von Gittern erfolgt auf der Grundlage der Modaltheorie und ergibt die Beugungseffizienz in Abhängigkeit von der Tiefe der Gitterlinien [2].

Zur Herstellung der Resistoriginale werden optisch ebene Substrate aus Glas, Quarz oder Kupfer durch Aufschleudern von Resist beschichtet. Die Belichtung erfolgt entweder lithografisch über Masken in einem Belichter vom Typ MJ4 der Fa. Karl Süss oder durch Direktschreiben mit dem Laserscanner. Hochfrequente sin-Gitter mit Gitterkonstanten < 1 μm wurden im Interferenzfeld zweier kohärenter Teilbündel eines Argonionenlasers belichtet. Mit einer sich anschließenden Entwicklung mit genau abgestimmten Parametern wird die Intensitätsverteilung des Belichtungsfeldes in eine Reliefstruktur des Resistes übertragen. Dieses Relief muß unter Erhalt der Ebenheit der Fläche und der Form der Strukturen in ein strahlungsfestes Metall übertragen werden.

Die Strahlungsbilanz der Oberfläche eines metallischen optischen Bauelements unter Laserbelastung wird durch die Absorption α und die Wärmeleitfähigkeit κ bestimmt. Die Temperatur der Oberfläche ergibt sich für metallische Werkstoffe aus der Wärmeleitungsgleichung zu $T(t) = \alpha\, I_0\, 2\sqrt{t}\, /\, \sqrt{(\pi\, \kappa\, \rho\, C)}$; darin bedeuten:

t = Bestrahlungszeit, I_0 = Strahlintensität, C = spezifische Wärme, ρ = Dichte.

Der Faktor $F = \sqrt{(\pi\, \kappa\, \rho\, C)}\, /\alpha$ enthält die Materialkonstanten und ist ein Maß für die Strahlungsfestigkeit eines Metalls. Bezogen auf Kupfer mit F=1 ergibt Silber 2.5, Nickel dagegen nur 0.6. Aus diesen Überlegungen und auch aus wirtschaftlichen Gründen wurde eine elektrolytische Abformmethode in Kupfer gewählt. Zur Abformung wurden eine Replikationstechnik für Strukturen < 1μm und eine Aufbautechnik für Strukturen > 1μm entwickelt.

Die mikrogalvanische Replikation geht von der strukturierten Resistschicht aus, deren Oberfläche mit einer Leitschicht aus Silber oder Kupfer durch thermisches Bedampfen oder Sputtern versehen wurde. Diese Leitschicht wird kontaktiert und das Hologramm wird in einem sauren Kupferelektrolyten als Kathode geschaltet. Die Anode besteht aus phosphatisiertem Kupfer. Die Leitschicht wird elektrolytisch auf eine Dicke von 200 - 500 μm verstärkt. Diese Replikation wird unter Erhalt der optischen Ebenheit auf einen kühlbaren Kupferspiegel gelötet. Nach der Trennung der Replikation vom Original liegt die invertierte diffraktive Struktur in Kupfer vor. Bild 1 zeigt wurde. Der Abstand der Schreibschritte beträgt 3 μm. Diese Abformtechnik eignet sich besonders gut für Strukturen $\leq$ 1μm. Oberflächenprüfungen ergaben, daß die Strukturen bis zu < 50 nm fehlerfrei übernommen werden.

Die mikrogalvanische Aufbautechnik geht von einer Resistschicht aus, die auf einem ultrapräzisionsbearbeiteten Kupferspiegel aufgebracht wurde. Über eine Maske wird diese Resistschicht in einem Kontaktbelichter belichtet und anschließend bis auf die Kupferoberfläche durchentwickelt. Im Kupferbad erfolgt eine galvanische Abscheidung von Kupfer an den

Resists liegt die abgeformte Oberflächenstruktur in Kupfer vor. Dieser Zyklus kann für mehrere Maskenbelichtungen wiederholt werden. Auf diese Weise wurden sowohl 8-stufige Hologrammstrukturen in Kupfer hergestellt mit Stufenhöhen von 1.33 μm als auch Gitter mit Tiefen wahlweise von 250 nm bis 4 μm. Bild 1 zeigt das Höhenprofil eines binären Kupfergitters, gemessen mit einem konfokalen Mikroskop. Auf Bild 2 ist die Kupferreplikation eines Hologramms abgebildet, das mit einem Laser in Resist geschrieben wurde.

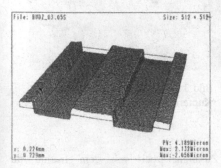

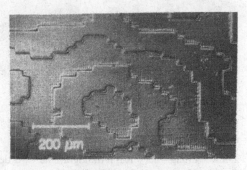

Bild 1: Höhenprofil eines Kupfergitters
(Konfokales Mikroskop, H.J. Jordan, ITO)

Bild 2 : Oberfläche eines Kupferhologramms
(Auflichtmikroskop

Mit den beschriebenen Verfahren wurden Gitter zur Strahlauskopplung für Diagnosezwecke (Gitterkonstante 100 μm, Tiefe 0.3 μm, Auskoppelverhältnis 1:10⁻³) und Gitter zur Strahlteilung 1:1 zum Betreiben von 2 Bearbeitungsstationen hergestellt. Bild 3 a zeigt eine Intensitätsverteilung zum Laserhärten mit einem stuhlförmigen Profil, das auf dem Werkstück eine gleichmäßige Temperaturverteilung erzeugt. Auf Bild 3 b ist das unter diesen Vorgaben hergestellte Kupferhologramm auf einem kühlbaren Träger dargestellt und Bild 3 c zeigt das Ergebnis des Einsatzes des kühlbaren Kupferhologramms mit 1.5 KW Laserleistung als Plexiglaseinbrand.

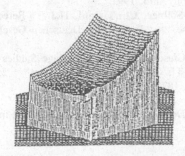

Bild 3 a: Stuhlförmige Intensitätsverteilung auf dem Werkstück zum martensitischen Laserhärten

918

Bild 3 b: kühlbares Kupferhologramm
als fertiges Bauteil

Bild 3 c : Härteprofil, erzeugt mit
Kupferhologramm, 1.5 kW CO_2-Laser

Ergebnisse:

Die Prüfung der Kupferhologramme wurde in Zusammenarbeit mit dem Institut für Strahlwerkzeuge, Universität Stuttgart vorgenommen. Die Absorption wurde kalorimetrisch nach ISO/TC -Standard gemessen [3]. Für s- und p-Polarisation ergab sich für 5 °ein Wert von $\alpha = 0.7$ %. Zur Prüfung der Strahlungsfestigkeit wurden im gleichen Institut Belastungstests durchgeführt. Die gekühlten Bauelemente wurden unter Einhaltung der Anwendungsgeometrie (P-Polarisation, 45° Einfallswinkel) bis zur Erwärmung auf 70 ° C mit ca 13 kW/ cm^2 belastet. Oberflächeninspektionen vor und nach der Belastung mit dem Nomarski-Mikroskop zeigten keine Veränderungen der Oberfläche. Die Beugungseffizienz betrug 85% für das 8-stufige Hologramm, für Gitter mit der Tiefe 300 nm wie berechnet 0.1%. Die Ebenheit ohne Strahlungsbelastung beträgt $\lambda/5$ der Anwendungswellenlänge, die Tests unter Strahlungsbelastung werden z. Z. durchgeführt.

Referenzen

[1 **Haupt, C. , Pahlke, M., Jäger, E., Tiziani, H. J.**:
" Design of diffractive optical elements for CO_2 - laser material processing", SPIE Vol. 1718 Workshop on Digital Holography, 175, 1992

[2] **Budzinski, Ch., Hembd-Söllner, Ch.,Tiziani, H.J.** :" Berechnung, Herstellung und Test von Gittern zur Strahldiagnose,"96. Tagung der Deutschen Gesellschaft für angewandte Optik 1995, Binz/Rügen

[3] **Plass, W., Krupka,R., Giesen, A.**:" Laser Damage Studies of Metal Mirrors and ZnSe-Optics by Long Pulse- and TEA-CO_2-Lasers at 10.6 μm" , 25th Annual Boulder Damage Symposium , SPIE Vol. 2114, 187, 1993

Die Arbeiten werden von der Deutschen Forschungsgemeinschaft im Rahmen des SFB 349 gefördert.

Diffractive Optics for CO_2 Laser up to 5kW

M. Pahlke, C.Haupt, L. Niessen, H. J. Tiziani
Institut für Technische Optik
Universität Stuttgart
Pfaffenwaldring 9
70569 Stuttgart

1. Introduction

Laser material processing by CO_2-lasers became a standard technology in industrial production. For cutting, welding and drilling the focal spot is a suitable intensity distribution, because a power density of about 10^8 W/cm² is necessary for these processes (1). For some surface treatment processes power densitys of 10^4W/cm² are required. These processes can be optimized by special intensity distributions. Marking can be optimized by replacing the masks by optical components that creat the mark. Different technics are developed to create these intensity distribution. The high intensity of the laser radiation requires reflective optical components. We have realized a reflective diffractive element which reconstructs on optimized intensity distribution for surface hardening (2) by a TLF5000 Laser.

The development in fabrication of diffractive reflective elements for high power laser radiation can be devided into two streams. One technic is based on electro-plating with Copper (3), the other based on microlithographic technology (4), which is the technology we used to manufacture our element.

The diffractive element had to be designed for a optical system with an aperture radius of 20mm. It should satisfy the quality reqirements for reflective optical components, especially with regard to surface flatness and absorption of laser radiation. Additional we had to minimize the mass of the device, because the diffractive element is a part of the moving segment in a laser processing machine.

2. Micro heat sink

The quality of a mirror depends on the flatness of its surface. The absorption of the laser radiation deforms the surface of the mirror. The water pressure and temperature differences between the cooling water and the mirror surface will also cause distortions of the mirror. In VLSI-technology and in diode laser applications heat sinkings with a high heat capacity are developed (5). These coolers consist of a large number of small rectangular grooves in a Silicon waver with a typical width of 50μ and a depth of 300-350μm. These grooves are covered with a Pyrex plate which is anodically bonded to the Silicon waver. With such device a heat transfer of 1kW/cm² can be achieved. This heat transfer depends on the temperature difference between the mirror surface and the cooling water. This temperature difference causes deformation of the surface. On the other hand we need a heat sink for power densities of 50W/cm² and a minimal surface deformation. Moreover we can achieve massreduction by using this technology.

The heat sink was designed that the whole mirror surface can be cooled. We use a double finger-like structure in the Pyrex plate for water source and drain. Source and drain are connected by the grooves in the Silicon waver. So we acieved a effective lenth of the grooves of 4mm. The Pyrex plate was structured by ultrasonic drilling.

To test the heat sink a mirror was fabricated. The surface was coated with protected silver, because the absorption of a high performance coating is too low to test the heat sinks properties. The flatness of the mirror surface was meassurement at an Fizeau-interferometer at our institute. The PV-value of the flatness was 1.3μm. The absorbtion was 0.8%. The surface deformation of the mirror in the beam of aTrumpf TLF 5000 with a diameter of 15mm was 50nm. The mass of the test element was 60g without mounting.

3. Calculation and fabrication of the device

To avoid energy losses diffractive elements had to be realized as a phase only devices. The diffractive structure was calculated as a Fourier element - because a focussing mirror is available in our machine - by an Gerchberg-Saxton algorithm (6). This algorithm delivers high efficient diffractive structures also for complex intensity distributions with an acceptable computational

afford, because only a small part of the diffractive structure had to be calculated. The result is a phase distribution with values between 0 and 2π, which is quantized in our case to 8 values using a hardclip. The repetition of this structure yields the diffracting surface. The necessary calculation time depends on the size of the element and the CPU-type. A standard calculation by an I486DX 50 CPU needs a few hours.

The diffractive structure was manufactured by lithographic technology. The three masks for the diffractive structure are fabricated at the Laser Photoplotter at our institute. The basic structures of the element is 4μm. Therefore the possitioning accuracy of 200nm of the xy-Unit of Photo-

Fig 1: Calculated phase distribution to reconstruct an optimized intensity distribution for surface hardening

plotter is sufficient. The mask for the heat sink was also manufactured at this Photoplotter.

The grooves of heat sink were etched in the back side of a standard 4'' <110> Silicon waver using KOH. The 8-level diffractive structure was manufactured in three ion etching processes. Then the Pyrex plate was structured by ultrasonic drilling and bonded with the Silcon waver. To get good reflection properties the diffractive structure was coated with super enhanced Gold at the LZH in Hannover.

The quality of the diffractive structure depends on several important facts. The depth of the diffracting structures had to be manufactured with an accuracy better than 50nm of the structure depth. Deviations will cause a miss matching of the phase, which reduces the efficiency of the element and increases errors in the reconstruction. Alignment errors have only small effects on the optical properties, but in the high power laser beam these fabrication errors may be the starting-point of the destruction in such an element. During the anodic bonding of the Silicon waver and the Pyrex plate the errors from both surfaces are added on the silicon waver, because of the elasticity of the silicon waver.

4. Characterizing the element

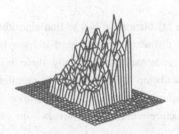

Fig. 2: Computersimulation of optical reconstruction. The diffraction efficiency achieves 90%.

Before the fabrication of the diffractive element the quality of the reconstruction is tested in a computer simulation. For a 8-level diffractive structure the simulation predicts a diffraction efficiency of 90% which could be increased by a further lithograpic processes, so the element will have 16 or more levels. The simulation shows noise over the whole reconstruction, which is characteristic for that calculation algorithm. There exist proposals for modification of the algorithm, but they have no effect or they reduce the efficiency.

The measurement of the surface flatness on the diffractive structure is difficult, because the structures are too small and the fringe density between the discrete phaselevels cannot be resolved. We expect a deformation comparable to that of the mirror.

The absorption was meassured after ISO-DIS 11/551 and (7). The optical component has an area without any diffractive structure, to compare the effects of the diffractive structure on the absorption. The unstructured surface has an absorption of 0.5%, the absorption of the diffractive structure was meassured to 2%.

The diffraction efficiency was measured to 88% which is in correspondance to the computer simulation, because the absorption of the element was not taken into respect at the computer simulation. The reconstruction of the diffractive element shows the principle characteristics of the optimized intensity distribution for surface hardening. The reconstruction is very sensitive on inclination and small deviations will introduce noticable errors in the reconstruction. On the other hand this components are insensitve against translation of the component and the change of the reconstructing intensity distribution.

Fig. 3: Diffractive element

922

5. Conclusions

A diffractive element for CO_2-laser material processing was manufactured in Silicon. A efficient numerical algorithm was used to calculate the diffracting structure, which was fabricated in a lithographic process in silicon. A high efficient heat sink is used for cooling the diffractive element. The diffractive stucture was coated with super enhanced Gold. The absorption of the diffractive structure was four times higher than on a nun-structured area on the same waver. The diffraction efficiency for the 8-level Phase structure reached 88% which corresponds with our simulation.

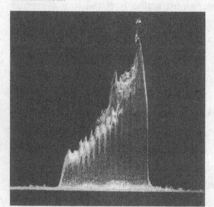

Fig. 4 Optical reconstruction in Plexiglass

With reguard to calculation we had to find algorithms which compute diffractive structures that have less errors in the reconstruction. Especially there is an interest in calculation of binary and multilevel diffractive structures because the quantization of calculated structure introduces errors on the reconstruction.

The quality of the optical surface has to be improoved. We need a better surface for the mirror, and a more acurate alignment in the lithographic process. For that purpose the reproducibility of the work of our photoplotter had to be increased.

Our experiments had shown that our diffractive element resists to radiation of a 5kW CO_2-Laser. The quality of the reconstruction with the 5kW Laser does not differ from the reconstruction at low power.

Diffractive elements are an expansion of the set of optical components. Further development of fabrication technology will decrease fabrication time and costs.

6. Acknowledments

We thank the DFG for the financial support of our work in the SFB 349. The lithograhic processes were perfomed at the IMIT, Villingen-Schwenningen. The optical components were tested at the IFSW, University Stuttgart.

7. Literature

1 Hügel, H., Strahlwerzeug Laser, Teubner, Stuttgart 1992

2 Burger, D., Optimicrung der Strahlqualität beim Laserhärten, LASER 89, Springer Verlag, Berlin 1990

3 Budzinski, Ch., Tiziani, H. J., Strahlungsfeste diffraktive Optiken durch mikrogalvanische Abformung, Laser und Optoelektronik, 1/1995

4 Pahlke, M., Haupt, C., Tiziani, H.J., Diffraktive Elemente für die Lasermatrialbearbeitung, LASER 93, Springer Verlag, Berlin 1994

5 Tuckerman D.B., Pease R.F.W. . High-performance heat sinking for VLSI , IEEE Electron Device Letters 2 126 129 , 1981

6 Gerchberg R.W., Saxton W.O., A practical algorithm for the determination of phase from image and diffraction plane pictures, Optik / Vol.35, No.2 / pp.237-246 , 1972

7 Serchinger R.W., Giessen, A., Absorptionsmessungen an optischen Komponenten und zwei Stahllegierungenbei 10,6μm, LASER 89, Springer , Berlin 1990

Laser in der Mikrostrukturtechnik
Laser in Microsrtucture Technology

H. Opower, H. Hügel, W. Pompe

Einleitung

Der Kongress "Laser in der Mikrostrukturtechnik" wurde 1995 als ein neues Gebiet
aufgenommen. Die damit erfaßten Prozesse reichen von den Phänomenen der primären
Wechselwirkung über die Dünnschichtbildung mit lasererzeugten Partikelströmen bis zur
lateralen Feinstrukturierung. Bei letzterer spielen neben den direkten Abtragungen
insbesondere auch photochemische Methoden eine wichtige Rolle. Lasereinsätze zur
Feinanalyse von Gittereigenschaften und Strukturen gehören selbstverständlich ebenfalls
in das Gebiet der Mikrostrukturtechnik.

Auf Grund der Neuheit der Verfahren befindet sich die industrielle Einführung noch im
Anfangsstadium. Erst in einigen Jahren können nachhaltige Einflüsse vorzugsweise in
Spezialbereichen der Halbleiterfertigung erwartet werden. Die wissenschaftlich
technischen Erkenntnisse erfahren aber gerade in der gegenwärtigen Zeit mit den
grundlegenden Institutsarbeiten einen stürmischen Fortschritt. Die ausgezeichneten
Beiträge des Jahres 1995 lassen eine spannende Entwicklung erwarten, so daß der
nächste Kongreß bereits in wesentlich größerem Umfang und mit schon starker
Einbindung der Industrie geplant werden kann.

Hans Opower

Introduction

The congress G "Lasers in microstructuring" is a new arrangement within the general scope of the conference. The topics to be presented in this congress comprise primary interaction phenomena, thin film formation by laser generated particle beams and precise ablations for lateral structuring including direct writing as well as photochemical processes. Obviously special methods of laser diagnostics play an important role for the analysis of microstructures.

Due to the present state of this new branch of laser technology the industrial application is yet in its infancy. But despite the fact, that large scale impacts to industry may be expected in the course of the coming years, the exceedingly rapid progress and the excellent results, which have been presented by prominent scientists of several research institutes will form a strong basis for the technical application in the near future. As a consequence even the next congress might be characterized by a considerable increase of contributions and especially by a numerous participation of industrial companies.

Hans Opower

Bedeutung der Mikrostrukturtechnik – Perspektiven für Laseranwendungen

H.-J. Warnecke, A. Gillner

Fraunhofer-Gesellschaft München

Fraunhofer-Institut für Lasertechnik Aachen

1. Einleitung

Die Lasertechnik hat sich in den vergangenen Jahren in vielen Bereichen der Technik eine Stellung erhalten, die für Nutzer und Anwender ein fester Teil von Produkt- und Fertigungsplanung geworden ist. Dies gilt sowohl für den Laser als Element in technischen Produkten, wie z. B. in der Unterhaltungsindustrie (CD-Spieler) oder Kommunikationsindustrie als auch als System zur Materialbearbeitung und Anwendung, wie z. B. in der Medizin oder in der Produktionstechnik. In der Fertigungstechnik ist vor allen Dingen das Schneiden mit Laserstrahlung ein eingeführtes Verfahren, für das eine Vielzahl von Anwendungen und Bearbeitungssystemen gibt. Heute vereinzelt, in Zukunft jedoch in stärkerem Maße, wird der Laser in der Herstellung von Mikroprodukten eine wichtige Rolle spielen. Ursache hierfür sind die herausragenden Eigenschaften des Werkzeugs Laserstrahl, die es erlauben, Energie auf kleinstem Volumina und in kürzesten Zeiten in einem Werkstoff zu deponieren. Damit können Vorgänge und Effekte hervorgerufen werden, die jenseits der Gleichgewichtsprozesse ablaufen und völlig neue Produkteigenschaften und Qualitäten erlauben. Darüber hinaus ist der Laser für eine Vielzahl von Anwendungen einsetzbar, ohne das Fertigungssystem selbst hierfür zu ändern. Damit eignet sich dieses Werkzeug insbesondere für die Herstellung von Produkten mit kleinen Losgrößen und für den Einsatz bei kleinen und mittelständischen Unternehmen, für die die hohen Investitionskosten einer Mikroelektronik nicht zu tragen sind.

Als Resultat aus den physikalischen Eigenschaften und Qualitätsmerkmalen der Laserstrahlung kann der Laser für die Bearbeitung beinahe jedes Werkstoffs mit sehr hohen Prozeßgeschwindigkeiten eingesetzt werden. In der Materialanalyse eignet sich die Laserstrahlung für eine Vielzahl von Meß- und Prüfverfahren mit größter Genauigkeit bei Auflösung im Nanometer-Bereich. Diese Vorzüge prädestinieren den Laserstrahl als Werkzeug für Sensorik und Mikrosystemtechnik, die nach verschiedenen Prognosen in den kommenden Jahrzehnten ein erhebliches Wachstum aufweisen werden (Bild 1). Insbesondere die Sensorik wird durch Anfragen aus der Automobilindustrie einen exponentiellen Markzuwachs erfahren /1/.

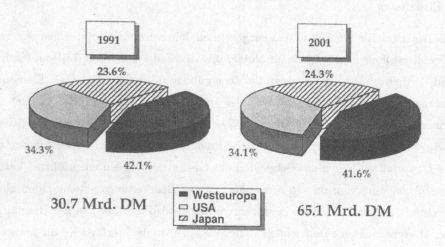

Bild1: Marktentwicklung für Sensoren, gegliedert nach Herstellerregionen /1/

Dabei werden sich die Laseranwendungen in Sensorik und Mikrostrukturierung nicht nur auf die klassischen Bereiche der Mikrotechnik, wie Resist-Belichtung und Mikrostrurierung beschränken. Vor allem in der Montage der Systeme werden die Vorteile der Lasertechnik

- minimale Energiedeposition,

- sehr kleine Bearbeitungsgeometrien,

- hohe Bearbeitungeschwindigkeit

zu neuartigen Fertigungsverfahren und Konstruktionen führen. Technologien wie das Mikrokunststoffschweißen, laserunterstütztes Silicon-Fusion Bonding und Laserstrahl-Mikrojustieren eröffnen neue Möglichkeiten des Mikrosystementwurfs und können bei konsequenter Anwendung zu einem festen Bestandteil von Laseranwendungen führen.

Gleichermaßen stürmisch ist mit einer Entwicklung von Mikrosystemen für die minimalinvasive Therapie und Chirurgie zu rechnen. Für beide Bereiche kann die Laserstrahlung geeignete Werkzeuge zur Herstellung, Integration und Prüfung bereitstellen, die sowohl den Anforderungen des Produktes als auch den Anforderungen an eine kostengünstige Fertigung gerecht werden.

Ein Beispiel für ein Mikrosystem aus dem Automobilbereich ist in Bild 2 dargestellt /2/. Es handelt sich dabei um einen Luftmassensensor, der hybrid aus anisotrop geätzten Silizium, einem metallischen Träger und verschiedenen Funktionsschichten aufgebaut ist. Ein Teil der Fertigungsverfahren, die heute aus der Mikroelektronik übernommen sind, lassen sich mit Laserverfahrenabdecken, wie zum Beispiel das selektive Beschichten mit piezoelektrischen Funktionsschichten. An anderer Stelle, zum Beispiel beim Fügen und Justieren der Einzelkomponenten eröffnet die Lasertechnik eine Reihe von Fertigungsvereinfachungen.

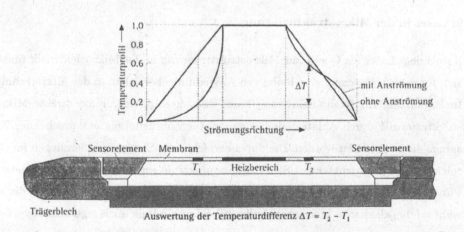

Bild 2: Querschnitt eines Luftmassensensors auf Si-Basis /2/

Vor diesem Hintergrund kann sich der Markt für Lasersysteme, der in den vergangenen Jahren eine gewisse Stagnation zeigte, wieder verstärkt entwickeln und in dem Bereich der Mikrotechnik neue Marktsegemente gewinnen. Ursache hierfür ist vor allen Dingen die verstärkte Entwicklung von Hochleistungs-Diodenlasern und den daraus resultierenden Derivaten, wie diodengepumpte Festkörperlaser und All-Solid-State-Laser vom Infraroten bis zum UV /3/. Vor allem UV-Laser werden im Bereich der Mikrostrukturierung neben dem bereits besetzten Terrain der UV-Belichtung verstärkt Anwendungen finden.

Dabei resultieren die Anforderungen aus den herrschenden Entwicklungstrends in der Herstellung von mikroelektronischen Bauelementen und Mikrosystemen. Beispielhaft hierfür sei die Entwicklung von dynamischen Speicherbausteinen genannt, die bis zum Jahr 2000 die Gigabitmarke überschreiten werden. Die dabei geforderten minimalen Strukturgrößen liegen weit unterhalb eines Mikrometers. Gleichzeitig steigt durch die erhöhte Integration die Anzahl der Maskenebenen auf über 30 /4/. Zusätzlich werden heute noch überwiegend durchgeführte Batchprozesse durch Single-Wafer-Prozesse abgelöst, die weitgehend auf Flüssigverfahren verzichten. All diese Anforderungen können durch den Einsatz von UV-Lasern erfüllt werden, mit denen sowohl Belichtungsaufgaben mit hohen Taktfrequenzen als auch nachfolgende Bearbeitungsschritte durchgeführt werden können.

2. Laser in der Mikrostrukturierung

Neben dem Laser als Quelle zur Mikrostrukturierung in der Mikroelektronik finden sich für dieses Werkzeug eine Reihe von Anwendungsbereichen in der Mikrotechnik. In Ergänzung zur Resist-Belichtung kann der Laser dabei für die direkte Mikrostrukturierung durch Ablation und Ätzen unter Zuhilfenahme entsprechender Zusatzmedien verwendet werden. Die dabei erreichbaren Strukturgrößen liegen im Bereich weniger Mikrometer mit Strukturgenauigkeiten unterhalb eines Mikrometers. Für die Herstellung von Funktionsschichten gibt es eine Reihe von Verfahren, die sowohl an die klassische Mikroelektronik angelehnt sind als auch eigenständige Neuentwicklungen für Anwendungen des Lasers sind. Zuletzt findet sich der Laser als vielfältiges Werkzeug in der Aufbau- und Verbindungstechnik mit als klassisch anzusehenden Verfahren der Fügetechnik, der Strukturierung und der Prüftechnik. Die Materialpalette reicht dabei von Metallen über Halbleiter zu den Keramiken und die Elektriker, wobei hier vor allen Dingen die Keramiken zu nennen sind. In verstärktem Maße kommen auch Kunststoffe zum Einsatz, für die es eine Reihe von Bearbeitungsverfahren mit Laserstrahlung gibt, die sich von den konventionellen wesentlich durch Geschwindigkeit und Qualität unterscheiden.

Einige der Laserverfahren in der Mikrostrukturtechnik, die in Zukunft verstärkt zum Einsatz kommen, werden im folgenden anhand von Beispielen kurz vorgestellt.

2.1 Photolithographie mit UV-Strahlquellen

Hochintegrierte mikroelektronische Speicherbausteine mit Kapazitäten oberhalb 100 Mbit verlangen eine hohe Beweglichkeit der Elektronen, um kurze Schaltzeiten zu ermöglichen. Für den Aufbau auf Silizium-Basis werden hierfür extrem kleine Strukturen im Bereich unterhalb eines 1 µm benötigt. Die in CMOS-Technologie hergestellten Bauelemente sind aus einer Vielzahl von Einzelstrukturen zusammengesetzt, wobei diese über verschiedene Maskentechniken und Zwischenbeschichtungen erzeugt werden. Die kleinsten dabei auftretenden Geometrien liegen im Bereich weniger 100 nm. Die Strukturierung dieser Bauteile erfolgt dabei in der Regel durch Belichtung mit UV-Licht. Neben dem Resist ist für die Genauigkeit der Belichtung die Linienbreite der Lichtquelle maßgeblich. Um kurze Belichtungszeiten zu ermöglichen, sind hierfür Strahlquellen notwendig, die bei geringer Linienbreite eine möglichst hohe Strahlungsleistung aufweisen. Modifizierte Excimerlaser werden hierfür in Einzelfällen bereits eingesetzt und zeigen vor allen Dingen für eine erhöhte Integrationsdichte ein erhebliches Marktpotential /5/.

Nach dem Belichten der Resist-Schicht wird diese abgelöst und das darunter liegende Oxid mit entsprechenden Mitteln naßchemisch strukturiert. Der abschließende Schritt ist die Silizium-Ätzung, die die Basis für weiterführende Beschichtungen ist. Die mit UV-Lasern erzielbaren Geometrien liegen heute in der Regel im Bereich 0,1 bis 0,3 µm und können mit fortschreitender Maskierungstechnik und erweiterten Resist-Technologien in den Größenbereich unter 100 nm reduziert werden /6/. Ein Beispiel für eine solche Struktur ist in Bild 3 dargestellt.

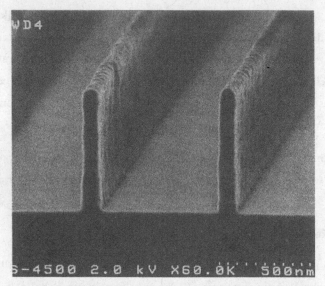

*Bild 3: Nanometer-Strukturen in Photoresist, hergestellt
durch UV-Belichtung /6/*

Zum Erreichen solcher Geometrien reichen einfache Abbildungs- und Fokussierungs-
verfahren nicht mehr aus. In der Proximity-Technik haben sich zur Erzeugung von
Strukturen unterhalb 100 nm Verfahren wir die Phase-Edge-Maskierung durchgesetzt. Die damit erzielbaren Aspektverhältnisse liegen im Bereich 10-20 und sind somit vergleichbar mit anisotropen Ätzverfahren wie z.B. das reaktive Ionenstrahlät-
zen.

2.2 Laserstrukturierung durch Ablation

Für großflächigere Strukturierungen im nicht-elektronischen Bereich sind Laserverfahren verfügbar, die auf der Basis eines thermischen Direktabtrages bzw. unter
Verwendung zusätzlicher Ätzmedien eingesetzt werden können. Dadurch, daß sie an
bereits vorstrukturierten bzw. bestückten Bauteilen eingesetzt werden können, erlauben sie eine nachträgliche Strukturierung bzw. Personalisierung des Bauteils.
Für dieses Verfahren eignen sich sowohl Nd:YAG-Laser, bei denen die Strukturierung mit Genauigkeiten ⪯ 0,1 mm erfolgen kann als auch Excimerlaser, die einen
Strukturabtrag in Größenordnungen < 10 μm mit Genauigkeiten < 1 μm erlauben.

Bei Metallen, Keramiken und Halbleitern erfolgt der Materialabtrag thermisch durch Verdampfen des Werkstoffes und Austrieb infolge Dampfexpansion. Bei Kunststoffen kann ein Teil des Abtragprozesses über die direkte Dissoziation der Makromoleküle erfolgen, da die Photonenenergie der UV-Laser zum Teil über der Bindungsenergie der Kunststoffverbindungen liegt.

Allen Verfahren gemeinsam ist, daß sie an Atmosphäre eingesetzt werden können und somit eine hohe Flexibilität hinsichtlich Materialwahl und Bearbeitungsgeometrie aufweisen. Durch eine Steuerung der Strahlführung lassen sich dreidimensionale Bauteile mit der gleichen Genauigkeit bearbeiten wie ebene mit großer Wiederholpräzision. Auch schwer bearbeitbare Werkstoffe werden durch eine geeignete Prozeßführung mit dem Laser strukturierbar. So wird zum Beispiel bei der Bearbeitung von Diamant der thermisch bedingte Phasenübergang zwischen kristallinem Diamant und Graphit für die Strukturierung ausgenutzt /7/. Der umgewandelte Bereich wird nach der Laserbearbeitung chemisch herausgelöst. Damit steht ein Verfahren zur Verfügung, mit dem der zukünftig sehr vielversprechende Mikroelektronik-Werkstoff Diamant auf kostengünstige Weise in der Bauteilherstellung und in der Kalibrierung an teilgefertigten Komponenten strukturiert werden kann.

3. Mikrostrukturierung durch Laserstrahlbeschichten

Ebenfalls an bereits vorprozessierten Bauteilen können die laserbasierten Beschichtungsverfahren eingesetzt werden. Vorteil dieser Verfahren gegenüber konventionellen CVD- und PVD-Verfahren ist die geringe Substratbelastung durch den in der Regel Wegfall einer notwendigen Substratheizung. Bei den Verfahren Laser-CVD und -PLD sind die Prinzipien denen konventioneller Verfahren, wie CVD und PVD angelehnt. Der Unterschied besteht lediglich darin, daß das zu beschichtende Bauteil nicht integral erhitzt, sondern nur lokal bearbeitet wird. Mit dem Laser-CVD-Verfahren können Metalle, Halbleiter und Isolatoren mit ähnlichen Qualitäten, wie im klassichen CVD, aufgetragen werden. Das PLD-Verfahren zeichnet sich durch hohe Schichtraten und Schichtqualitäten aus, mit dem Metalle, Metallverbindungen, Oxide und Halbleiter aufgebracht werden können. Beispielhaft seien hier Aktorschichten aus PZT und Supraleiter (YBCO-Schichten) genannt.

Anwendungen finden diese Verfahren in der Mikrosystemtechnik, wo mikrostrukturierte Sensor- und Aktorelemente mit Funktions- und Kontaktierungsschichten versehen werden müssen. Ein Beispiel hierfür sind Drucksensoren für die minimalinvasive Diagnostik, die mit einer Anzahl monolithisch hergestellter Druckdosen versehen sind. Als Sensorschicht kommen piezoelektrische Dünnschichten zum Einsatz, die durch PLD-Verfahren aufgebracht werden können. Bei diesem Verfahren wird mit einem Laserstrahl, in der Regel Excimer-Laser, ein Target verdampft, das die gleiche Stöchiometrie wie die zu erzeugende Funktionsschicht aufweist. Der Materialdampf in Richtung zu beschichtendes Substrat und kondensiert darauf. Durch Bewegung von Target und Substrat können sowohl lokale als auch großflächige Beschichtungen auf ganzen Wafern mit Beschichtungsraten von 20 $\mu m/cm^2/min$ erzielt werden. Gegenüber konventionellen Verfahren zeichnet sich das PLD Beschichten in der Regel durch eine geringere thermische Substratbelastung aus /8/.

Durch Steuerung der Verfahrensparameter können solche Schichten ein weites Spektrum möglicher Anwendungen durchlaufen. Als amorphe Struktur zeigen sie eine hohe Durchschlagsfestigkeit und können als Isolator in mikroelektronischen Schaltungen dienen. Die Grenzfeldstärke beträgt bei diesen Schichten ca. 40 kV/mm. Das gleiche Material kann bei veränderter Prozeßführung Sensor- und Aktoreigenschaften aufweisen. In Bild 4 sind Aufnahmen der unterschiedlich ausgebildeten Schichtstrukturen dargestellt.

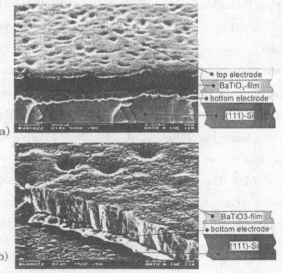

Bild 4:
Durch PLD hergestellte
Bariumtitanat-Schicht mit
a : amorpher Struktur;
b: kristalliner Perovskit-Struktur

Tetragonal orientiertes Bariumtitanat mit einer Perovskitstruktur ist ferroelektrisch, d. h. zeigt eine Speicherwirkung und ist auch als Sensor z.B. in pyrolelektrischen Detektoren verwendbar. Ähnliche Eigenschaften zeigen PLD-Schichten auf der Basis Bleizirkontitanat, die über eine hohe piezoelektrische Striktion verfügen und somit als Aktor eingesetzt werden können. Durch die geringen Schichtdicken können mit diesem Verfahren Mikroaktoren erzeugt werden, die in Optoelektronik und Mikromechanik Einsatz finden.

Durch Einsetzen geeigneter Masken und Einstellung anderer Bearbeitungsparameter (höhere Bearbeitungsenergiedichte) wird die zuvor aufgebrachte Dünnschicht mit dem Laser abgetragen. Die Abtragsleistungen liegen dabei im Bereich einiger 100 nm pro Puls, so daß eine Mehrlagenstrukturierung als Funktionsschicht in einer einzigen Bearbeitungsstation und in kurzen Bearbeitungszyklen erfolgen kann. Somit eignet sich das Verfahren vor allem für Kleinserien, bei denen die Bearbeitung im Batchbetrieb bei großen Stückzahlen wirtschaftlich nicht sinnvoll ist.

4. Laser in der Mikrosystemtechnik

Neben dem Einsatz in der Mikroelektronik wird der Laser vor allen Dingen in der Mikrosystemtechnik ein entscheidendes Werkzeug zur Durchsetzung dieser Systeme in marktgängige kostengünstige Produkte sein. Mikrosysteme, zu diesen können auch im weitesten Sinne Sensoren gezählt werden, werden in der Regel in geringen Stückzahlen (100 000 bis 1 000 000, z.T. auch geringer) benötigt. Die derzeitigen Herstellungsverfahren für Mikrosysteme sind denen der Mikroelektronik entlehnt und nur bei hohen Stückzahlen kostengünstig. Auch die LIGA-Technik die große Möglichkeiten bezüglich Mikrostrukturierung und Präzision bietet, kann hier keine entscheidenden Vorteile leisten, da die zur Herstellung der Spritzgußteile notwendigen Werkzeuge teuer sind und sich nur bei großen Stückzahlen amortisieren. Beispiele solcher Mikrosysteme sind in einer Vielzahl industrieller Einsatzbereich zu finden, wie z. B. in der Kommunikationstechnik, wo optische Duplexer als hybride Bauteile auf einer durch LIGA-Technik hergestellten mikrooptischen Bank aufgebaut sind. Laserverfahren können hier in einer Vielzahl von Anwendungen, z. B. zur Laserstrukturierung oder zum Laserfügen von Glasfasern, Linsen und Teilerplatten eingesetzt werden.

Darüber hinaus zeigen die Laserverfahren ein großes Potential in der Aufbau- und Verbindungstechnik von Mikrosystemen. Aufgeteilt auf einzelnen Verfahrensschritte finden sich hier Laserverfahren in der Chipvereinzelung, z. B. durch Laserschneiden, in der Kontaktierung einzelner Bauelemente, z. B. durch laserunterstütztes Bonding, und das bekannte Laserlöten. In der Hybridintegration werden diese Verfahren in einem größeren Bereich eingesetzt, wobei vor allen Dingen die Fügeverfahren Laserlöten, Lasermikroschweißen, laserunterstütztes Kleben und auch das Lasermikrobiegen zum Einsatz kommen. Vor allen Dingen hier werden die Vorteile der Lasertechnik: geringste Energieeinbringung bei kleinstem Volumen sichtbar, da ein Justieren und thermisches Nachbehandeln der gefügten Struktur in der Regel nicht notwendig ist.

4.1 Laser zur Mikrokontaktierung

Für die Mikrokontaktierung bei niedrigen Temperaturen kann das lokale Beschichten mit Laserstrahlung aus der Flüssigphase eingesetzt werden. Bei diesem Verfahren wird ein flüssiger Precursor auf das zu beschichtende Bauteil aufgebracht und an den zu beschichtenden Stellen mit dem Laser durchstrahlt. Infolge lokaler Temperaturerhöhung scheidet sich aus dem Precursor der Metallkomplex ab und auf dem Substrat nieder. Die Beschichtungstemperaturen liegen bei diesem Prozeß unterhalb 300 ° C, so daß der Prozeß auch für die Beschichtung von technischen Kunststoffen geeignet ist. Die Strukturgrößen werden lediglich durch die Fokussierung des Laserstrahls beeinflußt und liegen in der Regel zwischen 5 und 100 μm . In Bild 5 ist eine durch Laserabscheidung hergestellte Leiterbahn auf Silizium dargestellt.

Die erzielbaren Leitfähigkeiten einer Einzelschicht mit einer Dicke von 200 nm betragen 10 % der Bulkleitfähigkeit von Gold und eignen sich somit für die Signalübertragung. Durch Mehrfachbeschichtungen sind Schichtdicken bis zu 2 μm erzielbar, so daß auch Leistungselemente in der Sensorik mit diesem Verfahren kontaktiert werden können. Gegenüber allen anderen Beschichtungsverfahren eignet sich das laserunterstützte CLD-Beschichten für die Erzeugung lokaler leitfähiger Strukturen auf 3dimensionalen Mikrobauteilen, wie z. B. Sensoren und Mikrosystemen in hybrider Bauweise.

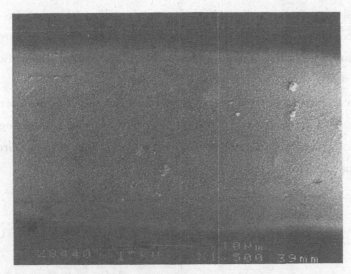

Bild 5: Leiterbahn auf Silizium-Wafer, mit Laserstrahlung aus einem flüssigen Precursor abgeschieden

4.2 Mikrointegration durch Laserstrahlfügen

Ein Anwendungsfeld mit festem Marktsegment für den Laser ist das Mikrofügen. Neben Anwendungen der Feinwerktechnik, wo das Laserschweißen bereits vielfach in der Fertigung zu finden ist, kann das Lasermikrolöten als lokal wirkendes Verfahren zur Kontaktierung von Elektronikbauteilen in hybrider Bauweise genannt werden /9/. Im Vergleich zu allen anderen Lötverfahren wird beim Lasermikrolöten nicht das Gesamtbauteil erhitzt, sondern lediglich die zu fügenden Kontaktstellen. Für einige Applikationen, bei denen temperaturempfindliche Bauteile auf bereits vorbestückte Platinen aufgebracht werden, finden sich bereits erste industrielle Applikationen. Zukünftig wird vor allen Dingen dieser Bereich verstärkt von Laserdioden dominiert sein, bei denen die Laserdioden direkt als Strahlquelle zur Anwendung kommen. Somit können Laserlötanlagen zu einer wirtschaftlichen Alternative gegenüber konventionellen thermischen Lötanlagen werden /10/.

Ebenfalls durch Laserdioden lassen sich Fügeaufgaben bewältigen, die vor allen Dingen im Bereich der Mikrosystem anzutreffen sind. Häufig sind in diesen Systemen Zwei-Schichtverbünde, bestehend aus Silizium und Glas oder Silizium bzw. Metall und Kunststoff herzustellen. Zur Zeit werden diese Verbünde geklebt bzw. im Fall von Silizium und Glas durch anodisches Bonden miteinander verbunden. Im ersten Fall setzt die Genauigkeit der Kleberzuführung den einzusetzenden Strukturgrößen eine Grenze. Beim anodischen Bonden lassen sich die Strukturen durch die hohe Temperaturbelastungen nicht beliebig verkleinern.

In beiden Fällen können Laserverfahren aufgrund ihrer lokalen Wärmeeinbringung Abhilfe schaffen. Grundprinzip dieser Verfahren ist, daß einer der beiden Fügepartner vom Bearbeitungslaserstrahl durchstrahlt wird und die Wärme an der Grenzfläche beider Fügepartner frei wird /11/.

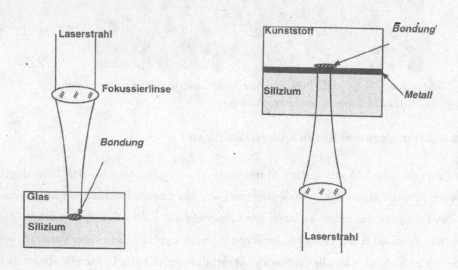

Bild 6: Fügeanordnungen zum Verbinden von Mehrschichtverbünden mit Laserstrahlung

Auf diese Art und Weise lassen sich z. B. Silizium-Glas-Verbünde erzeugen, die mit einem Minimum an Energieeinbringung verbunden sind. Anwendungen solcher Verbünde finden sich im Verschließen von Silizium-Kavitäten durch Glasdeckel zur Erzeugung von Druckmeßdosen oder im Verbinden von Kunststoff-Liga-Bauteilen auf Metallen und strukturierten Silizium-Trägern. Neben einer Verringerung der Bauteilbelastung ist mit diesem Verfahren auch eine Verkürzung der Prozeßzeiten verbunden. Die zu fügenden Bauteile werden in einem üblichen Mask-Aligner zueinander justiert und nach Abschluß des Justagevorganges kraftschlüssig miteinander verbunden. Aushärtezeiten, wie sie beim Kleben notwendig sind, entfallen, so daß nachfolgende Bearbeitungsschritte sofort einsetzen können.

5. Laser in der Mikroanalyse

Neben den verschiedenen Bearbeitungsverfahren, für die der Laser in der Mikrostrukturtechnik eingesetzt werden und zu konkreten Problemlösungen in Mikroelektronik, Mikrosystemtechnik und Sensorik beitragen kann eignet sich Laserstrahlung in besonderem Maße für die Analyse und Bewertung von Fertigungsschritten und Bauteilqualitäten. Die optische Vermessung geometrischer Größen ist hierfür ein bereits eingeführtes Verfahren und in Form interferometrischer Meßtechnologien, z.B. durch Speckle-Interferometrie, Doppelpulsholographie und Triangulation für die Oberflächenprüfung einsetzbar. Die Analyse vergrabener Strukturen kann bei entsprechender Wellenlänge mit Durchstrahlungsmethoden erfolgen. Für solche Aufgaben werden in der Regel Kurzpulslaser eingesetzt, die eine Zeitauflösung im Picosekundenbereich erlauben. Damit können Hochfrequenzbauteile hinsichtlich ihrer Eigenschaften und internen Modenverteilung analysiert werden /12/.

Bild 7: Picosekunden-Mapping eines Hochfrequenzbauteiles

Durch Abscannen bzw. Integralbelichtung des Bauteils lassen sich auch größere Strukturen analysieren und Rückschlüsse auf die Auslegung und Betriebsweise der Komponenten ziehen.

6. Trends in der Mikrostrukturtechnik

Durch Anwendungen der beschriebenen Lasertechnologie im Zusammenspiel mit verbesserten Verfahren der Mikrostrukturierung aus der Mikroelektronik können viele der zukünftigen Anforderungen aus der Mikrostrukturtechnik, Mikrosystemtechnik und Sensorik befriedigt werden. Als Kernanforderungen sind hier vor allen Dingen die hohe Integrationsdichte mit Strukturgrößen $\leq 1\,\mu m$ bei hohem Durchsatz und geringem Ausschuß zu nennen. Höhere Flexibilität und kundenspezifische Lösungen erfordern ein Einzel-Wafer-Prozessing bei ständig wachsendem Wafer-Durchmesser.

Vor allen Dingen die Mikrosystemtechnik mit einer Vielzahl aktiver und passiver Funktionen erfordert eine Veränderung im Aufbau hin zu einer hybriden oder teilhybriden Integration. Hierfür sind geeignete Strukturierungs-, Beschichtungstechniken sowie Integrationstechnologien, wie z. B. das Mikrofügen, notwendig. Der Laser kann für alle drei Kernanforderungen geeignete Verfahren liefern.

Eine Strukturierung von Schichten kann auf photolythischen, ablativen und chemisch katalythischen Prozessen bis hinunter zu 100 nm Strukturgröße erfolgen. Die durch Laserverfahren hergestellten Schichten zeigen ähnliche Eigenschaften und Schichtraten, wie die konventionellen Beschichtungsverfahren, und die Integration aller Bauteile kann durch das Mikrostruktruieren und Mikrofügen sowohl bei Werkstoffen gleicher Art als auch Werkstoffen unterschiedlicher Art mit hohen Geschwindigkeiten und Genauigkeiten erfolgen.

Zusammen mit neuentwickelten Methoden des Bauteilhandlings, der Mikropositionierung, einer auf Mikrobauteile abgestimmten Greifertechnik sowie einer erhöhten Flexibilität in der Fertigung von Mikrobauteilen durch modifizierte Fertigungstechniken kann mit dem gezielten Einsatz der zur Verfügung stehenden Laserverfahren ein neues Feld der Produktionstechnik, der Mikro-Produktionstechnik eröffnet werden. Mit diesem Ansatz lassen sich vorhandene Entwürfe für Mikrosysteme entsprechend ihrem Nutzen auch bei kleinen und mittleren Stückzahlen in marktgängige Produkte umwandeln.

7. Literatur

[1] Sensormärkte 2001
     Intechno Consulting Basel 1993

[2] J. Marek, M. Möllendorf
     Mikrotechniken im Automobil, microelectronik, Heft3/1995

[3] V. Krause, H.-G. Treusch , E. Beyer, P. Loosen,
     High power laser diodes as a beam source for materials processing,
     Laser Treatment of Materials, DGM-Verlag 1992

[4] M.D. Levenson
     Extending optical lithography to the gigabit era,
     Solid State Technology, Vol 38, No. 2

[5] U.C. Böttiger, T. Fischer, A. Grassmann,
     European Deep-UV Lithography competitive by Lambda 248 LITHO Laser
     Lambda Physik Industrial Report No. 9, January 1995

[6] J. C. Langston, G. T. Dao
     Extending optical lithography to 0.25 µm and below
     Soild State Technology, Vol. 38, No. 3

[7] J.D. Hunn, C. P. Christensen
     Ion beam and laser-assisted micromachining of single-crystal diamond
     Solid State Technology, Vol. 37, No. 12

[8] A. Gillner, M. Wehner, E.W. Kreutz, H. Frerichs, M. Mertin. D. Offenberg
     Lasergestützte Schichtabscheidung zur Herstellung funktionaler Schichten in
     Mikrosystemtechnik und Mikroelektronik,
     This Proceedings

[9] P. Horneff, H.-G. Treusch, E. Beyer, G. Herziger, D. Knödler, W. Möller,
     Temperaturgeregeltes Mikrolöten
     Laser in der Fertigung, Proceedings Laser '91, Springer Verlag

[10] G. Herziger, P. Loosen
     Neue Laserstrahlquellen für die industrielle Fertigungstechnik
     Proceedings Laser '95, Wagenbach Verlag

[11] A. Gillner,
     Laseranwendungen in Sensorik und Mikrosystemtechnik
     Laser Magazin 3/95

[12] J. Bell, H. Roskos
     Pulses map and verify integrated electronics
     Opto & Laser Europe, No. 17, February 1995

Herstellung mikromechanischer Bauteile mit Excimer-Laserstrahlung

K. Dickmann und A. Eschenburg
Labor für Lasertechnik, Fachbereich Physikalische Technik, FH Münster
Stegerwaldstraße 39, D-48565 Steinfurt

Innerhalb der Mikrosystemtechnik werden Herstellungsverfahren für mikroelektronische und mikromechanische Komponenten benötigt. Dabei können für die Herstellung von elektronischen und elektrischen Komponenten etablierte Standardverfahren aus der Mikroelektronik übernommen werden. Für die Herstellung mikromechanischer Bauteile könen je nach Material konventionelle Verfahren nur bedingt eingesetzt werden. Als innovatives Verfahren hat auch die Lasertechnik an Interesse gewonnen

1 Stand der Technik

Bei Verfahren, die auf der isotropen und anisotropen Ätztechnik beruhen, wird geeignetes Material auf ein Substrat aufgebracht und mit Fotolack überzogen. Anschließend erfolgt die Belichtung sowie das teilweise Entfernen des Materials durch Ätzen. Dieser Vorgang wir so oft wiederholt, bis die gewünschte Struktur fertiggestellt ist. Durch den Einsatz von Opferschichten, die nach der Strukturierung vollständig entfernt werden, können auch frei schwebende Bauteile erzeugt werden /1/.

Das LIGA-Verfahren kann als konsequente Weiterentwicklung der Ätztechnik gesehen werden. Der zentrale Schritt des Verfahrens ist die Röntgentiefenlithografie, bei der Mikrostrukturen mit einem Aspektverhältnis von bis zu 100:1 durch die Belichtung einer PMMA-Schicht mit kollimierter Röntgenstrahlung erzielt werden. Durch eine nachfolgende galvanische Abformung können Mikrostrukturen aus Ni, Au, Cu, sowie aus Ni-Co- und Ni-Fe-Legierungen hergestellt werden. Ähnlich wie beim Ätzen können auch hier durch den Einsatz von Opferschichten teilweise freistehende Strukturen erzeugt werden /2/. In einem neueren „Laser-Liga-Verfahren" /3/ wird die Polymerschicht auf einem Substrat direkt durch Excimer-Laserstrahlung strukturiert und anschließend wie beim konventionellen Liga-Verfahren abgeformt.

Mit Laserstrahlung werden heute im industriellen Maßstab Mikrobohrungen mit einem Durchmesser von wenigen 10 μm in verschiedenste Materialien eingebracht. Zur Herstellung von beliebigen Mikrogeometrien wird heute noch überwiegend im Forschungsstadium der Einsatz des Excimerlasers untersucht. So wurde beispielsweise in /4/ ein lasergeschnittenes Mikroflügelrad aus Keramik mit einem Durchmesser von 250 μm vorgestellt.

2 Versuchstechnik zur Mikrofertigung

Als Strahlquelle für die Versuche wird ein KrF-Excimer-Laser (λ=248 nm) eingesetzt. Der Laser hat einen Strahlquerschnitt von 25*8 mm² und erreicht bei einer Pulsfrequenz von 200 Hz eine mittlere Leistung von 40 W. Die Pulsenergie beträgt max. 200 mJ. Für die Positionierung der Substrate wird ein 3-Achsen-Positioniersystem mit einer Positioniergenauigkeit von ± 1 μm bei einem Stellweg von 100 mm eingesetzt.

944

Zur Untersuchung der Einsatzmöglichkeiten in der Mikrobearbeitung wurde im Gegensatz zu /4/ das Masken-Abbildungsverfahren mit einer afokalen Teleskopanordnung gewählt (siehe Bild 1). Es wurde mit

einem Abbildungsverhältnis 1:20 und einem 2,5-mm Pinhole zur Verbesserung der Abbildungsqualität gearbeitet. Abgebildet wurden Masken in der Form der zu erzeugenden Mikrostrukturen. Die Masken bestehen aus einem Quarzglas, auf dem eine Chromschicht aufgebracht wurde, um die Laserstrahlung der gewünsch-

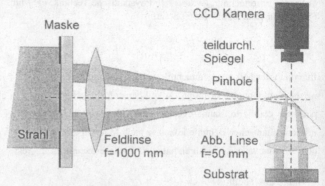

Bild 1 : Prinzip des optischen Aufbaus zur Mikrostrukturierung

ten Geometrie entsprechend abzuschatten. Durch das hohe Abbildungsverhältnis sind die Energiedichten der Laserstrahlung am Ort der Maske so gering, daß ein dünne Chromschicht (ca. 100 µm) ausreicht. Die Masken wurden mit Hilfe von konventionellen Ätztechniken hergestellt.

3 Bearbeitungsergebnisse

Unter praktischen Gesichtspunkten ist es von Interesse, wie minimale Strukturgröße und Tiefenschärfe am Bearbeitungsort voneinander abhängen. Die minimale Strukturgröße läßt sich nur mit Aufwand berechnen, da die Grenzen der Strahlenoptik im vorliegenden Fall bereits überschritten worden sind. Für eine theoretische Betrachtung müßten demnach sämtliche im Strahl vorkommenden Modi aus der Gauß'schen Strahlentheorie betrachtet werden. Daher wurde die minimale Strukturgröße, die für die Berechnung der Tiefen-

schärfe bekannt sein muß, experimentell bestimmt. Für diese Bestimmung wurden Abbildungen mit einer speziellen Testmaske eingebracht und anschließend vermessen. Aus der Vermessung der erzielten Abtragsgeometrie bestimmt sich die minimale Strukturgröße unter den vorliegenden Versuchsbedingungen zu ca. 2.3 µm.

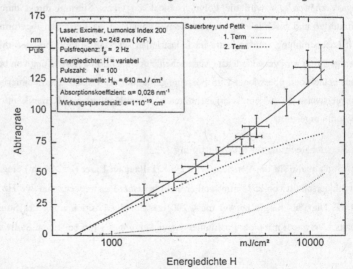

Bild 2: Abtragsraten in Saphir; Experiment und Theorie /5/

Aus der minimalen Strukturgröße r_{min} kann mit dem Pinholeradius r_P und den Abmessungen des teleskopischen Aufbaus (f_2=Brennweite der Abbildungslinse) die Tiefenschärfe z_t berechnet werden zu :

$$z_t = \frac{r_{min}}{r_p}\left(\sqrt{2}-1\right)f_2 .$$

Beim vorliegenden Aufbau ergibt sich eine Tiefenschärfe von ca. 40 µm. Wie aus der Formel ersichtlich ist, kann die Tiefenschärfe insbesondere durch den Radius des Pinholes beeinflußt werden.

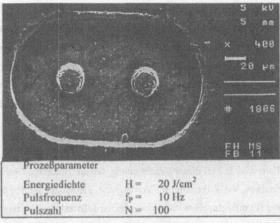

Prozeßparameter		
Energiedichte	H =	20 J/cm²
Pulsfrequenz	f_P =	10 Hz
Pulszahl	N =	100

Gleichzeitig wird aber der Energieverlust **Bild 3:** REM-Aufnahme eines Mikrogetriebegehäuses in Saphir
am Pinhole erhöht, so daß die Größe des Pinholes an die entsprechende Bearbeitungsaufgabe angepaßt werden muß.

Mit Hinblick auf eine gute Bearbeitbarkeit mit Excimer-Laserstrahlung wurde für erste Strukturuntersuchungen Polymid als Versuchswerkstoff gewählt. Anschließend wurden die Ergebnisse für spätere praktische Anwendungen auf Saphir übertragen. Während Abtragsuntersuchungen an Polyimid in der Vergangenheit umfangreich vorgestellt wurden, wurde das Abtragsverhalten von Saphir experimentell ermittelt und auch theoretisch beschrieben /5/. Bild 2 zeigt die gute Übereinstimmung des experimentellen Resultats mit einem angepaßten Modell von /6/.

Als beispielhaftes mit der vorgestellten Anordung hergestelltes Bauteil für Anwendungen in der Mikrosystemtechnik zeigt Bild 3 ein Mikrogetriebegehäuse aus Saphir zur Aufnahme von Zahnrädern. Für den Tiefenabtrag von ca. 25 µm wurden bei einer Energiedichte von H=20 J/cm² ca. 100 Pulse benötigt. Die Abtragskanten haben sich nahezu senkrecht zur Oberfläche ausgebildet.

Zur weiteren Demonstration der Bearbeitungsmöglichkeiten wurden zu dem in Bild 3 abgebildeten Getriebegehäuse passende Mikrozahnräder aus Polyimid hergestellt. Die Bauteile (s. Bild 4) mit einem Durchmesser < 100 µm weisen an der Strahlaustrittsseite (oben in Bild 4) einen scharfkantigen Abtrag auf. Hingegen sind an der Strahleintrittsseite deutliche Kantenverundungen erkennbar. Dieser Effekt läßt sich durch die im Vergleich zur

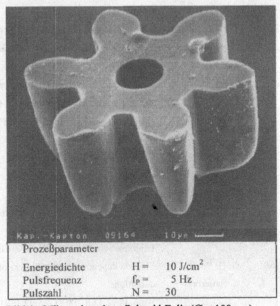

Prozeßparameter		
Energiedichte	H =	10 J/cm²
Pulsfrequenz	f_P =	5 Hz
Pulszahl	N =	30

Bild 4: Mikrozahnrad aus Polymid-Folie ($\varnothing$ < 100 µm)

vorhandenen Tiefenschärfe größere Materialstärke erklären, aber auch auf zusätzliche Beugungseffekte an der Maske zurückführen. Hierfür sind die an der Maske in Richtung der optischen Achse gebeugten Strahlen verantwortlich, während entgegengesetzt gebeugte Strahlen durch das Pinhole herausgefiltert werden. Diese Vorstellung wird dadurch bestätigt, daß die Flanken der inneren Zahnradbohrung im Gegensatz zu den Flanken der Außenkontur nahezu senkrecht sind. Um die Auswirkung der beugungsbedingten Intensitätsmodulation auf den Abtragsprozeß zu reduzieren, bietet sich das Arbeiten mit Energiedichten in der Nähe der Abtragsschwelle an.

4 Ausblick

Es war nicht Ziel dieser Untersuchung, funktionstüchtige Getriebe für den Einsatz in der Mikromechanik zu fertigen. Vielmehr stand im Vordergrund, die Möglichkeiten der Excimerlasertechnik als Fertigungsverfahren für mikromechanische Komponenten aufzuzeigen. Die repräsentativ dargestellten Bauteile stellen erzielte Minimalgrößen dar. Für Bauteile mit größerer Geometrie für praktische Anwendungen in der Mikrosystemtechnik läßt sich dieses Verfahren ebenfalls einsetzen, soweit eine damit verbundene reduzierte Energiedichte oberhalb der materialspezifischen Abtragsschwelle liegt. Die vorgestellten Ergebnisse lassen sich auf eine Vielzahl weiterer Materialien übertragen, vor allem keramische- und Halbleiter-Materialien.

5 Danksagung

Die Arbeiten wurden vom Land NRW im Rahmen eines Verbundforschungsvorhabens „Mikrosystemtechnik" sowie eines gleichnamigen Forschungsschwerpunktes an der FH Münster unterstützt. Die Autoren bedanken sich bei Prof. Dr. B. Lödding, Leiter des Labors für Metallographie im Fachbereich Physikalische Technik der FH Münster.

6 Literatur

/1/	S. Büttgenbach	Mikromechanik, B.G. Teubner Verlag
/2/	N.N.	Broschüre : - LIGA - Bewegliche Mikrostrukturen, Kernforschungszentrum Karlsruhe
/3/	N.N.	Laser Liga : Universelles Verfahren für mikrooptische und mikromechanische Anwendungen (BMFT-Vorhaben, FKZ 13N6164)
		Broschüre : Materialbearbeitung mit Excimerlasern, VDI TZ, Düsseldorf, 3.4.1995, S. 28-32
/4/	M. Gonschoir,	Oberflächenmodifikation von Keramik mi Excimerlaserstrahlung
	H. Kappel	Tagung : „Anwendung der Lasertechnik", 11. Int. Mittweidaer Fachtagung, 24.-26.11.1994
/5/	R. Börger	Oberflächenbearbeitung durch Excimer-Laser-Ablation und ihre Charakterisierung mittels Kraftmikroskopie,
		Diplomarbeit, Westfälische Wilhelms Universität Münster, FH Münster, 1995
/6/	G. H. Pettit	Fluence-dependent transmission of polyimide at 248 nm under laser ablation
	R. Sauerbrey	conditions
		Appl. Phys. Lett. 58(8), 25.02.1991, 793-795

Mikrostrukturierung von Metallen und Halbleitern durch ultrakurze Laserimpulse

J.Jandeleit[1], G.Urbasch[1], D.Hoffmann[1], H.G.Treusch[2] und E.W.Kreutz[1]

[1] Lehrstuhl für Lasertechnik der Rheinisch-Westfälischen Technischen Hochschule,
Steinbachstraße 15, D-52074 Aachen, FRG

[2] Fraunhofer-Institut für Lasertechnik, Steinbachstraße 15, D-52074 Aachen, FRG

1. Einleitung

Intensive, ultrakurze Laserpulse im ps-Bereich können zur präzisen Mikrostrukturierung unterschiedlichster Werkstoffe verwendet werden /1/. Ein Vorteil der kurzen Pulsdauer liegt in der geringeren thermischen Belastung des Materials aufgrund der reduzierten Wärmeeindringtiefe. Die durch kurze Pulse erreichbaren hohen Leistungsdichten können die Verdampfungsrate beim thermischen Abtragen steigern und den Schmelzanteil verringern, was zu einer höheren Präzision der Bearbeitung führen kann.

Das Abtragen von Kupfer und Silizium wird an Hand von Bohrungen charakterisiert. Für Kupfer wird die Abhängigkeit des Bohrlochdurchmessers und der Bohrlochtiefe von der Leistungsdichte und der Pulsanzahl, sowie die Bohrgeschwindigkeit bestimmt. Für einkristallines Silizium wird der Wirkungsgrad beim Bohren in Abhängigkeit von der Leistungsdichte und der verwendeten Laserwellenlänge gemessen. Die erzeugten Strukturen werden mit Lichtmikroskopie und REM analysiert. Der Einfluß der Laserwellenlänge und des laserinduzierten Plasmas auf das Abtragen und die Strukturierbarkeit werden diskutiert. Die Untersuchungen führen zu einem besseren Verständnis der grundlegenden Wechselwirkungsmechanismen ultrakurzer Pulse mit Festkörpern.

2. Experimenteller Aufbau

Zur Erzeugung der ultrakurzen Pulse wird ein modengekoppelter Nd:YLF-Laser mit nachgeschaltetem regenerativen Verstärker verwendet. Die Pulsdauer beträgt t_p = 30 ps und die maximale Pulsenergie E = 1 mJ. Die Grundwellenlänge von λ = 1053 nm und die frequenzverdoppelte Strahlung der Wellenlänge λ = 527 nm stehen zur Verfügung. Der Laserstrahl wird durch eine Strahlaufweitung und eine Pockelszelle geführt, und dann auf die Werkstückoberfläche fokussiert. Einzelne Pulse werden mit der Pockelszelle separiert. Die Repetitionsrate beträgt f_p = 1 kHz. Die Positionierung der Proben erfolgt über eine PC-gesteuerte x-y-Verschiebeeinheit.

3. Experimentelle Ergebnisse und Diskussion

Der Einfluß der Pulsdauer auf den Bohrprozeß wird an einem Vergleich der Bearbeitungsergebnisse von Bohrungen, die mit unterschiedlich langen Pulsen hergestellt wurden, ersichtlich. Abbildung 1 zeigt zwei REM-Aufnahmen von Bohrungen in Kupfer, die mit einer Pulsdauer von t_p = 30 ns und t_P = 30 ps hergestellt wurden. Der Bohrlochdurchmesser d_m beträgt in beiden Fällen 20 µm. Bei der mit ns-Pulsen erzeugten Bohrung treten in einem großen Bereich der Bohrlochumgebung Schmelzablagerungen auf. Das Material wird überwiegend im schmelzförmigen Zustand ausgetrieben. Demgegenüber ergibt sich bei der mit ps-Pulsen hergestellten Bohrung ein wohldefinierter schmelzförmiger Bohrlochrand. Die Schmelzablagerungen in der Bohrlochumgebung sind stark reduziert. Die Verwendung von ps-Pulsen ergibt demnach eine Verringerung der Schmelzablagerungen, was im Vergleich zu den ns-Pulsen zu einer höheren Bearbeitungspräzision in Form von geringerer Konizität der Bohrung und einer definierteren Bohrlochgeometrie führen kann.

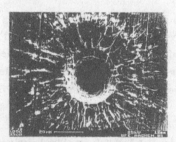 

t_p = 30 ns, I = 3.4x10^{11} W/cm² t_p = 30 ps, I = 1.1x10^{12} W/cm²

Abb. 1: Makroaufnahmen von Bohrungen in Kupfer, d_m = 20 µm, B.tiefe = 8 µm, 2 Pulse

Zur Charakterisierung des Wechselwirkungsprozesses der kurzen Pulse mit Metallen wurde die Bohrgeschwindigkeit v_d in unterschiedlich dicken Kupferfolien bestimmt und mit der berechneten Bohrgeschwindigkeit gemäß /2/

$$v_d = \frac{A \cdot I}{\rho \left(c\, T_v + L_m + L_v \right)} \tag{1}$$

verglichen. Hierbei bedeuten A Absorptionsgrad (A{Cu, 527 nm} = 0.38), I Leistungsdichte, ρ spez. Gewicht, c spez. Wärme, T_v Verdampfungstemperatur, L_m Schmelzwärme, L_v Verdampfungswärme. Abbildung 2 zeigt die Bohrgeschwindigkeiten in unterschiedlich dicken Kupferfolien als Funktion der Leistungsdichte. Bis zu einer Leistungsdichte von I = $1 \cdot 10^{12}$ W/cm² steigt die Bohrgeschwindigkeit nahezu linear mit zunehmender Leistungsdichte an. Die Übereinstimmung mit der berechneten Geschwindigkeit ist gut. Oberhalb von I = $1 \cdot 10^{12}$ W/cm² zeigen die Bohrgeschwindigkeiten ein Sättigungsverhalten. In diesem Bereich muß sich der Oberflächenzustand des Materials während der Einwirkungsdauer verändern, sodaß es zu einer Änderung der Energieeinkopplungesmechanismen kommt.

Abbildung 2 zeigt ebenfalls die zugehörigen Bohrlochdurchmesser d_m als Funktion der Leistungsdichte im Vergleich zum berechneten Bohrlochdurchmesser. Der berechnete Durchmesser ergibt

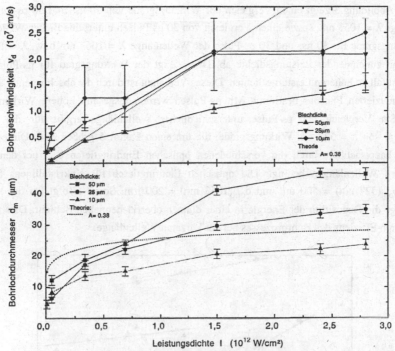

Abb. 2: Bohrgeschwindigkeit und Bohrlochdurchmesser in Kupfer
$\lambda = 527$ nm, $t_p = 40$ ps, $r_f = 7.25$ µm

sich unter der Annahme eines Gaußförmigen Strahlprofils, eindimensionaler Wärmeleitung und einer Oberflächenquelle zu

$$d_m = r_f \cdot \sqrt{2 \ln\left(\frac{4\, A\, I\, \sqrt{t_p}}{T_v\, \rho\, c\, \sqrt{\pi\, \kappa}}\right)}. \tag{2}$$

Hierbei bedeuten κ Temperaturleitfähigkeit und r_f Fokusradius. Unterhalb einer Leistungsdichte von $I = 1 \cdot 10^{12}$ W/cm² erreichen die experimentellen Werte nicht den theoretisch berechneten Bohrlochdurchmesser. Die gesamte Energie wird in diesem Bereich wahrscheinlich am Bohrlochgrund absorbiert. Oberhalb von $I = 1 \cdot 10^{12}$ W/cm², in dem Bereich, wo die Bohrgeschwindigkeiten nicht weiter ansteigen, übersteigen die experimentell bestimmten Bohrlochdurchmesser den theoretischen Wert. Somit kommt es in diesem Bereich zu einer verstärkten Energieeinkopplung in die Bohrlochwand. Dieses Verhalten kann zum Beispiel durch ein oberflächennahes Plasma erklärt werden, welches bei den hohen Leistungsdichten entsteht und die Energie über die Plasmaoberfläche homogen in den Bohrlochgrund und die Bohrlochwand einkoppelt.

Die Wechselwirkung ultrakurzer Laserpulse mit Halbleitern wird an einkristallinem Silizium untersucht. Der Wirkungsgrad beim Bohren η_v wird als Funktion der Leistungsdichte gemäß

$$\eta_v = \frac{abgetragenes\ Volumen \cdot Verdampfungsenergiedichte}{Laserpulsenergie}$$

bestimmt. Abbildung 3 zeigt einen Vergleich der Wirkungsgrade von 30 ns- und 30 ps-Pulsen der Wellenlänge λ = 1053 nm, sowie einen Vergleich von 30 ps-Pulsen unterschiedlicher Wellenlänge. Die Wirkungsgrade für 30 ns- und 30 ps-Pulse der Wellenlänge λ = 1064 nm bzw. λ = 1053 nm nehmen mit zunehmender Leistungsdichte ab. Dabei zeigt der Wirkungsgrad für ns-Pulse einen größeren Abfall zu höheren Leistungsdichten. Dieses Verhalten ist durch die abschirmende Wirkung des laserinduzierten Plasmas bestimmt. Mit ps-Pulsen werden insgesamt höhere Wirkungsgrade erreicht. Ein Vergleich der ps-Pulse unterschiedlicher Wellenlänge ergibt für die kürzere Wellenlänge von λ = 527 nm Wirkungsgrade, die um einen Faktor 5 größer sind. Dieser Unterschied ist hauptsächlich durch die verschiedenen optischen Eindringtiefen δ_{opt} bei den beiden verwendeten Wellenlängen bedingt. Die optischen Eindringtiefen für einkristallinens Silizium betragen δ_{opt}(527 nm) = 500 nm und δ_{opt}(1053 nm) = 200 µm. Somit erfolgt bei der kürzeren Wellenlänge die Deposition der Energie in einer dünnen oberflächennahen Schicht. Dies führt zu einer höheren Effizienz des Bohrprozesses bei der kürzeren Wellenlänge.

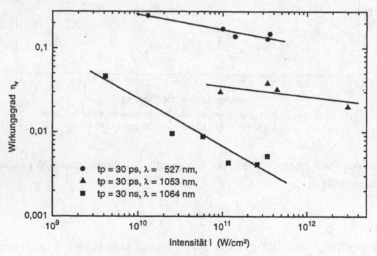

Abb. 3: Wirkungsgrad als Fkt. der Leistungsdichte für Silizium

4. Zusammenfassung

Für Kupfer und einkristallines Silizium wurde das Abtragen bei der Bestrahlung mit 30 ps-Pulsen untersucht. Durch die Verwendung von ps-Pulsen werden die Schmelzablagerungen in der Bohrlochumgebung reduziert. Die Bohrgeschwindigkeit ist zu hohen Leistungsdichten begrenzt, in diesem Bereich ergibt sich eine erhöhte Energieeinkopplung in die Bohrlochwand. Im Vergleich zu ns-Pulsen konnte ein höherer Wirkungsgrad beim Bohren durch ps-Pulse erzielt werden.

5. Referenzen

/1/ W.Kautek, J.Krüger, SPIE Proc. 2207 (1995) 600.
/2/ M.von Allmen, J.Appl.Phys. 47 (1976) 5460.

Mikrostrukturierung sprödharter Werkstoffe mit Excimer-Laserstrahlung

H.K. Tönshoff, H. Kappel, L. Overmeyer
Laser Zentrum Hannover e.V.
D-30419 Hannover

Einleitung

Excimer-Laser haben sich im Bereich der Forschung in einer Vielzahl von Anwendungen etabliert. Die kurze Pulsdauer von wenigen Nanosekunden ermöglicht eine Materialbearbeitung mit sehr geringer thermischer Beanspruchung. Der Excimer-Laser besitzt somit die Voraussetzung thermoschockempfindliche Werkstoffe, wie Keramik und Glas, schädigungsarm zu bearbeiten. Bei der Kunstoffbearbeitung dominiert infolge der hohen Photonenenergie der kurzwelligen elektromagnetischen Strahlung der nichtthermische photochemische Abtrag. Durch diese Eigenschaften hat der Excimer-Laser auch in verschiedenen Bereichen der industriellen Fertigung Einzug gehalten [GEDR92]. Meist beschränken sich die Anwendungen noch auf verfahrenstechnisch einfache Prozesse, die ausschließlich auf ein spezielles Bauteil und einen Werkstoff abgestimmt sind. Auf das Potential des Lasers zur flexiblen Fertigung präziser Mikrobauteile wird jedoch nicht zurückgegriffen [TÖNS94]. Nachfolgend wird eine für diese Aufgaben entwickelte Bearbeitungsstation vorgestellt und die Strukturierungsmöglichkeiten anhand von zwei Beispielen erläutert. Ergänzend wird eine weitere UV-Laserstrahlquelle, ein frequenzvervierfachter Nd:YAG-Laser, vorgestellt, mit dem Strukturgrößen im Submikrometerbereich zu erzielen sind.

Anlagentechnik

Die oben genannten Eigenschaften prädestinieren den Excimer-Laser zur Bearbeitung sprödharter Werkstoffe. Darüber hinaus kann die UV-Strahlung mit optischen Systemen auf sehr geringe laterale Flächen gebündelt werden. Über die Kombination von Maskenprojektion und teleskopischer Abbildung, z.B. nach Kepler, können minimale Strahlabmessungen am Werkstück von ca. 5x5 µm² erzeugt werden. Flächige Strukturen werden durch Variation des optischen Strahlengangs, der Maskengeometrie oder die gleichzeitige Relativbewegung zwischen Werkstück und Werkzeug hergestellt. Sollen mit diesem Verfahren Mikrostrukturen mit hoher Präzision und Reproduzierbarkeit geschaffen werden, ergibt sich daraus die Forderung nach einer flexiblen Bearbeitungsstation mit einer Auflösung und Genauigkeit im Mikrometerbereich. Da entsprechende Systeme kommerziell nur eingeschränkt verfügbar sind, wurde am Laser Zentrum Hannover eine Werkzeugmaschine zur Mikrostrukturierung mit Excimer-Laserstrahlung (*ELPECµ; Excimer Laser Precision Engineering Center*) entwickelt und gebaut, Bild 1. Wesentliche Bestandteile der Bearbeitungsstation sind die 8-Achsen CNC-Steuerung, zwei

Bild 1: Mikrobearbeitungsstation für Excimer-Laser

die flexiblen Masken zur Strahlformung und eine CAD/CAM-Software, die u.a. d
vollautomatische Steuerung des Excimer-Lasers erlaubt. Zur Strahlformung wird eine numerisc
gesteuerte Maske eingesetzt, bei der vier unabhängig voneinander bewegbare Schneiden d
Excimer-Rohstrahl begrenzen. Auf diese Weise kann der Strahlquerschnitt und damit d
Abbildungsfläche am Werkstück während der Bearbeitung stufenlos variiert werden. Nich
rechteckige Strahlquerschnitte können durch vorgefertigte Revolvermasken im Strahlengan
positioniert und ebenfalls während des Bearbeitungsprozesses bewegt werden [TÖNS93].

Der wachsende Bedarf an miniaturisierten Bauteilen im Bereich der Halbleiter- un
Mikrosystemtechnik hat in den letzten Jahren dazu geführt, daß neben dem Excimer-Laser ander
UV-Laserstrahlquellen für den industriellen Einsatz qualifiziert wurden. In diesem Zusammenhan
sind vor allem der frequenzvervierfachte Nd:YAG-Laser (λ=266 nm; f_p=1-15.000 Hz) und de
frequenzverdoppelte Ar^+-Laser (λ=244-257 nm; kontinuierliche Strahlung) hervorzuheben. Di
Strahlung dieser Systeme wird auf das Werkstück fokussiert und erreicht bei Objektiven mit hoh
numerischer Apertur einen Fokusdurchmesser im Submikrometerbereich. In Verbindung m
zahlreichen Optimierungen, die auch nach der Frequenzkonversierung noch eine hoh
Strahlqualität sicherstellen, konnte mit einer numerischen Apertur von 0,9 ein Abtragsdurchmesse
von 0,18 µm erzielt werden. Ein Einsatzgebiet dieser Laserstrahlquelle ist z.B. die Reparatur vo
Photolithographiemasken.

Laser- und Prozeßdaten	
Laser : KrF-Excimer Wellenlänge : λ = 248 nm Pulsdauer : t$_l$ = 25 ns (FWHM)	Pulsfrequenz : f$_p$ = 100 Hz Werkstoff : AlN

Vorschubgeschwindigkeit : v$_f$ = 1,2 mm/min Abbildungsmaßstab : m = 2000 : 50 Strahlfleck □ : 15 x 15 µm^2 Anzahl der Durchläute : 20	Vorschubgeschwindigkeit : v$_f$ = 0,6 mm/min Abbildungsmaßstab : m = 2000 : 25 Strahlfleck □ : 8 x 8 µm^2 Anzahl der Durchläufe : 30

3 02561-31 He

© LZH 1994

Bild 2: Schneiden von Mikroflügelrädern aus Keramik

Laser : Excimerlaser Material : Glas (FG 375) Wellenlänge : λ = 248 nm Energiedichte : H = 10 J/cm^2 Pulszahl : n = 35999

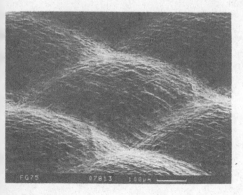

Bearbeitungszeit : t$_{ges}$ = 18 min mittlere Frequenz : f$_m$ = 33,3 Hz

Anwendungen

Die beiden nachfolgenden Applikationen sind mit Excimer-Laserstrahlung und Verwendung ~
ELPECµ gefertigt worden. Bild 2 zeigt Mikroflügelräder, die direkt aus Aluminiumni~
ausgeschnitten wurden. Das Bauteil in der linken Bildhälfte weist einen Gesamtdurchmesser v
250 µm bei einer Materialstärke von 70 µm auf [HESS94], [KAPP94]. Über eine Verdopplu~
des Abbildungsverhältnisses wurde der Durchmesser noch einmal halbiert. Eine weit~
Anwendung ist die Erzeugung dreidimensionaler Strukturen, hier am Beispiel von Mikrolin~
dargestellt, Bild 3. Die Oberflächen besitzen noch eine verfahrensbedingte Restrauheit, die du~
einen nachfolgenden Glättungsschritt auf optische Qualität verringert werden kann. I
Anwendungen dieser Mikrolinsen liegen vor allem in der Telekommunikation und Medizintech~
[OVER95].

Zusammenfassung

Excimer-Laser und frequenzkonvertierte Laser eignen sich in hohem Maße zur Mikrobearbeitu~
Mit dem Excimer-Laser können Präzisionsbauteile aus Polymeren, Keramiken u
Glaswerkstoffen hergestellt werden, die völlig neue Möglichkeiten in der Prototypen- u
Kleinserienferigung eröffnen. Darüber hinaus verhilft die Entwicklung weiterer UV-Las~
strahlquellen zu vielversprechenden Strukturierungsmöglichkeiten im Submikrometer-Bereich.

Literaturverzeichnis

GEDR92   Gedrat, O.: Strukturierung technischer Keramik mit Excimer-Laserstrahlung, Dr.-In~
         Diss. Universität Hannover,1992

HESS94   Hesse, D.: Abtragen und Schneiden von Kohlefaserverbundwerkstoffen mit Excime~
         Laserstrahlung, Dr.-Ing. Diss. Universität Hannover,1994

KAPP94   Kappel, H., Gonschior, M.: Oberflächenmodifikation und Mikrobearbeitung n~
         Excimer-Laserstrahlung, 11. Int. Mittweidaer Fachtagung, Mittweida, 11/1994

OVER95   Overmeyer, L., Mommsen, J., Alvensleben v., F.: Neues Fertigungsverfahren für d~
         Mikrooptik und Mikrosensorik, Jahrbuch für Optik und Feinmechanik, 1995

TÖNS93   Tönshoff, H.K., Hesse, D., Mommsen, J.: Micromachining Using Excimer Lase~
         Annals of the CIRP Vol. 42/1/1993

TÖNS94   Tönshoff, H.K., Hesse, D., Kappel, H., Mommsen, J,: Excimer Laser Systems, LAN~
         '94, Vol. II, 12.-13. Okt. 1994, Erlangen

Lateral Structuring of Thin Semiconductor Films by Laser Interference Techniques

M. Stutzmann, C. E. Nebel, G. Groos, J. Zimmer, B. Dahlheimer, and M. Kelly
Walter Schottky Institut, Technische Universität München,
Am Coulombwall, 85748 Garching

Abstract

The use of single Nd:YAG laser pulses for lateral structuring of Si and GaAs in the sub-µm range is described. Using threshold phenomena such as recrystallization or melting, structural sizes as small as $\lambda/10$ can be realized over large areas by exposure to a periodic interference pattern. Structural and electronic properties of the obtained samples as well as possible applications (e.g. in photovoltaics and integrated optics) are also briefly mentioned.

Introduction

Lasers have a wide range of potential applications in modern semiconductor physics, including laser recrystallization, thermal processing, deposition and etching, structuring and, of course, characterization [1,2]. Most of these applications deal with large area processes or with laser writing techniques, which are limited in lateral resolution by the wavelength, λ, of the laser source. With the increasing importance of nanometer-scale semiconductor devices, however, there is also an increasing need for lateral structuring methods with a resolution of 100 nm or less. For example, such a resolution is necessary to process 2-dimensional electron gases prepared by hetero-epitaxy further to 1-dimensional quantum wires or 0-dimensional quantum dots. Today most of this nanostructring is done by electron beam or focused ion beam lithography, which have the disadvantage of being serial and thus rather time-consuming processes. An alternative approach for the realization of periodic structures is holographic lithography using UV lasers for the preparation of etching masks. In particular in combination with spatial period division and other frequency doubling schemes, structures with 100 nm period have been demonstrated [3,4]. On the other hand, the use of UV lasers for such purposes is less desirable in view of a number of practical problems, such as the small coherence length of excimer lasers, safety considerations, requirements for optical components, etc. More importantly, the specific optical properties of common semiconductor materials can be exploited much better for excitation wavelengths in the range 700 - 350 nm (2 - 4 eV).

In order to reconcile the wavelength requirement of $\lambda \approx 500$ nm with structure sizes of ≤ 100 nm, we make use of the contrast enhancement provided by direct structural modification of semiconductors exposed to intense (pulsed) laser radiation. The basic processes to be considered are shown schematically in Fig. 1. In the simplest case, a one-dimensional sinusoidal variation of the light intensity (and thus of the temperature rise due to absorption in the semiconductor film) is provided by the interference of two coherent laser beams brought together on the sample surface with an angle α between both beams. The period Λ of the intensity grating is then given by

$$\Lambda = \frac{\lambda}{2 \sin \alpha/2} \quad , \quad \frac{\lambda}{2} \leq \Lambda < \infty$$

where λ is the laser wavelength, and the spatial period Λ can be changed over a large range by adjusting the angle α and/or the wavelength λ. Thus, it is possible to optimize the absorption properties of the laser radiation (determined by λ) independent of the spatial period Λ. Contrast enhancement can occur either because of a threshold phenomenon, such as recrystallization of an amorphous layer or local melting, which requires the local intensity to exceed a certain limit, or because of a nonlinear dependence of a process on the local temperature. A typical example for the second possibility is diffusion of impurities or constituent atoms [5], which usually has an exponential dependence on the local temperature :

$$D(x) = D_0 \, exp \left(-\frac{E_{act}}{kT(x)} \right)$$

The main aim of the investigations described below is to explore the limitations of laser-induced sub-μm structuring via these contrast-enhancing methods for different semiconductor systems. Depending on the specific applications in mind, not only the structural, but also electronic properties of the laser-treated samples are of interest. Most experimental results deal with the recrystallization of amorphous Si on glass, but a few results pertaining to crystalline Si and GaAs will also be mentioned. Finally, the relevance of lateral structuring for future applications will be discussed briefly.

Experiment

Fig. 2 shows the principal components of the laser system employed in our experiments. The basic laser is an injection-seeded, Q-switched Nd:YAG laser emitting pulses of about 8 ns duration with a repetition rate of 30 Hz and pulse power of 1 J. The fundamental frequency is then doubled, tripled and quadrupled with KD*P crystals. The third harmonic (355 nm, 250 mJ/pulse) is used to pump an optical parametric

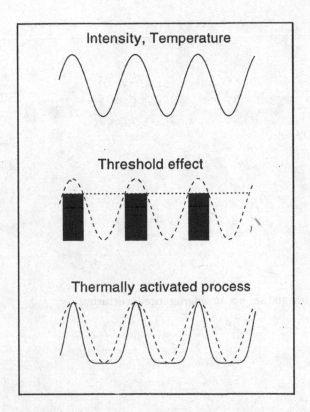

Fig. 1: Spatial variation of the intensity and temperature caused by the exposure to a laser interference pattern. The use of threshold phenomena and thermally activated processes for contrast enhancement are shown schematically.

oscillator operating between 430 and 2000 nm with a typical power of 50 mJ/pulse. All experiments reported here were performed with the second harmonic of the ND:YAG laser at 532 nm. Using $\lambda/10$ optical components for the laser interference stage, homogeneous laser structuring can be achieved typically on areas of about 3 mm x 3 mm.

The laser-treated samples are characterized by a number of methods, including x-ray diffraction, microscopy (optical, SEM, AFM), Raman scattering, infrared reflectivity and Hall measurements.

Results and Discussion

As an example for contrast enhancement via a threshold phenomenon, we show in Fig. 3 an AFM analysis of a 50 nm thick amorphous Si film on a glass substrate. The

958

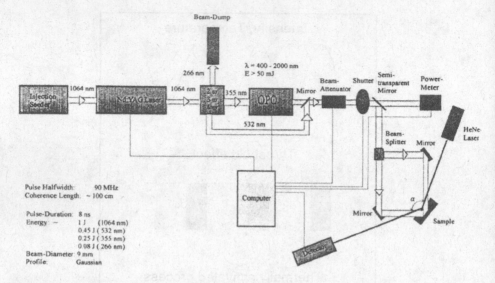

Fig. 2: Experimental set-up for interference structuring

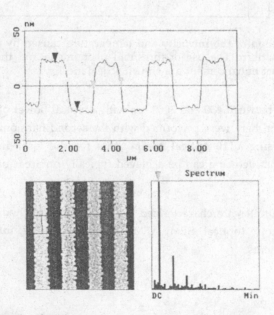

Fig. 3: Atomic Force Microscope analysis of a 50 nm thick amorphous Si film recrystallized with an interference period of 2.5 μm.

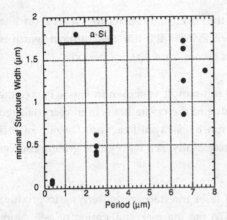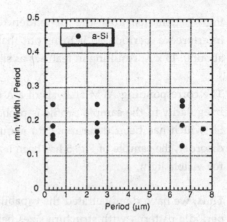

Fig. 4: Dependence of the minimal structure width of recrystallized a-Si on the interference period, Λ.

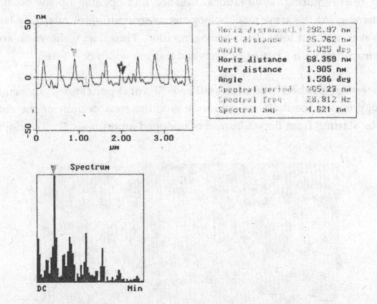

Fig. 5: AFM analysis of a recrystallized sample with a period Λ = 400 nm. The width of the obtained stripes is about 70 nm.

film was partially recrystallized by laser interference with a period of 2.5 μm, followed by a preferential etch of the remaining amorphous tissue in KOH. The AFM picture clearly shows the step like structure of the crystallized stripes as expected from Fig. 1. The width of these stripes can be changed continuously by varying the intensity of the laser pulse used for the recrystallization. For the 2.5 μm period stripes a width of 400 nm or less can be produced routinely.

Fig. 4 demonstrates the dependence of the minimal structure width on the interference period, Λ. It turns out that down to $\Lambda = 400$ nm the minimal width is about $0.15 \times \Lambda$, resulting in features as small as 70 nm for a 400 nm period.

The corresponding AFM analysis of a $\Lambda = 400$ nm sample is shown in Fig. 5. Part of the irregularity in the sample profile is probably due to the crystallite size in recrystallized Si, which has been determined to about 70 nm by x-ray diffraction. Despite of this disorder, the sample of Fig. 5 has been used successfully as a waveguide grating coupler for visible light.

Thus, we have demonstrated the capability of laser interference patterning to produce periodic patterns with structure sizes below 100 nm via recrystallization of a-Si. Non-periodic structures could be produced in a similar way with the use of holograms. An attractive feature of the method is that structuring can be done with a single shot of ns-duration, thus requiring no vibrational isolation and opening up the possibility of in-situ patterning. At the same time, homogeneous structuring of relatively large areas is possible even without laser beam homogenization. Thus, Fig. 6 shows an area of about 1 mm^2 structured with a period of 7 µm and a stripe width of ≈ 1 µm.

An interesting property of recrystallized Si for applications in semiconductor technology is the possibility to achieve a simultaneous doping of the recrystallized regions by starting from doped instead of undoped amorphous Si . The doping level of

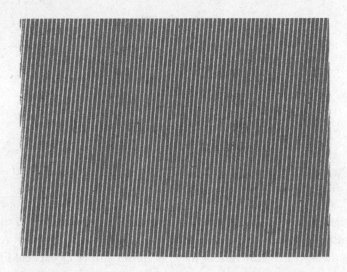

Fig. 6: Optical microscope picture of recrystallized a-Si on glass (period 7 µm). The total area is about 1 mm^2.

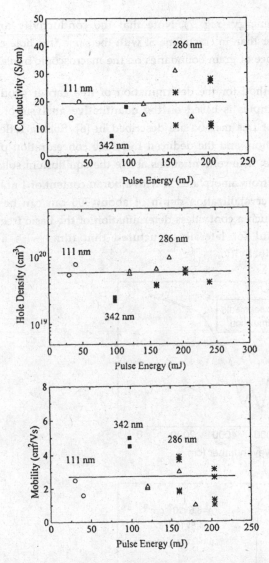

Fig. 7: Hall properties (conductivity, free hole density, and mobility) of boron-doped (1000 ppm) amorphous Si films with different thicknesses recrystallized by a single laser pulse (at 532 nm) as a function of pulse energy.

the recrystallized regions is then determined by the dopant content of the starting amorphous film and the efficiency of dopant activation. For the case of crystallization by a single 8 ns pulse at 532 nm, the latter efficiency turns out to be close to unity. It is possible to obtain highly doped microcrystalline layers with conductivities above 20 S/cm. The transport characteristics are rather insensitive to the sample thickness and the laser pulse energy (Fig. 7). Note that the conductivity and the mobility are significantly lower than in crystalline Si with the same free hole concentration. This is due to the influence of grain boundaries on the macroscopic transport properties.

A contactless method for the determination of free carrier profiles in only partially recrystallized samples is based on the quantitative analysis of infrared reflectivity spectra. Details of this method are described in [6]. Such a reflectivity spectrum, its computer simulation, and the deduced free hole concentration profiles are shown in Fig. 8. In this case, hole concentrations above the equilibrium solubility of boron were obtained starting from amorphous Si with a boron content of 1 at. % and a thickness of 1 μm. Also, the crystallization depth of about 300 nm can be estimated from the quantitative fit. Such a contactless determination of the basic free carrier properties is particularly useful for laterally structured thin films with a highly anisotropic macroscopic conductivity.

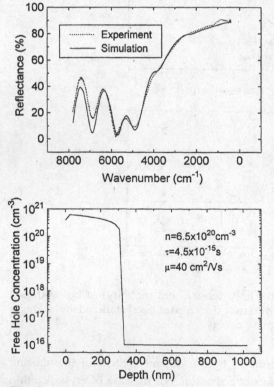

Fig. 8: Infrared reflectance (experiment and simulation) of a 1000 nm thick a-Si sample doped with 1 at.% boron on glass. The sample was recrystallized with 10 shots of energy 140 mJ/cm². The lower part of the figure shows the deduced depth profile of the free carrier concentration.

In addition to lateral structuring by recrystallization, we are also investigating the possibility of local laser-induced diffusion e.g. of doping atoms as a further way of contrast enhancement in direct laser structuring. Significant diffusion due to a single laser pulse can only occur when the sample surface is essentially molten. As shown by the reflectivity transients in Fig. 9, melting of crystalline Si by a 8ns pulse at 532 nm requires an energy density of approximately 100 mJ/cm^2, and resolidification occurs in about 20 - 30 ns. In GaAs, where the absorption depth of 532 nm light is about five times smaller, already about 30 mJ/cm^2 are sufficient to cause melting, and the resolidification times are between 30 and 100 ns.

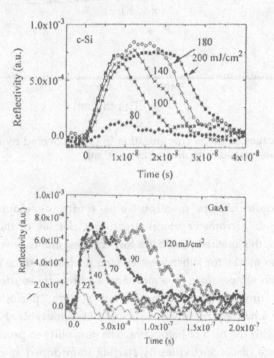

Fig. 9: Reflectivity transients due to melting of crystalline Si and GaAs surfaces after laser irradiation at 532 nm with different pulse energy densities.

As an example of laser pulse induced diffusion, Fig. 10 shows SIMS profiles of Si in GaAs subjected to one or three laser shots with an energy of 60 mJ. The Si source was in this case a thin amorphous Si film deposited on the surface by plasma-assisted CVD. The observed diffusion depth of about 100 nm agrees well with the absorption depth of the laser radiation. Lateral structuring with a period of 400nm is detectable in such samples by optical diffraction and AFM analysis, but further details still need to be investigated.

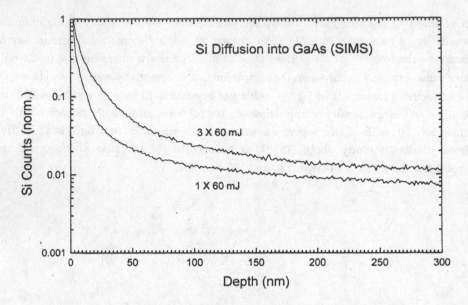

Fig. 10: SIMS profiles of the Si concentration in GaAs covered by a thin a-Si layer after one and three laser pulses with an energy of 60 mJ.

Finally, we would like to mention some future developments and possible applications of sub-micrometer lateral structuring, notably of thin amorphous silicon films. First off all, this method provides a low cost, fast and easy way to produce optical gratings, but also masks for subsequent etching processes, since amorphous Si can be deposited at room temperature on almost any substrate by sputtering or plasma-CVD. Using the 355 nm laser radiation of our system, periods of 200 nm with correspondingly smaller structure sizes should be achievable. Non-periodic patterns can be realized with the help of holograms. The possibility to produce doped structures with a wide range of conductivities by starting from doped amorphous Si is also an attractive variant. Currently we investigate the possibility to incorporate regular arrays of highly doped Si dots with a period of about 400 nm and a diameter around 100 nm as point contacts in large area amorphous Si solar cells. Such dot patterns, produced by rastering the solar cell substrate with single shots of three or four interfering beams, could replace the expensive structured transparent conducting oxide currently used to optimize light trapping and carrier collection in advanced a-Si cells.[7] Another interesting application is that of a grating coupler in integrated optics. As a first example, we have used the structure shown in Fig. 5 to couple green and red HeNe laser beams into an SiO_2 waveguide at almost normal incidence. Although the structured layer is only 20 nm thick, coupling efficiencies of several percent have been achieved. Interesting is here the possibility to combine such a thin a-Si layer with SiO_2

or oxynitride based integrated optics, so that coupler gratings can be produced after deposition at any position and with variable periods by simple laser exposure.

Acknowledgements:

We gratefully acknowledge financial support by Deutsche Forschungsgemeinschaft (SFB 348) and by Bundesministerium für Bildung, Wissenschaft, Forschung und Technologie under contract 0329617. We also would like to thank M. Heintze for providing some of the samples used in this study.

References:

[1]   Laser Annealing of Semiconductors, edited by J. M. Poate, J. W. Mayer
       (Academic Press, New York, 1982)

[2]   Semiconductor Processing and Characterization with Lasers, Mat. Science
       Forum Vol. 173 - 174 (1995)

[3]   D. C. Flanders, A. M. Hawrylak, H. I. Smith, J. Vac. Sci. Technol. 16, 1949 (1979)

[4]   E. H. Anderson, C. M. Horwitz, H. I. Smith, Appl. Phys. Lett. 49, 834 (1983)

[5]   Laserinduzierte Herstellung und Charakterisierung niedrigdimensionaler
       GaAs/AlGaAs-Strukturen
       PhD thesis by K. F. Brunner, Technische Universität München (1993)

[6]   C.N. Waddell, W.G. Spitzer, G.K. Hubler, J.E. Fredrickson, J. Appl. Phys. 53, 5851
       (1982)

[7]   M. Heintze, P.V. Santos, C.E. Nebel, M. Stutzmann, Appl. Phys. Lett. 64, 3148
       (1994)

Femotosecond-Pulse Laser Micromachining of Metal Layer Composites

Jörg Krüger, Wolfgang Kautek*
Laboratory for Laser and Chemical Thin Film Technology
Federal Institute for Materials Research and Testing (BAM), D-12200 Berlin, Germany

Markus Staub, Günther G. Scherer
Department of General Energy Technology, Paul Scherrer Institute (PSI), CH-5232 Villigen PSI, Switzerland

Abstract:

Femtosecond-pulse lasers allowed precise micromachining of platinum-gold layer composite model electrodes for polymer electrolyte fuel cells (PEFC). In contrast, conventional nanosecond-pulse lasers caused serious melting, splashing and debris-contamination of the catalytical Pt top layer. The heat affected zone (HAZ) determined the quality of the ablation process. It could be reduced from >1 μm typical for ns pulse durations to values <20 nm by application of sub-picosecond laser pulses with intensities ~10^{13} W cm^{-2} at ~615 nm.

1. Introduction:

Laser-micromachining of metal layer composites covers cutting, scribing, and drilling. It involves vaporization and heat spreading into the non-illuminated substrate regions. Conventional nanosecond pulse lasers (Nd:YAG, excimer) hit intrinsic precision limits because heat-affected zones (HAZ) of 1-10 μm prevent etching precision. The HAZ depends on the thermal diffusion coefficient D of the substrate and the laser pulse duration τ according to HAZ $\approx (D\tau)^{0.5}$. The light absorption process and the removal of material take place coincidently and lead to plasma shielding. Nanosecond-pulse laser ablation of metals like gold results in extensive splashing and melting. The ns-laser ablation mechanism is determined by the HAZ and is practically independent of the optical properties of the metal: even though the penetration depth of light (α^{-1}) is extremely small in d-metals (10 -100 nm), melting and plastification was observed within a HAZ of several μm's [1,2]. The ablation threshold of Ni and Au films on thermally insulating substrates with 14 ns pulses at 248 nm depends not only on the thermal properties of the metal but also on the film thickness when it is less than the HAZ [3]. With 0.5 picosecond pulses, on the other hand, the HAZ becomes less than 0.1 μm, and, therefore, no influence of the film thickness ≥0.1 μm on the threshold could be observed [4].

Polymer Electrolyte Fuel Cells (PEFC) are one of the most promising energy conversion systems of the future [5,6]. In a PEFC a proton-conducting polymer membrane is sandwiched between two porous gas diffusion electrodes [7]. Typically, such activated electrodes consist of an agglomeration of carbon particles covered by nanoparticles of platinum. Geometric areas of such electrodes are not well defined. Therefore, we intended to micromachine a model structure of parallel lines with typical widths of several μm's on top of a solid conductive, inert and catalytically inactive substrate by femtosecond-pulse laser ablation. The upper surface was to be covered by a thin catalytic platinum layer. For sufficient gas flow, the interstitial grooves should exhibit

---

* Corresponding author

aspect ratios >1. Such a microstructured electrode could be pressed onto the membrane surface by a defined compaction pressure.

2. Experimental

1 μm platinum films were sputter-deposited on polycrystalline gold sheets ($10\times10\times1$ mm^3). Nanosecond-pulse laser experiments were performed with a Nd:YAG laser (1064 nm) and Excimer lasers (193 nm and 248 nm). The femtosecond-pulses were generated and amplified by means of a colliding pulse modelocked dye laser and a dye amplifier, respectively [1]. This generator/amplifier system yielded pulses with a duration of ~300 fs, a centre wavelength of 615 nm, and a single pulse energy of 200 μJ. The amplified spontaneous emission (ASE) could be supressed to at least a background level <3 %. The energy stability was better than ±10% from the average. Pulse energies were determined by means of a pyroelectric detector. The beam was focused by a 58 mm lense to a spot of 2×10^{-5} cm^2. The specimen for microstructuring was mounted on a motorised x-y-z-translation stage and positioned perpendicular to the direction of laser incidence. The ablation depth was determined with an optical microscope (1 μm scale). Scanning electron micrographs were obtained on a scanning electron microscope equipped with a cold-field electron emission cathode at an accelerating voltage of 20 kV.

3. Results and Discussion

Excimer laser ablation (193 and 248 nm) of gold substrates can produce narrow grooves (<10 μm, Fig. 1a,b). However, thick melt films are squeezed and droplets are thrusted out of the ditches, and resolidify at the edges and on the remaining plains. Near infrared ns Nd:YAG laser ablation (1064 nm) resulted in even larger melt extrusions of the order of the groove width (~30 μm, Fig. 1c). These melt features and redeposited droplets are still too bulky for a defined model groove electrode structure. Moreover, this debris would almost completely cover up any catalytic Pt top layer [7].

Table 1: Reflectivity R [%] and Absorption Coefficients α [cm^{-1}] [8]

	275 nm		600 nm		670 nm		1070 nm	
	R	α	R	α	R	α	R	α
Au	27	5.8×10^5	79	6.8×10^5	95	9.6×10^5	98	8.3×10^5
Pt	43	9.0×10^5	59	7.6×10^5	59	6.9×10^5	77	7.9×10^5

The absorption coefficient α, and the penetration depth of light (α^{-1}) are not responsible for the differences of the micromaching results: in the wavelengths range employed here, α is practically constant if we consider low energy values (Table 1). Reflectivities on the other hand, increase from <27% in the UV to 98% in the near IR. That means that about 35 times less of the incoming IR light can penetrate the surface as compared to the UV light. At comparable penetration depths, energy volume densities, $(1-R)\alpha F$, differ by the same factor as the absorption. Therefore, local temperatures are much less under IR illumination, and the vaporization process is not as efficient resulting in a higher percentage of melt. These facts suggest the surprising conclusion that Nd:YAG processing can produce similar metal micromachining qualities when sufficient fluences are employed.

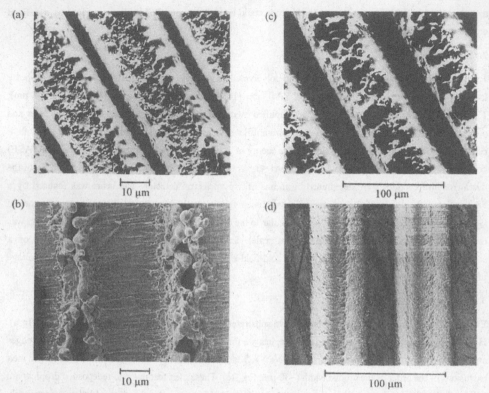

Fig. 1 Pulse laser ablation of gold. (a) ns-laser (Excimer, 193 nm); (b) ns-laser (Excimer, 248 nm), (c) ns-laser (Nd:YAG, 1064 nm), (d) fs-laser (dye, 615 nm, $F = 1.5$ J cm^{-2}, 0.5 μm s^{-1}).

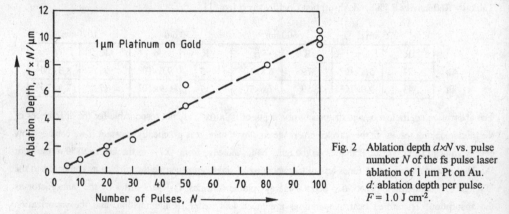

Fig. 2 Ablation depth $d \times N$ vs. pulse number N of the fs pulse laser ablation of 1 μm Pt on Au. d: ablation depth per pulse. $F = 1.0$ J cm^{-2}.

Femtosecond-pulse laser ablation, in contrast, generates microstructures without any melt, droplet, and debris formation (Fig. 1d). That means that the HAZ can be neglected [1,2]. The lateral resolution is limited by ripple formation due to laser interferences between incident and surface scattering fields leading to spatially periodic power deposition and solid/melt structures with a period related to the wavelength ($\lambda \approx 0.6$ μm). The ablation rate is independent of the number of laser pulses and the cavity depth (Fig. 2) at $F = 1.0$ J cm^{-2}. The ablation threshold of a platinum film on a gold substrate is $F_{th}(Pt) \approx 0.1$ J cm^{-2}, and deviates from the bulk gold value of $F_{th}(Au) \approx 0.25$ J cm^{-2} (Fig. 3) [2]. This difference can be used for an intrinsic etch stop when fluences between $F_{th}(Pt)$ and $F_{th}(Au)$ are applied. 100 pulses with $F = 0.16$ J cm^{-2} somewhat greater than $F_{th}(Pt)$ were not sufficient to remove the 1 μm thick Pt layer (Fig. 4a). A fluence near $F_{th}(Au)$, $F = 0.23$ J cm^{-2} resulted in preferable ablation of the Pt films (Fig. 4b). Energy dispersive x-ray analysis showed a centre spot of ~13 μm open Au substrate [9].

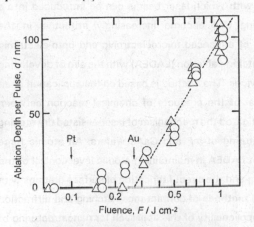

Fig. 3   Ablation depth per pulse vs. laser fluence of the fs pulse laser ablation of Au (Δ) and 1 μm Pt on Au (O). The points represent experimental data averaged over 100 pulses.

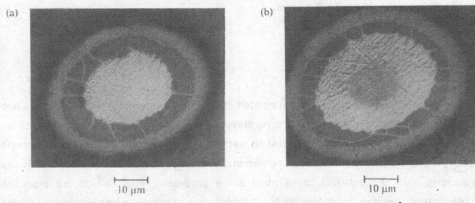

Fig. 4   Femtosecond-pulse laser ablation of 1 μm Pt on Au. 100 pulses; (a) $F = 0.16$ J cm^{-2}; (b) $F = 0.23$ J cm^{-2}.

References:

[1]   W. KAUTEK and J. KRÜGER, SPIE Proceedings, Vol. 2207 (1994) 600.

[2]   J. KRÜGER and W. KAUTEK, SPIE Proceedings, Vol. 2403 (1995) in press.

[3]   E. MATTHIAS, M. REICHLING, J. SIEGEL, O.W. KÄDING, S. PETZOLDT, H. SKURK, P. BIZENBERGER, and E. NESKE, Appl. Phys. A 58 (1994) 129.

[4]   S. PREUSS, E. MATTHIAS, and M. STUKE, Appl. Phys. A 59 (1994) 79.

[5]   K. KORDESCH, Ber. Bunsenges. Phys. Chem. 94 (1990) 902.

[6]   K. STRASSER, Ber. Bunsenges. Phys. Chem. 94 (1990) 1000.

[7]   G.G. SCHERER, Ber. Bunsenges. Phys. Chem. 94 (1990) 1008.

[8]   Taschenbuch für Physiker und Chemiker, eds. J. D'Ans and E. Lax, Springer, Berlin (1943).

[9]   J. KRÜGER and W. KAUTEK, to be published.

Excimer Laser Microstructuring of InP and GaAs Semiconductor Compounds

J.J. Dubowski, M. Bielawski, M. Fallahi and B. Mason
Institute for Microstructural Sciences
National Research Council of Canada, OTTAWA, Ontario K1A 0R6, Canada

ABSTRACT

Due to unique features and the ease with which laser beams can be introduced into a processing chamber, laser beam processing has become increasingly attractive in the investigation of methods for the fabrication of advanced microelectronic and opto-electronic systems. We have studied laser-assisted dry etching ablation (LADEA) with the aim of developing *in-situ* fabrication of novel opto-electronic devices. The method is based on the application of an excimer laser ($\lambda = 308$ nm) for the removal of the products of chemical reaction between chlorine and a semiconductor. We have investigated the mechanism of laser-assisted dry etching and implementation of LADEA for the structuring of InP and GaAs wafers. An atomic force microscopy study has shown the feasibility of LADEA in maintaining atomic level control of the surface morphology. The anisotropy of etching and the direct fabrication of surface gratings with sub-half-μm periods have been analyzed. The methods of contact mask etching and diffraction image printing have been examined and the applicability of this approach for nanostructuring of III-V semiconductors has been discussed.

1. INTRODUCTION

The fabrication of devices and integrated structures for advanced microelectronics and optoelectronics require numerous processing steps. Of particular importance in this respect are methods of dry etching which, in contrast to wet etching, are capable of realizing anisotropic etching. Being compatible with a vacuum environment, dry etching offers the possibility of *in-situ* processing with the potential to eliminate some processing steps which are prone to contamination. Reactive ion etching (RIE), in which beams of energetic ions are used to desorb the products of reaction from the surface of a wafer kept in a reactive atmosphere, is a method of fabrication commonly applied by the microelectronic industry. It has, however, been demonstrated that RIE induces structural damage which may extend to more than 20 nm in depth of a processed sample [1-5]. Surface damage becomes an especially important issue when the fabrication of sub-micrometer size devices and nanostructural systems such as quantum wires and dots is considered.

Laser-assisted dry etching has been suggested as an alternative method of structuring microelectronic materials. For instance, the direct generation of three-dimensional structures in Si has been demonstrated [6] with the use of laser controlled microreactions of this material in a chlorine ambient. The application of pulsed lasers operating in an ultraviolet wavelength range offers the possibility of digital etching with a rate of material removal near one atomic layer per pulse [7,8]. The application of low-energy photons is especially attractive for obtaining a less damaged or even damage-free surface structure of a processed wafer. We have recently demonstrated [9] that laser dry etching leads to structures of superior quality when compared to those obtained by a standard RIE.

The mechanisms of laser dry etching of Si [10], GaAs [11-13], InP [14-16] have been investigated by numerous research groups. Results have been obtained for laser fluences from about 10 mJ/cm^2 to more than 1 J/cm^2. The full advantage of anisotropic structuring attainable with laser-driven etching may be realized if a processed wafer is kept at a sufficiently low temperature. Consequently, the suppression of spontaneous chemical reactions is expected between a reactive gas and a semiconductor. This problem has been addressed in a recent study of laser-induced photoetching of GaAs and AlGaAs with Cl$_2$ physisorbed on cryogenically cooled samples [17].

The ability to control the surface morphology which is desirable for a particular application is one of the most critical issues concerning laser microstructuring of semiconductors. Although a significant effort in this respect has been documented by experimental results [e.g., 11-16], no comprehensive picture has been available. Some results concerning the application of laser dry etching for device fabrication have been published [17], however a demonstration of the advantage of this technique for the fabrication of an operating device has yet to be proven. We have applied laser-assisted dry etching ablation (LADEA) for microstructuring of III-V semiconductors. Preliminary results concerning micropatterning of InP have been published elsewhere [18-21]. In this paper we investigate both the mechanism of laser dry etching of InP and GaAs and the potential of LADEA for nanostructure fabrication.

2. EXPERIMENTAL DETAILS

The LADEA setup consists of a quartz chamber equipped with a fused silica window of high optical quality to make possible a low-distortion transmission of the laser beam to a sample. Chlorine diluted in helium (1:10) has been used as an etching gas. The chamber was evacuated by a turbomolecular pump and a liquid nitrogen trap which was used to collect chlorine. The base pressure in the system was 5x10^{-6} mbar. An almost three orders of magnitude better vacuum has been obtained in the chamber when compared to that obtained in our previous system equipped with a charcoal filter [21]. This enabled us to better control the amount of Cl_2 injected on the sample and to conduct experiments at lower Cl_2/He pressures. Studies were carried out at a dynamic pressure of Cl_2/He up to 3×10^{-3} mbar using a constant flow of the gas mixture to the chamber. Etching was realized with an XeCl excimer laser (λ = 308 nm, τ = 10 ns) operating at a repetition rate of 5 Hz. The laser fluence was measured with the use of a Joulmeter and a diaphragm placed in the plane of a sample. The diameter of the diaphragm was equal to the diameter of a laser etched crater. The estimated error of the laser fluence measurements was ± 20 %, but the relative error of the fluence measured in one series of experiments was not worse than ± 5 %. The samples etched were S-doped (001) InP and Si-doped (001) GaAs. No special cleaning procedure was applied to the samples before mounting them on a stainless steel holder. The experiments were carried out on samples kept at room temperature. The laser beam irradiated samples at normal incidence except when it has been otherwise indicated.

3. RESULTS AND DISCUSSION

3.1 Etch rate in LADEA

Figure 1 shows the etching gas pressure, p, dependence of the InP etch rate, R, obtained for laser fluences of 60 and 65 mJ/cm² and averaged over 600 pulses for each point. The etch rate is negligible at pressures below 1×10^{-4} mbar and it rapidly increases at p ≈ 1.5×10^{-3} mbar to about 0.8 Å/pulse and 1.4 Å/pulse for laser fluences of 60 and 65 mJ/cm², respectively. We note that such dependence was not observed in our previously used experimental system [21].

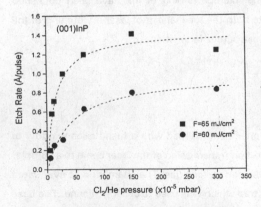

Fig. 1. Etch rate dependence on gas pressure for two laser fluences. Results obtained for 600 pulses delivered to each site.

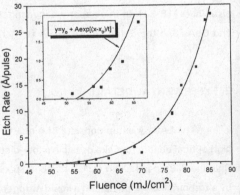

Fig. 2. Etch rate of (001) InP as a function of the applied laser fluence. The results of fitting are shown by a solid line.

Clearly, the etch rate measured for the same concentration of the etching gas did not show any decrease with decreasing Cl_2/He pressure down to about 1.34×10^{-2} mbar [the system base pressure was about 6.7×10^{-3} mbar (5 mTorr)]. This difference appears to be due to higher concentrations of chlorine and/or water vapour in the chamber. The increase of the etch rate with etching gas pressure, as it is shown in Fig. 1, suggests that for $p < 1 \times 10^{-3}$ mbar the etching is source-limited. An additional mechanism also seems to play a role since the etch rate increases with laser fluence even in the saturation pressure regime. Qualitatively, this logarithmic increase of the etch to a fluence-dependent saturation value is similar to that reported in literature [15,16]. However, substantially higher etch rates have been observed in our experiments. This is due to the fact that our experiments have been carried out on wafers kept at room temperature where a higher sticking coefficient of chlorine is expected on the etched surfaces. Thus, a thicker chlorinated layer could be formed between two consecutive laser pulses. An additional enhancement in the etch rate may result from a strong absorption of 308 nm radiation by Cl_2 [22]. This could lead to the photodissociation of Cl_2 into more reactive Cl atoms. As will be discussed below, we have observed etch rates in excess of 25 Å/pulse for laser fluences which were well below those required to melt InP.

The dependence of the InP etch rate on laser fluence, F, obtained for $p = 1 \times 10^{-3}$ mbar is shown in Fig. 2. Each point in this figure has been obtained for 600 laser pulses delivered at a site. A threshold fluence of ~ 48 mJ/cm^2 required to initiate the etching is characteristic of laser induced thermal desorption [23,24]. Threshold fluence values of ~ 20 mJ/cm^2 [15] and 50 mJ/cm^2 [16] have been observed for laser etching of InP wafers kept at ~ 440 K. We note, however, that the threshold fluence required to initiate the etching and/or surface damage strongly depends on the ambient in which the laser irradiation takes place. A significant shift towards lower laser fluences capable of inducing surface defects in InP and GaAs has been observed with *in-situ* photoluminescence [25] when samples were transferred from a vacuum to an atmospheric environment. The composition of a reactive atmosphere also appears to be responsible for the slightly higher etch rates of InP observed for $F \leq 50$ mJ/cm^2 in our previous experiments [21].

An exponential dependence of R on F is expected if the process is due to laser-induced thermal desorption [23,24]. Indeed, such dependence is observed in the whole range of fluences studied in this work, i.e., for 45 mJ/cm^2 $< F <$ 90 mJ/cm^2. The results shown in Fig. 2 have been fitted with a curve: $R = R_o + A\exp[(x-x_o)/t]$, which is shown by a solid line. The fitting parameters are: $R_o = -0.56$, $A = 0.001$, $x_o = -1.48$ and $t = 8.43$.

The results presented in Figs. 1 and 2 can be qualitatively understood in the frame of laser-induced heating of InP, formation of $InCl_x$ and desorption of reaction products from the InP surface. Experiments of InP etching in Cl_2 atmosphere [26,27] indicate that at temperatures below 430 K the surface is covered with $InCl_3$. At higher temperatures, the conversion of $InCl_3$ into a mixture of P_y ($1 \leq y \leq 4$) and $InCl_x$ ($0 \leq x \leq 2$), with InCl dominating at temperatures above

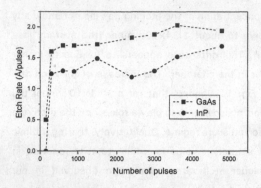

Fig. 3. Etch rate dependence on the number of laser pulses for InP and GaAs.

520 K, has been observed. In a vacuum, InP evaporates at approximately 600 K, with preferential loss of P from the surface and formation of an In rich surface. With the laser used in our experiment, a fluence of about 75-80 mJ/cm² is required to reach a 600 K surface temperature (T_s). However, as has been confirmed by Auger electron spectroscopy [28], the preferential loss of P appears not to be of concern for fluences as high as 120 mJ/cm². This can be explained by the formation of a thick $InCl_x$ layer at the surface of etched InP. The thickness, as well as the chemical composition of this layer, depend on the laser fluence applied for the processing. For laser fluences below 55-60 mJ/cm² ($T_s < 500$ K), the surface is expected to be covered with $InCl_3$ and at low pressures of the etching gas ($p \leq 1.0 \times 10^{-3}$ mbar) one laser pulse is sufficient for the complete removal of this layer. For higher fluences, the formation of $InCl_x$ ($0 < x \leq 2$) takes place at the end of each laser pulse. The thickness of this $InCl_x$ layer increases with laser fluence due to a temperature enhanced diffusion of chlorine into InP. Also, the diffusion of chlorine may be promoted by a laser-induced electromigration [8]. The etch rate will increase with laser fluence as long as the thickness of the chlorinated layer increases and the fluence is sufficient to carry out ablation. If the reaction product, which is formed between two laser pulses, is not removed with one pulse then a slow build-up of it may take place [29]. This hypothesis is supported by the etch rate dependence on the number of pulses, N, for InP and GaAs which is shown in Fig. 3. The experiment was carried out at 1.0×10^{-3} mbar and for the laser fluence of 60 mJ/cm². It can be seen that the etch rate of InP obtained with the first 300 pulses is about 1.25 Å/pulse, and it increases to about 1.7 Å/pulse for 5000 pulses. The small value of R for a low number of pulses (N < 300) suggests a different from the bulk mechanism of material removal. This may be due to the presence of a native oxide layer on the surface and/or the requirement of achieving a critical concentration of surface defects which would increase the coupling of an incident laser beam to the material. The results indicate a similar etching mechanism for both InP and GaAs. We note that under similar experimental conditions slightly higher etch rates are obtained for GaAs which is most likely due to a higher surface temperature obtained with the same laser fluence. In the case of GaAs, the most stable products of chemical reaction with Cl_2 are $GaCl_x$ [27].

3.2 Surface morphology of LADEA-processed samples

The use of laser dry etching for device fabrication requires a surface with a smoothness which would be acceptable for a particular application. The surface morphology of LADEA-processed InP samples was found to depend on the depth of an etched crater [19-21]. Figure 4 shows AFM images of 80 nm deep craters obtained with etching rates of 40, 2.0 and 1.0 Å/pulse. It can be seen (Fig. 4a) that the surface of a crater etched at the rate of 40 Å/pulse ($F \approx 90$ mJ/cm^2) has a strongly corrugated surface morphology. The average period and maximum amplitude of the corrugations are about 260 nm and 75 Å, respectively. However, the surface exhibits macro-undulations as is indicated by the scan shown on the right side of the AFM image. The spatial and/or temporal (pulse-to-pulse) nonuniformity of the laser beam could lead to such nonplanar etching. It appears that the formation of a surface layer mixed of InCl$_3$ and InCl$_x$ ($0 < x \leq 2$), which takes place at temperatures induced with such laser fluence ($T_s \approx$ 700 K), is even more important. The etching with lower laser fluences, i.e., $F \approx 65$ mJ/cm^2 (R = 2.0 Å/pulse) produces superior quality surfaces as Fig. 4b illustrates. The surface structure consists of corrugations with a maximum amplitude of about 75 Å. The etched surface is more planar indicating a more uniform process of etching. The corrugations in this case are less regular and no specific periodicity can be assigned to them. An AFM image in Fig. 4c shows the surface morphology of a sample etched with the laser fluence of about 60 mJ/cm^2 (R = 1.0 Å/pulse). It can be seen that the surface is smooth almost on the atomic scale, with the maximum amplitude of the roughness being about 6 Å. This compares to the roughness amplitude of about 3 Å for an unetched surface. It is likely that even better surface morphology is attainable with LADEA. For instance, our system is not equipped with a loadlock and it has been found [17] that the presence of residual water vapor in a system without the loadlock assembly influences the quality of laser etched semiconductor surfaces. The increase of the etch rate with an increasing number of laser pulses (see Fig. 3) may also lead to changes in surface morphology. These problems should be addressed in future investigations.

3.3 Laser direct etching of sub-μm gratings

The surface corrugations such as those shown in Figs. 4a and b are due to an inhomogeneous deposition of laser energy on the sample surface. It has been indicated in the literature [30,31] that the nature of this inhomogeneous deposition is due to the interference between the laser light scattered on the surface and the primary laser beam. If a linearly polarized laser beam irradiates the sample at an angle Θ with respect to the normal, damage in the form of gratings with periods $\Lambda_1 = \lambda/[n(1 \pm \sin\Theta)]$ and $\Lambda_2 = \lambda/n\cos\Theta$ (n is the index of refraction of a medium where the interference takes place) can occur [30]. For p-polarized light, as Θ is

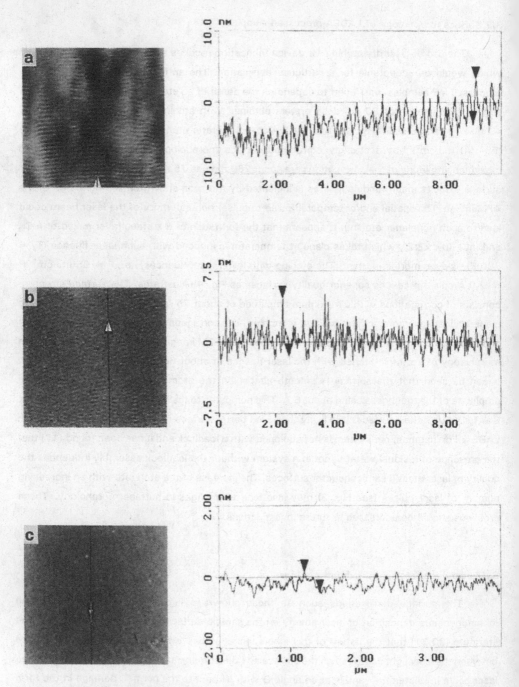

Fig. 4. AFM images and associated cross scans which illustrate the surface morphology of 80 nm deep craters etched at the rate of a) 40 Å/pulse, b) 2.0 Å/pulse and c) 1.0 Å/pulse.

increased above 30°, the damage pattern is dominated by a single set of fringes running perpendicular to the polarization with a spacing of $\Lambda = \Lambda_2$. It is interesting to note that Λ_2-type fringes have been observed for lower laser fluences than Λ_1-type [31].

In an attempt to examine if the laser formed corrugation could be applied for integrated reflectors in laser diodes, we have analyzed the surface morphology of samples etched with various laser fluences for a normal incidence of the laser beam [20,21]. Etching with rates of up to 2 Å/pulse and for normal laser beam incidence did not produce gratings that would be uniform over reasonable areas. Here, we investigate the formation of gratings with laser fluences about 120 mJ/cm² for the irradiation angles: $\Theta = 0$ and 45°. It has been found that for the laser beam at normal incidence, it was necessary to develop some surface roughness before a well developed and oriented grating could be observed. Typically, it was necessary to remove about a 0.5 μm thick layer of material. An AFM 10 μm x 10 μm image shown in Fig. 5 was obtained from the surface of a 2.5 μm deep crater. A well oriented and uniform grating can be seen. It was difficult however to obtain gratings that were developed uniformly at the same depth even though their orientation was preserved over reasonable areas, typically 100 x 100 μm. The period of the grating shown in Fig. 5 is: $\Lambda_1 = 288$ nm. This implies that the index of refraction of an $InCl_x$ layer, which is formed during laser interaction with the sample, is about 1.07. A Fourier analysis has also revealed in this structure the presence of a second order grating with $\Lambda = 144$ nm.

An example of a grating obtained with a laser beam irradiating the sample at $\Theta = 45°$ is shown in Fig. 6. The structure was developed at the bottom of a 3000 Å deep crater which was formed with 3000 pulses. This means that the formation of the grating proceeded at the rate of about 1 Å/pulse and implies that if the etching mechanism in this case is similar to that discussed in Sec. 3.1, about 50% of the initial laser fluence was reflected from the surface. This result confirms a conclusion about lower laser fluences required to form Λ_2-type gratings. The period of the grating shown in Fig. 6 is equal to 352 nm. We deduce that the refractive index

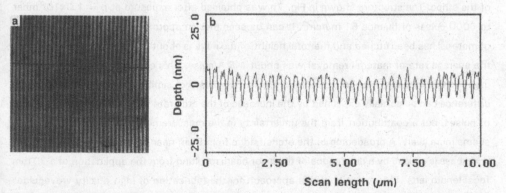

Fig. 5. An AFM image of the grating structure obtained at the bottom of a 2.5 μm deep crater etched in (001) InP. The period of the grating is 288 nm.

Fig. 6. An SEM image (45° tilt) of the grating structure obtained at the bottom of a 3000 Å deep crater etched in InP. The laser irradiation angle was $\Theta = 45°$ and the electric field vector E was in the plane of a sample. The period of the grating is 352 nm.

of $InCl_x$, in this case, is about 1.25, a significantly larger value than that obtained for the grating in Fig. 5, but near the value of n = 1.2 obtained for gratings produced at normal incidence with $F < 70$ mJ/cm² [20]. It appears from this study that the direct laser formation of Λ_2-type gratings has the potential in the fabrication of grating structures as an alternative and/or complementary method to conventional holographic etching which involves exposure to a photoresist.

3.3 Contact mask etching

The process of LADEA has been examined for etching of ridges and for microstructuring of InP with the use of Cr masks deposited directly on wafers and with Si masks which were placed on top of processed wafers. Figure 7a shows a scanning electron micrograph (SEM) of the surface of an initial structure which was used in this study. A DEKTAK scan which is included in the top part of this figure provides details of the structure. It consists of a 70 nm thick and ~ 5 μm wide Cr stripe. As a result of the preparation procedure, which involved a wet chemical etching, two V-shape grooves with depth of 0.1 and 0.2 μm can be seen on both sides of the stripe. The structure shown in Fig. 7b was obtained after exposure at p = 1.0x10⁻³ mbar to 6000 pulses of fluence 61 mJ/cm². It can be seen that an approximately 1.1 μm thick layer of material has been etched and the total height of the ridge is about 1.3 μm. This indicates that the average rate of material removal was about 1.8 Å/pulse. This etch rate is almost twice as great as that measured for craters obtained with 600 pulses of similar fluence (see Fig. 2). The difference can be partially explained by the increase of the etch rate with the increasing number of pulses, but a contribution from the uncertainty in the measurements of laser fluence seems more likely. A broadening of the etched ridge from 5 μm near the top to an almost 15 μm wide base is caused by a divergence of the laser beam resulting from the application of a 30 mm focal length lens. The potential of this approach for the fabrication of high quality waveguides and laser ridge structures is obvious. A protective passivation layer, which is required to prevent

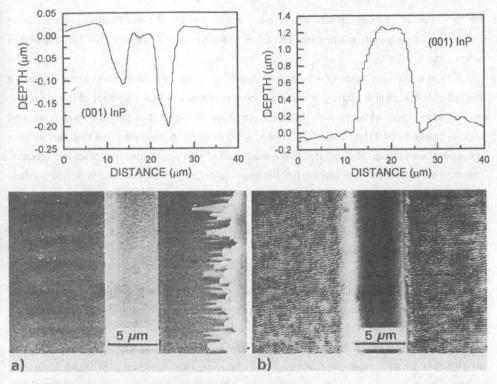

Fig. 7. SEM micrographs and DEKTAK profiles of an initial structure consisting of a Cr coated ridge (a), which was developed to a greater depth upon exposure to 6000 laser pulses at $p_{Cl/He} = 1.0 \times 10^{-3}$ mbar (b).

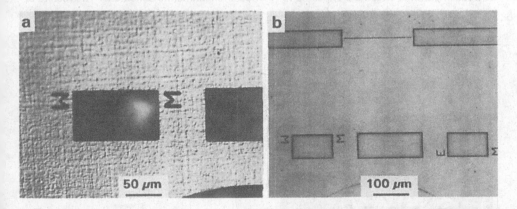

Fig. 8. Optical micrographs showing a) a fragment of the 2 μm thick Si mask, b) the resulting microstructure obtained in InP upon exposure to 540 laser pulses.

the formation of surface defects which are a major source of non-radiative recombination centers, could be grown *in-situ* since LADEA is compatible with vacuum thin film deposition techniques.

With a Si mask, made of a 2-μm-thick membrane, we have been able to demonstrate a method of contact mask etching of various microstructures in InP. A fragment of the mask and the etched structure obtained with this mask are shown in Fig. 8. The structure was fabricated upon exposure to 540 laser pulses of fluence $\sim$ 80 mJ/cm^2. A horizontal line that can be seen in the upper part of Fig. 8b illustrates that a resolution of $\sim$ 1 μm is attainable with this process. This method of contact mask etching has the potential for large scale processing due to the ease with which the mask can be transferred from wafer to wafer without breaking the vacuum.

3.4 Diffraction image printing

We have investigated a near-field diffraction for the intentional projection of regular patterns on etched samples. The patterns could extend over reasonable surface areas due to the coherence length of the excimer laser radiation which was more than 100 μm. Masks with arrays of various apertures which were made of 50 μm thick Au membranes have been applied for this purpose [21]. An SEM micrograph of a fragment of the structure obtained with a mask consisting of 40 μm wide slits is shown in Fig. 9. The structure was obtained after exposure to 1000 pulses at 66 mJ/cm^2. Well defined fringes of 0.3 μm to 1.8 μm widths can be seen in this picture. The frequency of fringes depends on a distance between a diffracting edge of the mask and the sample. This distance can be estimated from the calculations of the intensity of near-field diffraction patterns which are expected for particular experimental conditions. The results of

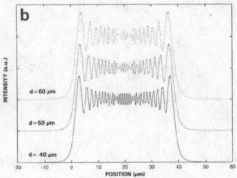

Fig. 9. An SEM micrograph of the structure obtained with a mask consisting of 40 μm wide slits (a), and the simulated near-field diffraction patterns obtained for various distances between the mask and a wafer (b).

of the calculations for a 40 μm wide slit are shown in Fig. 9b. They indicate that the number of etched grooves increases from 21 to 31 as the distance between a diffracting edge and the substrate is reduced from 60 to 40 μm. A total of 27 grooves is observed in the etched structure which indicates that the distance between a diffracting edge and the sample was about 50 μm.

A fragment of the structure etched in (001) InP with a mask consisting of an array of 12 μm x 12 μm windows is shown by an AFM image in Fig. 10. The structure consists of an 8 x 8 array of dots formed by cones which are 1200 Å high and 0.5 μm wide. A cross section AFM scan which is included in Fig. 10 indicates that a device quality surface has been achieved during this process.

It is known [e.g., 32] that the interference pattern obtained with a pair of narrow slits consists of fringes that have more uniform amplitudes and widths than those obtained with a single slit. Thus, with a suitable mask and high-power UV laser the diffraction image printing method should enable the fabrication of linear gratings as well as wire and dot-like structures with a significantly better contrast than that obtained in this experiment.

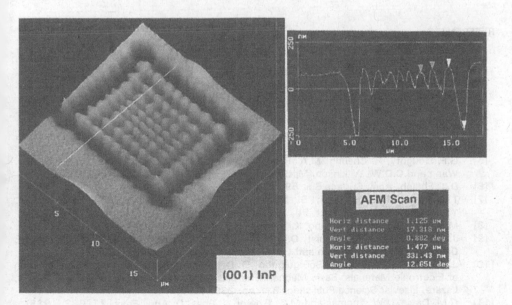

Fig. 10.   An AFM image of an 8 x 8 array of dot-like structure obtained in InP. The structure has been etched as a result of the projection of a diffraction image on the InP wafer held in a Cl₂/He atmosphere.

4. CONCLUSIONS

We have reviewed a method of excimer laser (308 nm) assisted dry etching ablation (LADEA) of InP and GaAs in a 10% Cl_2/He atmosphere. The mechanism of material removal has been discussed based on knowledge of Cl_2 reaction with these materials [26,27] and the laser induced surface temperature. It has been shown that the surface morphology of material etched at rates ≤ 1 Å/pulse can be controlled at the near-atomic level. The feasibility of LADEA has been demonstrated for the direct formation of sub-half-μm period gratings. In particular, high quality gratings have been formed on surfaces irradiated with the laser beam at 45°. A method of diffraction image printing has been proposed for structuring of materials. It has been argued that LADEA has the potential for the fabrication of nanostructural devices.

ACKNOWLEDGEMENTS

The authors wish to acknowledge the assistance provided by Michel Julier and Erwin Sproule in the LADEA and AES experiments, respectively.

REFERENCES

[1]   M. Watt, C.M. Sotomayor-Torres, R. Cheung, C.D.W. Wilkinson, H.E.G. Arnot and S.P. Beaumont, J. Mod. Optics **35**, 365 (1988).
[2]   D.G. Lishan, H.F. Wong, D.L. Green, E.L. Hu, J.L. Merz and D. Kirillov, J. Vac. Sci. Technol. **B7**, 556 (1989).
[3]   T.R. Hayes, U.K. Chakrabarti, F.A. Baiocchi, A.B. Emerson, H.S. Luftman and W.C. Dautremont-Smith, J. Appl. Phys. **68**, 785 (1990).
[4]   S.J. Pearton, A. Katz and U.K. Chakrabarti, Proc. Third Int. Conf. on Indium Phosphide and Related Materials, Cardiff, Wales, UK, April 8-11, 1991, p. 252.
[5]   G.F. Doughty, R. Cheung, M.A. Foad, M. Rahman, N.I. Cameron, N.P. Johnson, P.D. Wang and C.D.W. Wilkinson, Mat. Res. Soc. Symp. Proc., vol. 236, 223 (1992).
[6]   D.J. Ehrlich, Appl. Surface Sci. **69**, 115 (1993).
[7]   T. Meguro, M. Hamagaki, S. Modaressi, T. Hara, Y. Aoyagi, M. Ishii and Y. Yamamoto, Appl. Phys. Lett. **56**, 1552 (1990).
[8]   M. Ishii, T. Meguro, T. Sugano, K. Gamo and Y. Aoyagi, Appl. Surf. Sci. **86**, 554 (1995).
[9]   J.J. Dubowski, B. Rosenquist, D.J. Lockwood, H.J. Labbé, A.P. Roth, C. Lacelle, M. Davies, R. Barber, B. Mason and G.I. Sproule, J. Appl. Phys. **78**, (1995).
[10] J. Boulmer, J.-P. Budin, B. Bourguignon, D. Débarre and A. Desmur, in: "Laser Ablation of Electronic Materials: Basic Mechanisms and Applications", Ed. E. Fogarassy and S. Lazare, Elsevier Science Publishers B.V., p. 239 (1992).
[11] G.M. Davis, D.W. Thomas and M.C. Gower, J. Phys. D: Appl. Phys. **21**, 683 (1988).
[12] P.A. Maki and D.J. Ehrlich, Appl. Phys. Lett **55**, 91 (1989).
[13] F. Foulon, M. Green, F.N. Goodall and S.De Unamuno, J. Appl. Phys. **71**, 2898 (1992).
[14] J. Meiler, R. Matz and D. Haarer, Appl. Sur. Sci. **43**, 416 (1989).
[15] V.M. Donnelly and T.R. Hayes, Appl. Phys. Lett. **57**, 701 (1990).
[16] R. Heydel, R. Matz and W. Göpel, Appl. Surface Science **69**, 38 (1993).

[17]   M.C. Shih, M.B. Freiler, R. Scarmozzino and R.M. Osgood, Jr., J. Vac. Sci. Technol. **B13,** 43 (1995).
[18]   J.J. Dubowski, Trans. Mat. Res. Soc. Jpn., **17,** 333 (1994).
[19]   J.J. Dubowski, Mater. Sci. Forum, vol. 173-174, 73 (1995).
[20]   J.J. Dubowski, A. Compaan and M. Prasad, Appl. Surface Sci., **86,** 548 (1995).
[21]   M. Prasad, J.J. Dubowski and H.E. Ruda, Proc. SPIE, **2403,** 414 (1995).
[22]   G.E. Gibson and N.S. Bayliss, Phys. Rev. **40,** 188 (1933).
[23]   T.J. Chuang, Surface Science Reports, vol. **3,** 1-105 (1983).
[24]   G.W. Tyndall and C.R. Moylan, Appl. Phys. **A50,** 609 (1990).
[25]   M. Julier, J.J. Dubowski and M. Bielawski, to be published.
[26]   S.C. McNevin, J. Vac. Sci. Technol. **B4,** 1203 (1986).
[27]   S.C. McNevin, J. Vac. Sci. Technol. **B4,** 1216 (1986).
[28]   G.I. Sproule, private communication.
[29]   B. Li, U. Streller, H.-P. Krause, I. Twesten, N. Schwentner, V. Stepanenko and Yu. Poltoratskii, Appl. Surface Sci. **86,** 577 (1995).
[30]   J.F. Young, J.S. Preston, H.M. van Driel and J.E. Sipe, Phys. Rev. **B27,** 1155 (1983).
[31]   S.E. Clark and D.C. Emmony, Phys. Rev. **B40,** 2031 (1989)].
[32]   G.O. Reylolds, J.B. DeVelis, G.B. Parrent, Jr. and B.J. Thompson, *Physical Optics Notebook: Tutorials in Fourier Optics,* SPIE Optical Engineering Press, Bellingham, 1989.

Novel Laser Techniques for Microoptical Compounds Fabrication

Dr. Sci. Prof. V. Veiko*,
Dr. Prof. S. Metev**, Dr. Prof. G. Sepold**
*Saint-Petersburg University of Fiine Mechanics and Optics,
Saint-Petersburg, Russia.
**Bremen Institute of Applied Beam Technology (BIAS),
Bremen, Germany

1. INTRODUCTION

One of the most stable tendencies of modern techniques - it is tendency to miniaturization. Semiconductor transistors invention led to new directions in information processing and to the appearance of microelectronics.

The recent progress in semiconductor lasers and fiber optics put us to the threshold of the same jump in optics. The microoptics development is stimulated now by the next reasons [1]:

- development of semiconductor lasers, lines and matrix with power of 1-100 W at room temperature and their applications in industry, medicine, ecology, etc. (instead of He-Ne, YAG:Nd, and other types of lasers),

- broad development of fiber-optical systems in medicine and in industry for delivery of laser radiation and as optical tools,

- microsystem technique progress, which often includes micro-opto-electro-mechanical components in an integrated unit,

- integrated optical systems development which consist of a number of special micro-optical components (MOC) like wave-guides, optical interconnections, geodesical lenses, special diffraction elements, etc.,

- information technique for optical communication, cable television, optical recording, data storage, optical diagnostics in complicated technological processes and constructions, in medicine, biology and environmental applications broad propagation, which now connected more and more with fiber-optical systems.

All these fields of laser applications demand a new, much more variable component base, which could not be fabricated by traditional ways of optical technology (mechanical polishing technique). The main disadvantages of optical technology such as small productivity, difficulties of control and automation put it at the edge of art. That is why so many new methods appear nowaday for the new optical components producing: ion and molecular exchange (diffusion methods), photolithography, beam (laser, electron, ion) techniques and so on. All of them have their own advantages and disadvantages.

Laser techniques [2-5] attract more and more researchers last time due to next main advantages: high resolution of elements size, possibility of complete automation of the process, possibilities of aspherical elements production and high productivity.

2. PRINCIPLES OF LASER EQUIPMENT FOR MOC FABRICATION.

All the versions of laser techniques for MOC fabrication based on thermal action of laser radiation on glass or glass-like materials. Some of them will be considered below. But it significant to note there, that in all cases the complete temperature distribution $T(x, y, z, t)$ is the responsible parameter of treatment.

On the other hand T is the function of heating conditions and depends only on radiation intensity distribution $q(r, t)$. As a result it is enough to provide the control on laser radiation intensity in time $q(t)$ (in simplest case by duration of action), and spatial distribution $q(r)$. It is the first principle of laser installation for MOC fabrication.

The second principle is an automatic feed-back, which can be realized by the optical control of the process. Idea is in continious observing the process of laser fabrication of MOC by the beam of He-Ne or semiconductor laser. The action of technological laser can be stopped at the moment, when the optical parameters of MOC reach the values of desired ethalon parameters.

Scheme of laser set, based on these principles is shown in Figure 1. The main elements of it are:

1) continuous CO_2-laser, stabilized in power;

2) optical system (O) for control on parameters, consisting of optical modulator,which control by the duration of action; scanning system for controlling cross-section energy distribution and focusing lens;

3) system for observing of MOC optical parameters during the process of fabrication by the characteristics of wavefront of radiation from additional source (He-Ne or semiconductor laser, fiber-optics source, etc.) which is going through the MOC (PR);

4) sample of glass material (S) in necessitive handling (H);

5) computer for control the work of the whole complex (PC).

In other cases various methods of this process control were realized:

1) by the temperature in zone under action - with micropirometer and computer;

2) by the form of refractive zone which is projected by CCD-camera. The form of power distribution defined by $q(r)=const$, $q_0 r^2$, $q_0(1+ar)$ or more complicated expression depends on the law of moving of scanning system, which is controlled by computer.

The main parameters of laser set are:
- frequency of scanning - 0.1 - 150 Hz;
- size of processed area - from 50 μm to 50 mm;
- power of radiation - 2-60 W;
- duration of action - 0.1-600 s.

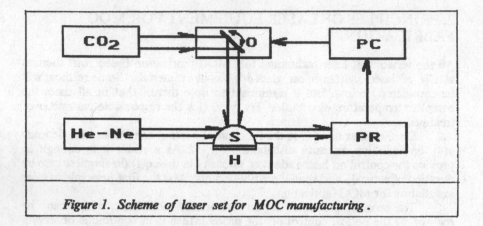

Figure 1. Scheme of laser set for MOC manufacturing.

3. MAIN MOC LASER FABRICATION TECHNIQUES.

3.1. Soft laser heating of glass and glass-like materials.

There is an extensive area of the soft laser heating (SLH) of glass and glass - like materials. Processes of the local deformation, microstructure modification and extratempering as well as the laser melting and fusion of the optical glasses, densification of the porous glasses, amorphization of glass - ceramics and ceramics take place in this area.

SLH of optical glasses propers of the few connected phenomena. There is a thermal expansion and structure modification are responsible for the volumes optical properties changing (density, refractive index (RI) ,etc.) on the first stage. And surface tensions forces - for the shape of the surfaces in the irradiated zone. This regime ,when the surface temperature T_s less than 600 - 800 °C and heating rate less than 50 K/s and cooling rate less than 100 K/s takes place. When high - speed cooling to the initial temperature T_0 occurs, microstructure alteration haven't time to relax. Because of that the sample has a residual deformation of the profile on the irradiated (10 - 100 μm) and the opposite (1-20 μm) sides. Both surfaces become convex so that positive optical components forms.

Our measurements show that focal lengths of produced by this technique miniature optical components were from 20 to 62 mm, and numerical aperture increased with the time from 0.01 up to 0.04. So the lenses have a positive optical force although the less value of RI due to the big influence of profile deformation.

Other case when we have deal with the light intensity about $q=(4-5) \cdot 10^6$ W/m². This regime is characterized by more values of heating rate, and the surface temperature of glass plate quickly reaches the values more than 900 - 1000° C. Viscosity of glass is hardly decreases and the processes of viscous flowing begin to play an important role in the forming of refractive

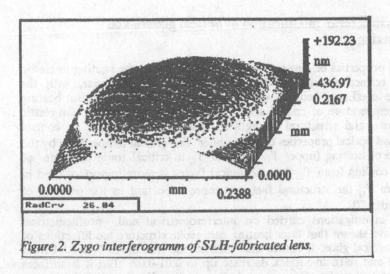

Figure 2. Zygo interferogramm of SLH-fabricated lens.

surfaces. The irradiated surface shrinks down its initial level, although it has a convexity, connected with thermal expansion. Back surface becomes a drop-shaped due to the action of gravitation forces. When the temperature of glass reaches the values of 1300 - 1500° C, an irradiated surface becomes plane or concave due to the hard viscosity decreasing.

The thickness of glass plate plays an important role for refractive surfaces forming. If the thickness of plate more than 4-6 r_0 (where r_0 - radius of laser treatment zone (LTZ)), the temperature on the opposite side is too small for quick flowing because we get the big temperature gradient between the surfaces. If we will increase the power of irradiation, we can really reach the evaporation threshold or exceed the critical heating rate.

At the expense of viscous flowing the partial relaxation of stresses takes place during the heating process. During the fast cooling the residual stresses appears again causing the tempering effect. In our investigations we found, that the maximum values of cooling rate were about 200 - 250 K/s. If the cooling rate being maded more than this value, the destruction of local zone of laser treatment can occurs.

The volume alteration has more value than in thermoexpansion regime, but the RI alterations not exceed its values. The numerical aperture increases during the heating up to 0.04, the focal length of perturbed zones in the case of viscous flowing changes more slowly. The typical value of focal length produced by viscous - flowing method is about 50 - 100 mm.

Both regimes of SLH - technique is just suitable for small-aperture microlenses and microlens arrays fabrication. And its parameters is good correspond to requirements for laser beam cross-section distribution improving. Zygo - interferogramm of one of the SLH - fabricated lens is given in Figure 2. One can see that it has a good optical quality.

3.2. Microstructural modification of optical glasses and extratempering.

The optical properties of glasses change reversely during the heating in elastic zone. The refractive index in this zone changes, in particular, with the temperature coefficient because of changing the glass density . But heating up to the temperature of critical zone (zone of glassing) cause a non-elastic deformations and structural changes in glass. It is possible to control structural and optical properties of glass under the critical temperature by the temperature of heating (upper T_U or lower T_L in critical zone) and rate of cooling. In cooling from T_L the mechanical factor is more important, and in cooling from T_U the structural factor is more important in the creating of refractive index [7].

Our investigations carried on interferometrical and profilemetrical methods have shown that laser heating can evoke structure modifications of monolithic optical glass. It causes the volume increase up to 0.03-0.04 from initial value and refractive index decrease up to 0.01-0.02. But it is surfaces deformations that make greater contribution to the alterations of the optical properties in the LTZ. Only in case when irradiation acts on cylindrical samples from lateral side takes place another situation. Acting from this side on the necessary part of the cylindrical sample (rod) it is possible to accumulate corresponding phase changing along the axis of the rod compare to the off - axis ways. By this way of preliminary irradiation was realized the compensation of thermolenses in laser active glass elements - rods with diameter 0.5-1.0 cm [7].

Other opportunity of extratempering is to use these alterations of RI for laser non-mechanical aspherization (compensation of spherical aberrations) of lenses.

3.3. Local laser melting and fusion techniques.

Local laser melting and fusion (LLMF) techniques based on dependence of glass viscosity η from temperature

$$\eta(T) = B \cdot \exp(-W/_{kT}), \qquad (1)$$

where W - activation energy of viscosity, B - constant depending on the material properties.

In the soft heating zone one can observe the softening penetration, diffusion and fusion of two different kinds of glass. The depth of melting and fusion are the functions of time. Different hydrodynamical processes take place at the melted (softened) area under the action of surface tension, external mechanical forces (from special handling, see below) and from gravitation forces sometime. This technique is especially useful for fiber-tip components fabrication [8]. By this way it is possible to create achromatic assemblies, non-spherical optical components and optical components with high numerical aperture. By this technique was realized MOC from fused silica and lead glass, for example.

3.4. Amorphization of glass – ceramics and ceramics.

Usually glass-ceramics and ceramics (GCC) are not suitable materials for optical components. The reason of that is the high degree of structure crystallization, which is impossible to avoid by the traditional technology of GCC making. As a result GCC materials have a big scattering in optical diapason and can not be used for refractive elements production. Naturally with increasing of cooling rate (after the heating up to the temperature of crystallization T_{cr}) the degree of crystallization will decrease. And from the $|dT/dt| > |dT/dt|_{cr}$ at cooling (see Figure 3) it is possible to reach the quazi-amorphous state for GCC. Suitable cooling rate can be reached (at thermoconductivity typical for glasses) in natural conditions of cooling the plots of laser heating with characteristic size about 1 mm. The proof of that is the experiment on measuring the degree of GCC crystallization before and after the laser heating carried out by X - rays analysis . Experiments [6] show that the amorphous GCC is a good optical material, it has a high optical transparency and small scattering.

There is increasing of substance volume due to densities differences between the amorphous and crystalline phases. As a result of surface tensions and gravity forces action the amorphizated zone is acquiring the shape of double - convex lens. Geometrical profile of the lens is close to spherical one. It is stated that the microlenses could be useful for optical signals processing of objects with spatial frequencies until 15 mm^{-1}. It is possible to produce micro-optical arrays with different spatial period, focal length and size of each element.

3.5. Densification of porous glasses and ceramics.

The main mechanism of porosity alteration is considered to be the pores flowing under the action of surface tension forces at the decrease of the silicate matrix viscosity of porous glass during the process of heating. This process of densification will continue until porous glass (PG) is cooled or gets monolithic structure with minimum surface square. Local laser action at wavelength λ = 10.6 μm leads to heating and softening of PG. At the temperature of "sintering" T_S (T_S is a function of heating rate) forces of surface tension in pores become more than viscosity friction. In result matrix of porous glass flows and fills the pores. At the T_S this process is spontaneos and it goes without external forces.It is finished when all of the pores disappear ($\sigma \to 0$), or material is cooled (viscosity η strives for ∞). The real structure of refractive index distribution in this zone corresponding to density structure is more complicated because it has the axes and radial gradients.

Thermodensification for the main kinds of porous glasses begins at the laser irradiation intensity q ~ $5 \cdot 10^5$ W/m^2 , surface temperature being 700 - 1000 $^{\circ}$C. The densification rate (surface setting down) may reach mm/s values, heating rate for such regimes doesn't exceed permissible dT/dt ~ 10^3 K/s. At $q > q_{ev} = 8 \cdot 10^7$ W/m^2 porous glass starts to evaporate, and if dT/dt > 10^3-10^4 K/s, thermomechanical destruction of glass can occur.

Kinetics of porous glass heating and sintering in general depends on its water contents, because its absorption changes in the band of part - transparency and hence the heating rate also changes. If before laser treatment porous glass hasn't been annealed, it may start boiling due to the presence of water vapour which fails leaving the sample because of high rate of heating. Thus a fine dispersive silica structure appears and it scatters light greatly.

A number of different MOC was fabricated using thermodensification of porous glass. Among them concave and convex lenses, microobjectives and microtelescopes, cylindrical lenses and double - cylindrical systems, soft microdiaphragmes, etc. [2] [Table 1].

3.6. Thermostable glass evaporation

When the light intensity $q > q_{ev}$, evaporation of glass begins. However, due to the fast heating ($dT/dt > 5 \cdot 10^3$ K/s) to the point of evaporation high temperature gradients appear. As a result thermal stresses arise in the LTZ. They do not exceed the destruction limit for thermostable quartz glass. Other glasses need to be heated preliminary in a furnace.

An important advantage of quartz is also a possibility of removal with the temperature close to the softening point (viscosity about 10^9 Puaz at the temperature T = 1940 °C). Due to this fact a liquid bath practically doesn't form on a quartz glass surface. It allows to remove the quartz glass layer corresponding to the laser irradiation distribution. Thus the accuracy of laser evaporation depends only from the method of measuring.

Local laser evaporation process was successfully used for cylindrical lenses fabrication [5].

Beside above-mentioned laser technologies it is possible to use other laser-assisted physical and chemical processes. Among them laser sintering of powders, laser diffusion in solids and liquids, laser deposition and laser decomposition from gases and liquids, etc. But this new possibilities for microoptics fabrication needs in fundamental investigation yet.

4. COMPARISION OF DIFFERENT LASER TECHNIQUES FOR MOC FABRICATION.

It is not a time to make a complete comparison of so new methods of MOC fabrication as laser technologies. Now it is possible to say , that it needs in the new approaches in calculation, measurements and applications due to its specific pecularities:

1) small sizes and gradients of refractive index in working area;

2) possibility to control upon an optical parameters during the processing and;

3) possibility to form various non-spherical shape of optical element surface.

Only it is clear that it present laser methods more suitable for fabrication energetical adjustment (non-imaging) components (see Table 1) due to simplicity of control of its optical parameters (usually only one - numerical aperture) by optical feed-back.

Most important thing is a physical limitations of laser technologies.

Table 1   Comparison of different laser techniques
                              for MOC fabrication.

Type of laser technique	Characteristics of method										
	Thermo-stresses	bub-bles	surface quality	tempo-ral stabi-lity	possi-bility of mea-suring during fabri-cation	possi-bility of comp-lete auto-mation	limit on optical charac-teris-tics NA, F, D	number of useful optical materi-als	produc-tivity	sum balls	place
Densification of porous glass	3	6	5	6	2	3	3	6	4	38	5-6
Amorphization of glass-ceramics	4	5	4	2	6	5	4	5	3	38	5-6
Structure modification of optical glasses	5	3	3	4	4	4	5	3	2	33	3
Extra tempering of optical glasses	6	2	2	5	5	6	6	4	1	37	4
Melting of optical glasses	2	4	1	3	1	1	2	1	5	20	1
Evaporation of thermostable glasses	1	1	6	1	3	2	1	2	6	23	2

Remark: The number in the columns represent in order of preferential use; the best technique have the least sum.

First of them are thermal stresses. This reason limits the maximal size of elements, especially in (soft) heating processes. Appearance of thermal stresses need special thermal regime of treatment MOC before, during and after laser irradiation. If somebody breaks out an optimal conditions of treatment and goes out of optimal heating-cooling zone., big thermal stresses or thermomechanical cracking arise. The second one from main limitations is a bubbles appearance.

Other problems of laser-fabricated MOC is more or less usual . It is quality of surface, accuracy of wave-front profile, temporal stability and so on. Comparison of different methods of laser technologies for MOC fabrication is given in Table 1.

5. COMBINED LASER-ASSISTED TECHNIQUES.

As one can see from the Table 1, the most suitable technique for current moment are evaporation and melting. Evaporation (etching) technique provides good level of most parameters excluding from surface quality. It is connected with the kinetics of machining. The new conception to exclude this disadvantage was suggested to combine laser evaporation with further smoothing of surface. The best solution of the irregularities of surface decreasing is a thermal treatment by the high concentrate source, like plasma - jet, electron-beam, laser-beam, etc. The theoretical consideration of hydrodynamical task gives the criteria which should be satisfied:

$$V_{sc} \leq V_m = h/\eta \cdot dT/dz , \qquad (2)$$

where : V_{sc} is a scanning speed of heat source, V_m - a speed of liquid crest moving, h- the crest highest, η - viscosity, dT/dz - temperature gradient at the crest highest direction.

If $V_{sc} > V_m$ liquid crest goes to the side of preform analogously to the process of semiconductors cleaning from defects and impurities by the zone "melting" methods. The heat smoothing of laser-prepared preform gave a very good results. The photograph of cylindrical non-spherical lenses one can see at Figure 3.

Some melting technique limitation can be overcomed by using of combination of laser heating or evaporation with mechanical forces action (stretching, twisting , etc.) and mechanical movement like rotation of samples. It gives the possibilities to improve the quality and accuracy of MOC.Very useful application of this version of laser-assisted technique for fiber-tip optical components fabrication are shown in Figure 4 [8].

6. SOME APPLICATIONS OF LASER-FABRICATED MOC.

As it clear from above-mentioned laser techniques can be used for fabrication different MOC and their assemblies.

The main task which was formulated and realized is the following:

1. Fiber-optical tools (FOT) development. A number of FOT is necessary now for focusing , directing and scattering components for medical and technological applications. They provide the desirable numerical aperture NA=d/2f (where d is linear aperture, f is focal length of MOC), different directions of irradiation and various intensity distributions. The main type of produced fiber-tip optical components are presented at Table 2 and photos in Figure 4.

2. Creation of MOC for adjustment of fiber optics, the main parameter of which is NA. By the laser densification of the porous glass after the acid treatment spherical and cylindrical MOC with d = 0.5 - 3 mm and NA = 0.27 were maded. On the same glasses after additional alkalic treatment NA could be increased till 0.37. Double - sides elements and systems maded by the laser action on the both sides of the plate permit to increase NA till 0.47. MOC maded by laser amorphization of GC [6] (CT-50-1, thickness h = 0.3 - 0.6 mm) have diameter of d = 0.4 - 1.2 mm, f = 12 - 14 mm and numerical aperture NA = 0.015 - 0.05.

3. Creation of new type of MOC - fiber concentrators. There are two types of them: a) laser amorphization, focal length f = 1-3 mm [6] ; b)

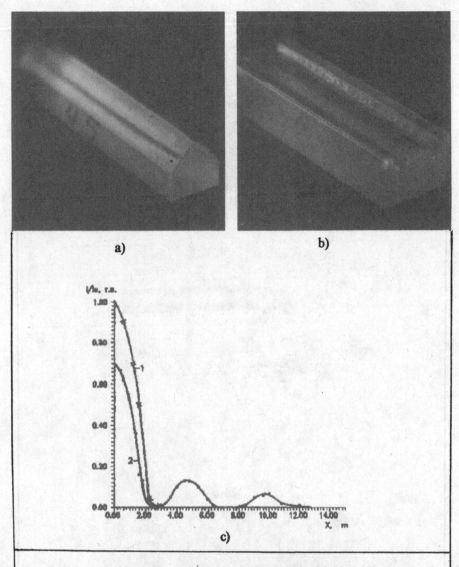

a)

b)

c)

Figure 3. Cylindrical non-spherical lemses for laser diodes collimation produced by laser-assisted technique (a,b) and intensity distributions at focal plane of these lenses (c).

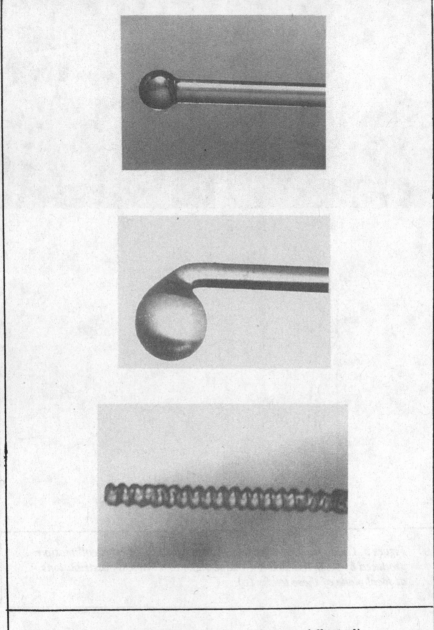

Figure 4. Some examples of laser-produced focusing (elliptical), side-fiber and scattering components.

Table 2.

#	Type – sketch	Name
	I. Focusing end-lenses	
1		arc-end
2		drop-end
3		spherical (bulb) end
4		elliptical end
	II. Side fiber components	
5		focusing side fiber
6		double-spherical side fiber
7		spherical side fiber
	III. Focons components	
8		focon
9		lens-ended focon
	IV. Scattering components	
10		all-side scatterer
11		half-space scatterer
12		point scatterer

Table 3.

N	Application	Form (construction)	Main optical parameters			
			diameter d, mm	focal length f, mm	numerical apertura NA	focal point ρ, μm
1	**Fiber optics**					
	-fiber connector		0.5-2.0	1.5-4.0	0.1-0.2	10
	1. fiber					
	2. fiber-handler					
	3. MOC lens with f'=h					
	- fiber-tip-concentrator		1.5-2.5	1.5-10	0.1-0.5	150-200
	1. fiber					
	2. Al_2O_3 -tube 2,3					
	3. MOC-lens					
2	**Diode lasers optics**					
	- cylinderical collimator (non spherical)		0.5-6.0	1.0-6.0	0.25-0.5	3-10
	- double-sided system for symmetrization and collimation		0.5-50	0.5-10	0.4-0.8	-
3	**Laser optics**					
	- soft-diafragms		0.05-5.0	beam strength $10 \div 10^3$ W/sm^2		
	- lens arrays and matrixes		0.3-5.0	1.5-20	step 0.5-1.0 mm	

N	Application	Form (construction)	Main optical parameters			
			diameter d, mm	focal length f', mm	numerical apertura NA	focal point ρ, μm
4	Imaging optic					
	- objective		0.5-3.0	2.5-15	0.08-0.20	8-15
	1. porous glass		1.0-10	20-200	0.01-0.04	
	2. MOC-lens					
	3. optic glass		1.0-2.0	2.0-8.0	0.2-0.4	5-10
	- double components		1-10	1-10	0.1-1.0	10
5	Integral optics					
	Plane surface and internal optics guides		width 0.3-1.0 mm depth 0.05-0.2 mm $\Delta n = n_2 - n_1 = 0.1$-$0.2$			
	1. glass-like material					
	2. optic guide					
	- Y-multiplexer					
	1. glass-like material					
	2. optic guide					

fabricated using optical glass (rods) by two - sided evaporation, $d = 1\text{-}2$ mm, $f = 0.3$ mm [7].

4. Creation of systems for symmetrization and collimation of laser diodes irradiation. They demand non - spherical astigmatic optics. One of them contained double - sided cylindrical lens system and provided energy transfer effectivity about 0.6. Some photos of new laser etched and fire-smoothed lenses are shown at Figure 3.

Typical examples of MOC fabricated by laser-assisted technique is given at Table 3.

The future progress in laser-assisted techniques of MOC fabrication will give a possibility to create optical elements with the new required parameters.

7. REFERENCES.

1. OE-REPORTS, SPIE PUBLISHING, NN 128-129 Aug.-Sept. 94.
2. V.P.Veiko, "Laser technology of microoptical components with gradient refraction index," Proceedings of SPIE, vol.1352, "Laser surface microprocessing", pp.92-99, 1989.
3. D.J.Shaw, T.A.King, "Densification of Sol-Gel Silica Glass by Laser Irradiation," Proceedings of SPIE, vol.1328, 46, 990.
4. V.P.Veiko, E.B.Yakovlev, V.V.Frolov, V.A.Chuiko, A.K.Kromin, M.O.Abbakumov, A.T.Shakola, P.A.Fomichev, "Laser Heating and Evaporation of Glass and Glass-borning Materials and Its Application for Creating MOC," Proceedings of SPIE, vol.1544, 1991, pp. 152 - 163.
5. J.Bartley, W.Goltoos, "Laser ablation of refractive micro-optic lenslet arrays," Proceedings of SPIE, v.1544, pp.140- 145.
6. V.P.Veiko, K.G.Predko, V.P.Volkov, P.A.Skiba, "Laser Formation of Micro-Optical Elements Based on Glass-Ceramics Materials," Proceedings of SPIE, vol. 1751, 43, 1992.
7. A.A.Mak, V.M.Mit'kin, G.P.Petrovsky, "Formation of Gradient Refractive Index of Glass by Laser Radiation," Dokladi Ac. Sci. USSR, vol. 287, N 4, pp. 845-849, 1986 (in Russian).
8. V.P. Veiko, V.G.Artjushenko , Yu.D.Berezin, V.A.Chuiko, V.P. Chulkov, A.K. Kromin, S.V. Kukhtin, S.A. Rodionov, "Innovative laser technologies for fiber tools fabrication" ,Proceedings of SPIE, vol.2396, 19, 1995.

Mikrostrukturierung von Keramiken im Direct-Write-Verfahren mit Festkörperlasern

A. Raiber, F. Dausinger, H. Hügel
Universität Stuttgart, Institut für Strahlwerkzeuge
Stuttgart / D

1. Einführung

Trotz interessanter Werkstoffeigenschaften haben nicht oxidische Keramiken wie Si_3N_4, SiC und AlN in der Mikrosystemtechnik bisher keine große Verbreitung gefunden. Der Grund ist ihre schwere Bearbeitbarkeit mit den klassischen formgebenden oder mechanischen Verfahren, die die erreichbare minimale Strukturgröße auf 500 µm beschränken. Gebräuchliche Techniken zur Herstellung von Mikrostrukturen, wie z.B. das Ätzen, sind für Keramiken problematisch, da sie nicht selektiv und ausreichend formgenau wirken. Dagegen können beim Einsatz des Maskenprojektionsverfahrens zusammen mit einem Excimerlaser gute Strukturierungsergebnisse erzielt werden.

Eine Alternative zum Maskenprojektionsverfahren stellt das Direktschreiben mit stark fokussierter Laserstrahlung dar. Als Strahlquellen bieten sich hier Festkörperlaser von hoher Strahlqualität an. Verfahrensbedingt ist der Ausnutzungsgrad der angebotenen Strahlenergie sowie die Flexibilität beim Direktschreiben wesentlich höher als bei der Maskenprojektion. Die Stärken des Maskenprojektionsverfahrens dagegen liegen in der Simultanbearbeitung von flächig angeordneten Strukturen.

Dieser Beitrag untersucht die Möglichkeit der Strukturierung von Si_3N_4-, SiC- sowie AlN-Keramiken mit kurzgepulsten Festköperlasern im Direct-Write-Vefahren. Neben den technologischen Daten werden zwei Verfahren zur Erhöhung der Strukturgenauigkeit vorgestellt.

2. Anlagentechnik

Die Strukturierung erfolgt durch das Verfahren des Werkstücks unter einem feststehenden Laserstrahl. Dazu verfügt die Anlage über einen luftgelagerten xy-Tisch mit Drehachse sowie einer z-Achse zur Fokusnachführung für dreidimensionale Bearbeitung. Die Struktur entsteht durch Pulsüberlappung in Vorschubrichtung und seitliche Überlappung einzelner Bahnen. In die Anlage sind ein lampengepumpter Nd:YAG-Laser sowie ein diodengepumpter Nd:YLF-Laser integriert. Die Daten der Laser und der Fokussierung zeigt Tabelle 1.

Laser		Nd:YLF	Nd:YAG
Wellenlänge	nm	1047	1064
Mode		TEM_{00}	TEM_{00}
Pulsenergie	mJ	0,200	2,06
Pulsdauer	ns	10	100
Pulsleistung (100 %)	kW	20	20,6
Pulsfrequenz	kHz	1 bis 50	1 bis 50
F-Zahl		5	5
Brennweite	mm	100	58

Tabelle 1: Daten der Laser und Fokussierung

3. Einflußfaktoren auf die Strukturgenauigkeit und Verbesserungsansätze

Die Genauigkeit der erzeugten Struktur wird durch die Positionierung und den Ablationsprozeß bestimmt. Um die mechanischen Einflüsse der Positionierung auf die Konturschärfe und den Strukturgrund zu minimieren muß die Anlage über eine gute Wiederholgenauigkeit und gute Gleichlaufeigenschaften verfügen.

Bei der Laserablation von nicht oxidischen Keramiken entstehen bei der Zersetzung flüssige Phasen, die in der Struktur verbleiben. Abhängig von der Prozeßführung lagern sich weitere Reaktionsprodukte durch Rekombination und Redeposition in der Struktur ab. Zur Entfernung bzw. Minimierung dieser Ablagerungen wurden zwei Verfahren entwickelt, das chemische Post-Processing sowie das In-situ-Processing. Beim Post-Processing wird die komplett bearbeitete Struktur in ein Ätzbad getaucht. Nach einer Neutralisation erfolgt das Reinigen der Struktur bzw. das Vereinzeln von Strukturteilen nach dem Feinschneiden in einem Ultraschallbad. Die Verfahrensparameter sind abhängig von der verwendeten Keramik. In Tabelle 2 sind die Ätzsubstanzen und -bedingungen aufgeführt.

Für Si_3N_4 und SiC können während der Laserbearbeitung durch In-situ-Processing die Ablagerungen minimiert werden.

Material	Ätzsubstanz	Ätzbedingungen
Si_3N_4	NaOH-Schmelze	350...400 °C, 1..60s
SSiC(α-SiC)	modifizierte Murakami-Lösung (H_2O+KOH+$K_3Fe(CN)_6$)	siedend, 1...10 min
AlN	verdünnte HF	siedend, 1...120 s

Tabelle 2: Verfahrensparameter für das Post-Processing

Dazu wird eine Gasglocke über der Bearbeitungsstelle mitgeführt, die mit Frigen (R22) gespült wird. Die Strukturen sind danach ohne Ätzprozeß direkt im Ultraschallbad zu reinigen oder zu vereinzeln.

4. Direct-Write-Verfahren

Beim Direct-Write-Verfahren werden komplexe Mikrostrukturen durch Einbringung von Nuten oder Schnitten in den Werkstoff erzeugt.

Die maximale Abtragleistung mit den obengenannten Lasern wird bei einer Pulsintensität von ca. 800 bis 1000 MW/cm² erreicht (Bild 1). Trotz einer deutlichen Verringerung der Abtragleistung sollte im Hinblick auf eine gute Strukturgenauigkeit die Pulsüberlappung Ü mindestens 0,9 betragen.

Als ein Kriterium für die Strukturgenauigkeit ist in Bild 2 die Standardabweichung der Abtragtiefe bei einer Pulsfrequenz von 1 kHz gezeigt. Man erkennt, daß der günstigste Bereich der Intensität zwischen 2000 und 3000 MW/cm² liegt und nicht mit dem der maximalen Abtragleistung zusammenfällt. Eine weitere Verbesserung der Genauigkeit ist durch Vorbehandlung der Keramiksubstrate zu erzielen. Die Standardabweichung reduziert sich beim Übergang von unpolierten Oberflächen (R_Z=4 µm) auf polierte (R_Z=0,76 µm) um ca. 30%.

Bild 3 zeigt die Entwicklung der Flankenwinkel in einer 200 µm dicken Si_3N_4-Platte in Abhängigkeit von der Streckenenergie und der Pulsdauer bei konstanter Pulsspitzenleistung von 16 kW. Die

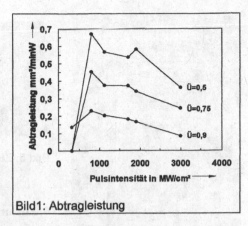

Bild1: Abtragleistung

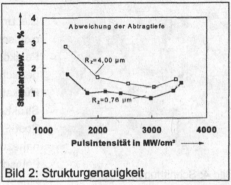

Bild 2: Strukturgenauigkeit

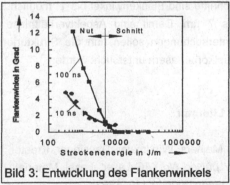

Bild 3: Entwicklung des Flankenwinkels

Vermessung der Winkel wurde nach der DIN-Norm 2310 T5 durchgeführt. Bei der Herstellung von Nuten bewirkt die kürzere Pulsdauer deutlich steilere Flanken während beim Schneiden nahezu kein Einfluß zu erkennen ist. Der Übergang von einer Nut zum Schnitt findet unabhängig von der Pulsdauer bei gleicher Streckenenergie statt.

Die Schnittfugenbreiten liegen bei 200 µm dicken Keramik-Platten bei 5 bis 7 µm. Dies entspricht einem Aspektverhältnis von ca. 1:30. Bild 4 zeigt einen Querschliff durch zwei Schnittspalte in Si_4N_4, die eine Zunge von 40 µm Breite stehen lassen. Es ist keine Werkstoffschädigung zu erkennen. Bild 5 zeigt die komplett ausgeschnittenen Strukturen.

Bild 5: Zungenstrukturen

5. Zusammenfassung

Die Untersuchungen haben gezeigt, daß sich nicht oxidische Keramiken mit kurzgepulsten Festkörperlasern im infraroten Wellenlängenbereich schädigungsarm strukturieren lassen. Rückstände und Ablagerungen, die die Strukturgenauigkeit beeinflussen, können durch das Post-Processing bzw. durch das In-situ-Processing nahezu vollständig beseitigt werden. Beim Feinschneiden sind dadurch die Mikrostrukturen in einem Ultraschallbad vereinzelbar. Für Nuten und

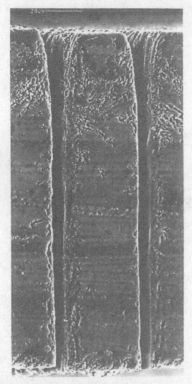

Bild 4: Schnittfugen in Si₃N₄

Schnitte sind Flankenwinkel bis 1° möglich. Die kleinsten Schnittspaltbreiten liegen bei 5 bis 7 μm. Damit sind Aspektverhältnisse von 1:30 zu erzielen. In weiterführenden Untersuchungen sollen nun die Strukturierungsmöglichkeiten mit frequenzvedoppelten Festkörperlasern untersucht werden.

6. Literatur

E. Meiners, Wiedmaier, Dausinger, Krastel, Masek, Kessler: *Micro Machining of ceramics by pulsed Nd:YAG-Laser*. Laser Materials Processing, ICALEO ´91, San José 1991 - Orlando, Florida: LIA, 1992

P. Burck: *Chemisch unterstützter Feinabtrag nicht oxidischer Keramiken mit gepulstem Nd:YAG-Laser*. In: VDI-Broschüre zur Abschlußpräsentation des BMFT-Verbundprojekts: *Abtragen und Bohren mit Festkörperlasern*. Düsseldorf, 1993

Diese Arbeit wurde im Teilvorhaben 13 N 6156 des BMBF-Verbundprojekts „Grundlagen der präzisen, optischen Behandlung von Festkörpern" durchgeführt. Die Autoren sind für den Inhalt verantwortlich.

Laserbearbeitung in der Medizin- und Mikrosystemtechnik

A. Schüßler[1], T. Haas[1], M. Kohl[2], P. Schloßmacher[1] und R. Trapp[3]
Forschungszentrum Karlsruhe
Institut für Materialforschung I[1], Institut für Mikrostrukturtechnik[2], Hauptabteilung
Ingenieurtechnik[3], D-76021 Karlsruhe

NiTi-Formgedächtnislegierungen (FGL) finden gegenwärtig starke Beachtung in Mikroaktorik und Medizintechnik. Für Mikroaktoren sind die hohen Kräfte bzw. Stellwege interessant, beim Bau von mikrochirurgischen Instrumenten nutzt man das pseudoelastische Verhalten und die gute Biokompatibilität der Legierungen aus. Die zunehmende Miniaturisierung und multifunktionale Auslegung von endoskopischen Instrumenten und die Entwicklung von Aktorkonzepten auf der Basis von FGL stellt hohe Anforderungen an die Bearbeitung dieser Werkstoffe im Submillimeterbereich. Die Möglichkeiten zur Herstellung miniaturisierter Bauteile sind allerdings noch unzureichend. NiTi kann galvanisch nicht abgeschieden werden und über das chemische oder elektrolytische Ätzen können lediglich Aspektverhältnisse von maximal 2:1 realisiert werden. NiTi, insbesondere die pseudoelastischen Legierungen, lassen sich wegen der Rückfederung nur schwierig spanend bearbeiten. Der Eintrag von Wärme oder hohe mechanische Spannungen können die Funktionseigenschaften verändern und müssen vermieden werden. Problematisch ist auch das Fehlen geeigneter Fügetechniken für NiTi-Komponenten. Mit den heute noch überwiegend angewendeten kraftschlüssigen Techniken, wie Quetschen, Nieten und Klemmen lassen sich nur wenige Anwendungen abdecken, konventionelle Schweißtechniken, wie das Widerstandsstoßschweißen und das Reibschweißen sind in der Regel auf Makroabmessungen und einfache Geometrien beschränkt. Verfahren zum Kleben sind aufgrund ungenügender Langzeitsicherheit ebenfalls wenig aussichtsreich.

Durch Laserfeinschneiden lassen sich aus NiTi-Röhrchen bzw. Folienmaterial Teile für Endoskope Stents und Katheter (Bild 1), sowie für Aktoranwendungen Miniaturfedern mit definierten Kraft-Weg-Kennlinien (Bild 2) herstellen. Die starke Fokussierung des Laserstrahls auf Durchmesser um 30 µm und das Austreiben der Schmelze mit Argon-Hochdruckgas hält die thermische Belastung der Bearbeitungsflächen beim Schneiden gering, insbesondere im Vergleich zu spanabhebenden Verfahren. Durch kalorimetrische Messungen (Differential Scanning Calorimetry, DSC) an Miniaturkomponenten konnten wir nachweisen (1), daß beim Feinschneiden filigraner Strukturen der Formgedächniseffekt erhalten bleibt.

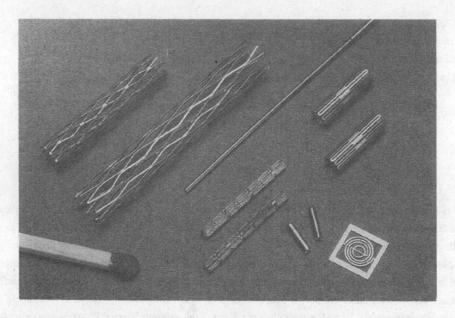

<u>Bild 1:</u> Lasergeschnittene NiTi-Komponenten <u>Bild 2:</u> Kraft-Weg-Kennlinien einer NiTi-Spiralfeder

Für das Verschweißen filigraner mikrochirurgischer Komponenten wurde das Fügen mit Nd:YAG-Lasern entwickelt. Die hier beschriebenen materialkundlichen Untersuchungen wurden an einer kommerziellen NiTi-Legierung (alloy SE, Fa. NDC, USA) mit 51,5 Atom-% Ni durchgeführt, Ergebnisse für eine 49,5% Ni-Legierung sind in (2) veröffentlicht. Das Material lag in Form gewalzter Bleche in den Stärken 0,5 mm und 0,17 mm vor und wurde vor dem Schweißen einer Glühung (550°C/5min) zur Einstellung des pseudoelastischen Verhaltens unterzogen. Eine beiderseitige Spülung der Bleche mit Argon verhinderte die Oxidation der sauerstoffaffinen Legierung beim Fügen.

Das Spannungs-Dehnungsverhalten von NiTi-Legierungen ist stark temperaturabhängig, da sich in Abhängigkeit von der Prüftemperatur (hier -40 °C, +20 °C, +110 °C) die Gefügezustände Martensit, pseudoelastischer Zustand und Austenit einstellen. Die Zugfestigkeit der geschweißten Verbindungen beträgt unabhängig von der Prüftemperatur etwa 820 MPa gegenüber 1050 MPa im Ausgangsmaterial (Bild 3). Für den Standardeinsatz bei Raumtemperatur (pseudoelastischer Zustand) wurde aus 10 Versuchen eine mittlere Zugfestigkeit von 823 MPa mit einer Standardabweichung von +/-53 MPa ermittelt. Dieser Wert entspricht 80% des Ausgangsmaterials. Wichtiger noch als die hervorragende Festigkeit der Verbindungen ist das Ergebnis, daß die Laserschweißnähte unter Bruchverzweigung und plastischer Einschnürung (ε = 8,4%) versagen und somit eine hohe Sicherheit bieten. Die Untersuchung der Mikrostruktur im Transmissions-Elektronenmikroskop zeigt an einigen Stellen im Schweißgefüge Martensitlatten (Bild 4), die erst

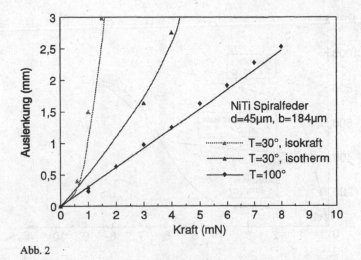

Abb. 2

ab 400 MPa, der Plateauspannung der Umwandlung vom Austenit in spannungsinduzierten Martensit (vgl. Bild 3), entstehen können. Im Schweißgefüge kann also spannungsinduzierter Martensit gebildet werden. Dadurch besteht die Möglichkeit, Spannungsspitzen in der Schweißnaht abzubauen.

Das pseudoelastische Verhalten lasergeschweißter Verbindungen wurde durch Be- und Entlastung bei konstanter Prüftemperatur (20 °C) ermittelt. Bei der Belastung erfolgt die spannungsinduzierte Martensitbildung, während der Entlastung die Rückumwandlung in den Austenit unter einer ausgeprägten Spannungshysterese (Bild 5). Die nach Durchlaufen der Belastungs/Entlastungszyklen

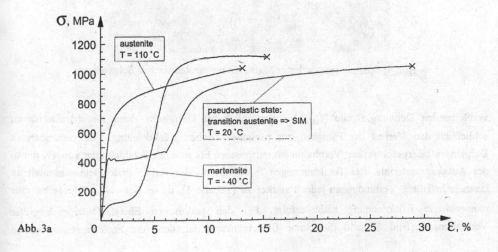

Abb. 3a

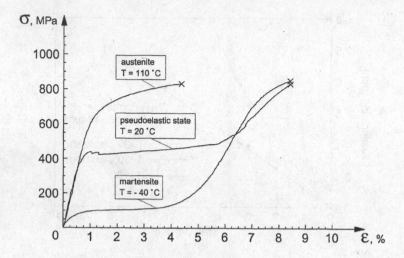

Bild 3: Zugspannungs-Dehnungskurven des Ausgangsmaterials (a) und lasergeschweißter Verbindungen (b)

Bild 4: Spannungsinduzierter Martensit im austenitischen Schweißgut

.verbleibenden Dehnungsanteile (ε_{irrev}) sollen möglichst klein sein, denn sie charakterisieren schließlich den Verlust der Fähigkeit zur pseudoelastischen Rückdehnung. Die verbleibenden Dehnungen lasergeschweißter Verbindungen entsprechen bis zu einer Totaldehnung von 4% denen des Ausgangsmaterials. Bei Totaldehnungen > 4% nimmt der verbleibende Dehnungsanteil an lasergeschweißten Verbindungen jedoch stärker zu (Tabelle 1), da im Schweißgefüge selbst eine nennenswerte Rückdehnung nicht erfolgt. Für den praktischen Einsatz lasergeschweißter Verbindungen (Bild 6) stellt dies keine Einschränkung dar, denn die Funktionseigenschaften

	ε_t	ε_{irrev}
Ausgangs-material	7%	0,05%
	6%	0,04%
	5%	0,03%
	4%	0,02%
geschweißte Verbindung	7%	0,15%
	6%	0,12%
	5%	0,08%
	4%	< 0,05%

Tabelle 1: Irreversible Dehnung nach verschiedenen Totaldehnungen für Ausgangsmaterial und lasergeschweißte Verbindungen

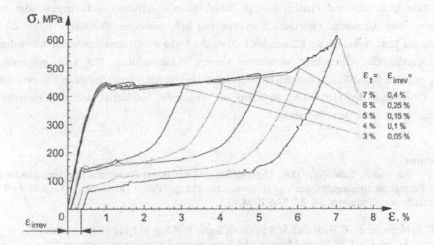

Bild 5: Pseudoelastische Rückdehnung geschweißter Verbindungen

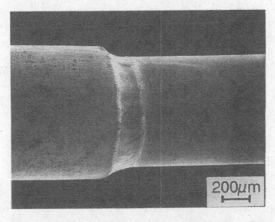

Bild 6: Mikroschweißnaht an einem mikrochirurgischem Instrument

(Formgedächtnis und Pseudoelastizität) werden vom gesamten Bauteil getragen. Durch die hohe Fokussierbarkeit und Einschweißtiefe beim Fügen mit dem Nd:YAG-Laser gelingt es, das Materialvolumen, welches nicht aktiv zur Funktion beiträgt, auf sehr kleine Abmessungen zu beschränken. In Verbindung mit der hohen Versagenssicherheit lasergeschweißter Verbindungen läßt dieses Ergebnis eine weite Verbreitung des Laserschweißens in der Medizintechnik aussichtsreich erscheinen.

Während artgleiche Verbindungen durch Laserschweißen hergestellt werden können, bereitet das Verbinden von NiTi und Titanlegierungen mit artfremden Materialien, wie etwa rostfreien Stählen erhebliche Probleme. Bei der metallurgischen Verbindung dieser Werkstoffe entstehen versprödend wirkende intermetallische Phasen, so daß Schmelzschweißverfahren nicht angewendet werden können. Eine Alternative bietet das Hartlöten mit Silberbasisloten (T: 800 - 1000 °C) unter Schutzgas bzw. Vakuum. So können auch Flußmittel, die Probleme hinsichtlich Korrosion und Biokompatibilität verursachen, vermieden werden. Laserstrahlung läßt sich problemlos in entsprechende Behandlungskammern einkoppeln und besitzt den Vorteil der guten Fokussierbarkeit. Die Erwärmung der Materialien wird auf die Fügestelle beschränkt und ein unerwünschter Wärmeeintrag in die Bauteile reduziert.

Literatur
(1) T. Haas und A. Schüßler, "DSC-Untersuchungen an laserstrukturierten und lasergeschweißten NiTi-Formgedächtnislegierungen", Hauptversammlung der Deutschen Gesellschaft für Materialkunde, Göttingen, 24.-27. Mai 1994.

(2) P. Schloßmacher, T. Haas and A. Schüßler, "Laser Welding of NiTi Shape Memory Alloys", 1st Int. Conf. on Shape Memory and Superelastic Technologies, Monterey, California, March 6-10 1994.

Zur Wechselwirkung von ps-Pulsen mit Metallen

Bernd Hüttner, Gernot Rohr

DLR - Institut für Technische Physik

Pfaffenwaldring 38-40, 70569 Stuttgart

Einleitung

Laser mit Pulszeiten im ps-Bereich und darunter besitzen ein großes Potential für viele Anwendungen (Strukturierung, laserablatierte Schichten), dessen Ausschöpfung allerdings wesentlich von dem theoretischen Verständnis der bei der Laserabsorption ablaufenden Prozesse bedingt wird.

Hier stellt sich nun die Frage, inwieweit können theoretische Ansätze, welche die Wechselwirkung von ns-Pulsen mit Metallen erfolgreich beschreiben, auch für kürzere Pulse angewendet werden, oder sind prinzipielle Ergänzungen oder gar neue Ansätze erforderlich.

Wir glauben, daß diese oder ähnliche Fragen der Ursprung einer gewissen Renaissance von älteren ad hoc Modifikation der Wärmeleitungsgleichung /1,2/ sind, die oft mit dem, u. E., irreführenden Begriff 2. Schall verbunden ist.

Theorie

Mathematisch gesehen ist die Wärmeleitungsgleichung (WLG) eine parabolische Differentialgleichung, die einen Markov-Prozeß, d.h. einen Prozeß bei dem Vergangenheit und Zukunft unabhängig voneinander sind, beschreibt. Physikalisch impliziert die WLG eine vollständige Kopplung zwischen dem Elektronen- und dem Phononensystem zu *allen* Zeiten. Für Metalle folgt daraus, daß die WLG eine Gleichung für die Elektronen mit dem Wärmediffusionskoeffizienten der Phononen ist. Dieser Ansatz beschreibt die Temperaturverteilung im Metall korrekt, solange die Laserpulszeit sehr viel größer als die Energierelaxationszeiten für die Elektronen und Phononen ist. Im ps-Bereich ist dies a priori nicht mehr zu erwarten und man hat deshalb ein gekoppeltes Differentialgleichungssystem für die Elektronen und Phononen /3/ zu betrachten.

Der Grad der Entkopplung wird durch die Elektron-Phonon-Wechselwirkung bestimmt. Dies bedeutet, daß die guten Leiter früher entkoppeln und bis zu größeren Zeiten durch zwei getrennte Temperaturen für die Elektronen und die Phononen beschrieben werden müssen. Zusätzlich werden wir dem Wärmestrom, der durch die Elektronen geleistet wird, noch ein "Gedächtnis" geben, so, daß der Wärmestrom am Orte z zur Zeit t nicht gänzlich von früheren Zeiten unabhängig ist. Als "natürliche" Gedächtnisrelaxationszeit wählen wir die Elektronenenergierelaxationszeit. Dies ergibt mit der Memoryfunktion g(t-t')

$$q(z,t) = - \int_0^t g(t-t') \ \lambda \ gradT(z,t') \ dt' \ .$$

Wählt man nun für g(t-t') eine Delta-Funktion, so gelangt man zur parabolischen WLG, im Grenzfall langer Relaxationszeiten zur hyperbolischen Telegraphengleichung. Mit einem exponentiellen Gedächtnis besitzt man einen allgemeineren Ansatz, der beide Grenzfälle enthält.

Setzt man dies in die Kontinuumsgleichung für die Energie ein und differenziert nach der Zeit, so findet man eine Differentialgleichung 2. Ordnung in der Zeit und zusätzlich einen Term proportional zur Zeitableitung der absorbierten Leistungsdichte. Nach einigen Umformungen ist folgendes Gleichungssystem zu lösen

$$\frac{\partial^2 T_e}{\partial t^2} + \nu \ \frac{\partial T_e}{\partial t} - \nu k_e \ \frac{\partial^2 T_e}{\partial z^2} = \frac{\nu}{c_e} [I(z,t) + \tau_e \frac{\partial I(z,t)}{\partial t}] - \nu^2 (T_e - T_{ph})$$

$$\frac{\partial T_{ph}}{\partial t} = \varphi \cdot (T_e - T_{ph})$$

wobei $\nu = \tau_e / \tau_L$ und $\varphi = \tau_{ph} / \tau_L$ durch das Verhältnis der Energierelaxationszeit der Elektronen bzw. der der Phononen zur Laserzeit definiert sind. k_e und c_e stehen für den Wärmeleitungskoeffizient bzw. für die spezifische Wärme der Elektronen. Die absorbierte Energie pro Volumen und Zeit ist gegeben als

$$I(z,t) = I_{abs} \cdot f(t) \cdot \alpha \cdot \exp(-\alpha z)$$

mit dem Absorptionskoeffizient α und $f(t) = (t/\tau_L) \cdot \exp(-t/\tau_L)$ als zeitliche Verteilung der Laserleistungsdichte.

Die Berechnung erfolgte mittels der Laplace-Transformation und numerischer Rücktransformation.

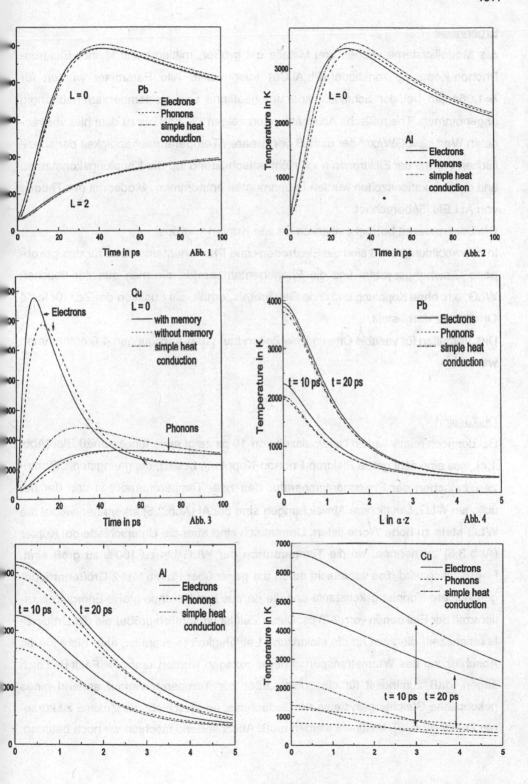

Abb. 1

Abb. 2

Abb. 3

Abb. 4

Ergebnisse

Als Modellsysteme wurden drei Metalle mit großer, mittlerer und kleiner Elektron-Phonon-Kopplungskonstante (Pb,Al,Cu) ausgewählt. Alle Parameter wurden für $\lambda=1.064$ µm bei der Schmelztemperatur bestimmt und als temperaturunabhängig angenommen. Theoretische Abschätzungen zeigen /3/, daß bis zu dem hier verwendeten Wert $I_{abs}=5GW/cm^2$ der Einfluß der linearen Temperaturabhängigkeit der spezifischen Wärme der Elektronen noch nicht entscheidend ist. Die Kopplungskonstanten und die Relaxationszeiten wurden Experimenten entnommen /4/ oder mit der Theorie von ALLEN /5/ berechnet.

Die Laserpulszeit beträgt einheitlich für alle Kurven $\tau_L=10ps$.

In den Abbildungen 1-3 sind die Elektronen- und Phononentemperatur für das gekoppelte Gleichungssystem und die Phononentemperatur, die man aus der üblichen WLG, d.h. ohne Kopplung und ohne Gedächtnis, erhält, als Funktion der Zeit für feste Orte $L=\alpha\cdot z$ dargestellt.

Das Verhalten für variable Orte und fixe Zeiten kann den Abbildungen 4-6 entnommen werden.

Diskussion

Bei der noch relativ langen Laserpulszeit von 10 ps zeigt erwartungsgemäß Blei (Abb. 1,4), was eine sehr große Elektron-Phonon-Kopplung besitzt, die geringsten Differenzen zwischen der Phononentemperatur des zwei Temperaturmodells und der der üblichen WLG. Deutlichere Abweichungen sind bei Al (Abb.2,5) erkennbar, wobei die WLG stets zu hohe Werte liefert. Dramatisch sind aber die Unterschiede bei Kupfer (Abb.3,6) zu nennen, wo die Temperaturen der WLG bis zu 100% zu groß sind. Physikalisch wird dies verursacht durch die gegenüber Pb um fast 2 Größenordnungen kleinere Kopplungskonstante und die daraus resultierende große Energierelaxationszeit der Elektronen von (2-3) ps. Diese Zeit ist wesentlich größer als die Druderelaxationszeit, die zwar für die elektrische Leitfähigkeit relevant ist, aber nicht für die Abschätzung des Wärmetransportes herangezogen werden kann. Als Fazit läßt sich sagen, daß zumindest für die guten Leiter der Temperaturverlauf anhand eines gekoppelten Gleichungssystems mit Gedächtnis, es verschiebt die Maxima zu kürzeren Zeiten (Abb. 3), bestimmt werden muß. Abschließend möchten wir noch betonen, daß bei höheren Leistungsdichten die Temperaturabhängigkeit der spezifischen Wärme der Elektronen und damit verbunden die der Energierelaxationszeit explizit berücksichtigt werden muß, was zu einem Wettbewerb zwischen dem stärkeren Aufheizen der Elektronen und einer einsetzenden Kühlung durch thermische Emission führen wird.

Literatur

/1/ C. Cattaneo - Compt. Rend. 247, (1958), 431
/2/ P. Vernotte - Compt. Rend. 252, (1961), 3154
/3/ S. Anisimov et al. - Effects of high-power radiation on metals, Moscow, Nauka Publishing House, 1970
/4/ S. D. Brorson et al. - Phys. Rev. Lett. 64, (1990) 2172
/5/ P. B. Allen - Phys. Rev. Lett. 59 (1987) 1460

Micromachining of Metals by Laser Chemical Processing in Liquids

R. Nowak, S. Metev, K. Meteva, G. Sepold

BIAS, Bremer Institut für angewandte Strahltechnik,

Klagenfurterstraße 2, D-28359 Bremen, Germany

I. Introduction

Laser radiation provides a unique tool for various micromachining methods using either physical (1) or chemical (2) processes resulting in material removal. Physical processes for material removal, however, are often accompanied by redeposition of evaporated mater near the laser interaction region. Especially in the case of metals a temporary molten phase seems to be unavoidable (1). Its resolidification results in a reduction of the accuracy of the treatment (surface roughness, deviation from the desired geometry). In such cases material removal by localized chemical etching reactions can significantly increase the accuracy of laser micromachining, providing some additional means for direct 3D-micromachining of metals. In comparison to laser-induced physical processes for material removal, however, chemical processes in many cases are expected to be of lower processing speeds, as the amount of the removed material depends not only on laser power but also on the rate of mass transport of incoming and outgoing reactands as well as on the rate of elementary chemical reactions. A high density of reactands, as in liquids, and a pronounced exchange of reactands, i. e. induced by convection, are prerequisites for high chemical reaction rates. In addition, processing in an aqueous environment is accompanied by an efficient cooling of the sample's surface.

The aim of this contribution is to present some recent results of our investigation on laser-induced wet chemical etching of metals using a focussed cw-Nd:YAG- and cw-Ar-laser. While our previous work (3, 4) was devoted mainly to investigations on thin metal films, this work will concentrate on laser micromachining of Ti and stainless steel 304 foils.

II. Experimental Method

A detailed description of the experimental system used in the course of this study can be found elsewhere (3). In brief, the beam of either a cw-Ar- (< 1.5 W, 514 nm, TEM_{00}) or cw-Nd:YAG-laser (<10 W, 1.06 µm, TEM_{00}) was focussed to an estimated focal spot diameter of about 1 µm or 25 µm, respectively, on thin metal foils immersed in the liquid etchant (5M phosphoric or 1.8M sulfuric acid). Samples were translated using a computer-controlled xy-translation stage at speeds ranging from about 1 µm/s to about 2 mm/s. Etched structures were investigated using optical or scanning electron (SEM) microscopy. Depths and widths of etched grooves were determined by stylus profilometry and SEM investigations.

III. Results

a. Titanium

It is well-known that Ti is passivated at room temperature by a thin oxide layer resulting in an apparant electrochemical potential similar to that of Pt (5). Thus, a corrosion rate of Ti in phosphoric acid less than about 0.1 mm/year at room temperature and at a wide range of concentrations (5) is observed. At elevated temperatures Ti becomes unstable in phosphoric acid and dissolves at a rate of about 2 nm/s in 5M H_3PO_4 at 90 ^oC as was measured in our laboratory. This peculiarity offers a prerequisite for localized activation of chemical etching reactions. Consequently, a temperature increase of the surface induced by the incident laser radiation results in chemical etching reactions which are approximately confined in the laser spot. Focussing laser radiation on the sample surface results in etching of small circular holes in 5M H_3PO_4 at rates up to about 10 μm/s and a laser power of about 1W. Since in deionized water the samples showed no indications of melting of the surface when irradiated under the same experimental conditions, the etching mechanism is believed to be exclusively due to chemical reactions. In addition, etch rates and diameter of etched holes were found to depend strongly on laser power, as can be expected for thermally activated chemical reactions. Volumetric etch rates up to 10^6 μm³/s were obtained at a laser power of 5 W.

Fabrication of microstructures with aspect ratios larger than one requires a detailed consideration of the influence of heat diffusion on geometry. The geometry of etched grooves and cuts in Ti foils strongly depends on laser power and scanning velocity. Investigations on etching of grooves in thin foils revealed etched widths approximately 1 to 2 times larger than the etched depths at scanning speeds up to 100 μm/s and laser powers up to 5 W. In this parameter range an approximately linear dependence of etched depths on laser power was found. For example 140 μm wide grooves of 70 μm depth were produced at a scanning velocity of 20 μm/s and 4 W using the Nd:YAG-laser. On the other hand, thermally activated chemical etching reactions can result in a feature size much smaller than the laser spot diameter. Previous investigations on laser-induced chemical etching of lines in Ti thin films resulted in etched line widths approximately two times smaller than the laser spot diameter (3). Precise cutting of 25 μm thick foils in phosphoric acid was obtained at scanning speeds up to 50 μm/s. Cut angles of about 70 to 80° were measured. However, the width of the etched cuts was about two times larger than the samples thickness.

Aspect ratios of etched grooves larger than two were obtained upon multiple scanning the same path. Fig. 1 shows some results on the influence of the number of scans across the same path on the width and depth of etched grooves. Obviously in this case multiple scanning results in a relatively unchanged width of approximately the same size as the laser spot diameter while the depth steadily increases. It is interesting to note, that etching of grooves in phosphoric acid under the given experimental conditions is exclusively due to chemical etching reactions. Repeating the experiments in deionized water under the same conditions resulted in no visible change of the sample's surface.

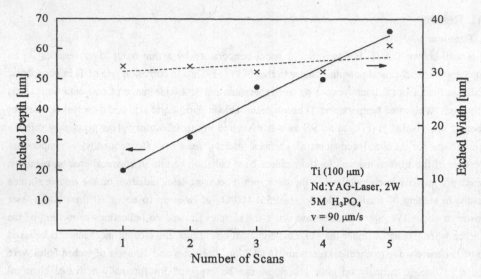

Fig. 1: Influence of multiple scanning on depth and width of etched grooves in Ti foil (thickness 100 µm).

Fig. 2 shows a SEM microphotograph of a small spiral which has been produced in a 50 µm thick Ti foil using a Nd:YAG-laser power of 4.2 W, 5M phosphoric acid and a scanning velocity of about 20 µm/s. Again repeating the experiment with deionized water instead of the acid resulted in no visual change of the metal foil revealing the pure chemical mechanism. From SEM investigations a roughness of the etched side walls of less than 0.5 µm has been estimated.

Fig. 2: Microspiral fabricated from Ti by Nd:YAG-laser-induced wet chemical etching in H_3PO_4

b. Stainless Steel 304

Investigations on electrochemical enhanced laser-induced etching of stainless steel 304 were performed in 5M phosphoric acid and 1.8M sulfuric acid using an Ar-laser. Experiments on drilling of small holes were carried out in the range of anodic passivation of the metal at electrical background current densities less than 1 $\mu A/cm^2$. It was found that in sulfuric acid laser-induced electric current densities of the order of 10 A/cm^2 resulted upon illumination at a laser power of 0.6 W. The diameter of etched holes was approximately 100 μm using a focal spot diameter of about 100 μm. Several mechanisms may contribute to this result. One mechanism is based on a laser-induced break-down of the passivation layer because of thermal effects. Another mechanism probably could origin in a localized shift of the electrical potential in the laser interaction zone. Currently the detailed mechanisms of electrochemical enhanced laser-induced etching are being under detailed investigations.

III. Conclusions

Laser-induced wet chemical etching was shown to be a clean and precise method for microstructuring of metals without any significant thermal load of the workpiece. Obtained roughness of etched side walls were less than about 0.5 μm. Etching of grooves was obtained at speeds up to about 100 $\mu m/s$ at laser powers of about 4 W and a wavelength of 1.06 μm. Cutting of thin metal foils resulted at maximum velocities of about 50 $\mu m/s$ for 25 μm thick foils. Enhancement of electrochemical etching reactions by laser irradiation in the passivation region was demonstrated.

IV. Acknowledgements

The authors gratefully acknowledge A. Klett for his technical assistance and the German Ministry for Education and Science under contract No. 13N6158 for financial support.

V. References

[1]   S. M. Metev and V. P. Veiko, *Laser-assisted Microtechnology*, Springer, Heidelberg, 1994
[2]   D. Bäuerle, *Laser Chemical Processing*, Springer, Heidelberg, 1986
[3]   R. Nowak, S. Metev and G. Sepold, Materials and Manufacturing Processes 9, 429(1994)
[4]   R. Nowak, S. Metev and G. Sepold, *High Power Lasers and Applications* SPIE 2207, 633(1994)
[5]   DECHEMA Corrosion Handbook Vol. 12, ed. by G. Kreysa and R. Eckermann, VCH Weinheim, 1993, pp. 273-276

Laserunterstütztes Ätzen zur 3D-Strukturierung von Silizium
Laser-Supported Etching of 3D Structures in Silicon

A. Schumacher, M. Alavi, B. Schmidt
Institut für Mikro- und Informationstechnik
D-78052 Villingen-Schwenningen

Einführung

Bei der Herstellung mikromechanischer Bauelemente durch 3D-Strukturierung von Silizium werden üblicherweise anisotrope naßchemische oder ionenstrahlgestützte Trockenätzverfahren eingesetzt. In beiden Fällen wird die Form der Mikrostrukturen durch eine geeignete photolithographisch strukturierte Maskierschicht auf der Oberfläche des Siliziumsubstrats festgelegt. Zudem ist beim anisotropen Naßätzen das maximal realisierbare Verhältnis von Tiefe zu Breite, das Aspektverhältnis, grundsätzlich durch ätzresistente {111}-Kristallebenen begrenzt.

In dieser Arbeit werden laserunterstützte Naß- und Trockenätzprozesse zur 3D-Strukturierung von Silizium untersucht sowie ihre Anwendungsmöglichkeiten in der Mikromechanik demonstriert. Die Untersuchungen wurden mit einem Mikrobearbeitungssystem bestehend aus einem kontinuierlich angeregten frequenzverdoppelten Nd:YAG-Laser mit Güteschalter (TEM$_{00}$, Wellenlänge: 532 nm, mittlere Leistung: bis 8 W), einer computergesteuerten xy-Positioniervorrichtung (Verfahrweg: 100 mm × 100 mm, Positioniergenauigkeit: 1 μm) und einem Kleinrechner durchgeführt.

Laserunterstütztes anisotropes Naßätzen

Eine Vielfalt neuer Strukturen mit definierten Berandungsflächen und variierbaren Aspektverhältnissen wird möglich, wenn Photolithographie und Ätzen mit einer vorangehenden Lasermikrobearbeitung kombiniert werden. Mit Hilfe eines fokussierten Laserstrahls werden die Kristallordnung im Silizium und damit auch die ätzresistenten {111}-Ebenen lokal zerstört. Das nachfolgende anisotrope Ätzen beseitigt diese zerstörten Kristallbereiche und kommt erst bei Erreichen von unversehrten {111}-Ebenen zum Stillstand. Somit können völlig neuartige

Strukturen mit höherem Aspektverhältnis realisiert werden. Im Falle von <110>-Silizium beispielsweise entstehen bei einer Ausrichtung der Zerstörungszonen in [1$\overline{1}$0]-Richtung teilweise geschlossene Mikrokanäle. Bei einem bestimmten Abstand zwischen zwei benachbarten, in [1$\overline{1}$0]-Richtung ausgedehnten Schmelzzonen in einem <110>-Siliziumwafer erhält man nach dem Ätzprozeß freitragende und monolithisch mit dem Substrat verbundene

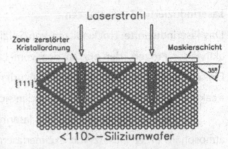

Abb. 1: Schematische Darstellung des Verfahrens zur Herstellung von freitragenden Balkenstrukturen in Silizium.

Balkenstrukturen mit dreieckiger Querschnittsform und Dicken bis zu einigen 100 μm. Die auf diese Weise realisierten Balken werden an ihrer Unterseite durch {111}-Ebenen begrenzt, welche zur Substratoberfläche einen Winkel von etwa 35 Grad bilden (Abb. 1) [ALAVI93].

Nach diesem Prinzip lassen sich auch weitaus komplexere Strukturen für Anwendungen in der Mikrosensorik formen. Mit Hilfe eines CAD-Systems und durch Variation der Laserstrahlparameter können exakte geometrische Vorgaben, sowohl in der Breite als auch in der Tiefe, auf das Siliziumsubstrat übertragen werden. Durch lokale Zerstörung der Kristallordnung innerhalb einer definierten Fläche mit dem fokussierten Laserstrahl und nachfolgen-

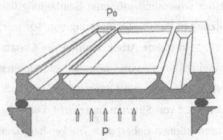

Abb. 2: Schematische Darstellung der durch laserunterstütztes anisotropes Ätzen hergestellten Basisstruktur für einen Drucksensor.

dem Ätzprozeß wurde eine monolithisch integrierte Balken-Membran-Struktur hergestellt, die als Basisstruktur für einen Drucksensor dient [SCHUM94]. Die in Abb. 2 dargestellte Anordnung besteht aus einer dünnen Membran, die auf ihrer Rückseite über Träger mit einer Balkenstruktur verbunden ist. Infolge der Durchbiegung der Membran unter Druckeinwirkung erfährt der Balken über die Trägerstützen eine Deformation in Längsrichtung (Kniehebeleffekt). Die dadurch verursachte Änderung des Spannungszustandes der Balkenstruktur kann mit Hilfe von sensitiven Schichten in elektrische Meßgrößen umgewandelt und zur Erfassung des Druckes verwendet werden.

Da bei der Laserbearbeitung des Siliziumwafers eine darüber befindliche Maskierschicht lokal mit zerstört wird, kann auf den Photolithographieschritt ganz verzichtet werden, wenn keine extremen Anforderungen an die Strukturgenauigkeit vorliegen. Das bedeutet, daß mit dem laserunterstützten anisotropen Naßätzen relativ schnell Prototypen hergestellt werden können ("Rapid Prototyping").

Laserinduziertes Trockenätzen

Das laserinduzierte Trockenätzen von Silizium in Chlorgas ist ein thermisch/photochemisch aktivierter Prozeß, der nur an den vom Laser bestrahlten Bereichen stattfindet. So ist der Einsatz von photolithographisch strukturierten Maskierschichten nicht erforderlich. Da die Reaktionsprodukte gasförmig sind, finden sich auf der Substratoberfläche keine Ätzrückstände.

In den letzten Jahren war das laserinduzierte Trockenätzen von Silizium in Chlorgas-atmosphäre unter Einsatz von Excimerlasern (Impulsbetrieb) oder Edelgas-Ionen-Lasern (cw-Betrieb) Gegenstand zahlreicher Untersuchungen. Entscheidend für einen wirtschaftlichen und damit industriellen Einsatz des laserinduzierten Trockenätzens als ein seriell durchführbares Verfahren ist in erster Linie neben der erzielbaren Bearbeitungsqualität die maximal realisier-bare Ätzrate. Je nach der verwendeten Laserintensität wurden beim Argon-Ionen-Laser Volu-menätzraten von 15 $\mu m^3/s$ (anisotropes Ätzen [ARRONE89]) bis 7500 $\mu m^3/s$ (isotropes Ätzen [BLOOM91]) erzielt. Excimerlaser liefern eine homogen verteilte Strahlungsintensität innerhalb einer wesentlich größeren Bearbeitungsfläche, so daß die maximal realisierbare Volumen-ätzrate in der Größenordnung von 10^5 $\mu m^3/s$ liegt [BOULMER92].

In dieser Arbeit wurden die Einsatz-möglichkeiten von frequenzverdoppelten Nd:YAG-Lasern mit Güteschalter zum Trok-kenätzen von Silizium untersucht. Diese Sy-steme können neben dem Trockenätzprozeß für zahlreiche andere Anwendungen in der Mikrotechnik, wie beispielsweise zur Chipse-parierung oder in der Mikroverbindungs-technik (Laserlöten) eingesetzt werden. Zur Durchführung der Untersuchungen zum la-serinduzierten Trockenätzen von Silizium

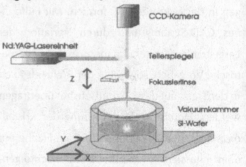

Abb. 3: Schematische Darstellung der Pro-zeßkammer, montiert auf dem xy-Tisch, mit Strahlquelle, Fokussierungs- und Beobach-tungsoptik.

wurde das oben beschriebene Lasermikrobearbeitungssystem mit einer Prozeßkammer und einer Vakuum- und Gassteuerung ausgestattet. Abb. 3 zeigt schematisch diese Prozeßkammer mit der Strahlquelle und der Fokussierungs- und Beobachtungsoptik.

Bei den folgenden Experimenten wurde bei einer Pulsfrequenz von 50 kHz und einem Fokusdurchmesser von etwa 25 μm gearbeitet. Es wurden <100>-Siliziumwafer mit einer Phosphordotierung von 4 ... 7,2 Ωcm verwendet. Durch Variation der Bearbeitungsparameter kann das Ätzergebnis in weiten Bereichen beeinflußt werden. Mit steigender Laserleistung nimmt die Ätzrate zu (Abb. 4), wobei die entstehenden Strukturen sowohl in ihrer Tiefe als auch in der Breite anwachsen. Deren Verhältnis, die Anisotropie, bleibt nahezu unverändert. Der Ätzboden ist in diesem Bereich sehr glatt. Bei einer Leistungsdichte von 680 kW/cm^2

wurde ein Volumenabtrag von 40.000 $\mu m^3/s$ erzielt. Bei weiterer Leistungserhöhung schmilzt das Substrat lokal, die Ätzrate nimmt nicht weiter zu, und der Ätzboden wird uneben. Die Probenvorschubgeschwindigkeit wirkt sich in weiten Bereichen nicht auf die Volumenabtragsrate aus. Abb. 5 zeigt in einer REM-Aufnahme den Querschnitt einer Ätzgrube, hergestellt in 250 mbar Chlorgas bei etwa 340 kW/cm² Laserleistungsdichte.

Die Anwendungsmöglichkeiten des laserinduzierten Trockenätzens liegen in der maskenlosen, zerstörungs- und rückstandsfreien Mikrostrukturierung von bereits prozessierten Siliziumwafern zur Prototypen- und Kleinserienfertigung bzw. zur kundenspezifischen Modifizierung von mikromechanischen Bauteilen.

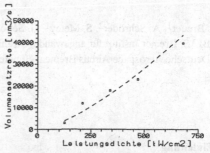

Abb. 4: Volumenabtragsrate von <100>-Silizium in Abhängigkeit von der Laserleistungsdichte (Chlorgasdruck: 200 mbar).

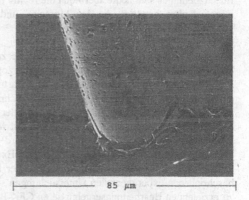

Abb. 5: Querschnitt einer Grube, hergestellt durch laserinduziertes Trockenätzen in 250 mbar Chlorgas und bei 340 kW/cm² Laserleistungsdichte.

Danksagung

Die vorliegende Arbeit wurde im Rahmen des AiF-Vorhabens (Arbeitsgemeinschaft industrieller Forschungsvereinigungen e. V.) Nr. 9017 durchgeführt, welches aus Mitteln des Bundesministers für Wirtschaft gefördert wurde.

Literatur

[ALAVI93]   M. Alavi, Th. Fabula, A. Schumacher, H.-J. Wagner, Monlithic microbridges in silicon using laser machining and anisotropic etching, Sensors and Actuators A, 37–38 (1993) 661–665.

[ARNONE89] C. Arnone, G. B. Scelsi, Anisotropic laser etching of oxidized (100) silicon, Appl. Phys. Lett. 54(3) (1989) 225–227.

[BLOOM91] T. M. Bloomstein, D. J. Ehrlich, Laser desorption and etching of three-dimensional microstructures, 1991 IEEE, pp. 507–511.

[BOULMER92] J. Boulmer, J.-P. Budin, B. Bourguignon, D. Débarre, A. Desmur, Laser ablation and laser etching, Laser Ablation of Electronic Materials – Basic Mechanism and Applications, E. Fogarassy, S. Lazare (Editors), 1992, Elsevier Science Publishers B. V., pp. 239–253.

[SCHUM94] A. Schumacher, M. Alavi, Th. Fabula, B. Schmidt, H.-J. Wagner, Monolithic bridge-on-diaphragm microstructure for sensor applications, Proc. Micro System Technologies '94, Berlin, 1994, pp. 309–316.

Einfluß der Excimerlaserstrahlung auf das Fügeverhalten von Faserverbundwerkstoffen

J. Breuer[1], A. Schröder[2], S. Metev[1], G. Sepold[1]

[1]BIAS, Bremer Institut für angewandte Strahltechnik; Klagenfurter Str. 2; 28359 Bremen
[2]Deutsche Aerospace Airbus Bremen; Hünefeldstr. 1-5; 28199 Bremen

Einleitung

Faserverstärkte Werkstoffe (CFK) werden insbesondere aufgrund ihrer Vorteile hinsichtlich Gewichtseinsparung und konstruktiver Gestaltungsmöglichkeiten gegenüber Leichtmetallen in zunehmendem Maße, beispielsweise in der Flugzeugindustrie, als Strukturwerkstoff eingesetzt. Generelles Einsatzhemmnis sind heute aber noch die relativ hohen Bauteilkosten. Deshalb werden u.a. geeignete Fügetechniken benötigt, die den Montage- und Reparaturaufwand zu verringern helfen. Ansatzpunkte dafür bietet die Klebtechnologie.

Prinzipiell müssen CFK-Werkstoffe, bevor sie zu einem langzeitstabilen Verbund verklebt werden können, einem Vorbehandlungsverfahren unterzogen werden. Standardverfahren sind das sog. "Peel-Ply"-Abziehen - als Decklage wird vor dem Härtungsprozeß ein später abzuziehendes Zusatzgewebe aufgelegt - und, häufig auch als zusätzlicher Schritt, das mechanische Anschleifen. Ein großflächiges Anschleifen wird ebenfalls für Reparaturen oder zum Ausbessern schadhafter Peel-Ply-Flächen angewandt. Diese Vorgehensweisen sind sehr aufwendig, zum Teil auch schlecht reproduzierbar, weshalb nach Verfahrensalternativen gesucht wird. Günstige Voraussetzungen für einen Einsatz als alternatives Vorbehandlungswerkzeug liefern Excimerlaser aufgrund ihrer charakteristischen Eigenschaften, der hohen Photonenenergien und der kurzen Impulsdauern /1,2,3/. In den im folgenden beschriebenen experimentellen Untersuchungen sollte deshalb geklärt werden, ob die mit dem Laser zu erzielenden Bearbeitungsergebnisse an CFK-Proben für Klebverbindungen brauchbar sind und ob diese insbesondere Festigkeitsansprüchen der Luftfahrtindustrie genügen.

Beschreibung des lasergestützten Vorbehandlungsverfahrens

Die Untersuchungen zum lasergestützten Vorbehandeln von Faserverbundwerkstoffen wurden vor dem Hintergrund durchgeführt, daß die Beeinflussung der Bauteiloberfläche so gezielt erfolgen muß, daß zwar Matrixharz in ausreichender Menge abgetragen wird um einen sauberen Neuaufbau des Verbundes zu erzeugen, andererseits aber kein Angriff auf die Fasern selbst erfolgt. Wegen der kurzen Impulsdauern der Excimerlaser von 10 - 40 ns ist die Abtragstiefe je Impuls in der Regel auf weniger als 1 μm beschränkt. Dies stellt eine günstige Voraussetzung für die Steuerbarkeit des Prozesses dar. Wenn sich zusätzlich chemische Veränderungen einstellen, können diese auf eine dünne Oberflächenschicht beschränkt werden.

Für die nachfolgend beschriebenen Untersuchungen stand ein Excimerlaser, der mit der Wellenlänge $\lambda = 308$ nm betrieben wurde, zur Verfügung. Probenwerkstoff war das CFK-Epoxidsystem C6376-HTA der Fa. Ciba Composites. Durch geeignete Strahlformung wurde auf der Probenoberfläche ein rechteckiges Bestrahlungsfeld von ca. 5x5mm² erzeugt. Da für die Bindefestigkeitsuntersuchungen, die in Anlehnung an die für die Luftfahrt relevante Prüfnorm prEN2243-1 erfolgten, größere bestrahlte Zonen erforderlich waren, wurden die Proben auf einem in zwei Achsrichtungen verfahrbaren Probentisch aufgespannt und konnten entsprechend abgerastert werden. Zur qualitativen Beurteilung des Bestrahlungsergebnisses dienten zunächst mikroskopische Aufnahmen. Um chemische Änderungen an den Werkstoffoberflächen festzustellen, wurden XPS-Analysen angefertigt. Für Bindefestigkeitsuntersuchungen wurden jeweils zehn Klebproben mit zwei unterschiedlichen Bestrahlungsparametern hergestellt.

Mikroskopische Untersuchungen

Zur Beurteilung struktureller Änderungen, die durch das Laserbestrahlen initiiert worden sind, wurden rasterelektronenmikroskopische Aufnahmen von den Bestrahlungszonen angefertigt, Bild 1.

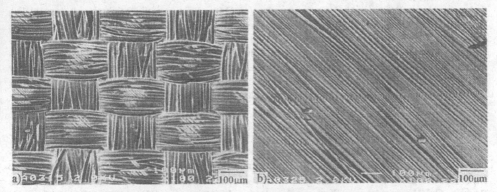

Bild 1: Gesteuerter Werkstoffabtrag an CFK-Laminaten nach Peel-Ply-Abzug (REM-Aufnahmen)
Bestrahlungsparameter: $\lambda = 308$ nm, Energiedichte: 0,08 J/cm^2; f = 2 Hz
a) 10 Impulse; b) 100 Impulse

In den Aufnahmen von Proben, die zum einen mit 10 Impulsen und zum anderen unter sonst gleichen Bedingungen mit 100 Impulsen beaufschlagt worden sind, ist ein aus morphologischer Sicht gleichförmiger Harzabtrag zu erkennen. Während bei der Bestrahlung mit niedriger Impulszahl die Oberfläche noch größtenteils mit Harz bedeckt ist, liegen bei der hohen Impulsanzahl die Fasern vollständig frei. Zum Teil ist Matrixharz auch zwischen den Fasern entfernt worden. Die Fasern selbst sind aber, wie auch durch zusätzliche Schrägschliffe bestätigt werden konnte, nicht beschädigt. Schädigungen bilden sich erst bei wesentlich höheren Impulsanzahlen aus. Die homogene Strahlverteilung des Excimerlaserlichts in der Bestrahlungsebene führte dazu, daß sich ein bis in die Randbereiche gleichförmiges Bearbeitungsfeld ausbildete. Mit Hilfe zusätzlicher Profiltiefenmessungen wurde ermittelt, daß die Einstrahlung des Laserlichtes beispielsweise mit einer Energiedichte von 1 J/cm^2 eine atmosphärenunabhängige durchschnittliche Abtragsrate von 0,6 µm pro Impuls ergibt. Eine Variation der Energiedichte zeigte bei gleichen Impulsanzahlen in weitem Rahmen keine qualitativen, sondern im wesentlichen quantitative Unterschiede im Abtragsvolumen. Dies bedeutet, das gewünschte Abtragsergebnis ist in relativ breitem Rahmen sowohl durch Wahl der Energiedichte als auch durch die Impulsanzahl einstellbar.

Photoelektronenspektroskopische Analysen

Um zu zuordnungsfähigen Aussagen über eventuelle chemische Änderungen zu gelangen, wurden jeweils getrennt an C-Fasern (HTA, Fa. Akzo) und am Epoxidharz (6376, Fa. Ciba Geigy) Bestrahlungsuntersuchungen vorgenommen und die Proben anschließend mit Hilfe der Photoelektronenspektroskopie (XPS) analysiert. Dabei sollte festgestellt werden, ob die Excimerlaserbehandlung bei der hier eingestellten Wellenlänge von 308 nm chemische Änderungen an den einzelnen Werkstoffkomponenten bewirkt /4/. Beispielsweise könnten unkontrollierte Zersetzungen an den C-Fasern oder im Matrixharz einen Verlust an Festigkeit im späteren Verbund zur Folge haben. Die Ergebnisse der XPS-Messungen an bestrahlten Fasern und an bestrahltem Harz zeigten jedoch, daß durch die Laserbehandlung mit unterschiedlichen Bedingungen insgesamt nur sehr geringfügige Änderungen in deren chemischer Zusammensetzung initiiert worden sind. So blieben sowohl bei der C-Faser als auch beim Matrixharz die C- und O-Anteile nahezu konstant. Aus sog. Fit-Prozeduren (d. h. Entfaltung des C_{1s}-Signals) war ableitbar, daß der Sauerstoff auf der C-Faseroberfläche hauptsächlich einfach gebunden in Form von C-O-Gruppen vorlag /4/. Dagegen zeigten die Reinharzproben zusätzlich einen geringen Anteil höher oxidierten Oberflächenkohlenstoffs (C=O-Gruppen), der sich allerdings auch nach dem Bestrahlen nicht wesentlich veränderte. Lediglich ein geringfügig erhöhter Stickstoffanteil bei der mit einer hohen Impulszahl durchgeführten Laserbehandlung des Matrixharzes deutete auf eine beginnende Zersetzung hin. Insgesamt kann man also bei einer Vorbehandlung unter den vorliegenden Bedingungen von einem nahezu reinen Abtag der Oberflächenverbindungen ohne signifikante chemische Veränderungen in der resultierenden Oberflächenstruktur ausgehen.

Bindefestigkeitsuntersuchungen

Für die Bindefestigkeitsuntersuchungen sind einschnittig überlappende Prüfkörper hergestellt worden. Aufbau und Härtung der Laminate erfolgte unter luftfahrttechnischen Standardbedingungen, die Laservorbehandlung mit Gesamtenergiedichten von 1,65 und 6,51 J/cm^2 (10 und 50 Impulse). Das Peel-Ply-Gewebe wurde umittelbar vor dem Bestrahlen abgezogen. Die späteren Fügeflächen der Platten sind in einem Randstreifen von 12,5 mm Breite (entsprach dem Überlappungsbereich der Proben) durch "Abrastern" über eine Länge von ca. 300 mm erzeugt worden. Anschließend wurden daraus 25 mm breite Streifen geschnitten, so daß aus je zwei dieser Platten zehn Proben hergestellt werden konnten. Als Klebstoff wurde FM300 (175°C-System, Fa. Cyanamid) verwendet. Als Referenz dienten der Peel-Ply-Abzug und das Anschleifen. Für die Beurteilung der Bindefestigkeiten und des Bruchbildes der Proben wurde sowohl der ungealterte Zustand als auch das Verhalten nach Auslagerung in verschiedenen Medien herangezogen.

Die unter den verschiedenen Bedingungen erzielten Festigkeiten sind in einem Diagramm eingetragen worden, Bild 2.

Die dem Diagramm zu entnehmenden Bindefestigkeitswerte sind sowohl bei den Standardverfahren als auch bei der Laservorbehandlung als hoch einzustufen. Die Festigkeitswerte nach dem Schleifen und nach der Laserbehandlung liegen in gleicher Größenordnung und sind höher als die nach alleinigem PP-Abzug als Vorbehandlung. Nach Feuchtwarmauslagerung liegt der Festigkeitsabfall nach PP-Abzug und nach Laserbehandlung in gleicher Größenordnung, der zulässige Maximalwert von 20% gegenüber den Ausgangswerten wird jedoch nicht überschritten, das Bruchbild ist zu 100% kohäsiv und liegt im Klebstoff. Bei der stärkeren Laservorbehandlung ist der Festigkeitsabfall nach Auslagerung höher als bei der Referenz Schleifen. Hier ist ein etwas höherer Anteil an Adhäsionsbruch an der Grenzfläche Klebstoff/Laminat festgestellt worden. Auffallend ist, daß die Proben nach Lagerung in dem sehr aggressiven Medium Skydrol/Wasser und anschließender Prüfung bei Raum-

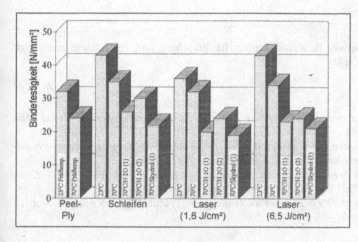

Bild 2:
Ergebnisse von Bindefestig-
keitsuntersuchungen am
Epoxidharzsystem C6376-
HTA

temperatur sowohl beim Schleifen als auch bei der Laservorbehandlung noch Festigkeitswerte liefern, die für die Flugzeugindustrie akzeptabel sind.

Zusammenfassung

Abgesehen von der Tatsache, daß für die Laservorbehandlung aufgrund der aufwendigen und zeitin-
tensiven Prüftechnologie nur eine sehr begrenzte Parametervariation vorgenommen werden konnte,
zeigen die erzielten Ergebnisse zusammengefaßt, daß mit der konventionell eingesetzten Schleifvor-
behandlung und mit einer Laservorbehandlung, die also ohne Optimierung erfolgte, sowohl im un-
gealterten Zustand als auch nach verschiedenen Arten der Auslagerung bereits vergleichbare Festigkeiten erzielbar sind. Als positiver Aspekt für die Weiterentwicklung der Laservorbehandlungstech-
nologie ist anzumerken, daß mit Bearbeitungsparametern innerhalb eines relativ großen Spektrums
das Harz reproduzierbar abgetragen werden konnte, ohne einen nennenswerten Einfluß auf die Fasern zu nehmen. Dies spiegelte sich auch in XPS-Messungen wider, bei denen keine wesentlichen
Änderungen in der chemischen Zusammensetzung von Harz und C-Fasern erkennbar waren.

Es kann aufgrund der bisher erzielten Ergebnisse davon ausgegangen werden, daß mit der Excimer-
laservorbehandlung von CFK-Verbundwerkstoffen langzeitstabile Klebverbindungen erzielt werden
können. Es ist ein schonendes und kontrollierbares Abtragsverhalten einstellbar. Erst bei einer sehr
intensiven Behandlung waren Schädigungen und damit vermutlich Schwächungen im späteren Verbund erkennbar. Unter Berücksichtigung der allgemein bekannten Vorteile eines lasergestützten Verfahrens können die vorgestellten Untersuchungsergebnisse somit als sinnvolle Grundlage für weitere
erfolgversprechende Schritte in der Verfahrensentwicklung zu einer fortschrittlichen Vorbehand-
lungstechnologie angesehen werden. Allerdings sind vor einer produktionstechnischen Anwendung
der Technologie noch eine Reihe grundsätzlicher Fragen zu klären. Diese betreffen beispielsweise die
Optimierung der Laser- und der Prozeßparameter, die Minimierung von Kontaminationsrückständen
oder eine zusätzliche Aktivierung der Substratoberflächen. Können diese Punkte zufriedenstellend
gelöst werden, so steht ein Vorbehandlungsverfahren zur Verfügung, das automatisier- und reproduzierbar sowohl auf kleine Flächen beschränkt als auch großflächig mit akzeptabler Bearbeitungsge-
schwindigkeit einsetzbar ist.

Danksagung

Die Untersuchungen wurden vom Bundesministerium für Forschung und Technologie unterstützt. Dafür sei an dieser Stelle gedankt.

Literatur

/1/ S. Metev, G. Sepold: *Excimer Lasers and New Trends in Laser Microtechnology;* Verlag Meisenbach, Bamberg (1993)

/2/ M. Wehner, W. Barkhausen, K. Wissenbach: *Klebeflächenvorbehandlung von Faserverbundwerkstoffen mit Laserstrahlung;* VDI-Z 134 (1992), 1, S. 70-72

/3/ J. Breuer, S. Metev, G. Sepold: *Photolytic Surface Modification of Polymers with UV-Laser Radiation;* J. Adhesion Sci. Technol. Vol. 9, No. 3, pp. 351-363 (1995)

/4/ D. Briggs: *Applications of XPS in Polymer Technology;* in: Practical Surface Analysis, Vol. 1,; John Wiley & Sons, pp. 437-483 (1990)

Lasergestützte Aufbau und Verbindungstechnik für Mikrosysteme

A. Gillner, M. Niessen, F. Legewie

Fraunhofer-Institut für Lasertechnik Aachen Steinbachstraße 15, 52074 Aachen

1 Einleitung

Die höhere Integrationsdichte industrieller Produkte mit Strukturgrößen unterhalb 1 µm erfordert Fertigungsverfahren, die den dabei gestiegenen Anforderungen an Genauigkeit und Reproduzierbarkeit gerecht werden. Vor allem in der Sensorik werden Strukturen im µm-Bereich benötigt, um die Forderungen z.B. aus der Automobilindustrie nach immer kleineren und leichteren Sensoren erfüllen zu können. Eine im Vergleich hierzu noch größere Integrationsdichte verlangt die Mikrosystemtechnik, bei der elektronische, optische, mechanische und vielfach auch fluidische Komponenten zu einem Gesamtsystem vereinigt werden und das als Sensor, Aktor und Übertragungsglied in vielfältiger Weise fungieren kann /1/. Im Gegensatz zur Mikroelektronik, bei der die einzelnen Komponenten in monolithischer Bauweise integriert werden können, erfordert die Heterogenität der Mikrosystemtechnik Verfahren, die einen hybriden Aufbau mit annähernd gleicher Genauigkeit ermöglichen.

Verfahren	Bearbeitungs-geometrie	Laserart	Geschwindigkeit, Prozeßzeit	Anwendung
Laserstrahl-schweißen	0.05 - 5 mm	CO_2-Laser Nd:YAG-Laser	max 60 m/min	Packaging Kontaktierung Housing
Laserstrahllöten	0.02 - 2 mm	Nd:YAG-Laser	50 - 500 ms	Kontaktierung
Laserstrahl-bohren	0.005 - 0.5 mm	Nd:YAG-Laser Excimer-Laser	0.1 - 10 s	Düsen Multi-Layer-Kontaktierungen
Laserstrahl-schneiden	0.01 - 0.1 mm (Schnittspalt)	Nd:YAG-Laser	> 5 m/min	Vereinzeln MCM-Module
Laserstrahl-beschichten	lokal und großflächig	VIS- und Excimer-Laser	> 1 µm/min Schichtrate	Leiterbahnen, Sensorschichten
Laserstrahl-strukturieren	0.01 - einige mm	Nd:YAG-Laser Excimer-Laser	100 nm/Puls bis 10 g/s	Faseraufnahmen Kavitäten

Tabelle 1: Übersicht über Laserverfahren in der Mikrosystemtechnik

Die Lasertechnik bietet hierzu Technologien zum Mikrofügen, Beschichten und Strukturieren, die flexibel auf die Anforderungen der jeweiligen Komponenten abgestimmt werden können und die ein Höchstmaß an Präzision erlauben. In Tabelle 1 sind einige der Verfahren für den Einsatz in der Mikrosystemtechnik und deren Anwendungsfelder aufgelistet.

2 Mikrofügen mit Laserstrahlung

Bei der Herstellung von Bauelementen von Sensoren und Mikrosystemen mit optischen und elektronischen Hybridbauteilen werden Fügegenauigkeiten z.T. im Submikrometerbereich verlangt. Beispiele hierfür sind vor allem in der Optoelektronik zu finden, bei denen Einzelelemente diskret aufgebaut sind und die mit konventionellen Fügeverfahren, d.h. Schweißen, Löten oder Kleben miteinander verbunden sind /2/. Insbesondere bei optoelektronischen Bauelementen, bei denen in Hybridbauweise alle Elemente, wie Laser, Empfänger Linsen und Fasern miteinander vereint werden hängt die Leistungsfähigkeit eines solchen Elementes von der Genauigkeit ab, mit der die Einzelbauteile gefügt werden können. Bei der Einkopplung des Laserlichtes in die Monomodefaser genügen wenige Mikrometer lateraler Versatz, um Einkoppelverluste von bis zu 20 dB zu bewirken /3/. Durch eine Justage der optischen Bauelemente während bzw. nach dem Fügeprozeß kann dieser Anteil auf 6 db reduziert werden. Die Folge sind geringere notwendige Laserleistung und eine größere Lebensdauer der optoelektronischen Bauelemente (OEICs). Eine weitere Reduzierung der justagebedingten Einkoppelverluste kann durch die Verwendung von Fügemethoden erfolgen, bei dem der Energieeintrag minimal ist und thermische Verzüge vermieden werden können. Hier bieten sich die Laserverfahren Laserschweißen und Laserlöten an, bei denen im Mikrobereich Energiemengen von $E < 1$ J/mm^3 deponiert werden und eine dementsprechend geringe Aufheizung des Werkstücks bewirken. Als Alternative könnten Klebeverbindungen dienen, die jedoch die Qualität optischer Bauteile nachteilig beeinflussen, da sich beim Aushärten Lösungsmittel auf den Bauelementen niederschlagen.

Als eine Besonderheit der Fügetechnologien kann das Bonden mit Laserstrahlung bezeichnet werden. Dabei werden ähnlich den konventionellen Bondverfahren Werkstoffverbünde erzeugt, bei dem die Materialien nicht durchmischt, sondern über Adhäsion und Ionenbindung miteinander verbunden werden. In der Mikrosystemtechnik werden solche Werkstoffverbünde im Bereich der Sensorherstellung benötigt, um z.B. einen empfindlichen Silizium-Sensor mit frei tragenden Membranen und schwingungsfähigen Biegebalken von den zum Teil rauhen

Umgebungsbedingungen zu trennen. Als Standard dieser Werkstoffverbünde hat sich die Verbindung von Glas mit Silizium herauskristallisiert. Mit entsprechend ausgewählten Gläsern können dabei die mechanischen und thermischen Werkstoffeigenschaften beider Materialien angepaßt werden, so daß der Sensor in weiten Temperaturbereichen und unter mechanischer Belastung eingesetzt werden kann /4/. Die Verbindung der beiden Werkstoffe erfolgt in der Regel über anodisches Bonden, bei dem über ein angelegtes Feld ein Ionenaustausch initiiert und damit eine Ionenbindung erzeugt wird. Mit dem Bonden durch Laserstrahlung steht ein alternatives Verfahren zur Verfügung, das beide Partner kraftschlüssig miteinander verbindet und das sich auch unter den hohen Anforderungen einer CMOS-Linie einsetzen läßt.

In Bild 2 sind zwei typische Fügegeometrien für die Verbindung von Silizium mit Glas bzw. Kunststoff dargestellt. Bild 2a zeigt dabei eine in Silizium geätzte Kavität, z.B. eine Druckmeßdose, die über ein Pyrex-Glas-Plättchen verschlossen wird. In Bild 2b ist Kunststoff-Mikrospritzgußteil dargestellt, das auf einen Silizium-Träger aufgebracht werden muß. Solche Mikro-Spritzgußteile werden vermehrt durch LIGA-Technik hergestellt. Ein Beispiel hierfür ist ein Tintenstrahldruckkopf, dessen Mikro-Tintenkavitäten und Auslassdüsen bei einer Geometrie von < 10 µm ohne Veränderung der Bauteilform gefügt werden müssen.

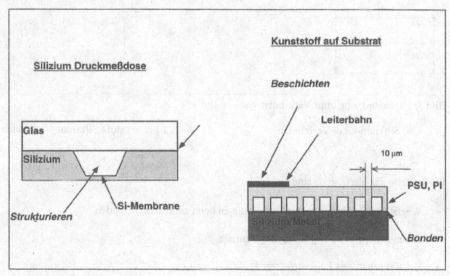

a) Verschließen einer Siliziumkavität   b) LIGA-Kunststoffbauteil auf
                                                              Silizumträger

Bild 2: Fügegeometrien von Werkstoffverbünden in der Mikrosystemtechnik
Das Fügen von LIGA-Bauteilen erfolgt zur Zeit in der durch Kleben, wobei auf die Auswahl des jeweiligen Klebstoffes besondere Sorgfalt verwendet werden muß. Insbesondere bei sehr kleinen

Geometrien und dünnflüssigen Klebstoffen können durch die Kapillarkräfte im LIGA-Bauteil Öffnungen verschlossen werden und somit zum Versagen des Bauteils führen.

Für beide Geometrien und Werkstoffverbünde kann das Laserstrahlfügen eingesetzt werden. Grundprinzip dabei ist jeweils, daß einer der beiden Fügepartner durchstrahlt und der jeweils andere Partner in der Fügezone aufgeschmolzen wird. Durch Vermischung oder Adhäsion erfolgt dann ein Fügen des Werkstoffverbundes. In Bild 3 sind die zwei unterschiedlichen Fügeanordnungen schematisch dargestellt.

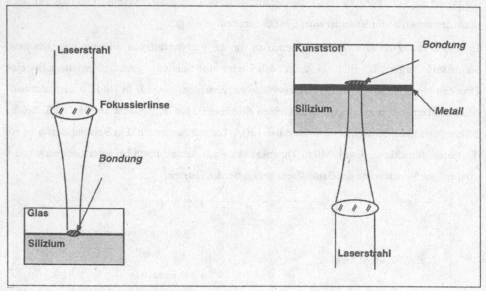

Bild 3: Fügeanordnung zum Verbinden von

    a) Silizium-Glas-Verbünden   b) Kunststoff-Silizium-Verbünden

Die Vorteile des Verfahrens sind:

- Wegfall mechanischer Spannvorrichtungen beim anodischen Bonden

- Geringe thermische Belastung der Bauteile

- Lokales selektives Fügeverfahren

- Tauglichkeit für CMOS-Linien-Fertigung

- Keine zusätzlichen Fügehilfsstoffe

Bei der Anordnung zum Verbinden des Silizium-Glas-Verbundes wird der Glasträger mit nur einer geringen Absorption von einem Nd:YAG-Laser durchstrahlt. Die Absorption der Strahlung erfolgt ausschließlich im Silizium, wodurch das Silizium aufschmilzt und die Energie durch Wärmeleitung teilweise an das Glas übertragen wird. In Bild 4 ist das Ergebnis einer Punktverbindung zwischen Silizium und Glas dargestellt.

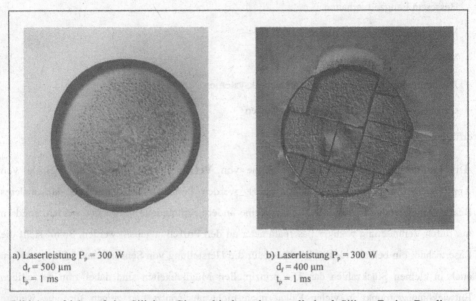

a) Laserleistung P_p = 300 W
d_f = 500 µm
t_p = 1 ms

b) Laserleistung P_p = 300 W
d_f = 400 µm
t_p = 1 ms

Bild 4: Aufsicht auf eine Silizium-Glasverbindung, hergestellt durch Silicon-Fusion-Bonding

Die Fügeverbindung wurde mit einem Nd:YAG-Laser bei einer Pulsleistung von 300 W und einer Pulsdauer von 1 ms hergestellt. In der Verbindung in Bild 4a wurde durch eine entsprechende Strahlformung eine qualitativ hochwertige Fügezone erzeugt. Demgegenüber zeigt Bild 4b Muschelbrüche und starke Anschmelzungen im Silizium. Ursache hierfür sind eine zu hohe Intensität und entsprechend hohe Gradienten, die über Anpassung von Laserleistung und Strahlgeometrie gesteuert werden können. Eine andre Art der Energieanpassung zeigt Bild 3b, bei dem der Werkstoffverbund von der Silizumseite durchstrahlt wird, was zu einer gleichmäßigeren Temperaturerhöhung führt. Insbesondere bei Kunststoffen mit einem kleinen Prozeßfenster, d. h. geringem Temperaturbereich der schmelzflüssigen Phase ist eine solche Anordnung zu wählen, da Temperaturspitzen infolge ungleichmäßiger Intensitätsverteilungen zur thermischen Degradation des Kunststoffes führen können.

3 Zusammenfassung

Mit dem Einsatz der Laserstrahlung als Werkzeug in der Herstellung von Sensoren und Mikrosystemen können die bestehenden Anforderungen aus der Fertigung und den Anwenderforderungen hinsichtlich

- Hoher Fügegenauigkeit

- Besserem Langzeitverhalten

- Vereinfachten Montageplätzen

- Verbesserten Bauteilspezifikationen

- Abstimmbaren Mikro- und Feinwerktechniksystemen

- Geringeren Montage- und Produktionskosten

erfüllt werden.

Die Lasertechnik bietet hierfür eine Reihe von Verfahren, mit denen eine Vielzahl von Applikationen und Problemstellungen gelöst werden können. Dabei muß für ein anderes Bearbeitungsverfahren häufig nicht einmal eine andere Strahlquelle verwendet werden, sondern nur durch Veränderung weniger Laserparameter an den Prozeß angepaßt werden. Somit stellt die Lasertechnik ein beinahe ideales Werkzeug für die Herstellung von Sensoren und Mikrosystemen auch in kleinen Stückzahlen dar. Die prinzipiellen Möglichkeiten sind dabei zum derzeitigen Entwicklungszeitpunkt in vielen Fällen noch nicht ausgenutzt und zeigen für eine weitere Erhöhung der Integrationsdichte ein erhebliches Entwicklungspotential.

4 Literatur

[1] J. Marek, M. Möllendorf,
     Mikrotechniken im Automobil, Mikroelektronik 9 (1995), Nr. 3

[2] VDI/VDE Technologiezentrum GmbH (Hrsg)
     Förderschwerpunkt Mikrosystemtechnik, Zweiter Erfahrungsbericht
     Berlin 1993

[3] H. Gruhl, AVT bei Laser-Einmodenfaserkopplung
     Workshop AVT in der Photonik, VDI/VDE, Berlin 1993

[4] W. Menz, P. Bley,
     Mikrosystemtechnik für Ingenieure, VCH-Verlag 1993

Deposition of Elements from Fe(CO)$_5$ Induced with Laser Radiation

S.A.Mulenko[1], A.N.Pogorelyi[1], R.Alexandrescu[2], I.Voicu[2]
[1]Institute of Metal Physics NAS of Ukraine, Kiev, 252142, UKRAINE
[2]Institute of Atomic Physics, Bucharest MG-6, ROMANIA

Deposition of elements induced with laser radiation of volatile
metal carbonyls has a great interest owing to the application
this method in solid state microelectronics and integrated optics.
This method gives the possibility for maskless technology in manufacturing of microelectronical structures on a substrate surface
directly. Laser chemical vapor deposition (LCVD) processes from
organometallic compounds may be driven at different wavelengthes
and irradiation regimes: photochemical [1,2] or thermochemical [3].
To control deposition processes of elements from gas-phase induced with laser radiation it is necessary to know the dissociation
mechanism of precursor molecules as in the volume above the substrate surface as on it and to have kinetic data about deposition
processes of elements which are forming thin films on the substrate surface. In the present work we have studied deposition processes of elements from iron pentacarbonyl - Fe(CO)$_5$ induced with
Ar$^+$-laser radiation (λ_L= 488 nm) and KrF-laser radiation (λ_L =
248 nm). The experimental arrangement used to deposit of elements
from Fe(CO)$_5$ vapors by means of laser radiation is described in
[3]. Laser radiation of 488 nm wavelength was focused on SiO$_2$ or
Si substrate surface to have been placed into the deposition cell
with Fe(CO)$_5$ vapors at 5 torr. The cell befor filling with vapors
was pumped up to 10^{-3} torr. The Ar$^+$-laser power density on the
substrate surface was 10^2 W/cm^2 with power $\sim$ 60 mW. The KrF-laser
power density on SiO$_2$ substrate surface was 3x10^6 W/cm^2 with pulse length $\sim$10 ns, repetition frequency 1 Hz and pulse energy $\sim$
4 mJ. The thickness of deposited films having been received at
different laser exposure time ($\tau_{exp.}$) or number of pulses (N),
was measured with optical method taking into account the interference efect in thin films. The element analysis with AES was made
with 20 Å step into film depth at different laser exposure time
using JEOLJAMP-10 S AUGER MICROPROBE. The chemical composition of
films was analysed with XPS using VG Escalab MK 11 spectrometer.
So it was possible to receive kinetic data about deposited films
under certain conditions of laser radiation by means of this ex-

perimental arrangement. The temperature calculation of the substrate at the laser spot was made according the expression from [4]:

$$\Delta T(\mathcal{T}_{exp.}) = \frac{P_{abs.}}{4\pi K \times L} \ln\left[1 + (\frac{4 \times \mathcal{T}_{exp.}}{r_o^2})\right], \qquad (1)$$

where $P_{abs.}$-the absorbed power (36 mW for SiO_2 and about 60 mW for Si substrate); K-the thermal conductivity; χ-the coefficient of temperature conductivity; L-the substrate thickness (0.02 cm for SiO_2 and 0.04 cm for Si). The maximum temperature rise at the SiO_2 surface at $\mathcal{T}_{exp.}$ = 20 min. is about 110 ^oC and ~2 ^oC for the Si surface. The radius of the deposited film spot on the substrate surface is about 0.027 cm (r_o). Such heating of the substrate was not enough for thermochemical dissociation of $Fe(CO)_5$ molecules as thermodissociation reaction takes place at the temperature above 380 ^oC [5]. The maximum temperature rise at the substrate surface while using KrF-laser was about 750 ^oC [6]. So there is thermochemical dissociation of adsorbed molecules which takes place simultaneously with photochemical one under KrF-laser radiation. Photochemical dissociation resulting from simple one-photon absorption at 488 nm or at 248 nm wavelength may be expressed:

$$Fe(CO)_5 + E_{h\nu} \longrightarrow Fe(CO)_x + CO_{5-x}, \qquad (2)$$

where x=3-dissociation process under Ar^+-laser radiation; x=1-dissociation process under KrF-laser radiation. It is seen from dependences shown in Fig.1 that deposition rate of elements (V) on SiO_2 substrate (curve 1) in the initial part is essentially higher ($V_1 \simeq 1.5$ Å/s) than while using Si substrate (curve 2, $V_2 \simeq 0.8$ Å/s). In both cases a marked decrease in deposition rate was recorded for a longer exposure time with Ar^+-laser radiation ($V_1 \simeq V_2 \simeq$ 0.1 Å/s for $\mathcal{T}_{exp.}$ =20 min.). The higher deposition rate of elements on SiO_2 substrate may be explained that the sticking coefficient of $Fe(CO)_x$ molecules is more for SiO_2 substrate than for Si one. Deposition process of elements as the result of photochemical dissoci-

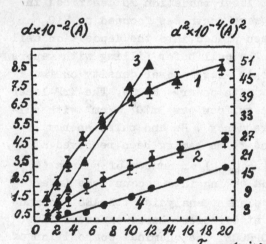

Fig.1. Dependences of film thickness (d) as a function of exposure time of Ar^+-laser radiation on SiO_2 substrate (curve 1) and on Si substrate (curve 2); d^2-dependences for SiO_2 substrate (line 3) and for Si substrate (line 4).

ation of Fe(CO)$_5$ molecules consists of two stages in conditions of our experiment. The first stage is stipulated of primary photoproduced molecules in the volume of laser caustic and diffusion of these molecules to the substrate surface. The second one is stipulated of reactions between adsorbed primary photoproduced molecules and precursor Fe(CO)$_5$ molecules. Dependences confirming of diffusion mechanism for film growth:

$$d^2 = D \times \tau_{exp.} \qquad (3)$$

are presented in Fig.1 (line 3 for SiO$_2$ substrate and line 4 for Si substrate). Where d is the film thickness and D is the diffusion coefficient (7×10^{-14} cm^2/s for SiO$_2$ substrate and 1.3×10^{-14} cm^2/s for Si substrate). The film growth while deposition of elements using KrF-laser radiation submits to the diffusion law at number of pulses less than 300 too. The investigation was being carried on with AES method at different exposure time of Ar$^+$-laser radiation to receive quantitative data about the element consistence of films deposited on Si substrate. Auger analysis of Si substrate surface after deposition of elements under Ar$^+$-laser radiation reveals the presence of iron, carbon and oxygen only. This analysis is able to provide the depth-profiling kinetics of elements such as Fe, O and C (Fig.2). The same element distribution are through deposited films at other exposure Ar$^+$-laser times. The main elements in deposited films are Fe, O, C provided atomic ratios C/Fe and O/Fe are less than unity. The C atom concentration increases up to 20% at the substrate surface and than decreases to zero at the film depth of 30% its thickness. Such behavious of C atoms may be explained by the very weak disolubility of C atoms in iron: while CO dissociation on the surface of growing film the process of the displacement for C atoms is developing as to the substrate surface and as to the film surface. It was found from the comparison of AES and XPS data that ratios C/Fe and O/Fe are

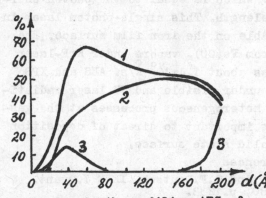

Fig.2.The depth-profiling AES of Fe (1),O (2),C (3) atoms through the film thickness while deposition of elements from Fe(CO)$_5$ induced with Ar$^+$-laser radiation at $\tau_{exp.}$=5 min.

always less than unity and are being changed in the range of 30% while using Ar^+-laser radiation. While using KrF-laser radiation for deposition of films the sharp increase of ratios Fe/O and Fe/C in films takes place (about 5 times) at increased number of pulses from 10 to 300. This is the result of a high enough surface temperature to promote rapid desorption of CO group from the surface, preventing their possible readsorption in the deeper layers and efficient desorption of dissociated carbon, preventing to some extent non-volatile carbide formation of iron (Fe_xC), which is not sufficient while using Ar^+-laser radiation in comparison with oxidized phase Fe_xO. It is clear from kinetic data of element content of films having been obtained at different exposure times that the main phase was Fe_xO through the film depth. This phase are forming owing to the oxidation of atomic iron with the atomic oxygen having been formed while CO dissociation on the film surface. Dissociation of adsorbed CO molecules on the surface is owing to that Fe atoms as atoms of transition metals act as catalyzer of dissociation reaction CO molecules[7]. As it is known[7], chemisorbed CO molecules on iron surface form $CO(\alpha_3)$ molecular state with low stretching vibrational frequency 1200 cm^{-1} of the α_3 state. This frequency on the iron film surface corresponds to a C-O bond dissociation energy of 4.4-4.9 eV which is equal to the photon energy of KrF-laser at 248 nm wavelength. This single-photon laser-induced CO dissociation is possible on the iron film surface. The deposition rate of elements from $Fe(CO)_5$ vapors under KrF-laser radiation on SiO_2 substrate was about 1.4×10^9 Å/s. AES and XPS analysises of deposited films under visible and UV laser radiation gave the information about heterogeneous processes at the gas-solid state interface. This is important to direct of deposition processes from gas-phase on solid state surface.

References

1. J.D.Ehrlich, R.M.Osgood,JR., and T.F.Deutsch, IEEE J.Quant. Electr., v.QE 16, No.11, 1980, p.1233.
2. R.Alexandrescu, A.Andrei, I.Morjan, S.Mulenko, M.Stoica and I.Voicu, Thin Solid Films, v.218, 1992, p.68.
3. S.A.Mulenko, A.I.Chaus, R.Alexandrescu, I.Voicu, Proc.10 Intern. Congress "LASER-91", Berlin Heidelberg New York London Paris Tokyo Hong Kong Barcelona Budapest, Springer-Verlag, 1991,P.768.
4. W.W.Duley. Laser Processing and Analyses of Materials, N.Y.and London, Plenum Press, 1983, P.502.
5. V.G.Syrkin, V.N.Babin. Gas Grows of Metals, M., Nauka, 1986, P.190.
6. X.Xu and J.I.Steinfeld, Appl.Surf.Science, v.45, 1990, p.281.
7. D.W.Moon, S.L.Bernasek, J.P.Lu, J.L.Gland and D.J.Dwyer, Surf. Science, v.184, No.1,2, 1987, p.90.

Tiefenstrukturierung von Silizium mit Nd:YAG-Q-Switch-Laser

K. Dickmann und A. Niessner
Labor für Lasertechnik, Fachbereich Physikalische Technik, FH Münster
Stegerwaldstr. 39, D-48565 Steinfurt

Das Abtragen mit Laserstrahlung wurde in den letzten Jahren umfangreich untersucht und wird bereits in einer Reihe verschiedener Anwendungen industriell umgesetzt. Zu den Standardmaterialien bisheriger Untersuchungen zählen vor allem Metalle, Keramik und Kunststoffe. Zunehmendes Interesse gilt auch dem Laserabtrag von Silizium, dem Schlüsselwerkstoff für die Halbleiter- und Mikrosystemtechnik. Hier bietet das Laserverfahren im Vergleich zu konventionellen Ätzverfahren durch den Verzicht von Masken eine hohe Geometrieflexibilität.

1 Stand der Technik

Das Abtragen mit Laserstrahlung wird heute bereits industriell für verschiedene Anwendungen genutzt. Dabei wird bzgl. der Verfahren der Massenabtrag mit unbestimmter Geometrie (z.B. für das Auswuchten bzw. Tarieren) /1,2,3/ vom Formabtrag mit definierter Geometrie (z.B. Formerstellung) /3,4,10/ unterschieden. Weitere Anwendungen unter Verwendung des CO_2 - Lasers werden in der Literatur /1-4,11/ erläutert.

Aufgrund seiner gegenüber dem CO_2-Laser i.a. besseren Fokussierbarkeit sowie der für das Absorptionsverhalten vieler Werkstoffe günstigeren Wellenlänge finden sich auch für den Nd:YAG-Laser in industriellen Anwendungen zum Abtragen mit Laserstrahlung ständig neue Einsatzgebiete /3,4,7/. Untersuchungen z.B. in /5,6,8/ haben Abtragsergebnisse mit Nd:YAG-Laser an Stahl und Keramik ermittelt und damit die hohen erreichbaren Qualitäten sowie die Wirtschaftlichkeit nachgewiesen. Dabei wird darauf verwiesen, daß durch gepulste Q-Switch-Laserstrahlung eine qualitative Verbesserung des Abtragsvorganges aufgrund des steigenden Anteils an verdampftem Material bei gleichzeitiger Verringerung der Wärmebelastung erreicht wird. Vergleichende Untersuchungen in /7/ haben die Vorteile einer Scanneroptik gegenüber Lineartischen bei der Ausbildung von Vertiefungen belegt.

Im Vergleich zu umfangreichen Abtragsuntersuchungen an verschiedenen Metallen und Keramik /4-7/ wurde der Materialabtrag mit Laserstrahlung an Silizium bisher nur in ersten Ansätzen verfolgt /9/. Etablierte Verfahren zur Tiefenstrukturierung von Silizium für vielfältige Anwendungen in der Halbleiter- und Mikrosystemtechnik beziehen sich vor allem auf chemische Ätzverfahren.

2 Anlagentechnik und Versuchsdurchführung

Für die Abtragsuntersuchungen wurde ein Q-Switch-Nd:YAG-Laser mit einer maximalen (cw-) Ausgangsleistung von 90 W verwendet. Die Relativbewegungen bei der Bearbeitung werden durch zwei Scannerspiegel mit maximaler Scangeschwindigkeit von 1,5 m/s realisiert. Durch den integrierten Q-Switch können Pulsfrequenzen bis 20 kHz erzeugt werden. Die von der Pulsfrequenz abhängige Pulsdauer (250 ns ...1 µs) führt bei Pulsenergien (3 mJ...15 mJ) zu hohen Pulsspitzenleistungen von einigen 10 kW. Dadurch ist bei dem Abtragsprozeß ein hoher Sublimationsanteil gewährleistet, der im Vergleich zum Schmelzabtrag zu einer erhöhten Bearbeitungsqualität führt. Der nicht vermeidbare Schmelzanteil wird durch den entstehenden Dampfdruck ausgetrieben.

Die folgenden Untersuchungen wurden ohne Zusatzgas unter Normalatmosphäre durchgeführt. Für den Materialabtrag an Silizium wurden einzelne Abtragsspuren mit einem Überlappungsgrad von >60% nebeneinander gelegt. Die Tiefenstrukturierung erfolgte durch Mehrfachscannen. Das Prinzip der Versuchsanordnung sowie die Abtragsstrategie sind schematisch in Bild 1 dargestellt.

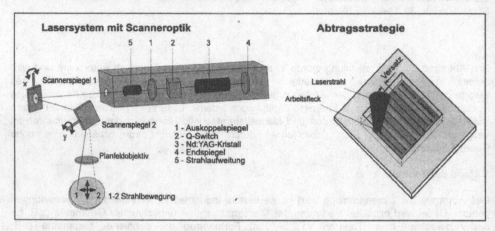

Bild 1 : Verwendete Anlagentechnik und Abtragsstragie

3 Experimentelle Ergebnisse

Die Abtragsrate wurde unter Variation der Prozeßparameter Q-Switch-Pulsfrequenz und Laserleistung (Lampenstrom I_L) untersucht. Zur Auswertung wurden mit einem laseroptischen 3D-Profilometer (Typ UBM) Abtragsvolumina bestimmt und daraus die Abtragsrate [mm^3/min] abgeleitet. Bild 2 zeigt ein repräsentatives Abtragsprofil. Bei kleinen Pulsfrequenzen unterhalb von 2 kHz zeigte eine Variation der Laserleistung keinen nennenswerten Einfluß auf die Abtragsrate (siehe Bild 3). Erst im mittleren Frequenzbereich nimmt die Abtragsrate erwartungsgemäß mit steigender Laserleistung (Lampenstrom) zu.

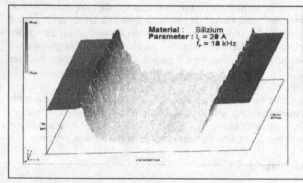

Bild 2 :   Repräsentatives Abtragsprofil in Silizium
(laseroptisch abgetastet)

Bzgl. der Q-Switch-Pulsfrequenz zeigte sich mit Hinblick auf eine möglichst hohe Abtragsrate ein eindeutiges Optimum, das sich mit steigender Laserleistung in Richtung höhere Pulsfrequenz verschiebt.

Laserparameter	
Laser :	Nd:YAG
max. mittlere Leistung :	P_{Lmax}= 90 W
Betriebsart :	Q-Switch
Verfahrensparameter	
Geschwindigkeit	v = 400 mm/s
Versatz	w = 0,05 mm
Modenblende	b = 3,2 mm
Lampenstrom	I_L = variabel
Frequenz	f_P = variabel

Tab. 1 : Bearbeitungsparameter

Das ausgeprägte Frequenzoptimum bei konstanter Laserleistung läßt sich dadurch erklären, daß bei kleinen Pulsfrequenzen zwar die Spitzenleistung eines Einzelimpulses höhere Werte annimmt, jedoch die geringe Pulsfolgefrequenz zu einer entsprechend geringen Abtragsleistung führt. Hingegen reduziert sich bei hohen Pulsfrequenzen die Pulsspitzenleistung auf Werte, die schließlich unter der Abtragsschwelle liegen.

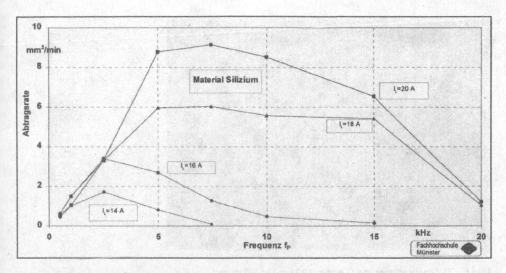

Bild 3 : Abtragsrate in Abhängigkeit von Q-Switch-Pulsfrequenz und Laserleistung
(Lampenstrom)

Die Untersuchung der erzielten Bearbeitungsqualität zeigte, daß ebenfalls im Maximum der Abtragskurve auch geringe Rauheiten am Abtragsgrund auftraten. Ein repräsentatives Abtragsresultat zeigt die REM-Aufnahme in Bild 4. Rißbildungen konnten nicht festgestellt werden.

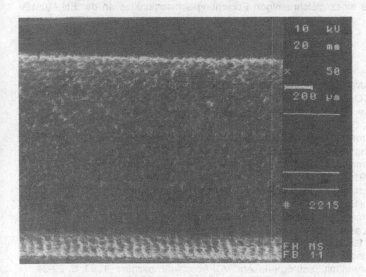

Bild 5 zeigt eine beispielhafte Tiefenstrukturierung (Tiefe = 120μm) in Silizium anhand einfacher positiv und negativ eingebrachter Dreiecksgeometrien.

Während für konventionelle Ätzverfahren Masken erforderlich sind, bietet das vorgestellte rechnergestützte Laserverfahren den Vorteil einer erheblich höheren Geometrieflexibilität.

Bild 4 : Blick auf die Oberfläche eines Silizium-Tiefenabtrages

Ein wirtschaftlicher Einsatz bietet sich jedoch aufgrund der geringen erzielten Abtragsraten von <9 mm³/min nur für die Prototypen- oder Einzelfertigung sowie Abtragsprozesse in der Serienfertigung mit geringen Abtragsvolumina an.

1040

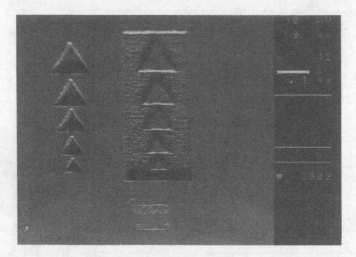

Bild 5 : Tiefenstrukturierung (Tiefe = 120 µm) in Silizium

Danksagung : Die Autoren bedanken sich bei Herrn Prof. Dr. B. Lödding, Leiter des Labors „Metallographie" im Fachbereich Physikalische Technik / FH Münster, der die REM-Untersuchung von laserbearbeiteten Proben durchgeführt hat.
Die Arbeiten wurden vom Land NRW im Rahmen eines Verbundforschungsvorhabens „Mikrosystemtechnik" sowie eines gleichnamigen Forschungsschwerpunktes an der FH Münster unterstützt.

Literatur

[1]   Tönshoff,H.K.;F.v. Alvensleben: „Verfahren zur Qualitätskontrolle für den Materialabtrag mit CO_2-Laser", Laser u. Optoelektronik (2)1990; S. 62-66
[2]   Haferkamp,H. ; Vinke,T.; Engel, K. : „ Abtragen mit CO_2-Laserstrahlung, Laser Magazin 6/88; S. 24-27
[3]   Tönshoff,H.K. ; Stürmer, M.;et al.:"Strukturieren technischer Oberflächen mittels Laserstrahlung", Laser u. Optoelektronik (2)1993; S. 56-61
[4]   Tönshoff,H.K. ; Stürmer, M. :„Laserfräsen- Formabtrag mit Hochleistungslaser", Laser Magazin 6/1991; S. 16-24
[5]   Burck,P. : „ Chemisch unterstützter Feinabtrag nichtoxidischer Keramiken „ Abtragen und Bohren mit Festkörperlasern" VDI-Abschlußbroschüre 1993 ; S. 43-50
[6]   Tönshoff,H.K. ; Stürmer, M. :" Definierter Materialabtrag metalischer Werkstoffe mit gepulsten Nd:YAG-Lasern" , in „Abtragen und Bohren mit Festkörperlasern" VDI-Abschlußbroschüre 1993 ; S.51- 57
[7]   Treusch, H.G. ; Läßinger, B. : „Formabtrag mit cw-Q-switch-Nd:YAG-Lasern" in „Abtragen und Bohren mit Festkörperlasern" VDI-Abschlußbroschüre 1993; S. 29-34
[8]   Tönshoff,H.K. ; Stürmer, M. : „Untersuchungen von Stahl mit Nd:-YAG-Lasern" , Laser u. Optoelektronik 22(5)1990; S. 59-63
[9]   Neske, E. ; Knoll,G. : „Strukturierung von Einzel- und Mehrfachschichten mit Nd:YAG-Laser" in „Abtragen und Bohren mit Festkörperlasern" VDI-Abschlußbroschüre 1993 ; S. 59-66
[10] Eberl, G. : „Laserbearbeitung für den Formenbau" ; Laser 3 (1990), S. 24-27
[11] Eberl, G.; et al.: „Laserspanen - eine neue Technologie zum Abtragen" ; Laser und Optoelektronik 25(3) 1993; S.80-86

Molekülbildung und Abscheideverhalten bei der Excimerlaserablation von Cu

L. Pfitzner, R. König

Max-Born-Institut für nichtlineare Optik und Kurzzeitspektroskopie

Rudower Chaussee 6

D - 12489 Berlin - Adlershof

Die beim Ablationsprozess von Cu mit KrF-Laserstrahlung erzeugte Materialwolke wurde im Nanosekunden-Bereich durch laserinduzierte Fluoreszenz untersucht. Neben ablatierten Cu-Atomen kommt es zur Bildung von Cu_2 bzw. -bei zusätzlicher Injektion oxidhaltiger Gase- von CuO.

Dieses frei ins Vakuum propagierende CuO kann auf geeigneten geheizten Substraten (z.B. MgO) zur Erzeugung dünner Schichten abgeschieden werden [1]. Hierbei ist das CuO-System von besonderem Interesse für die Herstellung dünner, supraleitender YBCO-Schichten, da die Güte der CuO-Schichten entscheidenden Einfluß auf die Eigenschaften des Supraleiters hat [2].

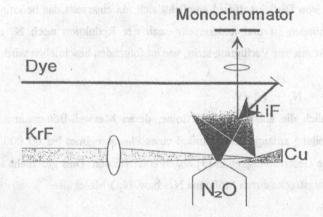

Bild 1: Versuchsaufbau

Der Ablationsprozess des Kupfers ist mit der verwendeten Strahlung (248 nm, 30 nsec) thermischer Natur, d.h. es werden durch den Laserpuls Elektronen im Material erzeugt, deren Temperatur die des Festkörpers übersteigt. Während der vergleichsweise langsamen Relaxation kommt es danach zu einem Energieübertrag auf das Festkörpergitter, so daß auch nach Ende des Laserpulses (~100 nsec) ein Materialaustrag beobachtet werden kann.

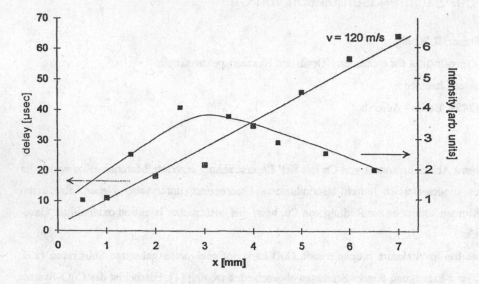

Bild 2: Delay der CuO-Wolke als Funktion des Ortes sowie CuO-₂ ?-Intensit. ⸱ ⸱n dem genannten Ort zu dem genannten Delay

Zur Erzeugung von CuO wird eine synchron zum Ablationslaser gepulste N_2O-Düse verwendet. Die Verwendung von Distickstoffoxid empfiehlt sich, da einerseits das benötigte Radikal relativ schwach gebunden ist und andererseits nach der Reduktion noch N_2 als Stoßpartner für schnelle Cu-Atome zur Verfügung steht, wie im folgenden beschrieben wird. Damit es zu einer Reaktion

$$Cu + N_2O \rightarrow CuO + N_2 + 1.1\ eV$$

kommen kann, müssen nämlich die ablatierten Cu-Atome, deren Maxwell-Boltzmannsche (MB-)-Geschwindigkeitsverteilung anfangs unter Einfluß eines Plasmaregimes bei 60 000 K liegt, zunächst auf thermische Geschwindigkeiten [3] abgebremst werden. Dies geschieht auf den ersten Zentimetern der Flugstrecke durch Stöße mit N_2 - bzw. N_2O-Molekülen .

Das entstehende CuO hat dann ebenfalls nur niedrige Geschwindigkeiten, die einer MB-Temperatur von 300 K entsprechen [Bild 2]. Aufgrund dieses Verhaltens haben weder die Ablationsfluenz noch der Nachweisort Einfluß auf die gemessene Temperaturverteilung. Der Einfluß der Fluenz beschränkt sich auf die Anzahl der zur Verfügung stehenden Reaktions-partner und geht bei 9 J/cm² in den Sättigungsbereich über. Bei Fluenzen oberhalb 9 J/cm² besitzt das über der Probe entstehende Plasma ein Transparenz von unter 33%, so daß bei einer Erhöhung der Fluenz der Energieeintrag in das Material nicht erhöht wird.

Betrachtet man einen Konkurenzprozess zur Bildung von CuO, nämlich die Dimerisierung zu Cu_2, stellt man fest, daß hier keine Abremsung der Cu-Atome nötig ist. So hat Cu_2 eine Translationsenergieverteilung von 1250 K und wird derart in angeregten Zuständen gebildet, daß bis zu Verzögerungs-Zeiten von 1 µsec nur spontane Emission sichtbar ist. Durch Vergleich berechneter mit gemessenen Spektren stellt man eine Rotationstemperaturverteilung fest, die von 450 K in der Nähe des Ablationsortes (~1 cm) auf 300 K in etwa 8 cm Entfernung gefallen ist [Bild 3].

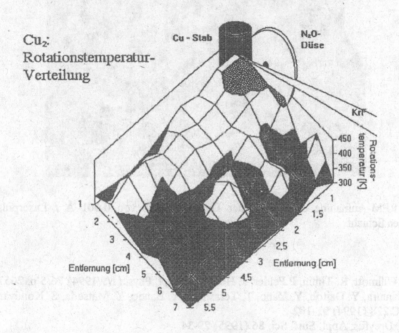

Bild 3: Rotationstemperaturen von Cu_2 als Funktion des Ortes

Für die Erzeugung von supraleitenden Schichten ist das beschriebene Verfahren der Laserverdampfung mit anschließender Abscheidung sehr elegant, da die Stöchiometrie des abgeschiedenen Materials über die Stöchiometrie des Targets bzw. der kontrollierten Zuführung von Reaktionspartnern in weiten Grenzen frei gewählt werden kann.
Es gibt jedoch noch Probleme mit Tröpfchenbildung auf den abgeschiedenen Oberflächen. Durch Wahl einer niedrigen Abscheiderate scheint dieses Problem überwindbar [Bild 4].

Nachteilig bei der Herstellung elektronischer Bauteile ist die z. Zt. noch notwendige hohe Temperatur des Substrates (~300-600°C). Die Randbedingungen, unter denen auf diese Temperaturen verzichtet werden kann, sind eine aktuelle Fragestellung.

Bild 4: REM-Aufnahme einer mit einer Depositionsrate von 0.001 Å / Laserpuls abge-schiedenen Schicht

[1] P.R. Willmott, R. Timm, P.Felder, J. Huber; J. Appl. Phys. (1/9/1994) 76/5 p. 2657-69
[2] K. Shimura, Y. Daitoh, Y. Yano, T. Terashima, Y. Bando, Y. Matsuda, S. Komiyama; Physica C 228 (1994) 91-102
[3] R.W. Dreyfus; Appl. Surf. Sci. 86 (1995) 29-34

Patterning of Si(100) in Microwave-Assisted LDE (Laser Dry Etching)

W. Pfleging[1], M. Wehner[2], F. Lupp[3], and E. W. Kreutz[1],

[1] Lehrstuhl für Lasertechnik der Rheinisch-Westfälischen Technischen Hochschule Aachen, Steinbachstraße 15, D-52074 Aachen, FRG

[2] Fraunhofer-Institut für Lasertechnik, Steinbachstraße 15, D-52074 Aachen, FRG

[3] Siemens AG, ZPL 1 TW 54, Postfach 830953, D-81709 München, FRG

1. Introduction

Patterning of silicon is necessary in a number of microelectronic and micromechanic applications. Plasma processes, such as reactive ion etching, play an important role in VLSI and nanotechnology applications. The charged particles impinging onto the semiconductor surface may cause radiation damage, such as crystal defects of different dimensionality. In plasma etching processes resist patterning and a large number of processing steps are needed to perform etching. The laser-induced dry etching (LDE) process offers the possibilities of direct pattern transfer by mask projection, anisotropic etching with negligible damage, and etching with very high selectivity. Modifying the LDE-process by applying a microwave-excited gas results in more radicals and an increased etch rate. Some authors use Cl_2 as the processing gas during the microwave-assisted laser dry etching (MALDE) with excimer laser radiation (248 nm) [1, 2]. In this case the etch rate of Si(100) is below 1 Å/pulse.

Employing a microwave discharge allows to use excited CF_4 and non-excited CCl_4 as processing gases. CCl_4 and CF_4 gases don't etch Si spontaneously, and with excimer laser radiation (248 nm) only for CCl_4 a gas phase dissociation occurs [3]. The microwave discharge generates the gas-phase F radicals necessary for etching. By adding non-excited CCl_4 to the processing gas, the etch rate increases due to a chemical process in the gas phase. The formation of reactive gas-phase species in the MALDE process is investigated with quadrupole mass spectroscopy. The etching mechanisms of the MALDE are investigated by analysing ex-situ the reaction products on the etched surfaces with X-ray photoelectron spectroscopy (XPS). The geometries of etched features are measured with stylus profilometry.

2. Experimental details

The apparatus has been described previously [4,5]. The samples (p(100)-type Si, 5Ω cm; p(111)-type Si, 1-30 Ω cm) are mounted in a vacuum chamber with a base pressure of 1×10^{-4} mbar. CF_4 is excited by a 2.45 GHz microwave source (microwave power 675 - 790 W) to generate F atoms and CF_X fragments, which are introduced into the reaction chamber through a quarz tube. Pulsed excimer laser radiation (wavelength 248 nm) with a pulse length of 25 ns is used. A slit mask 0.3-0.5 mm in width is imaged onto the sample surface by a quartz lense (demagnification 3:1). The laser entrance window is protected by Ar purging gas. Without excitation CCl_4 is fed through an additional inlet. Thus a gas mixture of excited CF_4 and non-excited CCl_4 is used as the processing gas. The gas flow of CF_4 is

20-40 sccm and that of CCl$_4$ is 0-20 sccm. The gas flows are regulated by mass flow controllers and a butterfly exhaust valve. The total pressure is kept at 1-1.5 mbar. Typical processing times are 15-20 min at laser pulse repetition rates of 10-20 Hz. The etched profiles are characterized by stylus profilometry. The etching mechanisms are investigated by analysing ex-situ the reaction products on the etched surfaces with angle-resolved XPS (300 W MgK$_\alpha$ radiation). Quadrupol mass spectroscopy is used to examine the gas-phase reaction between the microwave-excited CF$_4$ and the nonactivated CCl$_4$.

3. Results and discussion

Because of the large distance between the plasma discharge region and the sample surface (20 cm) the etching process occurs in an after-glow region. While the excited CF$_4$ processing gas impinges onto the Si surface, the laser radiation (248 nm) initiates or supports etching. Figure 1 reveals that etch rates of the MALDE process with excited CF$_4$ gas are below 1 Å/pulse [6]. Similar results are obtained with Cl$_2$ as processing gas [1].

A modification of the MALDE process is obtained by adding non-activated CCl$_4$ as an additional processing gas. Figure 1 shows the enhanced etch rates obtained with a CF$_4$/CCl$_4$ gas mixture ratio of 0.25 compared to the MALDE process with only microwave excited CF$_4$ as processing gas. A maximum etch rate is found [4,5] when the flow rate of CCl$_4$ reaches approximately 20 % of the total gas flow (CF$_4$ + CCl$_4$). The spontaneous etch process (i.e., the thermally driven etch process without laser radiation) is of course also present. However, since the investigations are carried out without an additional mask or polymer to protect the Si surface against spontaneous etching, both the surface and the etch features are affected. The measured etch depth, thus, is due to the laser-induced process.

The spontaneous etch rate is investigated as a function of temperature in the range 298-623 K for a gas mixture ratio of CCl$_4$/[CF$_4$+CCl$_4$] = 0.20. For example, a Si etch rate of 0.5 μm/min is obtained at room temperature and 2.6 μm/min for T = 623 K.

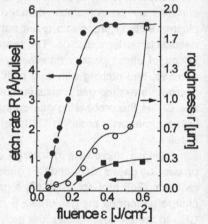

Fig. 1: Etch rate of silicon in MALDE ([CCl$_4$]/[CF$_4$+CCl$_4$] = 0.25) as a function of fluence with (●)and without (■) CCl$_4$; roughness (○) of etched area for case (●) for an etching time of 15 min (CF$_4$ gas flow 30 sccm, CCl$_4$ gas flow 10 sccm, total pressure 1mbar).

The thermal evaporation cannot explain the etch rates obtained during a single laser pulse, as evidenced from a simple evaporation model, so that a non-pyrolytic contribution to etching is implied [5].

The etch rate for p-Si(111) and p-Si(100) is nearly the same (0.9 μm/min≡ 15Å/pulse) as found under the conditions CF$_4$: 32 sccm; CCl$_4$: 8 sccm; pressure: 1 mbar, laser fluences: 0.24 J/cm^2, repetition rate 10 Hz. This investigations are performed with a polymer film to protect the Si surface which is not exposed to the laser beam against spontaneous etching. Investigations with stylus profilometry show that the surface morphology of the etched area depends on the laser fluence (Fig. 1). For fluences below 0.3 J/cm^2 smooth surfaces (< 0.3

µm roughness for an etching time of 15min) are obtained while for fluences above 0.3 J/cm^2 an additional chemical reaction and/or partial surface melting occur on the Si surface. For fluences above 0.53 J/cm^2 Si surface melting occurs.

The different etching mechanisms as function of CF_4/CCl_4 gas mixture ratio are investigated by XPS on etched Si surfaces. Figure 2 reveals the Si 2p core level spectrum for different gas mixtures. For non-excited CCl_4 as processing gas mainly $SiCl_2$ is formed on the Si surface [7] which is exposed to the excimer laser radiation. This result is similar to that of LDE of Si with Cl_2 as processing gas [1]. For microwave excited CF_4 as processing gas only Si and a small amount of SiO_2 (natural oxide layer resulting from exposure to air after MALDE) are detected. For different gas mixtures SiF_x ($3 \leq x \leq 4$) compounds are formed on the Si surface [8]. For a gas mixture ratio $[CCl_4]$/$[CF_4+CCl_4]$ = 0.2 (Fig. 2) a SiF_4 fluorosilyl species at 104.9 eV is detected.

Without laser radiation ClF^+ species are detected by mass spectroscopy. The QMS signal of ClF^+ as a function of the gas flow of CCl_4 shows a similar maximum to the etch rate curve [5]. It is possible that ClF^+ is only a fragment of a larger molecule, e.g., ClF_3 which might be formed in the after glow of the microwave discharge.

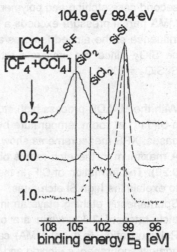

Fig. 2: Si 2p core level binding energies; XPS measurements are performed *ex-situ* on etched Si surfaces for different gas mixture ratios.

The MALDE process might be modified by polymerization of CCl_x fragments following the reaction $CCl_4 \xrightarrow{h\nu} CCl_3 + Cl$ [3]. For an increasing amount of CCl_4 the polymerization of CCl_x species and surface passivation via Si-Cl formation supresses the etch process. For a small amount of CCl_4 the etching process with "ClF_3" formation dominates the MALDE process via the polymerization/passivation effect. The F atoms from the microwave discharge seem to be converted to ClF_3 by a titration reaction with Cl in the gas phase. Other authors [9] have studied etching with simultaneous energetic bombardement of F atoms and Cl_2 molecules on a Si surface. They observed an increasing F atom etch rate and concluded that the formation of ClF_3 is responsible for the increased etch rate. It is observed, that ClF_3 etches Si spontaneously [10,11], and that the etch rate for ClF_3 is much higher than that for ClF [12].

Figure 3 reveals an example of an large area (4 mm^2) etched feature in Si(100). The angle of the side walls is less than 25°. The etch depth is of about 15 µm with an etch rate of 1µm/min. To avoid spontanous etching on the areas which are not exposed to the excimer laser radiation a protective polymer film has to be deposited onto the Si surface. This is realized by adding non-excited MMA ($C_5H_8O_2$) to the processing gas. Chlorine and F atoms are required for the formation of a non-volatile polymer on the Si surface. The XPS C1s-line

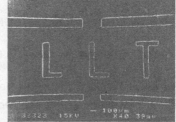

Fig. 3: SEM of a large area etched feature in Si(100) for $[CCl_4]$/$[CF_4+CCl_4]$ = 0.2.

shape of the polymer is very similar to that of PMMA. The polymerization can be performed either before the etching process or during the etching. In the first case the resulting polymer can be ablated with a few number of laser pulses below a fluence of 0.2 J/cm^2. In the second case etching and polymer formation compete, and etching will be stopped when the MMA partial pressure exceeds a threshold value. More details about this process and the influence of the etched features are described elsewhere [5]. Another contact mask could be SiO_2, since the etch rate ratio for the MALDE (gas mixture ratio: 0.20) is of about $Si:SiO_2 = 200:1$.

4. Summary

With the MALDE process high etch rates of up to 1 μm/min are obtained for p-Si(100) and p-Si(111) at room temperature by using excited CF_4 and non-excited CCl_4 as processing gases. XPS measurements show that the formation of $SiCl_2$ and F atom etching compete. A maximum in the etch rate is observed for a suitable $CF_4:CCl_4$ ratio ($CCl_4/[CF_4+CCl_4]$ = 0.20). The formation of ClF_3 in the MALDE process is proposed as a possible mechanism to explain the high Si etch rates.

Spontaneous etching plays an important role in MALDE. For temperatures up to 350 °C etch rates of 2.6 μm/min are otained via thermal evaporation without laser radiation. Polymerization (e.g., of MMA) can be used to protect the Si surface and the sidewalls against spontaneous etching and to make the process more anisotropic.

For applications in microsystem technologies (e.g., sensors or actuators), large areas (> 4mm^2) can be etched by mask projection techniques.

5. Acknowledgments

We are grateful to our colleague A. Vörckel for stimulating discussions. We also thank F.J. Schroeteler from Forschungszentrum Jülich (ISI: Institut für Schicht- und Ionentechnik) for his support in sample preparation and stylus profilometry.

6. References

[1] W. Sesslemann, E. Hudeczek, and F. Bachmann, J. Vac. Sci. Technol. B. 7 (1989) 1284.

[2] Y. Horiike, N. Hayasaka, M. Sekine, T. Arikado, M. Nakase, and H. Okano, Appl. Phys. A 44 (1987) 313.

[3] D.J. Ehrlich and J.Y. Tsao (Eds.), "Laser Microfabrication: Thin Film Processes and Lithography" (Academic Press, Inc., San Diego, 1989).

[4] W. Pfleging, D.A. Wesner, and E.W. Kreutz, Appl. Surf. Science, in press

[5] W. Pfleging, M. Wehner, F. Lupp, and E.W. Kreutz, SPIE Proceedings 2403 (1995) 387.

[6] E.W. Kreutz, H. Frerichs, M. Mertin, D.A. Wesner, W. Pfleging, Appl. Surf. Science 86 (1995) 266.

[7] C. C. Cheng, K. V. Guinn, V. M. Donnelly, and I. P. Herman, J. Vac. Sci. Technol. A 12 (1994) 2630.

[8] G. S. Oehrlein, J. Vac. Sci. Technol. A 11 (1993) 34.

[9] J. W. Coburn, J. Vac. Sci. Technol. A 12 (1994) 617.

[10] Y. Saito, O. Yamaoka, and A. Yoshida, J. Vac. Sci. Technol. B 9 (1991) 2503.

[11] Y. Saito, O. Yamaoka, and A. Yoshida, J. Vac. Sci. Technol. B 10 (1992) 175.

[12] D. E. Ibbotson, J. A. Mucha, and D. L. Flamm, J. Appl. Phys. 56 (1984) 2939.

Cleaning of Copper Traces on Circuit Boards with Excimer Laser Radiation

D. A. Wesner[1], M. Mertin[1], F. Lupp[2], and E. W. Kreutz[1]

[1]Lehrstuhl für Lasertechnik der Rheinisch-Westfälischen Technischen Hochschule Aachen, Steinbachstraße 15, D-52074 Aachen, FRG

[2]Siemens AG, ZPL 1 TW 54, Postfach 83 09 53, D-81709 München, FRG

1. Introduction

"Laser dry cleaning" of metals in air using pulsed, UV excimer laser radiation offers several advantages compared to more conventional techniques requiring solvent baths or low pressure plasmas [1]. In this method the short laser pulse length, combined with the high optical absorptivity for the laser radiation of most metals, ensures that the thermally affected zone remains spatially limited, thus avoiding possible damage to neighboring regions. The high photon energy (5 eV for the KrF-excimer laser) is large enough to sever surface chemisorption bonds, permitting contaminant removal via a non-thermal process. There is no exposure to energetic ions, and the cleaning effect can be laterally confined to selected areas of the surface by using projection mask techniques. Sample handling is relatively simple, since only a line of sight to the surface is required, and no transfer into vacuum is necessary. Environmental problems associated with hydrocarbon solvents are also avoided. This process is potentially useful in microsystem technology for applications requiring area-selective cleaning of bonding surfaces on complex modules containing sensitive structures that cannot tolerate thermal stress, solvents, or energetic ions.

We have applied laser dry cleaning to Cu surfaces having a contaminant layer consisting of mainly Cu_2O and a carbon species. Such passivation layers [2] form on Cu surfaces after long exposure to air at room temperature or after shorter exposures at elevated temperatures. Since even a relatively thin Cu_2O overlayer significantly affects the surface energy of Cu [3], it is desirable to reduce the contamination layer thickness before bonding or soldering to several nm or less. We evaluate laser dry cleaning by comparing the chemical, optical and morphological properties of oxidized Cu traces on circuit boards before and after cleaning.

1050

2. Experimental details

A KrF-excimer laser having a pulse width of about 20 ns and a pulse energy ≤300 mJ is used. The laser and associated beam-forming optics (Fig. 1) yield fluences at the sample ≤2 J/cm² over areas about 0.05-0.15 cm² in size at pulse repetition rates near 1 s⁻¹. Typically ≤10³ pulses are necessary for cleaning.

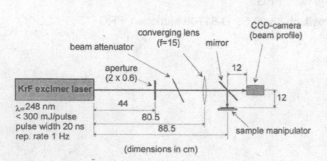

Fig. 1: Schematic diagram of the apparatus (dimensions in cm).

Commercially available printed circuit board material having a Cu coating about 50 μm in thickness is oxidized by exposure to air for 30 min at temperatures of 200 or 240°C. As confirmed by chemical analysis (see below), this yields a dark red oxide overlayer containing Cu_2O. Several analytical methods are used before and after cleaning. Samples are transferred into an ultrahigh vacuum chamber and chemically analyzed with X-ray photoelectron spectroscopy (XPS), a sensitive probe of the stoichiometry and valence state in the near-surface region (sampling depth ~10 nm) [4]. Optical and scanning electron microscopies and optical profilometry are used to determine the surface morphology. Ellipsometry (wavelength 628 nm, 70° angle of incidence) is used to determine the oxide overlayer thickness, using the known optical constants (n, k) of overlayer and substrate [5].

3. Results and discussion

Repeated irradiation (≤5 pulses) at fluences per pulse in the range 0.6-1.5 J/cm² visibly affects the dark red Cu_2O overlayer, which is initially 15-25 nm thick. Cleaned regions are clearly visible as lighter areas in comparison to the darker, untreated areas surrounding them. Micrographs of the surface show that the gross morphology (e.g., scratches) is unchanged in cleaned regions, suggesting that large amounts of material have not been removed. For fluences ≥1.2 J/cm² small droplets appear, indicating that the surface has been partially melted by the high-fluence pulses. The color change caused by irradiation (dark red changing to bright gold/yellow) is what one would expect for the removal of the surface oxide layer, leaving the metallic Cu substrate behind.

The chemical change in cleaned regions is revealed by XPS measurements of the Cu Auger transitions and the C1s core levels (Fig. 2). For reference the results for an unirradiated sample are plotted at the bottom. Using the Cu Auger line shape, it is possible to distinguish

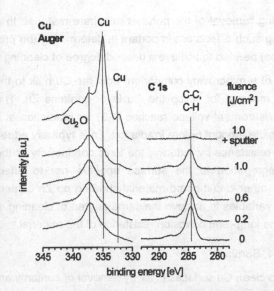

Fig. 2: XPS spectra of oxidized Cu traces on a circuit board before and after cleaning with 5 KrF excimer laser pulses at various fluences. Spectra for the Cu Auger line (left) and the C1s core level (right) are plotted. The topmost spectra were obtained after *in-situ* sputter-etching of the region irradiated at a fluence of 1 J/cm^2.

between metallic Cu (valence 0) and the first Cu oxidation state in Cu_2O (valence +1) [6]. The Cu Auger line shape for the untreated sample (Fig. 2) shows that the surface is totally oxidized to Cu_2O to within at least the XPS information depth. The growth of metallic Cu peaks near 332 and 335 eV binding energy for fluences above 0.6 J/cm^2 shows that the Cu_2O oxide overlayer thickness has been reduced to the order of the XPS information depth or less: spectral structures from both overlayer and metallic substrate are simultaneously present. Cleaning removes not only a good part of the oxide layer, but also reduces the carbon contamination (Fig. 2). The C1s peak near 284.6 eV binding energy is characteristic for a carbonaceous overlayer containing mainly C-C and C-H bonds [7]. For com-

parison, the topmost spectra In Fig. 2 show the result after sputter-etching and *in-situ* analysis of the sample that was cleaned with 1 J/cm^2 pulses. The Cu_2O and C overlayer is almost completely removed, leaving metallic Cu.

The cleaning effect relies on the greater optical absorptivity for the laser radiation of the contamination layer in comparison to the Cu substrate. The optical reflectivity of an oxidized Cu trace for specularly reflected light at $\lambda=248$ nm increases from about 7% to 28% after cleaning, compared to a value of about 36% for a reference Cu sample. This implies a strong decrease in the optical absorptivity with cleaning. Contaminant removal can in principle occur at laser fluences below the material removal threshold for Cu, and the process tends to be self-limiting, holding damage to neighboring regions to a minimum.

In general, qualitatively similar cleaning effects can be achieved with either many pulses at a low fluence or fewer pulses at a high fluence [1]. At either high or low fluence, the Cu_2O overlayer thickness as a function of number of cleaning pulses first decreases, reaches a minimum, and then for a large number of pulses increases. This is observed in ellipsometric measurements of the Cu_2O overlayer thickness and is attributable to a reoxidation of the surface, as confirmed by XPS measurements. Irradiation with many pulses at low fluence (in contrast to a few at high fluence) can also adversely affect areas of the circuit board adjacent to the metal traces, e.g., by causing removal of the polymer substrate material. In a practical application of laser dry cleaning such effects are important in determining the process variables (number of pulses, fluence) needed to achieve a desired degree of cleaning.

Irradiation can also induce segregation of a minor alloy component from the Cu bulk to the surface, as observed for circuit board material for which the Cu layer contains Zn. The atomic ratio (Zn:Cu) within the XPS measurement volume reaches (1:3) for irradiation at a fluence of 1 J/cm^2, compared to a negligible amount before irradiation. Zn is typically added to Cu in order to increase its corrosion resistance by reducing the ionic conductivity of the passivating Cu_2O overlayer [8]. Its segregation to the surface appears not to affect significantly the cleaning process, since other circuit board material showing no Zn content in the Cu layer required similar process variables to achieve the same degree of cleaning. It could, however, conceivably influence the long-term corrosion resistance of the material.

4. Summary

Pulsed excimer laser radiation is used to clean Cu surfaces in air by removal of contaminant overlayers containing Cu_2O and C. Within limits, the process variables (number of pulses, pulse fluence) can be varied to achieve cleaning at low fluence with a large number of pulses, or at high fluence with fewer pulses. Secondary effects of the cleaning include localized melting at high fluence and segregation of Zn from the Cu bulk.

5. Acknowledgments

We are grateful to U. Giesen, G. Schriever, O. Treichel, F. Weller, and M. Wehner for technical assistance and for stimulating discussions.

6. References

[1]   Y. F. Lu, M. Takai, S. Komuro, T. Shiokawa, and Y. Aoyagi, Appl. Phys. A 59 (1994) 281; Y. F. Lu and Y. Aoyagi, Jpn. J. Appl. Phys. 33 (1994) L430.

[2]   T. L. Barr, J. Phys. Chem. 82 (1978) 1801.

[3]   K. V. Ravikumar, V. Dhir, and T. H. K. Frederking, Cryogenics 32 (1992) 430.

[4]   D. Briggs and M. P. Seah, Eds., "Practical Surface Analysis by Auger and X-ray Photoelectron Spectroscopy" (London, John Wiley & Sons, 1983).

[5]   C. G. Ribbing and A. Roos, in "Handbook of Optical Constants of Solids II", E. D. Palik, Ed. (Academic Press, Boston, 1991).

[6]   G. Schön, Surface Sci. 35 (1973) 96.

[7]   P. Swift, Surf. Interf. Anal. 4 (1982) 47.

[8]   N. W. Polan, F. J. Ansuini, C. W. Dralle, F. King, W. W. Kirk, T. S. Lee, H. Leidheiser, Jr., R. O. Lewis, and G. P. Sheldon, in "Metals Handbook, Ninth Edition, Volume 13, Corrosion", J. R. Davis, Ed. (American Society for Metals, Metals Park, Ohio, 1987).

Lasergestützte Schichtabscheidung zur Herstellung funktionaler Schichten in Mikrosystemtechnik und Mikroelektronik

A. Gillner, M. Wehner, E.W. Kreutz, H. Frerichs, M. Mertin, D. Offenberg
Fraunhofer-Institut für Lasertechnik Aachen
Lehrstuhl für Lasertechnik, RWTH Aachen

1 Einleitung

In Sensorik und Mikrosystemtechnik werden in vielen Anwendungen dünne Schichten benötigt, die Teil eines Sensors, einer Funktionsgruppe sind, als Abschirmung bzw. optisches Element dienen oder einfach nur Kontaktierungen zwischen einzelnen Baugruppen bzw. Komponenten darstellen. Häufig werden an die Erzeugung der Schichten in diesem Anwendungsfeld besondere Anforderungen gestellt, da als Substrat entweder dreidimensionale Bauteile zum Einsatz kommen oder bereits vorprozessierte Komponenten nachträglich mit Funktionsschichten versehen werden müssen. Konventionelle Beschichtungsmethoden können hierbei in den meisten Fällen nicht eingesetzt werden. Die Lasertechnik bietet hier Alternativen, um lokal mit hoher Präzision Funktionsschichten zu erzeugen, die mit konventionell erzeugten hinsichtlich Qualität und Gebrauchseigenschaften vergleichbar sind. In Tabelle 1 sind hierzu die gebräuchlichsten Verfahren und deren Einsatzgebiete zusammengestellt.

Verfahren	Substrate	Beschichtungs- werkstoffe	Anmerkungen
Laser-CVD	Metalle Keramiken Halbleiter	Metalle Keramiken	Vakuum- beschichtung
Laser-CLD	Kunststoffe, Keramiken Halbleiter	Metalle	3D-Beschichtung Atmosphäre
Laser-PVD	Metalle Keramiken Halbleiter	Keramiken Metalle	Vakuum- beschichtung

Tabelle 1: Laserverfahren zur Herstellung dünner Funktionsschichten

2 PLD-Verfahren zur Herstellung von Funktionsschichten

Zur Erzeugung von dielektrischen und ferroelektrischen Schichten kann das sogenannte PLD (Pulsed Laser Deposition) Verfahren eingesetzt werden /1/. Dabei wird das Beschichtungsmaterial von einem Excimer-Laser verdampft, wobei die Abtragsrate bei den meisten Werkstoffen bei einigen mg/cm^2/min liegt. Das verdampfte Material strömt vom Target ab und rekondensiert auf dem Substrat. Für eine reproduzierbare Schichtherstellung und gezielt einstellbare Schichteigenschaften sind alle relevanten Prozesse des Beschichtungsvorganges zu beachten /2/. Im einzelnen sind dies:

1. Materialabtrag

2. Materialtransfer durch Plasma/Dampf-Fcrmation

3. Rekondensation/Schichtwachstum

Für die Analyse der Einzelprozesse stehen eine Reihe von Methoden, wie z. B. XPS-Oberflächenanalytik, Massenspektrometrie, Plasmaspektrometrie, Ellipsometrie, zur Verfügung.

2.1 Anlagentechnik

Die Beschichtung mit dem PLD-Verfahren erfolgt bei Prozeßgasdrücken zwischen 10^{-4} und 10^{-5} bar. Zur Vermeidung von Kontaminationen wird die Prozeßkammer zunächst auf 10^{-8} bar evakuiert und anschließend mit Sauerstoff bis auf 10^{-4} bar gefüllt. Entsprechend Bild 1 wird das Target über ein Fenster mit Laserstrahlung beaufschlagt.

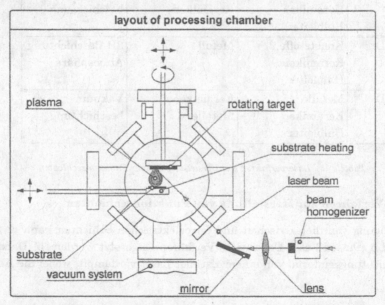

Bild 1: Schematische Darstellung des Anlagenaufbaus für den PLD-Prozeß

Dabei ist von eminenter Bedeutung, daß das Targetmaterial ohne Unterschiede bei verschiedenen Targetwerkstoffen instantan verdampft. Um dies zu erreichen, werden für das PLD-Verfahren Excimer-Laser verwendet, die bei Pulsdauern von 20 - 30 ns Pulsleistungen von $P_L \geq 2$ MW aufweisen. Die Strahlung wird zur Vermeidung von Intensitätsunterschieden auf dem Target homogenisiert und anschließend mit einer Bearbeitungsoptik in die Bearbeitungskammer fokussiert. Gute Ergebnisse in der Schichtherstellung ergeben sich bei Energiedichten von 2 - 5 J/cm^2.

Das zum Beschichten verwendete Substrat besteht aus einem mit Titan/Platin beschichteten Silizium-Wafer, der auf Temperaturen bis zu 900°C geheizt werden kann.

Zur Untersuchung der Abströmcharakteristik kann der Abstand zwischen Target und Substrat variiert sowie durch Drehen und Schwenken des Substrates bzw. des Targets Ungleichmäßigkeiten in der Dampfdichte ausgeglichen werden.

2.2   Bearbeitungsergebnisse

Durch Variation von Bearbeitungsenergiedichte, Substrattemperatur und anderen Prozeßparametern können unterschiedliche Schichtparameter eingestellt werden. Bei Substrattemperaturen oberhalb 570°C stellt sich eine kristalline Schicht mit einer ausgeprägten Dielektrizitätskonstante ein. Eine Steigerung der Substrattemperatur führt zu Werten von $E_r \geq 800$ bei einer Schichtdicke von 500 mm (Bild 2).

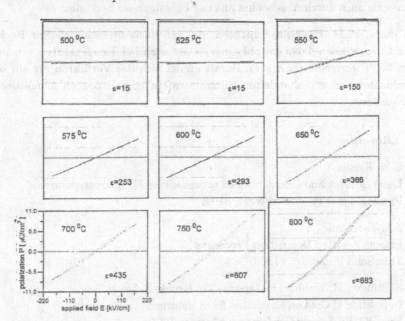

Bild 2:   Abhängigkeit der elektrischen Polarisation von der Substrattemperatur

Die beschichteten Proben werden an der Oberfläche mit einem Metallkontakt versehen und mit einem Sawyer-Tower-Meßverfahren vermessen. Unterhalb von 500°C scheidet sich die Schicht amorph ab und kann als Isolator verwendet werden. Die gemessene Durchbruchsfeldstärke beträgt dabei $E_{crit} = 10^8$ V/m. Neben der Substrattemperatur beeinflußt die verwendete Energiedichte die resultierenden Schichteigenschaften. Mit einer Erhöhung der Energiedichte können die gleichen Eigenschaften erzielt werden wie in Bild 1 dargestellt. Ursache hierfür ist die höhere Teilchenenergie des verdampften Materials, die durch Flugzeitmessungen nachgewiesen wurde.

3 Zusammenfassung

Durch Laserverfahren können Funktionsschichten bei niedrigen Temperaturen auf beinahe beliebige Substrate abgeschieden werden. Die Vorteile der Laser- verfahren sind :

- stöchiometrische Schichterzeugung
- Abscheidung und Strukturierung mit einem Werkzeug
- lokale Abscheidung auch auf vorprozessierten Bauteilen
- hohe Abscheideraten
- Hohe Flexibilität in Schicht- und Substratwahl

Leitende Metallschichten lassen sich dabei sowohl durch Laser-CVD aus der Dampfphase als auch durch Abscheiden aus der Flüssigphase herstellen /3/.

Die durch PLD erzeugten Schichten auf der Basis ferroelektrischer Bariumtitanstrukturen können dabei sowohl amorph als auch bei Temperaturen von nur wenig über 500°C ferroelektrisch sein. Damit eignet sich das Verfahren für ein selektives Beschichten von vorbehandelten Siliziumwafern zum Erzeugen monolitischer Sensorbausteine.

4 Literatur

/1/ E.W. Kreutz,
Laser Sources and Laser-induced processes for 3D microstructures
Proceedings of Photonics West, SPIE

/2/ J.A. Thornton
Plasma Assisted Deposition Processes
Thin Solid Films, 107 (1983)

/3/ S. Pflüger, M. Wehner, F. Jansen, Th. Kruck, F. Lupp
Deposition of Gold on Polyimide from Solutions
Proc. EMRS-Spring-Conference, Elsevier 1995

Lasergestützte Epitaxie von AlN- und GaN-Schichten

R.J. Dieter, J. Krampe, H. Schröder, H.-J. Stähle
Deutsche Forschungsanstalt für Luft- und Raumfahrt, Institut für Technische Physik
Pfaffenwaldring 38-40
D-70569 Stuttgart

I. Einleitung

Die Nitride der dritten Hauptgruppe (AlN, GaN, InN) sind in jüngster Zeit als neue Materialien für Halbleiterbau-elemente in den Mittelpunkt des Interesses gerückt. Ihre günstigen optischen, thermischen und akustischen Eigenschaften [1] ermöglichen die Herstellung optoelektronischer Bauelemente für den blauen bis ultravioletten Spektralbereich und von elektronischen Bauelementen, die auch bei hohen Betriebstemperaturen funktionsfähig sind.

Die Epitaxie dieser Materialien zeigt allerdings einige Schwierigkeiten. Zum einen lassen sich die Nitride nicht als Volumenkristall züchten, so daß keine geeigneten Nitrid-Substrate zur Verfügung stehen. Deshalb muß auf Fremdsubstrate zurückgegriffen werden - meist Saphir, Silizium oder SiC -, die sich im Vergleich zu den Nitriden zum Teil deutlich in Gitterkonstante und thermischer Ausdehnung unterscheiden. Dies erzeugt Spannungen und Versetzungen in den aufgebrachten Schichten. Weiterhin zeigen die bislang hergestellten Schichten Stickstoff-Fehlstellen, was sich in einer erhöhten n-Hintergrunddotierung äußert. Diese Hintergrunddotierung der Schichten muß bei der p-Dotierung der Materialien erst kompensiert werden.

Von den Gruppe III-Nitriden existieren zwei verschiedene Kristallmodifikationen [1]: die thermodynamisch stabile Wurtzit-Konfiguration, und die kubische Zinkblende-Struktur. Welche Struktur sich ausbildet, läßt sich über die Wahl des Substrats und der Wachstumsparameter beeinflussen.

Gruppe III-Nitride wurden bislang mit verschiedenen Methoden hergestellt, z. B. CVD [2], Plasma-unterstützte CVD [3], MOVPE [4], plasmaunterstützte MBE [5], PLD [6-9]. Lasergestützte Methoden bieten dabei den Vorteil, daß die Nitridbildung vom Schichtwachstum getrennt werden kann. Bei der herkömmlichen PLD wird von gepreßten nitridischen Targets ausgegangen. Dies hat den Nachteil, daß das pulverförmige Ausgangsmaterial aufgrund der unvermeidlichen Oxidation immer einen relativ hohen Sauerstoffanteil von einigen Prozent aufweist, und daß sich nicht alle Materialien, z.B. GaN, zu einem festen Target pressen lassen.

In dieser Arbeit soll daher von einem metallischen Target (Al, Ga, In) in Stickstoff-Atmosphäre ablatiert werden. Die Ausgangskomponenten (Metalle, N_2) sind in hoher Reinheit verfügbar, was eine geringe Verunreinigung der Schichten gewährleistet. Da die Bildung der Nitride nicht wie bei den anderen Methoden an der Substratoberfläche, sondern am Target bzw. in der Plasmafackel stattfindet, können Wachstumsparameter (Substrattemperatur, Wachstumsrate) und die Parameter für die Nitridbildung (über Variation der Laserparameter) getrennt voneinander optimiert werden.

II. Experimentelles

Für die Experimente wurden ein ns-Pulslaser (Nd:YAG, 8 ns Pulsbreite, 10 Hz Repetitionsrate P_{max}=10 W) mit Frequenzverdopplung und -vervierfachung (Modell BMI 502), und ein im Institut entwickelter diodengepumpter ps-

Laser (Nd:YAG, 25 ps Pulsbreite, 3 kHz Repetitionsrate, mittl. Leistung 1.5 W) verwendet. Eine genaue Beschreibung des ps-Lasers findet sich in [9].

Die Ablationsexperimente wurden einerseits in einer UHV-Kammer mit einem Basisdruck von ca. 10^{-10} mbar, andererseits in einer HV-Kammer (Basisdruck ca. 10^{-6} mbar), durchgeführt. Die UHV-Kammer verfügt dabei über eine Schleuse, so daß eine Kontamination der Reaktionskammer mit Umgebungsluft ausgeschlossen ist. Die Metalltargets werden in der UHV-Kammer vom Laser mit einem Scanner abgerastert, in der HV-Kammer kommt ein rotierendes Target zur Anwendung.

Der Stickstoff wird durch ein auf das Target gerichtetes Rohr mit einem bestimmten Fluß kontiniuerlich in die Kammer eingebracht. Eine Turbomolekularpumpe pumpt das Gas ab. Bei den beim Wachstum typischen Gasflüssen von 2 sccm bis 5 sccm wird während des Prozesses ein Druck von ca 10^{-3} mbar bis 10^{-2} mbar erreicht. Der Abstand zwischen Substrat und Target beträgt 5 cm. Die Wachstumstemperaturen in der HV-Kammer betrugen 700 °C, in der UHV-Kammer zwischen 700 °C und 1000 °C. Die Experimente mit dem ns-Laser betrugen die Pulsenergien ca. 100 mJ bei einem Durchmesser des Strahls auf dem Target von 2 mm, bei den Experimenten mit dem ps-Laser betrug die Pulsenergie ca. 0.5 mJ, bei einem Strahldurchmesser von etwa 200 µm. Als Substratmaterial wurde Saphir (0001) und Silizium (001) verwendet.

Die abgeschiedenen Schichten wurden mit Rasterelektronenmikroskopie, EDX (Energy-Dispersive X-ray Analysis) und Röntgendiffraktometrie charakterisiert, um Auskuft über Oberflächenmorphologie, Stöchiometrie und Kristallinität zu erhalten.

III. Ergebnisse und Diskussion

AlN-Schichten wurden sowohl mit dem ns- als auch mit dem ps-Laser hergestellt. Dabei zeigt sich, daß mit beiden Pulslängen im Rahmen der Meßgenauigkeit der EDX-Analyse ein stöchiometrischer AlN-Film erreicht werden kann. Der notwendige Stickstoff-Druck beträgt 10^{-2} mbar. Die abgeschiedenen Schichten sind transparent. Eine glatte Morphologie ist erst bei Wachstumstemperaturen oberhalb von 900 °C erreichbar. Bei tieferen Temperaturen erhält man aufgrund der geringeren Oberflächenbeweglichkeit dreidimensionales Inselwachstum. Während bei den mit dem ns-Laser abgeschiedenen Filmen Spratzer auf der Oberfläche gefunden werden, erscheinen die mit dem ps-Laser hergestellten Schichten nahezu frei von großen Spratzern (Siehe Abb. 1).

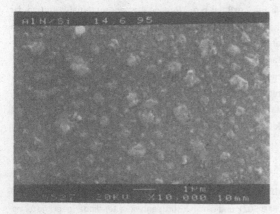

Abbildung 1: Oberflächenmorphologie einer AlN-Schicht. Substrat: Si. Wachstumstemperatur 900 °C. Abscheidung mit ps-Laser.

Die erkennbare Rauhigkeit ist auf noch nicht optimale Wachstumsbedingungen zurückzuführen. Messungen mit dem Röntgendiffraktometer zeigen orientiertes, aber nicht einkristallines Schichtwachstum (Abb. 2).

Wegen seines niedrigen Schmelzpunktes von 29.8 °C muß das Ga in einem horizontalen Tiegel für die Ablation bereitgestellt werden. Aus apparativen Gründen konnten die Experimente nur in der HV-Kammer mit dem ns-Laser gefahren werden. Als Tiegelmaterial dient heißgepreßtes Bornitrid. Durch die Strahlungswärme des geheizten Substrates liegt das Material flüssig vor. Wegen der stets glatten Flüssigkeitsoberfläche wird auf eine Targetrotation oder Abrastern verzichtet.

Ga zeigt bei den mit dem Nd:YAG-Laser zugänglichen Wellenlängen 1064 nm, 532 nm und 266 nm eine deutlich höhere Absorption als Al (siehe Abb. 3).

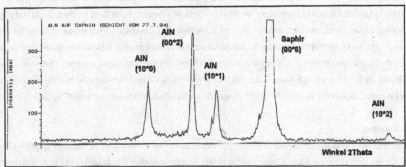

Abbildung 2: Diffraktogramm einer AlN-Schicht. Substrat Saphir. Neben dem Substratreflex sind die Peaks des AlN in der (00*2) Orientierung und der (10*0)-Orientierung (mit Reflexen höherer Ordnung) erkennbar.

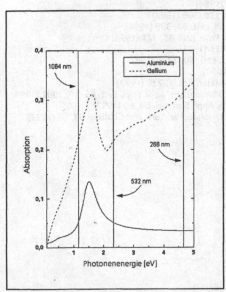

Abbildung 3: Berechnete Absorption von Ga und Al.

Aus diesem Grund wird erheblich mehr Material verdampft als im Fall des Al. Dies führt dazu, daß in der Plasmafackel größere Aggregate entstehen, so daß zuviel Ga im Vergleich zur N_2-Konzentration in der Gasphase angeboten wird, was nur durch noch kürzere Laserpulse günstig beeinflußt werden kann. Dadurch wird nur an der Oberfläche der Ga-Tröpfchen Stickstoff chemisch gebunden. Diese sind als Kristallite auf den Ga-Tröpfchen sichtbar. Der N_2-Partialdruck kann nicht beliebig erhöht werden, ohne daß durch Stöße im Gasraum die Abscheiderate drastisch sinkt.

Die vorliegenden Experimente zeigen, daß sowohl bei AlN als auch bei GaN durch Laserbeschuß von Metalltargets in Stickstoffatmosphäre eine stöchiometrische Nitridbildung möglich ist. Die Frage des Mechanismus der Nitrierung läßt sich aus den vorliegenden Daten nicht abschließend beantworten. Fest steht, daß die verwendeten Photonenenergien der Nd:YAG-Strahlung nicht zur Zerlegung des Stickstoffs ausreichen, da die Bindungsenergie des Stickstoff-Moleküls wesentlich größer ist. Allerdings können vibronische oder Rotationszustände angeregt werden. Bei Experimenten, bei denen der Stickstoffeinlaß nicht auf die Plasmafackel gerichtet ist, zeigen sich keine oder nur sehr geringe Konzentrationen von chemisch gebundenem Stickstoff in der Schicht. Wir schließen daher, daß die in der Plasmafackel entstehenden angeregten, teilweise ionisierten, energiereichen Teilchen Stickstoffmoleküle effektiv angreifen können und zur Nitridbildung führen.

Bei Pulsdauern von nur wenigen ps, wie in unserem Fall, ist die Zeit zur Bildung einer Metallschmelze zu kurz. Die Nitrierung an der heißen Metalloberfläche, wie aus der Literatur im Falle der Nitrierung von Ti bekannt [11,12], schließen wir daher aus.

IV. Zusammenfassung

In der vorliegenden Arbeit konnte gezeigt werden, daß die Bildung von AlN und GaN mittels Lasergestützter Epitaxie möglich ist. Es konnten AlN-Schichten guter Qualität hergestellt werden. Beim GaN ist die Schichtbildung noch nicht befriedigend, wobei allerdings durch den Einsatz von ps-Pulsen eine drastische Verbesserung erwartet werden kann. Gegenüber der herkömmmlichen PLD bietet die Ablation von Metalltargets in N_2-Atmosphäre den Vorteil, daß von hochreinen Ausgangsmaterialien ausgegangen werden kann. Im Gegensatz zu anderen Methoden, bei denen die Nitridbildung am Substrat erfolgt, können bei der hier verwendeten Abscheidetechnik, bei der die Nitride am Target gebildet werden, durch die zusätzlichen Freiheitsgrade die Wachstumsparameter weiter optimiert werden.

V. Danksagung

Wir danken G. Rohr und B. Hüttner für Rechnungen zur Absorption in Metallen, M. Schulze für XPS-Messungen, G. Schiller für die Diffraktometer-Messungen. Diese Arbeit wurde teilweise durch Mittel des BMVg unterstützt.

VI. Literatur

[1] Properties of Group III Nitrides, ed. J.H. Edgar, emis Dataview Series No. 11
[2] Y. Chubachi, K. Sato, K. Kojima, Thin Solid Films **122**, 259 (1984)
[3] W. Zhang, Y. Someno, M. Sasaki, T. Hirai, J. Cryst. Growth **130**, 308 (1993)
[4] A. Saxler, P. Kung, C.J. Sun, E. Bigan, M. Razhegi, Appl. Phys. Lett. **64**, 339 (1994)
[5] K.S. Stevens, A. Ohtani, M. Kinniburgh, R. Beresford, Appl. Phys. Lett. **65**, 321 (1994)
[6] M.G. Norton, P.G. Kotula, C.B. Carter, J. Appl. Phys. **70**, 2871 (1991)
[7] K. Seki, X. Xu, H. Okabe, J.M. Frye, J.B. Halpern, Appl. Phys. Lett. **60**, 2234 (1992)
[8] P. Bhattacharya, D.N. Bose, Jap. J. Appl. Phys. **30**, L1750 (1991)
[9] S.R. Nishitani, S. Yoshimura, H. Kawata, M. Yamaguchi, J. Mater. Res. **7**, 725 (1992)
[10] D.R. Walker, C.J. Flood, H.M. van Driel, U.J. Greiner, H.H. Klingenberg, Appl. Phys. Lett. **65** (16),1992 (1994)
[11] E.W. Kreutz, M. Krösche, H. Sung, A. Voss, K. Wissenbach, Appl. Surf. Sci. **54**, 69 (1992)
[12] E.W. Kreutz, M. Krösche, H. Sung, a. Voss, A. Jürgens, T. Leyendecker, Surf. and Coatings **53**, 57 (1992)

Laser-Induced Direct Patterning of Non-Metallic Substrates

Z. Illyefalvi-Vitéz[1], Z. Kocsis[1], J. Pinkola[1], M. Ruszinkó[1], P.J. Szabó[2]

Department of Electronics Technology (1)

Department of Electrical Engineering Materials (2)

Technical University of Budapest, Goldman t. 3., H-1521 Budapest, Hungary

1. INTRODUCTION

Laser-induced direct patterning is a process where laser radiation is used to affect a controlled area - a pattern - on the surface of a substrate. Materials modification as well as subtractive and additive processing of the surface structure using laser driven physical and/or chemical effects are of interest. The category of laser-induced direct patterning includes the process of **laser direct writing** where the surface pattern is generated by the controlled movement of a focused laser spot.

Three kinds of laser-induced direct patternings have been investigated as follows:

1. **Patterned modification of** the surface layer of insulating **aluminium-nitride (AlN) substrates** into conductive aluminium by laser-induced thermal decomposition. This decomposition is accompanied by a material removal and a redeposition process as well. In the present paper this type of processing is explained in detail.

2. **Patterned surface cladding of substrates** by applying screen-printed thick film paste on the entire surface, and firing the pattern selectively by a focused laser beam. The controlled movement of the laser spot defines the circuitry by sintering the powder, while the paste can be removed from the inactivated areas. A subsequent firing at the usual temperature can be used to improve the film parameters. The process can be an efficient method for the rapid prototyping of thick film interconnects.

3. **Laser-induced backward transfer** (LIBT) of copper from a black oxidized copper foil to a glass substrate by laser direct writing, where the transparency of the applied glasses for the wavelength of Nd:YAG lasers is utilized. The process can be used for marking of microscopic slides, or for producing simple interconnect patterns.

2. LASER PROCESSING OF AlN SUBSTRATES

Aluminum nitride has recently been introduced as a substrate and packaging material for multi-chip and similar microelectronic modules. This material is suitable for replacing alumina as a substrate for silicon chips, because it combines a high thermal conductivity with a thermal expansion coefficient close to that of silicon. AlN may also compete with beryllia since it is nontoxic and could be competitive in price. Compatible technologies developed for aluminum nitride ceramics improve the applicability of this material.

When a material surface is irradiated by a laser beam, the absorbed part of the beam can generally be considered as a source of thermal energy that rises the temperature of the surface. This temperature rise can cause physical and/or chemical changes which are utilized for processing. In the case of AlN physical processes dominate: over 2500 degree C AlN **decomposes** and **removes**, some part of the aluminum, however, **redeposits** and forms a conductive layer. Problems caused by the increased conductivity in the trimming cuts of thick-film metal resistors on AlN substrates were studied and published in [1].

The laser affected volume of material depends on the parameters of the radiation, the material surface characteristics and the environmental conditions. The most significant factor is the temperature distribution caused by the penetrating heat. A theoretical analysis on heat penetration, and some practical results on the example of laser processing of aluminum nitride substrates were provided by [2].

3. ANALYSIS OF THE LASER ABLATION OF AlN

The above mentioned analysis proved that the z_d **decomposition depth** of AlN at the end of a laser pulse of t_p width can be calculated on the basis of the energy balance for a unit area of the surface:

$$z_d = \frac{\alpha I t_p}{\rho C T_d} - \frac{\pi K T_d}{4 \alpha I} \, ,$$

where I is the intensity of the laser pulse, α is the absorption coefficient of the surface, K is the thermal conductivity, ρ is the density, C is the specific heat capacity, and T_d is the temperature of decomposition.

Q-switched Nd:YAG pulsed laser was used for the experimental investigation. The laser worked in multi-mode with a focal spot diameter of about 100 μm. The laser power was adjusted by the pumping current, and the processing intensity was calculated from the measured average power, the frequency and the above data. The laser beam scanned over a surface of 5 by 0.5 mm, and shot according to a net with the spacing of 5 μm and 25 μm in X and Y directions, respectively. Supposing a complete ablation, the calculated and measured **decomposition depth** can be compared, the results are plotted in Figure 1.

At higher intensities the calculated and measured values of the decomposition depth show very good coincidence, within 15%. At low intensities, however,

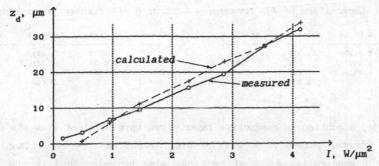

Fig.1. *Comparison of calculated and measured decomposition depths*

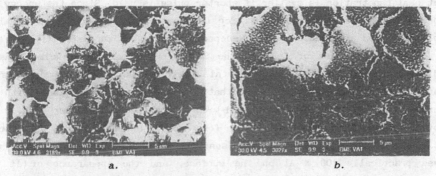

a.   b.

Fig.2. *The surface of AlN ceramic: a.before; b.after laser processing*

where AlN only slightly or does not decompose, material removal was still observed. The depth of the laser affected area is in the range of the calculated value of the melting depth ($z_m = 0.45$ μm), that proves the idea that the liquid phase always plays an important role in material removal.

4. THE EFFECT OF SHORT PULSES: MORE Al REDEPOSITS ON AlN

The **material characteristics** of the same processed surface were analyzed by scanning electron microscope including energy dispersive X-ray composition analysis. As an effect of laser processing, the morphology of the surface showed a significant change: its appearance became fused, cracked and much smoother, with metallic luster (Figure 2.). With X-ray qualitative material analysis the presence of O, the decrease of the amount of N and a small decrease of the Al content were detected on the laser processed areas in comparison with the original surface (Table 1.), which proves the decomposition of AlN and the simultaneous ablation, oxidization and redeposition of Al. By using the 200 ns pulses of Q-switched Nd:YAG laser for the processing of AlN, on the previous samples the presence of Al can be detected, but it does

Table 1. Composition of AlN ceramic: a.before; b.after laser processing

Element	K Ratio	Weight %	Atomic %
N K	0.0237	12.399	21.088
O K	0.0057	2.591	3.858
AlK	0.9706	85.010	75.054
Total		100.000	100.000

Element	K Ratio	Weight %	Atomic %
N K	0.0127	5.629	9.430
O K	0.0426	14.225	20.865
AlK	0.9447	80.146	69.704
Total		100.000	100.000

a.   b.

not supply a continuous conductive layer. From this point of view the application of shorter pulses is more favorable . From the theoretical analysis of [2] the L **penetration depth of heat** can also be determined in the form of $L = (4\kappa t_p)^{1/2}$, where $\kappa = K/\rho C$ is the thermal diffusivity. This equation shows that heat penetration is smaller if shorter pulse is used. Since the energy of shorter pulses is mainly utilized for decomposition, the substrate remains cooler, and, as a consequence, the redeposition of Al is more effective.

The 13 ns pulses of KrF and the 50 ns pulses of XeCl excimer laser radiation was used for the experimental investigation of the process. As a characteristic example, it was found that 20 laser pulses of KrF laser with an energy density of about 1,25 J/cm^2 and with a repetition rate lower than 10 Hz provide near optimum condition for the formation of aluminum layers with sheet resistance of 0.2-0.4 ohm. In comparison with Table 1. the material analysis gives practically 100 % Al on the surface, and the appearance of it is metallic, smooth and shining, as it is shown in Figure 3.

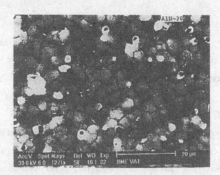

Fig.3. The surface of AlN after processing with KrF excimer laser pulses

REFERENCES

1. Takahashi,M.; Kurihara,Y.; Yamada,K.; Kanai,K.; Kurihara,K.: *Changes of Characteristics in AlN Ceramics Surface by Laser Irradiation*, Electronics and Communications in Japan, Part 2, Vol.73, No.8, 1990, pp.105-13.
2. Illyefalvi-Vitéz,Zs.; Pinkola,J.; Ruszinkó,M.: *Analysis of Laser Induced Surface Effects on AlN Substrates*, in Kinzy,W.J.; Kurzweil,K; Harsányi,G.; Mergui,S. (editors): Multi-Chip-Module Sensor Technologies, Kluwer Academic Publishers, Dordrecht, (being published in 1996)

Printed in the United States
By Bookmasters

Printed in the United States
By Bookmasters